GEOMETRIC FORMULAS

Triangles

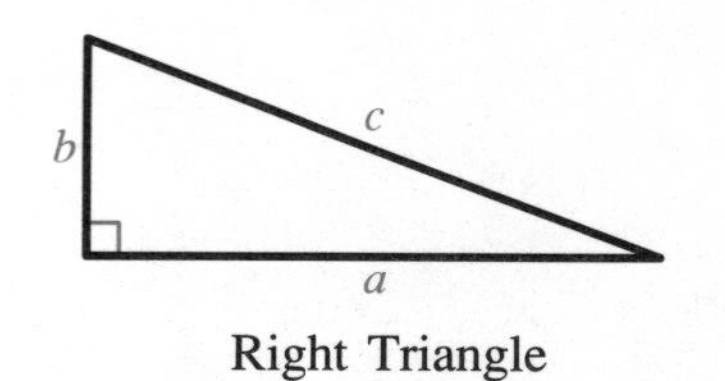

Right Triangle

Pythagorean Theorem $a^2 + b^2 = c^2$

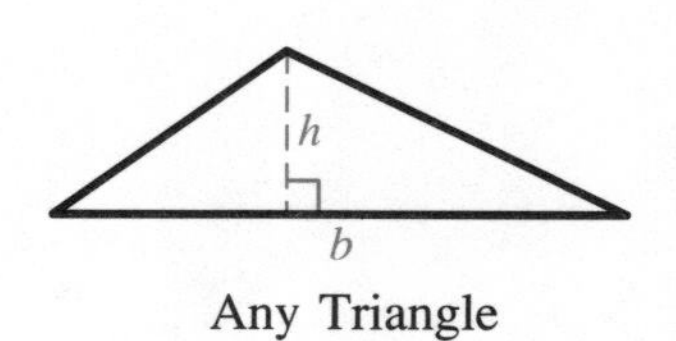

Any Triangle

Area $A = \frac{1}{2}bh$

Circles

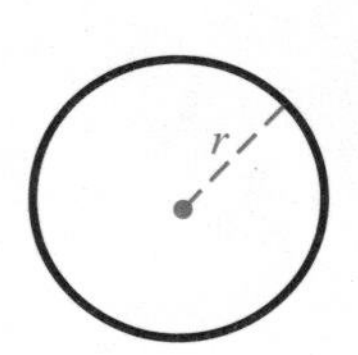

Area $A = \pi r^2$

Circumference $C = 2\pi r$

Cylinders

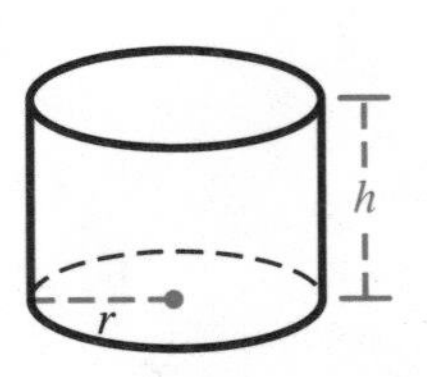

Surface Area $S = 2\pi r^2 + 2\pi rh$

Volume $V = \pi r^2 h$

Cones

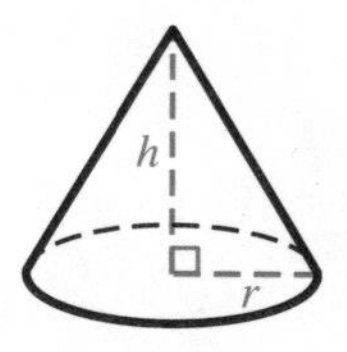

Surface Area $S = \pi r^2 + \pi r\sqrt{r^2 + h^2}$

Volume $V = \frac{1}{3}\pi r^2 h$

Spheres

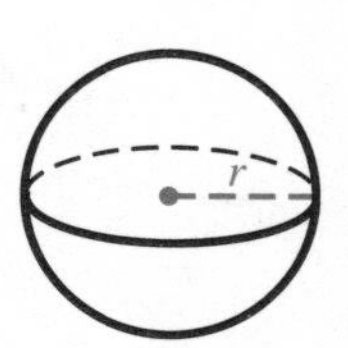

Surface Area $S = 4\pi r^2$

Volume $V = \frac{4}{3}\pi r^3$

College Algebra

Second Edition

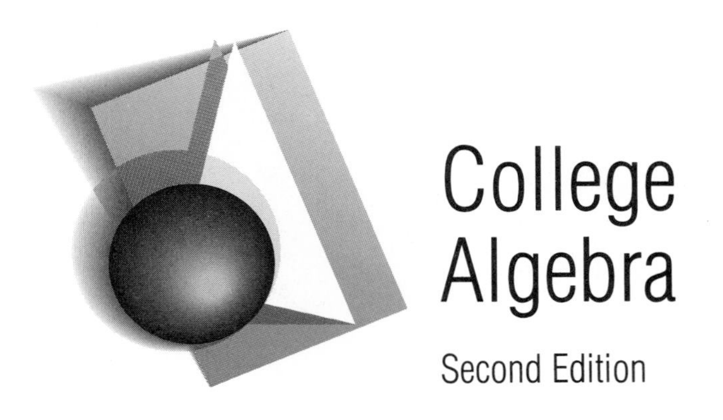

College Algebra

Second Edition

Stanley I. Grossman

University of Montana and University College London

SAUNDERS COLLEGE PUBLISHING
A Harcourt Brace Jovanovich College Publisher

Fort Worth Philadelphia San Diego New York Orlando Austin
San Antonio Toronto Montreal London Sydney Tokyo

THIS BOOK IS PRINTED ON **ACID-FREE, RECYCLED** PAPER 

Text typeface: Times Roman
Compositor: York Graphic Services
Acquisitions Editor: Robert B. Stern
Developmental Editor: Sandra Kiselica
Copy Editor: York Production Services
Art Director: Christine Schueler
Art Assistant: Caroline McGowan
Text Designer: Rebecca Lemna
Cover Designer: Lawrence R. Didona
Text Artwork: York Production Services
Layout Artist: York Production Services
Production Manager: Bob Butler

Cover Credit: © 1982 Barbara Kasten/Courtesy John Weber Gallery

Printed in the United States of America

ISBN 0-03-052168-8
Library of Congress Catalog Card Number:
91-25214

Preface

During the past few years the study of algebra has become increasingly important for students in almost every academic discipline for at least two reasons. First, physical, biological, economic, and social relationships are described in terms of algebraic equations. Second, algebra is an indispensable prerequisite for other mathematical topics such as finite mathematics, discrete mathematics, statistics, and calculus.

Many different kinds of students study algebra in college, but they fall mainly into two groups: recent high school graduates and returning students who may have been out of school for many years. Many students in both groups look upon algebra with less than unbridled enthusiasm. The comments "I never liked math" and "I haven't seen any math for years" are heard often by every algebra instructor.

College Algebra, Second Edition, contains all the algebraic material that is needed to meet the requirements for other college disciplines and to satisfy prerequisites for further mathematical study. I have taught these topics many times at the University of Montana — a university with large numbers of both recent high school students and nontraditional students (the average age of our students is 27). Consequently, I wrote this book with the needs and concerns of these students in mind.

First, I have made the material more accessible by providing large numbers of examples with every step included. There should be no mystery in algebraic computations and this book has none. Second, I have provided many realistic applications so that students will understand that they are studying algebra because it is useful to them, not because it is a requirement for a degree. Finally, I have attempted to make the material interesting. Solving cubic equations might seem unexciting to some, but reading how Renaissance men competed, sometimes violently, to solve them should add spice to the subject.

Above all, it is my goal that students will both learn to appreciate the value of algebra and, most importantly, gain the confidence to succeed in their mathematical studies. I really believe that every university student can learn mathematics if he or she works hard doing problems and is given sufficient help and encouragement.

Changes in the Second Edition

I have made a number of changes which, I believe, will make this book more effective as a teaching tool. Here are the major ones:

1. Captions: Captions have been added to every example and figure in the text. This will make it easier for students to determine exactly what is being done in each example and will be very helpful when tackling the problem sets. The captions on figures clearly describe exactly what each figure illustrates.

2. Readiness Checks: Each problem set (except two) begins with up to five ''Readiness Check'' problems. These *true-false* and *multiple-choice* problems are intended to test students' understanding of the material they have just read and do not involve difficult computations. The student who can solve these is ready to tackle the exercise set that follows. Answers to the Readiness Check problems are given at the bottom of the last page on which they occur in each section — not at the back of the book.

3. New Exercises: The second edition has approximately 4,500 exercises — about 1,000 more than in the first edition. In addition, about 600 drill problems from the first edition have been replaced here.

4. New Graphs: This edition has almost 1,000 figures, nearly double the number of the previous one. These are used in new ways as well. See the discussion under ''Figures'' on page viii.

5. New Topics: The following are new to this edition:

- **a.** Expanded coverage of the conic sections (four sections — Sections 5.2 to 5.5 — instead of one).
- **b.** Factoring with negative exponents, page 53.
- **c.** Horner's Method, pages 253–255.
- **d.** Upper and Lower Bound Theorem for Zeros of a Polynomial, page 262.
- **e.** The Bisection Method for determining the zeros of a polynomial, Section 4.5.
- **f.** Equations Involving Exponential and Logarithmic Functions, Section 6.5.
- **g.** Systems of Nonlinear Equations, Section 7.3.
- **h.** Introduction to Linear Programming, Section 7.6.

6. Use of the Graphing Calculator: Many students now have access to calculators that can, if used properly, provide accurate graphs of a great number of functions. In Appendix A I have shown students how to use their calculators effectively to draw graphs of functions, sketch conic sections, find zeros of polynomials, and solve other types of equations. Example 12 on page A.22 shows the limitations of such calculators by discussing a polynomial whose graph *cannot* accurately be sketched on a calculator. The appendix is written to be used with any calculator now available.

In fourteen sections within Chapters 2, 3, 4, 5, and 6, I have added problems that are intended for solution on a graphing calculator. I made the deliberate choice to limit the use of the graphing calculator to those sections where it is appropriate, rather than to integrate it throughout the text. This gives the instructor an option. I stress that ownership of a graphing calculator is *not required* for use with this book.

Organization

Chapter 1 is introductory and is intended as a review of the most important topics in first year algebra. Chapter 2 deals with equations and inequalities in one variable. As elsewhere in the text, the emphasis is on applications. Section 2.2 discusses applications of linear equations and Section 2.6 contains applications of quadratic equations.

Chapter 3 introduces functions and their graphs. The chapter begins with a thorough discussion of straight lines, followed by the general definition of a function. A wide variety of common functions are described and graphed. Section 3.5 contains a detailed description of how to shift and reflect known graphs to obtain new ones.

Chapter 4 discusses properties of polynomials and describes a number of ways to find their zeros. The new Section 4.5 presents the bisection method which is a simple procedure for obtaining crude approximations to zeros. The much more efficient Newton's method is described in Section 4.6.

Chapter 5 contains a discussion of rational functions and their graphs and goes on in Sections 5.3, 5.4, and 5.5 with descriptions of ellipses, parabolas, and hyperbolas. Each of these three sections concludes with illustrations of how these basic conic sections arise in real world applications.

Chapter 6 presents an introduction to exponential and logarithmic functions. Section 6.6 contains a large and diverse number of realistic applications of exponential growth and decay.

Chapters 7 and 8 contain an introduction to systems of equations and matrices. A brief introduction to linear programming can be found in the new Section 7.6.

The final chapter (Chapter 9) contains topics in discrete mathematics including counting techniques, probability, mathematical induction, the binomial theorem, and an introduction to sequences and series. Finally, Appendix A describes how to obtain accurate graphs and solve equations on a graphing calculator.

Features

Examples As a student, I learned algebra from seeing examples and doing exercises. There are 519 examples in this book. The examples include all the necessary steps so that students can see clearly how to get from "A" to "B." In many instances explanations are highlighted by colored notes to make steps easier to follow.

Exercises The text includes approximately 4,500 exercises — both drill and applied problems. More difficult problems are marked with an asterisk (*) and a few especially difficult ones are marked with two (**). In my opinion, exercises provide the most important tool in any undergraduate mathematics textbook. *If you don't work problems, you won't learn the mathematics*. Or, to quote a button popular at mathematics meetings,

MATH IS NOT A SPECTATOR SPORT.

Readiness Check Problems Each problem set but two contains multiple-choice and true-false questions that require relatively little computation. Answers to these problems appear on the bottom of the page on which the last such problem in the set appears. They are there to test whether the student understands the basic ideas in the section, and they should be done before tackling the more standard problems that follow.

Chapter Review Exercises At the end of each chapter I have provided a collection of review exercises. Any student able to do these exercises can feel confident that he or she understands the material in the chapter.

Chapter Summary Outlines A summary of the most important facts discussed in each chapter appears at the end of that chapter. Students should find these summaries useful, especially when studying for a test.

Applications We study algebra because of its great utility. This book has a large number and variety of applications. A list of these applications appears on pages xvii–xix. New to this edition are extensive examples of the conic sections (ellipses, parabolas, and hyperbolas) in the real world. These can be found beginning on pages 299, 308, and 318.

Figures There are approximately 1,000 figures in this book. Many of these appear in a standard way—as graphs of functions. Others are used in the problem sets to help the student understand how graphs change as functions change. Examples of this can be found on pages 168–169, 179–181, 198, 200, 212–213, 289, 302, 310–311, 333, and 351 among others. There are also a number of new figures attached to applied problems to help the student visualize what he or she is expected to solve.

Use of the Graphing Calculator Students with access to a calculator that can draw graphs can learn how to use it more effectively by consulting Appendix A at the back of the book. For more details on this feature, see item #6 on page vi.

Warnings An important part of the teaching process is helping students to avoid making mistakes, especially those that are commonly made. In thirty places in the book I provide warnings that illustrate common errors. Each warning illustrates the mistake and shows how it can be avoided.

Use of the Calculator Virtually all college and university students own or have access to a hand calculator. Problems that were computational monstrosities 20 years ago have become fairly easy with the aid of a calculator. I have used the calculator in many examples and have suggested its use in a number of exercises. Examples and problems that require the use of a calculator are marked with a 🖩.

Precalculus Many students studying algebra will go on to study calculus. In several places in the book I have provided examples that do arise in calculus and have labelled them as such. Some of these appear on pages 61–64, 136, 144, 169, 178, and 182.

Discrete Mathematics Algebra is a prerequisite for discrete mathematics and many students will go on to take that course. Many topics in discrete mathematics are introduced in Chapter 7, 8, and 9. A student learning this material should have no difficulty continuing into a more advanced discrete math course.

Focuses I believe that studying algebra can be fun. To make it more fun, students can read about real, interesting people from the history of mathematics who were interested in far more than a succession of dry formulas in the sketches that are headed "Focus on . . ." In these focuses students can learn, for example, about the first known instance of irrational behavior (p. 5) and the nasty scheming of Cardano (p. 264). One focus beginning on page 379 contains no history but, rather, asks students to think about the assumptions inherent in a mathematical model.

Accuracy The success of a mathematics textbook largely depends on its accuracy. Galleys and page proofs were carefully checked for accuracy by me and four other mathematicians: Lynne Kotrous and Ray Plankinton at Central Community College, Platte Campus, in Nebraska, Paul Allen at the University of Alabama and Bruce Sisko at Belleville Area College. George Bradley at Duquesne University wrote many of the replacement drill problems and provided solutions to them. Lynne Kotrous, Ray Plankinton, and I solved all the odd-numbered problems in the book. Finally, all three of us proofread the typeset answers to make sure they were accurate.

The result is a book and answer section that is as clean as is within human ability to compile. However, if you do find an error in an answer or in the text, please send it to the publisher or to me; it will be corrected in the next printing.

Chapter Interdependence The following chart indicates chapter interdependence—that is, which later chapters depend on the student's having mastered earlier material.

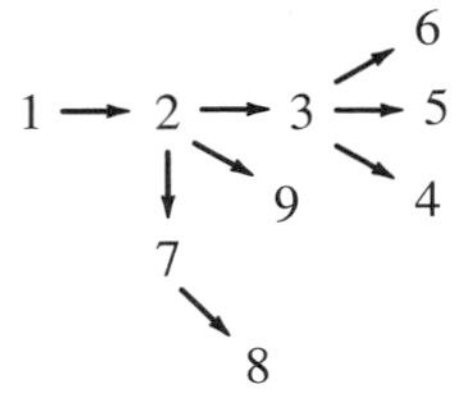

1 → 2 → 3 → 5
3 → 6
3 → 4
2 → 9
2 → 7
7 → 8

Supplements

The answers to most odd-numbered problems appear at the back of the book. In addition, the following instructional aids are available from the publisher:

A **Student Solutions Manual** prepared by George Bradley and Daniel Barbush at Duquesne University contains chapter summaries and detailed solutions for all odd-numbered problems.

An **Instructor's Manual With Transparency Masters** also prepared by George Bradley and Daniel Barbush provides solutions for all the even-numbered problems and transparency masters of key figures from the text.

A **Computerized Test Bank,** of 900 multiple choice and open ended questions, prepared by Jan Wynn at Brigham Young University. Available for IBM, Macintosh and Apple II computers. A **Printed Test Bank** of these questions is also available.

AT Software is a computer software package referenced to *College Algebra*. Both interactive and tutorial, this software is available for use with IBM and compatible computers.

Videotaped Lectures prepared by Pat Stanley and Becki Bergs at Ball State University covering the first seven chapters of *College Algebra*. These lectures are referenced directly to the sections of this text.

A **Graphing Calculator Supplement** is available for purchase. Written by James Angelos of Central Michigan University, it explains how to use the Casio and TI graphing calculators and uses examples from this text.

An additional **Graphing Calculator Supplement,** written by Iris Fetta of Clemson University, also available for purchase, explores the use and value of the TI-81 graphing calculator.

Acknowledgements

Most of us really don't know what a book is like until it's been used in class and we get comments on how it works. I am grateful to the following individuals for their helpful comments on this second edition:

Mickie Ahlquist, Casper College
Paul Allen, University of Alabama
Daniel Anderson, University of Iowa
Ruth Berger, Memphis State University
John Bruha, University of Northern Iowa
Gary Crown, Wichita State University
Lucy Dechene, Fitchburg State College
Kenneth Dodaro, Florida State University
Iris Fetta, Clemson University
Donald Goldsmith, Western Michigan University
H. T. Mathews, University of Tennessee, Knoxville
Pamela Matthews, Mount Hood Community College
Philip Montgomery, University of Kansas
Janina Udrys, Schoolcraft College
Ron Virden, Lehigh County Community College
Peter L. Waterman, Northern Illinois University
Carroll Wells, Western Kentucky University
Jan E. Wynn, Brigham Young University

The following reviewers made important contributions to the reliability and teachability of the first edition.

James Arnold, University of Wisconsin at Milwaukee
Daniel Barbush, Duquesne University

George Bradley, Duquesne University
Edgar Chandler, Paradise Valley College
Charles Cook, Tri-State University
Susan Danielson, University of New Orleans
Joe Diestel, Kent State University, Main Campus
Milton Eisner, Mount Vernon College
Susan Foreman, Bronx Community College
Jim Gussett, Longwood College
Barney Herron, Muskegon Community College
Lynne Kotrous, Central Community College, Platte Campus
David Logothetti, Santa Clara University
Reginald Luke, Middlesex Community College
Lyle Oleson, University of Wisconsin
Ray Plankinton, Central Community College, Platte Campus
Jean Rubin, Purdue University
Robert Sharpton, Miami Dade Community College, South Campus
Joseph Stokes, Western Kentucky University
George Szoke, University of Akron

I would like to thank and give credit to the following sources for the use of their original figures: On page 33 (cartoon) by permission of Universal Press Syndicate; on page 93 (cartoon), by permission of Johnny Hart and Creators Syndicated Inc.; on page 104 (cartoon), reprinted by permission of NEA, Inc.; and on page 183, reprinted by permission of the Wall Street Journal, © Dow Jones and Company, Inc., 1983. All Rights Reserved.

Some of the material in this book — especially in Chapter 9 — first appeared in *Mathematics for the Biological Sciences* (New York: Macmillan, 1974) written by James E. Turner and me. I am grateful to Professor Turner for permission to use this material.

I wrote a great deal of this book while I was a research associate at University College London. I am grateful to the Mathematics Department at UCL for providing office facilities, mathematical suggestions, and, especially, friendship, during my annual visits there.

The book was produced by York Production Services in York, Pennsylvania. I wish to acknowledge the spectacular job done by York, in general, and by my Production Editor Kirsten Kauffman, in particular.

Special thanks are due to the editorial and production staff at Saunders for the care and skill they brought to this process. Finally, I owe much to my Acquisitions Editor, Bob Stern, and my Developmental Editor, Sandi Kiselica, who provided much encouragement and help in determining the final form this book was to take.

Stanley I. Grossman

Contents

Index of Applications

Biological Sciences

Business and Economics

Physical Sciences

General Topics

To Kerstin, Erik, and Aaron

One learns by doing the thing; for though you think you know it, you have no certainty until you try.

SOPHOCLES

Chapter 1

Basic Concepts of Algebra

1.1 History, Motivation, and a Dissenting View

People have been studying and using algebra for a long time. How did algebra originate? What is it good for? We will give partial answers to these questions in this section.

Until the middle of the 19th century, when its basic concepts became more abstract, algebra was essentially the science of equations. Linear equations were being solved in Egypt and Mesopotamia in the second millenium before Christ. In fact, the inhabitants of the fertile land between the Tigris and the Euphrates rivers had solved certain types of quadratic equations and even some cubic equations.

Early in the development of algebra, every problem was written out in words. This was the nature of ancient Egyptian, Mesopotamian, and Arabic algebra. The first use of abbreviations to represent variables and powers is credited to the Hellenistic Greek, Diophantus, in his book *Arithmetica* (ca. A.D. 250). Diophantus's work was much more sophisticated than work done by earlier civilizations. For these achievements he is often referred to as the "father of algebra."

From Diophantus onward, algebraic techniques and symbolism were developed until the seventeenth century when René Descartes wrote *La Geometrie*. This book has been called "the earliest mathematical text that a present day student of algebra can follow without encountering difficulties in notation."† It is essentially Descartes's notation that we use today.

The study of algebra is not without its detractors. The following article by Jerry Zezima appeared in the Stamford (Connecticut) *Advocate* on September 5, 1985.‡

> Does anyone here remember algebra class? Do you remember what an utter waste of time it was? Or how you would have gladly submitted to Chinese water torture than be forced to solve just one more problem?

---

†Carl B. Boyer, *A History of Mathematics*, Wiley, New York, 1968, p. 371.

‡Copyright 1985, *The* (Stamford) *Advocate;* used with permission.

Of course you do. It all comes back to you every year at this time, I'll wager, for this would have been the beginning of school, the beginning of another long, miserable, thoroughly intolerable year of algebra.

Algebra class was a menacing and mystifying place where logic was nowhere to be found, where irrelevance and insignificance were the order of the day. Algebra class also was where we were introduced to some of the strangest people in the annals of American education — people like Jim, the protagonist of Problem No. 5 on Page 19.

Memory fails me on the particulars of Jim's dilemma, thank heavens, but it went something like this: "I am one-third as young as my father was two years ago. If the age of my father is seven times that of half my age last year, how old am I?"

Here it is easy to see the root of the problem. We are dealing with a guy who knows he is one-third as young as his father was two years ago. Furthermore, he knows that the age of his father is seven times that of half his age last year. Yet, given all this information, irrelevant and confusing though it may be, he cannot even figure out how old he is at this very moment!

This naturally begs the question: What kind of moron is Jim, anyway? The answer, of course, is that Jim is the kind of moron found only in algebra books. Brilliant men like Einstein reside in science books; brave men like Columbus are found in history books. But only fat-headed fools like Jim can find sanctuary between the covers of an algebra book.

Not only can you find Jim there, in Problem No. 5 on Page 19, you can also find the Moores and the Smiths, who populate Problem No. 12 on Page 26.

It seems that the Moores, who live in New York, are leaving for Boston at 8 A.M. They will average 50 miles an hour over the 200-mile journey. The Smiths, on the other hand, are leaving Boston for New York at 9 A.M. averaging 55 miles an hour.

Question: At what time will the Moores and the Smiths pass each other on the highway?

Answer: Who cares? Any motorist who spends his time craning to see when his friends will pass him on the other side of the highway instead of watching his own side of the road, as he should be doing, is a menace to society and ought to have his license revoked.

But does that ever occur to the people who dream up these problems? Apparently not. They're too busy baffling us with the antics of ninnies like Mrs. Fisher, who, in Problem No. 15 on Page 32, has gone to the store to buy some coffee.

Of course, it is impossible for Mrs. Fisher to simply grab a can of coffee, bring it to the counter and pay for it. Instead, Mrs. Fisher must decide how much regular coffee at $2.52 a pound and how much decaffeinated coffee at $2.37 a pound she has to buy to equal six and a half pounds of coffee at $2.43 a pound.

Why are people like Mrs. Fisher allowed to exist? Why isn't teaching algebra a felony? Why do grown men and women — who themselves were once victimized by the likes of Jim in Problem No. 5 on Page 19 — persist in polluting young minds with this kind of tommyrot?

Obviously because the adults who concoct or condone these conundrums are not half as sensible as the children who know full well that they are sheer folly. Or, to put it into language we can all relate to, the aforementioned adults are not half as sensible as said children were five years ago. If the children are three times as sensible as the adults were two years ago, and the adults are not two-thirds as sensible as the children are today, how sensible are the adults?

Fortunately, you don't have to be a whiz at algebra to figure that one out.

Jerry Zezima is partially right. High school students spend lots of time in algebra classes solving problems with dubious practicality. It is easy to look at the three problems in the article and say "who cares?"

But algebra hasn't survived thousands of years in order to answer silly questions. Algebra *is* important because it provides us with techniques for working with equations that occur in virtually everything we do. Here are some equations that frequently occur.

Distance formula: $d = vt$

where d is the distance traveled, v is the velocity (assumed to be constant), and t is the elapsed time.

Simple interest formula: $I = Prt$

where I is the interest earned, P is the amount invested (principal), r is the interest rate, and t is the amount of time (usually the number of years) the money was invested.

Ideal-gas law: $P = \dfrac{nRT}{V}$

where P is the pressure of a gas, T is its temperature, V is its volume, n is the amount of gas (in moles), and R is a constant.

Algebra is used whenever we attempt to solve problems involving these or other equations. It might not be useful to know when the Moores and the Smiths meet, but it is often important to determine when the Moores can expect to get home, or to know how much gas is used by the Smiths, or by a commercial aircraft, or by a space shuttle. By solving appropriate algebraic equations, we can compute compound interest, the value of an annuity, the age of a fossil (using carbon dating), the temperature of a cooling liquid, the approximate population of India in the year 2000, the frictional force on a sliding object, the magnitude of a star, or the intensity of an audible sound.

Algebra (together with trigonometry) is also needed in the study of calculus, which some of you will go on to take.

In the chapters that follow, we will describe most of the algebraic techniques used to solve the kinds of problems we've mentioned. After reading this book, you will, we expect, recognize the importance of algebra in almost every other subject you study.

1.2 The Real Number System

The collection of **real numbers,** denoted by $\mathbb{R}$, consists of the natural numbers, integers, rational numbers, and irrational numbers. Real numbers can be represented on a **number line** in such a way that each point on the line corresponds to exactly one real number and each real number corresponds to one point on the line. The number 0 (zero) is placed. Then the positive real numbers are placed at regular intervals to the right of 0, and the negative real numbers are placed at regular intervals to the left of zero.

The **natural numbers** (also called **positive integers** or **counting numbers**) are the numbers of counting: 1, 2, 3, 4, . . . (the three dots indicate that the string of numbers goes on indefinitely). The number 2 is placed 1 unit to the right of 1 on the number line, the number 3 is placed one unit to the right of 2, and so on. The natural numbers are denoted by the symbol N.

The **integers** consist of the natural numbers, their negatives (called the **negative integers**), and the number 0. The collection of integers is denoted by the symbol Z.

In Figure 1 we represent the integers 0, ± 1, ± 2, ± 3, and ± 4 on the number line.

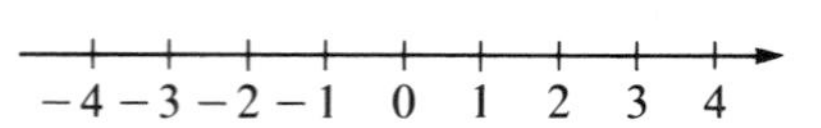

Figure 1 The part of the number line that includes the integers from -4 to 4.

A **rational number** is a real number that can be written as the quotient of two integers, where the integer in the denominator is not zero.

Definition of a Rational Number

$$r = \frac{m}{n} \quad \text{where } m \text{ and } n \text{ are integers and } n \neq 0$$

Every integer n is also a rational number because $n = n/1$. The collection of rational numbers is denoted by the symbol Q (for *quotient*).

EXAMPLE 1 ***Six Rational Numbers***

The following are rational numbers:

(a) $\frac{1}{2}$ (b) $-\frac{3}{4}$ (c) -5 (d) $0 = \frac{0}{1}$ (e) $-\frac{127}{105}$

(f) $0.721 = \frac{721}{1000}$

As in part (f), any terminating decimal is a rational number.

All rational numbers can be represented in an infinite number of ways. For example

$$\frac{1}{2} = \frac{2}{4} = \frac{4}{8} = \frac{3}{6} = \frac{125}{250} = \cdots$$

Usually, however, a rational number is written as m/n, where m and n have no common factors. That is, we will write the rational number in lowest terms.

Any real number that is not rational is called **irrational.** Examples of irrational numbers are $\pi = 3.141592654\ldots$ and $\sqrt{2} = 1.414213562\ldots$ A proof that $\sqrt{2}$ is irrational is given in Section 1.6.

There is an interesting graphical way to depict $\sqrt{2}$. In Figure 2 we draw a right triangle with each leg having length 1. Then, by the Pythagorean theorem

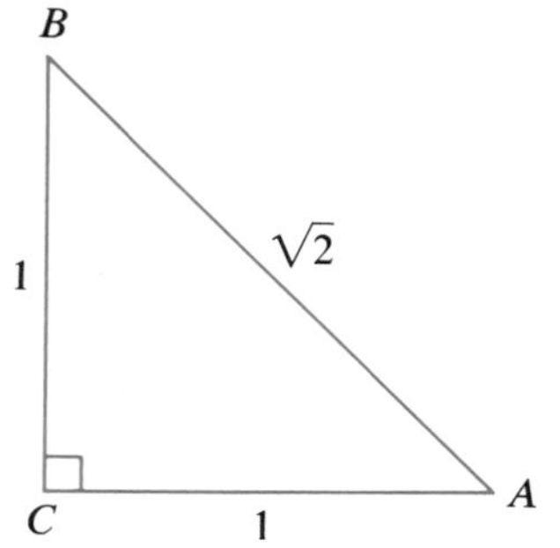

Figure 2 The right (isosceles) triangle with two sides of length 1. The hypotenuse has length $\sqrt{2}$.

$$\begin{aligned}
\overline{AB}^2 &= \text{square of the length of the hypotenuse} \\
&= \overline{AC}^2 + \overline{BC}^2 = 1 + 1 \\
\overline{AB}^2 &= 2 \\
\overline{AB} &= \sqrt{2}
\end{aligned}$$

That is, if a right triangle has two sides each of length 1, then the length of the hypotenuse is $\sqrt{2}$.

Repeating Decimals

Every rational number can be written as a **repeating decimal,** but no irrational number can be written in this way. For example, $\frac{1}{3} = 0.33333\ldots = 0.\overline{3}$, $\frac{3}{11} = 0.272727\ldots = 0.\overline{27}$, $\frac{5}{4} = 1.25000\ldots$ and $\frac{2}{7} = 0.285714285714\ldots = 0.\overline{285714}$ are examples of rational numbers written as repeating decimals. The overbars indicate that the numbers underneath repeat infinitely.

FOCUS ON
The Discovery of Irrational Numbers

According to some historians of mathematics, the discoverer of irrational numbers (then called **incommensurable ratios**) was Hippasus of Metapontum, a follower of Pythagoras, who lived in the fifth century B.C. At the time of his discovery, Hippasus and fellow Pythagoreans were at sea. His colleagues were very upset that someone could produce a number that contradicted the Pythagorean belief that all phenomena on earth could be reduced to natural numbers or their ratios. As a result, Hippasus was thrown overboard. This is the first known instance of irrational behavior.

EXAMPLE 2 *Writing a Repeating Decimal as a Fraction*

What rational number, written as a fraction, is represented by the repeating decimal $0.23232323\ldots = 0.\overline{23}$.

SOLUTION Let $r = 0.23232323\ldots$.
Then

$$100r = 23.232323\ldots$$

To multiply by 100, move the decimal point two places to the right

We subtract:

$$\begin{array}{rl} 100r = & 23.2323\ldots \\ -\quad r = & 0.2323\ldots \\ \hline 99r = & 23 \end{array}$$

Therefore $r = \frac{23}{99}$.

You should check this result on a calculator.

Note In the first step, we multiplied by $100 = 10^2$ because two digits were repeating.

On the number line, rational numbers are depicted by subdividing an interval. For example, $\frac{8}{3} = 2\frac{2}{3}$ is two thirds of the way from 2 to 3. Irrational numbers are placed between appropriate rational approximations. For example, since $\sqrt{2} \approx 1.141421$, $\sqrt{2}$ appears between 1.4142 and 1.4143. Some rational and irrational numbers are shown on the number line in Figure 3.

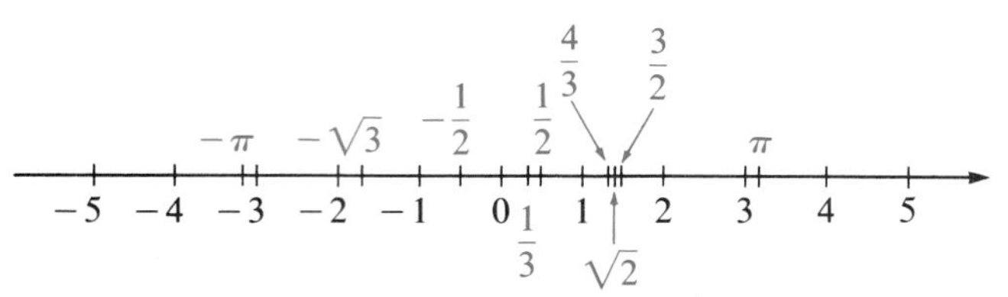

Figure 3 Some rational and irrational numbers depicted on the number line.

Sets of Real Numbers

We will often be interested in **sets** of numbers. A set of numbers is any well-defined† collection of numbers. For example, the integers and rational numbers are each sets of numbers. The collection of "large numbers" does not constitute a set since it is not well defined. There is no universal agreement about whether a given number is or is not in this collection.

The numbers in a set are called **members** or **elements** of that set. If x is an element of the set A, we write $x \in A$ and read this notation as "x is an element of A" or "x belongs to A."

The elements of a set can be written with braces, { }. For example, the set A of integers between $\frac{1}{2}$ and $\frac{11}{2}$ can be written as

$$A = \left\{x \in \mathbb{R}\text{: } x \text{ is an integer and } \frac{1}{2} < x < \frac{11}{2}\right\}.$$

This notation is read as "A is the set of real numbers x such that x is an integer and x is between $\frac{1}{2}$ and $\frac{11}{2}$." Alternatively, we can write the same set as

$$A = \{1, 2, 3, 4, 5\}$$

NOTE The symbol $\notin$ is read "is not an element of." For example, if A is the set described above, then $0 \notin A$.

Problems 1.2

Readiness Check

I. Which of the following represents a natural number?

a. 0 b. −7 c. π d. $\frac{6}{3}$

II. Which of the following represents an integer?

a. 0 b. −7 c. π d. 0.1111 . . .
e. $\frac{27}{9}$

III. Which of the following represents a rational number?

a. 23 b. $\frac{17}{41}$ c. $\frac{-362}{471}$ d. 0.254
e. 0.7367367367 . . . f. 23.4261457323232 . . .

Answers to Readiness Check

I. d II. a, b, and e III. All of them

† A set of numbers is **well defined** if we can determine with certainty whether or not a number belongs to the set.

In Problems 1–6 write each repeating decimal as the quotient of two integers in lowest terms.

1. $0.222222\ldots = 0.\overline{2}$
2. $0.323232\ldots = 0.\overline{32}$
3. $0.147147147\ldots = 0.\overline{147}$
4. $0.231723172317\ldots = 0.\overline{2317}$
5. $2.123123123\ldots = 2.\overline{123}$ [Hint: First write $0.\overline{123}$ as a fraction. Then add 2.]
6. $12.301301301\ldots = 12.\overline{301}$

† In Problems 7–14 use a calculator to write each fraction as a repeating decimal.

7. $\frac{5}{6}$ 8. $-\frac{5}{11}$ 9. $\frac{5}{12}$ 10. $\frac{7}{15}$
11. $\frac{143}{1111}$ * 12. $\frac{5}{14}$ 13. $\frac{57}{32}$ 14. $\frac{27}{26}$
15. Find two irrational numbers a and b such that $a + b$ is rational.
16. Is it true that the sum of two rational numbers is rational?
17. (a) Show that $\frac{1}{9} = 0.\overline{1}\overline{1}$, and $\frac{1}{11} = 0.\overline{09}$.
    (b) Show that $\frac{1}{27} = 0.\overline{037}$ and $\frac{1}{37} = 0.\overline{027}$.
    ** (c) Find a general pattern that includes the examples given in parts (a) and (b).

† This symbol is used throughout the book to indicate that a calculator is used or needed in the problem.

1.3 Properties of Real Numbers

Real numbers have many important properties. We list several in Table 1. Here, a, b, and c denote real numbers.

Table 1
Basic Properties of Real Numbers

Name of Property	Statement of Property	Illustration
Additive closure	$a + b$ is real	since 2 is real and 3 is real, $2 + 3$ is real
Multiplicative closure	ab is real	since 2 is real and 3 is real, $2 \cdot 3$ is real
Additive identity	$a + 0 = 0 + a = a$ 0 is called the **additive identity**	$2 + 0 = 0 + 2 = 2$
Multiplicative identity	$1 \cdot a = a \cdot 1 = a$ 1 is called the **multiplicative identity**	$2 \cdot 1 = 1 \cdot 2 = 2$
Associative law for addition	$a + (b + c) = (a + b) + c$	$2 + (3 + 7) = (2 + 3) + 7$
Associative law for multiplication	$a(bc) = (ab)c$	$2(3 \cdot 7) = (2 \cdot 3)7$
Additive commutativity	$a + b = b + a$	$2 + 3 = 3 + 2$
Multiplicative commutativity	$ab = ba$	$2 \cdot 3 = 3 \cdot 2$
Additive inverse	For every a there is a real number, denoted by $-a$, called the **additive inverse** of a such that $a + (-a) = (-a) + a = 0$. $-a$ is called the **negative** of a	$2 + (-2) = (-2) + 2 = 0$
Multiplicative inverse	If $a \neq 0$, there is a number denoted by $a^{-1} = \frac{1}{a}$, called the **multiplicative inverse** or **reciprocal** of a such that $a\left(\frac{1}{a}\right) = \left(\frac{1}{a}\right)a = 1$	$2\left(\frac{1}{2}\right) = \left(\frac{1}{2}\right)2 = 1$

Left distributive law	$a(b + c) = ab + ac$	$2(3 + 4) = 2 \cdot 3 + 2 \cdot 4$
Right distributive law	$(a + b)c = ac + bc$	$(2 + 3)4 = 2 \cdot 4 + 3 \cdot 4$
Additive substitution law	If $a = b$, then $a + c = b + c$	$2 \cdot 3 = 6$ so $2 \cdot 3 + 5 = 6 + 5$
Multiplicative substitution law	If $a = b$, then $ac = bc$	$2 \cdot 3 = 6$ so $(2 \cdot 3)5 = 6 \cdot 5$

We can use these laws to prove many other properties of real numbers. Here are two of them.

Reduction Laws

Additive Reduction Law If a, b, and c are real numbers and $a + c = b + c$, then $a = b$.

Multiplicative Reduction Law If $ac = bc$ and $c \neq 0$, then $a = b$.

Here is an outline of the proof of the additive reduction law. We start with (as given)

$$a + c = b + c$$

Then

$(a + c) + (-c) = (b + c) + (-c)$	Additive substitution law
$a + (c + (-c)) = b + (c + (-c))$	Associative law for addition
$a + 0 = b + 0$	Additive inverse
$a = b$	Additive identity

You are asked to prove the multiplicative reduction law in Problem 51.

We now turn to properties of negatives.

Properties of the Additive Inverse — Rules of Signs

Property	***Description***	***Illustration***
(a) $-a = (-1)a$	To obtain the negative or additive inverse of a number, multiply the number by -1.	$-2 = (-1)2$
(b) $-(-a) = a$	The negative of the negative of a number is the original number.	$-(-2) = 2$
(c) $-ab = (-a)b$ $-ab = a(-b)$	The negative of the product of two numbers is the product of one of the numbers and the negative of the other.	$2(-3) = (-2)3$ $= -(2 \cdot 3)$

(d) $(-a)(-b) = ab$	The product of the negatives of two numbers equals the product of the numbers.	$(-2)(-3) = 2 \cdot 3$
(e) $(-a) + (-b) = -(a + b)$	The sum of the negatives equals the negative of the sum.	$(-2) + (-3) = -(2 + 3)$

The proofs of these properties are left as exercises.

EXAMPLE 1 ***Multiplying Two Negative Numbers***

$$(-2)(-5) = 2 \cdot 5 = 10 \qquad (-a)(-b) = ab \quad \blacksquare$$

EXAMPLE 2 ***Multiplying Three Negative Numbers***

$$(-4)(-3)(-7) = [(-4)(-3)](-7) = 12(-7) \overset{a(-b) = -ab}{=} -(12 \cdot 7) = -84$$

We emphasize the following:

The product of two negative numbers is a positive number.
The product of three negative numbers is a negative number.

In general, the product of an even number of negative numbers is positive while the product of an odd number of negative numbers is negative.

We now define the difference and quotient of real numbers.
The **difference** of a and b is defined by

$$a - b = a + (-b)$$

If $b \neq 0$, the **quotient** $\frac{a}{b}$ is defined by

$$\frac{a}{b} = a \div b = a\left(\frac{1}{b}\right)$$

Properties of Quotients — Rules of Fractions

Let a, b, c, and d be real numbers with $b \neq 0$ and $d \neq 0$.

Property	***Illustration***
(a) If $a = b$, then $\frac{a}{d} = \frac{b}{d}$	$6 = 2 \cdot 3$ so $\frac{6}{3} = \frac{2 \cdot 3}{3}$
(b) $\frac{a}{b} = \frac{c}{d}$ if and only if $ad = bc$	$\frac{4}{6} = \frac{2}{3}$ since $4 \cdot 3 = 2 \cdot 6$

(c) $\dfrac{a}{b}+\dfrac{c}{b}=\dfrac{a+c}{b}$	$\dfrac{2}{7}+\dfrac{3}{7}=\dfrac{2+3}{7}=\dfrac{5}{7}$
(d) $\dfrac{ad}{bd}=\dfrac{a}{b}$ Division property	$\dfrac{6}{21}=\dfrac{2\cdot 3}{7\cdot 3}=\dfrac{2}{7}$ [Note that $2\cdot 21=6\cdot 7=42$]
(e) $\dfrac{a}{b}+\dfrac{c}{d}=\dfrac{ad+bc}{bd}$ and $\dfrac{a}{b}-\dfrac{c}{d}=\dfrac{ad-bc}{bd}$	$\dfrac{2}{3}+\dfrac{5}{7}=\dfrac{2\cdot 7+3\cdot 5}{3\cdot 7}$ $=\dfrac{14+15}{21}=\dfrac{29}{21}$
(f) $a\cdot\dfrac{c}{b}=\dfrac{a\cdot c}{b}=\dfrac{a}{b}\cdot c$	$3\cdot\dfrac{4}{5}=\dfrac{3\cdot 4}{5}=\dfrac{3}{5}\cdot 4$
(g) $\dfrac{a}{b}\cdot\dfrac{c}{d}=\dfrac{ac}{bd}$	$\dfrac{2}{3}\cdot\dfrac{5}{7}=\dfrac{2\cdot 5}{3\cdot 7}=\dfrac{10}{21}$
(h) $\dfrac{a/b}{c/d}=\dfrac{a}{b}\cdot\dfrac{d}{c}=\dfrac{ad}{bc}$	$\dfrac{3/5}{6/11}=\dfrac{3}{5}\cdot\dfrac{11}{6}=\dfrac{33}{30}$ From (d) $=\dfrac{3\cdot 11}{3\cdot 10}=\dfrac{11}{10}$

NOTE Property (a) follows from the multiplicative substitution law on page 8; simply set $c=\dfrac{1}{d}$. Then $ac=bc$, so $a\cdot\dfrac{1}{d}=b\cdot\dfrac{1}{d}$.

EXAMPLE 3 *Dividing by a Negative Number*

It follows from the division property that

$$\frac{1}{-3}=\frac{(-1)1}{(-1)(-3)}=\frac{-1}{3}=-\frac{1}{3}$$

In general,

$$\frac{1}{-a}=\frac{-1}{a}=-\frac{1}{a} \tag{1}$$

EXAMPLE 4 *Adding Fractions*

Compute $\dfrac{3}{4}+\dfrac{5}{6}$ and reduce the answer to lowest terms.

SOLUTION We do this in two ways.

First way: We use Property (e):

$$\frac{3}{4}+\frac{5}{6}=\frac{3\cdot 6+4\cdot 5}{4\cdot 6}=\frac{18+20}{24}=\frac{38}{24}=\frac{19\cdot 2}{12\cdot 2}\overset{\text{Property (d)}}{=}\frac{19}{12}$$

Second way: We observe that in Property (e) the denominator of the result is bd—the product of b and d. We seek the smallest number that is a multiple of b and d, in this case 4 and 6. The **least common multiple** (LCM) of 4 and 6 is 12. We then use Property (d) to write each fraction with 12 in the denominator and then use Property (c) to add them:

$$\frac{3}{4}+\frac{5}{6}=\frac{3\cdot 3}{4\cdot 3}+\frac{5\cdot 2}{6\cdot 2}=\frac{9}{12}+\frac{10}{12}=\frac{19}{12}$$

Multiply 4 by 3 to get 12   Multiply 6 by 2 to get 12

NOTE We can also write this answer as $1\frac{7}{12}$. Usually, however, improper fractions will be left as the quotient of two integers.

Fractions cause difficulty for many students. Here are two more examples showing how they are added, subtracted, multiplied, and divided.

EXAMPLE 5 *The Reciprocal of a Fraction*

Compute $\dfrac{1}{7/15}$.

SOLUTION

$$\frac{1}{7/15}=\frac{1/1}{7/15}\overset{\text{Property (h)}}{=}\frac{1}{1}\cdot\frac{15}{7}=\frac{15}{7}$$

In general,

The Reciprocal of a Fraction

$$\frac{1}{a/b}=\frac{b}{a} \qquad (2)$$

EXAMPLE 6 *Simplifying a Compound Fraction*

Simplify the **compound fraction** $\dfrac{\dfrac{2}{3}+\dfrac{3}{4}}{\dfrac{5}{6}-\dfrac{3}{8}}$.

SOLUTION We first add the fractions in the numerator and those in the denominator separately:

The LCM of 3 and 4 is 12:

$$\frac{2}{3}+\frac{3}{4}=\frac{2\cdot 4}{3\cdot 4}+\frac{3\cdot 3}{4\cdot 3}=\frac{8}{12}+\frac{9}{12}=\frac{17}{12}$$

The LCM of 6 and 8 is 24 (not 48):

$$\frac{5}{6}-\frac{3}{8}=\frac{5\cdot 4}{6\cdot 4}-\frac{3\cdot 3}{8\cdot 3}=\frac{20}{24}-\frac{9}{24}=\frac{11}{24}$$

Thus

$$\frac{\frac{2}{3}+\frac{3}{4}}{\frac{5}{6}-\frac{3}{8}}=\frac{17/12}{11/24}\underset{\text{Property (h)}}{=}\frac{17}{12}\cdot\frac{24}{11}=\frac{17\cdot 24}{12\cdot 11}=\frac{17\cdot 2\cdot 12}{11\cdot 12}=\frac{17\cdot 2}{11}=\frac{34}{11}$$

WARNING Two errors are frequently made by beginning algebra students (and, unfortunately, by more advanced students as well).

COMMON ERROR 1 It is true that

$$\frac{a}{b}+\frac{c}{b}=\frac{a+c}{b} \quad \text{Property (b)}$$

This is correct

But

$$\frac{a}{b}+\frac{a}{d} \text{ Does not equal } \frac{a}{b+d}$$

For example,

$$\frac{2}{3}+\frac{2}{5}=\frac{2\cdot 5+2\cdot 3}{3\cdot 5}=\frac{10+6}{15}=\frac{16}{15} \quad \text{Correct}$$

but

$$\frac{2}{3}+\frac{2}{5} \text{ Does not equal } \frac{2}{3+5}=\frac{2}{8}=\frac{1}{4}$$

COMMON ERROR 2 It is true that

$$\frac{ad+bd}{cd}=\frac{(a+b)d}{cd}=\frac{a+b}{c} \quad \text{Multiplicative reduction law}$$

But

$$\frac{a+bd}{cd} \text{ Does not equal } \frac{a+b}{c}$$

That is, you can divide the d's from the numerator and denominator only when *each term* in the numerator and denominator has the factor d. For example,

$$\frac{4 \cdot 5 + 7 \cdot 5}{3 \cdot 5} = \frac{4 + 7}{3} = \frac{11}{3} \quad \text{Correct}$$

but

$$\frac{39}{15} = \frac{4 + 7 \cdot 5}{3 \cdot 5} \quad \text{Does not equal} \quad \frac{4 + 7}{3} = \frac{11}{3}$$

The general rule is *you can cancel factors, not terms.* ■

Problems 1.3

Readiness Check

I. Which of following properties is illustrated by

$$2 + (3 + 7) = (2 + 3) + 7?$$

a. The associative law of addition
b. The commutative law of addition
c. The left distributive law
d. The additive closure law

II. Which of the following illustrates the closure law of multiplication if x is a real number?
a. $x + 4 = 4 + x$ b. If $3x = 6$, then $x = 2$
c. $-7x \in \mathbb{R}$ d. $1x = x$

III. Which of the following is the additive inverse of -5?
a. $-\frac{1}{5}$ b. $\frac{1}{5}$ c. 0   d. 5   e. -5

IV. Which of the following is justified by the multiplicative reduction law if $3x = 3(2)$?
a. $3x = 6$ b. $x = 2$ c. $3x + 2 = 3(2) + 2$
d. $x = 3$

V. When $\frac{a}{5}$ is added to $\frac{b}{7}$, the smallest possible denominator of the sum is ______. [Assume that a is not a multiple of 5 and b is not a multiple of 7.]
a. 5   b. 7   c. 35   d. 2   e. None of the above

In Problems 1–20 write the law or laws that justify the given statement. Each letter represents a real number.

1. $6 + y = y + 6$
2. $5x = x \cdot 5$
3. $5(uv) = (5u)v$
4. $5 + 0 = 5$
5. $z \cdot 1 = z$
6. $ab + (-ab) = 0$
7. $x + (2y + 3z) = (x + 2y) + 3z$
8. $z\left(\frac{1}{z}\right) = 1,\ z \neq 0$
9. $(xyz)\left(\frac{1}{xyz}\right) = 1,\ xyz \neq 0$
* 10. $\frac{xz - xw}{x} = z - w$
11. $3(u + w) = 3u + 3w$
12. $(3 + r)s = 3s + rs$
13. If $x = y$, then $x + 2 = y + 2$.
14. If $x = y$, then $-x = -y$.
15. If $x = y$, then $5x = 5y$.
16. If $\frac{x}{2} = \frac{y}{2}$, then $x = y$.
17. If $x - 3 = y - 3$, then $x = y$.
18. If $x + 4z = y + 4z$, then $x = y$.
19. If $a = 7$, then $a + 3 = 10$.
20. If $xy = z$ and $x = 4$, then $4y = z$.

In Problems 21–50 perform the indicated operation and express the answer in lowest terms.

21. $\frac{2}{3} + \frac{5}{3}$
22. $\frac{1}{2} + \frac{5}{2}$
23. $\frac{1}{4} + \frac{1}{3}$
24. $\frac{2}{3} + \frac{3}{5}$
25. $\frac{5}{7} - \frac{3}{7}$
26. $\frac{5}{7} - \frac{8}{7}$
27. $\frac{4}{5} - \frac{3}{4}$
28. $\frac{7}{8} - \frac{5}{6}$
29. $3 \cdot \frac{7}{4}$

Answers to Readiness Check

I. a   II. c   III. d   IV. b   V. c

30. $5 \cdot \frac{8}{5}$
31. $\frac{7}{2} \cdot 4$
32. $\frac{9}{5} \cdot 6$
33. $\frac{3}{4} \cdot \frac{4}{5}$
34. $\frac{4}{9} \cdot \frac{9}{4}$
35. $\frac{7}{3} \cdot \frac{6}{7}$
36. $\frac{11}{4} \cdot \frac{13}{11}$
37. $\frac{3}{4} \cdot \frac{5}{7}$
38. $\frac{3}{5} \cdot \frac{10}{21}$
39. $\frac{3}{4} \cdot \frac{2}{5}$
40. $\frac{2}{3}\left(\frac{3}{7} - \frac{2}{3}\right)$
41. $\frac{3/5}{4/7}$
42. $\frac{2/3}{1/5}$
43. $\frac{1}{8/15}$
44. $\frac{-1}{3/4}$
45. $\frac{1}{2} + \frac{1}{3} + \frac{1}{4}$
46. $\frac{2}{3} - \frac{1}{2} + \frac{3}{7}$
47. $\frac{3}{4} + \frac{2}{5} - \frac{5}{6}$
48. $\frac{6}{5} - \frac{2}{7} - \frac{3}{8}$
49. $\dfrac{\frac{2}{3} + \frac{1}{4}}{\frac{3}{5} - \frac{1}{6}}$
50. $\dfrac{\frac{3}{7} + \frac{2}{9}}{\frac{5}{21} - \frac{2}{3}}$

Prove each of the properties in Problems 51–62 by using the laws described in this section.

51. If $ac = bc$ and $c \neq 0$, then $a = b$. [Hint: Mimic the proof of the additive reduction law, but substitute the multiplicative inverse for the additive inverse.]
52. $a \cdot 0 = 0$. [Hint: $a \cdot 0 = a \cdot (0 + 0) = a \cdot 0 + a \cdot 0$. Now use the additive inverse and the substitution laws.]
53. The additive inverse is unique. [Hint: Assume that $a + b = 0 = a + c$. Show that $b = c$.]
54. $(-a) = (-1)a$. [Hint: $a + (-1)a = 1 \cdot a + (-1)a = [1 + (-1)]a = 0 \cdot a$. Then use the results of Problems 52 and 53.]
55. $-(-a) = a$. [Hint: $(-1)(-1) = 1$]
56. $a(-b) = -ab$
57. $(-a)(-b) = ab$
58. $(-a) + (-b) = -(a + b)$
59. $a(b - c) = ab - ac$
60. The multiplicative inverse of a nonzero number is unique.
61. If $ab = 0$, then $a = 0$, $b = 0$, or both. [Hint: If $a \neq 0$ and $b \neq 0$, then each has a multiplicative inverse, and you can show that $ab = 0$ implies that $1 = 0$.]
62. $(a + b)(c + d) = ac + ad + bc + bd$

In Problems 63–68, prove that each statement is false by providing a counterexample.

63. $a - b = b - a$
64. $\frac{a}{b} = \frac{b}{a}$
65. $\frac{a + b}{c + d} = \frac{a}{c} + \frac{b}{d}$
66. $(a + b)(c + d) = ac + bd$
67. The negative numbers are closed under subtraction; that is, if a and b are negative, then $a - b$ is negative.
68. The negative numbers are closed under multiplication.

1.4 Order, Inequalities, and Absolute Value

Order and Inequalities

Consider the numbers a and b in the graph of Figure 1. The number line can

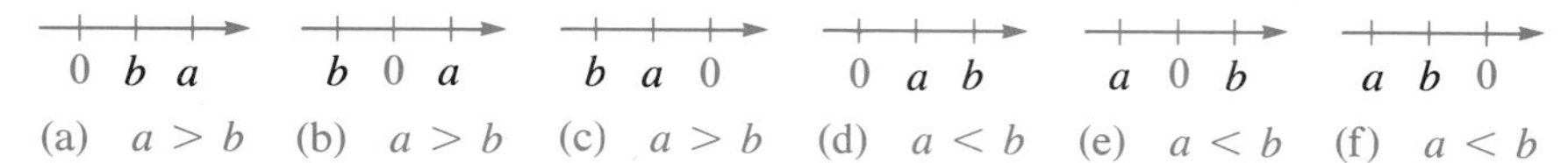

Figure 1 The inequalities $a > b$ and $a < b$ can be illustrated on a number line.

be used to illustrate our sense of order. We put the number a to the right of the number b if a is greater than b. We then write this inequality as $a > b$. Similarly, if $b > a$, then a is to the left of b, and we write the inequality as $a < b$. We use the notation $a \leq b$ to indicate that a is less than or equal to b; that is, $a < b$ or $a = b$. Finally, we write $a \geq b$ to indicate that a is greater than or equal to b.

EXAMPLE 1 ***Illustrating Inequalities on a Number Line***

The following inequalities are illustrated on the number line in Figure 2.

(a) $2 < 3$ (b) $\frac{1}{3} < \frac{1}{2}$ (c) $\frac{4}{3} < \sqrt{2} < \frac{3}{2}$

(d) $-3 < 1$ (e) $-3 < -2 < 0$ (f) $-4 < -\pi < -3$

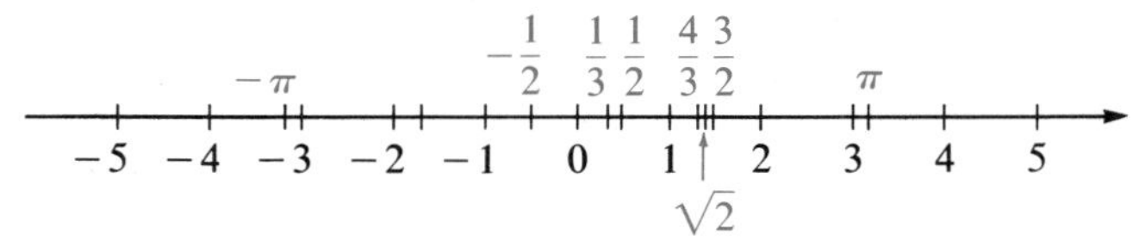

Figure 2 Some real numbers on a number line.

The notation $a < c < b$ indicates that c lies between a and b on the number line, as in parts (c), (e), and (f).

Summary of Notation

Symbolic Representation	*Number Line Location*	*Common Terminology*
$a > 0$	a is to the right of zero on the number line.	a is positive.
$a < 0$	a is to the left of zero on the number line.	a is negative.
$a > b$	a is to the right of b on the number line.	a is greater than b and $a - b$ is positive; that is, $a - b > 0$.
$a < b$	a is to the left of b on the number line.	a is less than b and $a - b$ is negative; that is, $a - b < 0$.

Inequalities have some useful properties

Properties of Inequalities

Property	*Description*	*Illustration*
(a) If $a < b$, then $a + c < b + c$	Adding a number to both sides of an inequality preserves the inequality.	Since $2 < 3$, $2 + 5 < 3 + 5$ ($7 < 8$); $2 - 6 < 3 - 6$ ($-4 < -3$)

(b) If $a < b$ and $b < c$, then $a < c$	This property is called the **transitive law.**	Since $2 < 3$ and $3 < 5$, we have $2 < 5$
(c) If $a < b$ and $c > 0$, then $ac < bc$	Multiplying an inequality by a positive number preserves the inequality.	Since $2 < 3$, we have $2 \cdot 4 < 3 \cdot 4$ $8 < 12$
(d) If $a < b$ and $c < 0$, then $ac > bc$	Multiplying an inequality by a negative number *reverses* the sense of the inequality.	Since $2 < 3$, we have $2(-4) > 3(-4)$ $-8 > -12$
(e) If a and b are both positive or both negative and $a < b$, then $\frac{1}{a} > \frac{1}{b}$	Taking reciprocals reverses an inequality of numbers with the same sign.	$2 < 3$ so $\frac{1}{2} > \frac{1}{3}$ $-3 < -2$ so $-\frac{1}{3} > -\frac{1}{2}$

EXAMPLE 2 ***Illustration of Four Properties of Inequalities***

Suppose that $0 < x < y$.

(a) $x + z < y + z$ for every real number z
(b) $5x < 5y$
(c) $-3x > -3y$
(d) $\frac{1}{x} > \frac{1}{y}$

Absolute Value

The **absolute value** of a number a is the distance on the number line between that number and zero and is written $|a|$. See Figure 3. Thus 2 is 2 units from zero, so $|2| = 2$. The number -3 is 3 units from zero, so $|-3| = 3$.

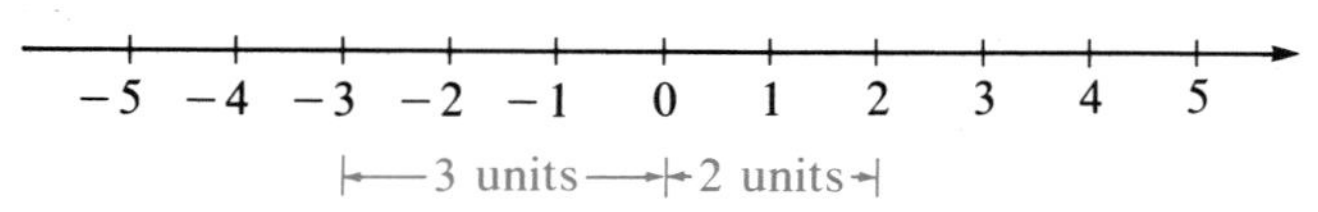

Figure 3 2 is 2 units from 0, and -3 is 3 units from 0, so $|2| = 2$ and $|-3| = 3$.

We may define

Definition of Absolute Value

If a is a real number, then the absolute value of a is given by

$$|a| = a \qquad \text{if } a \geq 0 \qquad (1)$$

$$|a| = -a \qquad \text{if } a < 0 \qquad (2)$$

Put another way, let n be a *nonnegative* number. Then the absolute value of n is n, and the absolute value of $-n$ is also n.

EXAMPLE 3 *Illustrations of Absolute Value*

$$|5| = 5 \quad \text{and} \quad |-5| = -(-5) = 5 \qquad \text{From (2)}$$

We stress that

The absolute value of a number is a nonnegative number.

Properties of Absolute Value

$$|a| \geq 0 \text{ for every real number and } |a| = 0 \text{ if and only if } a = 0 \qquad (3)$$

$$|-a| = |a| \qquad (4)$$

$$|ab| = |a||b| \qquad (5)$$

$$|a + b| \leq |a| + |b| \quad \text{Triangle inequality} \qquad (6)$$

EXAMPLE 4 *Illustration of the Properties of Absolute Value*

We illustrate the properties of absolute value.

(a) $|-6| = |6| = 6$ Illustration of (4)
(b) $|(-2)(3)| = |-6| = 6 = 2 \cdot 3 = |-2||3|$ Illustration of (5)
(c) $|2 + 3| = 5 = 2 + 3 = |2| + |3|$ Illustration of triangle inequality (6) when $a > 0$ and $b > 0$
(d) $|-2 - 3| = |-5| = 5 = 2 + 3 = |-2| + |-3|$ Illustration of triangle inequality (6) when $a < 0$ and $b < 0$
(e) $|2 + (-3)| = |2 - 3| = |-1| = 1 < 2 + 3 = |2| + |-3|$ Illustration of triangle inequality when $a > 0$ and $b < 0$
(f) $|-2 + 3| = |1| = 1 < 2 + 3 = |-2| + |3|$ Illustration of triangle inequality when $a < 0$ and $b > 0$

Parts (c)–(f) illustrate a method of proof of the triangle inequality. (See Problems 45–49.)

Distance Between Two Numbers

The **distance** on the number line between a and b is given by $|a - b|$. This is illustrated in Figure 4. Note that by property (4) of absolute value $|b - a| = |-(a - b)| = |a - b|$.

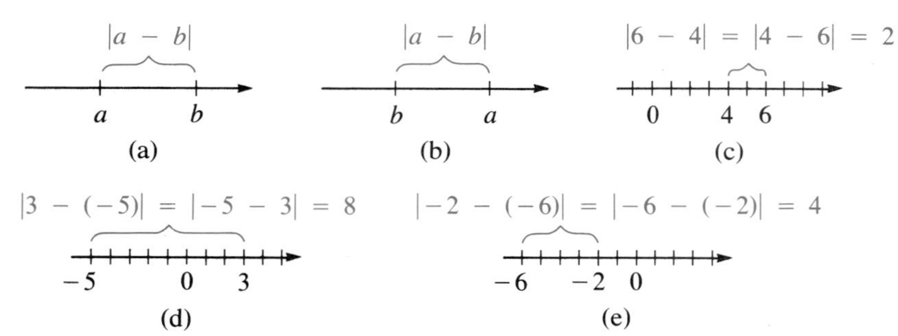

Figure 4 Illustration on a number line of the difference between two numbers.

EXAMPLE 5 ***Finding the Distances Between Four Pairs of Numbers***

Find the distance on the number line between the following pairs of numbers:
(a) 1, 4 (b) −2, 5 (c) 4, −8 (d) −12, −5

SOLUTION

(a) $|1 - 4| = |-3| = 3$
(b) $|-2 - 5| = |-7| = 7$
(c) $|4 - (-8)| = |12| = 12$
(d) $|-12 - (-5)| = |-12 + 5| = |-7| = 7$

These distances are illustrated in Figure 5.

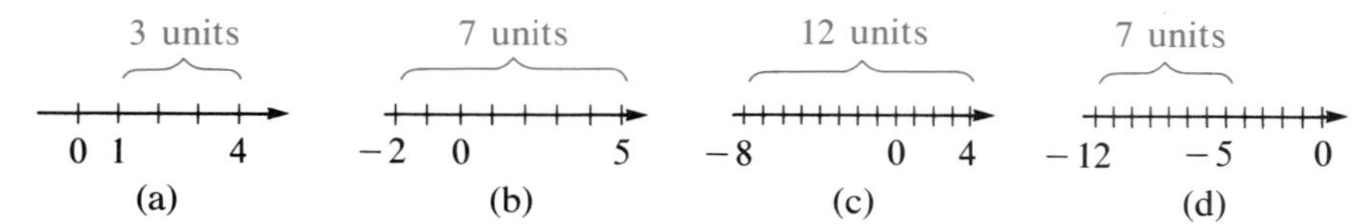

Figure 5 The distance between four pairs of numbers.

One other property of absolute value is sometimes useful. If $a > 0$, then $|a| = a$ and $|a|^2 = a^2$. If $a < 0$, then $|a| = -a$ and $|a|^2 = (-a)(-a) = a^2$. Thus, in either case,

$$|a|^2 = a^2 \tag{7}$$

Problems 1.4

Readiness Check

I. Which of the following is true about the two numbers a and b on the graph below?

a. b is a negative number and a is a positive number.
b. $a > b$ c. $b > a$
d. No relationship can be determined until values of a and b are known.

II. Suppose that $y < 0$. Then it must be true that ______.
a. $\frac{1}{y} > 0$ b. $\frac{1}{y} < 0$ c. $-y < 0$
d. $-y > 0$ e. $y + 2 < 0$ f. $y - 2 < 0$
g. $3 - y > 0$ h. $3 - y < 0$

III. Suppose that $x > 2$. Then ______.
a. $3x > 6$ b. $-3x > -6$ c. $3x < -6$
d. $-3x < -6$ e. $\frac{1}{x} > \frac{1}{2}$ f. $\frac{1}{x} < \frac{1}{2}$

IV. Which of the following is the distance between a and b on the number line below?

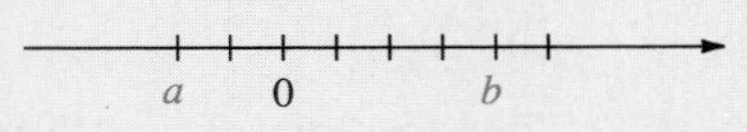

a. $|-2 + 4|$ b. $|4 - (-2)|$ c. $|4 - 2|$
d. $|2 - 4|$

V. Which of the following is equal to $|-3| - |-5|$?
a. -2 b. 5 c. 2 d. -8

In Problems 1–10 determine whether $<$ or $>$ belongs in the space provided.

1. -3 ______ 2
2. 4 ______ -6
3. -3 ______ -8
4. $\frac{\pi}{3}$ ______ 1
5. $\frac{2}{3}$ ______ 0.6
6. $\frac{1}{19}$ ______ $\frac{1}{20}$
7. $-\frac{1}{3}$ ______ $-\frac{1}{2}$
8. $\frac{1}{5}$ ______ $-\frac{1}{2}$
9. $-\frac{1}{3}$ ______ $\frac{1}{2}$
10. 3π ______ 9

In Problems 11–16 compute each value.

11. $|7| - |3|$
12. $|3| - |-2|$
13. $|4| - |-9|$
14. $|10| - |12|$
15. $|\sqrt{2} - 4|$
16. $|\pi - 7|$

In Problems 17–22 two numbers are depicted on a number line. For each graph, which of the following conclusions can be drawn?

a. x is negative. b. y is negative. c. x is positive.
d. y is positive. e. $x < y$ f. $y < x$
g. No relationship can be determined until values of x and y are given.

17.

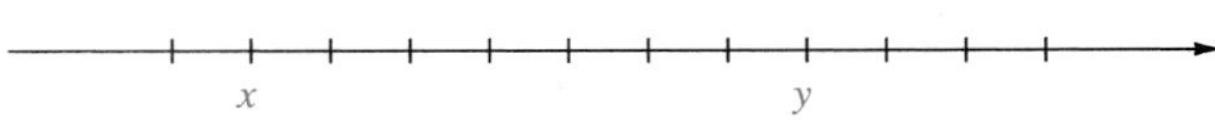

18.

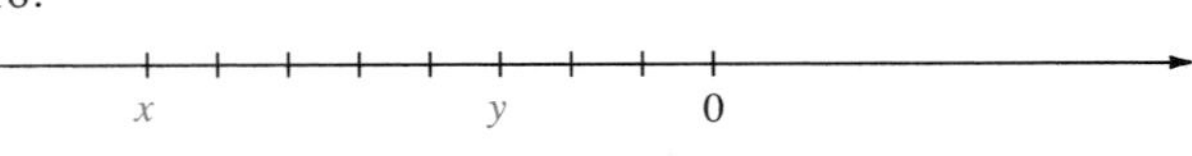

19.

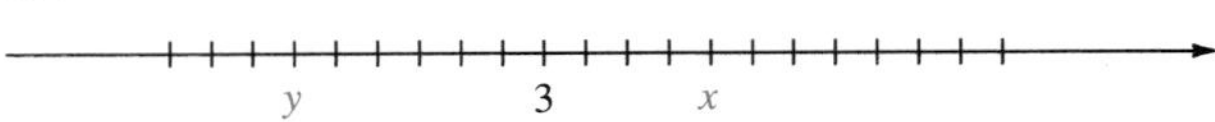

20.

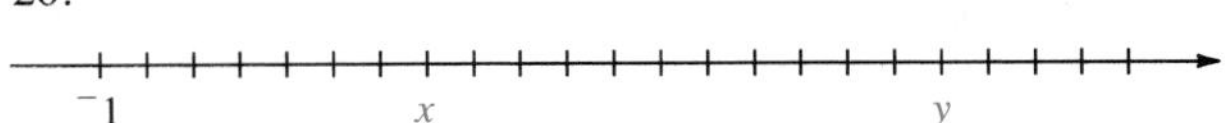

21.

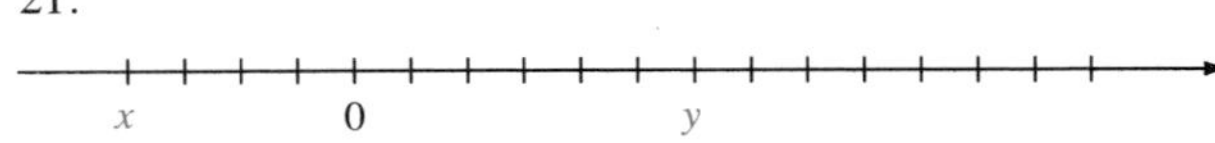

22.

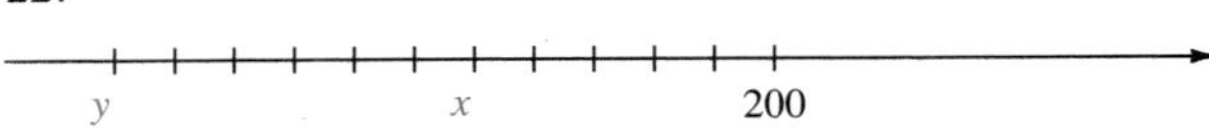

In Problems 23–32 state why the given conclusion is valid.

23. If $a > b$, then $2a > a + b$.
24. If $a < 2b$, then $b > a - b$.
25. If $w > 3z$, then $2w > 6z$.

Answers to Readiness Check

I. b II. b, d, f, g III. a, d, f IV. b V. a

26. If $w + 5 < z - 5$, then $w + 10 < z$.
27. If $y < 3x$, then $-9x < -3y$.
28. If $w - 2 < z - 5$, then $10 - 2z < 4 - 2w$.
29. If $a - b < c - d$, then $b - a > d - c$.
30. If $0 < x < 2y$, then $\frac{1}{x} > \frac{1}{2y}$.
31. If $0 < \frac{3}{x} < \frac{y}{4}$, then $\frac{4}{y} < \frac{x}{3}$.
32. If $0 < x < 1$, then $\frac{1}{x} > 1$.

In Problems 33–42 find the distance on the number line between the two numbers.

33. 2, 7
34. −1, 9
35. 2, −7
36. −2, −7
37. −5, 0
38. −3, 0
39. 1.6, 3.8
40. −1.6, 3.8
41. 1.6, −3.8
42. −1.6, −3.8
43. Show that $|-a| = |a|$.
44. Show that $|ab| = |a||b|$. [Hint: Consider the separate cases $a > 0$ and $b > 0$, $a < 0$ and $b > 0$, $a < 0$ and $b < 0$, and $a > 0$ and $b < 0$. What happens if $a = 0$ or $b = 0$?]
45. If $a > 0$ and $b > 0$, show that $|a + b| = |a| + |b|$.
46. If $a < 0$ and $b < 0$, show that $|a + b| = |a| + |b|$.
47. If $a < 0$ and $b > 0$, show that $|a + b| < |a| + |b|$.
48. If $a > 0$ and $b < 0$, show that $|a + b| < |a| + |b|$.
49. If $a = 0$ or $b = 0$, show that $|a + b| = |a| + |b|$.

1.5 Integral Exponents

In this section we discuss the basic algebraic notion of taking powers. We begin with a definition.

Definition of a^n Where n Is a Positive Integer

Let a be a real number, and let n be a positive integer ($n > 0$). Then

$$a^n = \underbrace{a \cdot a \cdot a \cdot \cdots \cdot a}_{n \text{ factors}} \tag{1}$$

That is, a^n is the product of n factors, each of which is equal to a. In this setting, the number n is called an **exponent,** and the number a is called the **base** of the exponent.

EXAMPLE 1 ***Evaluating a^n Where n Is a Positive Integer***

(a) $6^2 = 6 \cdot 6 = 36$

(b) $\left(\frac{1}{2}\right)^3 = \frac{1}{2} \cdot \frac{1}{2} \cdot \frac{1}{2} = \frac{1}{8}$

(c) $(-2)^5 = (-2)(-2)(-2)(-2)(-2) = -32$

(d) $5^1 = 5$, since there is now only one factor in the product

We now define a^n where n is a negative integer.

Definition of a^n Where n Is a Negative Integer

If n is a positive integer and $a \neq 0$, then

$$a^{-n} = \frac{1}{a^n} \tag{2}$$

EXAMPLE 2 ***Evaluating a^{-n} Where n Is a Positive Integer***

(a) $5^{-2} = \dfrac{1}{5^2} = \dfrac{1}{25}$

(b) $(-2)^{-3} = \dfrac{1}{(-2)^3} = \dfrac{1}{-8} = -\dfrac{1}{8}$

(c) $\left(\dfrac{1}{2}\right)^{-3} = \dfrac{1}{\left(\dfrac{1}{2}\right)^3} = \dfrac{1}{\left(\dfrac{1}{8}\right)} = 8$

(d) $\left(\dfrac{5}{7}\right)^{-1} = \dfrac{1}{\left(\dfrac{5}{7}\right)} = \dfrac{7}{5}$

Properties of Integral Exponents

Suppose that $a \neq 0$ and m and n are integers (positive or negative).

Property	***Illustration***
(a) $a^1 = a$	$5^1 = 5$
(b) $a^{-n} = \dfrac{1}{a^n}$	$2^{-3} = \dfrac{1}{2^3} = \dfrac{1}{8}$
(c) $a^m a^n = a^{m+n}$	$2^3 2^2 = 2^{3+2} = 2^5 = 32$
(d) $\dfrac{a^m}{a^n} = a^{m-n} = \dfrac{1}{a^{n-m}}$	$\dfrac{2^5}{2^3} = 2^{5-3} = 2^2 = 4$
(e) $a^0 = a^{n-n} = \dfrac{a^n}{a^n} = 1$	$3^0 = 3^{4-4} = \dfrac{3^4}{3^4} = \dfrac{81}{81} = 1$
(f) $(ab)^n = a^n b^n$	$(2 \cdot 4)^3 = 2^3 \cdot 4^3 = 8 \cdot 64 = 512$
(g) $\left(\dfrac{a}{b}\right)^n = \dfrac{a^n}{b^n}$	$\left(\dfrac{2}{3}\right)^3 = \dfrac{2^3}{3^3} = \dfrac{8}{27}$
(h) $(a^m)^n = a^{mn}$	$(2^3)^2 = 2^{3 \cdot 2} = 2^6 = 64$
(i) $a^{-1} = \dfrac{1}{a}$	$2^{-1} = \dfrac{1}{2}$
(j) $\left(\dfrac{a}{b}\right)^{-1} = \dfrac{1}{\left(\dfrac{a}{b}\right)} = \dfrac{b}{a}$	$\left(\dfrac{2}{3}\right)^{-1} = \dfrac{3}{2}$

NOTE The expression 0^0 is *not defined*. That is, no value is assigned to 0^0.

EXAMPLE 3 ***Using the Fact That $a^m a^n = a^{m+n}$***

Compute $4^2 4^3$.

SOLUTION $4^2 4^3 = 4^{2+3} = 4^5$. We can verify this by noting that $4^2 = 16$, $4^3 = 64$, and $4^2 4^3 = 16 \cdot 64 = 1024 = 4^5$.

EXAMPLE 4 ***Using the Fact That $a^m/a^n = a^{m-n}$***

Compute $\dfrac{3^7}{3^4}$.

SOLUTION $\dfrac{3^7}{3^4} = 3^{7-4} = 3^3 = 27$. Again, this can be verified by computing $3^7 = 2187$, $3^4 = 81$, and $\dfrac{2187}{81} = 27 = 3^3$. In this example it is much easier to simplify the exponents before doing any computations. ■

EXAMPLE 5 ***Using the Fact That $a^n/b^n = (a/b)^n$***

Compute $\dfrac{50^9}{25^9}$.

SOLUTION $\dfrac{50^9}{25^9} = \left(\dfrac{50}{25}\right)^9 = 2^9 = 512$. Here direct computations are horrendous. For example, $50^9 = 1{,}953{,}125{,}000{,}000{,}000$. ■

EXAMPLE 6 ***Using the Fact That $(a^m)^n = a^{mn}$***

Compute $(2^3)^4$.

SOLUTION $(2^3)^4 = 2^{3 \cdot 4} = 2^{12} = 4096$. ■

EXAMPLE 7 ***Using the Facts That $a^m a^n = a^{m+n}$ and $a^m/a^n = a^{m-n}$***

Simplify $\dfrac{5^{-2} \cdot 5^4}{5^8 \cdot 5^{-5}}$.

SOLUTION $\dfrac{5^{-2} \cdot 5^4}{5^8 \cdot 5^{-5}} = \dfrac{5^{-2+4}}{5^{8-5}} = \dfrac{5^2}{5^3} = 5^{2-3} = 5^{-1} = \dfrac{1}{5}$. ■

EXAMPLE 8 ***Using the Fact That $a^m a^n = a^{m+n}$***

Simplify $(3x^2y^3)(2x^4y)$.

$y = y^1$
↓

SOLUTION $(3x^2y^3)(2x^4y) = 3 \cdot 2x^{2+4}y^{3+1} = 6x^6y^4$. ■

EXAMPLE 9 ***Using the Fact That $(a^m)^n = a^{mn}$***

Simplify $(2u^2v^3)^4$.

SOLUTION $(2u^2v^3)^4 = 2^4(u^2)^4(v^3)^4 = 16u^8v^{12}$. ■

EXAMPLE 10 ***Using the Facts That $(a^m)^n = a^{mn}$ and $a^{-n} = 1/a^n$***

Simplify $(3a^2b^{-3})^2$.

SOLUTION $(3a^2b^{-3})^2 = 3^2(a^2)^2(b^{-3})^2 = 9a^4b^{-6} = \dfrac{9a^4}{b^6}$ ■

EXAMPLE 11 ***Showing That $(a/b)^{-1} = b/a$***

Simplify $\left(\dfrac{a}{b}\right)^{-1}$.

SOLUTION $$\left(\frac{a}{b}\right)^{-1} = \frac{1}{\left(\frac{a}{b}\right)} \overset{\text{Multiply top and bottom by } b}{\downarrow}{=} \frac{b}{b\left(\frac{a}{b}\right)} = \frac{b}{a}$$

We repeat a useful fact first given on page 21.

If $a \neq 0$ and $b \neq 0$, then $\dfrac{1}{\left(\frac{a}{b}\right)} = \dfrac{b}{a}$ and, in particular, $\dfrac{1}{\left(\frac{1}{a}\right)} = a$.

EXAMPLE 12 ***Illustrating That $(a/b)^{-1} = b/a$***

(a) $\left(\dfrac{1}{3}\right)^{-1} = \dfrac{1}{\left(\frac{1}{3}\right)} = 3$ (b) $\left(\dfrac{4}{7}\right)^{-1} = \dfrac{1}{\left(\frac{4}{7}\right)} = \dfrac{7}{4}$

(c) $\dfrac{\left(\frac{x}{y}\right)}{\left(\frac{y}{x}\right)} = \dfrac{x}{y} \cdot \dfrac{x}{y} = \dfrac{x^2}{y^2}$ ■

EXAMPLE 13 ***Using the Fact That $(a/b)^{-1} = b/a$***

Simplify $(5x^2y^{-3})^{-1}$.

SOLUTION $(5x^2y^{-3})^{-1} = \left(\dfrac{5x^2}{y^3}\right)^{-1} = \dfrac{y^3}{5x^2}$

or alternatively,

$$(5x^2y^{-3})^{-1} = 5^{-1}(x^2)^{-1}(y^{-3})^{-1} = 5^{-1}x^{-2}y^{(-3)(-1)} = \frac{1}{5} \cdot \frac{1}{x^2} \cdot y^3 = \frac{y^3}{5x^2}$$

EXAMPLE 14 ***Simplifying an Expression Involving Integer Exponents***

Simplify $\dfrac{(4a^2b^3)^{-2}(2a^3b)^3}{5\left(\dfrac{b}{a}\right)}$.

SOLUTION We do this in steps.

Step 1: $(4a^2b^3)^{-2} = \dfrac{1}{(4a^2b^3)^2} = \dfrac{1}{4^2(a^2)^2(b^3)^2} = \dfrac{1}{16a^4b^6}$

Step 2: $(2a^3b)^3 = 2^3(a^3)^3b^3 = 8a^9b^3$

Step 3: $\dfrac{1}{5\dfrac{b}{a}} = \dfrac{1}{5} \cdot \dfrac{1}{\dfrac{b}{a}} = \dfrac{1}{5} \cdot \dfrac{a}{b} = \dfrac{a}{5b}$

Thus

Step 4: $\dfrac{(4a^2b^3)^{-2}(2a^3b)^3}{5\left(\dfrac{b}{a}\right)} = \dfrac{8a^9b^3}{16a^4b^6} \cdot \dfrac{a}{5b} = \dfrac{8a^{10}b^3}{16 \cdot 5a^4b^7}$

$$= \frac{8}{80}\,\frac{a^{10}}{a^4}\,\frac{b^3}{b^7} = \frac{1}{10}a^6 \cdot \frac{1}{b^4} = \frac{a^6}{10b^4}$$

WARNING Here's another common algebraic mistake:

$$\text{Correct} \quad (a + b)^{-1} = \frac{1}{a + b}$$

but

$$(a + b)^{-1} \text{ is not equal to } a^{-1} + b^{-1}$$

For example,

$$(2 + 3)^{-1} = 5^{-1} = \frac{1}{5}$$

but

$$\frac{1}{5} = (2 + 3)^{-1} \text{ is not equal to } 2^{-1} + 3^{-1} = \frac{1}{2} + \frac{1}{3} = \frac{5}{6} \quad ■$$

Scientific Notation

In many applications, especially those involving very large or very small numbers, it is convenient to use scientific notation.

In **scientific notation** numbers are expressed in terms of powers of 10:

Scientific Notation

$$x = a \times 10^n$$

where $1 \le |a| < 10$ and n is an integer.

EXAMPLE 15 ***Illustrating Scientific Notation***

(a) $253 = 2.53 \times 100 = 2.53 \times 10^2$

(b) $32{,}584 = 3.2584 \times 10{,}000 = 3.2584 \times 10^4$

(c) $0.23 = \dfrac{2.3}{10} = 2.3 \times 10^{-1}$

(d) $0.000005 = \dfrac{5}{1{,}000{,}000} = 5 \times 10^{-6}$

(e) The mass of the earth is $5{,}983{,}000{,}000{,}000{,}000{,}000{,}000{,}000 = 5.983 \times 10^{24}$ kilograms. ■

EXAMPLE 16 ***Writing Two Expressions in Scientific Notation***

Evaluate (a) $(3 \times 10^4)^2$ (b) $(-2 \times 10^{-3})^5$

SOLUTION

(a) $(3 \times 10^4)^2 = 3^2 \times (10^4)^2 = 9 \times 10^8$

(b) $(-2 \times 10^{-3})^5 = (-2)^5 \times (10^{-3})^5 = -32 \times 10^{-15}$
$= (-3.2 \times 10^1) \times 10^{-15} = -3.2 \times 10^{-14}$ ■

EXAMPLE 17 ***Using a Calculator to Simplify an Expression Written in Scientific Notation***

Compute the following:

$$\frac{6.252 \times 10^{-4}}{9.8106 \times 10^{-9}}$$

Round the number multiplying a power of 10 to 4 decimal places.

SOLUTION

$$\frac{6.252 \times 10^{-4}}{9.8106 \times 10^{-9}} = \frac{6.252}{9.8106} \times \frac{10^{-4}}{10^{-9}} \approx 0.63727 \times \frac{10^{-4}}{10^{-9}}$$

(From a calculator: $\frac{6.252}{9.8106} \approx 0.63727$; Fact (d) on p. 21 applied to $\frac{10^{-4}}{10^{-9}}$)

$$= 0.63727 \times 10^{-4-(-9)}$$

$$= (6.3727 \times 10^{-1}) \times 10^5 = 6.3727 \times 10^4$$

CALCULATOR NOTE Your calculator should have a power function button, usually appearing as $\boxed{y^x}$ This is used to obtain powers.

EXAMPLE 18 *Obtaining Powers on a Calculator*

Obtain the following, using a calculator.
(a) $(2.3525)^4$ (b) $(135.23)^{-3}$ (c) $(0.2305)^5$

SOLUTION The following answers were obtained with a calculator displaying 10 significant figures. Here are the steps:

(a) 2.3525 [yˣ] 4 [=] 30.62799224 $(= 3.062799224 \times 10^1)$

Changes 3 to −3

(b) 135.23 [yˣ] 3 [+/−] [=] 0.000000404 $(= 4.04 \times 10^{-7})$
(some calculators will display the more accurate answer $4.04371795 \times 10^{-7}$)

(c) 0.2305 [yˣ] 5 [=] 0.00065066 $(= 6.5066 \times 10^{-4})$
(more precisely, $6.506608087 \times 10^{-4}$)

CALCULATOR NOTE Your calculator should have a key labeled [EE] or [EXP] that can be used to carry out computations in scientific notation.

EXAMPLE 19 *Computing Expressions Involving Scientific Notation on a Calculator*

Use a calculator to compute

$$(1.27 \times 10^4) \times (2.37 \times 10^3) \div (4.93 \times 10^{-2})$$

SOLUTION We enter the following keystrokes:

1.27 [EE] 4 × 2.37 [EE] 3 ÷ 4.93 [EE] 2 [+/−] [=]

The answer is displayed in one of the following ways (depending on the calculator used):

610527383.4
6.105273834^{08} (Which denotes 6.105273834×10^8)
6.1053 08 (On a calculator that carries only 4 decimal places in this mode)
6.105273834 E08

FOCUS ON The Magnitude of Exponents

Quantities that are initially bigger than 1 increase very rapidly as the exponent increases, and those that are between 0 and 1 decrease very rapidly. To illustrate this point, we retell an old fable. A brave soldier in the service of a king won many battles. The king offered him his choice of reward. The soldier, being mathematically inclined, asked for what seemed to the king to be a small gift. He requested that one grain of gold be given him on the first day of the month, two grains the second day, four the third day, eight the fourth day, and so on until the 30-day month was over. That is, he asked that the number of grains be doubled each day.

The king granted this wish with pleasure, thinking the soldier a fool—but not for long. On the tenth day the man received $2^9 = 512$ grains of gold and on the 20th, $2^{19} = 524{,}288$ grains. The soldier, alas, never lived until the 30th day, since before then he would have depleted the king's (and the world's) supply of gold. (On the 30th day he would have been owed $2^{29} = 536{,}870{,}912$ grains.)† The king found it more expedient to do away with the soldier. Did you think that all fairy tales had happy endings?

†There are 437.5 grains in 1 ounce. So 536,870,912 grains = 536,870,912/437.5 ounces ≈ 1,227,133.5 ounces = 1,227,135.5/16 pounds ≈ 76,696 pounds (≈ 34,862 kilograms) = 76,696/2000 tons ≈ 38.35 tons of gold.

Problems 1.5

Readiness Check

I. $\frac{x^2x^3}{x^4} =$ ______.

a. x^5 b. x c. $\frac{1}{x}$ d. $\frac{1}{x^2}$

II. $\frac{x^{-2}x^{-3}}{x^{-4}} =$ ______.

a. x^{-5} b. x c. $\frac{1}{x}$ d. x^2

III. Which of the following is equal to $7x^{-1} - y^{-1}$?

a. $-7x + y$ b. $\frac{1}{7x} - \frac{1}{y}$

c. $\frac{1}{7x - y}$ d. $\frac{7}{x} - \frac{1}{y}$

IV. Which of the following is equal to $(2x + y)^{-1}$?

a. $\frac{1}{2x + y}$ b. $\frac{2}{x} + \frac{1}{y}$

c. $\frac{2x + y}{xy}$ d. $\frac{-1}{2x} - \frac{1}{y}$

V. Which of the following is the scientific notation for 0.000781?

a. 7.81×10^4

b. 0.781×10^{-3}

c. 7.81×10^{-4}

d. 7.81×10^2

In Problems 1–30 write each expression as a rational number with no exponent.

1. 3^2
2. 2^3
3. 4^{-3}
4. 3^{-2}
5. $\left(\frac{1}{3}\right)^2$
6. $\left(\frac{1}{2}\right)^3$
7. $\left(\frac{1}{2}\right)^{-4}$
8. $\left(\frac{1}{3}\right)^{-2}$
9. $4^{-1}2^3$
10. $4^{-2}(-2)^2$
11. $(-3)^{-3}(3)^3$
12. $(1^22^3)^2$
13. 4^22^3
14. 4^32^{-6}
15. 2^23^{-2}
16. $(2^{-3})^2$
17. $(2^2)^{-3}$
18. 1^{-5}
19. 23^05^{-1}
20. $(3 \cdot 4)^{-2}$
21. $(-1)^{-1}$
22. $(-1)^{100}$
23. $(-1)^{101}$
24. $(-10)^0$
25. $(-10)^1$
26. $(-10)^2$
27. $(-10)^3$
28. $\left(\frac{5}{3}\right)^{-1}$
29. $\left(\frac{257}{1049}\right)^{-1}$
30. $(1.333\ldots)^{-2}$

In Problems 31–72 simplify the given expression.

31. $\frac{x^3}{x^5}$
32. $\frac{y^6}{y^5}$
33. $\frac{y^6}{y^4}$
34. $\frac{6z^8}{3z^3}$
35. $\left(\frac{2}{x}\right)^2$
36. $\left(\frac{1}{3a}\right)^2$
37. $\left(\frac{2b}{3}\right)^{-1}$
38. $\left(\frac{3c}{4d}\right)^{-1}$
39. $\left(\frac{7d}{5e}\right)^{-2}$
40. $\left(\frac{1}{4z^3}\right)^{-2}$
41. $\left(\frac{5x}{2y}\right)^{-3}$
42. $\frac{x^3x^4a^2}{ax}$
43. $\frac{5a^2b^3}{10ab^4}$
44. $\frac{u^3v^5}{u^7v^6}$
45. $\left(\frac{u^2v^2}{u^3v}\right)^{-1}$
46. $\frac{a^5d^3e^9}{a^4d^5e^7}$
47. $\left(\frac{x^3z^4w^3}{xz^5w^2}\right)^2$
48. $\frac{x^2x^{-3}y^4}{y^5}$
49. $\frac{a^2a^{-5}b^{-3}b^4}{ab^6b^{-2}}$
50. $\frac{u^4u^{-7}v^6v^{-8}v^{-3}}{u^{-4}u^3v^3v^{-1}v^6}$
51. $(abc)^2$
52. $\left(\frac{ab}{c}\right)^3$
53. $x^{-1}y(xy)^2$
54. $xy(xy)^3$
55. $(wz)^4w^{-2}z^{-3}$
56. $\left(\frac{x}{3y}\right)^2\frac{y}{x}$
57. $(ab)^{-1}(ab)^2$
58. $\frac{(u^6w^3)^5(u^6w^3)^{-4}}{u^6w^3}$
59. $\left(\frac{x}{2y^2}\right)^{-1}\left(\frac{y}{4x}\right)^2$
60. $\left(\frac{ab}{3c}\right)^{-2}\left(\frac{6c}{4a^2b}\right)(a^5bc^{-1})$
61. $\left(\frac{2x^{-1}y^{-2}}{3x^2}\right)^{-1}\left(\frac{6x}{5y}\right)^{-2}$

Answers to Readiness Check

I. b II. c III. d IV. a V. c

62. $\left(\dfrac{3d^{-2}e^{-1}f^{2}}{e^{3}f^{5}}\right)^{-2}\left(\dfrac{d^{2}ef^{3}}{e^{2}f^{-1}}\right)^{2}$
63. $\left(\dfrac{4x^{2}}{5a^{3}}\right)\left(\dfrac{9}{2x}\right)^{3}$
64. $\left(\dfrac{x^{0}y}{y^{0}x}\right)^{-1}$
65. $\left(\dfrac{x^{3}wz^{-5}}{4x^{2}w^{2}z^{-8}}\right)^{0}$
66. $\left(\dfrac{u^{0}v^{0}}{v}\right)^{-3}$
67. $\left(\dfrac{x^{2}y^{3}}{xy^{5}}\right)^{0}\left(\dfrac{xy^{5}}{x^{2}y^{3}}\right)$
68. $\left(\dfrac{x}{y^{2}}\right)\left(\dfrac{y^{2}}{z^{5}}\right)\left(\dfrac{z^{6}}{w^{7}}\right)\left(\dfrac{w^{8}}{u^{9}}\right)\left(\dfrac{u^{10}}{x^{2}}\right)$
69. $\left(\dfrac{w^{-2}v^{5}}{wv^{-1}}\right)^{0}\left(\dfrac{v}{w}\right)^{-1}$
70. $\dfrac{ab^{2}c^{3}}{bac}$
71. $\left(\dfrac{x^{2}y^{-2}z^{-3}}{z^{3}x^{4}y^{-1}}\right)^{-2}$
72. $\left[\left(\dfrac{a}{b}\right)\left(\dfrac{c}{d}\right)\left(\dfrac{a}{b}\right)^{-1}\left(\dfrac{c}{d}\right)^{-1}\right]^{-1}$

In Problems 73–87 write each number in scientific notation.

73. 365 74. 3,231,406 75. 521.236
76. 0.01605 77. 0.0000001 78. 1,000,000,000
79. The peak of Mt. Everest is at an elevation of 29,028 ft.
80. The area of the Pacific Ocean covers 64,186,300 square miles.
81. One **light year** (the distance light travels in one year) is approximately 5,878,499,830,120 miles.†
82. The U.S. gross national product in fiscal 1984 was $3,662,800,000,000.
83. The mass of an electron is 0.000000000000000000000000000000911 kilogram.
84. The mass of Mars is approximately 634,200,000,000,000,000,000,000 kilograms.
85. The mean distance from the earth to the sun is 92,900,000 miles.
86. **Avogadro's number,** N_0, the number of molecules per mole of substance, is 602,300,000,000,000,000,000,000.
87. Light travels in waves. Light waves at the center of the range visible to the human eye have a length of 0.000000555 meter.

Calculator Problems

In Problems 88–104 write the answer in scientific notation.

88. $(2.35)^2$ 89. $(-25.4)^3$ 90. $(537.2)^{-1}$
91. $(0.235)^{-4}$ 92. $(58.205)^{-4}$ 93. $(2{,}310{,}624)^5$
94. $(2.37\times10^2)\times(4.95\times10^5)$
95. $-(8.2403\times10^{-3})\times(4.106\times10^2)$
96. $\dfrac{7.6024\times10^6}{5.3109\times10^3}$ 97. $\dfrac{-6.105\times10^{-3}}{9.011\times10^4}$
98. $(5.29\times10^5)\times(8.07)\times10^6\div(5.33)\times10^5$
99. $\dfrac{3.2105\times10^{-4}}{-5.6615\times10^{-7}}\div7.1124\times10^2$
100. $\dfrac{3.2611\times10^8}{4.0923\times10^{12}}\times(9.2106\times10^{11})$
101. $(-1.2724\times10^{-7})\div[(3.4107\times10^{-8})\times(-8.2993\times10^{-6})]$
102. $(1.1)^{10}$ 103. $(1.01)^{100}$ 104. $(1.001)^{1000}$
105. Find the smallest integer n such that $\left(\dfrac{1}{2}\right)^n < 0.001$.
106. Find the smallest integer n such that $(0.9)^n < 0.00000001$.
107. Find the smallest integer n such that $(1.2)^n > 1000$.
108. Find the smallest integer n such that $(1.001)^n > 1{,}000{,}000{,}000$.
109. Which is bigger: 4^5 or 5^4?
110. Which is bigger: 50^{51} or 51^{50}?
111. Guess which is bigger: 1000^{1001} or 1001^{1000}.
112. It has been estimated that the number of grains of sand on the beach in Coney Island is about 10^{20}.‡ After how many days would the soldier in the fable retold in this section be due to receive this number of grains of gold?
113. The closest star to our sun is alpha centauri, 4.4 light years away. How far away is it in miles? [see Problem 81].

---

†Obtained by multiplying the speed of light (186,282.3976 mi/sec) × 365.24219879 (no. of days in a year) × 24 × 60 × 60.

‡See James R. Newman, *The World of Mathematics,* Vol. 3, Simon and Schuster, New York, 1956, p. 2007.

1.6 Integral Roots and Rational Exponents

*n*th Root

We know how to compute a^m, where m is an integer. To extend this definition to noninteger exponents, we need first to define the nth root of a number, where n is a positive integer.

Definition of an *n*th Root of a Number

Let a be a real number. Suppose that there is a real number b such that $b^n = a$. Then b is called an **nth root of a.** This is denoted by $a^{1/n}$ or $\sqrt[n]{a}$. We have

$$(a^{1/n})^n = \underbrace{a^{1/n}a^{1/n}\cdots a^{1/n}}_{n \text{ times}} = a$$

NOTE If $n = 2$, we call $a^{1/2}$ the **square root of a** and write

$$a^{1/2} = \sqrt{a}$$

Similarly, we call $a^{1/3}$ the **cube root of a** and write

$$a^{1/3} = \sqrt[3]{a}$$

In addition, $a^{1/4}$ is called the **fourth root of a,** $a^{1/5}$ is called the **fifth root of a,** and so on.

EXAMPLE 1 *Two Cube Roots and Two Square Roots*

(a) $8^{1/3} = \sqrt[3]{8} = 2$ because $2^3 = 8$.
(b) $(-8)^{1/3} = \sqrt[3]{-8} = -2$ because $(-2)^3 = -8$.
(c) $(-4)^{1/2} = \sqrt{-4}$ is not a real number because the square of a real number is nonnegative.
(d) $-4^{1/2} = -\sqrt{4} = -2$

In general,

If n is even and $a < 0$, then $\sqrt[n]{a} = (a)^{1/n}$ is not a real number.

EXAMPLE 2 *Four Roots That Are Not Real Numbers*

The following are not real numbers:
(a) $\sqrt{-2}$ (b) $(-3.1415)^{1/4}$ (c) $(-123.407)^{1/6}$ (d) $(-2.37 \times 10^8)^{1/2}$

NOTE Every positive real number has two square roots: one positive and one negative. When we write $a^{1/2} = \sqrt{a}$, we mean the positive square root of a. This is called the **principal square root.** If we want the negative square root, we shall write $-\sqrt{a}$. That is, if $a > 0$

The Two Square Roots of a Positive Number

$$a^{1/2} = \sqrt{a} > 0$$
$$-a^{1/2} = -\sqrt{a} < 0$$

For example, using this convention, $\sqrt{4} = 2$. The same rule applies to other even-numbered roots: fourth roots, sixth roots, and so on.

EXAMPLE 3 *Two Square Roots and Two Fourth Roots*

(a) $9^{1/2} = 3$ and $-9^{1/2} = -3$.
(b) $16^{1/4} = 2$ because $2^4 = 16$, and $-16^{1/4} = -2$.

Every real number has *exactly one* real cube root because the cube root of a positive number is positive and the cube root of a negative number is negative. The same holds for other odd-numbered roots: fifth roots, seventh roots, and so on.

We now define a^r where $r = m/n$ is a rational number.

Definition of a Rational Exponent

Suppose that m and n are integers, $n > 0$, $a \neq 0$, and $a^{1/n}$ exists. Then

$$a^{m/n} = (a^{1/n})^m = (\sqrt[n]{a})^m = \sqrt[n]{a^m}$$

Moreover, if $m/n > 0$, then $\quad 0^{m/n} = 0$

EXAMPLE 4 *Computing Five Rational Powers*

Compute each number, if it is a real number.

(a) $4^{3/2}$ (b) $(-8)^{5/3}$ (c) $(-4)^{5/4}$ (d) $9^{-5/2}$

(e) $\left(\dfrac{1}{10{,}000}\right)^{-5/4}$

SOLUTION

(a) $4^{3/2} = (4^{1/2})^3 = 2^3 = 8$

Alternatively: $4^{3/2} = (4^3)^{1/2} = 64^{1/2} = 8$. In most cases, it is computationally easier to take the root first and then raise to the integer power rather than the other way around.

(b) $(-8)^{5/3} = [(-8)^{1/3}]^5 = (-2)^5 = -32$

(c) $(-4)^{5/4}$ cannot be computed as a real number because $(-4)^{1/4}$ is not a real number

(d) $9^{-5/2} = (9^{1/2})^{-5} = (3)^{-5} = \dfrac{1}{3^5} = \dfrac{1}{243}$

(e) $\left(\dfrac{1}{10{,}000}\right)^{-5/4} = \left[\left(\dfrac{1}{10{,}000}\right)^{1/4}\right]^{-5} = \left(\dfrac{1}{10}\right)^{-5} = (10^{-1})^{-5} = 10^5 =$ 100,000

In Section 1.5 we gave a number of properties of integral exponents. The same properties hold for rational exponents.

Properties of Rational Exponents

Suppose that $a \neq 0$ and r and s are rational numbers. Assume further that a^r and a^s are defined.

Property	*Illustration*
(a) $a^0 = 1$	$3^0 = 1$
(b) $a^1 = a$	$5^1 = 5$
(c) $a^{-r} = \frac{1}{a^r}$	$4^{-3/2} = \frac{1}{4^{3/2}} = \frac{1}{8}$
(d) $a^r a^s = a^{r+s}$	$16^{1/4}16^{1/2} = 16^{1/4+1/2} = 16^{3/4} = 8$
(e) $\frac{a^r}{a^s} = a^{r-s}$	$\frac{(1/27)^{5/3}}{(1/27)^1} = \left(\frac{1}{27}\right)^{5/3-3/3} = \left(\frac{1}{27}\right)^{2/3} = \frac{1}{9}$
(f) $(ab)^r = a^r b^r$	$(4 \cdot 9)^{1/2} = 4^{1/2}9^{1/2} = 2 \cdot 3 = 6$
(g) $\left(\frac{a}{b}\right)^r = \frac{a^r}{b^r}$, $b \neq 0$	$\left(\frac{36}{9}\right)^{1/2} = \frac{36^{1/2}}{9^{1/2}} = \frac{6}{3} = 2$
(h) $(a^r)^s = a^{rs}$	$(16^{1/4})^3 = 16^{3/4} = 8$
(i) $a^{-1} = \frac{1}{a}$	$2^{-1} = \frac{1}{2}$; $\left(\frac{1}{3}\right)^{-1} = 3$
(j) $\left(\frac{a}{b}\right)^{-1} = \frac{b}{a}$	$\left(\frac{2}{3}\right)^{-1} = \frac{3}{2}$

The square root sign $\sqrt{}$, the cube root sign $\sqrt[3]{}$, and similar symbols are called **radicals.** Since $\sqrt{x} = x^{1/2}$, $\sqrt[3]{x} = x^{1/3}$, and so on, we can rewrite properties (f) and (g) in terms of radicals. We do this below for the square root radical.

Properties of Radicals

Property		*Illustration*
(k) $\sqrt{a}\sqrt{b} = \sqrt{ab}$	if $a \geq 0$ and $b \geq 0$	$\sqrt{3}\sqrt{12} = \sqrt{36} = 6$
(l) $\frac{\sqrt{a}}{\sqrt{b}} = \sqrt{\frac{a}{b}}$	if $a \geq 0$ and $b > 0$	$\frac{\sqrt{12}}{\sqrt{3}} = \sqrt{\frac{12}{3}} = \sqrt{4} = 2$

Properties (k) and (l) follow from (f) and (g) when $r = \frac{1}{2}$.

WARNING The properties of square root radicals hold only when a and b are nonnegative. Otherwise, you can end up with nonsensical results.

For example, by definition of the square root,

$$\sqrt{-1}\sqrt{-1} = -1$$

But, from (k),

$$\sqrt{-1}\sqrt{-1} = \sqrt{(-1)(-1)} = \sqrt{1} = 1$$

Thus $-1 = 1$. It is easy to derive incorrect formulas if you ignore the rules of algebra. ■

EXAMPLE 5 ***Simplifying $\sqrt{20}$***

Show that $\sqrt{20} = 2\sqrt{5}$.

SOLUTION $\sqrt{20} = \sqrt{4 \cdot 5} \overset{\text{From (k)}}{=} \sqrt{4}\sqrt{5} = 2\sqrt{5}$. ■

EXAMPLE 6 ***Simplifying $\sqrt[3]{24}$***

Show that $\sqrt[3]{24} = 2\sqrt[3]{3}$.

SOLUTION $\sqrt[3]{24} = \sqrt[3]{8 \cdot 3} \overset{\text{Property (f) for } r = \frac{1}{3}}{=} \sqrt[3]{8} \cdot \sqrt[3]{3} = 2\sqrt[3]{3}$.

Rationalizing the Denominator

To eliminate the square root in the expression $\dfrac{a}{\sqrt{b}}$, multiply the numerator and denominator by $\sqrt{b}$:

$$\frac{a}{\sqrt{b}} = \frac{a\sqrt{b}}{\sqrt{b}\sqrt{b}} = \frac{a\sqrt{b}}{b}$$

This process is called **rationalizing the denominator.**

EXAMPLE 7 ***Rationalizing the Denominator***

(a) $\dfrac{10}{\sqrt{5}} = \dfrac{10\sqrt{5}}{\sqrt{5}\sqrt{5}} = \dfrac{10\sqrt{5}}{5} = 2\sqrt{5}$

(b) $\dfrac{x}{\sqrt{y}} = \dfrac{x\sqrt{y}}{\sqrt{y}\sqrt{y}} = \dfrac{x\sqrt{y}}{y}$

(c) $\dfrac{6}{\sqrt{3u}} = \dfrac{6\sqrt{3u}}{\sqrt{3u}\sqrt{3u}} = \dfrac{6\sqrt{3u}}{3u} = \dfrac{2\sqrt{3u}}{u}$

WARNING $\sqrt{a} + \sqrt{b} \neq \sqrt{a+b}$ and $\sqrt{a} - \sqrt{b} \neq \sqrt{a-b}$.

Incorrect	Correct
(a) $2 + 3 = \sqrt{4} + \sqrt{9}$	$\sqrt{4} + \sqrt{9} = 2 + 3 = 5$
Incorrect $= \sqrt{4+9}$	
$= \sqrt{13} \approx 3.61$	
(b) $\sqrt{10} - \sqrt{6} = \sqrt{10-6}$	$\sqrt{10} - \sqrt{6} \approx 3.16227766 - 2.449489743$
Incorrect $= \sqrt{4} = 2$	$= 0.712787917$ ■

FOX TROT by Bill Amend

NOTE $2\sqrt{5}$ is considered to be a *simpler* way to write $\sqrt{20}$ because the number under the radical (the square root sign) is smaller (5 is smaller than 20). If you are asked to simplify $\sqrt{20}$, you should look for a factor of 20 that is a perfect square (here 4 is a factor of 20 and $2^2 = 4$). Similarly, to simplify $\sqrt[3]{24}$, we seek a factor of 24 that is a perfect cube. Also, $\sqrt{72x^4y} = \sqrt{(36x^4)2y} = \sqrt{36x^4}\sqrt{2y} = 6x^2\sqrt{2y}$ so $6x^2\sqrt{2y}$ is considered simpler than $\sqrt{72x^4y}$. Finally, $\dfrac{5\sqrt{2}}{2}$ is considered simpler than $\dfrac{5}{\sqrt{2}}$ because there is no radical in the denominator.

EXAMPLE 8 ***Simplifying an Expression Involving Square Roots***

Compute $3\sqrt{15}\sqrt{5} - 4\sqrt{27}$ and simplify your answer.

SOLUTION

$$\begin{aligned} 3\sqrt{15}\sqrt{5} &= 3\sqrt{15 \cdot 5} = 3\sqrt{75} = 3\sqrt{25 \cdot 3} = 3\sqrt{25}\sqrt{3} \\ &= 3 \cdot 5\sqrt{3} = 15\sqrt{3} \\ 4\sqrt{27} &= 4\sqrt{9 \cdot 3} = 4\sqrt{9}\sqrt{3} = 4 \cdot 3\sqrt{3} = 12\sqrt{3} \end{aligned}$$

Thus

$$3\sqrt{15}\sqrt{5} - 4\sqrt{27} = 15\sqrt{3} - 12\sqrt{3} = 3\sqrt{3} \quad ■$$

EXAMPLE 9 ***Simplifying an Expression Involving Rational Exponents***

Simplify $(16a^4b^{1/3}b^2)^{3/2}$.

SOLUTION

$$\begin{aligned} (16a^4b^{1/3}b^2)^{3/2} &\overset{\text{Use (d)}}{=} (16a^4b^{1/3+2})^{3/2} = (16a^4b^{7/3})^{3/2} \\ &= 16^{3/2}(a^4)^{3/2}(b^{7/3})^{3/2} \\ &\overset{\text{Use (h) twice}}{=} 64a^{(4)(3/2)}b^{(7/3)(3/2)} = 64a^6b^{7/2} \end{aligned}$$

EXAMPLE 10 ***Simplifying an Expression Involving Rational Exponents***

Simplify $\dfrac{(x+3)^{3/2}(y+2)^{-1/2}}{(y+2)^{5/2}(x+3)^{-1/2}}$.

SOLUTION

$$\begin{aligned}\frac{(x+3)^{3/2}(y+2)^{-1/2}}{(y+2)^{5/2}(x+3)^{-1/2}} &= \frac{(x+3)^{3/2}}{(x+3)^{-1/2}} \cdot \frac{(y+2)^{-1/2}}{(y+2)^{5/2}} \\ &= (x+3)^{3/2-(-1/2)}(y+2)^{-1/2-5/2} \\ &= (x+3)^{4/2}(y+2)^{-6/2} = (x+3)^2(y+2)^{-3} \\ &= \frac{(x+3)^2}{(y+2)^3}\end{aligned}$$

CALCULATOR NOTE The [y^x] button on your calculator can be used to calculate rational powers. To do so, you must first convert the exponent to a decimal. The easiest way to do this is illustrated in the following example:

EXAMPLE 11 ***Computing $20^{2/7}$ on a Calculator***

Compute $20^{2/7}$.

SOLUTION We carry out each step on a calculator.

Step 1: 2 [÷] 7 [=] [STO] Stores the number $\dfrac{2}{7} = 0.285714285$ in memory

Recall from memory
↓

Step 2: 20 [y^x] [RM] [=] computes 20 to the power in memory

Answer: 2.353546894

On some calculators you can avoid the storing step by using parentheses: 20 [y^x] [(] 2 [÷] 7 [)] [=]

Almost every scientific calculator can store at least one number in memory. The [STO] command used above stores the number in memory, and [RM] displays the number in memory. Your calculator may use some other labels, but the result is the same.

FOCUS ON

Prime Numbers, the Fundamental Theorem of Arithmetic, and the Irrationality of Roots

In Section 1.2 we said that $\sqrt{2}$ is irrational. In fact, it can be shown that if a is not a square (i.e., $a \neq 1, 4, 9, 16, 25, 36, \ldots$), then $\sqrt{a}$ is irrational. To do this, we briefly discuss a property of integers.

Let n be a positive integer greater than 1. Then n is a **prime number** if n has no factors except 1 and itself. The first 12 prime numbers are 2, 3, 5, 7, 11, 13, 17, 19, 23, 29, 31, and 37. A positive integer that is not prime is called a **composite number.** The following are composite numbers

$$6 = 2 \times 3 \tag{1}$$

$$20 = 5 \times 4 = 5 \times 2 \times 2 \tag{2}$$

$$375 = 125 \times 3 = 5 \times 5 \times 5 \times 3 \tag{3}$$

Every composite number can be written as a product of primes (some of which may be repeated as in (2) or (3)).

Fundamental Theorem of Arithmetic

Every integer greater than one can be expressed as a product of primes in one and only one way (except for the order of the factors). This product is called the **prime factorization** of n.

Suppose that a is a positive integer that is not the square of an integer. That is, $\sqrt{a}$ is not an integer. We show that $\sqrt{a}$ is irrational. Our proof is a proof by contradiction. That is, we assume to the contrary that $\sqrt{a}$ is a rational number and then obtain an impossible result.

Hypothesis

$\sqrt{a}$ is a rational number that can be written as $\sqrt{a} = \dfrac{m}{n}$ in lowest terms.

That is, m and n have no prime factors in common (otherwise these would cancel out and $\dfrac{m}{n}$ would not be in lowest terms). Also, $n \neq 1$ because $\sqrt{a}$ is not an integer. Then

$$a = (\sqrt{a})^2 = \left(\frac{m}{n}\right)^2 = \frac{m^2}{n^2} = \frac{m \cdot m}{n \cdot n}$$

We see that the prime factors of m^2 are the same as the prime factors of m, with each such factor repeated twice as often. The same holds for n. Therefore, since m and n have no common prime factors, the same holds for m^2 and n^2. Thus

$$a = \frac{m^2}{n^2} \text{ in lowest terms}$$

As an illustration, look at $r = \dfrac{12}{35} = \dfrac{2 \cdot 2 \cdot 3}{5 \cdot 7}$.

This fraction is given in lowest terms because the numerator and denominator have no common prime factors. Then

$$r^2 = \frac{12^2}{35^2} = \frac{144}{1225} = \frac{2 \cdot 2 \cdot 3 \cdot 2 \cdot 2 \cdot 3}{5 \cdot 7 \cdot 5 \cdot 7}$$

Again there are no common factors, so $r^2 = \dfrac{144}{1225}$ in lowest terms. The point is that since nothing cancels in $\dfrac{12}{35}$, nothing cancels in $\left(\dfrac{12}{35}\right)^2 = \dfrac{144}{1225}$.

Now, since $n \neq 1$, we conclude that $n^2 \neq 1$, so *a is not an integer*. But we started by assuming that a was a positive integer. This *contradiction* shows that our hypothesis is false. We conclude that $\sqrt{a}$ cannot be written as m/n. Thus $\sqrt{a}$ is irrational.†

Remark

A similar argument can be used to show that if $n > 1$ is an integer and if $a^{1/n}$ is not an integer, then $a^{1/n}$ is irrational.

For more on this interesting topic, read the article "Why Square Roots Are Irrational" by William C. Waterhouse in *The American Mathematical Monthly*, 93(3), March, 1986, p. 213.

† A different proof that $\sqrt{2}$ is irrational appears in Section 4.4.

Problems 1.6

Readiness Check

I. Which of the following is *not* a real number?
a. $(-3)^{1/5}$ b. $-4^{1/2}$ c. $0^{1/2}$ d. $(-8)^{1/4}$

II. $-16^{3/2} =$ ______.
a. 64 b. -64 c. $\dfrac{1}{64}$ d. -24
e. This is not a real number.

III. Which of the following is true?
a. $\sqrt{-4}\sqrt{-5} = \sqrt{20}$ b. $\sqrt{4} + \sqrt{5} = 3$
c. $\dfrac{\sqrt{36}}{\sqrt{6}} = \sqrt{6}$ d. $\sqrt{\sqrt[3]{x}} = \sqrt[5]{x}$

IV. $x^{1/2}x^{1/3} =$ ______.
a. $x^{5/6}$ b. $x^{1/6}$
c. $x^{2/3}$ d. $x^{3/2}$

V. Which of the following is equal to $6x^{2/3}$?
a. $(\sqrt[3]{6x})^2$ b. $(\sqrt{6x})^3$
c. $6\sqrt[3]{x^2}$ d. $6\sqrt{x^3}$

In Problems 1–50 compute the indicated value, if it is a real number.

1. $4^{5/2}$
2. $4^{-3/2}$
3. $4^{9/2}$
4. $8^{1/3}$
5. $8^{-1/3}$
6. $8^{2/3}$
7. $(-8)^{2/3}$
8. $(-8)^{-1/3}$
9. $8^{-2/3}$
10. $(-8)^{-2/3}$
11. $\left(-\dfrac{1}{64}\right)^{-2/3}$
12. $\left(-\dfrac{1}{64}\right)^{2/3}$
13. $27^{1/3}$
14. $(-27)^{1/3}$
15. $(-27)^{-2/3}$
16. $(27)^{-2/3}$
17. $100^{1/2}$
18. $-100^{1/2}$
19. $(-100)^{1/2}$
20. $(-9)^{3/2}$
21. $9^{5/2}$
22. $100^{-1/2}$
23. $100^{5/2}$
24. $100^{-5/2}$
25. $(-100)^{5/2}$
26. $1000^{1/3}$
27. $-(1000^{1/3})$
28. $1000^{-1/3}$
29. $(-1000)^{1/3}$
30. $(-1000)^{-1/3}$
31. $128^{1/7}$
32. $128^{-3/7}$
33. $128^{5/7}$
34. $(-128)^{-2/7}$
35. $(-128)^{6/7}$
36. $(-128)^{-9/7}$
37. $64^{-1/6}$
38. $(-64)^{5/6}$
39. $4^{1/2}4^{3/2}$
40. $2^{5/7}2^{2/7}$
41. $10^{1/3}10^{5/3}$
42. $\dfrac{12^{4/3}}{12^{1/3}}$
43. $\dfrac{(11)^{3/4}}{(11)^{-5/4}}$
44. $\dfrac{12^4}{6^4}$
45. $\dfrac{8^3}{24^3}$
46. $\dfrac{1.6^5}{0.8^5}$
47. $\dfrac{(3.15)^3}{(6.3)^3}$
48. $\dfrac{4^{1/2}4^{3/2}}{4^4 4^{-5/2}}$
49. $\dfrac{(-64)^{1/3}(-64)^{-2/3}}{(-64)^{1/3}(-64)^{2/3}}$
50. $\dfrac{7^{1/5}7^{-9/5}}{7^{11/5}7^{-14/5}}$

In Problems 51–93 simplify the given expression.

51. $\sqrt{108}$
52. $\sqrt{200}$
53. $\sqrt[3]{-5000}$
54. $\sqrt[4]{80}$
55. $\dfrac{100}{\sqrt{3}}$
56. $\dfrac{2\sqrt{10}}{\sqrt{7}}$
57. $\dfrac{\sqrt{0.1}}{\sqrt{0.4}}$
58. $\dfrac{30}{\sqrt[3]{10}}$ [Hint: Multiply and divide by $10^{2/3}$.]
59. $\dfrac{-5}{\sqrt[3]{-3}}$
60. $\dfrac{2}{\sqrt[4]{4}}$
61. $x^{3/2}x^{-3/2}$
62. $\dfrac{x^{1/2}x^{2/3}}{x^{3/4}}$
63. $\dfrac{y^{3/7}y^{8/7}}{y^{1/7}y^{10/7}}$
64. $a^2a^{1/2}$
65. $(x^2y^4)^{1/2}$
66. $\dfrac{x^{n/3}x^{n/2}}{x^{n/6}x^{2n/3}}$, n an integer
67. $\left(\dfrac{u}{2u}\right)^2$
68. $\left(\dfrac{4y}{\sqrt{y}}\right)^3$
69. $\left(\dfrac{a^{1/2}}{b^2}\right)^2\left(\dfrac{b^{3/2}}{a^{2/3}}\right)^3$
70. $[u^{1/2}v^{-1/3}u^{-1/4}v^{5/3}]^2$
71. $\left(\dfrac{y^{2/3}z^{1/4}}{z}\right)^{-1}$
72. $x^{1.23}x^{-0.6}x^{1.92}x^{-1}$
73. $\dfrac{y^{1.3}y^{-2.6}}{y^{-2.4}y^3}$
74. $\sqrt{x}\sqrt{y}$
75. $\dfrac{\sqrt{x}}{\sqrt{y}}$
76. $\dfrac{\sqrt{4x}}{\sqrt{9x}}$
77. $\sqrt[6]{x^4y^{12}}$
78. $\sqrt[3]{a^{12}/b^6}$
79. $\sqrt{xy^2}\sqrt{yx^2}$
80. $\sqrt[3]{x\sqrt[4]{x}}$
81. $\sqrt{\sqrt{y^{16}}}$
82. $\sqrt[3]{\sqrt{z^{12}}}$
83. $\dfrac{\sqrt{w\sqrt[3]{w}}}{\sqrt[4]{w}}$
84. $\sqrt{\dfrac{x^6y^4}{z^8}}$
85. $\sqrt{(x-1)^5y^2}$
86. $\sqrt[3]{x^6y^{12}}$
87. $\sqrt{\dfrac{x^{12}}{y^9}}$
88. $\sqrt{40x^3y^7z^2}\sqrt{5xy^2z^3}$
89. $\sqrt{8} + \sqrt{18}$
90. $4\sqrt{8} - 3\sqrt{18} + 2\sqrt{50}$
91. x^nx^{2n}
92. $\dfrac{z^{5a}}{z^{2a}}$
93. $\sqrt{\dfrac{y^{4k}y^{7k}}{y^{8k}}}$

In Problems 94–103, solve for x.

94. $9^x = 3$ [Hint: Ask yourself, 9 to what fractional power equals 3?]
95. $2^x = \dfrac{1}{4}$
96. $8^x = 2$
97. $16^{-x} = \dfrac{1}{2}$
98. $8^{2x} = \dfrac{1}{2}$
99. $8^{x+3} = 4$
100. $4^{-x+3} = 2$
101. $\left(\dfrac{1}{2}\right)^{3x/2} = 16$
102. $\left(\dfrac{1}{3}\right)^{-1/x} = \dfrac{1}{9}$
103. $(0.01)^{-1/x} = 1000$

Calculator Exercises

In Problems 104–114 find an approximation to each number on your calculator.

104. $\sqrt{5}$
105. $\sqrt[3]{10}$
106. $\sqrt[4]{8}$
107. $\sqrt[5]{-1000}$
108. $10^{2/3}$
109. $80^{-3/7}$
110. $50^{0.737}$
111. $(100{,}000)^{0.1182}$
112. $\left(\dfrac{1}{3}\right)^{1/3}$
113. $(0.235)^{1.46}$
114. π^π

In Section 6.1 we will derive the compound interest formula

$$A(t) = P(1 + r)^t$$

Here $A(t)$ denotes the amount of money in an account after t time periods (usually years) if P dollars is invested at an interest rate of r. For example, 6% corresponds to $r = 0.06$. In Problems 115–117 use this formula.

Answers to Readiness Check

I. d II. b III. c IV. a V. c

115. If $1000 is invested in a savings account paying 6% annual interest, how much money will be in the account after (a) 5 years? (b) $7\frac{1}{2}$ years? (c) $11\frac{3}{4}$ years?
116. If $4500 is invested in a savings account paying 7% interest compounded quarterly, how much money will be in the account after 3 years? [Hint: A time period is now $\frac{1}{4}$ year = 3 months so the quarterly interest is 0.07/4 and the number of time periods is $3 \cdot 4 = 12$.]
* 117. A bond pays 12% interest compounded monthly. How much will the bond be worth in 10 years if $5000 is invested now?
118. In the kinetic theory of gases, the **root mean square** (rms) speed of gas molecules is related to the average or typical speed of the molecules comprising the gas and is given by

$$v_{\text{rms}} = \sqrt{\frac{3P}{\rho}},$$

where P is the pressure (measured in atmospheres, atm) of the gas and ρ is its density (measured in kilograms per cubic meter).

Calculate the v_{rms} of hydrogen molecules at 0°C and 1 atm of pressure. (At that temperature, the density of hydrogen is 8.99×10^{-2} kg/m^3.)

119. In astronomy, the **luminosity** of a star is the star's total energy output. Loosely speaking, a star's luminosity is a measure of how bright the star would appear at the surface of the star. The **mass-luminosity relation** gives the approximate luminosity of a star as a function of its mass. It has been found experimentally that, approximately,

$$\frac{L}{L_0} = \left(\frac{M}{M_0}\right)^r,$$

where L and M are the luminosity and mass of the star and where L_0 and M_0 are the luminosity and mass of our sun; the exponent r depends on the mass of the star according to the following table.†

Mass ratio, $\frac{M}{M_0}$	r	Mass ratio, $\frac{M}{M_0}$	r
1.0–1.4	4.75	5–10	3.38
1.4–1.7	4.28	10–20	2.80
1.7–2.5	4.15	20–50	2.30
2.5–5	3.95	50–100	1.90

The mass of a given star is three times the mass of the sun. For that star compute the ratio L/L_0.

120. Answer the question of Problem 119 for a star that is 16 times as massive as the sun.

† These data are based on stellar models computed by D. Ezer and A. Cameron in their paper "Early and main sequence evolution of stars in the range 0.5 to 100 solar masses," *Canadian Journal of Physics*, 45(1967), 3429–3460.

1.7 Polynomials

You will encounter polynomials often in this text.

EXAMPLE 1 *Four Polynomials*

(a) $x - 3$ is a polynomial in x
(b) $x^2 - 7x + 6$ is a polynomial in x
(c) $4x^4 - 3x^3 + 2x^2 - 8x + 10$ is a polynomial in x
(d) $x^2 - 2xy + y^2$ is a polynomial in x and y

Definition of a Polynomial in One Variable *x*

A **polynomial in *x*** is an expression of the form

$$p(x) = a_n x^n + a_{n-1}x^{n-1} + \cdots + a_2x^2 + a_1x + a_0 \qquad (1)$$

Where $a_0, a_1, \ldots, a_n$ are real numbers and n is a nonnegative integer.

The numbers $a_0, a_1, \ldots, a_n$ are called the **coefficients** of the polynomial p. In equation (1) the number a_n is called the **leading coefficient** of p, and a_0 is called the **constant term.**

We assume in (1) that $a_n \neq 0$. Then p is a polynomial of **degree n.** We denote the degree of p by deg p. A polynomial of **degree 0** is a nonzero constant.

EXAMPLE 2 ***Five Polynomials in x and Their Degrees***

(a) $x^2 - 2x + 4$ is a polynomial of degree 2, or a **quadratic** polynomial.
(b) $2x^3 - 4x^2 + 5x - 7$ is a polynomial of degree 3, or a **cubic** polynomial.
(c) $5x - 6$ is a first-degree or **linear** polynomial.
(d) $5x^7 - 6x^4 + 3x - 2$ is a polynomial of degree 7.
(e) 3 is a polynomial of degree 0.

In this section we will discuss the addition, subtraction, and multiplication of polynomials. First we need a definition.

The **corresponding terms** of two polynomials are terms of the same degree.

EXAMPLE 3 ***Identifying Corresponding Terms of Two Polynomials***

Let $p(x) = x^2 + 3x + 5$ and $q(x) = x^3 - 2x^2 + 4x + 7$. Then the following pairs are corresponding terms:

x^2 and $-2x^2$ Two terms of degree 2

$3x$ and $4x$ Two terms of degree 1

5 and 7 Two constant terms (degree 0)

Note that no term in $p(x)$ corresponds to the x^3 term in $q(x)$.

We can add or subtract polynomials by adding or subtracting coefficients.

Addition Rule for Polynomials

To add (or subtract) two polynomials, combine corresponding terms by adding (or subtracting) their respective coefficients.

EXAMPLE 4 ***Adding and Subtracting Two Polynomials***

Let $p(x) = 2x^3 - 7x^2 + 4x - 5$ and $q(x) = 5x^2 - 2x + 6$. Find (a) $p(x) + q(x)$ and (b) $p(x) - q(x)$.

SOLUTION $q(x)$ has no x^3 term, so we add the term $0x^3$ to $q(x)$ in order to have a term that corresponds to the $2x^3$ term in $p(x)$.

(a)
$$\begin{aligned} p(x) + q(x) &= (2x^3 - 7x^2 + 4x - 5) + (0x^3 + 5x^2 - 2x + 6) \\ &= (2 + 0)x^3 + (-7 + 5)x^2 + (4 - 2)x + (-5 + 6) \\ &= 2x^3 - 2x^2 + 2x + 1 \end{aligned}$$

(b)
$$\begin{aligned} p(x) - q(x) &= (2x^3 - 7x^2 + 4x - 5) - (0x^3 + 5x^2 - 2x + 6) \\ &= (2 - 0)x^3 + (-7 - 5)x^2 + (4 - (-2))x + (-5 - 6) \\ &= 2x^3 - 12x^2 + 6x - 11 \end{aligned}$$

Rule for Multiplying a Polynomial by a Constant

To multiply a polynomial by a constant, multiply each term in the polynomial by that constant.

EXAMPLE 5 *Multiplying a Polynomial by 4*

Let $p(x) = 5x^4 - 2x^3 + 3x^2 - x + 7$. Find $4p(x)$.

SOLUTION By the rule above,

$$\begin{aligned} 4p(x) &= 4(5x^4 - 2x^3 + 3x^2 - x + 7) \\ &= 4 \cdot 5x^4 + 4 \cdot (-2)x^3 + 4 \cdot 3x^2 + 4 \cdot (-1)x + 4 \cdot 7 \\ &= 20x^4 - 8x^3 + 12x^2 - 4x + 28 \quad \blacksquare \end{aligned}$$

EXAMPLE 6 *Multiplying a Polynomial by −1*

Compute $-(x^3 - 4x^2 - 5x + 7)$.

SOLUTION

$$\begin{aligned} -(x^3 - 4x^2 - 5x + 7) &= -x^3 - (-4x^2) - (-5x) - (7) \\ &= -x^3 + 4x^2 + 5x - 7 \end{aligned}$$

WARNING When multiplying a polynomial by -1, that is, when finding the negative of a polynomial, it is necessary to change the sign of *every term* in the polynomial, as in Example 6.

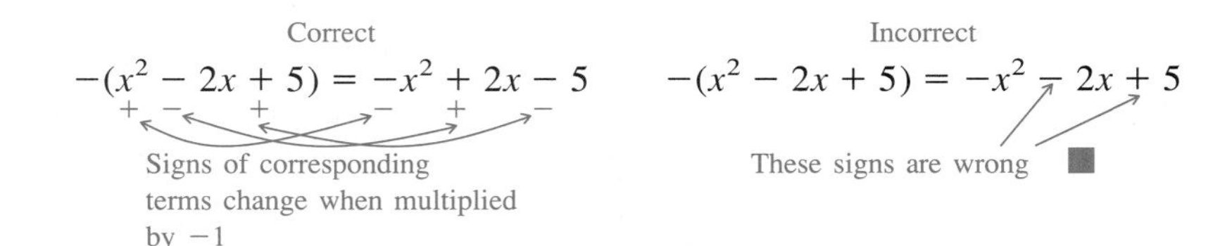

EXAMPLE 7 *Multiplying Polynomials by Constants and Adding Them*

Let $p(x) = 5x^4 - 2x^3 + 3x^2 - x + 7$ and $q(x) = x^5 - 2x^4 + 3x^2 - 5$. Compute $4p(x) - 3q(x)$.

SOLUTION We computed $4p(x)$ in Example 5. Then $3q(x) = 3x^5 - 6x^4 + 9x^2 - 15$. Hence,

$$\begin{aligned} 4p(x) - 3q(x) &= (20x^4 - 8x^3 + 12x^2 - 4x + 28) \\ &\quad - (3x^5 - 6x^4 + 9x^2 - 15) \\ &= (0 - 3)x^5 + (20 - (-6))x^4 + (-8 - 0)x^3 \\ &\quad + (12 - 9)x^2 + (-4 - 0)x + (28 - (-15)) \\ &= -3x^5 + 26x^4 - 8x^3 + 3x^2 - 4x + 43 \end{aligned}$$

To multiply two polynomials together, use the following rule (we first saw this rule applied to real numbers on p. 8):

Distributive Properties

Let a, b, and c denote polynomials. Then

$$a(b + c) = ab + ac \tag{2}$$

and

$$(a + b)c = ac + bc \tag{3}$$

EXAMPLE 8 ***Illustration of Distributive Property***

By equation (2),

$$\begin{aligned} 3x(5x^2 + 2) &= (3x)(5x^2) + (3x)(2) \\ &= 15x^3 + 6x \quad \blacksquare \end{aligned}$$

EXAMPLE 9 ***Multiplying Two Polynomials Together***

Let $p(x) = 2x^2 - 3x + 4$ and $q(x) = x^2 + 4x - 5$. Using the distributive property, find $p(x)q(x)$.

SOLUTION

$$\begin{aligned} p(x)q(x) &= (2x^2 - 3x + 4)(x^2 + 4x - 5) \\ &= (2x^2 - 3x + 4)x^2 + (2x^2 - 3x + 4)(4x) \\ &\quad + (2x^2 - 3x + 4)(-5) \quad \text{From (2)} \\ &= (2x^4 - 3x^3 + 4x^2) + (8x^3 - 12x^2 + 16x) \\ &\quad + (-10x^2 + 15x - 20) \quad \text{From (3)} \\ &= 2x^4 + 5x^3 - 18x^2 + 31x - 20 \end{aligned}$$

Product of first terms $2x^2$ and x^2

Sum of products $(2x^2)(4x) + (-3x)(x^2)$

Sum of products $2x^2(-5) + (-3x)(4x) + 4(x^2)$

Sum of products $(-3x)(-5) + 4(4x)$

Product of constant terms $4(-5)$

We can also perform this multiplication by arranging our work in rows, just as we would multiply two integers:

$$\begin{array}{rrrrrl} & & 2x^2 & - \ 3x & + \ 4 & \\ & & x^2 & + \ 4x & - \ 5 & \\ \hline & & -10x^2 & + 15x & - 20 & \leftarrow \text{Terms on top multiplied by } -5 \\ & +8x^3 & -12x^2 & + 16x & & \leftarrow \text{Terms on top multiplied by } 4x \\ +2x^4 & -3x^3 & +4x^2 & & & \leftarrow \text{Terms on top multiplied by } x^2 \\ \hline = 2x^4 & + 5x^3 & - 18x^2 & + 31x & - 20 & \leftarrow \text{Sum in each column} \end{array}$$

Example 9 illustrates three things that happen when we multiply two polynomials $p(x)$ and $q(x)$ together:

(a) If $\deg p = m$ and $\deg q = n$, then $\deg pq = m + n$.
(b) The leading coefficient of pq is the product of the leading coefficients of p and q.
(c) The constant term of pq is the product of the constant terms of p and q.

EXAMPLE 10 *Determining the Degree, the Leading Coefficient, and the Constant Term in the Product of Two Polynomials in x*

Consider the polynomials

$$p = 3x^5 + 2x^4 - x^3 + 5x^2 + 2x - 8$$

and

$$q = 9x^7 + 7x^6 + 11x^5 - 10x^4 + 3x^3 + 9x^2 - 8x + 3$$

Without multiplying through, find (a) the degree of the product polynomial, pq, (b) the leading coefficient of pq, and (c) the constant term of pq.

SOLUTION

(a) $\deg pq = \deg p + \deg q = 5 + 7 = 12$.
(b) The leading coefficient of $pq = 3 \times 9 = 27$.
(c) The constant term of $pq = (-8) \times 3 = -24$.

This means that the first term in pq is $27x^{12}$, and the last (constant) term is -24. ■

EXAMPLE 11 *Determining the Constant Term in the Product*

Let $p = x^5 + 3x^4 + 2x^2 - 5x$ and $q = x^3 + 3x^2 - 4x + 5$. Since the constant term of p is 0, we can conclude without multiplying through that the constant term of pq is also 0.

Polynomials in More Than One Variable

The **degree** of a **term** of a polynomial is the sum of the powers of the variables that appear in that term. The **degree** of a polynomial in more than one variable is the highest degree of its terms.

EXAMPLE 12 *Determining the Degree of a Polynomial in x and y*

Determine the degree of $x^3 + 3x^2y^2 + xy^5 - y^6 + 4x^4y^3$.

SOLUTION We first determine the degree of each term:

x^3	has degree 3	
$3x^2y^2$	has degree $2 + 2 = 4$	
xy^5	has degree $1 + 5 = 6$	$x = x^1$
$-y^6$	has degree 6	
$4x^4y^3$	has degree $4 + 3 = 7$	Highest degree

Thus the degree of the polynomial is 7.

A polynomial with only one term is called a **monomial.** If it has two terms, it is called a **binomial,** and if it has three terms, it is called a **trinomial.** The most commonly occurring binomials are the polynomials $x + y$ and $x - y$. We can use the distributive properties (2) and (3) to multiply these and other polynomials in more than one variable.

EXAMPLE 13 *Three Products Involving Binomials*

Compute
(a) $(x + y)^2$ (b) $(x + y)(x - y)$ (c) $(x + y)^3$

SOLUTION

(a)
$$(x + y)^2 = (x + y)(x + y) = (x + y)x + (x + y)y = x^2 + yx + xy + y^2$$
$xy = yx$ ↓
$$= x^2 + 2xy + y^2$$

(b)
$$(x + y)(x - y) = (x + y)x + (x + y)(-y) = x^2 + yx - xy - y^2 = x^2 - y^2$$

(c) From (a) ↓
$$\begin{aligned}(x + y)^3 = (x + y)(x + y)^2 &= (x + y)(x^2 + 2xy + y^2)\\ &= x(x^2 + 2xy + y^2) + y(x^2 + 2xy + y^2)\\ &= x^3 + 2x^2y + xy^2 + yx^2 + 2xy^2 + y^3\\ &= x^3 + 3x^2y + 3xy^2 + y^3\end{aligned}$$

The formulas in Example 13 come up very often. They, and a few others, are worth memorizing. The most important formulas are the following:

Important Binomial Multiplication Formulas

$$(x + y)^2 = x^2 + 2xy + y^2 \quad (4)$$
$$(x - y)^2 = x^2 - 2xy + y^2 \quad (5)$$
$$(x + y)(x - y) = x^2 - y^2 \quad (6)$$
$$(x + y)^3 = x^3 + 3x^2y + 3xy^2 + y^3 \quad (7)$$
$$(x - y)^3 = x^3 - 3x^2y + 3xy^2 - y^3 \quad (8)$$

EXAMPLE 14 *Squaring a Trinomial*

Compute $(x + y + z)^2$.

SOLUTION We write $(x + y + z)^2 = ((x + y) + z)^2$

$$\overset{\text{From 4}}{=} (x + y)^2 + 2(x + y)z + z^2$$

$$\overset{\text{Using (4) and (3)}}{=} x^2 + 2xy + y^2 + 2(xz + yz) + z^2$$

$$= x^2 + 2xy + 2xz + 2yz + y^2 + z^2$$

Multiplying Two Binomials

Consider the product

$$(ax + b)(cx + d) = ax(cx + d) + b(cx + d)$$
$$= acx^2 + adx + bcx + bd$$
$$= acx^2 + (ad + bc)x + bd$$

That is

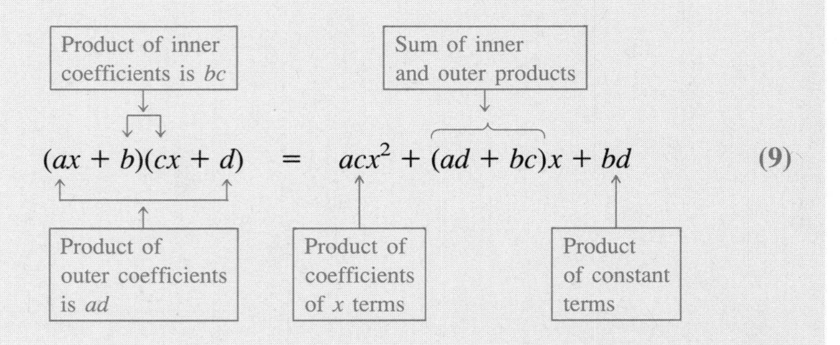

EXAMPLE 15 *Multiplying Two Binomials*

Compute the product $(2x + 3)(4x - 5)$.

SOLUTION Using (9), we obtain

$$(2x + 3)(4x - 5) = 2 \cdot 4x^2 + [2(-5) + 3 \cdot 4]x + 3(-5) = 8x^2 + 2x - 15$$

(Arrows label: $2 \cdot 4$ — Product of coefficients of x terms; $2(-5)$ — Outer product; $3 \cdot 4$ — Inner product; $3(-5)$ — Product of constant terms)

WARNING $(x + y)^2 \neq x^2 + y^2$
Many students make this mistake. Do not make it yourself.

Correct

$$(x + 5)^2 = x^2 + 2 \cdot 5x + 25$$
$$= x^2 + 10x + 25$$

Incorrect

$$(x + 5)^2 = x^2 + 25$$

Incorrect because the middle term $+10x$ is missing

Multiplying Algebraic Expressions

We can use the distributive properties of polynomial multiplication to multiply other kinds of expressions. An **algebraic expression** is an expression involving powers or roots.

EXAMPLE 16 *Multiplying Two Algebraic Expressions*

Multiply $(\sqrt{a} - b^{3/2})(a^{3/2} + b^{-1/2})$.

SOLUTION We proceed as we do when we multiply polynomials:

$$\begin{aligned}(\sqrt{a} - b^{3/2})(a^{3/2} + b^{-1/2}) &= (\sqrt{a} - b^{3/2})a^{3/2} + (\sqrt{a} - b^{3/2})b^{-1/2} \quad (\sqrt{a} = a^{1/2})\\ &= (a^{1/2} - b^{3/2})a^{3/2} + (a^{1/2} - b^{3/2})b^{-1/2}\\ &= a^{1/2}a^{3/2} - b^{3/2}a^{3/2} + a^{1/2}b^{-1/2} - b^{3/2}b^{-1/2}\\ &= a^2 - b^{3/2}a^{3/2} + a^{1/2}b^{-1/2} - b^1\\ &= a^2 - b^{3/2}a^{3/2} + \frac{a^{1/2}}{b^{1/2}} - b\end{aligned}$$

Product of first terms: $(\sqrt{a})(a^{3/2}) = a^{1/2}a^{3/2}$

Product of $-b^{3/2}$ and $a^{3/2}$

Product of $\sqrt{a} = a^{1/2}$ and $b^{-1/2}$

Product of last terms: $(-b^{3/2})(b^{-1/2})$

EXAMPLE 17 *Cubing an Algebraic Expression*

Compute $(x^{1/2} - x^{1/3})^3$.

SOLUTION Using formula (8), we obtain

$$\begin{aligned}(x^{1/2} - x^{1/3})^3 &= (x^{1/2})^3 - 3(x^{1/2})^2x^{1/3} + 3x^{1/2}(x^{1/3})^2 - (x^{1/3})^3\\ &= x^{3/2} - 3 \cdot x \cdot x^{1/3} + 3x^{1/2}x^{2/3} - x\\ &= x^{3/2} - 3x^{4/3} + 3x^{7/6} - x\end{aligned}$$

Problems 1.7

Readiness Check

I. $(3x + 2) - (4x - 6) =$ ______.
a. $7x + 8$ b. $-x - 4$ c. $7x - 4$ d. $-x + 8$

II. $(a - 2b)^2 =$ ______.
a. $a^2 + 2ab + 4b^2$ b. $a^2 - 4ab + 4b^2$
c. $a^2 + 4ab + 4b^2$ d. $4a^2 - 4ab + b^2$
e. $4a^2 - 2ab + b^2$ f. $a^2 - 4ab - 4b^2$

III. Which of the following is the first term of pq if $p(x) = 3x^2 - 7x + 1$ and $q(x) = -2x^3 + 5x$?
a. $6x^6$ b. $-6x^5$ c. $-5x$ d. $-2x^3$

IV. $(2x + 5)(2x - 5) =$ ______.
a. $4x^2 + 25$ b. $4x^2 + 20x + 25$
c. $4x^2 - 20x + 25$ d. $4x^2 - 20x - 25$
e. $4x^2 + 20x - 25$ f. $4x^2 - 25$

V. $(x + 3)(x - 2) =$ ______.
a. $x^2 + 3x - 6$ b. $-6x^2 - 5x - 6$ c. $x^2 - x + 6$
d. $x^2 + 5x - 6$ e. $x^2 + x - 6$

In Problems 1–6 determine the degree of the given polynomial.

1. $x^4 - 1$ 2. $x^2 - 3x + 2$ 3. 6
4. $5x - 6$
5. $-3x^6 + 11x^3 + 12x^2 - 17x$
6. $-8x^7 - 5x^3 + 2x - 4$

In Problems 7–15 let $p(x) = 2x^2 - 3x + 4$ and $q(x) = 3x^3 - x^2 + 5x - 3$. Compute the following:

7. $2p(x)$ 8. $4q(x)$ 9. $p(x) + q(x)$
10. $p(x) - q(x)$ 11. $q(x) - p(x)$ 12. $-2p(x) + 3q(x)$
13. $3q(x) - 4p(x)$ 14. $\deg pq$ 15. $p(x)q(x)$

In Problems 16–22 let $p(x) = x^4 - 3$ and $q(x) = x^7 - 2x + 3$. Compute the following:

16. $p(x) + q(x)$ 17. $3q(x)$ 18. $-8p(x)$
19. $q(x) - p(x)$ 20. $p(x) - q(x)$ 21. $\deg pq$
22. $p(x)q(x)$

In Problems 23–62 perform the multiplication and determine the degree of the product.

23. $(x + 2)(x + 4)$ 24. $x(x - 6)$
25. $(x + 5)(x - 5)$ 26. $(2x - 3)(x + 2)$
27. $(3x - 5)(-5x + 2)$ 28. $(x - a)(x + a)$
29. $(3x - 2)(4x + 1)$ 30. $(5x + 3)(x - 6)$
31. $(7x + 2)(7x + 1)$ 32. $(3x - 6)(2x + 5)$
33. $(8x + 1)(3x + 4)$ 34. $(3x + 6)\left(x - \frac{1}{2}\right)$
35. $(x - 3)(2x - 9)$ 36. $(3x - 5)(5x - 3)$
37. $(2 - x)(5 - 2x)$ 38. $(3 + 7x)(-2 + 5x)$
39. $(x^4 - 4)(x^4 + 4)$
40. $(x^2 - 4x + 3)(2x^2 - x + 2)$
41. $(-3x^2 - 4x + 2)(6x^2 - 3x + 2)$
42. $x^2(x^3 - 1)$ 43. $(x^3 - 1)(x^2 + 1)$
44. $(x^3 - 1)(x^3 + 1)$
45. $(2x^3 - 3x^2 + 4)(2x^2 - 3x)$
46. $(ax^2 + bx + c)(dx^2 + ex + f)$; $a, d \neq 0$
47. $(ax^3 + bx)(cx^3 + dx + e)$; $a, c \neq 0$
48. $(x^4 + 1)(x^5 - 1)$ 49. $x^{10}(x^{20} - 2)$
50. $(x^{10} + x^5 - 2)(x^{10} - x^5 - 2)$
51. $(x^4 - 2x^2 + 3)(x^4 + 2x^2 - 3)$
52. $(3x^8 - 2x^4 + 3x + 1)(6x^7 - 5x^2 + 6)$
53. $(x^2 - 4)(x^2 + 4)$ 54. $(x^n - 1)(x^n + 1)$, $n \neq 0$
55. $(3x - 2)^2$ 56. $(2x + 3y)^2$
57. $(x - 1)^3$ 58. $(x + 2)^3$
59. $(x + 1)(x - 2)(x + 3)$ 60. $(2x + 3)(x - 5)(4x + 1)$
61. $(x + 3)^4$ 62. $(3x - 1)^5$

In Problems 63–71 two polynomials are given. Without multiplying them through, determine (a) the degree of the product, (b) the leading coefficient of the product, and (c) the constant term of the product.

63. $p = 2x^3 - x^2 + 3$; $q = 3x^4 - 11x^2 + 2x + 4$
64. $p = x^7 - 2$; $q = 5x^8 + 4x^4 - 1$
65. $p = x^4 + x^2 + 1$; $q = x^5 + x^3 + x$
66. $p = (x - 1)^3$; $q = (x + 2)^2$
67. $p = (2x + 3)^2$; $q = (x - 5)^3$
* 68. $p = (x - 1)^{11}$; $q = (x + 1)^{10}$
* 69. $p = (2x - 3)^4$; $q = (x + 2)^4$
* 70. $p = (x + 1)^n$; $q = (x - 1)^m$, m odd
* 71. $p = (x + 1)^n$; $q = (x - 1)^m$, m even

In Problems 72–77 determine the degree of each polynomial.

72. $x^2y + yx^3$

Answers to Readiness Check

I. d II. b III. b IV. f V. e

73. $x^3y^5 - 3x^2y^4 + 8xy^6$
74. $x^2y + xy^2 + x^4 + y^3$
75. $xy + 3x^2y^2 - 5x^3 - y^3$
76. $6xyz + x^2 + y^2 + z^2$
77. $x^3z^5 - 2xy^2z^2 + 3x^5y - 8x^2y^3z^4$

In Problems 78–111 multiply through and simplify each expression.

78. $(x - 2y)^2$
79. $(2a - b)^2$
80. $(2u - v)^3$
81. $(3w - 4z)^3$
82. $\left(\frac{1}{x} + \frac{1}{y}\right)^2$
83. $\left(\frac{1}{x} - \frac{1}{y}\right)^2$
84. $(x + y + z)^2$
85. $(x + y - z)^2$
86. $(x - y - z)^2$
87. $(x - y + z)^2$
88. $(x + y + 1)(x + y - 1)$
89. $(x + y + z)(x + y - z)$
90. $(x^2 + y^2 + z^2)^2$
91. $(x^2 - y^2 + z^2)^2$
92. $(x^2 + y^2 + z^2)(x^2 + y^2 - z^2)$
93. $\left(x + \frac{1}{x} + y\right)^2$
94. $\left(x - y + \frac{1}{y}\right)\left(x + y + \frac{1}{y}\right)$
95. $(x + y + z + w)^2$
96. $(x + y + z - u - v)^2$
97. $(x + y + z + w)(x + y - z - w)$
98. $(\sqrt{x} + \sqrt{y})^2$
99. $(\sqrt{x} - \sqrt{y})^2$
100. $\left(x + \frac{1}{x}\right)^3$
101. $\left(w^2 - \frac{1}{w^2}\right)^3$
102. $(x^{2/3} + 3y^{1/3})^3$
103. $(3x^2 - 12y^2)(10x^3 + 4y)$
104. $(7w^8 - 2x^5)(w^3 + z^{10})$
105. $(a + b)(c + d)$
106. $(x - 3y)(x + 3y)$
107. $\left(z + \frac{1}{z}\right)\left(z - \frac{1}{z}\right)$
108. $\left(y^n - \frac{1}{y^n}\right)\left(y^n + \frac{1}{y^n}\right)$
109. $(x^3 + 2y + 3z^4)(4x - y^3 + 5z^2)$
110. $\left(x + \frac{1}{x} + \frac{1}{x^2}\right)\left(x - \frac{3}{x} + \frac{10}{x^2}\right)$
111. $\left(\frac{2}{x} - 3y\right)^3$

* 112. Consider the product $97 \times 89 = 8633$. Observe that $97 = 100 - 3$, $89 = 100 - 11$, and $3 \times 11 = 33$. Observe further that $3 + 11 = 14$ and $100 - 14 = 86$. Use your observations to formulate a general rule for multiplying certain pairs of two-digit numbers. Under what conditions will your rule work?

113. Use the rule derived in Problem 112 to find each product.
(a) 93×88 (b) 96×85

1.8 Factoring

It is often important to **factor** a given polynomial, that is, to write the polynomial as the product of two or more polynomials of smaller degree. Each term in the product is called a **factor** of the original polynomial.

EXAMPLE 1 ***Five Factored Polynomials***

Here are some factored polynomials:

$$x^2 - 4 = (x + 2)(x - 2)$$
$$x^2 - 3x - 10 = (x - 5)(x + 2)$$
$$6x^2 - 10x - 4 = 2(3x^2 - 5x - 2) = 2(x - 2)(3x + 1)$$
$$x^3 - 6x^2 + 11x - 6 = (x - 1)(x - 2)(x - 3)$$
$$x^3 + x^2 + 2x + 2 = (x + 1)(x^2 + 2)$$

Factoring out Common Factors

Recall the distributive properties on p. 40:

$$a(b + c) = ab + ac \quad \text{and} \quad (a + b)c = ac + bc$$

In this section we use these formulas in the reverse order:

$$ab + ac = a(b + c) \quad (1)$$

$$ac + bc = (a + b)c \quad (2)$$

If *each term* in a polynomial has the same common factor, then we use (1) or (2) to factor it out.

EXAMPLE 2 ***Factoring Out a Common Factor***

Each term in the polynomial $x^5 + 2x^2$ has the factor x^2. Thus

$$x^5 + 2x^2 = x^2 \cdot x^3 + x^2 \cdot 2 \overset{\text{From (1)}}{=} x^2(x^3 + 2)$$ ■

EXAMPLE 3 ***Factoring Out the Largest Common Factor***

Factor out the largest common factor in the expression

$$4x^5y^2 - 12x^3y + 8x^2y^3$$

SOLUTION
(a) 4, -12, and 8 contain the largest common factor 4.
(b) x^5, x^3, and x^2 contain the largest common factor x^2.
(c) y^2, y, and y^3 contain the largest common factor y.

Thus each term has the factor $4x^2y$ and

$$\begin{aligned} 4x^5y^2 - 12x^3y + 8x^2y^3 &= 4x^2y \cdot x^3y + 4x^2y(-3x) + 4x^2y(2y^2) \\ &= 4x^2y(x^3y - 3x + 2y^2) \end{aligned}$$

NOTE The terms in $x^3y - 3x + 2y^2$ have no common factors, so this is as far as we can go. ■

EXAMPLE 4 ***Factoring Out a Common Factor and Then Factoring Again***

Factor $(x^2 + y)(x - y) + (5x^2 - 4y)(x - y)$.

SOLUTION Each term has the factor $x - y$, so we may write

$$\begin{aligned} (x^2 + y)(x - y) + (5x^2 - 4y)(x - y) &\overset{\text{From (2)}}{=} [(x^2 + y) + (5x^2 - 4y)](x - y) \\ &= (6x^2 - 3y)(x - y) \end{aligned}$$

Since $6x^2$ and $-3y$ have the common factor 3, we can go further:

$$(6x^2 - 3y)(x - y) = [3 \cdot 2x^2 + 3(-y)](x - y) \overset{\text{From (1)}}{=} 3(2x^2 - y)(x - y)$$

Factoring Quadratics

The following formulas are important. They are derived from the multiplication formulas on p. 42.

Some Factored Quadratics

Quadratic	*Factored Form*	*Illustration*
Difference of two squares	$x^2 - y^2 = (x + y)(x - y)$ (Difference; Note opposite signs)	$x^2 - 9 = x^2 - 3^2 = (x + 3)(x - 3)$ (Opposite signs)
Square of the sum	$x^2 + 2xy + y^2 = (x + y)^2$ (Note plus signs)	$x^2 + 6x + 9 = (x + 3)^2$ (Plus signs)
Square of the difference	$x^2 - 2xy + y^2 = (x - y)^2$ (Note minus signs)	$x^2 - 6x + 9 = (x - 3)^2$ (Minus signs)
Product of two linear factors	$acx^2 + (ad + bc)xy + bdy^2 = (ax + by)(cx + dy)$	$6x^2 - 11x - 35 = (2x - 7)(3x + 5)$

EXAMPLE 5 ***Factoring a Difference of Two Squares***

Factor $x^2 - 25$.

SOLUTION $25 = 5^2$, so $x^2 - 25$ is a difference of two squares and

$$x^2 - 25 = x^2 - 5^2 = (x + 5)(x - 5) \quad ■$$

EXAMPLE 6 ***Factoring a Difference of Two Squares***

Factor $4a^2 - 9b^2$.

SOLUTION $4a^2 - 9b^2 = (2a)^2 - (3b)^2 = (2a + 3b)(2a - 3b)$

Recognizing a Perfect Square Trinomial

A trinomial that can be written as a square has one of two forms:

$$x^2 + 2xy + y^2 = (x + y)^2 \quad \text{or} \quad x^2 - 2xy + y^2 = (x - y)^2$$

(Same sign) (Same sign)

NOTE

(a) x^2 is a square with a plus sign in front of it.
(b) y^2 is a square with a plus sign in front of it.
(c) The middle term is twice the product of x and y (with a plus sign or a minus sign in front of it).

EXAMPLE 7 *Recognizing a Perfect Square Trinomial*

Factor $u^2 - 6u + 9$.

SOLUTION We check that the terms in (a), (b), and (c) of the Note on p. 48 are present: (a) $u^2 = (u)^2$; (b) $9 = 3^2$; (c) $-6u = -2(u)(3)$. Thus

$$u^2 - 6u + 9 = (u - 3)^2 = (u - 3)(u - 3)$$

WARNING Be Careful of Signs

$x^2 - 9$ can be factored, but $x^2 + 9$ cannot (we will see why not in Section 2.3).

$x^2 - 2x + 1 = (x - 1)^2$ can be written as a square, but $x^2 - 2x - 1$ cannot. ■

Before continuing, we note that some quadratic polynomials cannot be factored at all if we limit ourselves to real numbers. Such polynomials are called **irreducible.** We will say more about irreducible quadratics in Section 2.3.

Factoring Trinomials into a Product of Binomials

Consider the factoring

$$ax^2 + bx + c = (\blacksquare x + \blacksquare)(\blacksquare x + \blacksquare)$$

The product of these is c

The product of these is a

In order to factor $ax^2 + bx + c$, we look for the factors of a and factors of c so that the sum of outer and inner products is b.

EXAMPLE 8 *Factoring a Trinomial*

Factor $x^2 + 10x + 21$ into a product of linear factors with integer coefficients.

SOLUTION Suppose that

$$x^2 + 10x + 21 = (x + r)(x + s) = x^2 + (r + s)x + rs$$

Then, equating the coefficients of corresponding terms, we have

$$r + s = 10$$
$$rs = 21$$

r and s must be factors of 21. The factors of 21 are ± 1, ± 3, ± 7, ± 21. Only two of these add up to 10: $3 + 7$. Thus $r = 3$, $s = 7$, and

$$x^2 + 10x + 21 = (x + 3)(x + 7)$$

EXAMPLE 9 ***A Trinomial That Cannot Be Factored***

Factor $x^2 + 4x + 10$ into a product of linear factors with integer coefficients, if possible.

SOLUTION Suppose that

$$x^2 + 4x + 10 = (x + r)(x + s) = x^2 + (r + s)x + rs$$

Then

$$r + s = 4$$
$$rs = 10$$

The factors of 10 are ± 1, ± 10, ± 2, ± 5. No two of these numbers sum to 4 and have a product of 10, so we conclude that $x^2 + 4x + 10$ *cannot* be written as the product of linear factors with integer coefficients. ■

EXAMPLE 10 ***Factoring a Trinomial***

Write $2x^2 + x - 6$ as a product of linear factors with integer coefficients.

SOLUTION We write

$ps + rq$ = coefficient of $x = 1$

$$2x^2 + x - 6 = (px + r)(qx + s) = pqx^2 + (ps + rq)x + rs$$

$pq = 2$ $\quad$ rs = constant term $= -6$

Then $pq = 2$, so $p = 2$ and $q = 1$, or $p = 1$ and $q = 2$. Also, $rs = -6$, so r and $s = 1, -1, 2, -2, 3, -3, 6,$ or -6. Checking the various possibilities and using the fact that $ps + rq = 1$, we find that

$$2x^2 + x - 6 = (2x - 3)(x + 2)$$

Once you have factored a few quadratics, you will be able to check many of these possibilities in your head.

In Section 2.3 we will discuss a method for determining whether a given quadratic can be factored.

EXAMPLE 11 ***Factoring a Trinomial***

Factor $4x^2 - 19xy - 5y^2$.

SOLUTION We write

$$4x^2 - 19xy - 5y^2 = (ax + by)(cx + dy)$$

so $ac = 4$ and $bd = -5$. Then a and c are factors of 4, and b and d are factors of -5. Here are the possibilities:

$$(4x - y)(x + 5y) = 4x^2 + 19xy - 5y^2$$
$$(4x + y)(x - 5y) = 4x^2 - 19xy - 5y^2$$
$$(4x - 5y)(x + y) = 4x^2 - xy - 5y^2$$
$$(4x + 5y)(x - y) = 4x^2 + xy - 5y^2$$
$$(2x - y)(2x + 5y) = 4x^2 + 8xy - 5y^2$$
$$(2x + y)(2x - 5y) = 4x^2 - 8xy - 5y^2$$

Thus

$$4x^2 - 19xy - 5y^2 = (4x + y)(x - 5y)$$

Factoring Higher-Order Polynomials

It is generally very difficult to factor higher-order polynomials. We give two formulas here for factoring cubics. We will say more in Chapter 4.

Factoring the Sum or Difference of Cubes

Cubic	*Factored Form*	*Illustration*
Sum of two cubes	$x^3 + y^3 = (x + y)(x^2 - xy + y^2)$	$x^3 + 27 = x^3 + 3^3 = (x + 3)(x^2 - 3x + 9)$
Difference of two cubes	$x^3 - y^3 = (x - y)(x^2 + xy + y^2)$	$x^3 - 8 = x^3 - 2^3 = (x - 2)(x^2 + 2x + 4)$

EXAMPLE 12 *Factoring a Difference of Cubes*

Factor $8x^3 - 1$.

SOLUTION

$$8x^3 - 1 = (2x)^3 - 1^3 = (2x - 1)[(2x)^2 + (2x)(1) + 1^2]$$
$$= (2x - 1)(4x^2 + 2x + 1) \quad ■$$

EXAMPLE 13 *Factoring a Sum of Cubes*

Factor $x^3 + 64y^3$.

SOLUTION

$$x^3 + 64y^3 = x^3 + (4y)^3 = (x + 4y)[x^2 - x(4y) + (4y)^2]$$
$$= (x + 4y)(x^2 - 4xy + 16y^2)$$

Grouping

Sometimes a rearrangement of the terms of a polynomial will enable us to factor it.

EXAMPLE 14 ***Factoring by Grouping***

Factor $4xy - 6 + 2y - 12x$.

SOLUTION We write these terms in groups of two and see if each group has common linear factors:

First attempt:	$(4xy - 6) + (2y - 12x)$	Two groups
	$2(2xy - 3) + 2(y - 6x)$	Each group factored but no common linear factors
Second attempt:	$(4xy - 12x) + (2y - 6)$	Two groups (same terms written in a different order)
	$4x(y - 3) + 2(y - 3)$	Each group factored and each group has $y - 3$ as a factor

so

$$4xy - 6 + 2y - 12x = (4x + 2)(y - 3) = 2(2x + 1)(y - 3)$$ ■

EXAMPLE 15 ***Factoring by Grouping***

Factor $x^2 - 2x - y^2 + 1$.

SOLUTION We recognize that $x^2 - 2x + 1 = (x - 1)^2$. So we regroup to obtain

$$x^2 - 2x - y^2 + 1 = (x^2 - 2x + 1) - y^2 = (x - 1)^2 - y^2$$

Difference of squares
↓

$$= (x - 1 + y)(x - 1 - y)$$

Complete Factoring

Sometimes it is possible to factor more than once.

EXAMPLE 16 ***Factoring a Trinomial Completely***

Factor completely $x^4 - 5x^3 + 6x^2$.

SOLUTION We first factor out the largest common factor:

$$x^4 - 5x^3 + 6x^2 = x^2(x^2 - 5x + 6)$$

We observe that

$$x^2 - 5x + 6 = (x - 3)(x - 2)$$

Thus

$$x^4 - 5x^3 + 6x^2 = x^2(x - 3)(x - 2)$$

Factoring Expressions with Negative Exponents

EXAMPLE 17 *Factoring Out the Smallest (Most Negative) Exponent*

Factor the expression $4x^{-3} - 6x^{-4}$ completely by factoring out the term with the smallest exponent.

SOLUTION
(a) 4 and -6 contain the largest common factor 2.
(b) The smallest exponent is -4.

Therefore, we factor out $2x^{-4}$ to obtain

$$4x^{-3} - 6x^{-4} = 2x^{-4}(2x - 3)$$

Note that $x^{-4} \cdot x = x^{-4} \cdot x^1 = x^{-4+1} = x^{-3}$ ■

EXAMPLE 18 *Factoring a Difference of Squares with Negative Exponents*

Factor $4x^{-4} - 9x^{-6}$.

SOLUTION

$$4x^{-4} = (2x^{-2})(2x^{-2}) = (2x^{-2})^2$$

and

$$9x^{-6} = (3x^{-3})(3x^{-3}) = (3x^{-3})^2$$

Thus, we have a difference of squares and using the formula on p. 48, we have

$$4x^{-4} - 9x^{-6} = (2x^{-2})^2 - (3x^{-3})^2 = (2x^{-2} + 3x^{-3})(2x^{-2} - 3x^{-3})$$ ■

EXAMPLE 19 *Factoring an Expression with Negative Exponents*

Factor $4x^{-3} + 4x^{-5/6} + x^{4/3}$, $x > 0$.

SOLUTION Observe that

$$4x^{-3} = (2x^{-3/2})^2$$

$$x^{4/3} = (x^{2/3})^2$$

$$2(2x^{-3/2})(x^{2/3}) = 4x^{-9/6}x^{4/6} = 4x^{-(9/6)+(4/6)} = 4x^{-5/6}$$

Thus, if $a = 2x^{-3/2}$ and $b = x^{2/3}$

$$4x^{-3} + 4x^{-5/6} + x^{4/3} = a^2 + 2ab + b^2 = (a + b)^2 = (2x^{-3/2} + x^{2/3})^2$$

Problems 1.8

Readiness Check

I. Which of the following are factors of $4x^2 - 9y^2$?
a. $x - 3y$ b. $2x - 3y$ c. $2x - y$
d. $x + 3y$ e. $2x + 3y$ f. $2x + y$

II. ______ is a factor of $x^6 - 8$.
a. $x - 2$ b. $x - 4$ c. $x^2 - 2$
d. $x^2 - 4$ e. $x^2 + 4$

III. ______ is a factor of $y^9 + 1$.
a. $y + 1$ b. $y - 1$ c. $y^2 + 1$
d. $y^2 - 1$ e. $y^3 + 1$ f. $y^3 - 1$

IV. Which of the following is a linear factor of $2y^2 + 12y - 54$?
a. 2 b. $y + 6$ c. $y - 3$
d. $y^2 + 6y - 27$

V. Which of the following is the factorization of the difference of two squares?
a. $(x - y)^2$ b. $[(2a + b) - 3][(2a + b) + 3]$
c. $9x - 16y$ d. $(2x - 1)(3x + 2)$

In Problems 1–12 factor out the largest common factor in each expression.

1. $2x + 4$ 2. $3x^2 - 18$ 3. $10x + 25$
4. $6x - 2y^2$ 5. $8x + 12y$ 6. $x - 4x^2$
7. $ab + 2ac$ 8. $a^2b - 5a^3b^5$ 9. $8x^3 + 14x^2$
10. $15x^4y^7 + 25x^6y^3 - 30x^5y^4$
11. $(x + 2)(x - 3) + (x^2 + 1)(x + 2)$
12. $(a - 2b)(a + 3b)(b - a) - (4a + 5b)(a - b)(a - 2b)$

In Problems 13–66 write each expression as the product of factors with integer coefficients, if possible.

A. Differences of Squares

13. $x^2 - 1$ 14. $x^2 - 36$
15. $z^2 - w^2$ 16. $9x^2 - 16z^2$
17. $(x - 2)^2 - 9$ 18. $(x + 1)^2 - 16$
19. $(y + 4)^2 - (z - 2)^2$ 20. $81 - a^2$
21. $64 - 36b^4$ 22. $1 - w^2$
23. $4z^2 - 25$ 24. $x^{100} - y^{50}$
25. $w^6 - 9u^6$
* 26. $w^8 - 16$ [Hint: Factor twice.]

B. Square of Sum or Difference

27. $w^2 - 2w + 1$ 28. $z^2 + 4z + 4$
29. $q^2 - 6q + 9$ 30. $9x^2 + 6x + 1$
31. $16w^4 - 24w^2z + 9z^2$ 32. $16r^2 - 40rs + 25s^2$
33. $-x^2 + 4x - 4$ [Hint: First factor out -1.]
34. $2w^2 + 32w + 128$ [Hint: First factor out 2.]
35. $x^4 - 6x^2 + 9$ 36. $z^8 - 12z^4 + 36$
37. $16w^4 - 8w^2 + 1$ 38. $x^2y^2 - 6xyz + 9z^2$
39. $3r^2 - 30r + 75$ * 40. $4x^4 + 4x^2y + y^2 - 4z^2$
* 41. $a^6 + 2a^3b^3 + b^6 - 9$

C. Product of Two Linear Factors

42. $x^2 - 3x + 2$ 43. $x^2 + 7x + 10$
44. $x^2 - 7x + 10$ 45. $x^2 + 7x - 10$
46. $x^2 - 7x - 10$ 47. $x^2 + x - 6$
48. $x^2 - 15x + 56$ 49. $x^2 - 12x - 28$
50. $x^2 + 13x + 42$ 51. $x^2 - 13x + 42$
52. $x^2 + 2x + 4$ 53. $x^2 + 2x - 4$
54. $3x^2 - 12x + 12$
55. $-2x^2 + 6x - 4$ [Hint: First factor out -2.]
56. $4x^2 - 16x - 48$ 57. $5x^2 - 3x - 2$
58. $4x^2 + 8x + 3$ 59. $12x^2 + 13x - 14$
60. $21x^2 + 29x - 10$ 61. $x^2 - 3xy + 2y^2$
62. $x^2 - xy - 6y^2$ 63. $6x^2 - 13xy - 5y^2$
64. $35x^2 + 29xy + 6y^2$
65. $x^2 + (m - n)x - mn$, m and n are integers
66. $2x^2 - 2(m + n)x + 2mn$

In Problems 67–76 factor each cubic.

67. $x^3 - 27$ 68. $x^3 + 8$ 69. $x^3 + 64$
70. $8x^3 - 1$ 71. $8x^3 + 1$ 72. $27x^3 - y^3$
73. $x^3 + 27y^3$ 74. $(x + 2)^3 + (x - 5)^3$
75. $2x^3y^3 + 2y^3z^3$ 76. $64x^9y^6 - 0.001z^3w^6$

In Problems 77–82 factor by grouping.

77. $xy + y^2 - x - y$ 78. $wz - z^2 - 2z + 2w$
79. $2x^2y + 4x + 9xy^3 + 18y^2$ 80. $xz - 3xw + 2yz - 6yw$
81. $x^2 - 4xy - 4 + 4y^2$ 82. $x^2 - z^2 - 6xy + 9y^2$

Answers to Readiness Check

I. b, e II. c III. a, e IV. c V. b

In Problems 83–98, factor completely.

83. $2x^2 - 2$
84. $-4y^2 + 16$
85. $-6x^2 - 12x - 6$
86. $6p^2 + 5pq - 6q^2$
87. $-3x^2 + 21x - 30$
88. $x^3 + 13x^2 + 42x$
89. $x^5 - 8x^2$
90. $-2x^2y^2 - 2xy^2 + 6y^2$
91. $x^3y - x^2y^2 - 6xy^3$
92. $x^2z^2 - 4y^2z^2$
93. $(x + y)^2 - 4xy$
94. $(x - y)^2 + 4xy$
95. $(x^2 - y^2)^2 + 4x^2y^2$
96. $(x^2 + y^2)^2 - 4x^2y^2$
97. $(x + 1)(x^2 + 3x + 2) + (x + 1)(x^2 - 1)$
98. $(x - 3)(x^2 - 7x + 12) + (x^2 - 27)(x - 3)$

In Problems 99–104 factor the given expression completely by factoring out the term with the smallest exponent.

99. $x^{-2} - x^{-5}$
100. $x^{-3} + x^{-1}$
101. $6x^{-5} + 15x^{-3}$
102. $4x^{-2} - 2x^{-3} + 6x^{-4}$
103. $3x^2 - 6x^{-2} + 12x^{-1}$
104. $10x^3 - 5x^{-4} + 15x^{-1} + 20x$

In Problems 105–116 factor the given expression if possible.

105. $x^2 - x^{-2}$
106. $x^{-8} - x^{-6}$
107. $x^{-4} + 2x + x^6$
108. $9x^4 + x^{-4} - 6$
109. $25x^2 - 20x^{-2} + 4x^{-6}$
*110. $4x^{-1} + 24x^{-5/4} + 36x^{-3/2}$
111. $y^{-2} - 3y^{-1} + 2$
112. $w^{-2} - 2w^{-1} - 24$
113. $u^{-4} - 4u^{-2} - 5$
114. $y^{-6} - 9y^{-3} + 8$
115. $2z^{-2} - 7z^{-1} - 4$
116. $6p^{-2} - 11p^{-1} - 10$

1.9 Simplifying Rational Expressions

In this book we will, at various times, encounter rational expressions.

Definition of a Rational Expression

A **rational expression** is a sum, difference, product, or quotient of terms, each taking the form $p(x)/q(x)$, where p and q are polynomials.

EXAMPLE 1 *Five Rational Expressions*

The following are rational expressions:

(a) $\dfrac{1}{x}$ (b) $x + \dfrac{3}{x^2}$ (c) $\dfrac{x^2 - 1}{x^3 + x^2 - 2}$

(d) $\dfrac{\dfrac{x^2 + 3x + 5}{4x - 3}}{\dfrac{x^3 + 2}{x^5 + 5x - 6}}$ (e) $x^3 + \dfrac{5x^5 - 1}{2x + 3} - \dfrac{x}{x + 2}$

In this section we will show how rational expressions can be simplified. To do so, we use several rules.

Five Rules for Adding, Subtracting, Multiplying, and Dividing Rational Expressions

Let $a(x)$, $b(x)$, $c(x)$ and $d(x)$ be rational expressions with $b(x) \neq 0$ and $d(x) \neq 0$.

Rule 1. $\dfrac{a(x)d(x)}{b(x)d(x)} = \dfrac{a(x)}{b(x)}$ Common factor property or cancellation law

Rule 2. $\dfrac{a(x)}{b(x)} + \dfrac{c(x)}{b(x)} = \dfrac{a(x) + c(x)}{b(x)}$ Addition property

Rule 3. $\dfrac{a(x)}{b(x)} - \dfrac{c(x)}{b(x)} = \dfrac{a(x) - c(x)}{b(x)}$ Subtraction property

Rule 4. $\dfrac{a(x)}{b(x)} \cdot \dfrac{c(x)}{d(x)} = \dfrac{a(x)c(x)}{b(x)d(x)}$ Multiplication property

Rule 5. $\dfrac{a(x)/b(x)}{c(x)/d(x)} = \dfrac{a(x)}{b(x)} \cdot \dfrac{d(x)}{c(x)}$, Division property
if $b(x), c(x), d(x) \neq 0$

Simplification Using the Common Factor Property

If the numerator and denominator of an algebraic expression have a common factor, then divide each by that factor to obtain a simpler expression.

EXAMPLE 2 ***Simplifying a Rational Expression by Canceling Common Factors***

Simplify $\dfrac{12x^2}{20x}$.

SOLUTION We factor top and bottom:

$$\frac{12x^2}{20x} = \frac{3x \cdot \cancel{4x}}{\cancel{4x} \cdot 5} = \frac{3x}{5}$$ The $4x$ factor canceled (assuming that $x \neq 0$) ■

EXAMPLE 3 ***Simplifying by Canceling Common Factors***

Simplify $\dfrac{x^2 - 5x + 6}{x^2 + 2x - 15}$.

SOLUTION $$\frac{x^2 - 5x + 6}{x^2 + 2x - 15} = \frac{\cancel{(x - 3)}(x - 2)}{\cancel{(x - 3)}(x + 5)} = \frac{x - 2}{x + 5}$$

WARNING You can divide out factors. You cannot divide out terms.

Correct: $\dfrac{4(x + 1)}{4(x^2 + 3)} = \dfrac{x + 1}{x^2 + 3}$

Incorrect: $\dfrac{x + 4}{x^2 + 4} = \dfrac{x}{x^2}$ This is nonsense; for example, if $x = 2$, then $\dfrac{x + 4}{x^2 + 4} = \dfrac{2 + 4}{2^2 + 4} = \dfrac{6}{8} = \dfrac{3}{4}$, but $\dfrac{x}{x^2} = \dfrac{2}{4} = \dfrac{1}{2}$ ■

That is, $\dfrac{x + 4}{x^2 + 4} \neq \dfrac{x}{x^2}$

Simplification by Addition and Subtraction

If two (or more) rational expressions have the same denominator, then they can be added or subtracted using property 2 or 3.

EXAMPLE 4 *Adding Two Rational Expressions with the Same Denominator*

Add $\dfrac{x+2}{x^2+1} + \dfrac{2x-3}{x^2+1}$.

SOLUTION

$$\frac{x+2}{x^2+1} + \frac{2x-3}{x^2+1} \underset{\text{Addition property}}{=} \frac{(x+2)+(2x-3)}{x^2+1} = \frac{3x-1}{x^2+1} \quad \blacksquare$$

EXAMPLE 5 *Subtracting Two Rational Expressions with the Same Denominator*

Subtract $\dfrac{2x+7}{3x^4+8} - \dfrac{3x^3-4}{3x^4+8}$.

SOLUTION

$$\frac{2x+7}{3x^4+8} - \frac{3x^3-4}{3x^4+8} \underset{\text{Subtraction property}}{=} \frac{(2x+7)-(3x^3-4)}{3x^4+8} \underset{7-(-4)=11}{=} \frac{2x+11-3x^3}{3x^4+8}$$

Using the Multiplication and Division Properties

EXAMPLE 6 *Multiplying Two Rational Expressions and Simplifying*

Simplify $\dfrac{4}{x} \cdot \dfrac{x^3}{7}$.

SOLUTION

$$\frac{4}{x} \cdot \frac{x^3}{7} \underset{\text{Multiplication property}}{=} \frac{4x^3}{7x} = \frac{4x^2 \cdot \cancel{x}}{7 \cdot \cancel{x}} \underset{\text{Cancel common factors}}{=} \frac{4x^2}{7} \quad \blacksquare$$

EXAMPLE 7 *Multiplying Two Rational Expressions and Simplifying*

Simplify $\dfrac{4x^3}{y^5} \cdot \dfrac{y^7}{6x^4}$.

SOLUTION

Cancel common factors

$$\frac{4x^3}{y^5} \cdot \frac{y^7}{6x^4} = \frac{4x^3y^7}{6x^4y^5} = \frac{2 \cdot \cancel{2} \cdot \cancel{x^3} \cdot \cancel{y^5} \cdot y^2}{3 \cdot \cancel{2} \cdot \cancel{x^3} \cdot x \cdot \cancel{y^5}} \overset{\downarrow}{=} \frac{2y^2}{3x} \quad \blacksquare$$

EXAMPLE 8 *Using the Division Rule*

Simplify $\dfrac{1/x}{3/x^2}$.

SOLUTION

Division property; Common factor property

$$\frac{1/x}{3/x^2} \overset{\downarrow}{=} \frac{1}{x} \cdot \frac{x^2}{3} = \frac{x^2}{3x} = \frac{x \cdot \cancel{x}}{3 \cdot \cancel{x}} \overset{\downarrow}{=} \frac{x}{3}$$

The following useful rule can be used for dividing rational expressions:

To divide $\dfrac{a(x)}{b(x)}$ by $\dfrac{c(x)}{d(x)}$, multiply $\dfrac{a(x)}{b(x)}$ by $\dfrac{d(x)}{c(x)}$.

EXAMPLE 9 *Dividing Rational Expressions and Simplifying*

Simplify

$$\frac{\dfrac{x^2 + x - 6}{x^2 + 3x - 4}}{\dfrac{x^2 + 4x + 3}{x^2 + 9x + 20}}$$

SOLUTION Using the rule given above, we have

$$\frac{\dfrac{x^2 + x - 6}{x^2 + 3x - 4}}{\dfrac{x^2 + 4x + 3}{x^2 + 9x + 20}} = \frac{x^2 + x - 6}{x^2 + 3x - 4} \cdot \frac{x^2 + 9x + 20}{x^2 + 4x + 3}$$

Factor

$$\overset{\downarrow}{=} \frac{(x - 2)\cancel{(x + 3)}}{\cancel{(x + 4)}(x - 1)} \cdot \frac{\cancel{(x + 4)}(x + 5)}{\cancel{(x + 3)}(x + 1)}$$

Divide out like factors ↓ Multiply through ↓

$$= \frac{(x-2)(x+5)}{(x-1)(x+1)} = \frac{x^2+3x-10}{x^2-1}$$

Adding and Subtracting Rational Expressions

Consider the addition problem

$$\frac{1}{2} + \frac{1}{3} = \frac{3}{2 \cdot 3} + \frac{2}{2 \cdot 3} = \frac{3}{6} + \frac{2}{6} = \frac{5}{6}$$

The number 6 is the **least common multiple** (LCM) of the numbers 2 and 3.

We use the same idea to add or subtract rational expressions. The idea is to use the common factor property to multiply and divide each rational expression by some term so that each denominator equals the LCM of all the denominators of the given expressions. Then we add or subtract, using the addition or subtraction property.

EXAMPLE 10 *Adding Two Rational Expressions*

Add $\dfrac{3}{x-2} + \dfrac{2}{x+5}$.

SOLUTION The LCM of $x - 2$ and $x + 5$ is $(x - 2)(x + 5)$. Then

$$\frac{3}{x-2} + \frac{2}{x+5} = \frac{3}{x-2} \cdot \frac{x+5}{x+5} + \frac{2}{x+5} \cdot \frac{x-2}{x-2}$$

$$= \frac{3x+15}{(x-2)(x+5)} + \frac{2x-4}{(x+5)(x-2)}$$

$$= \frac{(3x+15)+(2x-4)}{(x+5)(x-2)} = \frac{5x+11}{(x+5)(x-2)}$$

Optional step ↓

$$= \frac{5x+11}{x^2+3x-10}$$ ■

EXAMPLE 11 *Adding Two Rational Expressions*

Combine terms: $x - \dfrac{1}{x}$.

SOLUTION $x = \dfrac{x}{1}$, so the LCM is x. Then

$$x - \frac{1}{x} = \frac{x}{1} \cdot \frac{x}{x} - \frac{1}{x} = \frac{x^2}{x} - \frac{1}{x} = \frac{x^2-1}{x}$$

EXAMPLE 12 ***Adding and Subtracting Rational Expressions***

Combine terms: $\dfrac{3}{x-1} + \dfrac{4}{(x-1)^2} - \dfrac{2}{(x-1)^3}$.

SOLUTION The LCM is $(x-1)^3$. Then

$$\begin{aligned}
&\frac{3}{x-1} + \frac{4}{(x-1)^2} - \frac{2}{(x-1)^3} \\
&\qquad = \frac{3}{x-1} \cdot \frac{(x-1)^2}{(x-1)^2} + \frac{4}{(x-1)^2} \cdot \frac{x-1}{x-1} - \frac{2}{(x-1)^3} \\
&\qquad = \frac{3(x-1)^2}{(x-1)^3} + \frac{4(x-1)}{(x-1)^3} - \frac{2}{(x-1)^3} \\
&\qquad = \frac{3(x-1)^2 + 4(x-1) - 2}{(x-1)^3} \\
&\qquad = \frac{3(x^2-2x+1) + 4x - 4 - 2}{(x-1)^3} \\
&\qquad = \frac{3x^2 - 6x + 3 + 4x - 6}{(x-1)^3} = \frac{3x^2 - 2x - 3}{(x-1)^3}
\end{aligned}$$

In this book, when we ask you to simplify a rational expression that is the sum of two or more terms, we will require that the answer be given as a rational function in *lowest terms*—that is, as a rational function $r(x) = p(x)/q(x)$, for which $p(x)$ and $q(x)$ have no common factors.

Complex Fractions

A **complex fraction** is a quotient in which there are fractions in the numerator or denominator or both.

To Simplify a Complex Fraction

Step 1 Find the LCM of the numerator and add the terms in the numerator to obtain a single rational function.
Step 2 Do the same for the terms in the denominator.
Step 3 Divide the result using the division property.

EXAMPLE 13 ***Simplifying a Complex Fraction***

Simplify $\dfrac{\dfrac{1}{x} - \dfrac{3}{2}}{\dfrac{2}{x-2} + \dfrac{5}{x}}$.

SOLUTION

Step 1: The LCM of the numerator is $2x$:

$$\frac{1}{x}-\frac{3}{2}=\frac{1}{x}\cdot\frac{2}{2}-\frac{3}{2}\cdot\frac{x}{x}=\frac{2-3x}{2x}$$

Step 2: The LCM of the denominator is $x(x-2)$:

$$\frac{2}{x-2}+\frac{5}{x}=\frac{2x}{x(x-2)}+\frac{5(x-2)}{x(x-2)}$$
$$=\frac{2x+5x-10}{x(x-2)}=\frac{7x-10}{x(x-2)}$$

Step 3:

Division property ↓ Cancel common factors ↓

$$\frac{\frac{1}{x}-\frac{3}{2}}{\frac{2}{x-2}+\frac{5}{x}}=\frac{\frac{2-3x}{2x}}{\frac{7x-10}{x(x-2)}}=\frac{2-3x}{2\not{x}}\cdot\frac{\not{x}(x-2)}{7x-10}=\frac{(2-3x)(x-2)}{2(7x-10)}$$

Optional step ↓

$$=\frac{-3x^2+8x-4}{14x-20}$$ ■

EXAMPLE 14 ***Simplifying an Expression That Comes Up in Calculus***

Simplify $\dfrac{\frac{1}{x+h}-\frac{1}{x}}{h}$.

SOLUTION

$$\frac{\frac{1}{x+h}-\frac{1}{x}}{h}=\frac{1}{h}\left(\frac{1}{x+h}-\frac{1}{x}\right)$$
$$=\frac{1}{h}\left(\frac{1}{(x+h)}\cdot\frac{x}{x}-\frac{1}{x}\cdot\frac{x+h}{x+h}\right)$$
$$=\frac{1}{h}\left[\frac{x-(x+h)}{x(x+h)}\right]=\frac{1}{\not{h}}\left[\frac{-\not{h}}{x(x+h)}\right]$$
$$=\frac{-1}{x(x+h)}$$

EXAMPLE 15 *Simplifying an Expression with Negative Exponents*

Simplify the expression

$$x(x+1)^{-2} + (3x-2)(x+1)^{-3} - 4(x+1)^{-5/2}$$

SOLUTION We first use the fact that $x^{-r} = \dfrac{1}{x^r}$.

$$x(x+1)^{-2} + (3x-2)(x+1)^{-3} - 4(x+1)^{-5/2} = \frac{x}{(x+1)^2} + \frac{3x-2}{(x+1)^3} - \frac{4}{(x+1)^{5/2}}$$

To make this easier, we first factor out the term in the denominator with the smallest exponent:

$$= \frac{1}{(x+1)^2}\left[x + \frac{3x-2}{x+1} - \frac{4}{(x+1)^{1/2}}\right]$$

The LCM of the denominator is $x + 1$:

$$= \frac{1}{(x+1)^2}\left[x\frac{x+1}{x+1} + \frac{3x-2}{x+1} - \frac{4(x+1)^{1/2}}{(x+1)^{1/2}(x+1)^{1/2}}\right]$$

$$= \frac{1}{(x+1)^2}\left[\frac{x^2+x}{x+1} + \frac{3x-2}{x+1} - \frac{4(x+1)^{1/2}}{x+1}\right]$$

$$= \frac{1}{(x+1)^2}\left[\frac{x^2+x+3x-2-4(x+1)^{1/2}}{x+1}\right]$$

$$= \frac{1}{(x+1)^2}\left[\frac{x^2+4x-2-4(x+1)^{1/2}}{x+1}\right]$$

$$= \frac{x^2+4x-2-4(x+1)^{1/2}}{(x+1)^3}$$ ■

EXAMPLE 16 *An Application in Electronics*

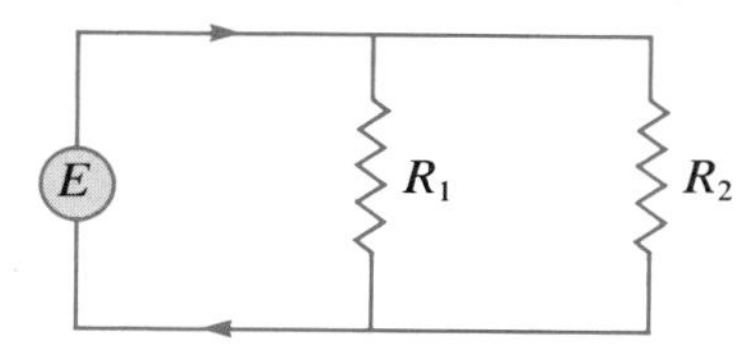

Figure 1 An electric circuit with two resistors.

Many electrical circuits contain **resistors.** A resistor opposes the current in the circuit. Resistance is measured in **ohms.** Consider the circuit in Figure 1. Here E stands for a source of **electromotive force** or **voltage.** The two resistors in the circuit are in parallel. It is a law of physics that when the resistors are in parallel the total resistance R in the circuit is given by

$$\frac{1}{R} = \frac{1}{R_1} + \frac{1}{R_2}$$

Write R in terms of R_1 and R_2.

SOLUTION

Multiply numerator and denominator by R_1R_2

$$R = \frac{1}{1/R} = \frac{1}{\dfrac{1}{R_1} + \dfrac{1}{R_2}} = \frac{R_1R_2}{\dfrac{1}{R_1} \cdot R_1R_2 + \dfrac{1}{R_2} \cdot R_1R_2} = \frac{R_1R_2}{R_2 + R_1} = \frac{R_1R_2}{R_1 + R_2}$$

Rational Expressions That Occur in Calculus

Some of the techniques that we used in this section can be used to rewrite certain expressions involving rational exponents.

EXAMPLE 17 ***Rationalizing the Denominator***

Rewrite $\dfrac{\sqrt{x} + 4}{\sqrt{x} - 4}$ so that there is no radical in the denominator.

SOLUTION We use the fact that $(a - b)(a + b) = a^2 - b^2$. If we multiply $\sqrt{x} - 4$ by $\sqrt{x} + 4$, we obtain

$$(\sqrt{x} - 4)(\sqrt{x} + 4) = (\sqrt{x})^2 - 4^2 = x - 16$$

Thus

$$\frac{\sqrt{x} + 4}{\sqrt{x} - 4} \cdot \frac{\sqrt{x} + 4}{\sqrt{x} + 4} = \frac{(\sqrt{x} + 4)^2}{x - 16} = \frac{(\sqrt{x})^2 + 2(4)\sqrt{x} + 4^2}{x - 16}$$

$$= \frac{x + 8\sqrt{x} + 16}{x - 16}$$

The process used in Example 17 is called **rationalizing the denominator.** See also page 32.

Rationalizing the Denominator

A rational expression of the form $\dfrac{a}{\sqrt{b} + c}$ or $\dfrac{a}{\sqrt{b} - c}$ can be written with no radical in the denominator by carrying out the following procedure:

$$\frac{a}{\sqrt{b} + c} = \frac{a}{\sqrt{b} + c} \cdot \frac{\sqrt{b} - c}{\sqrt{b} - c} = \frac{a(\sqrt{b} - c)}{(\sqrt{b})^2 - c^2} = \frac{a(\sqrt{b} - c)}{b - c^2}$$

Similarly,

$$\frac{a}{\sqrt{b} - c} = \frac{a}{\sqrt{b} - c} \cdot \frac{\sqrt{b} + c}{\sqrt{b} + c} = \frac{a(\sqrt{b} + c)}{b - c^2}$$

In calculus it is sometimes necessary to **rationalize the numerator.**

EXAMPLE 18 ***Rationalizing the Numerator***

Rationalize the numerator: $\dfrac{\sqrt{x+h}-\sqrt{x}}{h}$.

SOLUTION Using the same idea as in Example 17, we have

$$\begin{aligned}\frac{\sqrt{x+h}-\sqrt{x}}{h} &= \frac{\sqrt{x+h}-\sqrt{x}}{h}\cdot\frac{\sqrt{x+h}+\sqrt{x}}{\sqrt{x+h}+\sqrt{x}}\\ &= \frac{(\sqrt{x+h})^2-(\sqrt{x})^2}{h(\sqrt{x+h}+\sqrt{x})}\\ &= \frac{(x+h)-x}{h(\sqrt{x+h}+\sqrt{x})} = \frac{\cancel{h}}{\cancel{h}(\sqrt{x+h}+\sqrt{x})}\\ &= \frac{1}{\sqrt{x+h}+\sqrt{x}}\end{aligned}$$

Problems 1.9

Readiness Check

I. Which of the following is $\dfrac{2}{x-y}-\dfrac{7}{y-x}$ in lowest terms?

a. $\dfrac{9}{x-y}$ b. $\dfrac{9y-9x}{x^2-y^2}$ c. $\dfrac{-5}{x-y}$ d. $\dfrac{5y-5x}{x^2-y^2}$ e. $\dfrac{9}{y-x}$

II. $\dfrac{(x+2y)^2}{x^2-4y^2} =$ ______.

a. $x+2y$ b. $\dfrac{x+2y}{x-2y}$ c. $\dfrac{x-2y}{x+2y}$ d. x^2+4y^2 e. 1

III. Which of the following is $\dfrac{m^{-1}+z^{-1}}{m^{-1}-z^{-1}}$?

a. -1 b. $\dfrac{z+m}{z-m}$ c. z d. $\dfrac{m-z}{m+z}$

IV. Which of the following is the least common multiple of the denominators used to compute $\dfrac{3}{x^2-4x+4}-\dfrac{5}{x^2-4}$?

a. $(x-2)(x+2)$ b. $(x-2)^3(x+2)$ c. $(x-2)^2(x+2)^2$ d. $(x-2)^2(x+2)$

V. Which of the following is $\dfrac{(x+3)^2-12x}{(x-3)^2}$ in reduced form?

a. $-(1+12x)$ b. $-6x$ c. 1 d. $6x$ e. $-12x$ f. $1-12x$

In Problems 1–70 simplify, if possible, the given rational expression.

1. $\dfrac{6x}{3}$
2. $\dfrac{5}{10x}$
3. $\dfrac{2x+4}{6x^2+8}$
4. $\dfrac{3z}{6z^3}$
5. $\dfrac{12y^4}{24y^7}$
6. $\dfrac{4xz}{2xz}$
7. $\dfrac{x^2y^2}{x^3y^3}$
8. $\dfrac{4s+3}{12s^2+9s}$
9. $\dfrac{3}{x}\cdot\dfrac{x}{6}$

Answers to Readiness Check

I. a II. b III. b IV. d V. c

10. $\dfrac{y}{16} \cdot \dfrac{4}{3y^2}$

11. $\dfrac{64}{x^3} \cdot \dfrac{x^5}{12}$

12. $\dfrac{3}{s^2} \cdot \dfrac{s^3+1}{6}$

13. $\dfrac{x}{x+1} \cdot \dfrac{(x+1)^2}{x^4}$

14. $\dfrac{a^n}{b^n} \cdot \dfrac{b^{n-1}}{a^{n-1}}, \; a, b \neq 0$

15. $\dfrac{x/(x+1)}{x/(x+2)}$

16. $\dfrac{(x+1)/x}{(x+1)^2/(x+2)}$

17. $\dfrac{z/(z+1)}{(z+1)/z}$

18. $\dfrac{(w+2)/(w+3)}{(w+3)/(w+4)}$

19. $\dfrac{(w^2+1)/w}{w+1/w}$

20. $\dfrac{x^2-1}{(x-1)^2}$

21. $\dfrac{x^2+1}{(x+1)^2}$

22. $\dfrac{x^2-3x+2}{x^2-6x+8}$

23. $\dfrac{x^2-2x-3}{x^2+x-12}$

24. $\dfrac{y^2+3y-18}{y^2-6y+9}$

25. $\dfrac{z^2+2z-8}{z^2-5z-36}$

26. $\dfrac{w^2+4w-21}{w^2+8w+7}$

27. $\dfrac{6x^2-5x+1}{4x^2-4x+1}$

28. $\dfrac{y^2-3y+2}{6y^2-8y-8}$

29. $\dfrac{36z^2-24z-5}{36z^2-72z+35}$

30. $\dfrac{x^2-y^2}{(x-y)^2}$

31. $\dfrac{x^3-y^3}{(x-y)^3}$

32. $\dfrac{x^3+y^3}{(x+y)^3}$

33. $\dfrac{x^2-9}{x^2-16} \div \dfrac{x^2+3x-4}{x^2+8x+16}$

34. $\dfrac{\dfrac{x^2+2x-15}{x^2-2x-24}}{\dfrac{x^2-7x+6}{x^2+9x+20}}$

35. $\dfrac{(4z-12)/(z^2-4z+3)}{(5z+25)/(z^2+8z+15)}$

36. $\dfrac{(z^2+2z+1)/(z^2+4z+3)}{(z^2-6z-7)/(z^2+9z+8)}$

37. $2+\dfrac{y}{2}$

38. $6-\dfrac{5}{z}$

39. $\dfrac{1}{2}+2x$

40. $\dfrac{3}{s}+\dfrac{7}{s}$

41. $\dfrac{x-3}{x^2}+\dfrac{2x-5}{x^2}$

42. $\dfrac{5}{x^3}-\dfrac{7x^2+5}{x^3}$

43. $\dfrac{\frac{1}{2}}{1}-\dfrac{\frac{1}{2}}{x-1}$

44. $\dfrac{1}{x}+\dfrac{2}{x^2}$

45. $\dfrac{5}{z}-\dfrac{z}{4}$

46. $\dfrac{1}{z^2}-\dfrac{1}{z-1}$

47. $\dfrac{2}{3s}-\dfrac{4}{5s}$

48. $\dfrac{3}{2x}+\dfrac{x}{5x^2}$

49. $\dfrac{7}{2y^2}+\dfrac{8}{3y^2}$

50. $\dfrac{1}{x-1}-\dfrac{1}{x-2}$

51. $\dfrac{12}{x-b}+\dfrac{3}{x-a}$

52. $\dfrac{1}{x-1}+\dfrac{1}{x-2}+\dfrac{1}{x-3}$

53. $\dfrac{2}{x-3}-\dfrac{4}{x+5}$

54. $\dfrac{1}{x^2-1}+\dfrac{3}{x-1}$

55. $\dfrac{2x}{x^2-4}+\dfrac{5}{x-2}-\dfrac{3}{x+2}$

56. $\dfrac{6}{x^2-5x+4}-\dfrac{3}{x-1}$

57. $\dfrac{4}{x+y}-\dfrac{3}{y-x}+\dfrac{2}{x^2-y^2}$

58. $\dfrac{2x}{x^3}-\dfrac{4}{x^5}+\dfrac{9}{x^6}$

59. $\dfrac{3x}{(x-2)^2}-\dfrac{4}{x-2}$

60. $\dfrac{1}{x+2}+\dfrac{5x+3}{(x+2)^2}-\dfrac{6x^2+3x-2}{(x+2)^3}$

61. $\dfrac{3}{y-3}-\dfrac{7}{y+6}+\dfrac{2y-3}{y^2+3y-18}$

62. $\dfrac{-2}{z+4}-\dfrac{3}{z-2}+\dfrac{7z-5}{z^2+2z-8}$

63. $\dfrac{2+\dfrac{1}{x}}{3-\dfrac{1}{x}}$

64. $\dfrac{\dfrac{1}{x}-\dfrac{3}{x-1}}{\dfrac{2}{x-1}+\dfrac{4}{x}}$

65. $\dfrac{\dfrac{3}{x}-\dfrac{5}{y}}{\dfrac{6}{y}+\dfrac{2}{x}}$

66. $\dfrac{\dfrac{4}{x-2}+\dfrac{3}{x+5}}{(2x-5)/(x^2+3x-10)}$

67. $\dfrac{\dfrac{1}{x}-\dfrac{3}{x^2}+\dfrac{7x}{x^3}}{\dfrac{-4}{x}+\dfrac{3x-2}{x^2}+\dfrac{3x^2-5x+2}{x^3}}$

68. $\dfrac{\dfrac{1}{x-1}+\dfrac{3}{(x+2)^2}-\dfrac{x}{x+3}}{\dfrac{4}{x-5}+\dfrac{2}{x+2}}$

69. $\dfrac{\dfrac{1}{(x+h)^2}-\dfrac{1}{x^2}}{h}$

70. $\dfrac{\dfrac{1}{(x-h)^3}-\dfrac{1}{x^3}}{-h}$

In Problems 71–76 simplify each expression containing negative exponents by using the procedure of Example 15.

71. $x^{-1} + x^{-3}$
72. $(x + 2)^{-1/2} + x(x + 2)^{-3/2}$
73. $(4x + 3)(x + 5)^{-1/3} + (2x - 7)(x + 5)^{-4/3}$
74. $x(2x - 5)^{-2} - 5(2x - 5)^{-1} + (x + 1)(2x - 5)^{-3}$
75. $x^{-3} + x^{-4} - x^{-5} - x^{-6}$
76. $(x + 2)(x^2 + 3x + 10)^{-3/2} - (4x - 1)(x^2 + 3x + 10)^{-1/2} + x^2(x^2 + 3x + 10)^{-1}$

In Problems 77–88 rationalize the denominator.

77. $\dfrac{3}{\sqrt{x}}$
78. $\dfrac{5}{\sqrt[3]{x}}$
79. $\dfrac{\sqrt{5}}{\sqrt[4]{x+1}}$
80. $\dfrac{1}{\sqrt{7} - \sqrt{3}}$
81. $\dfrac{1}{\sqrt{5} + \sqrt{3}}$
82. $\dfrac{1}{\sqrt{x} - 1}$
83. $\dfrac{-1}{2 - \sqrt{2x}}$
84. $\dfrac{3 - 2\sqrt{x}}{1 + 4\sqrt{x}}$
85. $\dfrac{\sqrt{x} - \sqrt{y}}{\sqrt{x} + \sqrt{y}}$
86. $\dfrac{2\sqrt{x} - 3\sqrt{y}}{5\sqrt{x} + 2\sqrt{y}}$

* 87. $\dfrac{2}{1 - \sqrt[3]{x}}$ [Hint: Use the formula for the difference of two cubes on p. 51.]

* 88. $\dfrac{\sqrt[3]{x} + \sqrt[3]{y}}{\sqrt[3]{x} - \sqrt[3]{y}}$

In Problems 89–93 rationalize the numerator.

89. $\dfrac{\sqrt{x + h + 4} - \sqrt{x + 4}}{h}$
90. $\dfrac{\sqrt{(x + h)^2 + 2} - \sqrt{x^2 + 2}}{h}$
91. $\dfrac{\dfrac{1}{\sqrt{x+2}} - \dfrac{1}{\sqrt{x}}}{2}$
* 92. $\dfrac{\sqrt[3]{x + h} - \sqrt[3]{x}}{h}$
93. $\dfrac{\dfrac{1}{\sqrt{(x+h)^2}} - \dfrac{1}{x}}{h}$
94. What is the total resistance in the circuit of Figure 1 if $R_1 = 5$ ohms and $R_2 = 3$ ohms?
95. Answer the question of Problem 94 if $R_1 = 2 \times 10^{-3}$ ohms and $R_2 = 5 \times 10^{-4}$ ohms.

Summary Outline of Chapter 1

- **Real Numbers $\mathbb{R}$**

  The **natural numbers,** denoted by $\mathbb{N}$, are 1, 2, 3, . . . p. 3
  The **integers,** denoted by $\mathbb{Z}$, are $0, \pm 1, \pm 2, \ldots$ p. 4
  The **rational numbers,** denoted by $\mathbb{Q}$, are all numbers that can be written p. 4

$$r = \frac{m}{n}$$

  where m and n are integers and $n \neq 0$.
  The **irrational numbers** consist of all real numbers that are not rational. p. 4

- **Rules of Signs:** Let a and b be real numbers

  $-a = (-1)a$ p. 8
  $-(-a) = a$ p. 8
  $(-a)(-b) = ab$ p. 9
  $-a + (-b) = -(a + b)$ p. 9

- **Rules of Fractions**

  $\dfrac{a}{b} + \dfrac{c}{b} = \dfrac{a + c}{b}$ p. 10

  $\dfrac{ad}{bd} = \dfrac{a}{b}$ p. 10

$\frac{a}{b} + \frac{c}{d} = \frac{ad + bc}{bd}$ p. 10

$\frac{a}{b} \cdot \frac{c}{d} = \frac{ac}{bd}$ p. 10

- **Properties of Inequalities**

If $a < b$, then $a + c < b + c$ p. 15

If $a < b$ and $b < c$, then $a < c$ p. 16

If $a < b$ and $c > 0$, then $ac < bc$ p. 16

If $a < b$ and $c < 0$, then $ac > bc$ p. 16

If $ab > 0$ and $a < b$, then $\frac{1}{a} > \frac{1}{b}$ p. 16

- **Properties of Absolute Value**

$|a| = a$ if $a \geq 0$ p. 17

$|a| = -a$ if $a < 0$ p. 17

$|-a| = |a|$ p. 17

$|ab| = |a||b|$ p. 17

$|a + b| \leq |a| + |b|$ p. 17

$|a|^2 = a^2$ p. 18

- **Properties of Exponents:** $a \neq 0$ and m and n are integers

$a^n = \underbrace{a \cdot a \cdot a \cdot \cdots \cdot a}_{n \text{ times}}$ if $n > 0$ p. 20

$a^{-n} = \frac{1}{a^n}$ $\quad a^{1/n} =$ the nth root of a $\quad a^{m/n} = (a^{1/n})^m$ p. 20

$a^0 = 1 \quad a^1 = a \quad \frac{a^m}{a^n} = a^{m-n} = \frac{1}{a^{n-m}}$ p. 21

$(ab)^n = a^n b^n \quad \left(\frac{a}{b}\right)^n = \frac{a^n}{b^n} \quad (a^m)^n = a^{mn}$ p. 21

$a^{-1} = \frac{1}{a} \quad \left(\frac{a}{b}\right)^{-1} = \frac{1}{(a/b)} = \frac{b}{a}$ p. 21

for $a > 0$, $a^{1/2} = \sqrt{a} > 0$ and $-a^{1/2} = -\sqrt{a} < 0$ p. 30

If $a, b > 0$, $\sqrt{ab} = \sqrt{a}\sqrt{b}$ and $\frac{\sqrt{a}}{\sqrt{b}} = \sqrt{\frac{a}{b}}$ p. 31

- **Properties of Polynomials**

A **polynomial** in x is an expression of the form

$$p(x) = a_n x^n + a_{n-1} x^{n-1} + \cdots + a_2 x^2 + a_1 x + a_0$$ p. 37

where $a_0, a_1, \ldots a_n$ are real numbers called the **coefficients** of the polynomial p.
To add (or subtract) two polynomials, add (or subtract) their corresponding coefficients p. 37

If a, b, and c are polynomials, then
$a(b + c) = ab + ac$ and $(a + b)c = ac + bc$ p. 40

$(x + y)^2 = x^2 + 2xy + y^2$ p. 42

$(x - y)^2 = x^2 - 2xy + y^2$ p. 42

$(x + y)(x - y) = x^2 - y^2$ p. 42

$(x + y)^3 = x^3 + 3x^2y + 3xy^2 + y^3$ p. 42

$(x - y)^3 = x^3 - 3x^2y + 3xy^2 - y^3$ p. 42

$x^3 + y^3 = (x + y)(x^2 - xy + y^2)$ p. 51

$x^3 - y^3 = (x - y)(x^2 + xy + y^2)$ p. 51

- **Rules for Simplifying Rational Expressions**

$\frac{a(x)d(x)}{b(x)d(x)} = \frac{a(x)}{b(x)}$ p. 55

$\frac{a(x)}{b(x)} + \frac{c(x)}{b(x)} = \frac{a(x) + c(x)}{b(x)}$ p. 56

$\frac{a(x)}{b(x)} - \frac{c(x)}{b(x)} = \frac{a(x) - c(x)}{b(x)}$ p. 56

$\frac{a(x)}{b(x)} \cdot \frac{c(x)}{d(x)} = \frac{a(x)c(x)}{b(x)d(x)}$ p. 56

$\frac{a(x)/b(x)}{c(x)/d(x)} = \frac{a(x)}{b(x)} \cdot \frac{d(x)}{c(x)}$ p. 56

Review Exercises for Chapter 1

1. Write 0.424242 . . . as the quotient of two integers in lowest terms.
2. Do the same for 3.215215215 . . .
3. Write $\frac{5}{6}$ as a repeating decimal.
4. Write $-\frac{3}{7}$ as a repeating decimal.

In Exercises 5–10 perform the indicated operation and give the answer in lowest terms.

5. $\frac{1}{5} - \frac{2}{7}$
6. $\frac{3}{5} + \frac{17}{40}$
7. $\frac{4}{3} \cdot \frac{15}{16}$
8. $\frac{5/7}{10/9}$
9. $\frac{3}{8}\left(\frac{4}{3} - \frac{11}{6}\right)$
10. $\dfrac{\frac{5}{9} + \frac{1}{4}}{\frac{3}{8} - \frac{17}{16}}$
11. Find the distance between -2 and 5.
12. Find the distance between -3 and -10.

In Exercises 13–20 write each expression as a rational number with no exponent.

13. 5^2
14. $(-3)^3$
15. $\left(\frac{1}{2}\right)^4$
16. 1.7^0
17. $3^{-1}9^3$
18. 10^{-4}
19. $15^2 \cdot 15^{-4}$
20. $\left(\frac{493}{1085}\right)^{-1}$

In Exercises 21–27 simplify the given expression. No answer should be left with a negative exponent.

21. $\frac{w^4}{w^2}$
22. $\left(\frac{2}{5c}\right)^2$
23. $\left(\frac{4}{5d^2}\right)^{-1}$
24. $(a^2b)(ab^2)$
25. $\left(\frac{3x}{4y^2}\right)^2\left(\frac{2x}{y}\right)^{-1}$
26. $\left(\frac{2u^2v}{3v^3u}\right)^{-2}$
27. $\left(\frac{4xy^{-3}z^2}{3x^2y^{-5}z}\right)^{-1}$

In Exercises 28–31 write each number in scientific notation.

28. 23,651
29. 0.0003729
30. The mass of a proton is 0.0000000000000000000000000016726871 kilogram.
31. The distance from the planet Pluto to the sun is approximately 3,666,100,000 miles.

In Exercises 32–35 use a calculator to estimate the given value.

32. $(5.67)^4$
33. $(13.203)^{-3}$
34. $(4.205 \times 10^5) \times (8.802 \times 10^6)$
35. $\frac{9.256 \times 10^5}{4.213 \times 10^8}$

In Exercises 36–47 compute the indicated value, if it exists. Use a calculator only where indicated.

36. $9^{5/2}$
37. $9^{-3/2}$
38. $16^{-1/4}$

39. $\left(\frac{1}{16}\right)^{-1/2}$ 40. $(1000)^{4/3}$ 41. $17^{3/4}17^{5/4}$

42. $\frac{18^{5/8}}{18^{-3/8}}$ 43. $\frac{25^{3/2}25^{1/2}}{25^{9/2}25^{-11/2}}$ 44. $\sqrt{12}$

45. $\sqrt[3]{-14}$ 46. $(2.315)^{5/7}$ 47. $(14.218)^{-0.237}$

In Exercises 48–57 simplify the given expression.

48. $\sqrt{150}$ 49. $\sqrt[3]{270}$ 50. $\frac{7}{\sqrt{2}}$

51. $\frac{\sqrt{80}}{\sqrt{5}}$ 52. $x^{7/2}x^{-5/2}$ 53. $\frac{x^{3/4}x^{1/5}}{x^{3/5}}$

54. $(x^9y^6)^{2/3}$ 55. $\left(\frac{2w}{3w}\right)^4$ 56. $\left(\frac{u^{2/3}}{v^2}\right)^3\left(\frac{v^{3/2}}{u^{5/6}}\right)$

57. $\frac{x^{-2.3}x^{1.5}}{x^{4.6}x^{-3.2}}$

In Exercises 58–63 let $p(x) = 4x^2 - 5x + 3$ and $q(x) = x^3 - 2x^2 + 5x - 1$. Compute the following:

58. $p(x) + q(x)$ 59. $p(x) - q(x)$
60. $q(x) - p(x)$ 61. $3p(x) - 4q(x)$
62. deg pq 63. $p(x)q(x)$

In Exercises 64–74 perform the multiplication and determine the degree of the product.

64. $(x + 5)(x - 2)$ 65. $(x - 7)(x - 7)$
66. $(3x + 7)x$ 67. $(x + 8)(x - 8)$
68. $(x^2 + 3)(x^2 - 3)$ 69. $(x^2 - 2x + 2)(x - 4)$
70. $(x^3 - x)(x + 1)$ 71. $(3x^2 - 2x + 4)(2x + 3)$
72. $(4x - 2)^2$ 73. $(x + 2)^3$
74. $(x - 2)^4$

In Exercises 75–82 multiply and simplify each expression.

75. $(4x - y)^2$ 76. $(3x + 2y)^3$
77. $\left(\frac{2}{x} - \frac{3}{y}\right)^2$ 78. $(\sqrt{x} + 2\sqrt{y})^2$
79. $(x + 2y - 3z)^2$ 80. $(4x^2 - 3y^3)(2x + y^5)$
81. $(x^{3/4} - y^{1/2})(x^{1/4} + y^{3/2})$ 82. $\left(2 - \frac{5}{y}\right)^3$

In Exercises 83–105 factor each expression completely.

83. $3x^2 + 12$ 84. $4x^2 - 12y$
85. $4x^2 - 16y^2$ 86. $x^2 - 100$
87. $4x^2 - 9w^2$ 88. $w^2 - 12w + 36$
89. $z^2 + 14z + 49$ 90. $x^2 - 4xy + 4y^2$
91. $4r^2 - 12rs + 9s^2$ 92. $v^2 - v - 2$
93. $x^2 + 3x - 10$ 94. $y^2 - 17y + 70$
95. $4z^2 - 3z - 10$ 96. $5x^2 + 3x - 2$
97. $x^3 - 1000$ 98. $8x^3 + 27y^3$
99. $x^3y^3 - 8$ 100. $xy - y^2 - 3y + 3x$
101. $x^2 + 9y^2 - 16 + 6xy$ 102. $3x^2 - 48$
103. $x^3 + 5x^2 - 14x$ 104. $y^2z^2 - 6zy^2 + 9y^2$
105. $(x + 2)(x^2 - x - 6) + (x + 2)(x^2 - 4)$

In Exercises 106–123 simplify the given rational expression.

106. $\frac{10x}{5}$ 107. $\frac{x^3y^5}{x^2y^7}$

108. $\frac{x/(x + 3)}{x/(x + 4)}$ 109. $\frac{(x + 3)/(x + 4)}{(x + 4)/(x + 2)}$

110. $\frac{x^2 - 2x + 1}{x^2 + 9x - 10}$ 111. $\frac{3x^2 - x - 10}{x^2 - 4}$

112. $\frac{4x^2 - y^2}{(2x - y)^2}$

113. $\frac{(x^2 + 3x + 2)/(x^2 + 7x + 12)}{(x^2 - x - 6)/(x^2 + 2x - 3)}$

114. $\frac{(x^2 - 9)/(x^2 - 16)}{(x^2 - 7x + 12)/(x^2 - 5x + 4)}$

115. $\frac{1}{3} + x$ 116. $\frac{4}{x} - \frac{7}{x^2}$

117. $\frac{5}{x^2} + \frac{x^2}{5}$ 118. $\frac{3}{x} - \frac{4}{x - 2}$

119. $\frac{3x}{x - 2} + \frac{4}{x + 3}$ 120. $\frac{4x}{x - 1} - \frac{x^2}{(x - 1)^2}$

121. $\frac{1}{x} + \frac{2}{x + 1} + \frac{3}{x - 5}$ 122. $\frac{\frac{4}{x} - \frac{5}{x + 1}}{3 + \frac{4}{x^2}}$

123. $\frac{\frac{5}{x + 2} - \frac{6}{x - 2}}{\frac{2x + 3}{x^2 - 4}}$

In Exercises 124 and 125 rationalize the denominator.

124. $\frac{10}{3 - \sqrt{x}}$ 125. $\frac{\sqrt{x} + \sqrt{y}}{2\sqrt{x} - 3\sqrt{y}}$

Chapter 2

Equations and Inequalities in One Variable

2.1 Linear Equations

One of the central ideas in algebra is that of solving an equation. If x stands for an unknown quantity and is the only unknown quantity, then an equation involving x is called an **equation in one variable.**

EXAMPLE 1 *Four Equations in One Variable*

The following are equations in one variable.

(a) $2x = 4$
(b) $x - 7 = 5x + 2$
(c) $x^2 + 4x + 3 = 0$
(d) $\dfrac{x}{2} = \dfrac{5}{x}$

In this section we discuss a special type of equation in one variable called a linear equation.

Definition of a Linear Equation

A **linear equation** is an equation in one variable that can be written in the form

$$ax + b = 0 \tag{1}$$

where a and b are real numbers and $a \neq 0$.

Two equations are called **equivalent** if they have the same solutions.

We can solve a complicated-looking linear equation by using substitution laws from Chapter 1 to obtain an easier, equivalent equation. We then solve this equivalent equation.

We repeat here these useful laws (see pp. 8 and 9).

$c \neq 0$

Rules for Obtaining Equivalent Equations

If $a = b$, then $a + c = b + c$. Additive substitution law (2)

If $a = b$, then $ac = bc$. Multiplicative substitution law (3)

If $a = b$ and $c \neq 0$, then $\frac{a}{c} = \frac{b}{c}$ Division property (4)

EXAMPLE 2 ***Solving a Linear Equation by Using the Additive Substitution Law and the Division Property***

Solve the linear equation $x - 7 = 5x + 2$.

SOLUTION

$$x - 7 = 5x + 2$$

$$-x + x - 7 = -x + 5x + 2 \quad \text{From (2)}$$

$$-7 = 4x + 2$$

$$-7 - 2 = 4x + 2 - 2 \quad \text{From (2)}$$

$$-9 = 4x$$

$$-\frac{9}{4} = \frac{4}{4}x = x \quad \text{From (4)}$$

The solution is

$$x = -\frac{9}{4}$$

NOTE The equation can be written in the form (1). Since $4x = -9$, we find that $4x + 9 = 0$, which is equation (1) with $a = 4$ and $b = 9$. Note, too, that the equations $x - 7 = 5x + 2$ and $4x + 9 = 0$ are equivalent because each has the unique solution $x = -\frac{9}{4}$.

Check $x - 7 = 5x + 2$

$$-\frac{9}{4} - 7 \stackrel{?}{=} 5\left(-\frac{9}{4}\right) + 2 \quad \text{Substitute } -\tfrac{9}{4} \text{ for } x$$

$$-\frac{9}{4} - \frac{28}{4} \stackrel{?}{=} -\frac{45}{4} + \frac{8}{4}$$

$$-\frac{37}{4} = -\frac{37}{4} \quad \text{The answer is correct.}$$

Consider the linear equation

$$ax + b = 0, \qquad a \neq 0$$

Then

$$ax + b - b = 0 - b \qquad \text{From (2)}$$
$$ax = -b$$
$$\frac{ax}{a} = -\frac{b}{a} \qquad \text{From (4)}$$
$$x = -\frac{b}{a} \qquad (5)$$

We have shown that

The Solution to a Linear Equation

If $a \neq 0$, the linear equation $ax + b = 0$ has the unique solution $x = -\dfrac{b}{a}$.

EXAMPLE 3 ***Solving a Linear Equation on a Calculator***

Solve the equation $2.562x + 3.841 = 12.339$.

SOLUTION We use a calculator:

$$2.562x + 3.841 = 12.339$$
$$2.562x = 12.339 - 3.841 = 8.498$$

Round to 3 decimal places ↓

$$x = \frac{8.498}{2.562} = 3.316939891 \approx 3.317$$

Check

$$(2.562)(3.317) + 3.841 = 8.498154 + 3.841$$
$$= 12.339154$$
$$\approx 12.339$$

(The answer is not exact because of the rounding done above.)

WARNING Equations that look equivalent are not always equivalent. For example, the equation

$$x + 1 = 4 \qquad (6)$$

has the single solution $x = 3$. However, if we multiply both sides of (6) by x, we obtain

$$x(x + 1) = 4x \qquad (7)$$

Two solutions to (7) are $x = 3$ and $x = 0$. Thus the equations (6) and (7) are *not* equivalent because (7) has two solutions but (6) has only one. The solution $x = 0$ to (7) is called an **extraneous solution** because it is not a solution to equation (6).

Rules (3) and (4) on p. 71 give us equivalent equations when we multiply or divide by nonzero constants. If we multiply (or divide) by variables, we might gain (or lose) solutions. Put another way, whenever both sides of an equation are multiplied by an expression containing x (or some other variable), there is the possibility that **extraneous solutions** will occur. These are solutions to the new equation that are not solutions to the original equation. ■

One way to avoid the extraneous solution $x = 0$ is to insist that when we multiply both sides of an equation by x, that x not be allowed to take the value 0. If we do not restrict x to be nonzero, then we will always get $x = 0$ as one solution. For example

$$
\begin{aligned}
x^2 - 3x + 5 &= x^2 + 2x && \text{Original equation}\\
x(x^2 - 3x + 5) &= x(x^2 + 2x) && \text{Multiply both sides by } x\\
x^3 - 3x^2 + 5x &= x^3 + 2x^2 && \text{Multiply through}
\end{aligned}
$$

Evidently, $x = 0$ is a solution to the last equation, but it is not a solution to the original equation. In fact, the only solution to the original equation is $x = 1$.

We can get any number a to be an extraneous solution to an equation by multiplying both sides of the equation by $x - a$.

In Section 2.5 we will see another way that extraneous solutions can arise (see p. 116).

Many equations that are not linear reduce to linear equations.

EXAMPLE 4 *An Equation with Quadratic Terms That Reduces to a Linear Equation*

Solve the equation

$$(4x + 5)(3x - 2) = (2x + 1)(6x + 5) \tag{8}$$

SOLUTION We multiply through (see Section 1.7):

$$
\begin{aligned}
12x^2 + 7x - 10 &= 12x^2 + 16x + 5\\
7x - 10 &= 16x + 5 && \text{Subtract } 12x^2 \text{ from both sides (by Rule (2))}\\
-10 - 5 &= 16x - 7x && \text{Subtract 5 and } 7x \text{ from both sides}\\
-15 &= 9x\\
x &= -\frac{15}{9} = -\frac{5}{3}
\end{aligned}
$$

To check this answer, we substitute $-\frac{5}{3}$ for x in both sides of equation (8):

Check $$\left[4\left(-\frac{5}{3}\right)+5\right]\left[3\left(-\frac{5}{3}\right)-2\right]=\left(-\frac{20}{3}+5\right)(-5-2)$$
$$=\left(-\frac{5}{3}\right)(-7)=\frac{35}{3}$$

These are equal so answer is correct

$$\left[2\left(-\frac{5}{3}\right)+1\right]\left[6\left(-\frac{5}{3}\right)+5\right]=\left(-\frac{10}{3}+1\right)(-10+5)$$
$$=\left(-\frac{7}{3}\right)(-5)=\frac{35}{3}$$ ■

EXAMPLE 5 *An Equation with No Solution*

Solve the equation

$$\frac{3}{x(x-3)}-\frac{1}{x}=\frac{1}{x-3}$$

SOLUTION We first multiply both sides by the LCM (see p. 59) $x(x-3)$:

$$\cancel{x(x-3)}\cdot\frac{3}{\cancel{x(x-3)}}-\cancel{x}(x-3)\cdot\frac{1}{\cancel{x}}=x\cancel{(x-3)}\cdot\frac{1}{\cancel{x-3}}$$
$$3-(x-3)=x$$
$$3-x+3=x$$
$$6=2x$$
$$x=3$$

Thus, *if* our problem had a solution, then the solution would be $x = 3$.

Check $\frac{3}{3(3-3)}-\frac{1}{3}=\frac{1}{3-3}-\frac{1}{3}$; but $\frac{1}{3-3}=\frac{1}{0}$ is *not defined* because we cannot divide by 0. We conclude that *the problem does not have a solution*. This example illustrates the importance of checking your answer. The apparent solution $x = 3$ is an extraneous root. ■

EXAMPLE 6 *Another Equation with No Solution*

Solve

$$(x+3)(x+2)=(x+4)(x+1)$$

SOLUTION Multiplying through, we obtain

$$x^2+5x+6=x^2+5x+4$$
$$6=4 \quad \text{Subtract } x^2+5x \text{ from both sides}$$

The equation $6 = 4$ is false. We therefore conclude that the original equation has no solution.

TAUTOLOGY

Identity and Conditional Equation

An equation that holds for *all* values for which the expressions in the equation are defined is called an **identity.** An equation that has a solution and is not an identity is called a **conditional equation.** For example, the equation in Example 4 is a conditional equation because it holds only for $x = -\frac{5}{3}$.

EXAMPLE 7 *An Equation That Is Also an Identity*

Solve for x:

$$(x + 3)^2 = (x + 4)(x + 2) + 1 \tag{9}$$

SOLUTION Multiplying through on both sides, we obtain

left side	*right side*
$(x + 3)^2 = x^2 + 6x + 9$	$(x + 4)(x + 2) + 1 = x^2 + 6x + 8 + 1$
	$= x^2 + 6x + 9$

Thus equation (9) becomes

$$x^2 + 6x + 9 = x^2 + 6x + 9$$

This equation holds for *every* real number x. Thus equation (9) is an identity. ■

EXAMPLE 8 *Solving an Equation in x and y by Writing x in Terms of y*

Solve the following equation for x:

$$3(2x + 5y) + 7xy = -8y \tag{10}$$

SOLUTION We multiply through:

$$6x + 15y + 7xy = -8y$$

Then

$$6x + 7xy + 15y = -8y \quad \text{Rearrange terms}$$

$$x(6 + 7y) + 15y = -8y \quad \text{Factor to obtain a linear equation in } x$$

$$x(6 + 7y) = -8y - 15y = -23y$$

$$x = \frac{-23y}{6 + 7y}$$

NOTE Equation (10) is an equation in two variables. However, when we ask you to solve for x, we are asking you to treat x as if it were the only variable. When we do this, we treat the other variable (or variables) as constant.

Problems 2.1

Readiness Check

I. Which of the following equations is equivalent to $3x - 2 = 4x + 5$?
a. $x^2 - 2x = -9x$ b. $3x + 1 = 4x + 6$
c. $-2x + 1 = 15$ d. $x = -7$

II. Which of the following has no real solution?
a. $3x + 9 = 2x + 1$
b. $(x - 1)(x + 2) = x^2 - 1$
c. $\dfrac{7}{x-3} + 2 = \dfrac{2x}{x+3}$
d. $(x + 1)^2 = x^2 + 2x + 5$

III. Which of the following is an identity?
a. $\dfrac{x^2 - 9}{x + 3} = x + 3$
b. $7x + 8 = (x + 5)(x + 2) - (x^2 + 2)$
c. $3x - 7 = 4x + 8$ d. $\dfrac{x+2}{x} + 1 = \dfrac{2}{x}$

IV. Which of the following is the solution to $2x + 3 = 5x - 7$?
a. $\dfrac{10}{3}$ b. 5 c. 2 d. $-\dfrac{7}{2}$
e. $\dfrac{3}{5}$ f. $-\dfrac{2}{7}$

V. Which of the following is the solution for $\dfrac{1}{x-3} + 4 = \dfrac{5}{3-x}$?
a. 2 b. $\dfrac{3}{2}$
c. -4 d. 3
e. There are no solutions.

In Problems 1–30 find all solutions to the given equation. Check your answer.

1. $4z + 2 = 7$
2. $3w - 4 = 5$
3. $-2x + 7 = 7$
4. $2x + 5 = x - 3$
5. $3(1 - 4y) = -6(7y - 2)$
6. $(w + 1)^2 = (w + 1)(w + 2)$
7. $(z - 2)^2 = (z - 1)(z - 5)$
8. $(a + 2)(a - 5) = (a + 4)(a - 1)$
9. $(x - 6)(x - 2) = (x + 5)(x + 1)$
10. $(2u + 5)^2 = (4u + 1)(u - 3)$
11. $(4v + 1)(2v - 3) = (8v - 2)(v + 2)$
12. $\dfrac{y-3}{y+2} = \dfrac{4}{5}$
13. $\dfrac{z-2}{z+1} = 2$
14. $\dfrac{1}{x+1} = 2$
15. $\dfrac{1}{x} + 1 = 2$
16. $\dfrac{1}{z} - \dfrac{3}{z} = 4$
17. $\dfrac{3}{w} + \dfrac{2}{w} - \dfrac{8}{w} = 2$
18. $\dfrac{5}{7/x} = \dfrac{3}{5}$
19. $y = 3 - \dfrac{4}{2/y}$
20. $\dfrac{2}{z} - \dfrac{z}{2} = 1 - \dfrac{z}{2}$
21. $\dfrac{4}{m-1} = \dfrac{3}{m+2}$
22. $\dfrac{7}{x-2} + \dfrac{5}{x-3} = 0$
23. $\dfrac{s-2}{s+4} = \dfrac{s+3}{s-8}$
24. $\dfrac{2y+5}{3y-2} = \dfrac{4y+3}{6y-1}$
25. $(x - 1)^3 - (x + 2)^3 + 9x^2 = 4$
26. $p^3 - (p - 2)^3 = (2p + 1)(3p - 4)$
27. $\dfrac{1}{x-1} + \dfrac{1}{x+1} = \dfrac{1}{x^2-1}$
28. $\dfrac{3w+2}{w-2} + \dfrac{5w-3}{w+2} = \dfrac{8w^2+3w+2}{w^2-4}$
29. $\dfrac{2a}{a+5} - \dfrac{a}{a+3} = 1$
30. $\dfrac{6}{q+1} + \dfrac{q}{q+2} = 1$

Calculator Problems

In Problems 31–40 use a calculator to solve the given equation. In your answer, round to the same number of decimal places as given in the problem.

31. $1.207x = 4.103$
32. $2113.1y = 5.2038$

Answers to Readiness Check

I. c, d (a also has the root 0) II. d (it reduces to $1 = 5$) III. b IV. a V. b

33. $12.3156z + 15.2178 = 18.1432$
34. $3 \times 10^{-3}q = 5 \times 10^{-4}$
35. $8.34 \times 10^{-7}p = 5 \times 10^{-9}$ [Round to 2 decimal places.]
36. $5.12 \times 10^{8}m + 4.36 \times 10^{12} = 9.77 \times 10^{12}$
37. $\dfrac{4.106}{x} = -10.321$
38. $\dfrac{43.29y - 57.02}{23.75y + 14.64} = \dfrac{37.26}{49.8}$
39. $\dfrac{-5.7}{4w} + \dfrac{8.2}{5w} = 6$ [Round to 3 decimal places.]
40. $(q - 1.06)(8q + 3.24) = (2q - 6.33)(4q + 9.61)$

In Problems 41–50 explain why the given equation has no real solution.

41. $2(x + 3) = 2x - 1$
42. $(2x + 4)(3x - 1) = 6x^2 + 10x + 1$
43. $(x + 5)(x - 3) = (x + 1)^2$
44. $x(x + 7) = (x + 5)(x + 2)$
45. $(x + 2)(x + 3) = (x + 1)(x + 5) - x$
46. $\dfrac{3}{x} = 0$
47. $\dfrac{5}{y + 2} + 6 = 6$
48. $\dfrac{6}{z + 5} - \dfrac{2}{z + 5} = \dfrac{5}{z + 5}$
49. $\dfrac{2}{q + 1} - \dfrac{6}{q(q + 1)} = \dfrac{2}{q}$
50. $-\dfrac{2}{p + 3} + \dfrac{5}{p - 2} = \dfrac{10}{p^2 + p - 6}$

In Problems 51–56 show that the given equation is an identity. That is, show that the equation holds for every real number for which both sides are defined.

51. $(x - 1)^2 - 1 = x(x - 2)$
52. $\dfrac{x^2 - 4}{x + 2} = x - 2$
53. $(x + 5)^2 = (x + 3)(x + 7) + 4$
54. $(x - 2)^2 = (x + 2)(x - 6) + 16$
55. $\dfrac{5}{x} - \dfrac{x}{7} = -\dfrac{(x^2 - 35)}{7x}$
56. $(x + 1)^3 = (x^2 + 1)(x + 3) + 2x - 2$

In Problems 57–68 solve for the given variable (as in Example 8). You may assume, wherever necessary, that a variable is nonzero.

57. $y = 4z^3w$; solve for w.
58. $2y^2 + 6 = 5yz + 2$; solve for z.
59. $abc = bcd$; solve for d.
60. $2x + xy = 3y$; solve for x.
61. $\dfrac{z}{4} - xy + \dfrac{3}{x} = 10$; solve for y
62. $p(2 + 4q) - 3s(q - 5) = rq$; solve for q.
63. $\dfrac{az + b}{c} = 1 - b$; solve for z.
64. $\dfrac{1}{x - a} - \dfrac{1}{b - x} = 4$; solve for b.
65. $\dfrac{3}{x + a} - \dfrac{2}{x - b} = 3$; solve for a.
66. $\dfrac{pq}{y} + r = \dfrac{q + 2}{y}$; solve for y.
67. $\dfrac{1}{b + q} + \dfrac{1}{q} = 0$; solve for q.
68. $\dfrac{1}{b + q} + \dfrac{1}{q} = 0$; solve for b.
69. Explain why the equations $x = 1$ and $x^2 = 1$ are not equivalent.
70. Are the equations $x^2 = 1$ and $x^3 = 1$ equivalent? Explain.
71. Are the equations $x^3 = 8$ and $x^5 = 32$ equivalent? Explain.

* 72. (a) Show that the following equation has a solution:
$$x^2 - 4x - 5 = x^2 - 1$$
(b) What is wrong with the following argument that seems to show that the equation has no solution?
$$\begin{aligned} x^2 - 4x - 5 &= x^2 - 1 \\ (x - 5)(x + 1) &= (x - 1)(x + 1) \\ x - 5 &= x - 1 \\ -5 &= -1 \end{aligned}$$

* 73. What is wrong with the following proof that $0 = 1$? Let $x = 5$. Then
$$\begin{aligned} x &= 5 \\ 3x - 2x &= 15 - 10 \\ 3x + 10 &= 2x + 15 \\ -3x - 10 &= -2x - 15 \\ x^2 - 3x - 10 &= x^2 - 2x - 15 \\ (x - 5)(x + 2) &= (x - 5)(x + 3) \\ x + 2 &= x + 3 \\ x &= x + 1 \\ 0 &= 1 \end{aligned}$$

2.2 Applications of Linear Equations

Linear equations arise in an astonishingly wide variety of applications. We begin with a simple example from marketing.

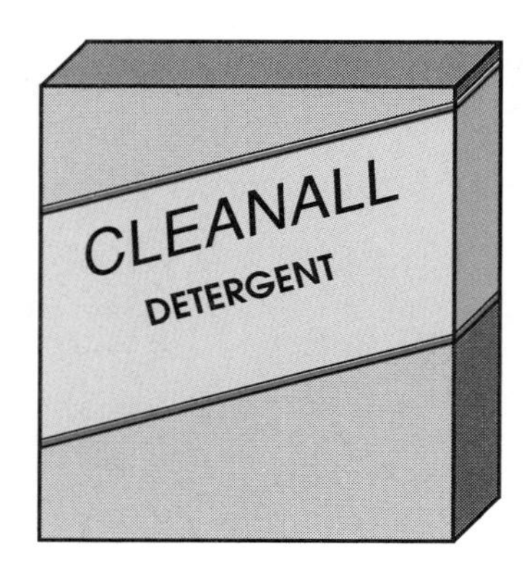

EXAMPLE 1 *A Marketing Problem*

The price of a 64-oz box of Cleanall detergent just increased by 16% and is now \$3.19. What was the original price?

SOLUTION Let p denote the original price (in dollars). Then 16% of the original price is $0.16p$. We have

$$\begin{aligned} \text{new price} &= \text{old price} + \text{price increase} && \text{Price equation} \\ 3.19 &= p + 0.16p && \text{The price equation written in symbols} \\ 3.19 &= p(1 + 0.16) = 1.16p && \text{Factor and combine terms} \\ p &= \frac{3.19}{1.16} = 2.75 \end{aligned}$$

so

$$\text{original price} = \$2.75$$

Check $(2.75)(0.16) = \$0.44$ and $2.75 + 0.44 = \$3.19$.

The procedure used in the last example can be used in many different types of problems. We can describe it as a five-step process.

Process for Solving Applied or Verbal Problems

Step 1 Read and reread the problem carefully. Note what information is given. Determine what quantity is sought.

Step 2 Assign a letter to one unknown quantity. Write other quantities in the problem in terms of this one quantity.

Step 3 Find an algebraic equation that relates the variables in the problem.

Step 4 Solve the equation for the unknown quantity.

Step 5 Check your answer in the words of the problem, not in your equation.

WARNING If you set up an equation incorrectly, you will obtain an incorrect answer, and your incorrect answer will check in your incorrect equation. Thus, you will believe that you have the correct answer when in fact you do not. This points out the importance of Step 1: Read the problem carefully! ■

Simple Interest Problems

The **simple interest formula** is an example of a linear equation. This formula is used to determine the interest on money that is invested or loaned over a period of time.

Simple Interest Formula

$$I = Prt \tag{1}$$

where I is the interest earned, P is the **principal** or the amount invested, r is the **rate** of interest (almost always a number between 0 and 1), and t is the **time** the investment is held (usually measured in years).

EXAMPLE 2 *Computing Simple Interest*

\$1500 is invested for 3 years, and \$360 in simple interest is earned. What is the interest rate?

SOLUTION Here $P = \$1500$, $t = 3$ years, and $I = \$360$. The unknown quantity is r. Since $I = Prt$, $I/Pt = Prt/Pt = r$, and

$$r = \frac{I}{Pt} = \frac{360}{(1500)(3)} = \frac{360}{4500} = 0.08$$

Thus

$$r = 0.08 = 8\%$$

Check $I = Prt = (1500)(0.08)(3) = \$360$ ■

EXAMPLE 3 *Computing the Amount Invested and the Effective Interest Rate*

A designer invests \$20,000 in two certificates of deposit (CD). The first CD pays 11% simple interest, and the second pays 12.5%. At the end of one year, she has earned \$2320 in interest. (a) How much did she invest in each CD? (b) What was the **effective interest rate** on the total investment?

SOLUTION (a) Let A denote the amount of money (in dollars) invested in the 11% CD. Then since the total amount invested is \$20,000, a total of \$$(20{,}000 - A)$ was invested in the 12.5% CD.

The facts of the problem are summarized in the table below.

	P **Amount Invested (in dollars)**	***r*** **Interest Rate**	***t*** **Time**	***Prt*** **Simple Interest Earned**
11% CD	A	0.11	1	$0.11A$
12.5% CD	$20{,}000 - A$	0.125	1	$0.125(20{,}000 - A)$
Total	20,000		1	\$2320

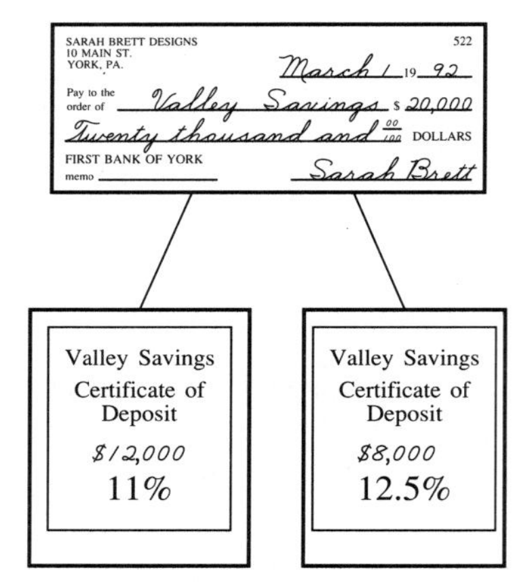

$20,000 is invested in two CDs, one at a rate of 11% and the other at 12.5%.

From the last column

$$0.11A + 0.125(20{,}000 - A) = 2320$$

$$0.11A + 2500 - 0.125A = 2320 \qquad (0.125)(20{,}000) = 2500$$

$$-0.015A = 2320 - 2500 = -180$$

$$A = \frac{-180}{-0.015} = 12{,}000$$

and

$$20{,}000 - A = 20{,}000 - 12{,}000 = 8000$$

Thus $12,000 was invested at 11% and $8000 was invested at 12.5%.

Check ($12,000)(0.11)(1) + ($8000)(0.125)(1) = $1320 + $1000 = $2320.

(b) Again, $I = Prt$ and $r = I/Pt$.

Here $I = 2320$ (the total interest), $P = 20{,}000$ (the total investment), and $t = 1$ year, so

$$r = \frac{2320}{(20{,}000)(1)} = \frac{2320}{20{,}000} = 0.116$$

Thus the effective interest earned on the total investment was 11.6%. Put another way, the **average simple interest** earned was 11.6%.

Velocity-Time-Distance Problems

Other linear problems derive from the **distance formula.** The distance formula is used to determine how far one has traveled over a certain period of time at a given (constant) velocity.

The Distance Formula

$$s = vt \qquad (2)$$

where s is the distance traveled, v is the average velocity, and t is the elapsed time. That is

$$\text{distance} = \text{velocity} \times \text{time}$$

Two other useful forms of the distance formula are obtained by dividing both sides of equation (2) by t or v:

$$v = \frac{s}{t}$$

or

$$\text{velocity} = \frac{\text{distance}}{\text{time}}$$

Also

$$t = \frac{s}{v}$$

or

$$\text{time} = \frac{\text{distance}}{\text{velocity}}$$

EXAMPLE 4 *Computing Average Velocity*

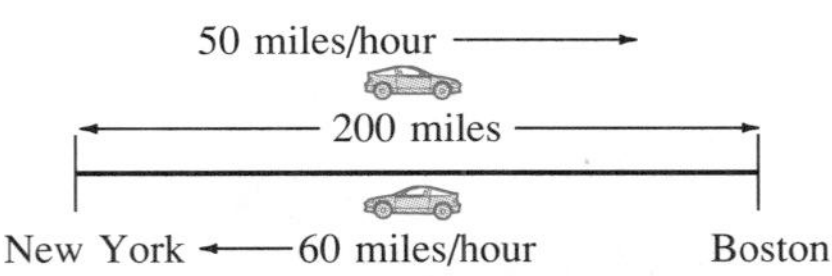

Figure 1 The car travels at 50 miles/hr from New York to Boston and at 60 miles/hr on the return trip from Boston to New York.

A car travels 200 miles from a suburb of New York to Boston at an average velocity of 50 miles/hour. It returns at an average velocity of 60 miles/hour. What is its average velocity over the entire trip? [See Figure 1]

SOLUTION Here,

$$\text{average velocity} = \frac{\text{total distance traveled}}{\text{total travel time}} = \frac{400 \text{ miles}}{\text{time (in hours)}}$$

We need to determine the time used in making the trip. From (2), $t = \dfrac{s}{v}$. Then

$$\text{time from New York to Boston} = \frac{200 \text{ miles}}{50 \text{ miles/hour}} = 4 \text{ hours}$$

$$\text{time from Boston to New York} = \frac{200 \text{ miles}}{60 \text{ miles/hour}} = \frac{200}{60} \text{ hours}$$

$$= \frac{10}{3} \text{ hours}$$

Thus

$$\text{total travel time} = 4 \text{ hours} + \frac{10}{3} \text{ hours} = \frac{22}{3} \text{ hours}$$

and

$$\begin{array}{l}\text{average velocity for}\\ \text{entire round trip}\end{array} = \frac{400 \text{ miles}}{22/3 \text{ hours}} = \frac{1200}{22} \text{ miles/hour}$$

$$= 54\frac{6}{11} \text{ miles/hour}$$

Note that the answer is *not* the 55 miles/hour (average of 50 and 60) that many people would guess.

EXAMPLE 5 *Computing the Distance Traveled Before Two Planes Pass Each Other*

The air distance from New York to London is 3458 miles. An American Airlines jet leaves New York bound for London at the same time as a British Airways plane leaves London for New York. The American Airlines jet flies at a constant velocity of 580 miles/hour while the British Airways jet flies at a constant velocity of 540 miles/hour. (a) How many miles from New York is the place where the two planes pass each other? (b) How many hours does each plane travel until it meets the other?

SOLUTION (a) Let s denote the distance traveled by the American Airlines jet (in miles) before the planes meet. Then since the total distance from New York to London is 3458 miles, the British Airways plane travels $3458 - s$ miles. The information in this problem is summarized in Figure 2 and the table below.

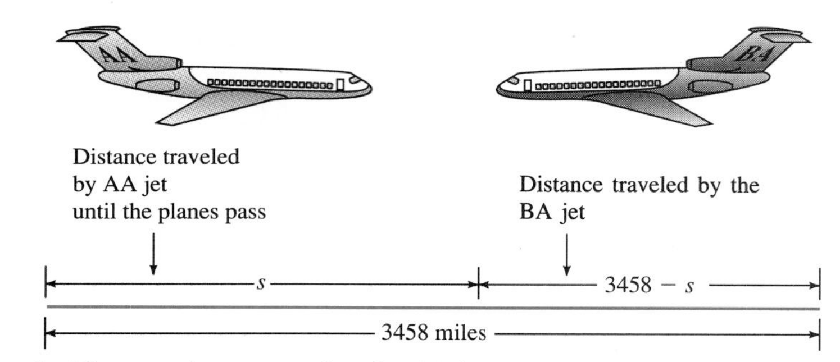

Figure 2 The two planes pass after the AA jet has traveled a total of s miles.

Jet	s Distance Traveled to Rendezvous (in miles)	v Velocity (in miles/hour)	$\frac{s}{v}$ Time Traveled to Rendezvous (in hours)	
American Airlines	s	580	$\frac{s}{580}$	These two numbers are equal because the planes leave and meet at the same time
British Airways	$3458 - s$	540	$\frac{3458 - s}{540}$	

Then, from the table

$$\frac{s}{580} = \frac{3458 - s}{540}$$

$$580 \cdot 540 \frac{s}{580} = 580 \cdot 540 \frac{3458 - s}{540} \quad \text{Multiply both sides by } 580 \cdot 540$$

$$540s = 580(3458 - s) = 2{,}005{,}640 - 580s$$

$$540s + 580s = 2{,}005{,}640 \quad \text{Add } 580s \text{ to both sides}$$

$$1120s = 2{,}005{,}640$$

$$s = \frac{2{,}005{,}640}{1120} \overset{\text{From a calculator}}{=} 1790.75$$

So the distance from New York is 1790.75 miles.

$$\text{(b) } t \text{ hours} = \text{elapsed time} \overset{\text{From the table}}{=} \frac{s}{580}$$

$$= \frac{1790.75}{580} = 3.0875 \text{ hours} = 3 \text{ hours } 5 \text{ minutes } 15 \text{ seconds}$$

Linear functions arise in a central way in business applications.

Break-Even Analysis

EXAMPLE 6 *The Total Revenue and Total Cost Functions*

The manager of a shoelace company has determined that the company operates according to the following conditions:

(a) It makes only shoelaces and sells them for 50¢ per pair.
(b) Ten people are employed, and they are paid 10¢ for each pair of shoelaces they make. Note that they are paid *only* for what they produce.
(c) Raw materials cost 24¢ for each pair of shoelaces made.
(d) The costs for equipment, plant, insurance, manager's salary, and fringe benefits (called **fixed costs**) amount to $2000 per month.

The manager of this shoelace factory is interested in how many pairs of shoelaces have to be made and sold to break even. This means that the manager has to determine **total revenue** and **total cost** at different levels of production. The information given above is sufficient to formulate equations for the total revenue and the total cost, and from these equations we can write an equation to determine the **profit** at different levels of production. From this equation we can determine when the profit is zero; that is, when the factory breaks even. Let's discuss this problem in more detail.

The **total revenue function** indicates how much money is brought into the firm each month by the sale of its product. The total revenue for a month is the amount taken in before any expenses are paid. If q pairs of shoelaces are sold in a month, the total revenue for the shoelace factory is $\$0.50q$. Thus the total revenue function is

$$R = 0.50q \quad \text{(In dollars)}$$

We can also write the equation for the **total cost function,** which measures the number of dollars the firm must pay to produce and sell its product. The total cost is composed of two parts: fixed costs and variable costs. **Fixed costs** are those that remain constant regardless of the number of units produced, such as those listed in (d) above. The fixed costs for the shoelace factory are given to be $2000 per month. Thus the cost for 1 month will always be at least $2000, even if no shoelaces are produced.

Variable costs are those directly related to the production of a commodity. The variable costs for producing q pairs of shoelaces are $0.10q$ for labor and $0.24q$ for raw materials. The total variable cost is $0.34q$. The total cost function is the sum of the fixed and variable costs. We have

$$\underset{\substack{\text{Fixed}\\ \text{costs}}}{} \qquad \underset{\substack{\text{Variable}\\ \text{costs}}}{}$$

$$C = \underset{\uparrow\ \text{Fixed costs}}{2000} + \underset{\uparrow\ \text{Variable costs}}{0.34q} \quad \text{(In dollars)}$$ ■

EXAMPLE 7 *Determining Total Revenue and Total Costs*

Find the total revenue and the total cost if the shoelace factory produces and sells

(a) 10,000 pairs of shoelaces in a month.
(b) 15,000 pairs of shoelaces in a month.

SOLUTION

(a) $R = 0.50q$. Thus, when $q = 10{,}000$,

$$R = (0.50)(10{,}000) = \$5000$$

Also

$$C = 2000 + (0.34)(\underset{\uparrow\ q = 10{,}000 \text{ here}}{10{,}000}) = \$5400$$

(b) Now $q = 15{,}000$, so

$$R = (0.50)(15{,}000) = \$7500$$

and

$$C = 2000 + (0.34)(15{,}000) = \$7100$$

Clearly the profit or loss for a month can be found by subtracting the total cost from the total revenue. In the previous example we see that producing and selling 10,000 pairs of shoelaces results in a loss of \$400 (\$5000 − \$5400), whereas producing and selling 15,000 pairs results in a profit of \$7500 − \$7100 = \$400. So the firm will "break even" at some point between 10,000 and 15,000 pairs of shoelaces.

If q units of a commodity are produced and sold, we find the total profit by subtracting the total cost from the total revenue.

Definition of the Total Profit Function

total profit (or loss) = total revenue − total cost

We define the total profit function

$$P = R - C$$

EXAMPLE 8 *Finding the Break-Even Point*

In Example 6, at what level of production will the firm break even?

SOLUTION At the **break-even point,** there is no profit or loss. That is, $P = 0$.

$$R = 0.50q \quad \text{and} \quad C = 2000 + 0.34q$$

so

$$\begin{aligned} P = R - C &= (0.50q) - (2000 + 0.34q) \\ &= 0.50q - 0.34q - 2000 \\ &= 0.16q - 2000 \underset{\text{Set } P = 0}{=} 0 \end{aligned}$$

Then

$$0.16q = 2000$$

and

$$q = \frac{2000}{0.16} = 12{,}500$$

That is, the break-even point is 12,500 pairs of shoelaces. If $q = 12{,}500$, then

$$R = 0.50(12{,}500) = 6250$$

and

$$C = 2000 + 0.34(12{,}500) = 2000 + 4250 = 6250$$

Since

$$P = 0.16q - 2000$$

we see that

if $q > 12{,}500$, the company makes a profit

and

if $q < 12{,}500$, the company loses money ■

EXAMPLE 9 *A Problem in Physiology: Pulmonary Ventilation*†

The rhythmic movement of air into and out of the lungs is called **pulmonary ventilation.** The amount of air ventilated by the lungs in 1 minute is referred to as **minute ventilation.**

Physiologists denote minute ventilation by $\dot{V}_E$. The V denotes volume (usually measured in liters), the E stands for "expired," and the dot over the V refers to rate per minute. Thus

$$\dot{V}_E = \text{the amount expired in 1 minute}$$

The quantity $\dot{V}_E$ depends on the amount of air ventilated per breath, referred to

† Adapted from *Sports Physiology,* 2nd ed. by Edward L. Fox, Saunders, Philadelphia, 1984, p. 165.

as the **tidal volume** (TV) and the number of breaths taken per minute (f = the **frequency**). We have

$$\dot{V}_E = TV \times f$$

Minute ventilation ↗ ($\dot{V}_E$), Tidal volume ↑ (TV), Frequency ↑ (f)

If your minute ventilation is 7.5 liters/min and you take 15 breaths per minute, what is your tidal volume?

SOLUTION Here $\dot{V}_E = 7.5$ liters/min and $f = 15$ breaths/min. Since $\dot{V}_E = TV \times f$,

$$TV = \frac{\dot{V}_E}{f} = \frac{7.5 \text{ liters/min}}{15 \text{ breaths/min}} = 0.5 \text{ liters of air per breath} \quad ■$$

EXAMPLE 10 *A Chemical Mixture Problem*†

In chemistry a **solution** is a homogeneous mixture of two or more substances. Simple solutions usually consist of one substance (the **solute**) dissolved in another substance (the **solvent**). Often the solvent is water. One common type of solute is an acid. A common acid is hydrogen chloride (HCl). The **concentration** of a solution is given as a percentage:

$$\begin{aligned} \text{concentration} &= \%\text{ solute} \\ &= \left(\frac{\text{grams of solute}}{\text{grams of solution}} \times 100\right)\% \end{aligned}$$

A chemist has two solutions of HCl. One has a concentration of 15%, and one has a concentration of 20%. How many grams of each solution must be uniformly mixed in order to obtain 50 grams of 18% solution?

SOLUTION *(NO PUN INTENDED)* The information in the problem is given in the table below. We let x denote the number of grams of the 15% solution used in the mixture.

Solution	Number of Grams in Solution	Concentration of HCl	Amount of HCl in Solution (in grams)
15% solution	x	15%	15% of $x = 0.15x$
20% solution	$50 - x$	20%	20% of $50 - x = 0.2(50 - x)$
Mixture	50	18%	18% of $50 = 9$

† This problem is adapted from *General Chemistry,* 2nd ed. by K. W. Whitten and K. D. Gailey, Saunders, Philadelphia, 1984, p. 67.

We depict this information in Figure 3.

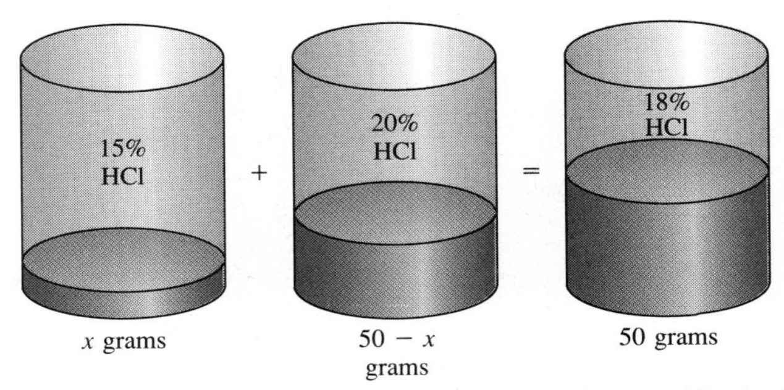

Figure 3 x grams of the 15% HCL concentration combined with $50 - x$ grams of the 20% HCL concentration yields 50 grams of the 18% HCL concentration.

From the last column in the table we have

$$\begin{aligned} 0.15x + 0.2(50 - x) &= 9 \\ 0.15x + 10 - 0.2x &= 9 \\ -0.05x &= 9 - 10 = -1 \\ x &= \frac{-1}{-0.05} = 20 \end{aligned}$$

so there are 20 grams of 15% solution and

$$50 - x = 50 - 20 = 30 \text{ grams of } 20\% \text{ solution}$$

Check

$$\begin{aligned} \text{concentration of mixture} &= \frac{\text{number of grams of HCl in mixture}}{\text{number of grams in mixture}} \times 100\% \\ &= \frac{(20)(0.15) + 30(0.2)}{50} \times 100\% \\ &= \frac{3 + 6}{50} \times 100\% = \frac{9}{50} \times 100\% \\ &= 0.18 \times 100\% = 18\% \end{aligned}$$ ■

EXAMPLE 11 *An Electric Circuit Problem*

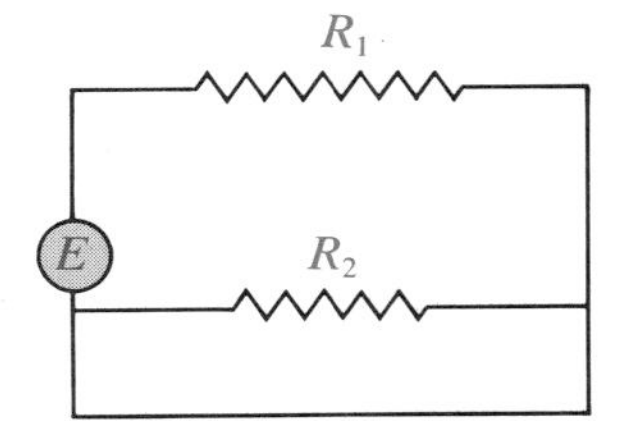

Figure 4 An electric circuit with two resistors.

An electric circuit contains two resistors in parallel (see Figure 4). The total resistance R (measured in ohms) in the circuit is given by

$$\frac{1}{R} = \frac{1}{R_1} + \frac{1}{R_2}$$

(a) Write R in terms of R_1 and R_2.
(b) Suppose $R_1 = 10$ ohms and $R = 8$ ohms. Find R_2.

SOLUTION (a) It is *not true* that $R = R_1 + R_2$. To see why, note that $\frac{1}{2} = \frac{1}{3} + \frac{1}{6}$ but $2 \neq 3 + 6$. To solve the problem, we must do a little work.

$$\frac{1}{R} = \frac{1}{R_1} + \frac{1}{R_2} \tag{3}$$

$$\frac{1}{R} = \frac{R_2 + R_1}{R_1 R_2} \quad \text{Combine fractions}$$

$$R = \frac{R_1 R_2}{R_1 + R_2} \quad \text{And then take reciprocals}$$

(b) Rewriting (3), we have

$$\frac{1}{R_2} = \frac{1}{R} - \frac{1}{R_1}$$

$$\frac{1}{R_2} = \frac{R_1 - R}{R R_1} \quad \text{Combine fractions}$$

$$R_2 = \frac{R R_1}{R_1 - R} = \frac{8(10)}{10 - 8} = \frac{80}{2} = 40 \quad \text{Take reciprocals and substitute the given values}$$

That is, $R_2 = 40$ ohms.

Check $\frac{1}{10} + \frac{1}{40} = \frac{4}{40} + \frac{1}{40} = \frac{5}{40} = \frac{1}{8}$. ■

EXAMPLE 12 ***Computing Completion Time for Two Crews Working Together***

Two crews work to pave a section of road. The first crew, working alone, could complete the job in 8 days. The second (larger) crew could complete the job in 6 days. How many days would it take to complete the job if the two crews worked together?

SOLUTION Let L denote the length (in miles) of the section of road. Then, crew 1 paves L miles in 8 days, so it can pave $\frac{L}{8}$ miles in 1 day; crew 2 paves L miles in 6 days, so it can pave $\frac{L}{6}$ miles in 1 day. Therefore, crew 1 and crew 2 working together, can in 1 day pave

$$\frac{L}{8} + \frac{L}{6} \overset{\text{LCM} = 24}{=} \frac{L}{8} \cdot \frac{3}{3} + \frac{L}{6} \cdot \frac{4}{4} = \frac{3L + 4L}{24} = \frac{7L}{24} \text{ miles}$$

Then

$$\frac{7L}{24} \text{ miles in 1 day}$$

means

$$L = \frac{7L}{24} \cdot \frac{24}{7} \text{ miles in } \frac{24}{7} \text{ days} \quad \text{Multiply both sides by } \frac{24}{7}$$

That is, the two crews working together can pave L miles in $\frac{24}{7}$ days (assuming an 8-hour work day, $\frac{24}{7}$ days $= 3\frac{3}{7}$ days ≈ 3 days 3.4 hours).

Problems 2.2

Readiness Check

I. The perimeter of a rectangular driveway is 106 feet, and the length is 23 feet longer than the width.

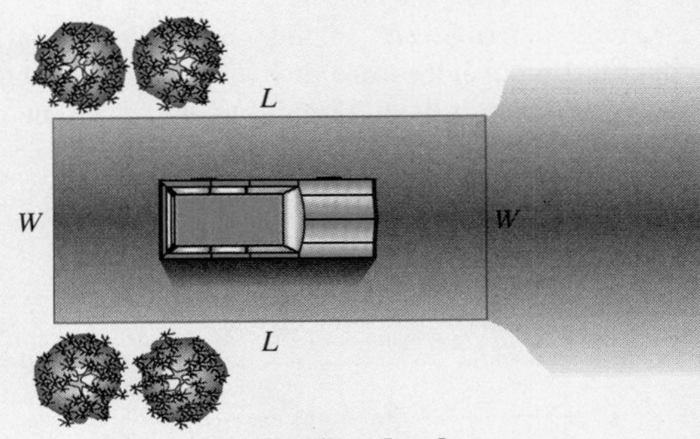

$$\text{Perimeter} = W + W + L + L$$
$$= 2W + 2L$$

Which of the following should be used to represent the length of the driveway if W represents the width?

a. $23 - W$ b. $23 + W$
c. $106 - W$ d. $106 + W$
e. $53 - W$ f. $53 + W$

II. Which of the following represents the second integer if n represents the first in finding the solutions to the problem below?

Find three consecutive integers such that three times the first diminished by the third is 4 less than twice the second.

a. $n + 1$ b. $n + 2$ c. $n - 1$ d. $2n - 4$

III. Which of the following is the number of gallons of 100% hydrochloric acid that must be added to 12 gallons of a 30% solution to obtain a 40% solution?

a. 4 gallons b. 2 gallons c. 1 gallon
d. $\frac{3}{2}$ gallons

IV. Which of the following is the equation needed to solve the following problem?

Susan is going to the local shopping mall. It takes 10 minutes for her to bike to the mall, but if she jogs the same route, it takes three times as long. If her rate of biking is 5 miles per hour more than twice her rate of jogging, how fast is she jogging? Here x denotes her rate of jogging.

a. $10(2x + 5) = 3x$ b. $20(x + 5) = 30x$

c. $2x + 5 = \frac{x}{3}$ d. $30x = 10(2x + 5)$

In Problems 1–10 a useful formula is given. Solve for the indicated quantity.

1. Circumference of a circle: $C = 2\pi r$; solve for r.
2. Area of a triangle: $A = \frac{1}{2}bh$; solve for b.
3. Volume of a right circular cylinder: $V = \pi r^2 h$; solve for h.
4. Simple interest: $I = Prt$; solve for t.
5. Ohm's law: $I = \frac{E}{R}$; solve for R.
6. Distance formula: $s = s_0 + v_0 t + \frac{1}{2}at^2$; solve for a.
7. Newton's law of universal gravitation: $F = G\frac{m_1 m_2}{r^2}$; solve for m_1.
8. Volume of a cone: $V = \frac{1}{3}\pi r^2 h$; solve for h.
9. Compound interest formula: $A = P\left(1 + \frac{r}{m}\right)^{mt}$; solve for P.
10. Resistance of a blood vessel of length l and radius r. (**Poiseuille's law**—c is a constant): $R = \frac{cl}{r^4}$; solve for l.

In many of the following problems a calculator will be useful.

11. The price of a bicycle increased by 23% to \$172.20. What was the original price?
12. The price of a Volga† automobile went from \$6000 to \$6400. What was the percentage increase?

Answers to Readiness Check

I. b, e II. a III. b IV. d

†The Volga is a Russian automobile.

13. A refrigerator was sold at a 15% discount off the original price. The sales price was \$528.70. What was the original price?
14. If the refrigerator of Problem 13 sells for \$447.84, what is the discount percentage?
15. \$20,000 was invested for 5 years and earned \$9400 simple interest. What was the interest rate?
16. John invested his life savings at $7\frac{1}{2}\%$ interest and earned \$3262.50 simple interest in 3 years. How much did he invest?
17. Mary invested \$5000 in a bank account that paid 6% and received \$1350 simple interest. How long was her money invested?
18. An investment banker invested \$50,000 in second mortgages that paid \$39,000 simple interest in 6 years. What was the average rate of interest?
19. A philanthropist wants to endow a chair of mathematics at her *alma mater*. The chair costs \$55,000 per year. She can invest money at $10\frac{1}{2}\%$ interest. If the chair is to be supported out of simple interest, how much (to the nearest dollar) must she invest?
20. A hairdresser invests \$15,000 in a bond fund that pays an average of 12% per year. He needs to earn \$10,000 in simple interest in order to open up his own salon. How long must he wait, assuming that his interest payments are not reinvested?
21. A teacher deposited \$8000 in two bank accounts, one paying $5\frac{1}{2}\%$ and one paying 6%. In 4 years she collected \$1851.60 in simple interest. (a) How much did she put in each account? (b) What was her effective interest rate on the total investment?
22. An engineer deposited \$4500 in a CD paying 7% and \$6000 in another CD. He earned a total of \$2421 in simple interest in 3 years. (a) What interest did the second CD pay? (b) What was the effective interest rate on the total investment?
23. Mr. Johnson invests a total of \$25,000 in a CD paying 8% and a mutual fund paying $9\frac{1}{2}\%$ simple interest. The annual simple interest from the mutual fund is \$60 more than the interest from the CD. To the nearest dollar, how much is invested in each?
24. A high school track star runs 400 meters in 50 seconds. What is her average velocity (in m/sec)?
25. The track star in Problem 24 runs 800 meters in 1 minute 52 seconds. What is her average velocity now?
26. A business person traveled to a city 89 miles away on a business trip. He drove from his house to a train station and arrived just in time to take an express train to the city. He drove an average of 40 miles/hour, and the train moved at an average velocity of 92 miles/hour. The total trip took him 1 hour 15 minutes. (a) How much time did he spend on the train? (b) How far does he live from the train station?

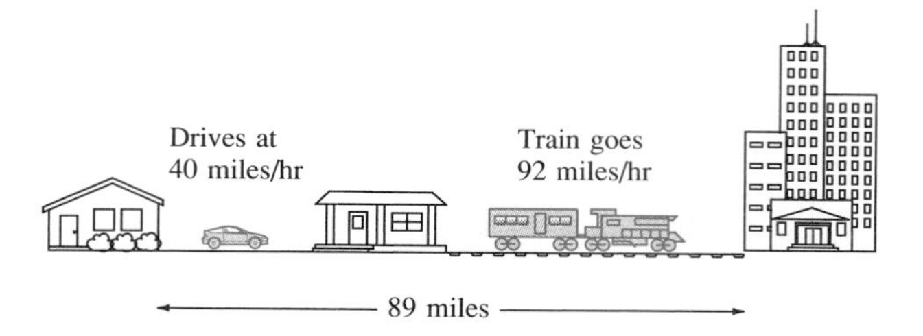

27. Two railroad termini are 600 miles apart. Two trains leave opposite termini at the same time. One travels at 60 miles/hour, and one travels at 75 miles/hour. After how many hours do the trains pass each other?

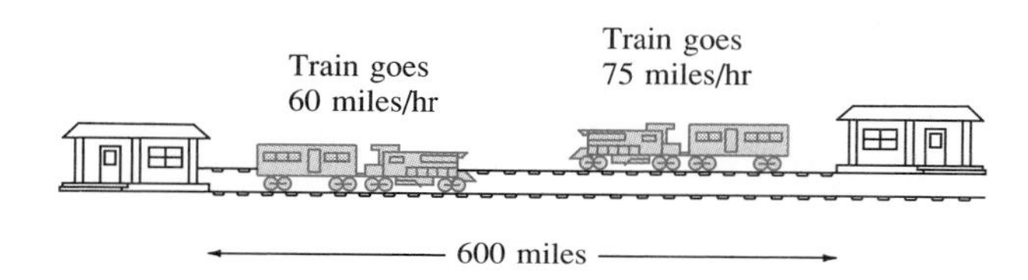

28. Suppose the price of a commodity is 40¢ each. If fixed costs are \$200 and the variable costs amount to 20¢ per item, find
    (a) The total revenue function.
    (b) The total cost function.
    (c) The total profit function.
    (d) The break-even point.
29. A product has a fixed cost of \$1650 and a variable cost of \$35 for each item produced during a given month.
    (a) Write the equation that represents total cost.
    (b) What will it cost to produce 215 items during the month?
30. The product of Problem 29 is sold for \$85 per item.
    (a) Write the equation representing the revenue function.
    (b) What will be the revenue from sales of 50 items?
    (c) What is the profit function for this product?
    (d) What is the profit on 50 items?
    (e) How many items must be sold in the month to avoid losing money?
31. In Example 9, if a person's minute ventilation is 6.8 liters/min and she takes 17 breaths per minute, what is her tidal volume?
32. In Example 9, if a person's tidal volume is 0.52 liter per breath and his minute ventilation is 7.02 liters/min, what is his average frequency (average breaths per minute)?
33. A chemist has two solutions of HCl: one with a 30% concentration and one with a 22% concentration. Approxi-

mately how many grams of each solution must be mixed to obtain 40 grams of a 25% solution?

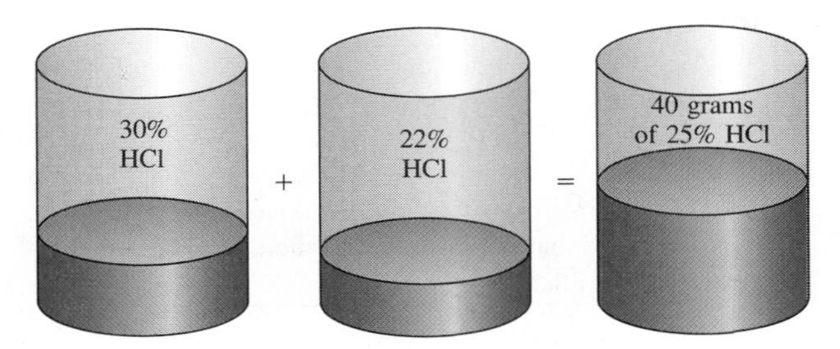

34. 100 grams of an 18.52% solution of sodium hydroxide (NaOH) were obtained by mixing 56 grams of a 15% solution and 44 grams of a second solution. What was the concentration of the second solution?

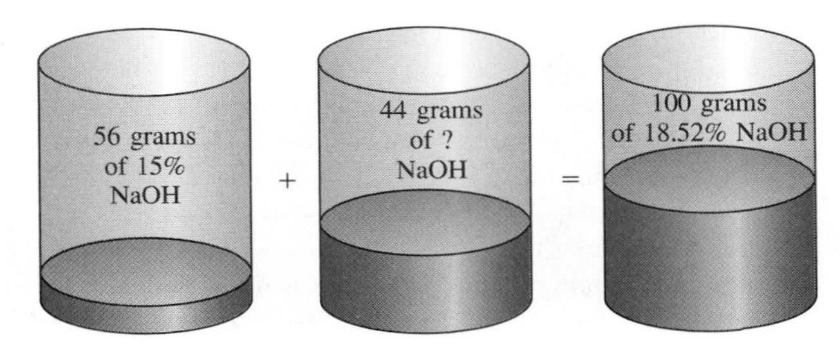

35. A quantity of 12% gold ore and a quantity of 18% gold ore together yield 20 ounces of pure gold. The weight of the 12% ore is 3 times the weight of the 18% ore. What is, approximately, the total weight of the ore?
36. According to **Okun's law,**† the growth rate of real output in an economy (y) is given by

$$y_n = g - 3(U_n - U_{n-1})$$

where the subscripts stand for the nth year, g is the growth rate of potential real output, and U denotes the unemployment rate. Solve the equation for U_{n-1} in terms of the other variables.
37. In a given economy $g = 3\% = 0.03$, $y_{1985} = 0.025$, and $U_{1985} = 7\frac{1}{2}\%$. What was the rate of unemployment in 1984? [Hint: First solve Problem 36.]
38. If an object of mass m kilograms is suspended from a height of h meters, then its **gravitational potential energy** (PE) is given by

$$\text{PE} = mgh$$

where $g \approx 9.8$ meters/sec^2 is the acceleration due to gravity. Solve for h in terms of PE, m, and g.

† See *Basic Economics,* 3rd ed. by Edwin G. Dolan, Dryden Press, Hinsdale, IL, 1983, p. 349.

39. A rock weighing 8 kilograms has the potential energy $361 \dfrac{\text{kg-m}^2}{\text{sec}^2}$. At what height is it held? [Hint: First solve Problem 38.]
40. If an object accelerates, then its velocity is given by

$$v = at + v_0$$

where v is the velocity, a is the acceleration of the object, t denotes time, and v_0 is the initial (starting) velocity of the object. Write t in terms of v, a, and v_0.
41. A vehicle accelerated from 20 feet per second to 140 feet per second. The acceleration is 30 feet per second per second (= 30 ft/sec^2). How long did the vehicle take to accelerate?
42. The formula relating temperature in degrees Fahrenheit (°F) to temperature in degrees Celsius (°C) is

$$°\text{F} = \frac{9}{5}°\text{C} + 32$$

Solve for °C in terms of °F.
43. The outside temperature is 80°F. What is the temperature in Celsius?
44. The temperature of a liquid gas is −150°F. What is its temperature in degrees Celsius?
45. The IQ (intelligence quotient) of a person is given by

$$\text{IQ} = \frac{\text{mental age}}{\text{actual age}} \times 100$$

Find the IQ of a 10-year-old child with a mental age of 13.
46. If an 18-year old has an IQ of 125, what is her mental age?
47. A person has a mental age of 20 and an IQ of 90. What is the actual age of the person?
48. According to **Ohm's law,** the voltage (E), current (I), and resistance (R) are related by the equation

$$E = IR$$

Here E is measured in volts, I in amperes, and R in ohms. Suppose that $E = 10$ volts and $I = 0.02$ ampere. Find R.
* 49. A major league baseball player bats .375 (= 37.5%) in his first 160 at bats. He then bats at a .285 average for the rest of the season. After how many at bats is his batting average exactly .325? [Hint: Batting average = hits/at bats.]
50. A coronary drug comes from a pharmaceutical company in concentrations of 20% (measured in mg/mL = milligrams per milliliter). The drug must be administered in an 8% concentration. How many milligrams (mg) of pure water must be mixed with 1.2 g (= 1200 mg) of the pharmaceutical solution to obtain the desired concentration?

51. The perimeter of a rectangular park is 200 feet. The length of the park is 20 feet longer than the width. What are the dimensions of the park?

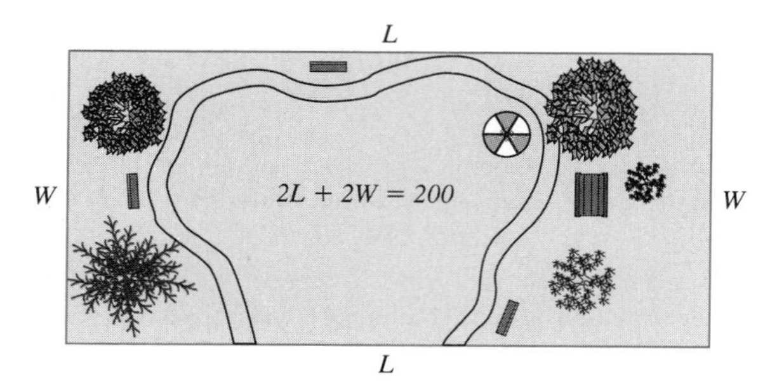

52. A student receives grades of 63, 77, and 89 on her first three algebra tests. What grade does she need on her fourth (and final) test in order to have an average of 80 (and get a B in the course)?
53. Coffee costing \$4.80 per pound is blended with coffee costing \$3.60 per pound. How many pounds of each must be mixed to obtain 25 pounds of a blend that costs an average of \$4.25 per pound?
54. The **octane rating** of gasoline is a number that is related to its ability to resist knocking. The principal ingredients in gasoline are *heptane,* which leads to knocking, and *isooctane,* which impedes knocking. If a brand of gasoline contains only heptane and isooctane and is 85% isooctane, then the octane rating of that brand is 85. In general, in determining octane ratings all brands of gasoline are treated as if they were mixtures of heptane and isooctane only. If two brands of 85 and 92 octane gasoline are blended to form 500 gallons of 90 octane gasoline, how many gallons of each brand were added?
55. In Problem 54, suppose that 150 gallons of 94 octane gasoline are combined with 250 gallons of Brand X gasoline to form 400 gallons of 89 octane gasoline. What is the octane rating of Brand X?
56. Salt is dissolved in 10 gallons of water to form a 12% saline solution. How many gallons of water must be evaporated to be left with a 15% solution?
* 57. John can dig a ditch in 3 hours. Mary can dig the same ditch in 4 hours. How long will it take the two of them working together to dig the ditch?
* 58. In Problem 57, Suzy can dig the same ditch in $2\frac{1}{2}$ hours. How long will it take all three of them to dig the ditch?
59. Resistors of 30 ohms and X ohms are connected in parallel in a circuit. The total resistance in the circuit is 20 ohms. Find X.
60. When three capacitors are connected in series in a circuit, the total capacitance C (measured in farads) is given by

$$\frac{1}{C} = \frac{1}{C_1} + \frac{1}{C_2} + \frac{1}{C_3}$$

(a) Write C in terms of C_1, C_2, and C_3.
(b) Find C if $C_1 = 0.1$ farad, $C_2 = 0.2$ farad, and $C_3 = 0.05$ farad.

Fun and Games

The following problems may be similar to problems you saw in high school. They have little practical significance, but they might be fun. Try them.

61. Find two consecutive integers whose sum is 115.
62. Find three consecutive even integers whose sum is 372.
63. Find four consecutive odd integers whose sum is 104.
64. Twice a certain number is 10 less than thrice the number. Find the number.
65. Sheila is 8 years older than Dot. In 4 years Sheila will be twice as old as Dot. How old is Dot?
66. In 3 years Horace will be 4 times as old as his son Scott was 2 years ago. In 12 years Horace will be 7 times as old as Scott was 5 years ago. How old are Horace and Scott?
67. Peter is meeting his girl friend Melissa at the airport. They haven't seen each other in 6 months. He is precisely 120 feet away from Melissa when he sees her walking toward him. They start running at the same moment, he at 15 feet/second, she at 12 feet/second (she is carrying luggage). How many feet does Peter run until they meet? How long does it take him?
68. A car enters a freeway driving 60 miles/hour. A second car enters the same freeway 15 minutes later and drives 70 miles/hour. How long will it take the second car to overtake the first?
69. Two trains initially 150 miles apart, move toward each other on the same track. Each train travels at 60 miles/hour. A fly starts on the front of the first train and flies at 80 miles/hour toward the second train. When it touches the second train, it immediately turns and flies toward the first train, without losing speed. The fly continues to fly back and forth in this manner until the trains collide and the fly is squashed. What is the total distance that the fly flies on its back and forth journeys? [Hint: This is not as hard as it seems.]
70. **A Number Trick**

Pick a positive number
Add 2
Square your answer
Add twice your number
Add 5
Take the square root
Subtract your number
The result is 3.

Explain how this trick works.

* 71. Devise a number trick whose result is always 10.

72. **The Pickle Problem.** Pickles are 99% water when fresh. 500 pounds of pickles were left to dry out overnight, with the result that they were only 98% water the next morning. How much did they weigh in the morning?

73. Look at the following cartoon.
    (a) Answer the question, assuming that the bird does *not* turn around.
    (b) What useful message does the cartoon convey?

2.3 Quadratic Equations

Up to now, we have been working mainly with linear equations. We now turn to quadratic equations.

Definition of a Quadratic Equation

A **quadratic equation** in standard form is an equation written in the form

$$ax^2 + bx + c = 0 \qquad (1)$$

where a, b, and c are real numbers and $a \neq 0$.

Roots

Numbers that satisfy the quadratic equation (1) are called **roots** or **zeros** or **solutions** of the equation. Three methods are used to find the roots of a quadratic equation. We discuss them one at a time.

Method 1: Factoring

If the expression $ax^2 + bx + c$ can be readily factored, as in Section 1.8, then the roots of (1) can be readily obtained. To do so, we need the following algebraic fact:

$$\text{If } u \text{ and } v \text{ are real numbers and } u \cdot v = 0, \text{ then } u = 0 \text{ or } v = 0. \qquad (2)$$

EXAMPLE 1 ***Solving a Quadratic Equation by Factoring: Two Real Roots***

Solve the quadratic equation $x^2 + 3x - 10 = 0$.

SOLUTION

$$x^2 + 3x - 10 = (x + 5)(x - 2) = 0 \quad \text{Factor}$$

$$x + 5 = 0 \qquad x - 2 = 0 \quad \text{Use fact (2)}$$

$$x = -5 \qquad x = 2$$

The solutions are $x = -5$ and $x = 2$.

Check $(-5)^2 + 3(-5) - 10 = 25 - 15 - 10 = 0$

$2^2 + 3(2) - 10 = 4 + 6 - 10 = 0$ ■

EXAMPLE 2 ***Writing a Quadratic Equation in Standard Form and Then Solving It by Factoring***

Solve the equation $x^2 + 3x - 5 = -x^2 + 2x + 1$.

SOLUTION We first write the equation in standard form:

$$2x^2 + x - 6 = 0 \quad \text{Gather terms}$$

$$(2x - 3)(x + 2) = 0 \quad \text{Factor—see Example 10 on p. 50}$$

$$2x - 3 = 0 \qquad x + 2 = 0 \quad \text{Use fact (2)}$$

$$x = \frac{3}{2} \qquad x = -2 \quad \text{These are the solutions}$$ ■

EXAMPLE 3 ***A Quadratic Equation Which Has Zero as a Root***

Solve $5x^2 - 20x = 0$.

SOLUTION

$$5x^2 - 20x = 0$$

$$5x(x - 4) = 0 \quad \text{Factor}$$

$$5x = 0 \qquad x - 4 = 0 \quad \text{Use fact (2)}$$

$$x = 0 \qquad x = 4 \quad \text{These are the solutions}$$ ■

EXAMPLE 4 ***Solving a Quadratic Equation by Factoring: One Double Root***

Solve the equation $x^2 - 8x + 16 = 0$.

SOLUTION

$$x^2 - 8x + 16 = 0$$

$$(x - 4)(x - 4) = 0 \quad \text{Factor}$$

$$x - 4 = 0 \quad \text{Use fact (2)}$$

$$x = 4 \quad \text{This is the only solution}$$

Note: $x = 4$ is called a **double root** of the quadratic equation.

Method 2: Completing the Square

In this method we use the following fact:

If $x^2 = d$ and $d \geq 0$, then

$$x = \pm\sqrt{d} \tag{3}$$

EXAMPLE 5 ***Solving a Quadratic Equation by Taking Positive and Negative Square Roots***

Solve $x^2 = 16$.

SOLUTION From (3), $x = \pm\sqrt{16} = \pm 4$. The roots are $x = 4$ and $x = -4$. ■

EXAMPLE 6 ***Solving a Quadratic Equation by Completing the Square***

Solve the quadratic equation $x^2 + 6x + 2 = 0$.

SOLUTION $(x + 3)^2 = x^2 + 6x + 9$. Thus if we add 9 to $x^2 + 6x$ we will have a "complete square."

$$x^2 + 6x + 2 = 0$$

$$x^2 + 6x = -2 \qquad \text{Subtract 2 from both sides}$$

$$x^2 + 6x + 9 = -2 + 9 \qquad \text{Add 9 to both sides}$$

$$x^2 + 6x + 9 = 7$$

$$(x + 3)^2 = 7$$

$$x + 3 = \pm\sqrt{7} \qquad \text{Use fact (3)}$$

$$x + 3 = \sqrt{7} \qquad x + 3 = -\sqrt{7} \qquad \text{Two possibilities}$$

$$x = -3 + \sqrt{7} \qquad x = -3 - \sqrt{7} \qquad \text{These are the two solutions}$$

Check $(-3 + \sqrt{7})^2 + 6(-3 + \sqrt{7}) + 2 = (9 - 6\sqrt{7} + 7) + (-18 + 6\sqrt{7}) + 2 = 0$

$(-3 - \sqrt{7})^2 + 6(-3 - \sqrt{7}) + 2 = (9 + 6\sqrt{7} + 7) + (-18 - 6\sqrt{7}) + 2 = 0$

Recall the formula (see p. 42)

$$(x + a)^2 = x^2 + 2ax + a^2$$

Then

$$\left(x + \frac{b}{2}\right)^2 = x^2 + 2\left(\frac{b}{2}\right)x + \left(\frac{b}{2}\right)^2$$

$$= x^2 + bx + \left(\frac{b}{2}\right)^2 \tag{4}$$

How to Complete the Square

1. Start with $x^2 + bx$.
2. Take half the coefficient of x (or the unknown): $\frac{1}{2}b$.
3. Square this number: $\left(\frac{b}{2}\right)^2$.
4. Add and subtract $\left(\frac{b}{2}\right)^2$ to the original expression.
5. Group the terms that belong together.
6. Factor the terms that form a square. You will obtain

$$x^2 + bx = x^2 + bx + \left(\frac{b}{2}\right)^2 - \left(\frac{b}{2}\right)^2$$
$$= \left[x^2 + bx + \left(\frac{b}{2}\right)^2\right] - \frac{b^2}{4} = \left(x + \frac{b}{2}\right)^2 - \frac{b^2}{4}$$

EXAMPLE 7 ***Solving a Quadratic Equation by Completing the Square: No Real Roots***

Solve the equation $x^2 + 2x + 5 = 0$.

SOLUTION $x^2 + 2x + 5 = 0$

$$x^2 + 2x + 1 - 1 + 5 = 0 \qquad \text{Add and subtract 1}$$
$$(x + 1)^2 + 4 = 0$$
$$(x + 1)^2 = -4 \qquad -4 \text{ is negative}$$

Thus there are *no real solutions* because the square of every real number is nonnegative.

In the next section we will discuss the case of no real roots in some detail.

Method 3: The Quadratic Formula

We now use the method of completing the square to derive a formula for solving quadratic equations. We start with

$$ax^2 + bx + c = 0, \qquad a \neq 0$$
$$x^2 + \frac{b}{a}x + \frac{c}{a} = 0 \qquad \text{Divide by } a$$
$$x^2 + \frac{b}{a}x = -\frac{c}{a} \qquad \text{Subtract } c/a \text{ from both sides}$$
$$x^2 + \frac{b}{a}x + \left(\frac{b}{2a}\right)^2 = -\frac{c}{a} + \left(\frac{b}{2a}\right)^2 \qquad \text{Add } \left(\frac{b}{2a}\right)^2 \text{ to both sides}$$
$$\left(x + \frac{b}{2a}\right)^2 = -\frac{c}{a} + \frac{b^2}{4a^2} \qquad \text{Factor using equation (4)}$$
$$\left(x + \frac{b}{2a}\right)^2 = -\frac{4a}{4a} \cdot \frac{c}{a} + \frac{b^2}{4a^2}$$

$$= \frac{b^2 - 4ac}{4a^2} \qquad \text{Combine terms}$$

$$x + \frac{b}{2a} = \pm\sqrt{\frac{b^2 - 4ac}{4a^2}} = \pm\frac{\sqrt{b^2 - 4ac}}{2a} \qquad \text{Fact (3)}$$

$$x + \frac{b}{2a} = \frac{\sqrt{b^2 - 4ac}}{2a} \qquad x + \frac{b}{2a} = -\frac{\sqrt{b^2 - 4ac}}{2a}$$

$$x = -\frac{b}{2a} + \frac{\sqrt{b^2 - 4ac}}{2a} \qquad x = -\frac{b}{2a} - \frac{\sqrt{b^2 - 4ac}}{2a}$$

These are the solutions

Quadratic Formula

A quadratic equation in standard form is

$$ax^2 + bx + c = 0, \qquad a \neq 0 \tag{5}$$

The solutions are

$$x = \frac{-b \pm \sqrt{b^2 - 4ac}}{2a} \tag{6}$$

The expression $b^2 - 4ac$ is called the **discriminant** of the quadratic equation (5).

The discriminant can be used to determine the number of real roots of the quadratic equation.

There are three cases to consider.

The Number of Real Roots of a Quadratic Equation

Case 1

If the discriminant is positive ($b^2 - 4ac > 0$), then the quadratic equation has *two distinct real roots:*

$$x = \frac{-b + \sqrt{b^2 - 4ac}}{2a} \quad \text{and} \quad x = \frac{-b - \sqrt{b^2 - 4ac}}{2a}$$

Case 2

If the discriminant is zero ($b^2 - 4ac = 0$), then the quadratic equation has *one real root,* called a **double root:**

$$x = -\frac{b}{2a}$$

Case 3

If the discriminant is negative ($b^2 - 4ac < 0$), then the quadratic equation has *no* real roots because no negative number has a real square root. In this case the quadratic polynomial $ax^2 + bx + c$ is said to be **irreducible.**†

† In this case the quadratic equation has two complex roots. Complex numbers are discussed in the next section.

EXAMPLE 8 ***Using the Quadratic Formula: Two Real Roots***

Solve the quadratic equation $5x^2 - 3x - 6 = 0$.

SOLUTION Here $a = 5$, $b = -3$, $c = -6$, and $b^2 - 4ac = (-3)^2 - 4(5)(-6) = 9 + 120 = 129 > 0$. Thus there are two roots:

$$x = \frac{-(-3) + \sqrt{129}}{2 \cdot 5} = \frac{3 + \sqrt{129}}{10} \quad \text{and} \quad x = \frac{3 - \sqrt{129}}{10}$$

Check $$5\left(\frac{3 + \sqrt{129}}{10}\right)^2 - 3\left(\frac{3 + \sqrt{129}}{10}\right) - 6$$

$$= \frac{5}{100}(9 + 6\sqrt{129} + 129) - \frac{9}{10} - \frac{3\sqrt{129}}{10} - 6$$

$5 \cdot 129 = 645$
↓

$$= \frac{45}{100} + \frac{645}{100} - \frac{90}{100} - \frac{600}{100} + \frac{30}{100}\sqrt{129} - \frac{3}{10}\sqrt{129} = 0$$

A similar check works for the other root. ■

EXAMPLE 9 ***Using the Quadratic Formula: One Double Root***

Solve the quadratic equation $9x^2 + 6x + 1 = 0$.

SOLUTION Here $a = 9$, $b = 6$, $c = 1$, and $b^2 - 4ac = 6^2 - 4 \cdot 9 \cdot 1 = 36 - 36 = 0$. Thus there is one double root

$$x = -\frac{b}{2a} = -\frac{6}{2 \cdot 9} = -\frac{6}{18} = -\frac{1}{3}$$

Check

$$9\left(-\frac{1}{3}\right)^2 + 6\left(-\frac{1}{3}\right) + 1 = 9\left(\frac{1}{9}\right) - 2 + 1 = 1 - 2 + 1 = 0.$$ ■

EXAMPLE 10 ***A Quadratic Equation with a Negative Discriminant: No Real Roots***

Solve the quadratic equation $x^2 + 3x + 10 = 0$.

SOLUTION Here $a = 1$, $b = 3$, $c = 10$, and $b^2 - 4ac = 9 - 4 \cdot 1 \cdot 10 = 9 - 40 = -31 < 0$. Thus the equation has no real roots.

WARNING Students commonly make two mistakes when using the quadratic formula. To avoid them, make sure you check two things:
Check 1: Divide both $-b$ and $\sqrt{b^2 - 4ac}$ by $2a$.
For example, $x^2 - 4x - 7 = 0$.

Correct	Incorrect
$x = \dfrac{4 \pm \sqrt{16 - 4(-7)}}{2}$	This term should have been divided by 2
$= \dfrac{4 \pm \sqrt{44}}{2}$	$x = \dfrac{4}{2} \pm 2\sqrt{11}$
$= \dfrac{4}{2} \pm \dfrac{2\sqrt{11}}{2}$	$= 2 \pm 2\sqrt{11}$
$= 2 \pm \sqrt{11}$	The error is to divide only one of the terms in the numerator by 2

Check 2: The quadratic formula works only when the quadratic equation is in the standard form $ax^2 + bx + c = 0$.

For example, $x^2 + 3x + 4 = 5$.

Correct	Incorrect
$x^2 + 3x + 4 = 5$	$x^2 + 3x + 4 = 5$
$x^2 + 3x - 1 = 0$	$x = \dfrac{-3 \pm \sqrt{9 - 4 \cdot 4}}{2}$
	$= \dfrac{-3 \pm \sqrt{-7}}{2}$ This is wrong: the quadratic formula cannot be used here
$x = \dfrac{-3 \pm \sqrt{13}}{2}$	So there are no real solutions ■

We have seen three methods for solving quadratic equations:

1. Factoring is the easiest one to use when it works—that is, when the quadratic polynomial is easily factored.
2. Completing the square is most useful as a theoretical tool (we used it to derive the quadratic formula). It is also used in other contexts. For example, we will complete squares in Section 3.1 in order to put equations of circles in standard form.
3. The quadratic formula is the most general and important method. It works in *all* cases and is easily programmed on a computer or programmable calculator.

The quadratic formula can be used to factor a quadratic.

EXAMPLE 11 *Using the Quadratic Formula to Factor a Quadratic*

Factor the quadratic trinomial $6x^2 + x - 15$

SOLUTION Suppose that a and b are roots of the quadratic equation $6x^2 + x - 15 = 0$. Then it follows that

$$6(x - a)(x - b) = 0$$

and so $6x^2 + x - 15 = 6(x - a)(x - b)$.

We will say more about this factorization in Section 4.2. From the quadratic formula, the roots of $6x^2 + x - 15 = 0$ are

$$x = \frac{-1 \pm \sqrt{1^2 - 4(6)(-15)}}{12} = \frac{-1 \pm \sqrt{361}}{12} = \frac{-1 \pm 19}{12}$$

and

$$x = \frac{-1 + 19}{12} = \frac{18}{12} = \frac{3}{2} \quad \text{or} \quad x = \frac{-1 - 19}{12} = \frac{-20}{12} = \frac{-5}{3}$$

Thus

$$6x^2 + x - 15 = 6\left(x - \frac{3}{2}\right)\left(x + \frac{5}{3}\right) \overset{6 = 2 \cdot 3}{=} \left[2\left(x - \frac{3}{2}\right)\right]\left[3\left(x + \frac{5}{3}\right)\right]$$

$$= (2x - 3)(3x + 5)$$

Sum and Product of Roots

The two roots of a quadratic equation are

$$r_1 = -\frac{b}{2a} + \frac{\sqrt{b^2 - 4ac}}{2a} \quad \text{and} \quad r_2 = -\frac{b}{2a} - \frac{\sqrt{b^2 - 4ac}}{2a}$$

Then

$$r_1 + r_2 = -\frac{b}{2a} + \frac{\sqrt{b^2 - 4ac}}{2a} - \frac{b}{2a} - \frac{\sqrt{b^2 - 4ac}}{2a} = -\frac{b}{a}$$

and, since $(x - y)(x + y) = x^2 - y^2$,

$$r_1 r_2 = \left[\frac{1}{2a}(-b + \sqrt{b^2 - 4ac})\right]\left[\frac{1}{2a}(-b - \sqrt{b^2 - 4ac})\right]$$

$$= \frac{1}{4a^2}[b^2 - (\sqrt{b^2 - 4ac})^2]$$

$$= \frac{1}{4a^2}[b^2 - (b^2 - 4ac)]$$

$$= \frac{4ac}{4a^2} = \frac{c}{a}$$

Thus

The Sum and Product of Roots of a Quadratic Equation

For the quadratic equation $ax^2 + bx + c = 0$

$$\text{the sum of the roots is } -\frac{b}{a} \tag{7}$$

$$\text{the product of the roots is } \frac{c}{a} \tag{8}$$

NOTE Formulas (7) and (8) are very useful because they provide an easy way to check whether computed solutions to a quadratic equation could be correct. It is very easy to compute the sum of the roots and check whether the sum is equal to $-\frac{b}{a}$. It is harder, but still not too difficult, to verify that the product is $\frac{c}{a}$.

EXAMPLE 12 ***Finding the Sum and Product of the Roots of a Quadratic Equation***

Find the sum and product of the roots of the quadratic equation

$$5x^2 - 3x - 6 = 0.$$

SOLUTION

Here $a = 5$, $b = -3$, and $c = -6$, so

$$\text{sum of roots} = -\frac{b}{a} = -\frac{(-3)}{5} = \frac{3}{5}$$

$$\text{product of roots} = \frac{c}{a} = -\frac{6}{5}$$

Check In Example 8 we computed the two roots

$$r_1 = \frac{3}{10} + \frac{\sqrt{129}}{10} \quad \text{and} \quad r_2 = \frac{3}{10} - \frac{\sqrt{129}}{10}$$

Then

$$r_1 + r_2 = \frac{6}{10} = \frac{3}{5}$$

and

$$\begin{aligned} r_1 r_2 &= \left(\frac{3}{10} + \frac{\sqrt{129}}{10}\right)\left(\frac{3}{10} - \frac{\sqrt{129}}{10}\right) \\ &= \left(\frac{3}{10}\right)^2 - \left(\frac{\sqrt{129}}{10}\right)^2 = \frac{9}{100} - \frac{129}{100} \\ &= -\frac{120}{100} = -\frac{6}{5} \end{aligned}$$

Problems 2.3

Readiness Check

I. Which of the following is the square that is completed when solving $x^2 + 6x - 10 = 0$?
a. $(x + 6)^2$
b. $(x - 6)^2$
c. $(x + 3)^2$
d. $(x - 3)^2$

II. Which of the following is the discriminant?
a. $-b \pm \sqrt{b^2 - 4ac}$
b. $b^2 - 4ac$
c. $\sqrt{b^2 - 4ac}$
d. $\dfrac{-b \pm \sqrt{b^2 - 4ac}}{2a}$

III. Which of the following is true of a quadratic equation if its discriminant is -1?
a. The equation has one real root.
b. -1 is the solution of the equation.
c. The equation has no real roots.
d. The equation has 2 real roots.

IV. Which of the following is the number to be added and subtracted to complete the square of the equation $x^2 - 7x = 4$?
a. $\dfrac{49}{4}$ b. $\dfrac{-7}{2}$ c. $\dfrac{7}{2}$ d. -14

V. Which of the following is the solution of $x^2 - 10x = 0$?
a. -10 b. 10 c. $-10, 10$ d. $0, 10$

In Problems 1–15 solve each quadratic equation by factoring.

1. $x^2 - 6x + 9 = 0$
2. $y^2 - 9 = 0$
3. $z^2 + 10z + 25 = 0$
4. $u^2 - 3u + 2 = 0$
5. $x^2 + x - 6 = 0$
6. $s^2 - \dfrac{1}{4} = 0$
7. $x^2 + 7x + 10 = 0$
8. $v^2 - 15v + 56 = 0$
9. $x^2 - 2x = 24$
10. $3p^2 - 12p = -12$
11. $-2n^2 + 6n - 4 = 0$
12. $5m^2 - 3m - 2 = 0$
13. $4x^2 + 8x + 3 = 0$
14. $7a^2 - 12a = 0$
15. $12y^2 + 16y - 3 = 0$

In Problems 16–24 complete the square by adding and subtracting an appropriate constant to the given expression.

16. $x^2 + 4x$
17. $y^2 - 3y$
18. $z^2 - 12z$
19. $r^2 + 7r$
20. $w^2 - \dfrac{9}{2}w$
21. $x^2 + 3.2x$
22. $r^2 - 1.076r$
23. $u^2 + 15.206u$
24. $p^2 - \dfrac{1}{6}p$

In Problems 25–36 find all real roots to the given equation by completing the square.

25. $x^2 + 2x - 2 = 0$
26. $y^2 - 5y + 1 = 0$
27. $x^2 + 4x - 3 = 0$
28. $w^2 + 3w + 5 = 0$
29. $u^2 - u - 1 = 0$
30. $v^2 - v = -1$
31. $r^2 - 11r = -7$
32. $4s^2 - 2s - 1 = 0$ [Hint: First divide by 4.]
33. $4s^2 - 2s + 1 = 0$
34. $x^2 + 15.36x - 18.45 = 0$
35. $y^2 - 0.0032y - 0.0156 = 0$
36. $z^2 + 748z + 1237 = 0$

In Problems 37–46 use the discriminant to determine the number of real roots of the given quadratic equation.

37. $x^2 - 3x + 1 = 0$
38. $y^2 - 6y - 12 = 0$
39. $v^2 + 6v + 9 = 0$

Answers to Readiness Check

I. c II. b III. c IV. a V. d

40. $u^2 - 6u + 9 = 0$
41. $z^2 - 5z + 10 = 0$
42. $3w^2 + 5w - 8 = 0$
43. $5r^2 + 12r = -10$
44. $3s^2 - 5s = 8$
45. $1.6t^2 - 8.9t + 7.4 = 0$
46. $\frac{t^2}{237} - \frac{t}{164} + \frac{14}{285} = 0$

In Problems 47–58 find all real roots to the given equation by using the quadratic formula.

47. $x^2 + 5x - 6 = 0$
48. $y^2 + 4y + 1 = 0$
49. $3z^2 + 4z - 10 = 0$
50. $2u^2 - 7u + 6 = 0$
51. $4v^2 - 6v = -5$
52. $\frac{w^2}{2} + w = 10$
53. $\frac{w^2}{2} + w + 10 = 0$
54. $\frac{1}{4}x^2 - \frac{1}{3}x - \frac{1}{6} = 0$
55. $\frac{1}{4}x^2 + \frac{1}{3}x - \frac{1}{6} = 0$
56. $3.16y^2 + 2.34y - 11.02 = 0$
57. $37{,}502u^2 - 15{,}106u - 23{,}208 = 0$
58. $10^{-4}w^2 + 2 \times 10^{-3}w + 10^{-2} = 0$

In Problems 59–68 find all real solutions (if any) to the given equation. Use any of the three methods.

59. $x^2 - 3x = 4x - 10$
60. $y - 2 = y^2 - 3$
61. $z^2 - 3z + 4 = 2z^2 + 5z - 7$
62. $2u^2 + 5u = u^2 + 8u + 4$
63. $w(w + 2) = (w - 1)^2$
64. $r^2 - 1 = (r - 1)^2$
65. $1 - x^2 = (x - 1)^2$
66. $(y + 3)(y - 5) = (2y - 4)(y + 7)$
67. $(y + 6)^2 = (y - 2)^2$
68. $(z + 1)^2 - 2 = (z - 2)^2 + 1$

In Problems 69–76 find the sum and product of the roots of each quadratic equation.

69. $2x^2 + 4x - 6 = 0$
70. $3y^2 - 8y + 3 = 0$
71. $7z^2 + 49z - 1 = 0$
72. $1.2u^2 - 0.8u + 2.4 = 0$
73. $v^2 - 50v + 137 = 0$
74. $w^2 - 0.123w + 0.962 = 0$
75. $r^2 + 8056r - 1137 = 0$
76. $q^2 - 80.57q + 99.02 = 0$

In Problems 77–87 find a value of c such that the given condition is satisfied.

77. $cx^2 + 8x + 24 = 0$; product of the roots is -3.
78. $cx^2 + 8x + 24 = 0$; product of the roots is -1.
79. $x^2 + cx - 24 = 0$; sum of the roots is 6.
80. $x^2 + cx - 24 = 0$; sum of the roots is -8.
81. $x^2 + 3x + c = 0$; product of the roots is -7.
82. $3x^2 - 2x - c = 0$; product of the roots is -14.
83. $cx^2 - 4x + 3 = 0$; sum of the roots is 5.
84. $cx^2 + 19x - 7 = 0$; sum of the roots is -11.
* 85. $x^2 - 9x + c = 0$; one root is double the other.
86. $cx^2 - 10x + 3 = 0$; one root is the reciprocal of the other.
87. $x^2 + 14x + c = 0$; one root is the negative reciprocal of the other.

In Problems 88–102 write a quadratic polynomial with leading coefficient 1 and that has the two given numbers as roots.

88. 1, 3
89. 1, -3
90. -1, 3
91. -1, -3
92. 0, 4
93. 6 (double root)
94. -3 (double root)
95. $-\frac{1}{2}$ (double root)
96. $\frac{1}{3}$ (double root)
97. $\frac{5}{2}$, $-\frac{3}{2}$
98. $-\frac{3}{7}$, $-\frac{4}{7}$
99. 1.206, -2.451
100. 26.8, -14.7
101. -2137, -4916
102. 0.003147, 0.006885

Solve each of the equations in Problems 103–107 on a calculator. Round to four decimal places.

103. $x^2 - 1.7x - 3.5 = 0$
104. $x^2 + 37.3x + 9.85 = 0$
105. $2.1x^2 + 3.2x - 5.7 = 0$
106. $x^2 + (3.2 \times 10^5)x - 4.7 \times 10^7 = 0$
107. $x^2 + 7.8 \times 10^{-3}x + 1.2 \times 10^{-6} = 0$

In Problems 108–113 use the quadratic formula to factor each expression.

108. $2x^2 + 3x - 2$
109. $2x^2 - 3x - 20$
110. $9x^2 + 9x + 2$
111. $10x^2 + 43x + 28$
112. $21x^2 - 11x - 2$
113. $15x^2 - 2x - 45$

2.4 Complex Numbers or What to Do if the Discriminant Is Negative

Consider the quadratic equation

$$ax^2 + bx + c = 0 \tag{1}$$

We saw in the last section that if the discriminant, $b^2 - 4ac$, is negative, then the equation has no real roots. The problem is that no negative number has a real square root. For example,

$$x^2 + 16 = 0$$

leads to

$$x^2 = -16$$

or

$$x = \pm\sqrt{-16}$$

But there is no real number whose square is -16. However, as we will soon see, the equation $x^2 + 16 = 0$ *does* have two roots.

In order to describe these roots, we must extend our number system. We do this by introducing a new kind of number.

> **Definition of the Imaginary Unit**
>
> The **imaginary unit** i is defined by
>
> $$i^2 = -1 \tag{2}$$

With this definition we write

$$i = \sqrt{-1} \tag{2'}$$

We interpret this to mean that i is the square root of -1. Using this definition, we can define the square root of a negative number.

> **Principal Square Root of a Negative Number**
>
> If $a > 0$ (so $-a < 0$), then the **principal square root** of $-a$ is given by
>
> $$\sqrt{-a} = \sqrt{a}i, \qquad a > 0 \tag{3}$$

WARNING Do *not* put the i under the radical (the square root sign) in (3). It is $\sqrt{a}i$, *not* $\sqrt{ai}$. ■

EXAMPLE 1 ***The Principal Square Roots of Four Negative Numbers***

(a) $\sqrt{-4} = \sqrt{4}i = 2i$

(b) $\sqrt{-16} = \sqrt{16}i = 4i$

(c) $\sqrt{-\frac{1}{9}} = \sqrt{\frac{1}{9}}i = \frac{1}{3}i$

(d) $\sqrt{-57} = \sqrt{57}i \approx 7.549834435i$

Definition of a Complex Number

A **complex number** is an expression that can be written in the form

$$a + bi \qquad (4)$$

where a and b are real numbers.

a is called the **real part** of the complex number.

b is called the **imaginary part** of the complex number.

EXAMPLE 2 ***Four Complex Numbers***

The following are complex numbers:

(a) $1 + i$ real part $= 1$; imaginary part $= 1$

(b) $2 - 7i$ real part $= 2$; imaginary part $= -7$

(c) 6 real part $= 6$; imaginary part $= 0$ A real number

(d) $3i$ real part $= 0$; imaginary part $= 3$ A pure imaginary number

FOCUS ON

The Word *Imaginary*

You should not be troubled by the term *imaginary*. It's just a name. The British mathematician Alfred North Whitehead, in the chapter on imaginary numbers in his *Introduction to Mathematics*, wrote:

> *At this point it may be useful to observe that a certain type of intellect is always worrying itself and others by discussion as to the applicability of technical terms. Are the incommensurable numbers properly called numbers? Are the positive and negative numbers really numbers? Are the imaginary numbers imaginary, and are they numbers?—are types of such futile questions. Now, it cannot be too clearly understood that, in science, technical terms are names arbitrarily assigned, like Christian names to children. There can be no question of the names being right or wrong. They may be judicious or injudicious; for they can sometimes be so arranged as to be easy to remember, or so as to suggest relevant and important ideas. But the essential principle involved was quite clearly enunciated in Wonderland to Alice by Humpty Dumpty, when he told her, apropos of his use of words, 'I pay them extra and make them mean what I like'. So we will not bother as to whether imaginary numbers are imaginary, or as to whether they are numbers, but will take the phrase as the arbitrary name of a certain mathematical idea, which we will now endeavour to make plain.*

Two complex numbers are **equal** if their real and imaginary parts are equal. That is,

$$a + bi = c + di \quad \text{if and only if} \quad a = c \quad \text{and} \quad b = d$$

We make two observations:

1. In part (c) of Example 2 we wrote the number 6 as a complex number. There is nothing special about the number 6, and we see that *every real number is a complex number*. Thus the definition of a complex number extends our number system.
2. The number $3i$ in part (d) of Example 2 is a complex number with real part zero. Such a number is called **pure imaginary.**

The Algebra of Complex Numbers

When we add, subtract, multiply, and divide complex numbers, we use the same rules we discussed in Chapter 1 for algebraic expressions.

Rules for Addition and Subtraction

$$(a + bi) + (c + di) = (a + c) + (b + d)i \qquad (5)$$

$$(a + bi) - (c + di) = (a - c) + (b - d)i \qquad (6)$$

EXAMPLE 3 ***Adding Two Complex Numbers***

Add $(2 + 5i) + (7 - 2i)$.

SOLUTION $(2 + 5i) + (7 - 2i) = (2 + 7) + [5 + (-2)]i = 9 + 3i$.

Use equation (5) ■

EXAMPLE 4 ***Subtracting Two Complex Numbers***

Compute $(-2 + 3i) - (5 + 8i)$.

SOLUTION $(-2 + 3i) - (5 + 8i) = (-2 - 5) + (3 - 8)i = -7 - 5i$.

Use equation (6)

Multiplication

The distributive properties discussed on p. 8 hold for complex numbers as well. We use them for complex number multiplication.

EXAMPLE 5 ***Multiplying Two Complex Numbers***

Compute $(3 + 4i)(5 + 2i)$.

SOLUTION

$$\begin{aligned}(3 + 4i)(5 + 2i) &= 3(5 + 2i) + (4i)(5 + 2i)\\ &= (3)(5) + (3)(2i) + (4i)(5) + (4i)(2i)\\ &= 15 + 6i + 20i + 8i^2\end{aligned}$$

But $i^2 = -1$, so $8i^2 = -8$, and we may continue:

$$= 15 + 26i - 8 = 7 + 26i \quad ■$$

EXAMPLE 6 *Obtaining Powers of i*

We can multiply i by itself repeatedly to obtain powers of i:

$$\begin{array}{ll} i^1 = i & i^5 = i \\ i^2 = -1 & i^6 = -1 \\ i^3 = -i & i^7 = -i \\ i^4 = 1 & i^8 = 1 \end{array}$$

etc.

For example,

$$i^{103} = i^{100}i^3 = (i^4)^{25}i^3 = 1^{25}i^3 = i^3 = -i \quad ■$$

EXAMPLE 7 *A Product of Two Complex Numbers That Is a Real Number*

Compute $(2 + 3i)(2 - 3i)$.

SOLUTION

$$\begin{aligned}(2 + 3i)(2 - 3i) &= 2(2 - 3i) + 3i(2 - 3i) \\ &= 4 - 6i + 6i - 9i^2 = 4 - 9(-1) \\ &= 4 + 9 = 13\end{aligned}$$

Definition of the Complex Conjugate

The **complex conjugate** of the number $a + bi$ is the number $a - bi$. It is denoted by $\overline{a + bi}$. That is,

$$\overline{a + bi} = a - bi$$

We compute

$$\begin{aligned}(a + bi)(a - bi) &= a^2 - abi + abi - b^2i^2 = a^2 - b^2(-1) \\ &= a^2 + b^2\end{aligned}$$

Thus, *when we multiply a nonzero complex number by its conjugate, we get a positive real number:*

The Product of a Complex Number and Its Conjugate Is a Real Number

$$(a + bi)(\overline{a + bi}) = (a + bi)(a - bi) = a^2 + b^2 \qquad (7)$$

Equation (7) enables us to factor expressions that we couldn't factor before. This is true because the product of complex conjugates is the sum of two squares.

EXAMPLE 8 ***Factoring a Sum of Squares***

Factor $x^2 + 16$.

SOLUTION In (7) set $x^2 = a^2$ and $16 = b^2$. Then

$$x = a, \qquad 4 = b$$

and

$$x^2 + 16 = (x + 4i)(x - 4i)$$

WARNING On p. 31 we discussed the rule

$$\sqrt{a}\sqrt{b} = \sqrt{ab} \qquad \text{if } a > 0,\ b > 0$$

This equation is *false* if $a < 0$ and $b < 0$. For example,

$$\sqrt{-1}\sqrt{-1} = i \cdot i = i^2 = -1$$

but

$$\sqrt{(-1)(-1)} = \sqrt{1} = 1$$

That is,

$$\sqrt{-1}\sqrt{-1} \neq \sqrt{(-1)(-1)} \quad \blacksquare$$

However, the following is true: If $a > 0$ and $b < 0$ or $a < 0$ and $b > 0$, then

$$\sqrt{ab} = \sqrt{a}\sqrt{b}$$

EXAMPLE 9 ***Simplifying a Radical***

$$\sqrt{-80} = \sqrt{5(-16)} = \sqrt{5}\sqrt{-16} = \sqrt{5}(4i) = 4\sqrt{5}i$$

Division of Complex Numbers

To divide two complex numbers, multiply the numerator and denominator by the conjugate of the denominator. This results in a positive real number in the denominator.

EXAMPLE 10 ***Dividing Two Complex Numbers***

Compute $\dfrac{4 - 5i}{2 + 3i}$.

SOLUTION

$$\frac{4 - 5i}{2 + 3i} = \frac{(4 - 5i)(2 - 3i)}{(2 + 3i)(2 - 3i)} \overset{\text{See Example 7}}{=} \frac{(8 - 15) + (-12 - 10)i}{13}$$

$$= \frac{-7 - 22i}{13} = -\frac{7}{13} - \frac{22}{13}i \quad \blacksquare$$

EXAMPLE 11 ***Dividing Two Complex Numbers***

Compute $\dfrac{3-7i}{2i}$.

SOLUTION The conjugate of $2i$ is $-2i$. We obtain

$$\frac{3-7i}{2i} = \frac{(3-7i)(-2i)}{(2i)(-2i)} = \frac{14i^2-6i}{-4i^2}$$
$$= \frac{-14-6i}{4} = -\frac{7}{2} - \frac{3}{2}i$$

Let z be a nonzero complex number with conjugate $\bar{z}$. Then

$$1 = z \cdot \frac{1}{z} \underset{\text{Multiply and divide by } \bar{z}}{=} z \cdot \frac{\bar{z}}{z\bar{z}} \tag{8}$$

The number $\dfrac{1}{z}$ is called the **inverse** of z. From (8), we see that the inverse of z is given by

$$\frac{1}{z} = \frac{\bar{z}}{z\bar{z}} \tag{9}$$

EXAMPLE 12 ***Computing the Inverse of a Complex Number***

Compute the inverse of $2-3i$.

SOLUTION $\overline{2-3i} = 2+3i$ and so, from (9), we obtain

$$\frac{1}{2-3i} = \frac{\overline{2-3i}}{(2-3i)(\overline{2-3i})} \underset{\text{From (7)}}{=} \frac{2+3i}{4+9} = \frac{2+3i}{13} = \frac{2}{13} + \frac{3}{13}i$$

Quadratic Equations with Complex Roots

In the last section we saw that the quadratic equation $ax^2 + bx + c = 0$ has two distinct real roots if the discriminant $b^2 - 4ac > 0$ and one double root if $b^2 - 4ac = 0$. We now discuss the case $b^2 - 4ac < 0$.

The Solutions to a Quadratic Equation with a Negative Discriminant

If the discriminant, $b^2 - 4ac$, is negative, then $4ac - b^2 > 0$, and two solutions to the quadratic equation (1) are

$$x = \frac{-b + \sqrt{4ac - b^2}\,i}{2a} \quad \text{and} \quad x = \frac{-b - \sqrt{4ac - b^2}\,i}{2a} \tag{10}$$

EXAMPLE 13 ***A Quadratic Equation with Complex Conjugate Roots***

Solve the quadratic equation $x^2 + 2x + 2 = 0$.

SOLUTION The discriminant is $4 - 4 \cdot 2 = -4 < 0$. Then the roots are

$$x = \frac{-2 \pm \sqrt{-4}}{2} = \frac{-2 \pm \sqrt{4}i}{2} = \frac{-2 \pm 2i}{2} = -1 \pm i$$

We check one of them:

$$\begin{aligned}(-1 + i)^2 + 2(-1 + i) + 2 &= 1 - 2i + i^2 - 2 + 2i + 2 \\ &= 1 - 2i - 1 - 2 + 2i + 2 = 0\end{aligned}$$ ■

EXAMPLE 14 ***Quadratic Equation With Complex Conjugate Roots***

Solve the quadratic equation $3x^2 + 5x + 4 = 0$.

SOLUTION $b^2 - 4ac = 5^2 - 4 \cdot 3 \cdot 4 = 25 - 48 = -23$. Thus solutions are

$$x = \frac{-5 + \sqrt{23}i}{6} \quad \text{and} \quad x = \frac{-5 - \sqrt{23}i}{6}$$

We may write these two solutions together as $x = \dfrac{-5 \pm \sqrt{23}i}{6}$

FOCUS ON
A Short History of Complex Numbers

Complex numbers came into use in the same way we have used them—as solutions to quadratic equations. One of the first to study quadratic equations was the Greek mathematician Diophantus of Alexandria, who lived about A.D. 250. Diophantus wrote a treatise called *Arithmetica,* in which he solved 189 problems involving quadratic equations, but did not describe a general method for solving such equations. In his solutions, he gave answers only when the roots were positive rational numbers. When roots were negative, irrational, or complex, he ignored them. In fact, when an equation had two positive rational roots, he gave only the larger of them.

GEROLAMO CARDANO
(New York Public Library Collection)

LEONHARD EULER
(Library of Congress)

The first real use of complex numbers as solutions to quadratic equations was given by the Italian mathematician Gerolamo Cardano (1501–1576). Cardano was born in Pavia in 1501 as the illegitimate son of a jurist and developed into a man of passionate contrasts. In his autobiography *De Vita Propria (Book of My Life),* he described himself as high-tempered, devoted to erotic pleasures, vindictive, humorless, and cruel. He gambled and played chess as an escape from "poverty, chronic illness and injustice." He commenced his turbulent professional life as a doctor, studying, teaching, and writing mathematics while practicing his profession. He once traveled as far as Scotland, and, upon his return to Italy, he successively held important chairs at the Universities of Pavia and Bologna. In 1570 he was imprisoned for heresy for casting the horoscope of Jesus. Shortly thereafter, he was hired as the personal astrologer of the Pope. (Not all mathematicians lead dull lives!) He died in Rome in 1576, by his own hand, one story says, so as to fulfill his earlier astrological prediction of the date of his death.

Cardano introduced complex numbers as a way to solve the following problem: Find two numbers whose sum is 10 and whose product is 40. If x and y denote the two numbers, we have

$$\begin{aligned} x + y &= 10 \quad \text{or} \quad y = 10 - x \\ xy &= 40 \\ x(10 - x) &= 40 \\ -x^2 + 10x &= 40 \\ 0 &= x^2 - 10x + 40 \\ x &= \frac{10 \pm \sqrt{100 - 160}}{2} = \frac{10 \pm \sqrt{-60}}{2} \\ &= \frac{10 \pm \sqrt{4(-15)}}{2} = \frac{10 \pm 2\sqrt{-15}}{2} \\ &= 5 \pm \sqrt{-15} \end{aligned}$$

Cardano obtained these two roots, but didn't really know what to do with them. He wrote, "putting aside the mental tortures involved," multiply $5 + \sqrt{-15}$ and $5 - \sqrt{-15}$ to obtain the product $25 - (-15) = 40$. He added, "So progresses arithmetic subtlety the end of which, as is said, is as refined as it is useless."

Complex numbers continued to plague mathematicians for the next three centuries. The German mathematician Gottfried Wilhelm Leibniz (1646–1716), codiscoverer (with Newton) of calculus, called the number i an "amphibean between being and nonbeing." In 1768 the great Swiss mathematician Leonhard Euler (1707–1783) wrote

> *Because all conceivable numbers are either greater than 0 or less than 0 or equal to 0, then it is clear that the square roots of negative numbers cannot be included among the possible numbers. Consequently we must say that these are impossible numbers. And this circumstance leads us to the concept of these numbers, which by their nature are impossible, and ordinarily are called imaginary or fancied numbers, because they exist only in the imagination.*

Actually, Euler made some use of complex numbers even though he called them "impossible." For example, in solving Cardano's problem he also obtained the "solution" $5 \pm \sqrt{-15}$. He then used this result to prove that the problem cannot be solved. Euler was also the first to use the letter i to denote $\sqrt{-1}$.

Complex numbers were not put on a concrete footing until the 19th century. By that time mathematicians and physicists had found a link between complex numbers and trigonometric functions. Today the study of complex numbers and complex functions is an important tool in solving a wide variety of problems involving cyclical or periodic activity. Such applications arise in all branches of science as well as in business and economics.

Problems 2.4

Readiness Check

I. Which of the following is true about $\sqrt{-16}$?
a. Its conjugate is $-4i$.
b. Its complex form is $4 + i$.
c. When it is written in $a + bi$ form, $b = -4$.
d. Its conjugate is -16.

II. Which of the following equations has solutions $\pm 7i$?
a. $x^2 = 49$ b. $x^2 - 7ix = 0$
c. $x^2 + 49 = 0$ d. $x^2 = -7ix$

III. Which of the following equations has the solutions $2 + i$ and $2 - i$?
a. $x^2 = 5 - 4x$ b. $x^2 - 4x + 5 = 0$
c. $x^2 - 2x + 1 = 0$ d. $x^2 + 4x - 2 = 0$

IV. $(1 + i)(1 - i) =$ ____.
a. 2 b. -2 c. $2i$ d. $-2i$

V. $\dfrac{1 + i}{1 - i} =$ ____.
a. 2 b. -2 c. 1 d. -1 e. i f. $-i$

In Problems 1–4 write each number in the form bi.

1. $\sqrt{-25}$ 2. $\sqrt{-100}$ 3. $\sqrt{-5}$ 4. $\sqrt{-80}$

In Problems 5–14 write the conjugate of each number.

5. $3 - 7i$ 6. $-2i$ 7. 5

Answers to Readiness Check

I. a II. c III. b IV. a V. e

8. $5 + 6i$
9. $-2 - \frac{1}{3}i$
10. $10 + \sqrt{53}i$
11. $\sqrt{73}i$
12. $\sqrt{73}$
13. $\frac{4 + \sqrt{2}i}{7}$
14. $\frac{2 - \sqrt{7}i}{12}$

In Problems 15–18 write each number in the form $a + bi$.

15. $3 - \sqrt{-64}$
16. $-5 + \sqrt{-4}$
17. $\frac{-4 + \sqrt{-64}}{8}$
18. $\frac{4 - \sqrt{-49}}{2}$

In Problems 19–52 carry out the indicated operations and write each answer in the form $a + bi$.

19. $(2 + 3i) + (4 + 5i)$
20. $(7 - 2i) + (8 + 5i)$
21. $\left(\frac{1}{3} + \frac{1}{5}i\right) + \left(\frac{1}{6} - \frac{1}{4}i\right)$
22. $(4 - i) - (2 + i)$
23. $(6 + \sqrt{3}i) - (6 - \sqrt{3}i)$
24. $(5 + \sqrt{2}i) + (5 - \sqrt{2}i)$
25. $(1 + i)(1 + 2i)$
26. $(1 + i)^2$
27. $(1 - i)^2$
28. $(3 - 4i)^2$
29. $(4 + 2i)^2$
30. $(3 + 4i)(3 - 4i)$
31. $(7 + 2i)(3 - 5i)$
32. $(4 - 3i)(7 + 5i)$
33. $\left(\frac{1}{2} + i\right)\left(-3 + \frac{1}{4}i\right)$
34. $(i + 5)(-3i + 7)$
35. $(2i - 3)(5 + 2i)$
36. i^5
37. i^{11}
38. i^{1002}
39. $(-i)^{15}$
40. $(1 + i)^3$
41. $(1 - i)^4$
42. $\frac{1}{i}$
43. $\frac{1}{1 + i}$
44. $\frac{1}{3 + 2i}$
45. $\frac{8 - 3i}{2i}$
46. $\frac{5 + i}{4i}$
47. $\frac{1 + i}{1 - i}$
48. $\frac{1 - i}{1 + i}$
49. $\frac{3 + 2i}{2 - 3i}$
50. $\frac{5 + i}{7 - 3i}$
51. $\frac{5 + 7i}{2 + 4i}$
52. $\frac{1}{3 - i} - \frac{i}{4 + 2i}$

In Problems 53–62 compute the inverse of each complex number.

53. $2i$
54. $-3i$
55. $1 - i$
56. $1 + 4i$
57. $2 - 5i$
58. $3 + 7i$
59. $4 - 6i$
60. $10 + 5i$
61. $a - bi,\ b \neq 0$
62. $a + bi,\ b \neq 0$

In Problems 63–66 solve for u and v, where u and v represent real numbers. [Hint: Equate real and imaginary parts.]

63. $u + v - 2i = 5 + 2vi$
64. $ui - 3v + 5 = 11 - 3i$
65. $2u - 3vi = 7ui + v - 2i$
66. $u + 2iv = 2u + iv + 3$

In Problems 67–78 find all solutions to the given quadratic equation.

67. $x^2 + 4 = 0$
68. $y^2 + 25 = 0$
69. $z^2 + z + 2 = 0$
70. $2u^2 + 2u + 1 = 0$
71. $p^2 - 6p + 10 = 0$
72. $3x^2 + 2x + 5 = 0$
73. $2y^2 - 3y + 5 = 0$
74. $8v^2 + 4v + 3 = 0$
75. $\frac{1}{2}x^2 + \frac{1}{3}x + \frac{1}{4} = 0$
76. $0.3y^2 + 0.4y + 0.26 = 0$
77. $12.72z^2 - 8.06z + 16.58 = 0$
78. $10^{-4}x^2 - 10^{-6}x + 10^{-5} = 0$
79. Show that if a complex number is equal to its conjugate, then the number is real.
80. Show that if a complex number is equal to the negative of its conjugate, then the number is pure imaginary.

* 81. Find the flaw in this "proof" that $i = 0$:

$$i = i$$
$$\sqrt{-1} = \sqrt{-1}$$
$$\sqrt{\frac{-1}{1}} = \sqrt{\frac{1}{-1}}$$
$$\frac{\sqrt{-1}}{\sqrt{1}} = \frac{\sqrt{1}}{\sqrt{-1}}$$
$$\frac{i}{1} = \frac{1}{i} = \frac{-(-1)}{i} = \frac{-i^2}{i}$$
$$i = -i$$
$$2i = 0$$
$$i = 0$$

2.5 Other Types of Equations

Many types of equations arise in applications. We discuss a few of them in this section.

Equations of Quadratic Type

An equation is of **quadratic type** if it can be written as the quadratic equation

$$ay^2 + by + c = 0 \tag{1}$$

EXAMPLE 1 ***A Fourth Degree Equation of Quadratic Type***

Solve the equation $x^4 - 13x^2 + 36 = 0$.

SOLUTION We rewrite the equation as

$$(x^2)^2 - 13x^2 + 36 = 0$$

This is an equation of the form (1) with $y = x^2$. We solve

$$\begin{aligned} y^2 - 13y + 36 &= 0 \qquad \text{Remember, } y = x^2 \\ (y-4)(y-9) &= 0 \\ y = 4 \quad &\text{or} \quad y = 9 \end{aligned}$$

We do not yet have a solution because we must solve for x. We have

$$x^2 = 4 \quad \text{and} \quad x^2 = 9$$

If $x^2 = 4$, then $x = \pm 2$. If $x^2 = 9$, then $x = \pm 3$. Hence four solutions are 2, −2, 3, −3.

We check two of them. You should check the other two.

$$2^4 - 13(2^2) + 36 = 16 - 13 \cdot 4 + 36 = 16 - 52 + 36 = 0$$

$$(-3)^4 - 13(-3)^2 + 36 = 81 - 13 \cdot 9 + 36 = 81 - 117 + 36 = 0 \quad \blacksquare$$

EXAMPLE 2 ***An Equation of Quadratic Type with Fractional Exponents***

Solve the equation $x^{1/3} + 2x^{1/6} - 2 = 0$. (2)

SOLUTION Let $y = x^{1/6}$. Then $y^2 = (x^{1/6})^2 = x^{1/3}$, and equation (2) can be rewritten as

$$y^2 + 2y - 2 = 0$$

From the quadratic formula,

$$\begin{aligned} y = \frac{-2 \pm \sqrt{4 - 4(-2)}}{2} &= \frac{-2 \pm \sqrt{12}}{2} = \frac{-2 \pm 2\sqrt{3}}{2} \\ &= -1 \pm \sqrt{3} \end{aligned}$$

But $y = x^{1/6}$, so $y^6 = (x^{1/6})^6 = x$. Hence two solutions are

$$x = (-1 + \sqrt{3})^6 \quad \text{and} \quad x = (-1 - \sqrt{3})^6$$

Using the $\boxed{y^x}$ key on a calculator, we obtain

$$(-1 + \sqrt{3})^6 = (0.7320508076)^6 = 0.1539030917$$
$$(-1 - \sqrt{3})^6 = (-2.7320508076)^6 = 415.8460969$$

Check We check one of the answers. You should check the other.

$$\begin{aligned}[(-1 + \sqrt{3})^6]^{1/3} + 2[(-1 + \sqrt{3})^6]^{1/6} - 2 &\\ &= (-1 + \sqrt{3})^2 + 2(-1 + \sqrt{3}) - 2\\ &= (1 - 2\sqrt{3} + 3) + (-2 + 2\sqrt{3}) - 2 = 0\end{aligned}$$

The two previous examples illustrate the following rule:

Substituting to Turn Certain Equations into Equations of Quadratic Type

We can write the equation

$$ax^{2s} + bx^s + c = 0$$

in the form

$$ay^2 + by + c = 0$$

by making the substitution $y = x^s$

In Example 1, $s = 2$, $2s = 4$, and $y = x^2$. In Example 2, $s = \frac{1}{6}$, $2s = \frac{1}{3}$, and $y = x^{1/6}$.

EXAMPLE 3 *Multiplying by the LCM to Obtain a Quadratic Equation*

Solve for x:

$$\frac{1}{x - 3} - \frac{x}{x + 2} = 6 \tag{3}$$

Here we must have $x \neq 3$ and $x \neq -2$.

SOLUTION The LCM of the denominators is $(x - 3)(x + 2)$, so we multiply both sides of (3) by $(x - 3)(x + 2)$:

$$\begin{aligned}\cancel{(x - 3)}(x + 2) \cdot \frac{1}{\cancel{x - 3}} - (x - 3)\cancel{(x + 2)} \cdot \frac{x}{\cancel{x + 2}} &= 6(x - 3)(x + 2)\\ x + 2 - x(x - 3) &= 6(x^2 - x - 6)\\ x + 2 - x^2 + 3x &= 6x^2 - 6x - 36\\ -x^2 + 4x + 2 &= 6x^2 - 6x - 36\\ 7x^2 - 10x - 38 &= 0 \quad \text{After simplification}\end{aligned}$$

From the quadratic formula

$$x = \frac{10 \pm \sqrt{100 - 4(7)(-38)}}{14}$$
$$= \frac{10 \pm \sqrt{1164}}{14} = \frac{10 \pm 2\sqrt{291}}{14}$$
$$= \frac{5 \pm \sqrt{291}}{7}$$

Check We check one of these with a calculator. You can check it directly, but that requires a great deal of algebraic computation.

(i) $x = \dfrac{5 + \sqrt{291}}{7} \approx 3.151246016$

(ii) $\dfrac{1}{x - 3} \approx \dfrac{1}{0.151246016} \approx 6.611744422$

(iii) $\dfrac{x}{x + 2} \approx \dfrac{3.151246016}{5.151246016} \approx 0.611744421$

(iv) $\dfrac{1}{x - 3} - \dfrac{x}{x + 2} \approx 6.000000001$

Our answer is correct. The extra 0.000000001 is roundoff error.

Equations Involving Radical Expressions

Sometimes it is possible to take a very complicated looking expression involving radicals and solve it by first raising both sides of the equation to an appropriate power.

EXAMPLE 4 *Squaring Both Sides to Obtain a Quadratic Equation*

Solve the equation $\sqrt{2x - 7} = x - 5$.

SOLUTION We can eliminate the square root by squaring both sides:

$$(\sqrt{2x - 7})^2 = (x - 5)^2$$
$$2x - 7 = x^2 - 10x + 25$$
$$x^2 - 12x + 32 = 0$$
$$(x - 4)(x - 8) = 0$$

Hence, $x = 4$ and $x = 8$ are the solutions. Or are they?

We have two answers, but one of them is not a solution: $x = 4$ cannot be a solution because $\sqrt{2x - 7} \geq 0$ for any real $x \geq 3.5$; but $x - 5 < 0$ if $x < 5$. In fact, if $x = 4$, $\sqrt{2x - 7} = \sqrt{2 \cdot 4 - 7} = \sqrt{1} = 1$ while $4 - 5 = -1$. The problem occurs because $(-\sqrt{2x-7})^2 = 2x - 7$, so when we square both sides we also solve the equation

$$-\sqrt{2x-7} = x - 5$$

(for which $x = 4$ *is* the solution).

As on p. 73, the number $x = 4$ is called an **extraneous solution** to the equation. We check the solution $x = 8$:

$$\sqrt{2x-7} = \sqrt{2 \cdot 8 - 7} = \sqrt{16-7} = \sqrt{9} = 3$$
$$x - 5 = 8 - 5 = 3$$

These are equal so $x = 8$ is a solution

Example 4 illustrates the importance of the following:

WARNING When solving equations involving radicals, make sure to check each solution. What looks like a solution might not be a solution at all. ■

A Note on Extraneous Solutions

Extraneous solutions often appear when both sides of an equation are squared. Here's a very simple example. We start with the equation $x = 2$.

$$x = 2 \quad \text{Given}$$
$$x^2 = 4 \quad \text{Square both sides}$$
$$x = \pm\sqrt{4} = \pm 2 \quad \text{The equation } x^2 = d \text{ has the solutions } x = \pm\sqrt{d}$$

The value $x = -2$, which is *not* a solution to the original equation $x = 2$, is an extraneous solution. It *is* a solution to the equation $x^2 = 4$, which is the square of the equation we originally wanted to solve.

EXAMPLE 5 ***Obtaining a Quadratic Equation from an Equation Involving Two Radicals***

Solve the equation

$$\sqrt{2x+7} - \sqrt{x+1} = 2 \qquad (6)$$

SOLUTION We put one radical on each side and then square:

$$\sqrt{2x+7} = \sqrt{x+1} + 2$$
$$(\sqrt{2x+7})^2 = (\sqrt{x+1} + 2)^2$$
$$2x + 7 = (x+1) + 4\sqrt{x+1} + 4 \quad (\sqrt{x+1})^2 = x + 1$$
$$x + 2 = 4\sqrt{x+1} \quad \text{Simplify}$$
$$(x+2)^2 = (4\sqrt{x+1})^2 \quad \text{Square again}$$
$$x^2 + 4x + 4 = 16(x+1) = 16x + 16$$
$$x^2 - 12x - 12 = 0$$
$$x = \frac{12 \pm \sqrt{12^2 - 4(-12)}}{2} = \frac{12 \pm \sqrt{192}}{2}$$
$$= \frac{12 \pm \sqrt{64 \cdot 3}}{2} = \frac{12 \pm 8\sqrt{3}}{2} = 6 \pm 4\sqrt{3}$$

Again there are two possible answers that must be checked.

Calculator check:

(a) $x = 6 + 4\sqrt{3} \approx 12.92820323$
$\sqrt{2x + 7} \approx \sqrt{32.85640646} \approx 5.732050808$
$\sqrt{x + 1} \approx \sqrt{13.92820323} \approx 3.732050808$
$\sqrt{2x + 7} - \sqrt{x + 1} = 5.732050808 - 3.732050808 = 2$

(b) $x = 6 - 4\sqrt{3} \approx -0.92820323$
$\sqrt{2x + 7} \approx \sqrt{5.143593539} \approx 2.267949192$
$\sqrt{x + 1} \approx \sqrt{0.071796769} \approx 0.267949192$
$\sqrt{2x + 7} - \sqrt{x + 2} = 2.267949192 - 0.267949192 = 2$

This time both possible solutions are indeed solutions. That is, the process of squaring did not, in this example, lead to extraneous solutions.

Higher-Order Polynomial Equations

Finding roots of polynomials of degree greater than 2 is much more difficult than solving quadratic equations. However, when the polynomial can be factored, then solutions can be found in the same way in which we solve a quadratic equation by factoring. More general techniques will be discussed in Chapter 4.

Recall two formulas from Section 1.8:

$$x^3 + y^3 = (x + y)(x^2 - xy + y^2) \tag{7}$$

$$x^3 - y^3 = (x - y)(x^2 + xy + y^2) \tag{8}$$

EXAMPLE 6 *Solving a Cubic Equation by Factoring*

Find all solutions to

$$x^3 - 8 = 0 \tag{9}$$

SOLUTION From (8),

$$(x^3 - 8) = (x^3 - 2^3) = (x - 2)(x^2 + 2x + 4) = 0$$

Thus $x - 2 = 0$ or $x^2 + 2x + 4 = 0$.

One solution is $x = 2$. To find the others, we use the quadratic formula:

$$x^2 + 2x + 4 = 0$$

$$\begin{aligned} x &= \frac{-2 \pm \sqrt{-12}}{2} = \frac{-2 \pm \sqrt{12}i}{2} \\ &= \frac{-2 \pm \sqrt{4 \cdot 3}i}{2} = \frac{-2 \pm 2\sqrt{3}i}{2} \\ &= -1 \pm \sqrt{3}i \end{aligned}$$

The three solutions to (9) are

$$x = 2, \quad x = -1 + \sqrt{3}i, \quad x = -1 - \sqrt{3}i$$

Check

(a) $2^3 - 8 = 0.$

(b) $(-1 + \sqrt{3}i)^3 - 8$

Formula (7) on p. 42 →
$$= (-1)^3 + 3(-1)^2(\sqrt{3}i) + 3(-1)(\sqrt{3}i)^2 + (\sqrt{3}i)^3 - 8$$
$$= -1 + 3\sqrt{3}i - 9i^2 + 3\sqrt{3}i^3 - 8$$
$i^3 = -i$ ↓
$$= -1 + 3\sqrt{3}i + 9 - 3\sqrt{3}i - 8 = 0$$

(c) Similarly $(-1 - \sqrt{3}i)^3 - 8 = 0.$

NOTE In this example we have found three cube roots of 8.

Problems 2.5

Readiness Check

I. Which of the following are equations of quadratic type?
a. $x^6 + x^4 - 2x^2 + 3 = 0$ b. $x^6 + x^3 - 8 = 0$
c. $x - x^{1/2} + 6 = 0$ d. $x^{1/2} + x^2 - 4 = 0$
e. $x^{3/8} - x^{3/4} + 25 = 0$ f. $x^5 - 2x^3 + 6 = 0$
g. $x^{10} - 6x^5 + 8 = 0$ h. $x^{10} - x^{15} + 2 = 0$

II. Which of the following equations can become quadratic equations when they are solved?
a. $\frac{1}{x+1} + \frac{x}{x-1} = 2$
b. $\frac{1}{x+1} - \frac{1}{x-3} = x$
c. $\frac{1}{x+5} + \frac{3}{(x+5)^2} = 2$
d. $\frac{1}{x+5} + \frac{3}{(x+5)^2} = 2x$
e. $\frac{1}{x+5} + \frac{3}{(x+5)^2} = 2x^2$
f. $\frac{1}{(x-2)^2} + \frac{5}{x+1} = 2$

III. Which of the following is true about the equation $\sqrt{x+5} + 2 = 0$?
a. It is a quadratic equation.
b. It has a solution of -1.
c. It has no extraneous solutions.
d. It has no solutions.

IV. To which of the following powers should both sides of $(x^2 - 2)^{1/4} = 8$ be raised to solve the equation?
a. $\frac{3}{4}$ b. 4 c. $\frac{1}{2}$ d. 2

V. Which of the following is an extraneous solution for $(x-1)^{1/2} + (x+2)^{1/2} = 3$?
a. -2 b. 2 c. -1
d. There is no extraneous solution.

In Problems 1–51 find all solutions to the given equation. Make sure to check your answer.

1. $\sqrt{x-3} = 4$
2. $\sqrt{y+1} = 3$
3. $\sqrt[3]{z+5} = 5$
4. $\frac{1}{\sqrt{x+2}} = 3$
5. $(y-1)^{3/2} = 27$
6. $x^4 - 2x^2 + 1 = 0$
7. $z^4 - 7z^2 + 12 = 0$
8. $u^4 - 1 = 0$
9. $v^4 - 81 = 0$
10. $w^4 - 8w^2 + 15 = 0$
11. $x^4 - 25x^2 + 144 = 0$
12. $y^4 + 3y^2 + 2 = 0$
13. $p^3 - 64 = 0$
14. $p^3 + 27 = 0$
15. $x^6 - 9x^3 + 8 = 0$
16. $y^6 - 28y^3 + 27 = 0$
17. $z^6 + 9z^3 + 8 = 0$
18. $x^3 + 1 = 0$
19. $y^8 - 17y^4 + 16 = 0$
20. $x^{2/3} - 3x^{1/3} + 1 = 0$
* 21. $w^{4/7} - 3w^{2/7} - 10 = 0$
22. $4z^{7/2} + 8z^{7/4} + 3 = 0$
23. $u^{0.2} + 13u^{0.1} + 42 = 0$
24. $\frac{3}{x-1} + \frac{3}{x-3} = 4$
25. $\frac{3}{y-2} = \frac{4}{y+1} + 5$
26. $\frac{y^2}{y+3} = 1 - y$
27. $\frac{1}{z} + z = 2$

Answers to Readiness Check

I. b, c, e, g II. a, c III. d IV. b V. d

28. $\dfrac{p+3}{p-1} + \dfrac{2p+2}{p+1} - 4 = 3$
29. $(x^2 - 3x - 2)^{1/3} = 2$
30. $(y^2 + 5y + 8)^5 = 32$ [Real roots only]
31. $\sqrt{3z + 4} = z$
32. $\sqrt{2w - 1} = w - 2$
33. $\sqrt{x + 9} = x - 3$
34. $\sqrt{v^2 - 3v + 9} = v + 3$
35. $\sqrt{3p + 3} = p + 1$
36. $\sqrt{4q - 3} = q + 2$
37. $\sqrt{3x + 4} = 2x - 1$
38. $\sqrt{y + 1} = y - 2$
39. $\sqrt{2z - 3} = z - 3$
40. $\sqrt{3x + 4} - \sqrt{x + 5} = -1$
41. $\sqrt{s + 1} + \sqrt{s + 2} = 1$
42. $\sqrt{w - 3} = 6 - \sqrt{2w + 2}$
43. $\sqrt{2x - 1} - \sqrt{x + 5} = 3$
* 44. $\sqrt{x + 2} + \sqrt{x + 3} = \sqrt{2x + 1}$
* 45. $\sqrt{2x + 1} + \sqrt{3x - 5} = \sqrt{4x + 7}$
46. $|x| = x^2 - 2$
47. $|x| = x^2 - 6$
48. $|x^2 + 4| = 5x$
* 49. $x^2 + 1 = |x^2 - 1|$
[Hint: Treat the cases $x^2 - 1 \geq 0$ and $x^2 - 1 < 0$.]
* 50. $|x + 1| = x^2 - 3$
[Hint: Treat the cases $x + 1 \geq 0$ and $x + 1 < 0$ separately.]
* 51. $|x^2 + 7x| = x + 7$

Solve Problems 52–57 on a calculator. Round each answer to four decimal places.

52. $x^4 + 5x^2 + 2 = 0$
53. $x^6 + 17x^3 - 35 = 0$
54. $x - \sqrt{x} - 2.7 = 0$
55. $x^{1/4} - 6.72x^{1/8} + 1.84 = 0$
56. $\dfrac{1}{x^2} - \dfrac{3.9}{x} - 8.8 = 0$ $\left[\text{Hint: Set } y = \dfrac{1}{x}.\right]$
57. $\dfrac{1}{x^4} - \dfrac{15}{x^2} + 4.3 = 0$
58. Find three cube roots of 1.
59. Find three cube roots of -1.
60. Find three cube roots of 1000.
61. Find three cube roots of 5.
62. Find three cube roots of -2.
* 63. What is wrong with the following "proof" that $-1 = 2$: Suppose A is a solution to $A = 1 + A^2$. But A cannot be 0; so we can divide by it:

$$1 = \frac{1}{A} + A$$

We substitute $\dfrac{1}{A} + A$ for 1 in the equation $A = 1 + A^2$ to obtain

$$A = \frac{1}{A} + A + A^2$$

so

$$0 = \frac{1}{A} + A^2 \quad \text{or} \quad A^2 = -\frac{1}{A} \quad \text{or} \quad A^3 = -1$$

Thus $A = \sqrt[3]{-1} = -1$, and the original equation $A = 1 + A^2$ becomes

$$-1 = 1 + (-1)^2 = 1 + 1 = 2$$

2.6 Applications of Quadratic and Other Equations

Quadratic equations and the kinds of equations we discussed in the last section arise frequently in a wide variety of applications to business and science. In this section we will cite a few of these applications.

Application in Business and Economics: Cost Functions

Suppose that a manufacturer produces an item in large quantities. Before any units are produced, however, it is necessary to pay certain **fixed costs.** These include rent or depreciation on the manufacturing plant, costs for tools, the monthly utility bills, and the wages of the employees. In addition, it is generally the case that the cost per unit decreases as the quantity produced increases. Thus, for example, it usually costs more to increase production from 1 unit to 2 units than from 101 units to 102 units. This fact is called **economy of scale.**

Economists model these phenomena by writing down a rule that gives the total cost C of producing q items (q stands for **quantity**). This rule is called a **total cost function.** One typical cost function is

$$C = aq^2 + bq + k, \qquad q \le M \tag{1}$$

where a, b, and k are real numbers. If $q = 0$ in (1), then $C = k$, so k represents the fixed costs. Usually, rule (1) applies only up to a certain maximum quantity M.

EXAMPLE 1 *Determining the Number of Units Produced Given the Total Cost*

The cost function for a firm was found by an economist to be†

$$C = 77 + 1.32q - 0.0002q^2, \qquad q \le 5000$$

where cost is measured in dollars.

(a) What are the fixed costs?
(b) If the total cost is \$818.52, how many units were produced?

SOLUTION

(a) If $q = 0$ is substituted into the cost function, we find that

$$\text{Cost of producing 0 units} = \$77$$

The fixed costs total \$77.

(b) We substitute $C = 818.52$ into the cost function:

$$\begin{aligned} 818.52 &= 77 + 1.32q - 0.0002q^2 \\ 0.0002q^2 - 1.32q + (818.52 - 77) &= 0 \\ 0.0002q^2 - 1.32q + 741.52 &= 0 \end{aligned}$$

$$\begin{aligned} q &= \frac{1.32 \pm \sqrt{(1.32)^2 - 4(0.0002)(741.52)}}{0.0004} \\ &= \frac{1.32 \pm \sqrt{1.7424 - 0.593216}}{0.0004} \\ &= \frac{1.32 \pm \sqrt{1.149184}}{0.0004} = \frac{1.32 \pm 1.072}{0.0004} \end{aligned}$$

or

$$q = \frac{2.392}{0.0004} = 5980 \quad \text{and} \quad q = \frac{0.248}{0.0004} = 620$$

Since q is restricted to be ≤ 5000, we throw out the root 5980. The answer is $q = 620$ units.

REMARK In real life, answers don't work out as precisely as they did in Example 1. Cost functions, and the answers they produce, are, at best, good approximations to reality. Certainty in business is achieved only in textbooks.

†This problem is adapted from *Statistical Cost Analysis* by J. Johnston, McGraw-Hill, New York, 1960, p. 65.

An Application in Physics: Falling Under the Influence of the Earth's Gravity

When an object falls, it falls because of the force of gravitational attraction. The acceleration under the pull of the earth's gravity has been measured to be approximately 32 feet/sec/sec (ft/sec^2) or 9.8 meters/sec/sec (m/sec^2). This means that if we neglect air resistance, a falling object will accelerate (increase its velocity) by 32 ft/sec or 9.8 m/sec each second. The following formula shows the height s of an object after t seconds:

Equation of Motion

$$s = -\tfrac{1}{2}gt^2 - v_0 t + s_0 \qquad (2)$$

where $g = 32$ ft/sec^2 $= 9.8$ m/sec^2, s is the height after t seconds, v_0 is the initial (starting) velocity, and s_0 is the initial height of the object. The minus signs indicate that the height s decreases as t increases (because the object is falling).

NOTE If s is measured in feet, then $g = 32$, $\tfrac{1}{2}g = 16$, and (2) becomes

$$s = -16t^2 - v_0 t + s_0 \qquad (3)$$

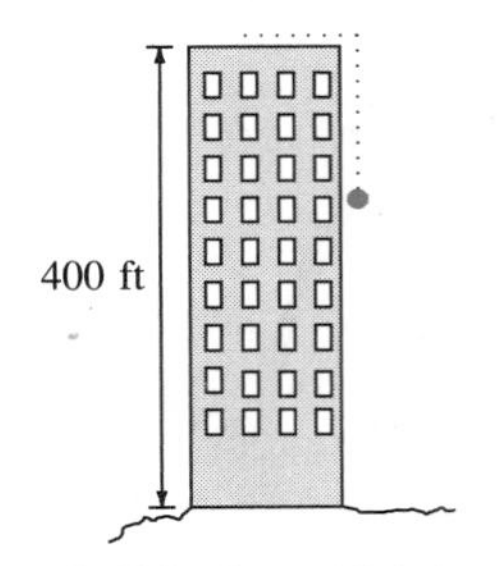

Figure 1 A ball is dropped from rest from a height of 400 ft.

EXAMPLE 2 *Determining Falling Time*

A ball is dropped from rest from the top of a building 400 feet high. When will the ball hit the ground? Neglect air resistance. [See Figure 1]

SOLUTION Here $s_0 = 400$. Since the ball is dropped from rest, its initial velocity is 0. Thus $v_0 = 0$, and (3) becomes

$$s = -16t^2 + 400$$

The ball hits the ground when $s = 0$. Thus we need to find t such that

$$\begin{aligned} 0 &= -16t^2 + 400 \\ 16t^2 &= 400 \\ t^2 &= \frac{400}{16} = 25 \\ t &= \pm 5 \end{aligned}$$

The value -5 seconds makes no sense in this problem. Thus the ball hits the ground after 5 seconds. ■

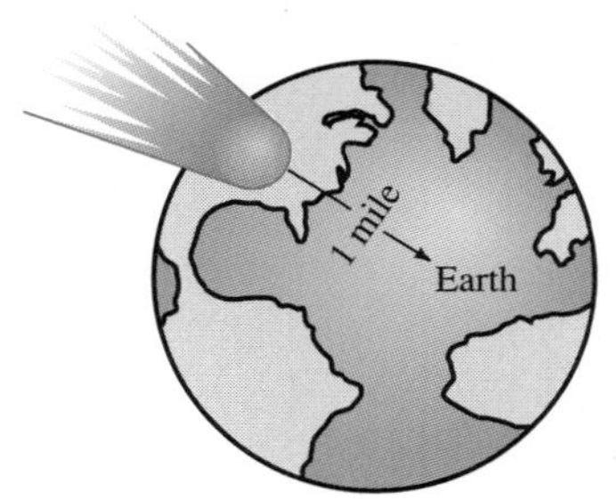

Figure 2 A meteorite traveling at 1000 ft/sec is 1 mile above the surface of the earth.

EXAMPLE 3 *Determining the Time Before Impact*

A meteorite enters the earth's atmosphere. When it is 1 mile above the surface of the earth, it is traveling at a rate of 1000 ft/sec. When will the meteorite hit the ground? Neglect air resistance. [See Figure 2]

SOLUTION Here $v_0 = 1000$ ft/sec and $s_0 = 1$ mile $= 5280$ ft. Thus, when the meteorite hits the ground, $s = 0$ and

$$0 = -16t^2 - 1000t + 5280 \tag{4}$$

$$16t^2 + 1000t - 5280 = 0$$

$$t = \frac{-1000 \pm \sqrt{(1000)^2 - 4(16)(-5280)}}{32}$$

$$= \frac{-1000 \pm \sqrt{1{,}337{,}920}}{32}$$

$$\approx \frac{-1000 \pm 1156.7}{32}$$

Two solutions of the quadratic equation (4) are, therefore,

$$t = \frac{-1000 + 1156.7}{32} \approx 4.9 \quad \text{and} \quad t = \frac{-1000 - 1156.7}{32} \approx -67.4$$

We reject the negative root. The answer is that the meteorite hits the ground approximately 4.9 seconds after it was 1 mile above the ground.

Compound Interest

In Section 2.2 we discussed the simple interest formula (see p. 79)

$$I = Prt \tag{5}$$

where I is the interest earned, P is the principal, r is the rate of interest, and t is the time the investment is held (usually measured in years).

Compound interest is interest paid on interest previously earned as well as on the original investment. Suppose that interest is paid annually. Then if P dollars are invested, the interest after one year ($t = 1$) is rP dollars, and the amount in the account is $P + rP$ dollars. After two years the interest is paid on $P + rP$ dollars (the P dollars originally invested plus the rP dollars earned after the first year). That is

$$\text{interest paid at end of second year} = r(P + rP) = rP(1 + r) \tag{6}$$

This means that

$$\begin{aligned} \begin{matrix}\text{total amount of investment}\\ \text{after 2 years}\end{matrix} &= P + \begin{matrix}\text{interest after}\\ \text{first year}\end{matrix} + \begin{matrix}\text{interest after}\\ \text{second year}\end{matrix} \\ &= P + rP + rP(1 + r) \\ &= P(1 + r) + rP(1 + r) = (P + rP)(1 + r) \\ &= P(1 + r)(1 + r) = P(1 + r)^2 \end{aligned} \tag{7}$$

After three years the investment is worth

$$\begin{aligned} \begin{matrix}\text{value at end of}\\ \text{second year}\end{matrix} + \begin{matrix}\text{interest at end}\\ \text{of third year}\end{matrix} &= P(1 + r)^2 + rP(1 + r)^2 \\ &= (P + rP)(1 + r)^2 \\ &= P(1 + r)(1 + r)^2 = P(1 + r)^3 \end{aligned}$$

If $A(t)$ denotes the value (amount) of our investment after t years, then we have the following:

Compound Interest Formula

$$A(t) = P(1 + r)^t \tag{8}$$

where P is the original principal, r is the rate of interest, t is the time the investment is held, and $A(t)$ is the total value of the investment after t years.

EXAMPLE 4 *Determining the Value of Investment in Three Years*

What is the value of a \$2000 investment after 3 years if it is invested at 12% interest compounded annually?

SOLUTION Here $P = 2000$, $r = 0.12$, and $t = 3$ in formula (8). Thus

$$A(3) = 2000(1 + 0.12)^3 = 2000(1.12)^3$$

Calculator
↓

$$= 2000(1.404928) \approx 2809.86$$

Thus the investment is worth \$2809.86 after 3 years. ■

EXAMPLE 5 *Determining the Rate of Interest*

Ms. Roschko invested \$1000 in January 1986 and \$1000 in January 1987. In January 1988 she had \$2325 in the account. If the rate of interest over the 2 years didn't change, and if it was paid and compounded annually, what was the rate of interest?

SOLUTION The \$1000 invested in 1986 earned interest for 2 years. Thus, from (8),

$$\text{value of \$1000 invested in January 1986} = 1000(1 + r)^2$$

The \$1000 invested in 1987 earned interest for 1 year:

$$\text{value of \$1000 invested in January 1987} = 1000(1 + r)$$

Then

$$\begin{aligned} &\text{total value of the 2 investments in 1988} \\ &\quad = 1000(1 + r) + 1000(1 + r)^2 \\ &\quad = 2325 \end{aligned} \tag{9}$$

Equation (9) is a quadratic equation. We solve it.

$$\begin{aligned}
1000(1 + r) + 1000(1 + r)^2 - 2325 &= 0 \\
(1 + r) + (1 + r)^2 - 2.325 &= 0 \quad \text{Divide by 1000} \\
1 + r + 1 + 2r + r^2 - 2.325 &= 0 \\
r^2 + 3r + 2 - 2.325 &= 0 \\
r^2 + 3r - 0.325 &= 0 \\
r &= \frac{-3 \pm \sqrt{9 - 4(-0.325)}}{2} \\
&= \frac{-3 \pm \sqrt{10.3}}{2}
\end{aligned}$$

We reject $\dfrac{-3 - \sqrt{10.3}}{2}$ because the rate of interest must be positive. Thus

$$r = \frac{-3 + \sqrt{10.3}}{2} \approx \frac{-3 + 3.20936}{2} = \frac{0.20936}{2} \approx 0.1047$$

Thus Ms. Roschko received a rate of interest of approximately 10.47% compounded annually over 2 years. ■

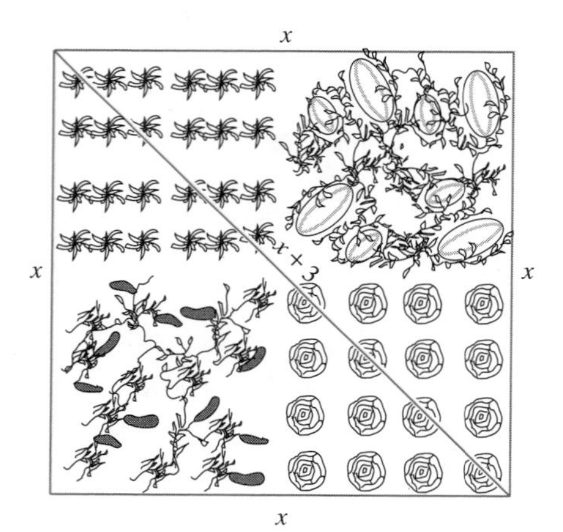

Figure 3 The diagonal of a square garden is 3 meters longer than one of its sides.

EXAMPLE 6 *A Problem in Geometry*

The diagonal of a square garden is 3 meters longer than one of its sides. Find the area of the garden.

SOLUTION We refer to Figure 3. By the Pythagorean theorem, the square of the hypotenuse is equal to the sum of the squares of the other two sides. In Figure 3 we have

$$\begin{aligned}
(x + 3)^2 &= x^2 + x^2 \\
x^2 + 6x + 9 &= 2x^2 \\
x^2 - 6x - 9 &= 0 \\
x &= \frac{6 \pm \sqrt{36 - 4(-9)}}{2} = \frac{6 \pm \sqrt{72}}{2} \\
&= \frac{6 \pm 6\sqrt{2}}{2} = 3 \pm 3\sqrt{2}
\end{aligned}$$

There seem to be two answers:

$$x = 3 + 3\sqrt{2} = 3(1 + \sqrt{2}) \text{ m} \quad \text{and} \quad \text{area} = x^2 \text{ m}^2 = 9(1 + \sqrt{2})^2 \text{ m}^2 \approx 52.5 \text{ m}^2$$

$$x = 3 - 3\sqrt{2} = 3(1 - \sqrt{2}) \text{ m}$$

But $3(1 - \sqrt{2}) < 0$, and the length of a side cannot be negative. Thus $x = 3(1 - \sqrt{2})$ is not an answer. ■

EXAMPLE 7 *A Division of Labor Problem*

Two trucks haul debris from a construction site. The site is cleared after 2 hours. If the larger truck worked alone, it could clear the site 3 hours earlier than the smaller one working alone. How long would it take each truck to clear the site if it worked alone?

SOLUTION The secret to solving problems of this type is to determine what part of the job can be completed *in 1 hour* by each participant. Suppose that it takes the larger truck x hours to clear the site. Then it would take the smaller truck $(x + 3)$ hours to complete the task. This means that in 1 hour the larger truck completes $\frac{1}{x}$ of the job. [For example, if $x = 5$ hours, then in 1 hour the truck would clear $\frac{1}{5}$ of the site.] We summarize the data in the following table:

	Time Necessary to Complete Task	Fraction Completed in 1 hour
Larger truck	x hours	$\frac{1}{x}$
Smaller truck	$x + 3$ hours	$\frac{1}{x+3}$
Two trucks working together	2 hours	$\frac{1}{2}$

The two trucks working together complete one half of the job in 1 hour. Thus we have

$$\frac{1}{x} + \frac{1}{x+3} = \frac{1}{2}$$

$$\frac{2x(x+3)}{x} + \frac{2x(x+3)}{x+3} = \frac{2x(x+3)}{2} \qquad \text{The LCD is } 2x(x+3) \text{ so we multiply both sides by this quantity}$$

$$2(x + 3) + 2x = x(x + 3)$$

$$4x + 6 = x^2 + 3x$$

$$x^2 - x - 6 = 0$$

$$(x - 3)(x + 2) = 0$$

Thus, $x = 3$ or $x = -2$. But $x = -2$ makes no sense in this problem. The answer is $x = 3$.

the larger truck takes 3 hours working alone to clear the site

the smaller truck takes 6 hours working alone to clear the site

Check: $\frac{1}{3} + \frac{1}{6} = \frac{2}{6} + \frac{1}{6} = \frac{3}{6} = \frac{1}{2}$.

Problems 2.6

Readiness Check

I. Which of the following equations might be used to find two consecutive odd integers whose product is 899?
   a. $x + x + 3 = 899$
   b. $x(x + 1) = 899$
   c. $x(x + 2) = 899$
   d. $x + x + 1 = 899$
   e. $(2x - 1)(2x + 1) = 899$

II. Which of the following equations might be used to find two even consecutive integers such that the sum of their reciprocals is 3 times the reciprocal of the larger integer?
   a. $x + x + 2 = 3(x + 2)$ b. $\frac{1}{2x + 1} = \frac{3}{x + 1}$
   c. $\frac{1}{x} + \frac{1}{x + 2} = \frac{3}{x + 2}$
   d. $x + x + 1 = 3(x + 1)$

III. An object is thrown straight down from the top of a 640-foot building with an initial velocity of 48 ft/sec. The equation that gives its height s after t seconds is ______.
   a. $s = -16t^2 + 48t + 640$
   b. $s = -16t^2 - 24t + 640$
   c. $s = -16t^2 - 48t + 640$
   d. $s = 16t^2 + 48t - 640$

IV. In Problem III, after how many seconds will the object hit the ground?
   a. 4 b. 5 c. 6 d. 8 e. 16

V. \$5000 invested now at 7% compounded annually will be worth \$______ in 10 years.
   a. $5000\ (1.07)^{10}$
   b. $5000\ (1.7)^{10}$
   c. $5000\ (1.1)^{7}$
   d. $5000\ (0.17)^{10}$
   e. $5000\ (0.11)^{7}$

In some of the problems that follow, the answers are not simple integers like 3, 10, or 45. We are not trying to make things difficult, but in real-life problems the answers almost never turn out to be arithmetically simple. In solving most of the problems, you will find a calculator useful. In problems where the answer is not an integer, but the answer must be an integer, round to the nearest integer.

1. The total cost function for a certain firm's product is $5q^2 - 200q + 1125$, $q \le 1000$. How many items can be produced for \$17,125?†
2. In Problem 1, how many items can be produced for \$30,000?
3. The total revenue received from selling q items of the product in Problem 1 is given by $R = -0.12q^2 + 100q$, $q \le 1000$. (A negative revenue means that selling costs exceed selling receipts.) If the firm received \$9694.72 in revenue, how many items were sold?
4. Answer the question in Problem 3 if \$12,000 in revenue was received.
5. A toy manufacturer finds that the cost in dollars of producing q copies of a certain doll is given by
$$C = 250 + 3q + 0.01q^2, \qquad q \le 300$$
The dolls can be sold for \$14 each.
   (a) Write the revenue function for this doll. [Hint: It is linear.]
   (b) Write the profit function. [Hint: Profit = revenue − costs; see p. 84.]
   (c) How many dolls were made and sold if the total profit was \$1406?
6. Answer question (c) in Problem 5 if the total profit is \$2000.
7. The cost of producing q color television sets is given by $C = 5000 + 250q - 0.01q^2$, $q \le 300$. How many sets were produced if the total cost was approximately \$19,466?
8. In Problem 7, the revenue received from selling q television sets is $R = 400q - 0.02q^2$. If \$48,502 was received, how many sets were sold?

Answers to Readiness Check

I. c, e II. c III. c IV. b V. a

† Problems 1–4 are adapted from *Mathematics for Economics and Business Analysis*, by Forsythe and Walter, Goodyear, Pacific Palisades, CA, 1976, p. 81.

9. Refer to Problems 7 and 8.
   (a) What is the profit received from producing and selling q television sets?
   (b) How many sets were manufactured and sold if the total profit was $18,010?

* 10. A manufacturer of kitchen sinks finds that if he produces q sinks per week, he has fixed costs of $1000, labor and material costs of $5 per sink, and advertising costs of $\$100\sqrt{q}$. If his average weekly total cost is $3831, how many sinks does he produce, on average, each week?

In Problems 11–16 determine how long, ignoring air resistance, it will take an object to hit the ground if it is dropped from the given height. [Hint: Use the values g = 32 feet/sec^2 = 9.8 meters/sec^2.]

11. 200 feet
12. 5000 feet
13. 200 meters
14. 1000 meters
15. 2500 meters
16. 5280 feet
17. A rock is thrown down from the top of a 500-ft building with an initial velocity of 50 ft/sec. When will the rock hit the ground?

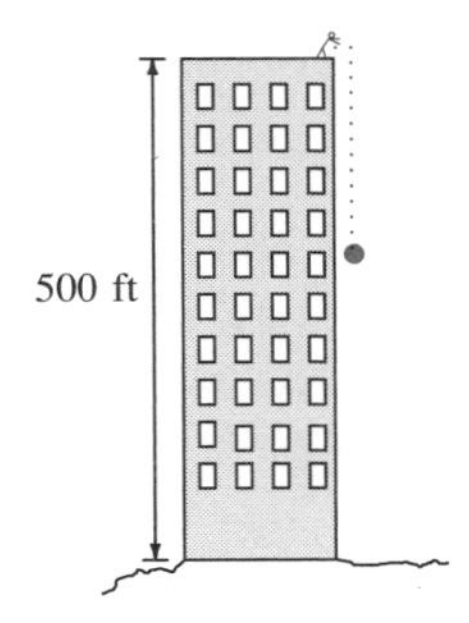

18. The acceleration due to gravity at the surface of Mars is approximately 3.92 meters/sec^2. How long will it take an object to hit the surface of Mars if it is dropped from a height of 250 meters?
19. Answer the question of Problem 18 if the object is dropped from a height of 1000 meters.
20. Mr. Newman invests $2000 in each of 2 consecutive years and, at the end of 2 years, he has $4500. What was his annual rate of return, assuming annual compounding?
21. On February 15, 1989, $10,000 was invested in a bond and $25,000 was invested in the same bond on February 15, 1990. The value of the bond on February 15, 1991 was $40,125. Assume that the rate of interest was constant over the 2-year period and that interest was paid annually. What was the rate of interest?

* 22. A person invested $1000 in a savings account paying interest semiannually. After 1 year, $1063.48 is in the account. What was the rate of interest? [Hint: Use the compound interest formula (8) with $t = 2$ time periods to determine the interest paid each 6 months.]

* 23. Answer the question in Problem 22 if an investment of $2500 is worth $2691 after 1 year.

24. One side of a child's rectangular bedroom is 4 feet longer than another side. The area of the bedroom is 96 square feet. What are the dimensions of the child's room?

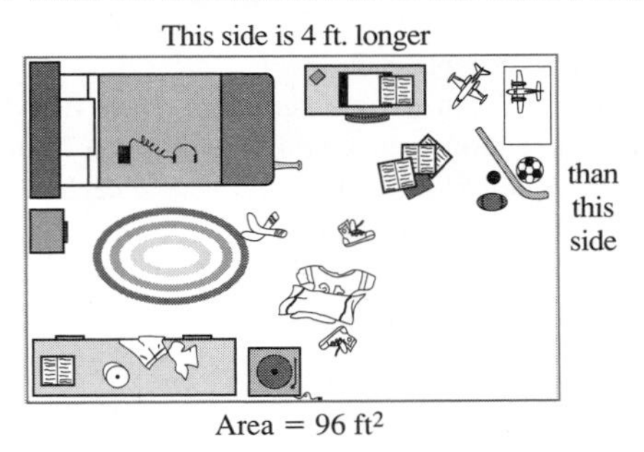

25. One side of a rectangular fenced-in courtyard is 3 feet longer than twice the length of the other side. The area of the courtyard is 275 square feet. Find its dimensions.

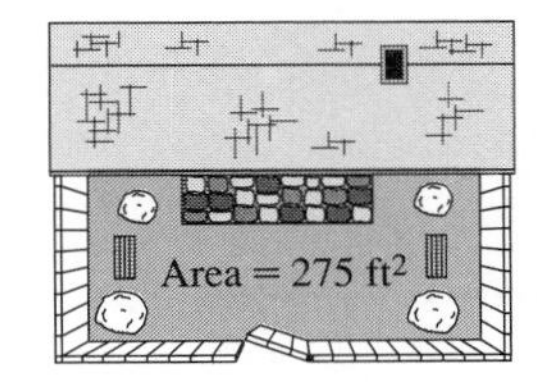

26. The sum of a number and its reciprocal is $\frac{65}{28}$. Find the number.
27. The product of two consecutive odd integers is 783. Find the integers.
28. One number is 6 less than another. The product of the two numbers is 187. What are the numbers?
29. The difference between a number and its reciprocal is $\frac{39}{40}$. Find the number.
30. The product of two consecutive even integers is 51,528. Find the integers.
31. Two resistors are connected in parallel in an electric circuit (see Example 11 on p. 87). The resistance of one is 2 ohms more than the other. The total resistance in the circuit is $\frac{40}{9}$ ohms. Find the resistance of each resistor.

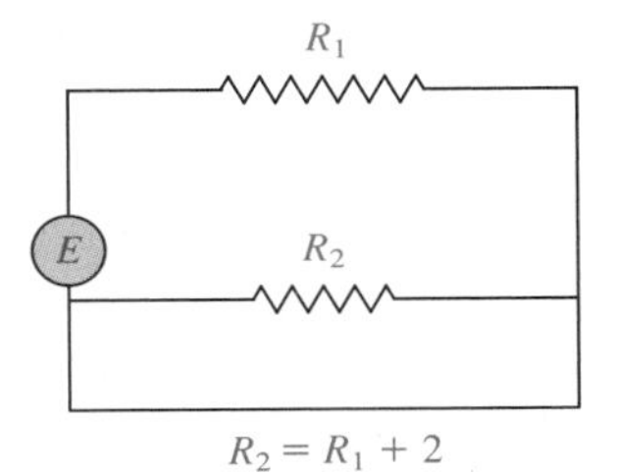

$R_2 = R_1 + 2$

32. Answer the question of Problem 31 if one resistance is 5 ohms more than the other and the total resistance is $\frac{100}{9}$ ohms.

33. The base of a triangle is 1 centimeter more than the height. The area is 15 square centimeters. What is the length of the base?
34. Working together, Jones and Smith can complete a job in 4 hours. Working alone, Smith takes 6 hours more than Jones to complete the job. Working alone, how long does each take to complete the job?
35. Jeff and Mary can mow a series of lawns in 8 hours. If Mary worked alone, she could finish the job 2 hours faster than Jeff could. How long would it take each to mow the lawns if they worked alone?

* 36. Two adjacent rectangular sports areas of the same size are to be fenced off with 310 feet of fencing. What are the dimensions of the field if the total area fenced off is 600 sq. ft.? [Hint: Write y in terms of x. Use the fact that the total amount of fencing used is 310 feet. There are two solutions.]

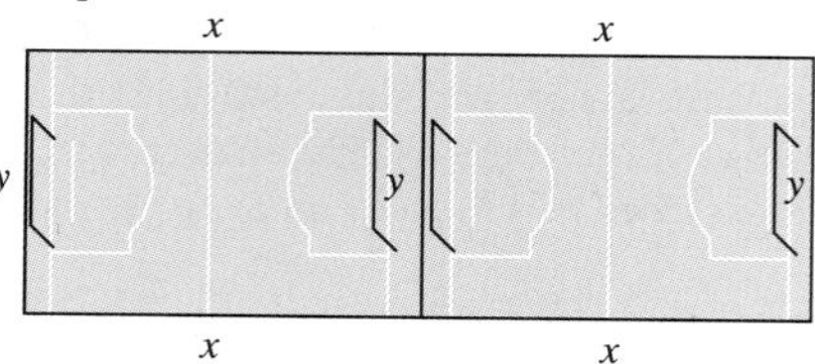

2.7 Linear Inequalities and Intervals

Intervals

Certain sets of real numbers are very important in applications.

(i) The **open interval** from a to b, denoted by (a, b), is the set of real numbers between a and b, *not including* the numbers a and b. We have

$$(a, b) = \{x: a < x < b\}$$

Note that $a \notin (a, b)$ and $b \notin (a, b)$. The members a and b are called **endpoints** of the interval. This is depicted in Figure 1.

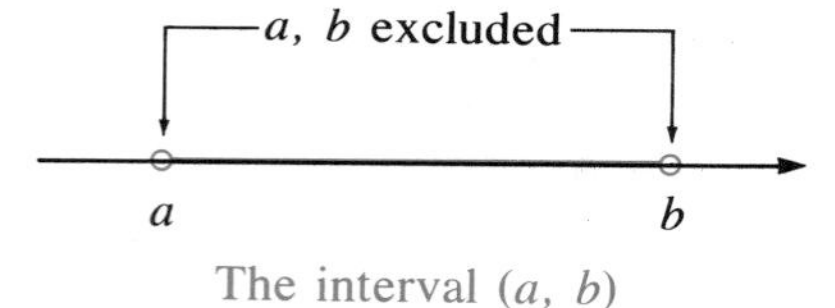

Figure 1 The open interval (a, b).

(ii) The **closed interval** from a to b, denoted by $[a, b]$, is the set of numbers between a and b, *including* the numbers a and b. We have

$$[a, b] = \{x: a \leq x \leq b\}$$

Note that $a \in [a, b]$ and $b \in [a, b]$.

As before, the numbers a and b are called **endpoints** of the interval. This situation is depicted in Figure 2.

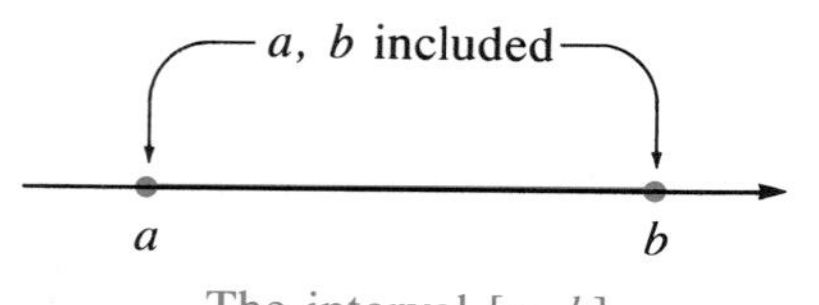

Figure 2 The closed interval $[a, b]$.

NOTE As in Figures 1 and 2 we use a solid dot • to indicate that an endpoint is included and an open dot ∘ to indicate that the endpoint is not included.

EXAMPLE 1 ***An Open Interval and a Closed Interval***

(a) $(-1, 5) = \{x: -1 < x < 5\}$
(b) $[0, 8] = \{x: 0 \leq x \leq 8\}$

Sometimes we will need to include one endpoint but not the other.

(iii) The **half-open interval** $[a, b)$ is given by

$$[a, b) = \{x: a \leq x < b\}$$

We include the endpoint $x = a$ but not the endpoint $x = b$. That is, $a \in [a, b)$ but $b \notin [a, b)$.

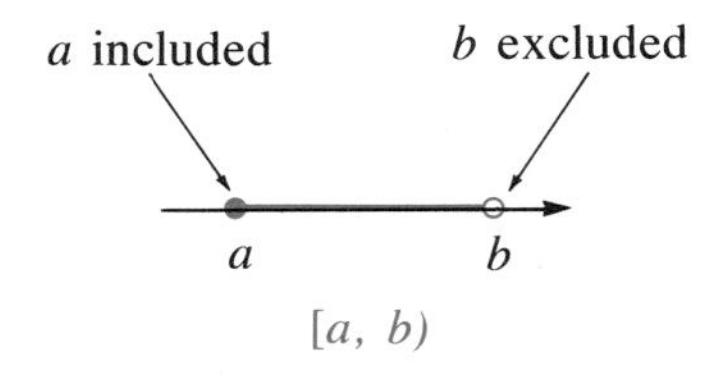

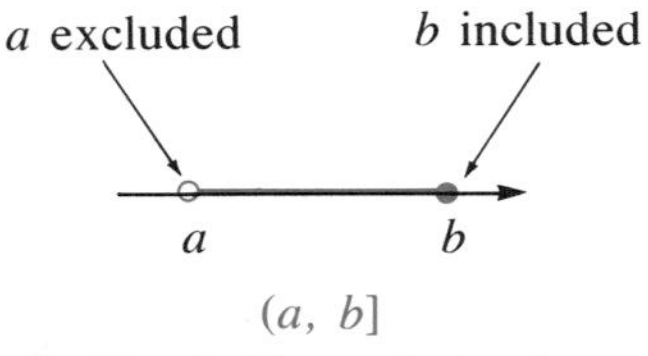

Figure 3 Two half-open intervals.

(iv) The half-open interval $(a, b]$ is given by

$$(a, b] = \{x: a < x \leq b\}$$

We have $a \notin (a, b]$ but $b \in (a, b]$. Here b is included, but a is not. See Figure 3.

EXAMPLE 2 ***A Half-Open Interval***

$[-\frac{5}{2}, 8) = \{x: -\frac{5}{2} \leq x < 8\}$. This is sketched in Figure 4.

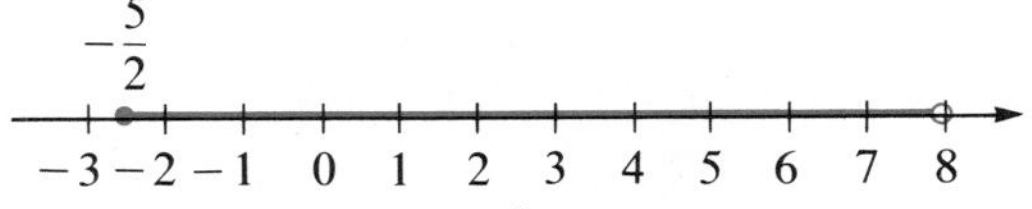

Figure 4 The interval $[-\frac{5}{2}, 8)$. ■

EXAMPLE 3 ***A Half-Open Interval***

$(3, 10] = \{x: 3 < x \leq 10\}$. This interval is sketched in Figure 5.

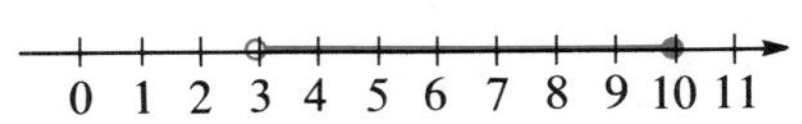

Figure 5 The interval $(3, 10]$.

Infinite Intervals

Intervals may be infinite in length. We define **infinite intervals** as follows:

Five Types of Infinite Intervals

$$[a, \infty) = \{x: x \geq a\}$$
$$(a, \infty) = \{x: x > a\}$$
$$(-\infty, a] = \{x: x \leq a\}$$
$$(-\infty, a) = (x: x < a\}$$
$$(-\infty, \infty) = \mathbb{R} = \text{the set of real numbers}$$

The symbols ∞ and $-\infty$, denoting infinity and minus infinity, respectively, are *not* real numbers and do not obey the usual laws of algebra, but they can be used for notational convenience. This idea is expressed in symbols as

$$[a, \infty) = \{x: a \leq x < \infty\} \quad \text{and} \quad (-\infty, a) = \{x: -\infty < x < a\}$$

EXAMPLE 4 ***Four Infinite Intervals***

(a) $(-\infty, 2) = \{x: x < 2\}$
(b) $[4, \infty) = \{x: x \geq 4\}$
(c) $(-\infty, -\frac{7}{3}] = \{x: x \leq -\frac{7}{3}\}$
(d) $(0, \infty) = \{x: x > 0\}$ = the positive real numbers

Intervals for which both endpoints are real numbers are called **bounded intervals.** Infinite intervals are also called **unbounded intervals.**

We summarize below the notation for bounded and unbounded intervals.

Bounded and Unbounded Intervals

Notation	*Type of Interval*	*Interval Written as an Inequality*	*Typical Graph*
(a, b)	open (bounded)	$a < x < b$	
$[a, b]$	closed (bounded)	$a \le x \le b$	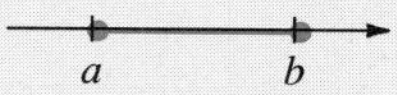
$[a, b)$	half-open (bounded)	$a \le x < b$	
$(a, b]$	half-open (bounded)	$a < x \le b$	
(a, ∞)	open (unbounded)	$x > a$	
$[a, \infty)$	half-open (unbounded)	$x \ge a$	
$(-\infty, a)$	open (unbounded)	$x < a$	
$(-\infty, a]$	half-open (unbounded)	$x \le a$	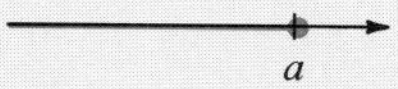
$(-\infty, \infty)$	The real number line (unbounded)		

Linear Inequalities

We now turn to the subject of linear inequalities. A **linear inequality in one variable** is an inequality that can be written in one of the following forms.

$$ax + b < c \tag{1}$$
$$ax + b \le c \tag{2}$$
$$ax + b > c \tag{3}$$
$$ax + b \ge c \tag{4}$$

Here a, b, and c are real numbers and $a \neq 0$. We can solve linear inequalities of the forms (1), (2), (3) or (4) by using the rules for dealing with inequalities discussed on pp. 15–16. By the **solution set** of an inequality, we mean the set of numbers that satisfy the inequality.

Linear inequalities can be solved by using one or more of the following properties:

Properties of Inequalities

	Property	*Description*	*Illustration*
(a)	If $a < b$, then $a + c < b + c$	Adding a number to both sides of an inequality preserves the inequality.	Since $2 < 3$, $2 + 5 < 3 + 5$ $2 - 6 < 3 - 6$

(b) If $a < b$ and $b < c$, then $a < c$	This property is called the **transitive law.**	Since $2 < 3$ and $3 < 5$, we have $2 < 5$
(c) If $a < b$ and $c > 0$, then $ac < bc$	Multiplying an inequality by a positive number preserves the inequality.	Since $2 < 3$, we have $2 \cdot 4 < 3 \cdot 4$
(d) If $a < b$ and $c < 0$, then $ac > bc$	Multiplying an inequality by a negative number *reverses* the sense of the inequality.	Since $2 < 3$, we have $2(-4) > 3(-4)$
(e) If a and b are both positive or both negative and $a < b$, then $\frac{1}{a} > \frac{1}{b}$	Taking reciprocals reverses an inequality of numbers with the same sign	$2 < 3$ so $\frac{1}{2} > \frac{1}{3}$ $-3 < -2$ so $-\frac{1}{3} > -\frac{1}{2}$

EXAMPLE 5 ***Illustration of Four Properties of Inequalities***

Suppose that $0 < x < y$.

(a) $x + z < y + z$ for every real number z — Property (a)
(b) $5x < 5y$ — Property (c)
(c) $-3x > -3y$ — Property (d)
(d) $\frac{1}{x} > \frac{1}{y}$ — Property (e) ■

EXAMPLE 6 ***Solving a Linear Inequality***

Solve the inequality $-3x + 5 \geq 12$.

SOLUTION

$$-3x + 5 \geq 12$$

$$-3x + 5 - 5 \geq 12 - 5 \quad \text{Subtract 5.}$$

$$-3x \geq 7$$

$$-\frac{1}{3}(-3x) \leq -\frac{1}{3}(7) \quad \text{Since } -\tfrac{1}{3} < 0 \text{, the inequality is reversed}$$

$$x \leq -\frac{7}{3}$$

The solution set is $\{x: x \leq -\frac{7}{3}\} = (-\infty, -\frac{7}{3}]$. It is sketched in Figure 6. ■

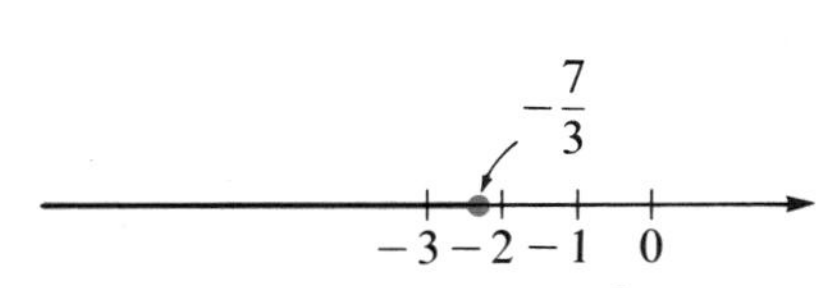

Figure 6 The interval $(-\infty, -\frac{7}{3}]$.

EXAMPLE 7 ***Solving Two Linear Inequalities***

Solve the inequalities $-3 < \dfrac{7-2x}{3} \le 4$.

SOLUTION There are two inequalities here. We solve both at the same time.

$$-3 < \frac{7-2x}{3} \le 4$$

$$3(-3) < 3\left(\frac{7-2x}{3}\right) \le 3(4) \qquad \text{Multiply by 3. This preserves the sense of the inequalities since } 3 > 0$$

$$-9 < 7 - 2x \le 12$$

$$-9 - 7 < 7 - 2x - 7 \le 12 - 7 \qquad \text{Add } -7$$

$$-16 < -2x \le 5$$

$$8 > x \ge -\frac{5}{2} \qquad \text{Multiply by } -\tfrac{1}{2} \text{ and reverse both inequalities}$$

or

$$-\frac{5}{2} \le x < 8$$

The solution set is $[-\frac{5}{2}, 8)$. This interval was sketched in Figure 4. Note that each of the steps in the computation served to simplify the term containing x.

Problems 2.7

Readiness Check

I. Which of the following numbers belongs to the solution set of the inequalities $-3 < x < 8$?
a. 0 b. -3 c. 8 d. -5

II. Which of the following does *not* represent the same set of real numbers?
a. $[-2, 1)$
b. $\{x: -\frac{3}{2} \le x - \frac{1}{2} < \frac{1}{2}\}$
c. 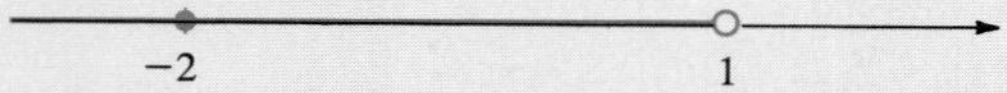

d. All the real numbers that are less than 1 but greater than or equal to -2.

III. Which of the following inequalities has a solution set represented by the interval $(-2, \infty)$?
a. $4x - 1 \ge -9$ b. $7 > 5 - x$
c. $2 - x < 0$ d. $8x + 16 < 0$

IV. Which of the following is the correct interval notation for the solution of $-3 < 3 - x \le 8$?
a. $[-6, 5)$ b. $(-6, 5]$ c. $[-5, 6)$
d. $(-5, 6]$

V. Which of the following is true about $0 < \dfrac{-2x}{3} < 4$?
a. The graph of its solution set contains its endpoints.
b. The solution set contains only negative numbers.
c. It represents an unbounded interval.
d. 0 is a solution of the inequality.

Answers to Readiness Check

I. a II. b III. b IV. c V. b

In Problems 1–12 a set of real numbers is depicted. Write an interval that describes the set.

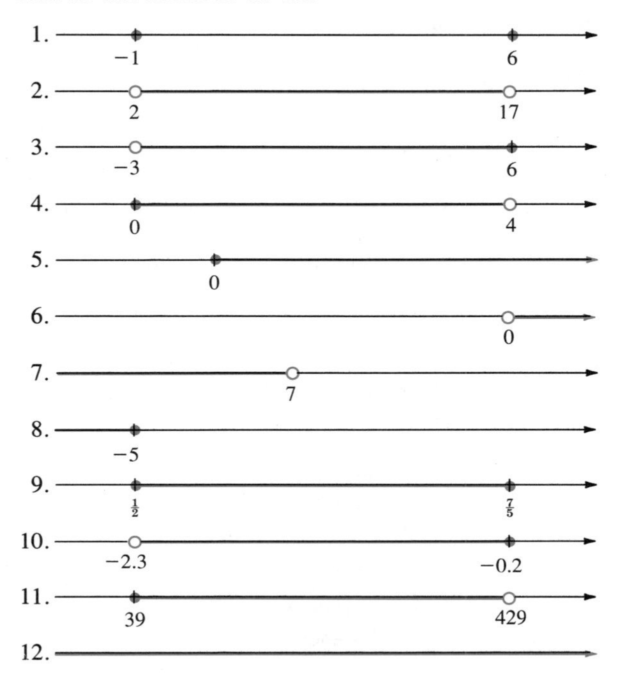

In Problems 13–28 represent each set of inequalities as an interval.

13. $1 < x < 2$
14. $-3 < x < 5$
15. $0 < x < 8$
16. $-5 \leq x \leq 4$
17. $-2 \leq x \leq 0$
18. $3.7 \leq x \leq 5.2$
19. $0 < x \leq 5$
20. $1 \leq x < \sqrt{2}$
21. $-1.32 \leq x < 4.16$
22. $x \geq 0$
23. $x < 0$
24. $x \leq 0$
25. $x \leq 2$
26. $x \geq -6$
27. $x < -5$
28. $x > 10^6$

In Problems 29–52 find the solution set of the given inequality (or inequalities).

29. $x - 4 < 7$
30. $x - 2 \leq 5$
31. $x - 4 > \frac{7}{2}$
32. $x - 2 \geq 5$
33. $-x + 2 \leq 3$
34. $-x + 4 > -5$
35. $2x - 7 \leq 2$
36. $-2x + \frac{4}{3} > 8$
37. $2 \geq 6x + 14$
38. $2 < 3x - 3$
39. $-2 < 4 - 8x$
40. $4 \leq 7x + 10$
41. $1 \leq x + 2 \leq 4$
42. $-2 < x - 1 \leq 3$
43. $-2 < -x + 3 \leq 7$
44. $\frac{1}{2} \leq 2x + 2 \leq 4$
45. $\frac{1}{5} \leq 2x + \frac{2}{5} < \frac{4}{5}$
46. $-1 < 2x - 2 < 4$
47. $-1 \leq 2x + 5 < 7$
48. $1 \leq 7x - 6 \leq 4$
49. $-2 < -2x - 4 \leq 10$
50. $2 < 3x + 4 \leq 7$
51. $-4 < \frac{2x - 4}{3} \leq 7$
52. $2 \geq \frac{4 - 2x}{5} > -4$

53. Solve the inequalities $a \leq \frac{bx + c}{d} < e$, with b and $d > 0$.

Calculator Problems

In Problems 54–61 use a calculator to find the solution set of each inequality.

54. $x - 3.16 < 2.52$
55. $x + 21.562 \geq 37.295$
56. $1.62x < 3.58$
57. $0.137x > 0.471$
58. $-7.85x \geq 14.962$
59. $-2 \times 10^{-5}x > 3 \times 10^{-4}$
60. $-14{,}385 < \frac{13x + 65.8}{23.97} \leq 2705$
61. $2 \times 10^{-6} \leq \frac{4 \times 10^{-5}x - 7.2 \times 10^{-7}}{3.1 \times 10^{-3}} \leq 6.5 \times 10^{-5}$

2.8 Other Inequalities

Inequalities Involving Absolute Value

In Section 1.4 we defined the absolute value of a real number as

$$|a| = a \quad \text{if} \quad a \geq 0 \tag{1}$$

$$|a| = -a \quad \text{if} \quad a < 0 \tag{2}$$

We also saw that $|a|$ represents the *distance* between the points on the number line representing a and 0.

In solving inequalities using absolute values, the following properties will be very useful.

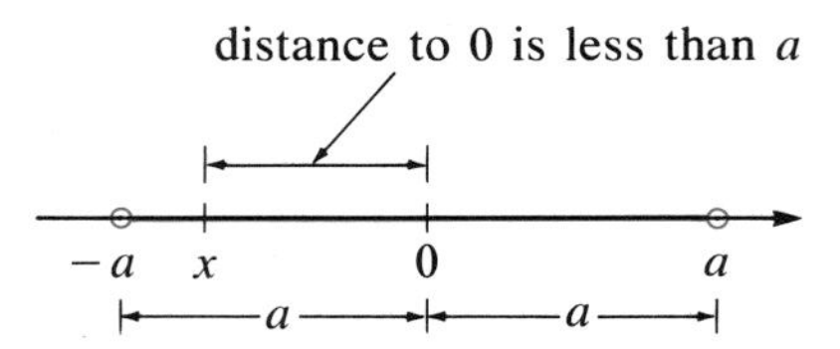

Figure 1 Illustration of the inequality $|x| < a$.

Two Properties of Absolute Value

If $a > 0$

$$|x| < a \text{ is equivalent to } -a < x < a \tag{3}$$

$$|x| \leq a \text{ is equivalent to } -a \leq x \leq a \tag{4}$$

That is, $\{x\colon |x| < a\} = \{x\colon -a < x < a\}$ (see Figure 1), which indicates that for any x in the open interval $(-a, a)$, the distance between the points representing x and zero is less than a.

EXAMPLE 1 ***Solving an Absolute Value Inequality***

Solve the inequality $|x| \leq 3$.

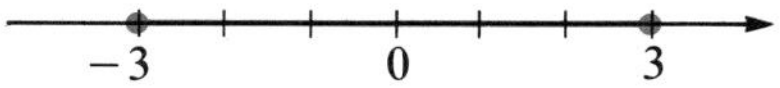

SOLUTION $|x| \leq 3$

$$-3 \leq x \leq 3 \qquad \text{From (4)}$$

The solution set is the closed interval $[-3, 3]$. It is sketched in Figure 2. ■

−3 0 3

Figure 2 Illustration of the inequality $|x| \leq 3$.

EXAMPLE 2 ***Solving an Absolute Value Inequality***

Solve the inequality $|x - 4| < 5$.

SOLUTION

$$|x - 4| < 5$$

$$-5 < x - 4 < 5 \qquad \text{From (3)}$$

$$-5 + 4 < x - 4 + 4 < 5 + 4 \qquad \text{Add 4 to each term}$$

$$-1 < x < 9$$

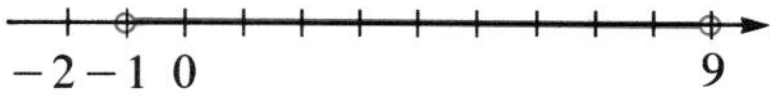

The solution set is the open interval $(-1, 9)$ and is sketched in Figure 3.

−2 −1 0 9

Figure 3 Illustration of the inequality $|x - 4| < 5$.

Here are two other useful properties for solving inequalities.

Two More Properties of Absolute Value

If $a > 0$,

$$|x| > a \text{ is equivalent to } x > a \text{ or } x < -a \quad (5)$$

$$|x| \geq a \text{ is equivalent to } x \geq a \text{ or } x \leq -a \quad (6)$$

These rules make sense because if $|x| > a$, then x is more than a units away from 0. If $x > 0$, then $x > a$. If $x < 0$, then $x < -a$. See Figure 4.

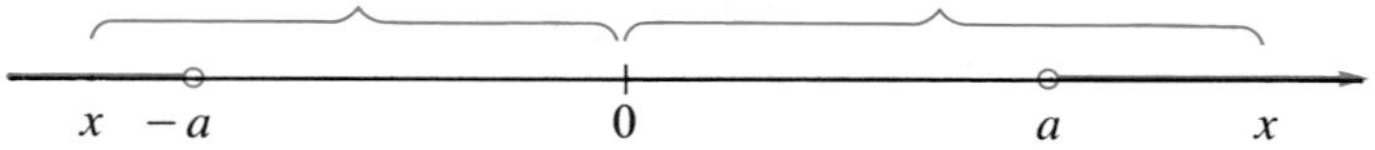

Figure 4 Illustration of the inequality $|x| > a$.

EXAMPLE 3 ***Solving an Absolute Value Inequality***

Solve the inequality $|x| > 3$.

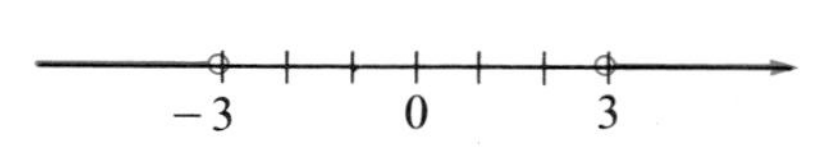

Figure 5 Illustration of the inequality $|x| > 3$.

SOLUTION $|x| > 3$

$$x > 3 \quad \text{or} \quad x < -3 \qquad \text{From (5)}$$

The solution set, $\{x\colon x < -3 \text{ or } x > 3\}$, is sketched in Figure 5. ■

EXAMPLE 4 ***Solving an Absolute Value Inequality***

Solve the inequality $|x + 2| \geq 8$.

SOLUTION $|x + 2| \geq 8$

$$x + 2 \geq 8 \quad \text{or} \quad x + 2 \leq -8 \qquad \text{From (6)}$$

$$x + 2 - 2 \geq 8 - 2 \quad \text{or} \quad x + 2 - 2 \leq -8 - 2$$

$$x \geq 6 \quad \text{or} \quad x \leq -10$$

The solution set, $\{x\colon x \leq -10 \text{ or } x \geq 6\}$, is sketched in Figure 6.

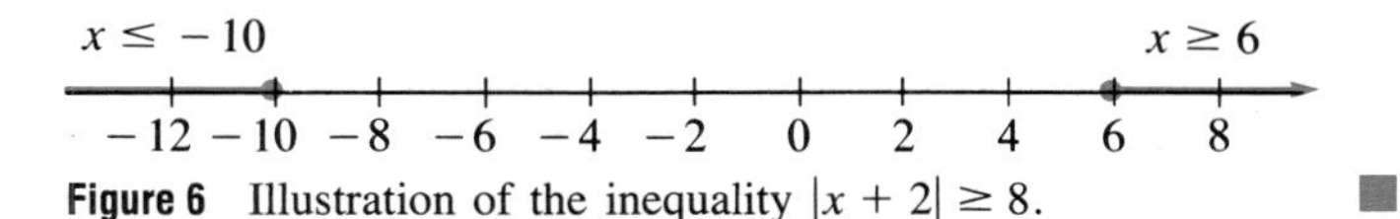

Figure 6 Illustration of the inequality $|x + 2| \geq 8$. ■

EXAMPLE ***Solving an Absolute Value Inequality***

Solve the inequality $|5 - 3x| \geq 1$.

SOLUTION $|5 - 3x| \geq 1$

$$5 - 3x \geq 1 \qquad \text{or} \qquad 5 - 3x \leq -1 \qquad \text{From (6)}$$

$$
\begin{aligned}
5 - 3x - 5 &\geq 1 - 5 &\quad \text{or} \quad& 5 - 3x - 5 \leq -1 - 5 \\
-3x &\geq -4 &\quad \text{or} \quad& -3x \leq -6 \\
\frac{-3x}{-3} &\leq \frac{-4}{-3} &\quad \text{or} \quad& \frac{-3x}{-3} \geq \frac{-6}{-3} \qquad \text{Divide by } -3 \text{ and reverse the inequalities} \\
x &\leq \frac{4}{3} &\quad \text{or} \quad& x \geq 2
\end{aligned}
$$

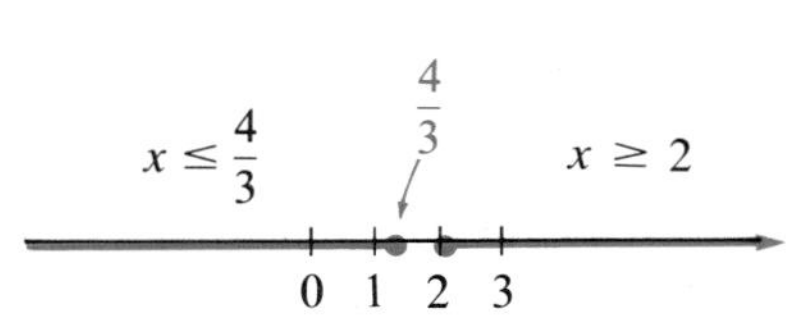

Figure 7 Illustration of the inequality $|5 - 3x| \geq 1$.

The solution set, $\{x: x \leq \frac{4}{3} \text{ or } x \geq 2\}$, is sketched in Figure 7.

Three Examples Useful in Calculus

Most calculus courses begin with a discussion of limits. The following three examples illustrate the kinds of inequalities that must be solved when proving facts about limits.

EXAMPLE 6 *An Absolute Value Inequality That Arises in Calculus*

Solve the inequality $|x - 3| < 0.01$

SOLUTION

$$
\begin{aligned}
|x - 3| &< 0.01 \\
-0.01 < x - 3 &< 0.01 \qquad \text{From (3)} \\
3 - 0.01 < x &< 3 + 0.01 \qquad \text{Add 3 to each term} \\
2.99 < x &< 3.01
\end{aligned}
$$

The solution set is the open interval (2.99, 3.01). ■

EXAMPLE 7 *A Basic Inequality That Arises in Calculus*

If a and δ are real numbers with $\delta > 0$, solve $|x - a| < \delta$ and represent the solution set graphically.

SOLUTION

$$
\begin{aligned}
|x - a| &< \delta \\
-\delta < x - a &< \delta \qquad \text{From (3)} \\
a - \delta < x &< a + \delta \qquad \text{Add } a \text{ to each term}
\end{aligned}
$$

Figure 8 The set of numbers that satisfy $|x - a| < \delta$.

The solution set, sketched in Figure 8, is the open interval $(a - \delta, a + \delta)$. ■

EXAMPLE 8 *Showing How One Inequality Leads to Another*

Show that $|(2x - 6) - 2| < 1$ whenever $|x - 4| < \frac{1}{2}$.

SOLUTION

$$|x - 4| < \frac{1}{2}$$

$$-\frac{1}{2} < x - 4 < \frac{1}{2} \qquad \text{From (3)}$$

$$4 - \frac{1}{2} < x < 4 + \frac{1}{2} \qquad \text{Add 4 to each term}$$

$$\frac{7}{2} < x < \frac{9}{2}$$

$$7 < 2x < 9 \qquad \text{Multiply each term by 2}$$

$$7 - 8 < 2x - 8 < 9 - 8 \qquad \text{Subtract 8 from each term}$$

$$-1 < 2x - 8 < 1$$

$$-1 < 2x - 6 - 2 < 1$$

$$-1 < (2x - 6) - 2 < 1$$

$$|(2x - 6) - 2| < 1 \qquad \text{From (3)}$$

Quadratic Inequalities

Quadratic Inequalities

A **quadratic inequality** in **standard form** is an inequality that can be written in one of the forms

$$\begin{array}{ll} ax^2 + dx + e > 0 & ax^2 + dx + e \geq 0 \\ ax^2 + dx + e < 0 & ax^2 + dx + e \leq 0 \end{array} \qquad (7)$$

where $a \neq 0$.

Since $a \neq 0$, we can divide each inequality by a and write the inequalities as

$$\begin{array}{ll} x^2 + bx + c > 0 & x^2 + bx + c \geq 0 \\ x^2 + bx + c < 0 & x^2 + bx + c \leq 0 \end{array} \qquad \textbf{(8)}$$

where $b = d/a$ and $c = e/a$.

We know from Section 2.3 that the quadratic equation $x^2 + bx + c = 0$ has two real roots, one real root, or no real roots. This leads to three possibilities when solving quadratic inequalities.

EXAMPLE 9 ***A Quadratic Inequality for Which the Related Quadratic Equation Has Two Real Roots***

Find the solution set of the inequality

$$x^2 < 2x + 8 \qquad (9)$$

SOLUTION We rewrite the inequality in the standard form

$$x^2 - 2x - 8 < 0 \quad \text{Subtract } 2x + 8 \text{ from both sides of (9)}$$

But $x^2 - 2x - 8 = (x + 2)(x - 4)$, so the inequality can be written

$$(x + 2)(x - 4) < 0$$

Evidently, the quadratic equation $x^2 - 2x - 8 = 0$ has the two real roots $x = -2$ and $x = 4$. These two numbers, called **dividing points,** divide the real numbers into three distinct intervals. Look at Table 1.

Table 1

Values of x	Test Point	Value of $x + 2$ at Test Point	Value of $x - 4$ at Test Point	Sign of $x + 2$	Sign of $x - 4$	Sign of $(x + 2)(x - 4)$
1. $x < -2$	-3	-1	-7	Negative	Negative	Positive
2. $-2 < x < 4$	0	2	-4	Positive	Negative	Negative
3. $x > 4$	5	7	1	Positive	Positive	Positive

In each range of values for x we choose a **test point** to make it easier to determine the signs of $x + 2$ and $x - 4$ over that range. As we will see in Chapter 4, a quadratic expression can change sign (from positive to negative or negative to positive) only by going through a zero. Thus, in any interval not containing a zero (-2 or 4), the sign of each factor, $(x + 2)$ and $(x - 4)$, remains the same. The sign of $(x + 2)(x - 4)$ is illustrated in Figure 9.

From Figure 9 we see that $(x + 2)(x - 4) < 0$ when $-2 < x < 4$. Thus our solution set is $(-2, 4)$.

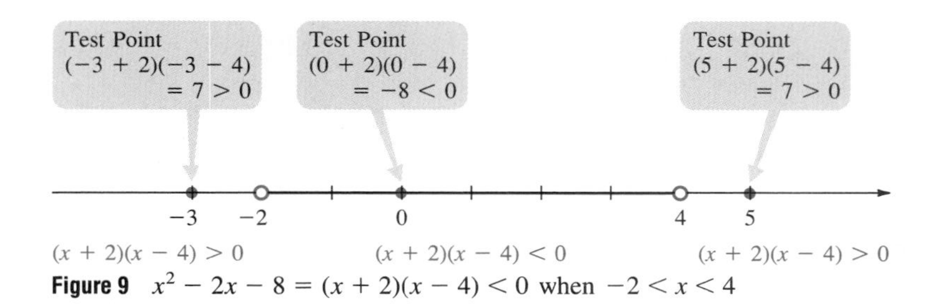

Figure 9 $x^2 - 2x - 8 = (x + 2)(x - 4) < 0$ when $-2 < x < 4$

WARNING In order to solve a quadratic inequality, it is first necessary to write the inequality in one of its four standard forms ((7) or (8)). Otherwise, it is very difficult to solve the inequality. ■

EXAMPLE 10 ***Two Real Roots***

Find the solution set of the inequality $x^2 - 2x - 8 \geq 0$.

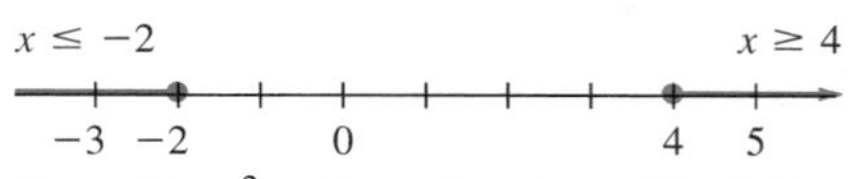

Figure 10 $x^2 - 2x - 8 = (x - 4)(x + 2) \ge 0$ when $x \le -2$ or $x \ge 4$

SOLUTION From Table 1 we see that $x^2 - 2x - 8 = (x - 4)(x + 2) > 0$ when $x < -2$ or $x > 4$. In addition, $x^2 - 2x - 8 = 0$ at $x = -2$ and $x = 4$. Since we have a $\ge$ equality, we must include these two points. Thus the solution set is

$$\{x: x \le -2 \text{ or } x \ge 4\}$$

This set is sketched in Figure 10. ■

EXAMPLE 11 *One Real Root*

Solve the inequality $6x < x^2 + 9$. **(10)**

SOLUTION We rewrite (10) as

$$x^2 - 6x + 9 > 0$$

or, factoring,

$$(x - 3)^2 > 0$$

But a square is nonnegative so $(x - 3)^2 \ge 0$; moreover, $(x - 3)^2 = 0$ only when $x - 3 = 0$ or $x = 3$. Thus the solution set is $\{x: x \ne 3\}$.

We add to this example three other cases:

The solution set of $(x - 3)^2 \ge 0$ is $\mathbb{R}$, the set of all real numbers because the square of every real number is nonnegative.
The solution set of $(x - 3)^2 \le 0$ is $x = 3$ (one point only).
The solution set of $(x - 3)^2 < 0$ is empty. There are no values of x for which $(x - 3)^2$ is negative. ■

EXAMPLE 12 *No Real Root*

Solve the inequality $x^2 + 2x + 2 \le 0$.

SOLUTION There are two ways to solve this problem.

A. Complete the Square

$$x^2 + 2x + 2 = x^2 + 2x + 1 + 1 = (x + 1)^2 + 1$$

But $(x + 1)^2 \ge 0$, so $(x + 1)^2 + 1 \ge 1$. This means that there is no x for which $x^2 + 2x + 2 \le 0$. The solution set is empty.

B. Use the Quadratic Formula The equation $x^2 + 2x + 2 = 0$ has no real roots because the discriminant $2^2 - 4(1)(2) = 4 - 8 = -4 < 0$. Because of the fact mentioned in Example 9, this means that $x^2 + 2x + 2$ is either always positive or always negative. Choosing the test point $x = 0$, we see that $0^2 + 2 \cdot 0 + 2 = 2 \ge 0$, so $x^2 + 2x + 2 > 0$ for all x and the solution set is empty.

Another Case The solution set of $x^2 + 2x + 2 \ge 0$ is $\mathbb{R}$, the set of all real numbers because $x^2 + 2x + 2 = (x + 1)^2 + 1 \ge 1$ for every real number x.

Rational Inequalities

A **rational inequality** is an equality that can be written in one of the forms

$$\frac{p(x)}{q(x)} > 0 \qquad \frac{p(x)}{q(x)} < 0 \qquad \frac{p(x)}{q(x)} \geq 0 \qquad \frac{p(x)}{q(x)} \leq 0$$

where p and q are polynomials.

EXAMPLE 13 *An Inequality with Linear Terms in the Denominator*

Solve the rational inequality

$$\frac{1}{x-4} < \frac{3}{x+2} \tag{11}$$

SOLUTION This problem is more difficult than the linear inequalities in Section 2.7. To solve it, we first write the equivalent inequalities

$$\frac{1}{x-4} - \frac{3}{x+2} < 0$$

$$\frac{(x+2) - 3(x-4)}{(x-4)(x+2)} < 0 \quad \text{Combine terms}$$

$$\frac{-2x+14}{(x-4)(x+2)} < 0$$

$$\frac{-2(x-7)}{(x-4)(x+2)} < 0$$

This inequality holds when either the numerator is positive and the denominator is negative or when the numerator is negative and the denominator is positive. The sign of the denominator is given in Figure 9 on p. 138. We summarize our information in Table 2 below.

Table 2

Interval	**Sign of** $-2(x-7)$	**Sign of** $(x-4)(x+2)$	**Sign of** $\frac{-2(x-7)}{(x-4)(x+2)}$
$(-\infty, -2)$	+	+	+
$(-2, 4)$	+	−	−
$(4, 7)$	+	+	+
$(7, \infty)$	−	+	−

$-2 < x < 4$ $x > 7$

−2 0 4 7

Figure 11 $\frac{1}{x-4} < \frac{3}{x+2}$ when $-2 < x < 4$ or $x > 7$

We see that $\frac{-2(x-7)}{(x-4)(x+2)} < 0$ if $x \in (-2, 4)$ or $x > 7$.

Thus, the solution set to inequality (11) is

$$\{x\colon -2 < x < 4 \text{ or } x > 7\}$$

This set is sketched in Figure 11. ■

EXAMPLE 14 *The Height of a Missile*

A missile is fired vertically from ground level with an initial velocity of 300 ft/sec. During what interval of time after the missile is launched will the missile be over 1350 ft above the ground?

SOLUTION According to the equation of motion described in p. 121, the height, s, of the missile after t seconds have elapsed is given by

$$s = -\tfrac{1}{2}gt^2 + v_0t + s_0 \tag{12}$$

Since s is measured in feet, $g \approx 32$ ft/sec^2 so $-\frac{1}{2}g = -16$, v_0 = initial velocity = 300 ft/sec, and s_0 = initial height = 0 (the missile starts from ground level). In (12) we write $+v_0t$ rather than $-v_0t$ because, initially, the object is rising so s is increasing. Then (12) becomes

$$s = -16t^2 + 300t$$

We seek values of t such that

$$\text{height} = -16t^2 + 300t > 1350$$

or

$$-16t^2 + 300t - 1350 > 0$$

From the quadratic formula, $-16t^2 + 300t - 1350 = 0$ when

$$t = \frac{-300 \pm \sqrt{300^2 - 4(-16)(-1350)}}{-32} = \frac{-300 \pm \sqrt{3600}}{-32}$$

$$= \frac{-300 \pm 60}{-32} \text{ or } t = 7.5 \text{ sec and } t = 11.25 \text{ sec}$$

We can consider values of only $t \geq 0$ (explain why).
Consider the following intervals:

$0 < t < 7.5$	$7.5 < t < 11.25$	$t > 11.25$
5 is a test point	10 is a test point	12 is a test point
$-16 \cdot 5^2 + 300(5) - 1350$	$-16 \cdot 10^2 + 300 \cdot 10 - 1350$	$-16 \cdot 12^2 + 300 \cdot 12 - 1350$
$= -250 < 0$	$= 50 > 0$	$= -54 < 0$

Thus the missile remains above 1350 ft in the interval

$$7.5 \text{ sec} < t < 11.25 \text{ sec}.$$

Problems 2.8

Readiness Check

I. Which of the following is graphed below?

a. $\{x: x \geq 7 \text{ or } x \leq -1\}$ b. $\{x: x \leq -1 \text{ and } x \geq 7\}$
c. $\{x: x \leq -1 \text{ or } x \geq 7\}$ d. $|x - 3| \geq 4$

II. Which of the following inequalities describes the numbers that are closer to -3 than to 4?
a. $-4 < x - 3 < 4$ b. $|x - 3| > |x + 4|$
c. $-3 < x + 4 < 3$ d. $|x + 3| < |x - 4|$

III. Which of the following inequalities has the solution set sketched below?

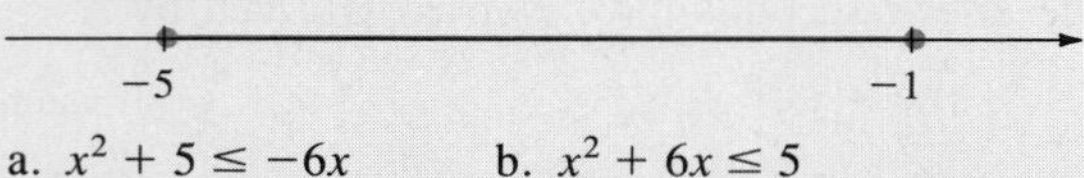

a. $x^2 + 5 \leq -6x$ b. $x^2 + 6x \leq 5$
c. $x^2 + 6x + 5 \geq 0$ d. $x^2 - 6x + 5 \geq 0$

IV. Which of the following is the solution set of $x^2 \geq 9x$?
a. $\{x: x \leq 0 \text{ or } x \geq 9\}$
b. $\{x: 0 \leq x \leq 9\}$
c. $\{x: x \leq -3 \text{ or } x \geq 3\}$
d. $\{x: -3 \leq x \leq 3\}$

V. Which of the following is the interval notation for the solution set of $\dfrac{1}{x} < \dfrac{1}{x+2}$?
a. $(-2, 0)$
b. $(-\infty, 2)$ or $(0, \infty)$
c. $(-\infty, 0)$
d. The solution set is empty.

In Problems 1–10 compute each value.

1. $|3| - |2|$
2. $|3| - |-2|$
3. $|2| - |3|$
4. $|-3| + |-2|$
5. $|10| - |12|$
6. $|-10| - |-12|$
7. $|\pi - 2|$
8. $|2 - \pi|$
9. $|\pi - 7|$
10. $|7 - \pi|$

In Problems 11–74 find the solution set of each inequality and sketch it on a number line.

11. $|x| < 1$
12. $|x| \leq 4$
13. $|x| \geq 4$
14. $|x| \leq 0$
15. $|x| \geq 0$
16. $|x| < 3$
17. $|x| \leq -1$
18. $|x| > -1$
19. $|x - 2| < 1$
20. $|x + 4| \leq 5$
21. $|x - 3| > 2$
22. $|x + 6| > 3$
23. $|4x + 1| < 5$
24. $|-x + 2| < 3$
25. $|5 - x| \geq 1$
26. $|2 - x| > 0$
27. $-1 + |-2x - 4| > 5$
28. $|3x + 4| - 2 \geq 4$
* 29. $|6 - 4x| \geq |x - 2|$ [Hint: Consider separate cases: $6 - 4x > 0$ and $x - 2 > 0$, and so on.]
30. $\left|\dfrac{8 - 3x}{2}\right| \leq 3$
31. $\left|\dfrac{3x + 17}{4}\right| > 9$
32. $|ax + b| < c,\ a > 0,\ c > 0$
33. $|ax + b| \geq c,\ a < 0,\ c > 0$
* 34. $x \leq |x|$
* 35. $|2x| > |5 - 2x|$
36. $x^2 - 3x - 4 > 0$
37. $x^2 > 7x - 10$
38. $x^2 \leq 7x - 10$
39. $s^2 < \dfrac{1}{4}$
40. $s^2 \geq \dfrac{1}{4}$
41. $w^2 - 4w \geq w$
42. $w^2 - 4w < w$
43. $x^2 > 4x - 4$
44. $x^2 \geq 4x - 4$
45. $x^2 \leq 4x - 4$
46. $x^2 < 4x - 4$
47. $x^2 < 4x - 9$
48. $x^2 \geq 4x - 9$
49. $x^2 > 4x - 9$
50. $x^2 \leq 4x - 9$
51. $9x - x^2 > 14$
52. $9x - x^2 \leq 14$
53. $6 - x^2 + x > 0$
54. $12 - x^2 - x \leq 0$
55. $x^2 + 2x - 7 \leq 0$ [Hint: Use the quadratic formula.]
56. $2x^2 + 3x - 2 < 0$
57. $3x^2 + 4x - 7 \geq 0$
58. $x(3x - 6) > 8$
59. $3x(6 - 5x) > 0$
60. $x^3 > x$ [Hint: Three test points are needed here and in Problems 61–63]
61. $x^3 < x^2$
62. $x^3 + x^2 \geq 2x$
63. $x^3 + 2x^2 \leq 15x$
64. $\dfrac{1}{x} > 3$

Answers to Readiness Check

I. a, c, d II. d III. a IV. a V. a

* 65. $2 < \dfrac{1}{5 - x}$

* 66. $\dfrac{4}{3x - 2} \leq -2$

* 67. $\dfrac{1}{x - 2} \geq \dfrac{2}{x + 3}$

* 68. $\dfrac{2}{x + 1} \leq -\dfrac{5}{x + 4}$

* 69. $\dfrac{x + 1}{x - 3} > -4$

* 70. $\dfrac{2 - x}{x + 4} \geq 1$

71. $\dfrac{2x - 4}{x + 1} \geq 0$

* 72. $\dfrac{4}{x^2 - 16} < 0$

* 73. $\dfrac{x - 3}{x^2 - 16} < 0$

** 74. $\dfrac{x + 4}{x^2 - 8x + 15} \geq 0$

75. Show that $|xy| = |x||y|$. [Hint: Deal with each of four cases separately: (1) $x \geq 0$, $y \geq 0$, (2) $x \geq 0$, $y \leq 0$, and so on.]

76. Show that if $x \geq 0$ and $y \geq 0$, then $|x + y| = |x| + |y|$.

77. If $x > 0$ and $y < 0$, show that $|x + y| < |x| + |y|$.

78. If $x < 0$ and $y < 0$, show that $|x + y| = |x| + |y|$.

79. Using Problems 76–78, prove the triangle inequality $|x + y| \leq |x| + |y|$.

* 80. Show that $\big||x| - |y|\big| \leq |x - y|$. [Hint: Write $x = (x - y) + y$ and apply the triangle inequality.]

* 81. Solve each inequality.
(a) $|2 - x| + |2 + x| \leq 10$
(b) $|2 - x| + |2 + x| > 6$
(c) $|2 - x| + |2 + x| \leq 4$
(d) $|2 - x| + |2 + x| \leq 3.99$

82. Use absolute value bars to translate each of the following statements into a single inequality.
(a) $x \in (-4, 10)$
(b) $x \notin (-3, 3)$
(c) $x \notin [5, 11]$

83. Write single inequalities that are satisfied in each case.
(a) All real numbers x that are closer to 5 than to 0.
(b) All real numbers y that are closer to -2 than to 2.

84. Show that

$$\frac{s + t + |s - t|}{2}$$

equals the larger of s and t.

85. Show that

$$\frac{s + t - |s - t|}{2}$$

equals the smaller of s and t.

86. (a) Show that

$$|A - B| \leq |A - W| + |W - B|$$

for all real numbers A, B, and W.
(b) Describe those situations in which the preceding less-than-or-equal statement is actually an equality.

87. For what choices of s is $3.72s > 4.06s$?

* 88. (a) Suppose that a and b are positive; show that

$$\sqrt{ab} \leq \frac{a + b}{2}$$

[Hint: Use the fact that $(x - y)^2 \geq 0$ for all real numbers x and y.]
(b) Use the inequality of part (a) to prove that among all rectangles with an area of 225 cm^2, the one with shortest perimeter is a square.
(c) Use the inequality of part (a) to prove that among all rectangles with perimeter of 300 cm, the one with largest area is a square.

* 89. An estimate for a particular number w is 1.3. Suppose that this estimate is accurate to one decimal place; that is, $|w - 1.3| \leq 0.05$. Observe that $(1.3)^2 = 1.69$, which rounds off to 1.7. Is the estimate 1.7 for w^2 also accurate to one decimal place? Describe the shortest interval that contains w^2.

* 90. The sides of a rectangular piece of paper (a page of this book, for example) are measured. Suppose that we measure to one-decimal-place accuracy and find that the rectangle is 16.1 cm by 23.4 cm; $16.1 \times 23.4 = 376.74$, which rounds off to 376.7. Does 376.7 cm^2 approximate the true area of the rectangle with one-decimal-place accuracy? Explain.

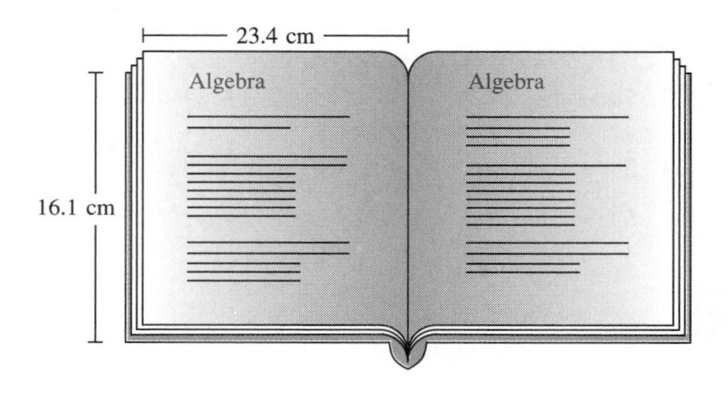

91. A cannonball is shot vertically into the air from ground level with an initial velocity of 120 ft/sec. During what interval of time after the cannonball is fired will the cannonball be at least 125 ft above the ground?

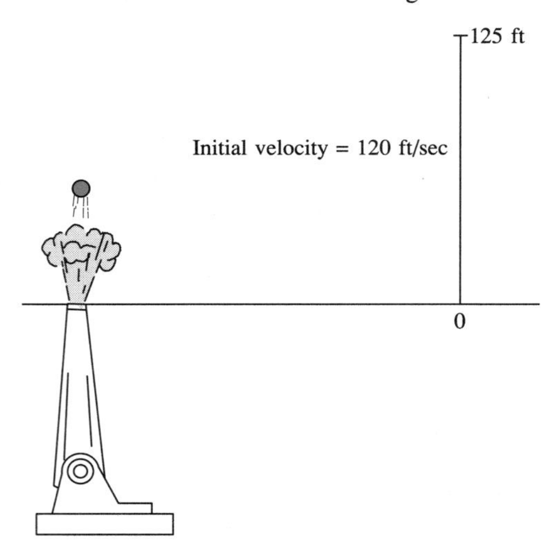

* 92. In Problem 91 find the time interval during which the cannonball will be below 125 ft but above the ground. [Hint: It does not remain above the ground indefinitely.]

In Problems 93–96 use a calculator to find the solution set of the given inequality.

93. $|1.6x - 2.7| \le 4.5$
94. $|37x + 59| < 1362$
95. $|326.5x + 242| > 3.8$
96. $|0.0316x - 0.1158| \ge 3.801$

In Problems 97–102 solve each inequality.

97. $|x - 3| < 0.001$
98. $|x + 5| < 0.01$
99. $|2x + 7| < 0.02$
100. $|3x + 4| < 0.005$
* 101. $|x^2 - 1| < 0.01$
* 102. $|x^2 - 2x| < 0.1$

103. Show that $|(3x + 10) - 1| < 1$ if $|x + 3| < \frac{1}{3}$.

104. Show that $|(2x + 1) - 5| < \frac{1}{2}$ if $|x - 2| < \frac{1}{4}$.

105. Show that $|(2x + 1) - 5| < 0.01$ if $|x - 2| < 0.005$.

* 106. Let ϵ represent a small, positive real number. Show that

$$|(2x + 1) - 5| < \epsilon \text{ if } |x - 2| < \frac{\epsilon}{2}.$$

* 107. Show that $|x^2 - 4| < 1$ if $|x - 2| < 0.2$.

In Problems 108–117 the graphs of 10 inequalities are given. Match each graph with one of the 10 inequalities given below.

(a) $|x - 5| \le 4$ (f) $x^2 \le x$
(b) $|2x + 6| \ge 3$ (g) $2x^2 \ge 4x$
(c) $|1 - x| \le 2$ (h) $x^2 + x \le 6$
(d) $|4x - 4| \ge 8$ (i) $x^2 - x \ge 6$
(e) $|3x + 7| \le 12$ (j) $2x^2 - 20 \le 3x$

108. 0, 2
109. -1, 3
110. $-\frac{19}{3}$, $\frac{5}{3}$
111. -2, 3
112. $-\frac{5}{2}$, 4
113. 1, 9
114. -3, 2
115. 0, 1
116. $-\frac{9}{2}$, $-\frac{3}{2}$
117. -1, 3

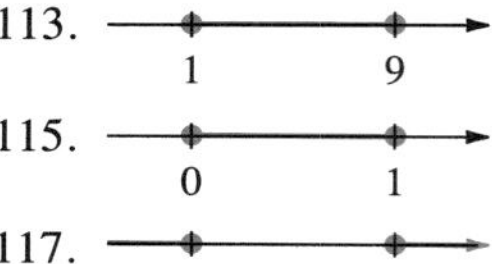

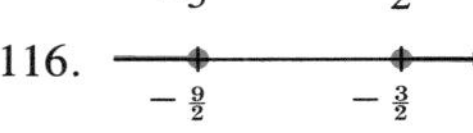

Graphing Calculator Problems

In Problems 118–129 solve each inequality by using your graphing calculator. Give all intervals with 1 decimal place accuracy. Before starting, read Example 8 in Appendix A.

118. $x^2 > 4x - 5$
119. $4 - 2x^2 < 3x$
120. $\dfrac{1}{1 - x} > 2.7$
121. $\dfrac{3 + x}{x - 2} > 14$
122. $\dfrac{5 - 2x}{3x + 7} < 10$
123. $x^3 + x^2 + x + 3 > 0$
124. $2x^3 - 4x^2 + 5x - 9 < 0$
125. $|2x - 3| > 1$
126. $|x - 2| > x + 1$
127. $x^2 > |x - 1|$
128. $|2x + 3| < |4 - x|$
129. $\dfrac{|x^2 - 2|}{x + 3} < |5 - 7x|$

Summary Outline of Chapter 2

- **Linear Equation**

  A linear equation is an equation of the form $ax + b = 0$ where $a \neq 0$. Its unique solution is $x = -b/a$. p. 70

- **Quadratic Equation**

  $ax^2 + bx + c = 0$ is a quadratic equation. p. 93

- **Solution by Factoring**

  If $x^2 + bx + c = (x - r)(x - s)$ then the solutions to $x^2 + bx + c = 0$ are $x = r$ and $x = s$. p. 93

- **Completing the Square**

  $x^2 + bx + c = \left(x + \frac{b}{2}\right)^2 + c - \frac{b^2}{4}$ p. 95

- **The Quadratic Formula**

  $x = \frac{-b \pm \sqrt{b^2 - 4ac}}{2a}$ are the solutions to $ax^2 + bx + c = 0$. p. 97

- If $ax^2 + bx + c = 0$, the sum of the roots is $-\frac{b}{a}$ and the product of the roots is $\frac{c}{a}$. p. 100

- **Discriminant**

  The expression $b^2 - 4ac$ is called the **discriminant** of $ax^2 + bx + c = 0$. If $b^2 - 4ac > 0$, there are two real solutions; if $b^2 - 4ac = 0$, there is one real solution; if $b^2 - 4ac < 0$, there are two complex conjugate solutions. p. 97

- **Imaginary Unit**

  The imaginary unit i is defined by $i^2 = -1$. p. 104
  If $a > 0$, then $\sqrt{-a} = \sqrt{a}i$. p. 104

- **Complex Number**

  A complex number is a number of the form $a + bi$ where a and b are real. p. 105

- **Algebra of Complex Numbers**

  $(a + bi) + (c + di) = (a + c) + (b + d)i$ p. 106
  $(a + bi) - (c + di) = (a - c) + (b - d)i$ p. 106
  $(a + bi)(a - bi) = a^2 + b^2$ p. 107
  $\overline{a + bi} = a - bi$ is the **complex conjugate** of $a + bi$. p. 107

- **Intervals**

  open interval $(a, b) = \{x: a < x < b\}$ p. 128
  closed interval $[a, b] = \{x: a \leq x \leq b\}$ p. 128
  half-open interval $[a, b) = \{x: a \leq x < b\}$ p. 128
  $[a, \infty) = \{x: x \geq a\}$ p. 129
  $(a, \infty) = \{x: x > a\}$ p. 129
  $(-\infty, a] = \{x: x \leq a\}$ p. 129
  $(-\infty, a) = \{x: x < a\}$ p. 129
  $(-\infty, \infty) = \mathbb{R}$ = the set of real numbers p. 129

- **Absolute Value Inequalities**

If $a > 0$, then

$|x| < a$ is equivalent to $-a < x < a$ p. 134

$|x| \leq a$ is equivalent to $-a \leq x \leq a$ p. 134

$|x| > a$ is equivalent to $x > a$ or $x < -a$ p. 135

$|x| \geq a$ is equivalent to $x \geq a$ or $x \leq -a$ p. 135

Review Exercises for Chapter 2

In Exercises 1–10 solve the given equation.

1. $x + 5 = 0$
2. $7x - 3 = 18$
3. $8x + 2 = -7$
4. $(z + 3)^2 = z^2 - 9$
5. $\dfrac{y + 1}{y - 3} = \dfrac{2}{7}$
6. $\dfrac{\sqrt{w}}{4} = 3$
7. $\dfrac{1}{x} - \dfrac{4}{x} = 7$
8. $\dfrac{3}{y - 2} = \dfrac{2}{y + 3}$
9. $\dfrac{x + 2}{x - 3} = \dfrac{x + 1}{x + 4}$
10. $\dfrac{3}{q - 1} + \dfrac{1}{q + 5} = 0$
11. Let $a = bc + d$; solve for c.
12. Let $\dfrac{ax + b}{c} = \dfrac{1}{2} + a$; solve for x.
13. Let $F = \dfrac{Gm_1m_2}{r^2}$; solve for r.
14. Let $K = \frac{1}{2}mv^2$; solve for v.
15. The price of a new compact car increased by $7\frac{1}{2}\%$ to \$5160. What was the original price of the car?
16. \$15,000 was invested for 4 years and earned simple interest of \$5100. What was the interest rate?
17. A product has a fixed cost of \$2300 and a variable cost of \$60 for each item produced in a given month.
    (a) Write the equation that represents total cost.
    (b) What will it cost to produce 135 items during the month?
18. The outside temperature is 30° Celsius. What is the temperature in °F?
19. A student receives grades of 61, 78, 83, and 82 on his first four tests. What grade does he need on his fifth test to have an average of 80?
20. Salt is dissolved in 20 gallons of water to form an 8% saline solution. How many gallons of water must be evaporated to be left with a 10% solution?

In Exercises 21–35 find all solutions to each quadratic equation.

21. $x^2 - 9 = 0$
22. $x^2 - x - 2 = 0$
23. $x^2 + 5x - 14 = 0$
24. $y^2 + 3y - 40 = 0$
25. $y^2 - 7y + 6 = 0$
26. $y^2 + 8y + 16 = 0$
27. $u^2 + 5u + 2 = 0$
28. $u^2 - 4u = 3$
29. $v^2 = 7 - 10v$
30. $w = 3w^2 - 1$
31. $5w^2 + 6w - 3 = 0$
32. $\dfrac{w^2}{4} - \dfrac{w}{3} + \dfrac{1}{9} = 0$
33. $z^2 + z + 1 = 0$
34. $z^2 + 4z + 5 = 0$
35. $z^2 + 2z + 3 = 0$

In Exercises 36–38 find the sum and the product of the roots of each quadratic equation.

36. $x^2 + 5x - 8 = 0$
37. $3y^2 + 10y - 4 = 0$
38. $\dfrac{z^2}{2} - z + 2 = 0$

In Exercises 39–43 compute the discriminant of each quadratic equation and determine the number of real roots.

39. $x^2 - 4x - 11 = 0$
40. $\dfrac{y^2}{2} - y + 1 = 0$
41. $w^2 - 0.4w + 0.04 = 0$
42. $z^2 + 3z - 1 = 0$
43. $v^2 + 4v + \sqrt{2} = 0$
44. Write each number in the form bi.
    (a) $\sqrt{-64}$ (b) $\sqrt{-75}$

In Exercises 45–54 write each number in the form $a + bi$.

45. $-5 + \sqrt{-9}$
46. $\dfrac{3 + \sqrt{-49}}{4}$
47. $(3 - 2i) + (4 - 6i)$
48. $(7 + 2i) - (-1 + 2i)$
49. $(2 + 5i)(6 - 3i)$
50. $(8 + 5i)\left(\dfrac{1}{2} - \dfrac{1}{3}i\right)$
51. i^7
52. $\dfrac{1}{3 - 2i}$
53. $\dfrac{2 + i}{2 - i}$
54. $\dfrac{1}{2 + i} - \dfrac{3i}{2 - 3i}$

In Exercises 55–64 find all solutions to the given equation.

55. $\sqrt{x + 2} = 7$
56. $(y + 3)^{2/3} = 9$
57. $z^4 - 8z^2 + 16 = 0$
58. $w^4 - 5w^2 + 4 = 0$
59. $u^4 - 2u^2 - 8 = 0$
60. $x^3 - 64 = 0$
61. $\dfrac{4}{y + 1} = \dfrac{7}{y - 3} - 2$
62. $\sqrt{w - 1} = w + 2$
63. $\sqrt{2v + 1} - \sqrt{v} = 1$
64. $\sqrt{v + 5} - \sqrt{v - 2} = 1$

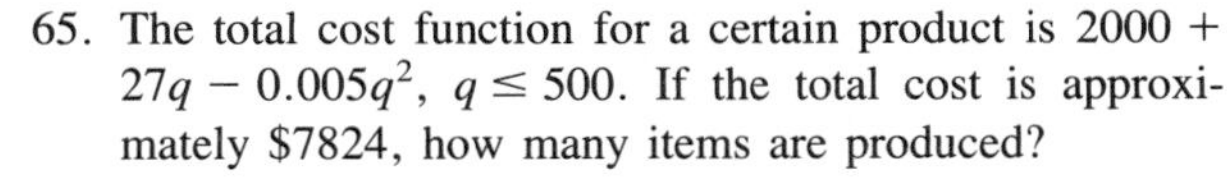

65. The total cost function for a certain product is $2000 + 27q - 0.005q^2$, $q \leq 500$. If the total cost is approximately \$7824, how many items are produced?
66. The total revenue received from selling q items of the product of Exercise 65 is $42q - 0.015q^2$, $q \leq 500$. If approximately \$8796 was received, how many units were sold?
67. Refer to Exercises 65 and 66.
    (a) What is the profit earned if q items are produced and sold?
    (b) How many items were produced and sold if the total profit is \$1776?
68. How long, ignoring air resistance, will it take for an object to hit the ground if it is dropped from a height of 350 feet?
69. A bolt is ejected from a missile with an initial velocity of 80 feet/second from a height of 15,000 feet. When will the bolt hit the ground? Assume that the bolt is sent straight down when it is ejected.

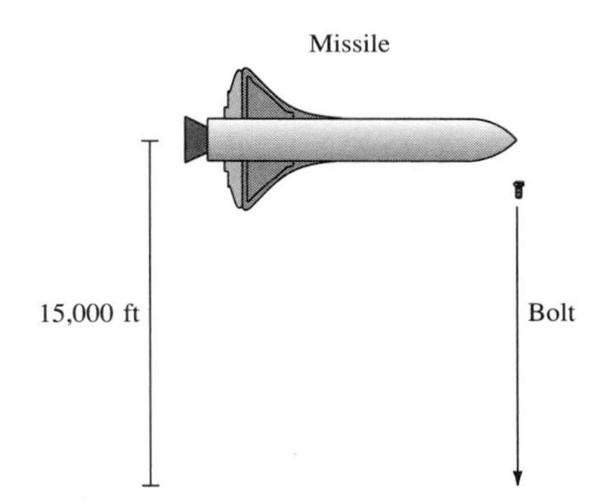

70. The product of two consecutive integers is 992. Find the integers.

In Exercises 71–76 write each set of numbers as an interval.

71. $3 \leq x \leq 7$
72. $-2 < x < 4$
73. $0 < x \leq 5$
74. $0 \leq x < 10$
75. $x \geq 4$
76. $x < 0$

In Exercises 77–104, find the solution set of the given inequality (or inequalities) and write your answers in interval notation.

77. $x - 3 < 4$
78. $x + 7 \geq 6$
79. $2x + 4 \leq 9$

80. $-7x + 3 > 8$
81. $-4 \leq x - 2 < 3$
82. $2 \leq 3x + 5 \leq 6$
83. $-2 < 5 - x < 4$
84. $|x| < 3$
85. $|x - 4| > 5$
86. $\left|x + \frac{1}{2}\right| \leq \frac{3}{4}$
87. $|2x - 5| < 3$
88. $|3x + 7| > -1$
89. $|4x - 9| < -2$
90. $|-3x + 2| \leq 5$
91. $|6 - 2x| > 1$
92. $\left|\frac{6 - 3x}{4}\right| \leq 7$
93. $\left|\frac{5x - 8}{3}\right| > 6$
94. $x^2 \leq 7x - 6$
95. $x^2 > 7x - 6$
96. $x^2 + 4 > 4x$
97. $x^2 + 6 < 7x$
98. $10 - x^2 > 3x$
99. $3x \geq 10 - x^2$
100. $x^2 + 11 > 6x$
101. $x^2 - 12 \leq x$
102. $\frac{1}{x} > 3$
103. $\frac{1}{x + 2} > \frac{2}{x - 1}$
104. $\frac{x + 3}{x - 5} \leq -3$

Chapter 3

Functions and Graphs

3.1 The Cartesian Coordinate System

In this section we describe the most common way of representing points in a plane: the **Cartesian coordinate system.**† To form the Cartesian coordinate system, we draw two mutually perpendicular number lines as in Figure 1: one horizontal line and one vertical line. The horizontal line is called the ***x*-axis,** and the vertical line is called the ***y*-axis.** The point at which the lines meet is called the **origin** and is labeled 0.

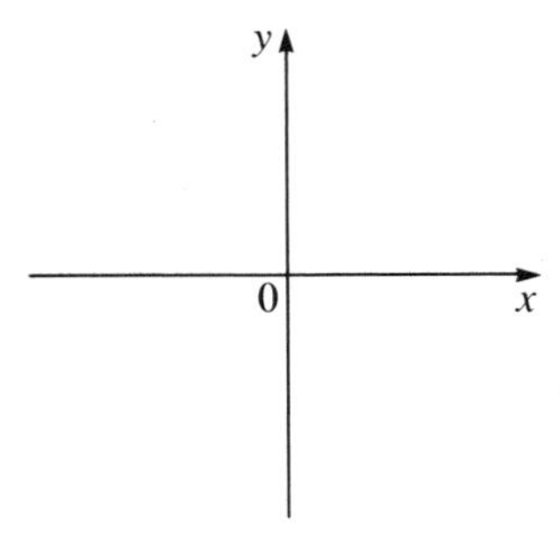

Figure 1 The x- and y-axes.

To every point in the plane, we assign an **ordered pair** of numbers. The first element in the pair is called the ***x*-coordinate,** and the second element of the pair is called the ***y*-coordinate.**

The x-coordinate measures the number of units from the y-axis to the point. Points to the right of the y-axis have positive x-coordinates, and those to the left have negative x-coordinates.

The y-coordinate measures the number of units from the x-axis to the point. Points above the x-axis have a positive y-coordinate, and those below have a negative y-coordinate. Figure 2 shows a typical point (a, b), where $a > 0$ and $b > 0$.

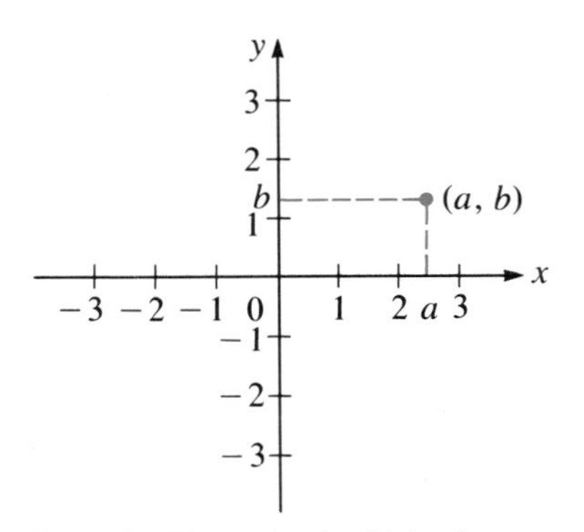

Figure 2 The point (a, b) in the xy-plane.

Two ordered pairs, or points, are **equal** if their first elements are equal and their second elements are equal. Note that $(1, 0)$ and $(1, 1)$ are *different*

†See the biographical sketch on p. 155.

because their second elements are different. Note too that (1, 2) and (2, 1) are different points.

In Figure 3, several different points are depicted.

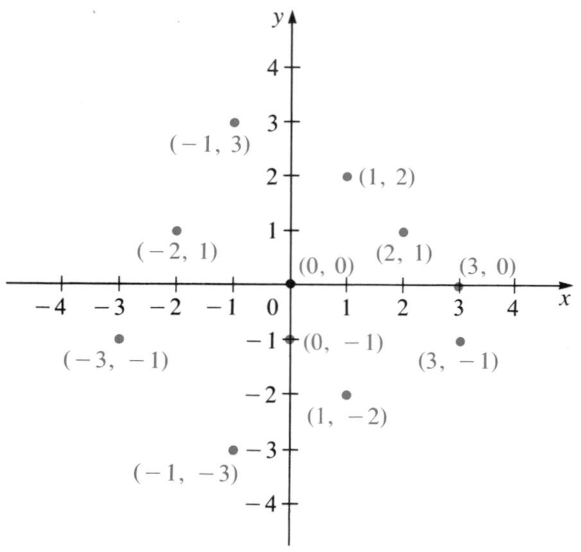

Figure 3 Eleven points in the xy-plane.

When points in the plane are represented by the Cartesian coordinate system, the plane is called the **Cartesian plane,** or the ***xy*-plane,** and is denoted by $\mathbb{R}^2$. We have

> **Definition of $\mathbb{R}^2$ (the *xy*-Plane)**
>
> $\mathbb{R}^2 = \{(x, y): x \in \mathbb{R} \text{ and } y \in \mathbb{R}\}$

Quadrants

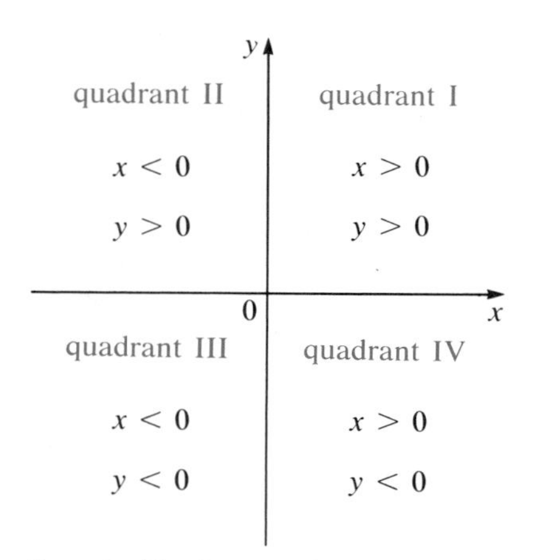

Figure 4 The four quadrants.

A glance at Figure 4 indicates that the x- and y-axes divide the xy-plane into four regions. These regions are called **quadrants** and are denoted as in the figure.

EXAMPLE 1 *The Quadrants of Four Points*

(a) (1, 3) is in the first quadrant because $1 > 0$ and $3 > 0$.
(b) $(-4, -7)$ is in the third quadrant because $-4 < 0$ and $-7 < 0$.
(c) $(-2, 5)$ is in the second quadrant.
(d) $(7, -3)$ is in the fourth quadrant.

We now give a formula for finding the distance between two points in the plane.

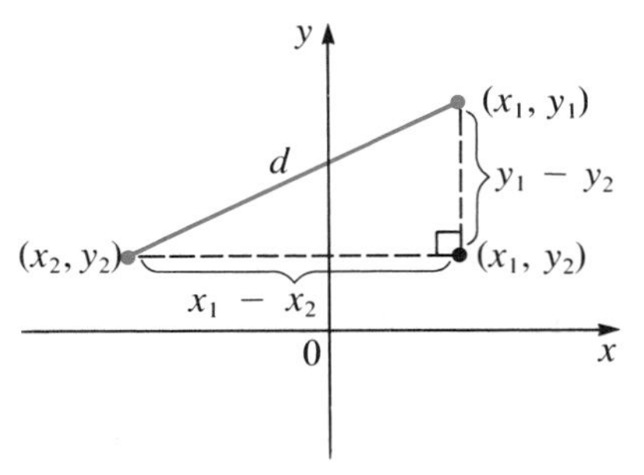

Figure 5 By the Pythagorean theorem, $(x_1 - x_2)^2 + (y_1 - y_2)^2 = d^2$.

Let (x_1, y_1) and (x_2, y_2) be two points in the xy-plane (see Figure 5).

Theorem: The Distance Formula

$$d = \sqrt{(x_1 - x_2)^2 + (y_1 - y_2)^2} \tag{1}$$

Proof of the Distance Formula

The triangle in Figure 5 is a right triangle. The lengths of the two legs are $|x_1 - x_2|$ and $|y_1 - y_2|$, and the length of the hypotenuse is d. The Pythagorean theorem states that for every right triangle

> The square of the hypotenuse equals the sum of the squares of the other two sides.

Thus

$$d^2 = |x_1 - x_2|^2 + |y_1 - y_2|^2 \tag{2}$$

But, $|a|^2 = a^2$. This means that

$$|x_1 - x_2|^2 = (x_1 - x_2)^2 \quad \text{and} \quad |y_1 - y_2|^2 = (y_1 - y_2)^2$$

So (2) becomes

$$d^2 = (x_1 - x_2)^2 + (y_1 - y_2)^2$$

and, taking the positive square root (since $d > 0$),

$$d = \sqrt{(x_1 - x_2)^2 + (y_1 - y_2)^2} \quad \blacksquare$$

EXAMPLE 2 *Finding the Distance Between Two Points*

Find the distance between the points (2, 5) and (−3, 7).

SOLUTION Let $(x_1, y_1) = (2, 5)$ and $(x_2, y_2) = (-3, 7)$. From (1),

$$d = \sqrt{(2 - (-3))^2 + (5 - 7)^2} = \sqrt{5^2 + (-2)^2} = \sqrt{29} \approx 5.385$$

Theorem: The Converse of the Pythagorean Theorem

If, in a triangle, the square of one side is equal to the sum of the squares of the other two sides, then the triangle is a right triangle.

EXAMPLE 3 *Determining Whether a Triangle is a Right Triangle*

Determine whether the triangle with vertices at (−3, 6), (−1, 2), and (5, 5) is a right triangle.

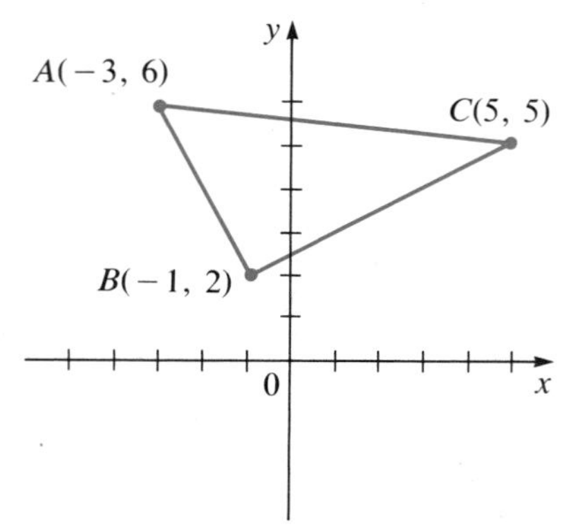

Figure 6 Triangle ABC is a right triangle with hypotenuse $\overline{AC}$.

SOLUTION The triangle is sketched in Figure 6. The three vertices are denoted by A, B, and C. We compute the square of the length of each side; $\overline{AB}$ denotes the length of AB. Then

$$\overline{AB}^2 = [-3 - (-1)]^2 + (6 - 2)^2 = (-2)^2 + 4^2 = 4 + 16 = 20$$
$$\overline{AC}^2 = (-3 - 5)^2 + (6 - 5)^2 = (-8)^2 + 1^2 = 64 + 1 = 65$$
$$\overline{BC}^2 = (-1 - 5)^2 + (2 - 5)^2 = (-6)^2 + (-3)^2 = 36 + 9 = 45$$

Since $65 = 45 + 20$, we see that triangle ABC is a right triangle with hypotenuse AC.

Using the distance formula, we can obtain an equation for the midpoint of a line segment. In Problem 32 in this section and Problem 66 in Section 3.2 you are asked to prove the following:

Theorem: The Midpoint Formula

$\left(\dfrac{x_1 + x_2}{2}, \dfrac{y_1 + y_2}{2}\right)$ is the midpoint of the line segment joining the points (x_1, y_1) and (x_2, y_2).

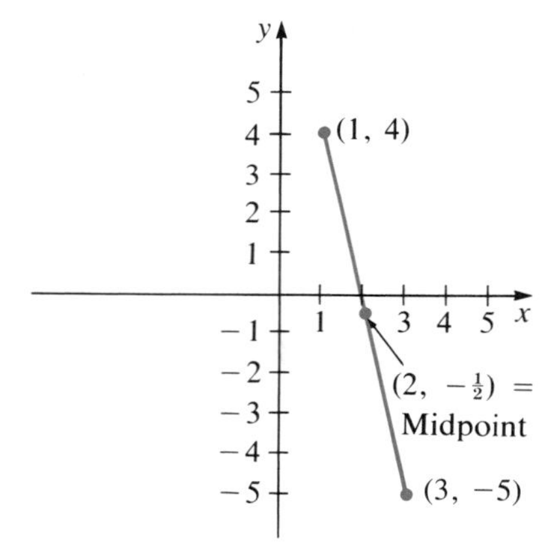

Figure 7 $(2, -\frac{1}{2})$ is the midpoint of the line joining (1, 4) and (3, −5).

EXAMPLE 4 *Finding the Midpoint of a Line Segment*

The midpoint of the line segment joining (1, 4) and (3, −5) is $\left(\dfrac{1+3}{2}, \dfrac{4-5}{2}\right) = \left(2, -\dfrac{1}{2}\right)$. This is illustrated in Figure 7.

Equations in Two Variables: Relations

Now that we have defined a coordinate system with two components, we will discuss **equations in two variables.** The set of ordered pairs that satisfy an equation in two variables is called a **relation.**

EXAMPLE 5 *Four Equations in Two Variables*

The following are equations in the two variables x and y:

(a) $y = 3x + 2$
(b) $x^2 + y^2 = 4$
(c) $y = \dfrac{1}{x}$
(d) $x^3y + \sqrt{x + y} = y$

Solution of an Equation in Two Variables

A **solution** to an equation in two variables is a point (x, y) whose coordinates satisfy the equation.

EXAMPLE 6 ***Solutions to Two Equations***

(a) (1, 5) is a solution to the equation $y = 3x + 2$ because

$$\underset{x}{\downarrow}\quad\quad\underset{y}{\downarrow}$$
$$3 \cdot 1 + 2 = 5$$

(b) $(\sqrt{3}, 1)$ and $(\sqrt{2}, -\sqrt{2})$ are solutions to the equation $x^2 + y^2 = 4$ because

$$(\underset{\underset{x}{\uparrow}}{\sqrt{3}})^2 + \underset{\underset{y}{\uparrow}}{1}^2 = 3 + 1 = 4$$

and

$$(\sqrt{2})^2 + (-\sqrt{2})^2 = 2 + 2 = 4$$

There's a cliché that states "a picture is worth a thousand words." In dealing with equations we can usually obtain answers more quickly and understand the answers better if we have a picture. Such a picture is called a graph.

Definition of a Graph

The **graph** of an equation in two variables is the set of all points in the *xy*-plane whose coordinates satisfy the equation.

In Sections 3.2 and 3.4 we will discuss equations whose graphs are straight lines. In Section 3.5 we will discuss more general equations and graphs. Now we consider the equation of a circle.

Circle

A **circle** is defined as the set of all points in a plane at a given distance from a given point. The given point is called the **center** of the circle, and the common distance from the center is called the **radius.**

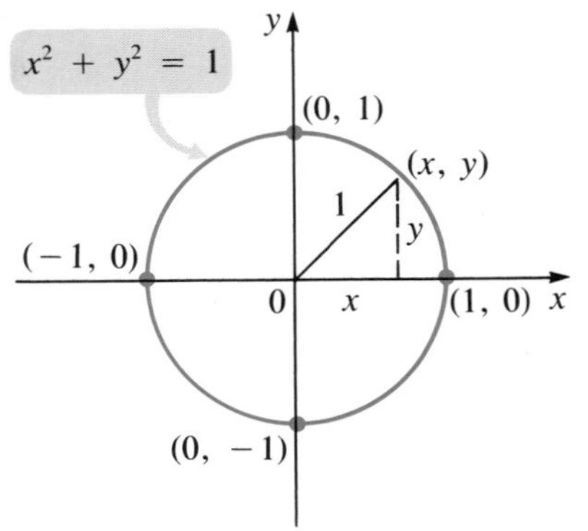

Figure 8 The unit circle.

EXAMPLE 7 ***The Unit Circle***

Find an equation of the circle centered at the origin with radius 1.

SOLUTION If (x, y) is any point on the circle, then the distance from (x, y) to $(0, 0)$ is 1. From (1) we have

$$\sqrt{(x - 0)^2 + (y - 0)^2} = 1 \tag{3}$$

Squaring both sides, we obtain

$$x^2 + y^2 = 1 \tag{4}$$

This circle is sketched in Figure 8. It is called the **unit circle** and is of central importance in the study of the trigonometric functions.

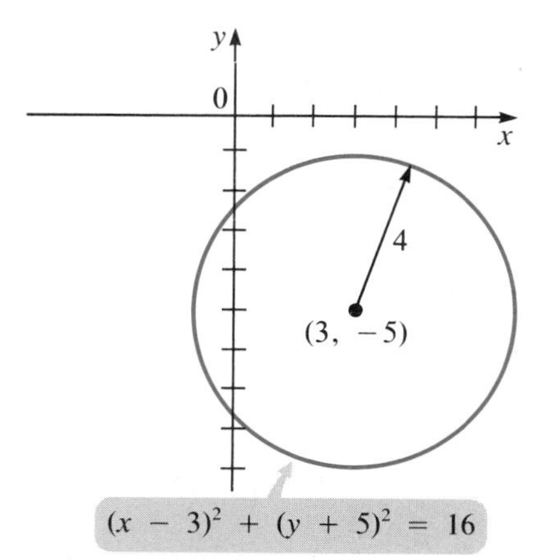

Figure 9 Circle centered at $(3, -5)$ with radius 4.

We now discuss circles with other centers and other radii.

EXAMPLE 8 *Finding an Equation of a Circle Given Its Center and Radius*

Find an equation of the circle centered at $(3, -5)$ with radius 4.

If (x, y) is on the circle, then the distance from (x, y) to $(3, -5)$ is 4, so that

$$\sqrt{(x-3)^2 + (y+5)^2} = 4 \quad \text{or} \quad (x-3)^2 + (y+5)^2 = 16$$

This circle is sketched in Figure 9.

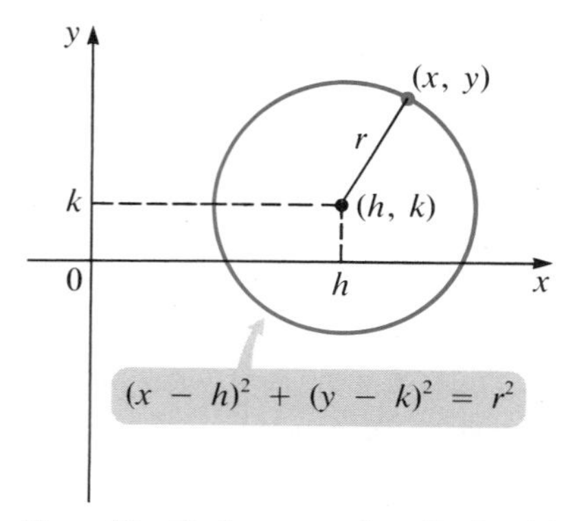

Figure 10 Circle centered at (h, k) with radius r.

Figure 10 shows the graph of the circle centered at (h, k) with radius r. The point (x, y) is on this circle if and only if its distance from (h, k) is r. We have the following:

Standard Equation of the Circle with Center at (h, k) and Radius r

$$(x-h)^2 + (y-k)^2 = r^2 \tag{5}$$

EXAMPLE 9 *Showing That a Second Degree Equation in Two Variables Represents a Circle*

Show that the equation $x^2 - 6x + y^2 + 2y - 17 = 0$ represents a circle; find its center and radius.

SOLUTION We use the technique of completing the square.† We have

$$x^2 - 6x + y^2 + 2y = 17$$

$$(x^2 - 6x + 9) + (y^2 + 2y + 1) = 17 + 9 + 1 \qquad \text{Add } 9 + 1 \text{ to both sides and group terms}$$

$$(x-3)^2 + (y+1)^2 = 27 \qquad \text{Factor}$$

This is the equation of a circle with center at $(3, -1)$ and radius $\sqrt{27} = 3\sqrt{3} \approx 5.196$. It is sketched in Figure 11.

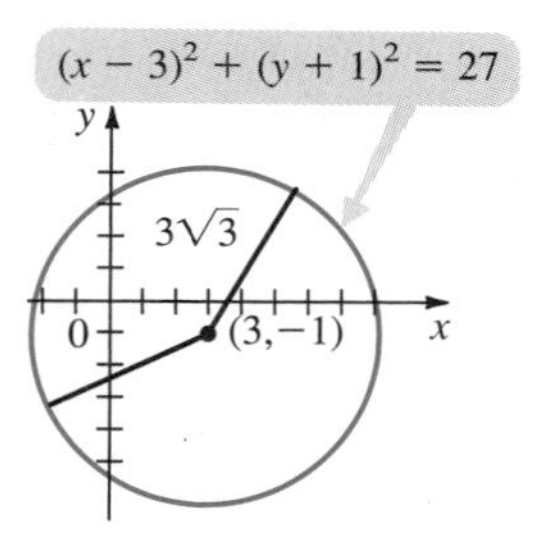

Figure 11 The circle centered at $(3, -1)$ with radius $3\sqrt{3}$.

We close this section by noting that the graphs of quadratic equations in two variables—that is, equations having the form

$$ax^2 + bxy + cy^2 + dx + ey + f = 0 \tag{6}$$

—are called **conic sections.** The equation of a circle is a special case of equation (6). Another special case is discussed in Example 1 in Section 3.5 (on p. 190). We will discuss conic sections in more generality in Chapter 5.

† Recall that this technique is sometimes used to solve the general quadratic equation $ax^2 + bx + c = 0$.

FOCUS ON

René Descartes (1596–1650)

The Cartesian plane is named after the great French mathematician and philosopher René Descartes. Born near the city of Tours in 1596, Descartes received his education first at the Jesuit school at La Flèche and later at Poitier, where he studied law. He had delicate health and, while still in school, developed the habit of spending the greater part of each morning working in bed. Later, he considered these morning hours the most productive period of the day.

At the age of 16, Descartes left school and moved to Paris, where he began his study of mathematics. In 1617, he joined the army of Maurice, Prince of Nassau. He also served with Duke Maximillian I of Bavaria and with the French army at the siege of La Rochelle.

Descartes was not a professional soldier, however, and his periods of military service were broken by periods of travel and study in various European cities. After leaving the army for good, he resettled in Paris to continue his mathematical studies and then moved to Holland, where he lived for 20 years.

Much stimulated by the scientists and philosophers he met in France, Holland, and elsewhere, Descartes later became known as the "father of modern philosophy." His statement "Cogito ergo sum" ("I think, therefore I am") played a central role in his philosophical writings.

Descartes's program for philosophical research was enunciated in his famous *Discours de la méthode pour bien conduire sa raison et chercher la vérité dans les sciences* (A Discourse on the Method of Rightly Conducting the Reason and Seeking Truth in the Sciences) published in 1637. This work was accompanied by three appendices: *La dioptrique* (in which the law of refraction—discovered by Snell—was first published), *Les météores* (which contained the first accurate explanation of the rainbow), and *La géométrie. La géométrie,* the third and most famous appendix, took up about a hundred pages of the *Discours*. One of the major achievements of *La géométrie* was that it connected figures of geometry with the equations of algebra. The work established Descartes as the founder of analytic geometry.

René Descartes
(David Smith Collection)

In 1649, Descartes was invited to Sweden by Queen Christina. He agreed, reluctantly, but was unable to survive the harsh, Scandinavian winter. He died in Stockholm in early 1650.

Problems 3.1

Readiness Check

I. Which of the following is true about the point $(0, b)$?
   a. It is located on the x-axis.
   b. It is located on the y-axis.
   c. It represents a vertical distance from the origin.
   d. If $b < 0$, its distance from the origin equals b^2.
   e. It is not possible to determine whether a, b, c, or d is the correct answer until the value of b is known.

II. Which of the following is the equation of a circle whose center is $(3, -4)$ and that passes through $(0, -2)$?
   a. $(x - 3)^2 + (y + 4)^2 = 45$
   b. $(x + 3)^2 + (y - 4)^2 = 45$
   c. $(x - 3)^2 + (y + 4)^2 = 13$
   d. $(x + 3)^2 + (y - 4)^2 = 40$

III. Which of the following are the vertices of a right triangle?
a. $(-4, 0)$, $(-1, -2)$, and $(1, 1)$
b. $(-3, -5)$, $(-1, -2)$, and $(-4, 2)$
c. $(-2, 1)$, $(-1, -2)$, and $(1, 1)$
d. $(-4, 2)$, $(-1, -2)$, and $(2, 2)$

IV. The midpoint of the line segment joining $(1, -2)$ and $(-2, 1)$ is ______.
a. $(0, 0)$ b. $\left(\frac{1}{2}, \frac{1}{2}\right)$ c. $\left(\frac{1}{2}, -\frac{1}{2}\right)$
d. $\left(-\frac{1}{2}, \frac{1}{2}\right)$ e. $\left(-\frac{1}{2}, -\frac{1}{2}\right)$ f. $(-1, -1)$

V. Which of the following is true about the Cartesian coordinate system?
a. The origin is the point of intersection of the x- and y-axes.
b. Every ordered pair is represented by a point on the x-axis.
c. The ordered pair (a, b) is located in quadrant IV if $a < 0$ and $b > 0$.
d. The vertical axis is called the x-axis.

In Problems 1–10 plot each point in the xy-plane. If the point is not on the x- or y-axis, determine the quadrant in which it lies.

1. $(3, -2)$ 2. $(4, 3)$ 3. $(2, 0)$
4. $(0, -5)$ 5. $(-4, -1)$ 6. $(-2, 3)$
7. $\left(\frac{1}{2}, \frac{1}{3}\right)$ 8. $\left(\frac{1}{3}, -\frac{3}{2}\right)$ 9. $\left(0, \frac{3}{4}\right)$
10. $\left(-\frac{2}{3}, \frac{7}{3}\right)$

In Problems 11–20 find the distance between the given points.

11. $(1, 3)$, $(4, 7)$ 12. $(-7, 2)$, $(4, 3)$
13. $(-9, 4)$, $(0, -7)$ 14. $\left(\frac{1}{2}, \frac{1}{3}\right)$, $\left(\frac{1}{3}, \frac{1}{2}\right)$
15. $(-3, -7)$, $(-1, -2)$ 16. (a, b), (b, a)
17. $(0, 0)$, $(-c, d)$
18. $(1.3, 4.5)$, $(6.2, 3.4)$
19. $(14.13, -2.16)$, $(11.19, 5.71)$
20. $(0.0135, 0.0146)$, $(0.0723, 0.0095)$

In Problems 21–24 determine whether the three given points are vertices of a right triangle.

21. $(0, 3)$, $(1, 4)$, $(5, 0)$
22. $(8, 3)$, $(2, 0)$, $(4, -4)$
23. $(-2, 8)$, $(1, 3)$, $(2, 7)$
24. $(-2, 6)$, $(3, 1)$, $(7, 2)$

In Problems 25–30 a triangle is sketched. Find the length of each side, and determine whether the triangle is a right triangle.

25.

26.

27.

Answers to Readiness Check

I. b II. c III. a IV. e V. a

28.

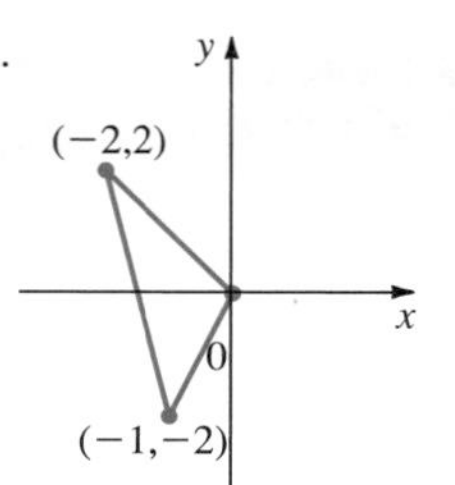

29.

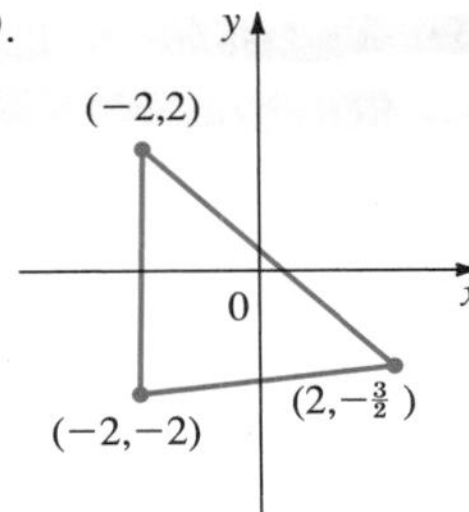

30.

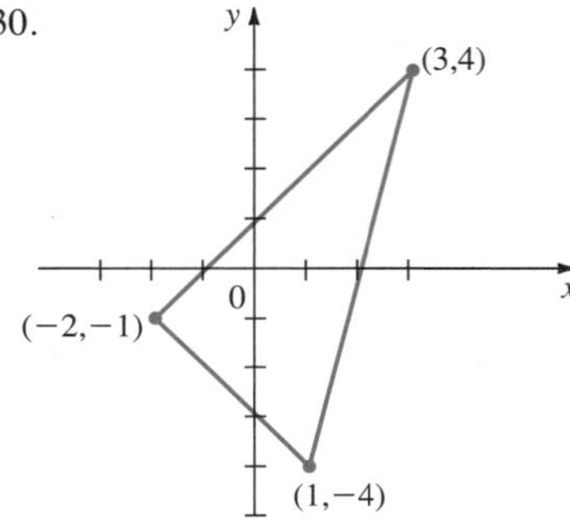

31. Find a number a such that the triangle with vertices at $(1, -1)$, $(2, 3)$, and $(a, 3)$ is a right triangle.

32. Show that the point $M\left(\frac{x_1 + x_2}{2}, \frac{y_1 + y_2}{2}\right)$ is halfway between the points $A(x_1, y_1)$ and $B(x_2, y_2)$. [Hint: Show that $\overline{AM} = \overline{MB} = \sqrt{\left(\frac{x_1 - x_2}{2}\right)^2 + \left(\frac{y_1 - y_2}{2}\right)^2}$.]

In Problems 33–38 find the midpoint of the line segment joining the two points.

33. $(4, -7)$, $(8, 5)$
34. $(3, 8)$, $(3, -4)$
35. $(5, 7)$, $(-2, 7)$
36. $(0, 10)$, $(3, -8)$
37. $(3.2, -1.2)$, $(4.6, 2.8)$
38. $(\sqrt{2}, 1)$, $(1, \sqrt{2})$

39. Show that the diagonals of a rectangle have equal lengths.

40. Show that the triangle with vertices at $(2, 5)$, $(3, 7)$, and $(0, 4)$ is isosceles (that is, two sides have equal lengths).

* 41. Three vertices of a square are $(6, 1)$, $(2, -1)$, and $(4, 5)$. Find the fourth vertex.

* 42. Show that the area of the parallelogram with vertices at $(0, 0)$, $P_1 = (x_1, y_1)$, $P_2 = (x_2, y_2)$, and $Q = (x_1 + x_2, y_1 + y_2)$ equals $|x_1y_2 - x_2y_1|$.

In Problems 43–51, find an equation of the circle with the given center and radius, and sketch its graph.

43. $(0, 2)$, $r = 1$
44. $(2, 0)$, $r = 1$
45. $(1, 1)$, $r = \sqrt{2}$
46. $(1, -1)$, $r = 2$
47. $(-1, 4)$, $r = 5$
48. $\left(\frac{1}{2}, \frac{1}{3}\right)$, $r = \frac{1}{2}$
49. $(\pi, 2\pi)$, $r = \sqrt{\pi}$
50. $(4, -5)$, $r = 7$
51. $(3, -2)$, $r = 4$

In Problems 52–55 an equation is given. Show that it is an equation of a circle, and find the circle's center and radius.

52. $x^2 + 4x + y^2 - 2y + 1 = 0$
53. $x^2 + y^2 - 6y + 3 = 0$
54. $x^2 + 6x + y^2 + 4y + 9 = 0$
55. $2x^2 + 2x + 2y^2 - y - \frac{61}{8} = 0$

** 56. Show that the equation $x^2 + ax + y^2 + by + c = 0$ is the equation of a circle if and only if $a^2 + b^2 - 4c > 0$.

* 57. Find an equation for the unique circle that passes through the points $(0, -2)$, $(6, -12)$, and $(-2, -4)$.

** 58. Show that the area A of a triangle with vertices $P_1 = (x_1, y_1)$, $P_2 = (x_2, y_2)$, and $P_3 = (x_3, y_3)$ is

$$A = \frac{1}{2}|x_1y_2 + x_2y_3 + x_3y_1 - x_1y_3 - x_2y_1 - x_3y_2|$$

59. Using the result of Problem 58, calculate the area of the triangle with the given vertices.
(a) $(2, 1)$, $(0, 4)$, $(3, -6)$
(b) $(4, 2)$, $(-1, -5)$, $(7, 3)$

3.2 The Slope and Equations of a Line

A **linear equation in two variables** or a **linear relation** is an equation of the form

$$ax + dy = c \tag{1}$$

where a and d are not both equal to zero. As we will soon see, the graph of a linear equation in two variables is a straight line.

Linear equations occur frequently in applications, some of which are discussed in Section 3.4. In this section we concentrate on properties of straight lines.

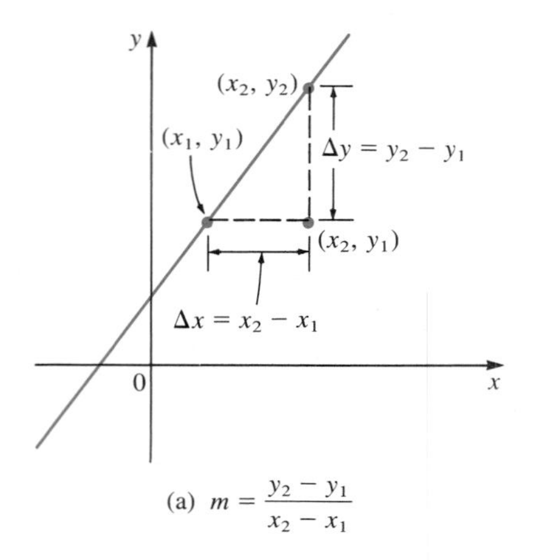

(a) $m = \dfrac{y_2 - y_1}{x_2 - x_1}$

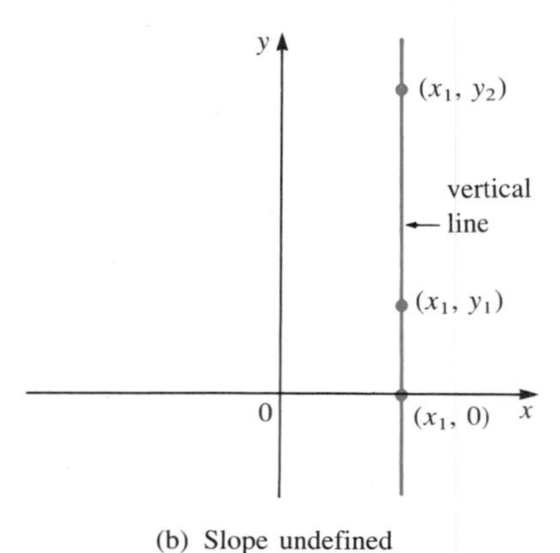

(b) Slope undefined

Figure 1 The slope of a line is $\dfrac{\Delta y}{\Delta x}$ unless the line is vertical.

We first discuss the *slope* of a line, which is a measure of the relative rate of change of the x- and y-coordinates of points on the line as we move along the line.

Slope

Let L denote a nonvertical line, and let (x_1, y_1) and (x_2, y_2) be two distinct points on the line. Then the **slope** of the line, denoted by m, is given by

Slope of a Line

$$m = \text{slope of } L = \frac{y_2 - y_1}{x_2 - x_1} = \frac{\Delta y}{\Delta x} \qquad (2)$$

Here Δy and Δx denote the changes in y and x, respectively (see Figure 1).† If L is vertical so that $x_2 = x_1$, then the slope is **undefined.**‡ If L is horizontal so that $y_2 = y_1$, then $\Delta y = 0$ and the slope is **zero.**

We have

Multiply numerator and denominator by -1

$$\frac{y_2 - y_1}{x_2 - x_1} \overset{\downarrow}{=} \frac{y_1 - y_2}{x_1 - x_2}$$

so, in (2), it doesn't matter which point we take first.

A problem remains. We must show that we get the same value for the ratio (2) no matter which two points are chosen. Look at Figure 2.

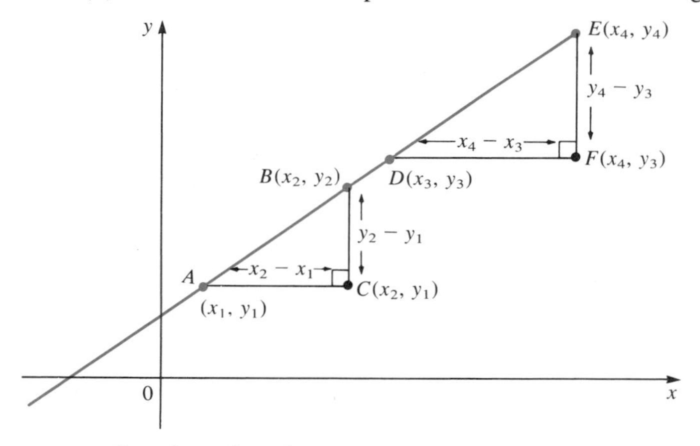

Figure 2 $\dfrac{y_2 - y_1}{x_2 - x_1} = \dfrac{y_4 - y_3}{x_4 - x_3}$ because triangles ABC and DEF are similar.

The two right triangles ABC and DEF are similar, so the ratios of corresponding sides are equal.† Thus $\overline{CB}/\overline{AC} = \overline{FE}/\overline{DF}$ or

† Δ is the capital Greek letter delta. Here Δx, read "delta x," denotes a change in x. It does not stand for the number Δ times the number x.

‡ In some books a line parallel to the y-axis is said to have an **infinite slope.**

$$m = \frac{y_2 - y_1}{x_2 - x_1} = \frac{y_4 - y_3}{x_4 - x_3} \tag{3}$$

Each ratio in (3) is equal to the slope.

Suppose $m > 0$ and $x_2 > x_1$. Then $x_2 - x_1 > 0$ and, from (2),

$$y_2 - y_1 = \underset{\substack{\downarrow \\ >0}}{m}\overbrace{(x_2 - x_1)}^{>0} > 0$$

so $y_2 > y_1$. That is, if m is positive, then y increases as x increases. Analogously, if $m < 0$ and if $x_2 > x_1$, then

$$y_2 - y_1 = \underset{\substack{\downarrow \\ <0}}{m}\overbrace{(x_2 - x_1)}^{>0} < 0$$

If m is negative, then y decreases as x increases. This means that the following are true.

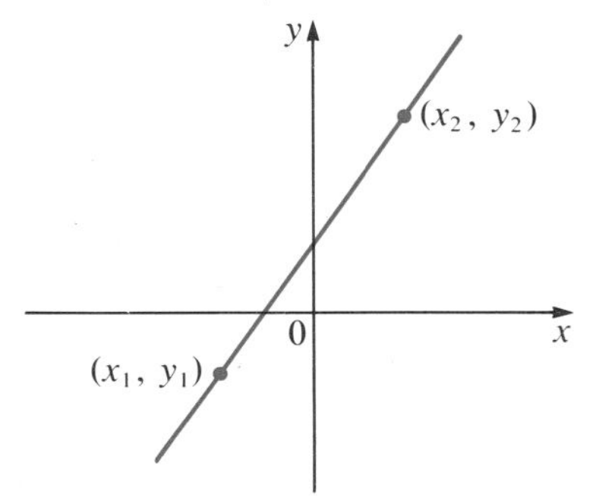

(a) Positive slope: $m > 0$

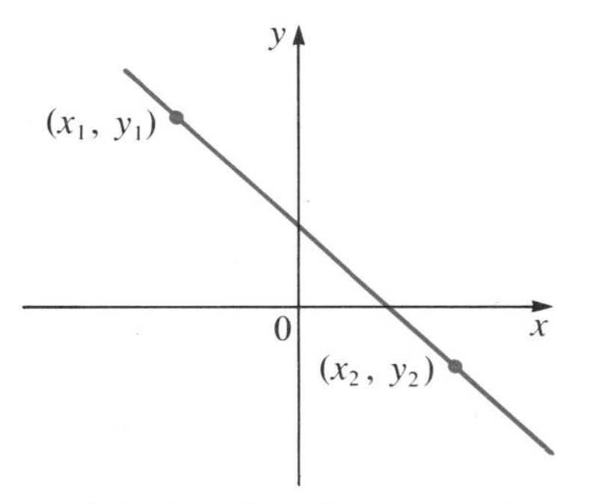

(b) Negative slope: $m < 0$

Figure 3 A line with a positive slope and a line with a negative slope.

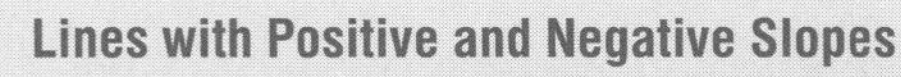

Lines with Positive and Negative Slopes

1. If $m > 0$, the graph of the line will rise as we move from left to right along the x-axis.
2. If $m < 0$, the graph of the line will fall as we move from left to right along the x-axis.

These facts are illustrated in Figure 3.

We will discuss two cases separately. In Figure 4(a) we have drawn the line $y = a$, which is horizontal. Here, as x changes, y does not change at all (since y is equal to the constant a). Therefore $\Delta y/\Delta x = 0/\Delta x = 0$.

Horizontal lines have a slope of zero.

In Figure 4(b) we have drawn the line $x = a$, which is vertical. Here, when y changes, x does not change at all and $\Delta x = 0$. In this case the slope is undefined, since division by zero is undefined.

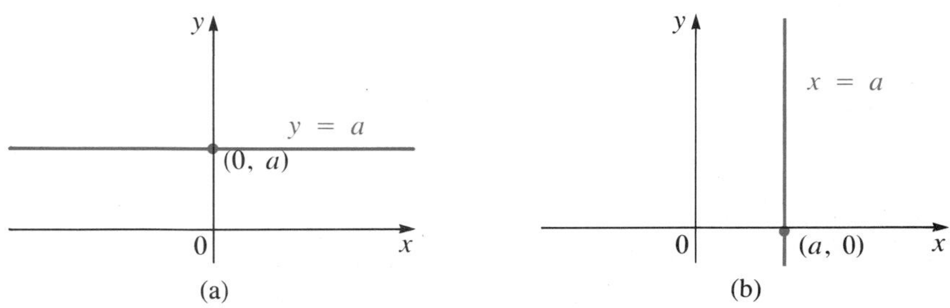

Figure 4 A horizontal line and a vertical line.

† Two triangles are similar if the measures of their angles are the same. The two triangles here are similar because $\angle EFD = \angle BCA$ (since both are right angles) and $\angle FDE = \angle CAB$ because, from geometry, when two parallel lines (DF and AC) are cut by a transversal (AE), the corresponding angles cut off by the transversal have equal measure.

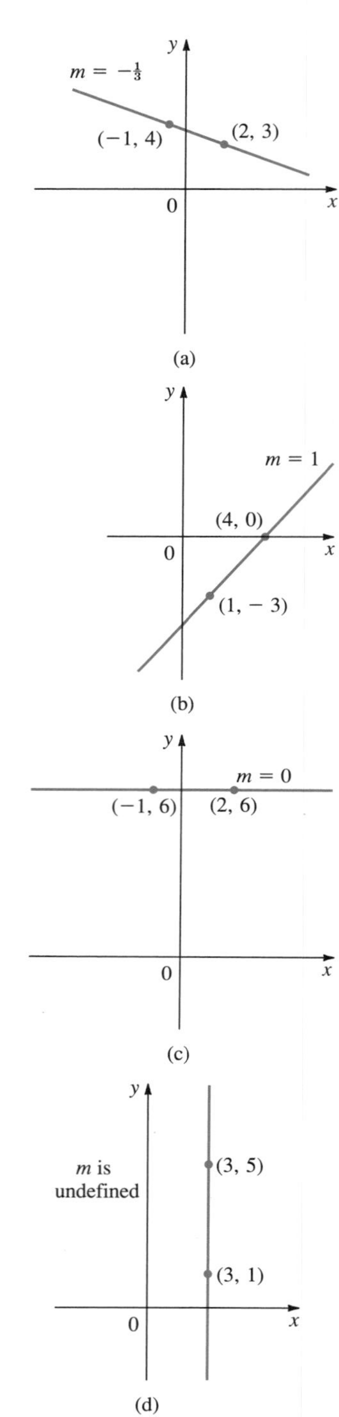

Figure 5 The slopes of four lines.

The slope of a vertical line is undefined.

EXAMPLE 1 *Computing the Slope*

Find the slope of the line containing each pair of points. Then sketch each line.

(a) (2, 3), (−1, 4) (b) (1, −3), (4, 0)
(c) (2, 6), (−1, 6) (d) (3, 1), (3, 5)

SOLUTION

(a) $m = \dfrac{\Delta y}{\Delta x} = \dfrac{4-3}{-1-2} = \dfrac{1}{-3} = -\dfrac{1}{3}$

(b) $m = \dfrac{\Delta y}{\Delta x} = \dfrac{0-(-3)}{4-1} = \dfrac{3}{3} = 1$

(c) $m = \dfrac{6-6}{-1-2} = \dfrac{0}{-3} = 0$. That is, as the x-coordinate changes, the y-coordinate does not vary. This line is horizontal.

(d) $m = \dfrac{5-1}{3-3} = \dfrac{4}{0}$, which is undefined. The line is vertical. The x-coordinate of each point has the constant value 3.

The lines are shown in Figure 5.

We now state two useful facts. Let L_1 and L_2 be two nonvertical lines with slopes m_1 and m_2, respectively.

Theorem

Two distinct nonvertical lines are parallel if and only if† they have the same slope. That is, if m_1 = slope of line L_1 and m_2 = slope of line L_2,

$$L_1 \text{ is parallel to } L_2 \text{ if and only if } m_1 = m_2. \tag{4}$$

EXAMPLE 2 *Two Lines with a Slope of 2 Are Parallel*

The line joining the points (1, −1) and (2, 1) is parallel to the line joining the points (0, 4) and (−2, 0) because the slope of each line is 2.

Theorem

Two nonvertical and nonhorizontal lines are perpendicular‡ if and only if their slopes are negative reciprocals of one another. That is,

$$L_1 \text{ is perpendicular to } L_2 \ (L_1 \perp L_2) \text{ if and only if } m_1 = -\frac{1}{m_2}. \tag{5}$$

---

† The words *if and only if* mean that each of the two statements implies the other. For example, (4) states that if L_1 is parallel to L_2, then $m_1 = m_2$, and if $m_1 = m_2$, then L_1 is parallel to L_2.

‡ Two lines are perpendicular if they meet at right angles.

The proof of this theorem is given at the end of the section.

NOTE If L_1 is vertical, then its slope is undefined. In that case a line perpendicular to L_1 is horizontal and has a slope of zero.

EXAMPLE 3 *Calculating the Slope of a Line Perpendicular to a Given Line*

Let the line L_1 contain the two points $(2, -6)$ and $(1, 4)$. Find the slope of a line L_2 that is perpendicular to L_1.

SOLUTION The slope of L_1 is $m_1 = \dfrac{4-(-6)}{1-2} = \dfrac{10}{-1} = -10$. Thus

$$m_2 = \frac{-1}{-10} = \frac{1}{10}$$

We make an important observation:

> The slope of a line is a measure of the steepness or inclination of the line; that is, the greater the slope (in absolute value), the steeper the line.

This important fact is illustrated in Figure 6.

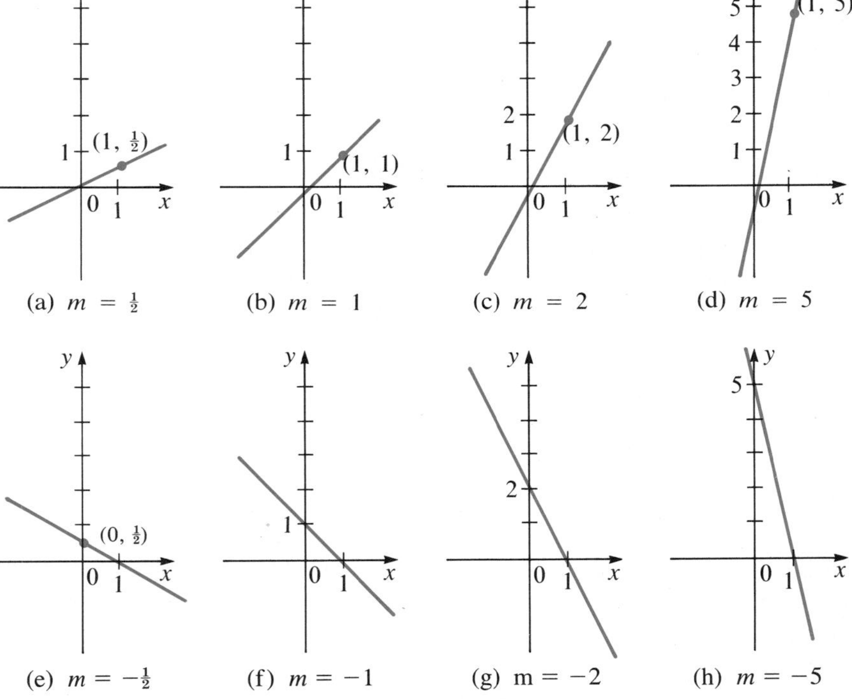

Figure 6 The greater the slope, the steeper the line.

Equation of a Line

An **equation** of a line is an equation in two variables that is satisfied by the coordinates of every point on the line and only the points on the line. If we know two points on a line, then we can find an equation of the line.

If a line is vertical, then it has the equation $x = a$, a constant.

If the line is not vertical, then it has slope given by (2):

$$m = \frac{y_2 - y_1}{x_2 - x_1}$$

or

$$y_2 - y_1 = m(x_2 - x_1) \quad \text{Multiply both sides by } x_2 - x_1 \qquad (6)$$

Let (x_1, y_1) be a point on a line with slope m. If (x, y) is any other point on the line, then, from (6), its coordinates must satisfy the following equation, called the **point-slope equation** of a line.

Point-Slope Equation of a Line

$$y - y_1 = m(x - x_1) \qquad (7)$$

EXAMPLE 4 *Finding a Point-Slope Equation of a Line*

Find a point-slope equation of the line passing through the points $(-1, -2)$ and $(2, 5)$.

SOLUTION We first compute

$$m = \frac{5 - (-2)}{2 - (-1)} = \frac{7}{3}$$

Thus, if we choose $(x_1, y_1) = (2, 5)$, a point-slope equation of the line is

$$y - 5 = \frac{7}{3}(x - 2)$$

Choosing $(x_1, y_1) = (-1, -2)$, we obtain another (equivalent) point-slope equation of the line:

$$y - (-2) = \frac{7}{3}(x - (-1))$$

or

$$y + 2 = \frac{7}{3}(x + 1)$$

Both equations yield the same graph, which is shown in Figure 7. Note that the two equations are really equivalent:

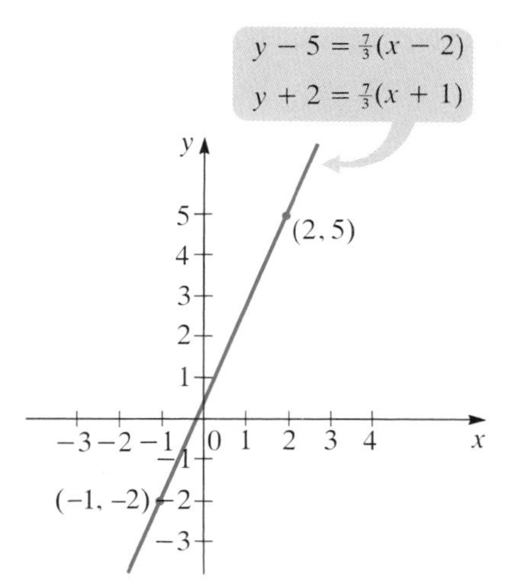

Figure 7 The line passing through $(-1, -2)$ and $(2, 5)$.

$$\begin{aligned} y - 5 &= \frac{7}{3}(x - 2) \\ y - 5 &= \frac{7}{3}x - \frac{14}{3} \\ y &= \frac{7}{3}x - \frac{14}{3} + 5 \\ y &= \frac{7}{3}x + \frac{1}{3} \end{aligned} \qquad \begin{aligned} y + 2 &= \frac{7}{3}(x + 1) \\ y + 2 &= \frac{7}{3}x + \frac{7}{3} \\ y &= \frac{7}{3}x + \frac{7}{3} - 2 \\ y &= \frac{7}{3}x + \frac{1}{3} \end{aligned}$$

As the last example shows, there are many equivalent point-slope equations of a line. In fact, there are an infinite number of them—one for each point on the line. A more commonly used equation of a line is given below.

Let m be the slope and let b be the y-intercept of a line. This is the y-value where the line crosses the y-axis (when $x = 0$). Then the **slope-intercept equation** of the line is the equation

Slope-Intercept Equation of a Line

$$y = mx + b \tag{8}$$

↑ Slope ↑ y-intercept

EXAMPLE 5 ***Finding the Slope-Intercept Equation of a Line***

Find the slope-intercept equation of the line passing through $(-1, -2)$ and $(2, 5)$.

SOLUTION In Example 4 we found the equation

$$y - 5 = \frac{7}{3}(x - 2)$$

As we saw in Example 4, this equation can be written in the form

$$y = \frac{7}{3}x + \frac{1}{3}$$

Note that when $x = 0$, $y = \frac{1}{3}$, so $\frac{1}{3}$ is the y-intercept. Thus the last equation is the slope-intercept equation of the line.

Consider the linear equation

$$ax + dy = c \tag{9}$$

If $d = 0$, then the equation becomes

$$ax = c \quad \text{or} \quad x = \frac{c}{a}$$

which is the equation of a vertical line (see Figure 4(b) with $\frac{c}{a}$ instead of a).

If $d \neq 0$, then we can divide both sides of the equation $ax + dy = c$ by d to obtain

$$\frac{a}{d}x + \frac{d}{d}y = \frac{c}{d}$$

$$\frac{a}{d}x + y = \frac{c}{d}$$

$$y = -\frac{a}{d}x + \frac{c}{d}$$

which is the slope-intercept equation of a line with slope $-\frac{a}{d}$ and y-intercept $\frac{c}{d}$. Conversely,

$$y = mx + b$$

can be written as

$$-mx + y = b$$

which is equation (9) with $a = -m$, $d = 1$, and $c = b$. Thus we have proved the following theorem:

Theorem

The graph of the linear equation

$$ax + dy = c$$

is a straight line, and every straight line is the graph of a linear equation.

This leads us to define a **standard equation** of a line.

Standard Equation of a Line

$$Ax + By = C \tag{10}$$

where A and B are not both equal to zero.

EXAMPLE 6 *Finding a Standard Equation of a Line*

Find a standard equation of the line passing through $(-1, -2)$ and $(2, 5)$.

SOLUTION In Example 5 we found that

$$y = \frac{7}{3}x + \frac{1}{3}$$

Then

$$3y = 7x + 1 \quad \text{Multiply through by 3}$$

$$3y - 7x = 1 \quad \text{Subtract } 7x \text{ from both sides}$$

A standard equation of the line is $-7x + 3y = 1$. Another standard equation is $7x - 3y = -1$. ■

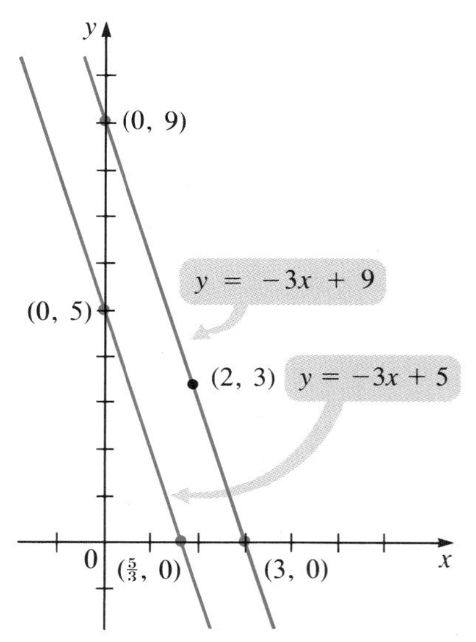

Figure 8 The lines $y = -3x + 9$ and $y = -3x + 5$ are parallel.

EXAMPLE 7 *Finding the Slope-Intercept Equation of a Line Parallel to a Given Line*

Find the slope-intercept equation of the line passing through the point (2, 3) and parallel to the line whose equation is $y = -3x + 5$.

SOLUTION Parallel lines have the same slope. The slope of the line $y = -3x + 5$ is -3 because the line is given in the slope-intercept form. Then, from (7), a point-slope equation of the line is

$$y - 3 = -3(x - 2)$$
$$y - 3 = -3x + 6$$
$$y = -3x + 9$$

This is the answer. Both lines are sketched in Figure 8. ■

EXAMPLE 8 *Finding the Slope-Intercept Equation of a Line Perpendicular to a Given Line*

Find the slope-intercept equation of the line passing through $(-1, 3)$ and perpendicular to the line $2x + 3y = 4$.

SOLUTION From $2x + 3y = 4$, we obtain

$$3y = -2x + 4$$

or

$$y = -\frac{2}{3}x + \frac{4}{3} \quad \text{Divide by 3}$$

This means that the slope of the line $2x + 3y = 4$ is $-\frac{2}{3}$, so the line whose equation we seek has the slope

$$m = \frac{-1}{-\frac{2}{3}} = \frac{3}{2}$$

We can use (7) to find the slope-intercept equation of the line passing through $(-1, 3)$ with slope $\frac{3}{2}$. Here's an alternative way to find it.

$$y = mx + b$$
$$3 = \frac{3}{2}(-1) + b$$
$$\frac{9}{2} = b$$

The equation is

$$y = \frac{3}{2}x + \frac{9}{2}$$

The two lines are sketched in Figure 9.

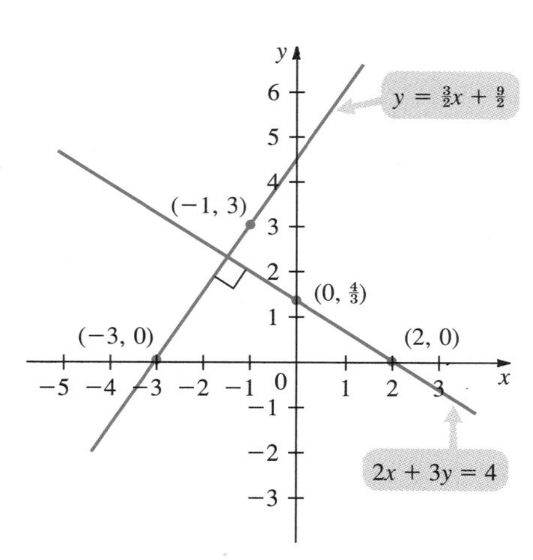

Figure 9 The lines $y = \frac{3}{2}x + \frac{9}{2}$ and $2x + 3y = 4$ are perpendicular.

Table 1 summarizes some facts about straight lines.

Table 1

Equation	Description of Line
$x = a$	*Vertical line*, x-intercept is a. No y-intercept (unless $a = 0$). Slope is undefined.
$y = b$	*Horizontal line*. No x-intercept (unless $b = 0$), y-intercept is b. Slope $= 0$.
$y - y_1 = m(x - x_1)$	*Point-slope form* of line with slope m passing through the point (x_1, y_1).
$y = mx + b$	*Slope-intercept form* of line with slope m and y-intercept b; x-intercept $= -b/m$ if $m \neq 0$.
$Ax + By = C$	*Standard form*. Slope is $-A/B$ if $B \neq 0$. x-intercept is C/A if $A \neq 0$, and y-intercept is C/B if $B \neq 0$.

There is another type of problem we will encounter.

EXAMPLE 9 *Finding the Point of Intersection of Two Lines*

Find the point of intersection, if one exists, of the lines $2x + 3y = 7$ and $-x + y = 4$.

SOLUTION The coordinates of the point of intersection, which we label (c, d), must satisfy both equations. For the first equation we have

$$y = -\frac{2}{3}x + \frac{7}{3} \qquad (11)$$

and for the second,

$$y = x + 4 \qquad (12)$$

The lines have different slopes ($-\frac{2}{3}$ and 1) and are therefore not parallel, *so they do have a point of intersection*. The point (c, d) is on both lines.

$$d = -\frac{2}{3}c + \frac{7}{3} \quad \text{From (11)} \qquad d = c + 4 \quad \text{From (12)}$$

Then

$$-\frac{2}{3}c + \frac{7}{3} = c + 4$$

$$c + \frac{2}{3}c = -4 + \frac{7}{3}$$

$$\frac{5}{3}c = -\frac{5}{3} \quad \text{so} \quad c = -1$$

Then $d = c + 4 = -1 + 4 = 3$, and the point of intersection is $(-1, 3)$. The two lines are graphed in Figure 10.

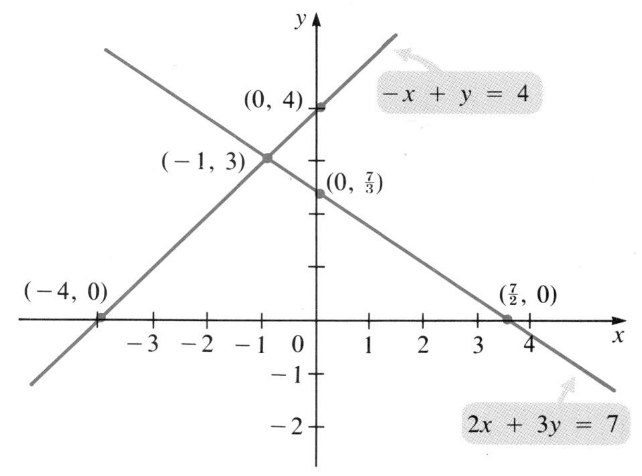

Figure 10 The lines $2x + 3y = 7$ and $-x + y = 4$ intersect at $(-1, 3)$.

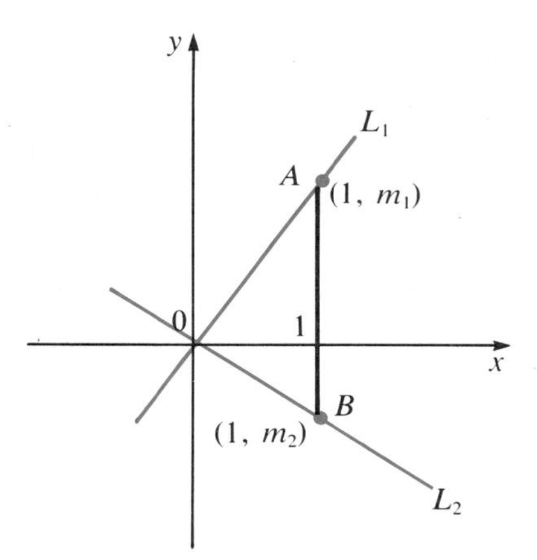

Figure 11 L_1 has slope m_1, and L_2 has slope m_2.

Proof of the theorem on page 160 (Optional)

We close this section with a proof that if neither L_1 nor L_2 is vertical, then $L_1 \perp L_2$ if and only if $m_1 = -\dfrac{1}{m_2}$. Look at Figure 11, where, to simplify the computations, we have assumed that the two lines intersect at the origin. The proof where they intersect at $(x^*, y^*) \neq (0, 0)$ is similar. We have

$$\begin{aligned} \text{slope-intercept equation for } L_1\text{:} \quad & y = m_1x + b_1 \\ \text{slope-intercept equation for } L_2\text{:} \quad & y = m_2x + b_2 \end{aligned}$$

But $b_1 = b_2 = 0$. Setting $x = 1$ in each equation, we find that $(1, m_1)$ is on L_1 and $(1, m_2)$ is on L_2. By the Pythagorean theorem and its converse, AOB is a right triangle if and only if

$$\overline{OA}^2 + \overline{OB}^2 = \overline{AB}^2$$

$$\begin{pmatrix}\text{distance between} \\ (0, 0) \text{ and } (1, m_1)\end{pmatrix}^2 + \begin{pmatrix}\text{distance between} \\ (0, 0) \text{ and } (1, m_2)\end{pmatrix}^2 = \begin{pmatrix}\text{distance between} \\ (1, m_1) \text{ and } (1, m_2)\end{pmatrix}$$

$$\begin{aligned} (1 + m_1{}^2) + (1 + m_2{}^2) &= (m_2 - m_1)^2 \\ 2 + m_1{}^2 + m_2{}^2 &= m_2{}^2 - 2m_2m_1 + m_1{}^2 \\ 2 &= -2m_2m_1 \\ -1 &= m_1m_2 \end{aligned}$$

If $m_2 = 0$, then L_2 is horizontal, and if L_1 were perpendicular to L_2, it would be vertical, which we have ruled out. Then $m_2 \neq 0$ and

$$m_1 = \frac{-1}{m_2}$$

Thus, triangle AOB is a right triangle if and only if $m_1 = -1/m_2$. ■

Problems 3.2

Readiness Check

I. Which of the lines below have a positive slope?
II. Which of the lines have a negative slope?
III. Which of the lines have a slope of zero?
IV. Which of the lines have an undefined slope?

a.
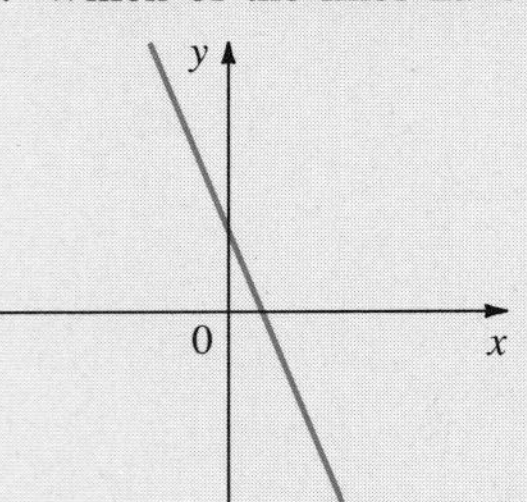

b.
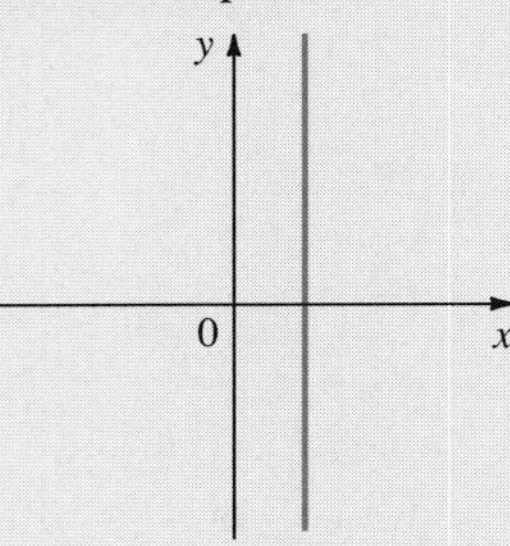

c.
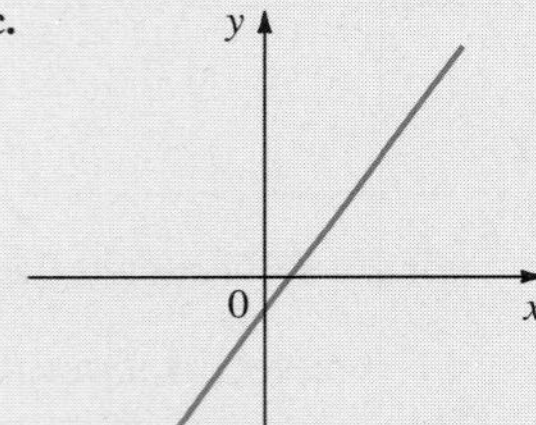

d.
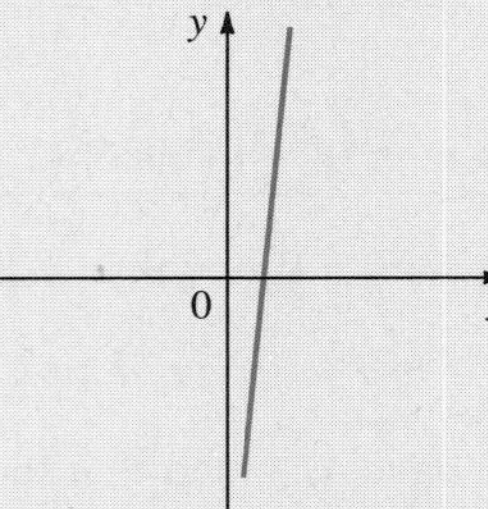

e.
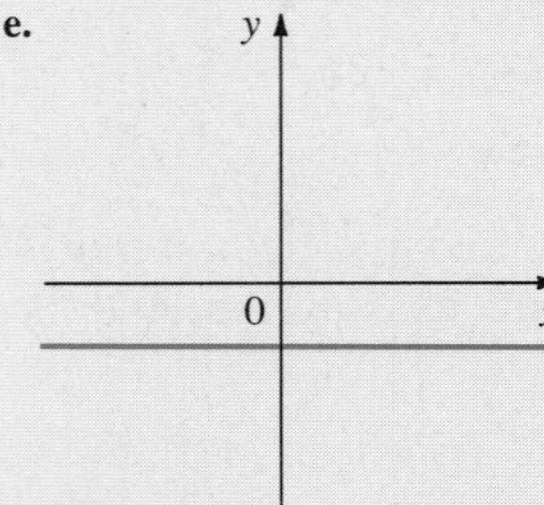

f.
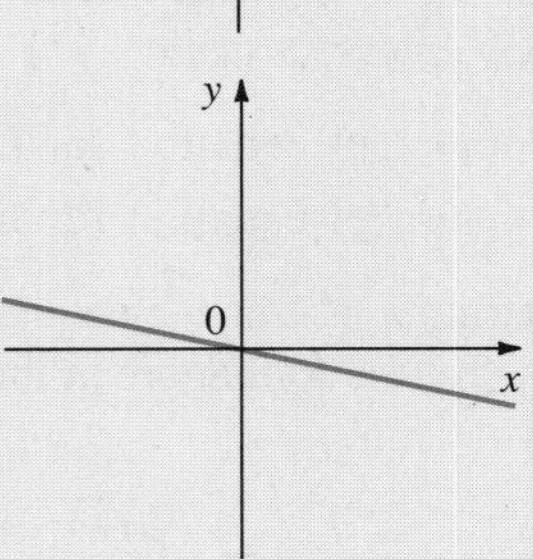

g.
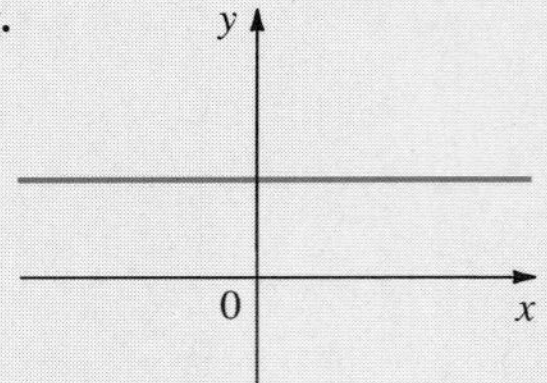

h.
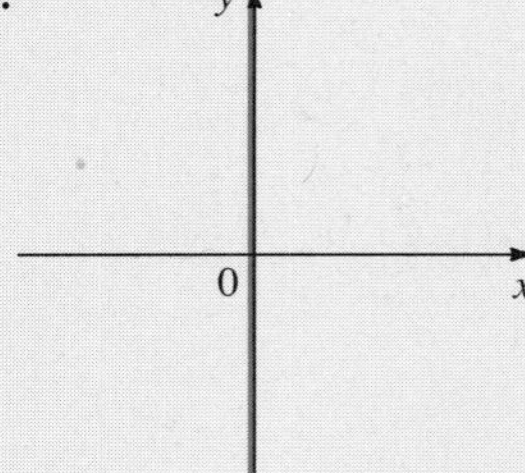

i.
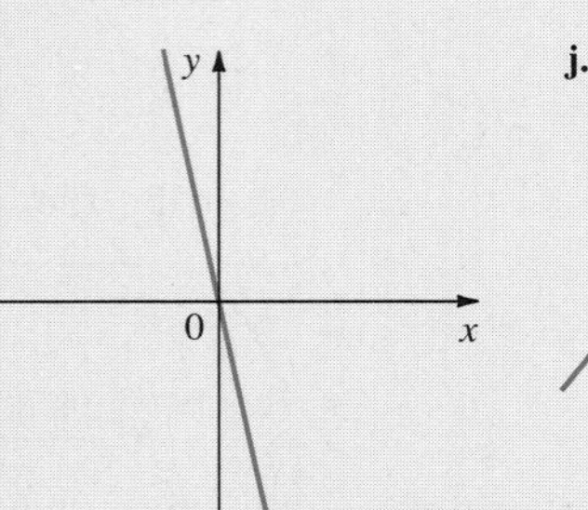

j.
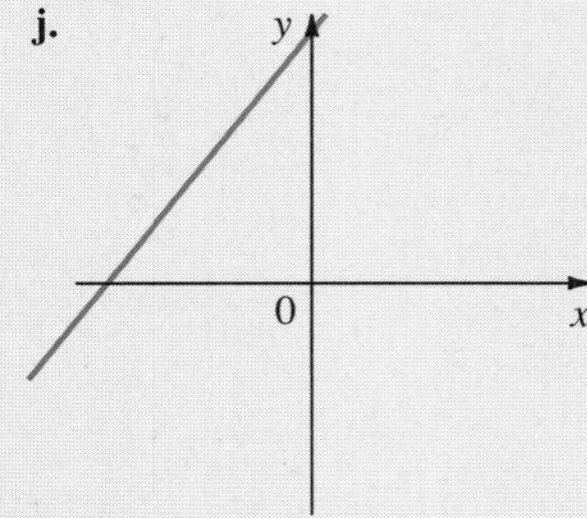

V. Which of the following is the equation of a line parallel to the x-axis?
a. $y = -1$
b. $y = x$
c. $x = -1$
d. $y = -x$

In Problems 1–11 find the slope of the line passing through the two given points. Graph the lines.

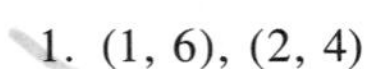

1. $(1, 6), (2, 4)$
2. $(-3, 4), (7, 9)$
3. $(-1, -2), (-3, -4)$
4. $(0, -3), (3, 0)$
5. $(-4, 3), (8, -2)$
6. $(1, 7), (-4, 7)$
7. $(2, -3), (5, -3)$
8. $(-2, 4), (-2, 6)$
9. $(0, a), (a, 0), a \neq 0$
10. $(a, b), (b, a), a \neq b$
11. $(a, b), (c, d), a \neq c$

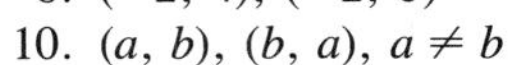
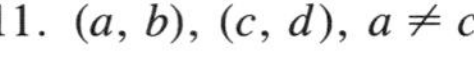

In Problems 12–19 determine the slope of each line.

12.
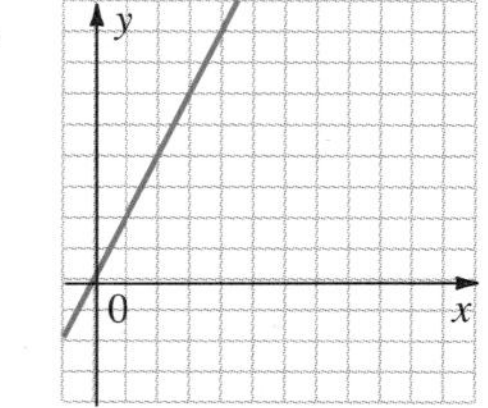

13.
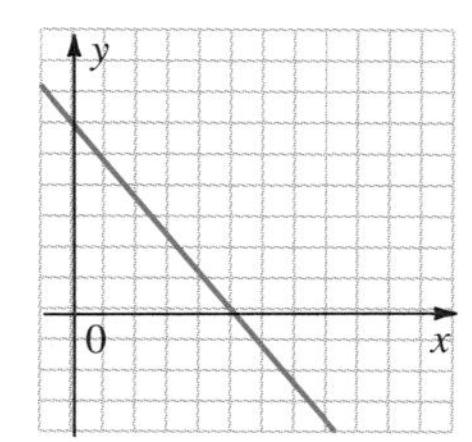

Answers to Readiness Check

I. c, d, j II. a, f, i III. e, g IV. b, h V. a

14. 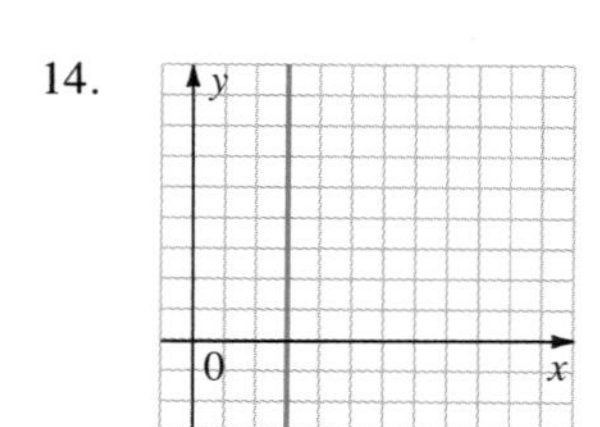

15. 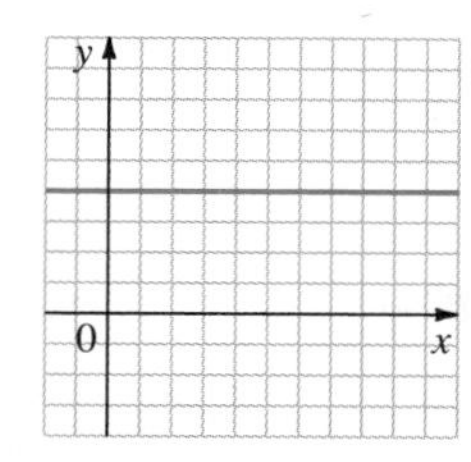

16. 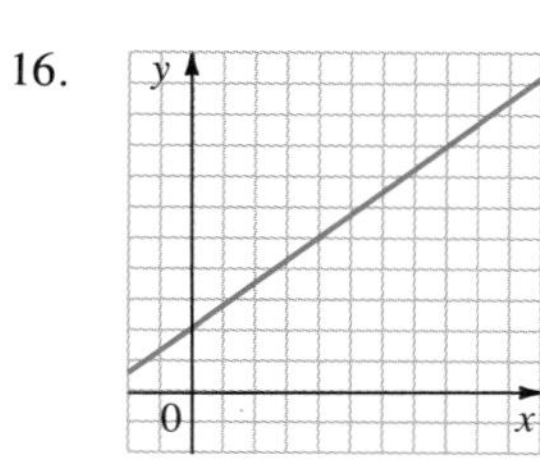

17. 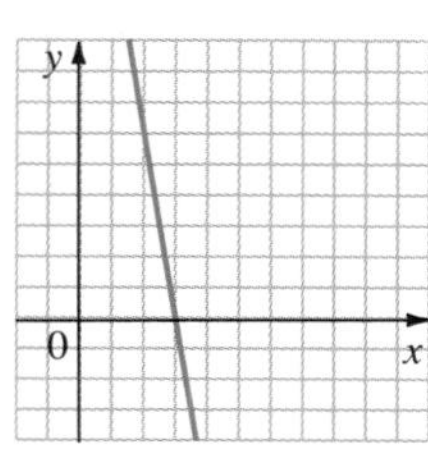

18. 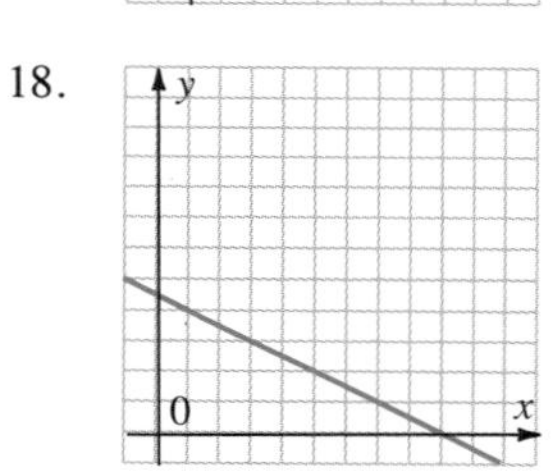

19. 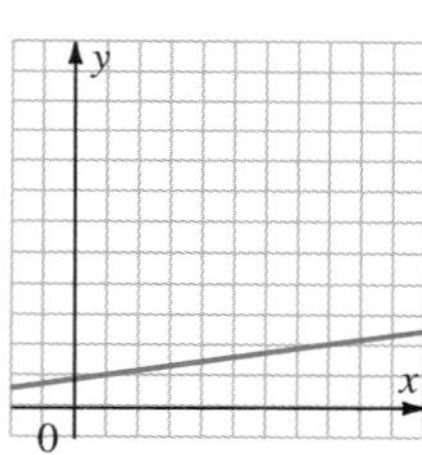

In Problems 20–28 two pairs of points are given. By calculating their slopes, determine whether the two lines containing these pairs of points are parallel, perpendicular, or neither. Graph the lines.

20. $(1, 8), (2, 9)$; $(1, 2), (0, 1)$
21. $(0, 5), (-5, 0)$; $(0, 2), (2, 0)$
22. $(2, 6), (8, -1)$; $(2, 4), (-2, 3)$
23. $(0, 5), (2, -1)$; $(0, 0), (-1, 3)$
24. $(5, 2), (1, 7)$; $(2, 5), (7, 1)$
25. $(1, -2), (2, 4)$; $(4, 1), (-2, 2)$
26. $(3, 2), (5, -2)$; $(0, 6), (-5, 6)$
27. $(3, 1), (3, 7)$; $(2, 4), (-1, 4)$
28. $(4, 3), (4, 1)$; $(-2, 4), (-2, 0)$

Problems That Arise in Calculus (Problems 29–39)

29. Find the slope of the line joining the points $(2, 4)$ and $(2.5, 6.25)$.
30. Find the slope of the line joining the points $(2, 4)$ and $(2.1, 2.1^2) = (2.1, 4.41)$.
31. Show that the slope of the line joining the points $(2, 4)$ and $(2 + h, (2 + h)^2)$ is $4 + h$.
32. Show that the slope of the line joining the points (x, x^2) and $(x + h, (x + h)^2)$ is $2x + h$.
33. Find the slope of the line joining the points $(2, \frac{1}{2})$ and $(2.5, 0.4)$.
34. Find the slope of the line joining the points $(2, \frac{1}{2})$ and $\left(2.1, \frac{1}{2.1}\right)$.
35. Show that the slope of the line joining the points $(2, \frac{1}{2})$ and $\left(2 + h, \frac{1}{2 + h}\right) = -\frac{1}{2(2 + h)}$.
36. Show that the slope of the line joining the points $\left(x, \frac{1}{x}\right)$ and $\left(x + h, \frac{1}{x + h}\right)$ is $-\frac{1}{x(x + h)}$.
37. Find the slope of the line joining the points $(4, 2)$ and $(4.1, \sqrt{4.1})$.
38. Show that the slope of the line joining the points $(4, 2)$ and $(4 + h, \sqrt{4 + h})$ is $1/(\sqrt{4 + h} + 2)$. [Hint: $(\sqrt{B} - \sqrt{A})(\sqrt{B} + \sqrt{A}) = B - A$ if $A, B \geq 0$.]
39. Suppose $a > 0$ and $a + h > 0$. Show that the straight line through $(a, \sqrt{a})$ and $(a + h, \sqrt{a + h})$ has slope $1/(\sqrt{a + h} + \sqrt{a})$.

In Problems 40–53, find a point-slope form, the slope-intercept form, and a standard form of the equation of the straight line when either two points on the line or a point and the slope of the line are given. Sketch the graph of the line in the xy-plane.

40. $(1, 2), (3, 6)$
41. $(3, -1), (2, -4)$
42. $(3, 7), m = \frac{1}{2}$
43. $(8, 3), m = 0$
44. $(2, -4)$, m undefined
45. $\left(3, -\frac{1}{2}\right), \left(\frac{1}{3}, 0\right)$
46. $(-2, -4), (3, 7)$
47. $(5, -1), (8, 2)$
48. $(7, -3), m = -\frac{4}{3}$
49. $(-5, 1), m = \frac{3}{7}$
50. $(a, b), (c, d)$
51. $(a, b), m = c$
52. $(1.602, 3.1527), m = -2.315$
53. $(0.0146, 0.0058), m = 12.611$
54. Find the slope-intercept equation of the line parallel to the line $2x + 5y = 6$ and passing through the point $(-1, 1)$.
55. Find the slope-intercept equation of the line parallel to the line $5x - 7y = 3$ and passing through the point $(2, 5)$.
56. Find a standard equation of the line perpendicular to the line $x + 3y = 7$ and passing through the point $(0, 1)$.
57. Find the slope-intercept equation of the line perpendicular to the line $2x - \frac{3}{2}y = 7$ and passing through the point $(-1, 4)$.
58. Find a standard equation of the line perpendicular to the line $ax + by = c$ and passing through the point (α, β). Assume that $a \neq 0$ and $b \neq 0$.

In Problems 59–65, find the point of intersection (if there is one) of the two lines.

59. $x + y = 5;\ 2x - y = 7$
60. $y - 2x = 4;\ 4x - 2y = 6$
61. $8x + 4y = 6;\ 12x + 6y = 9$
62. $8x - 4y = 6;\ 12x - 6y = 12$
63. $2x + 3y = 5;\ 3x - 4y = 10$
64. $3x + 4y = 5;\ 6x - 7y = 8$
65. $2.307x - 1.609y = 5.508;\ 0.1742x + 0.8196y = 163.24$
66. Show that the point $\left(\frac{x_1 + x_2}{2}, \frac{y_1 + y_2}{2}\right)$ is on the line segment joining the points (x_1, y_1) and (x_2, y_2). This fact, together with the result of Problem 32 in Section 3.1, proves that $\left(\frac{x_1 + x_2}{2}, \frac{y_1 + y_2}{2}\right)$ is the midpoint of the line segment joining (x_1, y_1) and (x_2, y_2).
* 67. Show that the line segment between (x_1, y_1) and (x_2, y_2) is divided in thirds by

$$Q = \left(\frac{2x_1 + x_2}{3}, \frac{2y_1 + y_2}{3}\right)$$

and

$$R = \left(\frac{x_1 + 2x_2}{3}, \frac{y_1 + 2y_2}{3}\right)$$

* 68. Consider the straight-line segment between $P_0 = (x_0, y_0)$ and $P_1 = (x_1, y_1)$. Show that the point

$$((1 - \lambda)x_0 + \lambda x_1, (1 - \lambda)y_0 + \lambda y_1),$$

where λ is a real number between 0 and 1, lies on that line segment.

Graphing Calculator Problems

In Problems 69–78 sketch the graph of each line on a graphing calculator for x in the given interval. In some cases it will be necessary to rewrite the equation of the line in its slope-intercept form. Before starting, decide on an appropriate range for the y-values.

69. $y = 2x - 5;\ [-3, 4]$
70. $y = 5 - 7x;\ [-2, 2]$
71. $y = 3.6x + 4;\ [-3, 3]$
72. $y = 37x + 105;\ [-4, 7]$
73. $y = -0.023x + 0.057;\ [-10, 10]$
74. $2x - 3y = 8;\ [-1000, 1000]$
75. $6x + 11y = 30;\ [-4, 0]$
76. $-237x + 108y = 1294;\ [-8, 10]$
77. $-0.002x + 0.004y = 0.009;\ [-0.23, 0.37]$
78. $\frac{y - 4}{x + 7} = 8;\ [200, 400]$

3.3 Functions

In Section 3.1 we discussed equations in two variables, and relations. We saw, in Section 3.2, that an equation of a straight line is a linear equation in two variables, or a linear relation. Functions constitute an important class of relations. We now define a function and give some examples of functions.

Definition of Function, Domain and Range, and Image

A **function** is a rule that assigns to each member of one set (called the **domain** of the function) a unique member of another set. The set of all assigned members is called the **range** of the function. If x in the domain is assigned the element y in the range, then y is called the **image** of x.

For many functions the domain and range are both equal to $\mathbb{R}$, the set of real numbers.

You have seen many examples of functions.

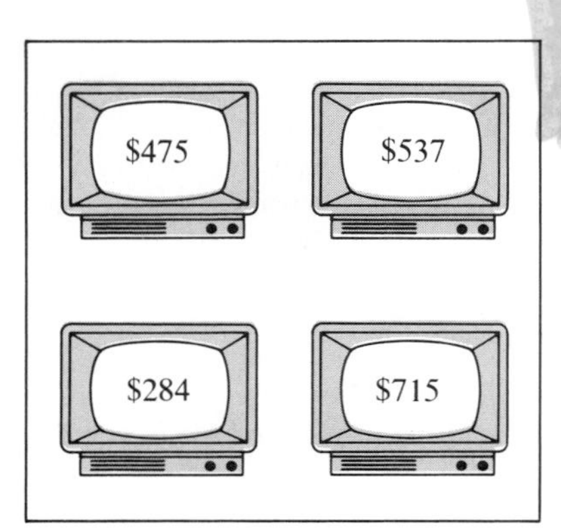

Figure 1 The prices of four TV sets.

EXAMPLE 1 ***The TV Price Function***

A certain appliance store sells TV sets. Each TV set is assigned a price as in Figure 1. This assignment defines a function because each TV set has exactly one price. The domain of the function is the set of TV sets in the store. The range is the set of prices of TV sets in the store. ■

EXAMPLE 2 *The Birthdate Function*

To each living person in the United States, assign his or her date of birth. This assignment defines a function because every person has exactly one date of birth. The domain of this function is the set of people living in the United States. The range of this function is the set of dates in history on which at least one person now living in the United States was born. ■

EXAMPLE 3 *The Squaring Function*

To each real number assign its square. For example, 4 is assigned to 2, 9 is assigned to -3, 1.69 is assigned to 1.3 and 2 is assigned to $\sqrt{2}$. This assignment defines a function because each real number has a unique square. The domain is the set of real numbers $\mathbb{R}$. The range is the set of nonnegative real numbers, denoted by $\mathbb{R}^+$.

NOTATION A function is often denoted by a single letter, usually f, g, or h. If x is a member of the domain of f, say, then $f(x)$ denotes the image of x. The symbol $f(x)$ is read "f of x." We often write $y = f(x)$,† read "y equals f of x," to indicate that to each value x in the domain we assign the unique image y in the range. When we write $y = f(x)$, x is called the **independent variable,** and y is called the **dependent variable.**

Domain of f Range of f

Figure 2 Illustration of domain and range.

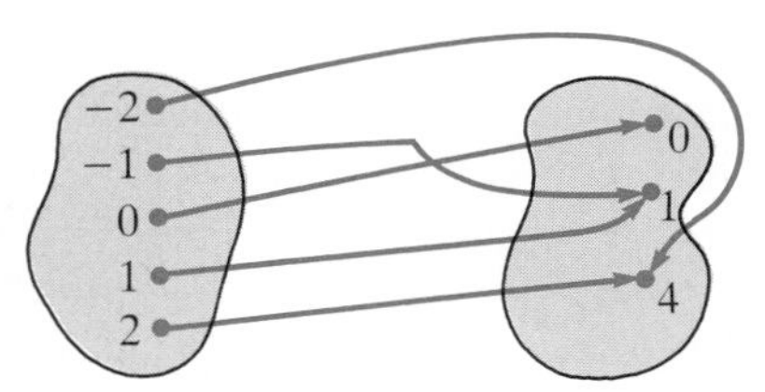

Figure 3 Some values of the squaring function $f(x) = x^2$.

EXAMPLE 4 *Using Function Notation*

(a) For the TV price function, if Brand X costs \$421, then

$$f(\text{Brand X}) = \$421$$

That is, the image of Brand X is \$421.

(b) For the squaring function, we have $f(2) = 2^2 = 4$, $f(-3) = (-3)^2 = 9$, $f(1.3) = 1.3^2 = 1.69$, and $f(\sqrt{2}) = (\sqrt{2})^2 = 2$. In general, $f(x) = x^2$ for every real number x.

To make this clearer, we write

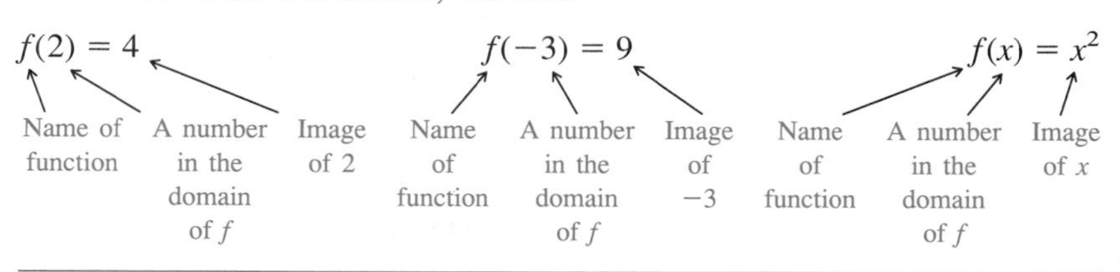

Functions are often depicted pictorially as in Figure 2. The diagram illustrates, for example, that the function f takes a value x in the domain and assigns to it a unique value $f(x)$ in its range.

In Figure 3 we depict some of the values taken by the squaring function.

† This notation was first used by the great Swiss mathematician Leonhard Euler (1707–1783) in the *Commentarii Academia Petropolitanae* (Petersburg Commentaries), published in 1734–1735.

EXAMPLE 5 *Determining Whether a Rule Is a Function*

Let D and R be two sets. In Figure 4 we show four rules that assign members of D to members of R. Which of the four rules represent functions with domain D?

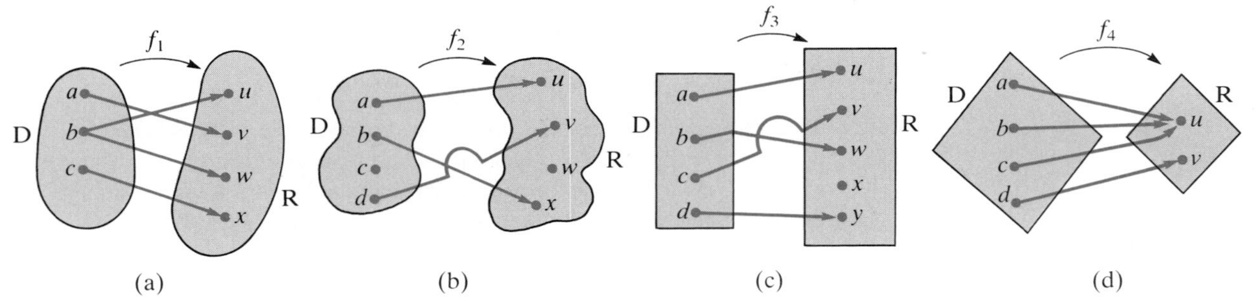

Figure 4 Four rules.

SOLUTION

(a) f_1 is not a function because it assigns two different members of R to one member (b) of D. That is, $f_1(b) = u$ and w. This contradicts the definition of a function. In order to be a function, f_1 must assign *exactly one* member of R to each member of D. If we delete the arrow from b to u or the arrow from b to w (but not both), then f will be a function with domain D.

(b) f_2 is not a function with domain D because f does not assign anything to c. In order to be a function with domain D, f must assign one member of R to *each* member of D.

(c) f_3 is a function because it assigns one member of R to each member of D. It is irrelevant that nothing in D is assigned to x. This may make x feel lonely, but it does not contradict the definition of a function. The range of f_3 is $\{u, v, w, y\}$.

(d) f_4 is a function because it assigns exactly one member of R to each member of D. The fact that u is assigned three times does not contradict the definition of a function.

For the remainder of this chapter we will look at functions for which the domains and ranges are sets of real numbers.

When the domain of a function is not given, we usually take the domain to be the set of all real values of x so that in the equation $y = f(x)$, $f(x)$ is a defined real number.

EXAMPLE 6 *A Linear Function*

Let $f(x) = 3x + 5$. Then since $3x + 5$ is defined for every real number x, the domain of f is $\mathbb{R}$, the set of all real numbers.

What is the range? To answer that question, we ask: For what values of y is $y = 3x + 5$ for some real number x? For example, for $y = 10$, we seek an x such that

$$\begin{aligned} 10 &= 3x + 5 \\ 10 - 5 &= 3x \\ 5 &= 3x \\ x &= \frac{5}{3} \end{aligned}$$

Thus $f(\frac{5}{3}) = 3 \cdot \frac{5}{3} + 5 = 5 + 5 = 10$ so 10 is the image of $\frac{5}{3}$ and 10 is in the range of f.

In fact, every real number y is in the range of f. If y is given, we seek an x such that

$$\begin{aligned} y &= 3x + 5 \\ y - 5 &= 3x \\ x &= \frac{y-5}{3} \end{aligned}$$

Then $f(x) = f\left(\frac{y-5}{3}\right) = 3\left(\frac{y-5}{3}\right) + 5 = y - 5 + 5 = y$ so y is the image of $\frac{y-5}{3}$, which means that y is in the range of f.

More generally, the slope-intercept equation of a nonvertical straight line represents a **linear function:**

$$y = f(x) = mx + b$$

EXAMPLE 7 ***A Square Root Function***

Let $s = g(t) = \sqrt{3t + 1}$.

(a) Evaluate $g(0)$, $g(1)$, $g(8)$, $g(100)$, and $g(t^2 + 2)$.
(b) Find the domain of g.
(c) Find the range of g.

SOLUTION

(a) $g(0) = \sqrt{3 \cdot 0 + 1} = \sqrt{1} = 1$
$g(1) = \sqrt{3 \cdot 1 + 1} = \sqrt{4} = 2$
$g(8) = \sqrt{3 \cdot 8 + 1} = \sqrt{25} = 5$
$g(100) = \sqrt{3 \cdot 100 + 1} = \sqrt{301} \approx 17.35$
$g(t^2 + 2) = \sqrt{3(t^2 + 2) + 1} = \sqrt{3t^2 + 7}$

(b) $\sqrt{t}$ is defined as a real number as long as $t \geq 0$. Then, since the range must be a set of real numbers, $\sqrt{3t+1}$ is defined if

$$\begin{aligned} 3t + 1 &\geq 0 \\ 3t &\geq -1 \\ t &\geq -\frac{1}{3} \end{aligned}$$

Thus domain of $g = \{t\colon t \geq -\frac{1}{3}\} = [-\frac{1}{3}, \infty)$.

(c) Suppose that s^* is in the range of g. Then there is a t such that $s^* = \sqrt{3t+1} \geq 0$. So every number in the range of g is nonnegative. Moreover, if $s^* \geq 0$, then there is a t such that $s^* = \sqrt{3t+1}$. To find that t, we work backward:

$$\begin{aligned} s^{*2} &= 3t + 1 \\ s^{*2} - 1 &= 3t \\ t &= \frac{s^{*2} - 1}{3} \end{aligned}$$

and

$$\sqrt{3t+1} = \sqrt{3\left(\frac{s^{*2}-1}{3}\right) + 1} = \sqrt{(s^{*2} - 1) + 1} = \sqrt{s^{*2}} = s^* \geq 0$$

This means that every nonnegative real number is the image of some real number t, so

$$\text{range of } g = \mathbb{R}^+ = \{s\colon s \geq 0\}$$

Here we used the letters t, s, and g instead of x, y, and f to denote the independent variable, the dependent variable, and the function. The symbol used to denote the function is unimportant. It's what the function *does* that is important.

In Figure 5 we depict some of the values taken by this square root function. ■

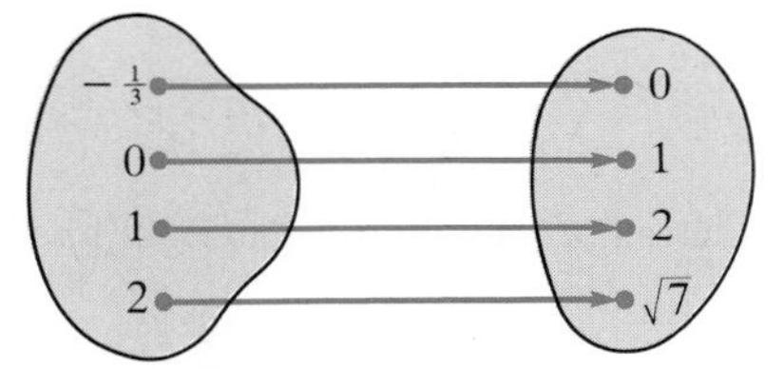

Figure 5 Some values of the function $g(t) = \sqrt{3t+1}$.

EXAMPLE 8 *A Quadratic Function*

Let $y = f(x) = x^2 + 4x + 5$.

(a) Evaluate $f(1)$, $f(-6)$, $f(4)$, and $f(x^3)$.
(b) Find the domain of f.
(c) Find the range of f.

SOLUTION

(a) $f(1) = 1^2 + 4 \cdot 1 + 5 = 10$
$f(-6) = (-6)^2 + 4(-6) + 5 = 17$
$f(4) = 4^2 + 4 \cdot 4 + 5 = 37$
$f(x^3) = (x^3)^2 + 4(x^3) + 5 = x^6 + 4x^3 + 5$

(b) The domain of $x^2 + 4x + 5$ is $\mathbb{R}$ because $x^2 + 4x + 5$ is defined for all real x.

(c) Suppose that y^* is in the range of f. Then, for some real number x,

$$y^* = x^2 + 4x + 5$$
$$= (x^2 + 4x + 4) + 1 = (x + 2)^2 + 1 \geq 1 \quad \text{since } (x + 2)^2 \geq 0$$

Therefore, $y^* \geq 1$. Thus every number in the range of f is greater than or equal to 1. Moreover, if $y^* \geq 1$, we can find an x such that $y^* = x^2 + 4x + 5 = (x + 2)^2 + 1$. We work backward:

$$y^* = (x + 2)^2 + 1$$
$$(x + 2)^2 = y^* - 1$$
$$x + 2 = \pm\sqrt{y^* - 1}$$
$$x = -2 \pm \sqrt{y^* - 1}$$

That is, if $y^* > 1$, there are two values of x for which $y^* = (x + 2)^2 + 1$. For example, if $y^* = 10$, then

$$x_1 = -2 + \sqrt{10 - 1} = -2 + 3 = 1 \quad \text{and} \quad x_2 = -2 - 3 = -5$$

satisfy

$$(x_1 + 2)^2 + 1 = (x_2 + 2)^2 + 1 = 10 \qquad \text{(check this)}$$

We have shown that

$$\text{range of } x^2 + 4x + 5 \text{ is } \{y\colon y \geq 1\} = [1, \infty) \quad ■$$

EXAMPLE 9 ***A Cubic Function***

Let $y = f(x) = x^3$.

(a) Compute $f(0)$, $f(2)$, $f(-2)$, $f(t^2)$, and $f\left(\frac{1}{t}\right)$.

(b) Find the domain of f.

(c) Find the range of f.

SOLUTION

(a) $f(0) = 0^3 = 0 \qquad f(2) = 2^3 = 8 \qquad f(-2) = (-2)^3 = -8$

$$f(t^2) = (t^2)^3 = t^6 \qquad f\left(\frac{1}{t}\right) = \left(\frac{1}{t}\right)^3 = \frac{1}{t^3}$$

(b) x^3 is defined for every real number x, so the domain is $\mathbb{R}$.

(c) If y^* is given, then $(\sqrt[3]{y^*})^3 = y^*$, so every real number is in the range. Note that the cube root of a negative number is negative, whereas the square root of a negative number is not defined.

WARNING Avoid the following common error in function evaluation. Let $f(x) = x^2 + 5x - 7$. Evaluate $f(2x)$.

Correct	Incorrect	
$f(2x) = (2x)^2 + 5(2x) - 7$	$f(2x) = 2x^2 + 5(2x) - 7$	$2x$ must be squared
$= 4x^2 + 10x - 7$	$= 2x^2 + 10x - 7$	$(2x)^2 \neq 2x^2$ ■

EXAMPLE 10 ***Domain Restricted by Division by Zero***

Let $y = f(x) = \dfrac{1}{x-3}$. Find (a) the domain of f and (b) the range of f.

SOLUTION

(a) $\dfrac{1}{x-3}$ is not defined if $x = 3$ because we cannot divide by 0. Thus,

$$\text{domain of } \frac{1}{x-3} \text{ is } \{x \colon x \neq 3\}$$

(b) Let $y^* = \dfrac{1}{x-3}$. We solve for x:

$$\begin{aligned} xy^* - 3y^* &= 1 && \text{Multiply by } (x-3) \\ xy^* &= 1 + 3y^* \\ x &= \frac{1 + 3y^*}{y^*} \end{aligned}$$

This is defined if $y^* \neq 0$. Thus

$$\text{range of } \frac{1}{x-3} \text{ is } \{y \colon y \neq 0\} \quad \blacksquare$$

EXAMPLE 11 ***A Constant Function***

Let $y = f(x) = 2$. Find

(a) $f(0)$, $f(-7)$, and $f(3.14)$. (b) domain of f. (c) range of f.

SOLUTION $f(x) = 2$ means that $f(x)$ is equal to 2 for every real number x. f is called a **constant function.** We have

(a) $f(0) = f(-7) = f(3.14) = 2$.
(b) domain of f is $\mathbb{R}$.
(c) range of f is 2 (this is the only image value). ■

EXAMPLE 12 ***A Function Defined in Pieces***

A news vendor buys an out-of-town newspaper from a distributor. He must pay 20¢ (= \$0.20) per paper if he buys 100 or fewer papers, but only 18¢ (= \$0.18) per paper if he buys more than 100. Then his total price, $p(q)$, is

$$p(q) = \begin{cases} 0.20q, & 0 \leq q \leq 100 \\ 0.18q, & q > 100 \end{cases}$$

Thus,

$$\text{domain of } p \text{ is } \{\text{nonnegative integers}\}$$

and

$$\text{range of } p \text{ is } \{0, 0.2, 0.4, \cdots, 20, 0.18(101), 0.18(102), \cdots\}$$

In all the examples considered so far, we gave you a function and asked questions about it. However, not everything that looks like a function is actually a function. We illustrate this in the next example. Remember that to have a function, there must be a unique value of $f(x)$ for each x in the domain of f.

EXAMPLE 13 *A Relation That Does Not Define a Function*

Consider the equation of a circle of radius 2 (see p. 154):

$$x^2 + y^2 = 4$$

Can we write y as a function of x? Let us try to solve for y:

$$y^2 = 4 - x^2$$
$$y = \pm\sqrt{4 - x^2}$$

That is, if x is given (and $x^2 < 4$), then there are two values of y that satisfy $x^2 + y^2 = 4$. For example, if $x = 1$, then $y = \sqrt{3}$ and $y = -\sqrt{3}$ satisfy $x^2 + y^2 = 4$. But, in order to have a function, there must be a *unique* y for every x in the domain. Thus the relation $x^2 + y^2 = 4$ does *not* define a function. ■

EXAMPLE 14 *A Relation That Does Define a Function*

Let

$$x^3 + y^3 = 4$$

We can solve for y:

$$y^3 = 4 - x^3$$
$$y = \sqrt[3]{4 - x^3}$$

If $4 - x^3 \geq 0$, then $y \geq 0$; and if $4 - x^3 < 0$, then $y < 0$. Thus, for every x there is a unique y such that $x^3 + y^3 = 4$. This defines a function with domain and range equal to $\mathbb{R}$.

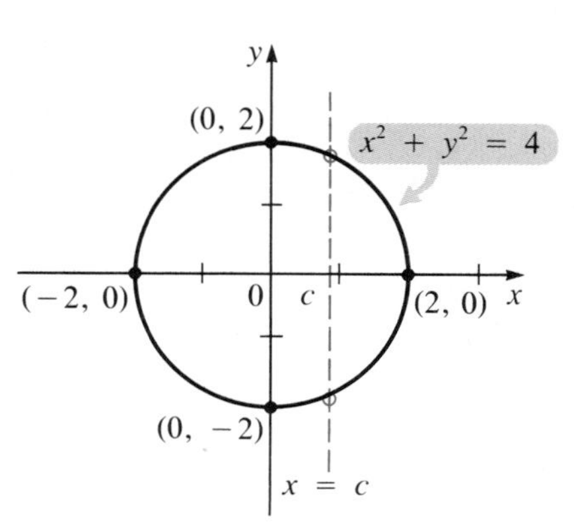

Figure 6 The equation $x^2 + y^2 = 4$ does not define a function because some vertical lines intersect its graph at two points.

The Vertical Lines Test

In Examples 13 and 14 we saw that an equation in two variables may or may not define a function. There is a graphical test to determine this. Consider the graph of $x^2 + y^2 = 4$ in Figure 6.

We draw the vertical line $x = c$ with c in the interval $(-2, 2)$. This line intersects the circle at *two* points. This means that for every c in $(-2, 2)$ there are two y's such that $c^2 + y^2 = 4$. Thus we cannot write y as a function of x.

Vertical Lines Test

A set of points in the Cartesian plane is the graph of a function if every vertical line in the plane intersects the set of points in at most one point.

We summarize the reasons why the domain of a function may be restricted.

Reasons Why the Domain of a Function May Not Be $\mathbb{R}$

(a) Division by zero is not allowed (see Example 10).
(b) Even roots (square roots, fourth roots, and so on) of negative numbers do not exist as real numbers (see Example 7).
(c) The domain is restricted by the nature of the problem under consideration (see Example 12).

You will see a fourth kind of restriction when we discuss logarithmic functions in Section 6.3.

In Section 3.5 we will discuss the graphs of functions.

EXAMPLE 15 ***An Example That Is Important in Calculus***

Let $f(x) = x^2$. Compute $\dfrac{f(x + \Delta x) - f(x)}{\Delta x}$ and simplify your answer. Assume that $\Delta x \neq 0$.

SOLUTION

$f(x + \Delta x) = (x + \Delta x)^2 = x^2 + 2x\Delta x + \Delta x^2$ and $f(x) = x^2$. Thus

Factor out Δx

$$\frac{f(x+\Delta x) - f(x)}{\Delta x} = \frac{(x^2 + 2x\Delta x + \Delta x^2) - x^2}{\Delta x} = \frac{2x\Delta x + \Delta x^2}{\Delta x} = \frac{\Delta x(2x + \Delta x)}{\Delta x}$$
$$= 2x + \Delta x$$

EXAMPLE 16 ***Another Example Useful in Calculus***

Let $f(x) = \dfrac{1}{x}$. Compute $\dfrac{f(x + \Delta x) - f(x)}{\Delta x}$ and simplify your answer. Assume that $\Delta x \neq 0$.

SOLUTION $f(x + \Delta x) = \dfrac{1}{x + \Delta x}$ and $f(x) = \dfrac{1}{x}$ so

Multiply numerator and denominator by $x(x + \Delta x)$

$$\frac{f(x + \Delta x) - f(x)}{\Delta x} = \frac{\dfrac{1}{x + \Delta x} - \dfrac{1}{x}}{\Delta x} = \frac{\dfrac{x(x + \Delta x)}{x + \Delta x} - \dfrac{x(x + \Delta x)}{x}}{\Delta x(x)(x + \Delta x)}$$
$$= \frac{x - (x + \Delta x)}{\Delta x(x)(x + \Delta x)} = \frac{-\Delta x}{\Delta x(x)(x + \Delta x)}$$
$$= \frac{-1}{x(x + \Delta x)}$$

Definition of Even and Odd Functions

A function $f(x)$ is called **even** if

$$f(-x) = f(x)$$

and **odd** if

$$f(-x) = -f(x)$$

EXAMPLE 17 *An Even Function and an Odd Function*

$f(x) = x^2$ is even because $f(-x) = (-x)^2 = x^2 = f(x)$.
$f(x) = x^3$ is odd because $f(-x) = (-x)^3 = (-1)^3x^3 = -x^3 = -f(x)$. ■

EXAMPLE 18 *Determining Whether a Function Is Even or Odd*

Determine whether each function is even, odd, or neither.

(a) $\sqrt[3]{x}$ (b) $\sqrt{x}$ (c) $x^4 + x^5$ (d) $x^4 - x^6$

SOLUTION

(a) $f(-x) = \sqrt[3]{-x} = \sqrt[3]{(-1)x} = \sqrt[3]{-1}\sqrt[3]{x} = (-1)\sqrt[3]{x} = -\sqrt[3]{x} = -f(x)$, so $\sqrt[3]{x}$ is odd.
(b) $f(-x) = \sqrt{-x}$ is not defined if $x > 0$, so $\sqrt{x}$ is neither even nor odd.
(c) If $f(x) = x^4 + x^5$, then $f(-x) = (-x)^4 + (-x)^5 = x^4 - x^5$. But $-f(x) = -x^4 - x^5$. Thus $f(-x) \neq f(x)$ and $f(-x) \neq -f(x)$, so $f(x) = x^4 + x^5$ is neither even nor odd.
(d) If $f(x) = x^4 - x^6$, then $f(-x) = (-x)^4 - (-x)^6 = x^4 - x^6 = f(x)$, so $f(x) = x^4 - x^6$ is even.

Problems 3.3

Readiness Check

I. Which of the following belongs to the range of $f(x) = -|x - 2|$?
a. 2 b. 1 c. -2
d. None of these because $f(x)$ is not a function.

II. Which of the following is the interval notation for the range of $f(x) = \sqrt{x + 2}$?
a. $[0, \infty)$ b. $[-2, \infty)$ c. $(-\infty, \infty)$ d. $(-\infty, 2]$

III. For which of the following is y a function of x?
a. $y = |x + 7|$
b. $x + y^2 = 4$
c. $y = f(x) = \begin{cases} x - 1, & \text{for } x \geq 2 \\ -4, & \text{for } x \leq 2 \end{cases}$
d. All of the above are functions.

IV. Which of the following is the graph of a function?

a.
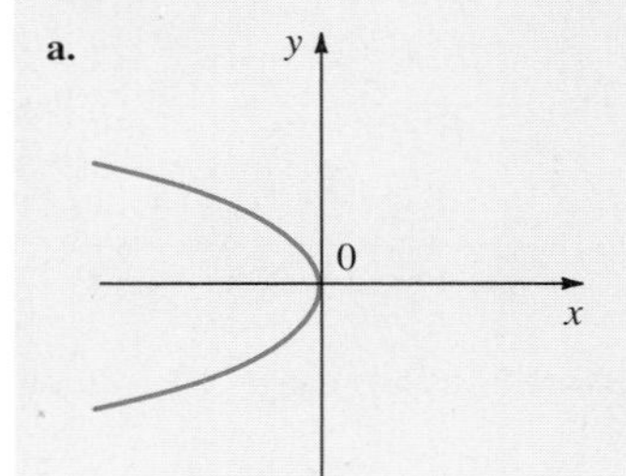

b.
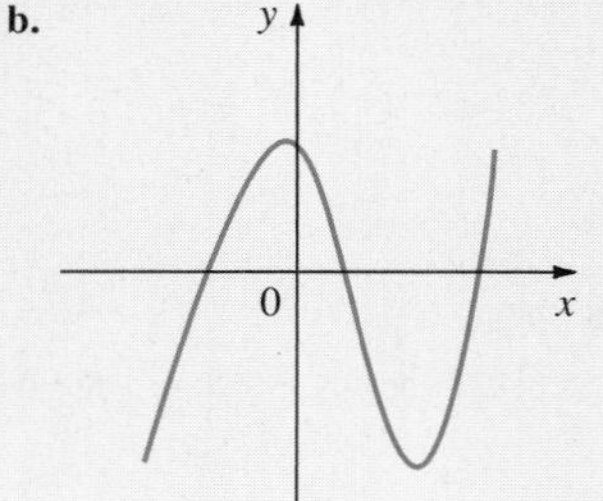

c.
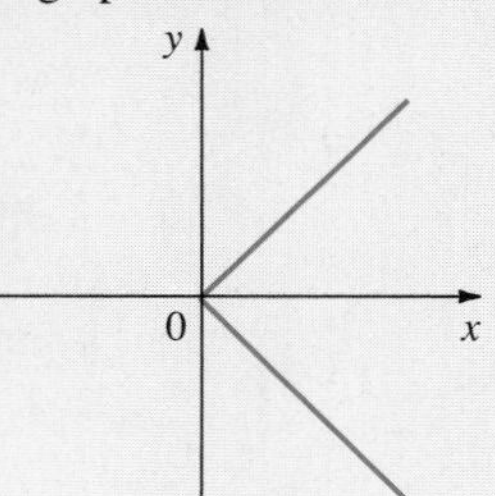

d.
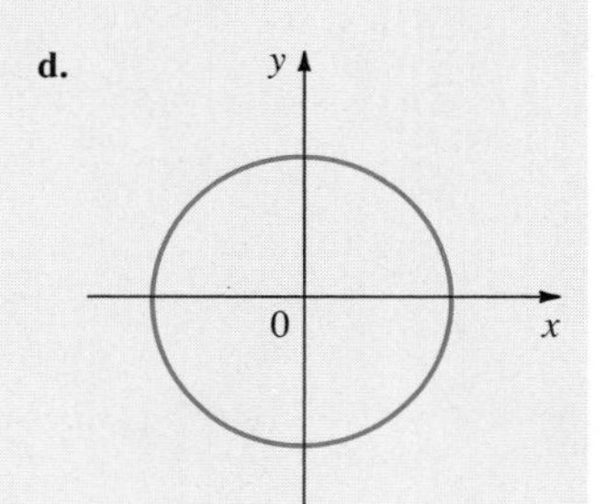

V. Which of the following has domain $\{x: x \geq 0\} = \mathbb{R}^+$?

a. $f(x) = x^2$ b. $f(x) = \dfrac{1}{x}$ c. $f(x) = \dfrac{1}{x^2}$ d. $f(x) = \sqrt{x}$

In Problems 1–10 evaluate the given function at the given values.

1. $f(x) = 1/(1 + x)$; $f(0)$, $f(1)$, $f(-2)$, $f(-5)$, $f(x^2)$, and $f(\sqrt{x})$
2. $f(x) = 1 + \sqrt{x}$; $f(0)$, $f(1)$, $f(16)$, $f(25)$, $f(y^4)$, and $f\left(\dfrac{1}{y}\right)$
3. $f(x) = 2x^2 - 1$; $f(0)$, $f(2)$, $f(-3)$, $f\left(\dfrac{1}{2}\right)$, $f(\sqrt{w})$, and $f(w^5)$
4. $f(x) = \dfrac{1}{2x^3}$; $f(1)$, $f(\frac{1}{2})$, $f(-3)$, $f(1 + t)$, and $f(t^2 - 2)$
5. $f(x) = x^4$; $f(0)$, $f(2)$, $f(-2)$, $f(\sqrt{5})$, $f(s^{1/5})$, and $f(s - 1)$
6. $g(t) = t/(t - 2)$; $g(0)$, $g(1)$, $g(-1)$, $g(3)$, $g(v^2)$, and $g(v + 5)$
7. $g(t) = \sqrt{t + 1}$; $g(0)$, $g(-1)$, $g(3)$, $g(7)$, $g(n^3 - 1)$, and $g\left(\dfrac{1}{w}\right)$
8. $h(z) = \sqrt[3]{z}$; $h(0)$, $h(8)$, $h(-\frac{1}{27})$, $h(1000)$, $h(x^{300})$, and $h(-8p)$
9. $h(z) = z^2 - z + 1$; $h(0)$, $h(2)$, $h(10)$, $h(-5)$, $h(n^2)$, and $h\left(\dfrac{1}{n^3}\right)$
10. $h(z) = z^3 + 2z^2 - 3z + 5$; $h(0)$, $h(1)$, $h(-1)$, $h(2)$, $h(t^{1/4})$, and $h(t + 1)$

In Problems 11–20 which of the rules depicted in the graphs are functions with each domain equal to D? If the rule is not a function, explain why it is not.

11.

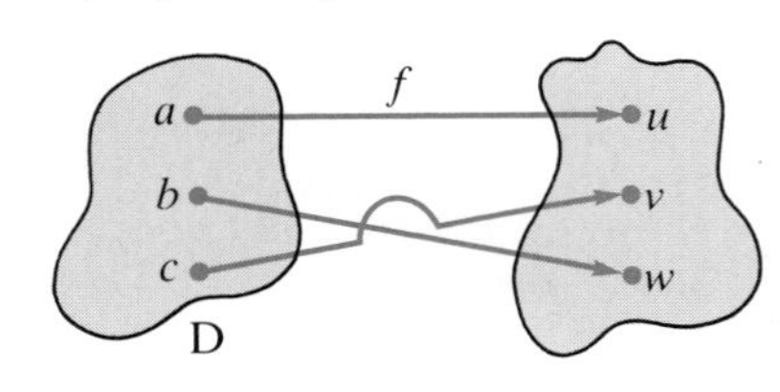

12.

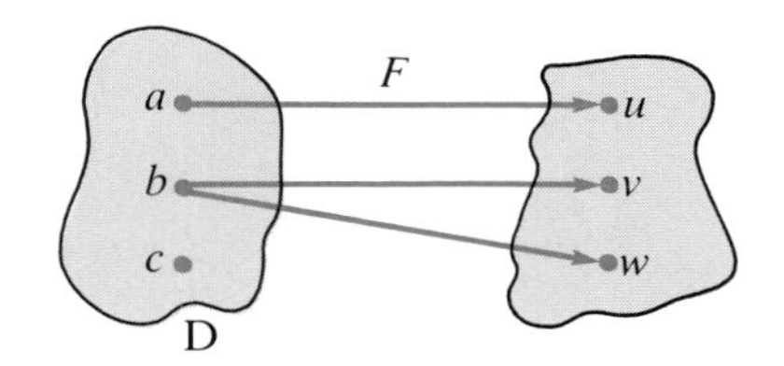

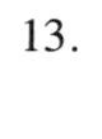

13.

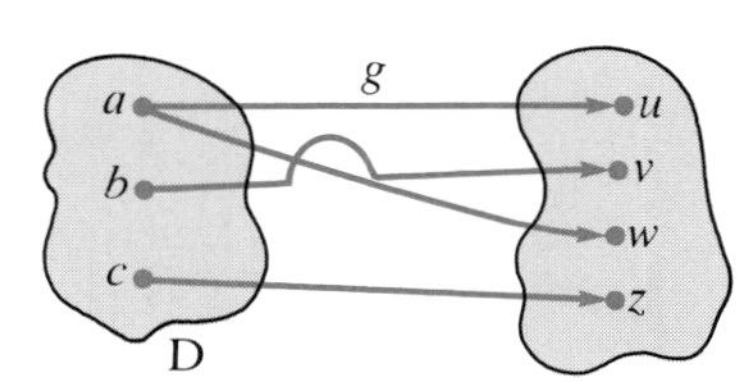

14.

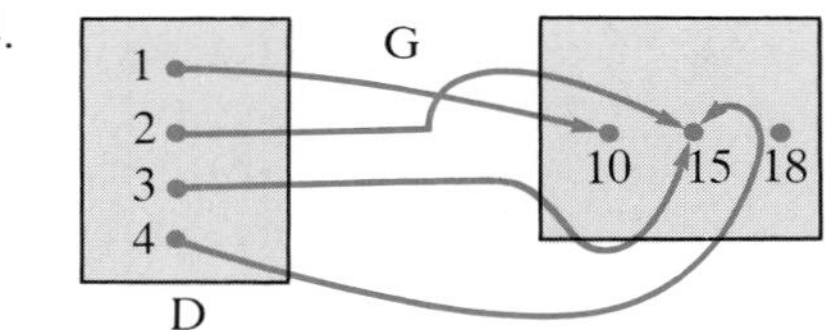

15.

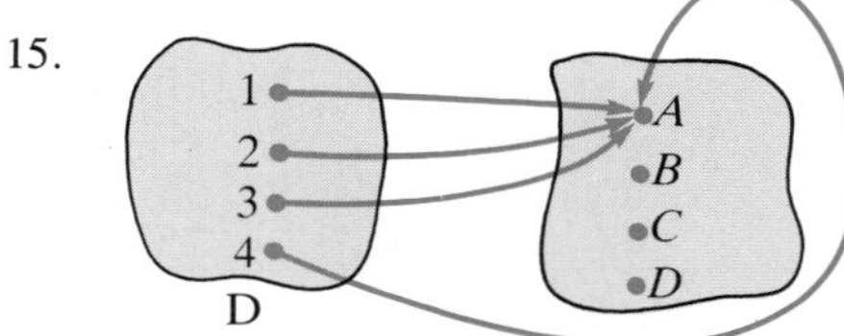

16.

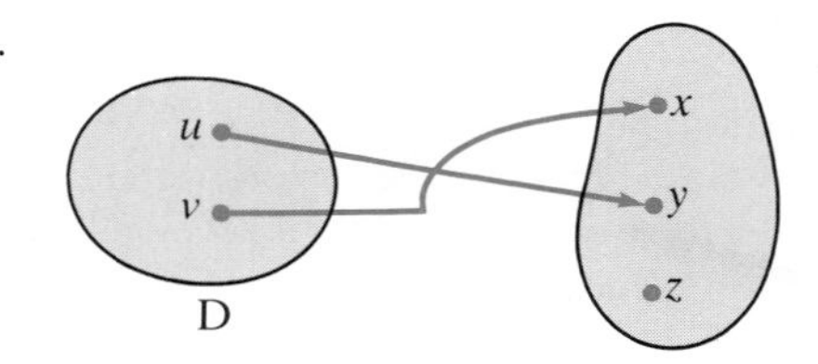

Answers to Readiness Check

I. c II. a III. a [The rule in c is not a function because 2 has two images.] IV. b V. d

17.

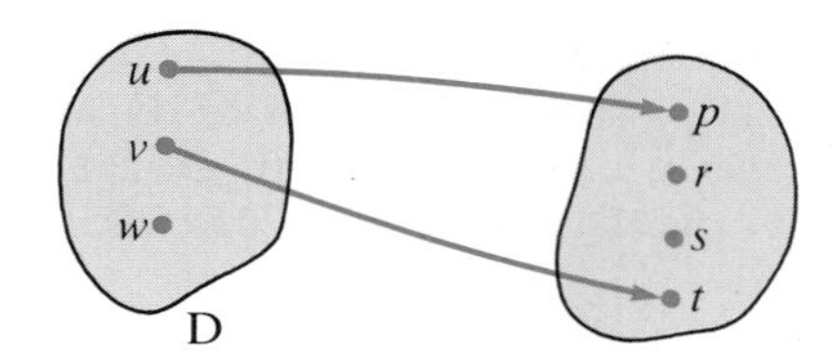

18.

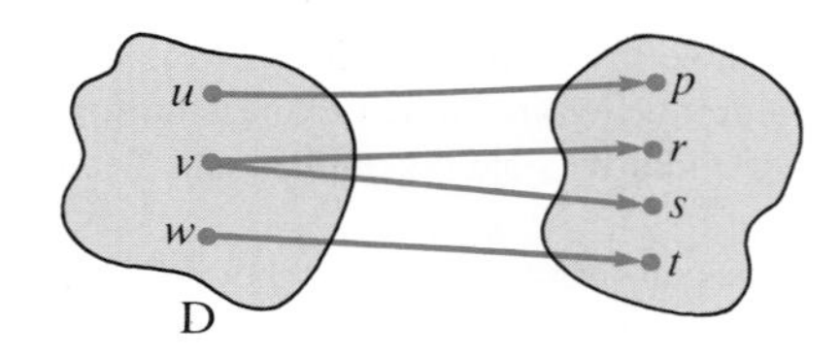

19.

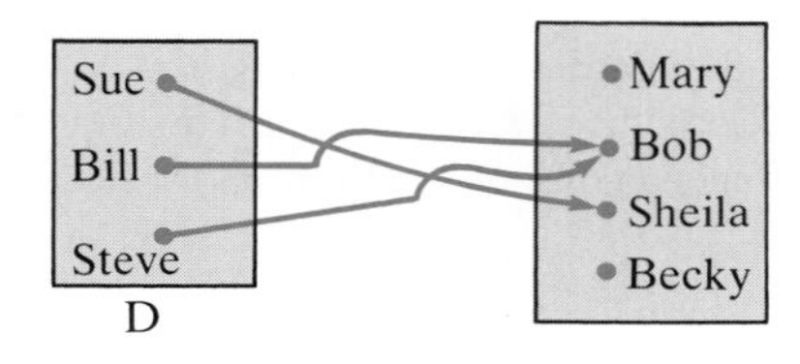

20.

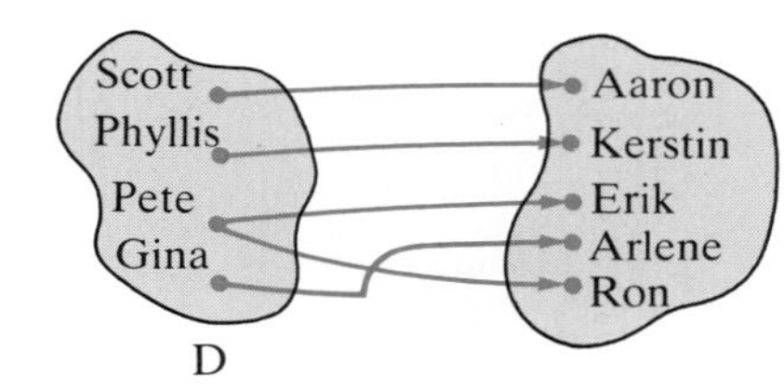

In Problems 21–32 an equation involving x and y is given. Determine whether y can be written as a function of x.

21. $2x + 3y = 6$
22. $\dfrac{x}{y} = 2$
23. $x^2 + 2y = 5$
24. $x - 3y^2 = 4$
25. $4x^2 + 2y^2 = 4$
26. $x^2 - y^2 = 1$
27. $\sqrt{x + y} = 1$
28. $y^2 + xy + 1 = 0$ [Hint: Use the quadratic formula.]
29. $y^3 - x = 0$
30. $y^4 - x = 0$
31. $y = |x|$
32. $y^2 = \dfrac{x}{x + 1}$
33. Explain why the equation $y^n - x = 0$ allows us to write y as a function of x if n is an odd integer but does not if n is an even integer. [Hint: First solve Problems 29 and 30.]

In Problems 34–43 the graph of an equation in two variables is given. Use the vertical lines test to determine whether the equation determines y as a function of x.

34.

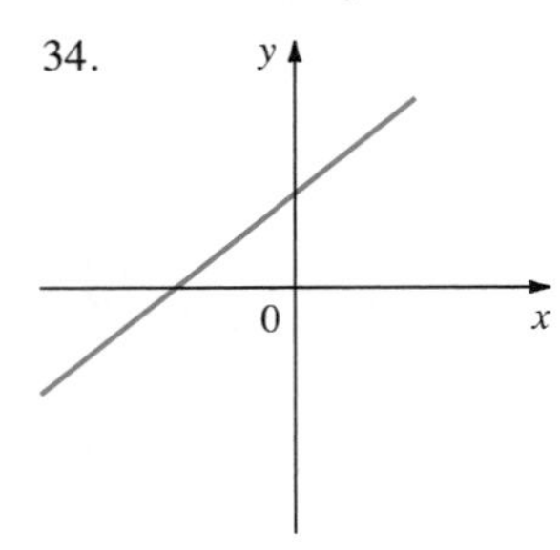

35.

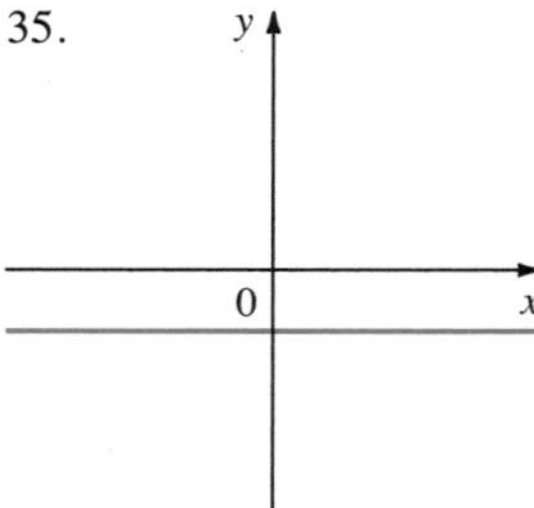

36.

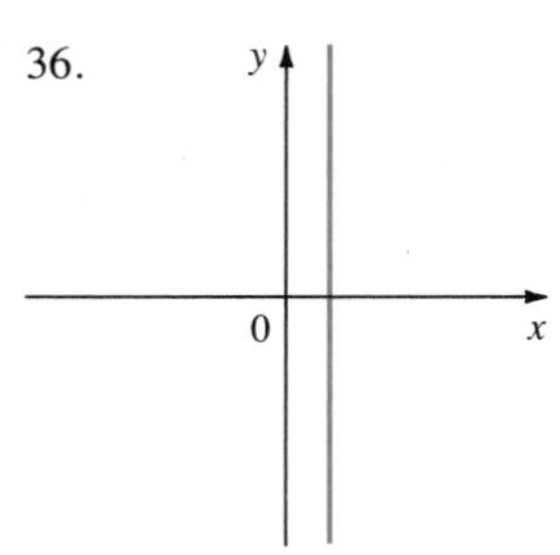

37.

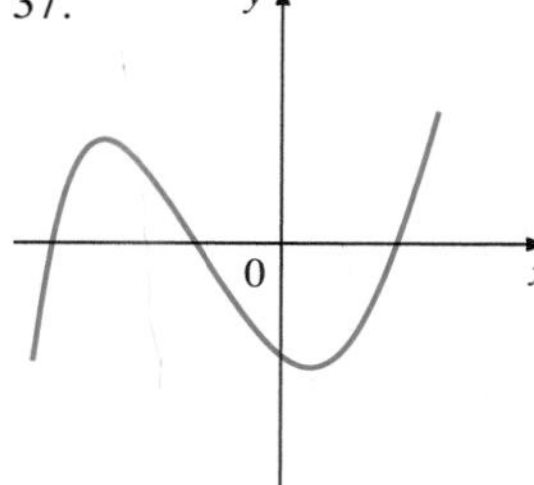

38.

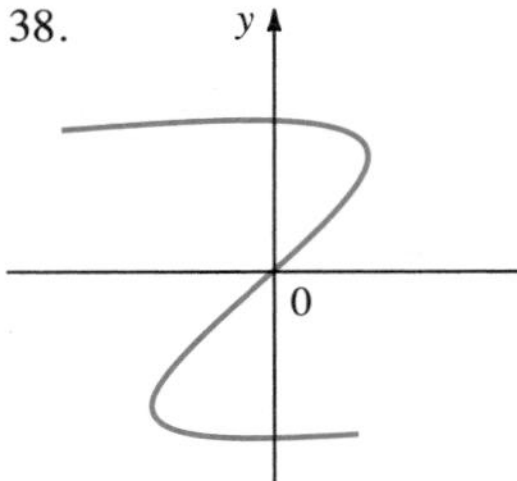

39.

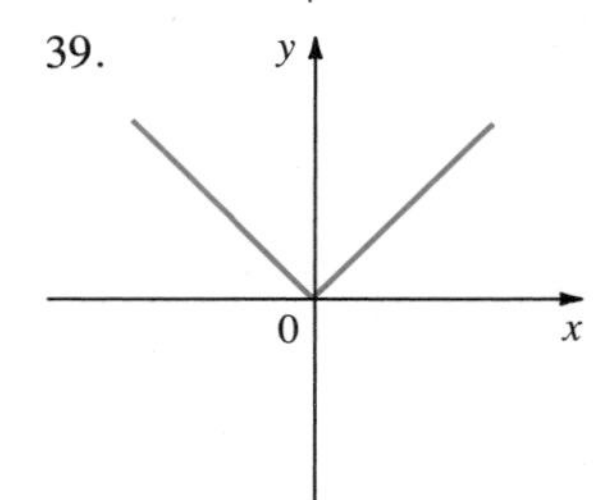

40.

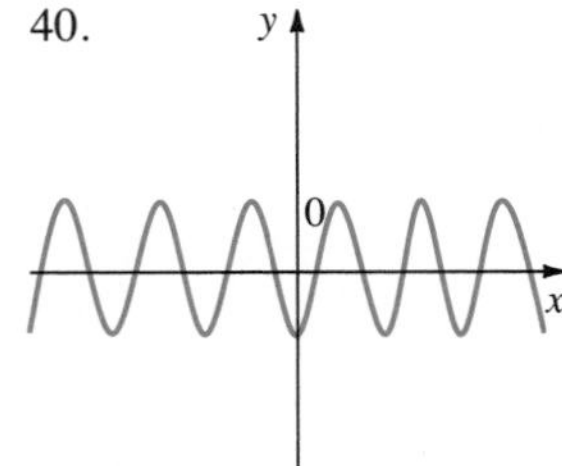

41.

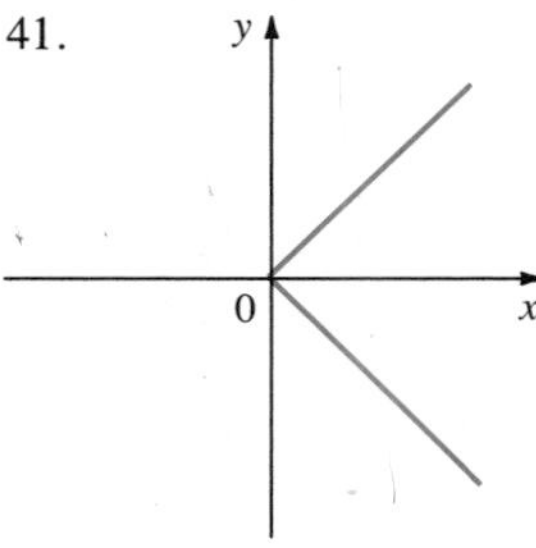

42.

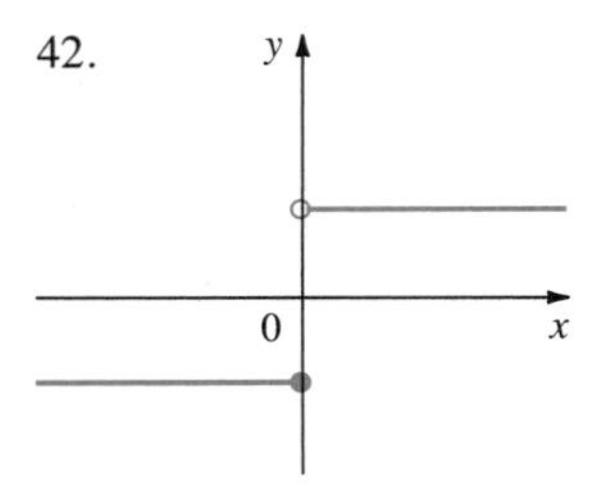

43.

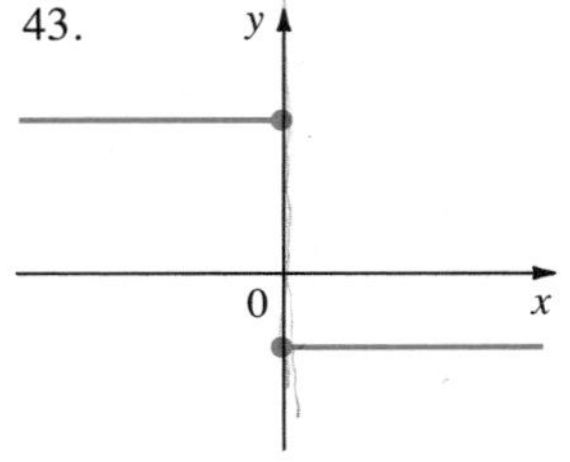

In Problems 44–57 find the domain and range of the given function.

44. $y = f(x) = 2x - 3$

45. $s = g(t) = 4t - 5$

46. $y = f(x) = 3x^2 - 1$

47. $v = h(u) = \dfrac{1}{u^2}$

48. $y = f(x) = x^3$

49. $y = f(x) = \dfrac{1}{x+1}$

50. $s = g(t) = t^2 + 2t + 1$

51. $y = f(x) = \sqrt{x^3 - 1}$

52. $v = h(u) = |u - 2|$

53. $y = f(x) = \dfrac{1}{|x|}$

54. $h(x) = \sqrt{4 - x^2}$

55. $y = \begin{cases} 2x, & x \geq 0 \\ -x, & x < 0 \end{cases}$

56. $y = \begin{cases} x, & x \geq 1 \\ 1, & x < 1 \end{cases}$

57. $y = \begin{cases} x^3, & x > 0 \\ x^2, & x \leq 0 \end{cases}$

58. The **greatest integer function** $g(x) = [x]$ is defined by $[x]$ = the largest integer less than or equal to x. Compute
(a) $[3.1]$ (b) $[\pi]$ (c) $[-2.7]$ (d) $[16]$
(e) $\left[\dfrac{17}{9}\right]$ (f) $\left[-\dfrac{10}{3}\right]$

59. Find the domain and range of $g(x) = [x]$.

In Problems 60–63 use a calculator to evaluate the given function at the given values.

60. $f(x) = 1.25x^2 - 3.74x + 14.38$; $f(2.34)$, $f(-1.89)$, $f(10.6)$

61. $g(t) = t^3 - 0.74t^2 + 0.756t + 1.302$; $g(0.18)$, $g(3.95)$, $g(-11.62)$

62. $h(z) = \dfrac{z+3}{z^2-4}$; $h(38.2)$, $h(57.9)$, $h(238.4)$

63. $f(x) = \dfrac{x - 1.6}{x + 3.4} + \dfrac{x^2 + 5.8}{6.2 - x^2}$; $f(5.8)$, $f(-23.4)$

64. Let $f(x) = \dfrac{1}{x-1}$. Find $f(t^2)$ and $f(3t + 2)$.

Two Exercises Useful in Calculus

65. Let $f(x) = x^3$. Find $f(x + \Delta x)$ and $\dfrac{f(x + \Delta x) - f(x)}{\Delta x}$, where x denotes an arbitrary real number. Simplify your answer. This computation is very important in calculus.

66. Let $f(x) = \sqrt{x}$. Show, assuming that $\Delta x \neq 0$, that

$$\frac{f(x + \Delta x) - f(x)}{\Delta x} = \frac{1}{\sqrt{x + \Delta x} + \sqrt{x}}$$

[Hint: Multiply and divide by $\sqrt{x + \Delta x} + \sqrt{x}$. That is, rationalize the numerator (see p. 64).]

67. Let $f(x) = |x|/x$. Show that $f(x) = \begin{cases} 1, & x > 0 \\ -1, & x < 0 \end{cases}$

Find the domain and range of f.

68. Describe a computational rule for expressing Fahrenheit temperature as a function of Centigrade temperature. [Hint: Pure water at sea level boils at 100°C and 212°F; it freezes at 0°C and 32°F. The graph of this function is a straight line.]

* 69. Consider the set of all rectangles whose perimeters are 50 cm. Once the width W of any one rectangle is measured, it is possible to compute the area of the rectangle. Verify this by producing an explicit expression for area A as a function of width W. Find the domain and the range of your function.

* 70. A spotlight shines on a screen; between them is an obstruction that casts a shadow. Suppose the screen is vertical, 20 m wide by 15 m high, and 50 m from the spotlight. Also suppose the obstruction is a square, 1 m on a side, and is parallel to the screen. Express the area of the shadow as a function of the distance from the light to the obstruction.

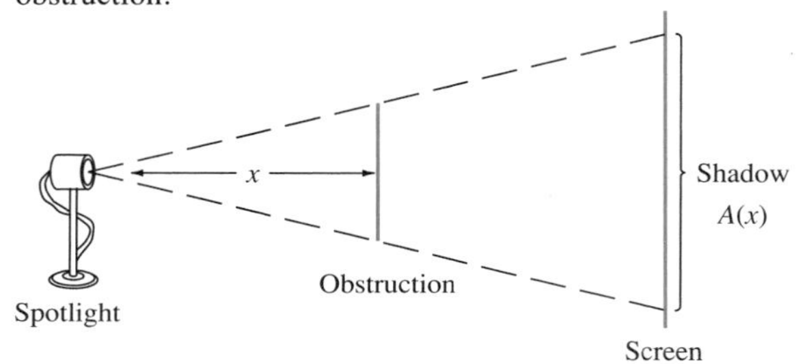

* 71. A baseball diamond is a square, 90 feet long on each side. Casey runs a constant 30 ft/sec whether he hits a ground ball or a home run. Today, in his first at-bat, he hit a home run. Write an expression for the function that measures his line-of-sight distance from second base as a function of the time t in seconds after he left home plate.

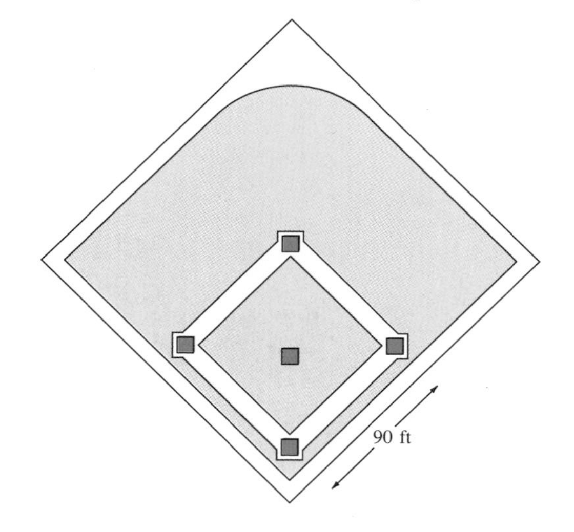

72. Let $f(x)$ be the fifth decimal place of the decimal expansion of x. For example, $f(\frac{1}{64}) = f(0.015624) = 2$, $f(98.786543210) = 4$, $f(-78.90123456) = 3$, and so on. Find the domain and range of f. [Note: Since, for example, $1.000\ldots = 0.9999\ldots$, assume, to avoid ambiguity, that every integer n is written as $1.000\ldots$, $2.000\ldots$, and so on.]

73. The Dow Jones closing averages for industrial stocks are given for the 3-month period from April 15 to July 15, 1983. Let April 18 (which was a Monday) be day 1 and July 15 be day 92. Let $A(t)$ be the closing average on day t. Find (a) $A(1)$, (b) $A(8)$, (c) $A(30)$, (d) $A(60)$, and (e) $A(88)$. [Hint: Each vertical bar represents a business day.]

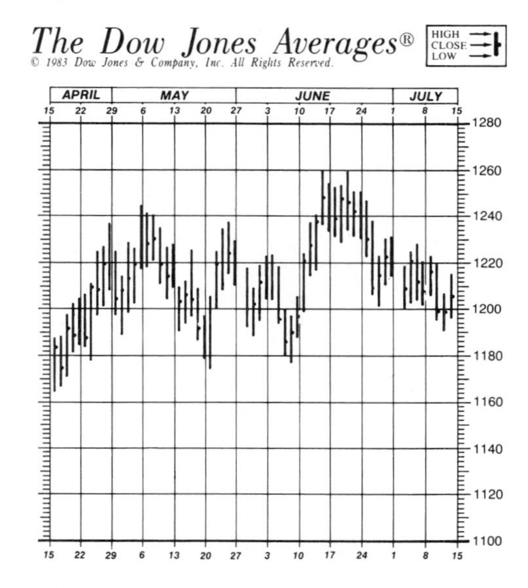

* 74. Alec, on vacation in Canada, found that he got a 25% premium on his U.S. money. When he returned, he discovered there was a 25% discount on converting his Canadian money back into U.S. currency. Describe each conversion function. Show that, after converting both ways, Alec lost money.

In Problems 75–92 determine whether the given function is even, odd, or neither.

75. $f(x) = x^2 - 1$
76. $f(x) = x^2 + x$
77. $f(x) = x^4 + x^2$
78. $f(x) = x^3 - x$
79. $f(x) = x^4 + x^2 + 2x$
80. $f(x) = \dfrac{1}{x}$
81. $f(x) = \dfrac{1}{x^2}$
82. $f(x) = \dfrac{1}{x+1}$
83. $f(x) = x^4 - x^2 + 2$
84. $f(x) = \dfrac{x}{x^3+1}$
85. $f(x) = \dfrac{x^2-2}{x^2+5}$
86. $f(x) = x^{1/3} - x^{1/5}$
87. $f(x) = x^{3/5}$
88. $f(x) = x^{17/4}$
89. $f(x) = \dfrac{x^3}{x^2+12}$
90. $f(x) = x^2 + x + 1$
91. $f(x) = x^3 + x + 1$
92. $f(x) = x^4 + 17x^2 - 5$

3.4 Applications of Linear Functions

Simple Interest Problems

Linear functions arise from the **simple interest formula.** This formula was given on page 79. We repeat it here.

Simple Interest Formula

$$I = Prt \tag{1}$$

where I is the interest earned, P is the **principal** or the amount invested, r is the **rate** of interest (almost always a number between 0 and 1), and t is the **time** the investment is held (usually measured in years).

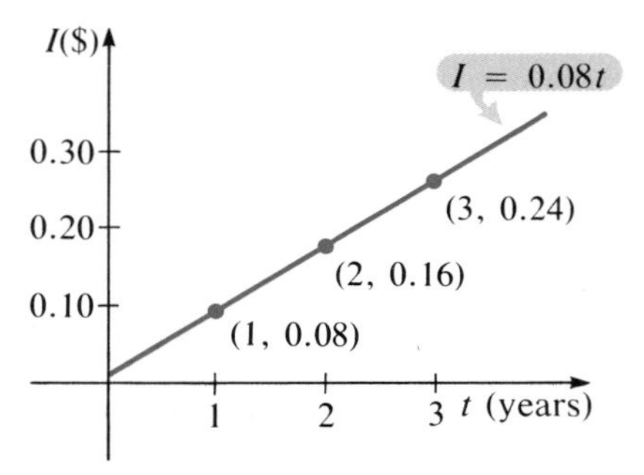

Figure 1 The graph of the equation $I = 0.08t$ is a straight line with slope 0.08.

EXAMPLE 1 *The Graph of I = rt Is a Straight Line*

Suppose that \$1 is invested for t years at an interest rate of r. Then, from (1) the simple interest paid is a function of t:

$$I = rt$$

The graph of this function is a straight line with slope r. For example, if $r = 8\% = 0.08$, we obtain the graph in Figure 1. ■

EXAMPLE 2 *Determining the Break-Even Point*

Consider the case of a shoelace manufacturer whose total revenue function is given by

$$R(q) = 0.50q$$

(that is, q is the number of items sold and revenue is 50¢ per item) and whose total cost function is

$$C(q) = 2000 + 0.34q$$

Profit, P, is given by

$$P(q) = R(q) - C(q) \tag{2}$$

Determine the **break-even point,** which is defined as the number of items sold for which profit is zero.

SOLUTION If $P(q) = 0$, then $R(q) = C(q)$. Since the graphs of R and C are straight lines, we need to find the point of intersection of the lines to find the value q for which $R = C$. Setting $R(q) = C(q)$, we have

$$0.50q = 2000 + 0.34q$$
$$0.16q = 2000$$

or $q = 2000/(0.16) = 12{,}500$. The functions R and C are sketched in Figure 2.

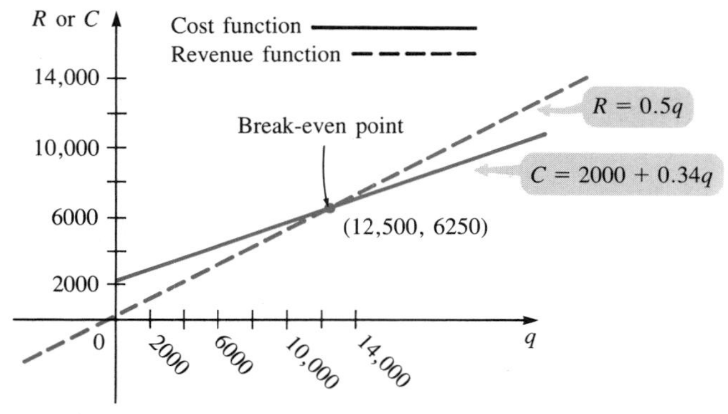

Figure 2 The break-even point is the point of intersection of the graphs of the cost and revenue functions.

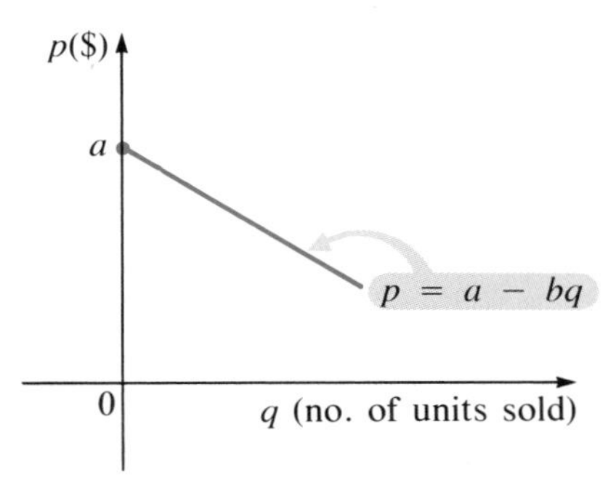

Figure 3 Typical demand function.

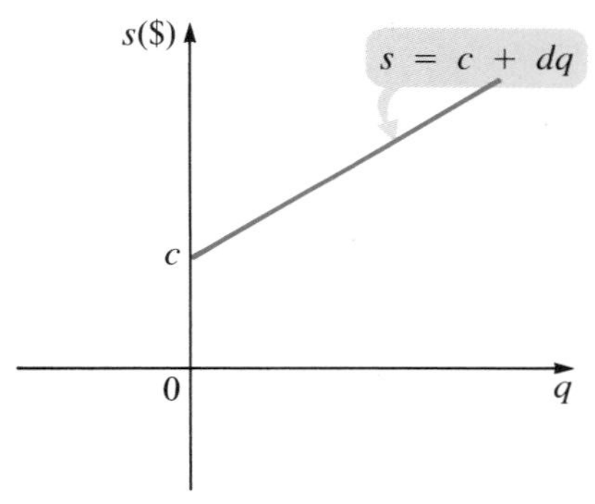

Figure 4 Typical supply function.

Demand and Supply Functions and Market Equilibrium

In economics a **demand function** expresses the relationship between the unit price that a product can sell for and the number of units, q, that can be sold for that price. Typically, the more units sold, the lower the price. A typical demand function is (see Figure 3)

$$p(q) = a - bq, \qquad b > 0 \tag{3}$$

A **supply function** expresses the relationship between the *expected* price, s, of a product and the number of units, q, the manufacturer will produce. It is reasonable to assume that as the expected price increases, the number of units the manufacturer will produce also increases, so q increases as s increases. A typical supply function is (see Figure 4)

$$s(q) = c + dq, \qquad d > 0 \tag{4}$$

The demand function gives price as a function of the number of units sold, and the supply function gives price as a function of the number of units produced. When the supply and demand are equal, we have achieved **market equilibrium.** The price at which the supply and demand curves intersect is called the **equilibrium price** (see Figure 5).

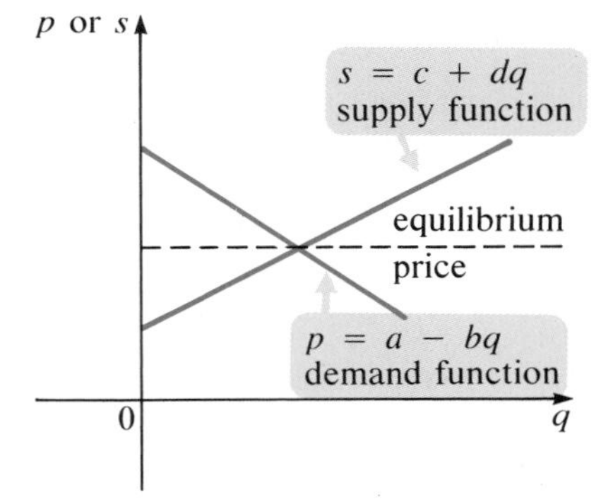

Figure 5 The equilibrium price is the price at the point of intersection of the supply and demand functions.

EXAMPLE 3 *Determining the Equilibrium Price*

A clothing manufacturer produces a certain type of western shirt. She estimates that her demand function is given by

$$p(q) = 35 - 0.02q \tag{5}$$

and her supply function is

$$s(q) = 20 + 0.01q \tag{6}$$

Determine the equilibrium price.

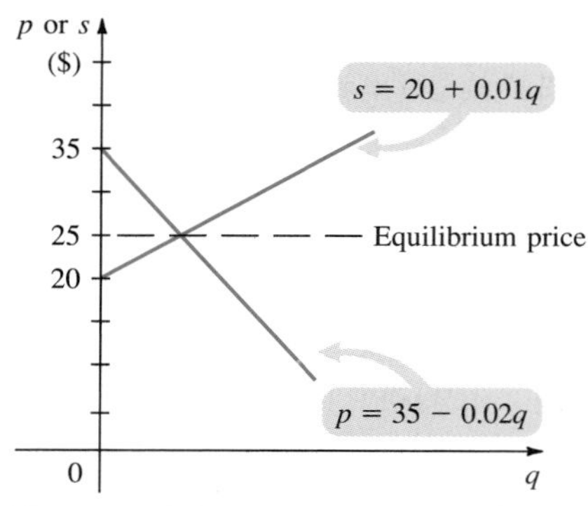

Figure 6 The equilibrium price is $25.

SOLUTION The two functions are sketched in Figure 6.

At equilibrium, $s(q) = p(q)$ or

$$35 - 0.02q = 20 + 0.01q$$
$$0.03q = 15$$
$$q = \frac{15}{0.03} = 500 \text{ units}$$

Then

$$p(500) = 35 - (0.02)500 = 35 - 10 = \$25$$

Check $s(500) = 20 + 0.01q = 20 + (0.01)500 = 20 + 5 = \$25$.

What does this answer mean? Suppose the price is higher than $25, say $30. The quantity supplied is obtained from (6):

$$30 = 20 + 0.01q$$
$$10 = 0.01q$$
$$q = \frac{10}{0.01} = 1000 \text{ shirts}$$

The quantity demanded is obtained from (5)

$$30 = 35 - 0.02q$$
$$0.02q = 5$$
$$q = \frac{5}{0.02} = 250 \text{ shirts}$$

Thus the manufacturer will have $1000 - 250 = 750$ shirts she can't sell. There is a surplus of shirts. The supply exceeds the demand, and the price will go down. On the other hand, suppose she charges less than the equilibrium price, say $22. Then

$$22 = 20 + 0.01q$$
$$2 = 0.01q$$
$$q = \frac{2}{0.01} = 200 \text{ shirts produced}$$

and

$$22 = 35 - 0.02q$$
$$0.02q = 13$$
$$q = \frac{13}{0.02} = 650 \text{ shirts demanded}$$

Thus there will be a deficit of $650 - 200 = 450$ shirts. The demand here exceeds the supply, and the price will go up. Only at market equilibrium will prices tend to remain stable. ■

EXAMPLE 4 *Athletic Performance as a Function of Altitude*†

One measure of the effectiveness of the cardiovascular system during exercise is the volume of oxygen, VO_2, carried in the blood. It has long been known that physical performance is reduced in altitudes above sea level. This is particularly evident at altitudes above 5000 feet. This reduction in performance is due to **hypoxia,** a decrease in the partial pressure of oxygen in the inspired air.

On the average, the maximum VO_2 for untrained individuals decreases about 3% for every 1000 feet above 5000 feet. For well-trained athletes, the rate of decrease in maximum VO_2 is less, but this decrease starts at sea level. The decrease is about 2% for every 1000 feet above sea level. If A denotes altitude, $\frac{A}{1000}$ represents the number of thousands of feet above sea level. In addition, if Pd denotes percent decrease, then

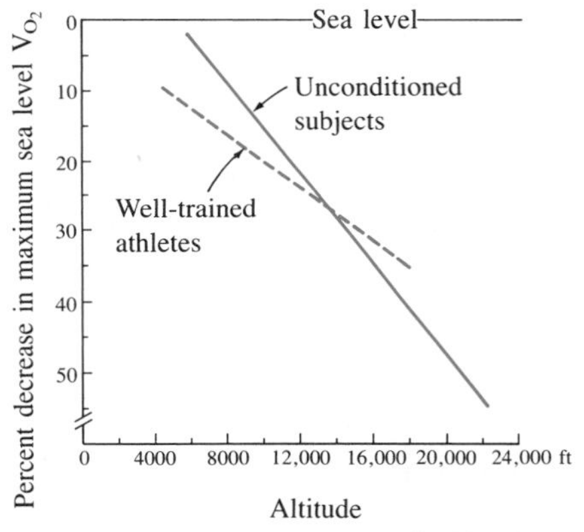

Figure 7 Percentage decrease in blood oxygen as a function of altitude for trained and untrained subjects.

for untrained individuals

$$Pd = -0.03\left(\frac{A}{1000} - \frac{5000}{1000}\right) = -0.00003(A - 5000),\ A \geq 5000 \qquad (7)$$

for trained athletes

$$Pd = -0.02\left(\frac{A}{1000}\right) = -0.00002A \qquad (8)$$

These graphs are sketched in Figure 7, where the vertical-axis values increase as we move downward to indicate that performance is impaired. Note that the numbers 0.00003 and 0.00002 represent the rates of decrease per foot.

At what altitude is the percentage decrease in performance the same for both trained and untrained individuals?

SOLUTION Equating the right-hand sides of (7) and (8), we obtain

$$-0.00003(A - 5000) = -0.00002A$$

$$-0.00003A + 0.15 = -0.00002A$$

$$0.15 = 0.00001A$$

$$A = \frac{0.15}{0.00001} = 15{,}000 \text{ feet}$$

At that altitude the percentage decrease for both trained and untrained individuals is

$$Pd = -0.00002(15{,}000) = 0.3 = 30\%$$

† Adapted from Edward L. Fox, *Sports Physiology,* 2d ed., Saunders, Philadelphia, 1984, pp. 194–195.

Problems 3.4

In Problems 1–10 sketch the linear function and determine its slope. What does the slope represent?

1. $I(t) = 0.12t$ 2. $I(t) = 500(0.07)t$
3. $I(t) = 1000(0.13)t$
4. $s(t) = 30t$ [Hint: In Problems 4, 5, and 6, t denotes time and s denotes distance.]
5. $s(t) = 70t + 100$ 6. $s(t) = 186{,}000t$
7. $p(q) = 42 - 0.025q$ 8. $p(q) = 500 - 0.15q$
9. $s(q) = 30 + 0.015q$ 10. $s(q) = 450 + 0.35q$
11. The demand and supply functions for a certain product are given in Problems 7 and 9. Determine the equilibrium price.
12. In Problem 11, what will be the quantities produced and demanded if the price is (a) \$3 above the equilibrium price and (b) \$2 below the equilibrium price?
13. The demand and supply functions for a certain compact car are given in Problems 8 and 10. Determine the equilibrium price.
14. In Problem 13, what will be the quantities produced and demanded if the price is (a) \$150 above the equilibrium price and (b) \$350 below the equilibrium price?
15. In the figure†, (a) find the slope of each line. (b) What does each slope represent?

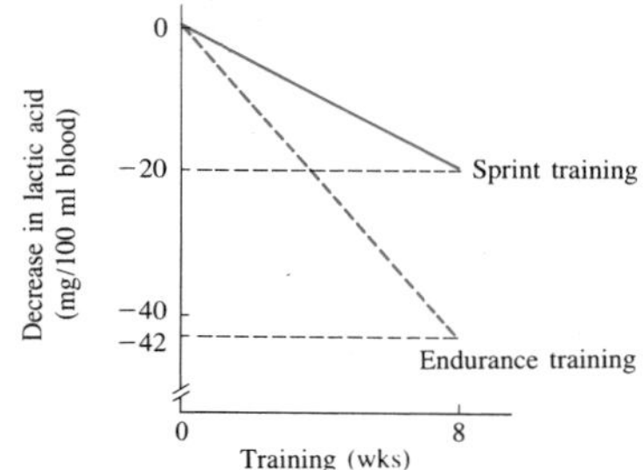

16. The graph‡ gives the supply curve for wheat in a given year. What is the slope of the curve?

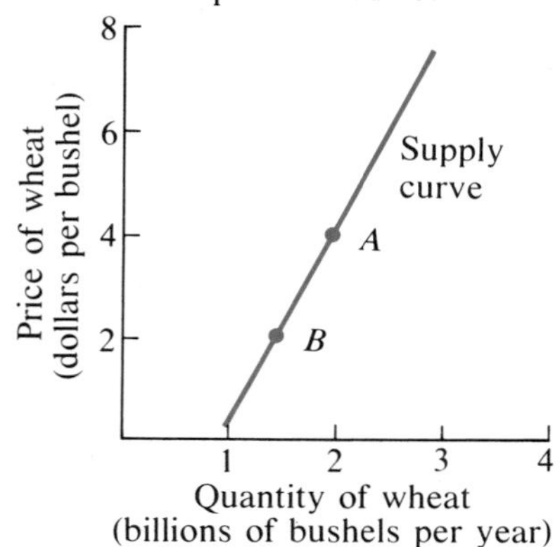

17. The demand curve for the wheat in Problem 16 is given below. (a) What is the slope of this curve? (b) What is the equilibrium price for wheat?

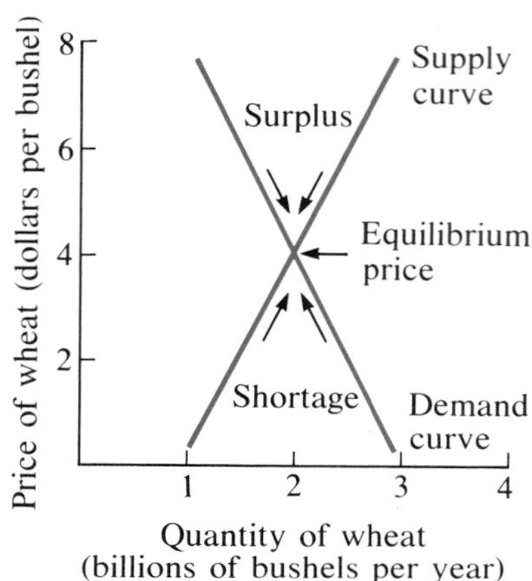

18. The price of a commodity is 50¢ each. If fixed costs are \$300 and the variable costs amount to 30¢ per item:
    (a) Find and sketch the total revenue function.
    (b) Find and sketch the total cost function.
19. The Thunder Power Company charges \$6 for the first 30 kilowatt-hours (kWh) or less each month, and 7¢ for each kilowatt-hour over 30.
    (a) Find a linear function that gives monthly cost as a function of the number of kilowatt-hours (kWh) of electricity used. Assume that at least 30 kWh will be used each month.
    (b) Graph the function.
    (c) What is the cost of 75 kWh in 1 month?
20. Some people paying federal income tax in the United States do not use tax tables. In 1980 they had to file a tax computation schedule (Schedule TC). Tax rates for single taxpayers in 1980 (before the tax tables were "simplified") who filed Schedule TC are given in the table.
    (a) Find a linear function that gives federal income tax due in 1980 as a function of income for single taxpayers earning between \$12,900 and \$15,000 per year.
    (b) Graph this function.
    (c) Determine the tax due on an income of \$14,000.

---

† See *Sports Physiology,* p. 237.

‡ See Dolan, *Economics,* p. 53.

From 1980 U.S. Federal Tax Tables

If the amount on Schedule TC, Part I, line 3, is:		Enter on Schedule TC, Part I, line 4:	
Not over $2,300		*-0-*	
Over—	**But not over—**		**of the amount over—**
$2,300	$3,400	14%	$2,300
$3,400	$4,400	$154 + 16%	$3,400
$4,400	$6,500	$314 + 18%	$4,400
$6,500	$8,500	$692 + 19%	$6,500
$8,500	$10,800	$1,072 + 21%	$8,500
$10,800	$12,900	$1,555 + 24%	$10,800
$12,900	$15,000	$2,059 + 26%	$12,900
$15,000	$18,200	$2,605 + 30%	$15,000
$18,200	$23,500	$3,565 + 34%	$18,200
$23,500	$28,800	$5,367 + 39%	$23,500
$28,800	$34,100	$7,434 + 44%	$28,800
$34,100	$41,500	$9,766 + 49%	$34,100
$41,500	$55,300	$13,392 + 55%	$41,500
$55,300	$81,800	$20,982 + 63%	$55,300
$81,800	$108,300	$37,677 + 68%	$81,800
$108,300	—	$55,697 + 70%	$108,300

21. (a) Find a linear function that gives tax due in 1980 as a function of income for single taxpayers earning between $28,800 and $34,100 a year.
    (b) Graph this function.
    (c) Determine the tax due on an income of $32,750.
22. (a) Find a linear function that gives tax due in 1980 as a function of income for single taxpayers earning over $108,300.
    (b) Graph this function.
    (c) What is the tax due on an income of $500,000?
23. What is the slope of the graph of the function that gives federal income tax in 1980 as a function of income for single taxpayers earning between $18,200 and $23,500 a year?
24. Answer the question of Problem 23 for single taxpayers earning between $55,300 and $81,800 a year.
25. The price of a gallon of gasoline is a linear function of the octane rating. If 85-octane gas costs $1.43 and 90-octane gas costs $1.49½, what is the price of 95-octane gas?
26. In Problem 19, we gave the following rates for users of electric power sold by the Thunder Power Company: $6 for the first 30 kWh or less each month; 7¢ for each kWh over 30. New rates are to go into effect: $4.50 for the first 20 kWh or less each month; 8¢ for each kWh over 20.
    Evidently, a consumer who uses less than 20 kWh each month will save money with the new rate. What is the break-even point? That is, what is the monthly level of usage (in kilowatt-hours) for which the bill will be the same under both the new and the old rates?

3.5 Graphs of Functions

Definition

The **graph** of the function f is the set of points

$$\{(x, f(x)): x \in \text{domain of } f\}.$$

We saw in Section 3.2 how to obtain the graph of a straight line. In this section we obtain the graphs of other types of functions.

EXAMPLE 1 ***Graph of $f(x) = x^2$***

Sketch the graph of $y = f(x) = x^2$.

SOLUTION In Example 3 on p. 171 we saw that $f(x) = x^2$ is a function with domain $\mathbb{R}$ and range $\mathbb{R}^+$. The graph of this function is obtained by plotting all points of the form $(x, y) = (x, x^2)$.† First we note that when $f(x) = x^2$,

† Of course, we can't plot *all* points (there are an infinite number of them). Rather, we plot some sample points and connect them to obtain the sketch of the graph.

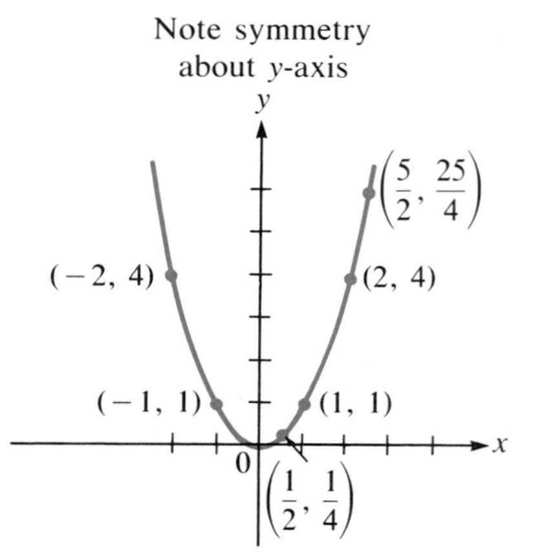

Figure 1 Graph of $f(x) = x^2$.

$f(-x) = (-x)^2 = x^2 = f(x)$. That is, f is even. Thus, it is necessary to calculate $f(x)$ only for $x \geq 0$. For every $x > 0$, there is a value of $x < 0$ that gives the same value of y. When a function is even, the graph of the function is **symmetric** about the y-axis. This means that the graph of $f(x)$ for x negative is the mirror image or reflection of the graph for x positive. Some values of $f(x)$ for $x \geq 0$ are shown in Table 1. The graph drawn in Figure 1 is a **parabola.**

Table 1

x	0	$\frac{1}{2}$	1	$\frac{3}{2}$	2	$\frac{5}{2}$	3	4	5
$f(x) = x^2$	0	$\frac{1}{4}$	1	$\frac{9}{4}$	4	$\frac{25}{4}$	9	16	25

■

EXAMPLE 2 *Graph of the Square Root Function*

Graph the function $f(x) = \sqrt{x}$.

SOLUTION First observe that f is defined only for $x \geq 0$. Note that $\sqrt{x} \geq 0$ for all $x \geq 0$. Table 2 gives values of $\sqrt{x}$. We plot these points and then join them to obtain the graph in Figure 2.

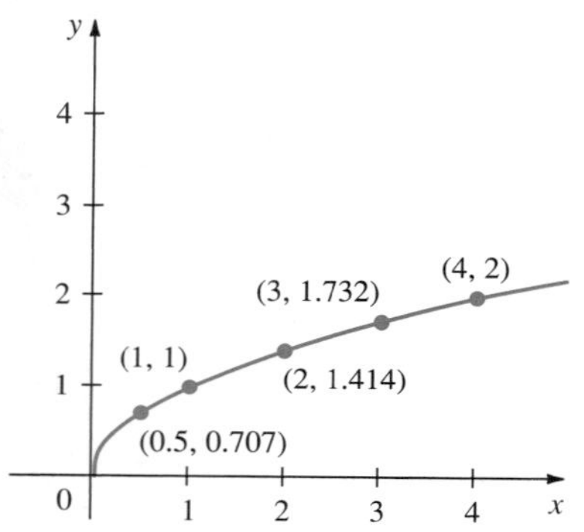

Figure 2 Graph of $f(x) = \sqrt{x}$.

Table 2

x	0	0.5	1	2	3	4	5	10	15	20	25
$f(x) = \sqrt{x}$	0	0.707	1	1.414	1.732	2	2.236	3.162	3.873	4.472	5

■

EXAMPLE 3 *Graph of $f(x) = x^3$*

Graph the function $f(x) = x^3$.

SOLUTION f is defined for all real numbers (domain of f is $\mathbb{R}$), and the cube of a negative number is negative, so we plot both positive and negative values for x. Some representative values are given in Table 3. We plot these points and join them to obtain the graph in Figure 3.

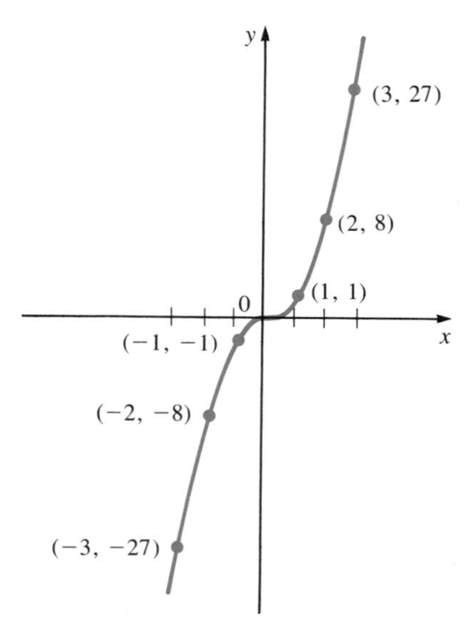

Figure 3 Graph of $f(x) = x^3$.

Table 3

x	0	$\frac{1}{2}$	$-\frac{1}{2}$	1	-1	2	-2	3	-3
$f(x) = x^3$	0	$\frac{1}{8}$	$-\frac{1}{8}$	1	-1	8	-8	27	-27

Since $f(-x) = (-x)^3 = -x^3 = -f(x)$, $f(x) = x^3$ is an odd function. In Figure 3, we say that the graph of x^3 for $x < 0$ is the graph of x^3 for $x > 0$ *reflected about the origin*, or *that the graph is symmetric about the origin.*

Before going further, we give a formal definition of symmetry.

Definition of Symmetry About the x-Axis, the y-Axis, and the Origin

(i) A graph is **symmetric about the x-axis** if whenever (x, y) is on the graph, $(x, -y)$ is also on the graph.
(ii) A graph is **symmetric about the y-axis** if whenever (x, y) is on the graph, $(-x, y)$ is also on the graph.
(iii) A graph is **symmetric about the origin** if whenever (x, y) is on the graph, $(-x, -y)$ is also on the graph.

The graph of $f(x) = x^2$ (Figure 1) is symmetric about the y-axis. The graph of $f(x) = x^3$ (Figure 3) is symmetric about the origin. The graphs in Figure 4 are symmetric about the x-axis.

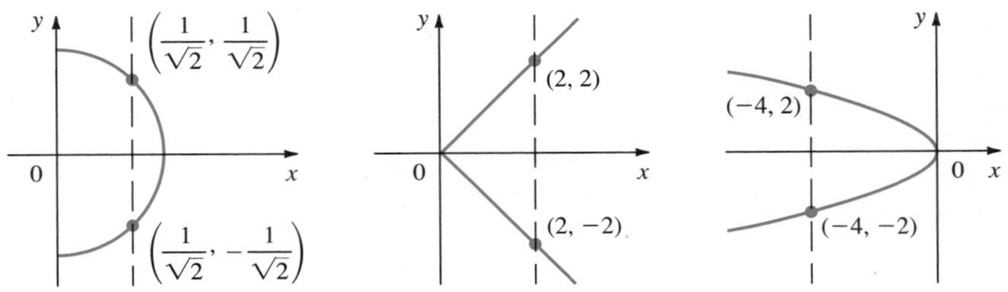

Figure 4 Three graphs that are symmetric about the x-axis.

Note that a graph that is symmetric about the x-axis is not the graph of a function because the vertical lines test is violated when $f(x) \neq 0$.

If $y = f(x)$ is a function, then it may be symmetric about the y-axis, symmetric about the origin, or neither.

Rules of Symmetry for a Function

(i) If $f(x)$ is an even function, then the graph of f is symmetric about the y-axis. To obtain the graph, first sketch it for $x \geq 0$, and then reflect the sketch about the y-axis.
(ii) If $f(x)$ is an odd function, then the graph of f is symmetric about the origin. To obtain the graph, first sketch it for $x \geq 0$, and then reflect the sketch about the origin (that is, reflect it first about the x-axis and then around the y-axis).

Suppose that f is an odd function. Then

Because f is odd
↓
$$f(0) = f(-0) = -f(0)$$

so

$$2f(0) = 0 \text{ or } f(0) = 0$$

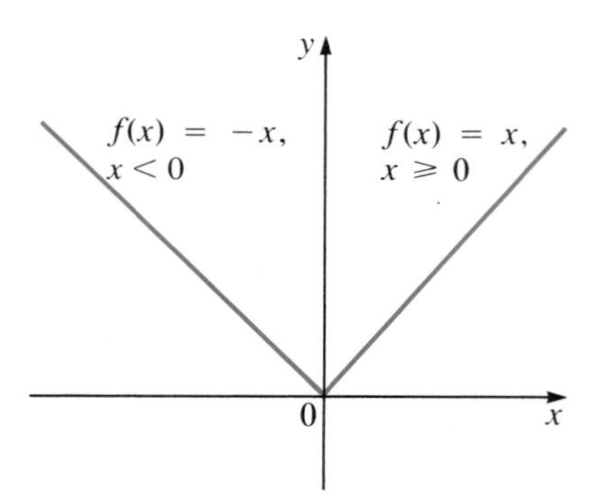

Figure 5 Graph of $f(x) = |x|$.

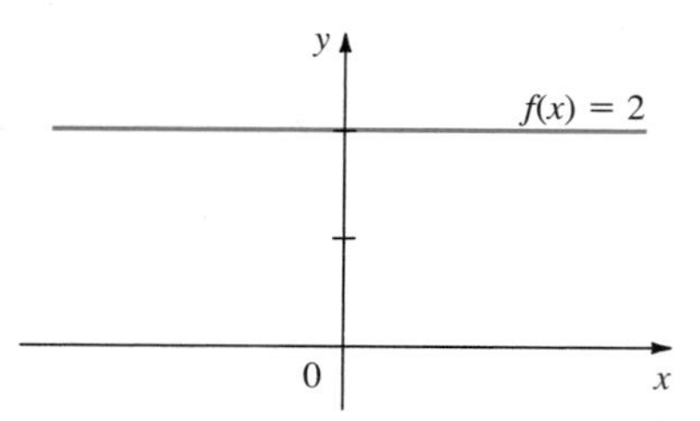

Figure 6 Graph of $f(x) = 2$.

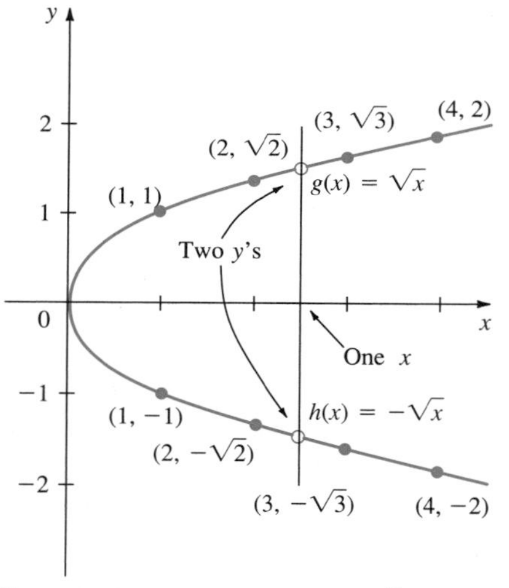

Figure 7 Graph of the equation $y^2 = x$.

That is,

> If f is an odd function, then the graph of f passes through $(0, 0)$.

EXAMPLE 4 *Graph of the Absolute Value Function*

Sketch the graph of $y = |x|$.

SOLUTION From p. 17 we know that

$$f(x) = |x| = \begin{cases} x, & x \geq 0 \\ -x, & x < 0 \end{cases}$$

Also, since $f(-x) = |-x| = |x| = f(x)$, f is an even function and its graph is symmetric about the y-axis. Thus we need only sketch $f(x) = |x| = x$ for $x \geq 0$ and reflect it about the y-axis. This is done in Figure 5. ■

EXAMPLE 5 *Graph of a Constant Function*

Sketch the graph of $f(x) = 2$.

SOLUTION The graph of f is the horizontal line given in Figure 6. Note that $f(-x) = 2 = f(x)$, so f is an even function and its graph is symmetric about the y-axis. ■

EXAMPLE 6 *An Equation That Does Not Determine a Function*

Consider the equation $y^2 = x$. The rule $f(x) = y$ where $y^2 = x$ does *not* determine y as a function of x, since for every $x > 0$ there are *two* values of y such that $y^2 = x$, namely, $y = \sqrt{x}$ and $y = -\sqrt{x}$. For example, if $x = 4$, then $y = 2$ and $y = -2$; both satisfy $y^2 = 4$. However, if we specify one of these values, say $g(x) = \sqrt{x}$, then we have a function. Here, domain of $g = \mathbb{R}^+$ and range of $g = \mathbb{R}^+$. We could obtain a second function, h, by choosing the negative square root. That is, the rule defined by $h(x) = -\sqrt{x}$ is a function with domain $\mathbb{R}^+$ and range $\mathbb{R}^-$ (the nonpositive real numbers). Note that both $g(x)$ and $h(x)$ pass through the origin.

We can look at these things in a different way. The graph of the equation $y^2 = x$ is given in Figure 7. The figure shows that for every positive x, there are two y's such that (x, y) is on the graph. This evidently violates the vertical lines test given on p. 177. This graph is symmetric about the x-axis. ■

EXAMPLE 7 *Graph of $f(x) = \dfrac{1}{x}$*

Sketch the graph of $f(x) = \dfrac{1}{x}$.

SOLUTION We first note that $f(0)$ is not defined.

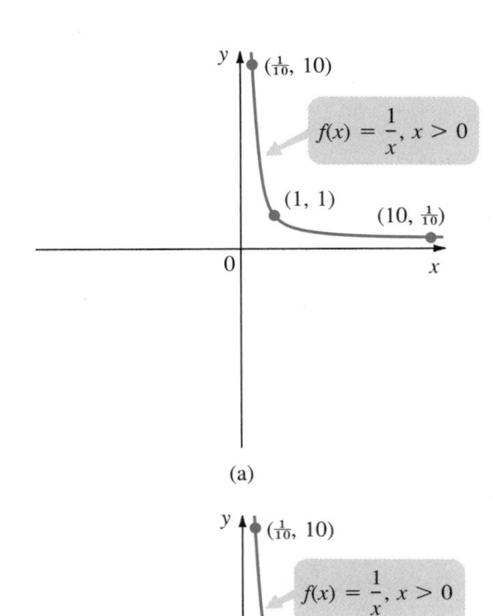

Figure 8 Graph of $f(x) = \dfrac{1}{x}$.

Second,

$$f(-x) = \frac{1}{(-x)} = -\frac{1}{x} = -f(x)$$

so f is an odd function, and its graph is symmetric about the origin. We therefore sketch the graph for $x > 0$ and then reflect it about the origin. Some values of $\dfrac{1}{x}$ for $x > 0$ are given in Table 4.

Table 4

x	1	2	10	100	$\frac{1}{2}$	$\frac{1}{10}$	$\frac{1}{100} = 0.01$	$\frac{1}{1000} = 0.001$
$f(x) = \dfrac{1}{x}$	1	0.5	0.1	0.01	2	10	100	1000

It is evident that $\dfrac{1}{x}$ gets very small as x gets large and $\dfrac{1}{x}$ gets very large as x gets small (we will discuss this type of behavior with more precision in Section 4.5). The graph of $f(x) = \dfrac{1}{x}$ for $x > 0$ is given in Figure 8(a), and the graph for $x \neq 0$ is given in Figure 8(b). ■

EXAMPLE 8 *Graph of $f(x) = \dfrac{|x|}{x}$*

Sketch the graph of $f(x) = \dfrac{|x|}{x}$.

SOLUTION

$$|x| = \begin{cases} x, & x \geq 0 \\ -x, & x < 0 \end{cases} \quad \text{so} \quad \frac{|x|}{x} = \begin{cases} 1, & x > 0 \\ -1, & x < 0 \end{cases}$$

We omit the value 0 from the last expression because we cannot divide by zero. Here the domain of f is $\{x \in \mathbb{R}: x \neq 0\}$ and the range of f is $\{-1, 1\}$. We sketch its graph in Figure 9.

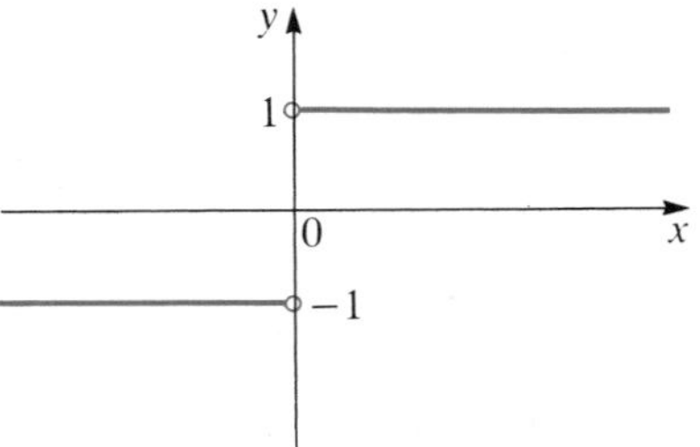

Figure 9 Graph of $f(x) = \dfrac{|x|}{x} = \begin{cases} 1, & x > 0 \\ -1, & x < 0 \end{cases}$.

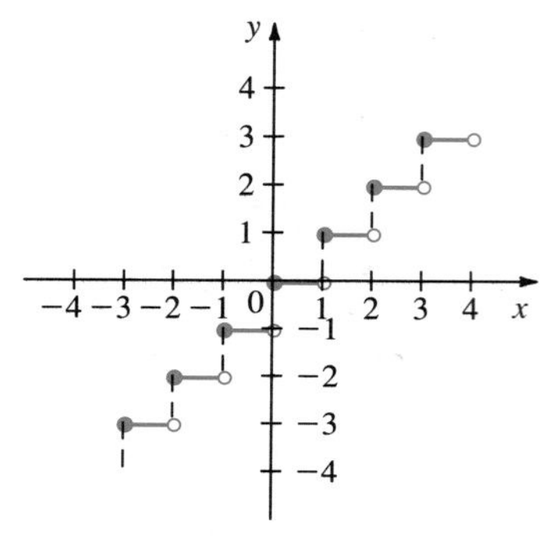

Figure 10 Graph of the greatest integer function $y = [x]$.

EXAMPLE 9 *The Greatest Integer Function*

The **greatest integer function** is defined by

$$f(x) = [x]$$

where $[x]$ is the greatest integer smaller than or equal to x. Thus $[3] = 3$, $[\frac{1}{2}] = 0$, $[2.16] = 2$, $[-5.6] = -6$, and so on. A graph of this function is given in Figure 10.

Shifting the Graphs of Functions

Although more advanced methods (involving calculus) are needed to obtain the graphs of most functions (without plotting a large number of points), there are some techniques that make it a relatively simple matter to sketch certain functions based on known graphs.

EXAMPLE 10 *Shifting a Graph Vertically*

Sketch the graphs of $y = f_1(x) = x^2 + 1$ and $y = f_2(x) = x^2 - 2$.

SOLUTION In Figure 11(a), we have used the graph of $y = x^2$ obtained in Figure 1. To graph $y = x^2 + 1$ in Figure 11(b), we add 1 unit to every y value obtained in Figure 11(a); that is, we shift the graphs of $y = x^2$ up 1 unit. Analogously, for Figure 11(c) we shift the graph of $y = x^2$ down 2 units to obtain the graph of $y = x^2 - 2$.

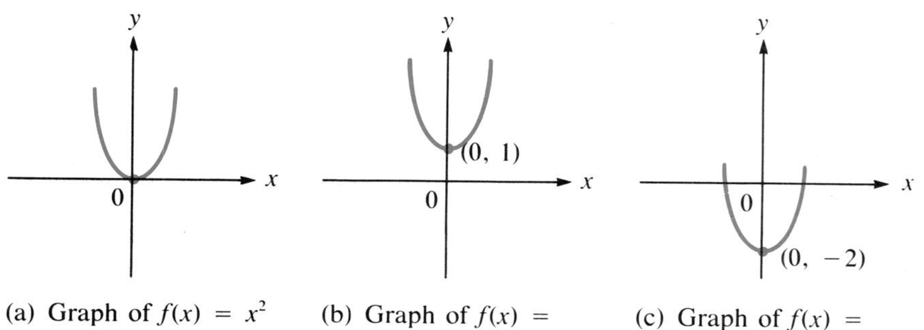

(a) Graph of $f(x) = x^2$ (b) Graph of $f(x) = x^2 + 1$ (c) Graph of $f(x) = x^2 - 2$

Figure 11 Vertical shifts of the graph of x^2.

The results of Example 10 can be generalized.

Vertical Shifts of Graphs

Let $y = f(x)$ and let $c > 0$.

(i) To obtain the graph of $y = f(x) + c$, shift the graph of $y = f(x)$ up c units.

(ii) To obtain the graph of $y = f(x) - c$, shift the graph of $y = f(x)$ down c units.

Table 5
Values taken by $(x - 1)^2$ are the values of x^2 taken one unit later

x	x^2	$(x - 1)^2$
−5	25	36
−4	16	25
−3	9	16
−2	4	9
−1	1	4
0	0	1
1	1	0
2	4	1
3	9	4
4	16	9

Table 6
Values taken by $(x + 2)^2$ are the values of x^2 taken two units earlier

x	x^2	$(x + 2)^2$
−5	25	9
−4	16	4
−3	9	1
−2	4	0
−1	1	1
0	0	4
1	1	9
2	4	16
3	9	25
4	16	36

EXAMPLE 11 ***Shifting a Graph Horizontally***

Sketch the graphs of $y = (x - 1)^2$ and $y = (x + 2)^2$.

SOLUTION Let us compare the functions x^2 and $(x - 1)^2$. For example, for the function $y = x^2$, $y = 0$ when $x = 0$, and for the function $y = (x - 1)^2$, $y = 0$ when $x = 1$. Similarly, $y = 4$ when $x = -2$ if $y = x^2$, and $y = 4$ when $x = -1$ if $y = (x - 1)^2$. By continuing in this manner, you can see that y values in the graph of $y = x^2$ are the same as y values in the graph of $y = (x - 1)^2$, except that they occur 1 unit to the right on the x-axis. Some representative values are given in Table 5. Thus, we find that the graph of $y = (x - 1)^2$ is the graph of $y = x^2$ *shifted 1 unit to the right*. This is indicated in Figure 12(b).

Similarly, in Figure 12(c) we find that the graph of $y = (x + 2)^2$ is the graph of $y = x^2$ *shifted 2 units to the left*. Some values are given in Table 6.

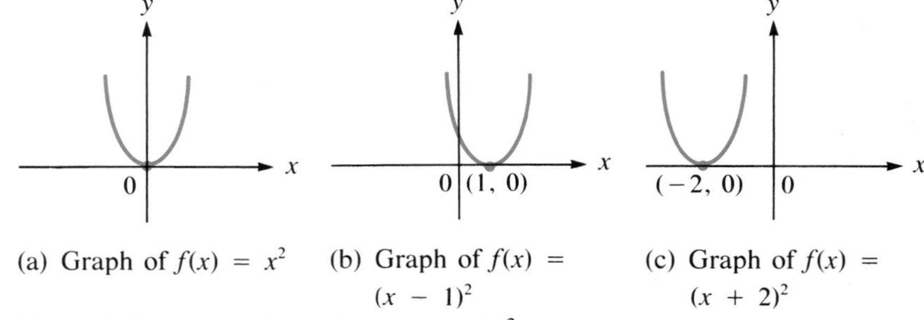

(a) Graph of $f(x) = x^2$ (b) Graph of $f(x) = (x - 1)^2$ (c) Graph of $f(x) = (x + 2)^2$

Figure 12 Horizontal shifts of the graph of x^2.

Horizontal Shifts of Graphs

Let $y = f(x)$ and let $c > 0$.

(iii) To obtain the graph of $y = f(x - c)$, shift the graph of $y = f(x)$ c units to the right.

(iv) To obtain the graph of $y = f(x + c)$, shift the graph of $y = f(x)$ c units to the left.

WARNING Be careful. $f(x + c) \neq f(x) + c$. For example, if $f(x) = x^2$, the graph of $f(x + 1) = (x + 1)^2 = x^2 + 2x + 1$ is the graph of x^2 shifted one unit to the left, but the graph of $f(x) + 1 = x^2 + 1$ is the graph of x^2 shifted 1 unit up. The graphs are not the same. ■

EXAMPLE 12 *Reflecting a Graph About the x-Axis*

Sketch the graph of $y = -x^2$.

SOLUTION To obtain the graph of $y = -x^2$, from the graph of $y = x^2$, note that each y value is replaced by its negative so that the graph of $y = -x^2$ is the graph of $y = x^2$ *reflected about the x-axis* (that is, turned upside down). See Figure 13. ■

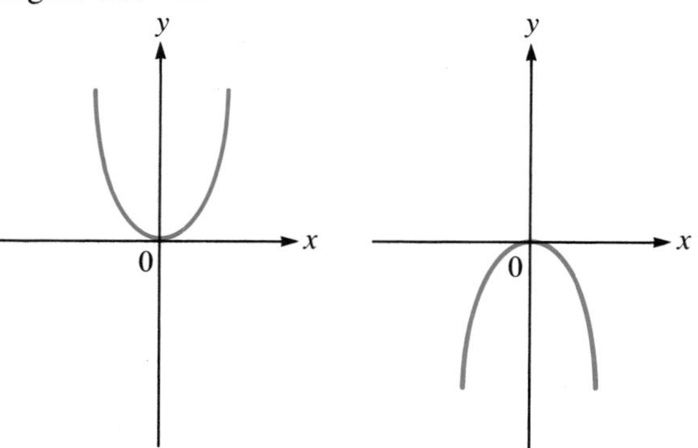

(a) Graph of $f(x) = x^2$ (b) Graph of $f(x) = -x^2$

Figure 13 Reflection of the graph of x^2 about the x-axis.

EXAMPLE 13 *Reflecting a Graph About the y-Axis*

Sketch the graph of $g(x) = \sqrt{-x}$.

SOLUTION $\sqrt{-x}$ is defined only when $x \le 0$ (so $-x \ge 0$). In Table 7 we provide sample values of $\sqrt{-x}$ for $x \le 0$.

Table 7

x	0	−0.5	−1	−2	−3	−4	−5	−10	−15	−20	−25
$g(x) = \sqrt{-x}$	0	0.707	1	1.414	1.732	2	2.236	3.162	3.873	4.472	5

These are the same values taken (in Table 2) by $f(x) = \sqrt{x}$ for $x \ge 0$. We see that $f(x) = \sqrt{x}$ and $g(x) = \sqrt{-x}$ take the same y values. $\sqrt{x}$ takes these values for $x \ge 0$ and $\sqrt{-x}$ takes them for $x \le 0$. The two graphs are given in Figure 14. The graph of $g(x) = \sqrt{-x}$ is *the reflection about the y-axis* of the graph of $f(x) = \sqrt{x}$.

Note that $g(x) = \sqrt{-x} = f(-x)$.

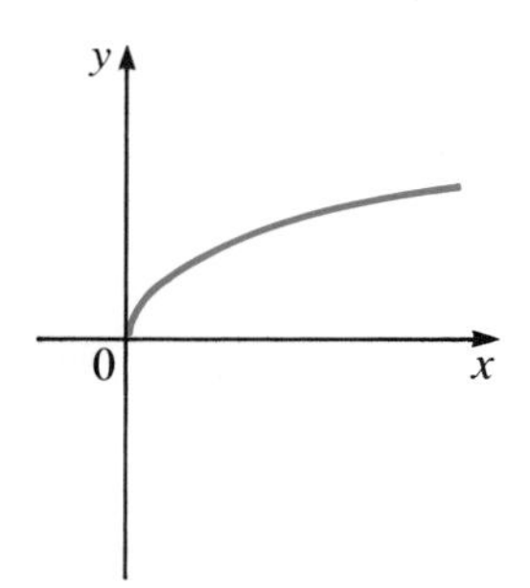

(a) Graph of $f(x) = \sqrt{x}$

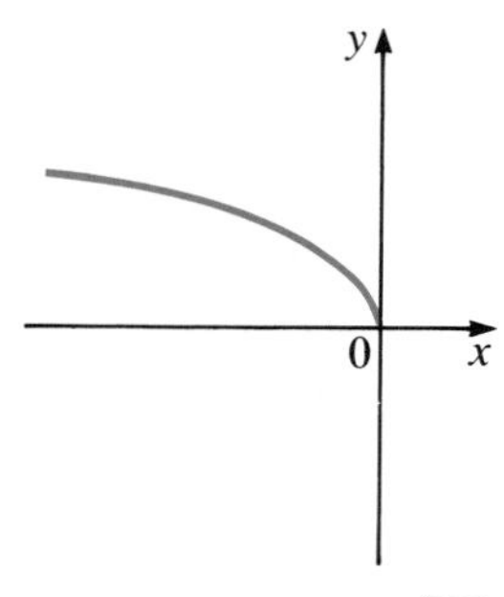

(b) Graph of $g(x) = \sqrt{-x}$

Figure 14 Reflection of the graph of $\sqrt{x}$ about the y-axis.

Reflection About the *x*-Axis and the *y*-Axis

The graph of $y = -f(x)$ is obtained by reflecting the graph of $y = f(x)$ about the x-axis.

The graph of $y = f(-x)$ is obtained by reflecting the graph of $y = f(x)$ about the y-axis.

WARNING Don't confuse $f(-x)$ and $-f(x)$. They are usually *not* the same. For example, if $f(x) = x^2$, then $f(-2) = (-2)^2 = 4$, but $-f(2) = -2^2 = -4$. If $f(x) = 2x + 3$, then $f(-5) = 2(-5) + 3 = -10 + 3 = -7$, but $-f(5) = -(2 \cdot 5 + 3) = -13$. ■

We can use our shifting results to sketch any quadratic function with leading coefficient ± 1.

EXAMPLE 14 *Shifting and Reflecting a Graph*

Sketch the graph of $f(x) = -x^2 + 4x - 9$.

SOLUTION

$$\begin{aligned} -x^2 + 4x - 9 &= -(x^2 - 4x + 9) && \text{Factor out } -1 \\ &= -(x^2 - 4x + 4 - 4 + 9) && \text{Complete the square by adding and subtracting 4} \\ &= -[(x^2 - 4x + 4) + 5] && \\ &= -[(x - 2)^2 + 5] && \text{Factor} \end{aligned}$$

Starting with the graph of $y = x^2$, we obtain the graph of $y = -x^2 + 4x - 9$ in three steps, as in Figure 15.

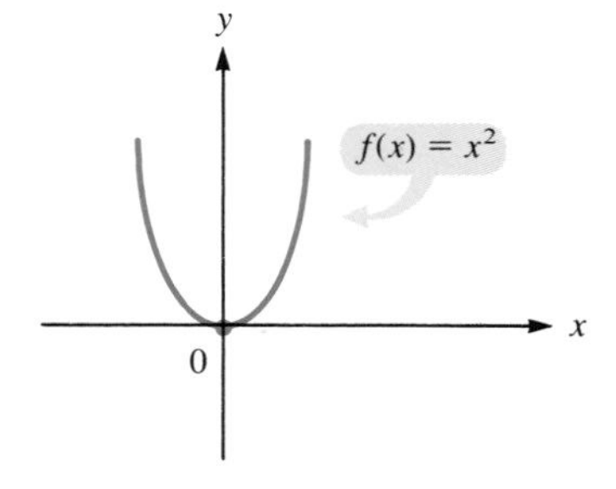

(a) Graph of $y = x^2$ (original graph).

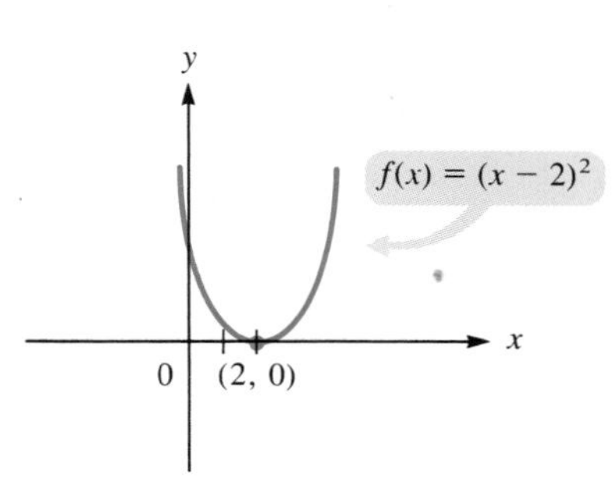

(b) Step 1: Shift graph of $f(x) = x^2$ right 2 units.

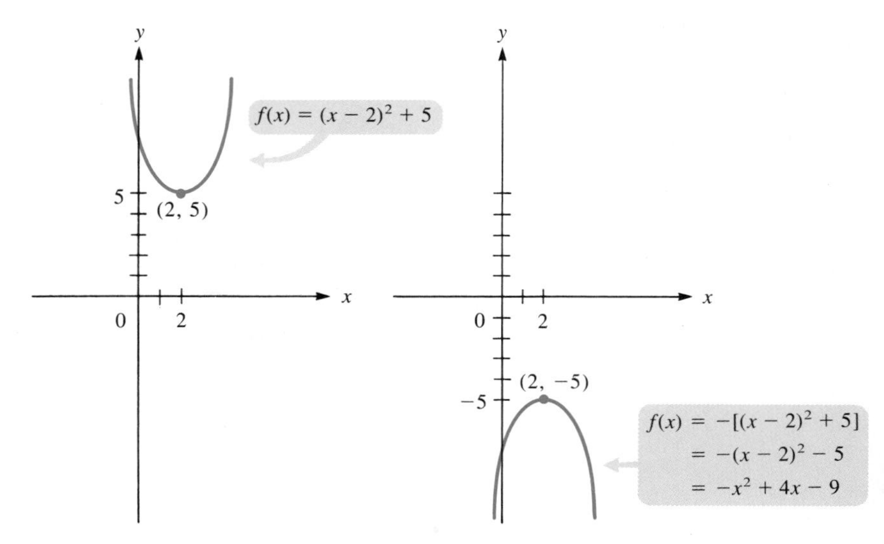

(c) Step 2: Shift graph of $f(x) = (x - 2)^2$ up 5 units.

(d) Step 3: Reflect graph of $f(x) = (x - 2)^2 + 5$ about the x-axis.

Figure 15 Shifting and reflecting the graph of x^2 to obtain the graph of $-x^2 + 4x - 9$.

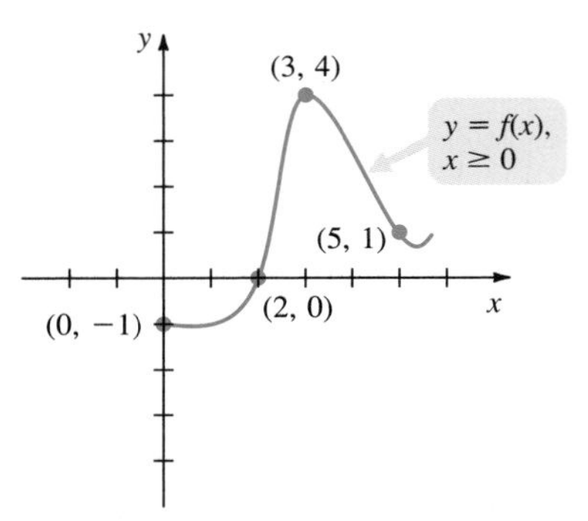

Figure 16 Graph of an even function for $x \geq 0$.

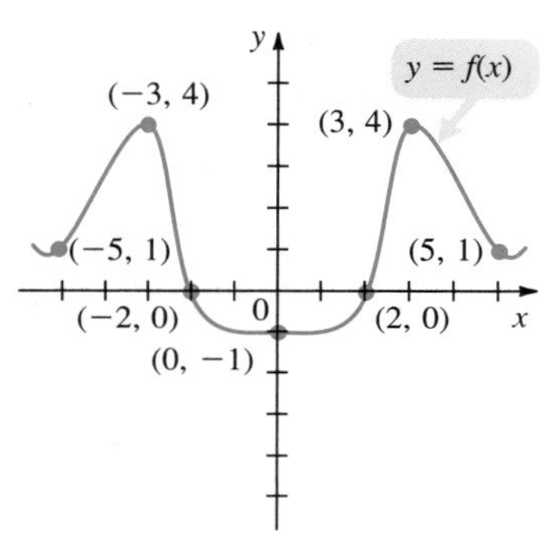

Figure 17 Graph of $f(x)$.

EXAMPLE 15 *Reflecting a Graph About the y-Axis*

The graph of an even function, for $x \geq 0$, is given in Figure 16. Complete the graph.

SOLUTION Since f is even, we may obtain its graph, for $x \leq 0$, by reflecting the graph for $x \geq 0$ about the y-axis. This is done in Figure 17. ■

EXAMPLE 16 *Reflecting and Shifting a Graph*

The graph of a certain function $f(x)$ is given in Figure 18(a). Sketch the graph of $-f(3 - x)$.

SOLUTION We do this in three steps:

(b) Reflect about the y-axis to obtain the graph of $f(-x)$ (Figure 18(b)).
(c) Shift to the right 3 units to obtain the graph of $f(-(x - 3)) = f(3 - x)$ (Figure 18(c)).
(d) Reflect about the x-axis to obtain the graph of $y = -f(3 - x)$ (Figure 18(d)).

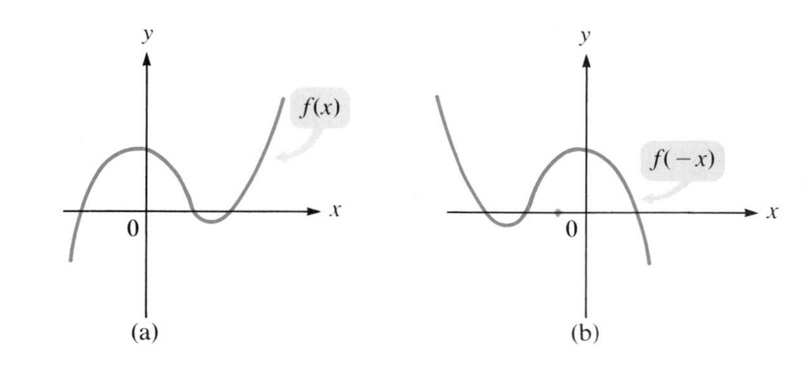

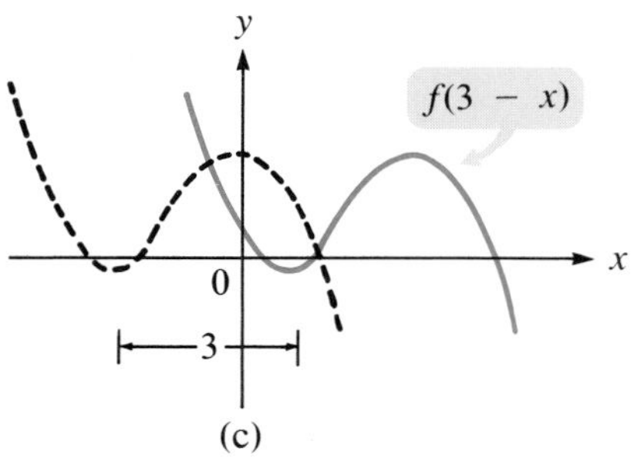

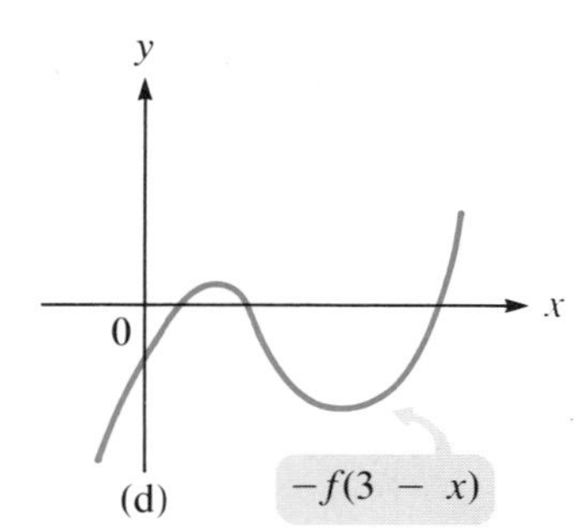

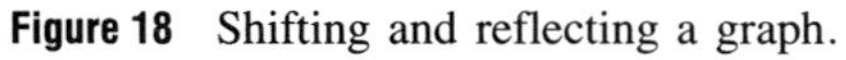

Figure 18 Shifting and reflecting a graph.

Problems 3.5

Readiness Check

I. Which of the following is true about an odd function f?
  a. Its graph is symmetric about the y-axis.
  b. Its graph is symmetric about the origin.
  c. $f(x) = f(-x)$
  d. If $(3, -2)$ is a point on the graph of f, then $(-3, -2)$ is also on the graph of $f(x)$.

II. Which of the following is graphed below?
  a. $f(x) = (x-2)^2 + 3$
  b. $g(x) = -(x+2)^2 - 3$
  c. $h(x) = -(x-2)^2 - 3$
  d. $k(x) = (x+2)^2 + 3$

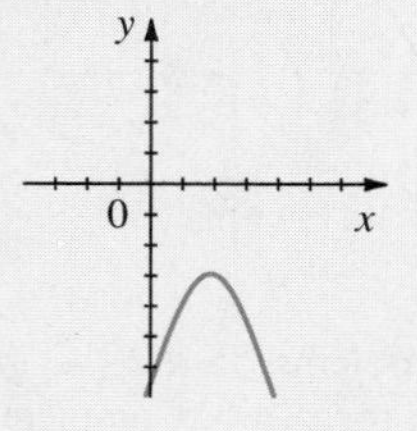

III. The graph of which of the following equations would be obtained by shifting $y = f(x) = \sqrt{x}$ three units to the left?
  a. $y = f(x) = \sqrt{x-3}$ b. $y = f(x) = \sqrt{x} - 3$
  c. $y = f(x) = \sqrt{x+3}$ d. $y = f(x) = \sqrt{x} + 3$

IV. Which of the following would have a graph that is the reflection of the graph of $f(x) = x^2 - 2x + 9$ about the x-axis?
  a. $f(x) = -x^2 + 2x - 9$ b. $f(x) = x^2 + 2x + 9$
  c. $f(x) = -(x-1)^2 + 10$ d. $f(x) = -[(x-1)^2 - 8]$

V. Which of the following is the function that results from shifting the graph of $f(x) = x^2 - 4x + 3$ up 2 units and to the left 3 units?
  a. $f(x) = x^2 + 8x + 6$ b. $f(x) = x^2 + 10x$
  c. $f(x) = x^2 - 3x + 2$ d. $f(x) = x^2 + 2x + 2$

In Problems 1–7 sketch the graph of the given function by plotting some points (if necessary) and then connecting them. Check first for symmetry about the y-axis and the origin. Use a calculator when marked.

1. $f(x) = x^4$ 2. $f(x) = x^2$
3. $f(x) = 5x^2$ 4. $f(x) = -2x^2$
5. $f(x) = \sqrt[3]{x}$ 6. $f(x) = x^3 + x$
7. $f(x) = \sqrt{x^2 + 1}$

In Problems 8–13 sketch the given function.

8. $y = x^2 + 3$ 9. $y = x^2 - 5$
10. $y = (x+3)^2$ 11. $y = (x-4)^2$
12. $y = 1 - x^2$ 13. $y = (x-1)^2 + 3$

In Problems 14–18 use the technique of Example 14 to sketch the given quadratic.

14. $y = x^2 - 4x + 7$ 15. $y = x^2 + 8x + 2$
16. $y = x^2 + 3x + 4$ 17. $y = -x^2 + 2x - 3$
18. $y = -x^2 - 5x + 8$

In Problems 19–24 use the graph in Figure 3 to obtain the graph of the given cubic.

19. $y = x^3 + 1$ 20. $y = x^3 - 2$
21. $y = (x+3)^3$ 22. $y = -x^3$
23. $y = -(x+2)^3$ 24. $y = 3 - (x+4)^3$

25. The graph of $f(x) = 1/x$ is given in Figure 8 (p. 193). Sketch the graph of
  (a) $\dfrac{1}{x+3}$ (b) $\dfrac{1}{x-4}$ (c) $3 + \dfrac{1}{x}$
  (d) $2 - \dfrac{1}{x}$ * (e) $\dfrac{5x-1}{x}$

In Problems 26–31 use the graph in Figure 5 (p. 192) to sketch the graph of the given function.

26. $y = |x| - 3$ 27. $y = |x| + 2$
28. $y = |x - 3|$ 29. $y = |x + 5|$
30. $y = -|x|$ 31. $y = 5 - |x + 4|$

In Problems 32–45 sketch the graph of each function defined in pieces.

32. $f(x) = \begin{cases} 2, & x \geq 0 \\ 1, & x < 0 \end{cases}$ 33. $f(x) = \begin{cases} 4, & x > 3 \\ -2, & x \leq 3 \end{cases}$

34. $f(x) = \begin{cases} 0, & x < 0 \\ 1, & 0 \leq x \leq 2 \\ 3, & x > 2 \end{cases}$

35. $f(x) = \begin{cases} x, & x < 0 \\ 2x, & x \geq 0 \end{cases}$

36. $f(x) = \begin{cases} x + 2, & -2 \leq x < 1 \\ 1 + 2x, & 1 \leq x \leq 5 \end{cases}$

Answers to Readiness Check

I. b II. c III. c IV. a V. d

37. $f(x) = \begin{cases} 1 + x, & -2 \le x \le 0 \\ 1 - x, & 0 < x \le 2 \end{cases}$

38. $f(x) = \begin{cases} 2x + 3, & x \ge 2 \\ -3x + 1, & x < 2 \end{cases}$

39. $f(x) = \begin{cases} x, & x \le 1 \\ x^2, & x > 1 \end{cases}$

40. $f(x) = \begin{cases} x, & x \le 2 \\ x^2, & x > 2 \end{cases}$

41. $f(x) = \begin{cases} x^2, & x \le 1 \\ x^3, & x > 1 \end{cases}$

42. $f(x) = \begin{cases} x^2 - 4, & x \ne 2 \\ 3, & x = 2 \end{cases}$

43. $f(x) = \begin{cases} \dfrac{x^2 - 1}{x - 1}, & x \ne 1 \\ 2, & x = 1 \end{cases}$

44. $f(x) = \begin{cases} \dfrac{x^2 - 4}{x + 2}, & x \ne -2 \\ 1, & x = -2 \end{cases}$

* 45. $f(x) = \begin{cases} \dfrac{x^3 - 1}{x - 1}, & x \ne 1 \\ 3, & x = 1 \end{cases}$

In each of Problems 46–54 the graph of a function is sketched. Obtain the graph of (a) $f(x - 2)$, (b) $f(x + 3)$, (c) $-f(x)$, (d) $f(-x)$, and (e) $f(2 - x) + 3$.

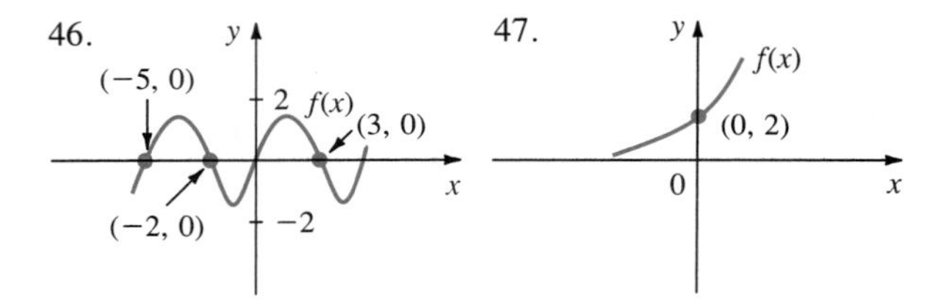

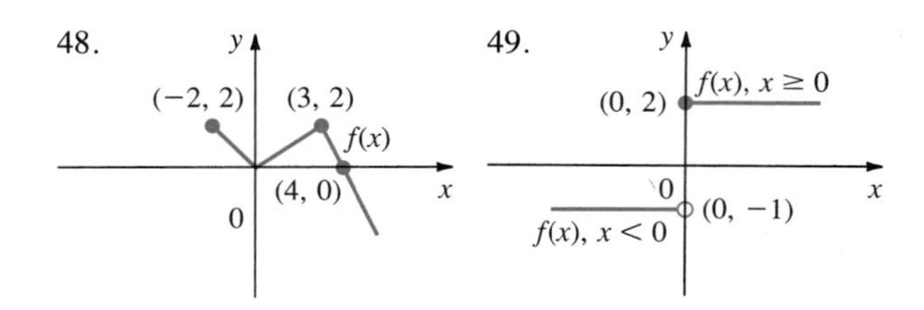

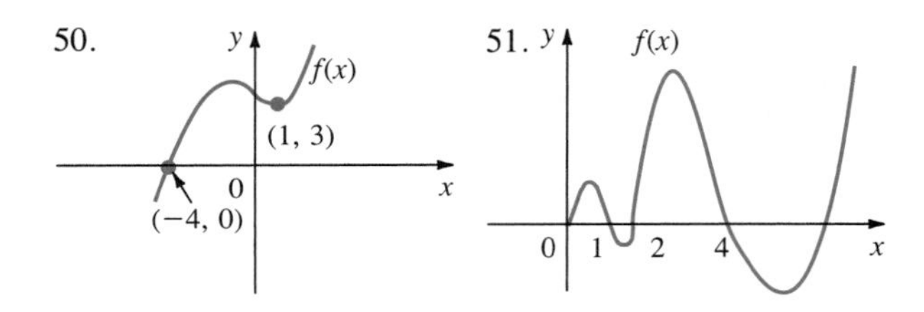

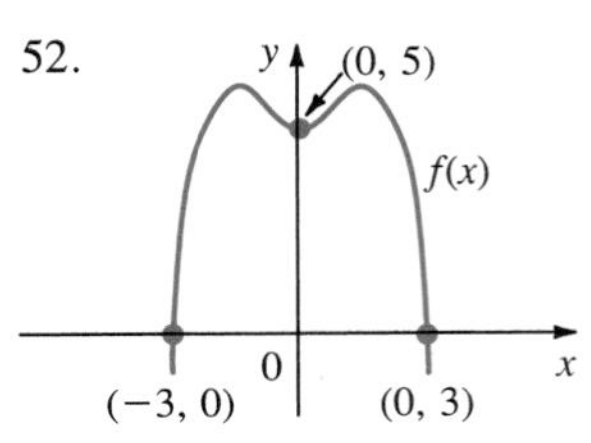

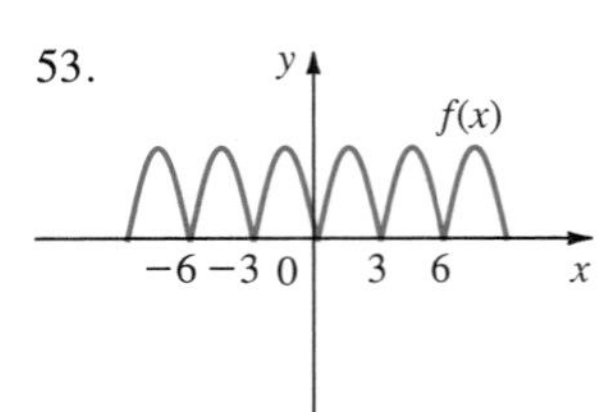

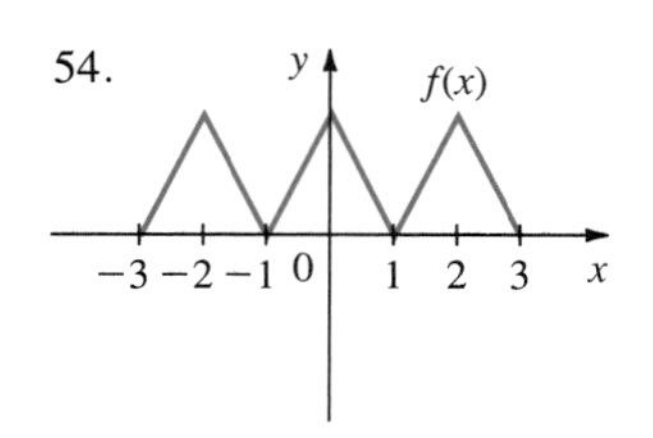

In Problems 55–62 the graph of an even or odd function is given for $x \ge 0$. Complete the graph of the function for $x < 0$.

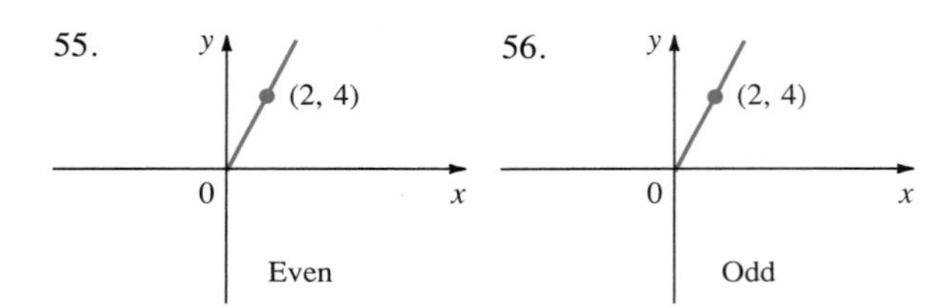

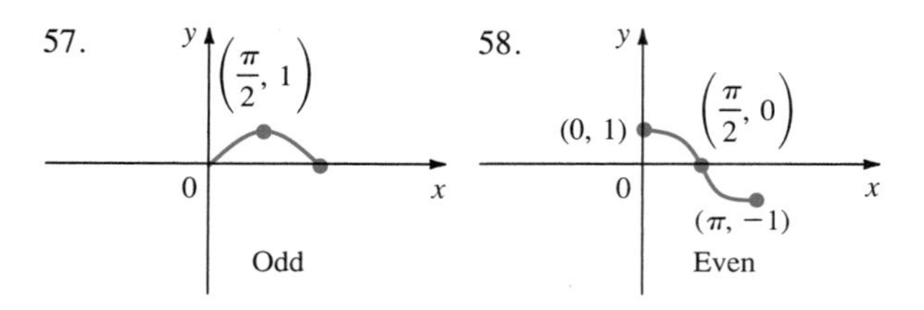

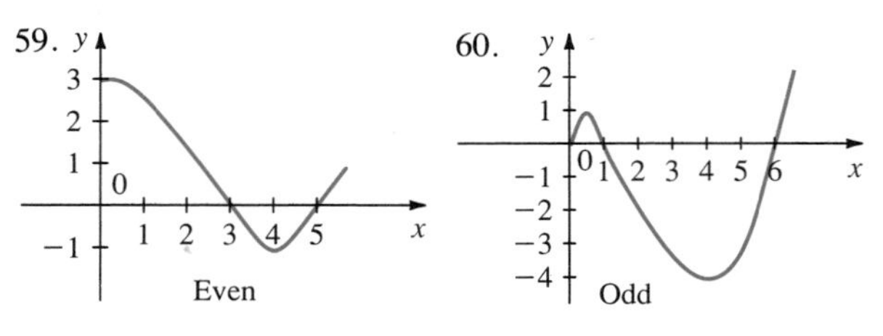

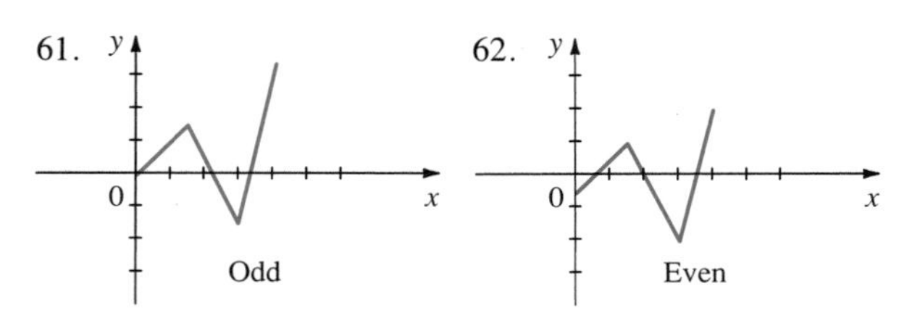

Graphing Calculator Problems

Before doing the following problems, read (or reread) Appendix A on using a graphing calculator.

63. Sketch the graph of $y = \sqrt[3]{x}$. Use the range values $-9 \le x \le 9$, $-2.5 \le y \le 2.5$, $x_{scl} = 2$, $y_{scl} = 1$.
64. Without clearing the graph of $\sqrt[3]{x}$, and without changing the range or scale values, sketch the graph of $y = \sqrt[3]{x-2}$.

In Problems 65–72 first clear previously sketched graphs, then sketch $y = \sqrt[3]{x}$ and, without clearing this graph, sketch the graph of the given function. It will be necessary to change the range values in some cases.

65. $y = \sqrt[3]{x+2}$
66. $y = \sqrt[3]{x} - 2$
67. $y = \sqrt[3]{x-2}$
68. $y = -\sqrt[3]{x}$
69. $y = \sqrt[3]{-x}$
70. $y = 3 - \sqrt[3]{x}$
71. $y = 2 + \sqrt[3]{x-1}$
72. $y = -1 + \sqrt[3]{2-x}$

In Problems 73–81 let $f(x) = -x^3 + 2x^2 + 5x - 6$. A graph of this function is given in Figure 3 in Appendix A for $-4 \le x \le 4$ and $-18 \le y \le 70$. Sketch each function after first changing the x- and y-range values in an appropriate way. (For example, if you shift 2 units to the right, then $-4 \le x \le 4$ is shifted to $-2 \le x \le 6$.)

73. $f(x-2)$
74. $f(x+3)$
75. $f(x) + 1$
76. $f(x) - 4$
77. $-f(x)$
78. $f(-x)$
79. $f(2-x)$
80. $-f(x) + 3$
81. $-4 + f(1+x)$

3.6 Operations with Functions

We begin this section by showing how functions can be added, subtracted, multiplied, and divided.

Definition of the Sum, Difference, Product, and Quotient of Two Functions

Let f and g be two functions. Then

(a) The sum $f + g$ is defined by

$$(f+g)(x) = f(x) + g(x)$$

(b) The difference $f - g$ is defined by

$$(f-g)(x) = f(x) - g(x)$$

(c) The product $f \cdot g$ is defined by

$$(f \cdot g)(x) = f(x)g(x)$$

(d) The quotient f/g is defined by

$$\left(\frac{f}{g}\right)(x) = \frac{f(x)}{g(x)}, \text{ whenever } g(x) \ne 0$$

The functions $f + g$, $f - g$, and $f \cdot g$ are defined for each x for which both f and g are defined. That is,

$$\begin{aligned}(\text{domain of } f+g) &= (\text{domain of } f-g) = (\text{domain of } f \cdot g)\\ &= \{x\text{: } x \in \text{domain of } f \text{ and } x \in \text{domain of } g\}.\end{aligned}$$

That is, in order for $(f+g)(x)$ to be defined, we must have both $f(x)$ *and* $g(x)$ defined.

Finally, f/g is defined whenever both f and g are defined and $g(x) \ne 0$ (so that we do not divide by zero). This last fact is very important.

$$(\text{domain of } f/g) = \{x\colon x \in \text{domain of } f,\ x \in \text{domain of } g \text{ and } g(x) \neq 0\}.$$

When taking the quotient of two functions, make sure you are not dividing by zero.

EXAMPLE 1 ***The Sum, Difference, Product, and Quotient of Two Functions***

Let $f(x) = \sqrt{1 + x}$ and $g(x) = 4 - x^2$. Find $(f + g)(x)$, $(f - g)(x)$, $(f \cdot g)(x)$, and $(f/g)(x)$ and determine the domain of each.

SOLUTION The domain of f is $\{x\colon x \geq -1\}$. Since the domain of g is $\mathbb{R}$, $f + g$, $f - g$, and $f \cdot g$ are defined for $\{x\colon x \geq -1\} = [-1, \infty)$. We have

(a) $(f + g)(x) = f(x) + g(x) = \sqrt{1 + x} + 4 - x^2$
(b) $(f - g)(x) = f(x) - g(x) = \sqrt{1 + x} - (4 - x^2) = \sqrt{1 + x} - 4 + x^2$
(c) $(f \cdot g)(x) = f(x)g(x) = \sqrt{1 + x}(4 - x^2)$
(d) $\left(\dfrac{f}{g}\right)(x) = \dfrac{f(x)}{g(x)} = \dfrac{\sqrt{1 + x}}{4 - x^2}$

Here the domain is different. The denominator $4 - x^2 = 0$ when $x = \pm 2$. The number -2 is not in the domain of f, so we don't worry about it. However, we must throw out 2. Thus

$$\text{domain of } \frac{f}{g} \text{ is } \{x\colon x \geq -1 \text{ and } x \neq 2\}$$

Composite Function

You will often need to deal with functions of functions. If f and g are functions, then their **composite function,** $f \circ g$, is defined as follows:

Definition of the Composite Function

$$(f \circ g)(x) = f(g(x))$$

and domain of $f \circ g$ is $\{x\colon x \in \text{domain of } g \text{ and } g(x) \in \text{domain of } f\}$.

That is, $(f \circ g)(x)$ is defined for every x such that $g(x)$ and $f(g(x))$ are defined. An illustration of the composition of two functions is given in the figure.

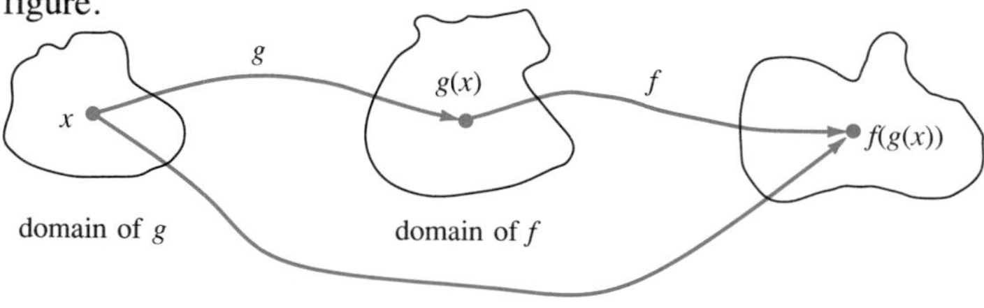

EXAMPLE 2 ***Finding the Composition of Two Functions***

Let $f(x) = \sqrt{x}$ and $g(x) = x^2 + 1$. Find $(f \circ g)(x)$ and $(g \circ f)(x)$ and determine their respective domains.

SOLUTION

$$(f \circ g)(x) = f(g(x)) = f(x^2 + 1) = \sqrt{x^2 + 1}$$

and

$$(g \circ f)(x) = g(f(x)) = g(\sqrt{x}) = (\sqrt{x})^2 + 1 = x + 1$$

Note that we must have $x \geq 0$ because $\sqrt{x}$ is defined (as a real number) only when $x \geq 0$. Now, the domain of f is $\mathbb{R}^+$, the domain of g is $\mathbb{R}$, and

$$\text{domain of } f \circ g \text{ is } \{x\colon g(x) = x^2 + 1 \in \text{domain of } f\}$$

But since $x^2 + 1 > 0$, $x^2 + 1 \in$ domain of f for every real x, so domain of $f \circ g$ is $\mathbb{R}$. On the other hand, domain of $g \circ f$ is $\mathbb{R}^+$ because f is defined only for $x \geq 0$.

WARNING It is *not* true, in general, that $(f \circ g)(x) = (g \circ f)(x)$. This is illustrated in Examples 2 and 3. ■

EXAMPLE 3 *Finding the Composition of Two Functions*

Let $f(x) = 3x - 4$ and $g(x) = x^3$. Find $(f \circ g)(x)$ and $(g \circ f)(x)$ and determine their respective domains.

SOLUTION

$$(f \circ g)(x) = f(g(x)) = f(x^3) = 3x^3 - 4$$

and

$$(g \circ f)(x) = g(f(x)) = g(3x - 4) = (3x - 4)^3$$

Here domain of $f \circ g$ = domain of $g \circ f = \mathbb{R}$. Note that the functions $f \circ g$ and $g \circ f$ are quite different. ■

EXAMPLE 4 *Finding a Function That Can Be Used to Obtain a Desired Composition*

Let $f(x) = 2x + 1$. Find a function $g(x)$ such that $(f \circ g)(x) = x^3$.

SOLUTION We must have $(f \circ g)(x) = f(g(x)) = 2g(x) + 1 = x^3$. Then $2g(x) = x^3 - 1$ and $g(x) = \dfrac{x^3 - 1}{2}$. ■

EXAMPLE 5 *A Technique Useful in Calculus*

Let $u(x) = \sqrt{x^3 + 5}$. Find three functions f, g, and h such that $u(x) = (f \circ g \circ h)(x)$.

SOLUTION We first observe that u is the square root of something. Since $(f \circ g \circ h)(x)$ is evaluated with f applied last, we set $f(x) = \sqrt{x}$. Next, to evaluate $x^3 + 5$, we first cube x and then add 5. Thus, if $h(x) = x^3$ and $g(x) = x + 5$, then

$$(g \circ h)(x) = g(h(x)) = g(x^3) = x^3 + 5$$

Finally,

$$(f \circ g \circ h)(x) = f((g \circ h)(x)) = f(x^3 + 5) = \sqrt{x^3 + 5}$$

So one solution is

$$f(x) = \sqrt{x}$$
$$g(x) = x + 5$$
$$h(x) = x^3$$

Problems 3.6

Readiness Check

I. Which of the following is true?
   a. The domain of $f \circ g$ is $\{x\colon x$ belongs to the domains of f and $g\}$.
   b. The domain of $f + g$ is the same as the domain of f/g for all f and g.
   c. The domain of a composite function cannot be determined until the composite function is written explicitly.
   d. The domain of $f + g$ is the same as the domain of $f - g$ for all f and g.

II. Which of the following has domain $= \{x\colon x \in \mathbb{R}, x > 0\}$ where $f(x) = \dfrac{-1}{x+1}$ and $g(x) = \sqrt{x}$?
   a. $f + g$ b. f/g c. $f \circ g$ d. $g \circ f$

III. Which of the following is true about $y = -8 + 6x - 2x^2$ if $f(x) = 2x^2 - x + 5$ and $g(x) = 5x - 3$?
   a. $y = (f + g)(x)$
   b. $y = (f - g)(x)$
   c. $y = (g - f)(x)$
   d. $y = (f + g)(-x)$

IV. Which of the following is the domain of $\dfrac{f}{g}$ if $f(x) = x^2 - 5x + 6$ and $g(x) = x - 3$?
   a. $\{x\colon x \in \mathbb{R}\}$
   b. $\{x\colon x \in \mathbb{R}$ and $x \neq 2, 3\}$
   c. $\{x\colon 2 \le x \le 3\}$
   d. $\{x\colon x \in \mathbb{R}$ and $x \neq 3\}$

In Problems 1–12 two functions, f and g, are given. Determine the functions $f + g$, $f - g$, $f \cdot g$, and f/g and find their respective domains.

1. $f(x) = 2x + 3$, $g(x) = -3x$
2. $f(x) = 2x - 5$, $g(x) = -3x + 4$
3. $f(x) = 4$, $g(x) = 10$
4. $f(x) = -3$, $g(x) = 0$
5. $f(x) = x$, $g(x) = \dfrac{1}{x}$
6. $f(x) = x^2$, $g(x) = x + 1$
7. $f(x) = \sqrt{x+1}$, $g(x) = \sqrt{1-x}$
8. $f(x) = x^3 + x$, $g(x) = \dfrac{1}{\sqrt{x+1}}$
9. $f(x) = 1 + x^5$, $g(x) = 1 - |x|$
10. $f(x) = \sqrt{1+x}$, $g(x) = \dfrac{1}{x^5}$
11. $f(x) = \sqrt[5]{x+2}$, $g(x) = \sqrt[4]{x-3}$
12. $f(x) = \dfrac{x+1}{x}$, $g(x) = \dfrac{x}{x-3}$

In Problems 13–24 find $f \circ g$ and $g \circ f$ and determine the domain of each.

13. $f(x) = 3x$, $g(x) = x - 2$
14. $f(x) = x - 3$, $g(x) = 2x + 1$
15. $f(x) = 5$, $g(x) = 8$
16. $f(x) = 0$, $g(x) = -2$
17. $f(x) = x$, $g(x) = \dfrac{1}{2x}$

Answers to Readiness Check

I. d II. b III. c IV. d

18. $f(x) = 2x^2, \qquad g(x) = x + 5$
19. $f(x) = 2x - 4, \qquad g(x) = x + 3$
20. $f(x) = \sqrt{x + 1}, \qquad g(x) = x^3$
21. $f(x) = \dfrac{x}{2 - x}, \qquad g(x) = \dfrac{x + 1}{x}$
22. $f(x) = |x|, \qquad g(x) = -x$
23. $f(x) = \sqrt{1 - x}, \qquad g(x) = \sqrt{x - 1}$
24. $f(x) = \begin{cases} x, & x \geq 0, \\ 2x, & x < 0, \end{cases}$

    $g(x) = \begin{cases} -3x, & x \geq 0 \\ 5x, & x < 0 \end{cases}$
25. Let $f(x) = 2x + 4$ and $g(x) = \frac{1}{2}x - 2$. Show that $(f \circ g)(x) = (g \circ f)(x) = x$. (When this occurs, we say that f and g are **inverse functions.**)
26. If $f(x) = -3x + 2$, find a function g such that $(f \circ g)(x) = (g \circ f)(x) = x$.
27. If $f(x) = x^2$, find two functions g_1 and g_2 such that $(f \circ g_1)(x) = (f \circ g_2)(x) = x^2 - 10x + 25$.
28. Let $h(x) = 1/\sqrt{x^2 + 1}$. Determine two functions f and g such that $f \circ g = h$.
29. Let $k(x) = (1 + \sqrt{x})^{5/7}$. Find the domain of k. Determine three functions f, g, and h such that $f \circ g \circ h = k$.
30. Let $h(x) = x^2 + x$, and let $f_1(x) = x^2 - x$, $g_1(x) = x + 1$, $f_2(x) = x^2 + 3x + 2$, and $g_2(x) = x - 1$. Show that $f_1 \circ g_1 = f_2 \circ g_2 = h$. This illustrates the fact that there is often more than one way to write a given function as the composition of two other functions.
31. Let f and g be the linear functions

    $$f(x) = ax + b, \qquad g(x) = cx + d$$

    Find conditions on a and b in order that $f \circ g = g \circ f$.

** 32. Each of the following functions satisfies an equation of the form $(f \circ f)(x) = x$ or $(f \circ f \circ f)(x) = x$ or $(f \circ f \circ f \circ f)(x) = x$, and so on. For each function, discover what type of equation is appropriate.
    (a) $A(x) = \sqrt[3]{1 - x^3}$
    (b) $B(x) = \sqrt[7]{23 - x^7}$
    (c) $C(x) = 1 - 1/x$, domain is $\mathbb{R} - \{0, 1\}$ (This is the set of all real numbers except 0 and 1.)
    (d) $D(x) = 1/(1 - x)$, domain is $\mathbb{R} - \{0, 1\}$
    (e) $E(x) = (x + 1)/(x - 1)$, domain is $\mathbb{R} - \{1\}$
    (f) $F(x) = (x - 1)/(x + 1)$, domain is $\mathbb{R} - \{-1, 0, 1\}$
    (g) $G(x) = (4x - 1)/(4x + 2)$,

    domain is $\mathbb{R} - \{-\frac{1}{2}, 0, \frac{1}{4}, \frac{1}{2}, 1\}$

* 33. A manufacturer of designer shirts determines that the demand function for her shirts is $x = D(p) = 400(50 - p)$, where p is the wholesale price she charges per shirt and x is the number of shirts she can sell at that price. Note that, as is common, the higher the price, the fewer shirts she can sell. Assume that the manufacturer's fixed cost is \$8000 and her material and labor costs amount to \$8 per shirt.
    (a) Determine the total cost function C as a function of p.
    (b) Determine the total revenue function R as a function of p.
    (c) Determine the profit function P as a function of p.
    (d) By completing the square, determine the price that yields the greatest profit. What is this maximum profit (or minimum loss)?
34. Let $f(x)$ be a function. Show that

    $$g(x) = \tfrac{1}{2}[f(x) + f(-x)]$$

    is an even function.
35. Let $f(x)$ be a function. Show that

    $$h(x) = \tfrac{1}{2}[f(x) - f(-x)]$$

    is an odd function.
36. Show that any function $f(x)$ whose domain is $\mathbb{R}$ can be written as the sum of an even function and an odd function. [Hint: Use the results of Problems 34 and 35.]

3.7 Inverse Functions

In Section 3.3 we defined a function as a rule: You give me an x in the domain and I'll give you $y = f(x)$. Sometimes it is necessary to reverse this procedure: You give me a y and I'll find the x for which $y = f(x)$.

Before giving a definition, we start with an example. Let $y = f(x) = 2x + 3$. What value of x leads to a given y? To answer this question, we solve the equation $y = 2x + 3$ for x.

$$y = 2x + 3$$
$$2x = y - 3$$
$$x = \frac{y - 3}{2}$$

Because it is customary (but not necessary) to write y as a function of x, interchange x and y in the last step to obtain the function $y = (x - 3)/2$.

Consider the two functions $f(x) = 2x + 3$ and $g(x) = (x - 3)/2$. We compute

$$(f \circ g)(x) = f(g(x)) = f\left(\frac{x-3}{2}\right) = 2\left(\frac{x-3}{2}\right) + 3 = x - 3 + 3 = x$$

and

$$(g \circ f)(x) = g(f(x)) = g(2x+3) = \frac{(2x+3)-3}{2} = \frac{2x}{2} = x$$

The functions $f(x) = 2x + 3$ and $g(x) = (x - 3)/2$ are said to be *inverses* of one another.

Definition of Inverse Functions

The functions f and g are **inverse functions** if the following conditions hold:

(i) For every x in the domain of g, $g(x)$ is in the domain of f and $f(g(x)) = x$.

(ii) For every x in the domain of f, $f(x)$ is in the domain of g and $g(f(x)) = x$.

In this case we write $f(x) = g^{-1}(x)$ or $g(x) = f^{-1}(x)$, and we say f is the inverse of g and g is the inverse of f.

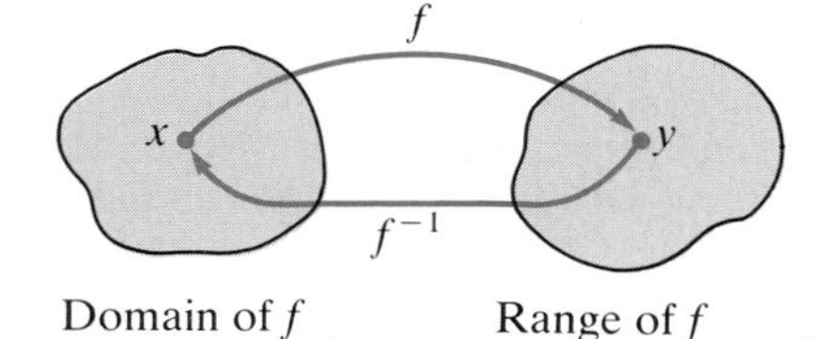

Figure 1 Pictorial representation of two inverse functions.

NOTE The domain of f^{-1} is the range of f, and the domain of f is the range of f^{-1}.

In the notation of Section 3.6, we may write

$$(f \circ f^{-1})(x) = x \quad \text{and} \quad (f^{-1} \circ f)(x) = x$$

In Figure 1 we give a pictorial representation of a function f and its inverse f^{-1}.

EXAMPLE 1 ***Two Inverse Functions***

$f(x) = x^3$ and $g(x) = \sqrt[3]{x} = x^{1/3}$ are inverses because

$$f(g(x)) = f(x^{1/3}) = (x^{1/3})^3 = x$$

and

$$g(f(x)) = g(x^3) = (x^3)^{1/3} = x$$

We do not have to worry about domains here because both x^3 and $\sqrt[3]{x}$ are defined for every real number x.

Computing an Inverse Function

To compute an inverse function, we use the following procedure:

(i) Replace $f(x)$ with y.
(ii) Interchange x and y.
(iii) Solve for y in terms of x, if possible.
(iv) The resulting y is equal to $f^{-1}(x)$.
(v) Set the domain of f^{-1} equal to the range of f.

We use the phrase *if possible* in step (iii) because it is not always possible to solve explicitly for y as a *function* of x.

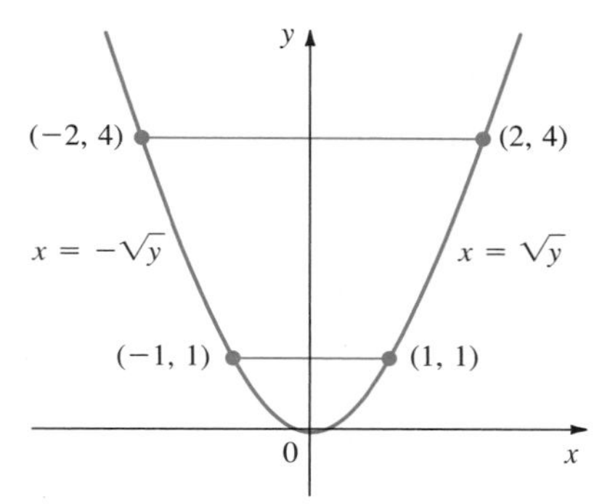

Figure 2 The graph of $y = x^2$. For each $y > 0$, there are two x's such that $y = x^2$.

EXAMPLE 2 ***A Function That Does Not Have An Inverse***

Let $y = x^2$. Then $x = \pm\sqrt{y}$. This means that each nonzero value of y comes from *two* different values of x. See Figure 2. We recall from Section 3.3 that a function $y = f(x)$ can be thought of as a rule that assigns to each x in its domain a *unique* value of y. Suppose we try to define the inverse of x^2 by taking the positive square root. If $f(x) = x^2$ and $g(x) = \sqrt{x}$ are inverses, then $g(f(x)) = x$ for every x in the domain of f. But $-2 \in$ domain of $f\,(= \mathbb{R})$, and $g(f(-2)) = g((-2)^2) = g(4) = \sqrt{4} = 2 \neq -2$. Thus $g(f(x)) \neq x$ for every x in the domain of f.

To avoid problems like the one just encountered, we make the following definition.

Definition of a One-to-One Function

The function $y = f(x)$ is **one-to-one on the interval $[a, b]$** if whenever $x_1, x_2 \in [a, b]$ and $f(x_1) = f(x_2)$, then $x_1 = x_2$. That is, each value of $f(x)$ comes from only one value of x. If $x_1 \neq x_2$ implies that $f(x_1) \neq f(x_2)$ for every x_1 and x_2 in the domain of f, we say that f is, simply, one-to-one.

The definition of a one-to-one function is illustrated in Figure 3.

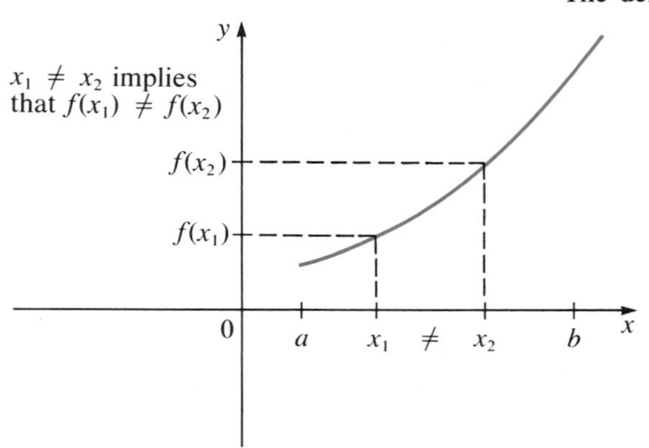

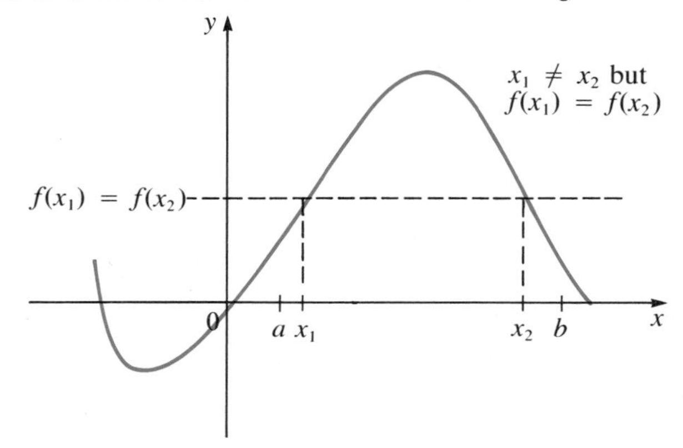

Figure 3 (a) one-to-one function (b) not a one-to-one function

It follows from the definition that

> If there are two distinct numbers x_1 and x_2 such that $f(x_1) = f(x_2)$, then f is *not* one-to-one.

In Section 3.3 we saw that an equation in x and y defined a function $y = f(x)$ if its graph satisfied the vertical lines test. We have a similar test for one-to-one functions.

Horizontal Lines Test

Let f be a function. If every horizontal line in the plane intersects the graph of f in at most one point, then f is one-to-one.

NOTE If f is a function, then the vertical lines test on p. 177 must also hold.

EXAMPLE 3 *Illustration of the Horizontal Lines Test*

We can see, in Figure 4, that $f(x) = x^2$ is not one-to-one, but $f(x) = x^3$ and $f(x) = \dfrac{1}{x}$ are one-to-one.

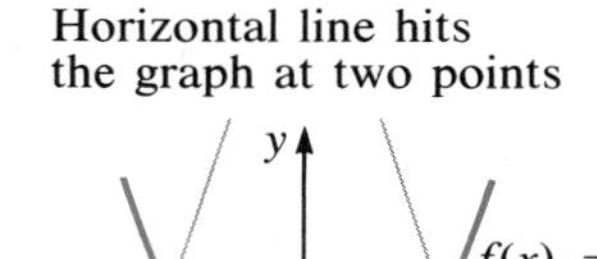

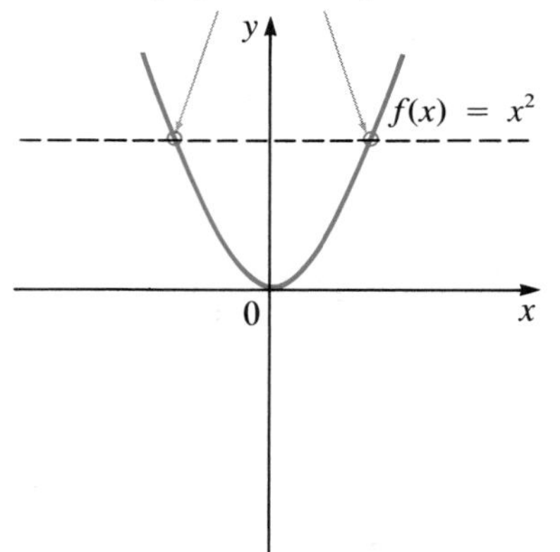

(a) $f(x) = x^2$ fails horizontal lines test

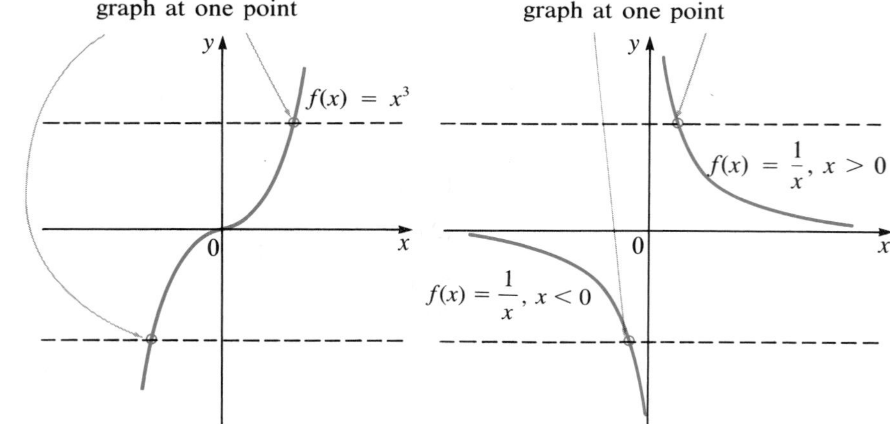

(b) $f(x) = x^3$ passes horizontal lines test

(c) $f(x) = \dfrac{1}{x}$ passes horizontal lines test

Figure 4 The horizontal line test applied to the graphs of three functions.

Why are we interested in one-to-one functions? Because if $y = f(x)$ is one-to-one, then every value of y comes from a unique value of x, so we can write x as a function of y and this new function is the inverse of y. We have shown the following:

Theorem

Every one-to-one function has an inverse.

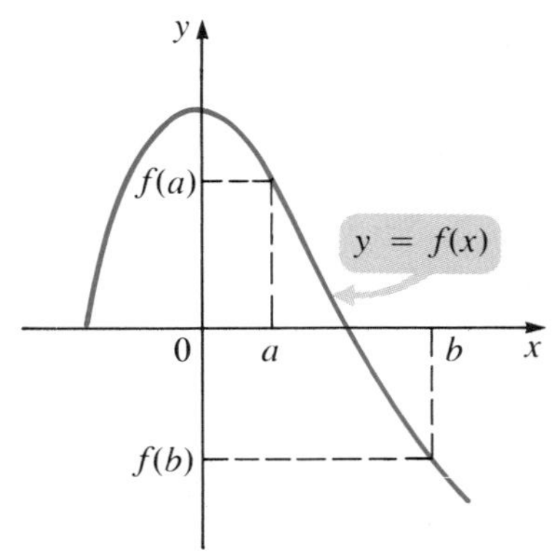

(a) $f(x)$ decreasing on $[a, b]$

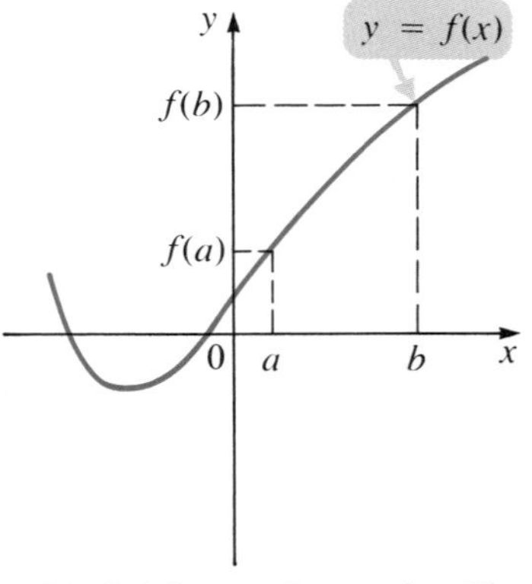

(b) $f(x)$ increasing on $[a, b]$

Figure 5

Increasing and Decreasing Functions

f is **increasing** on $[a, b]$ if, for every pair of numbers x_1, x_2 in $[a, b]$ with $x_2 > x_1$, $f(x_2) > f(x_1)$.

f is **decreasing** on $[a, b]$ if, for every pair of numbers x_1, x_2 in $[a, b]$ with $x_2 > x_1$, $f(x_2) < f(x_1)$.

These definitions are illustrated in Figure 5.

Theorem

If f is increasing or decreasing on $[a, b]$, then f is one-to-one on $[a, b]$.

Proof

Suppose f is increasing, x_1 and x_2 are in $[a, b]$ and $x_1 \neq x_2$. If $x_2 > x_1$, then $f(x_2) > f(x_1)$. If $x_1 > x_2$, then $f(x_1) > f(x_2)$. In either case, $f(x_1) \neq f(x_2)$, so f is one-to-one. The proof in the case f is decreasing is similar. ■

EXAMPLE 4 ***The Function $f(x) = x^2$ Is One-to-One When Its Domain Is Suitably Restricted***

From Figure 2 on p. 207, we see that $f(x) = x^2$ is decreasing on $(-\infty, 0]$ and increasing on $[0, \infty)$. As we have seen, $f(x) = x^2$ is not one-to-one and does not have an inverse. However, the function $f_1(x) = x^2$ with domain $[0, \infty)$ does have an inverse as does the function $f_2(x) = x^2$ on $(-\infty, 0]$. ■

EXAMPLE 5 ***Finding an Inverse Function***

Show that $f(x) = \sqrt{x}$ has an inverse given by $g(x) = x^2$, $x \geq 0$.

SOLUTION $y = \sqrt{x}$ is an increasing function for $x \geq 0$, so it has an inverse on $[0, \infty)$.

$$y = \sqrt{x}$$
$$x = \sqrt{y} \qquad \text{Interchange } x \text{ and } y$$
$$x^2 = y \qquad \text{Square both sides}$$

Thus $g(x) = f^{-1}(x) = x^2$, $x \geq 0$.

We need to have $x \geq 0$ to ensure that the domain of f^{-1} is equal to the range of f (the range of $f(x) = \sqrt{x}$ is $\mathbb{R}^+ = \{y\colon y \geq 0\}$).

Check $f(g(x)) = f(x^2) = \sqrt{x^2} = x \quad \text{if} \quad x \geq 0$

$$g(f(x)) = g(\sqrt{x}) = (\sqrt{x})^2 = x$$

NOTE The function $f(x) = x^2$ is not one-to-one and does not have an inverse, but the function $g(x) = x^2$, $x \geq 0$, which is a *different* function, does have the inverse $\sqrt{x}$.

A function changes if you change its domain.

EXAMPLE 6 *Finding an Inverse Function*

Show that $f(x) = \dfrac{1}{x}$ is one-to-one over its domain and compute its inverse.

SOLUTION $\dfrac{1}{x}$ is defined if $x \neq 0$. Suppose

$$\frac{1}{x_1} = \frac{1}{x_2} \qquad \text{Neither } x_1 \text{ nor } x_2 = 0$$

Then

$$x_2 = x_1 \qquad \text{Cross multiply}$$

Thus f is one-to-one. To find f^{-1}, we follow the steps on page 207:

$$y = \frac{1}{x} \qquad \text{Replace } f(x) \text{ with } y$$

$$x = \frac{1}{y} \qquad \text{Interchange } x \text{ and } y$$

$$y = \frac{1}{x} \qquad \text{Solve for } y \text{ in terms of } x$$

Thus $g(x) = f^{-1}(x) = \dfrac{1}{x}$. That is, the function $f(x) = \dfrac{1}{x}$, $x \neq 0$, is its own inverse.

We close this section by showing the relationship between the graph of a function f and the graph of its inverse function f^{-1}. The graphs of three functions and their inverses are sketched in Figure 6. It appears that the graphs of f and f^{-1} are symmetric about the line $y = x$. Let us see why this observation is true.

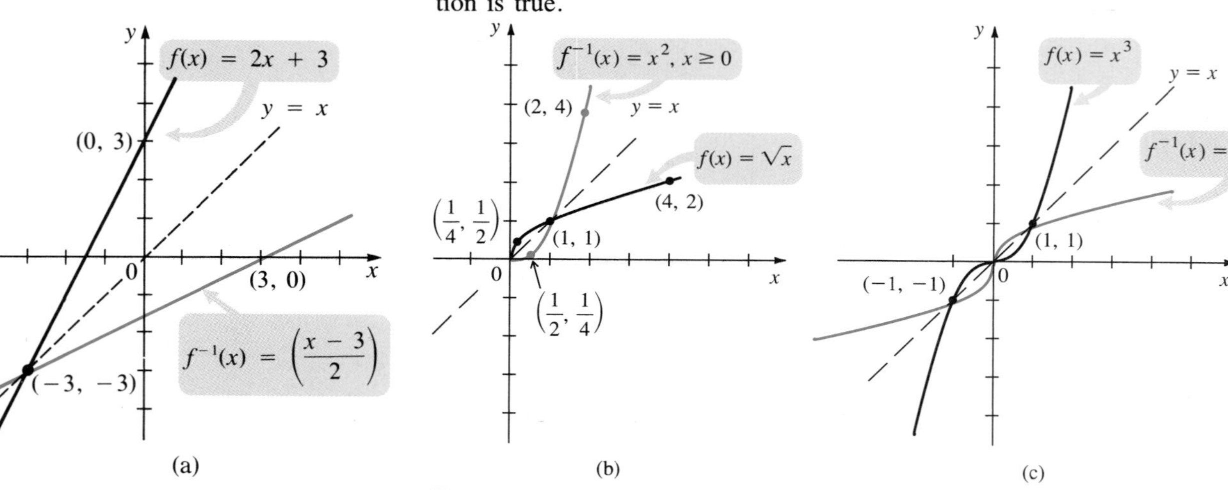

Figure 6 The graphs of three functions and their inverses.

Reflection Property

Suppose $y = f(x)$. Then, if f^{-1} exists, $x = f^{-1}(y)$. That is, (x, y) is in the graph of f if and only if (y, x) is in the graph of f^{-1}. In Figure 6(b), for example, $(\frac{1}{4}, \frac{1}{2})$, $(1, 1)$, and $(4, 2)$ are in the graph of $f(x) = \sqrt{x}$. The points $(\frac{1}{2}, \frac{1}{4})$, $(1, 1)$, and $(2, 4)$ are in the graph of $f^{-1}(x) = x^2, x \geq 0$. In Figure 6(b) if we fold the page along the line $y = x$, we find that the graphs of f and f^{-1} coincide. We see that

Reflection Property

The graphs of f and f^{-1} are **reflections** of one another about the line $y = x$.

This **reflection property,** as it is called, of the graphs of inverse functions enables us immediately to obtain the graph of f^{-1} once the graph of f is known.

Problems 3.7

Readiness Check

I. Which of the following is decreasing on $(-\infty, \infty)$?
a. $f(x) = -(x + 2)^2$ b. $f(x) = -|x - 1|$
c. $f(x) = -x^3$ d. $f(x) = \begin{cases} -2, & x \geq 0 \\ -|x|, & x < 0 \end{cases}$

II. Which of the following is a point on the graph of the inverse of the function graphed below?
a. $(2, 4)$
b. $(4, -2)$
c. $(-4, 2)$
d. $(-2, -4)$

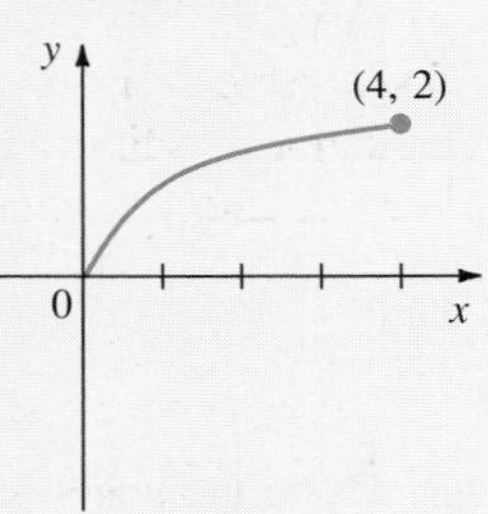

III. Which of the following is the inverse function for $y = f(x) = x^2 - 3$?
a. $f^{-1}(x) = x^2 + 3$
b. $f^{-1}(x) = \sqrt{x + 3}$
c. $f^{-1}(x) = \sqrt{x - 3}$
d. The function does not have an inverse on $(-\infty, \infty)$.

IV. Which of the following is *not* a one-to-one function?
a. $f(x) = |x + 1|$ b. $f(x) = x^3 + x$
c. $f(x) = 2x - 7$ d. $f(x) = \dfrac{1}{2x - 10}$

V. Which of the following is the inverse function of $g(x) = x^2 - 4, x \leq 0$?
a. $f(x) = -\sqrt{x + 4}$
b. $f(x) = \sqrt{x - 4}$
c. $f(x) = -\sqrt{x - 4}$
d. $g(x)$ does not have an inverse.

In Problems 1–24 show that the given function is one-to-one over its entire domain and find its inverse.

1. $f(x) = 2x - 1$
2. $f(x) = -7x + 4$
3. $f(x) = \frac{2}{3}x - \frac{1}{4}$
4. $f(x) = \dfrac{x + 5}{7}$
5. $f(x) = \dfrac{3 - 2x}{11}$
6. $f(x) = 3.862x - 1.803$
7. $f(x) = \dfrac{3}{x}$
8. $f(x) = \dfrac{-1}{2x}$
9. $f(x) = \dfrac{1}{x + 1}$
10. $f(x) = \dfrac{2}{x + 5}$
11. $f(x) = \dfrac{3}{4 - x}$
12. $f(x) = -x^3$

Answers to Readiness Check

I. c II. a III. d IV. a V. a

13. $f(x) = \dfrac{4}{2 + x^3}$ 14. $f(x) = \dfrac{3 - x^3}{7}$

* 15. $f(x) = \sqrt{x + 2}$ * 16. $f(x) = \sqrt{4x - 5}$

* 17. $f(x) = \sqrt{1 - 2x}$ 18. $f(x) = \sqrt[3]{x - 1}$

19. $f(x) = \dfrac{1}{\sqrt[3]{x - 7}}$ 20. $f(x) = x^5$

* 21. $f(x) = \dfrac{x}{x + 1}$ * 22. $f(x) = \dfrac{2x}{5 - x}$

23. $f(x) = (x + 3)^3$ 24. $f(x) = 1 - (x - 2)^3$

In Problems 25–38 determine intervals over which each function is one-to-one and find the inverse function over each such interval.

25. $f(x) = 1 + x^2$ 26. $f(x) = (1 + x)^2$

27. $f(x) = (4 - x)^2$ 28. $f(x) = 4 - (7 + x)^2$

29. $f(x) = x^4$ 30. $f(x) = \dfrac{1}{x^2}$

31. $f(x) = x^4 - 5$ * 32. $f(x) = \dfrac{x^2}{1 + x^2}$

* 33. $f(x) = x^2 - 7x + 6$ [Hint: Complete the square.]

34. $f(x) = x^2 + 2x + 2$

* 35. $f(x) = x^4 + 4x^2 + 4$

36. $f(x) = x^{10}$

37. $f(x) = |x|$

38. $f(x) = |x - 3|$

In Problems 39–41 show that each function has an inverse function. Do not try to compute it.

* 39. $f(x) = x + x^3$

* 40. $f(x) = 1 + x + x^3 + x^5 + x^7$

* 41. $f(x) = \dfrac{1}{x^3 + x + 1}$

In Problems 42–51 the graph of a function is given. Determine whether or not the function is one-to-one over the given interval.

42.

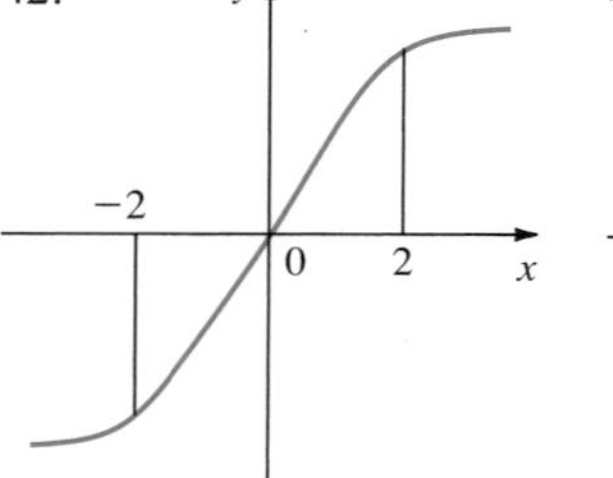

43.

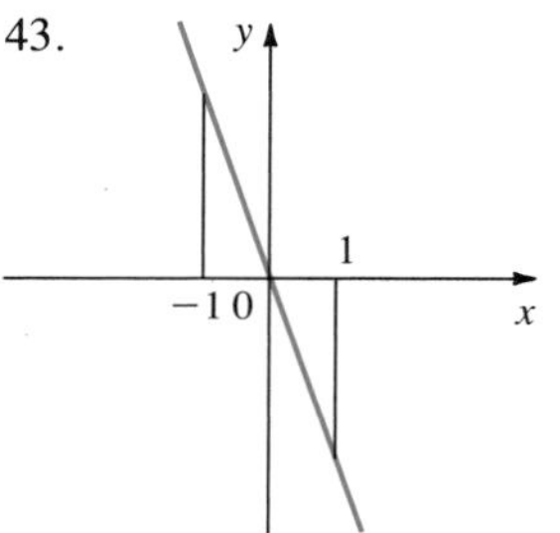

44.

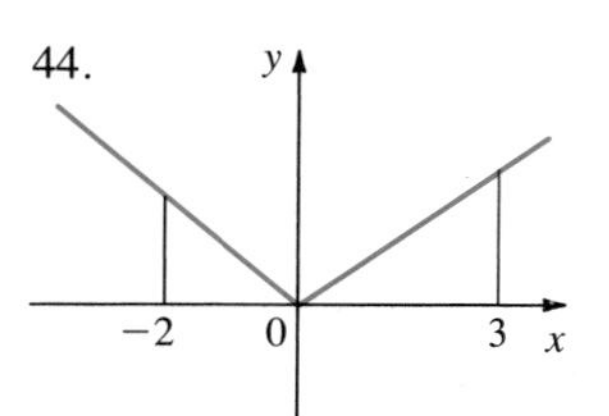

45.

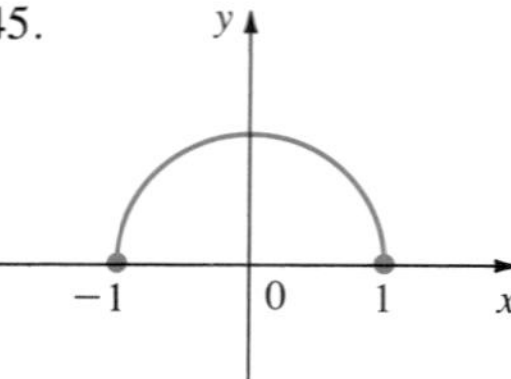

46.

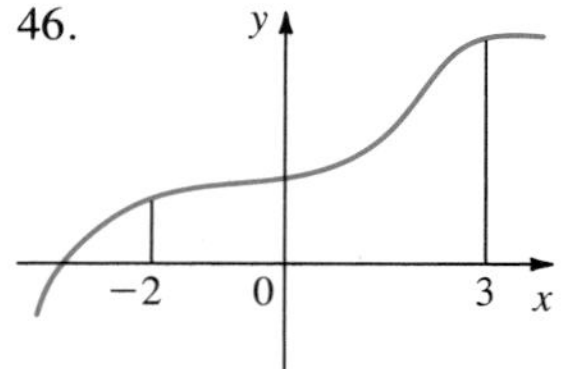

47.

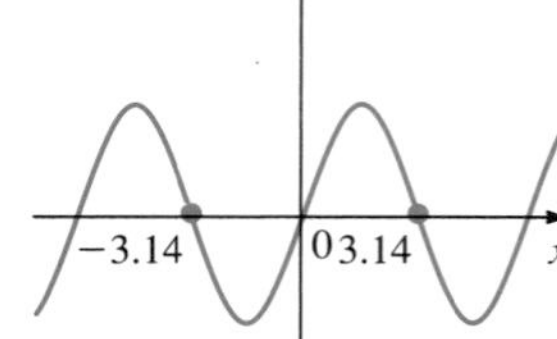

48.

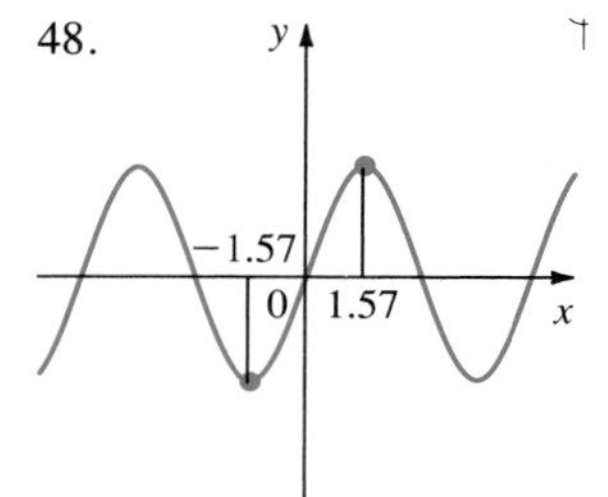

49.

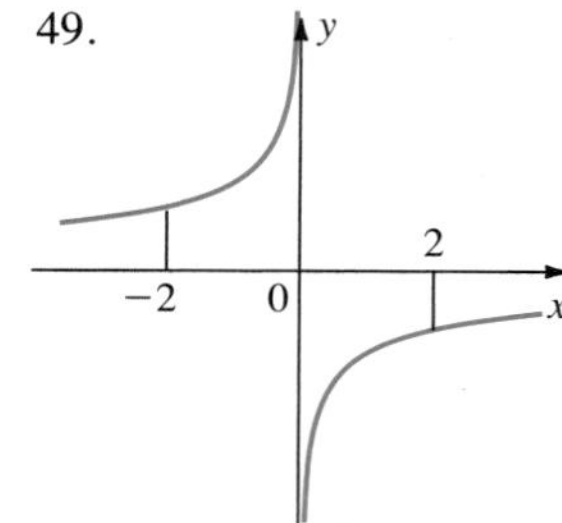

50.

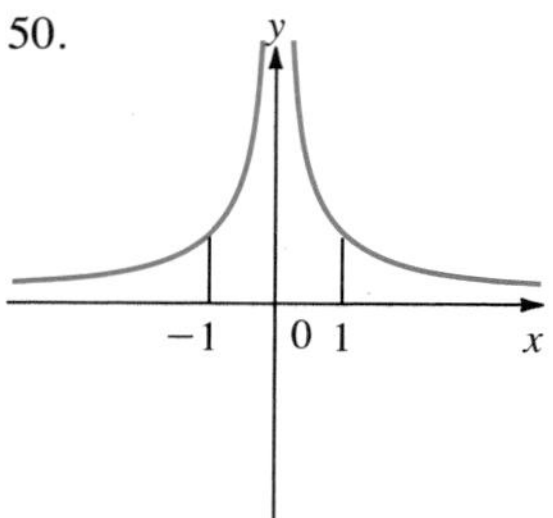

51.

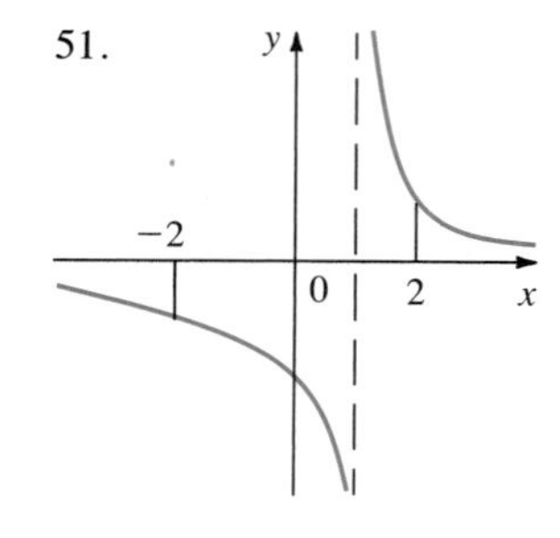

In Problems 52–56 the graph of a function over an interval $[a, b]$ is given. Sketch the graph of f^{-1} on the interval $[c, d] = [f(a), f(b)]$ or $[f(b), f(a)]$.

52.

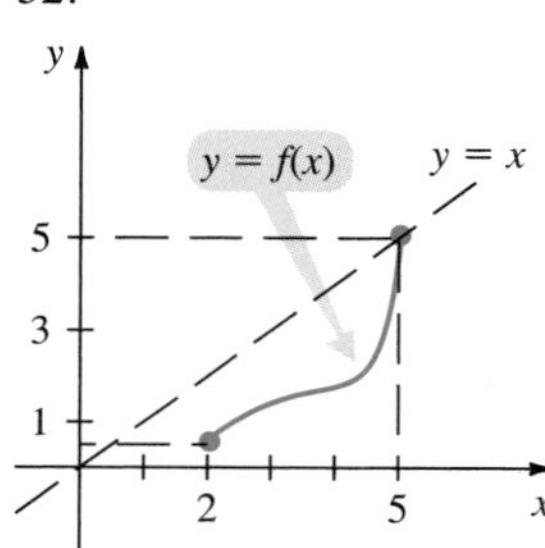

53.

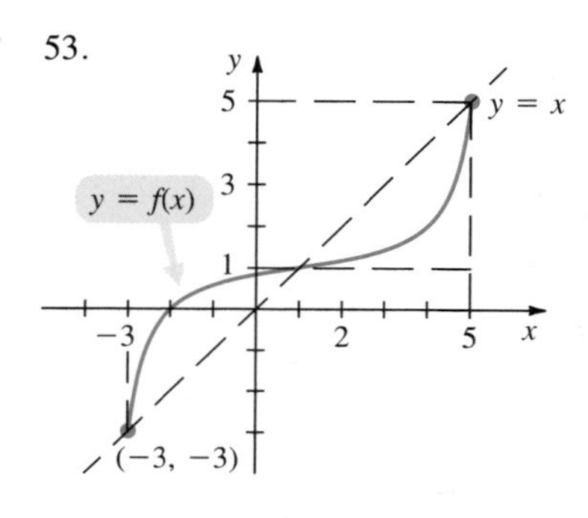

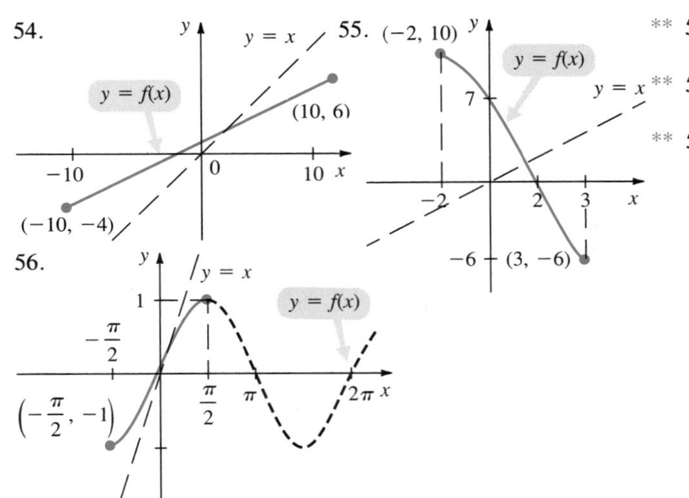

** 57. Prove that if a function is one-to-one, then its inverse is unique.

** 58. If f is one-to-one, show that the domain of f^{-1} is the range of f.

** 59. A function is called **continuous** on $[a, b]$ if its graph is unbroken—that is, if f is defined for all x in $[a, b]$ and if its graph has no gaps over $[a, b]$. The **intermediate value theorem** states that if f is continuous on $[a, b]$ and if c is any number with $f(a) < c < f(b)$ or $f(b) < c < f(a)$, then there is an $x \in (a, b)$ such that $f(x) = c$. That is, if f is continuous and assumes the values $f(a)$ and $f(b)$, then it assumes every value in between. Suppose that f is continuous and increasing with domain $[a, b]$. Prove the following:
(a) Range of $f = [f(a), f(b)]$.
(b) f^{-1} exists with domain $[f(a), f(b)]$.
(c) f^{-1} is an increasing function.

3.8 Direct and Inverse Proportionality

Consider the distance formula (p. 80)

$$s = vt \qquad \text{(distance = average velocity × time)}$$

If velocity is constant, then we say that distance s **varies directly** as t, or s is **directly proportional** to t.

Definition of Direct Variation

We say that y **varies directly as** x or is **directly proportional** to x or, simply, **proportional** to x if there is a number k such that

$$y = kx$$

k is called the **constant of proportionality.**

EXAMPLE 1 ***Distance Is Directly Proportional to Time***

A car is traveling at a constant 60 miles/hour. Then

$$s = 60t$$

Here s is directly proportional to t with constant of proportionality 60. ■

EXAMPLE 2 ***Simple Interest Is Directly Proportional to Time***

Suppose \$1000 is invested with 8% simple interest paid. Then, by the simple interest formula (see p. 183), the interest earned is

$$I = (1000)(0.08)t = 80t$$

Thus I is proportional to t with constant of proportionality 80.

EXAMPLE 3 *Kinetic Energy Is Directly Proportional to the Square of Velocity*

The kinetic energy K of a moving particle is given by

$$K = \tfrac{1}{2}mv^2$$

where v is the velocity of the particle and m is its mass. For an object of constant mass, we say that K **varies directly as the square** of velocity or that K is **directly proportional to v^2.** Here the constant of proportionality is $\frac{1}{2}m$. If, for example, $m = 10$ kilograms, then the constant of proportionality is 5. ■

EXAMPLE 4 *Hooke's Law*

Hooke's law† states that the force (in pounds) necessary to stretch a spring is proportional to the length of spring displaced; that is,

$$F = kx \tag{1}$$

The constant of proportionality in (1) is called the **spring constant.** A spring is stretched 10 inches by a force of 2 pounds (See Figure 1).

(a) What is the spring constant?
(b) What force is necessary to stretch the spring 15 inches?

Figure 1 A 2-pound force stretches a spring 10 inches.

SOLUTION

(a) From (1) and the information given, we have

$$F = kx$$
$$2 = k \cdot 10$$
$$k = \frac{1}{5}$$

Therefore, $F = \frac{1}{5}x$.

(b) $F = \frac{1}{5}(15) = 3$ pounds.

We now turn to another kind of variation.

> **Definition of Inverse Variation**
>
> We say that y **varies inversely** as x or is **inversely proportional** to x if there is a number k such that
>
> $$y = \frac{k}{x}$$

† Named for Robert Hooke (1635–1703), an English philosopher, physicist, chemist, and inventor.

EXAMPLE 5 ***Ideal Gas Law***

The **ideal gas law** states that, for an ideal gas,

$$PV = nRT \tag{2}$$

where P is the pressure of the gas (pounds per square inch = lb/in^2), V is its volume (in^3), n is the number of moles of the gas, T is its absolute temperature in Kelvin (Kelvin = degrees Celsius plus 273), and R is a constant. Suppose that the temperature of the gas is kept constant. Then (2) can be written

$$P = \frac{nRT}{V} = \frac{k}{V} \qquad \text{where } k = nRT \tag{3}$$

That is, pressure is inversely proportional to volume when the temperature is kept fixed. This fundamental law of physics is also known as **Boyle's law.** ■

EXAMPLE 6 ***Pressure Is Inversely Proportional to Volume***

The pressure of 100 in^3 of an ideal gas enclosed in an aluminum canister is 40 lb/in^2. The temperature of the gas does not vary.

(a) Find an equation relating volume and pressure.
(b) If the volume decreases to 30 in^3, what happens to the pressure?

SOLUTION

(a) In (3) we have

$$P = \frac{k}{V}$$

$$40 = \frac{k}{100}$$

$$k = 4000$$

So

$$P = \frac{4000}{V}$$

(b) $P = \dfrac{4000}{30} = \dfrac{400}{3} \approx 133.33$ lb/in^2

Definition of Joint Variation

We say that z **varies jointly** as x and y if there is a number k such that

$$z = kxy$$

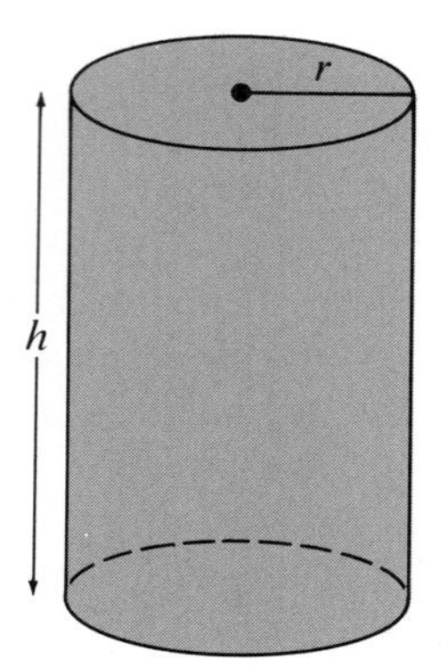

Figure 2 A cylinder with radius r and height h.

EXAMPLE 7 ***The Volume of a Cylinder Varies Jointly with the Height and the Square of the Radius***

The formula for the volume of a cylinder (see Figure 2) with radius r and height h is given by

$$V = \pi r^2 h$$

We see that V varies jointly as r^2 and h with constant of proportionality π. ■

EXAMPLE 8 ***Finding the Constant of Proportionality***

z varies directly with x and inversely with y^2. Suppose that $z = 5$ when $x = 2$ and $y = 4$. Find the constant of proportionality.

SOLUTION We have

$$z = k\frac{x}{y^2}$$

Inserting the given values, we have

$$5 = k\frac{2}{4^2} = k\frac{2}{16} = \frac{1}{8}k$$

Thus

$$k = 8 \cdot 5 = 40$$

Problems 3.8

Readiness Check

I. Which of the following is true about inverse variation if the constant of proportionality k is positive?
a. If one quantity increases, the other also increases.
b. $k = \frac{y}{x}$
c. $k = \frac{x}{y}$
d. If one quantity increases the other decreases.

II. Which of the following means the same as $B = \frac{ka^2c}{\sqrt{d}}$?
a. B is proportional to the product of the square of a and c and the square root of d.
b. B varies jointly as c and the square of a and inversely as the square root of d.
c. B is proportional to the square root of d and inversely to c and the square of a.
d. B varies inversely with the square root of d and directly with the square of a.

III. Which of the following is a statement of Newton's second law given below if the constant of proportionality is 9.8?
Newton's second law: The force on a body is directly proportional to its mass.
a. $F = 9.8/m$ b. $Fm = 9.8$ c. $F = 9.8m$
d. $F = m/9.8$

IV. If z varies directly as the square of w and $w = 3$ when $z = 90$, which of the following is w when $z = 125/2$?
a. 5/2 b. 25/4 c. 10 d. 100

In Problems 1–13 write a mathematical equation that represents the given information. Use the letter k to denote a constant of proportionality if no constant is given.

1. y is proportional to x with constant of proportionality 10.
2. t is proportional to s with constant of proportionality 15.
3. z is inversely proportional to w with constant of proportionality 3.3.
4. x is inversely proportional to t with constant of proportionality 0.2.
5. x is proportional to z^3.
6. r is inversely proportional to z^2.
7. w is proportional to $\sqrt{z}$.
8. w is proportional to $l^{3/2}$.

Answers to Readiness Check

I. d II. b III. c IV. a

9. y is inversely proportional to $v^{1/4}$.
10. x varies jointly with w^2 and z.
11. R varies jointly with x^3 and $\sqrt{z}$.
12. Q varies directly with r^2 and inversely with v.
13. w varies directly with u^2 and inversely with v^3.
14. A spring is stretched 15 in. by a force of 6 lb. How much will it be stretched by a force of 4 lb?
15. The cost C of producing a certain product is directly proportional to the square of the number, q, of items produced. It costs $12.80 to produce 80 items.
    (a) Write C as a function of q.
    (b) How much does it cost to produce 200 items?
16. The volume of a sphere is proportional to the cube of the radius. The volume of a sphere of radius 6 in. is 288π in^3.
    (a) What is the formula for the volume of a sphere?
    (b) What is the volume of a sphere of radius 4 in.?
17. The surface area of a sphere is proportional to the square of the radius. A sphere of radius 2 in. has a surface area of 16π in^2.
    (a) What is the formula for the surface area of a sphere?
    (b) What is the surface area of a sphere of radius 3 in.?
18. If y is proportional to x^2, what happens to y if x doubles?
19. If z is proportional to w^3, what happens to z if w triples?
20. If s is inversely proportional to t, what happens to s if t doubles?
21. If v is inversely proportional to u^2, what happens to v if u is halved?
22. According to **Kepler's third law,** the length of a planet's solar year (the time it takes to revolve about the sun) is proportional to $d^{3/2}$, where d is the distance from the planet to the sun. The planet earth is approximately 93 million miles from the sun, and its year is approximately 365 days. Mars is approximately 142 million miles from the sun. Estimate the length of the Martian year.

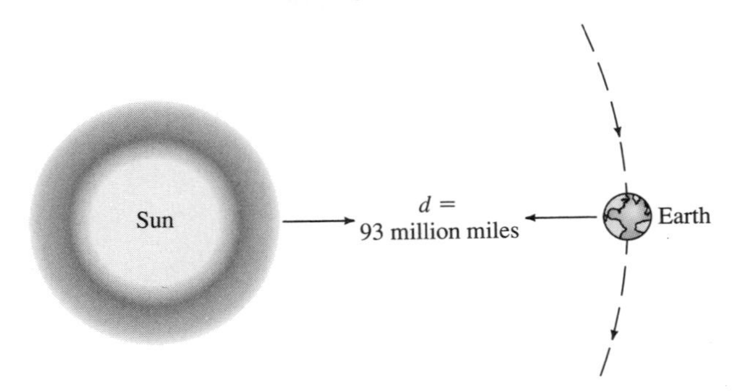

23. Estimate the length of the year on Mercury, which is approximately 36 million miles from the sun. [Hint: Use Kepler's third law.]
24. The pressure of 50 in^3 of an ideal gas enclosed in a balloon is 30 lb/in^2. The temperature of the gas is held constant. If the pressure decreases to 20 lb/in^3, what happens to the volume?
25. In Problem 24 what happens to the pressure if the volume decreases to 20 in^3?
26. In Problem 24 what must the pressure be increased to if the volume is to be cut in third?
27. The intensity I of illumination from a light source is inversely proportional to the square of the distance to the source. The intensity from a certain light is 100 foot-candles at a distance of 5 feet. What is the intensity at a distance of (a) 2 feet? (b) 10 feet?
28. The **centripetal force** (the "center-seeking" force) of an object revolving about a center varies inversely as the distance from the center and directly as the square of the velocity of the object. A car turns a corner of radius 80 feet at a speed of 100 ft/sec ($\approx$68 miles/hr). The centripetal force of the car is 15,000 lb. What is the centripetal force if the car goes around a curve of radius 60 feet at a speed of 60 ft/sec?

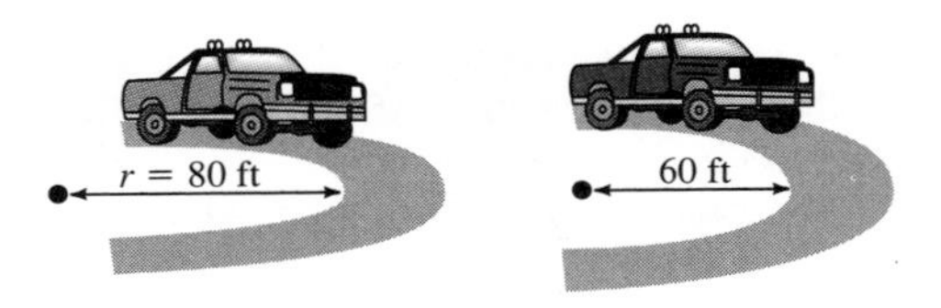

29. In Problem 28 what is the speed of the car if it generates a centripetal force of 20,000 lb as it goes around a curve of radius 75 feet?
30. The volume of a cone varies jointly as the height and the square of its radius. The volume of a cone with height 4 inches and radius 3 inches is 12π in^3.
    (a) Find the formula for the volume of a cone.
    (b) What is the volume of a cone of height 8 inches and radius 4 inches?
31. In Problem 30 what happens to the volume of a cone if its radius doubles and its height is cut in third?
32. According to **Poiseuille's law,**† the resistance R of a blood vessel varies directly as the length l and inversely as the fourth power of the radius r.
    (a) Write a formula giving R as a function of l and r.
    (b) What happens to R if both l and r double?
33. In Problem 32 what happens to r if both R and l double?
34. The frequency of vibration f of a violin string varies directly as the square root of the tension T on the string and inversely as the length l.
    (a) Write a formula giving f as a function of T and l.
    (b) What happens to f if both T and l double?
    (c) What happens to f if T doubles and l stays the same?

† Jean Louis Poiseuille (1799–1869) was a French physiologist.

Summary Outline of Chapter 3

- **Distance Formula:** The distance, d, between the points (x_1, y_1) and (x_2, y_2) is given by
$$d = \sqrt{(x_1 - x_2)^2 + (y_1 - y_2)^2}$$ p. 151
- **Midpoint Formula:** The midpoint of the line joining (x_1, y_1) and (x_2, y_2) is
$$\left(\frac{x_1 + x_2}{2}, \frac{y_1 + y_2}{2}\right)$$ p. 152
- **Graph:** The graph of an equation in two variables is the set of all points in the xy-plane whose coordinates satisfy the equation. p. 153
- **Circles:** The **unit circle** is centered at $(0, 0)$ with radius 1. Its equation is $x^2 + y^2 = 1$. p. 153
  The equation $(x - h)^2 + (y - k)^2 = r^2$ is the equation of a circle centered at (h, k) with radius r. p. 154
- **Linear Equation in Two Variables:** This is an equation of the form $ax + dy = c$, where a and d are not both equal to zero. p. 157
- **Slope:** The **slope,** m, of a line is given by $m = \dfrac{y_2 - y_1}{x_2 - x_1}$ where (x_1, y_1) and (x_2, y_2) are two points on the line and $x_1 \neq x_2$. p. 158
  If the line is vertical $(x_1 = x_2)$, then the slope is **undefined.** pp. 158, 160
  If the slope, m, of a line is positive, then the graph of the line will rise as x increases. p. 159
  If $m < 0$, then the graph of the line will fall as x increases. p. 159
  Horizontal lines have a slope of zero. p. 159
  Two lines are parallel if and only if their slopes are equal. p. 160
  Two nonvertical lines are perpendicular if and only if their slopes are negative reciprocals of one another $(m_1 = -1/m_2)$. p. 160
- **Equations of a Line:** Let (x_1, y_1) be a point on a line with slope m and y-intercept b.
  point-slope equation: $y - y_1 = m(x - x_1)$ p. 162
  slope-intercept equation: $y = mx + b$ p. 163
  standard equation: $Ax + By = C$ p. 164
- A **function,** f, is a rule that assigns to each member of one set (called the **domain** of the function) a unique member of another set (called the **range** of the function). We often write $y = f(x)$. p. 170
  Here y is called the **image** of x. The variable x is called the **independent variable,** and y is the **dependent variable.** p. 171
- **Vertical Lines Test:** An equation in two variables defines a function if every vertical line in the plane intersects the graph of the equation in at most one point. p. 177
- The function $f(x)$ is **even** if $f(-x) = f(x)$ or **odd** if $f(-x) = -f(x)$. p. 179
- The **graph** of a function f is the set of points $\{(x, f(x)): x \in \text{domain of } f\}$ p. 189
- If $f(x)$ is even, then its graph is symmetric about the y-axis. p. 191
- If $f(x)$ is odd, then its graph is symmetric about the origin. p. 191
- **Shifting Graphs**
  The graph of $f(x) + c$ is the graph of $f(x)$ shifted up c units if $c > 0$ or down $|c|$ units if $c < 0$. p. 194
  The graph of $f(x - c)$ is the graph of $f(x)$ shifted c units to the right if $c > 0$ or $|c|$ units to the left if $c < 0$. p. 195
  The graph of $-f(x)$ is the graph of $f(x)$ reflected about the x-axis. p. 196
  The graph of $f(-x)$ is the graph of $f(x)$ reflected about the y-axis. p. 196
- **Function Operations**
  Sum: $(f + g)(x) = f(x) + g(x)$ p. 201

Difference: $(f-g)(x)=f(x)-g(x)$ p. 201

Product: $(f\cdot g)(x)=f(x)g(x)$ p. 201

Quotient: $\left(\dfrac{f}{g}\right)(x)=\dfrac{f(x)}{g(x)}$ p. 201

Composition: $(f\circ g)(x)=f(g(x))$ p. 202

- **Inverse Function:** f and g are inverse functions if p. 206
  (i) for every $x\in$ domain of g, $g(x)\in$ domain of f and $f(g(x))=x$.
  (ii) for every $x\in$ domain of f, $f(x)\in$ domain of g and $g(f(x))=x$.
- **One-to-One Function:** $f(x)$ is **one-to-one** on $[a, b]$ if $f(x_1)=f(x_2)$ implies that $x_1=x_2$ whenever x_1 and x_2 are in $[a, b]$. p. 207
- A one-to-one function, f, has an inverse denoted by f^{-1}. p. 208
- **Horizontal Lines Test:** f is one-to-one on its domain if every horizontal line in the plane intersects the graph of f in at most one point. p. 208
- f is **increasing** on $[a, b]$ if $f(x_2)>f(x_1)$ when $x_2>x_1$. p. 209
- f is **decreasing** on $[a, b]$ if $f(x_2)<f(x_1)$ when $x_2>x_1$. p. 209
  If f is increasing or decreasing, then f is one-to-one. p. 209
- **Reflection Property:** The graphs of f and f^{-1} are reflections of one another about the line $y=x$. p. 211
- **Direction Variation:** y **varies directly** as x or y is **directly proportional** to x if $y=kx$ for some constant k (called the **constant of proportionality**). p. 213
- **Inverse Variation:** y **varies inversely** or is **inversely proportional** to x if $y=\dfrac{k}{x}$ for some constant k. p. 214
- **Joint Variation:** z **varies jointly** as x and y if $z=kxy$ for some constant k. p. 215

Review Exercises for Chapter 3

In Exercises 1–4 find the distance between the given points.

1. $(2, -1)$, $(3, 2)$
2. $(4, 0)$, $(0, 4)$
3. $(-1, 1)$, $(1, -1)$
4. $(2, 3)$, $(-4, -5)$
5. Find the midpoint of the line segment joining the points $(1, 2)$ and $(-5, 7)$.
6. Find the equation of the circle centered at $(2, -3)$ with radius 4.
7. Show that $x^2+4x+y^2-8y+16=0$ is the equation of a circle. Find its center and radius.

In Exercises 8–12, a linear function is given.

(a) Find the y value that corresponds to the given x value.
(b) Find the x- and y-intercepts (if any) of its graph.
(c) Sketch the graph.

8. $y=-2$; $x=5$
9. $x+y=2$; $x=-3$
10. $y=7x-4$; $x=2$
11. $3x+5y=15$; $x=-2$
12. $y=4$; $x=1$

In Exercises 13–18, find the slope-intercept equation of a straight line when either two points on it or its slope and one point are given. Also, find a standard form and a point-slope form of the equation of the line.

13. $(2, 5)$, $(-1, 3)$
14. $(-2, 4)$, $m=3$
15. $(4, -2)$, $(2, -4)$
16. $(-1, 4)$, $m=2$
17. $(1, 4)$, $(1, 7)$
18. $(3, -8)$, $(-8, -8)$
19. Find the slope-intercept equation of the line parallel to the line $2x+5y=4$ and that contains the point $(2, -3)$.
20. Find the slope-intercept equation of the line perpendicular to the line $3x+4y=5$ and that contains the point $(1, 7)$.
21. Sketch the demand function $p=80-0.04q$. What does the slope represent?
22. Sketch the supply function $s=50+0.02q$. What does the slope represent?
23. The demand and supply functions for a certain product are given in Exercises 21 and 22. Determine the equilibrium price.
24. In Exercise 23 describe what happens if the price charged is $5 more than the equilibrium price. What happens if it is $5 less?

In Exercises 25–33, determine whether the given equation defines y as a function of x, and if so, find its domain and range.

25. $4x - 2y = 5$
26. $\dfrac{x^2 - y}{2} = 4$
27. $\dfrac{y}{x} = 1$
28. $(x - 1)^2 + (y - 3)^2 = 4$
29. $y = \sqrt{x + 3}$
30. $3 = \dfrac{1 + x^2 + x^4}{2y}$
31. $y = \dfrac{x}{x^2 + 1}$
32. $y = \dfrac{x}{x^2 - 1}$
33. $y = \sqrt{x^2 - 6}$
34. For $y = f(x) = \sqrt{x^2 - 4}$, calculate $f(2)$, $f(-\sqrt{5})$, $f(x + 4)$, $f(x^3 - 2)$, and $f(-1/x)$.
35. If $y = f(x) = 1/x$, show that for $\Delta x \neq 0$,

$$\frac{f(x + \Delta x) - f(x)}{\Delta x} = -\frac{1}{x(x + \Delta x)}$$

In Problems 36–39 determine whether the given function is even, odd, or neither.

36. $f(x) = x^2 - 10$
37. $f(x) = \dfrac{1}{x^4}$
38. $f(x) = x^3 + 1$
39. $f(x) = \dfrac{x^3}{x^5 + x}$

In Problems 40–50, sketch the graph of the given function.

40. $f(x) = x^2 - 2$
41. $f(x) = x^3 + 1$
42. $f(x) = |x| - 1$
43. $f(x) = (x + 1)^2$
44. $f(x) = (x + 2)^3$
45. $f(x) = x^2 - 4x + 7$
46. $f(x) = x^2 + 8x + 20$
47. $f(x) = 5 - (x - 1)^3$
48. $f(x) = \dfrac{1}{x + 2}$
49. $f(x) = \begin{cases} 2, & x \geq 0 \\ 3, & x < 0 \end{cases}$
50. $f(x) = \begin{cases} 1, & x \leq 0 \\ x, & 0 < x \leq 2 \\ x^2, & x > 2 \end{cases}$

51. The graph of the function $y = f(x)$ is given in the figure below. Sketch the graph of $f(x - 3)$, $f(x) - 5$, $f(-x)$, $-f(x)$, and $4 - f(1 - x)$.

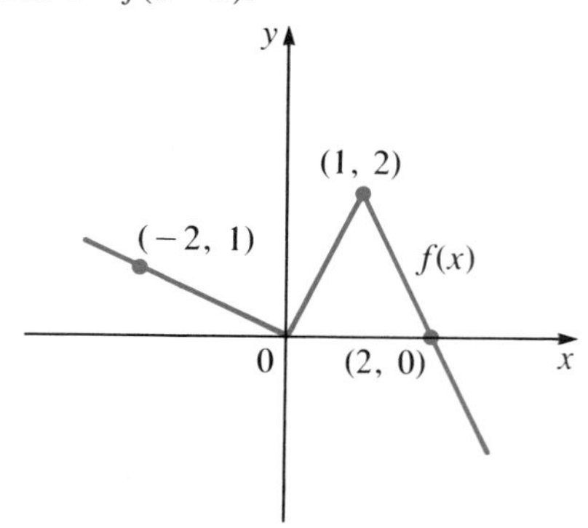

52. Repeat Exercise 51 for the function graphed in the figure below.

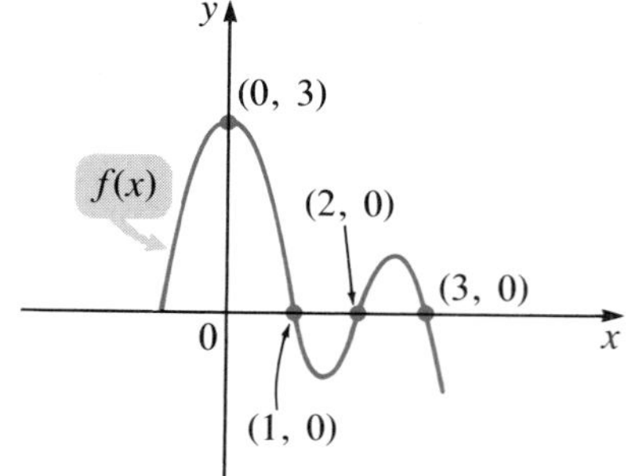

53. Let $f(x) = \sqrt{x + 1}$ and $g(x) = x^3$. Find $f + g$, $f - g$, $f \cdot g$, g/f, $f \circ g$, and $g \circ f$, and determine their respective domains.
54. Repeat Exercise 53 for $f(x) = 1/x$ and $g(x) = x^2 - 4x + 3$.

In Exercises 55–60 explain why the given function is one-to-one on its domain, and find its inverse.

55. $f(x) = 4x - 1$
56. $f(x) = -2x + 5$
57. $f(x) = \dfrac{2}{x}$
58. $f(x) = 3x^3 - 1$
59. $f(x) = \sqrt{x - 2}$
60. $f(x) = \sqrt[3]{x + 3}$

In Exercises 61 and 62 determine intervals over which each function is one-to-one, and determine the inverse function over each such interval.

61. $f(x) = 3 + 2x^2$
62. $f(x) = x^2 + 4x + 1$

63. The graph of a function over an interval $[a, b]$ is given. Sketch the graph of f^{-1} on the interval $[f(a), f(b)]$.

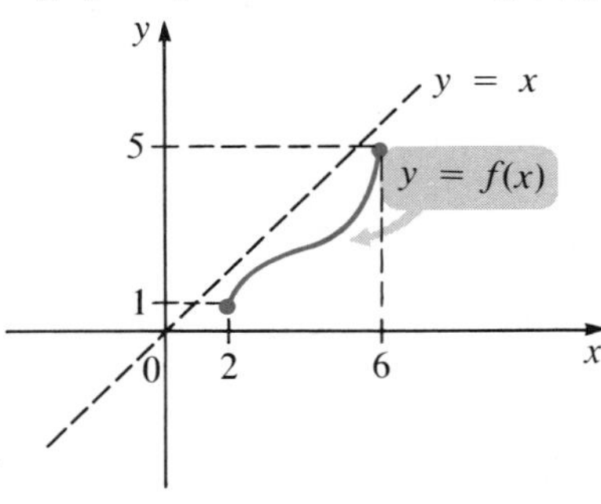

In Exercises 64–68 write a mathematical equation that represents the given information.

64. y is proportional to the square of x.
65. v is inversely proportional to u.
66. F is inversely proportional to the square root of d.
67. P varies directly as s and inversely as the cube of t.
68. z varies jointly with the square root of x and $y^{5/7}$.
69. A string is stretched 1 foot by a force of 10 lb. What force will stretch it 3 inches?
70. y varies inversely as x^2. If x doubles, what happens to y?

Chapter 4

Polynomials and Zeros of Polynomials

4.1 Quadratic Functions and Parabolas

In this section we begin a discussion of polynomial functions.

A **polynomial function** is a function that can be written as

$$p(x) = a_n x^n + a_{n-1} x^{n-1} + \cdots + a_2 x^2 + a_1 x + a_0 \tag{1}$$

where $a_n, a_{n-1}, \cdots, a_1$ and a_0 are real numbers and n is a positive integer. If $a_n \neq 0$, then the **degree** of the polynomial function is n. A nonzero constant function is said to have **degree zero.** A polynomial function of degree 2 is called a **quadratic** function. That is,

Definition of a Quadratic Function

$$p(x) = ax^2 + bx + c \tag{2}$$

where a, b, and c are real numbers and $a \neq 0$.

We have seen quadratic functions before. In Section 2.3 we solved quadratic equations. That is, we found zeros of quadratic functions.

In Section 3.5 we showed how to obtain the graph of a quadratic function of the form $y = x^2 + ax + b$ by completing the square and shifting or reflecting the graph of x^2 a number of times. We say more about the graphs of such functions now.

EXAMPLE 1 ***Graphing Two Parabolas***

Sketch the graphs of $f(x) = 2x^2$ and $f(x) = -2x^2$.

SOLUTION First we note that if $f(x) = 2x^2$, then

$$f(-x) = 2(-x)^2 = 2x^2 = f(x)$$

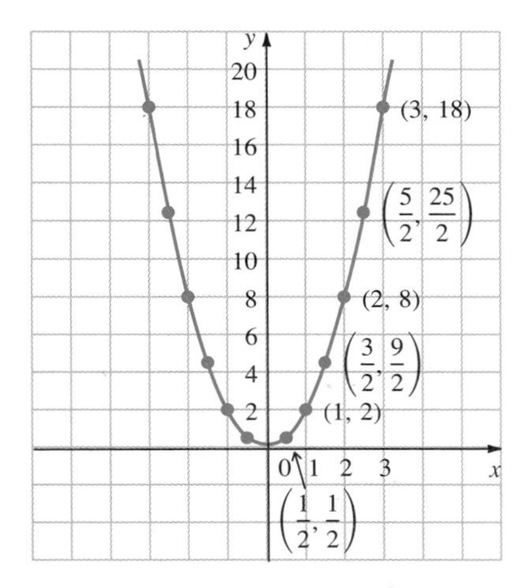

(a) Graph of $f(x) = 2x^2$

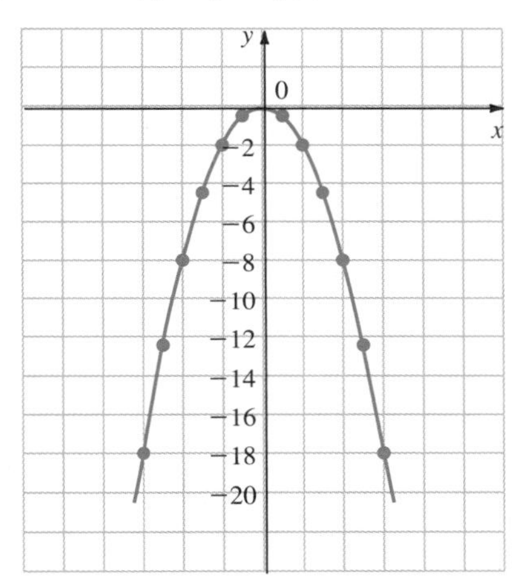

(b) Graph of $f(x) = -2x^2$

Figure 1 The graph of $-2x^2$ is the reflection about the x-axis of the graph of $2x^2$.

so f is an even function, and its graph is symmetric about the y-axis (see p. 191). We need only plot some points for $x \geq 0$, draw the graph for $x \geq 0$, and then reflect this graph about the y-axis. The graph is sketched in Figure 1(a). The graph of $f(x) = -2x^2$ is obtained by reflecting the graph of $f(x) = 2x^2$ about the x-axis (see p. 196) and is given in Figure 1(b).

The graphs of $\frac{1}{2}x^2$, x^2, $2x^2$, and so on, all have the same appearance. The only difference is in the scale. Graphs of $\frac{1}{2}x^2$, x^2, $2x^2$, $-\frac{1}{2}x^2$, $-x^2$, and $-2x^2$ are given in Figure 2.

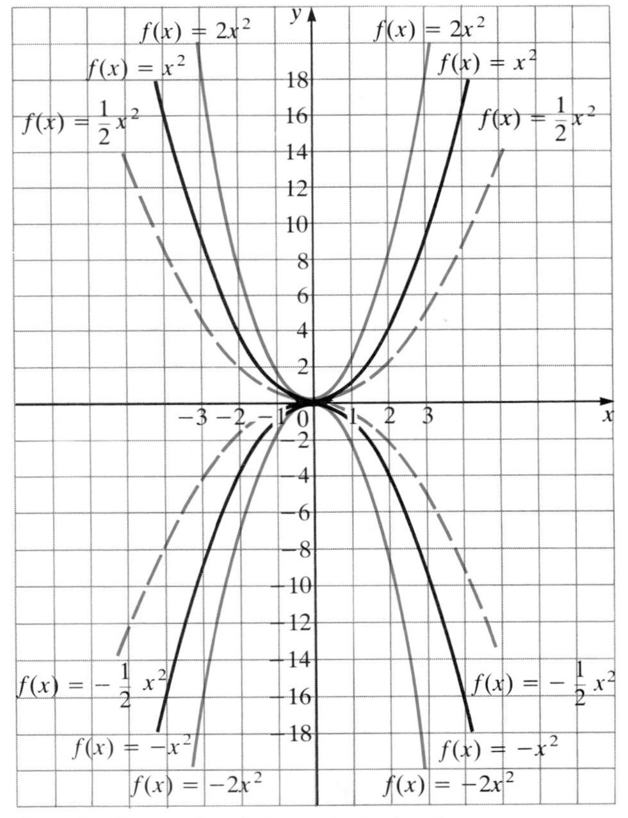

Figure 2 The graphs of six quadratic functions.

We observe that the graphs of $\frac{1}{2}x^2$, x^2, and $2x^2$ open upward, and the graphs of $-\frac{1}{2}x^2$, $-x^2$, and $-2x^2$ open downward. We have

> The graph of $f(x) = ax^2$ opens upward if $a > 0$ and opens downward if $a < 0$.

Using this information, we can sketch the graph of any quadratic function.

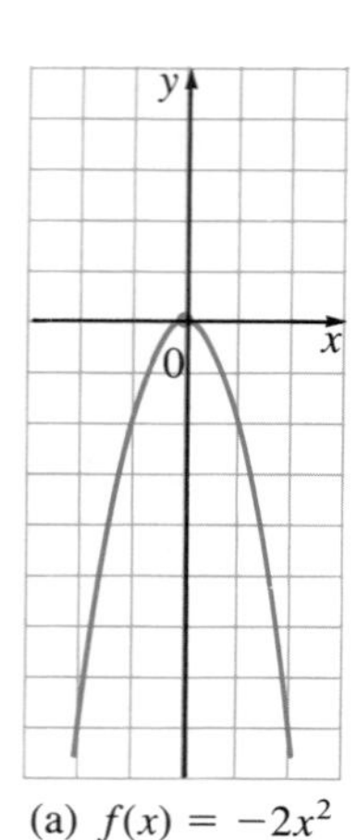

(a) $f(x) = -2x^2$

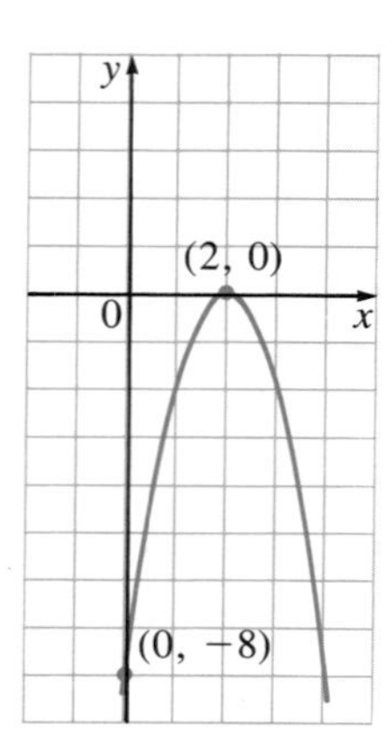

(b) $f(x) = -2(x - 2)^2$

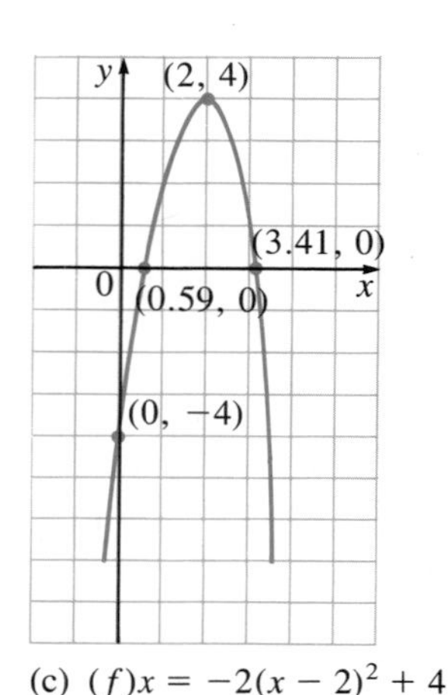

(c) $(f)x = -2(x - 2)^2 + 4$

Figure 3 The graph of $f(x) = -2x^2 + 8x - 4 = -2(x - 2)^2 + 4$ is obtained by shifting the graph of $-2x^2$ twice.

EXAMPLE 2 ***Graphing a Quadratic by Shifting a Known Graph***

Sketch the graph of the quadratic function

$$f(x) = -2x^2 + 8x - 4$$

SOLUTION

Step 1 Factor out -2:

$$-2x^2 + 8x - 4 = -2(x^2 - 4x + 2)$$

Step 2 Complete the square:

$$\begin{aligned}-2(x^2 - 4x + 2) &= -2[(x^2 - 4x + 4) - 4 + 2]\\ &= -2[(x - 2)^2 - 2]\\ &= -2(x - 2)^2 + 4\end{aligned}$$

Now, from the material in Section 3.5 (pp. 194 and 195), the graph of $-2(x - 2)^2$ is the graph of $-2x^2$ shifted 2 units to the right; the graph of $-2(x - 2)^2 + 4$ is the graph of $-2(x - 2)^2$ shifted 4 units up. Before drawing the graphs we note three other things:

(a) *y-intercept:* When $x = 0$, $y = -2 \cdot 0^2 + 8 \cdot 0 - 4 = -4$.
(b) *x-intercepts:* $y = -2x^2 + 8x - 4 = 0$ when

$$x = \frac{-8 \pm \sqrt{64 - 4(-2)(-4)}}{-4} \qquad \text{This is the quadratic formula}$$

$$= 2 \pm \frac{\sqrt{32}}{4} = 2 \pm \frac{\sqrt{16}\sqrt{2}}{4} = 2 \pm \sqrt{2}$$

$$\begin{aligned}2 + \sqrt{2} &\approx 3.41\\ 2 - \sqrt{2} &\approx 0.59\end{aligned} \qquad \text{These are the } x\text{-intercepts}$$

(c) *the vertex:* The point (2, 4) is called the **vertex** of the graph. Note that

$$-2(x - 2)^2 \le 0 \quad \text{and} \quad -2(x - 2)^2 = 0 \quad \text{when } x = 2$$

Further,

$$-2(x - 2)^2 + 4 \le 4 \quad \text{and} \quad -2(x - 2)^2 + 4 = 4 \quad \text{when } x = 2$$

Thus, the function $f(x) = -2x^2 + 8x - 4 = -2(x - 2)^2 + 4$ takes its **maximum** value 4 when $x = 2$.

With all this information, we can obtain an accurate picture of the graph. This is done in Figure 3(c).

The graph of $-2(x - 2)^2 + 4$ is called a **parabola** with **vertex** at (2, 4) and **axis** the vertical line $x = 2$. Note that the parabola is symmetric about its axis.

We now derive a formula for the standard form of a quadratic. Consider the quadratic

$$
\begin{aligned}
f(x) &= ax^2 + bx + c && (3)\\
&= a\left(x^2 + \frac{b}{a}x + \frac{c}{a}\right) && \text{Factor out the } a\\
&= a\left[\left(x^2 + \frac{b}{a}x + \frac{b^2}{4a^2}\right) - \frac{b^2}{4a^2} + \frac{c}{a}\right] && \text{Complete the square}\\
&= a\left[\left(x + \frac{b}{2a}\right)^2 + \frac{c}{a} - \frac{b^2}{4a^2}\right] && \text{Factor}\\
&= a\left(x + \frac{b}{2a}\right)^2 + c - \frac{b^2}{4a}
\end{aligned}
$$

If $h = -\dfrac{b}{2a}$ and $k = c - \dfrac{b^2}{4a}$, then the vertex is at (h, k) and (3) can be written in the **standard form**

$$f(x) = a(x - h)^2 + k$$

Note that

$$
\begin{aligned}
f(h) = f\left(-\frac{b}{2a}\right) &= a\left(\frac{-b}{2a}\right)^2 + b\left(\frac{-b}{2a}\right) + c\\
&= \frac{ab^2}{4a^2} - \frac{b^2}{2a} + c\\
&= c + \frac{b^2}{4a} - \frac{b^2}{2a} = c - \frac{b^2}{4a} = k
\end{aligned}
$$

That is, $f(h) = k$.

The curve $y = ax^2 + bx + c$ is called a **parabola with vertical axis.** Two typical parabolas with vertical axes are sketched in Figure 4.

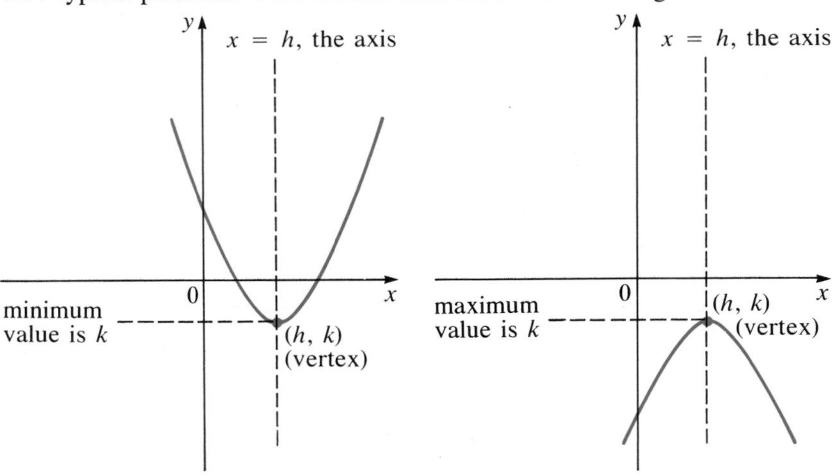

(a) Parabola with $a > 0$. (b) Parabola with $a < 0$.

Figure 4 Two parabolas with vertical axes.

We summarize our results.

To Sketch the Parabola $f(x) = ax^2 + bx + c$

Step 1 Determine whether it opens upward $(a > 0)$ or downward $(a < 0)$.

Step 2 Write the equation in the **standard form** $y = a(x - h)^2 + k$, where $h = \frac{-b}{2a}$ and $k = f(h)$. The point (h, k) is the **vertex.** The **axis** is the vertical line $x = h$. If $a > 0, f(x)$ takes its minimum value k when $x = h$. If $a < 0, f(x)$ takes its maximum value k when $x = h$.

Step 3 The y-intercept is the point $(0, c) = (0, f(0))$. This is the point at which the curve intersects the y-axis.

Step 4 Determine the x-intercepts (if any):

$$x = \frac{-b \pm \sqrt{b^2 - 4ac}}{2a}$$

If $b^2 - 4ac > 0$, there are two x-intercepts.

If $b^2 - 4ac = 0$, there is one x-intercept $\left(\frac{-b}{2a}, 0\right)$.

If $b^2 - 4ac < 0$, there is no x-intercept.
[In this case the quadratic is said to be **irreducible.**]

Step 5 Sketch the graph of $y = ax^2$ and

(a) translate this graph h units to the right if $h > 0$ or $|h|$ units to the left if $h < 0$;

(b) translate the curve $a(x - h)^2$ k units up if $k > 0$ or $|k|$ units down if $k < 0$.

In the graph in (b), plot the x- and y-intercepts and vertex. If there are no x-intercepts, then plot a few more points for accuracy.

We repeat the definition given in Step 4.

Irreducible Quadratic

The polynomial $ax^2 + bx + c$ is called **irreducible** if the quadratic equation $ax^2 + bx + c = 0$ has no real roots. This occurs if and only if $b^2 - 4ac < 0$.

EXAMPLE 3 *An Irreducible Quadratic*

Show that the quadratic $x^2 + 4x + 5$ is irreducible, and sketch its graph.

SOLUTION The discriminant of the quadratic equation $x^2 + 4x + 5 = 0$ is $4^2 - 4(1)(5) = 16 - 20 = -4 < 0$, so the equation has no real roots, and the quadratic is irreducible. We can see this in another way by completing the square:

$$x^2 + 4x + 5 = x^2 + 4x + 4 + 1 = (x + 2)^2 + 1 \geq 1$$

since $(x + 2)^2 \geq 0$ for every real number x. We see that the graph of $(x + 2)^2 + 1$ is obtained by shifting the graph of x^2 2 units to the left and 1 unit up so that its vertex is at $(-2, 1)$. This is done in Figure 5. It is evident from the graph that $x^2 + 4x + 5$ has no real zeros because it never crosses the x-axis.

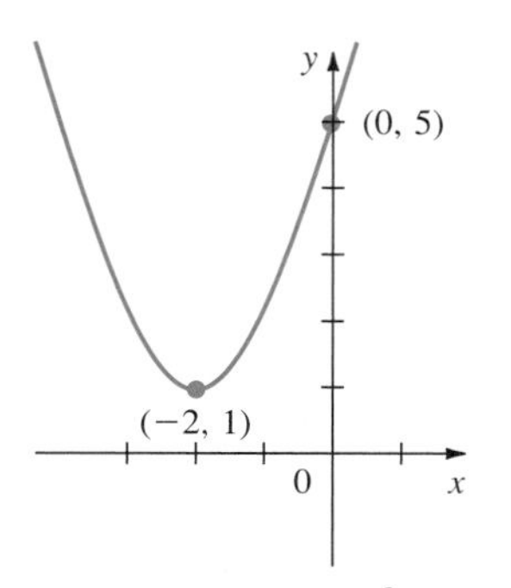

Figure 5 Graph of $x^2 + 4x + 5 = (x + 2)^2 + 1$.

Quadratic functions occur in a variety of applications. Here are two of them.

EXAMPLE 4 *Finding the Maximum Value of a Cost Function*

Find the maximum value taken by the cost function

$$C(q) = 77 + 1.32q - 0.0002q^2, \qquad 0 \leq q \leq 5000$$

Then sketch the curve.

SOLUTION In this parabola $a = -0.0002$, $b = 1.32$, and $c = 77$. Since $a < 0$, the curve opens downward and the vertex is a maximum. We have

$$h = \frac{-b}{2a} = \frac{-1.32}{-0.0004} = 3300$$

and

$$k = C(h) = C(3300) = 77 + 1.32(3300) - 0.0002(3300)^2 = 2255$$

So the maximum value taken by the cost function is \$2255.

We also compute the C-intercept,

$$C\text{-intercept} = C(0) = 77$$

and the q-intercepts:

$$q = \frac{-1.32 \pm \sqrt{(1.32)^2 - 4(-0.0002)(77)}}{-0.0004}$$

$$= \frac{-1.32 \pm \sqrt{1.804}}{-0.0004} \approx \frac{-1.32 \pm 1.343130671}{-0.0004}$$

$$q_1 = \frac{-1.32 + 1.343130671}{-0.0004} \approx -57.8$$

and

$$q_2 = \frac{-1.32 - 1.343130671}{-0.0004} \approx 6657.8$$

Both these values are outside the interval $0 \leq q \leq 5000$.

Also we compute

$$C(5000) = 77 + 1.32(5000) - 0.0002(5000)^2 = \$1677$$

The graph is sketched in Figure 6.

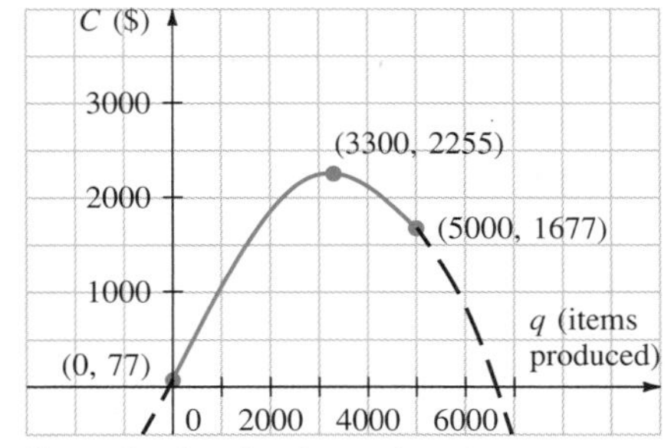

Figure 6 Graph of cost function $C(q) = 77 + 1.32q - 0.0002q^2$, $0 \leq q \leq 5000$.

EXAMPLE 5 *Sketching a Distance Function*

An object is thrown straight up from the ground into the air with an initial velocity of v_0 (measured in feet/second). Then, neglecting air resistance, the height s ft of the object at any time t sec (until it hits the ground) is given by

$$s(t) = -16t^2 + v_0 t \tag{4}$$

Suppose that $v_0 = 80$ ft/sec,

(a) After how many seconds does the object hit the ground?
(b) How high does the object get before it begins to fall?
(c) Sketch this function.

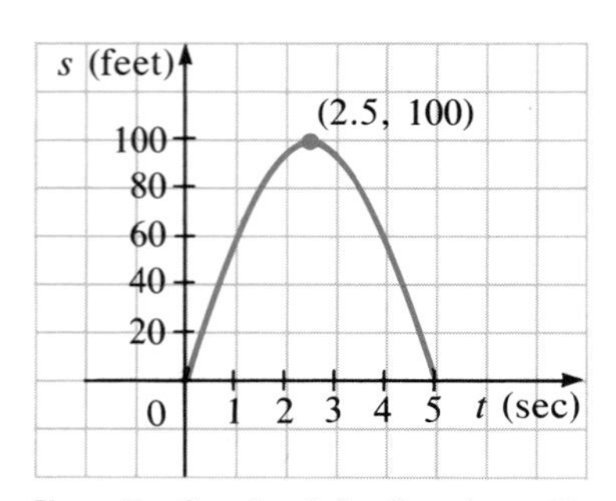

Figure 7 Graph of the function $s(t) = -16t^2 + 80t$ for $0 \le t \le 5$.

SOLUTION

(a) The object hits the ground when

$$s(t) = 0 = -16t^2 + 80t = t(-16t + 80)$$

One solution is $t = 0$. This is when the object is thrown. The other solution is obtained by setting

$$-16t + 80 = 0$$
$$t = 5 \text{ seconds}$$

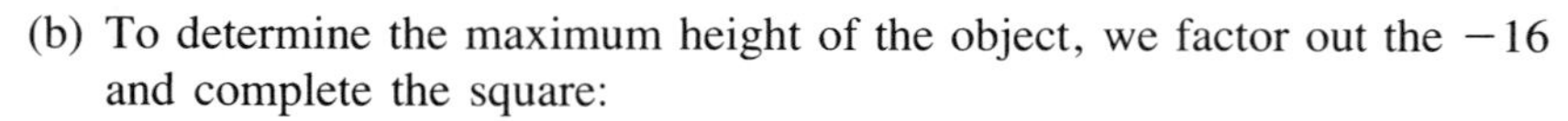

(b) To determine the maximum height of the object, we factor out the -16 and complete the square:

$$\begin{aligned} s(t) &= -16t^2 + 80t \\ &= -16(t^2 - 5t) \\ &= -16\left[\left(t - \frac{5}{2}\right)^2 - \frac{25}{4}\right] && \text{Complete the square} \\ &= -16\left(t - \frac{5}{2}\right)^2 + 100 && 16 \cdot \frac{25}{4} = 4 \cdot 25 = 100 \end{aligned}$$

Since $-16\left(t - \frac{5}{2}\right)^2 \le 0$ for every real number t, we see that the object reaches a maximum height of 100 feet after $\frac{5}{2} = 2.5$ seconds.

(c) The graph is sketched in Figure 7. Note that domain of $s = [0, 5]$.

We close this section with a different kind of problem.

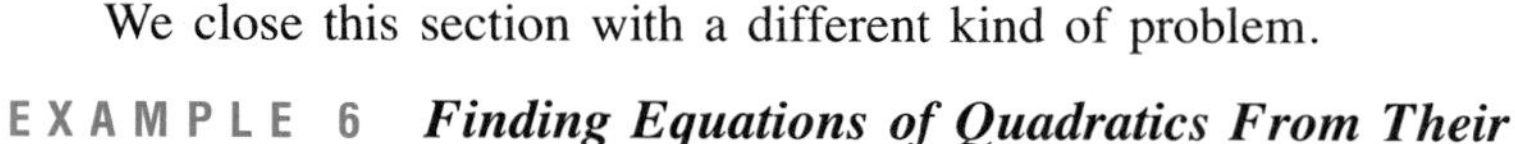

EXAMPLE 6 *Finding Equations of Quadratics From Their Graphs*

Find the equations in standard form of the two quadratics sketched in Figure 8.

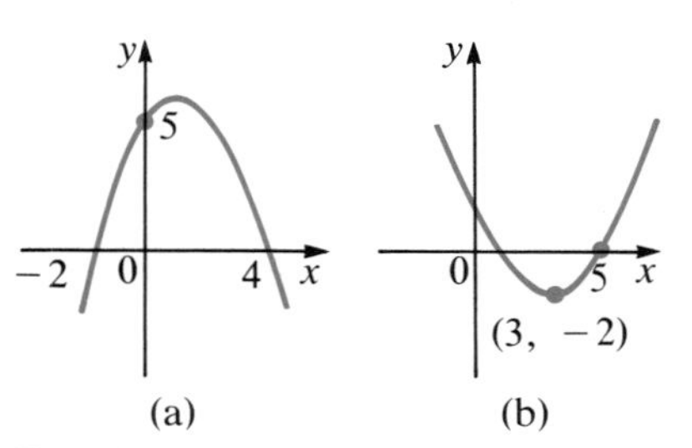

Figure 8 Graphs of two quadratics.

SOLUTION

(a) The x-intercepts are -2 and 4, so the quadratic has the form

$$y = a(x + 2)(x - 4) \quad \text{with } a < 0 \qquad \text{(since it opens downward)}$$

But the y-intercept is 5, so $(0, 5)$ is a point on the graph. Thus

$$5 = a(0 + 2)(0 - 4) = a(-8) \quad \text{and} \quad a = -\frac{5}{8}$$

so

$$y = -\frac{5}{8}(x + 2)(x - 4) = -\frac{5}{8}(x^2 - 2x - 8)$$
$$= -\frac{5}{8}x^2 + \frac{5}{4}x + 5$$

We write it in standard form by completing the square:

$$y = -\frac{5}{8}(x^2 - 2x - 8) = -\frac{5}{8}[(x - 1)^2 - 9] = -\frac{5}{8}(x - 1)^2 + \frac{45}{8}$$

The vertex is $\left(1, \frac{45}{8}\right)$.

(b) The vertex is $(3, -2)$, so the equation is

$$y = a(x - 3)^2 - 2$$

One x-intercept is $(5, 0)$. That is, when $x = 5$, $y = 0$, so we have

$$0 = a(5 - 3)^2 - 2$$
$$2 = 4a$$
$$a = \frac{1}{2}$$

and

$$y = \frac{1}{2}(x - 3)^2 - 2$$

Problems 4.1

Readiness Check

I. Which of the following has a maximum value of 2?
a. $y = -2x^2 + 12x - 16$ b. $y = x^2 - 6x + 11$
c. $y = -x^2 - 6x - 16$ d. $y = 3x^2 + 12x - 33$

II. Which of the following has a minimum on the line $x = 2$?
a. $y = 2x^2 + 8x + 1$ b. $y = x^2 - 4x + 1$
c. $y = -x^2 + 2$ d. $y = -x^2 + 2x - 3$

III. Which of the following is graphed below?
a. $y = x^2 - 4x$
b. $y = x^2 + 4x$
c. $y = -x^2 - 4x$
d. $y = -x^2 + 4x$

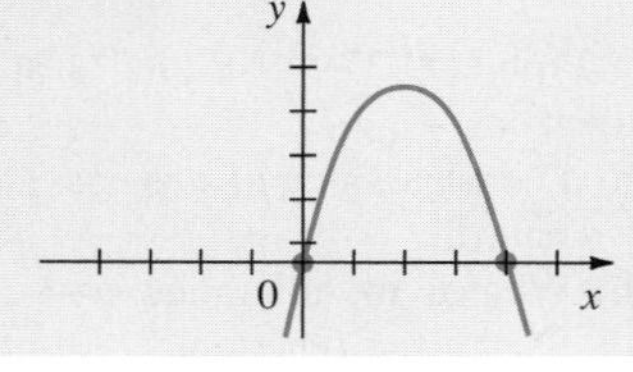

IV. If the discriminant for a quadratic equation is equal to zero, which of the following is true about its graph?
a. It has no x-intercepts.
b. It has two y-intercepts.
c. It has one x-intercept.
d. It has no y-intercept.

V. Which of the following is the minimum value for $y = 2x^2 - 12x + 8$?
a. -10
b. 3
c. 8
d. There is no minimum value.

Answers to Readiness Check

I. a II. b III. d IV. c V. a

In Problems 1–6 sketch the graph of $f(x) = ax^2$ for the indicated value of a.

1. $a = 3$
2. $a = \frac{1}{3}$
3. $a = -3$
4. $a = -\frac{3}{2}$
5. $a = 50$
6. $a = -\frac{1}{100}$

In Problems 7–18 write each quadratic in the form $y = a(x - h)^2 + k$. Then find the vertex and sketch the graph.

7. $y = x^2 + 2x$
8. $y = x^2 - 4x + 3$
9. $y = x^2 + 5x + 4$
10. $y = 2x^2 - 10x$
11. $y = 2x^2 + 4x - 5$
12. $y = 5x^2 + 10x + 5$
13. $y = -3x^2 + 9x + 12$
14. $y = -x^2 - x + 8$
15. $y = -7x^2 + 3x - 2$
16. $y = \frac{1}{2}x^2 + 2x - 3$
17. $y = \frac{x^2}{10} + \frac{x}{5} - 1$
18. $y = \frac{1}{4}x^2 + \frac{1}{8}x - \frac{1}{20}$

In Problems 19–30 sketch the given parabola. Indicate the vertex, axis, y-intercept, x-intercepts (if any), and the maximum or minimum value taken by the function.

19. $y = x^2 - 4$
20. $y = x^2 - 2x + 1$
21. $f(x) = x^2 - 5x + 4$
22. $f(x) = 9 - x^2$
23. $f(x) = 1 - 4x^2$
24. $f(x) = 2x^2 + 4$
25. $s(t) = 2t^2 + 4t - 6$
26. $C(q) = -\frac{1}{2}q^2 + 20q + 500$
27. $g(u) = 3u^2 + 6u + 3$
28. $h(v) = 4v^2 + 12v - 16$
29. $f(x) = \frac{1}{4}x^2 + x - 1$
30. $V(r) = 5r^2 - 10r + 20$

In Problems 31–36 write the equation of the parabola in the forms $y = ax^2 + bx + c$ and $y = a(x - h)^2 + k$.

31. 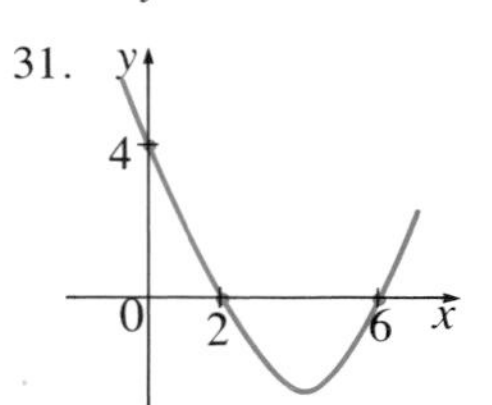

32. 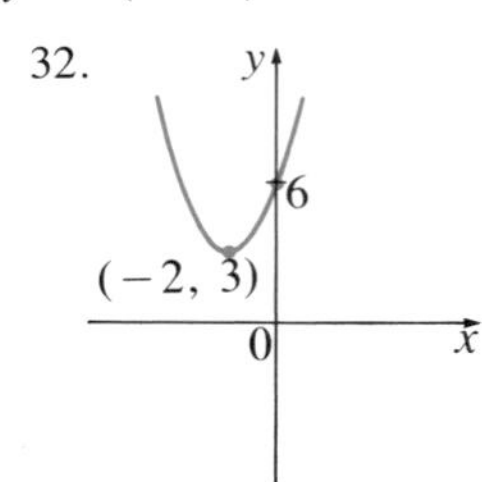

33. 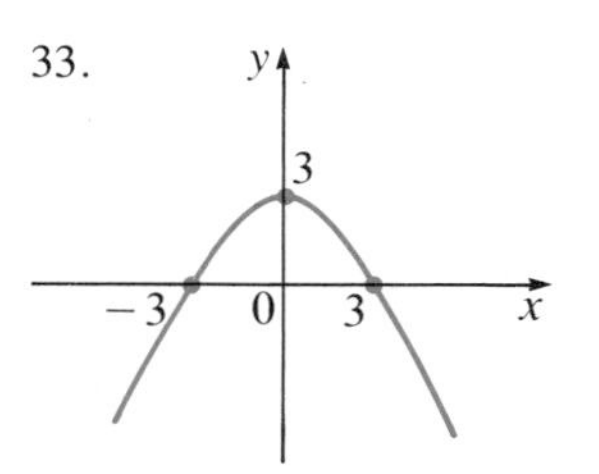

34. 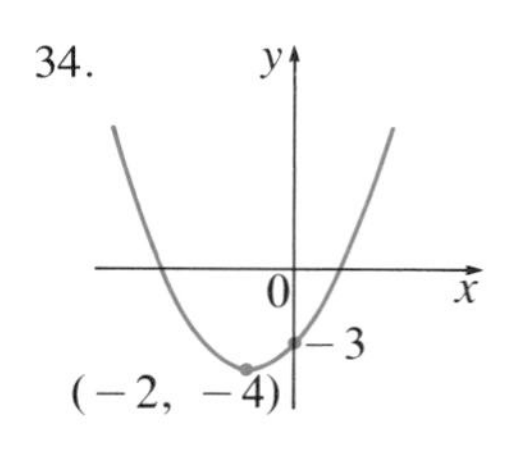

35. * 36.

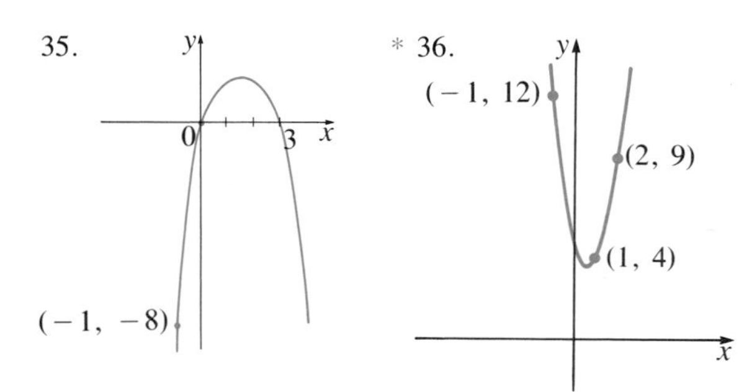

In Problems 37–40 a total cost or total revenue function is given.

(a) For what value of q is cost a minimum or revenue a maximum?
(b) What is the minimum cost or maximum revenue?
(c) Sketch the function.

37. $C(q) = 5q^2 - 200q + 3125,\ q \le 100$
38. $R(q) = -0.12q^2 + 150q,\ q \le 1000$
39. $C(q) = 5000 - 30q + 0.2q^2,\ q \le 300$
40. $R(q) = 40q - 0.02q^2,\ q \le 3000$

In Problems 41–46 an object is thrown straight up from the ground with an initial velocity of v_0 (see formula (12) on p. 141).
(a) Determine when the object hits the ground.
(b) Determine how high the object gets.

41. $v_0 = 40$ ft/sec
42. $v_0 = 200$ ft/sec
43. $v_0 = 5280$ ft/sec
44. $v_0 = 50$ m/sec [Hint: Use 4.9 ($= \frac{1}{2}(9.8)$ m/sec^2) instead of 16 in equation (12).]
45. $v_0 = 200$ m/sec
46. $v_0 = 1000$ m/sec
47. An object thrown straight up from the surface of Mars reaches a height of

$$h(t) = -1.96t^2 + v_0 t$$

where v_0, the initial velocity, is given in m/sec. How high will an object on Mars get if it is thrown up with a velocity of 100 m/sec?

Calculator Exercises

In Problems 48–52 find the vertex of each parabola. Determine the maximum or minimum value taken by each function.

48. $f(x) = 57x^2 - 103x + 87$
49. $f(x) = 2.37x^2 - 1.08x - 31.24$
50. $f(x) = \dfrac{x^2}{816} + \dfrac{x}{237} + \dfrac{23}{1056}$
51. $f(x) = 10^{-4}x^2 - 10^{-5}x + 2 \times 10^{-3}$
52. $f(x) = (3.15 \times 10^7)x^2 + (6.82 \times 10^5)x - 2.38 \times 10^9$

Graphing Calculator Exercises

In Problems 53–59 sketch each parabola $p(x)$ on a graphing calculator. Before starting, choose appropriate range values. From your graph determine the number of real zeros of the quadratic equation $p(x) = 0$. Before starting, read Appendix A on the use of a graphing calculator.

53. $p(x) = 5x^2$
54. $p(x) = -3x^2 + 5$
55. $p(x) = 5x^2 + 8x + 9$
56. $p(x) = -257x^2 + 362x + 2106$
57. $p(x) = 0.02x^2 - 0.036x - 0.071$
58. $p(x) = \dfrac{1}{5}x^2 + \dfrac{1}{9}x - \dfrac{1}{11}$
59. $p(x) = -12.6x^2 - 8.9x + 47.5$

4.2 Polynomial Functions of Degree Higher Than Two

In Section 4.1 we discussed quadratic functions. In this section we discuss the graphs of higher-order polynomial functions. Recall that a polynomial function takes the form

$$p(x) = a_n x^n + a_{n-1}x^{n-1} + \cdots + a_2x^2 + a_1x + a_0 \tag{1}$$

where $a_0, a_1, \cdots, a_n$ are real numbers, n is a positive integer, and $a_n \neq 0$.

Definition of a Power Function

A polynomial function that takes the form

$$p(x) = ax^n \tag{2}$$

is called a **power function.**

We now show how to graph power functions.

In Section 3.5 (Figures 1 and 3) we obtained the graphs of $f(x) = x^2$ and $f(x) = x^3$. We reproduce those graphs in Figure 1.

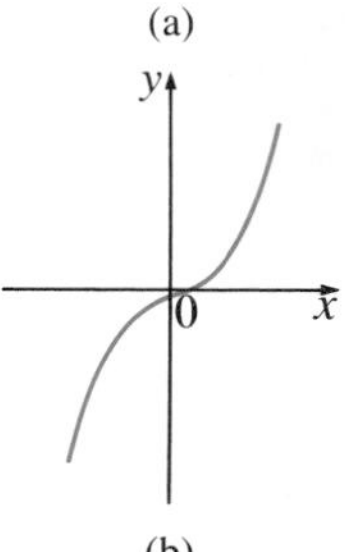

Figure 1 (a) Graph of $f(x) = x^2$.
(b) Graph of $f(x) = x^3$.

We saw in Section 4.1 that the graph of ax^2 for $a > 0$ is like the graph of x^2. The only difference is one of scale. The graphs of all functions ax^2 for $a > 0$ have the same appearance.

Similarly, the graph of ax^3 for $a > 0$ is very similar to the graph of x^3. Again, the only difference is one of scale. In Figure 2 we provide graphs of $\frac{1}{2}x^3$, x^3, and $2x^3$.

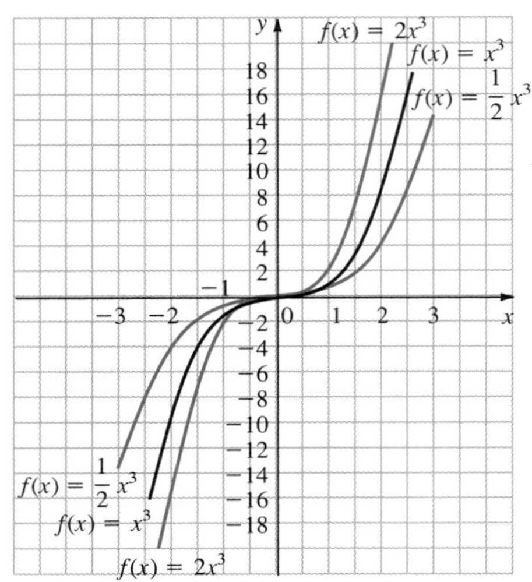

Figure 2 Graphs of $\frac{1}{2}x^3$, x^3, and $2x^3$.

If $a > 0$, then the graph of $-ax^2$ is the graph of ax^2 reflected around the x-axis, and the graph of $-ax^3$ is the graph of ax^3 reflected around the x-axis.

The Graph of ax^n

Suppose $a > 0$. The graph of $p(x) = ax^n$ looks like the graph of

$$p(x) = ax^2 \quad \text{if } n \text{ is even}$$
$$p(x) = ax^3 \quad \text{if } n \text{ is odd}$$

Table 1

x	x^2	x^4
0	0	0
0.5	0.25	0.0625
1	1	1
1.5	2.25	5.0625
2	4	16
2.5	6.25	39.0625
3	9	81

EXAMPLE 1 ***The Graphs of x^4 and $-x^4$***

Graph the functions x^4 and $-x^4$.

SOLUTION If $p(x) = x^4$, then $p(-x) = (-x)^4 = x^4 = p(x)$, so x^4 is even, and we need only plot the graph for $x \geq 0$ and reflect this graph about the y-axis. Values for x^4 are given in Table 1, and the graph is sketched in Figure 3(a). For comparison, the graph of $p(x) = x^2$ is also given. Note that the only important difference between the graphs is that x^4 increases much faster than x^2 for $x > 1$ and for $x < -1$. For $|x| < 1$, $x^4 < x^2$. The graph of $-x^4$ is given in Figure 3(b).

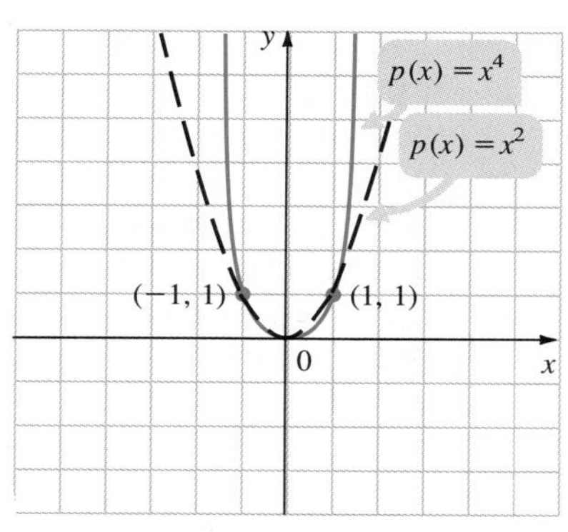

(a) Graph of $p(x) = x^4$

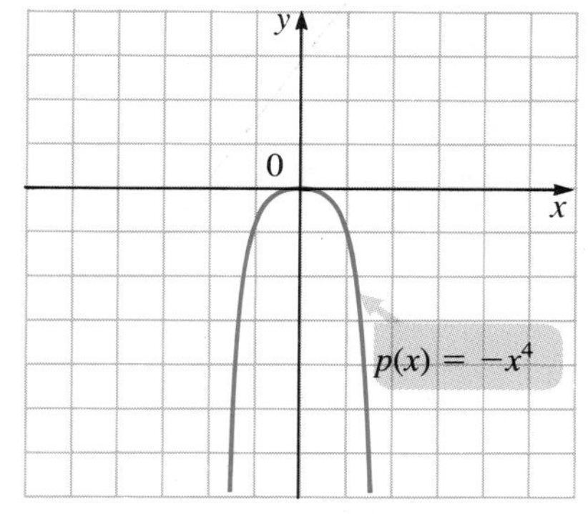

(b) Graph of $p(x) = -x^4$

Figure 3 The graph of $-x^4$ is obtained by reflecting the graph of x^4 about the x-axis.

Note that even though the graph of $y = x^4$ looks like the graph of $y = x^2$, the graph of $y = x^4$ is *not* a parabola. ■

EXAMPLE 2 *The Graphs of x^5 and $-x^5$*

Graph the functions x^5 and $-x^5$.

SOLUTION If $p(x) = x^5$, then $p(-x) = (-x)^5 = -x^5$ and x^5 is odd. Therefore we graph x^5 for $x \geq 0$ and reflect this graph about the origin. A list of sample values of x^5 is given in Table 2. Graphs of x^5 and x^3 are sketched in Figure 4(a). The graph of $-x^5$ is the graph of x^5 reflected around the x-axis. It is given in Figure 4(b).

Table 2

x	x^3	x^5
0	0	0
0.5	0.125	0.03125
1	1	1
1.5	3.375	7.59375
2	8	32
2.5	15.625	97.65625
3	27	243

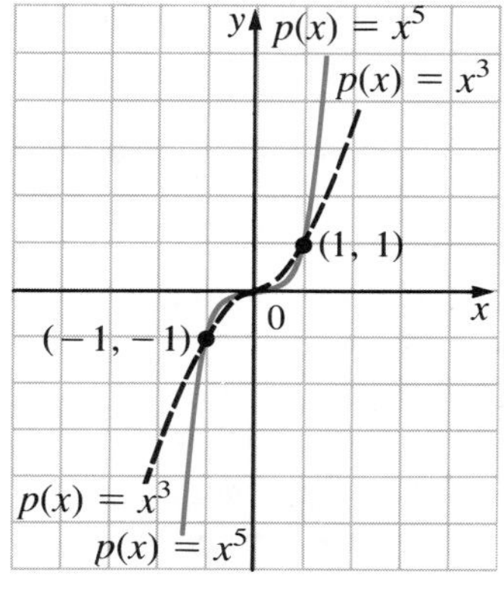

(a) Graph of $p(x) = x^5$

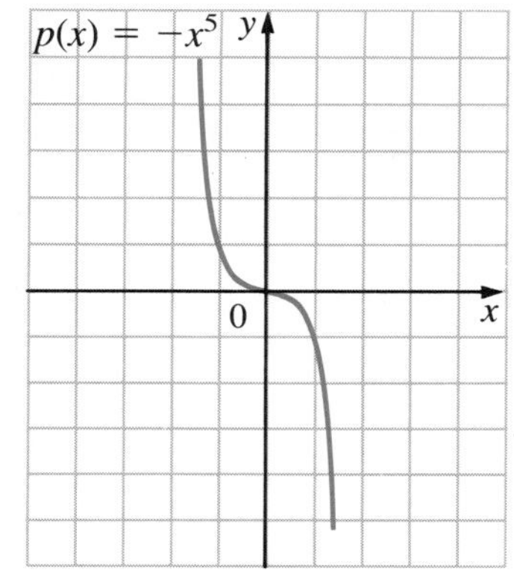

(b) Graph of $p(x) = -x^5$

Figure 4 The graph of $-x^5$ is obtained by reflecting the graph of x^5 about the x-axis.

We now discuss the graphs of some other polynomial functions. To do so, we first describe some properties of these functions.

Every polynomial function is **continuous** and **smooth.** The precise definition of these terms is given in a calculus course. However, it is easy to describe these properties geometrically.

A function f is **continuous** if it is defined at every real number and if its graph is unbroken. If f is not continuous, it is said to be **discontinuous.**

A function is **smooth** if it is continuous and if its graph contains only smooth, rounded turns, not sharp edges as in Figure 5(c).

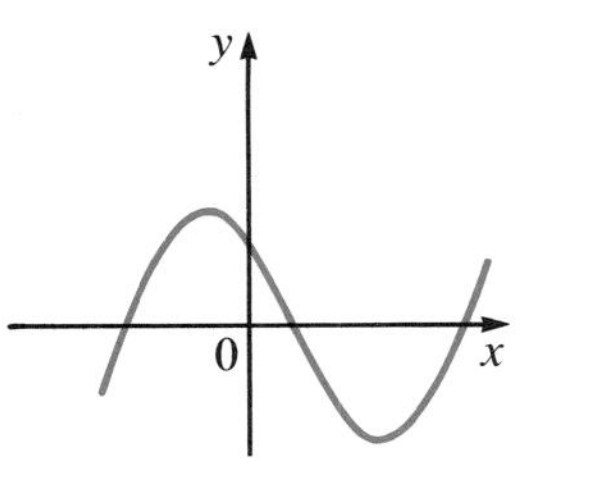

(a) Continuous, smooth curve.

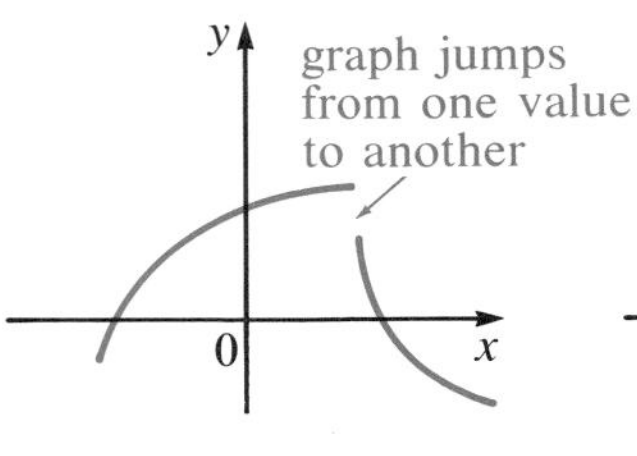

(b) Discontinuous curve.

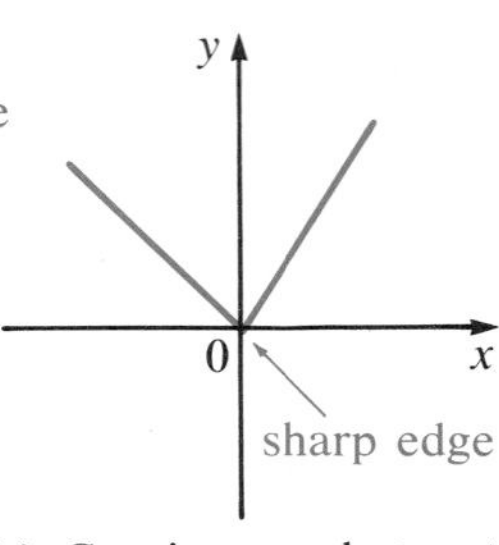

(c) Continuous, but not smooth curve.

Figure 5

Since a polynomial function is continuous and smooth, we can draw its graph by plotting a few points and connecting them, knowing that there will be no jumps, sharp edges, or gaps. In addition, there are other facts that can be very helpful.

Definition of Zero or Root

The number r is a **zero** or **root** of the polynomial $p(x)$ if $p(r) = 0$. In this book we will use the words **zero** and **root** interchangeably.

Consider the quadratic function

$$p(x) = x^2 - 3x + 2$$

$p(x)$ can be factored, and we find that its zeros are 1 and 2.

$$p(x) = x^2 - 3x + 2 = (x - 2)(x - 1)$$

Also, the points $(1, 0)$ and $(2, 0)$ are the x-intercepts of the graph of p. These facts can be generalized.

Facts About Real Zeros of Polynomials

Let $p(x)$ be a polynomial function. The following three statements are equivalent. That is, if one is true all are true.

(i) $p(r) = 0$; that is, r is a zero of p.
(ii) $x - r$ is a factor of p.
(iii) $(r, 0)$ is an x-intercept of the graph of p.

We will prove the equivalence of (i) and (ii) in the next section.

EXAMPLE 3 ***Real Zeros of a Fourth-Degree Polynomial***

Let $p(x) = x^4 + 3x^3 - 8x^2 - 22x - 24$. Verify that $x = 3$ and $x = -4$ are the only real zeros of p.

SOLUTION

$$p(3) = 3^4 + 3 \cdot 3^3 - 8 \cdot 3^2 - 22 \cdot 3 - 24$$
$$= 81 + 81 - 72 - 66 - 24 = 0$$

and

$$p(-4) = (-4)^4 + 3(-4)^3 - 8(-4)^2 - 22(-4) - 24$$
$$= 256 - 192 - 128 + 88 - 24 = 0$$

Thus $x - 3$ and $x - (-4) = x + 4$ are factors of p. In fact,

$$x^4 + 3x^3 - 8x^2 - 22x - 24 = (x - 3)(x + 4)(x^2 + 2x + 2)$$

Moreover, these are the only real zeros because if r were a different zero of p, it would be a zero of $x^2 + 2x + 2$. But $x^2 + 2x + 2$ has no real zeros since its discriminant, $2^2 - 4 \cdot 1 \cdot 2 = -4 < 0$. That is, $x^2 + 2x + 2$ is irreducible. Finally, the only two x-intercepts of p are $(3, 0)$ and $(-4, 0)$.

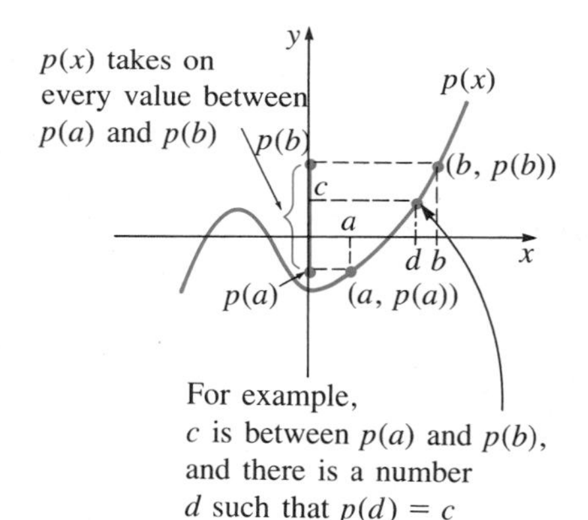

Figure 6 Illustration of intermediate value theorem.

How do we find the zeros of a polynomial? There are many ways to do so, and we will consider some of these in the next four sections. One very important result is the following.

> **Intermediate Value Theorem for Polynomial Functions**
>
> Let $p(x)$ be a polynomial function. Suppose p takes the value $p(a)$ at a and $p(b)$ at b. Then, $p(x)$ takes every value between $p(a)$ and $p(b)$ at values of x in $[a, b]$.

The theorem is illustrated in Figure 6. The theorem is especially useful because of the following fact, which is a direct consequence of the intermediate value theorem.

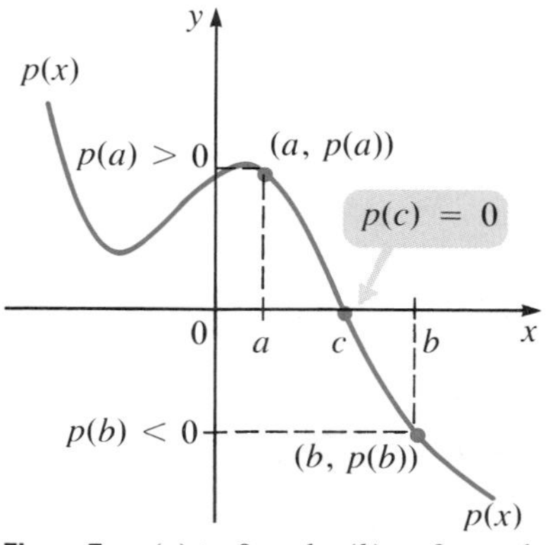

Figure 7 $p(a) > 0$ and $p(b) < 0$, so there is a number c between a and b such that $p(c) = 0$.

> **Locating Zeros of a Polynomial Function**
>
> Let $p(x)$ be a polynomial function and suppose that $a < b$. If $p(a) < 0$ and $p(b) > 0$ or if $p(a) > 0$ and $p(b) < 0$, then there is at least one number c in the interval (a, b) such that $p(c) = 0$.

This fact is illustrated in Figure 7.

NOTE The result states that there is at least one zero of $p(x)$ in (a, b). There may be more than one.

EXAMPLE 4 *Approximating the Real Zero of a Polynomial*

Approximate to four decimal places the only real zero of the cubic polynomial

$$p(x) = x^3 + x^2 + 7x - 3$$

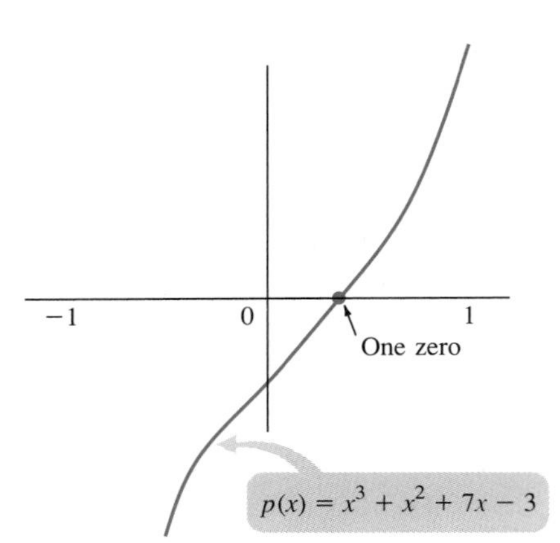

Figure 8 $p(x) = x^3 + x^2 + 7x - 3$ has one real zero in the interval (0, 1).

SOLUTION We observe that

$$p(0) = -3 \quad \text{and} \quad p(1) = 6$$

Thus p has a zero between 0 and 1. Also,

$$p(0.3) = (0.3)^3 + (0.3)^2 + 7(0.3) - 3 = -0.783$$

and

$$p(0.5) = (0.5)^3 + (0.5)^2 + 7(0.5) - 3 = 0.875$$

so there is a zero in the interval (0.3, 0.5). We can continue this ''narrowing down'' process. If we do so, we find that, to four decimal places, the zero is 0.3971. In Problem 40 you are asked to explain why $p(x)$ has only one real root. In Sections 4.5 and 4.6 we will discuss more efficient ways to approximate the zeros of a polynomial. In Figure 8 we provide a graph of $p(x)$ showing the one real zero.† ■

EXAMPLE 5 *Finding a Quadratic with Given Real Zeros*

Find a quadratic p with leading coefficient 1 and zeros 2 and -3.

SOLUTION Since 2 is a zero, $x - 2$ is a factor of p. Similarly, $x - (-3) = x + 3$ is also a factor. Since p is a quadratic, it has only two zeros, and so

$$p(x) = (x - 2)(x + 3) = x^2 + x - 6$$

Note that this is the only quadratic with the given zeros *and* leading coefficient 1. ■

EXAMPLE 6 *Finding a Cubic with Given Real Zeros*

Find a cubic polynomial p with real zeros -1, 1, and 2 such that $p(3) = 40$.

SOLUTION Three factors of p are $(x + 1)$, $(x - 1)$, and $x - 2$. Then p can be written

$$p(x) = a(x + 1)(x - 1)(x - 2) = a(x^2 - 1)(x - 2) = a(x^3 - 2x^2 - x + 2)$$

where a is the as yet unknown leading coefficient. Then

$$p(3) = a(3^3 - 2 \cdot 3^2 - 3 + 2) = 8a \underset{\uparrow\ p(3) = 40 \text{ is given}}{=} 40$$

†If you have access to a graphing calculator, you can approximate this zero fairly quickly. See Example 5 in Appendix A for details.

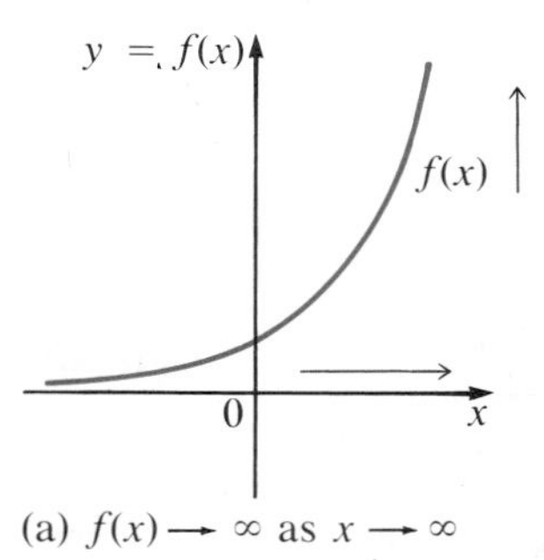

(a) $f(x) \to \infty$ as $x \to \infty$

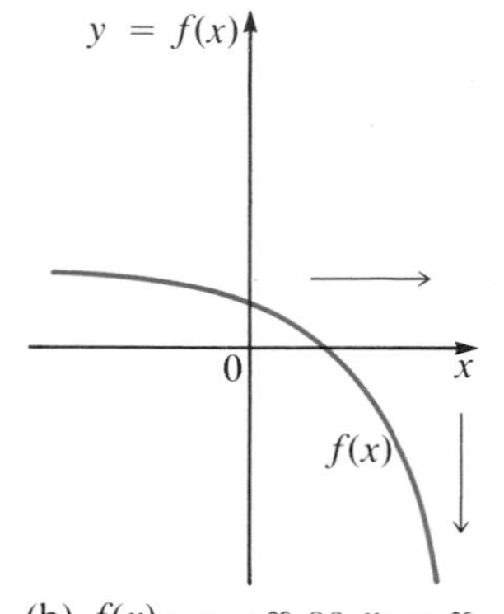

(b) $f(x) \to -\infty$ as $x \to \infty$

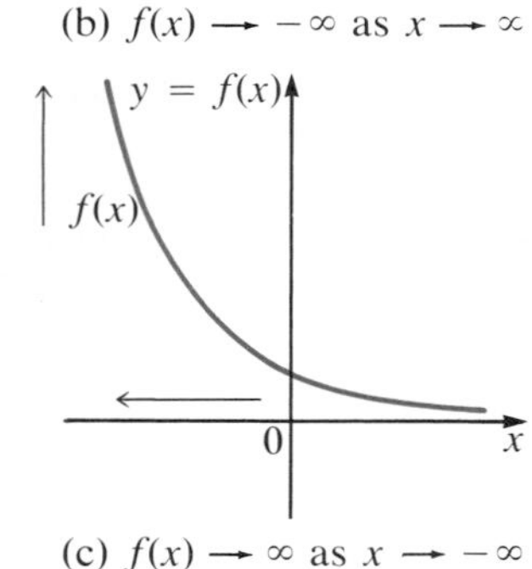

(c) $f(x) \to \infty$ as $x \to -\infty$

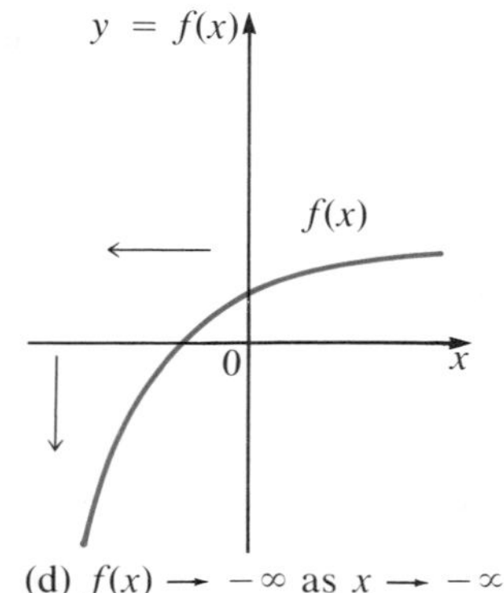

(d) $f(x) \to -\infty$ as $x \to -\infty$

Figure 9 Functions which approach ∞ or $-\infty$ as $x \to \infty$ or $-\infty$.

So $a = \dfrac{40}{8} = 5$ and

$$p(x) = 5(x^3 - 2x^2 - x + 2) = 5x^3 - 10x^2 - 5x + 10$$

Consider the polynomial

$$a_n x^n + a_{n-1}x^{n-1} + \cdots + a_1 x + a_0 \tag{3}$$

with $a_n \neq 0$. If $|x|$ is very large, then the term $a_n x^n$ is much larger in absolute value than the other terms in (3). For example, consider

$$x^4 - 10x^3 - 50x^2 - 1000$$

Suppose

$$x = 1{,}000{,}000 = 10^6$$

Then

$$x^4 = 10^{24}, \qquad |-10x^3| = 10 \cdot 10^{18} = 10^{19}$$
$$|50x^2| = 50 \times 10^{12} = 5 \times 10^{13} \quad \text{and} \quad |1000| = 10^3$$

Evidently, x^4 is much bigger than the other terms. Thus even though $x^4 - 10x^3 - 50x^2 - 1000$ is negative for some smaller values of x, it becomes positive and increasing as x increases. Using the symbols introduced below

$$x^4 - 10x^3 - 50x^2 - 1000 \to \infty \text{ as } x \to \infty$$

Here is what we mean by this use of the arrows:

Expression	Meaning	Graph
$f(x) \to \infty$ as $x \to \infty$	$f(x)$ increases without bound as x increases without bound	turns upward for large values of x
$f(x) \to -\infty$ as $x \to \infty$	$f(x)$ decreases without bound as x increases without bound	turns downward for large values of x
$f(x) \to \infty$ as $x \to -\infty$	$f(x)$ increases without bound as x decreases without bound	turns upward for negative values of x with $\|x\|$ large
$f(x) \to -\infty$ as $x \to -\infty$	$f(x)$ decreases without bound as x decreases without bound	turns downward for negative values of x with $\|x\|$ large

These kinds of behavior are illustrated in Figure 9.

The graph of $p(x)$ may turn many times, but from what we have seen, it will eventually resemble the graph of its first term, $a_n x^n$. That is, as x becomes large, the terms of $p(x)$ are negligible compared to the leading term $a_n x^n$. The graphs of $\pm x^4$ and $\pm x^5$ were given in Figures 3 and 4. From these we can conclude the following.

Leading Coefficient Test

Let the first term of the polynomial function (3) be $a_n x^n$. Then the behavior of $p(x)$ for x large is as in the table below.

n **Is Even or Odd**	**Sign of** a_n	**Behavior as** $x \to \infty$	**Behavior as** $x \to -\infty$	**Example**
n is even	$a_n > 0$	$p(x) \to \infty$	$p(x) \to \infty$	graph of x^2
n is even	$a_n < 0$	$p(x) \to -\infty$	$p(x) \to -\infty$	graph of $-x^2$
n is odd	$a_n > 0$	$p(x) \to \infty$	$p(x) \to -\infty$	graph of x^3
n is odd	$a_n < 0$	$p(x) \to -\infty$	$p(x) \to \infty$	graph of $-x^3$

There is one final fact that we need:

An nth-degree polynomial function has at most n zeros.

EXAMPLE 7 ***Sketching a Cubic by Using the Leading Coefficient Test***

Sketch the graph of

$$p(x) = -(x + 2)(x - 1)(x - 3) = -x^3 + 2x^2 + 5x - 6$$

SOLUTION The zeros are -2, 1, and 3. According to the preceding fact, these are the only zeros. Then, in any interval not containing a zero, the function is always positive or always negative (if it took both positive and negative values in an interval, then it would have a zero in that interval by the intermediate value theorem). To determine the sign of the function in an interval, we choose a **test point** in that interval. This is done in Table 3.

Table 3

Interval	Test Point in That Interval	Value of p at Test Point	Sign of p at That Point	Sign of p in the Interval
$(-\infty, -2)$	-3	24	+	+
$(-2, 1)$	0	-6	−	−
$(1, 3)$	2	4	+	+
$(3, \infty)$	4	-18	−	−

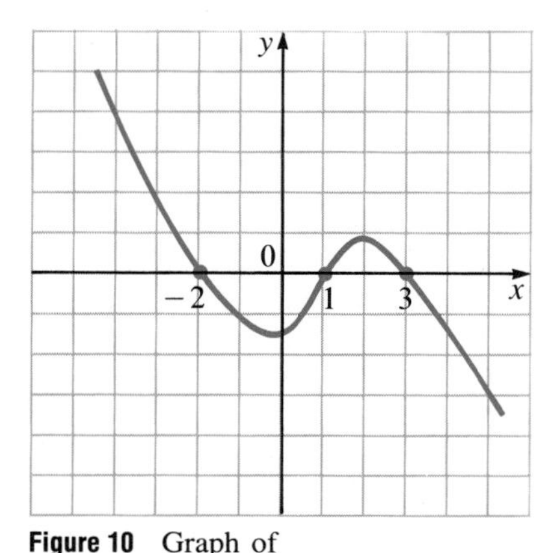

Figure 10 Graph of $p(x) = -x^3 + 2x^2 + 5x - 6 = -(x + 2)(x - 1)(x - 3)$.

Since the leading term is $-x^3$, $a_n = -1$ and n is odd, we see that

$$p(x) \to -\infty \quad \text{as } x \to \infty \qquad \text{and} \qquad p(x) \to \infty \quad \text{as } x \to -\infty$$

Finally, the y-intercept is $(0, -6)$ and the x-intercepts are $(-2, 0)$, $(1, 0)$, and $(3, 0)$. The graph is given in Figure 10.

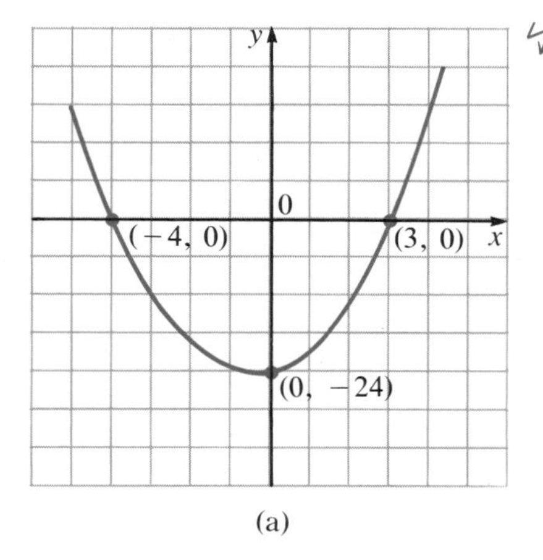

(a)

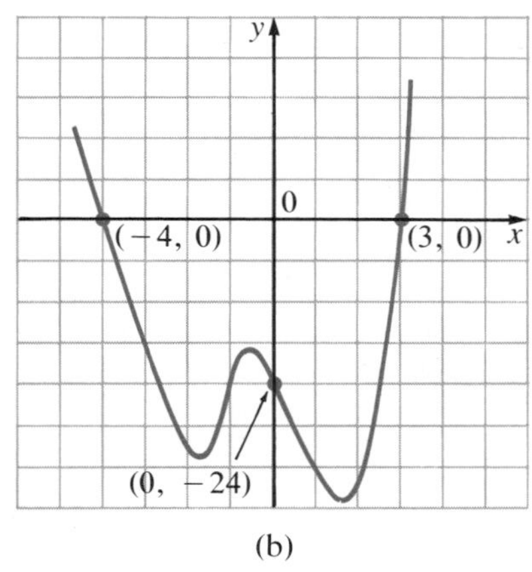

(b)

Figure 11 Two possible graphs for a fourth degree polynomial. The graph in (b) is the graph of $x^4 + 3x^3 - 8x^2 - 22x - 24$.

WARNING The sketching technique used in the last example is somewhat limited. To see why, consider the next example. ■

EXAMPLE 8 *Sketching a Fourth-Degree Polynomial*

Sketch the function

See Example 3 ↓

$$p(x) = x^4 + 3x^3 - 8x^2 - 22x - 24 = (x - 3)(x + 4)(x^2 + 2x + 2)$$

We have

$$p(x) > 0 \quad \text{if } x < -4$$
$$p(x) < 0 \quad \text{if } -4 < x < 3 \qquad \text{0 is a convenient test point}$$
$$p(x) > 0 \quad \text{if } x > 3$$

There are several possible graphs for $p(x)$. Two of them are given in Figure 11. The graph in Figure 11(b) is correct. The approximate places where the graph turns can be found by plotting a number of points. They can be obtained precisely using calculus methods. In general, more sophisticated methods for obtaining the shape of a curve are given in an introductory calculus course.

Obtaining the graph of a polynomial $p(x)$ is not too difficult if we can factor completely into a product of linear and irreducible quadratic factors. Unfortunately, to factor a polynomial requires that we find its zeros, and this is one of the most difficult things to do in algebra. In much of the rest of this chapter, we will discuss how to find the zeros of a polynomial. In the meantime, we summarize below a procedure for sketching the graph of a polynomial.†

To Sketch the Graph of a Polynomial $p(x)$

Step 1 Factor the polynomial as far as you can into linear and irreducible quadratic factors. The x-intercepts are the zeros of the linear factors.

Step 2 Construct a table of signs, showing where the polynomial is positive or negative.

Step 3 Compute the y-intercept. This is the point $(0, p(0))$.

Step 4 Determine the behavior of $p(x)$ for $|x|$ large by considering the term $a_n x^n$.

Step 5 If $p(x)$ was not factored into linear factors, then plot more points in order to determine the approximate shape of the curve (as in Figure 11).

† It is relatively easy to sketch the graphs of many polynomials if you have access to a graphing calculator. See Appendix A for details.

Problems 4.2

Readiness Check

I. Between which of the following is there a root of $p(x) = -x^3 + x^2 + x + 3$?
   a. -1 and 0
   b. 0 and 1
   c. 2 and 3
   d. 3 and 4

II. Which of the following is true of the left and right behavior of the graph of $p(x) = -2x^3 + 3x^2 - 5x + 1$?
   a. It is down on the left and up on the right.
   b. It is up on the left and down on the right.
   c. It is up on both the right and the left.
   d. It is down on both the right and the left.

III. Which of the following is true about the graph of $y = f(x) = x^4 - 81$?
   a. It looks like the graph of an even power function with a negative leading coefficient.
   b. As $x \to \pm\infty$, $f(x) \to -\infty$
   c. It is decreasing for $x > 0$.
   d. It has two real zeros.

IV. Which of the following is a polynomial whose graph has the same basic shape as $p(x) = -x^2$?
   a. $p(x) = 3x^4 - 2x^2 + 1$
   b. $p(x) = -3x^5 + 6x^3 + 7x$
   c. $p(x) = -2x^6 - 4$ d. $p(x) = x^4 + 2x^2 - 3$

In Problems 1–18 sketch the graph of the given function.

1. $p(x) = \frac{1}{3}x^3$
2. $p(x) = 3x^4$
3. $p(x) = \frac{-x^5}{10}$
4. $p(x) = \frac{-x^4}{3}$
5. $p(x) = x^3 + 2$
6. $p(x) = (x + 2)^3$
7. $p(x) = (x - 3)^3$
8. $p(x) = x^3 - 3$
9. $p(x) = x^4 + 1$
10. $p(x) = (x + 1)^4$
11. $p(x) = (x - 1)^4$
12. $p(x) = x^4 - 1$
13. $p(x) = -(x - 2)^3$
14. $p(x) = -(x + 3)^4$
15. $p(x) = 2(x - 1)^4 - 5$
16. $p(x) = 7 - (x + 3)^5$
17. $p(x) = (3x + 3)^4$
18. $p(x) = \frac{1}{4}(x - 2)^3 - \frac{1}{2}$

In Problems 19–32 (a) determine the zeros of each polynomial, (b) determine the sign of the polynomial between each pair of consecutive zeros, and (c) sketch the graph.

19. $p(x) = (x - 1)(x - 2)(x - 3)$
20. $p(x) = (x + 2)(x + 3)(x - 4)$
21. $p(x) = (x - 4)(x + 5)(x - 6)$
22. $p(x) = -(x + 1)(x - 3)(x + 2)$
23. $p(x) = -(x + 2)(x + 6)(x - 2)(x + 5)$
24. $p(x) = x(x^2 - 1)$
25. $p(x) = (x^2 - 1)(x^2 - 4)$
26. $p(x) = x^3 - 4x^2 + 3x$
27. $p(x) = x^3 - 9x^2$
28. $p(x) = x^3 - 5x^2 - 14x$
29. $p(x) = x^3 + 6x^2 + 8x$
30. $p(x) = (x^2 - x - 2)(x^2 + 2x - 15)$
31. $p(x) = x^2(x^2 - 3x + 2)$
32. $p(x) = (x - 1)^2(x - 2)^2$

In Problems 33–39 factor the polynomial into a product of linear and irreducible quadratic factors. Then plot some additional points to obtain its graph.

33. $p(x) = x^3 + x$
34. $p(x) = x^3 - x^2 + x$
35. $p(x) = x^3 + x^2 + 5x$
36. $p(x) = x^4 + 2x^2 + 1$
37. $p(x) = x^4 + 3x^2 + 2$
38. $p(x) = x^3 - 1$
39. $p(x) = x^3 + 1$

* 40. Let $p(x) = x^3 + x^2 + 7x - 3$.
   (a) Show that $p(x) < 0$ if $x < 0$.
   (b) Show that $p(x) > 0$ if $x > 0.4$.
   (c) Show that $p(x) < 0$ if $0 \le x < 0.39$.
   (d) Explain why $p(x)$ has exactly one real zero.

In Problems 41–44 show that each polynomial has a zero in the given interval, and estimate that zero to two decimal places.

41. $p(x) = 4x^3 - 5x^2 + 4x - 7$; $(1, 2)$
42. $p(x) = x^3 + 5x + 1$; $(-1, 0)$
43. $p(x) = 3x^5 + x^4 - 9x^2 + 3x - 4$; $(1, 2)$
44. $p(x) = x^4 - 2x^3 - 5x^2 - 4x - 8$; $(3, 4)$

Answers to Readiness Check

I. c II. b III. d IV. c

In Problems 45–51 find a polynomial p with leading coefficient 1 that has the given numbers as its only zeros.

45. $-4, 7$
46. $0, -2$
47. $0, 1, 2$
48. $-2, 2, 3$
49. $-1, 1, 1 + \sqrt{3}, 1 - \sqrt{3}$
50. $1, 2, 3, 4$
51. $\pm\sqrt{2}, \pm\sqrt{3}, \pm\sqrt{5}$
52. Find a polynomial of degree 5 with leading coefficient 1 for which -3 is the only zero.
53. Find a polynomial of degree n with leading coefficient 1 for which k is the only zero.
54. Find a quadratic p with zeros 2 and 5 such that $p(1) = 6$.
55. Find a cubic p with zeros -3, 0, and 2 such that $p(1) = 1$.
56. Find a cubic p with zeros 1, -3, and 5 such that $p(1) = 96$.
57. The demand function† for a certain product is given by

$$q(p) = 1000 - 5p - 0.1p^3, \qquad p \geq 0$$

where p is the price and q is the quantity demanded. Using the intermediate value theorem, estimate to two decimal places the price at which the demand would be zero.

58. If the price of equipment is fixed and l denotes the number of hours of labor used in production, then the output q for a certain product is given by

$$q(l) = 100 + 0.2l^2 - 0.001l^3, \qquad l \geq 0$$

This formula is accurate for small values of l but doesn't work if l is large, because $q(l) \to -\infty$ as $l \to \infty$. Find, to two decimal places, the value of l such that $q(l) = 0$.

59. In many applications in physics and statistics it is necessary to aproximate a function by polynomials. The **least squares polynomial approximation** is the polynomial that approximates the function in such a way that the square of the difference in the area between the function and the polynomial over a given interval is a minimum. One such approximation (over the interval $[-1, 1]$ makes use of **Legendre polynomials.** The third-degree Legendre polynomial is given by

$$P_3(x) = \frac{5x^3 - 3x}{2}.$$

Determine intervals over which $P_3(x)$ is positive and negative and graph it.

* 60. Answer the questions of Problem 59 for the fourth-degree Legendre polynomial

$$P_4(x) = \frac{35x^4 - 30x^2 + 3}{8}$$

* 61. If P_n denotes the nth-degree Legendre polynomial, then

$$(n + 1)P_{n+1}(x) + nP_{n-1}(x) = (2n + 1)xP_n(x)$$

Use this **recursion relation** and the information given in Problems 59 and 60 to determine $P_2(x)$. Graph this function.

* 62. Use the recursion relation in Problem 61 to determine $P_5(x)$ and $P_6(x)$.

63. A rectangular box is to be constructed with square ends in such a way that the length of the box is 2 feet longer than one of the sides of the square end.

s s s + 2

(a) Write the volume of the box as a function of s.
(b) Graph $V(s)$ for $s > 0$.

64. In Problem 63 determine s if the volume is (a) 10 ft^3 (b) 20 ft^3. Give both answers to 3 decimal place accuracy.

Graphing Calculator Exercises

In Problems 65–76 sketch the graph of each polynomial, and determine how many real zeros each one has. First determine appropriate range and scale values. Before starting, read (or reread) Example 1 in Appendix A.

65. $p(x) = x^3 + 2x^2 + x + 1$
66. $p(x) = 2x^3 + 10x^2 + 5x - 3$
67. $p(x) = -467x^3 - 506x^2 + 288x + 143$
68. $p(x) = -\frac{1}{2}x^3 + \frac{3}{5}x^2 - \frac{4}{3}x - \frac{10}{7}$
69. $p(x) = x^4 + x^3 + x^2 + x + 1$
70. $p(x) = x^4 - x^3 - x^2 - x + 1$
71. $p(x) = -2x^4 + 3x^3 - x + 5$
72. $p(x) = 7x^4 - 25x^3 - 11x^2 + 47x + 11$
73. $p(x) = -x^5 - 10x^4 + 6x^3 + 8x^2 - 12x + 1$
74. $p(x) = -x^5 - 2x^4 + 6x^3 + 8x^2 - 12x + 5$
75. $p(x) = 3x^5 - 50x^3 + 134x + 60$
76. $p(x) = x^7 - 6x^5 + 2x^3 + x^2 - x - 4$

† This and the next problem are adapted from *An Introduction to Mathematical Economics* by Achibald and Lipsey, Harper & Row, New York, 1976.

4.3 Division of Polynomials and Synthetic Division

In this section we show how one polynomial can be divided by another. Before doing so, we look at the division of one number by another.

EXAMPLE 1 *Dividing One Integer by Another*

Divide 6412 by 23.

SOLUTION The answer is found by the usual long division process.

```
      278      Quotient
   23)6412     23 goes into 64 two times; 2 × 23 = 46
      46
      181  ←   Subtract and bring down the 1
      161      23 goes into 181 seven times; 7 × 23 = 161
       202 ←   Subtract and bring down the 2
       184     23 goes into 202 eight times; 8 × 23 = 184
        18 ←   This is the remainder
```

Thus $\dfrac{6412}{23} = 278\dfrac{18}{23}$.

Check $23 \times 278\dfrac{18}{23} = \underbrace{23 \times 278 + 18}_{23\ \times\ \text{quotient}\ +\ \text{remainder}} = 6394 + 18 = 6412$

The process used in Example 1 can be used to divide one polynomial by another.

EXAMPLE 2 *Dividing Two Polynomials by Long Division*

Divide $2x^3 - 3x^2 + 6$ by $x - 2$.

SOLUTION We arrange our work in steps. The term $2x^3 - 3x^2 + 6$ is called the **dividend,** and the term $x - 2$ is called the **divisor.**

Step 1 Write the terms in both the dividend and divisor with exponents in descending order. If some power of x, x^k, is missing in the dividend, write $0x^k$.

$$x - 2\overline{)2x^3 - 3x^2 + 0x + 6}$$

Step 2 Divide the first term in the dividend by the first term in the divisor.

$$\begin{array}{r} 2x^2 \\ x - 2\overline{)2x^3 - 3x^2 + 0x + 6} \end{array} \qquad \frac{2x^3}{x} = 2x^2$$

Step 3 Multiply this new term by the divisor, and then subtract and bring down one more term from the dividend.

$$\begin{array}{rll} & 2x^2 & \\ x-2\,\big) & 2x^3 - 3x^2 + 0x + 6 & \\ & 2x^3 - 4x^2 & = 2x^2(x-2) \\ \hline & x^2 + 0x & -3x^2 - (-4x^2) = x^2 \end{array}$$

Step 4 Repeat steps 2 and 3. This time use the result of the last subtraction as the dividend. Continue until the degree of the remainder is less than the degree of the divisor.

$$\begin{array}{rrl} \text{Quotient} \rightarrow & 2x^2 + x + 2 & \\ x-2\,\big) & 2x^3 - 3x^2 + 0x + 6 & \\ & 2x^3 - 4x^2 \qquad\qquad & \text{Multiply} \\ \hline & x^2 + 0x \qquad & \text{Subtract} \\ & x^2 - 2x \qquad & \text{Multiply} \\ \hline & 2x + 6 & \text{Subtract} \\ & 2x - 4 & \text{Multiply} \\ \hline & 10 & \text{Subtract} \end{array}$$

The answer is

$$\begin{array}{l} \text{Dividend} \rightarrow \\ \text{Divisor} \rightarrow \end{array} \frac{2x^3 - 3x^2 + 6}{x - 2} = \overbrace{2x^2 + x + 2}^{\text{Quotient}} + \frac{10}{x - 2} \begin{array}{l} \leftarrow \text{Remainder} \\ \leftarrow \text{Divisor} \end{array}$$

Check

$$\begin{aligned} (x-2)\left[2x^2 + x + 2 + \frac{10}{x-2}\right] &= (x-2)(2x^2 + x + 2) + 10 \\ &= 2x^3 + x^2 + 2x - 4x^2 - 2x - 4 + 10 \\ &= 2x^3 - 3x^2 + 6 \quad \blacksquare \end{aligned}$$

EXAMPLE 3 *Dividing Two Polynomials by Long Division*

Divide $\dfrac{4x^5 + x + 8}{3 - 2x + x^2}$.

SOLUTION We proceed as in Example 2. Here the dividend is $4x^5 + x + 8$, and the divisor is $3 - 2x + x^2 = x^2 - 2x + 3$.

Step 1 $x^2 - 2x + 3\,\big)\overline{4x^5 + 0x^4 + 0x^3 + 0x^2 + x + 8}$

Step 2

$$\begin{array}{rl} & 4x^3 \\ x^2 - 2x + 3\,\big) & \overline{4x^5 + 0x^4 + 0x^3 + 0x^2 + x + 8} \end{array} \qquad \frac{4x^5}{x^2} = 4x^3$$

Step 3

$$\begin{array}{r} 4x^3 \\ x^2 - 2x + 3 \overline{)4x^5 + 0x^4 + 0x^3 + 0x^2 + x + 8} \\ \underline{4x^5 - 8x^4 + 12x^3} \\ 8x^4 - 12x^3 + 0x^2 \end{array} \quad = 4x^3(x^2 - 2x + 3)$$

Step 4

$$\begin{array}{rl} \text{Divisor} \quad 4x^3 + 8x^2 + 4x - 16 & \leftarrow \text{Quotient} \\ x^2 - 2x + 3 \overline{)4x^5 + 0x^4 + 0x^3 + 0x^2 + x + 8} & \\ \underline{4x^5 - 8x^4 + 12x^3} & \text{Multiply} \\ 8x^4 - 12x^3 + 0x^2 & \text{Subtract} \\ \underline{8x^4 - 16x^3 + 24x^2} & \text{Multiply} \\ 4x^3 - 24x^2 + x & \text{Subtract} \\ \underline{4x^3 - 8x^2 + 12x} & \text{Multiply} \\ -16x^2 - 11x + 8 & \text{Subtract} \\ \underline{-16x^2 + 32x - 48} & \text{Multiply} \\ \text{Remainder} \rightarrow -43x + 56 & \text{Subtract} \end{array}$$

NOTE We stop here because the degree of $-43x + 56$ is 1, the degree of the divisor is 2, and $1 < 2$. The answer is

$$\frac{\text{Dividend}}{\text{Divisor}} = \text{Quotient} + \frac{\text{Remainder}}{\text{Divisor}}$$

$$\frac{4x^5 + x + 8}{x^2 - 2x + 3} = 4x^3 + 8x^2 + 4x - 16 + \frac{-43x + 56}{x^2 - 2x + 3} \qquad (1)$$

Check

dividend = divisor × quotient + remainder   This is always true

$$4x^5 + x + 8 = (x^2 - 2x + 3)(4x^3 + 8x^2 + 4x - 16) - 43x + 56$$

This equation will hold if our answer is correct

$$\begin{aligned} &= x^2(4x^3 + 8x^2 + 4x - 16) - 2x(4x^3 + 8x^2 + 4x - 16) \\ &\quad + 3(4x^3 + 8x^2 + 4x - 16) - 43x + 56 \\ &= 4x^5 + 8x^4 + 4x^3 - 16x^2 - 8x^4 - 16x^3 - 8x^2 + 32x \\ &\quad + 12x^3 + 24x^2 + 12x - 48 - 43x + 56 \\ &= 4x^5 + (8 - 8)x^4 + (4 - 16 + 12)x^3 + (-16 - 8 + 24)x^2 \\ &\quad + (32 + 12 - 43)x + (-48 + 56) \\ &= 4x^5 + 0x^4 + 0x^3 + 0x^2 + x + 8 \\ &= 4x^5 + x + 8 \end{aligned}$$

In each of the last two examples we were able to write

dividend = divisor × quotient + remainder

where the quotient is a polynomial and the remainder is either zero or a polynomial of degree lower than that of the divisor. In general, we have the following:

Division Algorithm

If $p(x)$ and $d(x)$ are polynomials with $d(x) \neq 0$, and degree of $d(x) \leq$ degree of $p(x)$, then there exist unique polynomials $q(x)$ and $r(x)$ such that

$$\underset{\text{Dividend}}{p(x)} = \underset{\text{Divisor}}{d(x)}\,\underset{\text{Quotient}}{q(x)} + \underset{\text{Remainder}}{r(x)} \tag{2}$$

where either $r(x) = 0$ or degree of $r(x) <$ degree of $d(x)$. $q(x)$ is called the **quotient,** and $r(x)$ is called the **remainder** of the division $p(x)/d(x)$.

The division algorithm has three immediate consequences.

Quotient Theorem

If $p(x)$ and $d(x)$ are as in the division algorithm, then there exist unique polynomials $q(x)$ and $r(x)$ such that

$$\frac{p(x)}{d(x)} = q(x) + \frac{r(x)}{d(x)} \tag{3}$$

where $r(x) = 0$ or degree of $r(x) <$ degree of $d(x)$.

Proof of Quotient Theorem

From (2),

$$p(x) = d(x)q(x) + r(x)$$

so

$$\frac{p(x)}{d(x)} = \frac{d(x)q(x)}{d(x)} + \frac{r(x)}{d(x)} = q(x) + \frac{r(x)}{d(x)} \quad ■$$

NOTE

(a) In Example 2 we have $p(x) = 2x^3 - 3x^2 + 6$, $d(x) = x - 2$, $q(x) = 2x^2 + x + 2$, and $r(x) = 10$. Note that degree of $r(x) = 0$ (a constant) $<$ degree of $d(x) = 1$.

(b) In Example 3 we have $p(x) = 4x^5 + x + 8$, $d(x) = x^2 - 2x + 3$, $q(x) = 4x^3 + 8x^2 + 4x - 16$, and $r(x) = -43x + 56$. Here degree of $r(x) = 1 <$ degree of $d(x) = 2$.

Remainder Theorem

If the polynomial $p(x)$ is divided by $x - c$, then the remainder $r(x)$ is the constant $p(c)$.

Proof of Remainder Theorem

In (2), $d(x) = x - c$, so

$$\begin{aligned} p(x) &= d(x)q(x) + r \\ &= (x - c)q(x) + r \end{aligned}$$

and

$$p(c) = (c - c)q(c) + r = 0q(c) + r = r$$

Note that $r(x) = r$ is a constant because, by the remainder theorem, degree of $r(x) <$ degree of $d(x) = 1$. Constant functions have degree zero. ■

EXAMPLE 4 *Using the Remainder Theorem to Evaluate a Polynomial*

Use the remainder theorem to evaluate $p(3)$, where

$$p(x) = x^3 - 2x^2 + 5x - 8$$

SOLUTION We divide $p(x)$ by $x - 3$. The remainder is equal to $p(3)$.

$$\begin{array}{r|l} & x^2 + x + 8 \\ \hline x - 3 & x^3 - 2x^2 + 5x - 8 \\ & x^3 - 3x^2 \\ \hline & x^2 + 5x \\ & x^2 - 3x \\ \hline & 8x - 8 \\ & 8x - 24 \\ \hline & 16 \leftarrow \text{Remainder} \end{array}$$

Thus $p(3) = 16$.

Check $p(3) = 3^3 - 2 \cdot 3^2 + 5 \cdot 3 - 8 = 27 - 18 + 15 - 8 = 16$

Factor Theorem

$x - c$ is a factor of the polynomial $p(x)$ if and only if

$$p(c) = 0.$$

Proof of Factor Theorem

From the remainder theorem

$$p(x) = (x - c)q(x) + r$$

$x - c$ is a factor if and only if $r = 0$: But $r = p(c)$. Thus $x - c$ is a factor if and only if $p(c) = 0$. ■

EXAMPLE 5 ***Using the Factor Theorem***

Let $p(x) = x^3 - 3x^2 + 5x - 6$. We have

$$p(2) = 2^3 - 3 \cdot 2^2 + 5 \cdot 2 - 6 = 8 - 12 + 10 - 6 = 0$$

Thus $x - 2$ is a factor of $p(x)$. After dividing, we find that

$$x^3 - 3x^2 + 5x - 6 = (x - 2)(x^2 - x + 3)$$

Synthetic Division

We now give a procedure for dividing a polynomial by $x - c$. To illustrate it, we redo the division in Example 4 (division A). In order to subtract a polynomial, we change the sign (multiply by -1) and add. In B, we have changed the sign and added. In C, we repeat the computation in B, but write only the coefficients. For the divisor, we have written 3 in place of $x - 3$.

$$\text{A.}\quad \begin{array}{r|rrrrl}
 & x^2 & +\ x & +\ 8 & & \\
x - 3 & x^3 - 2x^2 & +\ 5x & -\ 8 & & \\
 & x^3 - 3x^2 & & & & \\
 & x^2 & +\ 5x & & & \text{Subtract} \\
 & x^2 & -\ 3x & & & \\
 & & 8x & -\ 8 & & \text{Subtract} \\
 & & 8x & -\ 24 & & \\
 & & & 16 & & \text{Subtract}
\end{array}$$

$$\text{B.}\quad \begin{array}{r|rrrrl}
 & x^2 & +\ x & +\ 8 & & \\
x - 3 & x^3 - 2x^2 & +\ 5x & -\ 8 & & \\
 & -x^3 + 3x^2 & & & & \text{Change the sign} \\
 & x^2 & +\ 5x & & & \text{Add} \\
 & -x^2 & +\ 3x & & & \text{Change the sign} \\
 & & 8x & -\ 8 & & \text{Add} \\
 & & -8x & +\ 24 & & \text{Change the sign} \\
 & & & 16 & & \text{Add}
\end{array}$$

$$\text{C.}\quad \begin{array}{r|rrrr}
 & & 1 & 1 & 8 \\
3 & 1 & -2 & 5 & -8 \\
 & \textcircled{-1} & 3 & & \\
 & & \textcircled{1} & \textcircled{5} & \\
 & & \textcircled{-1} & 3 & \\
 & & & \textcircled{8} & \textcircled{-8} \\
 & & & \textcircled{-8} & 24 \\
 & & & & 16
\end{array}$$

We now condense further. As we will soon see, the circled numbers are not needed to obtain the quotient and remainder. We throw them out and move the remaining numbers below the dividend to the same row:

$$\begin{array}{r|rrrr} & & 1 & 1 & 8 \\ \hline 3) & 1 & -2 & 5 & -8 \\ & & 3 & 3 & 24 \\ \hline & & & & 16 \end{array}$$

It is customary to move the quotient to the bottom row and put an open box around the remainder. Our quotient now looks like this:

$$\begin{array}{r|rrrr} 3) & 1 & -2 & 5 & -8 \\ & \downarrow & 3 & 3 & 24 \\ \hline & 1 & 1 & 8 & \boxed{16} \end{array}$$

The arrows indicate how each number is obtained:

(i) Bring down the 1.
(ii) Multiply $1 \times 3 = 3$ and insert 3 below -2.
(iii) Add $3 + (-2) = 1$.
(iv) Multiply $3 \times 1 = 3$ and insert 3 below 5.
(v) Add $3 + 5 = 8$.
(vi) Multiply $3 \times 8 = 24$ and insert 24 below -8.
(vii) Add $24 + (-8) = 16$. This is the remainder.

We read the answer as

$$\frac{x^3 - 2x^2 + 5x - 8}{x - 3} = x^2 + x + 8 + \frac{16}{x - 3}$$

The process used here is called **synthetic division.** Rather than write out a long, complicated-looking set of rules, we illustrate synthetic division with two additional examples.

EXAMPLE 6 *Dividing Using Synthetic Division*

Divide $2x^5 + 9x^4 - 4x^2 - 7$ by $x + 2$, using synthetic division.

SOLUTION

Step 1 Write the coefficients of the dividend with the powers of x in descending order. Put 0's in whenever a power of x is missing. Write c for the divisor $x - c$. (In this example $x - c = x - (-2) = x + 2$.) Draw a horizontal line, leaving room for the intermediate numbers above and the quotient and remainder below.

$x + 2 = x - (-2)$ Divisor; $2x^5$; $9x^4$; No x^3 term; $-4x^2$; No x term; -7

$$-2)\;2 \quad 9 \quad 0 \quad -4 \quad 0 \quad -7 \leftarrow \text{Dividend}$$

Intermediate values go here
← Horizontal line
Quotient goes here
↑ Remainder goes here

Step 2 Copy the leading coefficient below the line. This is the first term in the quotient since, from long division, $\dfrac{2x^5 + 9x^4 - 4x^2 - 7}{x + 2} = 2x^4 +$ other terms of lower degree $+ \dfrac{\text{remainder}}{x + 2}$.

$$\begin{array}{r|rrrrrr} -2 & 2 & 9 & 0 & -4 & 0 & -7 \\ & & & & & & \\ \hline & 2 & & & & & \end{array}$$

Step 3 Multiply 2 by -2 and write the product -4 above the line and under the 9 (the coefficient of x^4). Then add -4 to 9 and put the sum below the line.

$$\begin{array}{r|rrrrrr} -2 & 2 & 9 & 0 & -4 & 0 & -7 \\ & & -4 & & & & \\ \hline & 2 & 5 & & & & \end{array}$$

Note that we are adding, not subtracting

Step 4 Continue step 3. The next two numbers will be $-2 \cdot 5 = -10$ above the line and $0 + (-10) = -10$ below the line. Continue until you run out of numbers in the dividend line.

$$\begin{array}{r|rrrrrr} -2 & 2 & 9 & 0 & -4 & 0 & -7 \\ & & -4 & -10 & 20 & -32 & 64 \\ \hline & 2 & 5 & -10 & 16 & -32 & \boxed{57} \end{array}$$

Multiply by -2

The answer is

Remainder ↓

$$\frac{2x^5 + 9x^4 - 4x^2 - 7}{x + 2} = 2x^4 + 5x^3 - 10x^2 + 16x - 32 + \frac{57}{x + 2}$$

Check $(x + 2)(2x^4 + 5x^3 - 10x^2 + 16x - 32) + 57$
$= 2x^5 + 9x^4 - 4x^2 - 64 + 57 = 2x^5 + 9x^4 - 4x^2 - 7$ ■

EXAMPLE 7 *Using Synthetic Division to Evaluate a Polynomial*

Use the remainder theorem and synthetic division to compute $p(4)$, where $p(x) = x^4 - 3x^3 + 5x^2 - 4x + 2$.

SOLUTION $p(4)$ is the remainder when we divide $p(x)$ by $x - 4$. We do this by synthetic division.

$$\begin{array}{r|rrrrr} 4 & 1 & -3 & 5 & -4 & 2 \\ & & 4 & 4 & 36 & 128 \\ \hline & 1 & 1 & 9 & 32 & \boxed{130} \end{array}$$

Thus

$$x^4 - 3x^3 + 5x^2 - 4x + 2 = (x - 4)(x^3 + x^2 + 9x + 32) + 130$$

and

$$p(4) = 130$$

You should check this by computing $p(4)$ directly.

WARNING Synthetic division works only when dividing a polynomial by $x - c$ for some constant c. To divide by $ax - c$, when $a \neq 1$, first divide numerator and denominator by a, and then use synthetic division. For example,

$$\frac{x^3 + 2x^2 - 6}{2x + 5} = \frac{\frac{1}{2}x^3 + x^2 - 3}{x + \frac{5}{2}}$$

can be evaluated using synthetic division. Synthetic division cannot be used if the degree of the divisor is greater than 1. ■

Finding All Zeros of Some Cubic Polynomials

Consider the cubic equation

$$p(x) = x^3 + bx^2 + cx + d = 0 \tag{4}$$

If we can find one zero r, then $x - r$ is a factor of (4), and we can divide $p(x)$ by $x - r$ to get a quadratic polynomial. We can then obtain the remaining zeros using the quadratic formula.

EXAMPLE 8

Find all roots of $p(x) = x^3 - 2x^2 - 8x - 5 = 0$.

SOLUTION In order to start, we have to find one root. We try the simplest numbers 1 and -1 to begin:

$$p(1) = 1 - 2 - 8 - 5 = -14 \neq 0$$
$$p(-1) = -1 - 2 + 8 - 5 = 0$$

Thus -1 is a root and $x + 1$ is a factor of $p(x)$. We then divide $p(x)$ by $x + 1$, using synthetic division:

$$\begin{array}{r|rrrr} -1 & 1 & -2 & -8 & -5 \\ & & -1 & 3 & 5 \\ \hline & 1 & -3 & -5 & \boxed{0} \end{array}$$

So $x^3 - 2x^2 - 8x - 5 = (x + 1)(x^2 - 3x - 5)$. The roots of $x^2 - 3x - 5 = 0$ are

$$x = \frac{3 \pm \sqrt{9 + 20}}{2} = \frac{3 \pm \sqrt{29}}{2}$$

Thus the three roots of $x^3 - 2x^2 - 8x - 5 = 0$ are -1, $\dfrac{3 + \sqrt{29}}{2}$, and $\dfrac{3 - \sqrt{29}}{2}$.

NOTE In the next section we will show how to determine which numbers to "guess" when trying to find the rational roots of a polynomial.

Horner's Method

We have seen that synthetic division can be used to evaluate a polynomial (see Example 7). If we look carefully at this process, we can obtain a method for evaluating a polynomial on a calculator.

To see how this method arises, we evaluate $p(c)$ where $p(x) = 2x^3 + 5x^2 - 2x + 3$. Synthetic division yields the following:

$$\begin{array}{r|llll} c & 2 & 5 & -2 & 3 \\ & & 2c & (2c+5)c & [(2c+5)c-2]c \\ \hline & 2 & 2c+5 & (2c+5)c-2 & [(2c+5)c-2]c+3 \end{array}$$

Thus

$$p(c) = [(2c + 5)c - 2]c + 3$$

We can obtain this form in another way:

$$\begin{aligned} p(x) &= 2x^3 + 5x^2 - 2x + 3 \\ &= (2x^2 + 5x - 2)x + 3 && \text{Factor } x \text{ from each of the first three terms} \\ &= [(2x + 5)x - 2]x + 3 && \text{Factor } x \text{ from each of the first two terms} \end{aligned}$$

Thus

$$p(c) = [(2c + 5)c - 2]c + 3$$

For example,

$$\begin{aligned} p(4) &= [(2 \cdot 4 + 5)4 - 2]4 + 3 \\ &= [13 \cdot 4 - 2]4 + 3 = 50 \cdot 4 + 3 = 200 + 3 = 203 \end{aligned}$$

The process of writing a polynomial in this way by factoring out x repeatedly is called **Horner's method.** As we will see, it provides an easy way to evaluate a polynomial on a calculator.

EXAMPLE 9 *Evaluating a Polynomial by Using Horner's Method*

Use Horner's method to compute $p(5)$ and $p(-4.87)$ where

$$p(x) = 2x^4 - 3x^3 + 8x^2 - 10x + 2$$

SOLUTION We write

$$p(x) = 2x^4 - 3x^3 + 8x^2 - 10x + 2$$
$$= (2x^3 - 3x^2 + 8x - 10)x + 2 \quad \text{Factor } x \text{ from the first four terms}$$
$$= [(2x^2 - 3x + 8)x - 10]x + 2 \quad \text{Factor } x \text{ from the first three terms}$$
$$= \{[(2x - 3)x + 8]x - 10\}x + 2 \quad \text{Factor } x \text{ from the first two terms}$$

We now compute $p(5)$ on a calculator. First we store the number 5. Your calculator probably has a [STO] or an [x → M] key and a [RCL] (for *recall*) or [RM] (for *recall memory*) key. Here we denote these functions by [STO] and [RCL]. Then, to store 5, we press

[5] [STO]

The following sequence of key strokes yields $p(5)$

[2] [×] [RCL] [−] [3] [=]	$2x - 3 = 2 \cdot 5 - 3 = 7$
[×] [RCL] [+] [8] [=]	$(2x - 3)x + 8 = 7 \cdot 5 + 8 = 43$
[×] [RCL] [−] [10] [=]	$[(2x - 3)x + 8]x - 10 = 43 \cdot 5 - 10 = 205$
[×] [RCL] [+] [2] [=]	$\{[(2x - 3)x + 8] - 10\}x + 2 = 205 \cdot 5 + 2 = 1027$

Thus

$$p(5) = 1027$$

Check $p(5) = 2 \cdot 5^4 - 3 \cdot 5^3 + 8 \cdot 5^2 - 10 \cdot 5 + 2 = 1250 - 375 + 200 - 50 + 2 = 1027$

The same sequence of key strokes will give us $p(-4.87)$ if we store -4.87 instead of 5 at the beginning. The intermediate results are

-4.87	
-12.74	$2(-4.87) - 3$
70.0438	$[2(-4.87) - 3](-4.87) + 8$
-351.113306	$([2(-4.87) - 3](-4.87) + 8)(-4.87) - 10$
1711.9218	$\{([2(-4.87) - 3](-4.87) + 8)(-4.87) - 10\}(-4.87) + 2$

Thus $p(-4.87) = 1711.9218$†

Notice that in each step the five-step key sequence

[×] [RCL] [+] or [−] [NUMBER] [=]

is repeated.

† Actually, $p(-4.87) = 1711.92180022$, but the calculator used here displays only 10 digits.

Horner's Method for Evaluating a Polynomial on a Calculator

To find $p(c)$ on most calculators where

$$p(x) = a_n x^n + a_{n-1}x^{n-1} + a_{n-2}x^{n-2} + \cdots + a_1 x + a_0$$

use these steps

(i) Enter c by pressing c STO.

(ii) Evaluate the polynomial by using the following key sequence:

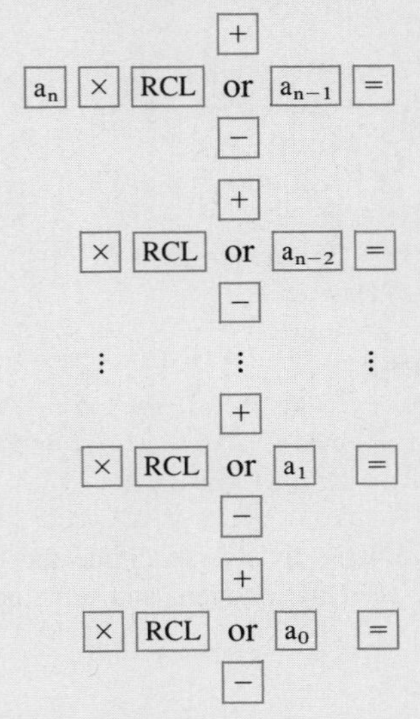

If $a_i = 0$, then skip the + or − a_i step.

Note that at every stage except for the first we use the same five-step key sequence

× RCL + or − a_i =

Problems 4.3

Readiness Check

I. Which of the following is the quotient for the synthetic division problem shown?

$$\begin{array}{r|rrrrr} -2 & 4 & 0 & -3 & -2 & +5 \\ & & -8 & +16 & -26 & +56 \\ \hline & 4 & -8 & +13 & -28 & +61 \end{array}$$

a. $4x^4 - 8x^3 + 13x^2 - 28x + 61$

b. $4x^3 - 8x^2 + 13x - 28 + \dfrac{61}{x-2}$

c. $4x^3 - 8x^2 + 13x - 28 + \dfrac{61}{x+2}$

d. $-4x^4 + 81x^3 - 13x^2 - 28x - 61$

II. Which of the following is true about the polynomial $p(x)$ used in the synthetic division problem shown?

$$\begin{array}{r|rrrrr} -1 & 2 & -1 & 0 & 0 & -3 \\ & & -2 & +3 & -3 & +3 \\ \hline & 2 & -3 & +3 & -3 & 0 \end{array}$$

a. $p(1) = 0$ b. $x + 1$ is a factor of $p(x)$.
c. $x - 1$ is a factor of $p(x)$. d. $p(x)$ is of degree 5.

III. Which of the following is true about $2x^3 - 7x^2 + 4$ divided by $1 - x + 2x^2$?

a. The remainder is $4x - 7$. b. The remainder is $-4x + 7$.
c. The quotient is of degree two.
d. The remainder is of degree zero.

In Problems 1–20 use long division to find the quotient and remainder when $p(x)$ is divided by $d(x)$.

1. $p(x) = x^2 - 3x + 2$; $d(x) = x - 1$
2. $p(x) = x^2 - 4x - 5$; $d(x) = x + 1$
3. $p(x) = 6x^2 - 7x - 20$; $d(x) = 2x - 5$
4. $p(x) = 6x^2 - 7x - 20$; $d(x) = 3x + 1$
5. $p(x) = 2x^2 + 7x - 4$; $d(x) = 2x + 1$
6. $p(x) = 2x^2 + 7x - 4$; $d(x) = 2x - 1$
7. $p(x) = 12x^2 + 4x + 40$; $d(x) = 3x - 5$
8. $p(x) = 4x^2 - 2x + 1$; $d(x) = 2x - 3$
9. $p(x) = 8x^2 + 4x - 1$; $d(x) = 4x + 2$
10. $p(x) = x^2 - x + 5$; $d(x) = 2x + 3$
11. $p(x) = x^3 + 4x^2 + 7x + 12$; $d(x) = x + 3$
12. $p(x) = -x^3 + 4x^2 + 7x + 12$; $d(x) = x - 1$
13. $p(x) = 2x^3 + 5x^2 + 3x + 35$; $d(x) = x - 2$
14. $p(x) = 2x^3 + 5x^2 + 3x + 35$; $d(x) = 2x + 7$
15. $p(x) = 2x^3 + x + 3$; $d(x) = 2x + 1$
16. $p(x) = x^3 - 1$; $d(x) = x^2 + x + 1$
17. $p(x) = x^4 + x^2$; $d(x) = x^3 + 1$
18. $p(x) = -x^5 - x^3 + 3x + 2$; $d(x) = x^3 - x + 5$
19. $p(x) = 4x^4 - 2x^3 + x^2 - x + 4$; $d(x) = x^2 - x + 5$
20. $p(x) = 2x^5 + 4x^4 - 3x + 2$; $d(x) = x^2 - 4$

In Problems 21–42 use synthetic division to divide the first polynomial by the second. Find the quotient and remainder.

21. $x^2 - 5x + 6$; $x - 3$
22. $2x^2 + 7x - 4$; $x + 4$
23. $5x^2 - 4x + 3$; $x + 5$
24. $x^3 - 1$; $x - 1$
25. $x^4 + 16$; $x + 2$
26. $x^3 - 3x^2 + 2x - 4$; $x + 2$
27. $2x^3 - 5x^2 + 3x - 4$; $x - 3$
28. $-3x^4 + x^2 - 2$; $x + 3$
29. $2x^5 - 1$; $x - 1$
30. $x^6 + 1$; $x + 1$
31. $-3x^4 - 2x^3 + 3$; $x + \frac{1}{2}$
32. $2x^3 - x + 8$; $x - 10$
33. $x^5 + x^4 + x^3 + x^2 + x + 1$; $x - \frac{1}{2}$
34. $-3x^6 - x^3 + 4$; $x + 1$
35. $x^{10} - 1$; $x - 1$
36. $x^{10} + 1$; $x + 1$
37. $x^4 + x^2 - 3$; $x + \frac{3}{4}$
38. $-x^3 + x$; $x - 5$
39. $-3x^4 + 3x^3 - 2x^2 - 2x + 5$; $x + 2$

* 40. $x^{100} - 1$; $x - 1$
* 41. $x^{200} + 1$; $x + 1$
42. $x^{10} - 1024$; $x - 2$

In Problems 43–50 use synthetic division to evaluate the given polynomial at the given point.

43. $p(x) = x^3 + 2x^2 - x + 3$; $x = -2$
44. $p(x) = x^4 + x^2 - 5$; $x = 3$
45. $p(x) = 5x^3 - 2x^2 + 4x + 9$; $x = -3$
46. $p(x) = 2x^4 + 3x^2 - 7x + 4$; $x = 2$
47. $p(x) = x^5 + 2x^3 - 5x + 1$; $x = 2$
48. $p(x) = x^3 - 2x^2 + x + 1$; $x = \frac{1}{2}$
49. $p(x) = 3x^4 + 2x^3 + 1$; $x = -\frac{1}{2}$
50. $p(x) = 2x^3 - x^2 - 1$; $x = \frac{2}{3}$

In Problems 51–57 three or four numbers are given. Find a third- or fourth-degree polynomial that has those numbers as zeros and that satisfies the given condition.

51. 2, 3, −1; $p(0) = 12$
52. −1, 1, 7; $p(0) = -3$
53. 2, i, $-i$; $p(0) = \frac{1}{4}$ [Hint: See Section 2.4.]
54. $\frac{1}{2}$, $\frac{1}{3}$, $\frac{1}{4}$; $p(0) = 5$
55. 3, −2, 0; $p(1) = 6$
56. 3, −2, 2, 4; $p(0) = -1$
57. 1, −2, 3, −4; $p(0) = 2$
58. Find a cubic polynomial with 3 as the only zero such that $p(0) = 4$.

* 59. Find a cubic polynomial with 1 and 2 as zeros such that $p(0) = 2$ and $p(3) = 14$.

* 60. Find a quartic (fourth-degree) polynomial with roots i, $-i$, $1 + i$, and $1 - i$ such that $p(0) = -7$.

Calculator Exercises

In Problems 61–66 use synthetic division and a calculator to evaluate the given polynomial at the given point. Give your answer to as many decimal places as are carried on your calculator.

61. $p(x) = x^2 + 3x - 4$; $x = 0.45$
62. $p(x) = x^3 + 2$; $x = -1.56$
63. $p(x) = 3x^3 - 2x^2 + x - 1$; $x = 4.58$

Answers to Readiness Check

I. c II. b III. b

64. $p(x) = 0.12x^3 - 2.35x^2 + 1.6x - 7.58$; $x = 8.27$
65. $p(x) = -0.01x^3 + 0.126x^2 - 5.84$; $x = -0.177$
66. $p(x) = 325x^4 + 87x^2 - 152x + 175$; $x = 23$

In Problems 67–76 use Horner's method to evaluate the given polynomial at the given value of x.

67. $p(x) = 4x^3 - 7x^2 + 5x + 3$; $x = 2.8$
68. $p(x) = 4x^3 - 7x^2 + 5x + 3$; $x = -14.6$
69. $p(x) = 4x^3 - 7x^2 + 5x + 3$; $x = 2.74 \times 10^{12}$
70. $p(x) = 4x^3 - 7x^2 + 5x + 3$; $x = 8.13 \times 10^{-5}$
71. $p(x) = -2x^4 - x^3 + 5x^2 + 12x - 8$; $x = 2$
72. $p(x) = -2x^4 - x^3 + 5x^2 + 12x - 8$; $x = -24.56$
73. $p(x) = -2x^4 - x^3 + 5x^2 + 12x - 8$; $x = 0.0182$
74. $p(x) = 8x^5 - 6x^3 + x^2 - 9$; $x = -7$
75. $p(x) = 8x^5 - 6x^3 + x^2 - 9$; $x = 4.3125 \times 10^{-7}$
76. $p(x) = 8x^5 - 6x^3 + x^2 - 9$; $x = 9.3621 \times 10^{12}$

In Problems 77–86 one zero of a cubic is given. Find the other zeros — real and/or nonreal.

77. $p(x) = x^3 + x^2 - 6x + 4$; $x = 1$
78. $p(x) = 2x^3 - 5x^2 - x + 6$; $x = 2$
79. $p(x) = -8x^3 + 12x^2 - 14x + 5$; $x = \frac{1}{2}$
80. $p(x) = x^3 - 7x^2 + 11x - 5$; $x = 5$
81. $p(x) = -2x^3 + 3x^2 - x - 6$; $x = -1$
82. $p(x) = 5x^3 + 4x^2 - 8x + 75$; $x = -3$
83. $p(x) = x^3 - 6x^2 + 16x - 32$; $x = 4$
84. $p(x) = -\frac{1}{2}x^3 + x^2 - 144$; $x = -6$
85. $p(x) = x^3 - 1.82x^2 - 5.14x + 2.100305$; $x = 0.37$
86. $p(x) = \frac{1}{4}x^3 + 1.5x^2 + 4.3x + 28.582$; $x = -6.2$

In Problems 87–90 two zeros of a fourth-degree polynomial are given. Find two more. [Hint: Use synthetic division twice.]

87. $p(x) = x^4 + x^3 - 13x^2 - x + 12$; $x_1 = 1$ and $x_2 = -1$
88. $p(x) = x^4 + 2x^3 - 13x^2 - 14x + 24$; $x_1 = -4$, $x_2 = 3$
89. $p(x) = 2x^4 + 3x^3 - 23x^2 - 27x + 45$; $x_1 = 3$, $x_2 = -3$
90. $p(x) = 8x^4 - 6x^3 + 9x^2 - 6x + 1$; $x_1 = \frac{1}{2}$, $x_2 = \frac{1}{4}$
91. Find a value of b such that $x - 2$ is a factor of $x^3 + 3x^2 + bx - 8$.
92. Find a value of b such that when $x^3 + bx^2 - 2x + 1$ is divided by $x + 3$, the remainder is 7.
93. Find a value of b such that when $2x^4 - bx + 4$ is divided by $x - 1$, the remainder is 2.
94. Find a value of b such that $x + 1$ is a factor of $3x^3 + bx^2 - 2b + 5$.
95. Show that $x^4 + x^2 + 1$ has no factor of the form $x - r$ for any real number r.

* 96. Show that if a, b, and d are *positive* real numbers, then $x - c$ is not a factor of $ax^4 + bx^2 + d$ for any real number c. That is, the polynomial has no real zeros.

97. Prove that $x - a$ is a factor of $x^n - a^n$ for any positive integer n and any real number a. [Hint: Use the factor theorem.]

* 98. If n is even, show that $x + a$ is a factor of $x^n - a^n$ for any real number a.

* 99. If n is odd, show that $x + a$ is a factor of $x^n + a^n$ for any real number a.

100. A rectangle with its base on the x-axis is inscribed under the parabola $y = 9 - x^2$, as shown below.

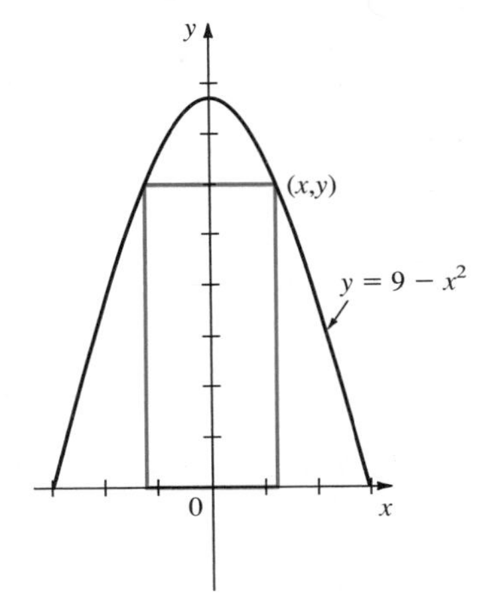

(a) Write the area of the rectangle, $A(x)$, as a function of x.
(b) Find the length and width of the rectangle that is formed when $x = 2$.
(c) Find the area of that rectangle.
* (d) Find the length and width of a second inscribed rectangle that has the same area as the area computed in part (*c*).

* 101. The cost function for processing q tons of coal at a certain plant is given by

$$C(q) = -0.001q^3 + 0.5q^2 + 80q + 10000; \quad 0 \le q \le 500.$$

(a) How much does it cost to process 300 tons?
(b) Find a second value for q that gives the same cost as in part (*a*). Round your answer to the nearest ton.

* 102. Answer the questions in Problem 101 if the cost function is given by

$$C(q) = -0.002q^3 + 1.4q^2 + 200q + 25{,}000; \quad 0 \le q \le 750$$

4.4 Zeros of Polynomials and Rational Zeros

In this section we say more about the zeros of polynomials.

The number c is a **simple zero** of the polynomial $p(x)$ if $x - c$ is a factor of $p(x)$ but $(x - c)^2$ is not. It is a **zero of multiplicity k** if $(x - c)^k$ is a factor of $p(x)$ but $(x - c)^{k+1}$ is not. Note that a simple zero is a zero of multiplicity 1.

EXAMPLE 1 *A Factored Polynomial and Its Zeros*

Let $p(x) = (x - 3)(x - 4)^3(x + 7)^6$.

Then 3 is a simple zero, 4 is a zero of multiplicity 3, and -7 is a zero of multiplicity 6. Note that the degree of $p(x) = 10 = 1 + 3 + 6 =$ multiplicity of 3 + multiplicity of 4 + multiplicity of -7.

In order to discuss the zeros of a polynomial, we need the following theorem, which is one of the most important theorems in algebra. Its proof is beyond the scope of this book.

Fundamental Theorem of Algebra

Every nonconstant polynomial function $p(x)$ with real or complex coefficients has at least one complex zero. That is, if $p(x)$ is of degree n with $n \geq 1$, then there is at least one number c, real or nonreal, such that $p(c) = 0$.

This theorem was first proved by the great German mathematician Carl Friedrich Gauss (1777–1855), when he was 22 years old. Gauss's proof depends on the theory of functions of a complex variable and cannot be given here. By proving this theorem, Gauss gave the concept of an imaginary number more legitimacy (see the focus on complex numbers on p. 110).

How many zeros does a polynomial have? We first must say how we determine the number of zeros of a polynomial. In Section 4.3 we showed that zeros of a quadratic could be complex numbers, so we should include these. What about $p(x) = (x - 4)^3 = (x - 4)(x - 4)(x - 4)$? We say that this polynomial has *three* zeros, all of which are equal. Put another way, count the zero c every time the factor $x - c$ occurs. If c is a zero of multiplicity k, then we count c as a zero k times.

We can use the fundamental theorem of algebra to determine the number of zeros of a polynomial. The next theorem follows from the fundamental theorem of algebra.

Theorem: The Number of Zeros of a Polynomial Function

Every polynomial $p(x)$ of degree $n > 0$ has, counting multiplicities and nonreal zeros, exactly n zeros.

This theorem tells us how many zeros a polynomial has. It does not tell us how to find them. We omit the proof of the next result, which follows from the fundamental theorem of algebra and facts about complex numbers.

Factoring a Polynomial

Any polynomial with real coefficients can be factored into a product of linear and irreducible quadratic factors. Some of the factors may be repeated. The two zeros of each irreducible quadratic are complex conjugate† zeros of the polynomial.

EXAMPLE 2 ***A Factored Polynomial and Its Zeros***

The polynomial $x^6 - 3x^5 - 2x^4 + 2x^3 + 12x^2$ can be factored as

$$p(x) = x^2(x^4 - 3x^3 - 2x^2 + 2x + 12) = x^2(x - 2)(x - 3)(x^2 + 2x + 2)$$

The roots of $x^2 + 2x + 2$ are $\dfrac{-2 \pm \sqrt{4 - 8}}{2} = \dfrac{-2 \pm \sqrt{4}\, i}{2} = -1 \pm i$. The six zeros of $p(x)$ are 0 (counted twice), 2, 3, $-1 + i$, and $-1 - i$.

We will not say much more about complex zeros in this section. Instead, we will concentrate on two questions.

I. How many real zeros does $p(x)$ have?
II. How can we find real zeros?

Consider the polynomial

$$p(x) = a_n x^n + a_{n-1}x^{n-1} + \cdots + a_1 x + a_0$$

The term a_0 is called the **constant term.** Note that $p(0) = a_0$, so 0 is a zero of $p(x)$ if and only if the constant term is zero.

If two consecutive coefficients of $p(x)$ have different signs, then there is a **variation of sign** in $p(x)$.

EXAMPLE 3 ***A Polynomial with Five Variations of Sign***

The polynomial

$$\begin{array}{cccccccccccccccc} p(x) = & x^7 & - & 4x^6 & - & 5x^5 & + & 10x^4 & - & x^3 & - & 2x^2 & + & 3x & - & 6 \\ & + & & - & & - & & + & & - & & - & & + & & - \\ & & ① & & & & ② & & ③ & & & & ④ & & ⑤ & \end{array}$$

has five variations of sign.

†Recall from p. 107 that $a + bi$ and $a - bi$ are complex conjugates and $(a + bi)(a - bi) = a^2 + b^2$.

EXAMPLE 4 ***Counting the Number of Variations in Sign for $p(-x)$***

Compute the number of variations in sign of $p(-x)$ where

$$p(x) = x^6 + 2x^5 - 3x^4 + x - 5$$

SOLUTION
$$\begin{aligned} p(-x) &= (-x)^6 + 2(-x^5) - 3(-x)^4 + (-x) - 5 \\ &= x^6 - 2x^5 - 3x^4 - x - 5 \\ &\quad + \quad - \quad - \quad - \quad - \end{aligned}$$

①

There is only one variation of sign in $p(-x)$. (There are three variations in sign in $p(x)$.)

Descartes' Rule of Signs
(The Number of Real Zeros)

Let $p(x)$ be a polynomial with nonzero constant term. Then

(a) The number of *positive* real zeros, P, of $p(x)$ is less than or equal to the number of variations of sign, V, in $p(x)$. The difference $V - P$ is an even integer.
(b) The number of *negative* real zeros, N, of $p(x)$ is less than or equal to the number of variations of sign, V_N, in $p(-x)$. The difference, $V_N - N$, is an even integer.

We omit the proof of Descartes' rule of signs.

NOTE When we say that $V - P$ is an even number, we mean that $P = V$ or $P = V - 2$ or $P = V - 4$ and so on. For example, if $V = 6$, then the polynomial has 6, 4, 2, or 0 positive real zeros. It *cannot* have more than 6 or 1, 3, or 5 positive real zeros.

EXAMPLE 5

What are the possible numbers of positive and negative zeros of $p(x) = x^6 + 2x^5 - 3x^4 + x - 5$?

SOLUTION From Example 4, $p(x)$ has 3 variations in sign, and $p(-x)$ has 1 variation in sign. Thus

number of positive zeros = 3 or 1 — 1 is 2 less than 3, and 2 is an even integer

number of negative zeros = 1 — There is no positive number an even number of units less than 1

Finding Rational Roots of a Polynomial

Although there is no general formula for finding all the real zeros of a polynomial, there is a way to find rational zeros, if any, when the coefficients of $p(x)$ are integers.

Rational Zeros Theorem

Let $p(x) = a_nx^n + a_{n-1}x^{n-1} + \cdots + a_1x + a_0$, where $a_n \neq 0$ and the coefficients $a_0, a_1, a_2, \cdots, a_n$ are all integers. Suppose $\frac{m}{k}$ is a rational zero of $p(x)$ and $\frac{m}{k}$ is reduced to lowest terms. Then

(i) m is a factor of a_0,
(ii) k is a factor of a_n.

The proof of this theorem is omitted.

EXAMPLE 6 ***Finding the Rational Zeros of a Polynomial***

Find all rational zeros of $p(x) = 27x^3 - 81x^2 + 27x + 10$.

SOLUTION $p(x)$ has two variations in sign and so has either two positive zeros or none. $p(-x) = -27x^3 - 81x^2 - 27x + 10$ has one variation in sign, and so $p(x)$ has exactly one negative zero. Here

$$a_n = 27 \quad \text{and} \quad a_0 = 10$$

so, if $\frac{m}{k}$ is a zero of $p(x)$ in lowest terms, then

$$m = \pm 1, \pm 2, \pm 5, \pm 10 \quad \text{Factors of 10}$$
$$k = \pm 1, \pm 3, \pm 9, \pm 27 \quad \text{Factors of 27}$$

The possibilities are given in the table below.

k \ m	1	2	5	10	-1	-2	-5	-10
1	1	2	5	10	-1	-2	-5	-10
3	$\frac{1}{3}$	$\frac{2}{3}$	$\frac{5}{3}$	$\frac{10}{3}$	$-\frac{1}{3}$	$-\frac{2}{3}$	$-\frac{5}{3}$	$-\frac{10}{3}$
9	$\frac{1}{9}$	$\frac{2}{9}$	$\frac{5}{9}$	$\frac{10}{9}$	$-\frac{1}{9}$	$-\frac{2}{9}$	$-\frac{5}{9}$	$-\frac{10}{9}$
27	$\frac{1}{27}$	$\frac{2}{27}$	$\frac{5}{27}$	$\frac{10}{27}$	$-\frac{1}{27}$	$-\frac{2}{27}$	$-\frac{5}{27}$	$-\frac{10}{27}$

There are 32 candidates for a zero. Each can be checked by synthetic division or direct substitution. We will not check all 32 numbers here. The correct one is $\frac{2}{3}$, as is verified by synthetic division:

$$\begin{array}{r|rrrr} \frac{2}{3} & 27 & -81 & 27 & 10 \\ & & 18 & -42 & -10 \\ \hline & 27 & -63 & -15 & \boxed{0} \end{array}$$

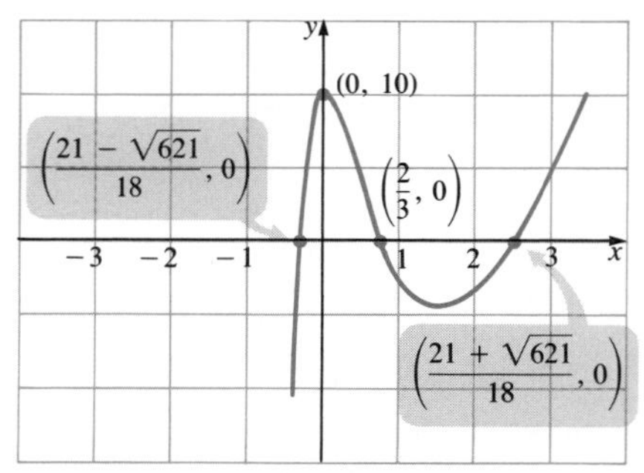

Figure 1 Graph of $p(x) = 27x^3 - 81x^2 + 27x + 10$.

We find that

$$\begin{aligned} 27x^3 - 81x^2 + 27x + 10 &= (x - \tfrac{2}{3})(27x^2 - 63x - 15) \\ &= 3(x - \tfrac{2}{3})(9x^2 - 21x - 5) \end{aligned}$$

The zeros of $9x^2 - 21x - 5$ are

$$x = \frac{21 \pm \sqrt{21^2 - 4(9)(-5)}}{18} = \frac{21 \pm \sqrt{621}}{18} \approx \frac{21 \pm 24.9199}{18}$$

or

$$x \approx 2.5511 \quad \text{and} \quad x = -0.2178$$

We see that $p(x)$ has two positive roots and one negative root. Only one root is rational. We sketch the graph of $p(x)$ in Figure 1. ■

EXAMPLE 7 *A Polynomial with No Rational Zeros*

Find all rational zeros of $x^5 + 3x^4 - 2x^3 + 4x^2 - 5x + 4$.

SOLUTION If $\frac{m}{k}$ is a rational root, then, since $a_0 = 4$ and $a_n = 1$, $m = \pm 1$, ± 2, or ± 4 and $k = \pm 1$.

There are six possible rational roots: 1, 2, 4, −1, −2, or −4. We check each one:

1)	1	3	−2	4	−5	4
		1	4	2	6	1
	1	4	2	6	1	5

2)	1	3	−2	4	−5	4
		2	10	16	40	70
	1	5	8	20	35	74

4)	1	3	−2	4	−5	4
		4	28	104	432	1708
	1	7	26	108	427	1712

−1)	1	3	−2	4	−5	4
		−1	−2	4	−8	13
	1	2	−4	8	−13	17

−2)	1	3	−2	4	−5	4
		−2	−2	8	−24	58
	1	1	−4	12	−29	62

−4)	1	3	−2	4	−5	4
		−4	4	−8	16	−44
	1	−1	2	−4	11	−40

We see that none of the six possibilities is a zero. We conclude that $p(x)$ has no rational zero.

Since $p(x)$ has four variations of sign, it has four, two, or no positive zeros. Also, $p(-x) = -x^5 + 3x^4 + 2x^3 + 4x^2 + 5x + 4$, which has one variation of sign, so $p(x)$ has exactly one negative zero, which must be an irrational number.

NOTE Example 7 is typical. Most polynomials, except those appearing in textbooks, do *not* have any rational zeros.

We note that approximating the zeros of a polynomial is not too difficult if you have access to a graphing calculator. To see how this can be done even when a polynomial has no rational zeros, look at Examples 5 and 6 in Appendix A.

In Example 2 we saw that two zeros of the polynomial are $-1 + i$ and $-1 - i$. Those numbers are complex conjugates of one another. This is no coincidence according to the following theorem.

Theorem: Complex Zeros of Polynomials with Real Coefficients Occur in Complex Conjugate Pairs

Let

$$p(x) = a_nx^n + a_{n-1}x^{n-1} + \cdots + a_1x + a_0$$

where $a_0, a_1, \ldots, a_{n-1}, a_n$ are real numbers. Then if $a + bi$ $(b \neq 0)$ is a zero of p, then $a - bi$ is also a zero of p.

EXAMPLE 8 *Finding a Polynomial with a Given Complex Zero*

Find a cubic polynomial p with real coefficients such that 3 and $1 + 2i$ are zeros and $p(0) = 30$.

SOLUTION Since $1 + 2i$ is a zero, so is its conjugate $1 - 2i$. If a denotes the leading coefficient of p, then

$$p(x) = a(x - 3)[x - (1 + 2i)][x - (1 - 2i)]$$

Now

$$\begin{aligned}[x - (1 + 2i)][x - (1 - 2i)] \\ &= x^2 - [(1 + 2i) + (1 - 2i)]x + (1 + 2i)(1 - 2i) \\ &= x^2 - 2x + 5 \qquad (a + ib)(a - ib) = a^2 + b^2\end{aligned}$$

Thus

$$p(x) = a(x - 3)(x^2 - 2x + 5) = a(x^3 - 5x^2 + 11x - 15)$$

But $p(0) = -15a = 30$, so $a = -2$ and

$$p(x) = -2(x^3 - 5x^2 + 11x - 15) = -2x^3 + 10x^2 - 22x + 30$$

There is a fairly easy way to prove that some polynomials have no rational zeros.

Theorem: Conditions That Guarantee That a Polynomial Has No Rational Zeros†

Let

$$p(x) = a_nx^n + a_{n-1}x^{n-1} + \cdots + a_1x + a_0$$

where all the coefficients are integers and $n \geq 2$.
If a_0, a_n, and $p(1)$ are all odd, then $f(x)$ has no rational zeros.

---

†This theorem and several other interesting results appear in ''Finding Rational Roots of Polynomials'' by Don Redmond in *The College Mathematics Journal,* Vol. 20, No. 2, March 1989, pp. 139–141. A proof is suggested in Problems 59–66.

EXAMPLE 9

Show that the following polynomial has no rational zeros:

$$x^5 + 18x^4 - 166x^3 + 323x^2 + 100x - 6225$$

SOLUTION $a_0 = -6225$, $a_n = a_5 = 1$, and

$$p(1) = 1 + 18 - 166 + 323 + 100 - 6225 = -5949$$

are all odd so, by the theorem above, p has no rational zeros. Note how much work is saved here by not having to check all the factors of 6225 (these include ± 3, ± 5, and ± 83).

Upper and Lower Bounds for the Zeros of a Polynomial

In Figure 1 on p. 260 we can see that the polynomial $p(x) = 27x^3 - 81x^2 + 27x + 10$ has no zero less than -1 or greater than 3. That is, 3 is an *upper bound* on the zeros of $p(x)$, and -1 is a *lower bound* on the zeros of $p(x)$. Without a graph, it is much more difficult to determine upper and lower bounds for the zeros of a polynomial. The following theorem makes this task a bit easier. We indicate why the theorem is true in Example 10.

Upper and Lower Bound Theorem

Let

$$p(x) = a_n x^n + a_{n-1}x^{n-1} + \cdots + a_1 x + a_0$$

with $a_n > 0$ and all coefficients real. Divide $p(x)$ by $x - c$ using synthetic division.

(i) c is an upper bound for the zeros of $p(x)$ if $c > 0$ and all the numbers in the bottom row of the synthetic division are nonnegative.
(ii) c is a lower bound for the zeros of $p(x)$ if $c < 0$ and the numbers in the bottom row alternate in sign (where 0 can be considered positive or negative, as required).

EXAMPLE 10 ***Finding Upper and Lower Bounds for the Zeros of a Polynomial***

Let $p(x) = 27x^3 - 81x^2 + 27x + 10$. We divide $p(x)$ by $x - 3$ using synthetic division to obtain

$$\begin{array}{r|rrrr} 3 & 27 & -81 & 27 & 10 \\ & & 81 & 0 & 81 \\ \hline & 27 & 0 & 27 & \boxed{91} \end{array} \leftarrow \text{All numbers are } \geq 0$$

All the numbers in the bottom row are nonnegative so that, as we already knew, 3 is an upper bound for the zeros.

We can see here why the upper and lower bound theorem works. The division tells us that

$$\frac{27x^3 - 81x^2 + 27x + 10}{x - 3} = 27x^2 + 27 + \frac{91}{x - 3}$$

or

$$p(x) = 27x^3 - 81x^2 + 27x + 10 = (27x^2 + 27)(x - 3) + 91$$

If $x > 3$, then all terms on the right are positive so that $p(x)$ could not be zero. That is, there are no zeros greater than 3.

When we divide $p(x)$ by $x + 1$ we obtain

$$\begin{array}{r|rrrr} -1 & 27 & -81 & 27 & 10 \\ & & -27 & 108 & -135 \\ \hline & 27 & -108 & 135 & \boxed{-125} \end{array}$$

Now $c = -1 < 0$ and the terms in the bottom row alternate in sign, so -1 is a lower bound for the zeros of $p(x)$. Again we can see why this must be true. Reading the results of the division, we can see that

$$\frac{27x^3 - 81x^2 + 27x + 10}{x + 1} = 27x^2 - 108x + 135 - \frac{125}{x + 1}$$

or

$$p(x) = 27x^3 - 81x^2 + 27x + 10 = \underbrace{(27x^2 - 108x + 135)}_{>0 \text{ if } x < -1}\underbrace{(x + 1)}_{<0} - \underbrace{125}_{<0}$$

If $x < -1$, then $27x^2 > 0$, $-108x > 0$, $135 > 0$, and $x + 1 < 0$ so $p(x) < 0$. That is, no number less than -1 can be a zero of $p(x)$ so -1 is a lower bound for the zeros of $p(x)$. ■

EXAMPLE 11 *Finding Upper and Lower Bounds for the Zeros of a Polynomial*

Find upper and lower bounds for

$$p(x) = x^5 - 2x^4 - 5x^3 + 2x^2 + 7x - 4$$

SOLUTION We see that $p(0) = -4$ and $p(1) = -1$. But $p(x) > 0$ for x large, so there must be a zero greater than 1 (explain why). Let us try 2:

$$\begin{array}{r|rrrrrr} 2 & 1 & -2 & -5 & 2 & 7 & -4 \\ & & 2 & 0 & & & \\ \hline & 1 & 0 & -5 & & & \end{array}$$

$\leftarrow$ We stop here because of a negative number in the bottom row

Next we try 3:

$$\begin{array}{r|rrrrrr} 3 & 1 & -2 & -5 & 2 & 7 & -4 \\ & & 3 & 3 & & & \\ \hline & 1 & 1 & -2 & & & \end{array}$$

$\leftarrow$ We stop again

Continuing, we try 4:

$$\begin{array}{r|rrrrrr} 4 & 1 & -2 & -5 & 2 & 7 & -4 \\ & & 4 & 8 & 12 & 56 & 252 \\ \hline & 1 & 2 & 3 & 14 & 63 & \boxed{248} \end{array}$$

$\leftarrow$ All numbers are nonnegative

By the upper and lower bound theorem, 4 is an upper bound for the zeros of $p(x)$.

To find a lower bound, we start with $c = -1$:

$$\begin{array}{r|rrrrrr} -1 & 1 & -2 & -5 & 2 & 7 & -4 \\ & & -1 & 3 & & & \\ \hline & 1 & -3 & -2 & & & \end{array} \leftarrow \text{Two consecutive negative numbers so we stop}$$

Next, we try -2:

$$\begin{array}{r|rrrrrr} -2 & 1 & -2 & -5 & 2 & 7 & -4 \\ & & -2 & 8 & -6 & 8 & -30 \\ \hline & 1 & -4 & 3 & -4 & 15 & \boxed{-34} \end{array} \leftarrow \text{Numbers alternate in sign}$$

We conclude that -2 is a lower bound for the zeros of $p(x)$.

Sometimes it is not very efficient to proceed consecutively as we did in Example 11. For example, if an upper bound is $c = 50$, then testing 2, 3, 4, . . . , 49 would take a long time. In many cases it is better to jump around a bit to get a larger than necessary upper bound and then go down from there.

FOCUS ON
The Battle to Find Zeros of Polynomials

One of the earliest problems in mathematics was to find the zeros of polynomials. Almost 4000 years ago, the Babylonians knew how to solve quadratic equations by completing the square. The Greeks also knew how to solve a number of quadratic equations. Euclid, for example, in Book II of his *Elements,* gave a geometric way to find the positive solution (there always is at least one) of the quadratic equation $x^2 - ax - b^2 = 0$, where $a > 0$. Many people dealt with quadratics and, as we mentioned on p. 110, Cardano dealt with complex numbers (which he claimed were "impossible") while discussing solutions of quadratic equations.

Finding zeros of higher-order polynomials proved to be much more difficult. For several hundred years, until about 1830, mathematicians sought general formulas, analogous to the quadratic formula, for finding solutions to cubic and higher-order equations.

Some significant progress on solving cubics was made by Arabic and Persian mathematicians in the eleventh century. One of the most important works in this period was the book *Algebra* by the great Persian poet Omar Khayyam (~1050–1122), known as the "tentmaker." Khayyam gave a geometric method, using intersecting conics, to find positive solutions to third-degree equations. His work generalized some earlier results, including work of Archimedes (who seemed to have dabbled in everything). But Khayyam believed, incorrectly, that algebraic solutions (that is, solutions using a formula) to cubic equations did not exist.

Probably the most spectacular mathematical achievement of the sixteenth century was the discovery, by Italian mathematicians, of the algebraic solution of cubic and quartic (fourth-degree) equations. The story of this discovery is very colorful. Briefly told, the facts seem to be these. About 1515, Scipione del Ferro (1465–1526), a professor of mathematics at the University of Bologna, solved algebraically the cubic equation $x^3 + mx = n$, probably basing his work on earlier Arabic sources. He did not publish his result, but revealed the secret to his pupil Antonio Fior. At about 1535, Nicolo Fontana of Brescia (1499?–1557), commonly referred to as Tartaglia (the stammerer) because of a childhood injury that affected his speech, claimed to have discovered an algebraic solution of the cubic equation $x^3 + px^2 = n$. Believing this claim to be a bluff, Fior challenged Tartaglia to a public contest of solving cubic equations. Tartaglia worked very hard and only a few days before the contest found an algebraic solution for cubics lacking a quadratic term. Entering the contest equipped to solve two types of cubic equations, whereas Fior could solve but one type, Tartaglia triumphed completely.

Word of Tartaglia's success reached Cardano, who immediately begged Tartaglia to reveal the secret. Tartaglia initially refused, but Cardano hinted that he could arrange a patron for the impoverished Tartaglia. Moreover, Cardano swore never to reveal the secret. Worn down, Tartaglia gave Cardano his formulas.

Cardano was not overburdened with scruples. In 1545 he published his famous work *Ars Magna,* which contained all of Tartaglia's formulas. Tartaglia was not pleased by this development. His vehement protests were met by Ludovico Ferrari, Cardano's most capable pupil, who argued that Cardano had received his information from del Ferro through a third party, and accused Tartaglia of plagiarism from the same source. There ensued an acrimonious dispute from which Tartaglia was perhaps lucky to escape alive.

Since the actors in this drama seem not always to have had the highest regard for truth, one finds a number of variations in the details of the plot.

The solution of the cubic equation $x^3 + mx = n$ given by Cardano in his *Ars Magna* is essentially the following. Consider the identity

$$(p - q)^3 + 3pq(p - q) = p^3 - q^3$$

If we choose p and q such that

$$3pq = m \quad \text{and} \quad p^3 - q^3 = n$$

then x is given by $p - q$. Solving the last two equations simultaneously for p and q we find that

$$p = \sqrt[3]{\left(\frac{n}{2}\right) + \sqrt{\left(\frac{n}{2}\right)^2 + \left(\frac{m}{3}\right)^3}}$$
$$q = \sqrt[3]{-\left(\frac{n}{2}\right) + \sqrt{\left(\frac{n}{2}\right)^2 + \left(\frac{m}{3}\right)^3}} \tag{1}$$

and x is thus determined. We have

Cardano's Formula for One Solution of $x^3 + mx = n$

$$x = \sqrt[3]{\sqrt{\left(\frac{n}{2}\right)^2 + \left(\frac{m}{3}\right)^3} + \frac{n}{2}} - \sqrt[3]{\sqrt{\left(\frac{n}{2}\right)^2 + \left(\frac{m}{3}\right)^3} - \frac{n}{2}} \tag{2}$$

For example, one solution to $x^3 + 6x = 2$ is obtained as follows:

$$n = 2, \quad \frac{n}{2} = 1, \quad m = 6, \quad \frac{m}{3} = 2,$$
$$\sqrt{\left(\frac{n}{2}\right)^2 + \left(\frac{m}{3}\right)^3} = \sqrt{1 + 8} = 3$$

and

$$x = \sqrt[3]{3 + 1} - \sqrt[3]{3 - 1} = \sqrt[3]{4} - \sqrt[3]{2}$$

To solve the general cubic $x^3 + ax^2 + bx + c = 0$, Cardano used the following trick: Substitute $y - \frac{a}{3}$ for x. Then (see Problem 69), the cubic becomes

$$y^3 + \left(b - \frac{a^2}{3}\right)y = \frac{ab}{3} - \frac{2a^3}{27} - c$$

One solution to this equation is given by (2) with $m = b - \frac{a^2}{3}$ and $n = \frac{ab}{3} - \frac{2a^3}{27} - c$. Then $x = y - \frac{a}{3}$ is a solution to the original equation.

Cardano's method led to difficulties. When the roots of a cubic are real and distinct, p and q (in (1)) will be complex. And Cardano stated that complex numbers are impossible.

A more useful formula was discovered in 1591 by Francois Viete (1540–1603) and published in 1615. But Viete's formula, too, gave only one solution. The first complete solution to the cubic equation was given by Leonhard Euler in 1732. Let w_1 and w_2 be the two complex cube roots of -1. [That is, w_1 and w_2 are solutions to $x^2 - x + 1 = 0$, which is obtained by writing $0 = x^3 + 1 = (x + 1)(x^2 - x + 1)$.] Then

Euler's Formula

Three solutions to $x^3 + mx = n$ are

$$x_1 = \sqrt[3]{p} - \sqrt[3]{q}, \quad x_2 = w_1\sqrt[3]{p} - w_2\sqrt[3]{q},$$
$$x^3 = w_2\sqrt[3]{p} - w_1\sqrt[3]{q}$$

where p and q are as in (1) and w_1 and w_2 are the two complex cube roots of -1.

It was not long after the cubic had been solved that an algebraic solution was discovered for the general quartic (or biquadratic) equation. In 1540, the Italian mathematician Zuanne de Tonini da Coi proposed a problem to Cardano that led to a quartic equation. Although Cardano was unable to solve the equation, his pupil Ferrari succeeded, and Cardano had the pleasure of publishing this solution also in his *Ars Magna.* For the next 300 years or so, many mathematicians attempted, without success, to find algebraic formulas for solving fifth- and higher-degree equations. They were unsuccessful because such formulas do not exist. In 1832, the young French mathematician Evariste Galois (1811–1832) proved that it is impossible to find an algebraic formula for solving all equations of degree n if $n \geq 5$. His paper was one of the first to use the modern notion of a *group.*

Galois had a tragic life. In 1830, he was sent to prison for his political activities in support of the revolution. He was released in 1832, just short of his twenty-first birthday. Shortly after his release, he was challenged to a duel over a woman. Realizing the night before that he might be killed, he wrote many of his mathematical thoughts in a letter to a friend. This letter contains some of the central ideas in modern group theory. The next day he was shot and killed.

Problems 4.4

Readiness Check

I. Which of the following is *not* a *possible* rational zero for $p(x) = 9x^5 + 36x^4 - 172x^3 + 122x^2 + 19x - 14$?
a. $-\frac{1}{3}$ b. -7 c. 2 d. 9

II. Which of the following are all of the possible rational zeros for $p(x) = 6x^3 - 5x^2 + 7$?
a. $\pm 7, \pm 6, \pm 5, \pm 3, \pm 2, \pm 1$
b. $\pm\frac{1}{6}, \pm\frac{1}{3}, \pm\frac{1}{2}, \pm 1, \pm\frac{7}{6}, \pm\frac{7}{3}, \pm\frac{7}{2}, \pm 7$
c. $\frac{1}{6}, \frac{1}{3}, \frac{1}{2}, 1, \frac{7}{6}, \frac{7}{3}, \frac{7}{2}, 7$
d. 7, 6, 5, 3, 2, 1

III. Which of the following is the leading coefficient a for the fourth-degree polynomial whose zeros are -2, 2, $-2i$, and $2i$ such that $p(0) = 4$?
a. $\frac{1}{4}$ b. -4 c. $-\frac{1}{4}$ d. 4

IV. Which of the following is another root of $p(x) = x^3 - x^2 + x$ if $\frac{1}{2} - \frac{\sqrt{3}i}{2}$ is one root?
a. $\frac{1}{2} + \frac{\sqrt{3}i}{2}$ b. $-\frac{1}{2} - \frac{\sqrt{3}i}{2}$
c. $\frac{\sqrt{3}i}{2}$ d. $\frac{\sqrt{3}}{2} - \frac{i}{2}$

V. Which of the following is the maximum number of positive real zeros based on Descartes' rule of signs for $p(x) = -7x^4 - x^3 + 4x^2 - 3x + 2$?
a. 1 b. 2
c. 3 d. 4

In Problems 1–8 determine the zeros of the polynomial and the multiplicity of each zero.

1. $p(x) = (x + 5)(x - 4)^2$
2. $p(x) = (x - 2)^3(x + 4)^5$
3. $p(x) = (x + 1)(x + 2)(x - 7)$
4. $p(x) = (x^2 - 1)^4$
5. $p(x) = (x^2 + 4)^5$
6. $p(x) = (x^2 - 6x + 8)^3$
7. $p(x) = (x^2 - 4)^3(x^2 - 3x - 18)^5$
* 8. $p(x) = (x^2 + 4x + 7)^4$

In Problems 9–20 use Descartes' rule of signs to determine the possible number of positive and negative zeros of each polynomial. Do not try to find these zeros.

9. $p(x) = x^2 + 2x + 3$
10. $p(x) = 5x^2 - 3x - 4$
11. $p(x) = 7x^3 + 1$
12. $p(x) = 2x^3 - x^2 - x - 4$
13. $p(x) = 2x^3 - 3x^2 + 4x - 1$
14. $p(x) = 8x^3 + x^2 + 3x + 10$
15. $p(x) = -3x^3 - 2x^2 - x - 8$
16. $p(x) = x^4 + 3x^3 - 2x^2 + 5x + 6$
17. $p(x) = 7x^5 + x^3 + 4x + 3$
18. $p(x) = x^5 - x^4 + 2x^3 - 3x^2 + 4x + 3$
19. $p(x) = 2x^5 + 4x^4 - 3x^3 - 2x^2 - 5x + 1$
20. $p(x) = 2x^5 - x^3 + x^2 - 3$
21. If $a > 0$ and $b > 0$, prove that $x^4 + ax^2 + b$ has no real zero.
22. If $a > 0$ and $b > 0$, prove that $x^3 + ax + b$ has exactly one real zero and this zero is negative.

In Problems 23–30 find all real zeros of each cubic. In each case at least one zero is rational. Factor each polynomial into a product of linear and irreducible quadratic factors.

23. $p(x) = x^3 - x^2 - 4x + 4$
24. $p(x) = x^3 + 2x^2 + 3x + 6$
25. $p(x) = x^3 - 2x^2 - x + 2$
26. $p(x) = x^3 - x^2 - 14x + 24$
27. $p(x) = 2x^3 + 11x^2 + 9x + 2$
28. $p(x) = 3x^3 - 4x^2 + 15x - 20$
29. $p(x) = 6x^3 + 13x^2 + 32x + 5$
30. $p(x) = 5x^3 - 2x^2 - 36x - 32$

In Problems 31–34 sketch the graph of the polynomial in

31. Problem 23
32. Problem 24
33. Problem 29
34. Problem 30
* 35. Find a real cubic polynomial with zeros -2 and i and leading coefficient 1.
* 36. Find a real cubic polynomial with zeros 1 and $1 - i$ with leading coefficient 1.
* 37. Find a real fourth-degree polynomial whose only zeros are 0 and $\pm i$ and leading coefficient 1.
* 38. Find a real cubic polynomial p with zeros 5 and $2 + 3i$ such that $p(0) = 9$.
* 39. Find a real fourth-degree polynomial with zeros 1, -1 and $-1 + 2i$ such that $p(0) = -20$.
* 40. Find a real fifth-degree polynomial with zeros 2, i and $1 + i$ with leading coefficient 1.

Answers to Readiness Check

I. d II. b III. c IV. a V. c

In Problems 41–54 determine upper and lower bounds for the real zeros of each polynomial and then find the rational zeros, if any.

41. $p(x) = x^4 + 3x^3 - x - 3$
42. $p(x) = x^5 - 3x^4 + x - 3$
43. $p(x) = x^4 + x^3 - 3x^2 - 4x - 4$
44. $p(x) = x^4 + x^3 + x - 1$
45. $p(x) = x^4 - 10x^2 + 9$
46. $p(x) = 4x^4 + 12x^3 + 19x^2 - 3x - 5$
47. $p(x) = 9x^4 + 8x^2 - 1$
48. $p(x) = x^4 - 4x^3 - 5x^2 + 36x - 36$
49. $p(x) = x^4 + 2x^3 + 6x^2 + 5x + 6$
50. $p(x) = x^5 - 5x^4 + 4x^3 - 8x^2 + 40x - 32$
51. $p(x) = x^5 - 3x^3 + 2x^2$
52. $p(x) = 16x^5 + 80x^4 - x - 5$
53. $p(x) = 6x^6 + x^5 - x^4 + 18x^2 + 3x - 3$
54. $p(x) = x^6 - 14x^4 + 19x^2 - 36$

Show that each polynomial in Problems 55–58 has no rational zero.

55. $p(x) = x^4 - 10x^3 + 27x^2 + 84x - 73$
56. $p(x) = 75x^7 - 46x^6 + 13x^5 + 44x^4 - 17x^3 + 88x^2 - 121x + 225$
57. $p(x) = x^{10} - 11x + 378{,}929$
58. $p(x) = 39x^8 + 76x^7 - 135x^6 + 238x^5 + 403x^4 + 290x^3 - 12x^2 + 833x + 2197$

In Problems 59–66 assume that $p(x) = a_nx^n + a_{n-1}x^{n-1} + \cdots + a_1x + a_0$ with rational zero $\dfrac{m}{k}$, expressed in lowest terms, and $n \geq 2$.

59. Use the rational zeros theorem to show that if a_0 and a_n are odd, then m and k are also odd.
60. Show that $k^np\left(\dfrac{m}{k}\right) = 0$. $\left[\text{Hint: } \dfrac{m}{k} \text{ is a zero of } p.\right]$
61. Show that $k^np\left(\dfrac{m}{k}\right) = a_nm^n + a_{n-1}m^{n-1}k + \cdots + a_1mk^{n-1} + a_0k^n = 0$.
62. Show that $f(1) = a_n + a_{n-1} + \cdots + a_1 + a_0$.
63. In Problem 59 show that n^rk^s is odd for any nonnegative integers r and s.
64. Using the results of Problems 61 and 62, show that

$$f(1) = f(1) - k^np\left(\frac{m}{k}\right) = a_n(1 - m^n) + a_{n-1}(1 - mk^{n-1}) + \cdots + a_1(1 - mk^{n-1}) + a_0(1 - k^n)$$

65. Using the results of Problems 63 and 64, show that if a_0 and a_n are odd, then $f(1)$ is even.
66. Using the results of Problem 65, show that if a_0, a_n, and $f(1)$ are odd, then p has no rational zero.
67. Prove that $\sqrt{5}$ is not a rational number.
68. * Prove that $\sqrt{p}$ is irrational for any prime number p.
69. Let $p(x) = x^3 + ax^2 + bx + c$. Show that

$$p\left(y - \frac{a}{3}\right) = y^3 + \left(b - \frac{a^2}{3}\right)y + \left(c + \frac{2a^3}{27} - \frac{ab}{3}\right)$$

70. Use the result of Problem 69 to write the equation $x^3 + 3x^2 + 6x - 4 = 0$ as $y^3 + my = n$ for some numbers m and n.
71. * Use the result of Problem 70 and Cardano's formula (2) to find one real solution of the equation

$$x^3 + 3x^2 + 6x - 4 = 0.$$

72. * Find one real solution to the cubic equation

$$x^3 - 6x^2 + 4x - 5 = 0.$$

Graphing Calculator Problems

In Problems 73–84 use your graphing calculator to find all zeros of each polynomial to within 2 decimal places of accuracy. Before beginning, read (or reread) Examples 5 and 6 in Appendix A. Also, see the note on accuracy on page A.9.

73. $p(x) = x^3 + 2x^2 + x + 1$
74. $p(x) = 2x^3 + 10x^2 + 5x - 3$
75. $p(x) = -467x^3 - 506x^2 + 288x + 143$
76. $p(x) = -\frac{1}{2}x^3 + \frac{3}{5}x^2 - \frac{4}{3}x - \frac{10}{7}$
77. $p(x) = x^4 + x^3 + x^2 + x + 1$
78. $p(x) = x^4 - x^3 - x^2 - x + 1$
79. $p(x) = -2x^4 + 3x^3 - x + 5$
80. $p(x) = 7x^4 - 25x^3 - 11x^2 + 47x + 11$
81. $p(x) = -x^5 - 10x^4 + 6x^3 + 8x^2 - 12x + 1$
82. $p(x) = -x^5 - 2x^4 + 6x^3 + 8x^2 - 12x + 5$
83. $p(x) = 3x^5 - 50x^3 + 134x + 60$
84. $p(x) = x^7 - 6x^5 + 2x^3 + x^2 - x - 4$
85. * Use the rational zeros theorem to prove that $\sqrt{2}$ is irrational. [Hint: $\sqrt{2}$ is a zero of $p(x) = x^2 - 2$.]

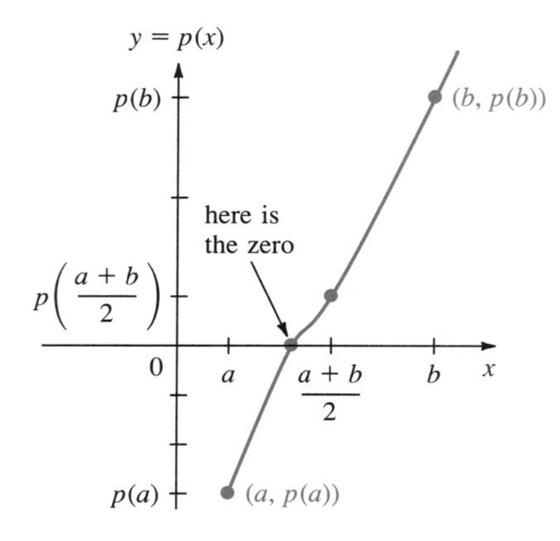

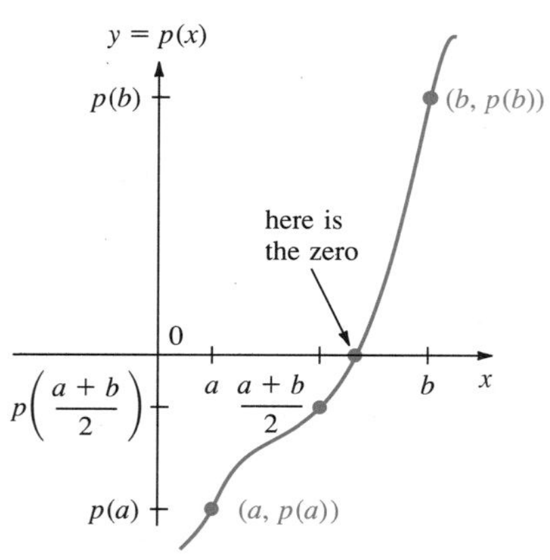

Figure 1 Illustration of the two possibilities in Step 3 of the Bisection Method.

4.5 The Bisection Method

In this section we discuss a simple method for approximating the real zeros of a polynomial. It is based on the intermediate value theorem (p. 235) and is similar to but more systematic than the trial-and-error method given on p. 236.

Here is the method.

The Bisection Method for Finding a Zero of a Polynomial Function $p(x)$

Step 1 Find two numbers a and b such that

$$p(a) < 0 \text{ and } p(b) > 0 \quad \text{or} \quad p(a) > 0 \text{ and } p(b) < 0$$

Then there is a zero in (a, b) by the Intermediate Value Theorem (p. 235). We assume here that $p(a) < 0$ and $p(b) > 0$. A similar process works in the other case.

Step 2 Compute p at the midpoint $\frac{a+b}{2}$. This is the first approximation to a zero.

Step 3 Case **(i)** If $p\left(\frac{a+b}{2}\right) > 0$, then there is a zero in $\left(a, \frac{a+b}{2}\right)$. See Figure 1(a).

Case **(ii)** If $p\left(\frac{a+b}{2}\right) < 0$, then there is a zero in $\left(\frac{a+b}{2}, b\right)$. See Figure 1(b).

Step 4 Continue to cut the interval containing a zero in half until the length of the interval is less than the accuracy you seek.

Step 5 Estimate the error. The maximum error of any approximation is less than half the length of the last interval used.

We illustrate the bisection method with an example.

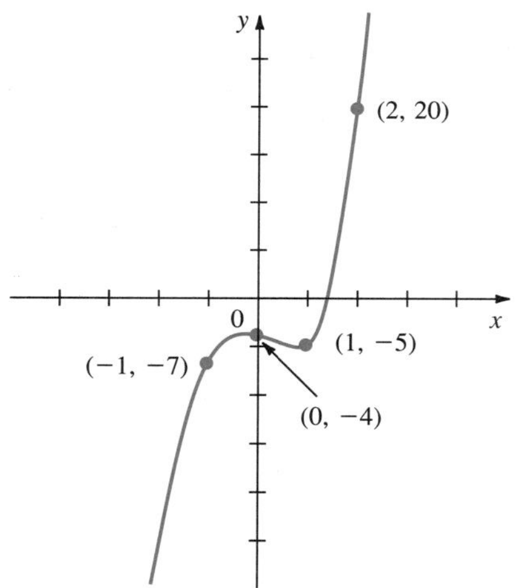

Figure 2 Graph of $p(x) = x^5 - 2x^2 - 4$.

EXAMPLE 1 ***Using the Bisection Method to Find Zeros of a Polynomial***

Approximate, with an error of less than 0.0005, all zeros of the polynomial

$$p(x) = x^5 - 2x^2 - 4$$

SOLUTION p has one variation in sign, so it has exactly one positive zero. Moreover, $p(-x) = -x^5 - 2x^2 - 4$ has no variations in signs so that p has no negative zeros. Thus, p has exactly one real zero. Since $p(1) = -5 < 0$ and $p(2) = 20 > 0$, the zero is in the interval $(1, 2)$. A graph of $p(x)$ is given in Figure 2. The midpoint of $(1, 2)$ is $\frac{1+2}{2} = 1.5$, and so 1.5 is our first approximation. The maximum error of this approximation is $\frac{2-1}{2} = 0.5$.

We then evaluate

$$p(1.5) = 1.5^5 - 2(1.5)^2 - 4 = -0.90625 < 0$$

Thus the zero is in (1.5, 2) and our next approximation is the midpoint of (1.5, 2), which is 1.75. The maximum error in the new approximation is $\frac{2 - 1.5}{2} = 0.25$. We continue in this fashion to obtain the values in Table 1. All numbers in the table are rounded to 5 decimal places before being used in further computations. We stop when the maximum error is less than 0.0005.

Table 1
Using the Bisection Method to Find a Zero of $p(x) = x^5 - 2x^2 - 4$

a	b	Midpoint (approximation to zero) $\frac{a+b}{2}$	Maximum Error in Approximation $\frac{b-a}{2}$	$p(a)$	$p(b)$	$p\left(\frac{a+b}{2}\right)$	New Interval (a, b)
1	2	1.5	0.5	−5	20	−0.90625	(1.5, 2)
1.5	2	1.75	0.25	−0.90625	20	6.28809	(1.5, 1.75)
1.5	1.75	1.625	0.125	−0.90625	6.28809	2.04971	(1.5, 1.625)
1.5	1.625	1.5625	0.0625	−0.90625	2.04971	0.43041	(1.5, 1.5625)
1.5	1.5625	1.53125	0.03125	−0.90625	0.43041	−0.27103	(1.53125, 1.5625)
1.53125	1.5625	1.54688	0.01563	−0.27103	0.43041	0.07125	(1.53125, 1.54688)
1.53125	1.54688	1.53907	0.00782	−0.27103	0.07125	−0.10189	(1.53907, 1.54688)
1.53907	1.54688	1.54298	0.00391	−0.10189	0.07125	−0.01574	(1.54298, 1.54688)
1.54298	1.54688	1.54493	0.00195	−0.01574	0.07125	0.02763	(1.54298, 1.54493)
1.54298	1.54493	1.54396	0.00098	−0.01574	0.02763	0.00602	(1.54298, 1.54396)
1.54298	1.54396	1.54347	0.00049 ↑ <0.0005	—	—	—	—

We see that the last approximation is correct to 0.0005 units, so the zero is within 0.0005 units of 1.54347.

In Example 2 in Section 4.6 we will show how to obtain this answer much faster using Newton's method. To 9 decimal places the zero is 1.543689013 (see p. 273).

The bisection method is useful for obtaining zeros of a function with one- or two-decimal place accuracy. A much faster and more efficient method is given in the next section.

We stress, too, that a reasonably good approximation to the zeros of a function can be obtained relatively quickly on a graphing calculator. Examples of this are given in Appendix A at the back of the book.

Problems 4.5

Readiness Check

I. If in the bisection method we obtain the interval $(2.6, 2.7)$ and find that $p(2.6) \approx 0.8$, $p(2.7) \approx -0.6$, and $p(2.65) \approx -0.2$, then the next interval to be used is

a. $(2.65, 2.7)$ b. $(2.6, 2.65)$
c. $(2.6, 2.625)$ d. $(2.625, 2.65)$

II. If in the bisection method we obtain the interval $(-4.77, -4.73)$ and find that $p(-4.77) \approx -0.28$, $p(-4.75) \approx -0.11$, and $p(-4.73) \approx 0.26$, then the next interval to be used is

a. $(-4.77, -4.74)$ b. $(-4.77, -4.75)$
c. $(-4.75, -4.73)$ d. $(-4.75, -4.74)$
e. $(-4.76, -4.75)$

III. If in the bisection method the interval $(2.163, 2.258)$ is obtained, the next approximation to the zero is 2.2105. The maximum error in this approximation is

a. 0.095
b. 2.2105
c. 0.0475
d. 0.02375

In Problems 1–8 use the bisection method to find all zeros to the given polynomial with an error less than 0.005.

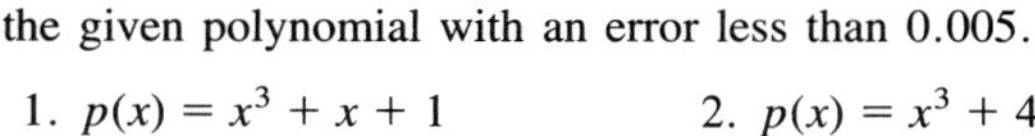

1. $p(x) = x^3 + x + 1$

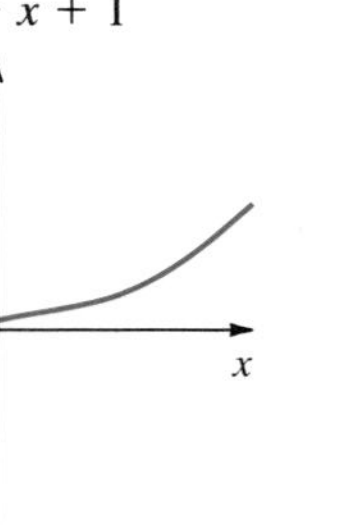

2. $p(x) = x^3 + 4x - 3$

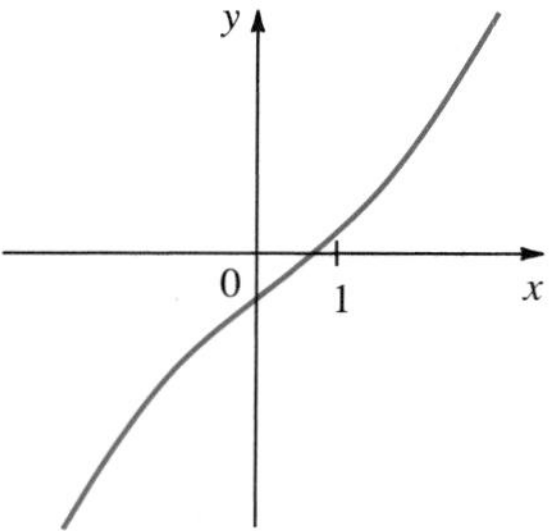

5. $p(x) = -5x^3 - 3x + 2$

6. $p(x) = -x^5 + x^3 + x - 100$

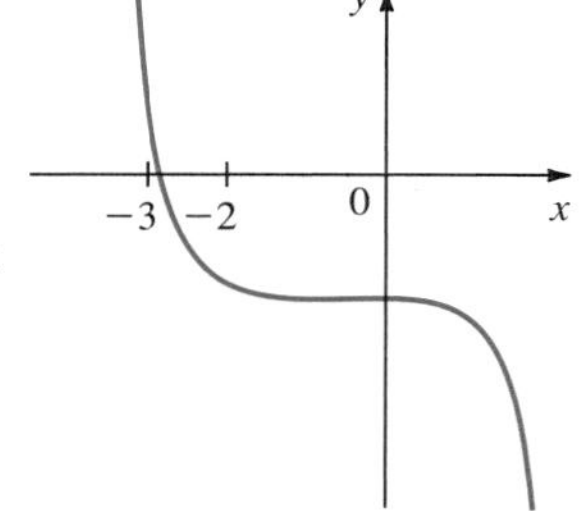

3. $p(x) = x^3 + 3x^2 + 6x + 2$

4. $p(x) = -x^3 - 2x + 8$

7. $p(x) = x^3 - 6x^2 - 15x + 4$

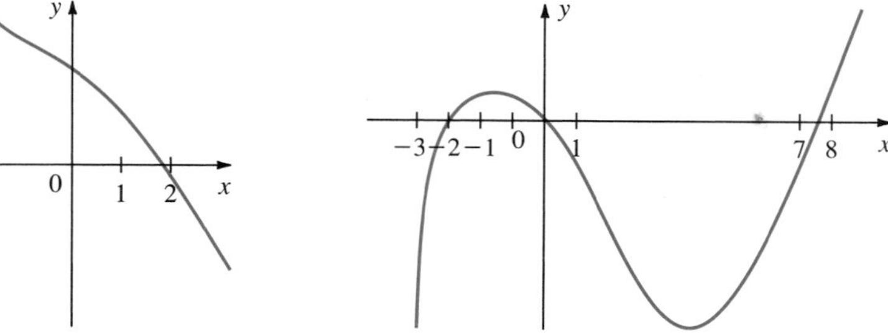

Answers to Readiness Check

I. b II. c III. c

8. $p(x) = x^4 - x - 1$

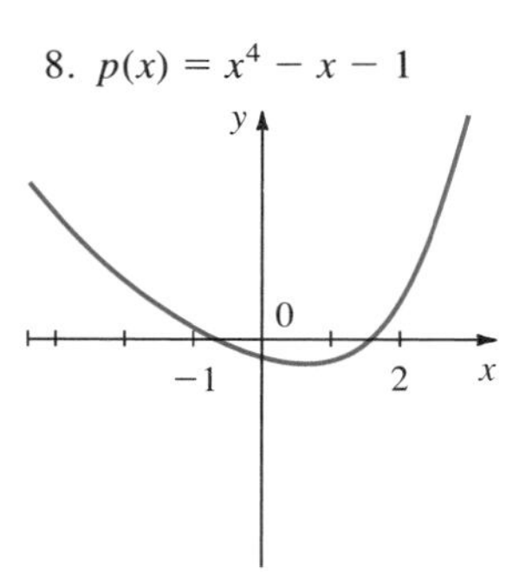

In Problems 9–13 use the bisection method to find one zero of the given polynomial in the given interval with an error less than 0.0005.

9. $x^3 - x^2 - 1$; $[1, 2]$
10. $x^4 - x^2 - 3x - 10$; $[2, 3]$
11. $3x^5 + 2x^4 + x^2 + 3$; $[-2, -1]$
12. $x^7 - x^2 - 12$; $[1, 2]$
13. $6x^4 - 2x^3 + 3x^2 - 4x - 2$; $[0, 1]$
* 14. Find all zeros of $x^3 + 14x^2 + 60x + 78$ with errors less than 0.01. [Hint: There are three of them.]
* 15. Find all zeros of $x^3 - 8x^2 + 2x - 14$ with errors less than 0.01.

4.6 Finding Zeros Numerically by Newton's Method (Optional)

We have spent a considerable amount of time trying to find zeros of polynomial functions. We were limited, however, to quadratic polynomials or higher-order polynomials with rational zeros. As we remarked on p. 261, most polynomials do not have rational zeros, although they do have real zeros.

For example, we know by Descartes' rule of signs that the polynomial $p(x) = x^5 - 2x^2 - 4$ has exactly one positive zero (there is one variation in sign). Moreover, since $p(1) = -5$ and $p(2) = 20$, the zero is between 1 and 2. How do we find it more precisely? One way is by trial and error, as in Example 4 on p. 236. Another is by using the bisection method described in Section 4.5. These methods are very slow. Fortunately, there is a very efficient method for estimating the real zeros of a polynomial, due to Sir Isaac Newton. We will describe this method here. The motivation for this method requires calculus, so all we can do is present it. First we need to introduce a new function.

Let $p(x) = a_nx^n + a_{n-1}x^{n-1} + \cdots + a_2x^2 + a_1x + a_0$ be a polynomial. Then the **derivative function,** $d(x)$, is defined by

Definition of the Derivative Function

$$d(x) = na_nx^{n-1} + (n-1)a_{n-1}x^{n-2} + \cdots + 2a_2x + a_1 \qquad (1)$$

Obtaining the Derivative Function

1. Start with a typical term in $p(x)$: a_kx^k.
2. Multiply this term by k and reduce the exponent by 1: ka_kx^{k-1}.
3. Take the sum of these new terms, omitting the constant term a_0.

EXAMPLE 1 *Finding the Derivative Function*

Let $p(x) = 3x^5 - 2x^4 + 7x^3 + 8x - 5$. Find $d(x)$.

SOLUTION $3x^5$ becomes $5 \cdot 3x^{5-1} = 15x^4$

$-2x^4$ becomes $4(-2x^{4-1}) = -8x^3$

$7x^3$ becomes $3(7x^{3-1}) = 21x^2$

$8x$ becomes $1(8x^{1-1}) = 8x^0 = 8$

-5 is eliminated

Thus

$$d(x) = 15x^4 - 8x^3 + 21x^2 + 8$$

Newton's Method for Finding Real Zeros of a Polynomial $p(x)$

Step 1 Find an interval $[a, b]$ that contains a zero of $p(x)$. To do so, find a and b such that either $p(a) > 0$ and $p(b) < 0$ or $p(a) < 0$ and $p(b) > 0$.

Step 2 Choose a number x_0 in $[a, b]$.

Step 3 Set $x_1 = x_0 - \dfrac{p(x_0)}{d(x_0)}$.

Step 4 Set $x_2 = x_1 - \dfrac{p(x_1)}{d(x_1)}$.

Step 5 Continue as in Steps 3 and 4. If x_k is chosen, then

$$x_{k+1} = x_k - \frac{p(x_k)}{d(x_k)}$$

The numbers $x_0, x_1, x_2, \ldots$ are called **iterates.**

Step 6 Stop when two successive iterates agree to the number of decimal places carried in the computation.

The iterates $x_0, x_1, x_2, x_3, \ldots$ form what is called a **sequence.** What is important here is that after a certain point, each iterate x_{n+1} is closer to the zero of $p(x)$ in $[a, b]$ than the iterate that preceded it. In fact, x_{n+1} will be accurate to twice as many decimal places as x_n. We illustrate all these things with an example.

EXAMPLE 2 *Using Newton's Method to Find a Zero*

Find the positive zero of $p(x) = x^5 - 2x^2 - 4$.

SOLUTION As we have already noted, $p(x)$ has exactly one positive zero in the interval $[1, 2]$. Now

$$d(x) = 5x^{5-1} - 2(2x^{2-1}) = 5x^4 - 4x$$

If we choose $x_0 = 1$, then we obtain the iterates in Table 1. We stop when two successive iterates are equal (to 1.543689013).

Table 1

n	x_n	$p(x_n) = x_n^5 - 2x_n^2 - 4$	$d(x_n) = 5x_n^4 - 4x_n$	$\frac{p(x_n)}{d(x_n)}$	$x_{n+1} = x_n - \frac{p(x_n)}{d(x_n)}$
0	1	−5	1	−5	6
1	6	7700	6456	1.192688972	4.807311029
2	4.807311029	2517.283395	2651.186564	0.949493117	3.857817911
3	3.857817911	820.7277108	1092.051311	0.751546839	3.106271072
4	3.106271072	265.9011430	453.0831992	0.586870454	2.519400618
5	2.519400618	84.80994240	191.3685285	0.443176017	2.076224600
6	2.0762246	25.95942134	84.60614435	0.306826667	1.769397932
7	1.769397932	7.081595821	41.93098229	0.168886952	1.600510980
8	1.60051098	1.379243697	26.40783562	0.052228577	1.548282402
9	1.548282402	0.102792706	22.53919173	0.004560620	1.543721781
10	1.543721781	0.000728097	22.22039135	0.000032767	1.543689014
11	1.543689014	0.000000037	22.21811162	0.000000001	1.543689013
12	1.543689013	0.00000000022	22.21811152	9.9×10^{-12}	1.543689013

Thus, to 9 decimal places, $s = 1.543689013$.

Check $s^5 - 2s^2 - 4 = 8.765951545 - 4.765951538 - 4 = 0.000000006$

In this example it took eleven steps to get the desired accuracy. We can reduce the work considerably if we do a bit more work to start. We have

$$p(1.5) = -0.90625 \quad \text{and} \quad p(1.6) = 1.36576$$

Thus we know that the zero is between 1.5 and 1.6. If we start with $x_0 = 1.5$, say, then we obtain the results in Table 2.

Table 2

n	x_n	$p(x_n) = x_n^5 - 2x_n^2 - 4$	$d(x_n) = 5x_n^4 - 4x_n$	$\frac{p(x_n)}{d(x_n)}$	$x_{n+1} = x_n - \frac{p(x_n)}{d(x_n)}$
0	1.5	−0.90625	19.3125	−0.046925566	1.546925566
1	1.546925566	0.072275308	22.44403321	0.003220246	1.543705320
2	1.543705320	0.000362333	22.21924607	0.000016307	1.543689013
3	1.543689013	0.000000006	22.21811152	3×10^{-10}	1.543689013

Here we got our answer after three steps.

One of the many nice things about Newton's method is that it is easily programmed on a computer or programmable calculator.

We cannot give a theorem here that guarantees that Newton's method will work. However, the following is true.

Theorem

If $p(a)$ and $p(b)$ have different signs and $d(x) \neq 0$ for x in $[a, b]$, then $p(x)$ has a unique zero in $[a, b]$.

The condition $d(x) \neq 0$ in $[a, b]$ is important because, in using Newton's method, we divide by $d(x)$. This condition alone doesn't guarantee that the iterates get closer and closer to the zero, but in most cases, it will be sufficient, especially if we start reasonably close to the zero.

Problems 4.6

In Problems 1–6 use Newton's method to find a zero of the given quadratic in the given interval. Compare each iterate with the correct answer obtained from the quadratic formula or factoring.

1. $x^2 - 3x - 10$; $[4, 6]$
2. $x^2 - 1$; $[\frac{1}{2}, \frac{3}{2}]$
3. $3x^2 - 5x - 2$; $[-1, 0]$
4. $6x^2 + 5x - 56$; $[-4, -3]$
5. $x^2 + x - 1$; $[0, 1]$
6. $x^2 + 3x - 8$; $[-5, -4]$
7. Use Newton's method to calculate the roots of $x^2 - 7x + 5 = 0$. It will be necessary to do two separate calculations, using two distinct intervals. Compare this result with the answers obtained by the quadratic formula.

In Problems 8–13 use Newton's method to find one zero of the given polynomial in the given interval.

8. $x^3 + x^2 + 5$; $[-3, -2]$
9. $x^3 - x^2 - 1$; $[1, 2]$
10. $x^4 - x^2 - 3x - 10$; $[2, 3]$
11. $3x^5 + 2x^4 + x^2 + 3$; $[-2, -1]$
12. $x^7 - x^2 - 12$; $[1, 2]$
13. $6x^4 - 2x^3 + 3x^2 - 4x - 2$; $[0, 1]$

* 14. Find all zeros of the polynomial $x^3 - 6x^2 - 15x + 4$. [Hint: Estimate, roughly, each of the three zeros and find intervals that contain them.]

* 15. Find all zeros of $x^3 + 14x^2 + 60x + 78$.

* 16. Find all zeros of $x^3 - 8x^2 + 2x - 15$.

17. Apply Newton's method to $p(x) = x^2 + 5x + 7$. What happens? Explain why the method is doomed to fail.
18. A rectangular box is to be constructed with square ends in such a way that the length of the box is 2 feet longer than one of the sides of the square end (see Problem 63 on page 241). Find, to the nearest hundredth of a foot, the dimensions of a box whose volume is 100 ft^3.
19. Answer the question of Problem 18 if the volume is 60 ft^3.
20. In Problem 101 on page 255 we gave the cost function

$$C(q) = -0.001q^3 + 0.5q^2 + 80q + 10000;$$
$$0 \le q \le 500$$

The cost was determined to be \$55,000. How many tons of coal were processed [to the nearest hundredth of a ton]? [Hint: There are two answers to this problem.]

21. Answer the question in Problem 20 if

$$C(q) = -0.002q^3 + 1.4q^2 + 200q + 25{,}000,$$
$$0 \le q \le 750$$

and the total cost is \$200,000.

Summary Outline of Chapter 4

- A **quadratic function** is a function of the form $p(x) = ax^2 + bx + c$. The graph of a quadratic function is a **parabola.** It opens upward if $a > 0$ and downward if $a < 0$. The vertex of the parabola is $\left(-\frac{b}{2a}, p\left(-\frac{b}{2a}\right)\right)$. pp. 225, 226

- The polynomial $ax^2 + bx + c$ is **irreducible** if the quadratic equation $ax^2 + bx + c = 0$ has no real roots (this occurs if $b^2 - 4ac < 0$). In this case its graph does not intersect the x-axis. p. 226
- The function $p(x) = ax^n$ is called a **power function.** If n is even, the graph resembles the graph of ax^2. If n is odd, the graph resembles the graph of ax^3. pp. 231, 232
- r is a **zero** or **root** of $p(x)$ if $p(r) = 0$. In this case $x - r$ is a factor of p, and $(r, 0)$ is an x-intercept of the graph of $y = p(x)$. p. 234
- If $p(a) > 0$ and $p(b) < 0$ (or vice versa), then $p(x)$ has a zero between a and b. p. 235
- $f(x) \to \infty$ as $x \to \infty$ means that $f(x)$ increases without bound as x increases without bound. p. 237
- **Division Algorithm:** If degree of the polynomial $d(x) <$ degree of the polynomial $p(x)$ and $d(x) \neq 0$, then there exist unique polynomials $q(x)$ and $r(x)$ such that $p(x) = d(x)q(x) + r(x)$. $q(x)$ is called the **quotient,** and $r(x)$ is called the **remainder.** Here degree of $r(x) <$ degree of $d(x)$. p. 245
- **Quotient Theorem:**

$$\frac{p(x)}{d(x)} = q(x) + \frac{r(x)}{d(x)}$$

  where $r(x) = 0$ or degree of $r(x) <$ degree of $d(x)$ p. 245
- **Remainder Theorem:** If $p(x)$ is divided by $x - c$, then the remainder $r = p(c)$. p. 245
- **Factor Theorem:** $x - c$ is a factor of the polynomial $p(x)$ if and only if $p(c) = 0$. p. 246
- **Synthetic division** is a process for quickly dividing a polynomial by $x - c$. pp. 247–250
- **Horner's method** is a technique for evaluating a polynomial quickly on a calculator by factoring out x repeatedly. pp. 251–253
- If $(x - c)^k$ is a factor of the polynomial $p(x)$ but $(x - c)^{k+1}$ is not, then c is a zero of $p(x)$ of **multiplicity *k*.** p. 256
- **Fundamental Theorem of Algebra:** Every polynomial $p(x)$ of degree n has at least one zero. In fact, counting multiplicities and nonreal zeros, it has exactly n zeros. p. 256
- **Descartes' Rule of Signs:** Let $p(x)$ be a polynomial with nonzero constant term.
  (i) The number of positive real zeros, P, is either equal to V, the number of variations of sign in $p(x)$, or is less than V, and $V - P$ is an even integer. p. 258
  (ii) The number of negative real zeros, N, is either equal to the number of variations of sign, V_N, in $p(-x)$, or is less, and $V_N - N$ is an even integer.
- Complex zeros of polynomials with real coefficients occur in complex conjugate pairs. p. 259
- **Rational Zeros Theorem:** Let $p(x) = a_n x^n + a_{n-1}x^{n-1} + \cdots + a_1 x + a_0$, where $a_n \neq 0$ and the coefficients $a_0, a_1, a_2, \cdots, a_n$ are all integers. Suppose $\frac{m}{k}$ is a rational zero of $p(x)$ and $\frac{m}{k}$ is reduced to lowest terms. Then p. 259
  (i) m is a factor of a_0,
  (ii) k is a factor of a_n.
- **Upper and Lower Bound Theorem:** Let p. 262

$$p(x) = a_n x^n + a_{n-1}x^{n-1} + \cdots + a_1 x + a_0$$

  with $a_n > 0$ and all coefficients are real. Divide $p(x)$ by $x - c$ using synthetic division.
  (i) c is an upper bound for the zeros of $p(x)$ if $c > 0$, and all the numbers in the bottom row of the synthetic division are nonnegative.
  (ii) c is a lower bound for the zeros of $p(x)$ if $c < 0$, and the numbers in the bottom row alternate in sign (where 0 can be considered positive or negative, as required).

- **The Bisection Method for Finding a Zero of a Function $p(x)$** p. 268

Step 1 Find two numbers a and b such that

$p(a) < 0$ and $p(b) > 0$ or $p(a) > 0$ and $p(b) < 0$

Then there is a zero in (a, b) by the intermediate value theorem (p. 235). We assume here that $p(a) < 0$ and $p(b) > 0$.

Step 2 Compute p at the midpoint $\frac{a+b}{2}$. This is your first approximation to the zero.

Step 3 (a) If $p\left(\frac{a+b}{2}\right) > 0$, then there is a zero in $\left(a, \frac{a+b}{2}\right)$.

(b) If $p\left(\frac{a+b}{2}\right) < 0$, then there is a zero in $\left(\frac{a+b}{2}, b\right)$.

Step 4 Continue to cut the interval containing a zero in half until the length of the interval is less than the accuracy you seek.

- **Newton's Method for Finding Real Zeros of a Polynomial $p(x)$** p. 272

Step 1 Find an interval $[a, b]$ that contains a zero of $p(x)$. To do so, find a and b such that either $p(a) > 0$ and $p(b) < 0$ or $p(a) < 0$ and $p(b) > 0$.

Step 2 Choose a number x_0 in $[a, b]$.

Step 3 Set $x_1 = x_0 - \frac{p(x_0)}{d(x_0)}$, where $d(x)$, the **derivative function,** is defined as on p. 271.

Step 4 Set $x_2 = x_1 - \frac{p(x_1)}{d(x_1)}$.

Step 5 Continue as in Steps 3 and 4. If x_k is chosen, then

$$x_{k+1} = x_k - \frac{p(x_k)}{d(x_k)}$$

The numbers $x_0, x_1, x_2, \ldots$ are called **iterates.**

Step 6 Stop when two successive iterates agree to the number of decimal places carried in the computation.

Review Exercises for Chapter 4

1. Sketch the graph of $f(x) = \frac{1}{4}x^2$.
2. Sketch the graph of $f(x) = -3x^2$.

In Exercises 3–5 write each quadratic in the form $y = a(x - h)^2 + k$, and find the vertex.

3. $y = x^2 + 4x$
4. $y = x^2 - 6x + 10$
5. $y = 2x^2 + 6x - 8$

In Exercises 6–8 sketch each parabola indicating the vertex, axis, y-intercept, x-intercepts (if any), and the maximum or minimum value taken by the function.

6. $f(x) = x^2 - 5x - 8$
7. $f(x) = 4 - x^2$
8. $f(x) = 2x^2 + 8x - 6$

In Exercises 9 and 10 a ball is thrown straight up from the ground with an initial velocity of v_0. Determine
(a) when the ball hits the ground and,
(b) how high it gets.

9. $v_0 = 80$ ft/sec
10. $v_0 = 400$ m/sec

In Exercises 11–19 sketch the graph of the given polynomial.

11. $p(x) = \frac{1}{2}x^3$
12. $p(x) = -3x^4$
13. $p(x) = (x + 2)^5$
14. $p(x) = 4 - x^3$
15. $p(x) = (x^2 - 1)(x - 2)$
16. $p(x) = x^3 - 3x^2 + 2x$
17. $p(x) = (x - 1)(x^2 - 3x - 4)$
18. $p(x) = x^3 - 8$
19. $p(x) = x^4 + 4x^2 + 4$

In Exercises 20–23 use long division to find the quotient and remainder when $p(x)$ is divided by $q(x)$.

20. $p(x) = x^2 - 4x + 5$; $q(x) = x + 4$
21. $p(x) = x^3 - x^2 + x + 1$; $q(x) = x^2 + 2x$
22. $p(x) = x^4 + x^3 - 7x^2 - 3x + 5$; $q(x) = -x^2 + 3x + 5$
23. $p(x) = x^5 - x^3 + 2x + 3$; $q(x) = x^3 - 4x^2 + 5$

In Exercises 24–27 divide the first polynomial by the second, using synthetic division. Find the quotient and remainder.

24. $x^3 - x^2 + 5$; $x - 2$
25. $3x^4 + 3x^3 - x^2 + 4x - 9$; $x + 1$
26. $x^3 + x^2 - 7x - 15$; $x - 3$
27. $4x^5 - x^3 + 2x - 1$; $x + 2$

In Exercises 28 and 29 use synthetic division to evaluate the given polynomial at the given point.

28. $p(x) = x^3 + 2x^2 - 5x + 6$; $x = -2$
29. $p(x) = 4x^4 - 3x^3 - x + 3$; $x = 5$
30. Find a cubic polynomial $p(x)$ with zeros 3, -2, and 2 such that $p(0) = 1$.

In Exercises 31–34 use Descartes' rule of signs to determine the possible number of positive and negative zeros of each polynomial.

31. $p(x) = 4x^3 - 7x^2 - 3x - 8$
32. $p(x) = x^4 + 6x^3 - 5x^2 + 8x + 9$
33. $p(x) = 4x^4 + 12x^3 - 11x^2 - 5x + 3$
34. $p(x) = -5x^5 - x^4 + x^3 + 4x^2 - 3x + 8$

In Exercises 35–38 find all three zeros of each cubic.

35. $p(x) = x^3 - 7x^2 + 14x - 8$
36. $p(x) = x^3 - x^2 - 18x + 18$
37. $p(x) = 2x^3 + x^2 + 2x + 1$
38. $p(x) = 5x^3 - x^2 - 11x + 6$

In Exercises 39–42 find all rational zeros of each polynomial.

39. $p(x) = x^4 - x^3 - 10x^2 + 4x - 24$
40. $p(x) = 4x^4 + 11x^3 + 3x^2 + x + 14$
41. $p(x) = x^5 - 4x^4 + 4x^3 - 16x^2 - 5x + 20$
42. $p(x) = x^6 + 7x^3 - 8$

In Exercises 43–46 use the bisection method to find, with a maximum error less than 0.005, a zero of the given polynomial in the given interval.

43. $x^3 + 2x - 2$; $[0, 1]$
44. $2x^3 - 3x^2 + 5x + 6$; $[-1, 0]$
45. $x^5 + 3x^3 + x - 10$; $[1, 2]$
46. $-x^4 + 3x^3 + 2x^2 + x + 8$; $[3, 4]$

In Exercises 47–50 use Newton's method to find, to eight decimal places, a zero of the given polynomial in the given interval.

47. The polynomial of Exercise 43.
48. The polynomial of Exercise 44.
49. The polynomial of Exercise 45.
50. The polynomial of Exercise 46.

Chapter 5

Rational Functions and Conic Sections

5.1 Rational Functions and Their Graphs

A **rational function** is a function of the form $f(x) = \dfrac{p(x)}{q(x)}$ where $p(x)$ and $q(x)$ are polynomials.

EXAMPLE 1 *Four Rational Functions*

The following are rational functions:

(a) $f(x) = \dfrac{x+3}{x-5}$ (b) $f(x) = \dfrac{1}{x}$

(c) $f(x) = \dfrac{3+x^2}{2-x^3}$ (d) $f(x) = \dfrac{x^3+x^2-3x+5}{x^2-4x+6}$

In this section we sketch the graphs of several rational functions. Before giving general rules, we begin with an example.

EXAMPLE 2 *The Graph of* $\dfrac{1}{x}$

Sketch the graph of $f(x) = \dfrac{1}{x}$.

SOLUTION We obtained this graph on p. 193. We say more about it here. $f(-x) = \dfrac{1}{-x} = -\dfrac{1}{x} = -f(x)$. Thus $f(x)$ is odd, and we need sketch only the graph for $x > 0$ and reflect this graph about the origin (see p. 191). $f(0)$ is not defined, and for every $x > 0$, $\dfrac{1}{x} > 0$. We write $x \to 0^+$ to indicate that x approaches 0 from the right (the *positive* side —that's why we use a +). As

Table 1

x	$1/x$
1	1
0.5	2
0.2	5
0.1	10
0.01	100
0.0001	10,000
10^{-10}	10^{10}

$x \to 0^+$, $\dfrac{1}{x}$ gets larger and larger, as indicated in Table 1. We have

$$\frac{1}{x} \to \infty \qquad \text{as } x \to 0^+$$

Also, reversing the columns in the table (which we can do because $1/(1/x) = x$), we see that

$$\frac{1}{x} \to 0^+ \qquad \text{as } x \to \infty\dagger$$

If $x_1 > x_2 > 0$, then $\dfrac{1}{x_1} < \dfrac{1}{x_2}$. For example, $3 > 2 > 0$ and $\dfrac{1}{3} < \dfrac{1}{2}$. Thus $\dfrac{1}{x}$ is a decreasing function. This is all we need to sketch the graph in Figure 1. The line $y = 0$ (the x-axis) is called a **horizontal asymptote** for $f(x)$. The line $x = 0$ (the y-axis) is called a **vertical asymptote** for $f(x)$.

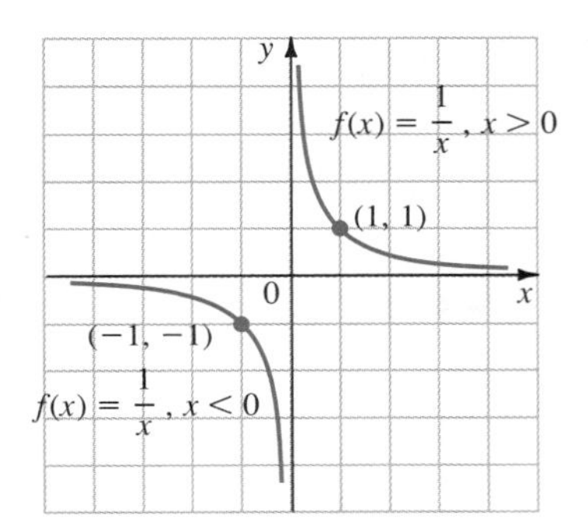

Figure 1 Graph of $f(x) = \dfrac{1}{x}$.

In Table 1 we saw that $\dfrac{1}{x} \to 0$ as $x \to \infty$. In fact, the following is true:

Behavior of $\dfrac{k}{x^n}$ as x Gets Large in Absolute Value

If k is a real number and n is a positive real number, then

$$\frac{k}{x^n} \to 0 \text{ as } |x| \to \infty$$

EXAMPLE 3 ***Behavior of k/x^n as $|x| \to \infty$***

(a) $\dfrac{3}{x^3} \to 0 \qquad$ as $x \to \infty$

(b) $\dfrac{-5}{x} \to 0 \qquad$ as $x \to \infty$

(c) $\dfrac{4.6}{x^2} \to 0 \qquad$ as $x \to -\infty$

You can verify that these are true by looking at values of each expression for $|x|$ large.

A line is called an **asymptote** to a curve if the curve gets closer and closer to the line as x increases without bound, decreases without bound, or approaches a fixed value. In Figure 1 we have two special kinds of asymptotes.

† See page 237 for an explanation of this notation.

Definition of Vertical and Horizontal Asymptotes

The vertical line $x = c$ is a **vertical asymptote** of the graph of f if $f(x) \to \infty$ or $f(x) \to -\infty$ as $x \to c^+$ or $x \to c^-$.

The horizontal line $y = c$ is a **horizontal asymptote** of the graph of f if $f(x) \to c$ as $x \to \infty$ or $x \to -\infty$.

NOTE $x \to c^-$ means that x gets closer and closer to c from the left (the *negative* side).

When does a graph have a vertical asymptote?

Vertical Asymptote Theorem

The line $x = c$ (c a real number) is a vertical asymptote of $f(x) = \dfrac{p(x)}{q(x)}$ if $q(c) = 0$ but $p(c) \neq 0$.

EXAMPLE 4 *The Graph of a Rational Function with Two Vertical Asymptotes*

Sketch the graph of $f(x) = \dfrac{1}{x^2 - x - 2}$.

SOLUTION We factor the denominator: $x^2 - x - 2 = (x - 2)(x + 1)$. The lines $x = 2$ and $x = -1$ are vertical asymptotes. What happens for x near -1 or 2? We know that $\dfrac{1}{(x-2)(x+1)}$ will get large as x gets close to either of these values. The only question is whether $f(x) \to \infty$ or $f(x) \to -\infty$. The answers are given in Table 2.

Table 2

Interval	Test Point	Sign of $x + 1$	Sign of $x - 2$	Sign of $(x + 1)(x - 2)$	Sign of $f(x) = \dfrac{1}{(x+1)(x-2)}$
$(-\infty, -1)$	-3	$-$	$-$	$+$	$+$
$(-1, 2)$	0	$+$	$-$	$-$	$-$
$(2, \infty)$	3	$+$	$+$	$+$	$+$

Thus

$$\frac{1}{(x+1)(x-2)} \to \infty \qquad \text{as } x \to -1^-$$

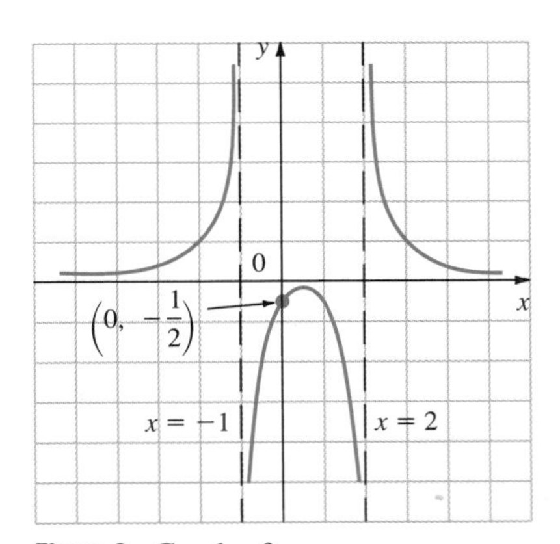

Figure 2 Graph of $f(x) = \dfrac{1}{x^2 - x - 2} = \dfrac{1}{(x-2)(x+1)}$.

$$\frac{1}{(x+1)(x-2)} \to -\infty \qquad \text{as } x \to -1^+$$
$$\frac{1}{(x+1)(x-2)} \to -\infty \qquad \text{as } x \to 2^-$$
$$\frac{1}{(x+1)(x-2)} \to \infty \qquad \text{as } x \to 2^+$$

Also, $f(0) = \dfrac{1}{-2} = -\dfrac{1}{2}$. Finally,

$$\frac{1}{x^2 - x - 2} \to 0 \quad \text{as } x \to \infty \quad \text{or } x \to -\infty$$

since the x^2 term is much bigger than the other terms when $|x|$ is large (see p. 237). Putting all this information together, we obtain the graph in Figure 2. ■

EXAMPLE 5 *The Behavior of Three Rational Functions as* $x \to \infty$

What happens as $x \to \infty$ for each rational function?

(a) $f(x) = \dfrac{3x^2 + 2x - 5}{x^3 + 10x^2 + 5x + 1}$ (b) $g(x) = \dfrac{2x^2 - 3x + 4}{5x^2 + 2x + 7}$

(c) $h(x) = \dfrac{x^3 - x^2 + x + 3}{10x^2 - 4x + 6}$

SOLUTION

(a) We divide numerator and denominator by x^3, the highest power of x in the denominator:

$$f(x) = \frac{3x^2 + 2x - 5}{x^3 + 10x^2 + 5x + 1} = \frac{\dfrac{3}{x} + \dfrac{2}{x^2} - \dfrac{5}{x^3}}{1 + \dfrac{10}{x} + \dfrac{5}{x^2} + \dfrac{1}{x^3}} \to \frac{0}{1} = 0$$
$$\text{as } |x| \to \infty$$

We conclude that $f(x) \to 0$ and $y = 0$ is a horizontal asymptote for f.

(b) We divide numerator and denominator by x^2:

$$g(x) = \frac{2x^2 - 3x + 4}{5x^2 + 2x + 7} = \frac{2 - \dfrac{3}{x} + \dfrac{4}{x^2}}{5 + \dfrac{2}{x} + \dfrac{7}{x^2}}$$

We see that $g(x) \to \frac{2}{5}$ as $|x| \to \infty$. The line $y = \frac{2}{5}$ is a horizontal asymptote for g.

(c) We have

$$h(x) = \frac{x^3 - x^2 + x + 3}{10x^2 - 4x + 6} = \frac{x - 1 + \dfrac{1}{x} + \dfrac{3}{x^2}}{10 - \dfrac{4}{x} + \dfrac{6}{x^2}}$$

Thus, for $|x|$ large,

$$h(x) \approx \frac{x-1}{10}$$

and $\dfrac{x-1}{10} \to \infty$ as $x \to \infty$. Similarly $\dfrac{x-1}{10} \to -\infty$ as $x \to -\infty$. Thus h has no horizontal asymptote.

We generalize the result of the last example.

Horizontal Asymptote Theorem

Let

$$f(x) = \frac{p(x)}{q(x)} = \frac{a_n x^n + a_{n-1}x^{n-1} + \cdots + a_1 x + a_0}{b_m x^m + b_{m-1}x^{m-1} + \cdots + b_1 x + b_0}, \quad a_n \neq 0 \text{ and } b_m \neq 0$$

(i) If $n < m$, then $y = 0$ is a horizontal asymptote for f. **(1)**

(ii) If $n = m$, then $y = \dfrac{a_n}{b_m}$ is a horizontal asymptote for f. **(2)**

(iii) If $n > m$, then f has no horizontal asymptote. **(3)**

EXAMPLE 6 ***The Graph of a Rational Function with One Vertical and One Horizontal Asymptote***

Sketch the graph of $f(x) = \dfrac{x}{x+1}$.

SOLUTION We have the following:

Vertical asymptote: $x = -1$ is a vertical asymptote.
Horizontal asymptote: $y = 1$ is a horizontal asymptote by (2).
x- and y-intercept: $y = 0$ when $x = 0$, so the curve passes through $(0, 0)$.

Sign Table

Interval	Test Point	Sign of x	Sign of $x + 1$	Sign of $\dfrac{x}{x+1}$
$(-\infty, -1)$	-2	$-$	$-$	$+$
$(-1, 0)$	$-\frac{1}{2}$	$-$	$+$	$-$
$(0, \infty)$	1	$+$	$+$	$+$

Special feature: If $x > 0$, then $\dfrac{x}{x+1} < 1$, since the denominator is larger than the numerator. If $x < -1$, then $\dfrac{x}{x+1} > 1$, since both x and $x + 1$ are negative and $|x| > |x + 1|$.

For example, $\dfrac{-3}{-3+1} = \dfrac{-3}{-2} = \dfrac{3}{2}$ and $\dfrac{-1.1}{-1.1+1} = \dfrac{-1.1}{-0.1} = 11$. The graph is given in Figure 3. Note that the graph of f never crosses the horizontal asymptote $y = 1$. For if it did, we would have

$$y = \frac{x}{x+1} = 1 \quad \text{or} \quad x = x + 1 \quad \text{or} \quad 0 = 1$$

which is false.

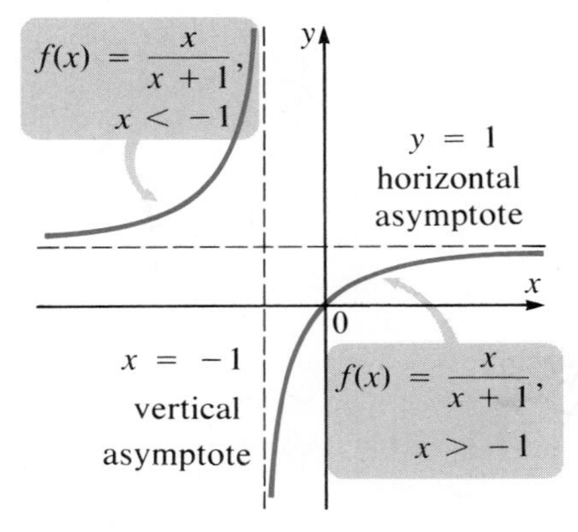

Figure 3 Graph of $f(x) = \dfrac{x}{x+1}$. ■

EXAMPLE 7 *A Rational Function with a Hole in Its Graph*

Sketch the graph of $f(x) = \dfrac{x^2 - 4}{x - 2}$.

SOLUTION We first note that if $x \neq 2$, then

$$\frac{x^2 - 4}{x - 2} = \frac{(x + 2)(x - 2)}{x - 2} = x + 2$$

Thus the graph of $f(x)$ is the same as the graph of the line $y = x + 2$, except the value $x = 2$ $(y = 4)$ must be omitted because $f(2)$ is not defined (division by zero is not permitted). The graph is given in Figure 4.

We see that when $p(x)$ and $q(x)$ have a common factor, we can divide by that common factor and then graph the resulting, simpler function. However, we stress that if $f(x) = \dfrac{p(x)}{q(x)}$ and $p(c) = q(c) = 0$, then $f(c)$ is not defined at $x = c$. Moreover, if the graph of f does not have a vertical asymptote at c, then the graph has a "hole" at c, as in Figure 4.

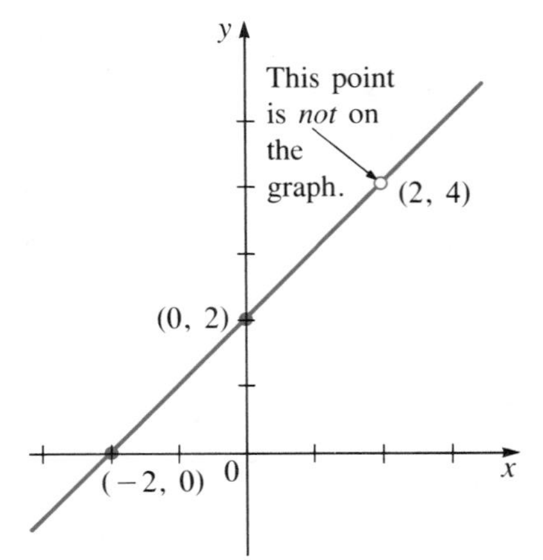

Figure 4 $f(x) = x + 2$, $x \neq 2$.

Sketching the Graph of a Rational Function $f(x) = \dfrac{p(x)}{q(x)}$

Step 1 Factor $p(x)$ and $q(x)$ if possible. Divide by common linear factors. If $x - c$ is a common factor, delete the point $(c, f(c))$ from the resulting graph.

Step 2 Check for symmetry.
If $f(-x) = f(x)$, the graph is symmetric about the y-axis.
If $f(-x) = -f(x)$, the graph is symmetric about the origin.

Step 3 Look for vertical asymptotes. If $q(c) = 0$ and $p(c) \neq 0$, then $x = c$ is a vertical asymptote. Draw it as a dashed line.

Step 4 Find the horizontal asymptotes (if there are any) by using the horizontal asymptote theorem. If $y = c$ is a horizontal asymptote, draw the dashed line $y = c$.

Step 5 Find all x-intercepts, if possible, by determining the zeros of $p(x)$.

Step 6 Find the y-intercept, if one exists, by evaluating $f(0)$. If $f(0)$ is defined (that is, if $q(0) \neq 0$), then the y-intercept is $(0, f(0))$.

Step 7 Let $x = c_1, x = c_2, \ldots, x = c_k$ be the vertical asymptotes. Determine the sign of $f(x)$ in each interval $(-\infty, c_1)$, (c_1, c_2), $\ldots$, (c_{k-1}, c_k), (c_k, ∞). For each c, determine whether $f(x) \to \infty$ as $x \to c^-$ or $f(x) \to -\infty$ as $x \to c^-$ and whether $f(x) \to \infty$ as $x \to c^+$ or $f(x) \to -\infty$ as $x \to c^+$.

Step 8 Look for special features that tell you about the behavior of $f(x)$ as $x \to \pm\infty$ (as in Example 6).

Step 9 Check to see if the graph crosses its horizontal or oblique asymptotes (see Examples 6, 9, and 10).

Step 10 Put everything together and sketch the curve. Plot a few points, if necessary. It is especially useful to plot some points before and after each x-intercept and each vertical asymptote.

EXAMPLE 8 *Sketching a Shifted Rational Function*

Sketch the curve $f(x) = \dfrac{x - 3}{x^2 - 6x + 13} = \dfrac{x - 3}{(x - 3)^2 + 4}$.

SOLUTION We first sketch $g(x) = \dfrac{x}{x^2 + 4}$ and then shift the graph 3 units to the right.

Step 1 Neither term can be factored.

Step 2 If $g(x) = \dfrac{x}{x^2 + 4}$, then

$$g(-x) = \frac{-x}{(-x)^2 + 4} = -\frac{x}{x^2 + 4} = -g(x)$$

so we have symmetry about the origin.

Step 3 $x^2 + 4 \neq 0$ for all real x, so there are no vertical asymptotes.

Step 4 $\dfrac{x}{x^2 + 4} \to 0$ as $x \to \infty$ or $x \to -\infty$ by (1), so $y = 0$ is a horizontal asymptote.

Steps 5, 6 (0, 0) is the only x- and y-intercept for $\dfrac{x}{x^2 + 4}$. $f(x) = \dfrac{x - 3}{(x - 3)^2 + 4} = 0$ when $x = 3$ so (3, 0) is the x-intercept. $f(0) = \dfrac{-3}{13}$, so the y-intercept is $\left(0, \dfrac{-3}{13}\right)$. Observe that $f(x) < 0$ for $x < 3$ and $f(x) > 0$ for $x > 3$.

Step 7 No vertical asymptotes.

Steps 8, 9 $\left|\dfrac{x}{x^2 + 1}\right| < 1$ for all x, so the graph stays between the lines $y = -1$ and $y = 1$.

Step 10 We plot a few points for x between -3 and 3.

The graphs of $f(x)$ and $g(x)$ are given in Figure 5.

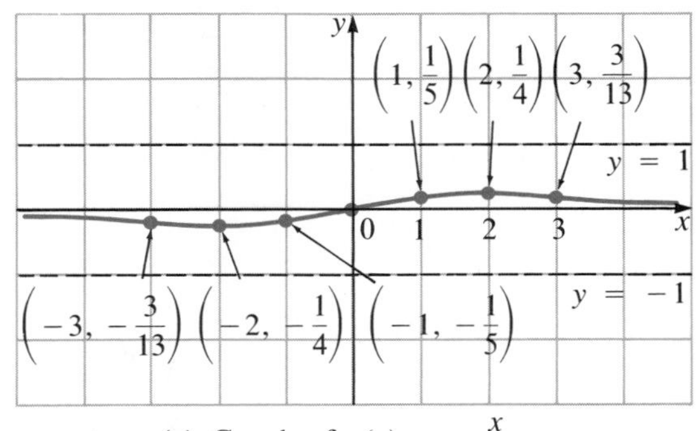

(a) Graph of $g(x) = \dfrac{x}{x^2 + 4}$

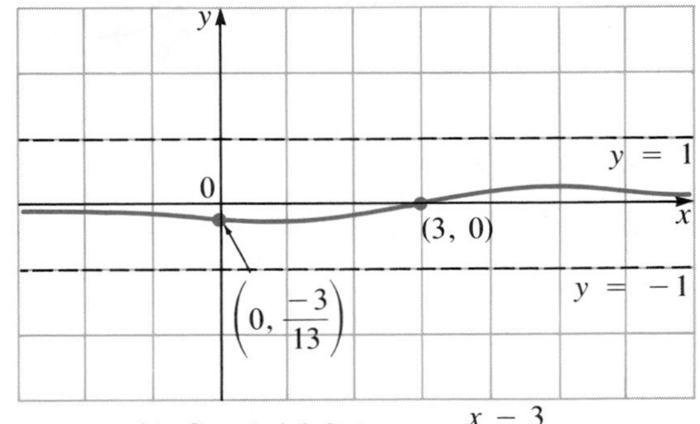

(b) Graph of $f(x) = \dfrac{x - 3}{(x - 3)^2 + 4}$

Figure 5 The graph of $\dfrac{x - 3}{(x - 3)^2 + 4}$ is obtained by shifting the graph of $\dfrac{x}{x^2 + 4}$ 3 units to the right. ■

EXAMPLE 9 ***A Rational Function with Two Vertical Asymptotes and an Oblique Asymptote***

Sketch the graph $f(x) = \dfrac{x^3 - 3x^2 - 10x}{x^2 - 9}$.

SOLUTION $\dfrac{x^3 - 3x^2 - 10x}{x^2 - 9} = \dfrac{x(x - 5)(x + 2)}{(x - 3)(x + 3)}$. There are no common factors and there is no symmetry.

Vertical asymptotes: $x = 3$ and $x = -3$.

Horizontal asymptotes: None (by (3)). Moreover, $f(x) \to \infty$ as $x \to \infty$ and $f(x) \to -\infty$ as $x \to -\infty$.

x-intercepts: $(0, 0)$, $(-2, 0)$, $(5, 0)$.

y-intercept: $(0, 0)$.

Sign Table

Interval	Test Point	Sign of $(x+2)(x)(x-5)$	Sign of $(x+3)(x-3)$	Sign of $\frac{x^3-3x^2-10x}{x^2-9}$
$(-\infty, -3)$	-4	$-$	$+$	$-$
$(-3, -2)$	$-\frac{5}{2}$	$-$	$-$	$+$
$(-2, 0)$	-1	$+$	$-$	$-$
$(0, 3)$	1	$-$	$-$	$+$
$(3, 5)$	4	$-$	$+$	$-$
$(5, \infty)$	6	$+$	$+$	$+$

We further observe that

Divide using long division

$$\frac{x^3 - 3x^2 - 10x}{x^2 - 9} = x - 3 - \frac{x + 27}{x^2 - 9}$$

But $\dfrac{x + 27}{x^2 - 9} \to 0$ as $x \to \pm\infty$ (by (1)) so

$$\frac{x^3 - 3x^2 - 10x}{x^2 - 9} \approx x - 3 \text{ for } |x| \text{ large}$$

The line $y = x - 3$ is called an **oblique asymptote** for the graph.

Does the graph ever cross the oblique asymptote? To check, we set

$$\frac{x^3 - 3x^2 - 10x}{x^2 - 9} = x - 3$$

$$\begin{aligned} x^3 - 3x^2 - 10x &= (x - 3)(x^2 - 9) = x^3 - 3x^2 - 9x + 27 \\ -10x &= -9x + 27 \\ -x &= 27 \\ x &= -27 \end{aligned}$$

Thus the graph crosses its oblique asymptote at the single point $(-27, -30)$. The graph is given in Figure 6.

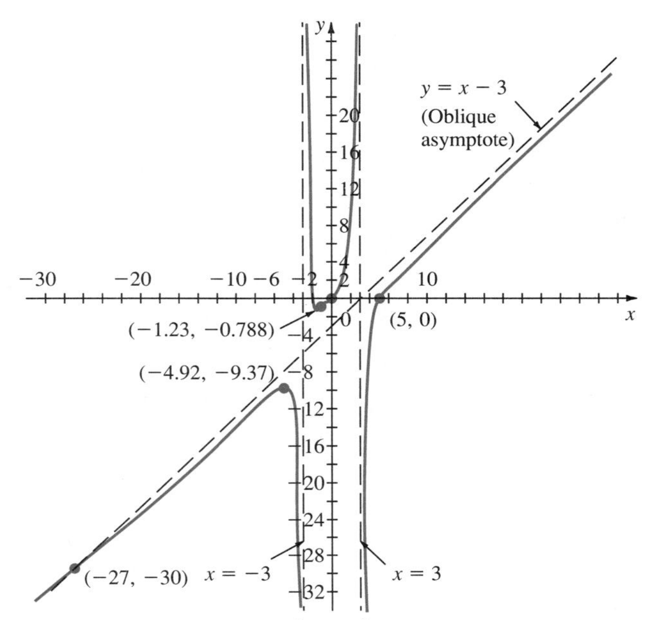

Figure 6 Graph of $f(x) = \dfrac{x^3 - 3x^2 - 10x}{x^2 - 9}$. Note the oblique asymptote $y = x - 3$.

The graph crosses this line at the point $(-27, -30)$. ■

EXAMPLE 10 *A Rational Function with an Oblique Asymptote and a Vertical Asymptote*

Sketch the graph of $f(x) = \dfrac{x^2 + 1}{x}$.

SOLUTION Symmetry: $f(-x) = \dfrac{(-x)^2 + 1}{-x} = \dfrac{x^2 + 1}{-x} = -f(x)$, so symmetric about the origin.

Vertical asymptote: $x = 0$.

Horizontal asymptote: None.

No x- or y-intercepts.

Special feature: $\dfrac{x^2 + 1}{x} = x + \dfrac{1}{x}$. For large $|x|$, $x + \dfrac{1}{x} \approx x$, so the line $y = x$ is an asymptote. For $x > 0$, $x + \dfrac{1}{x} > x$, so the graph is above the line $y = x$. For $x < 0$, $x + \dfrac{1}{x} < x$, so the graph is below the line $y = x$.

The line $y = x$ is an oblique asymptote. The graph is sketched in Figure 7. Note that the graph does not cross its oblique asymptote. If it did, we would have $\dfrac{x^2 + 1}{x} = x$ or $x^2 = x^2 + 1$ or $0 = 1$.

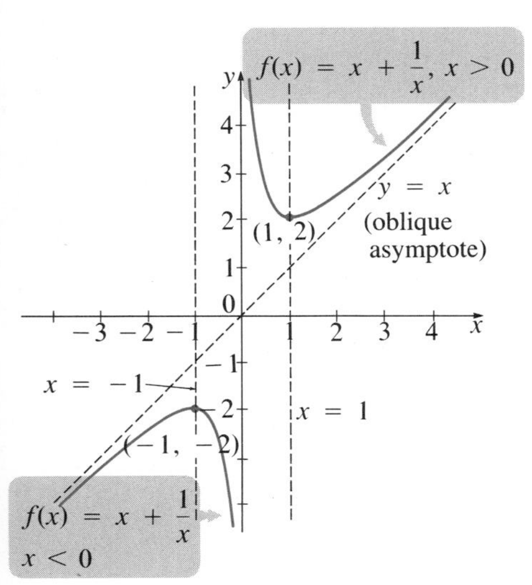

Figure 7 Graph of $f(x) = \dfrac{x^2 + 1}{x} = x + \dfrac{1}{x}$.

One Further Note About Asymptotes

Let

$$f(x) = \frac{p(x)}{q(x)} = \frac{a_n x^n + a_{n-1}x^{n-1} + \cdots + a_1 x + a_0}{b_m x^m + b_{m-1}x^{m-1} + \cdots + b_1 x + b_0}, \text{ where } a_n, b_m \neq 0$$

Then

(i) If $n < m$, the line $y = 0$ is a horizontal asymptote to the graph of $f(x)$ (as in Example 4).

(ii) If $n = m$, then the line $y = \frac{a_n}{b_m}$ is a horizontal asymptote (as in Example 6).

(iii) If $n = m + 1$ (that is, the degree of $p(x)$ is one more than the degree of $q(x)$), then, by the quotient theorem,

$$\frac{p(x)}{q(x)} = \left(\frac{a_n}{b_m}x + c\right) + \frac{r(x)}{q(x)}$$

where degree of $r(x) <$ degree of $q(x)$. The line $y = \frac{a_n}{b_m}x + c$ is an oblique asymptote (see Examples 9 and 10).

We close this section by noting that graphs of rational functions can be obtained on graphing calculators. See Examples 2 and 3 in Appendix A.

Problems 5.1

Readiness Check

I. Which of the following is a horizontal asymptote for $f(x) = \frac{x^2 - 3x + 2}{4 - x^2}$?

a. $y = 1$ b. $y = 2$ c. $y = -1$ d. $y = 0$

II. Which of the following is a vertical asymptote for $f(x) = \frac{1 - 2x^3}{x^3 + x}$?

a. $x = 0$ b. $x = -2$ c. $x = -1$

d. There is no vertical asymptote.

III. Which of the following has an oblique asymptote?

a. $f(x) = \frac{x^2 - 7x + 12}{x^3}$ b. $f(x) = \frac{9 - x^2}{x^2 - 7x + 12}$

c. $f(x) = \frac{x^3}{x^2 - 7x + 12}$

d. None of these curves has an oblique asymptote.

IV. Which of the following has the line $y = -x$ as an oblique asymptote?

a. $f(x) = \frac{x - 1}{x^2 + 1}$ b. $f(x) = \frac{3x^2 + 4x - 1}{5 - 3x}$

c. $f(x) = \frac{x^2 - 9}{3 - x}$ d. $f(x) = \frac{6 - x^2}{x}$

V. Which of the following has asymptote $y = 0$?

a. $f(x) = \frac{x^2 + 1}{x^4 - 1}$ b. $f(x) = \frac{x^4 - 1}{x^2 + 1}$

c. $f(x) = \frac{x^3 - x^2 + 1}{x^3}$ d. $f(x) = \frac{x^2 - x - 2}{2}$

Answers to Readiness Check

I. c II. a III. c IV. d V. a

In Problems 1–20 determine the vertical and horizontal asymptotes, if any, for each function.

1. $f(x) = \dfrac{1}{x^2 - 1}$

2. $f(x) = \dfrac{3}{x^2 + 9}$

3. $f(x) = \dfrac{x - 1}{x^2 + 5x + 6}$

4. $f(x) = \dfrac{x^2 + 2x + 1}{x^2 - 9}$

5. $f(x) = \dfrac{x + 2}{x^2 + 7x + 6}$

6. $f(x) = \dfrac{x^2 + 6x + 8}{x + 4}$

7. $f(x) = \dfrac{x^2 - 9}{(x - 9)^2}$

8. $f(x) = \dfrac{x^3}{1 - x^3}$

9. $f(x) = \dfrac{3x + 5}{x^2 - 2x}$

10. $f(x) = \dfrac{x^2}{x^3 - 3x^2 - 4x}$

11. $f(x) = \dfrac{3}{x - 2}$

12. $f(x) = \dfrac{x + 3}{2x + 4}$

13. $f(x) = \dfrac{x^2 + 2x + 1}{3x + 4}$

14. $f(x) = \dfrac{3x^2 - x + 2}{-7x^2 + x - 1}$

15. $f(x) = \dfrac{x^3 + 1}{1000x^2 + 50x + 100}$

16. $f(x) = \dfrac{5x^3 - 6x^2 + 5}{4x^3 + 6x + 1}$

17. $f(x) = \dfrac{(x + 1)(x - 3)(x + 5)}{(x - 1)(x + 2)(x - 4)}$

18. $f(x) = \dfrac{(x^2 - 1)^2}{2x^4 + 8}$

* 19. $f(x) = \dfrac{(x^2 - 1)^6}{1 - x^7}$

20. $f(x) = \dfrac{(x - 1)^3}{x^3 - 1}$

In Problems 21–26 determine oblique asymptotes for each function.

21. $f(x) = \dfrac{x^3 - 2}{x^2 + 2}$

22. $f(x) = \dfrac{x^3 + x^2 + x + 1}{x^2 + 4x + 7}$

23. $f(x) = \dfrac{2x^3 - 3x + 2}{1 - 5x^2}$

24. $f(x) = \dfrac{-6x^3 + 2x^2 + 3}{4x^2 - 3x + 1}$

25. $f(x) = \dfrac{x^5 - x^3 + 2x}{2x^4 - 3x^2 + 3}$

26. $f(x) = \dfrac{x^4 + x^3 + x^2 + x + 1}{3x^3 - 2x^2 + x - 4}$

In Problems 27–36 ten rational functions are given. Match each function with one of the ten graphs labeled (a)–(j).

27. $f(x) = \dfrac{1}{x + 2}$

28. $f(x) = \dfrac{x + 1}{x - 3}$

29. $f(x) = \dfrac{2}{1 - x^2}$

30. $f(x) = \dfrac{2}{x^2 + 3}$

31. $f(x) = \dfrac{x + 1}{x^2 + x - 2}$

32. $f(x) = \dfrac{x^2 + 1}{x - 1}$

33. $f(x) = \dfrac{x}{x^2 - 1}$

34. $f(x) = \dfrac{1}{(x - 1)^2}$

35. $f(x) = \dfrac{x^2 - 1}{(x - 1)^2}$

36. $f(x) = \dfrac{1}{x} + 2$

(a)
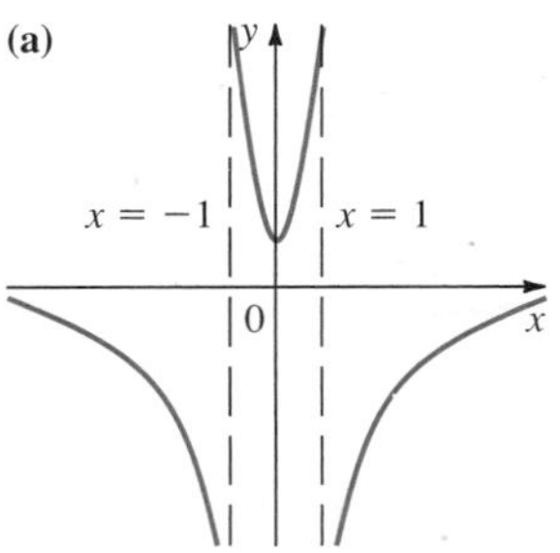

(b)
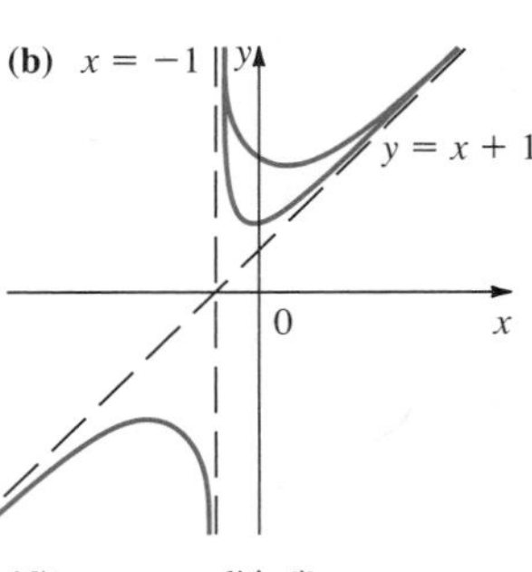

(c)
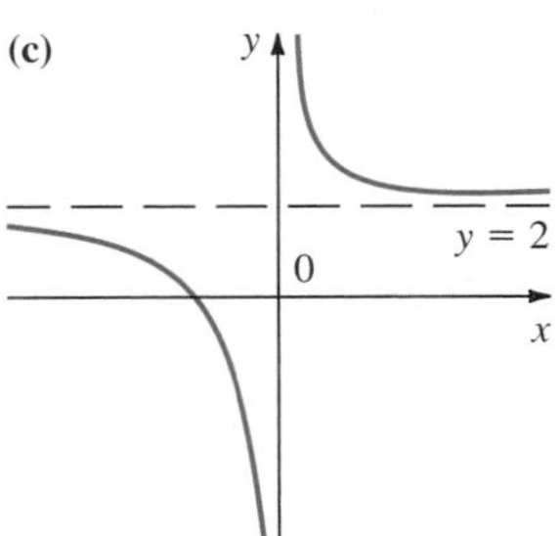

(d)
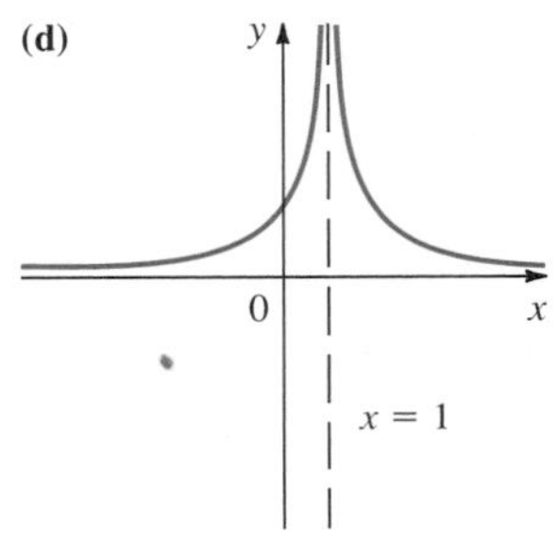

(e)
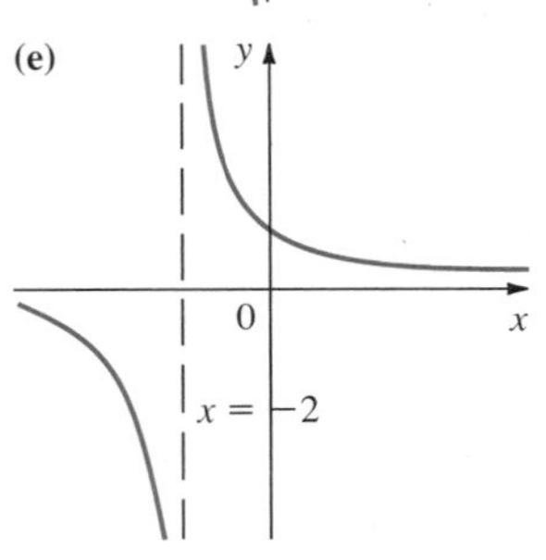

(f)
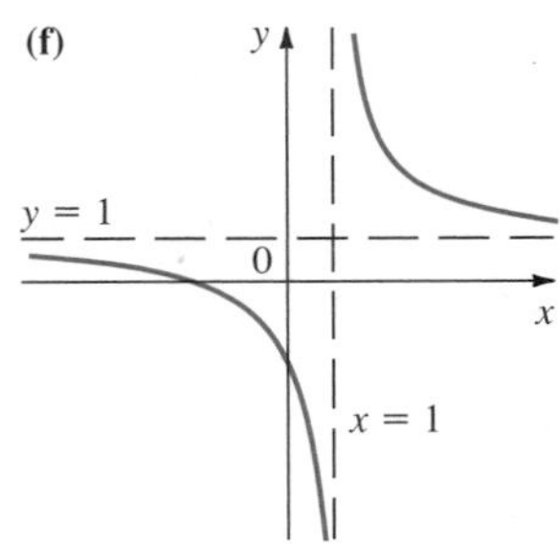

(g)
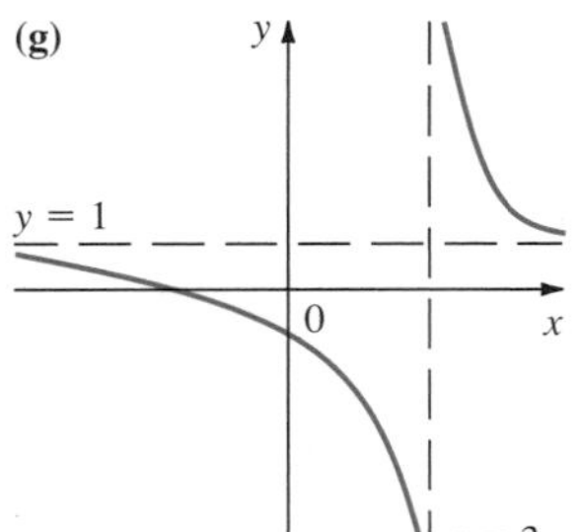

(h)
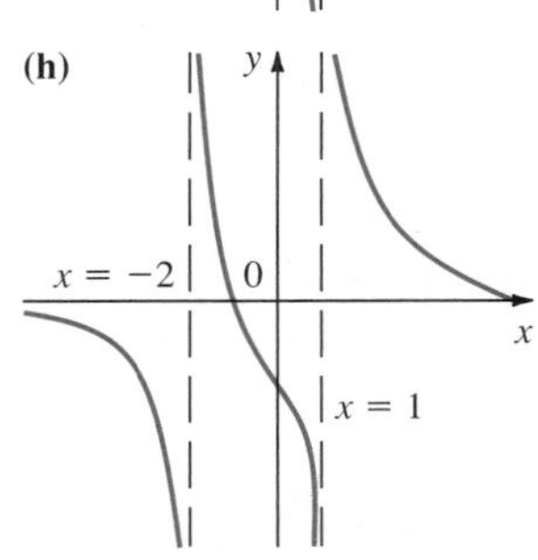

(i)
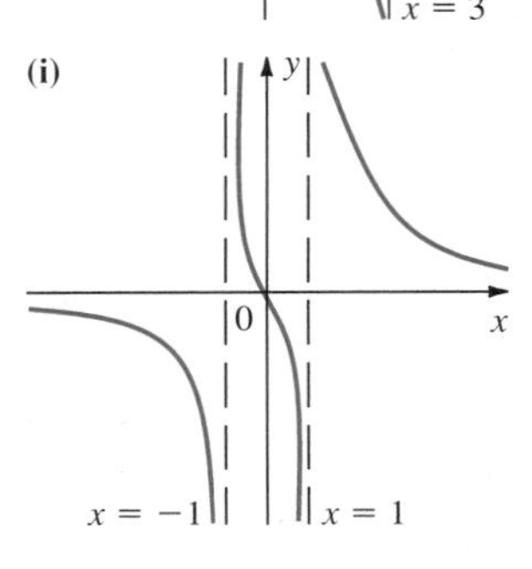

(j)
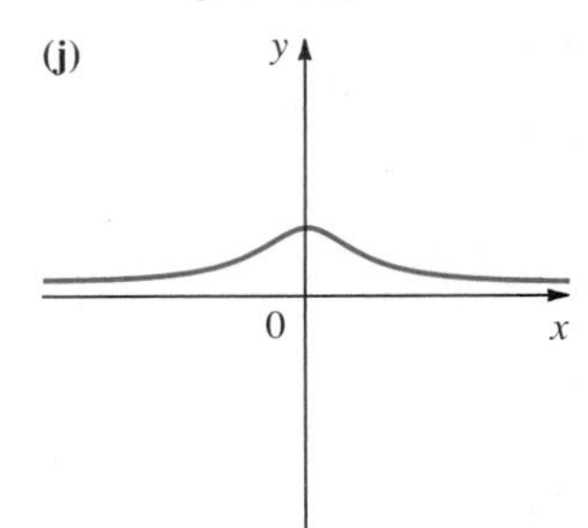

In Problems 37–70 sketch each curve.

37. $f(x) = \dfrac{1}{x-3}$

38. $f(x) = \dfrac{4}{x+5}$

39. $f(x) = \dfrac{1}{x^2}$

40. $f(x) = \dfrac{1}{(x+4)^2}$

41. $f(x) = \dfrac{1}{x^2} - 2$

42. $f(x) = \dfrac{1}{(x+3)^2} - 5$

43. $f(x) = \dfrac{1}{x^3}$

44. $f(x) = \dfrac{1}{(x-4)^3}$

45. $f(x) = \dfrac{1}{x^3} + 3$

46. $f(x) = \dfrac{3}{(x-2)^3} + 4$

47. $f(x) = \dfrac{1}{x^4}$

48. $f(x) = \dfrac{1}{x^5}$

49. $f(x) = \dfrac{1}{x - x^2}$

50. $f(x) = \dfrac{3x}{x^2 - 4}$

51. $f(x) = \dfrac{-2}{x^2 - 3x + 2}$

52. $f(x) = \dfrac{1}{(x^2 - 1)(x + 2)}$

53. $f(x) = \dfrac{2}{(x+5)(x-3)(x+7)}$

54. $f(x) = \dfrac{4}{x^2 + 1}$

55. $f(x) = \dfrac{4}{x^2 + 4x + 5}$ [Hint: Complete the square.]

56. $f(x) = \dfrac{x-1}{x-2}$

57. $f(x) = \dfrac{x-1}{(x-2)^2}$

58. $f(x) = \dfrac{x^2+1}{x^2-1}$

59. $f(x) = \dfrac{x^2-3}{x^2-3x+2}$

60. $f(x) = \dfrac{2x^3}{x^3 - x^2}$

61. $f(x) = \dfrac{(x+3)^2}{(x+3)^2 - 4}$

62. $f(x) = \dfrac{x^2-1}{x+2}$

63. $f(x) = \dfrac{x^2-1}{(x+2)^2}$

64. $f(x) = \dfrac{x^2-4}{x+2}$

65. $f(x) = \dfrac{x-3}{x^2+2x-15}$

66. $f(x) = \dfrac{x^3+2x^2}{x^2}$

67. $f(x) = \dfrac{x^2-1}{x^2}$

68. $f(x) = \dfrac{x^2-2x+3}{x}$

69. $f(x) = \dfrac{x^2-2x+4}{x-1}$ [Hint: Divide to obtain an oblique asymptote.]

70. $f(x) = \dfrac{x^2+6x+1}{x+3}$

71. According to **Newton's law of universal gravitation,** the gravitational force F between two particles of masses m_1 and m_2 is given by

$$F = G\frac{m_1 m_2}{r^2}$$

where G is a constant and r is the distance between the particles. Sketch this function if $Gm_1m_2 = 6$.

72. According to the **ideal gas law,** when temperature is constant, the pressure of an ideal gas is related to its volume by the formula $P = \dfrac{k}{V}$. Sketch this function for $k = 3.5$.

73. According to **Poiseuille's law,** the resistance R of a blood vessel is given by $R = \dfrac{kl}{r^4}$, where l is its length, r is its radius, and k is a constant. Suppose l is held fixed as r varies. Sketch this function when $kl = 0.75$.

Graphing Calculator Problems

In Problems 74–86 obtain the graph of each rational function on your graphing calculator. Before beginning, read (or reread) Examples 2 and 3 in Appendix A.

74. $f(x) = \dfrac{2}{x^2+5}$

75. $f(x) = \dfrac{-3}{x^2-5}$

76. $f(x) = \dfrac{x^2+2}{x^2-2}$

77. $f(x) = \dfrac{x^2-3}{x^2+3}$

78. $f(x) = \dfrac{2-x}{x^2+3x-5}$

79. $f(x) = \dfrac{15-27x}{32x-18}$

80. $f(x) = \dfrac{x^2-2x+7}{x^2-4x+12}$

81. $f(x) = \dfrac{x^2-2x+7}{x^2-4x-12}$

82. $f(x) = \dfrac{x^3+x-4}{x^2+15}$

83. $f(x) = \dfrac{x^3+x-4}{x^2-15}$

84. $f(x) = \dfrac{x^2-2x-8}{x^3+6}$

85. $f(x) = \dfrac{2x^3-2x^2+x+5}{x^3+x^2+x+3}$

86. $f(x) = \dfrac{4x^3+7x^2-12}{6-5x-8x^3}$

5.2 Introduction to the Conic Sections

The period from about 300 to 200 B.C. is known as the Golden Age of Greek mathematics because three of the world's greatest mathematicians lived during that period. The first of these, Euclid, invented much of the geometry that is studied in high schools today. The second, Archimedes, is considered by many to be the finest mathematician of any era. The third of the great Greek mathematicians, Apollonius of Perga, made many of the discoveries that are part of what we now call *analytic geometry*. The exact dates of Apollonius's life are not known, but he is believed to have lived from approximately 260 to 190 B.C.

Like most of the mathematicians of his day, Apollonius was an *applied* mathematician. He studied certain kinds of curves because they arose in practical ways. For example, he applied his work to the analysis of planetary motion and is considered to be the founder of Greek mathematical astronomy.

One of Apollonius's most important discoveries was that four different types of curves are obtained if a right circular cone is cut by a plane. These curves are circles, ellipses, parabolas, and hyperbolas. Because of the way they are formed, they are called **conic sections.** In Figure 1, we illustrate how the four curves can be obtained.

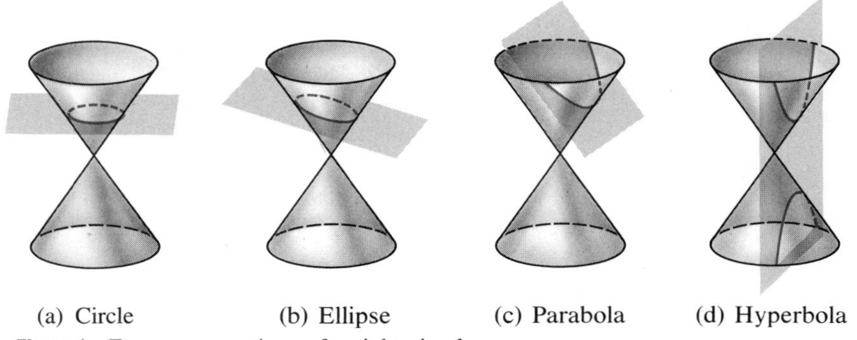

(a) Circle (b) Ellipse (c) Parabola (d) Hyperbola

Figure 1 Four cross sections of a right circular cone.

It turns out that each of the four curves can be written in the form

$$Ax^2 + Bxy + Cy^2 + Dx + Ey + F = 0 \qquad (1)$$

for certain constants A, B, C, D, E, and F.

We saw special cases of equation (1) earlier in this book. In Section 3.1, we discussed circles. In Example 9 on p. 154, we saw that the equation:

$$x^2 + y^2 - 6x + 2y - 17 = 0$$

is an equation of the circle of radius $\sqrt{27}$ centered at $(3, -1)$. This is equation (1) with $A = C = 1$, $B = 0$, $D = -6$, $E = 2$, and $F = -17$.

In Section 4.1 we graphed a number of parabolas. For instance, in Example 2 on p. 224 we showed that the equation $y = -2x^2 + 8x - 4$, which can be written as

$$-2x^2 + 8x - y - 4 = 0 \qquad (2)$$

is an equation of a parabola that opens downward with vertex at $(2, 4)$ and axis the line $x = 2$. Equation (2) is a special case of (1) with $A = -2$, $B = C = 0$, $D = 8$, $E = -1$, and $F = -4$.

In the rest of this chapter, we will discuss equations that can be written in the form (1) or some equivalent form. We also will show how these equations, and the curves they represent, arise in applications.

5.3 The Ellipse and Translation of Axes

Definition of an Ellipse

An **ellipse** is the set of points (x, y) such that the sum of the distances from (x, y) to two given points is fixed. Each of the two points is called a **focus** of the ellipse.

NOTE The plural of focus is foci. Thus we speak about the two foci of an ellipse.

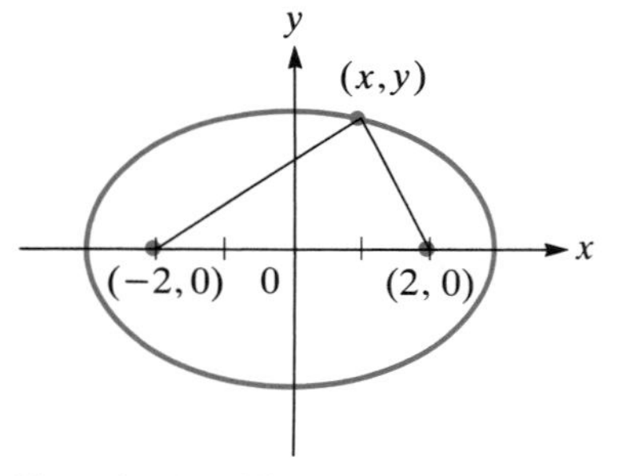

Figure 1 An ellipse with foci at $(-2, 0)$ and $(2, 0)$.

EXAMPLE 1 ***Finding an Equation of an Ellipse with Given Foci***

Find an equation of the ellipse with foci at $(-2, 0)$ and $(2, 0)$ such that the sum of the distances to the foci is 6.

SOLUTION As in Figure 1, let (x, y) denote a point on the ellipse. Then

$$[\text{distance from } (x, y) \text{ to } (2, 0)] + [\text{distance from } (x, y) \text{ to } (-2, 0)] = 6$$

or, from the distance formula (equation (1) on p. 151),

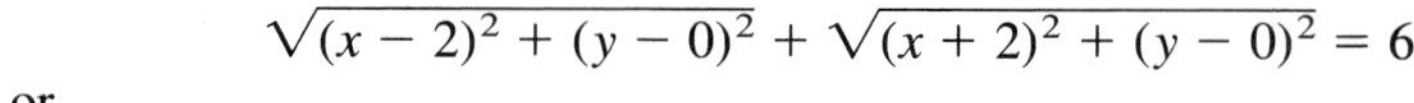

$$\sqrt{(x-2)^2 + (y-0)^2} + \sqrt{(x+2)^2 + (y-0)^2} = 6$$

or

$$\sqrt{(x-2)^2 + y^2} = 6 - \sqrt{(x+2)^2 + y^2}$$

$$(x-2)^2 + y^2 = 36 - 12\sqrt{(x+2)^2 + y^2} + (x+2)^2 + y^2 \qquad \text{We squared both sides}$$

$$x^2 - 4x + 4 + y^2 = 36 - 12\sqrt{(x+2)^2 + y^2} + x^2 + 4x + 4 + y^2$$

$$-4x = 36 - 12\sqrt{(x+2)^2 + y^2} + 4x \qquad \text{We subtracted } x^2 + y^2 + 4 \text{ from both sides}$$

$$36 + 8x = 12\sqrt{(x+2)^2 + y^2} \qquad \text{We rearranged terms}$$

$$3 + \frac{2}{3}x = \sqrt{(x+2)^2 + y^2} \qquad \text{We divided both sides by 12}$$

$$9 + 4x + \frac{4}{9}x^2 = (x+2)^2 + y^2 = x^2 + 4x + 4 + y^2 \qquad \text{We squared again}$$

$$\frac{5}{9}x^2 + y^2 = 5 \qquad \text{We combined terms}$$

$$\frac{x^2}{9} + \frac{y^2}{5} = 1 \quad \text{We divided by 5}$$

This is the **standard equation** of the ellipse with foci at $(-2, 0)$ and $(2, 0)$ and sum of distances equal to 6.

Suppose now that the foci, F_1 and F_2, of an ellipse are the points $(-c, 0)$ and $(c, 0)$. (See Figure 2.) If $P = (x, y)$ is on the ellipse, then by the definition of the ellipse, the distance from P to the first focus plus the distance from P to the second focus equals $2a$; that is,

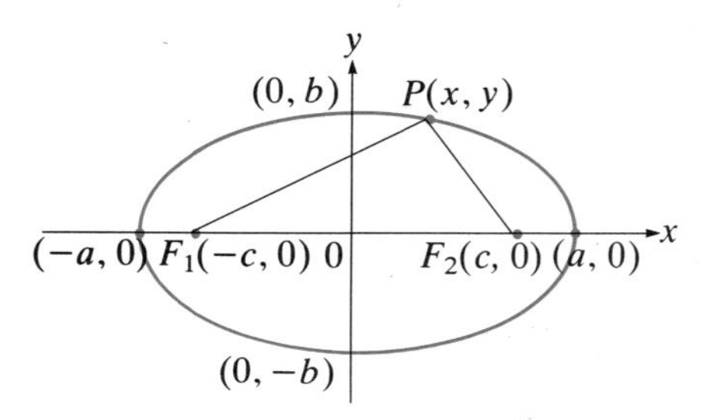

Figure 2 An ellipse with foci $(c, 0)$ and $(-c, 0)$ such that the sum of the distances from a point to each of the two foci is $2a$.

$$\overline{F_1P} + \overline{F_2P} = 2a \tag{3}$$

Since a straight line is the shortest distance between two points, we have

$$\overline{F_1F_2} = 2c < \overline{F_1P} + \overline{F_2P} = 2a$$

That is,

$$2a > 2c \quad \text{or} \quad a > c$$

($2c$ is the distance between the foci and $2a$ is the given sum of the distances.) Then

$$\sqrt{(x+c)^2 + y^2} + \sqrt{(x-c)^2 + y^2} = 2a \quad \text{From (3)}$$

or

$$\sqrt{(x+c)^2 + y^2} = 2a - \sqrt{(x-c)^2 + y^2}$$

$$(x+c)^2 + y^2 = 4a^2 - 4a\sqrt{(x-c)^2 + y^2} + (x-c)^2 + y^2 \quad \text{We squared both sides}$$

$$x^2 + 2xc + c^2 + y^2 = 4a^2 - 4a\sqrt{(x-c)^2 + y^2} + x^2 - 2xc + c^2 + y^2$$

$$4xc = 4a^2 - 4a\sqrt{(x-c)^2 + y^2} \quad \text{We subtracted } x^2 + c^2 + y^2 \text{ from both sides and simplified}$$

$$\sqrt{(x-c)^2 + y^2} = a - \frac{c}{a}x \quad \text{We divided by } 4a \text{ and rearranged terms}$$

$$(x-c)^2 + y^2 = a^2 - 2cx + \frac{c^2}{a^2}x^2 \quad \text{We squared again}$$

Then

$$x^2 - 2xc + c^2 + y^2 = a^2 - 2xc + \frac{c^2}{a^2}x^2$$

$$x^2\left(1 - \frac{c^2}{a^2}\right) + y^2 = a^2 - c^2$$

or

$$x^2\left(\frac{a^2 - c^2}{a^2}\right) + y^2 = a^2 - c^2$$

and after dividing both sides by $a^2 - c^2$, which is positive since $a > c$, we have

$$\frac{x^2}{a^2} + \frac{y^2}{a^2 - c^2} = 1$$

Finally, we define the positive number b by $b^2 = a^2 - c^2$ to obtain the standard equation of the ellipse.

Standard Equation of an Ellipse

$$\frac{x^2}{a^2} + \frac{y^2}{b^2} = 1 \qquad (4)$$

Here, since b^2 is defined by $b^2 = a^2 - c^2$, we have

$$a^2 = b^2 + c^2$$

where $(c, 0)$ and $(-c, 0)$ are the foci. Note that this ellipse is symmetric about both the x- and y-axes. Since $a > b$, the line segment from $(-a, 0)$ to $(a, 0)$ is called the **major axis,** and the line segment from $(0, -b)$ to $(0, b)$ is called the **minor axis.** The point $(0, 0)$, which is the intersection of the axes, is called the **center** of the ellipse. The points $(a, 0)$ and $(-a, 0)$ are called the **vertices** of the ellipse. In general, the vertices of an ellipse are the endpoints of the major axis.

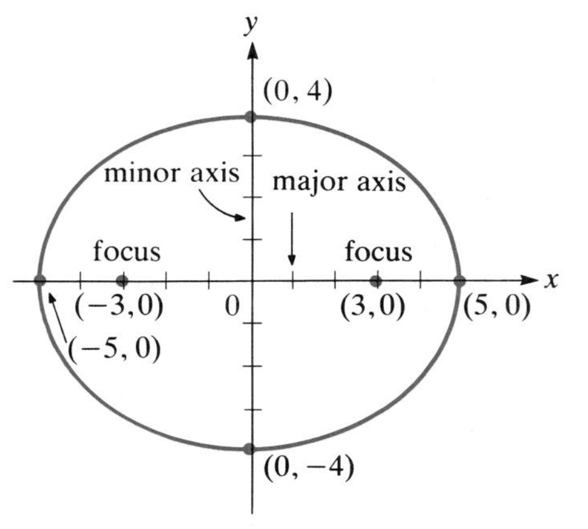

Figure 3 The ellipse $\frac{x^2}{25} + \frac{y^2}{16} = 1$.

EXAMPLE 2 *Finding the Standard Equation of an Ellipse*

Find the equation of the ellipse with foci at $(-3, 0)$ and $(3, 0)$ and with $a = 5$.

SOLUTION $c = 3$ and $a = 5$ so that $b^2 = a^2 - c^2 = 16$, and we obtain

$$\frac{x^2}{25} + \frac{y^2}{16} = 1$$

The ellipse is sketched in Figure 3.

We can reverse the roles of x and y in the preceding discussion. Suppose that the foci are at $(0, c)$ and $(0, -c)$ on the y-axis. Then if the fixed sum of the distances is given as $2b$, we obtain, using similar reasoning,

$$\frac{x^2}{a^2} + \frac{y^2}{b^2} = 1$$

where $a^2 = b^2 - c^2$. Now the major axis is on the y-axis, the minor axis is on the x-axis, and the vertices are $(0, b)$ and $(0, -b)$. In general, if the ellipse is given by (4), then we have the equations given in Table 1.

The last two entries in Table 1 are called **degenerate ellipses.** Degenerate ellipses have graphs containing one point or no point. For example, the equa-

Table 1 *Standard Ellipses*

Equation	Description	Picture
$\dfrac{x^2}{a^2} + \dfrac{y^2}{b^2} = 1,\ a > b$	Ellipse with major axis on x-axis; $a^2 = b^2 + c^2$	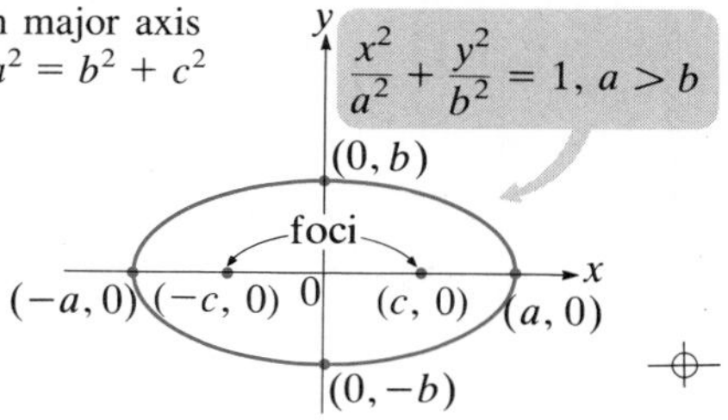
$\dfrac{x^2}{a^2} + \dfrac{y^2}{b^2} = 1,\ b > a$	Ellipse with major axis on y-axis; $b^2 = a^2 + c^2$	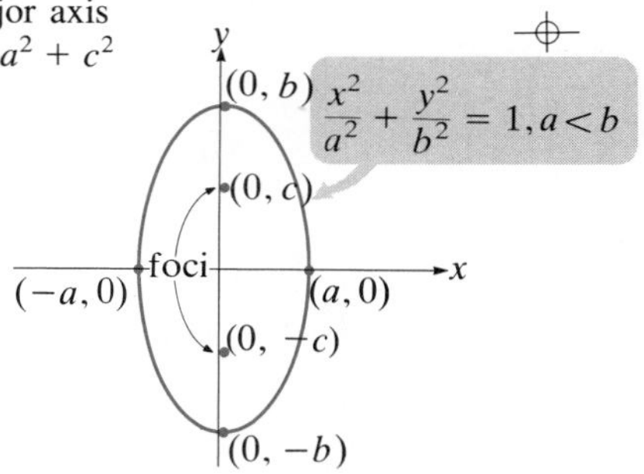
$\dfrac{x^2}{a^2} + \dfrac{y^2}{b^2} = 1,\ b = a$	Circle with radius a $(= b)$	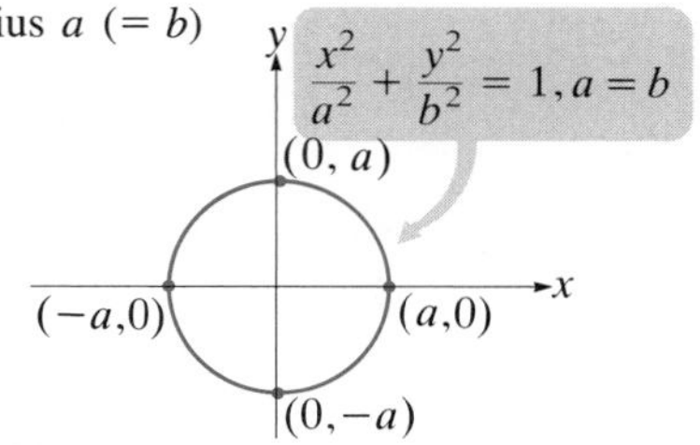
$\dfrac{x^2}{a^2} + \dfrac{y^2}{b^2} = 0$	Degenerate ellipse; single point $(0, 0)$	
$\dfrac{x^2}{a^2} + \dfrac{y^2}{b^2} = -1$	Degenerate ellipse; graph is empty	

tion $\dfrac{x^2}{4} + \dfrac{y^2}{9} = 0$ is satisfied only by the point $(0, 0)$. Similarly, no point lies on the graph of $\dfrac{x^2}{2} + \dfrac{y^2}{3} = -1$. The equations $\dfrac{x^2}{4} + \dfrac{y^2}{9} = 0$ and $\dfrac{x^2}{2} + \dfrac{y^2}{3} = -1$ are similar to the standard equation of an ellipse (with 0 or -1 in place of 1). That is one reason why they are called degenerate *ellipses*.

We note, too, that when $a = b$ in equation (4), we obtain

$$\frac{x^2}{a^2} + \frac{y^2}{a^2} = 1 \quad \text{or} \quad x^2 + y^2 = a^2$$

This is an equation of the circle centered at $(0, 0)$ with radius a (see equation (5) on p. 154). Thus a circle can be thought of as a special kind of ellipse. We discussed circles in Section 3.1, so we will say no more about them here.

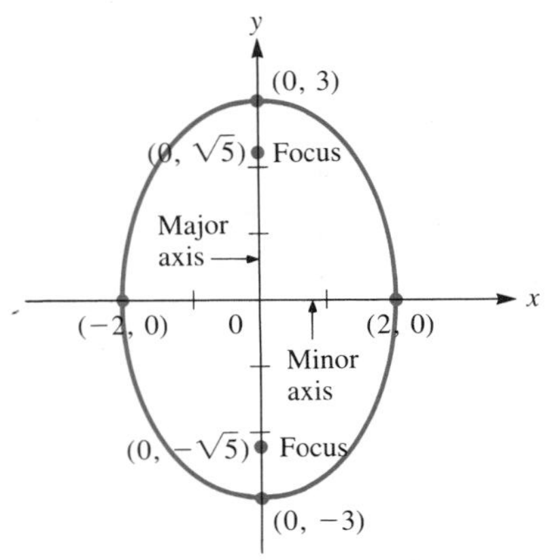

Figure 4 The ellipse $\dfrac{x^2}{4} + \dfrac{y^2}{9} = 1$.

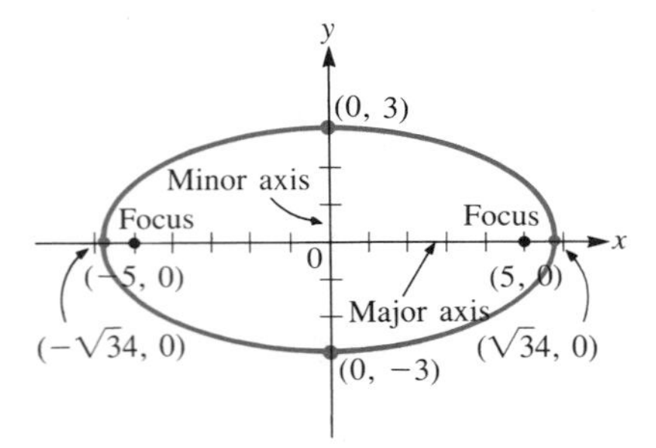

Figure 5 The ellipse $\dfrac{x^2}{34} + \dfrac{y^2}{9} = 1$.

EXAMPLE 3 ***Showing That a Second-Degree Curve Is the Equation of an Ellipse***

Discuss the curve $9x^2 + 4y^2 = 36$.

SOLUTION Dividing both sides by 36, we obtain

$$\frac{x^2}{4} + \frac{y^2}{9} = 1$$

Here $a = 2$, $b = 3$, and the major axis is on the y-axis. Since $c^2 = 9 - 4 = 5$, the foci are at $(0, \sqrt{5})$ and $(0, -\sqrt{5})$. This curve is sketched in Figure 4. ■

EXAMPLE 4 ***Finding the Equation of an Ellipse Given Its Foci and Minor Axis***

Find the equation of the ellipse with foci at $(-5, 0)$ and $(5, 0)$ and whose minor axis is the line segment extending from $(0, -3)$ to $(0, 3)$.

SOLUTION We have $c = 5$ and $b = 3$ so that

$$a^2 = b^2 + c^2 = 34$$

and the equation of the ellipse is

$$\frac{x^2}{34} + \frac{y^2}{9} = 1$$

It is sketched in Figure 5.

Definition of Eccentricity

The **eccentricity** e of an ellipse is defined by

$$e = \frac{c}{a} \quad \text{if} \quad a \geq b \tag{5}$$

and

$$e = \frac{c}{b} \quad \text{if} \quad b \geq a \tag{6}$$

NOTE If $a \geq b$, then the length of the major axis is $2a$; if $b \geq a$, the length of the major axis is $2b$. In both cases the distance between the foci is $2c$. Since $2c/2a = c/a$ and $2c/2b = c/b$, we have

Alternative Definition of Eccentricity

$$e = \frac{\text{distance between foci}}{\text{length of major axis}} = \frac{\text{distance between foci}}{\text{distance between vertices}} \tag{7}$$

The eccentricity of an ellipse is a measure of the shape of the ellipse and is always a number in the interval $[0, 1]$. If $e = 0$, then the ellipse is a circle, since in that case $c = 0$ so that $a^2 = b^2$, and the foci now coincide and are the center of the circle. As e approaches 1, the ellipse becomes progressively flatter and approaches the major axis: the straight-line segment from $(-a, 0)$ to $(a, 0)$ if $a > b$, and from $(0, -b)$ to $(0, b)$ if $b > a$. In general,

The larger the eccentricity, the flatter the ellipse.

In Example 2, $e = 3/5 = 0.6$. In Example 3, $c^2 = b^2 - a^2 = 5$ so that $e = \sqrt{5}/3 \approx 0.74536$. In Example 4, $e = 5/\sqrt{34} \approx 0.85749$.

In Figure 6, we draw four different ellipses showing how the ellipses get flatter as the eccentricity increases.

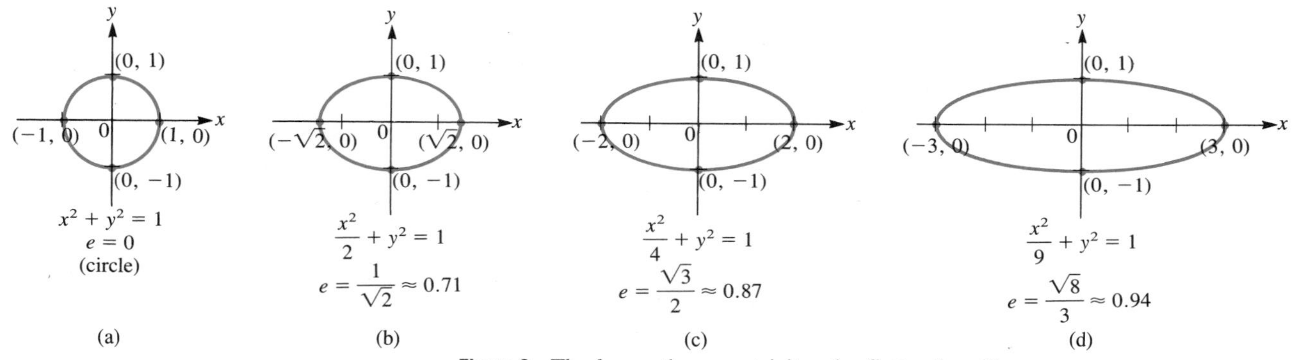

Figure 6 The larger the eccentricity, the flatter the ellipse.

Translation of Axes

We now turn to a different question. What happens if the center of the ellipse $(x^2/a^2) + (y^2/b^2) = 1$ is shifted from $(0, 0)$ to a new point (x_0, y_0) while the major and minor axes remain parallel to the x-axis and y-axis? Consider the equation

$$\frac{(x - x_0)^2}{a^2} + \frac{(y - y_0)^2}{b^2} = 1 \tag{8}$$

If we define two new variables by

$$x' = x - x_0 \quad \text{and} \quad y' = y - y_0 \tag{9}$$

then (8) becomes

$$\frac{(x')^2}{a^2} + \frac{(y')^2}{b^2} = 1 \tag{10}$$

This equation is the equation of an ellipse centered at the origin in the new coordinate system (x', y'). But $(x', y') = (0, 0)$ implies $(x - x_0, y - y_0) = (0, 0)$, or $x = x_0$ and $y = y_0$. That is, equation (8) is the equation of the "shifted" ellipse. See Figure 7. We have performed what is called a **translation of axes.** That is, we moved (or **translated**) the x- and y-axes to new positions so that they intersect at the point (x_0, y_0). If the original ellipse had its foci at $(-c, 0)$ and $(c, 0)$, then the translated ellipse has its foci at $(-c + x_0, y_0)$ and $(c + x_0, y_0)$. If the original ellipse had its foci at $(0, -c)$ and $(0, c)$, then the translated ellipse has its foci at $(x_0, -c + y_0)$, $(x_0, c + y_0)$.

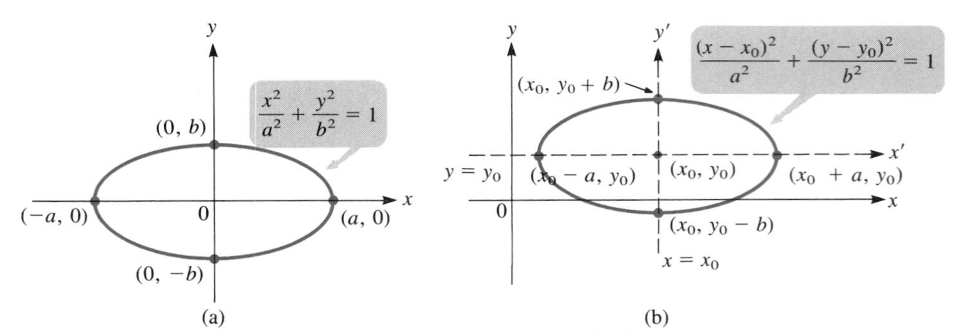

Figure 7 Translating an ellipse from center at $(0, 0)$ to center at (x_0, y_0).

EXAMPLE 5 *Finding the Equation of a Translated Ellipse*

Find the equation of the ellipse centered at $(4, -2)$ with foci at $(1, -2)$ and $(7, -2)$ and minor axis joining the points $(4, 0)$ and $(4, -4)$.

SOLUTION The points are sketched in Figure 8a. Here $c = (7 - 1)/2 = 3$, $b = [0 - (-4)]/2 = 2$, and $a^2 = b^2 + c^2 = 13$. Also, $(x_0, y_0) = (4, -2)$ so $x - x_0 = x - 4$ and $y - y_0 = y - (-2) = y + 2$. Thus, from (8), we have

$$\frac{(x - 4)^2}{13} + \frac{(y + 2)^2}{4} = 1$$

This ellipse is sketched in Figure 8b. Its major axis is on the line $y = -2$, and its minor axis is on the line $x = 4$. Its eccentricity is $3/\sqrt{13} \approx 0.83205$.

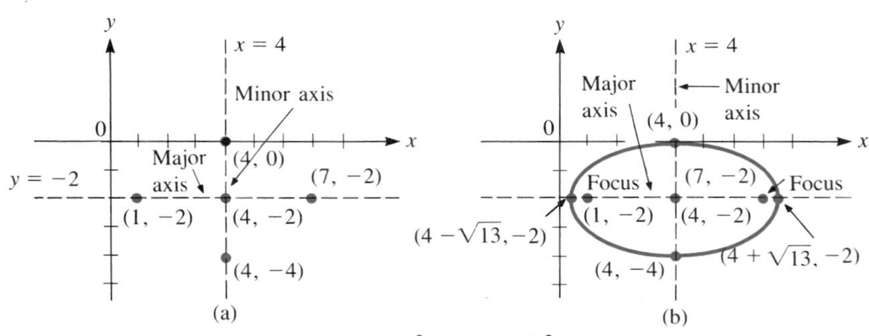

Figure 8 The translated ellipse $\frac{(x - 4)^2}{13} + \frac{(y + 2)^2}{4} = 1$. ■

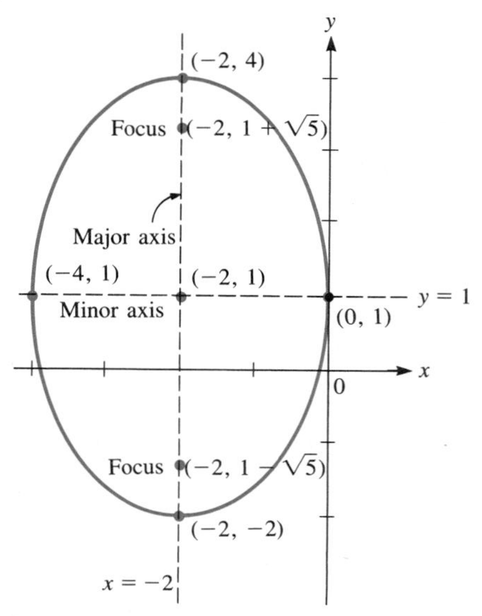

Figure 9 The translated ellipse $\dfrac{(x+2)^2}{4} + \dfrac{(y-1)^2}{9} = 1$.

EXAMPLE 6 ***Showing That a Second-Degree Equation Is the Equation of a Translated Ellipse by Completing Two Squares***

Discuss the curve given by

$$9x^2 + 36x + 4y^2 - 8y + 4 = 0$$

SOLUTION We write this expression as

$$9(x^2 + 4x) + 4(y^2 - 2y) = -4$$

Then after completing the squares, we obtain

$$9(\underset{\text{Added}}{x^2 + 4x + 4}) \underset{\text{Subtracted } 9\cdot 4}{- 36} + 4(y^2 - 2y \underset{\text{Added}}{+ 1}) \underset{\text{Subtracted } 4\cdot 1}{- 4} = -4$$

or

$$9(x + 2)^2 + 4(y - 1)^2 = 36$$

and, dividing both sides by 36,

$$\frac{(x+2)^2}{4} + \frac{(y-1)^2}{9} = 1$$

Since $x + 2 = x - x_0$ and $y - 1 = y - y_0$, $x_0 = -2$ and $y_0 = 1$. Thus this is the equation of an ellipse centered at $(-2, 1)$. Here $a = 2$, $b = 3$, and $c = \sqrt{b^2 - a^2} = \sqrt{5}$. The foci are therefore at $(-2, -\sqrt{5} + 1)$ and $(-2, \sqrt{5} + 1)$. The major axis is on the line $x = -2$, and the minor axis is on the line $y = 1$. The eccentricity is $\sqrt{5}/3 \approx 0.74536$. The ellipse is sketched in Figure 9.

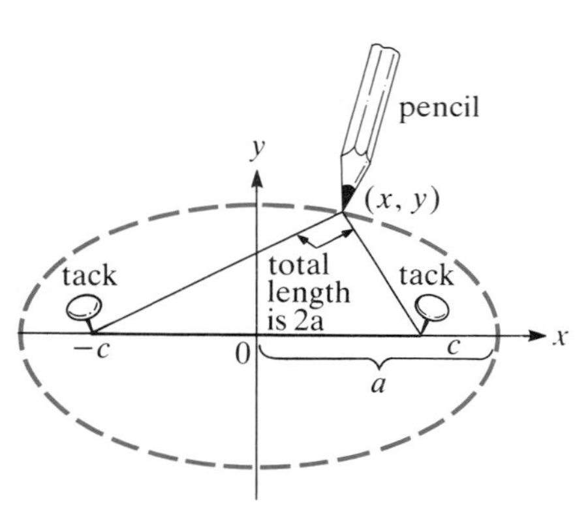

Figure 10 Drawing an ellipse.

How to Draw an Ellipse

Pick two positive numbers a and c with $a > c$. Place two tacks $2c$ units apart as in Figure 10, and attach a string of length $2a$ to the tacks. Then proceed as in the figure to obtain the ellipse whose equation is $\dfrac{x^2}{a^2} + \dfrac{y^2}{a^2 - c^2} = 1$.

FOCUS ON
The Ellipse in the Real World

Ellipses are all around you. To see one, take a circular glass of water (or any other liquid) and tilt it. The surface of the liquid now forms an ellipse. As another common example, hold a spherical ball in front of a light. The ball will cast an elliptical shadow (see Figure 11).

Ellipses are very important in astronomy. In 1609, the German astronomer Johannes Kepler (1571–1630) discovered that each planet follows an elliptical path with the sun at one focus. Except for Mercury and Pluto, the nearest and farthest planets

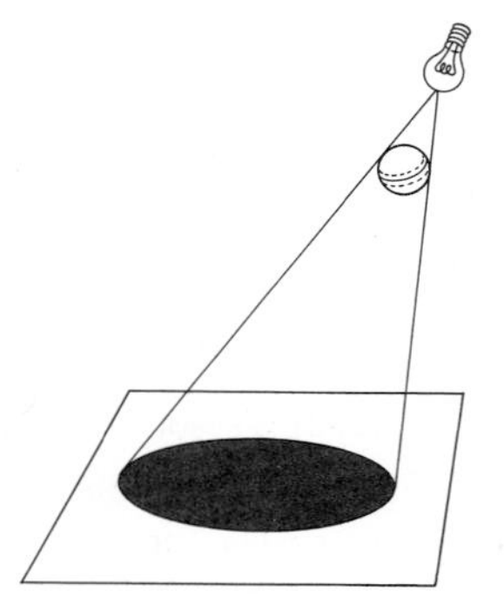

Figure 11 A ball casts an elliptical shadow.

from the sun, the orbits are nearly circular; that is, the eccentricities are close to 0. The table below gives the eccentricities of the nine planets.

Planet	Eccentricity, e
Mercury	0.2056
Venus	0.0068
Earth	0.0167
Mars	0.0934
Jupiter	0.0484
Saturn	0.0543
Uranus	0.0460
Neptune	0.0082
Pluto	0.2481

EXAMPLE 7

The comet Katounek orbits the sun in a very flat elliptical path. If 1 AU (astronomical unit) denotes the distance between the earth and the sun (1 AU $\approx$ 93 million miles), then the length of the minor axis of Katounek's orbit is 44 AU, while the length of the major axis is 3600 AU. Find the eccentricity of Katounek's orbit.

SOLUTION If the major axis is on the x-axis and the minor axis is on the y-axis, then

$$2a = 3600 \text{ AU}, \; 2b = 44 \text{ AU},$$
$$a = 1800 \text{ AU}, \; b = 22 \text{ AU}$$

and

$$c = \sqrt{a^2 - b^2} = \sqrt{(1800)^2 - (22)^2} = \sqrt{3{,}239{,}516} = 1799.865551$$

Then

$$e = \frac{c}{a} = \frac{1799.865551}{1800} \approx 0.999925$$

In Figure 12, we depict the very flat orbit of Katounek and use the orbit of Pluto as a reference.

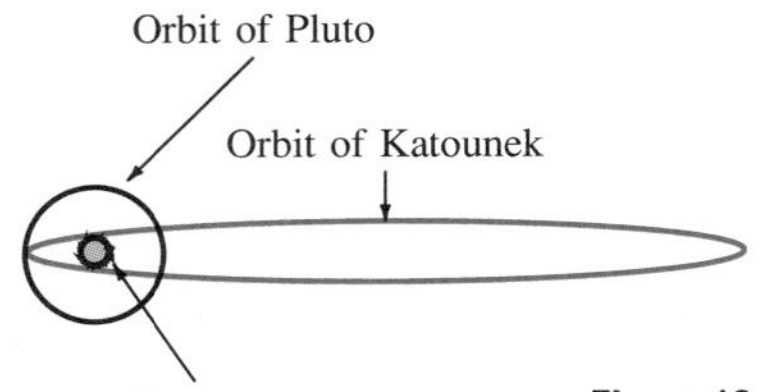

Figure 12 The orbit of Katounek.

The ellipse is used to design a more effective bicycle gear for racers. See Figure 13. The gear is designed to respond to the natural strengths and weaknesses of the racer's legs. At the top and bottom of the powerstroke, where the legs have the least leverage, the gear offers less resistance, but as the gear rotates, the resistance increases. This allows the legs to apply more power where it is most naturally available.

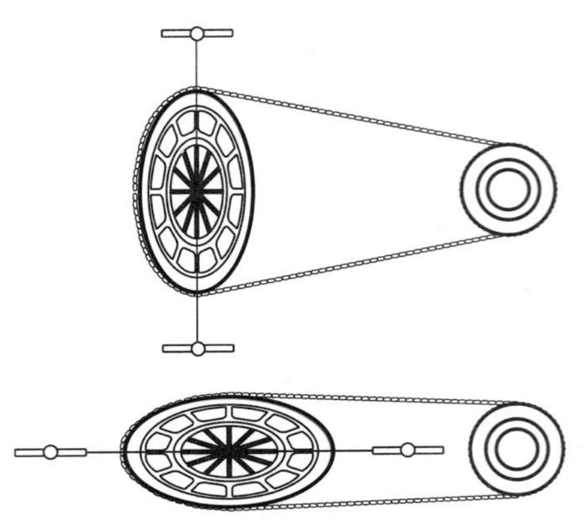

Figure 13 "Elliptical" bicycle gears give racers a significant advantage.

The Reflected-Wave Property of Ellipses and the Whispering-Gallery Effect

Let P be a point on an ellipse that is not a vertex (see Figure 14).

Figure 14 The reflected-wave property of an ellipse.

We draw the tangent line T that passes through P and draw lines joining P to each of the two foci of the ellipse. Let θ_1 denote the angle between T and PF_1, and let θ_2 denote the angle between T and PF_2. Then $\theta_1 = \theta_2$. This property is called the **reflected-wave property** of ellipses.

The reflected-wave property has been used to good effect by both Renaissance and modern architects. If the upper half of the ellipse in Figure 14 is rotated around the x-axis, then it will form a dome. In any room with an elliptical domed ceiling, a sound made at one focus will be reflected to the other focus where it will be heard very clearly. This phenomenon is known as the **whispering-gallery effect.** Some of the most famous rooms that exhibit the whispering-gallery effect include St. Paul's Cathedral in London (designed by the most famous of British architects, Sir Christopher Wren), the Caryatids room in the Louvre Museum in Paris, and the National Statuary Hall at the U.S. Capitol (the original House of Representatives) in Washington, D.C.

For more details on these and other interesting applications, see the excellent article, "The Standup Conic Presents: The Ellipse and Applications" by Lee Whitt in *The UMAP Journal,* Vol. 4, No. 2, 1983, pp. 157–186.

Problems 5.3

Readiness Check

I. Of the following, ______ is *not* an equation for an ellipse.

a. $x^2 + y^2 = 36$ b. $\left(\frac{x}{3}\right)^2 + \left(\frac{y}{4}\right)^2 = 1$

c. $16x^2 + 9y = 144$ d. $16x^2 + \frac{y^2}{9} = 12$

II. Among the ellipses satisfying the following equations, the one satisfying ______ is *not* a translation of the others.

a. $\frac{x^2}{25} + \frac{y^2}{16} = 1$

b. $25(x-1)^2 + 16(y-3)^2 = 400$

c. $\left(\frac{x}{5}\right)^2 + \frac{(y+3)^2}{16} = 1$

d. $16(x+7)^2 + 25(y-8)^2 = 16 \cdot 25$

III. The ellipse with foci $(-6, 0)$ and $(6, 0)$ and vertices $(-10, 0)$ and $(10, 0)$ can be drawn by attaching the ends of a string, ______ units in length, to thumbtacks at the foci and then tracing the outline with a pencil that moves in such a way as to keep the string taut.

a. 24 b. 20 c. 16 d. 12

IV. The ellipse with foci at $(-3, 0)$ and $(3, 0)$ and vertices $(-5, 0)$ and $(5, 0)$ has equation ______.

a. $\left(\frac{x}{5}\right)^2 + \left(\frac{y}{4}\right)^2 = 1$ b. $\left(\frac{x}{3}\right)^2 + \left(\frac{y}{4}\right)^2 = 1$

c. $\left(\frac{x}{3}\right)^2 + \left(\frac{y}{5}\right)^2 = 1$ d. $\left(\frac{x}{4}\right)^2 + \left(\frac{y}{3}\right)^2 = 1$

V. Which of the following is graphed below?

a. $x^2 + 9y^2 + 8x - 36y + 43 = 0$

b. $x^2 + 9y^2 - 8x + 36y + 43 = 0$

c. $9x^2 + y^2 - 72x + 4y - 139 = 0$

d. $9x^2 + y^2 + 72x - 4y - 139 = 0$

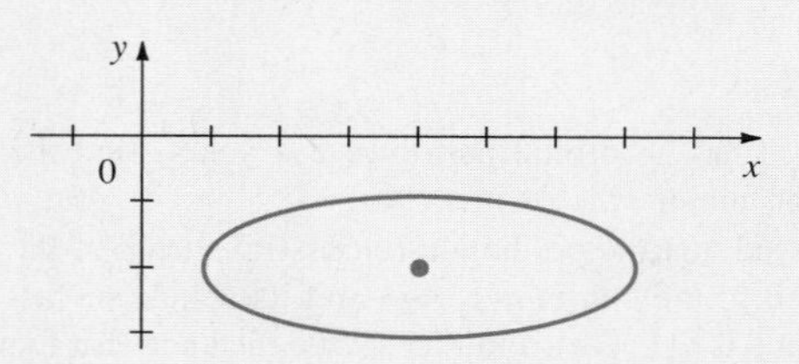

In Problems 1–18 the equation of an ellipse (or circle) is given. Find its center, foci, vertices, major and minor axes, and eccentricity. Then sketch it.

1. $\frac{x^2}{16} + \frac{y^2}{25} = 1$
2. $\frac{x^2}{25} + \frac{y^2}{16} = 1$
3. $x^2 + \frac{y^2}{9} = 1$
4. $\frac{x^2}{9} + y^2 = 1$
5. $x^2 + 4y^2 = 16$
6. $4x^2 + y^2 = 16$
7. $\frac{(x-1)^2}{16} + \frac{(y+3)^2}{25} = 1$
8. $\frac{(x+3)^2}{25} + \frac{(y-1)^2}{16} = 1$
9. $2x^2 + 2y^2 = 2$
10. $4x^2 + y^2 = 9$
11. $x^2 + 4y^2 = 9$
12. $4(x-3)^2 + (y-7)^2 = 9$
13. $4x^2 + 8x + y^2 + 6y = 3$
14. $x^2 + 6x + 4y^2 + 8y = 3$
15. $4x^2 + 8x + y^2 - 6y = 3$
16. $x^2 + 2x + y^2 + 2y = 7$
17. $3x^2 + 12x + 8y^2 - 4y = 20$
18. $2x^2 - 3x + 4y^2 + 5y = 37$
19. Find the equation of an ellipse with foci at $(0, 4)$ and $(0, -4)$ and vertices at $(0, 5)$ and $(0, -5)$.
20. Find the equation of the ellipse with vertices at $(2, 0)$ and $(-2, 0)$ and eccentricity 0.8.
21. * Find the equation of an ellipse with center at $(-1, 4)$ that is a translation of the ellipse of Problem 19.
22. * Find the equation of the "ellipse" centered at the origin that is symmetric with respect to both the x- and y-axes and that passes through the points $(1, 2)$ and $(-1, -4)$.
23. Show that the graph of the equation $x^2 + 2x + 2y^2 + 12y = c$ is:
    (a) An ellipse if $c > -19$.
    (b) A single point if $c = -19$.
    (c) Empty if $c < -19$.
24. * Find conditions on the numbers a, b, and c in order that the graph of the equation $x^2 + ax + 2y^2 + by = c$ be (a) an ellipse, (b) a single point, (c) empty.

Answers to Readiness Check

I. c II. b III. b IV. a V. b

Use a calculator to solve Problems 25–29.

25. The orbit of Halley's Comet is an ellipse with major axis approximately 36.2 AU and minor axis approximately 9.1 AU. Find the eccentricity of the orbit.

* 26. The major axis of the earth's orbit is approximately 185.5 million miles, and the eccentricity of the orbit is 0.0167. Find the largest and smallest distances between the earth and the sun. (The closest and farthest positions of the planet from the sun are called the **perihelion** and **aphelion,** respectively.) [Hint: Remember that the sun is at one focus of the orbit.]

27. A body orbits around the sun with major axis $2a$ measured in astronomical units (AU). Let T, measured in years, denote the period of the orbit. Then, according to **Kepler's third law,**

$$T^2 = a^3$$

If an asteroid has an orbital period of 8.4 years, find the length of the major axis of its orbit.

28. Kepler showed that the perihelion (closest) distance of the planet's orbit to the sun is $a(1 - e)$ and its aphelion (farthest) distance is $a(1 + e)$. Find these two distances for the asteroid of Problem 27 if the eccentricity of the orbit is 0.058.

29. The roof of a six-lane highway tunnel is constructed in the form of elliptical arches. Each of the six car lanes is 14 feet wide. Using the measurements in the figure, determine the vertical clearance in each lane; that is, determine how tall a truck can drive through without hitting any part of the roof. [Hint: First find the equation of the ellipse.]

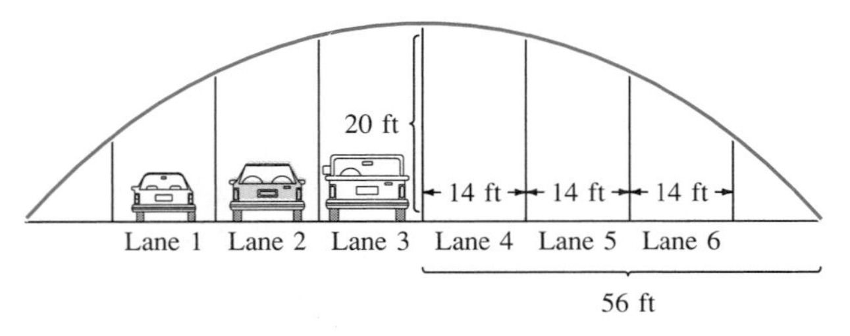

Eight ellipses are sketched in the right-hand column. Match each sketch with the equations given in Problems 30–37. Find the unmarked foci and vertices of each ellipse.

30. $(x - 3)^2 + 16(y - 2)^2 = 16$
31. $4x^2 + 25y^2 = 100$
32. $25x^2 + 21y^2 = 525$
33. $9(x + 2)^2 + 25(y - 1)^2 = 225$
34. $16(x + 2)^2 + 4(y + 1)^2 = 64$
35. $13x^2 + 4y^2 = 52$
36. $9x^2 + 16y^2 = 144$
37. $9(x - 3)^2 + 4(y - 1)^2 = 36$

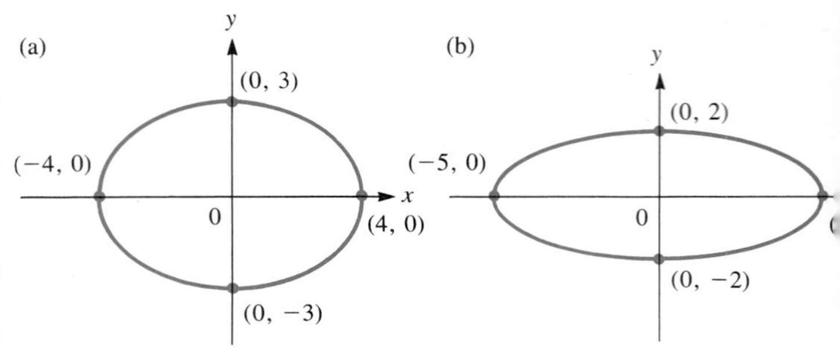

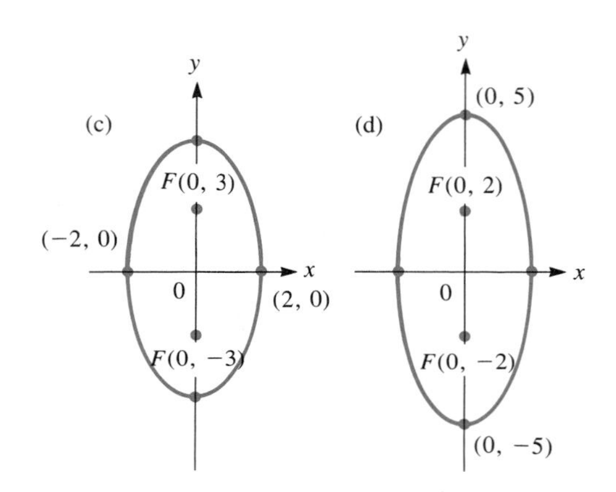

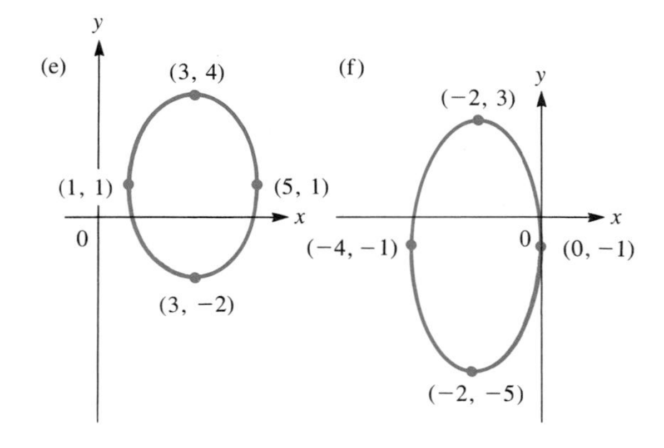

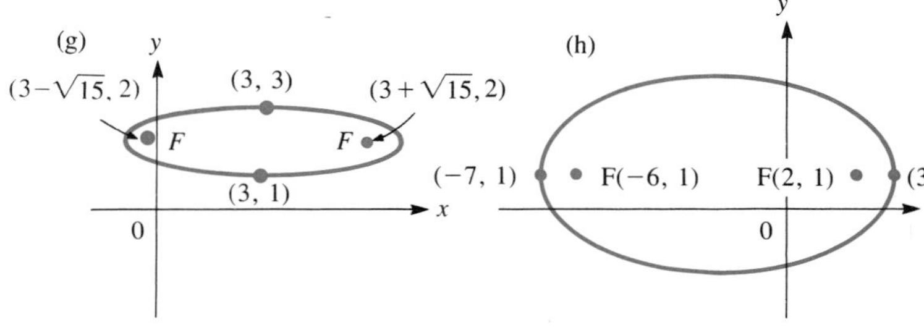

Graphing Calculator Problems

In Problems 38–47 obtain the graph of each ellipse on a graphing calculator. Before starting, read the material in Appendix A dealing with graphing conic sections. In particular, read Example 11 in that appendix.

38. $\dfrac{x^2}{10} + \dfrac{y^2}{37} = 1$
39. $\dfrac{x^2}{50} + \dfrac{y^2}{23} = 1$
40. $3x^2 + 8y^2 = 5$
41. $17x^2 + 10y^2 = 85$
42. $\dfrac{(x+3)^2}{20} + \dfrac{(y+1)^2}{30} = 1$
43. $\dfrac{3(x+4)^2}{7} + \dfrac{5(y+7)^2}{12} = 1$
44. $x^2 + 4x + y^2 + 10y = 18$
45. $x^2 - 7x + y^2 + 9y + 3 = 0$
46. $2x^2 + 3y^2 - 6y = 2$
* 47. $14x^2 + 23x + 27y^2 - 33y = 57$

5.4 The Parabola

In Section 4.1, we sketched a number of parabolas. In this section, we discuss parabolas in more generality. We begin with a definition.

Definition of a Parabola

A **parabola** is the set of points (x, y) equidistant from a fixed point and a fixed line that does not contain the fixed point. The fixed point is called the **focus,** and the fixed line is called the **directrix.**

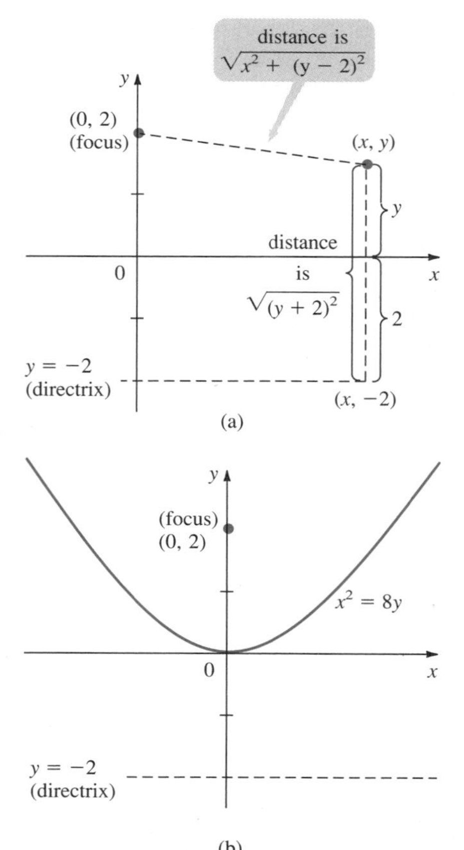

Figure 1 The parabola $x^2 = 8y$ has focus $(0, 2)$ and directrix the line $y = -2$.

EXAMPLE 1 ***Finding an Equation of a Parabola with Given Focus and Directrix***

Find an equation of the parabola whose focus is the point $(0, 2)$ and whose directrix is the line $y = -2$.

SOLUTION As in Figure 1(a), if (x, y) is a point on the parabola, then the distance from $(0, 2)$ to (x, y) is, from the distance formula (p. 151), equal to $\sqrt{(x-0)^2 + (y-2)^2}$. The distance between (x, y) and the line $y = -2$ is defined as the shortest distance from the point to the line. This is obtained by "dropping a perpendicular" from (x, y) to the line. Since the line $y = -2$ is horizontal, the perpendicular line will be vertical and will intersect $y = -2$ at the point $(x, -2)$. The distance between (x, y) and $(x, -2)$ is $\sqrt{(x-x)^2 + (y+2)^2} = \sqrt{(y+2)^2}$. Setting these two distances equal and squaring, we obtain

$$x^2 + (y-2)^2 = (y+2)^2$$
$$x^2 + y^2 - 4y + 4 = y^2 + 4y + 4$$
$$x^2 - 4y = 4y \qquad \text{Subtract } y^2 + 4 \text{ from both sides}$$
$$x^2 = 8y$$

This is an equation of the parabola we sought. Its vertex is at the origin. The parabola is sketched in Figure 1(b).

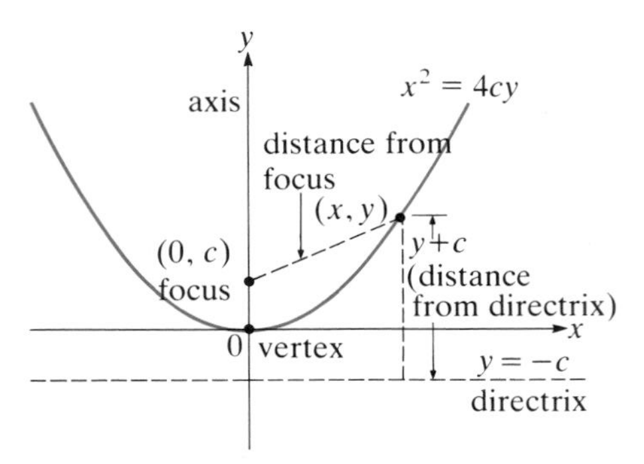

Figure 2 The parabola with focus $(0, c)$ and directrix the line $y = -c$.

We now calculate a more general equation of a parabola. We place the axes so that the focus is the point $(0, c)$ and the directrix is the line $y = -c$. (See Figure 2.) If $P = (x, y)$ is a point on the parabola, then, as in Example 1, we obtain (with c instead of 2)

$$\sqrt{x^2 + (y - c)^2} = \sqrt{(y + c)^2}$$

or squaring,

$$x^2 + (y - c)^2 = (y + c)^2$$

which reduces to

The Standard Equation of a Parabola

The standard equation of a parabola with vertex at the origin, focus at $(0, c)$, and directrix the line $y = -c$ is

$$x^2 = 4cy \qquad (1)$$

The parabola given by (1) is symmetric about the y-axis. This line is called the **axis** of the parabola. Note that the axis contains the focus and is perpendicular to the directrix. The point at which the axis and the parabola intersect is called the **vertex.** The vertex is equidistant from the focus and the directrix.

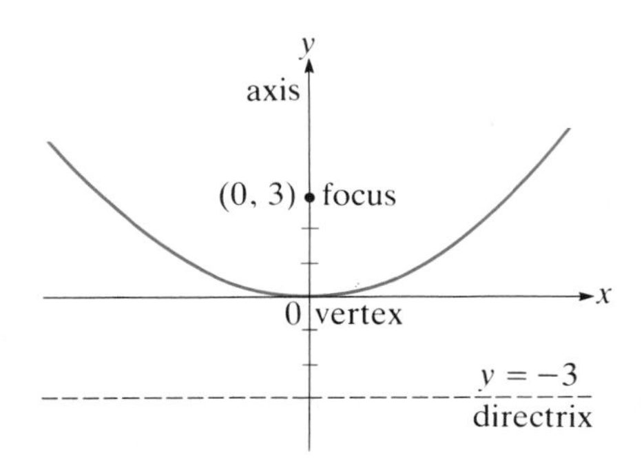

Figure 3 The parabola $x^2 = 12y$ has focus $(0, 3)$ and directrix the line $y = -3$.

EXAMPLE 2 *Sketching a Parabola*

Describe the parabola given by $x^2 = 12y$.

SOLUTION Here, as in equation (1), $4c = 12$, so that $c = 3$, the focus is the point $(0, 3)$, and the directrix is the line $y = -3$. The axis of the parabola is the y-axis, and the vertex is the origin. The curve is sketched in Figure 3. ■

EXAMPLE 3 *A Parabola That Opens Downward*

Describe the parabola given by $x^2 = -8y$.

SOLUTION Here $4c = -8$ so that $c = -2$, and the focus is $(0, -2)$, the directrix is the line $y = 2$, and the curve opens downward, as shown in Figure 4.

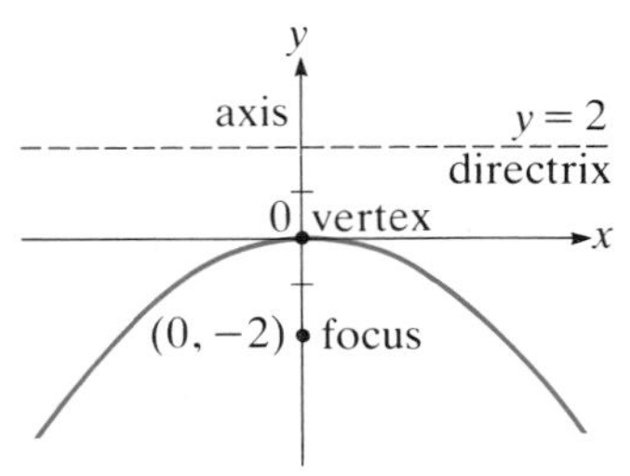

Figure 4 The parabola $x^2 = -8y$ opens downward.

In general,

Determining Whether a Parabola Opens Upward or Downward

The parabola described by $x^2 = 4cy$ opens upward if $c > 0$ and opens downward if $c < 0$.

As with the ellipse, we can exchange the role of x and y. We then obtain the following:

The Standard Equation of a Parabola

The standard equation of the parabola with vertex at the origin, focus at $(c, 0)$, and directrix the line $x = -c$ is

(2)

If $c > 0$, the parabola opens to the right; and if $c < 0$, the parabola opens to the left.

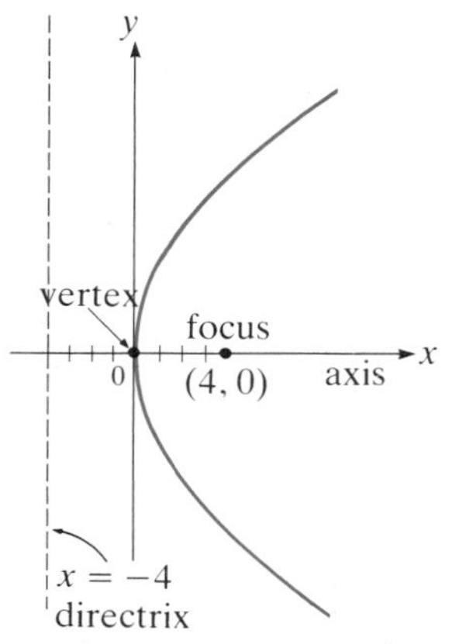

Figure 5 The parabola $y^2 = 16x$ opens to the right.

EXAMPLE 4 ***A Parabola That Opens to the Right***

Describe the parabola $y^2 = 16x$.

SOLUTION Here $4c = 16$ so that $c = 4$, and the focus is $(4, 0)$, the directrix is the line $x = -4$, the axis is the x-axis (since the parabola is symmetric about the x-axis), and the vertex is the origin. The curve is sketched in Figure 5. Note that it opens to the right.

In Table 1 below we list the **standard parabolas.**

Table 1
Standard Parabolas with Vertex at the Origin

Equation	Description	Picture
$x^2 = 4cy$	Focus: $(0, c)$ Directrix: $y = -c$ Axis: y-axis Vertex: $(0, 0)$ Curve opens upward if $c > 0$ and downward if $c < 0$	$x^2 = 4cy$; $(0, c)$; $y = -c$; $c > 0$; $c < 0$
$y^2 = 4cx$	Focus: $(c, 0)$ Directrix: $x = -c$ Axis: x-axis Vertex: $(0, 0)$ Curve opens to the right if $c > 0$ and to the left if $c < 0$	$y^2 = 4cx$; $x = -c$; $(c, 0)$; $c > 0$; $c < 0$
$x^2 = 0$	Degenerate parabola; graph is y-axis (one line)	
$y^2 = 0$	Degenerate parabola; graph is x-axis (one line)	
$x^2 = 1$	Degenerate parabola; graph consists of the two lines $x = 1$ and $x = -1$	
$y^2 = 1$	Degenerate parabola; graph consists of the two lines $y = 1$ and $y = -1$	

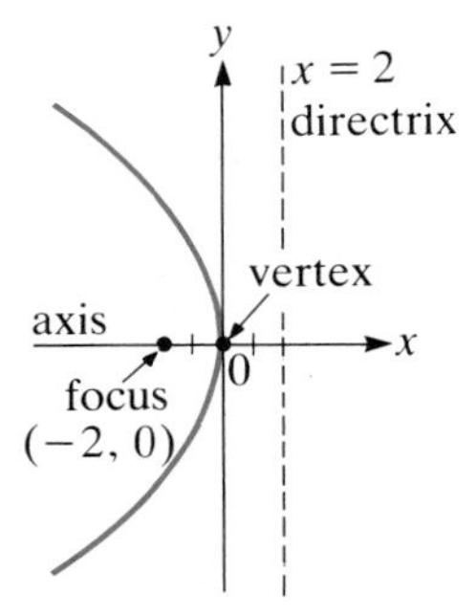

(a) $y^2 = -8x$

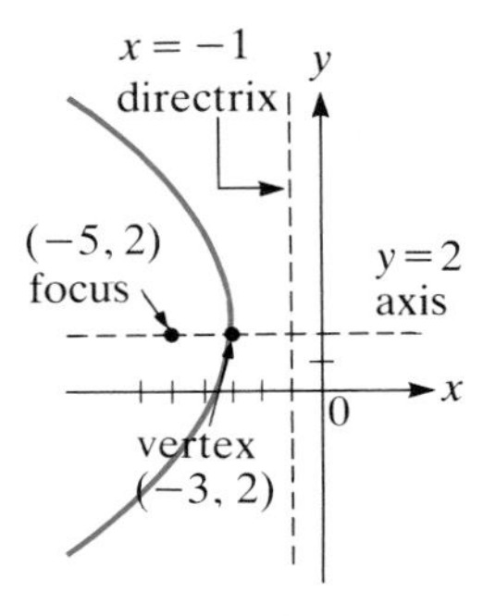

(b) $(y-2)^2 = -8(x+3)$

Figure 6 The parabola $(y-2)^2 = -8(x+3)$ is obtained by translating the parabola $y^2 = -8x$.

All the parabolas we have drawn so far have had their vertices at the origin. Other parabolas can be obtained by a simple translation of axes. The parabolas

$$(x - x_0)^2 = 4c(y - y_0) \tag{3}$$

and

$$(y - y_0)^2 = 4c(x - x_0) \tag{4}$$

have vertices at the point (x_0, y_0).

EXAMPLE 5 *A Translated Parabola*

Describe the parabola $(y - 2)^2 = -8(x + 3)$.

SOLUTION This parabola has its vertex at $(-3, 2)$. It is obtained by shifting the parabola $y^2 = -8x$ three units to the left and two units up (see Section 3.5). Since $4c = -8$, $c = -2$ and the focus and directrix of $y^2 = -8x$ are $(-2, 0)$ and $x = 2$. Hence after translation, the focus and directrix of $(y - 2)^2 = -8(x + 3)$ are $(-5, 2)$ and $x = -1$. The axis of this curve is $y = 2$. The two parabolas are sketched in Figure 6. ■

EXAMPLE 6 *Showing That a Second-Degree Equation Is the Equation of a Translated Parabola by First Completing the Square*

Describe the curve $x^2 - 4x + 2y + 10 = 0$.

SOLUTION We first complete the square:

$$x^2 - 4x + 2y + 10 = (x - 2)^2 - 4 + 2y + 10 = 0$$
$$(x - 2)^2 = -2y - 6$$
$$(x - 2)^2 = -2(y + 3)$$

This expression is the equation of a parabola with vertex at $(2, -3)$. Since $4c = -2$, $c = -\frac{1}{2}$, and the focus is $(2, -3 - \frac{1}{2}) = (2, -\frac{7}{2})$. The directrix is the line $y = -3 - (-\frac{1}{2}) = -\frac{5}{2}$, and the axis is the line $x = 2$. The curve is sketched in Figure 7.

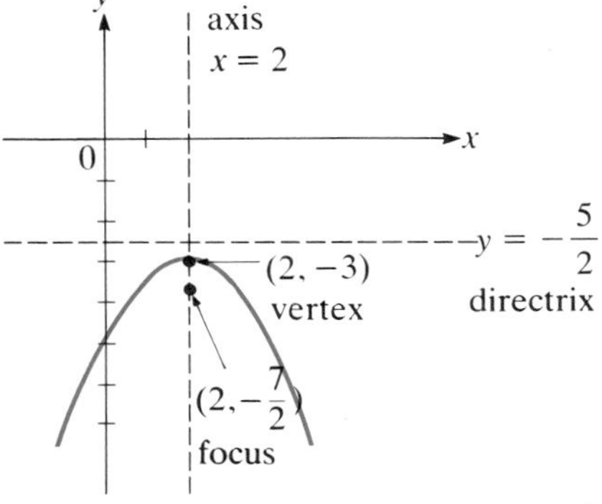

Figure 7 The translated parabola $(x - 2)^2 = -2(y + 3)$.

Translated Parabolas

The parabola $(x - x_0)^2 = 4c(y - y_0)$ has its vertex at (x_0, y_0) and opens upward if $c > 0$ or downward if $c < 0$.

The parabola $(y - y_0)^2 = 4c(x - x_0)$ has its vertex at (x_0, y_0) and opens to the right if $c > 0$ or to the left if $c < 0$.

The equation

$$Ax^2 + Dx + Ey + F = 0$$

can be written in the form (3) by first dividing by A and then completing the square.

The equation

$$Cy^2 + Dx + Ey + F = 0$$

can be written in the form (4) by first dividing by C and then completing the square.

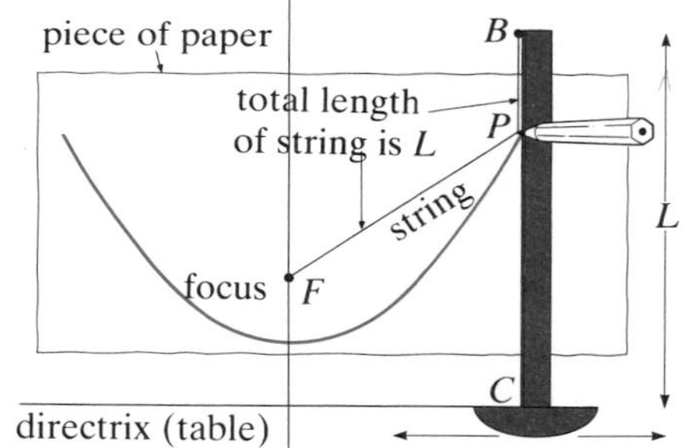

Figure 8 Kepler's method for drawing a parabola.

How to Draw a Parabola

Much of what we now know about the parabola was discovered by the great German physicist, astronomer, and mathematician Johannes Kepler (1571–1630). Kepler was the first to use the term *focus* (Latin for "hearthside") in the context of this section. He constructed parabolas using a table, a piece of string, a pencil, and the seventeenth-century version of a T-square. Place a piece of paper along a wall above the edge of a horizontal table as in Figure 8. The table's edge is the directrix of our parabola.

If the T-square has length L, then choose a string of length L. Pin one end to the focus F, and pin the other end to the top of the T-square. Slide a pencil along the string as in Figure 8, keeping the string taut. Then, as the T-square is moved along the side of the table, the pencil will trace out a parabola. The reason for this is that $\overline{FP} + \overline{PB} = \overline{CP} + \overline{PB} = L$ so that $\overline{FP} = \overline{CP}$. That is, the distance from a point P on the parabola to the focus equals the distance from the point to the directrix.

Figure 9 A light radio beam emanating from F is reflected parallel to the axis along the line L.

The Reflective Property of a Parabola

Consider the parabola with focus F sketched in Figure 9. Let $P(x, y)$ be a point on the parabola, T the line tangent to the parabola at P, and L the line passing through P that is parallel to the axis. Finally, let α denote the angle between T and L, and let β denote the angle between T and PF. Then $\alpha = \beta$. This means that a beam that starts at the focus F will be reflected off the parabola parallel to the axis. This **reflective property of the parabola,** as it is called, is useful in a wide variety of applications, as we will soon see.

The opposite effect is also seen: A beam coming in parallel to the axis will be reflected back to the focus. This can be illustrated in Figure 9 by reversing the arrows.

FOCUS ON
The Parabola in the Real World

Parabolas are all around us. The automobile headlight has the parabolic shape obtained by rotating a parabola around its axis. All the light emanating from a bulb placed at the focus is reflected parallel to the parabola's axis, that is, parallel to the ground. This follows from the reflective property just discussed (see Figure 10).

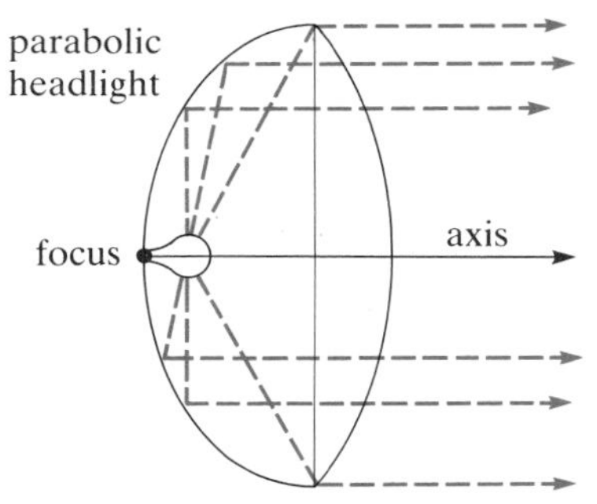

Figure 10 A parabolic headlight. Light rays from a bulb at the focus are reflected parallel to the axis.

Parabolic reflectors can be found in communications systems, electronic surveillance systems, radar systems, and telescopes. In radar, an electromagnetic beam is bounced off a target, and a parabolic reflector is used to collect and concentrate the deflected beam for signal processing. This works because the reflector reflects all returning signals back to the focus. In Figure 11, we show a parabolic reflector used on earth to track space probes.

Figure 11 Credit: Photo Researchers, Inc.

When a projectile is shot into the air, its path takes the shape of a parabola. This famous phenomenon was discovered by Galileo (1564–1642). Renaissance scientists were fascinated by this fact. The great artist Leonardo da Vinci drew the path of exploding mortar shells toward the end of the fifteenth century (Figure 12).

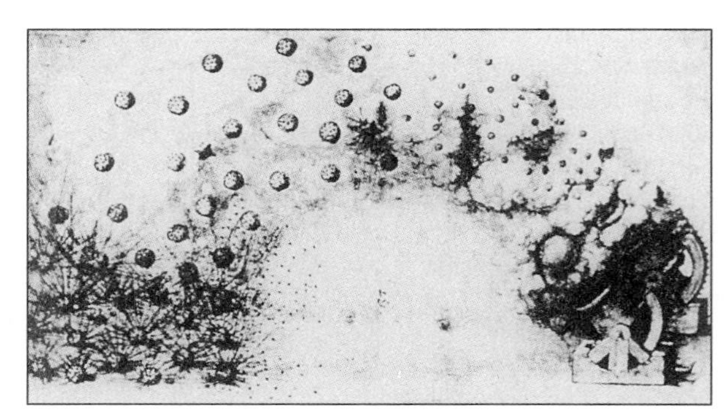

Figure 12 Cannon in action. The flight of exploding mortar shells, drawn by Leonardo da Vinci.

Sound waves can be transmitted effectively using parabolic reflectors. This is illustrated in Figure 13. One such "double reflector" device can be found at the Exploratorium in San Francisco. In this model, the reflectors have diameters of 8 feet and are placed facing each other about 50 feet apart.

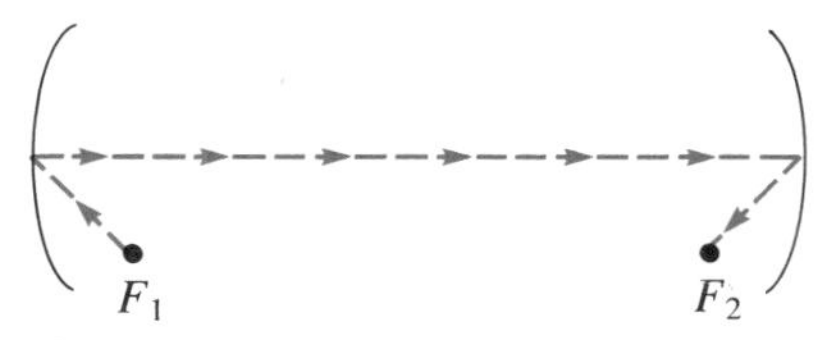

Figure 13 A sound made at focus F_1 will be reflected twice and clearly heard at focus F_2.

Parabolic reflectors also have been used in warfare and other unfortunate circumstances. The most famous example comes from Archimedes of Syracuse (287–212 B.C.), who seemed to have dabbled in everything. According to the Greek historian Plutarch, Syracuse was besieged by Romans led by their great general Marcellus. Archimedes helped save the city by designing, among other things, "burning" mirrors. These were parabolic mirrors capable of concentrating the rays of the sun onto attacking ships. Archimedes' clever devices also included a catapult that used the principle of the lever to hurl huge boulders. Because of Archimedes, Syracuse was able to hold out for nearly three years.

The story of the siege of Syracuse had an unhappy ending. After Marcellus failed to take Syracuse by a frontal siege, the city fell after a circuitous attack. A Roman soldier was sent by Marcellus to bring Archimedes to him. According to one version of the story, Archimedes had drawn a diagram in the sand and asked the soldier to stand away from it. This angered the soldier, who killed Archimedes with his spear.

The Greeks were fascinated by the parabola and sometimes suggested applications that were, to put it mildly, very unpleasant. The Greek Diocles, who lived in the second century B.C., wrote a book entitled *Burning Mirrors.* In his work, Diocles suggested that if victims were to be sacrificed in front of large crowds, parabolic mirrors could be used to provide a visible burning spot on the victim's body. It is not clear whether this idea was ever put into practice.

Parabolas are used in civil engineering. Bridges are built with twin parabolic cables that, ideally, will support a uniform horizontal load (see Figure 14).

These and many other applications of parabolas can be found in the fascinating paper, "The Standup Conic Presents: The Parabola and Applications," by Lee Whitt in *The UMAP Journal,* Vol. 3, No. 3, 1983, pp. 285–316.

Figure 14 In the ideal case, the main cable of a suspension bridge is parabolic. Credit: B.A. Lang Sr. Photo Researchers, Inc.

Problems 5.4

Readiness Check

I. The graph of $\frac{x}{-4} = \left(\frac{y}{3}\right)^2$ is a parabola opening ______.

a. to the right
b. to the left
c. upward
d. downward

II. The set of points $\{(a, b)\colon 4a = -b^2\}$ is a ______.

a. vertical line
b. horizontal line
c. single point
d. parabola

III. Answer True or False to each of the following assertions about the parabola satisfying

$$y = -(x - 1)^2.$$

a. The focus is (0, 0).
b. The vertex is (0, 0).
c. The vertex is (1, 0).
d. The focus is (1, 4).
e. The focus is $(1, -\frac{1}{4})$.
f. The directrix passes through the vertex.
g. The directrix passes through the focus.
h. The directrix is perpendicular to the axis of the parabola.

In Problems 1–18 the equation of a parabola is given. Find its focus, directrix, axis, and vertex. Determine whether it opens up, down, right, or left. Then sketch it.

1. $x^2 = 16y$
2. $y^2 = 16x$
3. $x^2 = -16y$
4. $y^2 = -16x$
5. $2x^2 = 3y$
6. $2y^2 = 3x$
7. $4x^2 = -9y$
8. $7y^2 = -20x$
9. $(x - 1)^2 = -16(y + 3)$
10. $(y - 1)^2 = -16(x - 3)$
11. $x^2 + 4y = 9$
12. $(x + 1)^2 + 25y = 50$
13. $x^2 + 2x + y + 1 = 0$
14. $x + y - y^2 = 4$
15. $x^2 + 4x + y = 0$
16. $y^2 + 4y + x = 0$
17. $x^2 + 4x - y = 0$
18. $y^2 + 4y - x = 0$
19. Find the equation of the parabola with focus (0, 4) and directrix the line $y = -4$.
20. Find the equation of the parabola with focus (−3, 0) and directrix the line $x = 3$.

Answers to Readiness Check

I. b II. d III. a. False b. False c. True d. False e. True f. False g. False h. True

21. Find the equation of the parabola obtained when the parabola of Problem 20 is shifted so that its vertex is at the point $(-2, 5)$.

** 22. Find the equation of the parabola with vertex at $(1, 2)$ and directrix the line $x = y$.

23. The parabola in Problem 20 is translated so that its vertex is now at $(3, -1)$. Find its new focus and directrix.

24. The **latus rectum** of a parabola is the chord passing through the focus that is perpendicular to the axis. Compute the length of the latus rectum of the parabola $x^2 = 6y$.

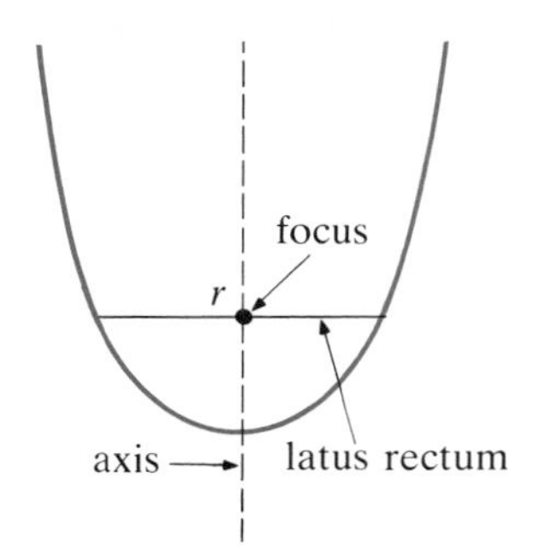

* 25. Show that if a parabola has the equation $x^2 = 4cy$ or $y^2 = 4cx$, then the length of its latus rectum is $|4c|$.

26. The tops of two towers of a suspension bridge (like the one in Figure 14) are 100 feet above water level and 375 feet apart. The lowest point of the parabolic cable connecting the two towers is 40 feet above the water. How high is a point on the cable that is 60 feet (horizontally) from one of the towers?

27. The receiver of a parabolic signal receptor is at the focus, which is 2 feet from the vertex. If the receptor is placed as in the figure, find an equation for the cross-sectional parabola that lies in the xy-plane.

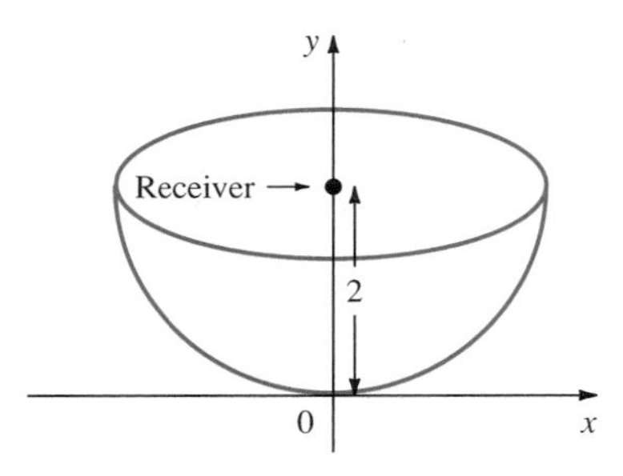

28. A missile is fired from ground level. Its path is a parabola opening downward. The missile reaches a height of 1200 meters and travels 15,000 meters (15 kilometers) horizontally. When the missile first reaches a height of 500 meters, how far is it (horizontally) from the firing site?

29. If an object is dropped from rest from an initial height of h_0 feet, then, after t seconds, its height, $h(t)$, is given by

$$h(t) = h_0 - 16t^2 \text{ feet}$$

(a) Graph this equation for $h_0 = 1000$ feet.
(b) When does the object hit the ground?

* 30. A bomber, flying at 350 miles per hour, releases a bomb from an altitude of 28,000 feet. How far does the bomb travel horizontally before it hits the ground?

The graphs of eight parabolas are given. Match the graphs with the equations given in Problems 31–38.

31. $x^2 + 10x + 3y + 13 = 0$
32. $x^2 - 4x - 5y - 11 = 0$
33. $4x - y^2 = 0$
34. $x^2 - 3y = 0$
35. $2x + y^2 = 0$
36. $y^2 + 4y + x = 0$
37. $y^2 - 2x - 4y + 6 = 0$
38. $x^2 + 4y = 0$

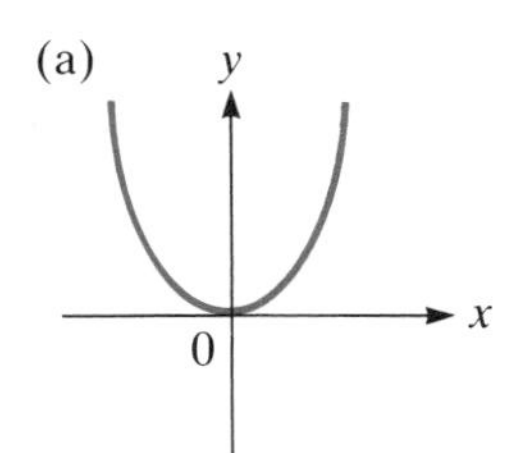

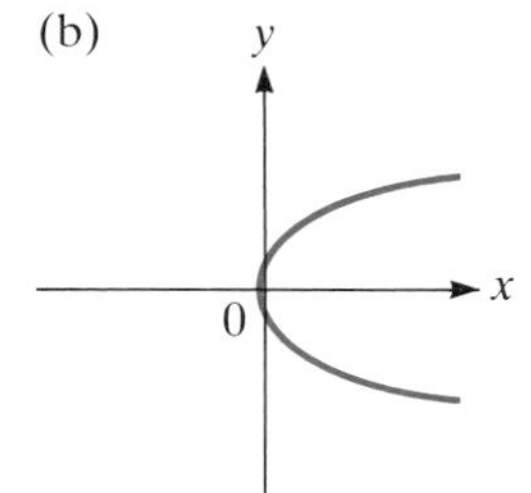

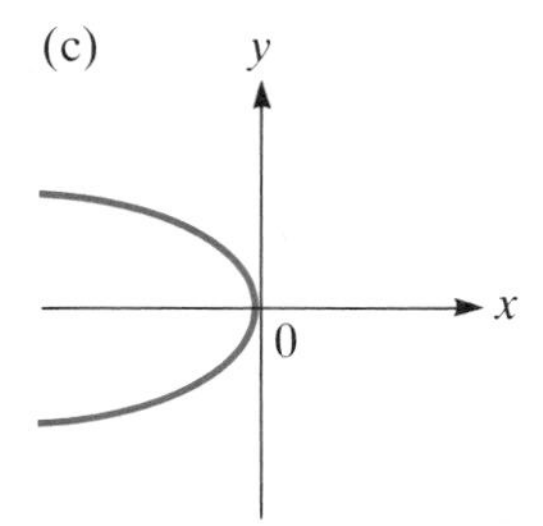

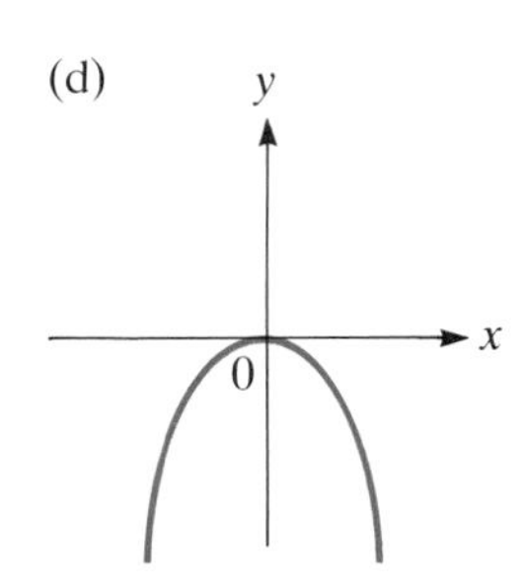

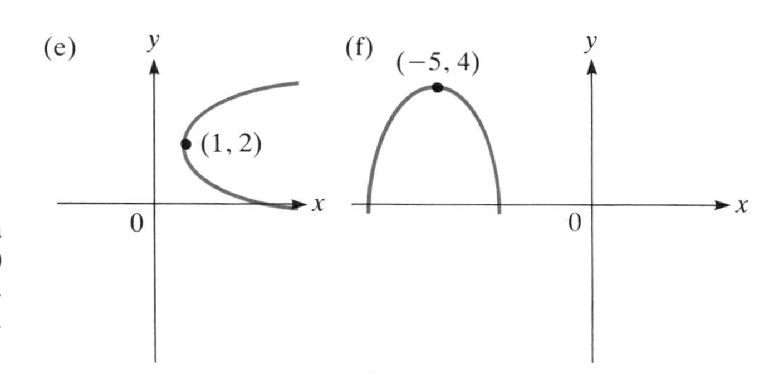

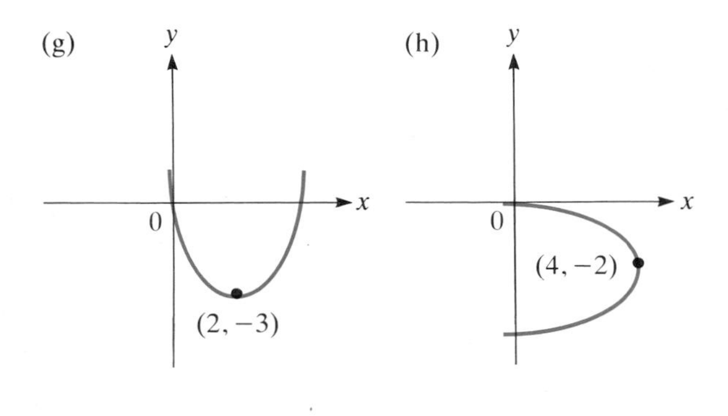

Graphing Calculator Problems

In Problems 39–48 obtain the graph of each parabola on a graphing calculator. Note that in order to graph the parabola $y^2 = 4cx$, it is necessary to graph *two* functions. Read Examples 9, 10, and 11 in Appendix A for more details.

39. $x^2 = 20y$
40. $y^2 = 20x$
41. $y^2 = -3x$
42. $x^2 = -4.7y$
43. $3x^2 = -17y$
44. $5y^2 = 23y$
45. $(x + 2)^2 = -5(y + 1)$
46. $(y - 3)^2 = 4(x + \frac{1}{2})$
47. $x^2 + 3x + 2y = 0$
48. $y^2 + 6y - 3x = 6$

5.5 The Hyperbola

Definition of a Hyperbola

A **hyperbola** is a set of points (x, y) with the property that the positive difference between the distances from (x, y), and each of two given (distinct) points is a constant. Each of the two given points is called a **focus** of the hyperbola.

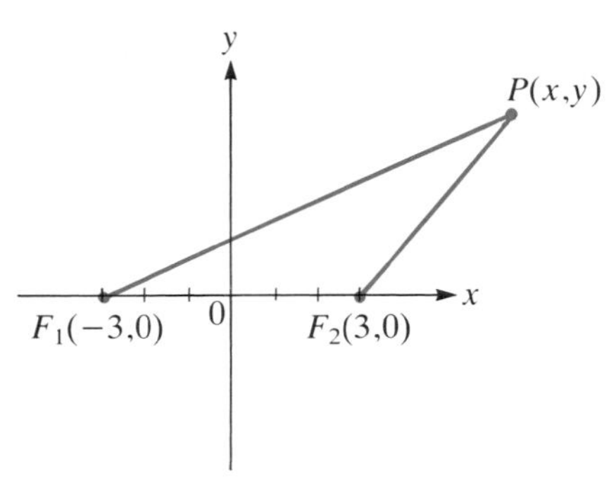

Figure 1 $\overline{PF_1} - \overline{PF_2} = 4$.

EXAMPLE 1 ***Finding the Equation of a Hyperbola with Given Foci***

Find an equation of the hyperbola whose foci are the points $(-3, 0)$ and $(3, 0)$ and in which the difference of the distances from a point (x, y) on the hyperbola to the foci is equal to 4.

SOLUTION If P is a point on the hyperbola and F_1 and F_2 denote the foci as in Figure 1, then the difference of the distances from P to the foci is

$$\overline{PF_1} - \overline{PF_2} = 4$$

Using the distance formula, we obtain

$$\sqrt{(x+3)^2 + y^2} - \sqrt{(x-3)^2 + y^2} = 4$$

$$\sqrt{(x+3)^2 + y^2} = \sqrt{(x-3)^2 + y^2} + 4$$

$$(x+3)^2 + y^2 = (x-3)^2 + y^2 + 8\sqrt{(x-3)^2 + y^2} + 16$$

We squared both sides

$$x^2 + 6x + 9 + y^2 = x^2 - 6x + 9 + y^2 + 8\sqrt{(x-3)^2 + y^2} + 16$$

We multiplied through

$$6x = -6x + 8\sqrt{(x-3)^2 + y^2} + 16$$

We subtracted $x^2 + 9 + y^2$ from both sides

$$12x - 16 = 8\sqrt{(x-3)^2 + y^2}$$

We rearranged terms

$$\frac{3}{2}x - 2 = \sqrt{(x-3)^2 + y^2} \qquad \text{We divided by 8}$$

$$\frac{9}{4}x^2 - 6x + 4 = (x-3)^2 + y^2 \qquad \text{We squared again}$$

$$\frac{9}{4}x^2 - 6x + 4 = x^2 - 6x + 9 + y^2$$

$$\left(\frac{9}{4} - 1\right)x^2 - y^2 = 5 \qquad \text{We combined terms}$$

$$\frac{5}{4}x^2 - y^2 = 5$$

$$\frac{x^2}{4} - \frac{y^2}{5} = 1 \qquad \text{We divided by 5} \qquad \textbf{(1)}$$

Equation (1) is the **standard equation** of the hyperbola. Here we assumed that $\overline{PF_1} > \overline{PF_2}$. If $\overline{PF_2} > \overline{PF_1}$, we obtain the same equation. You should verify this.

In order to help graph the hyperbola, we make several observations. First, from equation (1) we have

$$\frac{x^2}{4} = 1 + \frac{y^2}{5} \quad \text{or} \quad x^2 = 4 + \frac{4}{5}y^2$$

Since $y^2 \geq 0$ for all real numbers y, we have $x^2 \geq 4$ so that

$$x \geq 2 \quad \text{or} \quad x \leq -2$$

That is, no point on the hyperbola has an x-coordinate in the interval $(-2, 2)$. Second, if we replace x by $-x$ in (1), we obtain the same equation. This means that the graph is symmetric about the y-axis so that the hyperbola has two symmetric branches: one for $x > 0$ and one for $x < 0$. These correspond to the two cases $\overline{PF_1} - \overline{PF_2} = 4$ and $\overline{PF_2} - \overline{PF_1} = 4$.

We now solve (1) for y:

$$\frac{y^2}{5} = \frac{x^2}{4} - 1$$

$$y^2 = \frac{5}{4}x^2 - 5$$

$$y = \pm\sqrt{\frac{5}{4}x^2 - 5}$$

Suppose $|x|$ is large, then $\frac{5}{4}x^2 - 5 \approx \frac{5}{4}x^2$. To see this clearly, insert some numbers. For example, if $x = \pm 1000$, then $\frac{5}{4}x^2 - 5 = 1{,}249{,}995$ while $\frac{5}{4}x^2 = 1{,}250{,}000$. Thus, for x large

$$y \approx \pm\sqrt{\frac{5}{4}x^2} = \pm\frac{\sqrt{5}}{2}x$$

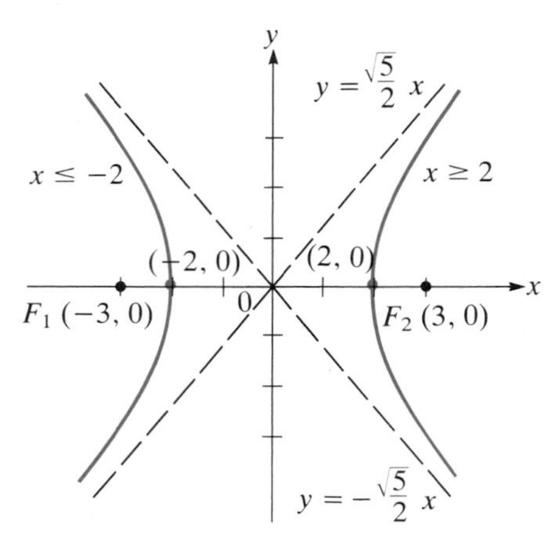

Figure 2 The hyperbola $\frac{x^2}{4} - \frac{y^2}{5} = 1$.

The lines $y = \frac{\sqrt{5}}{2}x$ and $y = -\frac{\sqrt{5}}{2}x$ are oblique asymptotes for the hyperbola (see p. 286). Putting this all together, we obtain the graph in Figure 2.

To calculate the equation of a more general hyperbola, we place the axes so that the foci are the points $(c, 0)$ and $(-c, 0)$, and the difference of the distances from a point (x, y) on the hyperbola to the foci is equal to $2a > 0$. In Figure 3, we assume that $\overline{PF_2} > \overline{PF_1}$. Then

$$\text{difference of distances} = \overline{PF_2} - \overline{PF_1} = 2a$$

so

$$\overline{PF_2} = 2a + \overline{PF_1}$$

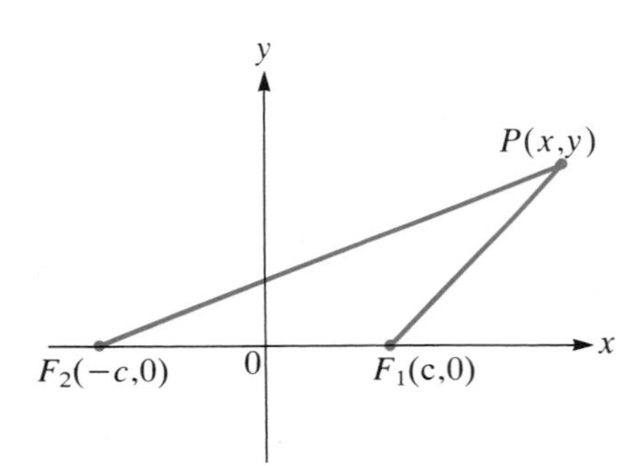

Figure 3 $\overline{PF_2} - \overline{PF_1} = 2a$.

But, as the shortest distance between two points is a straight line, we have

$$2a + \overline{PF_1} = \overline{PF_2} < \overline{PF_1} + \overline{F_1F_2} = \overline{PF_1} + 2c$$

Therefore,

$$2a + \overline{PF_1} < 2c + \overline{PF_1}$$

or

$$2a < 2c \quad \text{and} \quad c > a$$

If (x, y) is a point on the hyperbola,

$$\overline{PF_2} - \overline{PF_1} = 2a$$

or

$$\sqrt{(x+c)^2 + y^2} - \sqrt{(x-c)^2 + y^2} = 2a$$

(Again, we assumed that $\overline{PF_2} > \overline{PF_1}$ so that $\overline{PF_2} - \overline{PF_1}$ gives us a positive distance.) Then we obtain, successively,

$$\sqrt{(x+c)^2 + y^2} = \sqrt{(x-c)^2 + y^2} + 2a$$

$$(x+c)^2 + y^2 = (x-c)^2 + y^2 + 4a\sqrt{(x-c)^2 + y^2} + 4a^2 \qquad \text{We squared}$$

$$x^2 + 2cx + c^2 + y^2 = x^2 - 2cx + c^2 + y^2 + 4a\sqrt{(x-c)^2 + y^2} + 4a^2$$

$$4cx - 4a^2 = 4a\sqrt{(x-c)^2 + y^2} \qquad \text{We simplified}$$

$$\frac{c}{a}x - a = \sqrt{(x-c)^2 + y^2} \qquad \text{We divided by } 4a$$

$$\frac{c^2}{a^2}x^2 - 2cx + a^2 = (x-c)^2 + y^2 \qquad \text{We squared again}$$

$$\frac{c^2}{a^2}x^2 - 2cx + a^2 = x^2 - 2cx + c^2 + y^2$$

$$\left(\frac{c^2}{a^2} - 1\right)x^2 - y^2 = c^2 - a^2 \tag{2}$$

Finally, since $c > a > 0$, $c^2 - a^2 > 0$, and we can define the positive number b by

$$b^2 = c^2 - a^2$$

Then dividing both sides of (2) by $c^2 - a^2$, we obtain, since $\frac{c^2}{a^2} - 1 = \frac{c^2 - a^2}{a^2}$:

Standard Equation of a Hyperbola Centered at the Origin

$$\frac{x^2}{a^2} - \frac{y^2}{b^2} = 1 \tag{3}$$

A similar derivation shows that if we assume that $\overline{PF_1} > \overline{PF_2}$, we also obtain equation (3).

These two cases correspond to the right and left branches of the hyperbola sketched in Figure 4. Note that the hyperbola given by (3) is symmetric about

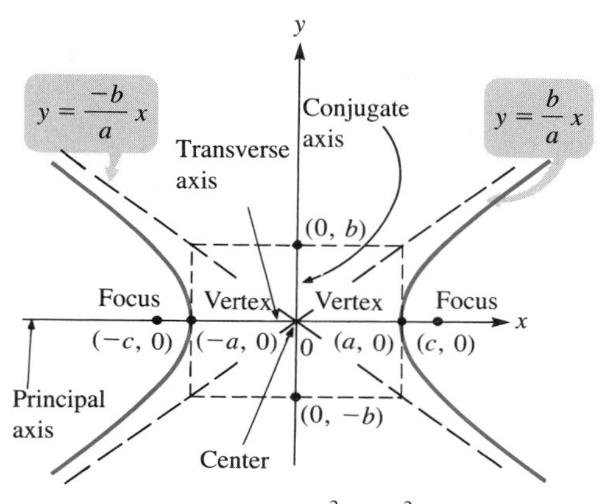

Figure 4 The hyperbola $\frac{x^2}{a^2} - \frac{y^2}{b^2} = 1$.

both the x-axis and the y-axis. The **principal axis** is the line containing the foci. The **vertices** of the parabola are the points of intersection of the hyperbola and its principal axis. The midpoint of the line segment joining the foci is called the **center** of the hyperbola. The **transverse axis** of the hyperbola is the line segment joining the vertices. The **conjugate axis** is the line joining the points $(0, -b)$ and $(0, b)$.

As in Example 1, we can write (3) as

$$y = \pm\sqrt{\frac{b^2}{a^2}x^2 - b^2} \approx \pm\sqrt{\frac{b^2}{a^2}x^2} = \pm\frac{b}{a}x \text{ for } |x| \text{ large}$$

Thus, the lines $y = \pm\frac{b}{a}x$ are oblique asymptotes to the hyperbola. To make it easier to sketch the hyperbola in Figure 4, we draw the rectangle with sides $x = \pm a$, $y = \pm b$. The lines that pass through the diagonals of this rectangle are the asymptotes of the hyperbola.

Finally, we note that the hyperbola given by (3) is called a hyperbola with **horizontal transverse axis.**

EXAMPLE 2 *Sketching a Hyperbola with a Horizontal Transverse Axis*

Discuss the curve given by $x^2 - 4y^2 = 9$.

SOLUTION Dividing by 9, we obtain $(x^2/9) - (4y^2/9) = 1$, or

$$\frac{x^2}{3^2} - \frac{y^2}{\left(\frac{3}{2}\right)^2} = 1$$

This is the equation of a hyperbola with $a = 3$ and $b = \frac{3}{2}$. Then $c^2 = a^2 + b^2 = 9 + 9/4 = 45/4$ so that the foci are $(\sqrt{45}/2, 0)$ and $(-\sqrt{45}/2, 0)$. The vertices are $(3, 0)$ and $(-3, 0)$, and the asymptotes are $y = \pm\frac{1}{2}x$. The curve is sketched in Figure 5.

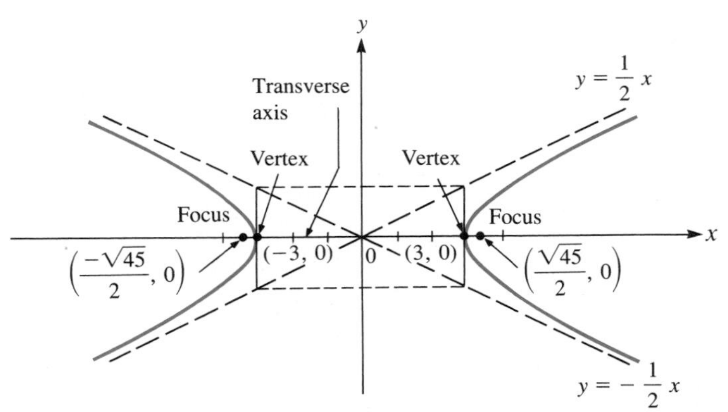

Figure 5 The hyperbola $\dfrac{x^2}{9} - \dfrac{y^2}{\frac{9}{4}} = 1$.

As with the ellipse and parabola, the roles of x and y can be reversed. The graph of the equation

$$\frac{y^2}{a^2} - \frac{x^2}{b^2} = 1 \tag{4}$$

is sketched in Figure 6. In (4) the transverse axis is on the y-axis. From (2) we have $c^2 = a^2 + b^2$, and the foci are at $(0, c)$ and $(0, -c)$. The vertices of the hyperbola given by (4) are the points $(0, a)$ and $(0, -a)$. The asymptotes are the lines $y = \pm(a/b)x$. This is a hyperbola with a **vertical transverse axis.**

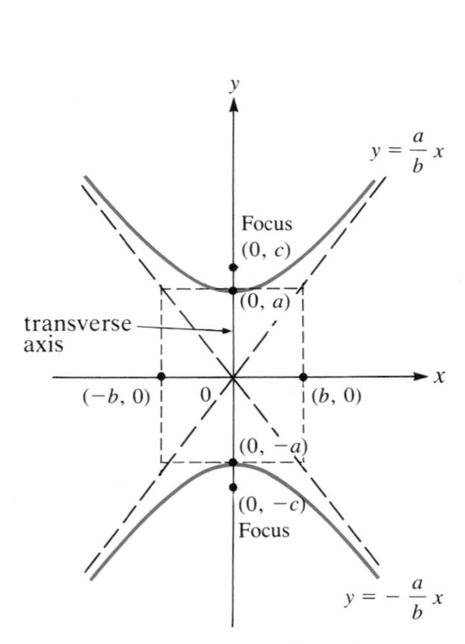

Figure 6 The hyperbola $\dfrac{y^2}{a^2} - \dfrac{x^2}{b^2} = 1$.

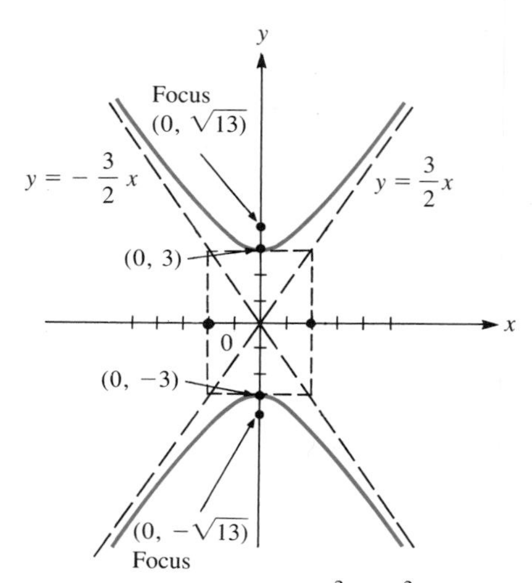

Figure 7 The hyperbola $\dfrac{y^2}{9} - \dfrac{x^2}{4} = 1$.

EXAMPLE 3 ***A Hyperbola with a Vertical Transverse Axis***

Discuss the curve given by $4y^2 - 9x^2 = 36$.

SOLUTION Dividing by 36, we obtain

$$\frac{y^2}{9} - \frac{x^2}{4} = 1$$

Hence $a = 3$, $b = 2$, and $c^2 = 9 + 4 = 13$ so that $c = \sqrt{13}$. The foci are $(0, \sqrt{13})$ and $(0, -\sqrt{13})$, and the vertices are $(0, 3)$ and $(0, -3)$. The asymptotes are the lines $y = \frac{3}{2}x$ and $y = -\frac{3}{2}x$. This curve is sketched in Figure 7.

The hyperbolas we have sketched to this point are **standard hyperbolas.** Our results are summarized in Table 1.

Table 1
Standard Hyperbolas Centered at the Origin

Equations	Description	Picture
$\dfrac{x^2}{a^2} - \dfrac{y^2}{b^2} = 1$	Hyperbola with foci at $(c, 0)$ and $(-c, 0)$, where $c^2 = a^2 + b^2$; transverse axis is on the x-axis; center at origin; asymptotes $y = \pm(b/a)x$; curve opens to right and left; horizontal transverse axis	See Figure 4
$\dfrac{y^2}{a^2} - \dfrac{x^2}{b^2} = 1$	Hyperbola with foci at $(0, c)$ and $(0, -c)$, where $c^2 = a^2 + b^2$; transverse axis is on the y-axis; center at origin; asymptotes $y = \pm(a/b)x$; curve opens at top and bottom; vertical transverse axis	See Figure 6
$\dfrac{x^2}{a^2} - \dfrac{y^2}{b^2} = 0$	Degenerate hyperbola; graph consists of two lines: $y = \pm(b/a)x$	y; $y = \frac{b}{a}x$; x; 0; $y = -\frac{b}{a}x$

The hyperbolas we have so far discussed have had their centers at the origin. Other hyperbolas can be obtained by a translation of the axes. The hyperbola

$$\frac{(x - x_0)^2}{a^2} - \frac{(y - y_0)^2}{b^2} = 1 \tag{5}$$

has its center at (x_0, y_0) and has a horizontal transverse axis.

The hyperbola

$$\frac{(y - y_0)^2}{a^2} - \frac{(x - x_0)^2}{b^2} = 1 \tag{6}$$

has its center at (x_0, y_0) and has a vertical transverse axis.

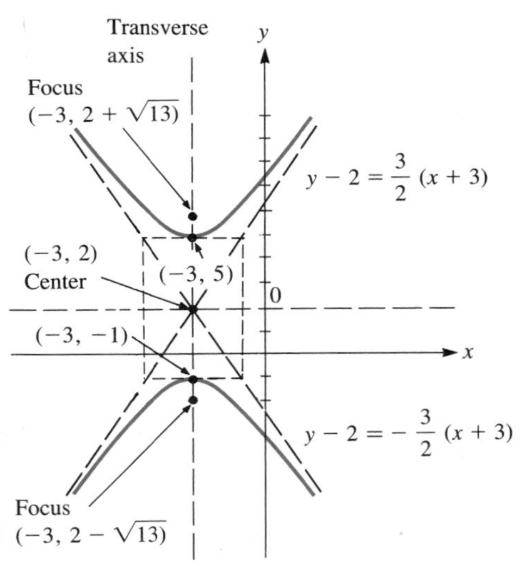

Figure 8 Graph of the hyperbola $\frac{(y-2)^2}{9}-\frac{(x+3)^2}{4}=1.$

EXAMPLE 4 *Translating a Hyperbola*

Describe the hyperbola $\frac{(y-2)^2}{9}-\frac{(x+3)^2}{4}=1.$

SOLUTION This curve is the hyperbola of Example 3 shifted three units to the left and two units up so that its center is at $(-3, 2)$. It is sketched in Figure 8. ■

EXAMPLE 5 *Determining the Nature of a Translated Conic by First Completing Two Squares*

Describe the curve $x^2-4y^2-4x-8y-9=0$.

SOLUTION We have

$$(x^2-4x)-4(y^2+2y)-9=0$$

or completing the squares,

$$(x-2)^2-4-4[(y+1)^2-1]-9=0$$
$$(x-2)^2-4(y+1)^2=9$$
$$\frac{(x-2)^2}{9}-\frac{4}{9}(y+1)^2=1$$

and

$$\frac{(x-2)^2}{9}-\frac{(y+1)^2}{\left(\frac{3}{2}\right)^2}=1$$

This is the equation of a hyperbola with center at $(2, -1)$ and horizontal transverse axis. Except for the translation, it is the hyperbola of Example 2 and is sketched in Figure 9.

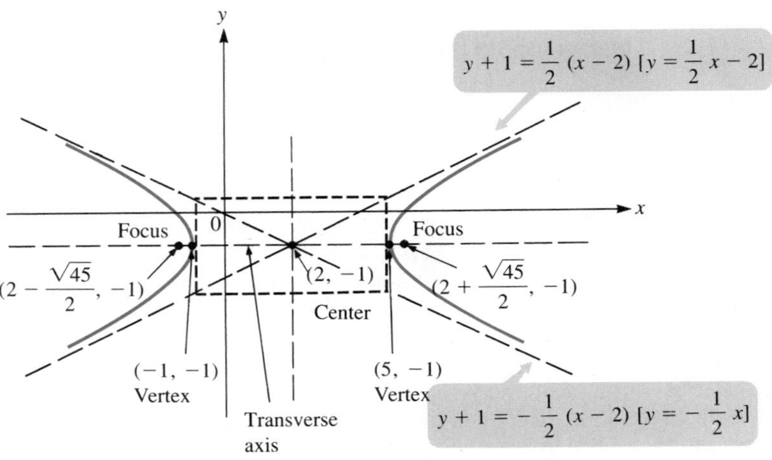

Figure 9 The translated hyperbola $\frac{(x-2)^2}{9}-\frac{(y+1)^2}{\left(\frac{3}{2}\right)^2}=1.$

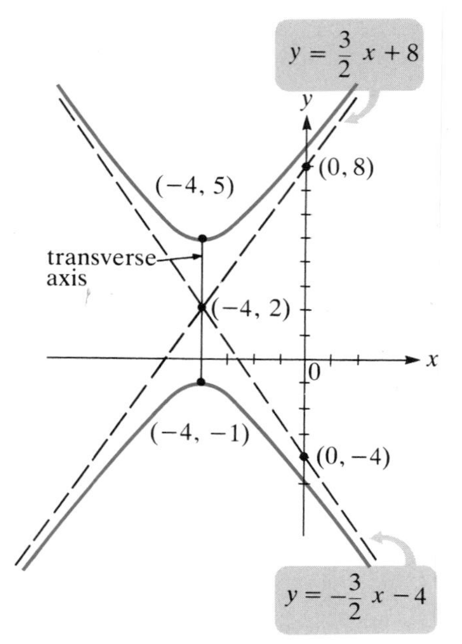

Figure 10 The translated hyperbola $\frac{(y-2)^2}{9} - \frac{(x+4)^2}{4} = 1$.

EXAMPLE 6 ***Finding the Standard Equation of a Translated Hyperbola***

Find the standard equation of the hyperbola with vertices at $(-4, -1)$ and $(-4, 5)$ and asymptotes the lines $y = \frac{3}{2}x + 8$ and $y = -\frac{3}{2}x - 4$.

SOLUTION As in Figure 10, the vertices lie on the vertical line $x = -4$, so the transverse axis is vertical. There are two ways to find the center. The simpler way is to observe that the center is the midpoint of the vertices. The x-coordinate is -4, and the y-coordinate (the average of -1 and 5) is $\frac{-1+5}{2} = 2$. Thus the center is at $(-4, 2)$. The other way is to find the point of intersection of the asymptotes. Now

$$2a = \text{distance between vertices} = 5 - (-1) = 6 \text{ so } a = 3$$

To find b, we observe that the asymptotes have the equations

$$y - 2 = \pm\frac{a}{b}(x + 4)$$

In our case, we must have $\frac{a}{b} = \frac{3}{2}$. Since $a = 3$, we see that $b = 2$. Therefore, the standard equation of the hyperbola is

$$\frac{(y-2)^2}{3^2} - \frac{(x+4)^2}{2^2} = 1 \quad \text{or} \quad \frac{(y-2)^2}{9} - \frac{(x+4)^2}{4} = 1$$

FOCUS ON

The Hyperbola in the Real World

Hyperbolas do not appear in physical constructions as often as ellipses and parabolas, but nevertheless they are useful. Hyperbolas do frequently appear as the graphs of important equations in physics, chemistry, biology, business, and economics. Examples include Ohm's law and supply and demand curves. In Einstein's theory of special relativity, an observer in an inertial reference frame sees a particle in a parallel force field follow a hyperbolic path in space time. The British physicist Ernest Rutherford (1871–1937) developed his now-accepted model of the atom by measuring the hyperbolic orbits of scattered positive-charged particles.

Capillary action is the elevation of the surface of liquids in fine tubes, and so on, due to surface tension and other forces. Suppose, as in Figure 11, that two pieces of glass are joined along one pair of edges and are slightly separated along the other pair. If the glass configuration is placed vertically in a dish of colored water, then capillary action will force the water to rise in such a way as to form a hyperbola. Try it.

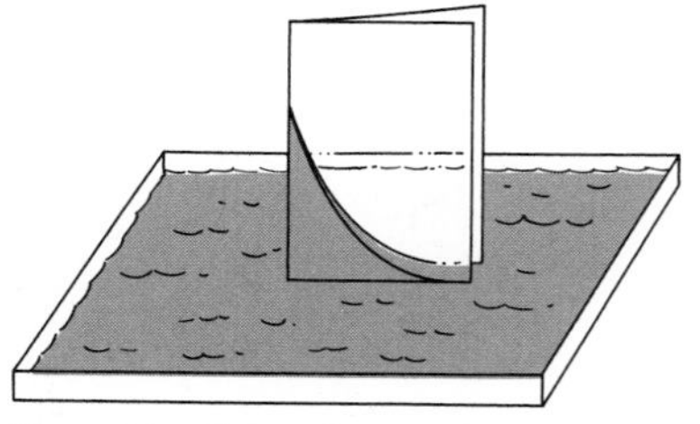

Figure 11 Colored water is drawn up by capillary action.

Hyperbolas are very useful in certain types of navigation. In order to explain why, we first do an example.

EXAMPLE 7

An explosion was heard on two ships 1 kilometer apart. Sailors on Ship B heard the explosion $1\frac{1}{2}$ seconds before those on Ship A. Relative to the two ships, where did the explosion occur?

SOLUTION The speed of sound in air (at 20°C) is approximately 340 meters/sec. In $1\frac{1}{2}$ seconds the sound traveled $1\frac{1}{2} \times 340 = 510$ meters. Therefore, the explosion took place at a point 510 meters closer to Ship B than to Ship A. In Figure 12, we draw a coordinate system and place A and B on the x-axis, equidistant from the origin. Since $\overline{AB} = 1$ km $= 1000$ m, the coordinates of A and B are $(-500, 0)$ and $(500, 0)$. The explosion took place at a

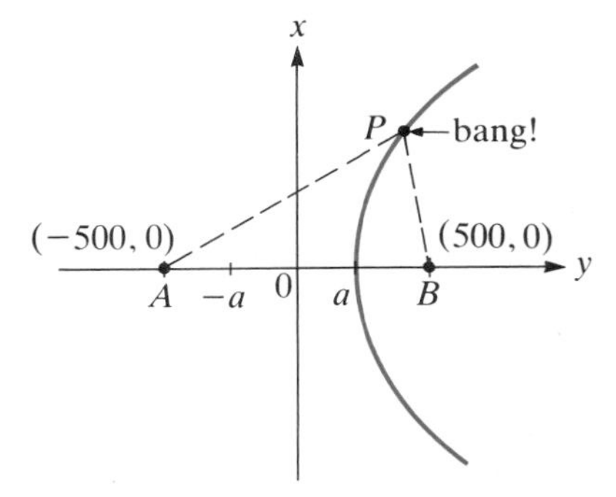

Figure 12 One branch of the hyperbola $\frac{x^2}{65{,}025} - \frac{y^2}{184{,}975} = 1$.

point P such that $\overline{PA} - \overline{PB} = 510$. Thus, P is on one branch of the hyperbola with foci $(-500, 0)$ and $(500, 0)$ such that the difference of the distances to the foci is $2a = 510$. Thus, $a = \frac{510}{2} = 255$, $c = 500$, $b^2 = c^2 - a^2 = 250{,}000 - 65{,}025 = 184{,}975$. The equation of this hyperbola is

$$\frac{x^2}{65{,}025} - \frac{y^2}{184{,}975} = 1$$

and the point P is on the branch of the hyperbola containing points that are closer to B than to A.

Of course, we have not located the point P precisely. However, if we have a third ship, Ship C, that hears the explosion, then we can obtain two more hyperbolas (one for Ships A and C and one for Ships B and C), and the point of the explosion is the single point at which the three hyperbolas intersect.

Because of the technique illustrated in the last example, the hyperbola is very useful in navigation — particularly in the LORAN (LOng RAnge Navigation) system. During World War II, LORAN served as a navigational aid for the strategic night bombing of Germany and for the long-range bombing of Japan from islands in the South Pacific. LORAN was used to draw highly accurate navigational maps, and one of these may have been used by the crew of the Enola Gay when it dropped the first atomic bomb over Hiroshima.†

The LORAN map in Figure 13 shows two sets of confocal hyperbolas (that is, hyperbolas with the same foci) with the foci

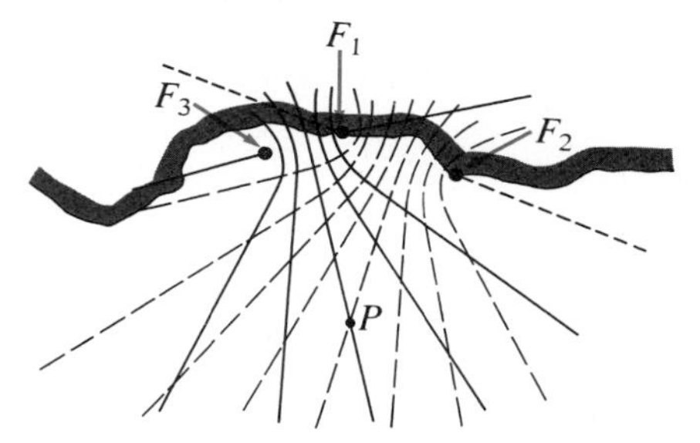

Figure 13 A LORAN map with hyperbolas of constant time difference. Two hyperbolas are needed to get a cross-fix at point P.

located at three radio broadcasting stations. As in Example 7, LORAN is based on the time difference between the reception of signals sent simultaneously from the stations in each broadcasting pair. A ship records the time difference as the signals from one pair (F_1 and F_2, say) arrive and determines its own position on one branch of a hyperbola. It uses the signals from the other pair (F_1 and F_3) to determine its position on one branch of a second hyperbola. The point of intersection of the two hyperbolas is the location of the ship.

LORAN has several advantages over other navigational systems. Radio waves are not affected by clouds or fog that do hamper celestial and visual navigation. Sun-spot activity and atmospheric storms can bend radio signals so that direction finders may be inaccurate. However, these storms do not seriously affect the *velocity* of the radio signals, so measurements based on time differences remain accurate.

Like the ellipse and parabola, the hyperbola has a useful **reflection property.** A light ray approaching (or leaving) one focus will be reflected toward (or away from) the other focus. This is illustrated in Figure 14, in which the direction of the arrows may be reversed.

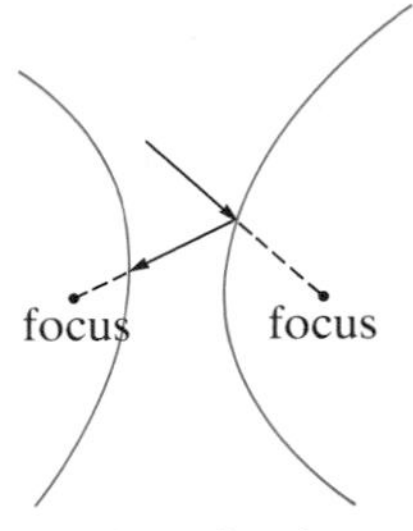

Figure 14 Illustration of the reflection property of the hyperbola.

†LORAN is now available in small, two-seat, and four-seat aircraft.

The reflection property is exploited in the design of telescopes. In 1672, the French sculptor and astronomer Guillaume Cassegrain designed a reflecting telescope with a large parabolic mirror and a smaller hyperbolic mirror both sharing a common focus. This is illustrated in Figure 15. The same principle is used in the 200-inch Hale telescope on Mount Palomar in California.

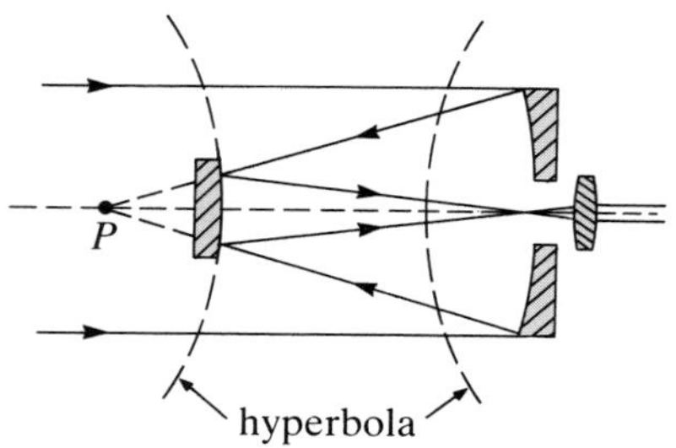

Figure 15 A Cassegrain telescope.

Hyperbolas appear in three-dimensional guises as well. In Figure 16, we provide a computer-drawn sketch of a **hyperboloid of one sheet.** This is a solid with the following properties: cross-sections (slices) parallel to the *xy*-plane are ellipses, while cross-sections perpendicular to the *xz*-plane are hyperbolas. The hyperboloid of one sheet is the design standard for all nuclear cooling towers.

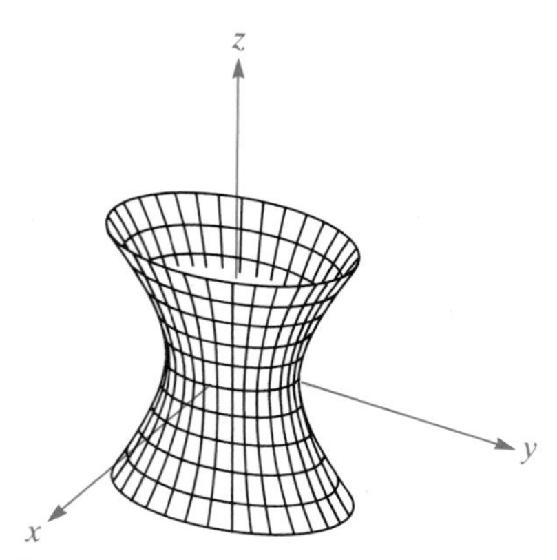

Figure 16 A hyperboloid of one sheet.

For more details of these and other interesting applications of the hyperbola, see Lee Whitt's delightful article, "The Standup Conic Presents: The Hyperbola and Applications," in *The UMAP Journal,* Vol. 5, No. 1, 1984, pp. 9–21.

Problems 5.5

Readiness Check

I. The graph of $\frac{y^2}{9} - \frac{x^2}{16} = 1$ is a hyperbola that opens ______.

a. to the left
b. to the left and right
c. upward
d. upward and downward

II. The transverse axis of the hyperbola satisfying $x^2 - y^2 = 1$ is ______.

a. the line segment between $(-\sqrt{2}, 0)$ and $(\sqrt{2}, 0)$
b. the line segment between $(-1, 0)$ and $(1, 0)$
c. the line $x = 0$
d. the line $y = 0$

III. The vertices of the hyperbola satisfying $\frac{x^2}{16} - \frac{y^2}{9} = 1$ are ______.

a. $(0, 0)$
b. $y = \frac{3x}{4}$ and $y = -\frac{3x}{4}$
c. $(-5, 0)$ and $(5, 0)$
d. $(-4, 0)$ and $(4, 0)$

IV. The asymptotes of the hyperbola satisfying $\frac{x^2}{16} - \frac{y^2}{9} = 1$ are ______.

a. $y = \frac{3x}{4}$ and $y = -\frac{3x}{4}$
b. $y = \frac{4x}{3}$ and $y = -\frac{4x}{3}$
c. $y = x$ and $y = -x$
d. the x-axis and the y-axis

V. Which of the following are the vertices of $9y^2 - x^2 + 36 = 0$?

a. $(0, \pm 2)$
b. $(0, \pm 3)$
c. $(\pm 6, 0)$
d. $(\pm 1, 0)$

Answers to Readiness Check

I. d II. b III. d IV. a V. c

In Problems 1–20 the equation of a hyperbola is given. Find its foci, transverse axis, conjugate axis, center, vertices, and asymptotes. Then sketch it.

1. $\frac{x^2}{16} - \frac{y^2}{25} = 1$
2. $\frac{y^2}{16} - \frac{x^2}{25} = 1$
3. $\frac{y^2}{25} - \frac{x^2}{16} = 1$
4. $\frac{x^2}{25} - \frac{y^2}{16} = 1$
5. $y^2 - x^2 = 1$
6. $x^2 - y^2 = 1$
7. $x^2 - 4y^2 = 9$
8. $4y^2 - x^2 = 9$
9. $y^2 - 4x^2 = 9$
10. $4x^2 - y^2 = 9$
11. $2x^2 - 3y^2 = 4$
12. $3x^2 - 2y^2 = 4$
13. $2y^2 - 3x^2 = 4$
14. $3y^2 - 2x^2 = 4$
15. $(x - 1)^2 - 4(y + 2)^2 = 4$
16. $\frac{(y + 3)^2}{4} - \frac{(x - 2)^2}{9} = 1$
17. $4x^2 + 8x - y^2 - 6y = 21$
18. $-4x^2 - 8x + y^2 - 6y = 20$
19. $2x^2 - 16x - 3y^2 + 12y = 45$
20. $2y^2 - 16y - 3x^2 + 12x = 45$

In Problems 21–30 find the standard equation of the indicated hyperbola.

21. foci: $(-4, 0)$, $(4, 0)$
vertices: $(-3, 0)$, $(3, 0)$
22. foci: $(-1, 0)$, $(1, 0)$
vertices: $(-\frac{1}{2}, 0)$, $(\frac{1}{2}, 0)$
23. foci: $(0, -3)$, $(0, 3)$
vertices: $(0, -2)$, $(0, 2)$
24. foci: $(0, -6)$, $(0, 6)$
vertices: $(0, -4)$, $(0, 4)$
25. foci: $(-1, 1)$, $(5, 1)$
vertices: $(0, 1)$, $(4, 1)$
26. foci: $(-3, -1)$, $(-3, 5)$
vertices: $(-3, 1)$, $(-3, 3)$
27. vertices: $(-2, 0)$, $(2, 0)$
asymptotes: $y = \pm x$
28. vertices: $(0, -3)$, $(0, 3)$
asymptotes: $y = \pm 2x$
29. vertices: $(1, 1)$, $(5, 1)$
asymptotes: $y = 2x - 5$, $y = -2x + 7$
30. vertices: $(-2, -1)$, $(-2, 9)$
asymptotes: $y = \frac{5}{2}x + 9$, $y = -\frac{5}{2}x - 1$
31. Find the standard equation of the hyperbola obtained by translating the hyperbola of Problem 21 so that its center is at $(4, -3)$.
32. Find the standard equation of the hyperbola obtained by translating the hyperbola of Problem 24 so that its center is at $(-3, 2)$.
33. The **eccentricity** e of a hyperbola is defined by

$$e = \frac{\text{distance between foci}}{\text{distance between vertices}}$$

Show that the eccentricity of any hyperbola is greater than 1.
34. For the hyperbola $x^2/a^2 - y^2/b^2 = 1$, show that $e = \sqrt{a^2 + b^2}/a = \sqrt{1 + (b/a)^2}$.

In Problems 35–44 find the eccentricity of the given hyperbola.

35. the hyperbola of Problem 1
36. the hyperbola of Problem 4
37. the hyperbola of Problem 5
38. the hyperbola of Problem 8
39. the hyperbola of Problem 9
40. the hyperbola of Problem 12
41. the hyperbola of Problem 13
42. the hyperbola of Problem 16
43. the hyperbola of Problem 19
44. the hyperbola of Problem 20

* 45. Find the equation of the hyperbola with center at $(0, 0)$ and axis parallel to the x-axis that passes through the points $(1, 2)$ and $(5, 12)$.

46. Find the equation of the curve having the property that the difference of the distances from any point on the curve to the points $(1, -2)$ and $(4, 3)$ is 5.

47. Show that the graph of the curve $x^2 + 4x - 3y^2 + 6y = c$ is (a) a hyperbola if $c \neq -1$, (b) a pair of straight lines if $c = -1$. (c) If $c = -1$, find the equations of the lines.

* 48. Find conditions relating the numbers a, b, and c such that the graph of the equation $2x^2 + ax - 3y^2 + by = c$ is (a) a hyperbola, (b) a pair of straight lines.

* 49. The speed of sound in sea water (at 25°C) is 1533 meters/second ($\approx$ 5030 ft/sec). Submarine A heard the sound of an exploding depth charge 2 seconds before Submarine B heard the sound. The submarines are 4 kilometers apart, and the depth charge was dropped by an enemy destroyer. Find an equation for the hyperbola that contains the point at which the destroyer dropped the charge. [Hint: Draw a coordinate system and put A and B on the x-axis, equidistant from the origin.]

** 50. In the coordinate system of Problem 49, assume that the positive x-axis points east. Submarine C, located 2 km due north of Submarine A, heard the sound of the depth charge 1 second after Submarine A heard it. Find the exact location of the destroyer at the moment it dropped the charge.

Graphing Calculator Problems

In Problems 51–60 obtain the graph of the given hyperbola on a graphing calculator. Before you begin, read Example 10 in Appendix A.

51. $\frac{x^2}{10} - \frac{y^2}{47} = 1$
52. $\frac{y^2}{43} - \frac{x^2}{19} = 137$
53. $3x^2 - 4y^2 = 7$
54. $12y^2 - 11x^2 = 34$
55. $-43x^2 + 19y^2 = 73$
56. $4(y + 2)^2 - 3(x - 7)^2 = 8$
57. $5(x + 1.9)^2 - 16(y + 1.1)^2 = 42$
58. $3x^2 - 4x - y^2 + 7y = 10$
59. $6x - 11x^2 + 4y^2 + 6y = 12$
60. $3.7x^2 - 4.9x + 12.9y - 6.3y^2 = 38.5$

Summary Outline of Chapter 5

- A **rational function** is a function of the form $f(x) = \dfrac{p(x)}{q(x)}$, where $p(x)$ and $q(x)$ are polynomials. p. 278
- The vertical line $x = c$ is a **vertical asymptote** of the graph of $y = f(x)$ if $f(x) \to \infty$ or $f(x) \to -\infty$ as $x \to c^+$ or $x \to c^-$. It is a vertical asymptote of $f(x) = \dfrac{p(x)}{q(x)}$ if $q(c) = 0$ but $p(c) \neq 0$. p. 280
- The horizontal line $y = c$ is a **horizontal asymptote** of the graph of $y = f(x)$ if $f(x) \to c$ as $x \to \infty$ or $x \to -\infty$. p. 280
- **Horizontal Asymptotes Theorem:** p. 282

  If $f(x) = \dfrac{p(x)}{q(x)} = \dfrac{a_nx^n + a_{n-1}x^{n-1} + \cdots + a_1x + a_0}{b_mx^m + b_{m-1}x^{m-1} + \cdots + b_1x + b_0}$, $a_n, b_m \neq 0$, then

  (i) if $n < m$, $y = 0$ is a horizontal asymptote for f.

  (ii) if $n = m$, then $y = \dfrac{a_n}{b_m}$ is a horizontal asymptote for f.

  (iii) if $n > m$, then f has no horizontal asymptote.
- An **ellipse** is the set of points (x, y) such that the sum of the distances from (x, y) to two given points is fixed. Each of the two points is called a **focus** of the ellipse. p. 292
- The **standard equation of the ellipse** is $\dfrac{x^2}{a^2} + \dfrac{y^2}{b^2} = 1$. p. 294
- If $a > b$, the line segment joining $(-a, 0)$ to $(a, 0)$ is the **major axis,** the line segment joining $(0, -b)$ to $(0, b)$ is the **minor axis,** and the points $(-a, 0)$ and $(a, 0)$ are **vertices** of the ellipse. If $b > a$, the major and minor axes are reversed. If $a = b$, the ellipse is a **circle.** The intersection of the axes is the **center** of the ellipse. pp. 294, 295
- The **eccentricity,** e, of an ellipse is given by $e = \dfrac{c}{a}$ if $a \geq b$ and $e = \dfrac{c}{b}$ if $b \geq a$, where $2c$ is the distance between the foci. p. 296
- A **translated ellipse** has the standard equation

$$\frac{(x - x_0)^2}{a^2} + \frac{(y - y_0)^2}{b^2} = 1$$

  p. 297
- A **parabola** is the set of points (x, y) equidistant from a fixed point called the **focus** and a fixed line (that does not contain the focus) called the **directrix.** p. 303
- The **standard equations** of a parabola are $x^2 = 4cy$ (which opens upward if $c > 0$ and downward if $c < 0$) and $y^2 = 4cx$ (which opens to the right if $c > 0$ and to the left if $c < 0$). pp. 304, 305
- The line about which a parabola is symmetric is the **axis** of the parabola. p. 304
- The point at which the axis and parabola intersect is the **vertex** of the parabola. p. 304
- A **translated parabola** takes the standard form $(x - x_0)^2 = 4c(y - y_0)$ or $(y - y_0)^2 = 4c(x - x_0)$. p. 306
- A **hyperbola** is a set of points (x, y) with the property that the positive difference between the distances from (x, y) and each of the two distinct points, called **foci,** is a constant. p. 311

- The **principal axis** of a hyperbola is the line containing the foci. The points of intersection of the principal axis and the hyperbola are the **vertices,** and the line segment joining the vertices is the **transverse axis.** The midpoint of the line segment joining the foci is the **center** of the hyperbola. p. 314

- The **standard equations** of a hyperbola are

$$\frac{x^2}{a^2} - \frac{y^2}{b^2} = 1 \quad \text{and} \quad \frac{y^2}{a^2} - \frac{x^2}{b^2} = 1$$

pp. 314–316

- A **translated hyperbola** takes the standard form

$$\frac{(x-x_0)^2}{a^2} - \frac{(y-y_0)^2}{b^2} = 1 \quad \text{or} \quad \frac{(y-y_0)^2}{a^2} - \frac{(x-x_0)^2}{b^2} = 1$$

p. 316

Review Exercises for Chapter 5

In Exercises 1–4 determine the vertical and horizontal asymptotes, if any, of the given rational function.

1. $f(x) = \dfrac{x}{x^2-4}$
2. $f(x) = \dfrac{x^2-1}{x^2+1}$
3. $f(x) = \dfrac{3x^2-4}{x^2-5x-4}$
4. $f(x) = \dfrac{3x^3-2x+4}{x^2-x}$

In Exercises 5–12 sketch the given rational function.

5. $f(x) = \dfrac{2}{x+3}$
6. $f(x) = -\dfrac{1}{(x+1)^2}$
7. $f(x) = \dfrac{1}{(x-2)^3} + 2$
8. $f(x) = \dfrac{x}{x-3}$
9. $f(x) = \dfrac{x-3}{x}$
10. $f(x) = \dfrac{1}{x^2-5x+6}$
11. $f(x) = \dfrac{x-2}{x^2-2x-8}$
12. $f(x) = \dfrac{5}{x^2+2}$

In Exercises 13–26, identify the type of conic. If it is an ellipse (or circle), give its foci, center, vertices, major and minor axes, and eccentricity. If it is a parabola, give its focus, directrix, axis, and vertex. If it is a hyperbola, give its foci, conjugate axis, transverse axis, center, vertices, asymptotes, and eccentricity. Finally, sketch the curve (if it is not degenerate).

13. $\dfrac{x^2}{9} + \dfrac{y^2}{16} = 1$
14. $\dfrac{x^2}{9} - \dfrac{y^2}{16} = 1$
15. $\dfrac{x^2}{9} - \dfrac{y}{16} = 0$
16. $\dfrac{y^2}{16} - \dfrac{x}{9} = 0$
17. $\dfrac{y^2}{9} - \dfrac{x^2}{16} = 1$
18. $\dfrac{x^2}{16} + \dfrac{y^2}{9} = 1$
19. $\dfrac{(x-1)^2}{4} + \dfrac{(y+1)^2}{9} = 1$
20. $\dfrac{(x+2)^2}{25} - \dfrac{(y+3)^2}{4} = 1$
21. $\dfrac{(x+2)^2}{25} + \dfrac{(y-5)^2}{25} = 0$
22. $x^2 + 2x + y^2 + 2y = 0$
23. $x^2 + 2x - y^2 + 2y = 0$
24. $x^2 + 2x - 2y = 0$
25. $4x^2 + 4x + 3y^2 + 24y = 5$
26. $-3x^2 + 6x + 2y^2 + 4y = 6$
27. Find the equation of an ellipse with foci at (3, 0) and (−3, 0) and eccentricity 0.6.
28. Find the equation of the parabola with focus (3, 0) and directrix the line $x = -4$.
29. Find the equation of the hyperbola with foci (0, 3) and (0, −3) and vertices (0, 2) and (0, −2).
30. Find the equation of the hyperbola centered at (0, 0) with vertices at (0, 3) and (0, −3) that is asymptotic to the lines $y = \pm 5x$.

Chapter 6

Exponential and Logarithmic Functions

6.1 Exponential Functions

In this chapter, we introduce some of the most important functions in mathematics. First, we review some facts that were discussed in Chapter 1.

Let a be a positive real number.

(a) If $x = n$, *a positive integer,* then

$$a^x = a^n = \underbrace{a \cdot a \cdot a \cdot \cdots \cdot a}_{n \text{ factors}} \quad \text{Equation (1) on p. 20}$$

(b) If $x = 0$, then

$$a^x = a^0 = 1 \quad \text{p. 21}$$

(c) If $x = -n$, *where n is a positive integer,* then

$$a^x = a^{-n} = \frac{1}{a^n} \quad \text{p. 20}$$

(d) If $x = 1/n$, *where n is a positive integer,* then

$$a^x = a^{1/n} = \text{the } n\text{th root of } a \quad \text{p. 29}$$

(e) If $x = m/n$ (*m and n are positive integers*), then

$$a^x = a^{m/n} = (a^{1/n})^m \quad \text{p. 30}$$

(f) If $x = -m/n$, $n \neq 0$, *a negative rational number,* then

$$a^x = a^{-m/n} = \frac{1}{a^{m/n}} \quad \text{Property (c) on p. 31}$$

Thus, a^x $(a > 0)$ is defined if x is a rational number. If x is not a rational number, then we have not as yet defined a^x. However, we can define an approximation to a^x by first approximating x as a decimal and then computing a to the power of this decimal. With the aid of a calculator, this is quite easily done.

EXAMPLE 1 *Approximating a^x Where x Is Irrational*

Use the procedure outlined above to approximate $4^{\sqrt{2}}$.

SOLUTION We find that $\sqrt{2} = 1.414213562$ . . . . Thus, $\sqrt{2}$ can be approximated, successively, by 1, 1.4, 1.41, 1.414, . . . , and, since each of these numbers is a rational number, we can compute 4^1, $4^{1.4}$, and so on. Some results are given in Table 1.

Table 1

r	1	1.4	1.41	1.414	1.4142	1.414213562
4^r	4	6.964404506	7.06162397	7.100890698	7.102859756	7.102993298

We can obtain this approximation on a calculator by the following key sequence:

[4] [y^x] [2] [$\sqrt{x}$] [=]

On a calculator carrying 10 digits, this results in the value 7.102993301.†

NOTE Some calculators require parentheses to obtain this number

[4] [y^x] [(] [2] [$\sqrt{x}$] [)] [=]

The procedure described above really provides us with the definition of a^x when $a > 0$ and x is irrational. We simply define a^x as the "limit" of a^r as r approximates x to more and more decimal places.

We can now define an exponential function.

Definition of an Exponential Function

Let $a \neq 1$ be a positive real number. Then the function f defined by $f(x) = a^x$ is called an **exponential function with base *a*.** Since $y = a^x$ is defined for every real number x, we see that domain of $f(x) = a^x$ is $\mathbb{R}$.

NOTE In an exponential function, the exponent is the variable. In the power function $y = x^n$, the base is the variable, and the exponent is constant.

EXAMPLE 2 *The Graph of 2^x*

Sketch the graph of the function $y = 2^x$.

---

† Some calculators that do not carry as many internal (that is, undisplayed) digits might give an answer that differs from this one in the last digit. The answer given here is correct.

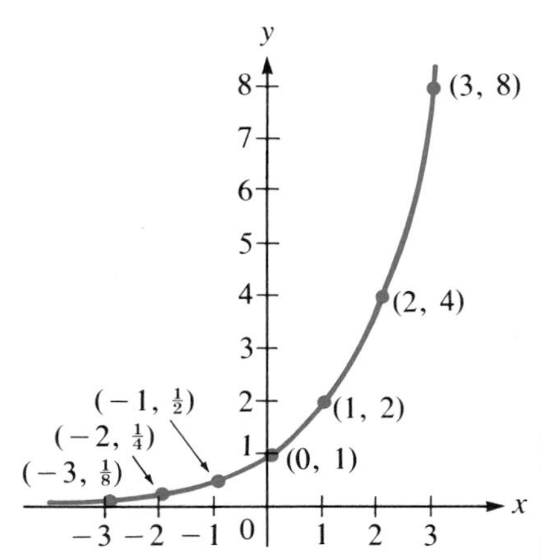

Figure 1 Graph of the exponential function $f(x) = 2^x$.

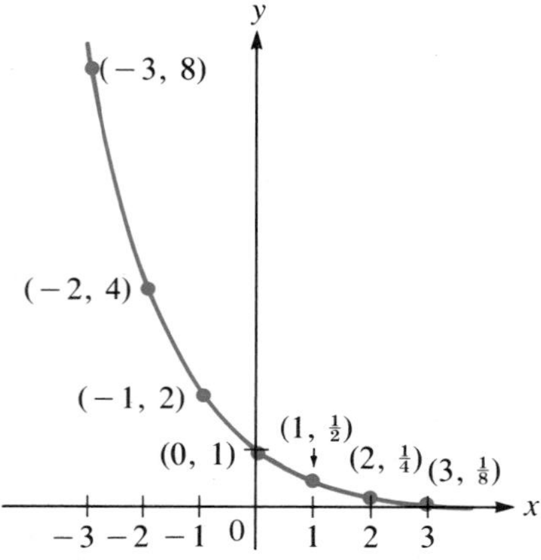

Figure 2 The graph of $f(x) = (\frac{1}{2})^x = 2^{-x}$ is the reflection about the y-axis of the graph of 2^x.

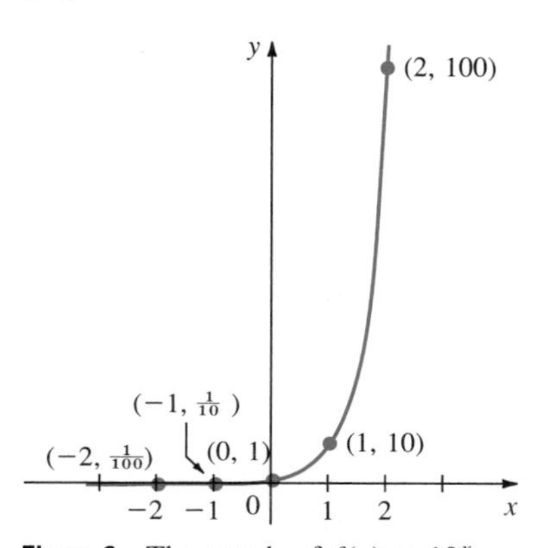

Figure 3 The graph of $f(x) = 10^x$.

SOLUTION We provide some values of 2^x in Table 2. We plot these

Table 2

x	-10	-5	-2	-1	0	$\frac{1}{2}$	1	$\frac{3}{2}$	2	3	5	10
2^x	0.001	0.03	0.25	0.5	1	1.4142	2	2.8284	4	8	32	1024

values and then draw a curve joining the points to obtain the sketch in Figure 1. ■

EXAMPLE 3 *The Graph of* $(\frac{1}{2})^x$

Sketch the graph of the function $y = (\frac{1}{2})^x$.

SOLUTION We see that $(\frac{1}{2})^x = 1/2^x = 2^{-x}$. Thus, if $f(x) = 2^x$, then $2^{-x} = f(-x)$, and the graph of $(\frac{1}{2})^x$ is the graph of 2^x reflected about the y-axis (see p. 196). The graph is given in Figure 2.

NOTE Let $f(x) = n^x$ and $g(x) = \left(\frac{1}{n}\right)^x$, where n is a positive integer. Then $f(-x) = n^{-x} = \frac{1}{n^x} = g(x)$ so, as in Figures 1 and 2, the graph of $\left(\frac{1}{n}\right)^x$ is the graph of n^x reflected about the y-axis.

EXAMPLE 4 *The Graph of* 10^x

Sketch the graph of $y = 10^x$.

SOLUTION We give some values of 10^x in Table 3 and draw the graph in Figure 3.

Table 3

x	-3	-2	-1	0	0.25	0.5	0.75	1	1.5	2	3
10^x	0.001	0.01	0.1	1	1.778	3.162	5.623	10	31.62	100	1000

■

EXAMPLE 5 *The Graph of* $(\frac{1}{10})^x$

Sketch the graph of $y = (\frac{1}{10})^x$.

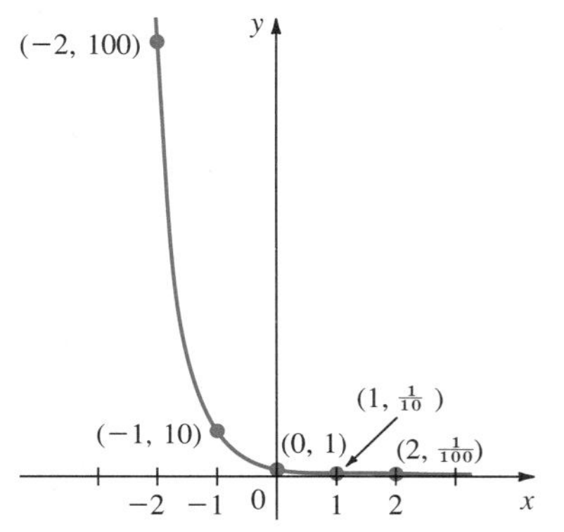

Figure 4 The graph of $f(x) = (\frac{1}{10})^x = 10^{-x}$ is the reflection about the y-axis of the graph of 10^x.

SOLUTION As in Example 3, we can obtain the graph by reflecting the graph of 10^x about the y-axis. We do this in Figure 4.

If you look at Figures 1 and 3, you may observe that the graphs of 2^x and 10^x are very similar. The only difference is that 10^x increases faster than 2^x as x increases and that 10^x decreases faster than 2^x when x is negative and x decreases. The functions $(\frac{1}{2})^x$ and $(\frac{1}{10})^x$ behave very similarly. These facts are not surprising after we observe that all exponential functions share some interesting properties. We cite some of these properties below. These all follow from the properties of rational exponents, given in Section 1.6.

Properties of Exponential Functions

Let $a > 0$ and let x and y be real numbers. Then

	Illustration
1. $a^x > 0$ and the range of $f(x) = a^x$ is $\{y: y > 0\}$.	
2. $a^{-x} = \dfrac{1}{a^x}$.	$2^{-3} = \dfrac{1}{2^3} = \dfrac{1}{8}$
3. $a^{x+y} = a^x a^y$.	$2^{3+4} = 2^3 2^4 = 8 \cdot 16 = 128$
4. $a^{x-y} = \dfrac{a^x}{a^y}$.	$2^{5-2} = \dfrac{2^5}{2^2} = \dfrac{32}{4} = 8$
5. $a^0 = 1$.	$2^0 = 1$
6. $a^1 = a$.	$2^1 = 2$
7. $a^{xy} = (a^x)^y = (a^y)^x$.	$2^{3 \cdot 2} = (2^3)^2 = 8^2 = 64$

8. If $a > 1$, a^x is an increasing function.†
9. If $0 < a < 1$, a^x is a decreasing function.
10. The graph of $y = a^x$ passes through the point (0, 1). That is, $f(0) = 1$.
11. **Exponentiation property:** If $x = y$, then $a^x = a^y$.
12. If $a > 1$, $a^x > 1$ if $x > 0$ and $0 < a^x < 1$ if $x < 0$.

Compound Interest Formulas

In Section 3.4 we discussed the simple interest formula (see p. 183)

$$I = Prt$$

where I is the interest earned, P is the principal, r is the rate of interest, and t is the time the investment is held (usually measured in years).

Compound interest is interest paid on interest previously earned as well as on the original investment. Suppose that interest is paid annually. Then if P dollars are invested, the interest after one year ($t = 1$) is rP dollars, and the amount in the account is $P + rP$ dollars. After two years the interest is paid on $P + rP$ dollars (the P dollars originally invested plus the rP dollars earned

---

†We discussed increasing and decreasing functions in Section 3.7 (p. 209).

after the first year). That is

$$\text{interest paid at end of second year} = r(P + rP) = rP(1 + r)$$

This means that

$$\begin{aligned}\begin{matrix}\text{total amount of investment}\\ \text{after 2 years}\end{matrix} &= P + \begin{matrix}\text{interest after}\\ \text{first year}\end{matrix} + \begin{matrix}\text{interest after}\\ \text{second year}\end{matrix}\\ &= P + rP + rP(1 + r)\\ &= P(1 + r) + rP(1 + r) = (P + rP)(1 + r)\\ &= P(1 + r)(1 + r) = P(1 + r)^2\end{aligned}$$

After three years the investment is worth

$$\begin{aligned}\begin{matrix}\text{value at end of}\\ \text{second year}\end{matrix} + \begin{matrix}\text{interest at end}\\ \text{of third year}\end{matrix} &= P(1 + r)^2 + rP(1 + r)^2\\ &= (P + rP)(1 + r)^2\\ &= P(1 + r)(1 + r)^2 = P(1 + r)^3\end{aligned}$$

If $A(t)$ denotes the value (amount) of our investment after t years, then we have

Compound Interest Formula

$$A(t) = P(1 + r)^t \qquad (1)$$

where P is the original principal, r is the rate of interest, t is the time the investment is held, and $A(t)$ is the total value of the investment after t years.

EXAMPLE 6 ***Computing the Value in Three Years with Annual Compounding***

What is the value of a \$2000 investment after 3 years if it is invested at 12% interest compounded annually?

SOLUTION Here $P = 2000$, $r = 0.12$, and $t = 3$ in formula (1). Thus

$$A(3) = 2000(1 + 0.12)^3 = 2000(1.12)^3$$

Calculator
↓

$$= 2000(1.404928) \approx 2809.86$$

Thus the investment is worth \$2809.86 after 3 years.

In practice, interest is compounded more frequently than annually. If it is paid m times a year, then in each interest period the rate of interest is r/m and in t years there are tm pay periods. Then, similar to formula (1), we have the

Compound Interest Formula:
Compounding m Times a Year

$$A(t) = P\left(1 + \frac{r}{m}\right)^{mt} \tag{2}$$

where

P is the original principal

r is the annual interest rate

t is the number of years the investment is held

m is the number of times interest is compounded each year

$A(t)$ is the amount (in dollars) after t years.

EXAMPLE 7 *Computing the Value in Five Years with Annual, Quarterly, Monthly, and Daily Compounding*

\$2000 is invested for 5 years at 6% interest. How much is the investment worth if interest is compounded (a) annually? (b) quarterly? (c) monthly? (d) daily?

SOLUTION

(a) Interest is compounded just once a year. Thus we can use either formula (1) or formula (2) with $m = 1$. Here $P = 2000$, $r = 0.06$, and $t = 5$. From (2), we have

$$\$A(5) = \$2000(1 + 0.06)^5 = \$2000(1.06)^5 = \$2000(1.338225578) = \$2676.45$$

Note that the total interest earned over the 5-year period is \$676.45.

(b) Interest is compounded 4 times a year. Thus $m = 4$ in (2), and we have

$$\$A(5) = \$2000\left(1 + \frac{0.06}{4}\right)^{(4)(5)}$$

(m indicated by arrows at the 4 and the exponent)

$$= \$2000(1.015)^{20} = \$2000(1.346855007) = \$2693.71$$

The total interest is now \$693.71.

(c) Here $m = 12$ (12 months in a year), so

$$\$A(5) = \$2000\left(1 + \frac{0.06}{12}\right)^{(12)(5)}$$

(m indicated by arrows at the 12 and the exponent)

$$= \$2000(1.005)^{60} = \$2000(1.348850153) = \$2697.70$$

The total interest is \$697.70.

(d) Here $m = 365$, so

$$\$A(5) = \$2000\left(1 + \frac{0.06}{365}\right)^{(365)(5)} = \$2000(1.000164384)^{1825}$$

(m indicated at the 365 and at the 365 in the exponent)

$$= \$2000(1.349825523) = \$2699.65$$

In this case, the total interest paid is \$699.65.

As the preceding example indicates, the more frequently interest is compounded, the more the investment increases in value. In Table 4, we show the value after 10 years and the interest earned on a \$1000 investment at 8% annual interest for different numbers of payment periods each year.

Table 4
Value of a \$1000 Investment Compounded m Times a Year for 10 Years at an Annual Rate of 8%

m **(number of times interest is compounded each year)**	**Value of \$1000 After 10 Years at 8% Interest (\$)**	**Total Interest Earned (\$)**
1 (annually)	2158.92	1158.92
2 (semiannually)	2191.12	1191.12
4 (quarterly)	2208.04	1208.04
12 (monthly)	2219.64	1219.64
52 (weekly)	2224.17	1224.17
365 (daily)	2225.35	1225.35
8,760 (hourly)	2225.53	1225.53
525,600 (each minute)	2225.54	1225.54

Table 4 is revealing. It suggests that though there is a considerable difference when we change from annual to semiannual compounding (a difference in this example of \$32.20), the difference becomes negligible as we increase the number of interest periods. For example, the difference between monthly compounding and hourly compounding is only \$5.89. The numbers in Table 4 suggest that, after a point, little is gained by increasing the number of interest periods per year.

Many bank advertisements contain statements like "our 8% savings plan carries an effective interest rate or yield of $8\frac{1}{3}$%." The **effective interest rate,** or **yield,** is the rate of simple interest received over a one-year period. For example, \$100 would be worth \$108 if that sum is invested for one year at 8%

interest compounded annually. But if it is compounded quarterly, for instance, then it is worth

$$\$100(1.02)^4 = \$108.24$$

after one year. Thus the interest paid is \$8.24, and the effective interest rate is $8.24\% \approx 8\frac{1}{4}\%$.

We see that for most problems there are two rates of interest: the *quoted* rate and the *effective* rate. The first of these is often called the **nominal** rate of interest. Thus, as we have seen, a nominal rate of 8% provides an effective rate of 8.24% when interest is compounded quarterly.

EXAMPLE 8 *Determining the Effective Rate of Interest*

If money is invested at a nominal rate of 15% compounded monthly, what is the effective rate of interest?

SOLUTION Starting with P dollars, there will be $P(1 + 0.15/12)^{12} \approx 1.161P$ dollars after 1 year. The increase is $0.161P = 16.1\%$ of P. Thus, P dollars will have grown by approximately 16.1% after 1 year. This is the effective rate of interest.

Table 5 gives the effective interest rates if a sum is invested at a nominal rate of 8% compounded m times a year.

Table 5

m (number of times interest is paid per year)	Effective Interest Rate (based on 8%) (%)
1	8.000
2	8.160
4	8.24322
8	8.28567
12	8.29995
24	8.31430
52	8.32205
365	8.32776
1,000	8.32836
10,000	8.32867
1,000,000	8.32871

Problems 6.1

Readiness Check

I. Which of the following is an exponential function?
   a. $y = x^{\sqrt{3}}$
   b. $y = \sqrt{2 + x^4}$
   c. $y = \sqrt{2}^x$
   d. $y = x^{-3}$

II. Which of the following is true about the graph of $y = 2^{-x}$?
   a. It is an increasing function.
   b. It has an x-intercept but no y-intercept.
   c. It has the same shape but does not decrease as fast as $y = (\frac{1}{10})^x$.
   d. As $x \to \infty$, $y \to -\infty$.

III. Which of the following functions would result if the graph of $f(x) = 2^x$ was shifted to the right 2 units and down 1 unit?
   a. $f(x) = 2^{x-1} + 2$
   b. $f(x) = 2^{x-2} - 1$
   c. $f(x) = 2^{x+1}$
   d. $f(x) = 2^{x-3}$

IV. Which of the following is true about the graph below if $a > 0$, $a \neq 1$, and $y = a^x$?
   a. It passes through (1, 0).
   b. Its range is all real numbers.
   c. $a > 1$
   d. $0 < a < 1$

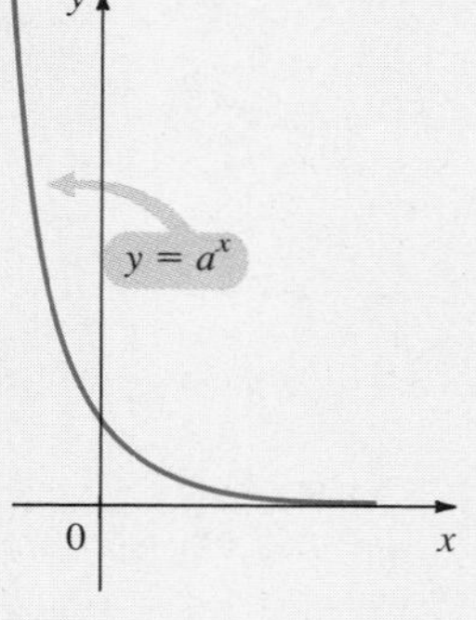

V. Which of the following is graphed below?
   a. $f(x) = 1 + 2^{x-2}$
   b. $f(x) = 1 - 2^{x+2}$
   c. $f(x) = 2^{x+2} + 1$
   d. $f(x) = 2^{x-2} - 1$

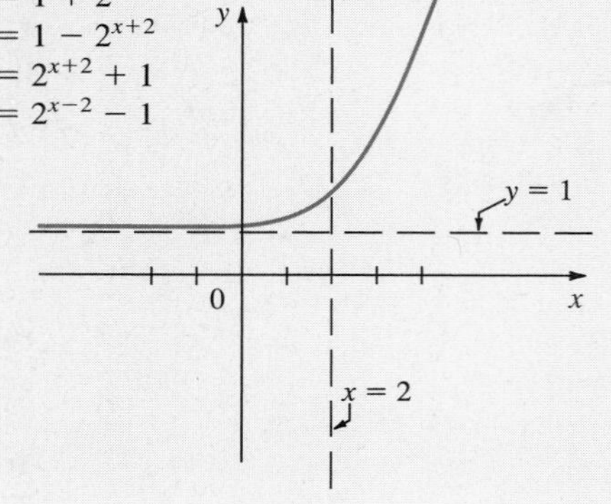

In Problems 1–14 draw a sketch of the given exponential function.

1. $y = 3^x$
2. $y = (\frac{1}{3})^x$
3. $y = (\frac{1}{5})^x$
4. $y = 5^x$
5. $f(x) = (7.2)^x$
6. $f(x) = (0.623)^x$
7. $f(x) = 3 \cdot 2^x$
8. $f(x) = 4 \cdot 10^x$
9. $y = -2 \cdot 10^x$
10. $y = 10 \cdot 2^x$
11. $y = 2^{x-1}$ [Hint: Shift a graph as in Section 3.5.]
12. $y = 3^{x-2}$
13. $y = 3 \cdot 10^{x+1} + 5$
14. $y = 4 \cdot 2^{1-x} - 1$

In Problems 15–22 use a calculator to estimate the given number to as many decimal places of accuracy as are carried on the machine.

15. $10^{2.2}$
16. $(3.8)^{4.7}$
17. $4^{-1.6}$
18. $(\frac{1}{2})^{5.1}$
19. $(0.35)^{0.42}$
20. $(53.21)^{-0.152}$
21. $3^{\sqrt{2}}$
22. $3^{\sqrt{2}}$

Answers to Readiness Check

I. c II. c III. b IV. d V. a

The remaining problems all require the use of a calculator.

In Problems 23–32 compute the value of an investment after t years and the total interest paid if P dollars is invested at a nominal interest rate of $r\%$ compounded m times a year.

23. $P = \$5000,\ r = 6\%,\ t = 4,\ m = 1$
24. $P = \$5000,\ r = 6\%,\ t = 4,\ m = 4$
25. $P = \$5000,\ r = 6\%,\ t = 4,\ m = 12$
26. $P = \$8000,\ r = 11\%,\ t = 4,\ m = 1$
27. $P = \$8000,\ r = 11\%,\ t = 4,\ m = 4$
28. $P = \$8000,\ r = 11\%,\ t = 4,\ m = 12$
29. $P = \$8000,\ r = 11\%,\ t = 4,\ m = 100$
30. $P = \$10{,}000,\ r = 8\frac{1}{2}\%,\ t = 6,\ m = 1$
31. $P = \$10{,}000,\ r = 8\frac{1}{2}\%,\ t = 6,\ m = 4$
32. $P = \$10{,}000,\ r = 8\frac{1}{2}\%,\ t = 6,\ m = 12$
33. Calculate the percentage difference in return on investment if P dollars is invested for 10 years at 6% compounded annually and quarterly.
34. As a gimmick to lure depositors, a bank offers 5% interest compounded daily in comparison with its competitor, who offers $5\frac{1}{8}\%$ compounded annually. Which bank would you choose?
35. Suppose the competitor in Problem 34 now compounds $5\frac{1}{8}\%$ semiannually. Which bank would you choose?
36. If \$20,000 is invested in bonds yielding 8% compounded quarterly, what will the bonds be worth in 9 years?
37. ** A certain government bond sells for \$750 and can be redeemed for \$1000 in 8 years. Assuming quarterly compounding, what is the nominal rate of interest paid?
38. A Roman deposited 1¢ in a bank at the beginning of the year A.D. 1. If the bank paid a meager 2% interest compounded quarterly, what would the investment be worth at the beginning of 1992?
39. Mrs. Jones has just invested \$400 in a 5-year term deposit (account A) paying 12% per year compounded twice per year, and she has invested another \$400 in a 5-year deposit (account B) at 11% per year compounded monthly.
    (a) Calculate the effective interest rate (as a percentage) for each account. Give answers correct to two decimal places.
    (b) Which investment is worth more after 5 years? by how much (to the nearest cent)?
40. (a) On November 1, 1975, Mr. Smith invested \$10,000 in a 10-year certificate that paid 11% interest per year compounded quarterly. When this matured on November 1, 1985, he reinvested the entire accumulated amount in Canada Savings Bonds with an interest rate of 7% compounded annually. To the nearest dollar, what was Mr. Smith's accumulated amount on November 1, 1990?
    (b) If Mr. Smith had made a single investment of \$10,000 in 1975 that matured in 1990 and had an effective rate of interest of 9%, would his accumulated amount be more or less than that in part (a)? by how much (to the nearest dollar)?

Four exponential functions are graphed below. Each is the graph of one of the functions given in Problems 41–44. Match each function with its corresponding graph.

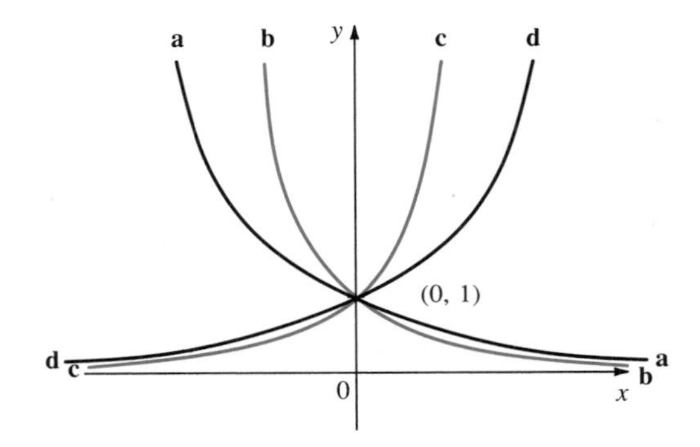

41. $f(x) = 3^x$
42. $f(x) = 5^x$
43. $f(x) = (\frac{1}{3})^x$
44. $f(x) = (\frac{1}{5})^x$

Graphing Calculator Problems

Before beginning, read Appendix A.

45. Use your calculator to sketch 2^x, 3^x, and 10^x on the same screen.
46. Sketch 2^{-x}, 3^{-x}, and 10^{-x} on the same screen.
47. Sketch 5^x and $(\frac{1}{5})^x$ on the same screen.
48. Sketch 2^x, 2^{x-1}, and $2^x - 1$ on the same screen.

In Problems 49–62 sketch each exponential function.

49. 3.7^x
50. 3.7^{x-2}
51. 3.7^{x+3}
52. $3.7^x + 5$
53. -3.7^x
54. 3.7^{-x}
55. 3.7^{2-x}
56. $4 - 3.7^{1-x}$
57. $4.2^{x/2}$
58. $3^{x/5}$
59. $-1.5^{1.5x}$
60. 2^{-2x}
61. $10^{x/100}$
62. $3^{-0.002x}$

6.2 The Natural Exponential Function

In this section, we introduce one of the most important functions in mathematics: the function e^x. To motivate the definition of the number e, we look again at the compound interest formula (equation (2) in Section 6.1)

$$A(t) = P\left(1 + \frac{r}{m}\right)^{mt} \tag{1}$$

In (1), we set $P = 1$, $r = 1$ (corresponding to 100% interest), and $t = 1$. Then (1) becomes

$$A(1) = \left(1 + \frac{1}{m}\right)^{m} \tag{2}$$

which is the value after one year of \$1 invested at 100% interest compounded m times a year. For example, if $m = 1$, then interest is paid once a year (annually) and

$$A(1) = \left(1 + \frac{1}{1}\right)^{1} = 2^1 = \$2$$

This is no surprise since the investor has the original \$1 plus \$1 interest (100% of \$1) paid at the end of the year. If $m = 2$, then

$$A(1) = \left(1 + \frac{1}{2}\right)^{2} = \left(\frac{3}{2}\right)^{2} = \frac{9}{4} = \$2.25$$

Again, this is reasonable. Now interest of 50% is paid twice a year. After six months the investor has $\$1 + \frac{1}{2}(\$1) = \$1.50$, and after one year she has \$1.50 plus 50% of \$1.50 or $\$1.50 + \$0.75 = \$2.25$.

In Table 1, we compute values of $\left(1 + \frac{1}{m}\right)^{m}$ for a number of values for m.

Table 1

m	$1/m$	$(1 + 1/m)^m$
1	1	2
2	0.5	2.25
5	0.2	2.48832
10	0.1	2.59374246
100	0.01	2.704813829
1,000	0.001	2.716923932
10,000	0.0001	2.718145927
100,000	0.00001	2.718268237
1,000,000	0.000001	2.718280469
1,000,000,000	0.000000001	2.718281827

It seems that the expression $(1 + 1/m)^m$ gets closer and closer to a fixed number. This number is denoted by e.

The Number e

Definition of e

The number $\boldsymbol{e}$ is defined to be the number approached by the expression $(1 + 1/m)^m$ as m increases without bound. To 10 decimal places

$$e \approx 2.7182818285$$

The number e was first discovered by the great Swiss mathematician Leonhard Euler (1707–1783), who described the number in 1728.†

The Function e^x

Once we know the number e, we can write the function $y = e^x$. This function arises in an astonishingly wide variety of applications. We will see how later in this section.

Values of e^x can be found on a scientific calculator in one of two ways:

To Find e^x on a Calculator

(a) If there is an [e^x] key, use it directly.
(b) If not, then there is a [ln] or [ln x] key. Press [INV] [ln x] or [2nd F] [ln x] to obtain e^x.

NOTE Sometimes the [e^x] key is called [exp x].

As we will see in Section 6.4, e^x is the inverse of a function called the natural logarithm function $\ln x$, so [INV] [ln x] gives us e^x.

EXAMPLE 1 *Computing e^x for Three Values of x*

Compute (a) e^2, (b) $e^{0.46}$, and (c) $e^{-3.14}$ on a calculator.

SOLUTION

(a) By pressing [2] [INV] [ln x], we obtain 7.389056099. We achieve the same result on a calculator with an [e^x] key by pressing [2] [e^x].
(b) [.] [4] [6] [INV] [ln x] yields 1.584073985.
(c) [3] [.] [1] [4] [+/−] displays −3.14, and then pressing [INV] [ln x] yields 0.043282797.

Alternatively, we could first compute $e^{3.14}$ and then use the reciprocal key [1/x] to compute $e^{-3.14} = 1/e^{3.14}$.

NOTE On some calculators [INV] [ln x] or [2nd F] [ln x] must be pressed *before* entering the number 2, 0.46 or −3.14.

† See the biographical sketch of Euler on p. 341.

In Table 1 at the back of the book, we provide values for e^x and e^{-x} $(= 1/e^x)$ for a wide range of numbers. However, each of these values can be obtained with more accuracy on a calculator.

The graph of $y = e^x$ is given in Figure 1(a). Since $2 < e < 3$, the graph of e^x is sandwiched between the graphs of 2^x and 3^x. We have

$$2^x < e^x < 3^x \qquad \text{if } x > 0$$
$$3^x < e^x < 2^x \qquad \text{if } x < 0$$

This is indicated in Figure 1(b).

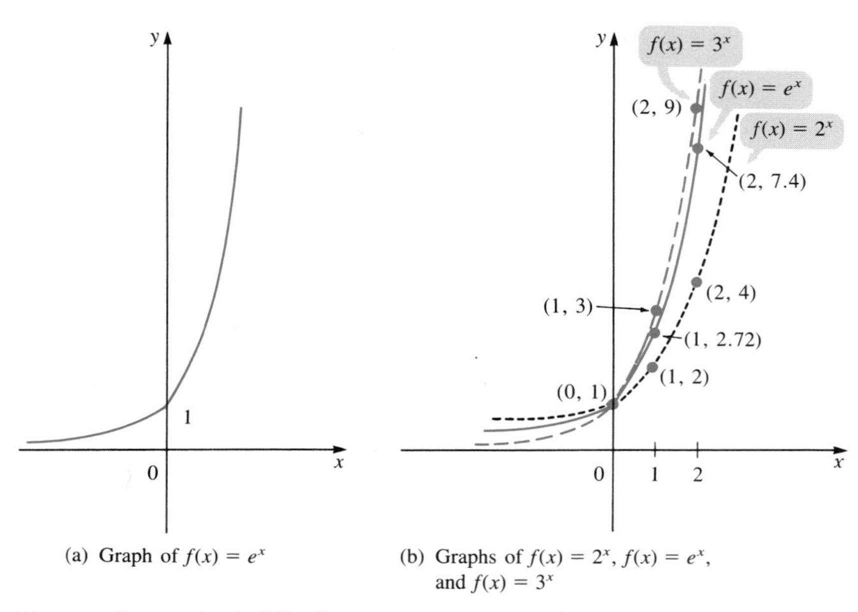

(a) Graph of $f(x) = e^x$

(b) Graphs of $f(x) = 2^x$, $f(x) = e^x$, and $f(x) = 3^x$

Figure 1 The graph of e^x lies between the graphs of 2^x and 3^x.

EXAMPLE 2 *Graphing a Shifted Exponential Function*

Sketch the graph of $f(x) = e^{x-1} - 2$.

SOLUTION We do this in two steps.

Step 1 The graph of e^{x-1} is the graph of e^x shifted 1 unit to the right (Figure 2(a)).

Step 2 The graph of $e^{x-1} - 2$ is the graph of e^{x-1} shifted 2 units down (Figure 2(b)).

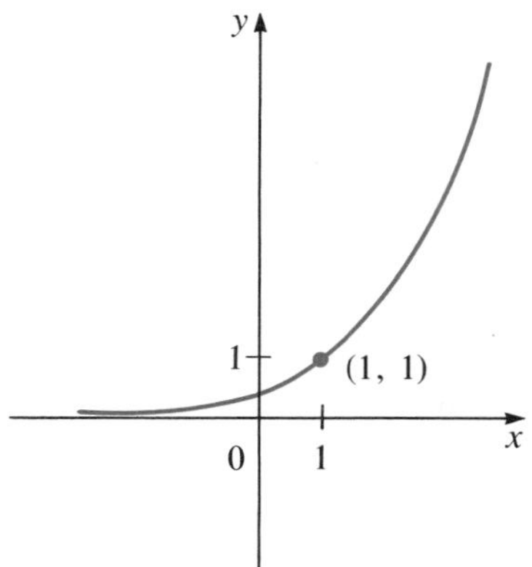

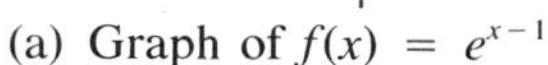
(a) Graph of $f(x) = e^{x-1}$

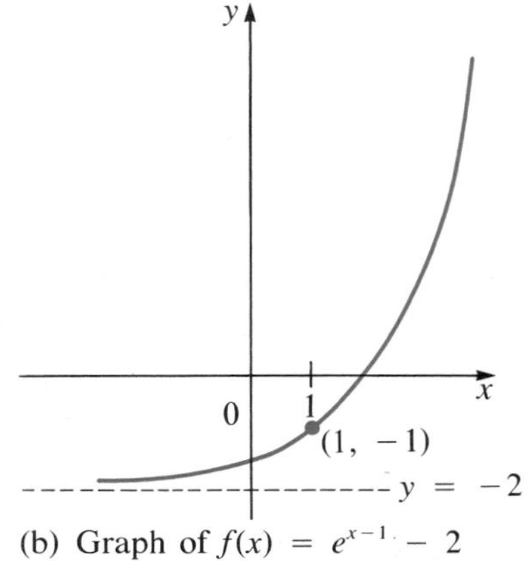

(b) Graph of $f(x) = e^{x-1} - 2$

Figure 2 The graph of $e^{x-1} - 2$ is obtained by shifting the graph of e^x 1 unit to the right and 2 units down.

Growth Rates of Power and Exponential Functions

Which grows faster: a power function or an exponential function? To get some idea, we give, in Table 2, values for x^2, x^5, x^{10}, and e^x.

Table 2
Some Approximate Values for x^2, x^5, x^{10}, and e^x

x	x^2	x^5	x^{10}	e^x
1	1	1	1	2.71828
10	100	100,000	10^{10}	22,026
50	2500	312,500,000	9.7656×10^{16}	5.1847×10^{21}
100	10,000	10^{10}	10^{20}	2.688×10^{43}
250	62,500	9.7656×10^{11}	9.5367×10^{23}	3.746×10^{108}
1000	1,000,000	10^{15}	10^{30}	1.9701×10^{434}

It seems that e^x grows much faster than the function x^{10}. In fact the following is true:

$$\frac{e^x}{x^n} \to \infty \text{ as } x \to \infty$$

It is difficult to prove this fact without using calculus. However, Table 2 certainly suggests that it is plausible.

Continuous Compounding of Interest

We return to equation (1):

$$A(t) = P\left(1 + \frac{r}{m}\right)^{mt}$$

What happens as the number of interest payment periods (m) increases without bound? That is, what happens if interest is compounded *continuously?* Let $k = m/r$. Then $m = kr$, $r/m = 1/k$, and

$$A(t) = P\left(1 + \frac{r}{m}\right)^{mt} = P\left(1 + \frac{1}{k}\right)^{krt} = P\left[\left(1 + \frac{1}{k}\right)^{k}\right]^{rt}$$

But, as k gets large, $(1 + 1/k)^k$ approaches e. Thus we have the following:

Formula for Continuous Compounding

If P dollars is invested at a rate of interest r **compounded continuously,** then, after t years, the investment is worth

$$A(t) = Pe^{rt} \tag{3}$$

EXAMPLE 3 ***The Value After Five Years with Continuous Compounding***

If \$2000 is invested at 6% interest compounded continuously, what is the investment worth after 5 years?

SOLUTION $A(5) = 2000e^{(0.06)5} = 2000e^{0.3}$
$= 2000(1.349858808) \approx \$2699.72$

NOTE In Example 7 in Section 6.1, we found that if interest were compounded daily, then the investment would be worth \$2699.65. The difference is 7¢. Continuous compounding sounds better but, as we see here, there really isn't much difference between frequent compounding and continuous compounding. ■

EXAMPLE 4 ***The Value After Ten Years with Continuous Compounding***

Suppose \$5000 is invested in a bond yielding $8\frac{1}{2}$% annually. What will the bond be worth in 10 years if interest is compounded continuously?

SOLUTION $A(t) = Pe^{rt} = 5000e^{(0.085)(10)} = 5000e^{0.85} = \$11,698.23$

In Table 5 on p. 331, we gave the effective interest rate of 8% interest compounded m times a year. We can extend the table to include continuous compounding by adding the result of the next example.

EXAMPLE 5 ***Computing the Effective Interest Rate with Continuous Compounding***

If money is invested at 8% compounded continuously, what is the effective interest rate?

SOLUTION One dollar invested will be worth $e^{r \cdot 1} = e^{0.08} = \$1.083287068$ after 1 year. The effective interest rate is 8.32871%.

Exponential Growth and Decay

We now give one of the reasons the exponential function is so important. Let $y = f(x)$ represent some quantity that is growing or declining, such as the volume of a substance, the population of a certain species, the value of an investment, or the mass of a decaying radioactive substance.

We define the **relative rate of growth** of y as follows:

$$\text{Relative rate of growth} = \frac{\text{actual rate of growth}}{\text{size of } f(x)}$$

When we say that a population is growing at 8% a year or an investment pays 8% interest, compounded continuously, or an ice cube is evaporating at a rate of 8% a minute, we are talking about the *relative* rate of growth.

The following remarkable fact is proved in a calculus course:

Constant Relative Rate of Growth = Exponential Growth

If the relative rate of growth of $y(t)$ is a constant, k, and if $y(t)$ is changing continuously, then

$$y(t) = y(0)e^{kt} \tag{4}$$

The value $y(0)$ of y at $t = 0$ is called the **initial value** of the quantity. If $k > 0$, then y is said to be **growing exponentially.** If $k < 0$, then y is said to be **decaying exponentially.**

These two ideas are illustrated in Figure 3.

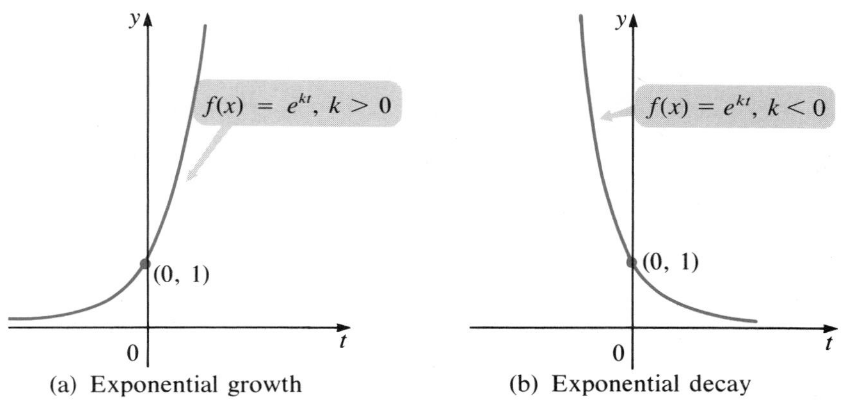

Figure 3 Graphs of exponential growth and decay.

EXAMPLE 6 *Population Growth*

A bacteria population is growing at a constant relative rate equal to 10% of its population each day. Its initial size is 10,000 organisms. How many bacteria are present after 10 days? after 30 days?

SOLUTION The relative rate of growth here is $10\% = 0.1$. Thus, if $P(t)$ denotes the population at time t, we have, from (4) with $k = 0.1$,

$$P(t) = P(0)e^{0.1t} \tag{5}$$

But the initial population is $P(0) = 10{,}000$, so (5) becomes

$$P(t) = 10{,}000e^{0.1t}$$

Now we can answer the questions:

$$\text{After 10 days,} \quad P(10) = 10{,}000e^{(0.1)(10)} = 10{,}000e^{1} \approx 10{,}000(2.7183) = 27{,}183$$

$$\text{After 30 days,} \quad P(30) = 10{,}000e^{(0.1)(30)} = 10{,}000e^{3} \approx 10{,}000(20.0855) = 200{,}855 \quad \blacksquare$$

EXAMPLE 7 ***Computing the Volume of a Melting Block of Ice***

A block of ice is melting at a constant relative rate and loses 5% of its volume each minute. If its initial volume is 20 cubic inches (in^3), what is its volume after (a) 5 minutes? (b) 20 minutes? (c) 1 hour?

SOLUTION The block is *losing* 5% of its volume each minute, so $k = -0.05$ in formula (4). If $V(t)$ denotes the volume after t minutes, then $V(0) = 20$ (this is given) and

$$V(t) = 20e^{-0.05t}$$

(a) $V(5) = 20e^{-0.05(5)} = 20e^{-0.25} \approx 15.576 \text{ in}^3$.
(b) $V(20) = 20e^{-0.05(20)} = 20e^{-1} \approx 7.358 \text{ in}^3$.

1 hour = 60 minutes
↓
(c) $V(60) = 20e^{-0.05(60)} = 20e^{-3} \approx 0.9957 \text{ in}^3$.

We will see many more examples of exponential growth and decay in Section 6.6, after we have discussed the natural logarithm function.

NOTE There may be some confusion between the relative growth rate and the percentage growth per unit time. To illustrate this difference, let us compute the growth in a population of 100,000 over one year when (a) the relative growth rate is 50% and (b) there is a 50% increase in the population each year:

(a) By (4) with $y(0) = 100{,}000$ and $k = 0.5$,

$$\text{population after 1 year} = y(1) = 100{,}000e^{0.5} \approx 164{,}872$$

which is an increase of 64,872 individuals.
(b) 50% of 100,000 = 50,000 so

$$\text{population after 1 year} = 100{,}000 + 50{,}000 = 150{,}000$$

Can you explain the difference between the increase of 64,872 in (a) and 50,000 in (b)? The answer is similar to the difference between simple and compound interest. A relative rate of growth implies that populations are changing continuously. For example, after 1 month (1/12 year) the population is

$$y\left(\frac{1}{12}\right) = 100{,}000e^{0.5(1/12)} \approx 104{,}255$$

The second month begins with a population of 104,255, which grows at a 50% rate. This is like "paying interest on interest." On the other hand, a growth rate of 50% per year means that each year the population increases by 50%. This is analogous to paying 50% simple interest at the end of each year. The two concepts are different.

FOCUS ON
Leonhard Euler (1707–1783)

The great Swiss mathematician Leonhard Euler (pronounced "Oiler") was born in Basel, Switzerland, in 1707. Euler's father was a clergyman who hoped that his son would follow him into the ministry. The father was adept at mathematics, however, and together with Johann Bernoulli, instructed young Leonhard in that subject. Euler also studied theology, astronomy, physics, medicine, and several Eastern languages.

In 1727, Euler applied to and was accepted for a chair of medicine and physiology at the St. Petersburg Academy. The day Euler arrived in Russia, however, Catherine I — founder of the Academy — died, and the Academy was plunged into turmoil. By 1730, Euler was pursuing his mathematical career from the chair of natural philosophy. Accepting an invitation from Frederick the Great, Euler went to Berlin in 1741 to head the Prussian Academy. Twenty-five years later, he returned to St. Petersburg, where he died in 1783 at the age of 76.

The most prolific writer in the history of mathematics, Euler found new results in virtually every branch of pure and applied mathematics. Although German was his native language, he wrote mostly in Latin and occasionally in French. His amazing productivity did not decline even when he became totally blind in 1766. During his lifetime, Euler published 530 books and papers. When he died, he left so many unpublished manuscripts that the St. Petersburg academy was still publishing his work in its *Proceedings* almost half a century later. Euler's work enriched such diverse areas as hydraulics, celestial mechanics, lunar theory, and the theory of music, as well as mathematics.

Euler had a phenomenal memory. As a young man, he memorized the entire *Aeneid* by Virgil (in Latin), and many years later could still recite the entire work. He was able to solve astonishingly complex mathematical problems in his head and is said to have solved, again in his head, problems in astronomy that stymied Newton. The French academician François Arago once commented that Euler could calculate without effort "just as men breathe, as eagles sustain themselves in the air."

Euler wrote in a mathematical language that is largely in use today. Among symbols first used by him are

- $f(x)$ for functional notation
- e for the base of the natural exponential and logarithm functions
- Σ for the summation sign
- i to denote $\sqrt{-1}$

Euler's textbooks were models of clarity. His texts included the *Introductio in analysin infinitorum* (1748), his *Institutiones calculi differentialis* (1755), and the three-volume *Institutiones calculi integralis* (1768–1774). This and others of his works served as models for many of today's mathematics textbooks.

It is said that Euler did for mathematical analysis what Euclid did for geometry. It is no wonder that so many later mathematicians expressed their debt to him.

Problems 6.2

Readiness Check

I. Which of the following is the definition of e?

a. The number that $(1+x)^{1/x}$ approaches as x approaches ∞.

b. The number that $\left(1-\frac{1}{x}\right)^x$ approaches as x approaches ∞.

c. The number that $\left(1+\frac{1}{x}\right)^x$ approaches as x approaches ∞.

d. The number that $(1-x)^{1/x}$ approaches as x approaches ∞.

II. Which of the following is true about $f(x) = e^{-x}$?

a. Its graph is decreasing.

b. Its graph is the reflection of $f(x) = e^x$ about the y-axis.

c. If $x > 0$, it takes negative values.

d. It is equal to $f(x) = \frac{-1}{e^x}$.

III. Which of the following is an increasing function?

a. $f(x) = e^{x-3}$ b. $f(x) = e^{-\pi x}$

c. $f(x) = 1 - e^{3x}$ d. $f(x) = -e^{3x}$

IV. Which of the following is true?

a. $2 < e^{-1} < 3$ b. $e > \pi$

c. $\sqrt{2} < e < \sqrt{3}$ d. $\frac{1}{3} < \frac{1}{e} < \frac{1}{2}$

V. Which function's graph is given in the figure below?

a. e^{-x} b. $e^x - 2$

c. $-e^x$ d. $-e^{-x}$

e. $e^{-x} - 2$

y
0
x
(0, −1)

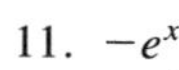

In Problems 1–10 use a calculator to compute e^x to six significant digits.

1. $e^{3.15}$ 2. $e^{-0.6}$ 3. $e^{\sqrt{3}}$ 4. $e^{12.02}$
5. $e^{29.4}$ 6. $e^{-4.17}$ 7. $e^{-15.9}$ * 8. e^e
* 9. e^π * 10. π^e

Match each function in Problems 11–20 with one of the ten graphs given below.

11. $-e^x$ 12. $-e^{-x}$ 13. $1 - e^x$ 14. e^{x+1}
15. e^{x-1} 16. $2 - e^{x-1}$ 17. $e^x - 2$ 18. $2e^x$
19. $-2e^x$ 20. $2 + e^{-x}$

(a)

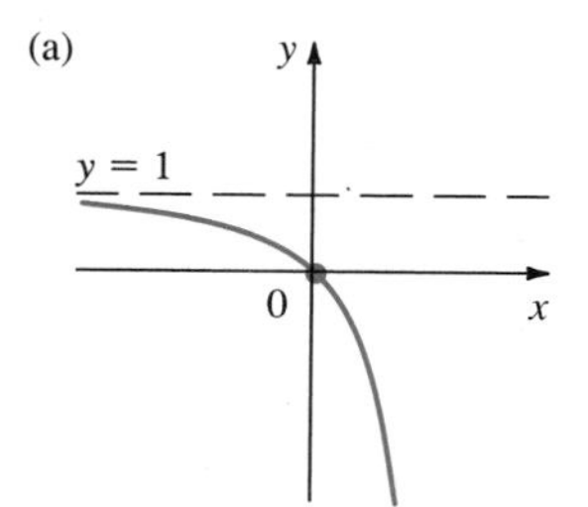

(b)

y
y = 2
0
x

(c)

y
(1, 1)
0
x

(d)

y
0
x
(0, −1)
y = −2

(e)

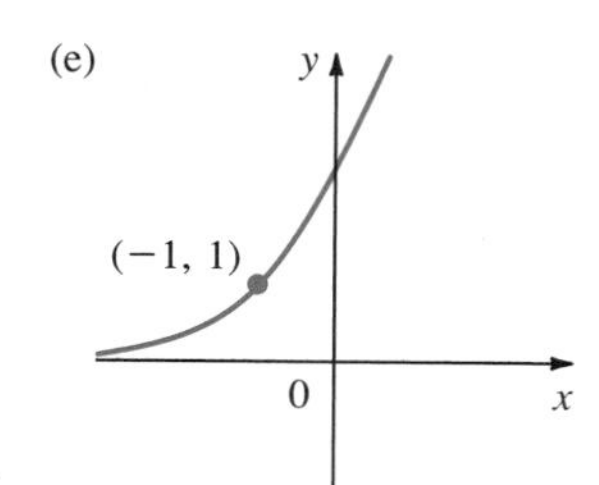

(f)

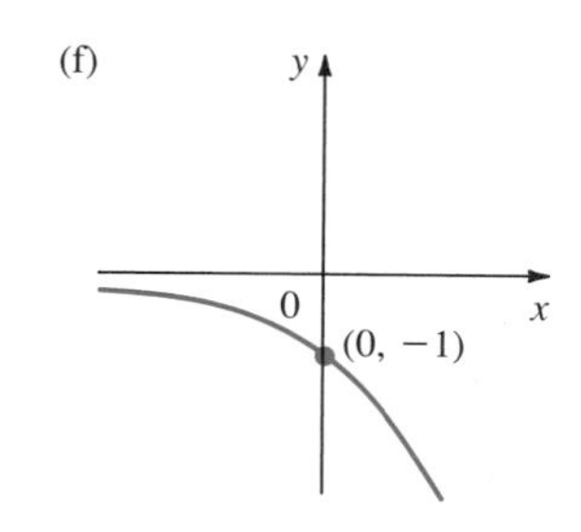

Answers to Readiness Check

I. c II. a, b III. a IV. d V. c

(g)

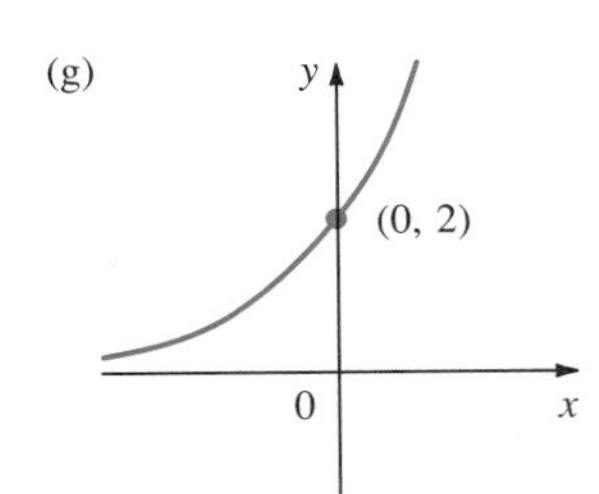

(h)

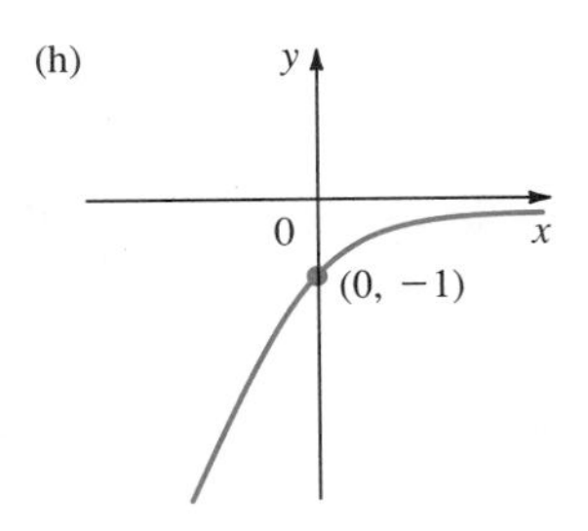

(i)

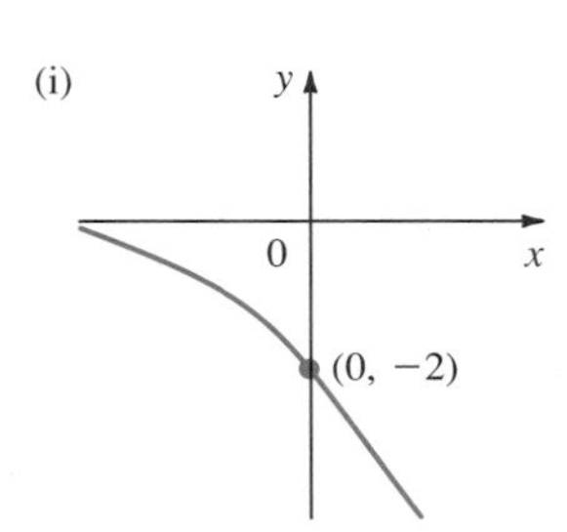

(j)

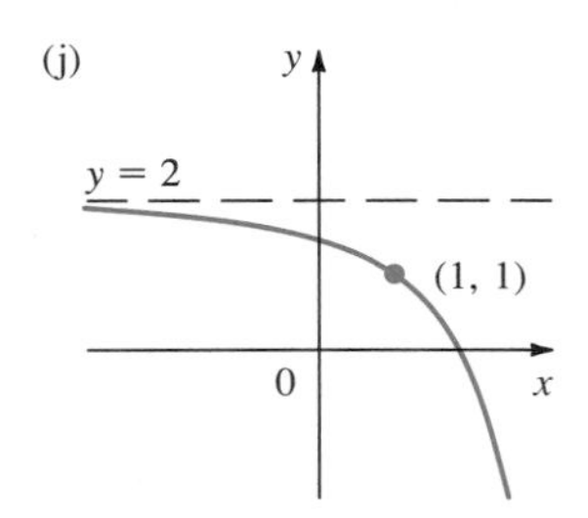

In Problems 21–28 sketch the graph of the exponential function.

21. e^{2x} 22. $e^{-x/2}$ 23. $3e^{1.5x}$ 24. $-e^x$
25. e^{x-2} 26. e^{x+3} 27. $e^x + 5$ * 28. $3 - e^{2-x}$

29. The exponential e^x can be estimated for x in $[-\frac{1}{2}, \frac{1}{2}]$ by the formula

$$e^x \approx \left(\left\{\left[\left(\frac{x}{5} + 1\right)\frac{x}{4} + 1\right]\frac{x}{3} + 1\right\}\frac{x}{2} + 1\right)x + 1$$

(a) Calculate an approximate value for $e^{0.13}$.
(b) Calculate an approximate value for $e^{-0.37}$.

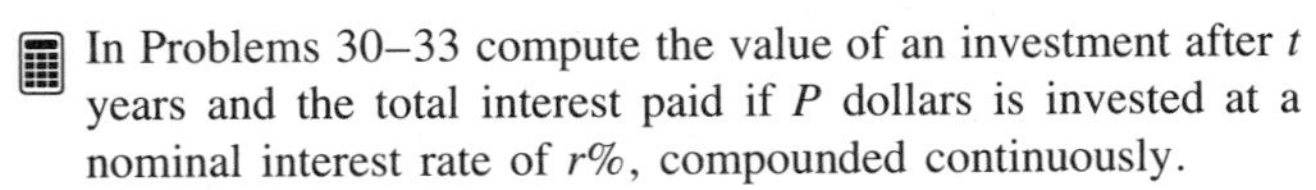

In Problems 30–33 compute the value of an investment after t years and the total interest paid if P dollars is invested at a nominal interest rate of $r\%$, compounded continuously.

30. $P = \$5000$, $r = 6\%$, $t = 4$
31. $P = \$9000$, $r = 10\%$, $t = 5$
32. $P = \$10,000$, $r = 6.5\%$, $t = 8$
33. $P = \$10,000$, $r = 13\%$, $t = 8$

Use a calculator to solve Problems 34–49.

34. A sum of \$5000 is invested at a return of 7% per year, compounded continuously. What is the investment worth after 8 years?
35. If \$10,000 is invested in bonds yielding 9% compounded continuously, what will the bonds be worth in 8 years?
36. An investor buys a bond that pays 12% annual interest compounded continuously. If she invests \$10,000 now, what will her investment be worth in (a) 1 year? (b) 4 years? (c) 10 years?
37. As a gimmick to lure depositors, a bank offers 5% interest compounded continuously in comparison with its competitor, which offers $5\frac{1}{8}\%$ compounded annually. Which bank would you choose?
38. If money is invested at 10% compounded continuously, what is the effective interest rate? [The effective interest rate is defined on p. 330.]
* 39. After how many years will the bond in Problem 36 be worth \$20,000? [Hint: We will see an easy way to solve this problem in Section 6.4. For now, use trial and error and give an answer to the nearest tenth of a year.]
40. A Roman deposited 1¢ in a bank at the beginning of the year A.D. 1. If the bank paid a meager 1% interest, compounded continuously, what would the investment be worth at the beginning of 1992?
41. Mrs. Jones has just invested \$400 in a 5-year term deposit (account A) paying 12% per year compounded twice per year, and she has invested another \$400 in a 5-year deposit (account B) at 11% per year compounded continuously.
(a) Calculate the effective interest rate (as a percentage) for each account. Give answers correct to two decimal places.
(b) Which investment is worth more after 5 years? by how much (to the nearest cent)?
42. The population of a certain city was 800,000 in 1985. If the population grows at a constant rate of 2% a year, what was the population in 1990? What will be the population in the year 2010? [Hint: Treat 1985 as year 0.]
43. The estimated world population in 1986 was 4,845,000,000. Assume that the population grows at a constant rate of 1% per year. Estimate the world population in (a) 1990, (b) 2000, and (c) 2010.
44. The mass of a radioactive substance is declining at a rate of 25% a week. The mass is initially 20 grams. What is the mass after (a) 2 weeks? (b) 10 weeks? (c) 1 year?
* 45. The temperature difference between a hot coal and the surrounding air declines 12% each minute. At noon, the temperature of the coal is 180°F, and the temperature of the air is 50°F. Assuming that the air temperature does not change, estimate the temperature of the coal at (a) 12:05 P.M., (b) 12:15 P.M., (c) 12:30 P.M.
46. The **hyperbolic sine** function, denoted by sinh x, is defined by

$$\sinh x = \frac{e^x - e^{-x}}{2}$$

Compute (a) sinh 0, (b) sinh 1, (c) $\sinh(-\frac{1}{2})$, (d) $\sinh(-4)$, (e) sinh 2.37.
47. (a) Show that sinh x is an odd function.
(b) Plot values of sinh x on graph paper for x between 0 and 5 in increments of 0.2.
(c) Sketch the graph of $y = \sinh x$.

48. The **hyperbolic cosine** function, denoted by $\cosh x$, is defined by

$$\cosh x = \frac{e^x + e^{-x}}{2}$$

Compute (a) $\cosh 0$, (b) $\cosh 1$, (c) $\cosh \frac{1}{2}$, (d) $\cosh 2.37$, (e) $\cosh 5$.

49. (a) Show that $\cosh x$ is an even function.
(b) Show that $\cosh 0 = 1$ and $\cosh x > 1$ if $x \neq 0$.
(c) Plot values of $\cosh x$ on graph paper for x between 0 and 5 in increments of 0.2.
(d) Sketch the graph of $y = \cosh x$.

* 50. The **hyperbolic tangent** function, denoted by $\tanh x$, is defined by

$$\tanh x = \frac{\sinh x}{\cosh x} = \frac{e^x - e^{-x}}{e^x + e^{-x}}$$

(a) Show that $\tanh x$ is defined for every real number.
(b) Show that $-1 < \tanh x < 1$ for every real number x.

Graphing Calculator Problems

Before beginning, read Example 4 in Appendix A.

51. Sketch $e^{x/3}$ and $e^{-x/3}$ on the same screen.
52. Sketch e^{2x} and $-e^{2x}$ on the same screen.
53. Sketch e^x, e^{2x}, and e^{3x} on the same screen.
54. Sketch e^{-x}, e^{-2x}, and e^{-3x} on the same screen.
55. Sketch $e^{x/2}$, $e^{(x-1)/2}$, and $e^{(x+1)/2}$ on the same screen.
56. Sketch $e^{1.5x}$, $e^{1.5x} + 3$, and $e^{1.5x} - 3$ on the same screen.

In Problems 57–73 obtain a sketch of each exponential function.

57. $e^{1.6x}$
58. $e^{1.6(x-2)}$
59. $-e^{1.6x}$
60. $e^{-1.6x}$
61. $2 - e^{1.6(1-x)}$
62. $e^{-0.23x}$
63. $e^{-3.5x}$
64. $e^{0.57x}$
65. $-e^{0.9(x-0.6)}$
66. $1 - 2e^{-\frac{1}{2}x}$
67. $-3 + 4e^{-0.8(x+0.4)}$
68. e^{x^2}
69. e^{-x^2}
70. $e^{1/x}$
71. $-e^{-1/x}$
72. $e^{\sqrt{x}}$
73. $e^{-\sqrt{x}}$

6.3 Logarithmic Functions

In Section 3.7, we discussed inverse functions. We proved that if a function f is increasing or decreasing, then f has an inverse. In Section 6.1, we saw that the function defined by $f(x) = a^x$ is increasing if $a > 1$ and decreasing if $0 < a < 1$. Putting these two facts together, we conclude that

The function $f(x) = a^x$ has an inverse for $a > 0$, $a \neq 1$

What does such an inverse look like? We know that $a^x > 0$ for every real x. Thus the domain of the inverse of a^x is the set of positive real numbers. Put another way, if $y > 0$ is given, we can find a unique x such that $a^x = y$.

EXAMPLE 1 ***Solving for the Exponent***

Solve for x: (a) $2^x = 8$, (b) $3^x = \frac{1}{9}$, (c) $(\frac{1}{2})^x = 8$.

SOLUTION (a) We see that $2^x = 8$ if and only if $x = 3$. Similarly, in (b), $3^x = \frac{1}{9}$ if and only if $x = -2$, and in (c), $(\frac{1}{2})^x = 8$ if and only if $x = -3$ $\left[(\frac{1}{2})^{-3} = \frac{1}{(\frac{1}{2})^3} = \frac{1}{(\frac{1}{8})} = 8\right]$.

We now reverse the roles of x and y† in $y = a^x$ and define an important new function, the inverse of $f(x) = a^x$.

† We reverse the roles of x and y so that we can write the inverse function with x as the independent variable and y as the dependent variable. This is the process we used to find inverse functions in Section 3.7.

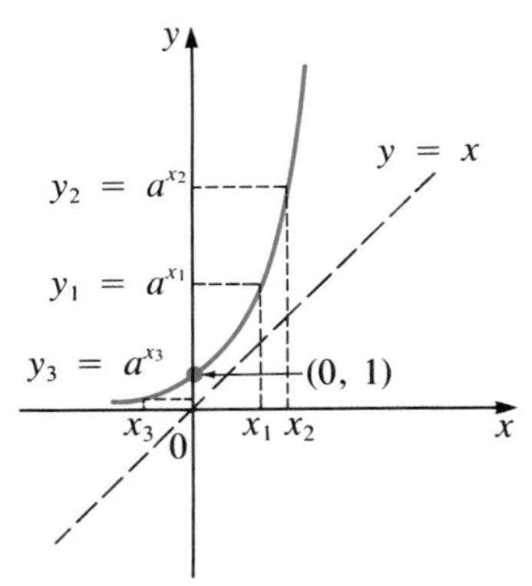

(a) Graph of $f(x) = a^x$

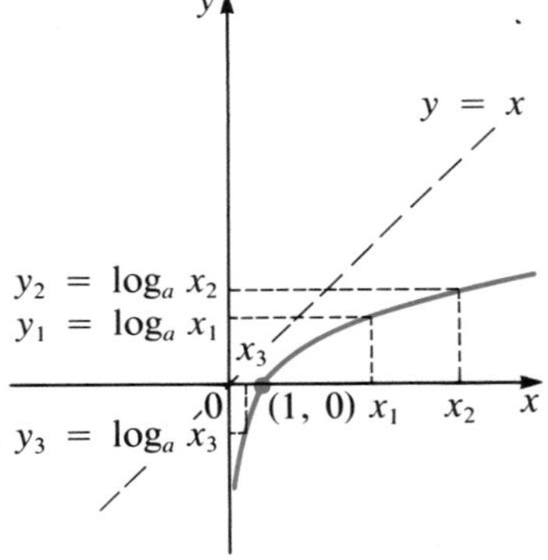

(b) Graph of $f(x) = \log_a x$

Figure 1 The graph of $\log_a x$ is the reflection about the line $y = x$ of the graph of a^x.

Definition of the Logarithm to the Base *a*

If $x = a^y$ with $a > 0$ and $a \neq 1$, then the **logarithm to the base *a*** of x is y. This is written

$$y = \log_a x \tag{1}$$

where $\log_a x$ is defined for $x > 0$.
This means that if

$$x > 0, \ a > 0, \quad \text{and} \quad a \neq 1$$

then

$$y = \log_a x \quad \text{if and only if} \quad x = a^y$$

By the reflection property discussed in Section 3.7 (p. 211), the graph of $y = \log_a x$ is the reflection of the graph of $y = a^x$ about the line $y = x$. In Figure 1, typical graphs for $y = a^x$ and $y = \log_a x$ are given for $a > 1$.

We stress the following facts about the logarithmic functions.

$$y = \log_a x \quad \text{is equivalent to} \quad x = a^y \tag{2}$$

$$y = a^x \quad \text{is equivalent to} \quad x = \log_a y \tag{3}$$

Think of $y = \log_a x$ as an answer to the question: To what power must a be raised to obtain the number x? This immediately implies that

$$a^{\log_a x} = x \qquad \text{for every positive real number } x \tag{4}$$

$$\log_a a^x = x \qquad \text{for every real number } x \tag{5}$$

EXAMPLE 2 ***Illustration of Properties of Logarithms***

(a) $2 = \log_3 9$ so $9 = 3^2$ From (2)

(b) $\frac{1}{16} = 2^{-4}$ so $-4 = \log_2 \frac{1}{16}$ From (3)

(c) $4^{\log_4 7} = 7$ From (4)

(d) $\log_{10} 10^{12} = 12$ From (5)

We also stress that $\log_a x$ is only defined for $x > 0$ because the equation $a^y = x$ has no solution when $x \leq 0$.† For example, $\log_2(-1)$ is not defined because there is no real number y such that $2^y = -1$. We have

$$\text{domain of } \log_a x = \{x: x > 0\}$$

EXAMPLE 3 ***Writing Equations Using Logarithmic Notation***

We rewrite the exponential equations solved in Example 1 using logarithmic notation: (a) $\log_2 8 = 3$, (b) $\log_3 \frac{1}{9} = -2$, and (c) $\log_{1/2} 8 = -3$. ■

† On p. 178, we discussed three reasons for restricting the domain of a function. The logarithmic function provides a fourth reason.

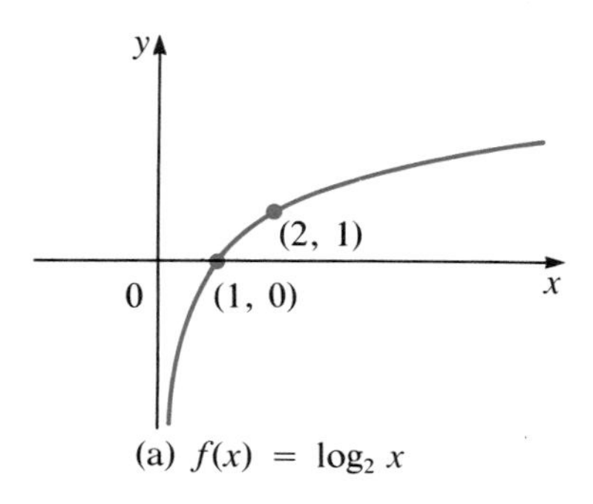

(a) $f(x) = \log_2 x$

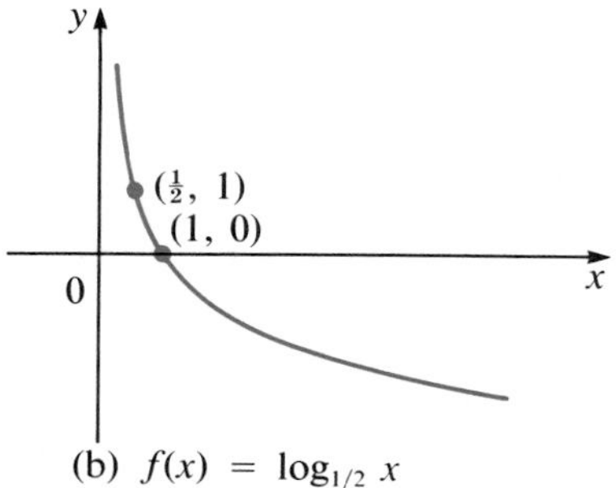

(b) $f(x) = \log_{1/2} x$

Figure 2 The graphs of $\log_2 x$ and $\log_{1/2} x$.

EXAMPLE 4 ***The Graphs of Two Logarithmic Functions***

The graphs of $y = \log_2 x$ and $y = \log_{1/2} x$ are given in Figure 2.

It follows from facts about the function a^x and the graph of $\log_a x$ that

$\log_a x$ is an increasing function if $a > 1$,
and a decreasing function if $0 < a < 1$.

EXAMPLE 5 ***Writing a Logarithmic Equation in Exponential Form***

Change to exponential form: (a) $\log_9 3 = \frac{1}{2}$ and (b) $\log_4 64 = 3$.

SOLUTION
(a) $\log_9 3 = \frac{1}{2}$ is equivalent to $9^{1/2} = 3$.
(b) $\log_4 64 = 3$ is equivalent to $4^3 = 64$. ■

EXAMPLE 6 ***Writing an Exponential Equation in Logarithmic Form***

Change to logarithmic form: (a) $5^4 = 625$ and (b) $(\frac{1}{2})^{-4} = 16$.

SOLUTION
(a) $5^4 = 625$ is equivalent to $\log_5 625 = 4$.
(b) $(\frac{1}{2})^{-4} = 16$ is equivalent to $\log_{1/2} 16 = -4$.

We note the following fact: $y = \log_a x$ defines a one-to-one function. That is,

$$\text{If} \quad \log_a x_1 = \log_a x_2, \quad \text{then } x_1 = x_2. \tag{6}$$

Properties of Logarithms

Logarithms have many useful properties. Four properties follow immediately from the definition of the logarithm. We have seen that $a^{\log_a x} = x$ and $\log_a a^x = x$ (equations (4) and (5)). Since $a^1 = a$, $1 = \log_a a$ from (2). Also, $a^0 = 1$, so $\log_a 1 = 0$.

Four Basic Properties of $\log_a x$

	Property	*Illustration*
1.	$a^{\log_a x} = x$	$2^{\log_2 27.3} = 27.3$
2.	$\log_a a^x = x$	$\log_5 5^{\sqrt{7}} = \sqrt{7}$
3.	$\log_a a = 1$	$\log_{12} 12 = 1$
4.	$\log_a 1 = 0$	$\log_{1/2} 1 = 0$

The next four properties are especially useful in computations.

Computational Properties of $\log_a x$

For $x > 0$, $w > 0$, and r a real number,

5. $\log_a xw = \log_a x + \log_a w$
6. $\log_a \dfrac{x}{w} = \log_a x - \log_a w$
7. $\log_a \dfrac{1}{x} = -\log_a x$
8. $\log_a x^r = r \log_a x$

NOTE Property 7 is a special case of both Property 6 and Property 8.

Proof of Property 5

Let $u = \log_a x$ and $v = \log_a w$. Then, from (2),

$$x = a^u \qquad \text{and} \qquad w = a^v$$

so

$$xw = a^u a^v \underset{\text{Property 3 on p. 327}}{=} a^{u+v}$$

Thus,

$$\log_a xw = \log_a a^{u+v} \underset{\text{Property 2}}{=} u + v = \log_a x + \log_a w \quad \blacksquare$$

Proof of Property 6

With u and v as in Property 5 above,

$$\frac{x}{w} = \frac{a^u}{a^v} \underset{\text{Property 4 on p. 327}}{=} a^{u-v}$$

so, from (2),

$$\log_a \frac{x}{w} = \log_a \frac{a^u}{a^v} = \log_a a^{u-v} = u - v = \log_a x - \log_a w \quad \blacksquare$$

Proof of Property 7

From Property 6 (already proven)

$$\log_a \frac{1}{x} = \log_a 1 - \log_a x \underset{\text{Property 4}}{=} 0 - \log_a x = -\log_a x \quad \blacksquare$$

Proof of Property 8

Let $u = \log_a x$. Then $x = a^u$, and

$$x^r = (a^u)^r \overset{\text{Property 7 on p. 327}}{=} a^{ur}.$$

Thus,

$$\log_a x^r = \log_a a^{ur} = ur = ru = r \log_a x \quad \blacksquare$$

EXAMPLE 7 ***Illustrating That $\log_a xw = \log_a x + \log_a w$***

Since $2^5 = 32$, we have $\log_2 32 = 5$. Also, $32 = 8 \cdot 4$ so, from Property 5,

$$\log_2 32 = \log_2 8 \cdot 4 = \log_2 8 + \log_2 4 = 3 + 2 = 5 \quad \blacksquare$$

EXAMPLE 8 ***Using the Fact That $\log_a x^r = r \log_a x$***

Compute $\log_5 \sqrt[3]{25}$.

SOLUTION $\sqrt[3]{25} = 25^{1/3}$ so, from Property 8,

$$\log_5 25^{1/3} = \frac{1}{3} \log_5 25 = \frac{1}{3} \log_5 5^2 = \frac{1}{3} \cdot 2 = \frac{2}{3} \quad \blacksquare$$

EXAMPLE 9 ***Using Properties of Logarithms to Compute New Logarithmic Values from Given Ones***

$\log_{10} 2 \approx 0.3010$ and $\log_{10} 3 \approx 0.4771$. Using these values approximate (a) $\log_{10} 6$, (b) $\log_{10} \frac{2}{3}$, (c) $\log_{10} 5$, (d) $\log_{10} 8$, (e) $\log_{10} 108$.

SOLUTION

(a) $\log_{10} 6 = \log_{10} 3 \cdot 2 \overset{\text{Property 5}}{=} \log_{10} 3 + \log_{10} 2$
$\approx 0.4771 + 0.3010 = 0.7781$

(b) $\log_{10} \frac{2}{3} \overset{\text{Property 6}}{=} \log_{10} 2 - \log_{10} 3 \approx 0.3010 - 0.4771 = -0.1761$

(c) $\log_{10} 5 = \log_{10} \frac{10}{2} = \log_{10} 10 - \log_{10} 2 \overset{\text{Property 3}}{=} 1 - \log_{10} 2$
$\approx 1 - 0.3010 = 0.6990$

(d) $\log_{10} 8 = \log_{10} 2^3 \overset{\text{Property 8}}{=} 3 \log_{10} 2 = 3(0.3010) = 0.9030$

(e) $\log_{10} 108 = \log_{10} 4 \cdot 27 = \log_{10} 2^2 \cdot 3^3 = \log_{10} 2^2 + \log_{10} 3^3$
$= 2 \log_{10} 2 + 3 \log_{10} 3 \approx 2(0.3010) + 3(0.4771)$
$= 2.0333$

NOTE Approximate values for $\log_{10} 2$ and $\log_{10} 3$ can be obtained from a calculator or from Table 3 at the back of the book.

EXAMPLE 10 *Solving an Equation Involving a Logarithm and an Exponent*

Find x if $x = 6^{-(1/2)\log_6 25}$.

SOLUTION

$$-\frac{1}{2}\log_6 25 \overset{\text{Property 8}}{=} \log_6 25^{-1/2} = \log_6 \frac{1}{25^{1/2}} = \log_6 \frac{1}{5}$$

Then

$$x = 6^{-(1/2)\log_6 25} = 6^{\log_6 1/5} \overset{\text{Property 1}}{=} \frac{1}{5} \quad ■$$

EXAMPLE 11 *Simplifying a Logarithmic Expression*

Write $\log_a \dfrac{x^4y^6}{\sqrt{x^{2/3}y^8}}$ in terms of $\log_a x$ and $\log_a y$.

SOLUTION

$$\begin{aligned}
\log_a \frac{x^4y^6}{\sqrt{x^{2/3}y^8}} &= \log_a x^4y^6 - \log_a \sqrt{x^{2/3}y^8} && \text{Property 6}\\
&= \log_a x^4 + \log_a y^6 - \log_a (x^{2/3}y^8)^{1/2} && \text{Property 5}\\
&= 4\log_a x + 6\log_a y - \tfrac{1}{2}\log_a x^{2/3}y^8 && \text{Property 8}\\
&= 4\log_a x + 6\log_a y - \tfrac{1}{2}(\log_a x^{2/3} + \log_a y^8) && \text{Property 5}\\
&= 4\log_a x + 6\log_a y - \tfrac{1}{2}(\tfrac{2}{3}\log_a x + 8\log_a y) && \text{Property 8}\\
&= 4\log_a x + 6\log_a y - \tfrac{1}{3}\log_a x - 4\log_a y\\
&= \tfrac{11}{3}\log_a x + 2\log_a y \quad ■
\end{aligned}$$

EXAMPLE 12 *Combining Logarithmic Terms*

Write the following expression as a single logarithm:

$$\tfrac{1}{2}\log_a x + 4\log_a y - 3\log_a z$$

SOLUTION

$$\begin{aligned}
\tfrac{1}{2}\log_a x + 4\log_a y - 3\log_a z &= \log_a x^{1/2} + \log_a y^4 - \log_a z^3 && \text{Property 8}\\
&= \log_a x^{1/2}y^4 - \log_a z^3 && \text{Property 5}\\
&= \log_a \frac{x^{1/2}y^4}{z^3} && \text{Property 6}
\end{aligned}$$

WARNING Two common errors are made by students who first study logarithms. Do not make them.

COMMON ERROR 1

$$\log_a (x + y) \neq \log_a x + \log_a y$$

Incorrect

$$\log_{10} 20 = \log_{10} (10 + 10)$$

This is the incorrect step ↓

$$= \log_{10} 10 + \log_{10} 10$$
$$= 1 + 1 = 2$$

Correct

$$\log_{10} 20 = \log_{10} 2 \cdot 10$$
$$= \log_{10} 2 + \log_{10} 10$$

See Example 9 ↓

$$\approx 0.3010 + 1 = 1.3010$$

In general, *the logarithm of the sum of two numbers cannot be simplified.* If you obtain the expression $\log_a (x + y)$, leave it alone. That's as far as you can go. No rule of logarithms will make the answer simpler.

COMMON ERROR 2

$$\frac{\log_a x}{\log_a y} \neq \log_a \frac{x}{y}$$

Incorrect

$$\frac{\log_{10} 20}{\log_{10} 10} = \log_{10} \frac{20}{10}$$
$$= \log_{10} 2 \approx 0.3010$$

This is the incorrect step

Correct

$$\frac{\log_{10} 20}{\log_{10} 10} \approx \frac{1.3010}{1} = 1.3010$$

$\log_{10} 20 = \log_{10} 10 + \log_{10} 2 \approx 1.3010$

■

Problems 6:3

Readiness Check

I. $2^{\log_2 16} =$ _____.
   a. 2 b. 4
   c. 16 d. 8 e. 32

II. Which of the following is true about $f(x) = \log_b x$?
   a. $f(x)$ is always positive.
   b. It is the inverse function of $f(x) = x^b$.
   c. If $b > 1$, its graph is decreasing.
   d. Its graph has an x-intercept but no y-intercept.

III. $\log_5 5^{125} =$ _____.
   a. 5 b. 125 c. $\frac{125}{5} = 25$
   d. $125 \cdot 5 = 625$ e. 3

IV. Which of the following is true of the graph of $f(x) = \log_2 (x - 1)$?
   a. It decreases from left to right.
   b. It passes through (2, 0).
   c. It is symmetric with respect to the x-axis.
   d. It is asymptotic to the negative y-axis.

Answers to Readiness Check

I. c II. d III. b IV. b

In Problems 1–10, change each equation to an exponential form. For example, $\log_9 3 = \frac{1}{2}$ is equivalent to $9^{1/2} = 3$.

1. $\log_{16} 4 = \frac{1}{2}$
2. $\log_2 32 = 5$
3. $\log_{1/2} 16 = -4$
4. $\log_3 \frac{1}{3} = -1$
5. $\log_{12} 1 = 0$
6. $\log_{10} 100 = 2$
7. $\log_{10} 0.001 = -3$
8. $\log_{1/10} 100 = -2$
9. $\log_e e^2 = 2$
10. $\log_{e^2} e = \frac{1}{2}$

In Problems 11–16, change each equation to a logarithmic form.

11. $2^5 = 32$
12. $(\frac{1}{2})^3 = \frac{1}{8}$
13. $(\frac{1}{3})^{-2} = 9$
14. $10^3 = 1000$
15. $10^{0.301029995} = 2$
16. $e^{0.69314718} = 2$

In Problems 17–27, solve for the unknown variable. (Do not use tables or a calculator.)

17. $y = \log_3 27$
18. $x = \log_4 16$
19. $u = \log_{1/2} 4$
20. $x = \dfrac{1}{3} \log_7 \dfrac{1}{7}$
21. $s = \log_{1/4} 2$
22. $v = \log_{81} 3$
23. $y = \log_e e^5$
24. $u = \log_{10} 0.01$
25. $y = 3 \log_5 \dfrac{1}{125}$
26. $y = \pi \log_\pi \dfrac{1}{\pi^4}$
27. $u = \log_3 \dfrac{1}{3^{4.5}}$

In Problems 28–31 a function is given. Match each function with one of the graphs given below.

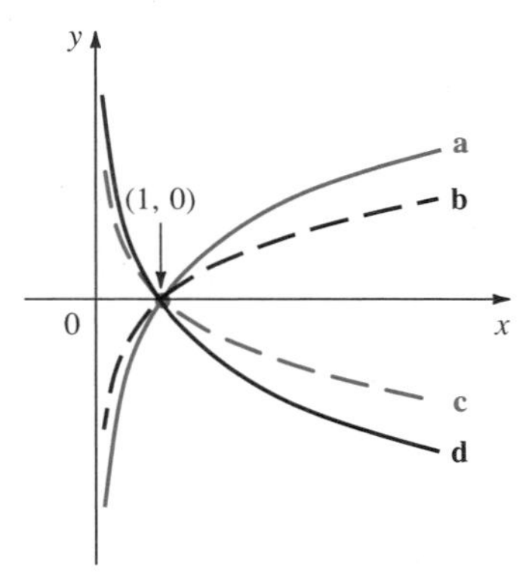

28. $\log_2 x$
29. $\log_3 x$
30. $\log_{1/2} x$
31. $\log_{1/3} x$

In Problems 32–36, use the fact that $\log_e 2 \approx 0.6931$ and $\log_e 6 \approx 1.7918$ to approximate each logarithm.

32. $\log_e 3$
33. $\log_e 8$
34. $\log_e \frac{1}{2}$
35. $\log_e 36$
36. $\log_e \frac{27}{16}$

In Problems 37–46, use the fact that $\log_{10} 3.45 \approx 0.5378$ to approximate each logarithm.

37. $\log_{10} 345$ [Hint: $345 = 3.45 \times 10^2$]
38. $\log_{10} 3{,}450{,}000$
39. $\log_{10} 0.345$ [Hint: $0.345 = 3.45 \times 10^{-1}$]
40. $\log_{10} 0.00000345$
41. $\log_{10} \sqrt{3.45}$
42. $\log_{10} \sqrt{34.5}$
43. $\log_{10} \dfrac{1}{\sqrt{3450}}$
44. $\log_{10} (345)^2$
45. $\log_{10} \left(\dfrac{1}{34.5}\right)^3$
46. $\log_{10} [(345)^{2/3}(3.45)^{3/4}]$

In Problems 47–54, write each expression in terms of $\log_a x$, $\log_a y$, and $\log_a z$.

47. $\log_a x^4y^3$
48. $\log_a \dfrac{x}{y^2}$
49. $\log_a \sqrt{x^4y^{3/2}}$
50. $\log_a \dfrac{xy^5}{z^2}$
51. $\log_a \dfrac{\sqrt{x}\sqrt[3]{y}}{\sqrt[5]{z}}$
52. $\log_a \dfrac{x^4}{y^7z^3}$
53. $\log_a \left(\dfrac{xy}{z}\right)^{4/5}$
54. $\log_a x^{20}y^{30}z^{-2}$

In Problems 55–68, write each expression as a single logarithm.

55. $\log_a x + \log_a \dfrac{1}{x}$
56. $2 \log_a u + 3 \log_a v^2 - 6 \log_a u^{1/2}$
57. $\log_a x + 2 \log_a y$
58. $3 \log_a x - \frac{1}{3} \log_a y$
59. $\log_a 4x + \log_a 5w$
60. $2 \log_a \dfrac{1}{x} - 3 \log_a \dfrac{1}{z}$
61. $\log_a x + 2 \log_a y + 3 \log_a z$
62. $3 \log_a u - 4 \log_a v + \dfrac{3}{5} \log_a w$
63. $\log_a (x^2 - 4) - \log_a (x + 2)$
64. $2 \log_a (w^2 - 1) - 3 \log_a (w + 1) - \log_a (w - 1)$
65. $\log_a (x^2 - 2y) + \log_a (y + 2)$
66. $\log_a (x - 1) + \dfrac{1}{2} \log_a (y - 2) + \dfrac{1}{3} \log_a (z - 3)$
67. $\log_a (x^2 - y^2) - 4 \log_a (x - y) + 3 \log_a (x + y)$
68. $2 \log_a (x^3 - y^3) - \log_a (x - y) - \log_a (x + y)$

6.4 Common and Natural Logarithms

Although any positive number ($\neq 1$) can be used as a base for a logarithm, two bases are used almost exclusively. The first of these is base 10. Logarithms to the base 10 are called **common logarithms.** The second, and more important, base for logarithms is the base e. This base is used in calculus. Logarithms to the base e are called **natural logarithms.**

Common Logarithms

Common logarithms are denoted by $\log x$. That is,

Definition of the Common Logarithm Function

$$\log x = \log_{10} x, \qquad x > 0 \tag{1}$$

Logarithms were discovered by the Scottish mathematician John Napier, Baron of Merchiston (1550–1617). In one of Napier's earliest notes, the following table appears:

	I	II	III	IIII	V	VI	VII	···
1	2	4	8	16	32	64	128	

You should note that the number represented by each Roman numeral above is the logarithm to the base 2 of the number beneath it. Alternatively, 2 to the power of the Roman numeral above is equal to the number below. In 1615, Henry Briggs (1561–1631) suggested to Napier that 10 be used as the base for the logarithm. Soon Napier discovered the importance of logarithms to the base 10 for computations since our numbering system is founded on the base 10. Because he foresaw the practical usefulness of logarithms in trigonometry and astronomy, he abandoned other mathematical pursuits and set himself the difficult task of producing a table of common logarithms—a task that took him 25 years to complete. When the tables were completed, they created great excitement on the European continent and were immediately used by two of the great astronomers of the day: the Dane Tycho Brahe and the German Johannes Kepler.

Until recently, common logarithms were frequently used for arithmetic computations. However, logarithms are now rarely used in this way because arithmetic computations can be carried out more accurately and much faster on a hand-held calculator. They are still used in certain scientific formulas. Most scientific calculators have a [log] or [log x] key for computing common logarithms.

FOCUS ON
Why Common Logarithms Were Once More Important Than They Are Today

Common logarithms were once very important in computations because base 10 is the basis for our number system. If one knows the common logarithms of all the numbers from 1 to 10, one can determine the common logarithm of every positive real number.

All algebra students 30 years ago had access to tables showing the common logarithms (usually to four decimal places) of all numbers from 1 to 10, in increments of 0.01. We give this in Table 3 at the back of the book. Thus one could look up, for example, log 2.73, log 1.02, and log 9.65. These values could then be used to perform some very difficult computations. One example is given in Example 2 below.

EXAMPLE 1 ***Obtaining Common Logarithms on a Calculator***

Use a calculator to compute (a) log 4.7 (b) log 0.02456 (c) log 10,584 and (d) log e.

SOLUTION Using the [log] key, we obtain

(a) $\log 4.7 = 0.672097857$
(b) $\log 0.02456 = -1.609771638$
(c) $\log 10{,}584 = 4.024649831$
(d) Using $e = 2.718281828$,† we have $\log e = 0.434294481$

EXAMPLE 2 ***Calculation Using Common Logarithms (An Example from the Past)***

Approximate

$$x = \frac{(395)^{1/3}(1280)^{3/4}}{(89)^{1/2}}$$

SOLUTION Taking common logarithms and using the rules of the last section, we have

$$\log x = \frac{1}{3}\log 395 + \frac{3}{4}\log 1280 - \frac{1}{2}\log 89$$

Now we use a table:

From Table 3 ↓

$$\log 395 = \log(3.95 \times 10^2) = \log 3.95 + \log 10^2 = 0.5966 + 2 = 2.5966$$
$$\log 1280 = \log(1.28 \times 10^3) = \log 1.28 + \log 10^3 = 0.1072 + 3 = 3.1072$$
$$\log 89 = \log(8.9 \times 10^1) = \log 8.9 + \log 10 = 0.9494 + 1 = 1.9494$$

---

†To get e on a calculator, press [1] [INV] [ln x] or [1] [2nd F] [ln x] (or [2nd F] [ln x] [1] or something similar to one of these).

Then
$$\log x = \frac{1}{3}(2.5966) + \frac{3}{4}(3.1072) - \frac{1}{2}(1.9494)$$
$$= 0.8655 + 2.3304 - 0.9747 = 2.2212$$

Thus

$$x \overset{\text{Property 1}}{=} 10^{\log x} = 10^{2.2212} = 10^2 10^{0.2212} = (100)10^{0.2212}$$

What is $10^{0.2212}$? If $y = 10^{0.2212}$, then $\log y = 0.2212$. We use the table again. We would find

$$\log 1.66 = 0.2201 \qquad \text{and} \qquad \log 1.67 = 0.2227$$

Now, 0.2212 is closer to 0.2201 than to 0.2227. Thus, $10^{0.2212} \approx 1.66$ and

$$x \approx (100)(1.66) = 166$$

Our answer is accurate as far as it goes. Using a calculator, the answer, correct to 10 digits, is

$$x = 166.4350685$$

Natural Logarithms

The second, and more important, base for logarithms is the base e. This is the base used in calculus. The number e was discussed in Section 6.2. Logarithms to the base e are called **natural logarithms** and are denoted by $\ln x$. That is,

Definition of the Natural Logarithm Function

$$\ln x = \log_e x, \qquad x > 0 \tag{2}$$

We stress that the function $y = \ln x$ is the inverse of the function $y = e^x$. It is the logarithmic function encountered in the overwhelming majority of applications. One reason for this is that $y = e^{kx}$ is the function that describes constant relative growth, as we saw in Section 6.2.

We can rewrite the four statements on p. 346 in terms of the natural logarithmic function:

Basic Properties of the Natural Logarithm Function

	Property	*Illustration*
(a)	$y = \ln x$ is equivalent to $x = e^y$	If $y = \ln 8$, then $8 = e^y$
(b)	$y = e^x$ is equivalent to $x = \ln y$	If $y = e^{-3}$, then $-3 = \ln y$
(c)	$e^{\ln x} = x$ for every positive real number x	$e^{\ln 17.4} = 17.4$
(d)	$\ln e^x = x$ for every real number x	$\ln e^{-8.2} = -8.2$

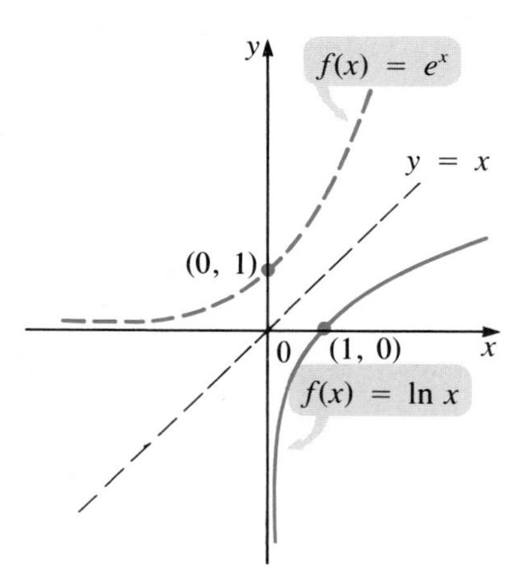

Figure 1 The graph of $f(x) = \ln x$ is the reflection about the line $y = x$ of the graph of e^x.

A graph of $\ln x$ is given in Figure 1. It is the reflection about the line $y = x$ of the graph of e^x.

EXAMPLE 3 *Finding the Domain of a Logarithmic Function*

Find the domain of each function.
(a) $f(x) = \ln(x - 2)$ (b) $f(x) = \ln(3 - x)$
(c) $f(x) = \ln(x^2 - 2x - 8)$

SOLUTION $\ln x$ is defined for $x > 0$.

(a) $\ln(x - 2)$ is defined if $x - 2 > 0$ or $x > 2$ so domain of $\ln(x - 2) = \{x: x > 2\}$
(b) We need $3 - x > 0$ or $3 > x$ so domain of $\ln(3 - x) = \{x: x < 3\}$
(c) In Example 9 in Section 2.8 (on p. 137) we found that

$$x^2 - 2x - 8 > 0 \quad \text{if} \quad x < -2 \text{ or } x > 4$$

Thus domain of $\ln(x^2 - 2x - 8) = \{x: x < -2 \text{ or } x > 4\}$ ■

EXAMPLE 4 *Sketching a Logarithmic Graph*

Sketch the graph of $f(x) = 2 - \ln(1 - x)$.

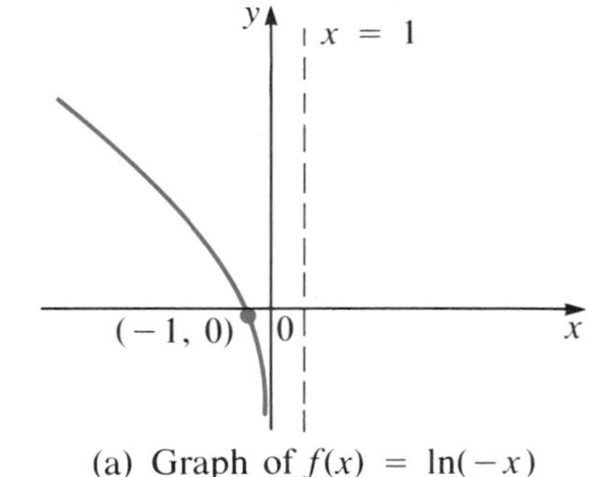

(a) Graph of $f(x) = \ln(-x)$

SOLUTION We do this in four steps.

Step 1 The graph of $y = \ln(-x)$ is given in Figure 2(a). It is the graph of $\ln(x)$ reflected about the y-axis (see p. 196).

Step 2 The graph of $y = \ln(1 - x) = \ln(-(x - 1))$ is the graph $y = \ln(-x)$ shifted 1 unit to the right (see p. 195). It is given in Figure 2(b).

Step 3 The graph of $y = -\ln(1 - x)$ is the graph of $y = \ln(1 - x)$ reflected about the x-axis (see p. 196) and is given in Figure 2(c).

Step 4 The graph of $y = 2 - \ln(1 - x)$ is the graph of $y = -\ln(1 - x)$ shifted up 2 units. It is given in Figure 2(d).

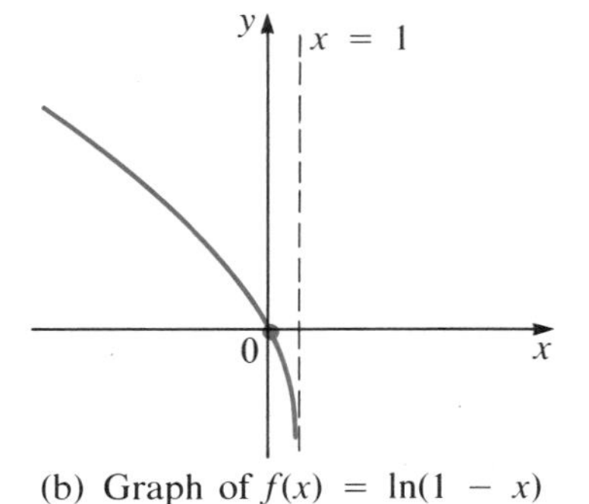

(b) Graph of $f(x) = \ln(1 - x)$

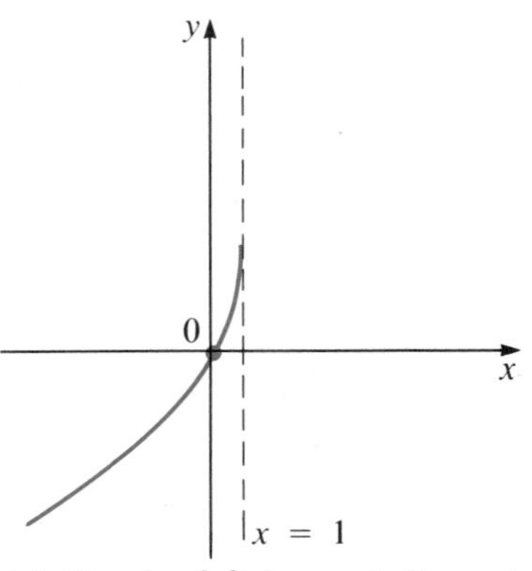

(c) Graph of $f(x) = -\ln(1 - x)$

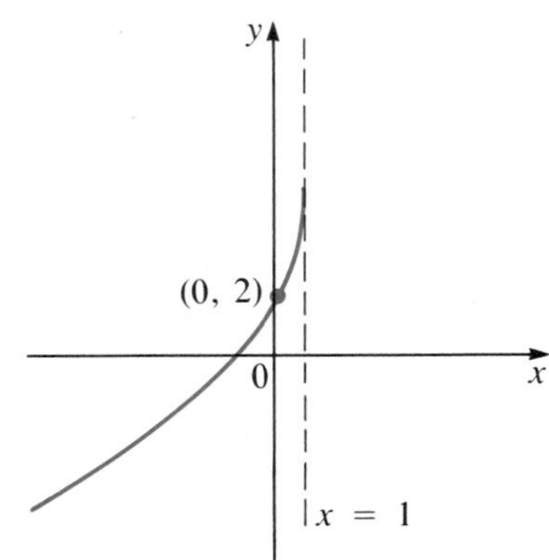

(d) Graph of $f(x) = 2 - \ln(1 - x)$

Figure 2 The graph of $2 - \ln(1 - x)$ is obtained from the graph of $\ln x$ by reflecting about the y-axis, shifting 1 unit to the right, reflecting about the x-axis, and shifting up 2 units (in that precise order).

Because ln x is so important, we repeat the rules of logarithms written in terms of ln x.

Properties of Logarithms for $\ln x = \log_e x$

Property	*Illustration*
1. $e^{\ln x} = x$	$e^{\ln 2} = 2$
2. $\ln e^x = x$	$\ln e^{1.7} = 1.7$
3. $\ln e = 1$	
4. $\ln 1 = 0$	
5. $\ln xw = \ln x + \ln w$	$\ln 6 = \ln 2 \cdot 3 = \ln 2 + \ln 3$
6. $\ln \frac{x}{w} = \ln x - \ln w$	$\ln 2 = \ln \frac{6}{3} = \ln 6 - \ln 3$
7. $\ln \frac{1}{x} = -\ln x$	$\ln \frac{1}{5} = -\ln 5$
8. $\ln x^r = r \ln x$	$\ln 81 = \ln 3^4 = 4 \ln 3$

EXAMPLE 5 *Using Properties of Logarithms to Compute New Logarithms from Given Ones*

Given that $\ln 2 \approx 0.6931$ and $\ln 5 \approx 1.6094$, compute (a) $\ln 10$, (b) $\ln \frac{2}{5}$, (c) $\ln 8$, and (d) $\ln \frac{16}{25}$.

SOLUTION

(a) $\ln 10 = \ln (2 \cdot 5) \overset{\text{Property 5}}{=} \ln 2 + \ln 5 \approx 0.6931 + 1.6094 = 2.3025$

(b) $\ln \frac{2}{5} \overset{\text{Property 6}}{=} \ln 2 - \ln 5 \approx 0.6931 - 1.6094 = -0.9163$

(c) $\ln 8 = \ln 2^3 \overset{\text{Property 8}}{=} 3 \ln 2 \approx 3(0.6931) = 2.0793$

(d) $\ln \frac{16}{25} = \ln 16 - \ln 25 = \ln 2^4 - \ln 5^2$

$= 4 \ln 2 - 2 \ln 5 \approx 4(0.6931) - 2(1.6094)$

$= -0.4464$ ■

EXAMPLE 6 *Combining Logarithmic Terms*

Write these as a single logarithm:

(a) $\ln (x - 1) - 2 \ln (x + 5)$

(b) $\ln x + 2 \ln (x + 1) + 3 \ln (x + 2)$

SOLUTION

(a) $\ln (x - 1) - 2 \ln (x + 5) \overset{\text{Property 8}}{=} \ln (x - 1) - \ln (x + 5)^2 \overset{\text{Property 6}}{=} \ln \dfrac{x - 1}{(x + 5)^2}$

$$\text{(b)}\ \ln x + 2\ln(x+1) + 3\ln(x+2) \overset{\text{Property 8}}{=} \ln x + \ln(x+1)^2 + \ln(x+2)^3$$

$$\overset{\text{Property 5}}{=} \ln[x(x+1)^2(x+2)^3]$$

All scientific calculators have [ln] or [ln x] buttons.

EXAMPLE 7 *Finding Values of ln x on a Calculator*

Use a calculator to compute (a) ln 4.7, (b) ln 0.02486, (c) ln 2486, (d) ln 10,584, and (e) ln 10.

SOLUTION

(a) $\ln 4.7 = 1.547562509$
(b) $\ln 0.02486 = -3.694495193$
(c) $\ln 2486 = 7.818430272$
(d) $\ln 10{,}584 = 9.267098706$
(e) $\ln 10 = 2.302585093$

In Table 1, we provide some values of ln x. Of course, these and other values can be obtained with greater accuracy on a calculator. A more complete table appears in Table 2 at the back of the book.

Table 1
Sample Values Taken by the Natural Logarithm Function ln x

x	0.0	0.1	0.5	1	2	5	10	50	100	1000	100,000	10^{50}
ln x	undefined	−2.3026	−0.6931	0	0.6931	1.6094	2.3026	3.9120	4.6052	6.9078	11.5129	115.1293

EXAMPLE 8 *The Growth Rate of ln x*

Suppose $y = 4 \ln x$.

(a) What happens to y if x doubles?
(b) What happens to y if x is halved?

SOLUTION

(a) If x doubles, then x becomes $2x$ and

$$y = 4\ln 2x = 4(\ln 2 + \ln x) = 4\ln 2 + 4\ln x$$

Thus, y increases by $4 \ln 2 \approx 2.77$.

(b) Now x is replaced by $\frac{x}{2}$ and

$$y = 4 \ln \frac{x}{2} = 4(\ln x - \ln 2) = 4 \ln x - 4 \ln 2$$

Thus, y is reduced by $4 \ln 2 \approx 2.77$.

Computing Logarithms to Other Bases

Sometimes we need to compute something like $\log_2 35$. The following three rules are very useful.

Change of Base Properties

Suppose $a > 0$, $b > 0$, and $x > 0$.

9. $\log_a b = \dfrac{1}{\log_b a}$ 10. $\log_a x = \dfrac{\ln x}{\ln a}$ 11. $\log_a x = \dfrac{\log x}{\log a}$

Proof of Property 9

Let $y = \log_a b$. Then

$a^y = b$	Definition of the logarithm
$\log_b a^y = \log_b b$	Take the logarithm to the base b of each side
$y \log_b a = 1$	Properties 8 and 3 on pp. 347 and 346
$y = \dfrac{1}{\log_b a}$	
$\log_a b = \dfrac{1}{\log_b a}$	Since $y = \log_a b$ ■

Proof of Property 10

Let $u = \log_a x$. Then

$a^u = x$	
$\ln a^u = \ln x$	Take the natural logarithm of each side
$u \ln a = \ln x$	Property 8
$u = \dfrac{\ln x}{\ln a}$	
$\log_a x = \dfrac{\ln x}{\ln a}$	Since $u = \log_a x$ ■

Proof of Property 11

Prove as for Property 10, but take common logarithms of each side instead. ■

EXAMPLE 9

Compute $\log_2 35$.

SOLUTION

Method 1 Using Property 10, we have

$$\log_2 35 = \frac{\ln 35}{\ln 2} \overset{\text{From a calculator}}{=} \frac{3.555348061}{0.69314718} = 5.129283017$$

Method 2 Using Property 11, we obtain

$$\log_2 35 = \frac{\log 35}{\log 2} \overset{\text{From a calculator}}{=} \frac{1.544068044}{0.301029995} = 5.129283017$$

Check If $\log_2 35 = 5.129283017$, then $2^{5.129283017} = 35$. We can verify on our calculator that this is true.

Problems 6.4

Readiness Check

I. Which of the following is a property of natural logarithms?
   a. $\ln x + \ln w = \ln (x + w)$ b. $\ln 0 = 1$
   c. $\ln (1/x) = 1/\ln x$ d. $\ln x^r = r \ln x$

II. Which of the following is equal to x for *every* real number x?
   a. $e^{\ln x}$ b. $\ln e^x$ c. $10^{\log x}$ d. $\ln x^e$

III. If $x = e^{4t}$, then $t =$ _____.
   a. $e^{(1/4)x}$ b. $\dfrac{\ln x}{\ln 4}$ c. $\dfrac{\ln x}{4}$ d. $\ln \dfrac{x}{4}$
   e. None of these.

IV. If $x = 5^{4u}$, then $u =$ _____.
   a. $\dfrac{\ln x}{4 \ln 5}$ b. $\dfrac{\ln x}{\ln 4 \ln 5}$
   c. $\ln x - \ln 4 - \ln 5 = \ln x - \ln 20$ d. $\dfrac{\ln \frac{x}{4}}{\ln 5}$
   e. $5 \ln \dfrac{x}{4}$

V. If $y = \ln (x + 5)$, then $x =$ _____.
   a. $e^y + 5$ b. $5e^y$ c. $e^y - \ln 5$
   d. $e^y - 5$ e. $\dfrac{e^y}{\ln 5}$

In Problems 1–12, solve for the unknown variable. Do not use a calculator.

1. $x = \log 1000$
2. $y = \log 0.001$
3. $z = \log 10^{20}$
4. $u = \log 10^{-8}$
5. $v = 10^{\log 3.4}$
6. $w = 10^{\log 10^{10}}$
7. $x = e^{\ln 0.235}$
8. $x = \ln e^{6.4}$
9. $x = \ln e^{\pi}$
10. $x = e^{2 \ln 3}$
11. $x = \ln e^{6.4}$
12. $x = \ln e^{-3.7}$

In Problems 13–22, write as a single logarithm.

13. $\ln (x + 3) + \ln x$
14. $\ln (x + 1) - \ln (x - 1) + \ln (x^2 - 1)$
15. $\ln x + \ln y + \ln z$
16. $3 \ln x - 4 \ln y$
17. $\dfrac{1}{2} \ln z + \dfrac{3}{4} \ln w - 2 \ln (x^2 + 1)$
18. $\ln 5 + \ln 2 + \ln 3$

Answers to Readiness Check

I. d II. b ($\ln x$ and $\log x$ are not defined for $x \le 0$) III. c IV. a V. d

19. $\ln 20 - \ln 2 - \ln 5$
20. $\ln \dfrac{1}{2} - 2 \ln \dfrac{1}{x}$
21. $3 \ln \dfrac{2}{w} - 4 \ln \dfrac{z}{3}$
22. $\dfrac{1}{2} \ln (x^2 - 1) - \dfrac{2}{3} \ln (x + 1) + \dfrac{3}{4} \ln (z^3 - 2)$

In Problems 23–35 find the domain of the given logarithmic function.

23. $f(x) = \ln (x + 1)$
24. $f(x) = \ln (x - 7)$
25. $f(x) = \ln (1 - x)$
26. $f(x) = \ln (x^2 + 1)$
27. $f(x) = \ln |x|$
28. $f(x) = \ln |x^2 - 1|$
29. $f(x) = \ln (x^2 - 1)$
30. $f(x) = \ln (x^2 - 2x + 1)$
31. $f(x) = \ln (x^2 - x - 12)$
32. $f(x) = \ln (x^2 + 4x - 21)$
33. $f(x) = \ln (x^2 + 4x + 10)$
34. $f(x) = \ln (-x^2 - 2x - 2)$
35. $f(x) = \ln \left(\dfrac{1}{x}\right)$

In Problems 36–44 nine functions are listed. Match each function with one of the graphs given below.

36. $\ln (-x)$
37. $\ln (x - 1)$
38. $\ln (x + 1)$
39. $\ln x + 1$
40. $\ln x - 1$
41. $\ln x$
42. $-\ln x$
43. $1 - \ln x$
44. $\ln x^2$

(a)

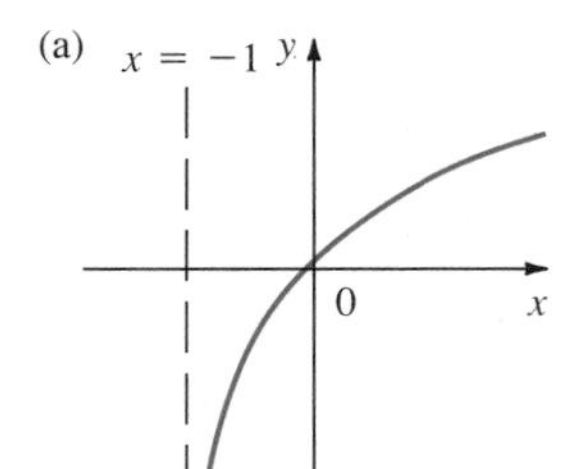

(b)

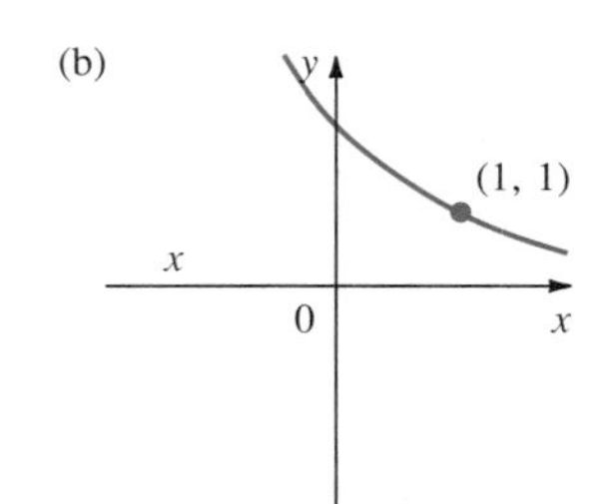

(c)

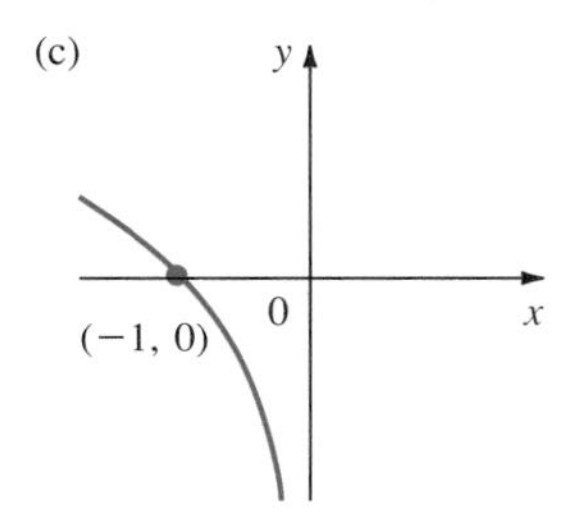

(d)

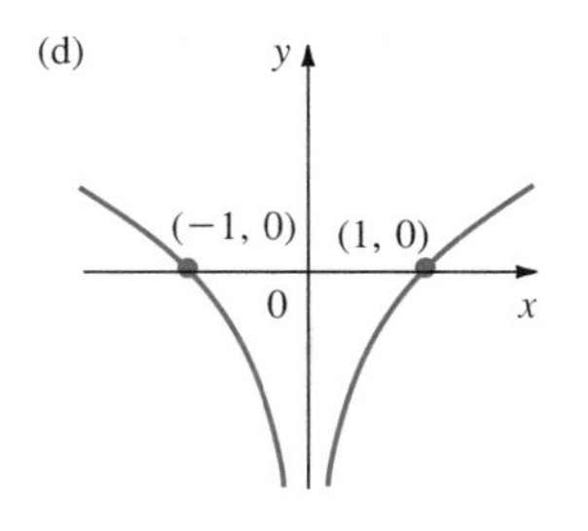

(e)

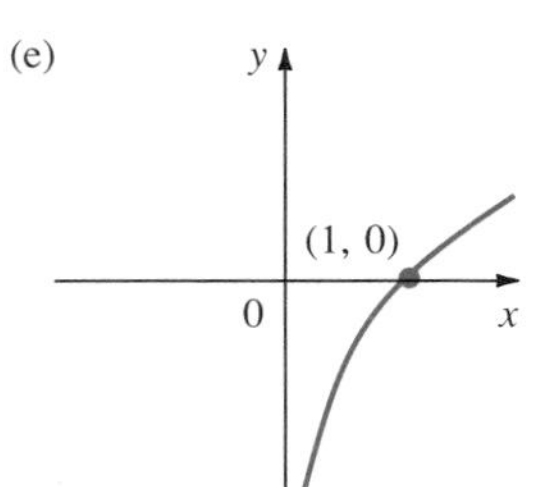

(f)

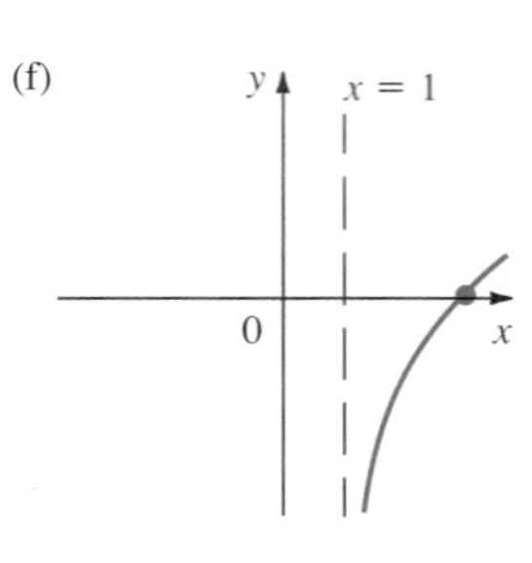

(g)

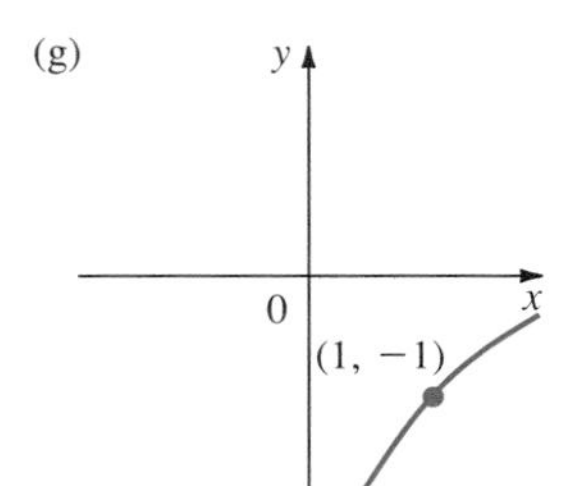

(h)

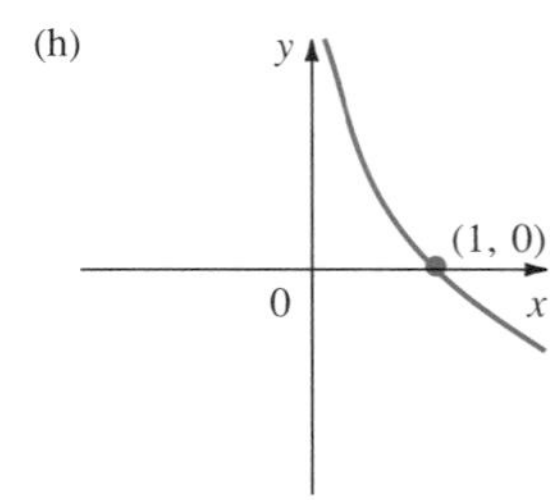

(i)

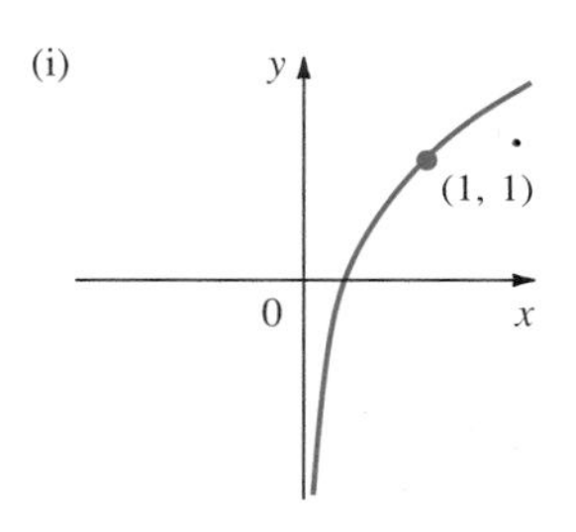

In Problems 45–54, solve for x. Do not use any table or calculator.

45. $\log x = 4 \log 2 + 2 \log 3$
46. $\log x = 4 \log \frac{1}{2} - 3 \log \frac{1}{3}$
47. $\log x = a \log b + b \log a$
48. $\log x = \log 1 + \log 2 + \log 3 + \log 4$
49. $\log x = \log 2 - \log 3 + \log 5 - \log 7$
50. $\ln x^3 = 2 \ln 5 - 3 \ln 2$
51. $\ln \sqrt[3]{x} = 2 \ln 3 + \ln 6$
52. $\ln x^2 - \ln x = \ln 18 - \ln 6$
53. $\ln x^{3/2} - \ln x = \ln 9 - \ln 3$
54. $\ln x^3 - \ln x - \ln 2x = \ln 32 - 2 \ln 4$

In Problems 55–57 do not use a calculator.

55. Given $\log 6.84 \approx 0.8351$. Approximate (a) $\log 684$ (b) $\log 6{,}840{,}000$ (c) $\log 0.0684$ (d) $\log 0.000000684$.
56. Given $\log 2.97 \approx 0.4728$. Approximate (a) $\log 2970$ (b) $\log (297 \times 10^3)$ (c) $\log 0.297$ (d) $\log (0.0297 \times 10^{-5})$.
57. Given $\log 9.52 \approx 0.9786$. Approximate (a) $\log 95.2$ (b) $\log 95{,}200$ (c) $\log 0.0952$ (d) $\log (95.2 \times 10^{-6})$.

In Problems 58–68, use the logarithms given in Problems 55–57 to approximate the common logarithm of the given expression.

58. $\dfrac{1}{684}$ 59. $\dfrac{1}{0.00952}$

60. $\dfrac{100}{2970}$ 61. $\sqrt{29{,}700}$

62. $\sqrt[3]{0.952}$ 63. $\dfrac{1}{\sqrt{68.4}}$

64. $(952)^{3/4}$ 65. $(68.4)(29.7)(95.2)$

66. $\dfrac{0.684}{0.0952}$ * 67. $\dfrac{\sqrt{6840}\sqrt[3]{95.2}}{(0.297)^5}$

* 68. $\dfrac{(29.7)^{3/8}(95{,}200)^{3/10}}{(68{,}400)^2}$

69. Let $y = 3 \ln x$. What happens to y if (a) x triples? (b) x is replaced by $\dfrac{x}{10}$?

70. Let $y = 6 \ln \dfrac{1}{\sqrt{x}}$. What happens to y if (a) x is multiplied by 4? (b) x is replaced by $0.2x$?

In Problems 71–86, find the given value on a calculator.

71. $\log 57.24$ 72. $\log 0.02315$
73. $\log 50002$ 74. $\log 0.00371496$
75. $\ln 1000$ 76. $\ln 37.85$
77. $\ln 0.0325$ 78. $\ln 18.18$
79. $\log \pi$ 80. $\ln \pi$
81. $\ln e^{\pi}$ * 82. $\ln \pi^e$
* 83. $\log \pi^{\pi}$ * 84. $\log e^{\pi}$
* 85. $\log \pi^e$ 86. $\ln \ln 10$

In Problems 87–94, use Property 10 or Property 11 to compute the given logarithm.

87. $\log_3 4$ 88. $\log_4 3$
89. $\log_6 25$ 90. $\log_{1/2} 5$
91. $\log_{1/4} \frac{1}{3}$ 92. $\log_{0.3} 57.4$
93. $\log_{28.2} 39.5$ 94. $\log_{6.4} 1753$

95. Given that $\log_2 5 \approx 2.3219$, approximate (a) $\log_5 2$ (b) $\log_5 4$ (c) $\log_5 \frac{1}{8}$.

96. Given that $\log_3 8 \approx 1.8928$, approximate (a) $\log_8 3$ (b) $\log_8 27$ (c) $\log_8 \frac{1}{9}$.

* 97. Natural logarithms can be approximated on a hand calculator even if the calculator does not have an [ln] key. If $\frac{1}{2} \le x \le \frac{3}{2}$ and if $A = \dfrac{x-1}{x+1}$, then a good approximation to $\ln x$ is

$$\ln x \approx \left[\left(\frac{3A^2}{5} + 1\right) \cdot \frac{A^2}{3} + 1\right] 2A, \qquad \frac{1}{2} \le x \le \frac{3}{2}$$

(a) Use this formula to calculate $\ln 0.8$ and $\ln 1.2$.
(b) Using the facts about logarithms, calculate (approximately)

$$\ln 2 = \ln\left(\frac{3}{2} \cdot \frac{4}{3}\right)$$

(c) Using the result of (b), calculate $\ln 3$ and $\ln 8$.

* 98. Show, using logarithms, that $a^b = b^a$, where $a = [1 + 1/n]^n$ and $b = [1 + 1/n]^{n+1}$. What does this result prove when $n = 1$?

* 99. The quantity $n! = n(n-1)(n-2) \cdot \cdots \cdot 3 \cdot 2 \cdot 1$ grows very rapidly as n increases. According to **Stirling's formula,** when n is large,

$$n! \approx \sqrt{2\pi n}\left(\frac{n}{e}\right)^n$$

Use Stirling's formula to estimate 100! and 200!. [Hint: Use common logarithms.]

100. Watch carefully. Suppose that $0 < A < B$. Because the logarithm is an increasing function, we have
(a) $\log A < \log B$; then
(b) $10A \cdot \log A < 10B \cdot \log B$,
(c) $\log A^{10A} < \log B^{10B}$,
(d) $A^{10A} < B^{10B}$.
On the other hand, we run into trouble with particular choices of A and B. For instance, choose $A = \dfrac{1}{10}$ and $B = \dfrac{1}{2}$. Clearly, $0 < A < B$, but $A^{10A} = \left(\dfrac{1}{10}\right)^1 = \dfrac{1}{10}$ is greater than $B^{10B} = \left(\dfrac{1}{2}\right)^5 = \dfrac{1}{32}$. Where was the first false step made?

101. In Problem 46 of Section 6.2 (p. 343), we defined the function

$$\sinh x = \frac{e^x - e^{-x}}{2}$$

(a) If $\sinh x = 2$, show that $e^x - e^{-x} = 4$.
(b) Show that $e^{2x} - 4e^x - 1 = 0$.
(c) Let $w = e^x$ to obtain the quadratic equation

$$w^2 - 4w - 1 = 0$$

and find two values of w that satisfy this equation.
(d) Show that only one of the values obtained in (c) is useful.
(e) Take a natural logarithm to find the unique value of x for which $\sinh x = 2$.

In Problems 102–104, use the technique of Problem 101 to solve for x.

102. $\sinh x = \frac{1}{2}$ 103. $\sinh x = 4$ 104. $\sinh x = -10$

In Problem 48 of Section 6.2 (p. 344) we defined the function

$$\cosh x = \frac{e^x + e^{-x}}{2}$$

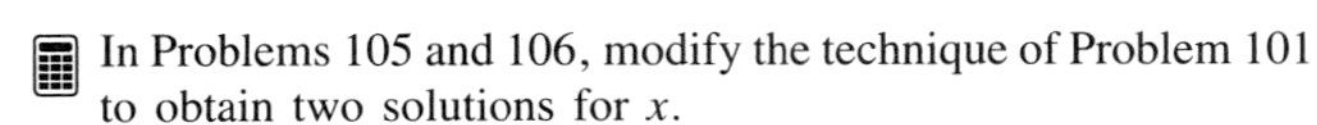

In Problems 105 and 106, modify the technique of Problem 101 to obtain two solutions for x.

105. $\cosh x = 2$ 106. $\cosh x = 5$

Graphing Calculator Problems

107. Sketch $\ln\left(\frac{x}{2}\right)$ and $-\ln\left(\frac{x}{2}\right)$ on the same screen.
108. Sketch $\ln 3x$ and $\ln(-3x)$ on the same screen.
109. Sketch $\ln x$, $\ln(x - 1.5)$, and $\ln(x + 1.5)$ on the same screen.
110. Sketch $\ln(1.7x)$, $\ln(1.7x) + 1$, and $\ln(1.7x) - 1$ on the same screen.

In Problems 111–125 obtain a sketch of the given function.

111. $\log 4x$
112. $4 \log x$
113. $\log\left(\frac{x-1}{2}\right)$
114. $\log 50(x - 1)$
115. $-3 \log(-x)$
116. $\ln(2 - x)$
117. $-2 + 3 \ln(x + 1)$
118. $2.3 + 12 \ln\left(\frac{x}{6} - 1\right)$
119. $\ln x^{10}$
120. $\ln(x^2 + 1)$
121. $\ln e^x$
122. $\ln \ln x$
123. $\log_4 x$ $\left[\text{Hint: } \log_4 x = \frac{\ln x}{\ln 4}.\right]$
124. $\log_{1/2} x$
125. $\log_2(x^2 + 1)$

6.5 Equations Involving Exponential and Logarithmic Functions

Exponential and logarithmic equations arise in a great variety of applications. We saw some of these already. In this section we see how to solve some types of these equations.

Exponential Equations

Definition

An **exponential equation** is an equation involving one or more exponential functions.

EXAMPLE 1 ***Solving an Exponential Equation***

Solve for x if

$$2^x = 8$$

SOLUTION Since $2^3 = 8$ (and 2^x is a one-to-one function), we conclude that $x = 3$.

It is rare that we can solve an exponential equation by inspection, as in Example 1. When we cannot, we can often solve it by taking logarithms, as the next three examples illustrate.

EXAMPLE 2 ***Solving an Exponential Equation by Taking Logarithms***

Find x if

$$5^x = 30$$

SOLUTION We take natural logarithms of both sides:

$$5^x = 30$$

$$\ln 5^x = \ln 30$$

$$x \ln 5 = \ln 30 \qquad \text{Since } \ln x^r = r \ln x$$

$$x = \frac{\ln 30}{\ln 5} \qquad \text{Divide by } \ln 5$$

$$x = \frac{3.401197382}{1.609437912} \approx 2.113 \qquad \text{Values obtained on a calculator}$$

We can also obtain this answer by taking common logarithms:

$$5^x = 30$$

$$\log 5^x = \log 30$$

$$x \log 5 = \log 30$$

$$x = \frac{\log 30}{\log 5} = \frac{1.447121255}{0.698970004} \approx 2.113$$

The reason we get the same answer in both cases is that by Property 10 on p. 358,

$$\frac{\log 30}{\log 5} = \frac{\log_{10} 30}{\log_{10} 5} = \frac{\dfrac{\ln 30}{\ln 10}}{\dfrac{\ln 5}{\ln 10}} = \frac{\ln 30}{\ln 5}$$

In the following examples we will take natural logarithms when logarithms are called for because, as we saw in Example 2, either logarithm leads to the same answer. [Sometimes, however, it is more convenient to use one form rather than the other, as the next example illustrates.]

EXAMPLE 3 ***Solving an Exponential Function***

Solve the following equation for r:

$$e^{2(r-4)} = 65$$

SOLUTION

$$e^{2(r-4)} = 65$$

$$\ln e^{2(r-4)} = \ln 65 \qquad \text{Take natural logarithms of both sides}$$

$$2(r-4) = \ln 65 \qquad \ln e^x = x \quad [\text{but } \log e^x \neq x\text{—that's why it is better to use } \ln x \text{ here}]$$

$$2r - 8 = \ln 65$$

$$2r = \ln 65 + 8$$

$$r = \frac{\ln 65 + 8}{2}$$

$$\approx \frac{4.17439 + 8}{2} \approx 6.087 \qquad \ln 65 \text{ obtained on a calculator}$$

NOTE In solving exponential equations involving the function e^x, we will always take natural logarithms because $\ln e^u = u$ for any real number [but $\log e^u = u \log e$—which is a more complicated expression].

EXAMPLE 4 *Solving an Exponential Equation*

Solve the following equation for u:

$$3^u = 5^{u-4}$$

SOLUTION

$$3^u = 5^{u-4}$$

$$\ln 3^u = \ln 5^{u-4} \qquad \text{Take natural logarithms on both sides}$$

$$u \ln 3 = (u-4)\ln 5 \qquad \ln x^r = r \ln x$$

$$u \ln 3 = u \ln 5 - 4 \ln 5 \qquad \text{Multiply through}$$

$$u \ln 5 - u \ln 3 = 4 \ln 5 \qquad \text{Rearrange terms}$$

$$u(\ln 5 - \ln 3) = 4 \ln 5 \qquad \text{Factor out the } u$$

$$u \ln \frac{5}{3} = 4 \ln 5 \qquad \ln x - \ln y = \ln \frac{x}{y}$$

$$u = \frac{4 \ln 5}{\ln \frac{5}{3}} \approx 12.6026 \qquad \text{Values obtained on a calculator}$$

Check You can check this answer on a calculator by using the $\boxed{y^x}$ key.

$$3^u \approx 3^{12.6026} = 1{,}030{,}310.972$$

$$5^{u-4} \approx 5^{8.6026} = 1{,}030{,}289.703$$

The slight difference in answers occurs because we rounded. If you store the value $u = \dfrac{4 \ln 5}{\ln \frac{5}{3}} \approx 12.60264041$, then on most calculators you will obtain the same value for 3^u and 5^{u-4}. ■

EXAMPLE 5 ***An Exponential Equation That Leads to a Quadratic Equation***

Solve the following equation for x:

$$e^x - 10e^{-x} = 3$$

SOLUTION

$$\begin{aligned}
e^x - 10e^{-x} &= 3 \\
e^x e^x - 10e^{-x}e^x &= 3e^x && \text{Multiply both sides by } e^x \\
e^{2x} - 10e^{-x+x} &= 3e^x && e^u e^v = e^{u+v} \\
e^{2x} - 10e^0 &= 3e^x \\
e^{2x} - 10 &= 3e^x && e^0 = 1 \\
e^{2x} - 3e^x - 10 &= 0
\end{aligned}$$

If we set $u = e^x$, then $u^2 = e^{2x} = (e^x)^2$, and we see that this is an equation of quadratic type (see p. 114):

$$\begin{aligned}
u^2 - 3u - 10 &= 0 && \text{Set } u = e^x \\
(u - 5)(u + 2) &= 0 && \text{Factor} \\
u = 5 \quad &\text{or} \quad u = -2 \\
e^x = 5 \quad &\text{or} \quad e^x = -2 && u = e^x
\end{aligned}$$

But $e^x > 0$, so the equation $e^x = -2$ is impossible, and we are left with

$$\begin{aligned}
e^x &= 5 \\
\ln e^x &= \ln 5 && \text{Take logarithms of both sides} \\
x &= \ln 5
\end{aligned}$$

Check

$$\begin{aligned}
e^x - 10e^{-x} &= e^{\ln 5} - 10e^{-\ln 5} \\
&= 5 - 10e^{\ln 1/5} && e^{\ln x} = x \text{ and } -\ln x = \ln \frac{1}{x} \\
&= 5 - 10 \cdot \frac{1}{5} && e^{\ln x} = x \\
&= 5 - 2 = 3
\end{aligned}$$

NOTE There is no way to solve the equation by first taking logarithms. We stress that $\ln(e^x - 10e^{-x}) \neq \ln e^x - \ln 10e^{-x}$.

EXAMPLE 6 *An Exponential Equation with Two Solutions*

Solve the following equation for x: $e^{2x} + 10e^{-2x} = 7$

SOLUTION We proceed as in Example 5:

$$
\begin{aligned}
e^{2x} + 10e^{-2x} &= 7 \\
e^{4x} + 10 &= 7e^{2x} && \text{Multiply both sides by } e^{2x} \\
e^{4x} - 7e^{2x} + 10 &= 0 \\
u^2 - 7u + 10 &= 0 && \text{Set } u = e^{2x} \\
(u - 2)(u - 5) &= 0 && \text{Factor} \\
(e^{2x} - 2)(e^{2x} - 5) &= 0 && u = e^{2x}
\end{aligned}
$$

$$
\begin{aligned}
e^{2x} &= 2 & \qquad e^{2x} &= 5 \\
2x &= \ln 2 & 2x &= \ln 5 \\
x &= \frac{1}{2}\ln 2 & x &= \frac{1}{2}\ln 5
\end{aligned}
$$

The two solutions are $x = \frac{1}{2}\ln 2$ and $x = \frac{1}{2}\ln 5$. ■

EXAMPLE 7 *An Exponential Equation with No Solution*

Solve the following equation for x: $e^x + e^{-x} + 2 = 0$

SOLUTION Since $e^x > 0$ and $e^{-x} > 0$ for every real number, there is no real x for which $e^x + e^{-x} = -2$, and the equation has no solution. We can see this in another way by proceeding as in Examples 5 and 6:

$$
\begin{aligned}
e^x + e^{-x} + 2 &= 0 \\
e^{2x} + 1 + 2e^x &= 0 && \text{Multiply by } e^x \\
u^2 + 2u + 1 &= 0 && \text{Set } u = e^x \\
(u + 1)^2 &= 0 && \text{Factor} \\
u + 1 &= 0 \\
u &= -1 \\
e^x &= -1 && u = e^x
\end{aligned}
$$

which is impossible, since $e^x > 0$ for every real number x.

Logarithmic Equations

Definition

A **logarithmic equation** is an equation involving one or more logarithmic functions.

EXAMPLE 8 *Solving a Logarithmic Equation*

Find x if

$$\ln x = 2$$

SOLUTION The expression $\ln x = \log_e x = 2$ means that

$$e^2 = x$$

That is, $x = e^2 \approx 7.389$.

In solving logarithmic equations, the following facts are useful. They are, essentially, the definition of $\log_a x$. The first property is equation (6) on p. 346.

$$\text{If } \log_a u = \log_a v \quad \text{then} \quad u = v \tag{1}$$

$$\text{If } u = \log_a v \quad \text{then} \quad v = a^u \tag{2}$$

In particular,

$$\text{if } u = \log v \quad \text{then} \quad v = 10^u \tag{3}$$

and

$$\text{if } u = \ln v \quad \text{then} \quad v = e^u \tag{4}$$

EXAMPLE 9 *Solving a Logarithmic Equation*

Find x if

SOLUTION

$$3 \ln (x - 2) = 1.5$$

$$\ln (x - 2) = \frac{1.5}{3} = 0.5$$

$$x - 2 = e^{0.5} \qquad \text{From (4)}$$

$$x = 2 + e^{0.5} \approx 3.6487 \quad ■$$

EXAMPLE 10 *Solving a Logarithmic Equation*

Solve for y:

$$2 = \log \frac{1}{y}$$

SOLUTION From Property 7 on p. 347, we have

$$2 = \log \frac{1}{y} = -\log y$$

$$\log y = -2$$

$$y = 10^{-2} = \frac{1}{100} \qquad \text{From (3)}$$

EXAMPLE 11 *Solving a Logarithmic Equation*

Solve the following equation for x:

$$\log(4x + 10) - \log(x - 2) = 1$$

SOLUTION

$$\log(4x + 10) - \log(x - 2) = 1$$

$$\log \frac{4x + 10}{x - 2} = 1 \qquad \text{Property 6 on p. 347}$$

$$\frac{4x + 10}{x - 2} = 10^1 = 10 \qquad \text{From 3}$$

$$4x + 10 = 10(x - 2)$$

$$4x + 10 = 10x - 20$$

$$6x = 30$$

$$x = 5$$

Check $4x + 10 = 30 > 0$ and $x - 2 = 3 > 0$ so that both $\log(4x + 10)$ and $\log(x - 2)$ are defined. This check must always be carried out. Otherwise, the answer might not make sense. Then, for $x = 5$,

$$\log(4x + 10) - \log(x - 2) = \log 30 - \log 3 = \log \frac{30}{3} = \log 10 = 1 \quad \blacksquare$$

EXAMPLE 12 *A Logarithmic Equation with No Solution*

Solve the following equation for x:

$$\ln(2x + 3) - \ln(x + 7) = 3$$

SOLUTION

$$\ln(2x + 3) - \ln(x + 7) = 3$$

$$\ln \frac{2x + 3}{x + 7} = 3 \qquad \text{Property 6 on p. 347}$$

$$\frac{2x + 3}{x + 7} = e^3 \qquad \text{From 4}$$

$$2x + 3 = e^3(x + 7) \qquad \text{Remember that } e^3 \text{ is a real number}$$

$$2x + 3 = e^3x + 7e^3$$

$$2x - e^3x = 7e^3 - 3 \qquad \text{Rearrange terms}$$

$$x(2 - e^3) = 7e^3 - 3$$

$$x = \frac{7e^3 - 3}{2 - e^3} \approx -7.608$$

Check If $x = -7.608$, then $2x + 3 \approx -12.2 < 0$ and $x + 7 = -0.6 < 0$ so neither $\ln(2x + 3)$ nor $\ln(x + 7)$ is defined, and $x \approx -7.6$ cannot be a solution to the original equation. We conclude that the equation has no solution.

One final note: there are many exponential and logarithmic equations that cannot be solved by the methods of this section. For example, consider the equation

$$e^x + 10^x = 8 \tag{5}$$

We see that $e^0 + 10^0 = 1 + 1 = 2 < 8$ and $e^1 + 10^1 = e + 10 \approx 12.718 > 8$. Thus there is one solution to (5) in the interval (0, 1). But how do we find it? None of the techniques discussed in this section will work. For example, we can try the following:

$$e^x = 8 - 10^x$$
$$\ln e^x = \ln (8 - 10^x)$$
$$x = \ln (8 - 10^x) \tag{6}$$

But equation (6) is no easier to solve than equation (5). In fact, it is worse because $\ln (8 - 10^x)$ is not defined if $8 - 10^x < 0$ or $10^x > 8$ or $x > \log 8 \approx 0.903$.

The best we can do is to approximate the solution by trial and error or by the bisection method discussed in Section 4.5—unless we have access to a graphing calculator. To four decimal places, the solution is $x \approx 0.7669$.

Problems 6.5

Readiness Check

I. What can be concluded from the fact that $\log x = 2.5$?
a. $x = e^{2.5}$ b. $x = 2.5^{10}$
c. $x = 10^{2.5}$ d. $x = \log 2.5$

II. What can be concluded from the fact that $\ln x = -0.27$?
a. $x = e^{-0.27}$ b. $x = -0.27^{10}$
c. $x = 10^{-0.27}$ d. $x = -\ln 0.27$

III. Which are the solutions to the equation $\ln x^2 = \ln x$?
a. $x = -1, 1$ b. $x = 0, 1$ c. $x = 1$
d. $x = 0$ e. There are no real solutions.

IV. Which are the solutions to $\log x^2 - \log (x^2 + 1) = 1$?
a. $x = 0$ b. $x = 1$ c. $x = 0, 1$
d. $x = \pm 1$ e. There are no real solutions.

In Problems 1–73 solve for the unknown variable. Use a calculator where necessary. Give each answer to 3-decimal-place accuracy.

1. $10^x = 100$
2. $e^x = e^3$
3. $4^{-x} = 4$
4. $e^x = \frac{1}{e}$
5. $\log_2 x = 8$
6. $\ln x = 2$
7. $\ln x = -1$
8. $\ln x = e$
9. $\log x = 1$
10. $\log x = -2$
11. $\log x = 0$
12. $\log x = 3$
13. $e^x = 10$
14. $e^x = 0.2$
15. $e^x = -1$
16. $4^x = 12$
17. $7^x = 14$
18. $2.7^x = 8.92$
19. $231^x = 8$
20. $0.019^x = 6$
21. $\ln x = 1.6$
22. $\log x = 0.23$
23. $\log x = -1.57$
24. $\ln x = -2.95$
25. $e^{x-1} = 2$
26. $e^{2x} = 5$
27. $e^{t/2} = 4$
28. $e^{t^2} = 16$
29. $10e^{-h/5} = 50$
30. $50e^{0.21t} = 100$
31. $\ln x + \ln 3 = 1$
32. $\ln x + \ln (x - 1) = \ln 6$
33. $\ln v + \ln \frac{1}{v} = \ln (v - 3)$
34. $\ln (2x - 3) = 4$
35. $2 \ln (0.172w) = 0.856$
36. $\frac{\ln v}{\ln 2} = \frac{\ln 3}{\ln 5}$

Answers to Readiness Check

I. c II. a III. c IV. e

37. $\ln(z-3) - \ln(z+4) = \ln 2$
38. $\ln(z-3) - \ln(z+4) = 3$
39. $e^x + 8e^{-x} = 6$
40. $e^{2x} + 3e^{-2x} = 4$
41. $e^{(1/2)x} - 6e^{-(1/2)x} + 1 = 0$
42. $e^{3x} - 14e^{-3x} + 5 = 0$
43. $e^x + 2e^{-x} + 3 = 0$
44. $e^{x/5} + 12e^{-x/5} + 7 = 0$
45. $z = 1.3^{\log_{1.3} 48}$
46. $p = e^{\log_e 37.4}$
47. $q = 10^{\log_{10} 0.0023}$
48. $r = 6^{\log_6 \sqrt{2}}$
49. $s = 3^{(1/2)\log_3 16}$
50. $t = 10^{-\log_{10} 4}$
51. $u = e^{-(1/2)\ln 100}$
52. $2\log_3 v = 4$
53. $3\log_5 2x = 1$
54. $y = e^{\log_e e}$
55. $\log_2 u^4 = 4$
56. $\log_y 64 = 3$
57. $\log_w 125 = -3$
58. $\log_q 32 = -5$
59. $5 \cdot 2^{x-1} = 20$
60. $e^x e^{x+1} = e^2$
61. $2\log_4 x + 3 = 0$
62. $\log_3(x-2) = 2$
63. $\log_{10}(x+5) = -3$
64. $3^{2y}\, 3^{\log_3 1/3} = 9$
65. $2 + \log_5 5^{1/x} = 5$
66. $10^{t^2+2t-8} = 1,\ t > 0$
67. $10^{t^2+2t-8} = 1,\ t < 0$
68. $\log_2 x + \log_2(x-1) = 3$
69. $\log_4 x - \log_4(x+1) = 1$
70. $\log 5x = \log(x^2+6)$
71. $\ln(x+3) = \ln(2x-5)$
72. $\log(\log(v+1)) = 0$
73. $10^{10^x} = 10$

Graphing Calculator Problems

In Problems 74–84 find all solutions (if any) to each equation by sketching the appropriate functions on your calculator. Give each answer to 2-decimal-place accuracy. Before beginning, read (or reread) Example 7 in Appendix A.

74. $e^x = x + 2$
75. $\ln x = \frac{1}{2}x - 1$
76. $x = e^{1/x}$
77. $e^x + \ln x = 5$
78. $e^x - \ln x = 2$
79. $e^{-x} + x = 1.5$
80. $xe^{-x} = 0.1$
81. $e^x = 2x^2 + 3x + 5$
82. $e^{x/5} = 10x^4 + 8x^3 + 20$
83. $e^{0.1x} = x^5 + 1000$
84. $5^x + 3^x = 100$

6.6 Applications of Exponential and Logarithmic Functions

In this section, we illustrate why logarithmic functions, especially the function $\ln x$, are so important by showing some applications. We begin by modifying problems we have seen before.

Interest Rate Problems

Recall the compound interest formulas discussed in Sections 6.1 and 6.2.

$$\text{Compounding } m \text{ times a year} \quad A(t) = P\left(1 + \frac{r}{m}\right)^{mt} \tag{1}$$

$$\text{Compounding continuously} \quad A(t) = Pe^{rt} \tag{2}$$

where P is the initial amount invested, r is the rate of interest, t is the amount of time (in years) the investment is held, m is the number of times that interest is compounded each year, and $A(t)$ is the value of the investment after t years.

EXAMPLE 1 ***Computing Doubling Time***

Money is invested at 8% compounded quarterly. How long does it take to double?

SOLUTION If P dollars is invested, then the investment has doubled when $A(t) = 2P$. From (1), with $r = 0.08$ and $m = 4$,

$$\begin{aligned} A(t) &= P\left(1 + \frac{0.08}{4}\right)^{4t} \\ 2P &= P(1.02)^{4t} \\ 2 &= (1.02)^{4t} && \text{Divide by } P \end{aligned}$$

We can now continue by taking logarithms. Logarithms to any base will work. We solve the problems using natural logarithms.

$$\begin{aligned} 2 &= (1.02)^{4t} && (3) \\ \ln 2 &= \ln\,(1.02)^{4t} && \text{Take natural logarithms of both sides} \\ \ln 2 &= 4t \ln\,(1.02) && \text{Use Property 8 on p. 347} \\ t &= \frac{\ln 2}{4 \ln\,(1.02)} \\ t &\approx \frac{0.69314718}{4(0.019802627)} \\ &\approx 8.750697195 && \text{From a calculator} \end{aligned}$$

That is, money doubles in about 8.75 years if it is invested at 8% interest compounded quarterly. ■

EXAMPLE 2 *Computing the Time It Takes for an Investment to Triple*

A bond pays 12% interest compounded continuously. If \$10,000 is initially invested, when will the bond be worth \$30,000?

SOLUTION From (2), we have

$$A(t) = Pe^{rt} = 10{,}000e^{0.12t} \qquad (4)$$

(with the annotation $0.12 = 12\%$ pointing to the exponent)

We seek a number t such that $A(t) = 30{,}000$. That is,

$$\begin{aligned} 30{,}000 &= 10{,}000e^{0.12t} \\ e^{0.12t} &= 3 && \text{Divide both sides by 10,000} \\ \ln e^{0.12t} &= \ln 3 && \text{Take natural logarithms of both sides} \\ 0.12t &= \ln 3 && \text{Property 2 on p. 356} \\ t &= \frac{\ln 3}{0.12} = \frac{1.098612289}{0.12} \approx 9.155 \text{ years} \end{aligned}$$

We have shown that an investment triples in approximately 9.155 years if it is invested at 12% interest compounded continuously.

In general, we can easily compute how long it takes for a sum to increase by a factor of k if it is invested at $r\%$ compounded continuously. Written as a decimal, $r\% = r/100$. Then, for P dollars to increase to kP dollars, we must find a t^* such that

$$kP = Pe^{(r/100)t^*}$$
$$k = e^{(r/100)t^*}$$
$$\ln k = rt^*/100$$

Multiplying an Investment by a Factor of *k*

$$t^* = \frac{\ln k}{r/100} = \frac{100 \ln k}{r} \text{ years} \tag{5}$$

is the amount of time over which money invested at $r\%$ compounded continuously will increase by a factor of k.

EXAMPLE 3 ***The Time for an Investment to Increase by a Factor of 4***

How long does it take for a sum of money to increase by a factor of 4 if it is invested at 13% compounded continuously?

SOLUTION Here $k = 4$ and $r = 13$, so from (5),

$$t^* = \frac{100 \ln 4}{13} \approx \frac{138.63}{13} \approx 10.66 \text{ years}$$

In particular,

Doubling Time

If money is invested at $r\%$ interest compounded continuously, it will double in $\frac{100 \ln 2}{r}$ years.

For example, money invested at 10% will double in $100(\ln 2)/10 = 10 \ln 2$ years. In Table 1, see the doubling time at various rates of interest.

Table 1

Interest Rate (%)	Doubling Time (years)	Interest Rate (%)	Doubling Time (years)
1	69.3	8	8.7
2	34.7	10	6.9
3	23.1	12	5.8
4	17.3	15	4.6
5	13.9	18	3.9
6	11.6	25	2.8
7	9.9	50	1.4

Exponential Growth and Decay

In Section 6.2, we used the formula of exponential or constant relative growth:

$$y(t) = y(0)e^{kt} \tag{6}$$

We use this formula in the examples below.

Population Growth

EXAMPLE 4 ***Population Growth***

A bacteria population grows exponentially. A population that is initially 10,000 grows to 25,000 after 2 hours. What is the population (a) after 5 hours? (b) after 24 hours?

SOLUTION From (6), we have

$$P(t) = P(0)e^{kt} \tag{7}$$

where $P(t)$ denotes the population after t hours. We are told that $P(0) =$ 10,000, so (7) becomes

$$P(t) = 10{,}000e^{kt} \tag{8}$$

This problem is different from the problems in Section 6.2 (and more realistic) because now the constant k is not given to us. However, we do have one more piece of information. We are told that $P(2) = 25{,}000$. Inserting $t = 2$ into (8), we obtain

$$\begin{aligned} P(2) = 25{,}000 &= 10{,}000e^{2k} \\ e^{2k} &= \frac{25{,}000}{10{,}000} = 2.5 \\ \ln e^{2k} &= \ln 2.5 \\ 2k &= \ln 2.5 \\ k &= \frac{\ln 2.5}{2} \end{aligned}$$

Then

$$P(t) = 10{,}000e^{kt} = 10{,}000e^{(\ln 2.5/2)t} = 10{,}000e^{(t/2)\ln 2.5}$$

Property 8 on p. 356 ↓ Property 1 ↓

$$= 10{,}000e^{\ln (2.5)^{t/2}} = 10{,}000(2.5)^{t/2}$$

[yˣ] key on a calculator ↓

(a) $P(5) = 10{,}000(2.5)^{5/2} \approx 98{,}821$
(b) $P(24) = 10{,}000(2.5)^{24/2} = 10{,}000(2.5)^{12} \approx 596{,}046{,}448$

NOTE We can compute $k = \dfrac{\ln 2.5}{2} = 0.458145365$, but it is not necessary in this problem. The value of k is interesting because it tells us that the population is growing at the very high rate of approximately 45.8% an hour. ■

EXAMPLE 5 *Estimating the Future Population of India*

The population of India was estimated to be 574,220,000 in 1974 and 746,388,000 in 1984. Assume that the relative growth rate is constant.

(a) Estimate the population in 1994.
(b) When will the population reach 1.5 billion?

SOLUTION
(a) From (7), we have

$$P(t) = P(0)e^{kt}$$

Treat the year 1974 as year zero. Then 1984 = year 10. We are told that

$$P(0) = 574{,}220{,}000 \qquad \text{and} \qquad P(10) = 746{,}388{,}000$$

Thus

$$P(t) = 574{,}220{,}000e^{kt}$$

$$P(10) = 746{,}388{,}000 = 574{,}220{,}000e^{10k}$$

$$e^{10k} = \frac{746{,}388{,}000}{574{,}220{,}000} = 1.299829334$$

$$\ln e^{10k} = \ln 1.299829334$$

$$10k = \ln 1.299829334$$

$$k = \frac{\ln 1.299829334}{10} \approx 0.026$$

so the population grew by about 2.6% a year between 1974 and 1984

The year 1994 is year 20. Thus

$$P(20) = 574{,}220{,}000e^{20k} = 574{,}220{,}000e^{(20 \ln 1.299829334)/10}$$

$$= 574{,}220{,}000e^{2 \ln 1.299829334} = 574{,}220{,}000e^{\ln (1.299829334)^2}$$

Property 1 →

$$= (574{,}220{,}000)(1.299829334)^2$$

$$= (574{,}220{,}000)(1.689556297) = 970{,}177{,}017$$

(b) We seek a number t such that $P(t) = 1{,}500{,}000{,}000$. That is,

$$P(t) = 1{,}500{,}000{,}000 = 574{,}220{,}000e^{kt}$$

$$e^{kt} = \frac{1{,}500{,}000{,}000}{574{,}220{,}000} = 2.612239211$$

$$\ln e^{kt} = \ln 2.612239211$$

$$kt = \ln 2.612239211$$

$$t = \frac{\ln 2.612239211}{k}$$

$$= \frac{\ln 2.612239211}{\ln 1.299829334/10} = \frac{10 \ln 2.612239211}{\ln 1.299829334}$$

$$= 36.6 \text{ years}$$

Then $\qquad \text{Year } 36.6 = 1974 + 36.6 = 2010.6$

We conclude that the population of India would reach 1.5 billion sometime in the year 2010 if its population growth rate continued at the same rate as it was between 1974 and 1984.

NOTE It is tempting to round 1.299829334 to 1.3. If we do this, we obtain

$$P(20) \approx (574{,}220{,}000)(1.3)^2 = (574{,}220{,}000)(1.69) = 970{,}431{,}800.$$

In doing this, we gained 314,783 people (970,431,800 − 970,177,017). This isn't a very large percentage of the total

$$\left(\frac{314{,}783}{970{,}177{,}017} \approx 0.0003 = 0.03\%\right),$$

but it's still a lot of people. This illustrates that you have to be careful when you round. In this example, since you are using a calculator anyway, there is no good reason to round until the end.

Newton's Law of Cooling

Newton's law of cooling states that the rate of change of the temperature of an object is proportional to the temperature difference between the body and its surrounding medium. That is, the temperature difference changes at a constant relative rate. An object that is hotter than its surroundings will cool off; one that is cooler will warm up. Thus the temperature difference will decrease exponentially as a function of time. Let $T(t)$ denote the temperature of the object at time t, and let T_S denote the temperature of the surroundings (T_S is assumed to be constant throughout). If the initial temperature is $T(0)$, then the initial temperature difference is $T(0) - T_S$. Thus, in equation (6) we may set $y(t) = T(t) - T_S$ to obtain

$$T(t) - T_S = (T(0) - T_S)e^{-kt}$$

or

$$T(t) = T_S + (T(0) - T_S)e^{-kt} \tag{9}$$

The minus sign in the exponent indicates that the temperature difference decreases as t increases.

EXAMPLE 6 ***Using Newton's Law of Cooling to Calculate the Temperature of Milk***

When a bottle of milk was taken out of the refrigerator, its temperature was 40°F. An hour later, its temperature was 50°F. The temperature of the air is 70°F, and this temperature does not change.

(a) What was the temperature of the milk after 2 hours?
(b) After how many hours was its temperature 65°F?

SOLUTION Here, $T(0) = 40$, $T_S = 70$, and $T(0) - T_S = -30$. Thus equation (9) becomes

$$T(t) = 70 - 30e^{-kt} \tag{10}$$

But $T(1) = 50$, so

$$\begin{aligned} 50 &= 70 - 30e^{-k} \\ 30e^{-k} &= 20 \\ e^{-k} &= \frac{20}{30} = \frac{2}{3} \\ \ln e^{-k} &= \ln \frac{2}{3} \\ -k &= \ln \frac{2}{3} \end{aligned}$$

Then (10) becomes

$$\begin{aligned} T(t) &= 70 - 30e^{(-k)t} \\ &= 70 - 30e^{t(\ln(2/3))} \overset{\text{Property 8}}{=} 70 - 30e^{\ln(2/3)^t} \\ &\overset{\text{Property 1}}{=} 70 - 30(\tfrac{2}{3})^t \end{aligned}$$

(a) $T(2) = 70 - 30(\frac{2}{3})^2 = 70 - 30(\frac{4}{9}) = 70 - \frac{40}{3} \approx 56.7°\text{F}$
(b) We seek t such that $T(t) = 65$. That is,

$$\begin{aligned} T(t) = 65 &= 70 - 30\left(\frac{2}{3}\right)^t \\ 30\left(\frac{2}{3}\right)^t &= 5 \\ \left(\frac{2}{3}\right)^t &= \frac{5}{30} = \frac{1}{6} \\ t \ln \frac{2}{3} &= \ln \frac{1}{6} && \text{Take natural logarithms of both sides} \\ t &= \frac{\ln \frac{1}{6}}{\ln \frac{2}{3}} = \frac{-1.791759469}{-0.405465108} \approx 4.42 \text{ hours} \\ &\approx 4 \text{ hours } 25 \text{ minutes} \end{aligned}$$

Carbon Dating

Carbon dating is a technique used by archaeologists, geologists, and others who want to estimate the ages of certain artifacts and fossils they uncover. The technique is based on certain properties of the carbon atom. In its natural state, the nucleus of the carbon atom ^{12}C has six protons and six neutrons. The **isotope** carbon-14, ^{14}C, is produced through cosmic-ray bombardment of nitrogen in the atmosphere. Carbon-14 has 6 protons and 8 neutrons and is **radioactive.** It decays by beta emission. That is, when an atom of ^{14}C decays, it gives up an electron to form a stable nitrogen atom ^{14}N. We make the assumption that the ratio of ^{14}C to ^{12}C in the atmosphere is constant. This assumption has been shown experimentally to be approximately valid, for although ^{14}C is being constantly lost through **radioactive decay** (as this process is often termed), new ^{14}C is constantly being produced. Living plants and animals do not distinguish between ^{12}C and ^{14}C, so at the time of death the ratio of ^{12}C to ^{14}C in an organism is the same as the ratio in the atmosphere. However, this ratio changes after death since ^{14}C is converted to ^{12}C but no further ^{14}C is taken in.

It has been observed that ^{14}C decays at a rate proportional to its mass and that its **half-life** is approximately 5580 years.† That is, if a substance starts with 1 gram of ^{14}C, then 5580 years later it would have $\frac{1}{2}$ gram of ^{14}C, the other $\frac{1}{2}$ gram having been converted to ^{14}N. Moreover, the relative rate of decay is constant.

EXAMPLE 7 *Estimating the Age of a Fossil*

A fossil is unearthed, and it is determined that the amount of ^{14}C present is 40% of what it would be for a similarly sized living organism. What is the approximate age of the fossil?

SOLUTION Let $M(t)$ denote the mass of ^{14}C present in the fossil. Since the ^{14}C decays at a constant relative rate, we have, from (6),

$$M(t) = M_0 e^{-kt} \tag{11}$$

The minus sign in the exponent indicates that the amount of ^{14}C present is decreasing as t increases.

Our first task is to determine k. We know that ^{14}C has a half-life of 5580 years. This means that M_0 grams initially ($t = 0$) decay to $\frac{1}{2}M_0$ grams 5580 years later ($t = 5580$), so

$$M(5580) = \frac{1}{2}M_0 = M_0 e^{-5580k}$$

$$e^{-5580k} = \frac{1}{2} \qquad \text{Divide by } M_0$$

† This number was first determined in 1941 by the American chemist W. S. Libby, who based his calculations on the wood from sequoia trees, whose ages were determined by rings marking years of growth.

$$\ln e^{-5580k} = \ln \frac{1}{2} \qquad \text{Take natural logarithms of both sides}$$

$$-5580k = \ln \frac{1}{2}$$

and

$$-k = \frac{\ln \frac{1}{2}}{5580}$$

Then

$$e^{-kt} = e^{(t \ln \frac{1}{2})/5580} \overset{\text{Property 8}}{=} e^{\ln (1/2)^{t/5580}} \overset{\text{Property 1}}{=} \left(\frac{1}{2}\right)^{t/5580}$$

and (11) becomes

$$M(t) = M_0\left(\frac{1}{2}\right)^{t/5580} \tag{12}$$

Now we are told that after t years (from the death of the fossilized organism to the present), $M(t) = 0.4M_0$, and we are asked to determine t. Then

$$0.4M_0 = M_0\left(\frac{1}{2}\right)^{t/5580}$$

$$0.4 = \left(\frac{1}{2}\right)^{t/5580} \qquad \text{Divide by } M_0$$

$$\ln 0.4 = \ln \left(\frac{1}{2}\right)^{t/5580} \qquad \text{Take natural logarithms of both sides}$$

$$\ln 0.4 = \frac{t}{5580} \ln \frac{1}{2}$$

$$5580 \ln 0.4 = t \ln \frac{1}{2} = t \ln 0.5$$

$$t = \frac{5580 \ln 0.4}{\ln 0.5} = \frac{5580(-0.916290731)}{-0.69314718} \approx 7376 \text{ years}$$

The carbon-dating method has been used successfully on numerous occasions. It was this technique that established that the Dead Sea scrolls were prepared and buried about 2000 years ago.

An Application of Common Logarithms: The pH of a Substance

The acidity of a liquid substance is measured by the concentration of the hydrogen ion $[H^+]$ in the substance. This concentration is usually measured in terms of moles per liter (mol/L). A standard way to describe this acidity is to define the pH of a substance by $\text{pH} = -\log [H^+]$.

Distilled water has an approximate H^+ concentration of 10^{-7} mol/L, so its pH is $-\log 10^{-7} = -(-7) = 7$.

A substance with a pH under 7 is termed an **acid,** and one with a pH above 7 is called a **base.**

It is standard to write the pH of a substance with two-decimal-place accuracy.

EXAMPLE 8 *Finding the pH of a Substance*

Find the pH of a substance whose hydrogen ion concentration is 0.05 mol/L.

SOLUTION We have

$$\begin{aligned} \text{pH} &= -\log (0.05) \\ &= -(-1.301029996) = 1.30 \quad \text{Two-decimal-place accuracy} \end{aligned}$$

■

EXAMPLE 9 *Finding the Hydrogen Ion Concentration Given the pH*

The pH of the juice of a certain type of lemon is 2.30. What is the hydrogen ion concentration for this kind of lemon juice?

SOLUTION We have

$$\begin{aligned} \text{pH} = -\log [H^+] &= 2.30 \\ \log [H^+] &= -2.30 \\ [H^+] &= 10^{-2.30} \quad \text{Definition of } \log x \\ [H^+] &= 0.005 \text{ mol/L} \quad \text{From a calculator} \end{aligned}$$

FOCUS ON
The Mathematical Model of Constant Relative Rate of Growth

In many of the examples given in this section, we assumed that the relative rate of growth was constant so that

$$y(x) = y(0)e^{kx}$$

Suppose that $k > 0$. Let's give it a value, say $k = 0.1$. This means that something—a population or investment, for example, is growing continuously at a rate of 10% per time period. Did you stop to consider the fact that this rate of growth *cannot possibly continue* for an indefinite period of time?

To illustrate this fact, suppose that a certain insect population grows at 10% a month and starts with 1000 individuals. Then the population after t months is

$$P(t) = 1000e^{0.1t}$$

The population after t months is given in Table 2. We see that after 240 months (20 years) there would be over 26 trillion insects.

Table 2

t	$P(t) = 1000e^{0.1t}$
10	2718
20	7389
30	20,086
60	403,429
120	162,754,791
240	2.649×10^{13}
480	7.017×10^{23}

After 40 years there would be over 7×10^{23} insects. That is an unimaginably large number. To give you some idea, let us do some calculations.

(1) The radius of the earth is approximately 4000 miles.
(2) The surface area of a sphere is given by $S = 4\pi r^2$ so surface area of the earth $\approx 4\pi(4000)^2 \approx 200{,}000{,}000$ square miles.
(3) There are 5280^2 square feet in a square mile so the surface area of the earth in square feet $\approx (200{,}000{,}000)(5280^2) \approx 5.6 \times 10^{15}$ square feet.
(4) We divide:

$$\frac{7 \times 10^{23} \text{ insects}}{5.6 \times 10^{15} \text{ sq ft}} \approx 125{,}000{,}000 \text{ insects per square foot.}$$

That is, 125 million insects occupying every square foot on earth! Where would they fit? What would they eat?

The situation is ridiculous. This should come as no surprise because as $t \to \infty$, $e^{kt} \to \infty$ if $k > 0$, so any population (or anything else) grows without bound if it is growing exponentially.

Since nothing on earth can possibly grow without bound, it seems that the model of this section is useless. This is not true. Many quantities grow exponentially—like continuously compounded interest, for example. But nothing can grow exponentially indefinitely. There are always limits to growth. Any population growing exponentially will eventually run out of space or food and suffer other effects from overcrowding. Then nature will force the growth rate to change.

But unlimited growth is not the only problem with this model. Is the assumption that growth continues at a constant relative rate a reasonable one? If a bond pays 7% compounded continuously for 20 years, then you are certain that, at least for 20 years, your money will grow exponentially at the constant rate of 7%.

However, for another kind of problem, the answer is likely to be no. In Example 5, we found that the population of India grew at an average rate of 2.6% a year from 1974 to 1984. In order to answer the questions in that example, we assumed that the growth rate would remain the same at least until the year 2010. This is very unlikely. There are many factors that influence population growth rates: climatic conditions such as drought might lead to insufficient food and deaths by famine; education in birth control might lead to smaller families; other factors may lead to greater population growth rates. You can undoubtedly think of a few.

So the assumption of indefinite growth at a constant rate is unrealistic for many different types of problems. Models based on this assumption may be very useful. But they are often useful only for limited periods of time. As with all mathematical models, you must question the validity of the assumptions inherent in the model before you make predictions based on the model.

Problems 6.6

Readiness Check

I. Which of the following is the concentration of hydrogen ions present in a certain substance if $\text{pH} = -\log\,[\text{H}^+]$ and its pH is 7.5?
a. $10^{7.5}$ b. $10^{-7.5}$ c. $\log 7.5$ d. $-\log 7.5$

II. Which of the following is true about $P(t) = P(0)e^{kt}$ if $k < 0$ and $t > 0$?
a. $P(0) < P(t)$
b. $P(t) < P(0)$
c. $P(t) = P(0)$
d. No determination can be made until t and $P(0)$ are known.

III. Which of the following is $A(t) = Pe^{rt}$ if $P = \$3000$, $r = 6.7\%$, and $t = 3$ months $= \frac{1}{4}$ year?
a. \$3050.67 b. \$3667.87 c. \$3801.97
d. \$3025.17

IV. Which of the following is true about $M(t) = M_0e^{kt}$ if $k > 0$?
a. $M(t)$ increases as t increases.
b. $M(t)$ decreases as t increases.
c. $M(t)$ remains constant as t increases.
d. The change in $M(t)$ cannot be determined until values of M_0, k, and t are given.

1. How long would it take an investment to increase by half if it is invested at 4% compounded monthly?
2. What must be the nominal interest rate in order that an investment triple in 15 years if interest is compounded semiannually?
3. How long will it take an investment to triple (assuming quarterly compounding) for each interest rate?
(a) 2% (b) 5% (c) 8% (d) 10% (e) 15%
4. Answer the questions of Problem 3 if interest is compounded monthly.
5. Answer the questions of Problem 3 if interest is compounded continuously.
6. A bank account pays 8.5% annual interest compounded monthly. How large a deposit must be made now in order that the account contain exactly \$10,000 at the end of 1 year?

Answers to Readiness Check

I. b II. b III. a IV. a

7. Answer the question in Problem 6 if interest is compounded continuously.
8. A doting father wants his newly born daughter to have what $10,000 would buy on the date of her birth as a gift for her 21st birthday. He decides to accomplish this by making a single initial payment into a trust fund set up specially for this purpose.
   (a) Assuming that the rate of inflation is to be 6% (effective annual rate), what sum will have the same buying power in 20 years as $10,000 does at the date of her birth?
   (b) What is the amount of the single initial payment into the trust if its interest is compounded continuously at a rate of 10%?
9. What is the most a banker should pay for a $10,000 note due in 5 years if he can invest a like amount of money at 9% compounded annually?
10. A sum of $20,000 is invested at a steady rate of return with interest compounded continuously. If the investment is worth $35,000 in 2 years, what is the annual interest rate?
11. A certain government bond sells for $750 and can be redeemed for $1000 in 8 years. Assuming continuous compounding, what is the rate of interest paid?
12. How long would it take an investment to increase by half if it is invested at 5% compounded continuously?
13. What must be the interest rate in order that an investment triple in 15 years if interest is continuously compounded?
14. A bacterial population grows at a constant relative rate. It is 7000 at noon and 10,000 at 3 P.M.
   (a) What is the population at 6 P.M.?
   (b) When will the population reach 20,000?
15. The estimated world population in 1986 was 4,845,000,000. Assume that the population grows at a constant rate of 1.9%. When will the world population reach 8 billion?
16. In Problem 15, at what constant rate would the population grow if it reached 6 billion in the year 2000?
17. According to the official U.S. census, the population of the United States was 226,549,448 in 1980 and 248,709,873 in 1990. Assume that population grows at a constant relative rate.
   (a) Estimate the population in the year 2000.
   (b) When will the population reach one-half billion?
18. The population of the United States was 76,212,168 in 1900 and 92,228,496 in 1910. If population had grown at a constant percentage until 1990, what would the population have been in 1980 and 1990? Compare this answer with the data given in Problem 17. To explain this discrepancy is a problem in history, not algebra.
19. The population of Australia was approximately 13,400,000 in 1974 and 16,643,000 in 1990. Assume constant relative growth.
   (a) Predict the population in the year 2000.
   (b) When will the population be 20 million?
20. The population of New York State was 17,558,165 in 1980 and 17,990,455 in 1990. Assume a constant relative growth in population.
   (a) Predict the population in 1995.
   (b) When will the population reach 18.5 million?
21. * The population of Florida was 9,747,197 in 1980 and 12,937,926 in 1990. Assuming that Florida continues its high growth rate, when will the population of Florida exceed the population of New York (see Problem 20)?
22. In 1920, the consumer price index (CPI) for perishable goods was 213.4, with 1913 assigned the base rate of 100.† Assuming that inflation was constant (that is, assume continuous compounding), find the rate of inflation of the price of perishable goods between 1913 and 1920.
23. In 1920, the CPI for construction materials was 262.0. Again, with CPI = 100 in 1913, find the rate of inflation of the price of construction materials between 1913 and 1920.
24. The average annual earnings of full-time employees in the United States was $4743 in 1960 and $7564 in 1970. Assuming a continuous increase in earnings at a constant rate during the period 1960–1970, what was the rate of increase in wages?
25. The CPI was 88.7 in 1960 and 116.3 in 1970 (1967 = 100). Assuming a continuous increase in prices at a constant rate between 1960 and 1970, find the rate of inflation.
26. The *real* increase in earnings is defined as the percentage increase in wages minus the rate of inflation. Using the data in Problems 24 and 25, determine the real percentage increase in the average U.S. worker's earnings between 1960 and 1970.
27. Assume that the figures in Problems 24 and 25 hold, except that the CPI in 1970 is unknown. What would the CPI be if it were known that workers between 1960 and 1970 experienced no gain or loss in real income?
28. When the air temperature is 60°F, an object cools from 170°F to 130°F in half an hour.
   (a) What will be the temperature after 1 hour?
   (b) When will the temperature be 80°F?
29. A hot coal (temperature 150°C) is immersed in a liquid (temperature −10°C). After 30 seconds the temperature of the coal is 60°C. Assume that the liquid is kept at −10°C.
   (a) What is the temperature of the coal after 2 minutes?
   (b) When will the temperature of the coal be 0°C?

---

†This means that an average item costing $1.00 in 1913 would cost 2.13\frac{4}{10}$ in 1920.

30. A fossilized leaf contains 65% of a "normal" amount of ^{14}C. How old is the fossil?
31. Forty percent of a radioactive substance disappears in 100 years.
 (a) What is its half-life?
 (b) After how many years will 90% be gone?
32. Salt decomposes in water into sodium $[Na^+]$ and chloride $[Cl^-]$ ions at a rate proportional to its mass. Suppose there were 35 kg of salt initially and 21 kg after 12 hours.
 (a) How much salt would be left after 1 day?
 (b) After how many hours would there be less than $\frac{1}{4}$ kg of salt left?
33. Radioactive beryllium is sometimes used to date fossils found in deep-sea sediment. The mass of radioactive beryllium satisfies equation (11) with $k = 1.5 \times 10^{-7}$. What is the half-life of beryllium?
34. In a certain medical treatment, a tracer dye is injected into the pancreas to measure its function rate. A normally active pancreas will secrete 4% of the dye each minute. A physician injects 0.4 gram of the dye, and 30 minutes later 0.15 gram remains. How much dye would remain if the pancreas were functioning normally?
35. Atmospheric pressure is a function of altitude above sea level and satisfies an equation of the form (6), where $P(a)$ denotes pressure at altitude a. The pressure is measured in millibars (mbar). At sea level ($a = 0$), $P(0)$ is 1013.25 mbar, which means that the atmosphere at sea level will support a column of mercury 1013.25 millimeters (mm) high at a standard temperature of 15°C. At an altitude of $a = 1500$ m, the pressure is 845.6 mbar.
 (a) What is the pressure at $a = 4000$ m?
 (b) What is the pressure at 10 km?
 (c) In California, the highest and lowest points are Mount Whitney (4418 m) and Death Valley (86 m below sea level). What is the difference in their atmospheric pressures?
 (d) What is the atmospheric pressure at Mount Everest (elevation 8848 m)?
 (e) At what elevation is the atmospheric pressure equal to 1 mbar?

* 36. A bacteria population is known to grow exponentially. The following data were collected:

Number of Days	Number of Bacteria
5	1054
10	2018
20	7405

 (a) What was the initial population?
 (b) If the present growth rate were to continue, what would be the population after 60 days?

* 37. A bacteria population is declining exponentially. The following data were collected:

Number of Hours	Number of Bacteria
12	5969
24	3563
48	1269

 (a) What was the initial population?
 (b) How many bacteria are left after 1 week?
 (c) When will there be no bacteria left (that is, when is $P(t) < 1$)?
38. What is the pH of a substance with a hydrogen ion concentration of 3.6×10^{-6}?
39. What is the pH of a substance with a hydrogen ion concentration of 0.6×10^{-7}?
40. Milk has a pH of 6.5. What is its hydrogen ion concentration?
41. Beer has a pH of 4.5. What is its hydrogen ion concentration?
42. Milk of magnesia has a pH of 10.5. What is its hydrogen ion concentration?
43. A general psychophysical relation was established in 1834 by the German physiologist Ernest Weber and given a more precise phrasing later by the German physicist Gustav Fechner. By the **Weber-Fechner law**, $S = c \log (R + d)$, where S is the intensity of a sensation, R is the strength of the stimulus producing it, and c and d are constants. The Greek astronomer Ptolemy catalogued stars according to their visual brightness in six categories or **magnitudes.** A star of the first magnitude was about $2\frac{1}{2}$ times as bright as a star of the second magnitude, which in turn was about $2\frac{1}{2}$ times as bright as a star of the third magnitude, and so on. Let b_n and b_m denote the apparent brightness of two stars having magnitudes n and m, respectively. Modern astronomers have established the Weber-Fechner law relating the relative brightness to the difference in magnitudes as

$$(m - n) = 2.5 \log \left(\frac{b_n}{b_m}\right)$$

 (a) Using this formula, calculate the ratio of brightness for two stars of the second and fifth magnitudes, respectively.
 (b) If star A is five times as bright to the naked eye as star B, what is the difference in their magnitudes?
 (c) How much brighter is Sirius (magnitude 1.4) than a star of magnitude 2.15?
 (d) The Nova Aquilae in a 2–3-day period in June 1918 increased in brightness about 45,000 times. How many magnitudes did it rise?

* (e) The bright star Castor appears to the naked eye as a single star but can be seen with the aid of a telescope to be really two stars whose magnitudes have been calculated to be 1.97 and 2.95. What is the magnitude of the two combined? [Hint: Brightnesses, but not magnitudes, can be added.]

* 44. The subjective impression of loudness can be described by a Weber-Fechner law. Let I denote the intensity of a sound. The least intense sound that can be heard is $I_0 = 10^{-12}$ watt/m^2 at a frequency of 1000 cycles/sec. (This value is called the **threshold of audibility.**) If L denotes the loudness of a sound, measured in decibels,† then $L = 10 \log (I/I_0)$.

(a) If one sound has twice the intensity of another, what is the ratio of the perceived loudness of the two sounds?

(b) If one sound appears to be twice as loud as another, what is the ratio of their intensities?

(c) Ordinary conversation sounds 6 times as loud as a low whisper. What is the actual ratio of intensity of their sounds?

Summary Outline of Chapter 6

- An **exponential function** is a function of the form $f(x) = a^x$, where $a > 0$, $a \neq 1$. p. 325
- **The number e** is the number approached by the expression $\left(1 + \dfrac{1}{m}\right)^m$ as $m \to \infty$. To 10 decimal places $e = 2.7182818285$. The function $y = e^x$ is the most important exponential function in applications. p. 335
- The function $y(t) = y(0)e^{kt}$ is the equation of **exponential growth** (if $k > 0$) or **exponential decay** (if $k < 0$). p. 339
- The **logarithm function** $y = \log_a x$ is the inverse of $y = a^x$. $y = \log_a x$ if and only if $x = a^y$. p. 345
- **Properties of Logarithmic Functions**

  $a^{\log_a x} = x \qquad \log_a a^x = x \qquad \log_a a = 1 \qquad \log_a 1 = 0$ pp. 345, 346

  $\log_a xw = \log_a x + \log_a w \qquad \log_a \dfrac{x}{w} = \log_a x - \log_a w$ p. 347

  $\log_a \dfrac{1}{x} = -\log_a x \qquad \log_a x^r = r \log_a x \qquad \log_a b = \dfrac{1}{\log_b a}$ p. 347

  $\log_a x = \dfrac{\log x}{\log a} = \dfrac{\ln x}{\ln a}$ p. 358
- **Common Logarithms** are logarithms to the base 10. p. 352
- **Natural Logarithms** are logarithms to the base e. p. 354
- **Properties of The Natural Logarithm** p. 356

  $e^{\ln x} = x \qquad \ln e^x = x \qquad \ln e = 1 \qquad \ln 1 = 0$

  $\ln xw = \ln x + \ln w \qquad \ln \dfrac{x}{w} = \ln x - \ln w \qquad \ln \dfrac{1}{x} = -\ln x \qquad \ln x^r = r \ln x$

Review Exercises for Chapter 6

In Exercises 1–8, draw a sketch of the given exponential or logarithmic function.

1. $y = 4^x$
2. $y = (\frac{1}{4})^x$
3. $y = 3 \cdot 5^x$
4. $y = -2 \cdot 2^{-x}$
5. $y = \log_2 (x + 3)$
6. $y = \log (1 - x)$
7. $y = 2 - \ln x$
8. $y = \ln (3 - x) - 1$

In Exercises 9–16, use a calculator to estimate the given number to as many decimal places of accuracy as your calculator carries.

9. $e^{1.7}$
10. $10^{3.45}$
11. $(\frac{1}{3})^{2.3}$
12. $2.4^{\sqrt{5}}$
13. $\log 28.4$
14. $\log 0.0032$
15. $\ln 1{,}000{,}000$
16. $\ln 0.00235$

† decibel (dB) = $\frac{1}{10}$ bel, named after Alexander Graham Bell (1847–1922), inventor of the telephone.

In Exercises 17–28, solve for the given variable.

17. $y = \log_2 16$
18. $y = \log_{1/3} 9$
19. $\frac{1}{2} = \log_x 5$
20. $y = e^{\ln 17.2}$
21. $\log x = 10^{-9}$
22. $\log_x 32 = -5$
23. $2 \ln (x + 3) = 4$
24. $e^{-2(x+1)} = 2$
25. $\ln x - \ln (x + 1) = \ln 2$
26. $3 \ln x = \ln 4 + \ln 2$
27. $\ln (y + 2) + \ln y = \ln 3$
28. $\ln z^3 - 2 \ln z = 1$
29. If $y = 3 \ln x$, what happens to y if x doubles?
30. If $y = -4 \ln x$, what happens to y if x is cut in half?
31. Convert each equation to an exponential form.
    (a) $\log_9 27 = \frac{3}{2}$
    (b) $\log_{1/2} 8 = -3$
32. Change each equation to a logarithmic form.
    (a) $5^3 = 125$
    (b) $3^{-2} = \frac{1}{9}$

In Exercises 33–38, write as a single logarithm.

33. $\log_3 x + \log_3 (x - 2)$
34. $\log x + \log y$
35. $\log (x + 1) - \log (x + 3) + \log z^2$
36. $3 \ln z - 4 \ln (x + 1) + 5 \ln (x^2 - 1)$
37. $3 \ln (x + 1) - \frac{1}{2} \ln (y + 4) + \frac{1}{3} \ln (z + 12)$
38. $\ln a - \ln b + \ln c - \ln d$

In Exercises 39–42 compute each logarithm.

39. $\log_2 7$
40. $\log_{1/2} \frac{1}{3}$
41. $\log_4 37$
42. $\log_{90} 2$

The remaining exercises all require the use of a calculator.

43. What is the simple interest paid on \$5000 invested at 7% for 6 years?
44. What is the simple interest paid on \$8000 invested at $7\frac{1}{2}$% for 12 years?

In Exercises 45–48, compute the value of an investment after t years and the total interest paid if P dollars is invested at a nominal rate of r% compounded m times a year.

45. $P = \$8000$, $r = 5\frac{1}{2}\%$, $t = 4$, $m = 1$
46. $P = \$10{,}000$, $r = 6\frac{1}{2}\%$, $t = 4$, $m = 2$
47. $P = \$6000$, $r = 10\%$, $t = 5$, continuously
48. $P = \$25{,}000$, $r = 7\frac{3}{4}\%$, $t = 15$, continuously
49. If money is invested at 8% compounded continuously, what is the effective interest rate?
50. What is the effective interest rate of 8% compounded monthly?
51. How long will it take an investment to double (assuming quarterly compounding) if the interest rate is (a) 3%? (b) $6\frac{1}{2}$%? (c) 8%? (d) 12%?
52. Answer the questions of Exercise 51 if money is compounded continuously.
53. The relative rate of growth of a population is 15%. If the initial population is 10,000, what is the population after 5 years? after 10 years?
54. In Exercise 53, how long will it take for the population to double?
55. When a cake is taken out of the oven, its temperature is 125°C. Room temperature is 23°C. The temperature of the cake is 80°C after 10 minutes.
    (a) What will be its temperature after 20 minutes?
    (b) How long will the cake take to cool to 25°C?
56. A fossil contains 35% of the normal amount of ^{14}C. What is its approximate age?
57. What is the half-life of an exponentially decaying substance that loses 20% of its mass in one week?
58. How long will it take the substance in Exercise 57 to lose 75% of its mass? 95% of its mass?

Chapter 7

Systems of Equations and Inequalities

In Section 3.2, we discussed the slopes and graphs of straight lines which, as we saw in Section 3.3, can be written as linear functions. In Section 3.4, we gave several applications of linear functions.

In a wide variety of problems, we obtain a number of linear equations in two or more variables. In this case, we have what is called a **system** of linear equations. In this chapter, we discuss systems of linear equations and show how they can be solved by using **matrices.** We also discuss the related topic of determinants.

7.1 Systems of Two Linear Equations in Two Unknowns

Consider the following system of two linear equations in two unknowns or variables:

$$\begin{aligned} 4x + 3y &= 1 \\ -2x + y &= 7 \end{aligned}$$

Each of these equations is an equation of a straight line. Using the material in Section 3.2, we find that the slope of the first line is $-\frac{4}{3}$. We see this by writing the equation in the slope-intercept form $y = -\frac{4}{3}x + \frac{1}{3}$. Similarly, the slope of the second line is 2. A **solution** to this system is a pair of numbers, denoted by (x, y), that satisfy the system. In order to satisfy each equation, (x, y) must be a point on each line. Since the slopes of the lines are not equal, the lines have exactly one point of intersection. You should verify that the only point that is on both lines is the point $(-2, 3)$. That is, $x = -2$, $y = 3$ is the unique solution to the system.

More generally, consider the system

$$\begin{aligned} a_{11}x + a_{12}y &= b_1 \\ a_{21}x + a_{22}y &= b_2 \end{aligned} \tag{1}$$

where a_{11}, a_{12}, a_{21}, a_{22}, b_1, and b_2 are given real numbers. Each of these equations is an equation of a straight line in the xy-plane.

The slope of the first line is $-a_{11}/a_{12}$; the slope of the second line is $-a_{21}/a_{22}$ (if $a_{12} \neq 0$ and $a_{22} \neq 0$). A **solution** to system (1) is a pair of numbers, denoted by (x, y), that satisfies (1). The questions that naturally arise are whether (1) has any solutions and, if so, then how many? We will answer these questions after looking at some more examples. In these examples, we will make use of two important facts about real numbers.

Let a, b, c, and d be real numbers.

If $a = b$ and $c = d$, then $a + c = b + d$. Addition of equals rule (2)

If $a = b$ and $c \neq 0$, then $ca = cb$. Multiplication rule (3)

The first rule states that if we add the corresponding sides of two equations together, we obtain a third, valid equation.

The second rule states that if we multiply both sides of an equation by a nonzero constant, we obtain a second, **equivalent** equation. The solutions of the two equations are the same.

EXAMPLE 1 *A System with a Unique Solution*

Consider the system

$$\begin{aligned} x - y &= 7 \\ x + y &= 5 \end{aligned} \tag{4}$$

From (2), we may add these equations together to obtain

$$2x = 12$$

$$x = 6 \quad \text{Multiply by } \tfrac{1}{2}$$

Then, from the second equation,

$$6 + y = 5$$

$$y = 5 - 6 = -1$$

Thus the pair $(6, -1)$ satisfies system (4), and the way we found the solution shows that it is the only pair of numbers to do so. That is, system (4) has a **unique solution.** In problems where there is a unique solution, it is not difficult to check the answer. In Figure 1, we see that the lines $x - y = 7$ and $x + y = 5$ intersect at $(6, -1)$.

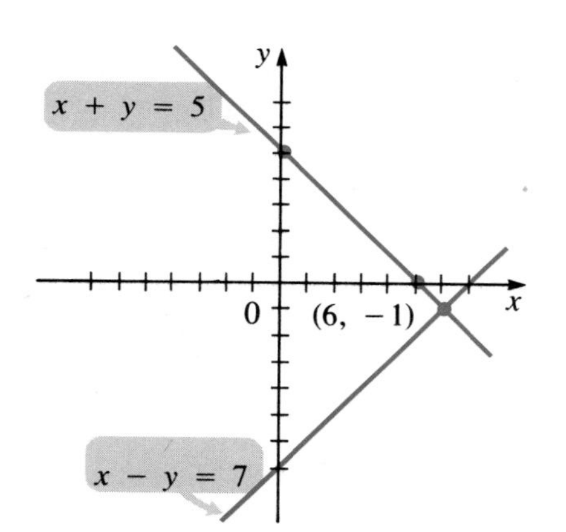

Figure 1 The system has a unique solution because the two lines intersect at a single point.

Check

$$x - y = 6 - (-1) = 7$$

$$x + y = 6 + (-1) = 5 \quad ■$$

EXAMPLE 2 *A System with an Infinite Number of Solutions*

Consider the system

$$\begin{aligned} x - y &= 7 \\ 2x - 2y &= 14 \end{aligned} \tag{5}$$

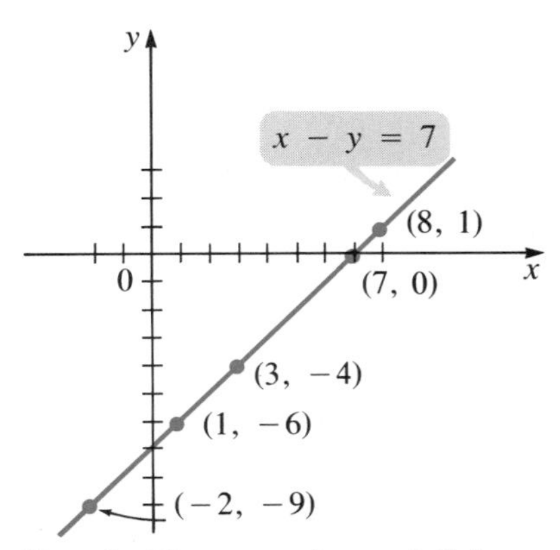

Figure 2 The system has an infinite number of solutions because the two lines coincide. Every point on the line $x - y = 7$ is a solution.

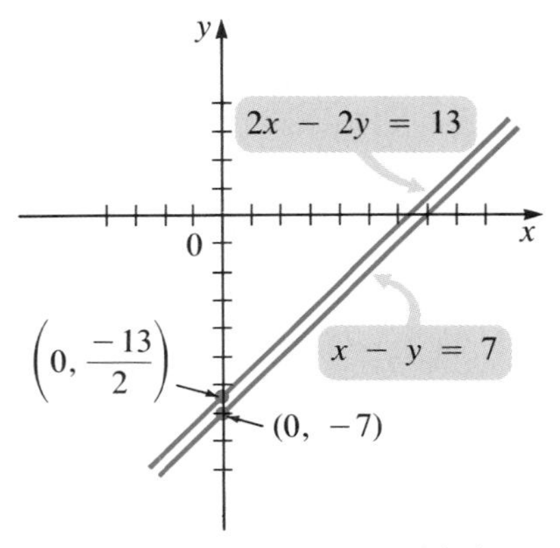

Figure 3 The system has no solution because the two lines have no point of intersection (they are parallel).

These two equations are **equivalent.** That is, they are equations of the same straight line. To see this, multiply the first by 2. (This is permitted by (3)). To obtain a description of the solutions, solve for y: $x - y = 7$, or $y = x - 7$. Thus the pair $(x, x - 7)$ is a solution to system (5) for any real number x; that is, system (5) has an **infinite number of solutions.** For example, the following pairs are solutions: $(7, 0)$, $(0, -7)$, $(8, 1)$, $(1, -6)$, $(3, -4)$, and $(-2, -9)$. The line is sketched in Figure 2. ■

EXAMPLE 3 *An Inconsistent System*

Consider the system

$$\begin{aligned} x - y &= 7 \\ 2x - 2y &= 13 \end{aligned} \tag{6}$$

Multiplying the first equation by 2 (which, again, is permitted by (3)) gives us $2x - 2y = 14$. This contradicts the second equation since $2x - 2y$ cannot be equal to both 13 and 14 at the same time. Thus, system (6) has *no* solution. In this case, the system is said to be **inconsistent.** We can see why this is true. Each of the equations in (6) is an equation of a straight line with slope 1. Thus the lines are parallel. The slope-intercept forms of the lines are $y = x - 7$ and $y = x - \frac{13}{2}$. Since the y-intercepts are different (-7 and $-\frac{13}{2}$), the lines are different and so have no point of intersection (parallel lines that are not identical never meet). Thus, system (6) has no solution. This is illustrated in Figure 3.

We summarize the results of Examples 1, 2, and 3. When we have a linear system of two equations in two unknowns, each equation is an equation of a straight line. There are only three possibilities: the lines are not parallel and intersect at one point only, the lines are the same, or the lines are parallel and different. That is,

Theorem

A system of two linear equations in two unknowns has no solution, exactly one solution, or an infinite number of solutions.

EXAMPLE 4 *A System Arising from a Manufacturing Problem*

The Sunrise Porcelain Company manufactures ceramic cups and bowls. For each cup or bowl, a worker measures a fixed amount of material and puts it into a forming machine, from which it is automatically glazed and dried. On the average, a worker needs 3 minutes to get the process started for a cup and 2 minutes for a bowl. The material for a cup costs 25¢, and the material for a bowl costs 20¢. How many of each can be manufactured in an 8-hour work day if a worker is working every minute and exactly $44 is spent on materials?

SOLUTION Let x denote the number of cups and y the number of bowls produced in an 8-hour day. Then, since there are 480 minutes in 8 hours, we obtain the following equations for x and y.

$$3x + 2y = 480 \quad \text{Time or labor equation}$$
$$0.25x + 0.20y = 44 \quad \text{Cost equation}$$

Multiplying the cost equation by 10, we obtain

$$2.5x + 2y = 440$$

Subtracting this from the labor equation, we have

$$0.5x = 40 \quad \text{or} \quad x = 80$$

We then have (From the labor equation ↓)

$$2y = 480 - 3x = 480 - 3(80) = 480 - 240 = 240$$

or

$$y = 120$$

Thus, 80 cups and 120 bowls can be manufactured in an 8-hour day. The labor and cost equations are sketched in Figure 4.

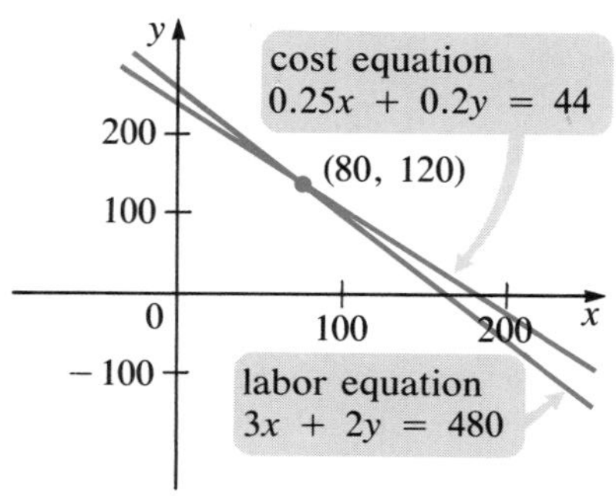

Figure 4 The solution is the point at which the cost and labor lines intersect.

Check

$$3x + 2y = 3 \cdot 80 + 2 \cdot 120 = 240 + 240 = 480$$
$$0.25x + 0.20y = 0.25(80) + 0.2(120) = 20 + 24 = 44$$

Problems 7.1

Readiness Check

I. Which of the following is *not* true about the solution for a system of two linear equations in two unknowns?
  a. It is an ordered pair that satisfies both equations.
  b. Its graph consists of the point(s) of intersection of the graphs of the equations.
  c. Its graph is the x-intercept of the graphs of the equations.
  d. If the system is inconsistent, there is no solution.

II. Which of the following is true of an inconsistent system of two linear equations?
  a. There is no solution.
  b. The graph of the system is on the y-axis.
  c. The graph of the solution is one line.
  d. The graph of the solution is the point of intersection of two lines.

III. Which of the following is true of the system of equations below?

$$3x - 2y = 8$$
$$4x + y = 7$$

  a. The system is inconsistent.
  b. The solution is $(-1, 2)$.
  c. The solution is on the line $x = 2$.
  d. The equations are equivalent.

IV. Which of the following is a second equation for the system whose first equation is $x - 2y = -5$ if there are to be an infinite number of solutions for the system?
  a. $6y = 3x + 15$ b. $6x - 3y = -15$
  c. $y = -\frac{1}{2}x + \frac{5}{2}$ d. $\frac{3}{2}x = 3y + \frac{15}{2}$

V. The graph of which of the following systems is a pair of parallel lines?
  a. $3x - 2y = 7$, $4y = 6x - 14$ b. $x - 2y = 7$, $3x = 4 + 6y$
  c. $2x + 3y = 7$, $3x - 2y = 6$ d. $5x + y = 1$, $7y = 3x$

Answers to Readiness Check

I. c II. a III. c IV. a V. b

In Problems 1–12, find all solutions (if any) to the given systems.

1. $2x + y = 4$
   $-x + 3y = 5$
2. $3x + y = 3$
   $-2x + 2y = 6$
3. $2x - y = -4$
   $3x + 2y = -6$
4. $2x - y = 1$
   $3x + 3y = 3$
5. $x - y = 2$
   $2x + 4y = -5$
6. $3x + y = 0$
   $2x - 3y = 0$
7. $4x - 6y = 0$
   $-2x + 3y = 0$
8. $5x + 2y = 3$
   $2x + 5y = 3$
9. $2x + 3y = 4$
   $3x + 4y = 5$
10. $ax + by = c$
    $ax - by = c$
11. $ax + by = c$
    $bx + ay = c$
12. $ax - by = c$
    $bx + ay = d$
13. Find conditions on a and b such that the system in Problem 10 has a unique solution.
14. Find conditions on a, b, and c such that the system in Problem 11 has an infinite number of solutions.
15. Find conditions on a, b, c, and d such that the system in Problem 12 has no solution.

In Problems 16–21, find the point of intersection (if there is one) of the two lines.

16. $3x - y = 2$
    $x + y = 4$
17. $3x + 2y = -3$
    $5x - y = 8$
18. $y - 2x = 4$
    $2x - 4y = -8$
19. $2x + 3y = 6$
    $4x + 6y = 8$
20. $3x + y = 4$
    $y - 5x = 2$
21. $3x + 2y = 4$
    $2x - 7y = 5$
22. A zoo keeps birds (two-legged) and beasts (four-legged). If the zoo contains 60 heads and 200 feet, how many birds and how many beasts live there?
23. A mutual fund has two investment plans. In plan A, 80% of one's money is invested in blue-chip stocks and 20% is invested in riskier "glamour" stocks. In plan B, 40% is invested in blue-chip stocks, and 60% is invested in glamour stocks. If the firm invests a total of \$3 million in blue-chip stocks and \$1 million in glamour stocks, how much money has been put in each of plans A and B?
24. The Atlas Tool Company manufactures pliers and scissors. Each pair of pliers contains 2 units of steel and 4 units of aluminum. Each pair of scissors requires 1 unit of steel and 3 units of aluminum. How many pairs of pliers and scissors can be made from 140 units of steel and 290 units of aluminum?
25. Answer the question in Problem 24 if each pair of scissors requires 2 units of aluminum and all other information is unchanged.
26. Answer the question in Problem 25 if only 280 units of aluminum are available and all other information is unchanged.
27. Ryland Farms in northwestern Indiana grows soybeans and corn on its 500 acres of land. During the planting season, 1200 hours of planting time will be available. Each acre of soybeans requires 2 hours, whereas each acre of corn requires 6 hours. If all the land and hours are to be utilized, how many acres of each crop should be planted?
28. Spina Food Supplies, Inc. manufactures frozen pizzas. Art Spina, President of Spina Food Supplies, personally supervises the production of both types of frozen pizzas produced by the company: Spina's regular and Spina's super deluxe. He currently has 150 lb of dough mix and 800 oz of topping mix available. Each regular pizza uses 1 lb of dough mix and 4 oz of topping, whereas each super deluxe uses 1 lb of dough and 8 oz of topping mix. How many of each type of pizza should he make in order to use all of his dough and topping mixes?

7.2 Higher-Order Systems: Gaussian Elimination

In this section, we describe a method for finding all solutions (if any) to a system of m linear equations in n unknowns. In doing so, we shall see that, as in the case of two equations in two unknowns, such a system has no solutions, exactly one solution, or an infinite number of solutions. Before launching into the general method, let us look at some examples.

EXAMPLE 1 ***A System of Three Equations in Three Unknowns with a Unique Solution***

Solve the system

$$\begin{aligned} 2x + 8y + 6z &= 20 \\ 4x + 2y - 2z &= -2 \\ 3x - y + z &= 11 \end{aligned} \tag{1}$$

SOLUTION Here we seek three numbers x, y, and z such that the three equations in (1) are satisfied. Our method of solution will be to simplify the equations so that solutions can be readily identified. We begin by dividing the first equation by the coefficient of x. That is, we divide the first equation by 2. This gives us

$$\begin{aligned} x + 4y + 3z &= 10 \\ 4x + 2y - 2z &= -2 \\ 3x - y + z &= 11 \end{aligned} \tag{2}$$

As we saw in the last section (addition of equals rule), adding two equations together leads to a third, valid equation. *We may use this new equation in place of either of the two equations used to obtain it.* We begin simplifying system (2) by multiplying both sides of the first equation in (2) by -4 and adding this new equation to the second equation. This gives us

$$\begin{array}{rl} -4x - 16y - 12z = -40 & \text{Multiply the first equation by } -4 \\ 4x + 2y - 2z = -\ 2 & \text{This is the second equation} \\ \hline -14y - 14z = -42 & \end{array}$$

The equation $-14y - 14z = -42$ is our new second equation, and the system is now

$$\begin{aligned} x + 4y + 3z &= 10 \\ -14y - 14z &= -42 \\ 3x - y + z &= 11 \end{aligned} \tag{3}$$

It is important to note that any solution to (1) is also a solution to (3), and vice versa. This is because of the two rules of Section 7.1. In this case, we say that systems (1) and (3) are **equivalent.** We then multiply the first equation by -3 and add it to the third equation.

$$\begin{aligned} x + 4y + 3z &= 10 \\ -14y - 14z &= -42 \\ -13y - 8z &= -19 \end{aligned} \tag{4}$$

Again, we note that systems (1) and (4) have the same solutions. Note that in the system above, the variable x has been eliminated from the second and third equations. Next we divide the second equation by -14.

$$\begin{aligned} x + 4y + 3z &= 10 \\ y + z &= 3 \\ -13y - 8z &= -19 \end{aligned}$$

We multiply the second equation by 13 and add it to the third equation.

$$\begin{aligned} x + 4y + 3z &= 10 \\ y + z &= 3 \\ 5z &= 20 \end{aligned}$$

We divide the third equation by 5.

$$\begin{aligned} x + 4y + 3z &= 10 \\ y + z &= 3 \\ z &= 4 \end{aligned} \tag{5}$$

We can now solve system (5). We have $z = 4$. Then

$$\begin{aligned} y + z &= 3 \quad \text{Second equation} \\ y + 4 &= 3 \\ y &= -1 \end{aligned}$$

$$\begin{aligned} x + 4y + 3z &= 10 \quad \text{First equation} \\ x + 4(-1) + 3(4) &= 10 \\ x - 4 + 12 &= 10 \\ x + 8 &= 10 \\ x &= 2 \end{aligned}$$

In each step, we obtained a system equivalent to system (1). This means that the solution(s) to (1) are the same as the solution(s) to (5). System (5) has the unique solution $x = 2$, $y = -1$, and $z = 4$, so this is the unique solution to the original system (1). We write this solution in the form $(2, -1, 4)$. The method we have used here is called **Gaussian elimination.**†

Check

$$\begin{aligned} 2x + 8y + 6z &= 2(2) + 8(-1) + 6(4) = 4 - 8 + 24 = 20 \\ 4x + 2y - 2z &= 4(2) + 2(-1) - 2(4) = 8 - 2 - 8 = -2 \\ 3x - y + z &= 3(2) - (-1) + 4 = 6 + 1 + 4 = 11 \end{aligned}$$

Before going on to another example, let us summarize what we have done in this example.

1. We divided to make the coefficient of x in the first equation equal to 1. (If this can't be done, then rearrange the equations so that the coefficient of x in the new first equation is not zero.)
2. We "eliminated" the x terms in the second and third equations. That is, we made the coefficients of these terms equal to zero by multiplying the first equation by appropriate numbers and then adding it to the second equation and the third equation, respectively.
3. We divided to make the coefficient of the y term in the second equation equal to 1 and then proceeded to use the second equation to eliminate the y term in the third equation.
4. We divided to make the coefficient of the z term in the third equation equal to 1. This gave us the value of z.
5. We used **back substitution** to obtain first the value of y and then the value of x.

---

† Named after the great German mathematician Karl Friedrich Gauss (1777–1855). See the biographical sketch on p. 400.

At every step, we obtained systems that were equivalent—that is, each system had the same set of solutions as the one that preceded it.

Before solving other systems of equations, we introduce notation that makes it easier to write down each step in our procedure. A **matrix** is a rectangular array of numbers. For example, the coefficients of the variables x, y, and z in system (1) can be written as the entries of a matrix A, called the **coefficient matrix** of the system:

$$A = \begin{pmatrix} 2 & 8 & 6 \\ 4 & 2 & -2 \\ 3 & -1 & 1 \end{pmatrix}$$

This is the coefficient matrix of our system

We will study properties of matrices in Chapter 8. We introduce them here for convenience of notation. Using matrix notation, system (1) can be represented as the **augmented matrix**

$$\left(\begin{array}{ccc|c} 2 & 8 & 6 & 20 \\ 4 & 2 & -2 & -2 \\ 3 & -1 & 1 & 11 \end{array}\right)$$

For example, the equation $2x + 8y + 6z = 20$ is represented by the row $(2 \quad 8 \quad 6 \mid 20)$. Note that each row of the augmented matrix corresponds to one of the equations in the system.

If we use this form, the solution to Example 1 looks like this.

This becomes a 1; Divide first row by 2; These become 0

$$\left(\begin{array}{ccc|c} 2 & 8 & 6 & 20 \\ 4 & 2 & -2 & -2 \\ 3 & -1 & 1 & 11 \end{array}\right) \longrightarrow \left(\begin{array}{ccc|c} 1 & 4 & 3 & 10 \\ 4 & 2 & -2 & -2 \\ 3 & -1 & 1 & 11 \end{array}\right)$$

Multiply first row by -4 and add it to second row; Multiply first row by -3 and add it to third row; This becomes 1

$$\longrightarrow \left(\begin{array}{ccc|c} 1 & 4 & 3 & 10 \\ 0 & -14 & -14 & -42 \\ 3 & -1 & 1 & 11 \end{array}\right) \longrightarrow \left(\begin{array}{ccc|c} 1 & 4 & 3 & 10 \\ 0 & -14 & -14 & -42 \\ 0 & -13 & -8 & -19 \end{array}\right)$$

Divide second row by -14; Multiply second row by 13 and add it to third row; This becomes 0; This becomes 1

$$\longrightarrow \left(\begin{array}{ccc|c} 1 & 4 & 3 & 10 \\ 0 & 1 & 1 & 3 \\ 0 & -13 & -8 & -19 \end{array}\right) \longrightarrow \left(\begin{array}{ccc|c} 1 & 4 & 3 & 10 \\ 0 & 1 & 1 & 3 \\ 0 & 0 & 5 & 20 \end{array}\right)$$

Divide third row by 5

$$\longrightarrow \left(\begin{array}{ccc|c} 1 & 4 & 3 & 10 \\ 0 & 1 & 1 & 3 \\ 0 & 0 & 1 & 4 \end{array}\right)$$

We write the system of equations represented by the last augmented matrix:

$$\begin{aligned} x + 4y + 3z &= 10 \\ y + z &= 3 \\ z &= 4 \end{aligned}$$

Again, we see that $z = 4$. Since $y + z = 3$, we have $y + 4 = 3$ so $y = -1$ and, as before, $x = 2$.

EXAMPLE 2 ***Three Equations in Three Unknowns: Infinite Number of Solutions***

Solve the system

$$\begin{aligned} 2x + 8y + 6z &= 20 \\ 4x + 2y - 2z &= -2 \\ -6x + 4y + 10z &= 24 \end{aligned}$$

SOLUTION We proceed as in Example 1, first writing the system as an augmented matrix.

$$\left(\begin{array}{ccc|c} 2 & 8 & 6 & 20 \\ 4 & 2 & -2 & -2 \\ -6 & 4 & 10 & 24 \end{array}\right)$$

$$\left(\begin{array}{ccc|c} 1 & 4 & 3 & 10 \\ 4 & 2 & -2 & -2 \\ -6 & 4 & 10 & 24 \end{array}\right)$$ Divide the first row by 2

$$\left(\begin{array}{ccc|c} 1 & 4 & 3 & 10 \\ 0 & -14 & -14 & -42 \\ 0 & 28 & 28 & 84 \end{array}\right)$$ Multiply the first row by -4 and add it to the second, and then multiply the first by 6 and add it to the third

$$\left(\begin{array}{ccc|c} 1 & 4 & 3 & 10 \\ 0 & 1 & 1 & 3 \\ 0 & 28 & 28 & 84 \end{array}\right)$$ Divide the second row by -14

$$\left(\begin{array}{ccc|c} 1 & 4 & 3 & 10 \\ 0 & 1 & 1 & 3 \\ 0 & 0 & 0 & 0 \end{array}\right)$$ Multiply the second row by -28 and add it to the third

This is equivalent to the system of equations

$$\begin{aligned} x + 4y + 3z &= 10 \\ y + z &= 3 \\ 0 &= 0 \end{aligned}$$

This is as far as we can go. There are now only two nonzero equations in the three unknowns x, y, and z and there are an infinite number of solutions.

To see this, let z be chosen arbitrarily. Then $y = 3 - z$ and $x = 10 - 4y - 3z$. For example, if $z = 0$, then

$$y = 3 - z = 3 - 0 = 3 \quad \text{and} \quad x = 10 - 4y - 3z = 10 - 4(3) - 3(0) = -2$$

For $z = 10$, we obtain

$$y = 3 - 10 = -7 \quad \text{and} \quad x = 10 - 4(-7) - 3(10) = 10 + 28 - 30 = 8$$

Thus, two solutions are $(-2, 3, 0)$ and $(8, -7, 10)$. We see that for every real number z, we get a different solution (x, y, z).

Elementary Row Operations and Row Reduction

We now introduce some terminology. We have seen that multiplying (or dividing) the sides of an equation by a nonzero number gives us a new, equivalent equation. Moreover, adding a multiple of one equation to another equation in a system gives us another valid equation. Finally, if we interchange two equations in a system of equations, we obtain an equivalent system. These three operations, when applied to the rows of the augmented matrix representation of a system of equations, are called **elementary row operations.**

To sum up, the following three elementary row operations can be applied to the augmented matrix representation of a system of equations to obtain the augmented matrix of an equivalent system.

The process of applying elementary row operations to simplify an augmented matrix is called **row reduction.**

Row Reduction

1. Replace a row with a nonzero multiple of that row.
2. Replace a row with the sum of the row and a multiple of some other row.
3. Interchange two rows.

NOTATION FOR ROW REDUCTION

1. $R_i \to cR_i$ stands for "replace the ith row by the ith row multiplied by c."
2. $R_j \to R_j + cR_i$ stands for "replace the jth row with the sum of the jth row and the ith row multiplied by c."
3. $R_i \rightleftarrows R_j$ stands for "interchange rows i and j."
4. $A \to B$ indicates that the augmented matrices A and B are equivalent; that is, the systems they represent have the same solution.

In Example 1, we saw that by using the elementary row operations (1) and (2) several times, we could obtain a system in which the solutions to the system were easily found. In the examples that follow, we will use our new notation to indicate the steps we are performing.

EXAMPLE 3 *An Inconsistent System*

Solve the system

$$\begin{aligned} 2x + 8y + 6z &= 20 \\ 4x + 2y - 2z &= -2 \\ -6x + 4y + 10z &= 30 \end{aligned} \tag{6}$$

SOLUTION We use the augmented-matrix form and proceed exactly as in Example 2 to obtain, successively, the following systems. (Note how, in each step, we use either elementary row operation 1 or 2.)

$$\left(\begin{array}{ccc|c} 2 & 8 & 6 & 20 \\ 4 & 2 & -2 & -2 \\ -6 & 4 & 10 & 30 \end{array}\right) \xrightarrow{R_1 \to \frac{1}{2}R_1} \left(\begin{array}{ccc|c} 1 & 4 & 3 & 10 \\ 4 & 2 & -2 & -2 \\ -6 & 4 & 10 & 30 \end{array}\right)$$

$$\xrightarrow[R_3 \to R_3 + 6R_1]{R_2 \to R_2 - 4R_1} \left(\begin{array}{ccc|c} 1 & 4 & 3 & 10 \\ 0 & -14 & -14 & -42 \\ 0 & 28 & 28 & 90 \end{array}\right)$$

$$\xrightarrow{R_2 \to -\frac{1}{14}R_2} \left(\begin{array}{ccc|c} 1 & 4 & 3 & 10 \\ 0 & 1 & 1 & 3 \\ 0 & 28 & 28 & 90 \end{array}\right)$$

$$\xrightarrow{R_3 \to R_3 - 28R_2} \left(\begin{array}{ccc|c} 1 & 4 & 3 & 10 \\ 0 & 1 & 1 & 3 \\ 0 & 0 & 0 & 6 \end{array}\right)$$

The last equation now reads $0x + 0y + 0z = 6$, which is impossible since $0 \neq 6$. Thus system (6) has *no* solution. As in the system of two unknowns, we say that the system is **inconsistent.**

Let us take another look at these three examples. In Example 1, the original coefficient matrix was

$$A_1 = \begin{pmatrix} 2 & 8 & 6 \\ 4 & 2 & -2 \\ 3 & -1 & 1 \end{pmatrix}$$

In the process of row reduction, A_1 was "reduced" to the matrix

$$B_1 = \begin{pmatrix} 1 & 4 & 3 \\ 0 & 1 & 1 \\ 0 & 0 & 1 \end{pmatrix}$$

In Example 2, we started with

$$A_2 = \begin{pmatrix} 2 & 8 & 6 \\ 4 & 2 & -2 \\ -6 & 4 & 10 \end{pmatrix}$$

and ended up with

$$B_2 = \begin{pmatrix} 1 & 4 & 3 \\ 0 & 1 & 1 \\ 0 & 0 & 0 \end{pmatrix}$$

In Example 3, we began with

$$A_3 = \begin{pmatrix} 2 & 8 & 6 \\ 4 & 2 & -2 \\ -6 & 4 & 10 \end{pmatrix}$$

and again ended up with

$$B_3 = \begin{pmatrix} 1 & 4 & 3 \\ 0 & 1 & 1 \\ 0 & 0 & 0 \end{pmatrix}$$

The matrices B_1, B_2, and B_3 are called the *row-echelon forms* of the matrices A_1, A_2, and A_3, respectively. In general, a matrix is in **row-echelon form** if the following three conditions hold:

1. All rows (if any) consisting entirely of zeros appear at the bottom of the matrix.
2. The first (starting from the left) nonzero number in any row not consisting entirely of zeros is 1.
3. If two successive rows do not consist entirely of zeros, then the first 1 in the lower row occurs farther to the right than the first 1 in the higher row.

EXAMPLE 4 *Five Matrices in Row-Echelon Form*

The following matrices are in row-echelon form.

$$\begin{pmatrix} 1 & 2 & -7 \\ 0 & 1 & 5 \\ 0 & 0 & 1 \end{pmatrix} \quad \begin{pmatrix} 1 & 6 & 2 & 4 \\ 0 & 1 & 9 & 3 \\ 0 & 0 & 0 & 1 \end{pmatrix} \quad \begin{pmatrix} 1 & 0 & 2 & 5 \\ 0 & 0 & 1 & 2 \end{pmatrix}$$

$$\begin{pmatrix} 1 & -1 \\ 0 & 1 \end{pmatrix} \quad \begin{pmatrix} 1 & 2 & 2 & 5 \\ 0 & 1 & 3 & 6 \\ 0 & 0 & 0 & 0 \end{pmatrix}$$ ■

EXAMPLE 5 *Three Matrices That Are Not in Row-Echelon Form*

The following matrices are *not* in row-echelon form:

(a) $\begin{pmatrix} 1 & 0 & 0 \\ 0 & 0 & 0 \\ 0 & 1 & 0 \end{pmatrix}$ Condition (1) is violated

(b) $\begin{pmatrix} 1 & 0 & 0 \\ 0 & 2 & 0 \\ 0 & 0 & 1 \end{pmatrix}$ Condition (2) is violated

(c) $\begin{pmatrix} 1 & 0 & 0 \\ 0 & 0 & 1 \\ 0 & 1 & 0 \end{pmatrix}$ Condition (3) is violated

---

As we saw in Examples 1, 2, and 3, there is a strong connection between the row-echelon form of a matrix and the existence of a unique solution to the system. In Example 1, the row-echelon form of the *coefficient matrix* (that is, the first three columns in the augmented matrix) had a 1 in each row, and there was a unique solution. In Examples 2 and 3, the row-echelon form of the coefficient matrix had a row of zeros, and the system had either no solution or an infinite number of solutions. This turns out always to be true in any system with the same number of equations as unknowns.

Gaussian Elimination

The process of solving a system of equations by reducing the coefficient matrix to its row-echelon form and then using back substitution is called **Gaussian elimination.**

We can use Gaussian elimination to solve a system of any number of equations in any number of unknowns. However, in our examples and exercises we will not discuss systems involving more than four unknowns.

EXAMPLE 6 *A System of Two Equations in Four Unknowns*

Solve the system

$$\begin{aligned} 2y - 2z + 4w &= 6 \\ x + 3y - 5z + w &= 4 \end{aligned}$$

SOLUTION We write this system as an augmented matrix and row-reduce.

$$\left(\begin{array}{cccc|c} 0 & 2 & -2 & 4 & 6 \\ 1 & 3 & -5 & 1 & 4 \end{array}\right) \xrightarrow{R_1 \rightleftarrows R_2} \left(\begin{array}{cccc|c} 1 & 3 & -5 & 1 & 4 \\ 0 & 2 & -2 & 4 & 6 \end{array}\right)$$

Interchange row 1 and row 2 to get a nonzero number in the upper left-hand corner

$$\xrightarrow{R_2 \to \frac{1}{2}R_2} \left(\begin{array}{cccc|c} 1 & 3 & -5 & 1 & 4 \\ 0 & 1 & -1 & 2 & 3 \end{array}\right)$$

This is as far as we can go. The coefficient matrix is in row-echelon form. There are an infinite number of solutions. The equations read

$$\begin{aligned} x + 3y - 5z + w &= 4 \\ y - z + 2w &= 3 \end{aligned}$$

Any two variables, z and w, say, can be chosen arbitrarily. Then

$$y = 3 + z - 2w \qquad \text{and} \qquad x = 4 - 3y + 5z - w$$

For example, we get one solution if we set $z = w = 0$. Then $y = 3$, $x = -5$, and $(x, y, z, w) = (-5, 3, 0, 0)$. If we set $z = 2$ and $w = -1$, then $y = 3 + 2 - 2(-1) = 7$ and $x = 4 - 3(7) + 5(2) - (-1) = -6$. Thus, another solution is $(x, y, z, w) = (-6, 7, 2, -1)$. We get one solution for every choice of the numbers z and w.

As you will see if you do a lot of system solving, the computations can become very messy. It is a good rule of thumb to use a calculator whenever the fractions become unpleasant.

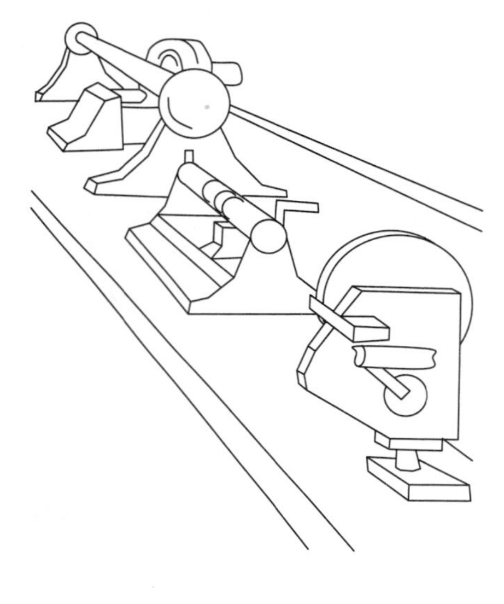

EXAMPLE 7 ***An Application of Systems to Manufacturing***

A manufacturing firm has discontinued production of a certain unprofitable product line, creating considerable excess production capacity. Management is planning to devote this excess capacity to three products, which we call products 1, 2, and 3. The available capacity on the machines used to produce these products is summarized in Table 1. The number of machine-hours required for each unit of the respective products is given in Table 2. How many units of each product should be manufactured in order to use all the available production capacity?

Table 1

Machine Type	**Available Time (in machine hours per week)**
Milling machines	1950
Lathes	1490
Grinders	2160

Table 2
Productivity (in Machine Hours per Unit)

Machine Type	**Product 1**	**Product 2**	**Product 3**
Milling machines	0.2	0.5	0.3
Lathes	0.3	0.4	0.1
Grinders	0.1	0.6	0.4

SOLUTION Let x, y, and z denote the number of units of each of the three products produced each week. Since each unit of product 1 requires 0.2 hour on a milling machine, the number of hours needed each week on the milling machines to produce x units is $0.2x$. Similarly, $0.5y$ and $0.3z$ represent the weekly requirements (in hours) on the milling machines to produce y units of product 2 and z units of product 3, respectively. Since 1950 hours are available on milling machines each week, we have (assuming that all capacity is to be used)

$$0.2x + 0.5y + 0.3z = 1950 \quad \text{Milling machine equation}$$

The equations for utilizing all the capacity of the other two machine types are obtained in a like manner:

$$\begin{aligned} 0.3x + 0.4y + 0.1z &= 1490 \quad \text{Lathe equation} \\ 0.1x + 0.6y + 0.4z &= 2160 \quad \text{Grinder equation} \end{aligned}$$

This is a system of three equations in three unknowns. To simplify matters algebraically, we first multiply each equation by 10 to eliminate the decimals. Then we row-reduce in the usual way.

$$\left(\begin{array}{ccc|c} 2 & 5 & 3 & 19{,}500 \\ 3 & 4 & 1 & 14{,}900 \\ 1 & 6 & 4 & 21{,}600 \end{array}\right) \xrightarrow{R_1 \to \frac{1}{2}R_1} \left(\begin{array}{ccc|c} 1 & \frac{5}{2} & \frac{3}{2} & 9750 \\ 3 & 4 & 1 & 14{,}900 \\ 1 & 6 & 4 & 21{,}600 \end{array}\right)$$

$$\xrightarrow[R_3 \to R_3 - R_1]{R_2 \to R_2 - 3R_1} \left(\begin{array}{ccc|c} 1 & \frac{5}{2} & \frac{3}{2} & 9750 \\ 0 & -\frac{7}{2} & -\frac{7}{2} & -14{,}350 \\ 0 & \frac{7}{2} & \frac{5}{2} & 11{,}850 \end{array}\right)$$

$$\xrightarrow{R_2 \to -\frac{2}{7}R_2} \left(\begin{array}{ccc|c} 1 & \frac{5}{2} & \frac{3}{2} & 9750 \\ 0 & 1 & 1 & 4100 \\ 0 & \frac{7}{2} & \frac{5}{2} & 11{,}850 \end{array}\right)$$

$$\xrightarrow{R_3 \to R_3 - \frac{7}{2}R_2} \left(\begin{array}{ccc|c} 1 & \frac{5}{2} & \frac{3}{2} & 9750 \\ 0 & 1 & 1 & 4100 \\ 0 & 0 & -1 & -2500 \end{array}\right)$$

$$\xrightarrow{R_3 \to -R_3} \left(\begin{array}{ccc|c} 1 & \frac{5}{2} & \frac{3}{2} & 9750 \\ 0 & 1 & 1 & 4100 \\ 0 & 0 & 1 & 2500 \end{array}\right)$$

We have

$$\begin{aligned} x + \tfrac{5}{2}y + \tfrac{3}{2}z &= 9750 \\ y + \quad z &= 4100 \\ z &= 2500 \end{aligned}$$

So

$$y = 4100 - z = 4100 - 2500 = 1600$$

and

$$\begin{aligned} x &= 9750 - \tfrac{5}{2}y - \tfrac{3}{2}z \\ &= 9750 - \tfrac{5}{2}(1600) - \tfrac{3}{2}(2500) = 2000 \end{aligned}$$

The solution is:

$$x = 2000 \text{ units of product } 1$$
$$y = 1600 \text{ units of product } 2$$
$$z = 2500 \text{ units of product } 3$$

must be produced in order to ensure full capacity.

Check

$$\begin{aligned} 0.2x + 0.5y + 0.3z &= 0.2(2000) + 0.5(1600) + 0.3(2500) \\ &= 400 + 800 + 750 = 1950 \\ 0.3x + 0.4y + 0.1z &= 0.3(2000) + 0.4(1600) + 0.1(2500) \\ &= 600 + 640 + 250 = 1490 \\ 0.1x + 0.6y + 0.4z &= 0.1(2000) + 0.6(1600) + 0.4(2500) \\ &= 200 + 960 + 1000 = 2160 \end{aligned}$$

FOCUS ON
Carl Friedrich Gauss, 1777–1855

The greatest mathematician of the nineteenth century, Carl Friedrich Gauss is considered one of the three greatest mathematicians of all time — the others being Archimedes and Newton.

Gauss was born in Brunswick, Germany, in 1777. His father, a hard-working laborer who was exceptionally stubborn and did not believe in formal education, did what he could to keep Gauss from appropriate schooling. Fortunately for Carl (and for mathematics), his mother, though uneducated herself, encouraged her son in his studies and took considerable pride in his achievements until her death at the age of 97.

Gauss was a child prodigy. At the age of 3, he found an error in his father's bookkeeping. A famous story tells of Carl, age 10, as a student in the local Brunswick school. The teacher there was known to assign tasks to keep his pupils busy. One day he asked his students to add the numbers from 1 to 100. Almost at once, Carl placed his slate face down with the words "There it is." Afterword, the teacher found that Gauss was the only one with the correct answer, 5050. Gauss had noticed that the numbers could be arranged in 50 pairs, each with the sum 101 ($1 + 100$, $2 + 99$, and so on), and $50 \times 101 = 5050$. Later in life, Gauss joked that he could add before he could speak.

When Gauss was 15, the Duke of Brunswick noticed him and became his patron. The duke helped him to enter Brunswick College in 1795 and, three years later, to enter the university at Göttingen. Undecided between careers in mathematics and philosophy, Gauss chose mathematics after two remarkable discoveries.

Carl Friedrich Gauss

First, he invented the method of least squares a decade before the result was published by Legendre. Second, a month before his nineteenth birthday, he solved a problem whose solution had been sought for more than two thousand years. Gauss showed how to construct, using compass and ruler, a regular polygon with the number of sides not a multiple of 2, 3, or 5. On March 30, 1796, the day of this discovery, he began a diary, which contained as its first entry rules for construction of a 17-sided regular polygon. The diary, which contains 146 statements of results in only 19 pages, is one of the most important documents in the history of mathematics.

After a short period at Göttingen, Gauss went to the University of Helmstädt and, in 1798 at the age of 20, wrote his now famous doctoral dissertation. In it he gave the first mathematically rigorous proof of the fundamental theorem of algebra—that every polynomial of degree *n* has, counting multiplicities, exactly *n* roots. Many mathematicians, including Euler, Newton, and Lagrange, had attempted to prove this result.

Gauss made a great number of discoveries in physics as well as in mathematics. For example, in 1801 he used a new procedure to calculate, from very little data, the orbit of the planetoid Ceres. In 1833, he invented the electromagnetic telegraph with his colleague Wilhelm Weber (1804–1891). While he did brilliant work in astronomy and electricity, however, it was Gauss's mathematical output that was astonishing. He made fundamental contributions to algebra and geometry. In 1811, he discovered a result that led to the development of complex variable theory by Cauchy. We encounter him here in the Gaussian elimination method. Students of numerical analysis study Gaussian quadrature—a technique for numerical integration in calculus.

Gauss became a professor of mathematics at Göttingen in 1807 and remained at that post until his death in 1855. Even after his death, his mathematical spirit remained to haunt nineteenth-century mathematicians. Often it turned out that an important new result was discovered earlier by Gauss and could be found in his unpublished notes.

In his mathematical writings, Gauss was a perfectionist and is probably the last mathematician who knew everything in his subject. Claiming that a cathedral was not a cathedral until the last piece of scaffolding was removed, he endeavored to make each of his published works complete, concise, and polished. He used a seal that pictured a tree carrying only a few fruit together with the motto *pauca sed matura* (few, but ripe). But Gauss also believed that mathematics must reflect the real world. At his death, Gauss was honored by a commemorative medal on which was inscribed "George V. King of Hanover to the Prince of Mathematicians."

Problems 7.2

Readiness Check

I. Which of the following systems has the coefficient matrix given at the right? $\begin{pmatrix} 3 & 2 & -1 \\ 0 & 1 & 5 \\ 2 & 0 & 1 \end{pmatrix}$

a. $3x + 2y = -1$, $y = 5$, $2x = 1$

b. $3x + 2z = 10$, $2x + y = 0$, $-x + 5y + z = 5$

c. $3x = 2$, $2x + y = 0$, $-x + 5y = 1$

d. $3x + 2y - z = -3$, $y + 5z = 15$, $2x + z = 3$

II. Which of the following is an elementary row operation?

a. Replace a row with a nonzero multiple of that row.
b. Add a nonzero constant to each entry in a row.
c. Interchange two columns.
d. Replace a row with a sum of the row and a nonzero constant.

III. Which of the following is true about the given matrix? $\begin{pmatrix} 1 & 0 & 0 & 3 \\ 0 & 1 & 1 & 2 \\ 0 & 0 & 0 & 3 \\ 0 & 0 & 0 & 0 \end{pmatrix}$

a. It is in row-echelon form.
b. It is not in row-echelon form because the fourth number in row 1 is not 1.
c. It is not in row-echelon form because the first nonzero entry in row 3 is 3.
d. It is not in row-echelon form because the last column contains a 0.

IV. Which of the following is true about the system given below?

$$\begin{aligned} x + y + z &= 3 \\ 2x + 2y + 2z &= 6 \\ 3x + 3y + 3z &= 10 \end{aligned}$$

a. It has the unique solution $x = 1$, $y = 1$, $z = 1$.
b. It is inconsistent.
c. It has an infinite number of solutions.

Answers to Readiness Check

I. d II. a III. c IV. b

In Problems 1–20, use Gaussian elimination to find all solutions (if any) to the given systems.

1. $\begin{aligned} x + y - z &= 1 \\ 2x - 2y + 4z &= 0 \\ -x + 2y + z &= -3 \end{aligned}$

2. $\begin{aligned} -2x + y + 6z &= 18 \\ 5x + 8z &= -16 \\ 3x + 2y - 10z &= -3 \end{aligned}$

3. $\begin{aligned} 2x + 4y - 6z &= 8 \\ -x + 3y + 2z &= 6 \\ x - y - z &= -2 \end{aligned}$

4. $\begin{aligned} 3x + 6y - 6z &= 9 \\ 2x - 5y + 4z &= 6 \\ 5x + 28y - 26z &= -8 \end{aligned}$

5. $\begin{aligned} x + 2y + 3z &= 1 \\ 2x - y + 4z &= 2 \\ 3x + y + z &= 5 \end{aligned}$

6. $\begin{aligned} x + y - z &= 7 \\ 4x - y + 5z &= 4 \\ 6x + y + 3z &= 18 \end{aligned}$

7. $\begin{aligned} x + y + z &= -2 \\ -2x + 2y + z &= -18 \\ 3x + 2y + 2z &= 6 \end{aligned}$

8. $\begin{aligned} x - 2y + 3z &= 0 \\ 4x + y - z &= 0 \\ 2x - y + 3z &= 0 \end{aligned}$

9. $\begin{aligned} x + y - z &= 0 \\ 4x - y + 5z &= 0 \\ 6x + y + 3z &= 0 \end{aligned}$

10. $\begin{aligned} 2y + 5z &= 6 \\ x - 2z &= 4 \\ 2x + 4y &= -2 \end{aligned}$

11. $\begin{aligned} x - 2y + 3z &= 4 \\ -2x + 4y - 6z &= 12 \end{aligned}$

12. $\begin{aligned} x + 2y - 4z &= 4 \\ -2x - 4y + 8z &= -8 \end{aligned}$

13. $\begin{aligned} x + 2y + z &= 4 \\ 2x + 5y + 3z &= 2 \end{aligned}$

14. $\begin{aligned} x + 2y - z + w &= 7 \\ 3x + 6y - 3z + 3w &= 21 \end{aligned}$

15. $\begin{aligned} 2x + 6y - 4z + 2w &= 4 \\ x - z + w &= 5 \\ -3x + 2y - 2z &= -2 \end{aligned}$

16. $\begin{aligned} x - 2y + z + w &= 2 \\ 3x + z - 2w &= -8 \\ y - z - w &= 1 \\ -x + 6y - 2z &= 7 \end{aligned}$

17. $\begin{aligned} x - 2y + z + w &= 2 \\ 3x + 2z - 2w &= -8 \\ 4y - z - w &= 1 \\ 5x + 3z - w &= -3 \end{aligned}$

18. $\begin{aligned} x - 2y + z + w &= 2 \\ 3x + 2z - 2w &= -8 \\ 4y - z - w &= 1 \\ 5x + 3z - w &= 0 \end{aligned}$

19. $\begin{aligned} x + 2y &= -1 \\ 3x + y &= 7 \\ 4x + 3y &= 6 \end{aligned}$

20. $\begin{aligned} x + 2y &= -1 \\ 3x + y &= 5 \\ 4x + 3y &= 8 \end{aligned}$

In Problems 21–30, determine whether the given matrix is in row-echelon form.

21. $\begin{pmatrix} 1 & 1 & 5 \\ 0 & 1 & 1 \\ 0 & 0 & 1 \end{pmatrix}$

22. $\begin{pmatrix} 2 & 0 & 0 \\ 0 & 1 & 0 \\ 0 & 0 & -1 \end{pmatrix}$

23. $\begin{pmatrix} 1 & 0 & 2 \\ 0 & 1 & 0 \end{pmatrix}$

24. $\begin{pmatrix} 1 & 0 & 0 & 10 \\ 0 & 0 & 1 & 1 \\ 0 & 0 & 0 & 1 \end{pmatrix}$

25. $\begin{pmatrix} 0 & 1 & 0 & 2 \\ 1 & 0 & -1 & 3 \\ 0 & 0 & 1 & 4 \end{pmatrix}$

26. $\begin{pmatrix} 1 & 5 & 1 & 2 \\ 0 & 1 & 3 & 4 \end{pmatrix}$

27. $\begin{pmatrix} 1 & -1 \\ 0 & 1 \\ 0 & 0 \end{pmatrix}$

28. $\begin{pmatrix} 1 & 0 & 0 \\ 0 & 0 & 1 \\ 0 & 0 & 1 \end{pmatrix}$

29. $\begin{pmatrix} 1 & 0 & 0 & 4 \\ 0 & 1 & 6 & 5 \\ 0 & 1 & 1 & 6 \end{pmatrix}$

30. $\begin{pmatrix} 1 & 7 & 2 & 1 \\ 0 & 1 & 3 & -4 \\ 0 & 0 & 1 & 8 \end{pmatrix}$

31. In Example 7, how many units of each product should be manufactured in order to use all the available production capacity for the data in Tables 3 and 4?

Table 3

Machine Type	Available Time (in machine hours per week)
Milling machines	1281
Lathes	942
Grinders	1185

Table 4
Productivity (in Machine Hours per Unit)

Machine Type	Product 1	Product 2	Product 3
Milling machines	0.2	0.5	0.4
Lathes	0.1	0.4	0.3
Grinders	0.3	0.3	0.5

32. The Robinson Farm in Illinois has 1000 acres to be planted with soybeans, corn, and wheat. During the planting season, Mrs. Robinson has 3700 labor-hours of planting time available to her. Each acre of soybeans requires 2 labor-hours, each acre of corn requires 6 labor-hours, and each acre of wheat requires 6 labor-hours. Seed to plant an acre of soybeans costs $12, seed for an acre of corn costs $20, and seed for an acre of wheat costs $8. Mrs. Robinson has $12,600 on hand to pay for seed. How should the 1000 acres be planted in order to use all the available land, labor, and seed money?
33. A traveler, just returned from Europe, spent $30 a day for housing in England, $20 a day in France, and $20 a day in Spain. For food, the traveler spent $20 a day in England, $30 a day in France, and $20 a day in Spain. The traveler spent $10 a day in each country for incidental expenses. The traveler's records of the trip indicate a total of $340 spent for housing, $320 for food, and $140 for incidental expenses while traveling in these countries. Calculate the number of days the traveler spent in each of the countries or show that the records must be incorrect because the amounts spent are incompatible with each other.
34. An intelligence agent knows that 60 aircraft, consisting of fighter planes and bombers, are stationed at a certain secret airfield. The agent wishes to determine how many of the 60 are fighter planes and how many are bombers. There is a type of rocket carried by both sorts of planes; the fighter carries six of these rockets, the bomber only two. The agent learns that 250 rockets are required to arm every plane at this airfield. Furthermore, the agent overhears a remark that there are twice as many fighter planes as bombers at the base (that is, the number of fighter planes minus twice the number of bombers equals zero). Calculate the number of fighter planes and bombers at the airfield or show that the agent's information must be incorrect because it is inconsistent.
35. An investor remarks to a stockbroker that all her stock holdings are in three companies, Eastern Airlines, Hilton Hotels, and McDonald's, and that 2 days ago the value of her stocks went down $350, but yesterday the value increased by $600. The broker recalls that 2 days ago the price of Eastern Airlines stock dropped by $1 a share. Hilton Hotels dropped $1.50, but the price of McDonald's stock rose by $0.50. The broker also remembers that yesterday the price of Eastern Airlines stock rose $1.50, there was a further drop of $0.50 a share in Hilton Hotels stock, and McDonald's stock rose $1. Show that the broker does not have enough information to calculate the number of shares the investor owns of each company's stock, but that when the investor says that she owns 200 shares of McDonald's stock, the broker can calculate the number of shares of Eastern Airlines and Hilton Hotels.
* 36. The activities of a grazing animal can be classified roughly into three categories: (1) grazing, (2) moving (to new grazing areas or to avoid predators), and (3) resting. The net energy gain (above maintenance requirements) from grazing is 200 calories per hour. The net energy losses in moving and resting are 150 and 50 calories per hour, respectively.
    (a) How should the day be divided among the three activities so that the energy gains during grazing exactly compensate for energy losses during moving and resting?
    (b) Is this division of the day unique?
* 37. Suppose that the grazing animal of Problem 36 must rest for at least 6 hours every day. How should the day be divided?
38. Suppose that an experiment has five possible (separate) outcomes with probabilities p_1, p_2, p_3, p_4, and p_5. If $p_1 = p_2 + p_3$, $p_3 + p_4 = 2p_2$, $p_2 + p_3 + p_4 = p_5$, and $p_1 + p_2 = p_5$, determine the probabilities of the five outcomes. [Hint: In any experiment, the sum of the probabilities of the outcomes is 1.]
39. Three chemicals are combined to form three grades of fertilizer. A unit of grade I fertilizer requires 10 kg of chemical A, 30 of B, and 60 of C. A unit of grade II requires 20 kg of A, 30 of B, and 50 of C. A unit of grade III requires 50 kg of A and 50 of C. If 1600 kg of A, 1200 of B, and 3200 of C are available, how many units of the three grades should be produced to use all available supplies?
40. Three species of squirrels have been introduced to an island with a total initial population of 2000. After 10 years, species I has doubled its population and species II has increased by 50%. Species III has become extinct. If the population increase in species I equals the increase in species II and if the total population has increased by 500, determine the initial populations of the three species.
41. A witch's magic cupboard contains 10 oz of ground four-leaf clovers and 14 oz of powdered mandrake root. The cupboard will replenish itself automatically provided she

uses up exactly all her supplies. A batch of love potion requires $3\frac{1}{13}$ oz of ground four-leaf clovers and $2\frac{2}{13}$ oz of powdered mandrake root. One recipe of a well-known (to witches) cure for the common cold requires $5\frac{5}{13}$ oz of four-leaf clovers and $10\frac{10}{13}$ oz of mandrake root. How much of the love potion and the cold remedy should the witch make in order to use up the supply in the cupboard exactly?

42. A factory for the construction of quality furniture has two divisions: a machine shop where the parts of the furniture are fabricated and an assembly and finishing division where the parts are put together into the finished product. Suppose there are 12 employees in the machine shop and 20 in the assembly and finishing division and that each employee works an 8-hour day. Suppose further that the factory produces only two products: chairs and tables. A chair requires $\frac{384}{17}$ hours of machine shop time and $\frac{480}{17}$ hours of assembly and finishing time. A table requires $\frac{240}{17}$ hours of machine shop time and $\frac{640}{17}$ hours of assembly and finishing time. Assuming that there is an unlimited demand for these products and that the manufacturer wishes to keep all employees busy, how many chairs and how many tables can this factory produce each day?

43. An ice cream shop sells only ice cream sodas and milk shakes. It puts 1 oz of syrup and 4 oz of ice cream in an ice cream soda and 1 oz of syrup and 3 oz of ice cream in a milk shake. If the store used 4 gal of ice cream and 5 qt of syrup in one given day, how many ice cream sodas and how many milk shakes did it sell on that day? [Hint: 1 qt = 32 oz; 1 gal = 128 oz.]

44. A farmer feeds his cattle a mixture of two types of feed. One standard unit of type A feed supplies a steer with 10% of its minimum daily requirement of protein and 15% of its requirement of carbohydrates. Type B feed contains 12% of the requirement of protein and 8% of the requirement of carbohydrates in a standard unit. If the farmer wishes to feed his cattle exactly 100% of their minimum daily requirement of protein and carbohydrates, how many units of each type of feed should he give a steer each day?

45. Answer the question of Problem 44 if the farmer wishes to satisfy 90% of the daily requirement of protein and 110% of the daily requirement of carbohydrates.

46. A large corporation pays its vice presidents a salary of $100,000 a year, 1000 shares of stock, and an entertainment allowance of $20,000. A division manager receives $70,000 in salary, 50 shares of stock, and $5000 for official entertainment. The assistant manager of a division receives $40,000 in salary but neither stock nor entertainment allowance. If the corporation pays out $1,600,000 in salaries, 1000 shares of stock, and $150,000 in expense allowances to its vice presidents, division managers, and assistant division managers in a year, how many vice presidents, division managers, and assistant division managers does the company have?

47. An automobile service station employs mechanics and station attendants. Each works 8 hours a day. An attendant pumps gas, whereas mechanics are expected to spend $\frac{3}{4}$ of their time repairing automobiles and $\frac{1}{4}$ of their time pumping gas. Suppose it takes $\frac{1}{10}$ hour to service an automobile that comes in for gas. If the service station owner wants to be able to sell gas to 320 cars a day and have 24 hours of mechanics' time available for repair work, how many attendants and how many mechanics should be hired?

48. Consider the system

$$\begin{aligned} 2x - y + 3z &= a \\ 3x + y - 5z &= b \\ -5x - 5y + 21z &= c \end{aligned}$$

Show that the system is inconsistent if $c \neq 2a - 3b$.

49. Consider the system

$$\begin{aligned} 2x + 3y - z &= a \\ x - y + 3z &= b \\ 3x + 7y - 5z &= c \end{aligned}$$

Find conditions on a, b, and c such that the system is consistent.

50. Solve the following system using a hand calculator and carrying 5 decimal places of accuracy.

$$\begin{aligned} 2y - z - 4w &= 2 \\ x - y + 5z + 2w &= -4 \\ 3x + 3y - 7z - w &= 4 \\ -x - 2y + 3z \qquad &= -7 \end{aligned}$$

51. Follow the directions of Problem 50 for the system

$$\begin{aligned} 3.8x + 1.6y + 0.9z &= 3.72 \\ -0.7x + 5.4y + 1.6z &= 3.16 \\ 1.5x + 1.1y - 3.2z &= 43.78 \end{aligned}$$

52. A system of equations is called **homogeneous** if the constant terms in the equations are all equal to zero. Explain why every homogeneous system of equations has solutions. [Hint: The systems in Problems 53–66 are all homogeneous.]

In Problems 53–66, find all solutions to the homogeneous systems.

53. $\begin{aligned} x + 2y &= 0 \\ 2x + 4y &= 0 \end{aligned}$

54. $\begin{aligned} 2x - y &= 0 \\ 3x + 4y &= 0 \end{aligned}$

55. $$\begin{aligned} x - 5y &= 0 \\ -x + 5y &= 0 \end{aligned}$$

56. $$\begin{aligned} x + y - z &= 0 \\ 2x - 4y + 3z &= 0 \\ 3x + 7y - z &= 0 \end{aligned}$$

57. $$\begin{aligned} x + y - z &= 0 \\ 2x - 4y + 3z &= 0 \\ -x - 7y + 6z &= 0 \end{aligned}$$

58. $$\begin{aligned} x + y - z &= 0 \\ 2x - 4y + 3z &= 0 \\ -5x + 13y - 10z &= 0 \end{aligned}$$

59. $$\begin{aligned} 2x + 3y - z &= 0 \\ 6x - 5y + 7z &= 0 \end{aligned}$$

60. $$\begin{aligned} 4x - y &= 0 \\ 7x + 3y &= 0 \\ -8x + 6y &= 0 \end{aligned}$$

61. $$\begin{aligned} x - y + 7z - w &= 0 \\ 2x + 3y - 8z + w &= 0 \end{aligned}$$

62. $$\begin{aligned} x - 2y + z + w &= 0 \\ 3x + 2z - 2w &= 0 \\ 4y - z - w &= 0 \\ 5x + 3z - w &= 0 \end{aligned}$$

63. $$\begin{aligned} -2x + 7w &= 0 \\ x + 2y - z + 4w &= 0 \\ 3x - z + 5w &= 0 \\ 4x + 2y + 3z &= 0 \end{aligned}$$

64. $$\begin{aligned} 2x - y &= 0 \\ 3x + 5y &= 0 \\ 7x - 3y &= 0 \\ -2x + 3y &= 0 \end{aligned}$$

65. $$\begin{aligned} x - 3y &= 0 \\ -2x + 6y &= 0 \\ 4x - 12y &= 0 \end{aligned}$$

66. $$\begin{aligned} x + y - z &= 0 \\ 4x - y + 5z &= 0 \\ -2x + y - 2z &= 0 \\ 3x + 2y - 6z &= 0 \end{aligned}$$

67. Show that the homogeneous system

$$\begin{aligned} a_{11}x + a_{12}y &= 0 \\ a_{21}x + a_{22}y &= 0 \end{aligned}$$

has an infinite number of solutions if and only if $a_{11}a_{22} - a_{12}a_{21} = 0$. Explain this result geometrically.

68. Show that a homogeneous system of m equations in n unknowns has an infinite number of solutions if $n > m$.

69. Consider the system

$$\begin{aligned} 2x - 3y + 5z &= 0 \\ -x + 7y - z &= 0 \\ 4x + 11y + kz &= 0 \end{aligned}$$

For what value of k will the system have nonzero solutions?

7.3 Systems of Nonlinear Equations

In Sections 7.1 and 7.2 we discussed systems in which every equation was linear. In a number of applications, however, it is necessary to solve systems in which one or more of the equations is nonlinear. For example, one application of integral calculus is to find the area enclosed by two intersecting curves. In order to do so, it is first necessary to find the points where the curves intersect.

In this section we provide methods for solving certain systems of nonlinear equations.

EXAMPLE 1 *A System Consisting of a Linear Equation and a Nonlinear Equation*

Find all the points in the xy-plane that satisfy the equations

$$y = x^2 - 2x + 5 \quad (1)$$

$$y = x + 9 \quad (2)$$

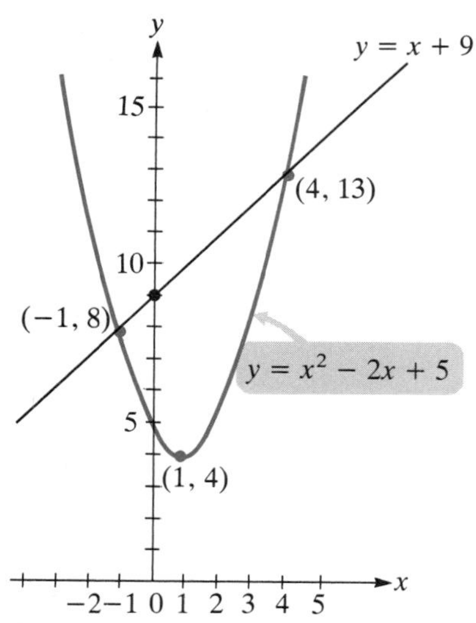

Figure 1 The line $y = x + 9$ intersects the parabola $y = x^2 - 2x + 5$ at the points $(-1, 8)$ and $(4, 13)$.

SOLUTION Since $y = x + 9$, we may substitute $x + 9$ for y in equation (1) to obtain

$$x + 9 = x^2 - 2x + 5$$

$$x^2 - 3x - 4 = 0 \quad \text{Combine terms}$$

$$(x - 4)(x + 1) = 0 \quad \text{Factor}$$

$$x - 4 = 0 \quad \text{or} \quad x + 1 = 0$$

$$x = 4 \quad \text{or} \quad x = -1$$

$$y = x + 9 = 4 + 9 = 13 \quad \text{or} \quad y = x + 9 = -1 + 9 = 8$$

The solutions are the points $(4, 13)$ and $(-1, 8)$.

Check $13 = 4^2 - 2 \cdot 4 + 5$ and $8 = (-1)^2 - 2(-1) + 5$
$13 = 16 - 8 + 5$ $\qquad 8 = 1 + 2 + 5$

How do we know that $(4, 13)$ and $(-1, 8)$ are the only solutions? To answer that question, we sketch the graphs of the two equations. Since $x^2 - 2x + 5 = (x - 1)^2 + 4$, the graph of $y = x^2 - 2x + 5$ is a parabola opening upward with vertex $(1, 4)$ (see pp. 223–226). It, and the graph of the straight line $y = x + 9$, are given in Figure 1.

From the graphs we seen that the parabola and the straight line have exactly two points of intersection. These are the points $(4, 13)$ and $(-1, 8)$, and we now know that these are the only solutions to the system (1), (2). ■

EXAMPLE 2 *The Points of Intersection of a Circle and an Ellipse*

Find all points of intersection of the circle and the ellipse:

$$x^2 + y^2 = 9 \tag{3}$$

$$4x^2 + y^2 = 16 \tag{4}$$

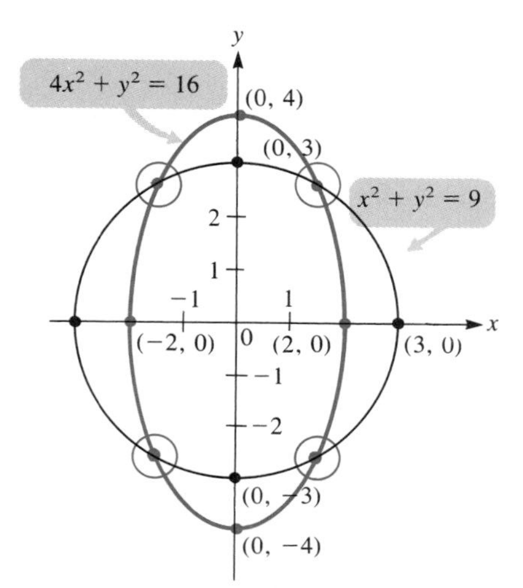

Figure 2 The circle and the ellipse intersect at four points (circled).

SOLUTION It is best to start by sketching the two curves. The circle (3) is centered at $(0, 0)$ with radius 3. The ellipse (4) passes through the points $(2, 0)$, $(-2, 0)$, $(0, 4)$, and $(0, -4)$. The curves are sketched in Figure 2, and we see that there are four points of intersection. From (3), we have

$$y^2 = 9 - x^2$$

and we substitute this into equation (4) to obtain

$$4x^2 + (9 - x^2) = 16$$

$$3x^2 = 7$$

$$x^2 = \frac{7}{3}$$

$$x = \pm\sqrt{\frac{7}{3}}$$

Then

$$y^2 = 9 - x^2 = 9 - \frac{7}{3} = \frac{20}{3}$$

and

$$y = \pm\sqrt{\frac{20}{3}}$$

The four points of intersection are

$$\left(\sqrt{\frac{7}{3}}, \sqrt{\frac{20}{3}}\right), \left(-\sqrt{\frac{7}{3}}, \sqrt{\frac{20}{3}}\right), \left(\sqrt{\frac{7}{3}}, -\sqrt{\frac{20}{3}}\right), \left(-\sqrt{\frac{7}{3}}, -\sqrt{\frac{20}{3}}\right)$$

You should check these answers. ■

EXAMPLE 3 *A System of Two Nonlinear Equations With No Real Solutions*

Find all real solutions to the system

$$4x^2 + 9y^2 = 36 \tag{5}$$
$$x^2 - 4y^2 = 16 \tag{6}$$

SOLUTION From (6) we have $x^2 = 16 + 4y^2$ and substituting this into (5) yields

$$\begin{aligned} 4(16 + 4y^2) + 9y^2 &= 36 \\ 64 + 25y^2 &= 36 \\ 25y^2 &= -28 \\ y^2 &= -\frac{28}{25} \end{aligned}$$

Evidently, since no real number has a negative square, the system has no real solutions. But it does have complex solutions. Using $i = \sqrt{-1}$ as in Section 2.4 (p. 104) we have

$$y = \pm\sqrt{\frac{28}{25}}i = \pm\frac{2}{5}\sqrt{7}i$$

Then

$$x^2 = 16 + 4y^2 = 16 + 4\left(\frac{-28}{25}\right) = \frac{400 - 112}{25} = \frac{288}{25}$$

and

$$x = \pm\sqrt{\frac{288}{25}} = \pm\frac{12}{5}\sqrt{2}$$

The system has four complex solutions:

$$x = \frac{12}{5}\sqrt{2},\ y = \frac{2}{5}\sqrt{7}i \qquad x = -\frac{12}{5}\sqrt{2},\ y = \frac{2}{5}\sqrt{7}i$$

$$x = \frac{12}{5}\sqrt{2},\ y = -\frac{2}{5}\sqrt{7}i \qquad x = -\frac{12}{5}\sqrt{2},\ y = -\frac{2}{5}\sqrt{7}i$$

Check

$$4x^2 + 9y^2 = 4\left(\frac{288}{25}\right) + 9\left(-\frac{28}{25}\right) = \frac{1152 - 252}{25} = \frac{900}{25} = 36$$

$$x^2 - 4y^2 = \frac{288}{25} - 4\left(-\frac{28}{25}\right) = \frac{288 + 112}{25} = \frac{400}{25} = 16$$

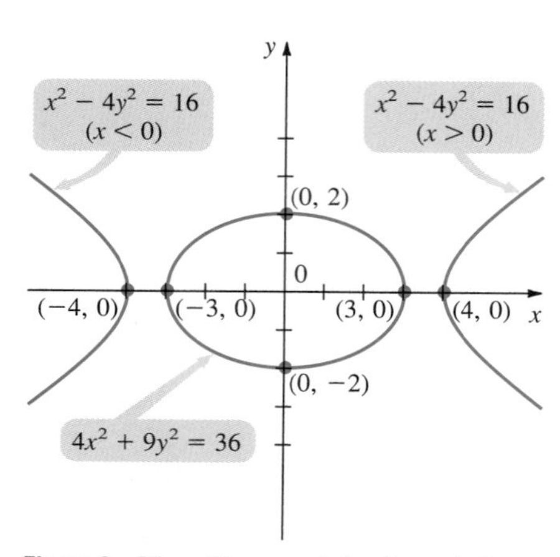

Figure 3 The ellipse and the hyperbola have no points of intersection.

We can see why there are no solutions in the xy-plane by drawing graphs. The ellipse $4x^2 + 9y^2 = 36$ is obtained by interchanging x and y in Figure 4 on p. 296, and the hyperbola $x^2 - 4y^2 = 9$ appears in Figure 5 on p. 315. The graph of the hyperbola $x^2 - 4y^2 = 16$ is obtained by changing 3 to 4. We superimpose these two sketches to obtain the graph in Figure 3. From the graphs it is evident that the two curves do not intersect so that the system (5), (6) has no solution of the form (x, y), where both x and y are real. In problems like this one, the answer to the question depends on how the question is phrased. If you are asked to find all points of intersection of the two curves, then the answer is "none." If you are asked to find all solutions, then the answer consists of the four solutions given above. ■

EXAMPLE 4 *The Points of Intersection of a Circle and a Parabola*

Find all points of intersection of the circle $x^2 + y^2 = 4$ and the parabola $y = x^2$.

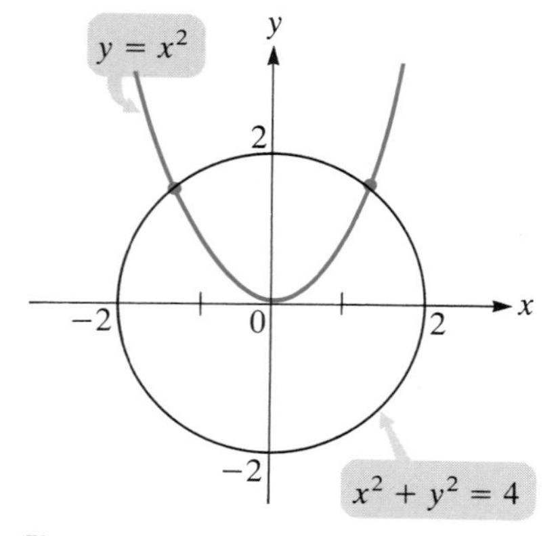

Figure 4 The circle and parabola intersect at two points.

SOLUTION We sketch the two curves in Figure 4 and observe that there are two points of intersection. To find them, we substitute x^2 for y in the equation $x^2 + y^2 = 4$ to obtain

$$x^2 + y^2 = 4$$
$$x^2 + (x^2)^2 = 4$$
$$x^4 + x^2 - 4 = 0 \quad \text{Simplify}$$

Setting $u = x^2$ leads to the quadratic equation (since $u^2 = (x^2)^2 = x^4$)

$$u^2 + u - 4 = 0$$

Then, from the quadratic formula on p. 97, we obtain

$$u = \frac{-1 \pm \sqrt{1 + 16}}{2} = \frac{-1 \pm \sqrt{17}}{2}$$

Thus $u = x^2 = \dfrac{-1 + \sqrt{17}}{2}$ or $u = x^2 = \dfrac{-1 - \sqrt{17}}{2}$. But x is a real number, so $x^2 \geq 0$ and the value $\dfrac{-1 - \sqrt{17}}{2}$ must be thrown out. Then

$$x^2 = \frac{-1 + \sqrt{17}}{2} \quad \text{and} \quad x = \pm\sqrt{\frac{-1 + \sqrt{17}}{2}}$$

For both values of x, $y = x^2 = \dfrac{-1 + \sqrt{17}}{2}$. Thus the two points of intersection are

$$\left(\sqrt{\frac{-1 + \sqrt{17}}{2}}, \frac{-1 + \sqrt{17}}{2}\right) \quad \text{and} \quad \left(-\sqrt{\frac{-1 + \sqrt{17}}{2}}, \frac{-1 + \sqrt{17}}{2}\right)$$
$$\approx (1.2496, 1.5616) \quad \text{and} \quad (-1.2496, 1.5616)$$

NOTE If we sought all *solutions* to the system $x^2 + y^2 = 4$, $y = x^2$ rather than all points of intersection, then we would keep the value $y = x^2 = (-1 - \sqrt{17})/2$ and obtain $x = \pm\sqrt{\dfrac{1 + \sqrt{17}}{2}}\,i$. ■

EXAMPLE 5 *Finding the Points of Intersection of Two Exponential Curves*

Find all points of intersection of the curves $y = e^x$ and $y = e^{2x} - 2$.

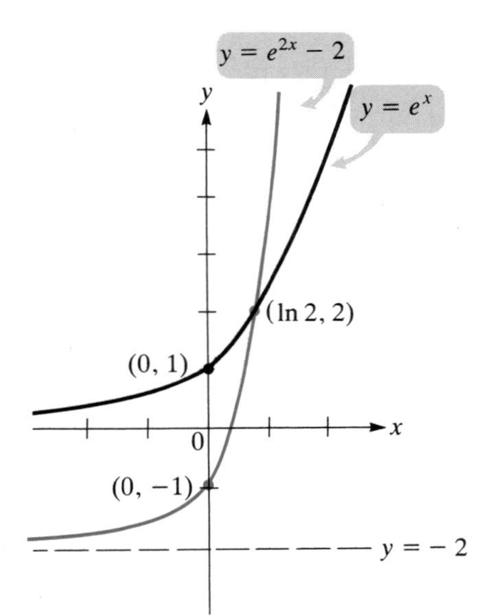

Figure 5 The curves $y = e^x$ and $y = e^{2x} - 2$ intersect at the single point (ln 2, 2).

SOLUTION Substituting e^x for y in the second equation, we obtain

$$e^x = e^{2x} - 2$$

Notice that $e^{2x} = (e^x)^2$ so if we set $u = e^x$, we obtain

$$e^x = e^{2x} - 2$$
$$u = u^2 - 2$$
$$u^2 - u - 2 = 0$$
$$(u - 2)(u + 1) = 0$$
$$u - 2 = 0 \quad \text{or} \quad u + 1 = 0$$
$$u = 2 \quad \text{or} \quad u = -1$$
$$e^x = 2 \quad \text{or} \quad e^x = -1 \qquad u = e^x$$

But $e^x = -1$ has no real solution since $e^x > 0$ for every real number x. Thus

$$y = e^x = 2 \quad \text{and} \quad x = \ln 2$$

Thus, the one point of intersection is $(\ln 2, 2) \approx (0.6931, 2)$. The two exponential curves and their point of intersection are sketched in Figure 5.

Problems 7.3

Readiness Check

I. How many real solutions are there for the following system?

$$x^2 + y^2 = 1$$
$$x^2 - y^2 = 1$$

a. 0 b. 1 c. 2 d. 3 e. 4

II. How many real solutions are there for the following system?

$$x^2 + y^2 = 1$$
$$x^2 - y^2 = 4$$

a. 0 b. 1 c. 2 d. 3 e. 4

III. What is the largest value that the number c can take in order that the following system has real solutions?

$$2x^2 + y^2 = 12$$
$$y^2 - x^2 = c$$

a. 3 b. 6 c. 12 d. 2 e. $\sqrt{12}$

In Problems 1–15 find all points of intersection of the two curves. Then sketch the curves.

1. $x^2 + y^2 = 4$, $y = x$
2. $x^2 + y^2 = 1$, $x + 2y = 0$
3. $x^2 + y^2 = 25$, $y = x - 1$
4. $x^2 + y^2 = 25$, $y = 2x - 2$
5. $y = x^2$, $y = 2x$
6. $y = x^2 - 2x - 3$, $y = 2(x + 1)$
7. $x^2 + 2y^2 = 17$, $2x^2 + y^2 = 22$
8. $-x + y^2 + 1 = 0$, $4x + 3y = 12$
9. $x^2 + y^2 = 4$, $x^2 - y^2 = 4$
10. $2x^2 + y^2 = 9$, $x^2 - y^2 = 3$
11. $x - 4y + y^2 = 2$, $-x + 2y = 2$
12. $-x + y^2 = 4$, $2x + y^2 = 1$
13. $-x^2 + 5y^2 = 0$, $4x^2 + 3y^2 = 0$
14. $5x^2 + 3y^2 = 17$, $-3x^2 + 2y^2 = 5$
15. $-xy = 15$, $3x + 4y = 3$

In Problems 16–40 find all *real* solutions to each system.

16. $y = x^2$, $x = y^2$
17. $y = 2x^2$, $x^2 + y^2 = 4$
18. $y = x^2$, $x^2 - 2y^2 = 1$
19. $xy = 4$, $x + y = 6$
20. $-4x^2 + 2y^2 = 7$, $2x^2 + y^2 = 3$
21. $\frac{1}{x} - \frac{3}{y} = 6$, $\frac{-4}{x} + \frac{2}{y} = 6$ [Hint: First make the substitutions $u = \frac{1}{x}$, $v = \frac{1}{y}$.]
22. $\frac{2}{x} - \frac{1}{y} = -3$, $\frac{5}{x} + \frac{7}{y} = 4$
23. $\frac{2}{x} - \frac{8}{y} = 5$, $\frac{-3}{x} + \frac{12}{y} = 8$
* 24. $\frac{2}{x} - \frac{8}{y} = 6$, $\frac{-3}{x} + \frac{12}{y} = -9$
25. $\frac{3}{x} + \frac{1}{y} = 0$, $\frac{2}{x} - \frac{3}{y} = 0$
26. $\frac{5}{x^2} + \frac{2}{y^2} = 3$, $\frac{2}{x^2} - \frac{3}{y^2} = -1$ [Hint: Make the substitutions $u = \frac{1}{x^2}$, $v = \frac{1}{y^2}$.]
27. $\frac{1}{x^2} + \frac{1}{y^2} = 4$, $\frac{1}{x^2} - \frac{1}{y^2} = 4$
28. $\frac{1}{x^2} + \frac{2}{y^2} = 17$, $\frac{2}{x^2} + \frac{1}{y^2} = 22$
29. $x^2 = y + 9$, $y = -x^2 - 2x + 3$
30. $4x^2 + 9y^2 = 36$, $4x^2 - 9y^2 = 36$

Answers to Readiness Check

I. c II. a III. c

31. $\frac{1}{x} - \frac{2}{y} + \frac{3}{z} = 11$
$\frac{4}{x} + \frac{1}{y} - \frac{1}{z} = 4$
$\frac{2}{x} - \frac{1}{y} + \frac{3}{z} = 10$

32. $\frac{-2}{x} + \frac{1}{y} + \frac{6}{z} = 18$
$\frac{5}{x} + \frac{8}{z} = -16$
$\frac{3}{x} + \frac{2}{y} - \frac{10}{z} = -3$

33. $y = e^x$
$y = e^{2x}$

34. $y = 2^x$
$y = 2^{2x} - 2$

* 35. $y = e^x + 1$
$y = e^{2x} - 11$

36. $y = e^{2x}$
$y = e^{4x} - 12$

37. $y = \ln x + 3$
$y = 2 \ln x + 1$

38. $y = 2 \log_2 x - 1$
$y = 4 \log_2 x + 3$

* 39. $x + y = 3$
$3^x 2^y = 8$
[Hint: First solve for y in terms of x. Then take natural logarithms. Finally, solve the resulting equation for x.]

* 40. $2x + 3y = 6$
$2^x 4^y = 20$

In Problems 41–46 show graphically that the given system has no real solutions. Then find all complex solutions.

41. $y = x^2 + 3$
$y = 1 - x^2$

42. $y = x^2 - 2x + 5$
$y = -x^2 - 2x + 1$

43. $x^2 + y^2 = 1$
$x^2 - y^2 = 4$

44. $x^2 + y^2 = 9$
$y^2 - x^2 = 16$

45. $x^2 + 4y^2 = 4$
$x^2 - 3y^2 = 9$

46. $y = x^2$
$4x^2 - y^2 = 5$

Solve Problems 47–53 by first setting up a system of nonlinear equations.

47. Find two numbers whose sum is 20 and whose product is 96.

48. Find two positive numbers whose difference is 8 and whose product is 65.

49. Find two positive numbers whose product is 48 and whose quotient is 3.

50. Find the lengths of the legs of a right triangle if its area is 12 and the length of the hypotenuse is $\sqrt{73}$.

51. Find the radius of each circle below if the total area enclosed by the two circles is $\frac{25}{2}\pi$ cm^2.

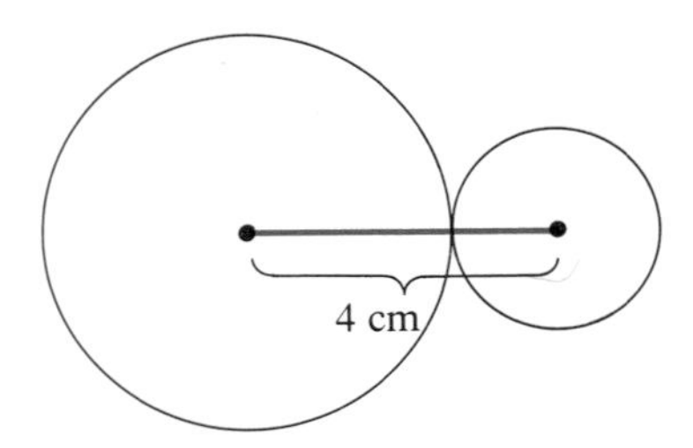

52. Mr. Jones buys a bond that pays him \$430 in simple interest at the end of one year. He then invests the same amount of money in a new bond that pays $1\frac{1}{2}$% more interest than the first bond and he receives \$505 in simple interest at the end of the second year. What was the rate of interest for each bond and how much did he invest?

53. A club in California chartered a plane for a trip to Hawaii. The total cost was \$36,000. After arrangements had been made, six additional people chose to go along, and the cost per person was reduced by \$10. How many people were originally scheduled to go on the trip, and what was the original cost per person?

7.4 Partial Fractions (Optional)

We have seen how to perform the following addition:

$$\frac{1}{x+3} + \frac{3}{x-2} = \frac{(x-2) + 3(x+3)}{(x+3)(x-2)} = \frac{4x+7}{x^2+x-6}$$

In calculus, it is important to be able to reverse the process:

$$\frac{4x+7}{x^2+x-6} = \frac{1}{x+3} + \frac{3}{x-2}$$

The terms $\frac{1}{x+3}$ and $\frac{3}{x-2}$ are called **partial fractions.**

In general, we wish to write a rational function as a sum of terms with the denominator in each term either linear or an irreducible quadratic. (Recall from p. 97 that the quadratic $ax^2 + bx + c$ is irreducible if $b^2 - 4ac < 0$.) In this section we will show how this is done using the material in Sections 7.1 and 7.2.

EXAMPLE 1 ***A Decomposition Where the Denominator Is the Product of Linear Terms***

Find the partial fraction decomposition of

$$\frac{1}{x^2 + 4x - 5}$$

SOLUTION We have $\dfrac{1}{x^2 + 4x - 5} = \dfrac{1}{(x + 5)(x - 1)}$. We write this as

$$\frac{1}{(x + 5)(x - 1)} = \frac{A}{x + 5} + \frac{B}{x - 1} \tag{1}$$

We must compute A and B. To do so, we multiply both sides of (1) by $(x + 5)(x - 1)$:

$$1 = A(x - 1) + B(x + 5) = Ax - A + Bx + 5B$$

or

$$(A + B)x + (-A + 5B) = 1 = 0 \cdot x + 1$$

In order that these be equal, the coefficients of x and 1 must be equal. On the left, the coefficients are $A + B$ and $-A + 5B$. On the right, the coefficients are 0 and 1. Thus, equating coefficients, we have

$$A + B = 0 \qquad \text{and} \qquad -A + 5B = 1$$

This is a system of two linear equations in the two unknowns A and B.

Since $A + B = 0$, $B = -A$, so $-A + 5B = -A - 5A = -6A = 1$. Then $A = -\frac{1}{6}$ and $B = -A = \frac{1}{6}$. We have

$$\frac{1}{(x + 5)(x - 1)} = \frac{-\frac{1}{6}}{x + 5} + \frac{\frac{1}{6}}{x - 1}$$

Check

$$\frac{-\frac{1}{6}}{x + 5} + \frac{\frac{1}{6}}{x - 1} = \frac{-\frac{1}{6}(x - 1) + \frac{1}{6}(x + 5)}{(x + 5)(x - 1)}$$

$$= \frac{-\frac{1}{6}x + \frac{1}{6} + \frac{1}{6}x + \frac{5}{6}}{(x + 5)(x - 1)} = \frac{1}{(x + 5)(x - 1)} \quad ■$$

EXAMPLE 2 ***An Alternative Way to Find the Coefficients***

Find the partial fraction decomposition of $\dfrac{x^2 + 4x + 1}{(x - 1)(x + 2)(x - 3)}$

SOLUTION We write

$$\frac{x^2+4x+1}{(x-1)(x+2)(x-3)} = \frac{A}{x-1} + \frac{B}{x+2} + \frac{C}{x-3} \tag{2}$$

We can compute A, B, and C as in Example 1, but there is an easier way.

Step A Multiply both sides of (2) by $x-1$, and then evaluate both sides at $x = 1$:

$$\frac{\cancel{(x-1)}(x^2+4x+1)}{\cancel{(x-1)}(x+2)(x-3)} = \cancel{(x-1)}\frac{A}{\cancel{x-1}} + (x-1)\frac{B}{x+2} + (x-1)\frac{C}{x-3}$$

$$\frac{x^2+4x+1}{(x+2)(x-3)} = A + (x-1)\frac{B}{x+2} + (x-1)\frac{C}{x-3}$$

Evaluate at $x = 1$ $$\frac{6}{3(-2)} = A + 0 + 0 \quad \text{or} \quad A = \frac{6}{-6} = -1$$

A shortcut is to cover the $x-1$ on the left side of (2) and evaluate the remaining factor at $x = 1$.

Step B Find B. Cover the $x+2$ on the left side of (2), and evaluate at $x = -2$:

$$B = \frac{x^2+4x+1}{(x-1)\boxed{(x+2)}(x-3)}\Bigg|_{\text{Evaluated at } x=-2} = \frac{(-2)^2+4(-2)+1}{(-2-1)(-2-3)} = \frac{-3}{15} = -\frac{1}{5}$$

($x+2$ factor covered)

Step C Find C. Cover the $x-3$ factor on the left side of (1), and evaluate at $x = 3$:

$$C = \frac{x^2+4x+1}{(x-1)(x+2)\boxed{(x-3)}}\Bigg|_{\text{Evaluated at } x=3} = \frac{9+12+1}{(3-1)(3+2)} = \frac{22}{10} = \frac{11}{5}$$

($x-3$ factor covered)

Thus $A = -1$, $B = -\frac{1}{5}$, $C = \frac{11}{5}$, and

$$\frac{x^2+4x+1}{(x-1)(x+2)(x-3)} = -\frac{1}{x-1} - \frac{\frac{1}{5}}{x+2} + \frac{\frac{11}{5}}{x-3}$$

This answer should be checked. The procedure used here always works to find the coefficient of $\dfrac{1}{x-c}$ if $x-c$ is a factor of the denominator but $(x-c)^2$ is not.

Before giving more examples, we provide a general procedure for decomposing a rational function into partial fractions.

Procedure for Obtaining the Partial Fraction Decomposition of $\dfrac{p(x)}{q(x)}$

Step 1 If degree of $p(x) \geq$ degree of $q(x)$, first divide to obtain

$$\frac{p(x)}{q(x)} = d(x) + \frac{r(x)}{q(x)}$$

where degree of $r(x) <$ degree of $q(x)$. Then find the decomposition of $\dfrac{r(x)}{q(x)}$.

Step 2 If the leading coefficient of $q(x)$ is $a_n \neq 1$, then factor a_n out so that the denominator function has leading coefficient 1.

Step 3 Factor the new $q(x)$ into linear and/or irreducible quadratic factors. This is the hardest step. Since the leading coefficient of $q(x)$ is 1, these factors will have the form $x - c$ or $x^2 + ax + b$.†

Step 4 For each factor of the form $x - c$, the partial fraction decomposition of $\dfrac{r(x)}{q(x)}$ contains a term of the form $\dfrac{A}{x - c}$.

Step 5 For each factor of the form $(x - c)^k$, $k > 1$, the partial fraction decomposition contains a sum of k terms:

$$\frac{A_1}{x - c} + \frac{A_2}{(x - c)^2} + \cdots + \frac{A_k}{(x - c)^k}$$

Step 6 For each factor of the form $x^2 + ax + b$, the partial fraction decomposition contains a term of the form

$$\frac{Ax + B}{x^2 + ax + b}$$

Step 7 For each factor of the form $(x^2 + ax + b)^k$, $k > 1$, the partial fraction decomposition contains a sum of k terms:

$$\frac{A_1x + B_1}{x^2 + ax + b} + \frac{A_2x + B_2}{(x^2 + ax + b)^2} + \cdots + \frac{A_kx + B_k}{(x^2 + ax + b)^k}$$

Step 8 Add the terms obtained in Steps 4, 5, 6, and 7 to obtain a rational function with denominator equal to $q(x)$.

Step 9 Equate the numerator in Step 8 to $p(x)$ or $r(x)$ to determine the coefficients (the A_k's and B_k's).

EXAMPLE 3 *A Partial Fraction Decomposition Where One Linear Term Is Raised to a Power Greater Than 1*

Find the partial fraction decomposition of $\dfrac{2x + 4}{(x - 5)(x + 3)^2}$

†If a quadratic factor is not irreducible but cannot be factored using integers, then treat it as if it were irreducible.

SOLUTION We now write (using Step 5)

$$\frac{2x+4}{(x-5)(x+3)^2} = \frac{A}{x-5} + \frac{B}{x+3} + \frac{C}{(x+3)^2} \tag{3}$$

Multiply both sides of (3) by $(x-5)(x+3)^2$:

$$\begin{aligned} 2x+4 &= A(x+3)^2 + B(x-5)(x+3) + C(x-5) \\ &= A(x^2+6x+9) + B(x^2-2x-15) + C(x-5) \\ &= (A+B)x^2 + (6A-2B+C)x + (9A-15B-5C) \end{aligned}$$

Equating the coefficients of x^2, x, and 1 in the numerators, we obtain

$$\begin{aligned} A + B \quad\quad &= 0 \\ 6A - 2B + C &= 2 \\ 9A - 15B - 5C &= 4 \end{aligned}$$

We solve this system by Gaussian elimination:

$$\left(\begin{array}{ccc|c} 1 & 1 & 0 & 0 \\ 6 & -2 & 1 & 2 \\ 9 & -15 & -5 & 4 \end{array}\right) \xrightarrow[R_3 \to R_3 - 9R_1]{R_2 \to R_2 - 6R_1} \left(\begin{array}{ccc|c} 1 & 1 & 0 & 0 \\ 0 & -8 & 1 & 2 \\ 0 & -24 & -5 & 4 \end{array}\right)$$

$$\xrightarrow{R_2 \to -\frac{1}{8}R_2} \left(\begin{array}{ccc|c} 1 & 1 & 0 & 0 \\ 0 & 1 & -\frac{1}{8} & -\frac{1}{4} \\ 0 & -24 & -5 & 4 \end{array}\right)$$

$$\xrightarrow{R_3 \to R_3 + 24R_2} \left(\begin{array}{ccc|c} 1 & 1 & 0 & 0 \\ 0 & 1 & -\frac{1}{8} & -\frac{1}{4} \\ 0 & 0 & -8 & -2 \end{array}\right)$$

$$\xrightarrow{R_3 \to -\frac{1}{8}R_3} \left(\begin{array}{ccc|c} 1 & 1 & 0 & 0 \\ 0 & 1 & -\frac{1}{8} & -\frac{1}{4} \\ 0 & 0 & 1 & \frac{1}{4} \end{array}\right)$$

The system now reads

$$\begin{aligned} A + B \quad\quad &= 0 \\ B - \frac{1}{8}C &= -\frac{1}{4} \\ C &= \frac{1}{4} \end{aligned}$$

so

$$B = -\frac{1}{4} + \frac{1}{8}C = -\frac{1}{4} + \frac{1}{8}\left(\frac{1}{4}\right) = -\frac{1}{4} + \frac{1}{32} = -\frac{7}{32}$$

and

$$A = -B = \frac{7}{32}$$

Thus, $$\frac{2x+4}{(x-5)(x+3)^2} = \frac{\frac{7}{32}}{x-5} - \frac{\frac{7}{32}}{x+3} + \frac{\frac{1}{4}}{(x+3)^2} \tag{4}$$

We want to check our answer. The easiest way is to insert some value for x (except 5 or -3) into each side of (4). If we get the same number on each side, then our partial fraction decomposition is almost certainly correct. (It could be wrong but yield the right answer in one special case, but this is extremely unlikely.) We check with a calculator. Let $x = 10$. Then

$$\frac{2x+4}{(x-5)(x+3)^2} = \frac{24}{5(13)^2} = \frac{24}{845} \approx 0.028402366$$

$$\frac{7}{32(x-5)} - \frac{7}{32(x+3)} + \frac{1}{4(x+3)^2} = \frac{7}{32(5)} - \frac{7}{32(13)} + \frac{1}{4(13)^2}$$
$$= \frac{7}{160} - \frac{7}{416} + \frac{1}{676} \approx 0.028402366$$

EXAMPLE 4 ***A Partial Fraction Decomposition with an Irreducible Quadratic in the Denominator***

Find the partial fraction decomposition of $\dfrac{x-2}{(x+4)(x^2+2x+2)}$

SOLUTION First we note that $x^2 + 2x + 2$ is irreducible. We then write

$$\frac{x-2}{(x+4)(x^2+2x+2)} = \frac{A}{x+4} + \frac{Bx+C}{x^2+2x+2} \tag{5}$$

We multiply both sides of (5) by $(x+4)(x^2+2x+2)$ to obtain

$$\begin{aligned} x - 2 &= A(x^2+2x+2) + (Bx+C)(x+4) \\ x - 2 &= Ax^2 + 2Ax + 2A + Bx^2 + 4Bx + Cx + 4C \\ 0x^2 + 1 \cdot x - 2 = x - 2 &\\ &= (A+B)x^2 + (2A+4B+C)x + (2A+4C) \end{aligned}$$

Equating coefficients of x^2, x, and 1, we obtain the system

$$\begin{aligned} A + B &= 0 \\ 2A + 4B + C &= 1 \\ 2A + 4C &= -2 \end{aligned}$$

We row-reduce:

$$\left(\begin{array}{ccc|c} 1 & 1 & 0 & 0 \\ 2 & 4 & 1 & 1 \\ 2 & 0 & 4 & -2 \end{array}\right) \xrightarrow[R_3 \to R_3 - 2R_1]{R_2 \to R_2 - 2R_1} \left(\begin{array}{ccc|c} 1 & 1 & 0 & 0 \\ 0 & 2 & 1 & 1 \\ 0 & -2 & 4 & -2 \end{array}\right)$$

$$\xrightarrow{R_2 \to \frac{1}{2}R_2} \left(\begin{array}{ccc|c} 1 & 1 & 0 & 0 \\ 0 & 1 & \frac{1}{2} & \frac{1}{2} \\ 0 & -2 & 4 & -2 \end{array}\right)$$

$$\xrightarrow{R_3 \to R_3 + 2R_2} \left(\begin{array}{ccc|c} 1 & 1 & 0 & 0 \\ 0 & 1 & \frac{1}{2} & \frac{1}{2} \\ 0 & 0 & 5 & -1 \end{array}\right)$$

$$\xrightarrow{R_3 \to \frac{1}{5}R_3} \left(\begin{array}{ccc|c} 1 & 1 & 0 & 0 \\ 0 & 1 & \frac{1}{2} & \frac{1}{2} \\ 0 & 0 & 1 & -\frac{1}{5} \end{array}\right)$$

We have

$$\begin{aligned} A + B \qquad &= 0 \\ B + \tfrac{1}{2}C &= \tfrac{1}{2} \\ C &= -\tfrac{1}{5} \end{aligned}$$

so

$$B = \tfrac{1}{2} - \tfrac{1}{2}C = \tfrac{1}{2} - \tfrac{1}{2}(-\tfrac{1}{5}) = \tfrac{1}{2} + \tfrac{1}{10} = \tfrac{3}{5} \quad \text{and} \quad A = -B = -\tfrac{3}{5}$$

Thus,

$$\frac{x-2}{(x+4)(x^2+2x+2)} = \frac{-\frac{3}{5}}{x+4} + \frac{\frac{3}{5}x - \frac{1}{5}}{x^2+2x+2}$$

Problems 7.4

Readiness Check

I. Which of the following is in the form of the partial fraction decomposition of $\dfrac{2x+3}{x^2-1}$?

a. $\dfrac{Ax+B}{x^2-1}$ b. $\dfrac{A}{x-1} + \dfrac{B}{x+1}$

c. $\dfrac{A}{x-1} + \dfrac{Bx+C}{x^2-1}$ d. $\dfrac{A}{x+1} + \dfrac{B}{x-1} + \dfrac{C}{x^2-1}$

II. Which of the following is the form of the partial fraction decomposition of $\dfrac{3x^2-2x+7}{x^2(x-5)}$?

a. $\dfrac{Ax+B}{x^2} + \dfrac{C}{x-5}$ b. $\dfrac{A}{x^2} + \dfrac{B}{x-5}$

c. $\dfrac{A}{x} + \dfrac{B}{x^2} + \dfrac{C}{x-5}$ d. $\dfrac{A}{x} + \dfrac{Bx+C}{x^2} + \dfrac{D}{x-5}$

III. Which of the following is the form of the partial fraction decomposition of $\dfrac{2x^2-4x+7}{(x-1)(x+3)}$?

a. $2 + \dfrac{A}{x-1} + \dfrac{B}{x+3}$ b. $\dfrac{A}{x-1} + \dfrac{B}{x+3}$

c. $\dfrac{A}{x-1} + \dfrac{B}{x+3} + \dfrac{Cx+D}{x^2+2x-3}$

d. $\dfrac{A}{x-1} + \dfrac{B}{x+3} + \dfrac{C}{x^2-2x+3}$

IV. Which of the following is the form of the partial fraction decomposition of

$$\frac{2x^2-5x+6}{(x+2)(x-3)^2(x^2-4x+10)}?$$

a. $\dfrac{A}{x+2} + \dfrac{B}{(x-3)^2} + \dfrac{C}{x^2-4x+10}$

b. $\dfrac{A}{x+2} + \dfrac{B}{(x-3)^2} + \dfrac{Cx+D}{x^2-4x+10}$

c. $\dfrac{A}{x+2} + \dfrac{Bx+C}{(x-3)^2} + \dfrac{Dx+E}{x^2-4x+10}$

d. $\dfrac{A}{x+2} + \dfrac{B}{x-3} + \dfrac{C}{(x-3)^2} + \dfrac{Dx+E}{x^2-4x+10}$

V. Which of the following should be done first in partial fraction decomposition?

a. Factor the numerator.
b. For each quadratic factor in the denominator, use a constant in the numerator.
c. Examine the degree of the numerator to see if it is less than the degree of the denominator.
d. Examine the denominator to see if it is irreducible.

Answers to Readiness Check

I. b II. c III. a IV. d V. c

In Problems 1–40, find the partial fraction decomposition of each rational expression.

1. $\dfrac{1}{x^2 - 1}$

2. $\dfrac{1}{x^2 - 9}$

3. $\dfrac{7x - 1}{x^2 - 2x - 3}$

4. $\dfrac{-2x + 4}{x^2 + x - 6}$

5. $\dfrac{2x^2 - 3x - 25}{x^2 - 4x + 3}$

6. $\dfrac{x^3}{x^2 - x - 42}$

7. $\dfrac{x + 2}{3x^2 - 12x - 15}$

$\left[\text{Hint: Write } \dfrac{x + 2}{3x^2 - 12x - 15} = \dfrac{(\frac{1}{3})(x + 2)}{x^2 - 4x - 5}.\right]$

8. $\dfrac{2}{4x^2 + 8x + 3}$

9. $\dfrac{1}{(x^2 - 4)(x - 1)}$

10. $\dfrac{2x^2 + 1}{x^3 - 6x^2 + 8x}$

11. $\dfrac{-3x^2 - 5x + 4}{(x + 1)(x + 2)(x + 3)}$

12. $\dfrac{x^3 + 3x^2 + 2x + 1}{(x - 1)(x - 2)(x - 3)}$

13. $\dfrac{x + 2}{x^3 - x}$

14. $\dfrac{x^2 + 1}{x^2(x + 3)}$

15. $\dfrac{3x^2 - 8x + 6}{(x - 2)^3}$

16. $\dfrac{x^2 + 1}{x(x - 3)^2}$

17. $\dfrac{x^4 + 5x^2 + 6}{x^2 + 1}$

18. $\dfrac{x^3 + x^2 - 2}{x^4}$

19. $\dfrac{2x + 5}{x^2 + 2x + 1}$

20. $\dfrac{2x^2 + 5}{(x - 1)(x^2 + 1)}$

21. $\dfrac{1 - x}{(1 + x)^3}$

22. $\dfrac{3x - 2}{(x - 2)(x^2 + 4x + 10)}$

23. $\dfrac{x + 4}{(x + 3)(x - 2)(x^2 + x + 1)}$

24. $\dfrac{5x + 1}{(x - 2)^2(x^2 + 4x - 10)}$

25. $\dfrac{1}{x^4 - 5x^2 + 4}$

26. $\dfrac{x^3 + x^2 - x - 1}{x^4 + 3x^2 + 2}$

27. $\dfrac{x}{x^4 - 18x^2 + 81}$

28. $\dfrac{x^2 + 1}{(x^2 - 1)^2}$

29. $\dfrac{x^4}{16x - x^3}$

30. $\dfrac{x^3 - x^2 + 5}{(x - 2)^2(x - 3)^2}$

31. $\dfrac{x^2 + 2}{x(x - 1)^2(x + 1)}$

32. $\dfrac{x^3}{x^4 - x^2}$

33. $\dfrac{-x^3 + x^2 + 1}{(x^2 + 3)(x^2 + 2)}$

* 34. $\dfrac{x^3 + x^2 + 3}{(x - 1)(x^2 + 2)^2}$

35. $\dfrac{x^4 + 1}{x^5 + x^3}$

36. $\dfrac{x^5}{x^4 + x^2}$

* 37. $\dfrac{5x^3 + 11x^2 - 18x + 17}{(x^2 + x + 1)(x^2 - 4x - 9)}$

* 38. $\dfrac{x^2 + 1}{x(x^2 + 16)^2}$

* 39. $\dfrac{x^3 + 1}{(x^2 + 4x + 13)^2}$

* 40. $\dfrac{2x^3 - 4x^2 + 70x - 35}{(x^2 - 2x + 17)(x^2 - 10x + 26)}$

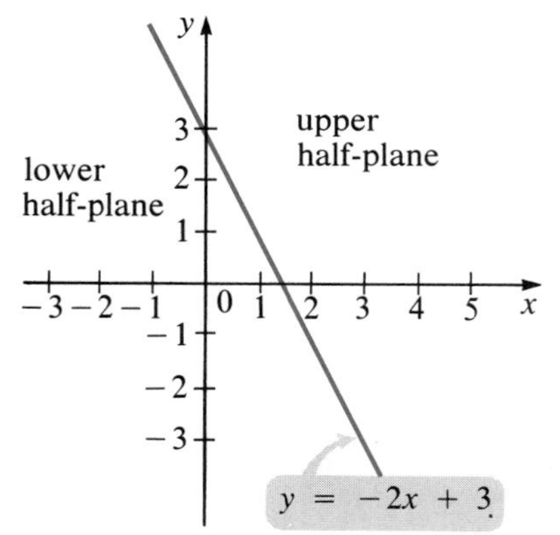

Figure 1 The line $y = -2x + 3$ divides the xy-plane into two half-planes.

7.5 Linear Inequalities in Two Variables

In Section 3.2, we saw how to find the equation and the graph of a straight line. In this section, we will show how to sketch linear inequalities in two variables.

Before citing general rules, we give two examples.

EXAMPLE 1 ***Sketching a > Inequality***

Sketch the set of points that satisfy the inequality $y > -2x + 3$.

SOLUTION We begin by drawing the graph of the line $y = -2x + 3$ in Figure 1. Since the line extends infinitely far in both directions, we can think of this line (or any other straight line) as dividing the xy-plane into two **half-planes.** In Figure 1 we have labeled these half-planes as *upper half-plane* and

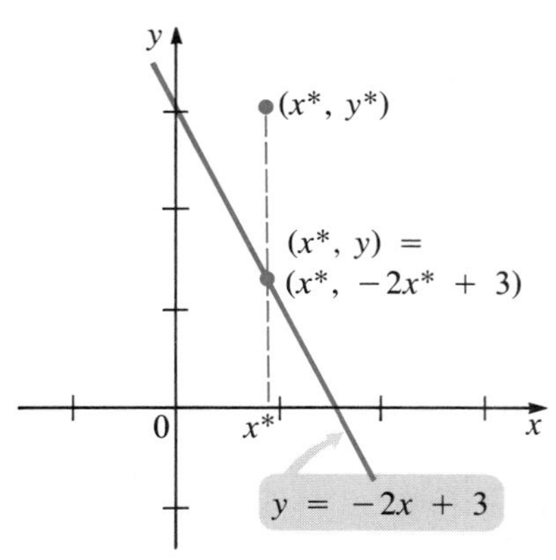

Figure 2 If $y^* > -2x^* + 3$, then the point (x^*, y^*) lies above the point $(x^*, -2x^* + 3)$ and is in the upper half-plane.

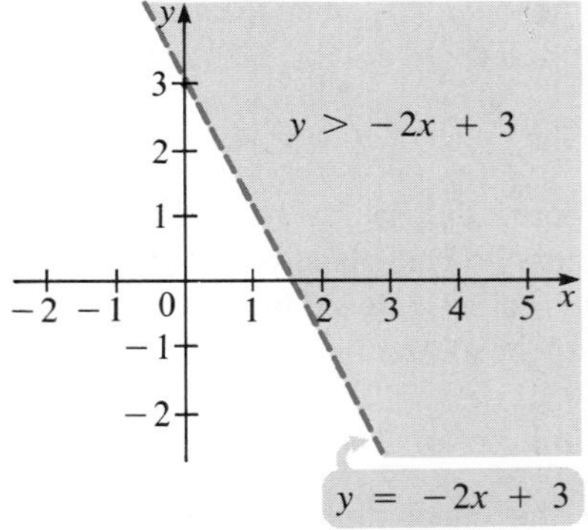

Figure 3 Every point in the shaded half-plane satisfies $y > -2x + 3$. The line $y = -2x + 3$ is dashed to indicate that the points on the line *do not* satisfy the inequality.

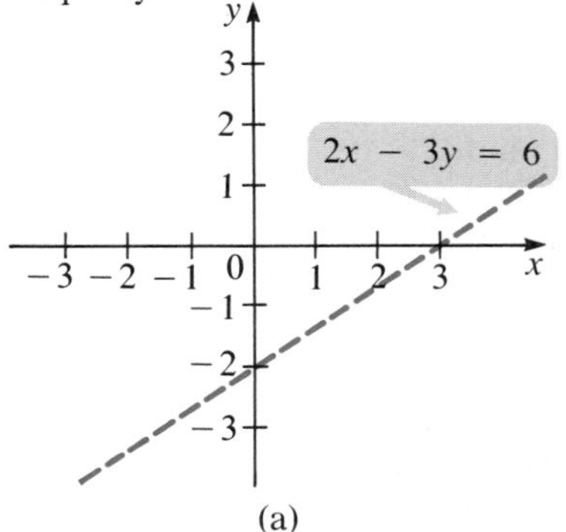

Figure 4 (a) The line $2x - 3y = 6$ is dashed because points on the line do not satisfy the inequality $2x - 3y < 6$.

lower half-plane. The set $L = \{(x, y): y = -2x + 3\}$ is the set of points on the line. We define two other sets by

$$A = \{(x, y): y > -2x + 3\} \qquad \text{and} \qquad B = \{(x, y): y < -2x + 3\}$$

For any pair (x, y) we have three possibilities: $y = -2x + 3$, $y > -2x + 3$, or $y < -2x + 3$. We see that every point in $\mathbb{R}^2$ is in exactly one of the sets L, A, or B.

We can see that A is precisely the upper half-plane in Figure 1. To see why, look at Figure 2. Let (x^*, y^*) be in A. Then, by the definition of A, $y^* > -2x^* + 3$, so the point (x^*, y^*) lies above the line $y = -2x + 3$. This follows because the y-coordinate of the point (x^*, y^*) is greater than (higher than) the y-coordinate of the point $(x^*, -2x^* + 3)$, which is on the line. Thus the set of points that satisfy $y > -2x + 3$ is precisely the upper half-plane shaded in Figure 3. The dashed line in the figure indicates that points on the line do *not* satisfy the inequality.

We now generalize this example.

A **linear inequality** in two variables is an inequality that can be written in one of the four forms

$$ax + by > c \tag{1}$$
$$ax + by \geq c \tag{2}$$
$$ax + by < c \tag{3}$$
$$ax + by \leq c \tag{4}$$

where a, b, and c are real numbers, and a and b are not both equal to zero.

NOTE Actually, there are only two distinct forms. For, if $ax + by < c$, then $-ax - by > -c$ (see Property (d), p. 131) and if $ax + by \leq c$, then $-ax - by \geq -c$.

There is a fairly easy method to use in graphing the set of points that satisfy one of these four inequalities. We illustrate this with an example.

EXAMPLE 2 *Sketching a < Inequality Using a Test Point*

Sketch the set of points that satisfy $2x - 3y < 6$.

SOLUTION In Figure 4(a), we first sketch the line $2x - 3y = 6$. Since no point on the line satisfies the given inequality, we draw a dashed line. As in Example 1, the set of points we seek is one of the two half-planes into which the xy-plane has been divided by the line. Which one? The simplest way to tell is to select a **test point,** such as $(0, 0)$. Now $2 \cdot 0 - 3 \cdot 0 = 0 < 6$, so $(0, 0)$ is in the set $\{(x, y): 2x - 3y < 6\}$. Thus the set we seek is the half-plane

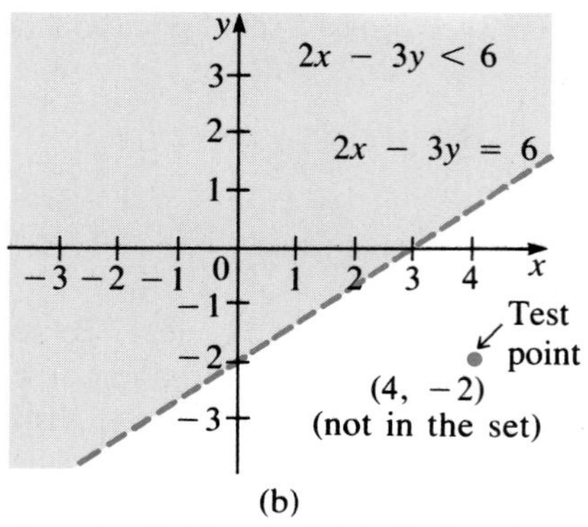

Figure 4 (b) The shaded half-plane is the solution set of the inequality $2x - 3y < 6$.

containing (0, 0), as indicated in Figure 4(b). Why did we choose (0, 0)? Because it is the easiest point to check. But any test point can be used. For example, let's try the point (4, −2). Then

$$2(4) - 3(-2) = 14 > 6$$

Thus the half-plane containing (4, −2) is *not* the half-plane we want. This leads us to the same graph as before.

We now state rules for sketching an inequality in one of the forms (1)–(4).

Procedure for Sketching the Set of Points Satisfying a Linear Inequality in Form (1), (2), (3), or (4)

1. Draw the line $ax + by = c$. Use a dashed line if equality is ruled out ((1) or (3)) and a solid line if it is not ruled out ((2) or (4)).
2. Pick any point in $\mathbb{R}^2$ not on the line, and use it as a test point. If the coordinates of the test point satisfy the inequality, then the set sought is the half-plane containing the test point. Otherwise, it is the other half-plane.

NOTE We have now seen that the set of points that satisfy a linear inequality is a half-plane. If equality is excluded, then the half-plane is called an **open half-plane.** If equality is included, then it is called a **closed half-plane.** These definitions are similar to the definitions of open and closed intervals.

EXAMPLE 3 *Sketching the Solution Set of Two Inequalities*

Sketch the set of points that satisfy the inequalities $x + 2y \geq 2$ and $-2x + 3y < 6$.

SOLUTION We begin by drawing, in Figure 5(a), the lines whose equations are given by $x + 2y = 2$ and $-2x + 3y = 6$. The coordinates (0, 0) satisfy the second inequality but not the first. This means that the half-plane $\{(x, y): -2x + 3y < 6\}$, which contains the point (0, 0), is the set of points below the line $-2x + 3y = 6$, whereas the half-plane $\{(x, y): x + 2y \geq 2\}$, which does not contain the point (0, 0), is the set of points on and above the line $x + 2y = 2$. Thus, the set of points that satisfy both inequalities is the intersection of these two half-planes. This solution set is sketched in Figure 5(b).

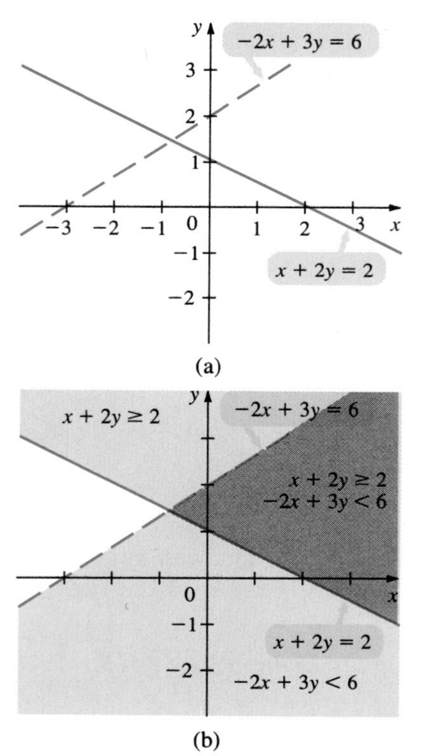

Figure 5 (a) The lines $-2x + 3y = 6$ and $x + 2y = 2$.
(b) The darker shaded region is the solution set of the two inequalities. The line $-2x + 3y = 6$ is dashed because it is not in the solution set.

Before continuing, we provide an easier way to sketch the set of points that satisfy two or more linear inequalities. It is easy to get confused when shading the sets of points that satisfy a number of linear equalities because the set you seek is the set that was shaded every time. (It is the set that satisfies every inequality.) To avoid this confusion, we will simply cross off those points that do *not* satisfy the inequality. The set that remains untouched will be the set that satisfies all the inequalities. Here is the procedure:

Procedure for Sketching the Solution Set of Points That Satisfy Two or More Linear Inequalities

For each linear inequality:

(a) Draw the bounding line (solid or dashed).
(b) Use a test point and cross off the side that does *not* satisfy the inequality.

The set of points that are never crossed off is the set we seek.

EXAMPLE 4 *Using the New Method to Sketch the Set of Points in Example 3*

Sketch the set of points that satisfy the linear inequalities $x + 2y \geq 2$ and $-2x + 3y < 6$.

SOLUTION We do this in four steps in Figure 6.

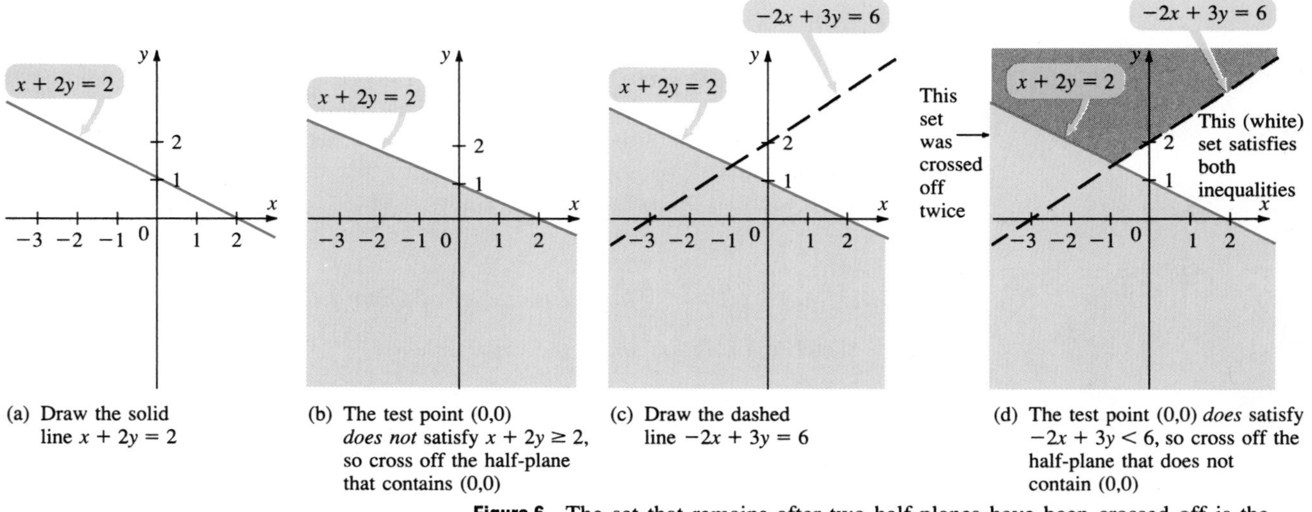

(a) Draw the solid line $x + 2y = 2$

(b) The test point (0,0) *does not* satisfy $x + 2y \geq 2$, so cross off the half-plane that contains (0,0)

(c) Draw the dashed line $-2x + 3y = 6$

(d) The test point (0,0) *does* satisfy $-2x + 3y < 6$, so cross off the half-plane that does not contain (0,0)

Figure 6 The set that remains after two half-planes have been crossed off is the set that satisfies both inequalities. ■

EXAMPLE 5 *An Empty Solution Set*

Sketch the set of points that satisfy the inequalities $x + y \leq 1$ and $2x + 2y \geq 6$.

SOLUTION We follow the procedure used in Example 4. The results are given in Figure 7.

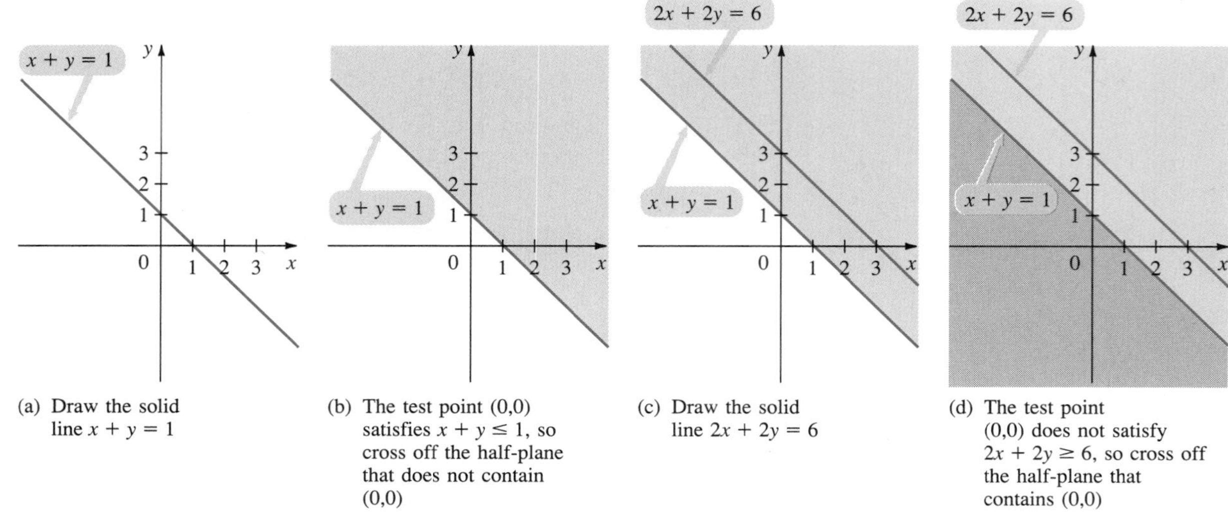

Figure 7 All points have been crossed off, so there are no points that satisfy both inequalities.

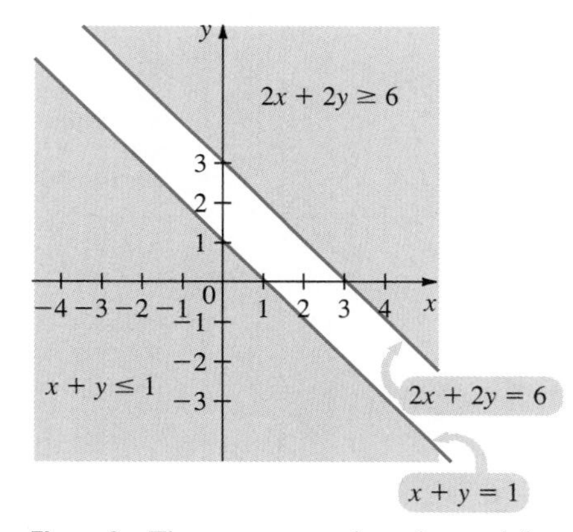

Figure 8 There are no points that satisfy both inequalities $2x + 2y \geq 6$ and $x + y \leq 1$.

We see that all points have been crossed off. That means that every point in the xy-plane fails to satisfy at least one of the inequalities, so there is no point that satisfies both of them. In this case, we say that the solution set is empty.

In Figure 8 we shade the set of points that satisfy each inequality. We again see that the solution set is empty. ■

EXAMPLE 6 *The Solution Set of Four Inequalities*

Sketch the solution set of the inequalities

$$\begin{aligned} -x + y &\leq 1 \\ x + 2y &\leq 6 \\ 2x + 3y &\geq 3 \\ -3x + 8y &\geq 4 \end{aligned}$$

SOLUTION The lines $-x + y = 1$, $x + 2y = 6$, $2x + 3y = 3$, and $-3x + 8y = 4$ are shown in Figure 9. The solution set of the four inequalities is the untouched region in the figure. Note that in this case the solution set is a four-sided region in the first quadrant with four "corners." Solution sets like this come up fairly frequently in linear programming problems (see the next section). Note that (0, 0) satisfies the first two inequalities but not the last two.

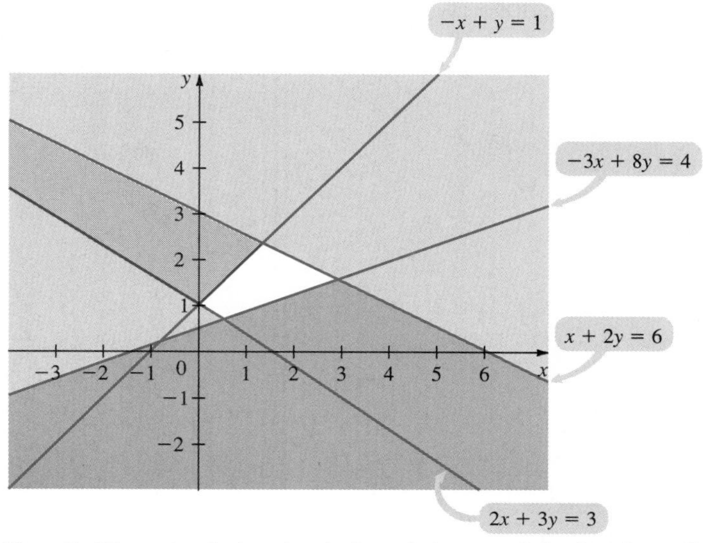

Figure 9 The untouched region is the solution set of the four inequalities $-x + y \le 1$, $x + 2y \le 6$, $2x + 3y \ge 3$, and $-3x + 8y \ge 4$. Note that (0, 0) satisfies the first two inequalities but not the last two.

Problems 7.5

Readiness Check

I. Which of the following describes the set of points that satisfies $3x - y \ge 4$?

a. The closed half-plane to the lower right of the boundary line.
b. The open half-plane to the lower right of the boundary line.
c. The open half-plane to the upper left of the boundary line.
d. The closed half-plane to the upper left of the boundary line.

II. Which of the following is the inequality whose solution is sketched below?

a. $2x + 3y > 6$
b. $2x + 3y \ge 6$
c. $2x + 3y < 6$
d. $2x + 3y \le 6$

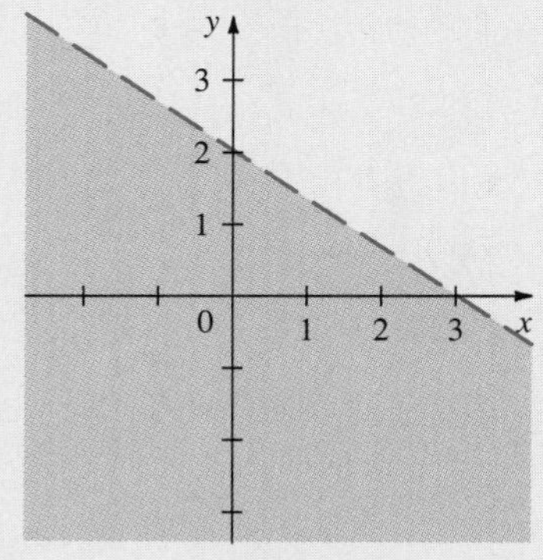

III. Which of the following is true of the set of points that satisfies the inequalities $x + 2y \ge 4$ and $-2x \ge 4y$?

a. The set is a strip between two closed half-planes.
b. The set is two closed half-planes.
c. The set includes the point (0, 0).
d. The set is empty.

Answers to Readiness Check

I. a II. c III. d

In the following problems, sketch the set of points that satisfy the given inequalities.

1. $x \le 3$
2. $y < 2$
3. $y \ge -4$
4. $x \le \frac{3}{2}$
5. $y > \frac{2}{3}$
6. $x + y > 2$
7. $x + y \le 2$
8. $2x - y < 4$
9. $2x - y \ge 4$
10. $y - 2x < 4$
11. $x - 2y > -4$
12. $3x - y < 3$
13. $y - 3x > -3$
14. $y - 3x \le 3$
15. $y + 3x \ge 3$
16. $y \le 2x - 5$
17. $y > 4x - 3$
18. $x < -2y + 7$
19. $y \le 4x - 3$
20. $3x + 4y \le 6$
21. $-3x + 4y > 6$
22. $3x - 4y \ge 6$
23. $-3x - 4y > 6$
24. $x - \dfrac{y}{2} > 4$
25. $\dfrac{x - y}{3} \le 2$
26. $\dfrac{x}{2} - \dfrac{y}{3} \ge 1$
27. $\dfrac{x}{3} + \dfrac{y}{2} < -1$
28. $\dfrac{x}{3} - \dfrac{y}{5} \ge \dfrac{1}{2}$
29. $-3 \le x < 0$
30. $1 < y \le 6$
31. $0 \le x \le 2,\ 0 \le y \le 3$
32. $-1 \le x < 4,\ -2 \le y < 2$
33. $-\frac{1}{2} < x < 1,\ \frac{1}{2} \le y < 2$
34. $|x| < 2,\ |y| < 3$

* 35. $|x - 1| < 4,\ |y + 2| \le 3$

36. $x + y \ge 1,\ 2x - 3y \le 6$
37. $x + y \le 1,\ 2x + 3y \ge 5$
38. $x - y \le 2,\ 2y - 3x > 6$
39. $x + 2y \le 2,\ 2x + 4y \ge 4$
40. $x + 2y < 2,\ 2x + 4y > 4$
41. $x + y \le 2,\ 5x + 2y \ge 4$
42. $3x - 4y \le 6,\ 2x + 3y > 3$
43. $-x + 2y \le 4,\ 3x + 2y \le 6$
44. $-x + 2y \le 4,\ 3x + 2y \le 6,\ x \ge 0,\ y \ge 0$
45. $x - y \le 2,\ x + 3y \ge 6,\ x \ge 0,\ y \ge 0$
46. $2x + y \ge 1,\ x + 2y \ge 1,\ x + y \le 3,\ x \ge 0,\ y \ge 0$

7.6 Introduction to Linear Programming

The problem of determining the maximum or minimum of a given function occurs in many applications of mathematics to business, economics, the biological sciences, and other disciplines. It is not difficult to think of examples of such problems. How can a businessman maximize profits or minimize costs? At what currency exchange rate will the balance of payments be most favorable? How can the food requirements of an animal be satisfied with a minimum expenditure of energy?

Maximization and minimization problems are often subject to constraints or limits on the variables. For example, a businessman is always limited by a finite supply of capital. Each of us could make virtually unlimited profits if we had unlimited sums to invest. A warehouse supervisor has limited space for storage. Biological variables may be constrained by physiological limits or by limits of resource availability. Some constraints are obvious by definition of the variables. A supermarket manager, for example, cannot order a negative number of pounds of tomatoes.

In this section we consider the special problem of maximizing or minimizing linear functions of several variables subject to linear constraints. Instead of giving a general definition of the problem at this point, we begin with an example.

EXAMPLE 1 ***Solving a Linear Programming Maximization Problem by a Graphical Method***

The Grant Furniture Company manufactures dining room tables and chairs. Each chair takes 20 board feet (bd ft) and 4 hours of labor. Each table requires 50 bd ft but only 3 hours of labor. The manufacturer has 3300 bd ft of lumber available and a staff able to provide 380 hours of labor. Finally, the manufacturer has determined that there is a net profit of $30 for each chair sold and $60 for every table sold. For simplicity, we assume that needed materials (such as nails or varnish) are available in sufficient quantities. How many tables and chairs should the company manufacture in order to maximize its profit, assuming that each item manufactured is sold?

The problem as stated seems difficult—there are lots of things going on. We begin simplifying the problem by putting all the information into Table 1.

Table 1
Data for the Grant Furniture Company

Raw Material	Amount Needed per Unit		Total Available
	Chair	*Table*	
Wood (board feet)	20	50	3300
Labor (hours)	4	3	380
Net unit profit (dollars)	30	60	

We now let x denote the number of chairs and y the number of tables produced by the company. Since it takes 20 bd ft of lumber to make one chair, it takes $20x$ bd ft of lumber to make x chairs. Similarly, it takes $50y$ bd ft of lumber to make y tables. Thus the data in the first line of Table 1 can be expressed algebraically by the linear inequality

$$20x + 50y \leq 3300 \quad \text{Lumber inequality}$$

Analogously, the linear inequality representing the information in the second line of Table 1 is

$$4x + 3y \leq 380 \quad \text{Labor inequality}$$

These two inequalities represent two of the **constraints** of this problem. They express, in mathematical terms, the obvious fact that raw materials and labor are finite (limited) quantities. There are two other constraints. Since the company cannot manufacture negative amounts of the two items, we must have

$$x \geq 0 \quad \text{and} \quad y \geq 0$$

The profit P earned when x chairs and y tables are manufactured is given (from the third line of Table 1) by

$$P = 30x + 60y \quad \text{Profit equation}$$

Putting all this information together, we can state our problem in a form called a **standard linear programming problem.**

Maximize

$$P = 30x + 60y \tag{1}$$

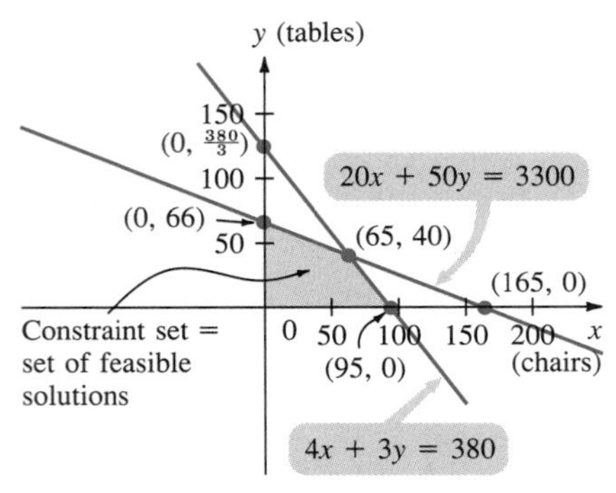

Figure 1 Solution set of the four inequalities $20x + 50y \leq 3300$, $4x + 3y \leq 380$, $x \geq 0$, and $y \geq 0$.

subject to the constraints

$$20x + 50y \leq 3300 \tag{2}$$

$$4x + 3y \leq 380 \tag{3}$$

$$x \geq 0, \ y \geq 0 \tag{4}$$

In this problem, the linear function given by (1) is called the **objective function.** Any point in the constraint set is called a **feasible solution.** Our problem is to find the point (or points) in the constraint set at which the objective function is a maximum. Our first method for solving this problem will employ techniques of the last section. We begin to find our solution by graphing the **constraint set,** which is the solution set of the inequalities (2), (3), (4). This is done in Figure 1. Consider the lines $30x + 60y = C$ for different values of the constant C. Some of the lines are sketched in Figure 2.

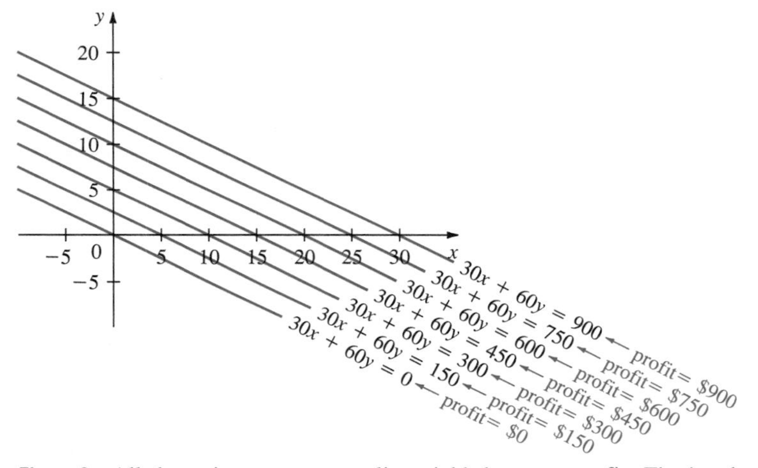

Figure 2 All the points on any one line yield the same profit. That's why the lines are called *constant profit* lines.

Each line $30x + 60y = C$ is called a **constant profit line** for this problem. To see why, consider the line $30x + 60y = 300$. For every point (x, y) lying both on this line and in the constraint set, the manufacturer makes a profit of \$300. Some points are (10, 0) (10 chairs and no tables), (6, 2) (6 chairs and 2 tables), and (0, 5) (no chairs and 5 tables). See Figure 3. From the manufacturer's point of view, these three points are equivalent because each leads to the same \$300 profit.

Now, consider the constant profit line $30x + 60y = 600$. This line lies to the right of the line $30x + 60y = 300$ and is a "nicer" line for the manufacturer because every point on it and in the constraint set gives a profit of \$600. Two such points are (20, 0) and (12, 4).

By now, the pattern may be getting clearer. All the constant profit lines are parallel (each has slope $-\frac{1}{2}$), and the profit increases as we move to the right from one line to the next. Each new line (to the right) leads to a higher profit. Our method now is to move to the right as much as we can while still

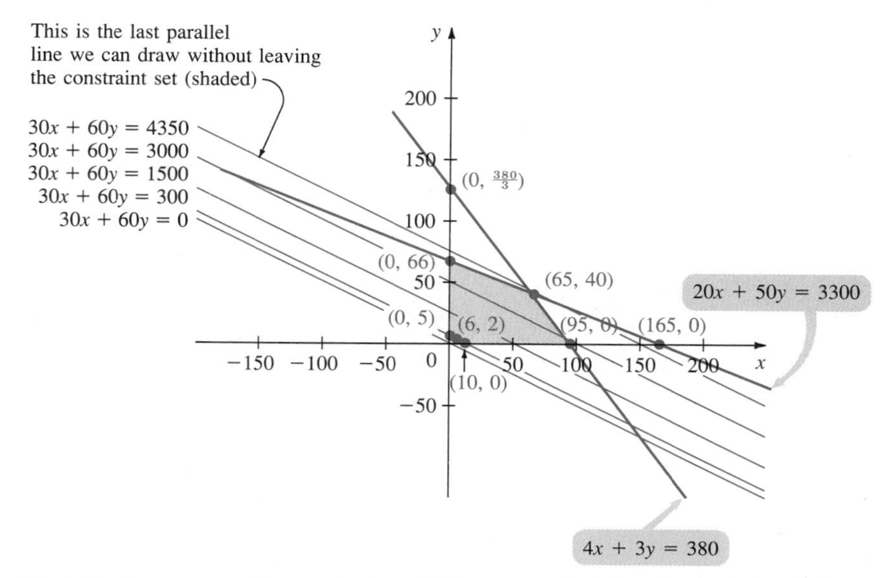

Figure 3 As we move the constant profit lines upward while still remaining in the constraint set, the profit increases. We do this until we can no longer move because any further upward movement will take us out of the constraint set.

remaining in the constraint set. From Figure 3, we see that the "last" constant profit line is the line that intersects the constraint set at the single point (65, 40). This means that the largest profit is earned when 65 chairs and 40 tables are manufactured. This yields a profit of $30 \cdot 65 + 60 \cdot 40 = \$4350$.

NOTE It may seem at first glance that the company can make more profit by putting as much as possible into the more profitable tables. From Figure 1, we see that as many as 66 tables (the largest value of y in the constraint set is 66) can be manufactured, which yields a profit of $60 \cdot 66 = \$3960$. On the other hand, the manufacture of 95 chairs and no tables gives a profit of \$2850. Thus the company does indeed do better by making 65 chairs and 40 tables.

The method used in Example 1 to solve the linear programming problem is called the **graphical method.** This method illustrates what is going on, but it is very impractical to use for two reasons: First, it is necessary to use very precise drawings to obtain the solution, and second, it can be used only with problems involving two variables because graphs in three dimensions are unwieldy and graphs in more than three dimensions cannot be drawn.

EXAMPLE 2 *Solving A Linear Programming Problem by the Corner Point Method*

We now introduce another method. To do so, we take another look at Example 1. The problem was to maximize

$$P = 30x + 60y \tag{5}$$

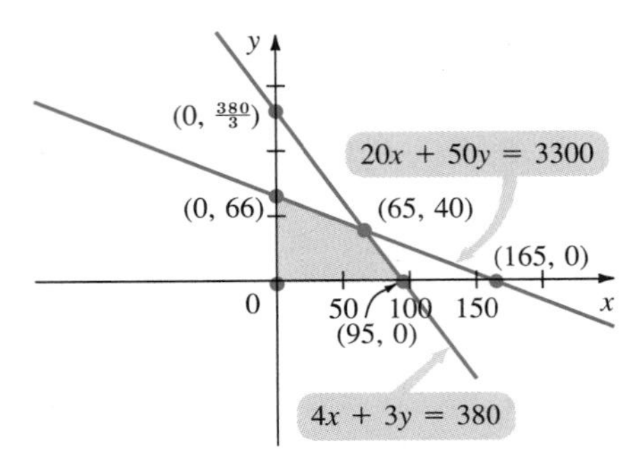

Figure 4 The constraint set redrawn.

subject to the constraints

$$20x + 50y \leq 3300 \tag{6}$$
$$4x + 3y \leq 380 \tag{7}$$
$$x \geq 0 \tag{8}$$
$$y \geq 0 \tag{9}$$

In Figure 4, we again sketch the constraint set of this problem. The constraint set is the intersection of four sets: $S_1 = \{(x, y): 20x + 50y \leq 3300\}$, $S_2 = \{(x, y): 4x + 3y \leq 380\}$, $S_3 = \{(x, y): x \geq 0\}$, and $S_4 = \{(x, y): y \geq 0\}$. Each of these four sets is the solution set of a linear inequality, and each is bounded by a straight line. The intersection of any two of these lines is a point in the plane and, if the point is in the constraint set, then the point is called a **corner point** of the constraint set. In Table 2 we list all the possible and actual corner points. A point is a *possible* corner point if it is the intersection of two of the lines that determine the constraint set. It is an *actual* corner point if it is in the constraint set; that is, it is an actual corner point if it is a feasible solution.

Table 2
Determining Corner Points

The Two Lines That Determine the Point	Possible Corner Point	Feasible Solution? (actual corner point?) (is it in the constraint set?)
$20x + 50y = 3300$ $y = 0$	(165, 0)	No (constraint (7) is violated since $4 \cdot 165 + 3 \cdot 0 = 660$, which is >380)
$4x + 3y = 380$ $y = 0$	(95, 0)	Yes
$x = 0$ $y = 0$	(0, 0)	Yes
$4x + 3y = 380$ $x = 0$	$(0, \frac{380}{3})$	No (constraint (6) is violated since $20 \cdot 0 + 50 \cdot \frac{380}{3} = \frac{19{,}000}{3}$, which is >3300)
$20x + 50y = 3300$ $x = 0$	(0, 66)	Yes
$20x + 50y = 3300$ $4x + 3y = 380$	(65, 40)	Yes

The following statement is true:

The Reason for Finding Corner Points

The maximum and minimum values of the objective function of any linear programming problem, if they exist, always occur at corner points of the constraint set.

We will indicate why this is true at the end of this section.

In our problem there are four actual corner points. So, in order to find a solution, we need only to evaluate the objective function at each corner point and choose the point that gives the maximum value. We do this in Table 3.

Table 3
Evaluation of the Objective Function at the Corner Points

Corner Point	Value of Objective Function $P = 30x + 60y$	
(65, 40)	$30 \cdot 65 + 60 \cdot 40 = 4350$	maximum value
(0, 66)	$30 \cdot 0 + 60 \cdot 66 = 3960$	
(95, 0)	$30 \cdot 95 + 60 \cdot 0 = 2850$	
(0, 0)	$30 \cdot 0 + 60 \cdot 0 = 0$	

Thus we see, as we saw in Example 1, that the maximum profit of $4350 is earned when 65 chairs and 40 tables are manufactured.

The Corner-Point Method

Step 1 Find all possible corner points. These are points of intersection of two of the lines making up the constraint set.

Step 2 Determine the actual corner points. These are the possible corner points that satisfy all the remaining constraints.

Step 3 Compute the value of f, the objective function, at each actual corner point.

Step 4 The maximum value of f is the largest number obtained in Step 3. The minimum value of f is the smallest number obtained in Step 3.

EXAMPLE 3 *Minimizing Water Costs Using the Corner-Point Method*

The water-supply manager for a midwest city must find a way to supply at least 10 million gal of potable (drinkable) water per day (mgd). The supply may be drawn from the local reservoir or from a pipeline to an adjacent town. The local reservoir has a daily yield of 5 mgd, which may not be exceeded. The pipeline can supply no more than 10 mgd because of its size. On the other hand, by contractual agreement it must pump out at least 6 mgd. Finally, reservoir water costs $300 for 1 million gal, and pipeline water costs $500 for 1 million gal. How can the manager minimize daily water costs?

SOLUTION Let x denote the number of reservoir gallons and y denote the number of pipeline gallons (in millions of gallons) pumped per day. Then the problem is:

Minimize

$$C = 300x + 500y$$

subject to

$$\begin{aligned} x + y &\geq 10 && \text{To meet the city water requirements} \\ x &\leq 5 && \text{Reservoir capacity} \\ y &\leq 10 && \text{Pipeline capacity} \\ y &\geq 6 && \text{Pipeline contract} \\ x &\geq 0 \\ y &\geq 0 \end{aligned}$$

The constraint set for this problem is sketched in Figure 5. From Figure 5 we can see that there are four corner points (see Table 4).

The minimum value of the constraint function over the corner points is 4200 at (4, 6). That is, the manager should draw 4 million gal per day from the reservoir and 6 million gal per day from the pipeline at a daily cost of $4 \cdot 300 + 6 \cdot 500 = \$4200$.

Table 4
Evaluating the Water Cost at Each Corner Point

Corner Point	Value of Objective Function $C = 300x + 500y$ at Corner Point	
(0, 10)	5000	
(5, 10)	6500	
(4, 6)	4200	minimum value
(5, 6)	4500	

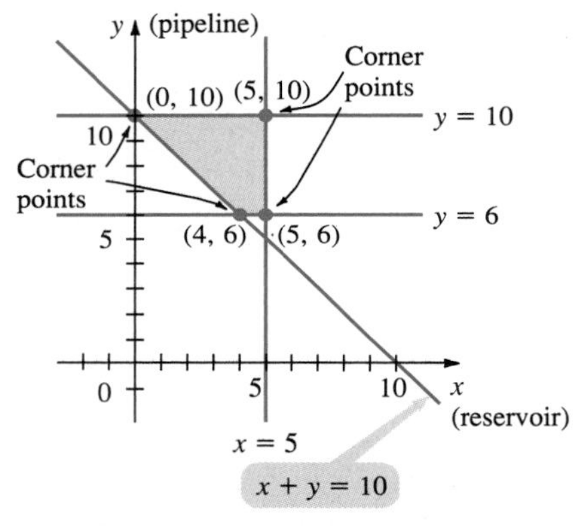

Figure 5 The constraint set for the water cost minimization problem has four corner points.

In the examples we considered in this section, there were two variables (which we denoted by x and y). The graphical method fails, as we have stated, if there are more than two variables. The corner-point method works with more than two variables, but the work required to compute the possible corner points can be tremendous.

We close this section by illustrating two of the difficulties that can arise when solving a linear programming problem.

EXAMPLE 4 *An Unbounded Linear Programming Problem*

Solve the following linear programming problem:

Maximize

$$f = 2x + 3y$$

subject to

$$\begin{aligned} x + y &\geq 5 \\ 6x + 2y &\geq 12 \\ x \geq 0,\ y &\geq 0 \end{aligned}$$

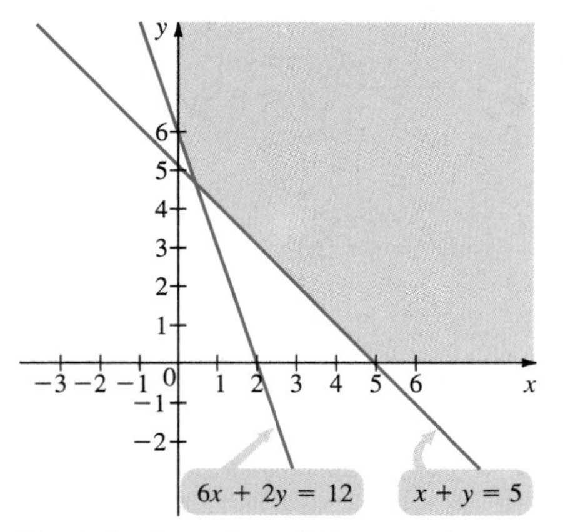

Figure 6 An unbounded constraint set: $f = 2x + 3y$ can be made as large as possible.

SOLUTION The constraint set is sketched in Figure 6. It is clear that x and y can take arbitrarily large values and still remain in the constraint set. Therefore f can take on arbitrary large values, and the problem has no solution. In this situation we say that the problem is **unbounded.** To see this more clearly, we evaluate, in Table 5, $f = 2x + 3y$ for several values of x and y in the constraint set.

Table 5
Values for $f = 2x + 3y$ for (x, y) in the Constraint Set

(x, y)	$f = 2x + 3y$
(10, 10)	50
(50, 50)	250
(100, 100)	500
(1000, 1000)	5000
(100,000, 100,000)	500,000

EXAMPLE 5 *An Infeasible Linear Programming Problem*

Solve the problem.
Maximize

$$f = 2x + 3y$$

subject to

$$\begin{aligned} x + \ y &\geq 5 \\ 2x + 3y &\leq 6 \\ x \geq 0,\ y &\geq 0 \end{aligned}$$

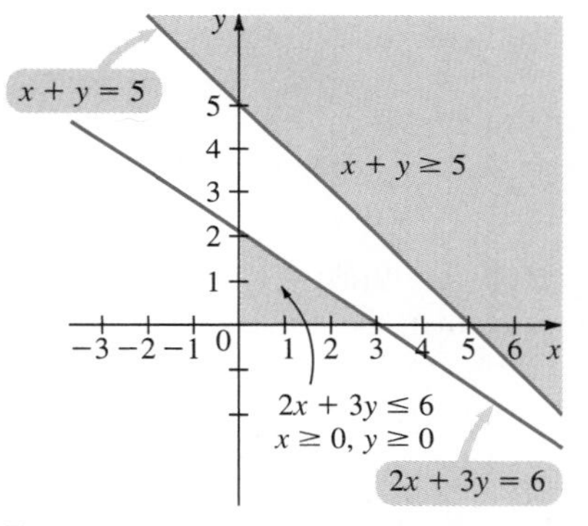

Figure 7 The constraint set is empty since no point satisfies all four inequalities.

SOLUTION The linear inequalities are sketched in Figure 7. It is evident that the constraint set is empty. Thus there are no feasible solutions, and we say that the problem is **infeasible.**

Why the Corner-Point Method Works

We now show that for a linear programming problem in two variables the maximum and minimum values of an objective function of the form $f = ax + by$ always occur at corner points of the constraint set.

In Figure 8 we draw a typical constraint set in the xy-plane. We will show that $f = ax + by$ takes its maximum value at one of the six corner points. A similar argument will show that f takes its minimum value at one of the corner points.

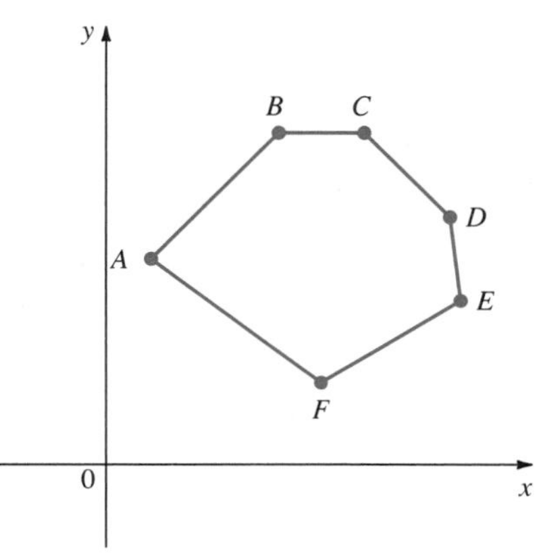

Figure 8 A constraint set with six corner points.

Suppose we are moving in the xy-plane along the line $y = mx + c$. Then

$$\begin{aligned} f = ax + \underset{\substack{\uparrow \\ y = mx + b}}{by} = ax + b(mx + c) &= ax + bmx + bc \\ &= (a + bm)x + bc \end{aligned}$$

Thus,

if $a + bm > 0$, then f increases as x increases
if $a + bm < 0$, then f decreases as x increases
if $a + bm = 0$, then f stays the same as x increases

In sum,

> As we move along a line in the xy-plane in a given direction, then f either always increases, always decreases, or stays the same.

Now let P be a point in the constraint set that is not a corner point (see Figure 9). We will show that there is at least one corner point at which f takes a value that is greater than or equal to the value of f at P.

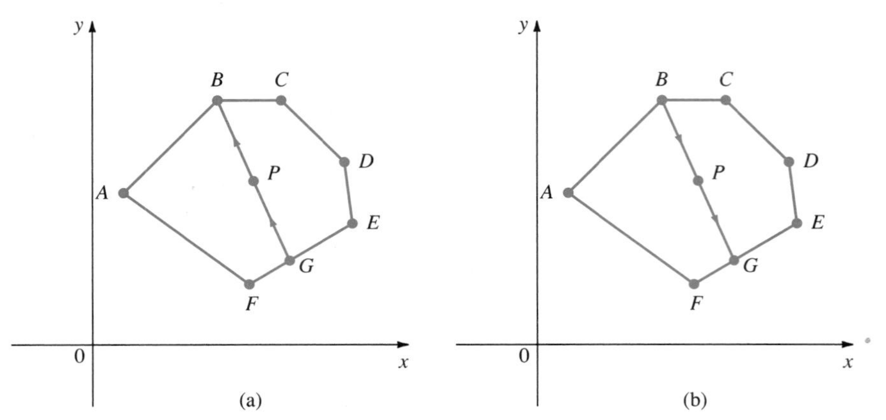

Figure 9 P is a point in the constraint set that is not a corner point.

We now pick a corner point, say B, draw a line from P to B, and extend the line in the other direction until it hits the boundary of the constraint set at a point that we label G.

Case 1 f increases as we move from P to B (Figure 9(a)). Then $f(B) > f(P)$. That is, the value of f at B is greater than the value of f at P.

Case 2 f increases as we move from P to G (Figure 9(b)). Then $f(G) > f(P)$. By the same reasoning, $f(E) \geq f(G)$ or $f(F) \geq f(G)$

so

$$\text{either } f(E) \geq f(G) > f(P) \quad \text{or} \quad f(F) \geq f(G) > f(P)$$

In either event, f takes a larger value at a corner point (E or F).

Case 3 f remains constant as we move from P to B. Then $f(B) = f(P)$. Thus, whenever P is a point in the constraint set, there is a corner point at which f takes at least as large a value. This means that f takes its maximum value at a corner point. ■

Problems 7.6

Readiness Check

I. Suppose we wish to maximize $f = 2x + 5y$ subject to a number of constraints. The corner points are (6, 0), (0, 3), (2, 4), and (5, 2). What is the maximum value for f?

a. 12 b. 15 c. 24 d. 30 e. 20

f. Cannot be determined from the information given

II. We wish to maximize $f = ax + by$ subject to being in the constraint set sketched in the figure to the right. The line $ax + by = c$ is drawn for some value of c. At which corner point is the maximum achieved?

a. A b. B c. C d. D e. E

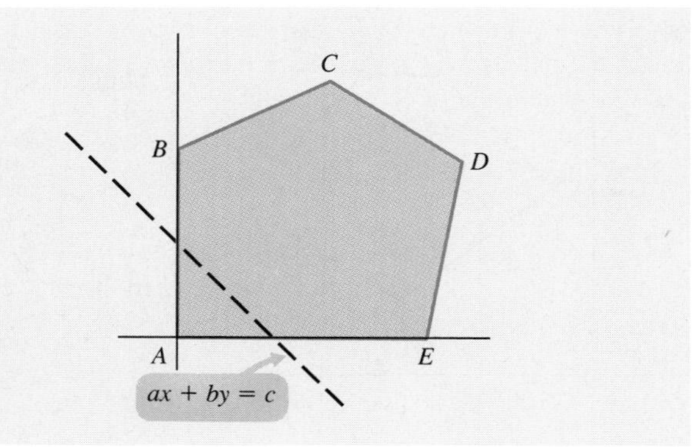

Answers to Readiness Check

I. c II. d

1. In Example 1, find the combination of tables and chairs that will maximize the manufacturer's profit if the profits are \$5 per chair and \$5 per table. Assume that all other data are unchanged.
2. Answer the question in Problem 1 if profits are \$8 per chair and \$2 per table.
3. Answer the question of Example 1 using the data in Table 6.

Table 6

Raw Material	**Amount Needed per Unit**		**Total Available**
	Chair	*Table*	
Wood (board feet)	30	40	11,400
Labor (hours)	4	6	1650
Net unit profit (dollars)	5	6	

4. Answer the question in Problem 3 if profits are \$2 per chair and \$8 per table and all other data are unchanged.
5. Answer the question in Problem 3 if profits are \$8 per chair and \$2 per table.
* 6. In Example 1, assume that a table and a chair require the same amount of wood and labor. If the unit profit is \$3 per chair and \$4 per table, show that if lumber and labor are limited, then the manufacturer can always maximize profit by manufacturing tables only.
7. In Example 3, how can the water manager minimize costs if there is no lower limit to the number of gallons that must pass through the pipeline?

In Problems 8–20, solve the given linear programming problem using the graphical or the corner-point method. Find the values of x and y at which the given objective function is maximized or minimized.

8. Maximize

$$f = 3x + 4y$$

subject to

$$\begin{aligned} x + y &\le 4 \\ 2x + y &\le 5 \\ x \ge 0,\ y &\ge 0 \end{aligned}$$

9. Maximize

$$f = 4x + 3y$$

subject to

$$\begin{aligned} x + y &\le 4 \\ 2x + y &\le 5 \\ x \ge 0,\ y &\ge 0 \end{aligned}$$

10. Maximize

$$f = x + y$$

subject to

$$\begin{aligned} 3x + 4y &\le 12 \\ 2x + y &\le 8 \\ x \ge 0,\ y &\ge 0 \end{aligned}$$

11. Maximize

$$f = 2x + 3y$$

subject to

$$\begin{aligned} x + y &\le 4 \\ 2x + 3y &\le 10 \\ 4x + 2y &\le 12 \\ x \ge 0,\ y &\ge 0 \end{aligned}$$

12. Maximize

$$f = 3x + 5y$$

subject to

$$\begin{aligned} 10x + y &\le 10 \\ x + 10y &\le 10 \\ 2x + 3y &\le 6 \\ x \ge 0,\ y &\ge 0 \end{aligned}$$

13. Maximize

$$f = 5x + 3y$$

subject to

$$\begin{aligned} 10x + y &\le 10 \\ x + 10y &\le 10 \\ 2x + 3y &\le 6 \\ x \ge 0,\ y &\ge 0 \end{aligned}$$

14. Maximize

$$f = 12x + y$$

subject to

$$\begin{aligned} 10x + y &\le 10 \\ x + 10y &\le 10 \\ 2x + 3y &\le 6 \\ x \ge 0,\ y &\ge 0 \end{aligned}$$

15. Maximize

$$f = x + 12y$$

subject to

$$\begin{aligned} 10x + y &\le 10 \\ x + 10y &\le 10 \\ 2x + 3y &\le 6 \\ x \ge 0,\ y &\ge 0 \end{aligned}$$

16. Minimize

$$g = 4x + 5y$$

subject to

$$\begin{aligned} x + 2y &\ge 3 \\ x + y &\ge 4 \\ x \ge 0,\ y &\ge 0 \end{aligned}$$

17. Minimize

$$g = 4x + 5y$$

subject to

$$\begin{aligned} x + 2y &\ge 4 \\ x + y &\ge 3 \\ x \ge 0,\ y &\ge 0 \end{aligned}$$

18. Minimize

$$g = 12x + 8y$$

subject to

$$\begin{aligned} 3x + 2y &\ge 1 \\ 4x + y &\ge 1 \\ x \ge 0,\ y &\ge 0 \end{aligned}$$

19. Minimize

$$g = 3x + 7y$$

subject to

$$\begin{aligned} 5x + y &\ge 1 \\ 2x + 3y &\ge 2 \\ x \ge 0,\ y &\ge 0 \end{aligned}$$

20. Minimize

$$g = 3x + 2y$$

subject to

$$\begin{aligned} x + 2y &\ge 1 \\ 2x + y &\ge 2 \\ 5x + 4y &\ge 10 \\ x \ge 0,\ y &\ge 0 \end{aligned}$$

21. Two foods contain carbohydrates and proteins only. Food I costs 50¢ per pound and is 90% carbohydrates (by weight). Food II costs $1 per pound and is 60% carbohydrates. What diet of these two foods provides at least 2 lb of carbohydrates and 1 lb of proteins at minimum cost? What is the cost per pound of this diet?
22. Spina Food Supplies, Inc. is a manufacturer of frozen pizzas. Art Spina, president of Spina Food Supplies, personally supervises the production of both types of frozen pizzas produced by the company: Spina's regular and Spina's super deluxe. Art makes a profit of $0.50 for each regular produced and $0.75 for each super deluxe. He currently has 150 lb of dough mix available and 800 oz of topping mix. Each regular pizza uses 1 lb of dough mix and 4 oz of topping, whereas each super deluxe uses 1 lb of dough and 8 oz of topping mix. Based on past demand, Art knows that he can sell at most 75 super deluxe and 125 regular pizzas. How many regular and super deluxe pizzas should Art make in order to maximize profits?
23. A predator requires 10 units of food A, 12 units of food B, and 12 units of food C as its average daily consumption. These requirements are satisfied by feeding on two prey species. One prey of species I provides 5, 2, and 1 units of foods A, B, and C, respectively. An individual prey of species II provides 1, 2, and 4 units of A, B, and C, respectively. To capture and digest a prey of species I requires 3 units of energy on the average. The corresponding energy requirement for species II is 2 units of energy. How many prey of each species should the predator capture to meet its food requirements with minimum expenditure of energy?
24. The Goody Goody Candy Company makes two kinds of gooey candy bars from caramel and chocolate. Each bar weighs 4 oz. Bar A has 3 oz of caramel and 1 oz of chocolate. Bar B has 2 oz of each. Bar A sells for 30¢, and bar B sells for 54¢. The company has stocked 90 lb of chocolate and 144 lb of caramel. How many units of each type of candy should be made in order to maximize the company's income?

Summary Outline of Chapter 7

- **Gaussian elimination** is a procedure used for reducing a linear system of equations to a simpler form in which solutions are readily obtainable. pp. 389–394
- A **matrix** is a rectangular array of numbers. p. 392
- **Elementary Row Operations** p. 394

  i. Replace a row with a nonzero multiple of that row, denoted by $R_i \to cR_i$.
  ii. Replace a row with the sum of the row and a multiple of some other row, denoted by $R_j \to R_j + cR_i$.
  iii. Interchange two rows, denoted by $R_i \rightleftarrows R_j$.

- An **inconsistent system** is a system with no solution. pp. 387, 395
- A matrix is in **row-echelon form** if (i) all rows (if any) consisting entirely of zeros appear at the bottom of the matrix, (ii) the first nonzero number in any row not consisting entirely of zeros is 1, and (iii) if two successive rows do not consist entirely of zeros, then the first 1 in the lower row occurs farther to the right than the first 1 in the higher row. p. 396

- **Procedure for Obtaining the Partial Fraction Decomposition of $\frac{p(x)}{q(x)}$:** p. 414

  Step 1 If degree of $p(x) \geq$ degree of $q(x)$, first divide to obtain

$$\frac{p(x)}{q(x)} = d(x) + \frac{r(x)}{q(x)}$$

  where degree of $r(x) <$ degree of $q(x)$. Then find the decomposition of $\frac{r(x)}{q(x)}$.

  Step 2 If the leading coefficient of $q(x)$ is $a_n \neq 1$, then factor a_n out so that the denominator function has leading coefficient 1.

  Step 3 Factor the new $q(x)$ into linear and/or irreducible quadratic factors. This is the hardest step. Since the leading coefficient of $q(x)$ is 1, these factors will have the form $x - c$ or $x^2 + ax + b$.

  Step 4 For each factor of the form $x - c$, the partial fraction decomposition of $\frac{r(x)}{q(x)}$ contains a term of the form $\frac{A}{x - c}$.

  Step 5 For each factor of the form $(x - c)^k$, $k > 1$, the partial fraction decomposition contains a sum of k terms:

$$\frac{A_1}{x - c} + \frac{A_2}{(x - c)^2} + \cdots + \frac{A_k}{(x - c)^k}$$

  Step 6 For each factor of the form $x^2 + ax + b$, the partial fraction decomposition contains a term of the form

$$\frac{Ax + B}{x^2 + ax + b}$$

  Step 7 For each factor of the form $(x^2 + ax + b)^k$, $k > 1$, the partial fraction decomposition contains a sum of k terms:

$$\frac{A_1x + B_1}{x^2 + ax + b} + \frac{A_2x + B_2}{(x^2 + ax + b)^2} + \cdots + \frac{A_kx + B_k}{(x^2 + ax + b)^k}$$

  Step 8 Add the terms obtained in Steps 4, 5, 6, and 7 to obtain a rational function with denominator equal to $q(x)$.

  Step 9 Equate the numerator in Step 8 to $p(x)$ or $r(x)$ to determine the coefficients (the A_k's and B_k's).

- A **linear inequality in two variables** takes one of the forms: p. 419

  $ax + by > c \qquad ax + by < c$

  $ax + by \geq c \qquad ax + by \leq c$

- The set of points in $\mathbb{R}^2$ that satisfy a linear inequality in two variables is a **half-plane.** pp. 418, 420

- A **standard linear programming problem** in two variables takes the form p. 425

  Maximize

$$f = ax + by \tag{1}$$

  subject to

$$\begin{aligned} a_{11}x + a_{12}y &\le b_1 \\ a_{21}x + a_{22}y &\le b_2 \\ \vdots \quad\quad \vdots \quad &\quad \vdots \\ a_{m1}x + a_{m2}y &\le b_m \\ x \ge 0,\ y &\ge 0 \end{aligned}$$

  where $b_1, b_2, \ldots, b_m \ge 0$

- The set of points that satisfy all the linear inequalities above is called the **constraint set** of the problem. p. 426
- The function f in (1) is called the **objective function.** p. 426
- Any point in the constraint set is called a **feasible solution.** p. 426
- The maximum and minimum values of the objective function are always taken at corner points. p. 428
- **The Corner-Point Method** p. 429

  Step 1 Find all possible corner points. These are points of intersection of two of the lines making up the constraint set.

  Step 2 Determine the actual corner points. These are possible corner points that satisfy all the remaining constraints.

  Step 3 Compute the value of f, the objective function, at each actual corner point.

  Step 4 The maximum value of f is the largest number obtained in Step 3. The minimum value of f is the smallest number obtained in Step 3.

- A linear programming problem for which the objective function can take arbitrarily large values is called **unbounded.** p. 430
- A linear programming program that has no feasible solutions is called **infeasible.** p. 431

Review Exercises for Chapter 7

In Exercises 1–15 find all solutions (if any) to the given systems.

1. $\begin{aligned} 3x + 6y &= 9 \\ -2x + 3y &= 4 \end{aligned}$

2. $\begin{aligned} 3x + 6y &= 9 \\ 2x + 4y &= 6 \end{aligned}$

3. $\begin{aligned} 3x - 6y &= 9 \\ -2x + 4y &= 6 \end{aligned}$

4. $\begin{aligned} x + 2y - z &= 6 \\ -x + y + 2z &= -1 \\ 2x + y + z &= 3 \end{aligned}$

5. $\begin{aligned} x + y + z &= 0 \\ 2x - y + 2z &= 0 \\ -3x + 2y + 3z &= 0 \end{aligned}$

6. $\begin{aligned} x + y + z &= 2 \\ 2x - y + 2z &= 4 \\ -x + 4y + z &= 2 \end{aligned}$

7. $\begin{aligned} x + y + z &= 4 \\ x - 2y + z &= 7 \\ -x + y + 3z &= 6 \end{aligned}$

8. $\begin{aligned} x + y + z &= 0 \\ 2x - y + 2z &= 0 \\ -x + 4y + z &= 0 \end{aligned}$

9. $\begin{aligned} 2x + y - 3z &= 0 \\ 4x - y + z &= 0 \end{aligned}$

10. $\begin{aligned} x + y &= 0 \\ 2x + y &= 0 \\ 3x + y &= 0 \end{aligned}$

11. $\begin{aligned} x + y &= 1 \\ 2x + y &= 3 \\ 3x + y &= 4 \end{aligned}$

12. $\begin{aligned} x + y + z + w &= 4 \\ 2x - 3y - z + 4w &= 7 \\ -2x + 4y + z - 2w &= 1 \\ 5x - y + 2z + w &= -1 \end{aligned}$

13. $$\begin{aligned} x + y + z + w &= 0 \\ 2x - 3y - z + 4w &= 0 \\ -2x + 4y + z - 2w &= 0 \\ 5x - y + 2z + w &= 0 \end{aligned}$$

14. $$\begin{aligned} x + y + z + w &= 0 \\ 2x - 3y - z + 4w &= 0 \\ -2x + 4y + z - 2w &= 0 \end{aligned}$$

15. $$\begin{aligned} x + z &= 0 \\ y - w &= 4 \end{aligned}$$

In Exercises 16–20 determine whether the given matrix is in row-echelon form.

16. $\begin{pmatrix} 1 & 0 & 0 & 0 \\ 0 & 1 & 0 & 3 \\ 0 & 0 & 1 & 3 \end{pmatrix}$ 17. $\begin{pmatrix} 1 & 8 & 1 & 0 \\ 0 & 1 & 5 & -7 \\ 0 & 0 & 1 & 4 \end{pmatrix}$

18. $\begin{pmatrix} 1 & 0 \\ 0 & 3 \\ 0 & 0 \end{pmatrix}$ 19. $\begin{pmatrix} 1 & 2 & 3 \\ 0 & 0 & 1 \\ 0 & 1 & 0 \end{pmatrix}$

20. $\begin{pmatrix} 1 & 1 & 1 & 1 \\ 0 & 1 & 1 & 1 \end{pmatrix}$

In Exercises 21 and 22, reduce the matrices to row-echelon form.

21. $\begin{pmatrix} 2 & 8 & -2 \\ 1 & 0 & -6 \end{pmatrix}$ 22. $\begin{pmatrix} 1 & -1 & 2 & 4 \\ -1 & 2 & 0 & 3 \\ 2 & 3 & -1 & 1 \end{pmatrix}$

In Exercises 23–28 find the partial fraction decomposition of each rational expression.

23. $\dfrac{1}{(x-2)(x+3)}$ 24. $\dfrac{x+18}{(x-6)(x+2)}$

25. $\dfrac{x^2+1}{(x+1)(x^2-4)}$ 26. $\dfrac{5x^2+x+12}{x(x^2+4)}$

27. $\dfrac{x^2-2x+4}{(x^2+2)(x^2-2x+3)}$ 28. $\dfrac{x^2+2}{(x-1)(x+2)(x-3)}$

In Exercises 29–33 find all points of intersection of the two curves.

29. $x^2 + y^2 = 9$, $y = -x$ 30. $x^2 + 3y^2 = 3$, $2x^2 + y^2 = 1$

31. $x^2 + 2y^2 = 14$, $x^2 - y^2 = 2$ 32. $x^2 + y^2 = 10$, $2x = 5$

33. $xy = 2$, $4x - 2y = 8$

In Exercises 34–38 find all solutions to each system.

34. $x = 2y^2$, $x^2 + y^2 = 20$ 35. $\dfrac{7}{x} + \dfrac{5}{y} = 4$, $-\dfrac{1}{x} + \dfrac{2}{y} = -3$ 36. $xy = 2$, $x + y = 3$

37. $y = e^{2x}$, $y = e^{4x}$ 38. $x = y^2 + 3$, $x = 1 - y^2$

In Exercises 39–48 sketch the set of points that satisfy the given inequality or inequalities.

39. $x \le 4$ 40. $x - y < 4$
41. $x + y \ge -1$ 42. $4x + 3y > 12$
43. $3x - 4y \le 12$ 44. $y \le 2x + 1$
45. $|x| > 2,\ |y| < 1$
46. $0 \le x \le 3,\ -1 < y < 4$
47. $x + y \le 1,\ 3x - 2y \le 6$
48. $2x - 2y > 1,\ x + 2y \ge 3$

In Exercises 49–54 solve the given linear programming problem by the graphical or corner-point method.

49. Maximize

$$f = 2x + 5y$$

subject to

$$\begin{aligned} 2x + y &\le 4 \\ x + 3y &\le 8 \\ x \ge 0,\ y &\ge 0 \end{aligned}$$

50. Minimize

$$g = x + 2y$$

subject to

$$\begin{aligned} x + y &\ge 3 \\ 2x + 3y &\ge 6 \\ x \ge 0,\ y &\ge 0 \end{aligned}$$

51. Minimize

$$g = 3x + 2y$$

subject to

$$\begin{aligned} x + 2y &\ge 4 \\ 2x + 4y &\ge 6 \\ 5x + y &\ge 10 \\ x \ge 0,\ y &\ge 0 \end{aligned}$$

52. Maximize

$$f = 5x + 3y$$

subject to

$$\begin{aligned} x + 10y &\le 10 \\ 2x + 5y &\le 5 \\ 3x + 2y &\le 6 \\ x \ge 0,\ y &\ge 0 \end{aligned}$$

53. Maximize

$$f = 2x + 3y$$

subject to

$$\begin{aligned} -x + y &\le 5 \\ 2x - 3y &\le 6 \\ x \ge 0,\ y &\ge 0 \end{aligned}$$

54. Minimize

$$f = 4x + 5y$$

subject to

$$\begin{aligned} x + 3y &\ge 3 \\ 3x + y &\ge 3 \\ x + y &\le 7 \\ x \ge 0,\ y &\ge 0 \end{aligned}$$

Chapter 8

Matrices and Determinants

8.1 Matrices

In Section 7.2, we introduced the notion of an augmented matrix in order to simplify the procedure for solving a system of equations. Matrices arise in many other situations as well. For example, matrices are often used to depict and work with data.

EXAMPLE 1 ***How a 5 × 5 Matrix Can Arise***

The table gives the distances in miles between the cities listed.

	Boston	New York	Chicago	Denver	San Francisco
Boston	0	208	980	2025	3186
New York	208	0	850	1833	3049
Chicago	980	850	0	1038	2299
Denver	2025	1833	1038	0	1270
San Francisco	3186	3049	2299	1270	0

We can write these numbers in a 5×5 rectangular array called a **5 × 5 matrix:**

$$
\begin{array}{c} \\ A = \end{array}
\begin{array}{c}
\begin{array}{ccccc} \text{Boston} & \text{New York} & \text{Chicago} & \text{Denver} & \text{San Francisco} \end{array} \\
\begin{pmatrix}
0 & 208 & 980 & 2025 & 3186 \\
208 & 0 & 850 & 1833 & 3049 \\
980 & 850 & 0 & 1038 & 2299 \\
2025 & 1833 & 1038 & 0 & 1270 \\
3186 & 3049 & 2299 & 1270 & 0
\end{pmatrix}
\end{array}
\begin{array}{l} \\ \text{Boston} \\ \text{New York} \\ \text{Chicago} \\ \text{Denver} \\ \text{San Francisco} \end{array}
$$

NOTE The labels are optional. They make the matrix easier to read.

The fourth row is the **row vector**

$$(2025 \quad 1833 \quad 1038 \quad 0 \quad 1270)$$

This row represents the distances from Denver to each of the five cities. The second column is the **column vector**

$$\begin{pmatrix} 208 \\ 0 \\ 850 \\ 1833 \\ 3049 \end{pmatrix}$$

This column gives the distances from each of the five cities to New York.

We now make these notions more precise.

Definition of an $m \times n$ Matrix

An $\boldsymbol{m \times n}$ **matrix** A is a rectangular array of mn numbers arranged in a definite order in m rows and n columns.†

$$A = \begin{pmatrix} a_{11} & a_{12} & \cdots & a_{1j} & \cdots & a_{1n} \\ a_{21} & a_{22} & \cdots & a_{2j} & \cdots & a_{2n} \\ \vdots & \vdots & & \vdots & & \vdots \\ a_{i1} & a_{i2} & \cdots & a_{ij} & \cdots & a_{in} \\ \vdots & \vdots & & \vdots & & \vdots \\ a_{m1} & a_{m2} & \cdots & a_{mj} & \cdots & a_{mn} \end{pmatrix}$$

The number a_{ij} appearing in the ith row and jth column of A is called the i, j **component** or **entry** of A. For convenience, the matrix A is written $A = (a_{ij})$. Usually, matrices will be denoted by capital letters.

If A is an $m \times n$ matrix with $m = n$, then A is called a **square matrix.** An $m \times n$ matrix with all components equal to zero is called the $\boldsymbol{m \times n}$ **zero matrix.**

EXAMPLE 2 *Five Matrices*

The following are $m \times n$ matrices for various values of m and n.

(a) $A = \begin{pmatrix} 1 & 3 \\ 4 & 2 \end{pmatrix}$, 2×2 (square)

(b) $A = \begin{pmatrix} -1 & 3 \\ 4 & 0 \\ 1 & -2 \end{pmatrix}$, 3×2

(c) $\begin{pmatrix} -1 & 4 & 1 \\ 3 & 0 & 2 \end{pmatrix}$, 2×3

(d) $\begin{pmatrix} 1 & 6 & -2 \\ 3 & 1 & 4 \\ 2 & -6 & 5 \end{pmatrix}$, 3×3 (square)

(e) $\begin{pmatrix} 0 & 0 & 0 & 0 \\ 0 & 0 & 0 & 0 \end{pmatrix}$, 2×4 zero matrix

---

† The term *matrix* was first used in 1850 by the British mathematician James Joseph Sylvestor (1814–1897) to distinguish matrices from determinants (which we will discuss in Section 8.4). In fact, matrix was intended to mean "mother of determinants."

Note, for example, that the 2,3 component of the matrix in (d) is 4 while the 3,1 component of the matrix in (b) is 1.

An $m \times n$ matrix is said to have the **size** $m \times n$. Two matrices $A = (a_{ij})$ and $B = (b_{ij})$ are **equal** if (1) they have the same size and (2) corresponding components are equal.

Row and Column Vectors

A matrix with one row and n columns is called an ***n*-component row vector** or, more simply, a **row vector.** A matrix with n rows and one column is called an ***n*-component column vector,** or **column vector.** Row and column vectors will usually be denoted by boldface, lowercase letters, such as **a**, **b**, **p**, **q**, **x**, or **y**.

EXAMPLE 3 *Four Row Vectors*

The following are row vectors.

(a) $(2 \quad 5)$
(b) $(-6 \quad 0 \quad 4)$
(c) $(5 \quad 0 \quad -2 \quad 3)$
(d) $(0 \quad 0 \quad 0)$, **zero vector** ■

EXAMPLE 4 *Four Column Vectors*

The following are column vectors.

(a) $\begin{pmatrix} 2 \\ 5 \end{pmatrix}$ (b) $\begin{pmatrix} -6 \\ 0 \\ 4 \end{pmatrix}$

(c) $\begin{pmatrix} 5 \\ 0 \\ -2 \\ 3 \end{pmatrix}$ (d) $\begin{pmatrix} 0 \\ 0 \\ 0 \end{pmatrix}$, **zero vector** ■

EXAMPLE 5 *Using a Vector to Keep Track of Order Quantities*

Suppose that the buyer for a manufacturing plant must order different quantities of steel, aluminum, oil, and paper. The buyer can keep track of the quantities to be ordered with a single column (or row) vector. The vector $\begin{pmatrix} 10 \\ 30 \\ 15 \\ 60 \end{pmatrix}$ indicates that 10 units of steel, 30 units of aluminum, 15 units of oil, and 60 units of paper would be ordered. ■

EXAMPLE 6 ***Using a Matrix to Keep Track of Order Quantities in Five Different Plants***

In Example 5 we saw how the vector $\begin{pmatrix} 10 \\ 30 \\ 15 \\ 60 \end{pmatrix}$ could represent order quantities for four different products used by one manufacturer. Suppose that there were five different plants. Then the 4×5 matrix

$$Q = \overset{\text{Plants}}{\begin{pmatrix} \overset{1}{10} & \overset{2}{20} & \overset{3}{15} & \overset{4}{16} & \overset{5}{25} \\ 30 & 10 & 20 & 25 & 22 \\ 15 & 22 & 18 & 20 & 13 \\ 60 & 40 & 50 & 35 & 45 \end{pmatrix}} \begin{matrix} \text{Steel} \\ \text{Aluminum} \\ \text{Oil} \\ \text{Paper} \end{matrix} \quad \text{Products}$$

could represent the orders for the four products in each of the five plants. We can see, for example, that plant 4 orders 25 units of aluminum, whereas plant 2 orders 40 units of paper.

Having defined matrices and seen how they could arise, we now turn to the question of adding matrices and multiplying them by real numbers. For historical reasons, numbers encountered when dealing with matrices are called **scalars.**

The study of vectors and matrices essentially began with the work of the Irish mathematician, Sir William Rowan Hamilton (1805–1865). Hamilton used the word *scalar* to denote a number that could take on "all values contained on the one *scale* of progression of numbers from negative to positive infinity," in a paper published in *Philosophy Magazine* in 1844.

Definition of Scalar Multiplication of a Matrix

To multiply a matrix by a scalar (real number), multiply each component of the matrix by that scalar.

EXAMPLE 7 ***Multiplying a Matrix by 2, −3, and 0***

Let $A = \begin{pmatrix} 1 & -3 & 4 & 2 \\ 3 & 1 & 4 & 6 \\ -2 & 3 & 5 & 7 \end{pmatrix}$. Then $2A = \begin{pmatrix} 2 & -6 & 8 & 4 \\ 6 & 2 & 8 & 12 \\ -4 & 6 & 10 & 14 \end{pmatrix}$

$-3A = \begin{pmatrix} -3 & 9 & -12 & -6 \\ -9 & -3 & -12 & -18 \\ 6 & -9 & -15 & -21 \end{pmatrix}$, and $0A = \begin{pmatrix} 0 & 0 & 0 & 0 \\ 0 & 0 & 0 & 0 \\ 0 & 0 & 0 & 0 \end{pmatrix}$ ■

EXAMPLE 8 *An Application of Multiplying a Matrix by a Scalar*

The sales of 4 products by a national retail chain in 3 different months is given in Table 1.

Table 1
Sales (in thousands of units)

	Product			
Month	*I*	*II*	*III*	*IV*
April 1991	6	19	14	46
May 1991	8	28	12	40
June 1991	4	26	17	55

We can represent these data in a 3×4 matrix S.

$$S = \begin{pmatrix} 6 & 19 & 14 & 46 \\ 8 & 28 & 12 & 40 \\ 4 & 26 & 17 & 55 \end{pmatrix}$$

Suppose now that in 1992, sales in all months and in all categories increased by 50%. Increasing by 50% is the same as multiplying by $1\frac{1}{2}$. Thus the new matrix representing 1992 sales of the four products in the 3 months is $1.5S$.

$$\begin{matrix} & & \text{Product} & & \\ & \text{I} & \text{II} & \text{III} & \text{IV} \end{matrix}$$

$$1.5S = \begin{pmatrix} 9 & 28.5 & 21 & 69 \\ 12 & 42 & 18 & 60 \\ 6 & 39 & 25.5 & 82.5 \end{pmatrix} \begin{matrix} \text{April 1992} \\ \text{May 1992} \\ \text{June 1992} \end{matrix}$$

For example, we see that whereas May 1991 sales of product III were 12,000 units (since these numbers represent thousands of units), sales of product III in May 1992 amounted to 18,000 units.

We now turn to the addition of matrices.

Definition of the Addition of Matrices

Let $A = (a_{ij})$ and $B = (b_{ij})$ be two $m \times n$ matrices. Then the sum of A and B is a matrix obtained by adding the corresponding components of A and B. That is

$$ij\text{th component of } A + B = (a_{ij} + b_{ij})$$

WARNING The sum of two matrices is defined only when both matrices have the same size. Thus, for example, it is not possible to add the matrices

$$\begin{pmatrix} 1 & 2 & 3 \\ 4 & 5 & 6 \end{pmatrix} \quad \text{and} \quad \begin{pmatrix} -1 & 0 \\ 2 & -5 \\ 4 & 7 \end{pmatrix}$$ ■

EXAMPLE 9 *Adding Two Matrices*

The following matrices can be added because they have the same size:

$$\begin{pmatrix} 2 & 4 & -6 & 7 \\ 1 & 3 & 2 & 1 \\ -4 & 3 & -5 & 5 \end{pmatrix} + \begin{pmatrix} 0 & 1 & 6 & -2 \\ 2 & 3 & 4 & 3 \\ -2 & 1 & 4 & 4 \end{pmatrix} = \begin{pmatrix} 2 & 5 & 0 & 5 \\ 3 & 6 & 6 & 4 \\ -6 & 4 & -1 & 9 \end{pmatrix}$$ ■

EXAMPLE 10 *Addition and Scalar Multiplication of Matrices*

Let $A = \begin{pmatrix} 1 & 2 & 4 \\ -7 & 3 & -2 \end{pmatrix}$ and $B = \begin{pmatrix} 4 & 0 & 5 \\ 1 & -3 & 6 \end{pmatrix}$. Calculate $-2A + 3B$.

SOLUTION

$$-2A + 3B = (-2)\begin{pmatrix} 1 & 2 & 4 \\ -7 & 3 & -2 \end{pmatrix} + (3)\begin{pmatrix} 4 & 0 & 5 \\ 1 & -3 & 6 \end{pmatrix}$$

$$= \begin{pmatrix} -2 & -4 & -8 \\ 14 & -6 & 4 \end{pmatrix} + \begin{pmatrix} 12 & 0 & 15 \\ 3 & -9 & 18 \end{pmatrix} = \begin{pmatrix} 10 & -4 & 7 \\ 17 & -15 & 22 \end{pmatrix}$$ ■

EXAMPLE 11 *An Application of Matrix Addition*

The retail chain of Example 8 has two stores in southern California. Sales of each store (in hundreds of units) in three different months are given in Table 2.

Table 2

Store A Sales (in hundreds of units)					Store B Sales (in hundreds of units)				
	Product					Product			
Month	***I***	***II***	***III***	***IV***	**Month**	***I***	***II***	***III***	***IV***
April 1992	7	12	2	28	April 1992	6	21	8	41
May 1992	5	14	8	17	May 1992	10	19	14	33
June 1992	6	9	5	33	June 1992	2	26	5	28

We can represent the sales in each store by a 3×4 matrix.

$$A = \begin{pmatrix} 7 & 12 & 2 & 28 \\ 5 & 14 & 8 & 17 \\ 6 & 9 & 5 & 33 \end{pmatrix}, \qquad B = \begin{pmatrix} 6 & 21 & 8 & 41 \\ 10 & 19 & 14 & 33 \\ 2 & 26 & 5 & 28 \end{pmatrix}$$

Then the matrix $A + B$ represents total sales for the two stores for each product in each month.

$$A + B = \begin{pmatrix} 13 & 33 & 10 & 69 \\ 15 & 33 & 22 & 50 \\ 8 & 35 & 10 & 61 \end{pmatrix}$$

Thus, for example, we find that 2200 units of product III were sold in May 1992 in stores A and B combined.

Problems 8.1

Readiness Check

I. Which of the following is true of the matrix
$$\begin{pmatrix} 1 & 2 & 3 \\ 7 & -1 & 0 \end{pmatrix}?$$
a. It is a square matrix.
b. If multiplied by the scalar -1, the product is $\begin{pmatrix} -1 & -2 & -3 \\ -7 & 1 & 0 \end{pmatrix}$.
c. It is a 3×2 matrix.
d. It is the sum of $\begin{pmatrix} 3 & 1 & 4 \\ 7 & 2 & 0 \end{pmatrix}$ and $\begin{pmatrix} -2 & 1 & 1 \\ 0 & 1 & 0 \end{pmatrix}$.

II. Which of the following is $2A - 4B$ if $A = (2 \quad 0 \quad 0)$ and $B = (3 \quad 1)$?
a. $(-8 \quad -4)$ b. $(5 \quad 0 \quad 1)$
c. $(16 \quad -4 \quad 0)$
d. This operation cannot be performed.

III. Which of the following is true when finding the difference of two matrices?
a. The matrices must have the same size.
b. The matrices must be square.
c. The matrices must both be row vectors or both be column vectors.
d. One matrix must be a row vector and the other must be a column vector.

IV. Which of the following would be the entries in the second column of matrix B, if
$$\begin{pmatrix} 3 & -4 & 0 \\ 2 & 8 & -1 \end{pmatrix} + B = \begin{pmatrix} 0 & 0 & 0 \\ 0 & 0 & 0 \end{pmatrix}?$$
a. $-2, -8, 1$ b. $4, -8$
c. $2, 8, -1$ d. $-4, 8$

V. Which of the following must be the second row of matrix B if $3A - B = 2C$ for
$$A = \begin{pmatrix} 1 & -1 & 1 \\ 0 & 0 & 3 \\ 4 & 2 & 0 \end{pmatrix} \text{ and } C = \begin{pmatrix} 1 & 0 & 0 \\ 0 & 1 & 0 \\ 0 & 0 & 1 \end{pmatrix}?$$
a. $-3 \quad 2 \quad 6$ b. $0 \quad -2 \quad 9$
c. $3 \quad -2 \quad 6$ d. $0 \quad 2 \quad -9$

In Problems 1–14 determine the size of the given matrix and indicate which matrices are square.

1. $\begin{pmatrix} 1 & 2 \\ 3 & 4 \end{pmatrix}$

2. $\begin{pmatrix} -1 & 2 \\ 3 & 1 \\ 1 & 6 \end{pmatrix}$

3. $\begin{pmatrix} 0 & 0 \\ 0 & 0 \end{pmatrix}$

4. $\begin{pmatrix} 0 & 0 & 0 \\ 0 & 0 & 0 \end{pmatrix}$

5. $\begin{pmatrix} 0 & 0 \\ 0 & 0 \\ 0 & 0 \end{pmatrix}$

6. $\begin{pmatrix} 1 & 0 & 0 \\ 0 & 1 & 0 \\ 0 & 0 & 1 \end{pmatrix}$

7. $\begin{pmatrix} 1 & 3 & 2 & 4 \\ 2 & 1 & 0 & 6 \end{pmatrix}$

8. $(1 \quad 0 \quad 2)$

9. $\begin{pmatrix} 1 \\ 0 \\ 2 \end{pmatrix}$

10. $\begin{pmatrix} 1 & 3 \\ 0 & 6 \\ 2 & 2 \\ 4 & 9 \end{pmatrix}$

11. $\begin{pmatrix} 3 & -6 & 2 \\ 1 & 7 & 2 \\ -1 & 4 & 6 \end{pmatrix}$

12. $\begin{pmatrix} a & b \\ c & d \end{pmatrix}$

13. $(a \quad b \quad c \quad d)$

14. $\begin{pmatrix} a \\ b \\ c \\ d \end{pmatrix}$

In Problems 15–19 determine whether the two matrices are equal.

15. $\begin{pmatrix} 1 & 0 \\ 0 & 1 \end{pmatrix}$ and $\begin{pmatrix} 0 & 1 \\ 1 & 0 \end{pmatrix}$

16. $\begin{pmatrix} 0 & 3 \\ 1 & 0 \end{pmatrix}$ and $\begin{pmatrix} 2-2 & 1+2 \\ 3-2 & -2+2 \end{pmatrix}$

17. $\begin{pmatrix} 0 & 0 & 0 \\ 0 & 0 & 0 \end{pmatrix}$ and $\begin{pmatrix} 0 & 0 \\ 0 & 0 \\ 0 & 0 \end{pmatrix}$

18. $\begin{pmatrix} 0 & 1 & 0 \\ 0 & 0 & 1 \\ 1 & 0 & 0 \end{pmatrix}$ and $\begin{pmatrix} 0 & 0 & 1 \\ 1 & 0 & 0 \\ 0 & 1 & 0 \end{pmatrix}$

Answers to Readiness Check

I. b II. d III. a IV. b V. b

19. $\begin{pmatrix} 1-2 & 3 & 1 \\ 0 & 4-5 & 3 \\ 2 & 6 & 2+3 \end{pmatrix}$ and $\begin{pmatrix} -1 & 1+2 & 1 \\ 5-5 & -1 & 5-2 \\ 1+1 & 6 & 5 \end{pmatrix}$

20. Let

$$A = \begin{pmatrix} 1 & 6 & -2 & 3 \\ 4 & 0 & 2 & 6 \\ -1 & 4 & 3 & 1 \end{pmatrix}$$

(a) Write the first row.
(b) Write the 1,3 component.
(c) Write the 3,2 component.
(d) Write the 2,4 component.

In Problems 21–30 perform the indicated computation, if possible.

21. $4\begin{pmatrix} 2 & 5 \\ 3 & -1 \end{pmatrix}$

22. $\begin{pmatrix} 2 & -4 \\ -3 & 0 \end{pmatrix} + \begin{pmatrix} 1 & 5 \\ 6 & 3 \end{pmatrix}$

23. $-\begin{pmatrix} 1 & 2 & 5 \\ 7 & -2 & 0 \end{pmatrix}$

24. $3\begin{pmatrix} 1 & 5 \\ 4 & -3 \\ 7 & 1 \end{pmatrix} - 2\begin{pmatrix} 7 & -4 \\ -2 & 5 \\ -3 & 2 \end{pmatrix}$

25. $\begin{pmatrix} 8 & -7 \\ 1 & 4 \\ 5 & 3 \end{pmatrix} + \begin{pmatrix} 2 & 0 & 5 \\ -3 & 5 & 2 \end{pmatrix}$

26. $\begin{pmatrix} 3 & 1 & 4 \\ 1 & 0 & 5 \\ -6 & 7 & 2 \end{pmatrix} + \begin{pmatrix} 7 & -2 & 3 \\ 5 & 0 & 6 \\ 0 & 1 & 2 \end{pmatrix}$

27. $4\begin{pmatrix} 0 & 3 & 2 \\ 5 & -9 & 0 \\ -8 & 5 & 3 \end{pmatrix} + 3\begin{pmatrix} -4 & 6 & 5 \\ 5 & 2 & 1 \\ 1 & 5 & -2 \end{pmatrix}$

28. $\begin{pmatrix} a & b \\ c & d \end{pmatrix} - \begin{pmatrix} e & f \\ g & h \end{pmatrix}$

29. $\begin{pmatrix} 5 & 6 & 2 \\ 3 & 4 & 1 \\ 0 & -7 & 2 \end{pmatrix} + \begin{pmatrix} 0 & 0 & 0 \\ 0 & 0 & 0 \\ 0 & 0 & 0 \end{pmatrix}$

30. $\begin{pmatrix} 1 & 6 \\ 2 & 3 \\ 4 & 7 \end{pmatrix} - \begin{pmatrix} 0 & 0 \\ 0 & 0 \\ 0 & 0 \end{pmatrix}$

In Problems 31–42 perform the indicated computation with

$$A = \begin{pmatrix} 1 & 3 \\ 2 & 5 \\ -1 & 2 \end{pmatrix}, B = \begin{pmatrix} -2 & 0 \\ 1 & 4 \\ -7 & 5 \end{pmatrix}, \text{ and } C = \begin{pmatrix} -1 & 1 \\ 4 & 6 \\ -7 & 3 \end{pmatrix}.$$

31. $2A$
32. $A - B$
33. $A + C$
34. $2C - 5A$
35. $0B$ (0 is the scalar zero.)
36. $-2A + 3C$
37. $A + B + C$
38. $C - A - B$
39. $2A - 3B + 4C$
40. $7C - B + 2A$
41. Find a matrix D such that $2A + B - D$ is the 3×2 zero matrix.
42. Find a matrix E such that $A + 2B - 3C + E$ is the 3×2 zero matrix.

In Problems 43–50 perform the indicated computation with

$$A = \begin{pmatrix} 1 & -1 & 2 \\ 2 & 4 & 5 \\ 0 & 1 & -1 \end{pmatrix}, B = \begin{pmatrix} 0 & 2 & 1 \\ 3 & 0 & 5 \\ 7 & -6 & 0 \end{pmatrix}, \text{ and}$$

$$C = \begin{pmatrix} 0 & 0 & 2 \\ 3 & 1 & 0 \\ 0 & -2 & 4 \end{pmatrix}.$$

43. $C - 2A$
44. $3A - C$
45. $A + B + C$
46. $2A - B + 2C$
47. $C - A - B$
48. $4C - 2B + 3A$
49. Find a matrix D such that $A + B + C + D$ is the 3×3 zero matrix.
50. Find a matrix E such that $3C - 2B + 8A - 4E$ is the 3×3 zero matrix.
51. Referring to Example 8, if sales increase 25% between 1991 and 1992, find a matrix that represents total sales (in thousands of units) of the four products in the months April, May, and June of 1992.
52. Answer the question of Problem 51 if sales *decrease* 40% in 1992.
53. Referring to Example 11, if sales of store A increase 30% in 1993 while sales in store B decrease 20%, find a matrix that represents combined sales of the four products in April, May, and June of 1993.

54. A polling organization gives the results of asking the preferences of a sample of 1089 voters in a coming election for governor in the table.

	Democratic Candidate	Republican Candidate	Other Candidates	Undecided
Democrats	415	91	6	77
Republicans	65	281	4	63
Independents	31	19	8	29

(a) Write these data as a matrix.
(b) What numbers form the fourth column?
(c) What numbers form the second row?
(d) What number is the 2,4 component?
(e) What number is the 4,2 component?

8.2 Matrix Products

In this section, we see how two matrices can be multiplied together. Quite obviously, we could define the product of two $m \times n$ matrices $A = (a_{ij})$ and $B = (b_{ij})$ to be the $m \times n$ matrix whose ijth component is $a_{ij}b_{ij}$. However, for just about all the important applications involving matrices, another kind of product is needed. Let us try to see why this is the case.

EXAMPLE 1 ***Multiplying a Demand Vector and a Price Vector***

Suppose that a manufacturer produces four items. The demand for the items is given by the demand vector $\mathbf{d} = (30 \quad 20 \quad 40 \quad 10)$ (a 1×4 matrix). The price per unit that the manufacturer receives for the items is given by the price vector $\mathbf{p} = \begin{pmatrix} \$20 \\ \$15 \\ \$18 \\ \$40 \end{pmatrix}$ (a 4×1 matrix). If the demand is met, how much money will the manufacturer receive?

SOLUTION Demand for the first item is 30, and the manufacturer receives \$20 for each of the first item sold. Thus, (30)(20) = \$600 is received from the sales of the first item. By continuing this reasoning, we see that the total amount of money received is

$$\begin{aligned}(30)(20) + (20)(15) + (40)(18) + (10)(40) &= 600 + 300 + 720 + 400 \\ &= \$2020\end{aligned}$$

We write this result as

$$(30 \quad 20 \quad 40 \quad 10)\begin{pmatrix} 20 \\ 15 \\ 18 \\ 40 \end{pmatrix} = 2020$$

That is, we multiplied a 4-component row vector and a 4-component column vector to obtain a scalar (real number).

In the last example, we multiplied a row vector by a column vector and obtained a scalar. In general, we have the following definition.

Definition of the Dot Product of Two Vectors

Let $\mathbf{p} = (p_1 \quad p_2 \quad \cdots \quad p_n)$ be an n-component row vector and $\mathbf{q} = \begin{pmatrix} q_1 \\ q_2 \\ \vdots \\ q_n \end{pmatrix}$ be an n-component column vector. Then the **dot product** (or **scalar product**) of $\mathbf{p}$ and $\mathbf{q}$, denoted by $\mathbf{p} \cdot \mathbf{q}$, is given by

$$\mathbf{p} \cdot \mathbf{q} = p_1q_1 + p_2q_2 + \cdots + p_nq_n \qquad (1)$$

We can write this as

Alternative Way to Write the Dot Product

$$\underset{1 \times n}{(p_1 p_2 \cdots p_n)} \underset{n \times 1}{\begin{pmatrix} q_1 \\ q_2 \\ \vdots \\ q_n \end{pmatrix}} = p_1q_1 + p_2q_2 + \cdots + p_nq_n \qquad (2)$$

This is a real number (a scalar)

WARNING When taking the dot product of $\mathbf{p}$ and $\mathbf{q}$, it is necessary that $\mathbf{p}$ and $\mathbf{q}$ have the same number of components. ■

EXAMPLE 2 ***Computing a Dot Product***

Let $\mathbf{a} = (2 \quad -3 \quad 4 \quad -6)$ and $\mathbf{b} = \begin{pmatrix} 1 \\ 2 \\ 0 \\ 3 \end{pmatrix}$. Compute $\mathbf{a} \cdot \mathbf{b}$.

SOLUTION Here $\mathbf{a} \cdot \mathbf{b} = (2)(1) + (-3)(2) + (4)(0) + (-6)(3)$
$= 2 - 6 + 0 - 18 = -22.$

We now define the product of two matrices.

Definition of the Product of Two Matrices

Let $A = (a_{ij})$ be an $m \times n$ matrix, and let $B = (b_{ij})$ be an $n \times p$ matrix. Then the **product** of A and B is an $m \times p$ matrix $C = (c_{ij})$, where

$$c_{ij} = (i\text{th row of } A) \cdot (j\text{th column of } B) \tag{3}$$

That is, the ijth element of AB is the dot product of the ith row of A and the jth column of B. If we write this out, we obtain

$$c_{ij} = a_{i1}b_{1j} + a_{i2}b_{2j} + \cdots + a_{in}b_{nj} \tag{4}$$

WARNING Two matrices can be multiplied together only if the number of columns of the first matrix is equal to the number of rows of the second. Otherwise, the vectors that are the ith row of A and the jth column of B will not have the same number of components, and the dot product in equation (3) will not be defined. To illustrate this, we write the matrices A and B:

jth column of B ↓

$$i\text{th row of } A \rightarrow \begin{pmatrix} a_{11} & a_{12} & \cdots & a_{1n} \\ a_{21} & a_{22} & \cdots & a_{2n} \\ \vdots & \vdots & & \vdots \\ a_{i1} & a_{i2} & \cdots & a_{in} \\ \vdots & \vdots & & \vdots \\ a_{m1} & a_{m2} & \cdots & a_{mn} \end{pmatrix} \begin{pmatrix} b_{11} & b_{12} & \cdots & b_{1j} & \cdots & b_{1p} \\ b_{21} & b_{22} & \cdots & b_{2j} & \cdots & b_{2p} \\ \vdots & \vdots & & \vdots & & \vdots \\ b_{n1} & b_{n2} & \cdots & b_{nj} & \cdots & b_{np} \end{pmatrix}$$

The shaded row and column vectors must have the same number of components. ■

EXAMPLE 3 *Computing the Product of Two 2 × 2 Matrices*

If $A = \begin{pmatrix} 1 & 3 \\ -2 & 4 \end{pmatrix}$ and $B = \begin{pmatrix} 3 & -2 \\ 5 & 6 \end{pmatrix}$, calculate AB and BA.

SOLUTION A is a 2×2 matrix and B is a 2×2 matrix, so $C = AB = (2 \times 2) \times (2 \times 2)$ is also a 2×2 matrix. If $C = (c_{ij})$, what is c_{11}? We know that

$$c_{11} = (\text{1st row of } A) \cdot (\text{1st column of } B)$$

Rewriting the matrices, we have

1st column of B ↓

$$\text{1st row of } A \rightarrow \begin{pmatrix} 1 & 3 \\ -2 & 4 \end{pmatrix} \begin{pmatrix} 3 & -2 \\ 5 & 6 \end{pmatrix}$$

Thus,

$$c_{11} = (1 \quad 3)\begin{pmatrix} 3 \\ 5 \end{pmatrix} = 3 + 15 = 18$$

Similarly, to compute c_{12} we have

$$\text{1st row of } A \rightarrow \begin{pmatrix} 1 & 3 \\ -2 & 4 \end{pmatrix}\begin{pmatrix} 3 & -2 \\ 5 & 6 \end{pmatrix} \quad \begin{matrix}\text{2nd column} \\ \text{of } B \\ \downarrow\end{matrix}$$

and

$$c_{12} = (1 \quad 3)\begin{pmatrix} -2 \\ 6 \end{pmatrix} = -2 + 18 = 16$$

Continuing, we find that

$$c_{21} = (-2 \quad 4)\begin{pmatrix} 3 \\ 5 \end{pmatrix} = -6 + 20 = 14$$

and

$$c_{22} = (-2 \quad 4)\begin{pmatrix} -2 \\ 6 \end{pmatrix} = 4 + 24 = 28$$

Thus,

$$C = AB = \begin{pmatrix} 18 & 16 \\ 14 & 28 \end{pmatrix}$$

Similarly, leaving out the intermediate steps, we see that

$$C' = BA = \begin{pmatrix} 3 & -2 \\ 5 & 6 \end{pmatrix}\begin{pmatrix} 1 & 3 \\ -2 & 4 \end{pmatrix} = \begin{pmatrix} 3 + 4 & 9 - 8 \\ 5 - 12 & 15 + 24 \end{pmatrix} = \begin{pmatrix} 7 & 1 \\ -7 & 39 \end{pmatrix}$$

WARNING Example 3 illustrates an important fact: *Matrix products do not, in general, commute;* that is, $AB \neq BA$ in general. It sometimes happens that $AB = BA$, but this will be the exception, not the rule. In fact, it may occur that AB is defined, whereas BA is not. Thus we must be careful of *order* when multiplying two matrices together. ■

EXAMPLE 4 *An Application of the Product of Two Matrices*

In Example 8 in the last section, we obtained the following matrix representation of the number of units (in thousands) of 4 items sold by a large retail chain in 3 successive months.

$$S = \begin{matrix} & \begin{matrix} \text{I} & \text{II} & \text{III} & \text{IV} \end{matrix} & \\ & \begin{pmatrix} 6 & 19 & 14 & 46 \\ 8 & 28 & 12 & 40 \\ 4 & 26 & 17 & 55 \end{pmatrix} & \begin{matrix} \text{April} \\ \text{May} \\ \text{June} \end{matrix} \end{matrix}$$

Table 1

Item	Profit (in hundreds of dollars)	Taxes (in hundreds of dollars)
I	2	1
II	3	2
III	5	3
IV	4	2

The gross (before-tax) profit made and the taxes paid (in hundreds of dollars) for each thousand units sold of each item are given in Table 1. How much profit was made and how much tax was paid on sales of the 4 items for each of the months of April, May, and June?

SOLUTION First, we write the information in Table 1 in matrix form to obtain what we may call the **profit-tax matrix,** P.

$$P = \begin{pmatrix} 2 & 1 \\ 3 & 2 \\ 5 & 3 \\ 4 & 2 \end{pmatrix} \begin{matrix} \text{I} \\ \text{II} \\ \text{III} \\ \text{IV} \end{matrix} \quad (\text{columns: Profit, Taxes})$$

Our problem now is really a problem in matrix multiplication. To see this, let us compute the profit in May, for example. In May, there were 8 units of product I sold and the profit per unit was 2, so the total profit on sales in May of product I was $(8)(2) = 16$. (Actually, the profit was $(8000)(\$200) = \$1,600,000$, but for simplicity, we will stick to the smaller numbers.) Similarly, the profit on product II in May was $(28)(3) = 84$. Thus, we see that the total profit from sales of all 4 products in May is

$$(8)(2) + (28)(3) + (12)(5) + (40)(4) = 320$$

But this is the dot product of the second (May) row of S and the first (profit) column of P. Analogously, we find that the total tax paid in June is equal to the dot product of the third (June) row of S and the second (tax) column of P. Therefore we compute

$$SP = \begin{pmatrix} 6 & 19 & 14 & 46 \\ 8 & 28 & 12 & 40 \\ 4 & 26 & 17 & 55 \end{pmatrix} \begin{pmatrix} 2 & 1 \\ 3 & 2 \\ 5 & 3 \\ 4 & 2 \end{pmatrix} = \begin{pmatrix} 323 & 178 \\ 320 & 180 \\ 391 & 217 \end{pmatrix} \begin{matrix} \text{April} \\ \text{May} \\ \text{June} \end{matrix} \quad (\text{columns: Profit, Taxes})$$

The last matrix tells us at a glance that gross profits and taxes in May were \$32,000,000 and \$18,000,000, so after-tax profit was \$14,000,000. ■

EXAMPLE 5 *First- and Second-Order Contact to a Contagious Disease*

Suppose that 3 persons have contracted a contagious disease. A second group of 6 persons is questioned to determine who has been in contact with the 3 infected persons. A third group of 7 persons is then questioned to determine contacts with any of the 6 persons in the second group. We define the 3×6 matrix $A = (a_{ij})$ by defining $a_{ij} = 1$ if the jth person in the second group has had contact with the ith person in the first group and $a_{ij} = 0$ otherwise. Similarly, we define the 6×7 matrix $B = (b_{ij})$ by defining $b_{ij} = 1$ if the jth person in the third group has had contact with the ith person in the second group and $b_{ij} = 0$ otherwise. These two matrices describe the *direct,* or *first-order, contacts* between the groups.

For example, we could have

$$A = \begin{array}{c} \\ 1 \\ 2 \\ 3 \end{array}\!\!\begin{array}{c} \text{Group 1} \downarrow \quad \text{Group 2} \\ \begin{pmatrix} 0 & 0 & 1 & 0 & 1 & 0 \\ 1 & 0 & 0 & 1 & 0 & 0 \\ 0 & 0 & 1 & 1 & 0 & 1 \end{pmatrix} \end{array} \quad \text{and} \quad B = \begin{array}{c} \\ 1 \\ 2 \\ 3 \\ 4 \\ 5 \\ 6 \end{array}\!\!\begin{array}{c} \text{Group 2} \downarrow \quad \text{Group 3} \\ \begin{pmatrix} 0 & 0 & 1 & 0 & 0 & 1 & 0 \\ 0 & 0 & 1 & 1 & 0 & 0 & 0 \\ 1 & 0 & 0 & 0 & 0 & 1 & 1 \\ 0 & 0 & 1 & 1 & 0 & 0 & 0 \\ 0 & 1 & 0 & 1 & 0 & 0 & 0 \\ 1 & 0 & 0 & 0 & 0 & 1 & 0 \end{pmatrix} \end{array}$$

(Columns of A numbered 1–6; columns of B numbered 1–7.)

In this case, we have $a_{24} = 1$, which means that the fourth person in the second group has had contact with the second infected person. Analogously, $b_{33} = 0$, which means that the third person in the third group has not had contact with the third person in the second group.

We may be interested in studying the *indirect,* or *second-order, contacts* between the 7 persons in the third group and the 3 infected people in the first group. The matrix product $C = AB$ describes these second-order contacts. The ijth component

$$c_{ij} = a_{i1}b_{1j} + a_{i2}b_{2j} + a_{i3}b_{3j} + a_{i4}b_{4j} + a_{i5}b_{5j} + a_{i6}b_{6j}$$

gives the number of second-order contacts between the jth person in the third group and the ith person in the infected group. With the given matrices A and B, we have

$$C = AB = \begin{pmatrix} 1 & 1 & 0 & 1 & 0 & 1 & 1 \\ 0 & 0 & 2 & 1 & 0 & 1 & 0 \\ 2 & 0 & 1 & 1 & 0 & 2 & 1 \end{pmatrix}$$

For example, the component $c_{23} = 2$ implies that there are two second-order contacts between the third person in the third group and the second contagious person. Note that the sixth person in the third group has had $1 + 1 + 2 = 4$ indirect contacts with the infected group. Only the fifth person has had no contacts.

The number 1 plays a special role in arithmetic. If x is any real number, then $1 \cdot x = x \cdot 1 = x$; that is, we can multiply any number on the right or the left by 1 and get the original number back. In matrix multiplication, a similar fact holds for square matrices.

The **main diagonal** of a square $(m \times n)$ matrix consists of the numbers that appear in the upper-left to lower-right diagonal of the matrix.

EXAMPLE 6 ***Three Main Diagonals***

The main diagonals of the square matrices given below are circled.

(a) $\begin{pmatrix} 2 & 3 \\ -1 & 4 \end{pmatrix}$ (b) $\begin{pmatrix} -2 & 2 & 4 \\ 1 & 0 & 5 \\ 3 & -2 & 6 \end{pmatrix}$ (c) $\begin{pmatrix} 5 & 2 & 1 & 6 \\ 2 & -3 & 2 & 1 \\ 4 & 1 & 7 & 2 \\ -3 & 2 & 0 & 5 \end{pmatrix}$

↖ Main diagonal ↖ Main diagonal ↖ Main diagonal

Identity Matrix

The $n \times n$ **identity matrix** I_n is the $n \times n$ matrix with 1's on the main diagonal and 0's everywhere else.

EXAMPLE 7 ***Two Identity Matrices***

$$I_3 = \begin{pmatrix} 1 & 0 & 0 \\ 0 & 1 & 0 \\ 0 & 0 & 1 \end{pmatrix} \quad \text{and} \quad I_5 = \begin{pmatrix} 1 & 0 & 0 & 0 & 0 \\ 0 & 1 & 0 & 0 & 0 \\ 0 & 0 & 1 & 0 & 0 \\ 0 & 0 & 0 & 1 & 0 \\ 0 & 0 & 0 & 0 & 1 \end{pmatrix}$$

It is not hard to show that I_n plays the role for $n \times n$ matrices that 1 plays for real numbers. Let A be an $n \times n$ matrix. Then the ijth component of $C = AI_n$ is, by (4) with $B = I_n$,

$$c_{ij} = a_{i1}\underset{=0}{b_{1j}} + a_{i2}\underset{=0}{b_{2j}} + \cdots + a_{ij}\underset{=1}{b_{jj}} + \cdots + a_{in}\underset{=0}{b_{nj}} = a_{ij} \tag{5}$$

In (5), the only nonzero term is $a_{ij}b_{jj} = a_{ij}$ (since $b_{jj} = 1$ and $b_{ij} = 0$ if $i \neq j$), so $c_{ij} = a_{ij}$. Thus $AI_n = A$. In a similar fashion, we can show that $I_nA = A$. In summary, we have

Multiplying a Square Matrix by the Identity Matrix

$$AI_n = I_nA = A \tag{6}$$

We will drop the subscript n since the size of the identity matrix we are using will always be obvious.

EXAMPLE 8 ***Multiplying a Square Matrix by the Identity Matrices***

We observe that

$$\begin{pmatrix} 2 & -3 & 1 \\ 4 & 0 & 6 \\ -2 & 4 & 5 \end{pmatrix}\begin{pmatrix} 1 & 0 & 0 \\ 0 & 1 & 0 \\ 0 & 0 & 1 \end{pmatrix} = \begin{pmatrix} 2 & -3 & 1 \\ 4 & 0 & 6 \\ -2 & 4 & 5 \end{pmatrix}$$

Matrices and Systems of Linear Equations

Any system of m linear equations in n unknowns can be written in the form $A\mathbf{x} = \mathbf{b}$, where A is an $m \times n$ matrix, $\mathbf{x}$ is an n-component column vector, and $\mathbf{b}$ is an m-component column vector.

EXAMPLE 9 ***Writing a System of Equations in Matrix Form***

Consider the system

$$\begin{aligned} 2x + 8y + 6z &= 20 \\ 4x + 2y - 2z &= -2 \\ 3x - y + z &= 11 \end{aligned}$$

(see Example 1 on p. 389). This can be written in the form $A\mathbf{x} = \mathbf{b}$, with

$$A = \begin{pmatrix} 2 & 8 & 6 \\ 4 & 2 & -2 \\ 3 & -1 & 1 \end{pmatrix}, \quad \mathbf{x} = \begin{pmatrix} x \\ y \\ z \end{pmatrix}, \quad \mathbf{b} = \begin{pmatrix} 20 \\ -2 \\ 11 \end{pmatrix}$$

That is, the system can be written as

$$\underset{A}{\begin{pmatrix} 2 & 8 & 6 \\ 4 & 2 & -2 \\ 3 & -1 & 1 \end{pmatrix}} \underset{\mathbf{x}}{\begin{pmatrix} x \\ y \\ z \end{pmatrix}} = \underset{\mathbf{b}}{\begin{pmatrix} 20 \\ -2 \\ 11 \end{pmatrix}}$$

Problems 8.2

Readiness Check

I. Which of the following is true of matrix multiplication of matrices A and B?
   a. It can be performed only if A and B are square matrices.
   b. Each entry c_{ij} is the product of a_{ij} and b_{ij}.
   c. $AB = BA$.
   d. It can be performed only if the number of columns of A is equal to the number of rows of B.

II. Which of the following would be the size of the product matrix AB when a 2×4 matrix A is multiplied by a 4×3 matrix B?
   a. 2×3 b. 3×2 c. 4×4
   d. This product cannot be found.

III. Which of the following is I_2?
   a. $\begin{pmatrix} 2 & 0 \\ 0 & 2 \end{pmatrix}$ b. $\begin{pmatrix} 1 & 0 \\ 0 & 1 \end{pmatrix}$
   c. $\begin{pmatrix} 0 & 1 \\ 1 & 0 \end{pmatrix}$ d. $\begin{pmatrix} 0 & 2 \\ 2 & 0 \end{pmatrix}$

IV. Which of the following is true of matrices A and B if AB is a column vector?
   a. B is a column vector.
   b. A is a row vector.
   c. A and B are square vectors.
   d. The number of rows in A must equal the number of columns in B.

V. Which of the following is true about a product AB if A is a 4×5 matrix?
   a. B must have 4 rows and the result will have 5 columns.
   b. B must have 5 columns and the result will be a square matrix.
   c. B must have 4 columns and the result will have 5 rows.
   d. B must have 5 rows and the result will have 4 rows.

Answers to Readiness Check

I. d II. a III. b IV. a V. d

In Problems 1–6 compute the dot product of the two vectors.

1. $(2 \quad 3), \begin{pmatrix} 4 \\ -2 \end{pmatrix}$
2. $(3 \quad -7), \begin{pmatrix} -4 \\ 0 \end{pmatrix}$
3. $(1 \quad 7 \quad 5), \begin{pmatrix} 2 \\ -3 \\ 4 \end{pmatrix}$
4. $(-3 \quad 1 \quad 7), \begin{pmatrix} -2 \\ 4 \\ 2 \end{pmatrix}$
5. $(2 \quad -3 \quad 1 \quad 4), \begin{pmatrix} 3 \\ 0 \\ 2 \\ 6 \end{pmatrix}$
6. $(-1 \quad 8 \quad 4 \quad 1), \begin{pmatrix} 5 \\ 0 \\ -2 \\ 4 \end{pmatrix}$
7. Let $\mathbf{a}$ be an n-vector. Show that $\mathbf{a} \cdot \mathbf{a} \geq 0$.
8. In Example 1 suppose that the demand vector is $\mathbf{d} = (25 \quad 45 \quad 20 \quad 30)$ and the price vector is $\mathbf{p} = \begin{pmatrix} \$12 \\ \$25 \\ \$16 \\ \$20 \end{pmatrix}$. If the manufacturer meets the demand, how much money will be received?
9. In Problem 8 let $\mathbf{c} = \begin{pmatrix} \$8 \\ \$15 \\ \$12 \\ \$14 \end{pmatrix}$ denote the costs of producing 1 unit of each of the 4 items. If the manufacturer meets the demand, what will be the cost of production?
10. In Problems 8 and 9 let **f** denote the profit vector—that is, **f** is a 4-component column vector showing the profit earned by selling 1 unit of each of the 4 items.
    (a) Write **f** in terms of **p** and **c**.
    (b) Write the total profit in terms of **f** and **d**.
    (c) Write the total profit in terms of **p**, **c**, and **d**.
    (d) Using the data of Problems 8 and 9, compute the total profit.

In Problems 11–25 perform the indicated computation, if possible.

11. $\begin{pmatrix} 3 & 4 \\ 4 & -1 \end{pmatrix}\begin{pmatrix} -2 & 1 \\ 0 & 3 \end{pmatrix}$
12. $\begin{pmatrix} 2 & 4 \\ 5 & 2 \end{pmatrix}\begin{pmatrix} 1 & -3 \\ 2 & 4 \end{pmatrix}$
13. $\begin{pmatrix} 1 & -1 \\ 1 & 1 \end{pmatrix}\begin{pmatrix} -1 & 0 \\ 2 & 3 \end{pmatrix}$
14. $\begin{pmatrix} 1 & -2 \\ 3 & 2 \\ -2 & 1 \end{pmatrix}\begin{pmatrix} 1 & 0 & 4 \\ 1 & 5 & -2 \end{pmatrix}$
15. $\begin{pmatrix} 2 & 4 & 3 \\ 1 & 0 & -2 \end{pmatrix}\begin{pmatrix} 1 & -2 \\ 4 & 3 \\ -3 & 0 \end{pmatrix}$
16. $\begin{pmatrix} 1 & 4 & -2 \\ 3 & -1 & 0 \end{pmatrix}\begin{pmatrix} 1 & 3 \\ 2 & -2 \\ 0 & 4 \end{pmatrix}$
17. $\begin{pmatrix} 1 & 3 & 7 \\ -3 & 2 & 0 \end{pmatrix}\begin{pmatrix} 4 & 2 \\ 1 & 5 \end{pmatrix}$
18. $\begin{pmatrix} 1 & 4 & -2 \\ 3 & 0 & 4 \end{pmatrix}\begin{pmatrix} 0 & 1 \\ 2 & 3 \end{pmatrix}$
19. $\begin{pmatrix} 1 & 2 & -2 \\ -3 & 0 & 2 \\ 2 & 3 & 1 \end{pmatrix}\begin{pmatrix} 3 & 0 & 1 \\ 2 & 1 & -3 \\ 4 & 0 & 2 \end{pmatrix}$
20. $\begin{pmatrix} 2 & -3 & 5 \\ 1 & 0 & 6 \\ 2 & 3 & 1 \end{pmatrix}\begin{pmatrix} 1 & 4 & 6 \\ -2 & 3 & 5 \\ 1 & 0 & 4 \end{pmatrix}$
21. $(1 \quad 4 \quad 0 \quad 2)\begin{pmatrix} 3 & -6 \\ 2 & 4 \\ 1 & 0 \\ -2 & 3 \end{pmatrix}$
22. $\begin{pmatrix} 3 & 2 & 1 & -2 \\ -6 & 4 & 0 & 3 \end{pmatrix}\begin{pmatrix} 1 \\ 4 \\ 0 \\ 2 \end{pmatrix}$
23. $\begin{pmatrix} 3 & -2 & 1 \\ 4 & 0 & 6 \\ 5 & 1 & 9 \end{pmatrix}\begin{pmatrix} 1 & 0 & 0 \\ 0 & 1 & 0 \\ 0 & 0 & 1 \end{pmatrix}$
24. $\begin{pmatrix} 1 & 0 & 0 \\ 0 & 1 & 0 \\ 0 & 0 & 1 \end{pmatrix}\begin{pmatrix} 3 & -2 & 1 \\ 4 & 0 & 6 \\ 5 & 1 & 9 \end{pmatrix}$
25. $\begin{pmatrix} a & b & c \\ d & e & f \\ g & h & j \end{pmatrix}\begin{pmatrix} 1 & 0 & 0 \\ 0 & 1 & 0 \\ 0 & 0 & 1 \end{pmatrix}$, where $a, b, c, d, e, f, g, h, j$ are real numbers.
26. Find a matrix $A = \begin{pmatrix} a & b \\ c & d \end{pmatrix}$ such that $A\begin{pmatrix} 2 & 3 \\ 1 & 2 \end{pmatrix} = \begin{pmatrix} 1 & 0 \\ 0 & 1 \end{pmatrix}$.
27. As in Example 5, suppose that two people have contracted a contagious disease. These people have contacts with a second group, who in turn have contacts with a third group. Let $A = \begin{pmatrix} 1 & 1 & 0 & 0 & 1 \\ 0 & 1 & 1 & 1 & 0 \end{pmatrix}$ represent the contacts

between the contagious group and the members of group 2, and let

$$B = \begin{pmatrix} 1 & 1 & 0 & 1 \\ 0 & 1 & 0 & 1 \\ 0 & 0 & 1 & 0 \\ 0 & 0 & 1 & 1 \\ 0 & 1 & 1 & 0 \end{pmatrix}$$

represent the contacts between groups 2 and 3.
(a) How many people are in group 2?
(b) How many are in group 3?
(c) Find the matrix of second-order contacts between groups 1 and 3.

28. Answer the questions of Problem 27 for $A = \begin{pmatrix} 1 & 1 & 1 & 1 & 1 & 0 \\ 0 & 0 & 1 & 1 & 0 & 1 \end{pmatrix}$ and

$$B = \begin{pmatrix} 0 & 0 & 1 & 0 & 0 \\ 0 & 1 & 1 & 0 & 1 \\ 1 & 0 & 1 & 0 & 1 \\ 0 & 1 & 1 & 1 & 0 \\ 0 & 0 & 0 & 0 & 0 \\ 1 & 0 & 1 & 1 & 0 \end{pmatrix}$$

29. An investor plans to buy 100 shares of telephone stock, 200 shares of oil stock, 400 shares of automobile stock, and 100 shares of airline stock. The telephone stock is selling for $46 a share, the oil stock for $34 a share, the automobile stock for $15 a share, and the airline stock for $10 a share.
(a) Express the numbers of shares as a row vector.
(b) Express the prices of the stocks as a column vector.
(c) Use matrix multiplication to compute the total cost of the investor's purchases.

30. A manufacturer of custom-designed jewelry has orders for 2 rings, 3 pairs of earrings, 5 pins, and 1 necklace. The manufacturer estimates that it takes 1 hour of labor to make a ring, $1\frac{1}{2}$ hours to make a pair of earrings, $\frac{1}{2}$ hour for each pin, and 2 hours to make a necklace.
(a) Express the manufacturer's orders as a row vector.
(b) Express the labor-hour requirements for the various types of jewelry as a column vector.
(c) Use matrix multiplication to calculate the total number of hours of labor it will require to complete all the orders.

31. A company pays its executives a salary and gives them shares of its stock as an annual bonus. Last year, the president of the company received $80,000 and 50 shares of stock, each of the three vice presidents was paid $45,000 and 20 shares of stock, and the treasurer was paid $40,000 and 10 shares of stock.
(a) Express the payments to the executives in money and stock by a 2×3 matrix.
(b) Express the number of executives of each rank by means of a column vector.
(c) Use matrix multiplication to calculate the total amount of money and the total number of shares of stock the company paid these executives last year.

32. A tourist returns from a European trip with the following foreign currency: 1000 Austrian schillings, 20 British pounds, 100 French francs, 5000 Italian lire, and 50 German marks. In American money, a schilling was worth $0.055, the pound $1.80, the franc $.20, the lira $0.001, and the mark $0.40.
(a) Express the quantity of each currency by a row vector.
(b) Express the value of each currency in American money by a column vector.
(c) Use matrix multiplication to compute how much the tourist's foreign currency was worth in American money.

33. A family consists of 2 adults, 1 teenager, and 3 young children. Each adult consumes $\frac{1}{5}$ loaf of bread, no milk, $\frac{1}{10}$ pound of coffee, and $\frac{1}{8}$ pound of cheese in an average day. The teenager eats $\frac{2}{5}$ loaf of bread, drinks 1 quart of milk, but no coffee, and eats $\frac{1}{8}$ pound of cheese. Each child eats $\frac{1}{5}$ loaf of bread, drinks $\frac{1}{2}$ quart of milk, but no coffee, and eats $\frac{1}{16}$ pound of cheese.
(a) Express the daily consumption of bread, milk, coffee, and cheese by the various types of family members using a matrix.
(b) Express the number of family members of the various types by a column vector.
(c) Use matrix multiplication to calculate the total amount of bread, milk, coffee, and cheese consumed by this family in an average day.

34. Sales, unit gross profits, and unit taxes for sales of a large corporation are given in the table below.

	Product					
	Sales				**Profit (in hundreds of dollars)**	**Taxes (in hundreds of dollars)**
Mo	***I***	***II***	***III***	**Item**		
Jan	4	2	20	I	3.5	1.5
Feb	6	1	9	II	2.75	2
Mar	5	3	12	III	1.5	0.6
Apr	8	2.5	20			

Find a matrix that shows total profits and taxes in each of the 4 months.

In Problems 35–40 write the given system in the form $A\mathbf{x} = \mathbf{b}$.

35. $\begin{aligned} 2x - y &= 3 \\ 4x + 5y &= 7 \end{aligned}$

36. $\begin{aligned} 3x - y + 2z &= 4 \\ 2x + 3y + 5z &= 0 \\ -3x + 4y - 10z &= -6 \end{aligned}$

37. $\begin{aligned} 3x + 6y - 7z &= 0 \\ 2x - y + 3z &= 1 \end{aligned}$

38. $\begin{aligned} 4x - y + z - w &= -7 \\ 3x + y - 5z + 6w &= 8 \\ 2x - y + z &= 9 \end{aligned}$

39. $\begin{aligned} 2x + 4z &= 9 \\ 3x + 2y &= 6 \\ -4y - 9z &= 3 \end{aligned}$

40. $\begin{aligned} 2x + 3y - z &= 0 \\ -4x + 2y + z &= 0 \\ 7x + 3y - 9z &= 0 \end{aligned}$

* 41. Let A, B, and C be 2×2 matrices. Show that

$$A(BC) = (AB)C$$

This is called the **associative law of matrix multiplication.** It holds whenever the products AB and BC are defined.

8.3 The Inverse of a Square Matrix

In this section, we define a kind of matrix central to matrix theory. We begin with a simple example. Let $A = \begin{pmatrix} 2 & 5 \\ 1 & 3 \end{pmatrix}$ and $B = \begin{pmatrix} 3 & -5 \\ -1 & 2 \end{pmatrix}$. It is not difficult to show that $AB = BA = I$, where $I = \begin{pmatrix} 1 & 0 \\ 0 & 1 \end{pmatrix}$ is the 2×2 identity matrix. The matrix B is called the *inverse* of A and is written A^{-1}. In general, we have the following.

Definition of the Inverse of a Matrix

Let A and B be square, $n \times n$ matrices and let I denote the $n \times n$ identity matrix. If

$$AB = BA = I \tag{1}$$

then B is called the **inverse** of A and is written as A^{-1}. We have

$$AA^{-1} = A^{-1}A = I \tag{2}$$

If A has an inverse, then A is said to be **invertible.**

NOTE 1 From this definition it follows that $(A^{-1})^{-1} = A$ if A is invertible because A is a matrix with the property that $A^{-1}A = AA^{-1} = I$.

NOTE 2 This definition does *not* state that every square matrix has an inverse. In fact, there are many square matrices that have no inverse. (See, for instance, Example 2 on p. 459.)

In the computation done above, we see that

$$\begin{pmatrix} 2 & 5 \\ 1 & 3 \end{pmatrix}^{-1} = \begin{pmatrix} 3 & -5 \\ -1 & 2 \end{pmatrix} \quad \text{and} \quad \begin{pmatrix} 3 & -5 \\ -1 & 2 \end{pmatrix}^{-1} = \begin{pmatrix} 2 & 5 \\ 1 & 3 \end{pmatrix}$$

Consider the system $$A\mathbf{x} = \mathbf{b}$$

and suppose that A is invertible. Then

$$\begin{aligned} A^{-1}A\mathbf{x} &= A^{-1}\mathbf{b} && \text{We multiplied on the left by } A^{-1} \\ I\mathbf{x} &= A^{-1}\mathbf{b} && A^{-1}A = I \\ \mathbf{x} &= A^{-1}\mathbf{b} && I\mathbf{x} = \mathbf{x} \end{aligned}$$

That is,

An Important Fact

If A is invertible, the system $A\mathbf{x} = \mathbf{b}$ has the unique solution $\mathbf{x} = A^{-1}\mathbf{b}$.

There are three basic questions that come to mind once we have defined the inverse of a matrix.

Question 1. Can a matrix have more than one inverse?
Question 2. What matrices do have inverses?
Question 3. If a matrix has an inverse, how can we compute it?

We answer all three questions in this section. The first one is the easiest. Suppose that B and C are two inverses for A. We can show that $B = C$. By equation (1) we have $AB = BA = I$ and $AC = CA = I$. Then $B(AC) = BI = B$ and $(BA)C = IC = C$. But $B(AC) = (BA)C$ by the associative law of matrix multiplication (see Problem 41 in Section 8.2). Hence $B = C$, and this means that A can have, at most, one inverse.

The other two questions are more difficult to answer. Rather than starting by giving you what seem to be a set of arbitrary rules, we first look at what happens in the 2×2 case.

EXAMPLE 1 ***Computing the Inverse of a 2 × 2 Matrix***

Let $A = \begin{pmatrix} 2 & -3 \\ -4 & 5 \end{pmatrix}$. Compute A^{-1} if it exists.

SOLUTION Suppose that A^{-1} exists. We write $A^{-1} = \begin{pmatrix} x & y \\ z & w \end{pmatrix}$ and use the fact that $AA^{-1} = I$. Then

$$AA^{-1} = \begin{pmatrix} 2 & -3 \\ -4 & 5 \end{pmatrix}\begin{pmatrix} x & y \\ z & w \end{pmatrix} = \begin{pmatrix} 2x - 3z & 2y - 3w \\ -4x + 5z & -4y + 5w \end{pmatrix} = \begin{pmatrix} 1 & 0 \\ 0 & 1 \end{pmatrix}$$

The last two matrices can be equal only if each of their corresponding components is equal. This means that

$$\begin{aligned} 2x \qquad - 3z \qquad &= 1 && (3) \\ 2y \qquad - 3w &= 0 && (4) \\ -4x \qquad + 5z \qquad &= 0 && (5) \\ - 4y \qquad + 5w &= 1 && (6) \end{aligned}$$

This is a system of four equations in four unknowns. Note that there are exactly two equations involving x and z only (equations (3) and (5)) and exactly two equations involving y and w only (equations (4) and (6)). We write these two systems in augmented matrix form.

$$\begin{array}{cc} x & z \end{array}$$
$$\left(\begin{array}{cc|c} 2 & -3 & 1 \\ -4 & 5 & 0 \end{array}\right) \tag{7}$$

$$\begin{array}{cc} y & w \end{array}$$
$$\left(\begin{array}{cc|c} 2 & -3 & 0 \\ -4 & 5 & 1 \end{array}\right) \tag{8}$$

Then row reduction applied to (7) results in

$$\left(\begin{array}{cc|c} 2 & -3 & 1 \\ -4 & 5 & 0 \end{array}\right) \xrightarrow{R_1 \to \frac{1}{2}R_1} \left(\begin{array}{cc|c} 1 & -\frac{3}{2} & \frac{1}{2} \\ -4 & 5 & 0 \end{array}\right) \xrightarrow{R_2 \to R_2 + 4R_1} \left(\begin{array}{cc|c} 1 & -\frac{3}{2} & \frac{1}{2} \\ 0 & -1 & 2 \end{array}\right)$$

$$\xrightarrow{R_2 \to -R_2} \left(\begin{array}{cc|c} 1 & -\frac{3}{2} & \frac{1}{2} \\ 0 & 1 & -2 \end{array}\right) \xrightarrow{R_1 \to R_1 + \frac{3}{2}R_2} \left(\begin{array}{cc|c} 1 & 0 & -\frac{5}{2} \\ 0 & 1 & -2 \end{array}\right)$$

Thus $x = -\frac{5}{2}$ and $z = -2$.
Similarly, we now reduce system (8) to obtain

$$\left(\begin{array}{cc|c} 2 & -3 & 0 \\ -4 & 5 & 1 \end{array}\right) \xrightarrow{R_1 \to \frac{1}{2}R_1} \left(\begin{array}{cc|c} 1 & -\frac{3}{2} & 0 \\ -4 & 5 & 1 \end{array}\right) \xrightarrow{R_2 \to R_2 + 4R_1} \left(\begin{array}{cc|c} 1 & -\frac{3}{2} & 0 \\ 0 & -1 & 1 \end{array}\right)$$

$$\xrightarrow{R_2 \to -R_2} \left(\begin{array}{cc|c} 1 & -\frac{3}{2} & 0 \\ 0 & 1 & -1 \end{array}\right) \xrightarrow{R_1 \to R_1 + \frac{3}{2}R_2} \left(\begin{array}{cc|c} 1 & 0 & -\frac{3}{2} \\ 0 & 1 & -1 \end{array}\right)$$

So $y = -\frac{3}{2}$ and $w = -1$.

Then $A^{-1} = \begin{pmatrix} x & y \\ z & w \end{pmatrix} = \begin{pmatrix} -\frac{5}{2} & -\frac{3}{2} \\ -2 & -1 \end{pmatrix}$.

Check

$$A^{-1}A = \begin{pmatrix} -\frac{5}{2} & -\frac{3}{2} \\ -2 & -1 \end{pmatrix}\begin{pmatrix} 2 & -3 \\ -4 & 5 \end{pmatrix} = \begin{pmatrix} 1 & 0 \\ 0 & 1 \end{pmatrix}$$

and

$$AA^{-1} = \begin{pmatrix} 2 & -3 \\ -4 & 5 \end{pmatrix}\begin{pmatrix} -\frac{5}{2} & -\frac{3}{2} \\ -2 & -1 \end{pmatrix} = \begin{pmatrix} 1 & 0 \\ 0 & 1 \end{pmatrix}$$

It is not necessary to solve systems (7) and (8) separately. Since the coefficient matrices in (7) and (8) are the same, we can perform the row reductions on the two augmented matrices simultaneously by considering the new augmented matrix

$$\left(\begin{array}{cc|cc} 2 & -3 & 1 & 0 \\ -4 & 5 & 0 & 1 \end{array}\right) \tag{9}$$

If A is invertible, then the system defined by (3), (4), (5), and (6) has a unique solution and, by what we said above, row reduction will result in

$$\left(\begin{array}{cc|cc} 1 & 0 & x & y \\ 0 & 1 & z & w \end{array}\right)$$

We now carry out the computation again, noting that the matrix on the left in (9) is A and the matrix on the right in (9) is I:

$$\left(\begin{array}{cc|cc} 2 & -3 & 1 & 0 \\ -4 & 5 & 0 & 1 \end{array}\right) \xrightarrow{R_1 \to \frac{1}{2}R_1} \left(\begin{array}{cc|cc} 1 & -\frac{3}{2} & \frac{1}{2} & 0 \\ -4 & 5 & 0 & 1 \end{array}\right)$$

$$\xrightarrow{R_2 \to R_2 + 4R_1} \left(\begin{array}{cc|cc} 1 & -\frac{3}{2} & \frac{1}{2} & 0 \\ 0 & -1 & 2 & 1 \end{array}\right)$$

$$\xrightarrow{R_2 \to -R_2} \left(\begin{array}{cc|cc} 1 & -\frac{3}{2} & \frac{1}{2} & 0 \\ 0 & 1 & -2 & -1 \end{array}\right)$$

$$\xrightarrow{R_1 \to R_1 + \frac{3}{2}R_2} \left(\begin{array}{cc|cc} 1 & 0 & -\frac{5}{2} & -\frac{3}{2} \\ 0 & 1 & -2 & -1 \end{array}\right)$$

Thus $x = -\frac{5}{2}$, $y = -\frac{3}{2}$, $z = -2$, $w = -1$, and $A^{-1} = \begin{pmatrix} -\frac{5}{2} & -\frac{3}{2} \\ -2 & -1 \end{pmatrix}$.

NOTE It is possible to prove that if A and B are square, $n \times n$ matrices such that $AB = I$, then $BA = I$ also. Thus it is not necessary to verify both that $AA^{-1} = I$ and $A^{-1}A = I$, as we did above. Verifying only one of these is enough. ■

EXAMPLE 2 *A 2 × 2 Matrix That Is Not Invertible*

Let $A = \begin{pmatrix} 1 & 2 \\ -2 & -4 \end{pmatrix}$. Calculate A^{-1} if it exists.

SOLUTION If $A^{-1} = \begin{pmatrix} x & y \\ z & w \end{pmatrix}$ exists, then

$$AA^{-1} = \begin{pmatrix} 1 & 2 \\ -2 & -4 \end{pmatrix}\begin{pmatrix} x & y \\ z & w \end{pmatrix} = \begin{pmatrix} x + 2z & y + 2w \\ -2x - 4z & -2y - 4w \end{pmatrix} = \begin{pmatrix} 1 & 0 \\ 0 & 1 \end{pmatrix}$$

This leads to the system

$$\begin{aligned} x \quad\quad + 2z \quad\quad &= 1 \\ y \quad\quad + 2w &= 0 \\ -2x \quad\quad - 4z \quad\quad &= 0 \\ - 2y \quad\quad - 4w &= 1 \end{aligned} \tag{10}$$

Using the same reasoning as in Example 1, we can write this system in the augmented matrix form $(A|I)$ and row-reduce.

$$\left(\begin{array}{cc|cc} 1 & 2 & 1 & 0 \\ -2 & -4 & 0 & 1 \end{array}\right) \xrightarrow{R_2 \to R_2 + 2R_1} \left(\begin{array}{cc|cc} 1 & 2 & 1 & 0 \\ 0 & 0 & 2 & 1 \end{array}\right)$$

This is as far as we can go. The last line reads $0 = 2$ or $0 = 1$, depending on which of the two systems of equations (in x and z or in y and w) is being solved. Thus, system (10) is inconsistent, and A is not invertible.

The last two examples illustrate a procedure that always works when you are trying to find the inverse of a matrix.

Procedure for Computing the Inverse of a Square Matrix *A*

Step 1 Write the augmented matrix $(A|I)$.

Step 2 Use row reduction to reduce the matrix A to the identity, if possible.

Step 3 Decide if A is invertible.

(a) If A can be reduced to the identity matrix I, then A^{-1} will be the matrix to the right of the vertical bar.

(b) If the row reduction of A leads to a row of zeros to the left of the vertical bar, then A is not invertible.

EXAMPLE 3 ***Calculating the Inverse of the General 2 × 2 Matrix***

Let $A = \begin{pmatrix} a & b \\ c & d \end{pmatrix}$. Show that if $ad - bc \neq 0$, then

$$A^{-1} = \begin{pmatrix} \dfrac{d}{ad - bc} & \dfrac{-b}{ad - bc} \\ \dfrac{-c}{ad - bc} & \dfrac{a}{ad - bc} \end{pmatrix}$$

SOLUTION

$$\begin{pmatrix} a & b \\ c & d \end{pmatrix} \begin{pmatrix} \dfrac{d}{ad - bc} & \dfrac{-b}{ad - bc} \\ \dfrac{-c}{ad - bc} & \dfrac{a}{ad - bc} \end{pmatrix} = \begin{pmatrix} \dfrac{ad - bc}{ad - bc} & \dfrac{-ab + ab}{ad - bc} \\ \dfrac{cd - cd}{ad - bc} & \dfrac{ad - bc}{ad - bc} \end{pmatrix} = \begin{pmatrix} 1 & 0 \\ 0 & 1 \end{pmatrix}$$

Similarly,

$$\begin{pmatrix} \dfrac{d}{ad - bc} & \dfrac{-b}{ad - bc} \\ \dfrac{-c}{ad - bc} & \dfrac{a}{ad - bc} \end{pmatrix} \begin{pmatrix} a & b \\ c & d \end{pmatrix} = \begin{pmatrix} 1 & 0 \\ 0 & 1 \end{pmatrix}$$

NOTE As we stated on p. 459, it is not necessary to verify both of these equations. One will do.

In Problem 19 on p. 464, you are asked to show that if $ad - bc = 0$, then A is not invertible because the row-echelon form of A has a row of zeros.

We summarize the result of the last example and Problem 19.

Procedure for Computing the Inverse of a 2 × 2 Matrix

Let $A = \begin{pmatrix} a & b \\ c & d \end{pmatrix}$.

1. A is invertible if and only if $ad - bc \neq 0$.
2. If $ad - bc \neq 0$, then

$$A^{-1} = \frac{1}{ad - bc}\begin{pmatrix} d & -b \\ -c & a \end{pmatrix} \tag{11}$$

The quantity $ad - bc$ is called the **determinant** of A and is abbreviated by det A. We will discuss determinants further in the next section.

EXAMPLE 4 *The Inverse of a 2 × 2 Matrix*

Let $A = \begin{pmatrix} 6 & -7 \\ 2 & 1 \end{pmatrix}$. Compute A^{-1} if it exists.

SOLUTION $ad - bc = (6)(1) - (-7)(2) = 6 + 14 = 20 \neq 0$. Thus, A^{-1} exists and

$$A^{-1} = \frac{1}{20}\begin{pmatrix} 1 & 7 \\ -2 & 6 \end{pmatrix}$$

Check

$$A^{-1}A = \frac{1}{20}\begin{pmatrix} 1 & 7 \\ -2 & 6 \end{pmatrix}\begin{pmatrix} 6 & -7 \\ 2 & 1 \end{pmatrix} = \frac{1}{20}\begin{pmatrix} 20 & 0 \\ 0 & 20 \end{pmatrix} = \begin{pmatrix} 1 & 0 \\ 0 & 1 \end{pmatrix}$$

As we noted in Examples 1 and 3, this proves that $AA^{-1} = I$ as well. ■

EXAMPLE 5 *The Inverse of a 3 × 3 Matrix*

Let $A = \begin{pmatrix} 2 & 8 & 6 \\ 4 & 2 & -2 \\ 3 & -1 & 1 \end{pmatrix}$ (see Example 1 in Section 7.2). Calculate A^{-1} if it exists.

SOLUTION We first put I next to A in an augmented matrix form

$$\left(\begin{array}{rrr|rrr} 2 & 8 & 6 & 1 & 0 & 0 \\ 4 & 2 & -2 & 0 & 1 & 0 \\ 3 & -1 & 1 & 0 & 0 & 1 \end{array}\right)$$

and then carry out the row reduction.

$$\xrightarrow{R_1 \to \frac{1}{2}R_1} \left(\begin{array}{ccc|ccc} 1 & 4 & 3 & \frac{1}{2} & 0 & 0 \\ 4 & 2 & -2 & 0 & 1 & 0 \\ 3 & -1 & 1 & 0 & 0 & 1 \end{array}\right)$$

$$\xrightarrow[R_3 \to R_3 - 3R_1]{R_2 \to R_2 - 4R_1} \left(\begin{array}{ccc|ccc} 1 & 4 & 3 & \frac{1}{2} & 0 & 0 \\ 0 & -14 & -14 & -2 & 1 & 0 \\ 0 & -13 & -8 & -\frac{3}{2} & 0 & 1 \end{array}\right)$$

$$\xrightarrow{R_2 \to -\frac{1}{14}R_2} \left(\begin{array}{ccc|ccc} 1 & 4 & 3 & \frac{1}{2} & 0 & 0 \\ 0 & 1 & 1 & \frac{2}{14} & -\frac{1}{14} & 0 \\ 0 & -13 & -8 & -\frac{3}{2} & 0 & 1 \end{array}\right)$$

$$\xrightarrow[R_3 \to R_3 + 13R_2]{R_1 \to R_1 - 4R_2} \left(\begin{array}{ccc|ccc} 1 & 0 & -1 & -\frac{1}{14} & \frac{4}{14} & 0 \\ 0 & 1 & 1 & \frac{2}{14} & -\frac{1}{14} & 0 \\ 0 & 0 & 5 & \frac{5}{14} & -\frac{13}{14} & 1 \end{array}\right)$$

$$\xrightarrow{R_3 \to \frac{1}{5}R_3} \left(\begin{array}{ccc|ccc} 1 & 0 & -1 & -\frac{1}{14} & \frac{4}{14} & 0 \\ 0 & 1 & 1 & \frac{2}{14} & -\frac{1}{14} & 0 \\ 0 & 0 & 1 & \frac{1}{14} & -\frac{13}{70} & \frac{1}{5} \end{array}\right)$$

$$\xrightarrow[R_2 \to R_2 - R_3]{R_1 \to R_1 + R_3} \left(\begin{array}{ccc|ccc} 1 & 0 & 0 & 0 & \frac{1}{10} & \frac{1}{5} \\ 0 & 1 & 0 & \frac{1}{14} & \frac{8}{70} & -\frac{1}{5} \\ 0 & 0 & 1 & \frac{1}{14} & -\frac{13}{70} & \frac{1}{5} \end{array}\right)$$

Since A has now been reduced to I, we have

$$A^{-1} = \begin{pmatrix} 0 & \frac{1}{10} & \frac{1}{5} \\ \frac{1}{14} & \frac{8}{70} & -\frac{1}{5} \\ \frac{1}{14} & -\frac{13}{70} & \frac{1}{5} \end{pmatrix} = \frac{1}{70}\begin{pmatrix} 0 & 7 & 14 \\ 5 & 8 & -14 \\ 5 & -13 & 14 \end{pmatrix}$$

Check

$$A^{-1}A = \frac{1}{70}\begin{pmatrix} 0 & 7 & 14 \\ 5 & 8 & -14 \\ 5 & -13 & 14 \end{pmatrix}\begin{pmatrix} 2 & 8 & 6 \\ 4 & 2 & -2 \\ 3 & -1 & 1 \end{pmatrix} = \frac{1}{70}\begin{pmatrix} 70 & 0 & 0 \\ 0 & 70 & 0 \\ 0 & 0 & 70 \end{pmatrix} = I$$

WARNING It is easy to make numerical errors in computing A^{-1}. Therefore, it is essential to check the computations by verifying that $A^{-1}A = I$. ■

EXAMPLE 6 *A 3 × 3 Matrix That Is Not Invertible*

Let $A = \begin{pmatrix} 1 & -3 & 4 \\ 2 & -5 & 7 \\ 0 & -1 & 1 \end{pmatrix}$. Calculate A^{-1} if it exists.

SOLUTION Proceeding as before we obtain, successively,

$$\left(\begin{array}{ccc|ccc}1 & -3 & 4 & 1 & 0 & 0\\ 2 & -5 & 7 & 0 & 1 & 0\\ 0 & -1 & 1 & 0 & 0 & 1\end{array}\right) \xrightarrow{R_2 \to R_2 - 2R_1} \left(\begin{array}{ccc|ccc}1 & -3 & 4 & 1 & 0 & 0\\ 0 & 1 & -1 & -2 & 1 & 0\\ 0 & -1 & 1 & 0 & 0 & 1\end{array}\right)$$

$$\xrightarrow[R_3 \to R_3 + R_2]{R_1 \to R_1 + 3R_2} \left(\begin{array}{ccc|ccc}1 & 0 & 1 & -5 & 3 & 0\\ 0 & 1 & -1 & -2 & 1 & 0\\ 0 & 0 & 0 & -2 & 1 & 1\end{array}\right)$$

This is as far as we can go. The matrix A *cannot* be reduced to the identity matrix, and we can conclude that A is *not* invertible. ■

EXAMPLE 7 *Using the Inverse to Solve a System of Equations*

Solve the system

$$\begin{aligned} 2x + 8y + 6z &= 20\\ 4x + 2y - 2z &= -2\\ 3x - y + z &= 11 \end{aligned}$$

SOLUTION We solved this problem in Example 1 in Section 7.2. Now we solve it by using inverses. We first write the system as $A\mathbf{x} = \mathbf{b}$, where

$$A = \begin{pmatrix}2 & 8 & 6\\ 4 & 2 & -2\\ 3 & -1 & 1\end{pmatrix}, \quad \mathbf{x} = \begin{pmatrix}x\\ y\\ z\end{pmatrix}, \quad \text{and} \quad \mathbf{b} = \begin{pmatrix}20\\ -2\\ 11\end{pmatrix}$$

We now proceed as on p. 457

$$\begin{aligned} A\mathbf{x} &= \mathbf{b}\\ A^{-1}A\mathbf{x} &= A^{-1}\mathbf{b}\\ I\mathbf{x} &= A^{-1}\mathbf{b}\\ \mathbf{x} &= A^{-1}\mathbf{b} \end{aligned}$$

Multiply both sides on the left by A^{-1}, which exists by the result of Example 5

But as we found in Example 5,

$$A^{-1} = \frac{1}{70}\begin{pmatrix}0 & 7 & 14\\ 5 & 8 & -14\\ 5 & -13 & 14\end{pmatrix}$$

Then, the solution is given by

$$\mathbf{x} = \begin{pmatrix}x\\ y\\ z\end{pmatrix} = A^{-1}\mathbf{b} = \frac{1}{70}\begin{pmatrix}0 & 7 & 14\\ 5 & 8 & -14\\ 5 & -13 & 14\end{pmatrix}\begin{pmatrix}20\\ -2\\ 11\end{pmatrix}$$

$$= \frac{1}{70}\begin{pmatrix}140\\ -70\\ 280\end{pmatrix} = \begin{pmatrix}2\\ -1\\ 4\end{pmatrix}$$

so $x = 2$, $y = -1$, and $z = 4$, as we found before.

Problems 8.3

Readiness Check

I. Which of the following is true?
  a. Every square matrix has an inverse.
  b. A square matrix has an inverse if its row reduction leads to a row of zeros.
  c. A square matrix is invertible if it has an inverse.
  d. A square matrix B is the inverse of A if $AI = B$.

II. Which of the following is true of a system of equations in matrix form?
  a. It is of the form $A^{-1}\mathbf{x} = \mathbf{b}$.
  b. If it has a unique solution, the solution will be $\mathbf{x} = A^{-1}\mathbf{b}$.
  c. It will have a solution if A is not invertible.
  d. It will have a unique solution.

III. Which of the following is invertible?
  a. $\begin{pmatrix} 1 & 3 \\ -3 & -9 \end{pmatrix}$ b. $\begin{pmatrix} 6 & -1 \\ 1 & -\frac{1}{6} \end{pmatrix}$
  c. $\begin{pmatrix} 2 & -3 \\ 1 & -1 \end{pmatrix}$ d. $\begin{pmatrix} 1 & 0 \\ 2 & 0 \end{pmatrix}$

IV. Which of the following is true of an invertible matrix A?
  a. The product of A and I is A^{-1}.
  b. A is a 2×3 matrix.
  c. $A = A^{-1}$.
  d. A is a square matrix.

V. Which of the following is true of the system

$$4x - 5y = 3$$
$$6x - 7y = 4?$$

  a. It has no solution because $\begin{pmatrix} 4 & -5 \\ 6 & -7 \end{pmatrix}$ is not invertible.
  b. It has the solution $(-1, -\frac{1}{2})$.
  c. If it had a solution, it would be found by solving $\begin{pmatrix} 4 & -5 \\ 6 & -7 \end{pmatrix}\begin{pmatrix} x \\ y \end{pmatrix} = \begin{pmatrix} 3 \\ 4 \end{pmatrix}$.
  d. Its solution is $\begin{pmatrix} 4 & -5 \\ 6 & -7 \end{pmatrix}\begin{pmatrix} 3 \\ 4 \end{pmatrix}$.

In Problems 1–15 calculate the inverse, if it exists.

1. $\begin{pmatrix} 1 & 3 \\ 2 & 5 \end{pmatrix}$
2. $\begin{pmatrix} -1 & 6 \\ 2 & -12 \end{pmatrix}$
3. $\begin{pmatrix} 0 & 1 \\ 1 & 0 \end{pmatrix}$
4. $\begin{pmatrix} 4 & -8 \\ -3 & 6 \end{pmatrix}$
5. $\begin{pmatrix} a & a \\ b & b \end{pmatrix}$
6. $\begin{pmatrix} 1 & 1 & 1 \\ 0 & 3 & 2 \\ 3 & 1 & 4 \end{pmatrix}$
7. $\begin{pmatrix} 1 & 2 & 3 \\ 2 & 1 & -1 \\ 3 & 2 & 3 \end{pmatrix}$
8. $\begin{pmatrix} 1 & 1 & 1 \\ 0 & 1 & 1 \\ 0 & 0 & 1 \end{pmatrix}$
9. $\begin{pmatrix} 1 & 3 & 0 \\ 4 & 2 & 3 \\ 6 & 8 & 3 \end{pmatrix}$
10. $\begin{pmatrix} 3 & 1 & 0 \\ 1 & -1 & 2 \\ 1 & 1 & 1 \end{pmatrix}$
11. $\begin{pmatrix} 2 & 1 & 0 \\ 0 & 1 & 3 \\ -2 & 3 & 2 \end{pmatrix}$
12. $\begin{pmatrix} 1 & 2 & 3 \\ 1 & 1 & 2 \\ 0 & 1 & 2 \end{pmatrix}$
13. $\begin{pmatrix} 1 & 0 & 1 & 2 \\ 0 & 1 & -1 & 0 \\ 2 & 1 & 2 & 1 \\ 1 & 2 & 1 & 0 \end{pmatrix}$
14. $\begin{pmatrix} 1 & 0 & 2 & 3 \\ -1 & 1 & 0 & 4 \\ 2 & 1 & -1 & 3 \\ -1 & 0 & 5 & 7 \end{pmatrix}$
15. $\begin{pmatrix} -1 & 1 & 2 & 3 \\ 3 & 2 & 1 & 0 \\ 1 & 1 & -1 & 1 \\ 3 & 4 & 2 & 4 \end{pmatrix}$
16. Show that if A and B are invertible matrices, then AB is invertible and $(AB)^{-1} = B^{-1}A^{-1}$.
17. Show that the matrix $\begin{pmatrix} 3 & 4 \\ -2 & -3 \end{pmatrix}$ is equal to its own inverse.
18. Let $A = \begin{pmatrix} 0 & b \\ c & d \end{pmatrix}$
    (a) Show that A^{-1} exists if and only if $bc \neq 0$.
    (b) If $bc \neq 0$, show that $A^{-1} = -\frac{1}{bc}\begin{pmatrix} d & -b \\ -c & 0 \end{pmatrix}$
* 19. Suppose that $ad - bc = 0$. Show that the row-echelon form of $A = \begin{pmatrix} a & b \\ c & d \end{pmatrix}$ has a row of zeros so that A is not invertible.

Answers to Readiness Check

I. c II. b III. c IV. d V. c

8.4 Determinants and Cramer's Rule

Consider the 2×2 system
$$\begin{aligned} ax + by &= e \\ cx + dy &= f \end{aligned} \tag{1}$$

We multiply the first equation by d and the second by $-b$. Then we add:

$$\begin{array}{r} adx + bdy = ed \\ -bcx - bdy = -bf \\ \hline (ad - bc)x + 0 = ed - bf \end{array}$$

If $ad - bc \neq 0$, then we can divide both sides of the equation by $ad - bc$ to obtain
$$x = \frac{ed - bf}{ad - bc}$$

We can use the first or second equation in (1) to find y:

$$y = \frac{e - ax}{b} \quad \text{if} \quad b \neq 0 \qquad \text{or} \qquad y = \frac{f - cx}{d} \quad \text{if} \quad d \neq 0$$

That is, if $ad - bc \neq 0$, then system (1) has a unique solution. In Problem 22 on p. 473 you are asked to show that if $ad - bc = 0$, then (1) has either no solution or an infinite number of solutions.

The quantity $ad - bc$ has a special name. First we note that system (1) can be written in the form
$$\begin{pmatrix} a & b \\ c & d \end{pmatrix}\begin{pmatrix} x \\ y \end{pmatrix} = \begin{pmatrix} e \\ f \end{pmatrix}$$

Then we define the **determinant of A,** denoted by $\det A$, by

Definition of the Determinant of a 2×2 Matrix A

$$\det A = ad - bc \tag{2}$$

We will often denote $\det A$ by
$$|A| = \begin{vmatrix} a & b \\ c & d \end{vmatrix}$$

EXAMPLE 1 ***Computing Three 2×2 Determinants***

Compute the determinant of each matrix.

(a) $A = \begin{pmatrix} 2 & 3 \\ -1 & 1 \end{pmatrix}$ (b) $B = \begin{pmatrix} 1 & 6 \\ 2 & 5 \end{pmatrix}$ (c) $C = \begin{pmatrix} 2 & -4 \\ 1 & -2 \end{pmatrix}$

SOLUTION

(a) $\det A = \begin{vmatrix} 2 & 3 \\ -1 & 1 \end{vmatrix} = 2 \cdot 1 - 3(-1) = 2 + 3 = 5$

(b) $\det B = \begin{vmatrix} 1 & 6 \\ 2 & 5 \end{vmatrix} = 1 \cdot 5 - 6 \cdot 2 = 5 - 12 = -7$

(c) $\det C = \begin{vmatrix} 2 & -4 \\ 1 & -2 \end{vmatrix} = 2(-2) - (-4)(1) = -4 + 4 = 0$

We saw that a system of two equations in two unknowns has a unique solution if and only if $ad - bc = \begin{vmatrix} a & b \\ c & d \end{vmatrix} \neq 0$. The following generalization is true.

A linear system of n equations in n unknowns has a unique solution if and only if the determinant of its coefficient matrix is not zero.

We now define the determinant of a square ($n \times n$) matrix when $n > 2$. We begin with an example. Consider the matrix

$$A = \begin{pmatrix} 3 & 5 & 2 \\ 4 & 2 & 3 \\ -1 & 2 & 3 \end{pmatrix}$$

To define the determinant of A, we first define what are called **cofactors** of A.

Step 1 Choose a component, say 5, in the first row of A, and cross off the row and column that contain 5.

$$\begin{pmatrix} \not{3} & \not{5} & \not{2} \\ 4 & \not{2} & 3 \\ -1 & \not{2} & 4 \end{pmatrix}$$

The 2×2 matrix that remains is called the **minor** of A corresponding to the component 5:

$$\text{minor of } A \text{ corresponding to } 5 = \begin{pmatrix} 4 & 3 \\ -1 & 4 \end{pmatrix}$$

Step 2 Take the determinant of this minor:

$$\begin{vmatrix} 4 & 3 \\ -1 & 4 \end{vmatrix} = 16 + 3 = 19$$

Step 3 Multiply this determinant by $+1$ or -1. Add the row number and the column number of the crossed out row and column. If this number is even, multiply by $+1$. If it is odd, multiply by -1. Here we crossed out row 1 and column 2. Since $1 + 2 = 3$ is odd, we multiply by -1.

$$A_{12} = -\begin{vmatrix} 4 & 3 \\ -1 & 4 \end{vmatrix} = -19$$

The number -19 is called the **cofactor** of A corresponding to the number 5. We denote it by A_{12} since 5 is in the first row and second column of A.

Rules for Finding Cofactors of a Square ($n \times n$) Matrix

Step 1 Choose a component a_{ij} in the matrix A. Cross off row i and column j of this matrix.

Step 2 Compute the determinant of the resulting $(n-1) \times (n-1)$ matrix.

Step 3 Multiply this determinant by $+1$ if $i+j$ is even and by -1 if $i+j$ is odd. The number obtained is the **cofactor** of A corresponding to the component a_{ij}. It is denoted by A_{ij}.

NOTE The signs of the cofactors follow a "checkerboard" pattern:

$$\begin{pmatrix} + & - & + & - & + \\ - & + & - & + & - \\ + & - & + & - & + \\ - & + & - & + & - \\ + & - & + & - & + \end{pmatrix}$$

↖ The sign of the cofactor of each component of the main diagonal is plus

EXAMPLE 2 *Finding a Cofactor*

Compute the cofactor of $\begin{pmatrix} 3 & 5 & 2 \\ 4 & 2 & 3 \\ \circled{-1} & 2 & 4 \end{pmatrix}$ corresponding to the circled component. That is, compute A_{31}.

SOLUTION The circled component is in row 3 and column 1, so we first cross off row 3 and column 1:

$$\begin{pmatrix} \not{3} & 5 & 2 \\ \not{4} & 2 & 3 \\ \not{-1} & \not{2} & \not{4} \end{pmatrix}$$

We next compute

$$\det \begin{pmatrix} 5 & 2 \\ 2 & 3 \end{pmatrix} = \begin{vmatrix} 5 & 2 \\ 2 & 3 \end{vmatrix} = 15 - 4 = 11$$

Since $1 + 3 = 4$ is even, we multiply 11 by $+1$ to find that $A_{31} = 11$.

Now we define the determinant of a square matrix.

Definition of the Determinant of a Square Matrix *A*

Multiply each component of A in the first row by its corresponding cofactor and add these products. The sum is called the **determinant** of A. It is denoted by $\det A$ or $|A|$.

NOTATION We stress that there are two commonly used ways to denote the determinant of matrix A: $\det A$ or $|A|$. Thus the determinant of matrix $A = \begin{pmatrix} 1 & 2 \\ 3 & 4 \end{pmatrix}$ is written

$$\det \begin{pmatrix} 1 & 2 \\ 3 & 4 \end{pmatrix} \quad \text{or} \quad \begin{vmatrix} 1 & 2 \\ 3 & 4 \end{vmatrix}$$

We will use these two notations interchangeably.

EXAMPLE 3 *Computing a 3 × 3 Determinant*

Compute the determinant of $A = \begin{pmatrix} 3 & 5 & 2 \\ 4 & 2 & 3 \\ -1 & 2 & 4 \end{pmatrix}$.

SOLUTION $\det A = 3A_{11} + 5A_{12} + 2A_{13}$. We already computed $A_{12} = -19$. We have

$$\begin{array}{l}\text{Minor corresponding to entry} \\ \text{in row 1 and column 1}\end{array} = \begin{pmatrix} \not{3} & \not{5} & \not{2} \\ \not{4} & 2 & 3 \\ -\not{1} & 2 & 4 \end{pmatrix} = \begin{pmatrix} 2 & 3 \\ 2 & 4 \end{pmatrix}$$

So

$$\overset{i + j = 1 + 1 = 2 \text{ (even)}}{A_{11} = \overset{\downarrow}{+}\begin{vmatrix} 2 & 3 \\ 2 & 4 \end{vmatrix} = 8 - 6 = 2}$$

Similarly,

$$A_{13} = \det \begin{pmatrix} \not{3} & \not{5} & \not{2} \\ 4 & 2 & \not{3} \\ -1 & 2 & \not{4} \end{pmatrix} = \begin{vmatrix} 4 & 2 \\ -1 & 2 \end{vmatrix} = 8 + 2 = 10$$

Here we again multiplied by $+1$ because $1 + 3 = 4$ is even. Thus

$$\det A = \begin{vmatrix} 3 & 5 & 2 \\ 4 & 2 & 3 \\ -1 & 2 & 4 \end{vmatrix} = 3A_{11} + 5A_{12} + 2A_{13}$$

$$= 3(2) + 5(-19) + 2(10) = 6 - 95 + 20 = -69 \quad ■$$

EXAMPLE 4 *Computing a 3 × 3 Determinant*

Calculate $\begin{vmatrix} 2 & -3 & 5 \\ 1 & 0 & 4 \\ 3 & -3 & 9 \end{vmatrix}$.

SOLUTION

$$\begin{vmatrix} 2 & -3 & 5 \\ 1 & 0 & 4 \\ 3 & -3 & 9 \end{vmatrix} = 2\begin{vmatrix} 0 & 4 \\ -3 & 9 \end{vmatrix} \overset{\substack{1+2=3\text{ (odd)} \\ \downarrow}}{-} (-3)\begin{vmatrix} 1 & 4 \\ 3 & 9 \end{vmatrix} + 5\begin{vmatrix} 1 & 0 \\ 3 & -3 \end{vmatrix}$$

$$= 2 \cdot 12 + 3(-3) + 5(-3) = 0$$

There is a simpler method for calculating 3×3 determinants. Note that

$$\begin{vmatrix} a_{11} & a_{12} & a_{13} \\ a_{21} & a_{22} & a_{23} \\ a_{31} & a_{32} & a_{33} \end{vmatrix} = a_{11}\begin{vmatrix} a_{22} & a_{23} \\ a_{32} & a_{33} \end{vmatrix} - a_{12}\begin{vmatrix} a_{21} & a_{23} \\ a_{31} & a_{33} \end{vmatrix} + a_{13}\begin{vmatrix} a_{21} & a_{22} \\ a_{31} & a_{32} \end{vmatrix}$$

$$= a_{11}(a_{22}a_{33} - a_{23}a_{32}) - a_{12}(a_{21}a_{33} - a_{23}a_{31}) + a_{13}(a_{21}a_{32} - a_{22}a_{31})$$

or

$$|A| = a_{11}a_{22}a_{33} + a_{12}a_{23}a_{31} + a_{13}a_{21}a_{32} - a_{13}a_{22}a_{31} - a_{12}a_{21}a_{33} - a_{11}a_{32}a_{23} \qquad (3)$$

We write A and adjoin to it its first two columns

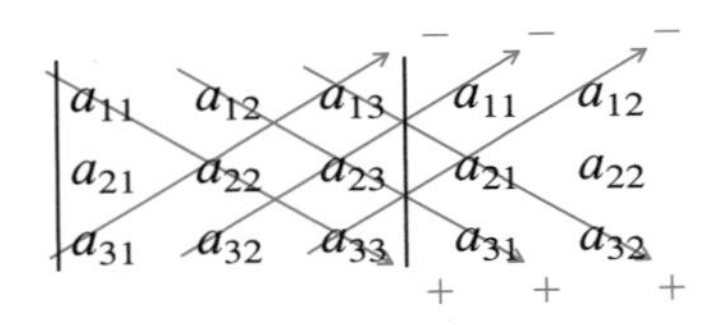

We then calculate the six products, change the signs of the products with arrows pointing upward, and add. This gives the sum in equation (3).

EXAMPLE 5 ***Computing a 3 × 3 Determinant by the New Method***

Calculate $\begin{vmatrix} 3 & 5 & 2 \\ 4 & 2 & 3 \\ -1 & 2 & 4 \end{vmatrix}$ by using this new method.

SOLUTION Writing $\left|\begin{matrix} 3 & 5 & 2 \\ 4 & 2 & 3 \\ -1 & 2 & 4 \end{matrix}\right|\begin{matrix} 3 & 5 \\ 4 & 2 \\ -1 & 2 \end{matrix}$ and multiplying as indicated, we obtain

$$|A| = (3)(2)(4) + (5)(3)(-1) + (2)(4)(2) - (-1)(2)(2) - 2(3)(3) - (4)(4)(5)$$

$$= 24 - 15 + 16 + 4 - 18 - 80 = -69$$

WARNING The method given above will *not* work for $n \times n$ determinants if $n \neq 3$. If you try something analogous for 4×4 or higher-order determinants, you will get the wrong answer. ■

Computations of 4×4 or higher-order determinants can be very tedious. There are methods for simplifying these computations. These methods are given in linear algebra textbooks. We will not discuss them here.

There is one other fact about determinants that is extremely useful. In Section 8.3 on p. 460 and in Problem 19 on p. 464 we showed that

$$A = \begin{pmatrix} a & b \\ c & d \end{pmatrix} \text{ is invertible if and only if } \det A = ad - bc \neq 0$$

That is, a 2×2 matrix is invertible if and only if its determinant is not zero. This is true for all square matrices.

Theorem

Let A be an $n \times n$ (square) matrix. Then A is invertible if and only if $\det A \neq 0$.

EXAMPLE 6 ***Determining Whether a Matrix Is Invertible by Looking at Its Determinant***

Determine whether or not each of the following matrices is invertible.

(a) $A = \begin{pmatrix} 3 & 5 & 2 \\ 4 & 2 & 3 \\ -1 & 2 & 4 \end{pmatrix}$ (b) $A = \begin{pmatrix} 2 & -3 & 5 \\ 1 & 0 & 4 \\ 3 & -3 & 9 \end{pmatrix}$

SOLUTION

(a) $\det A = -69 \neq 0$ so A is invertible. Example 3

(b) $\det A = 0$ so A is not invertible. Example 4

We now present another method for solving systems of equations. This one uses determinants.

Cramer's Rule

Consider the following system of three equations in three unknowns:

$$\begin{aligned} a_{11}x + a_{12}y + a_{13}z &= b_1 \\ a_{21}x + a_{22}y + a_{23}z &= b_2 \\ a_{31}x + a_{32}y + a_{33}z &= b_3 \end{aligned} \tag{4}$$

Let

$$D = \det \begin{pmatrix} a_{11} & a_{12} & a_{13} \\ a_{21} & a_{22} & a_{23} \\ a_{31} & a_{32} & a_{33} \end{pmatrix}, \quad D_1 = \det \begin{pmatrix} b_1 & a_{12} & a_{13} \\ b_2 & a_{22} & a_{23} \\ b_3 & a_{32} & a_{33} \end{pmatrix}$$

$$D_2 = \det \begin{pmatrix} a_{11} & b_1 & a_{13} \\ a_{21} & b_2 & a_{23} \\ a_{31} & b_3 & a_{33} \end{pmatrix}, \quad \text{and} \quad D_3 = \det \begin{pmatrix} a_{11} & a_{12} & b_1 \\ a_{21} & a_{22} & b_2 \\ a_{31} & a_{32} & b_3 \end{pmatrix}$$

If

$$A = \begin{pmatrix} a_{11} & a_{12} & a_{13} \\ a_{21} & a_{22} & a_{23} \\ a_{31} & a_{32} & a_{33} \end{pmatrix}$$

then $D = \det A$ and D_i, for $i = 1, 2,$ and 3, is obtained by replacing the ith column of A by the column vector $\mathbf{b} = \begin{pmatrix} b_1 \\ b_2 \\ b_3 \end{pmatrix}$ and then taking the determinant.

Cramer's Rule

If $D \neq 0$, system (4) has a unique solution given by

$$x = \frac{D_1}{D} \qquad y = \frac{D_2}{D} \qquad z = \frac{D_3}{D}$$

EXAMPLE 7 ***Solving a 3 × 3 System Using Cramer's Rule***

Solve the system

$$\begin{aligned} 2x + 4y - z &= -5, \\ -4x + 3y + 5z &= 14, \\ 6x - 3y - 2z &= 5 \end{aligned}$$

SOLUTION We have

$$D = \begin{vmatrix} 2 & 4 & -1 \\ -4 & 3 & 5 \\ 6 & -3 & -2 \end{vmatrix} = 112, \qquad D_1 = \begin{vmatrix} -5 & 4 & -1 \\ 14 & 3 & 5 \\ 5 & -3 & -2 \end{vmatrix} = 224$$

$$D_2 = \begin{vmatrix} 2 & -5 & -1 \\ -4 & 14 & 5 \\ 6 & 5 & -2 \end{vmatrix} = -112, \qquad D_3 = \begin{vmatrix} 2 & 4 & -5 \\ -4 & 3 & 14 \\ 6 & -3 & 5 \end{vmatrix} = 560$$

Therefore

$$x = \frac{D_1}{D} = \frac{224}{112} = 2, \qquad y = \frac{D_2}{D} = \frac{-112}{112} = -1, \qquad z = \frac{D_3}{D} = \frac{560}{112} = 5$$

NOTE Cramer's rule can be extended to a system of n equations in n unknowns with $n > 3$. To get D_1, for example, replace the first column of the coefficient matrix A by the column of constant terms in the system. However, it is usually simpler to solve an $n \times n$ system using Gaussian elimination if $n > 3$. There is less work involved.

FOCUS ON
Cramer's Rule

Cramer's rule is named for the Swiss mathematician Gabriel Cramer (1704–1752). Cramer published the rule in 1750 in his *Introduction to the Analysis of Lines of Algebraic Curves*. Actually, there is much evidence to suggest that the rule was known as early as 1729 to Colin Maclaurin (1698–1746), who was probably the most outstanding British mathematician in the years following the death of Newton. Cramer's rule is one of the most famous results in the history of mathematics. For almost 200 years it was central in the teaching of algebra and the theory of equations. Because of the great number of computations involved, the rule is used today less frequently. However, the result was very important in its time.

Problems 8.4

Readiness Check

I. Which of the following is the cofactor of 3 in $\begin{vmatrix} 1 & 2 & 3 \\ 2 & -2 & 1 \\ 4 & 0 & 2 \end{vmatrix}$?

a. 8 b. −8
c. 3 d. 6
e. −10 f. 0

II. Which of the following is 0 for all a and b?

a. $\begin{vmatrix} a & b \\ -b & a \end{vmatrix}$ b. $\begin{vmatrix} a & -b \\ -a & b \end{vmatrix}$ c. $\begin{vmatrix} a & a \\ b & -b \end{vmatrix}$

d. The determinants cannot be determined because values of a and b are not known.

III. Which of the following is 0?

a. $\begin{vmatrix} 1 & 2 & 3 \\ 1 & 2 & 4 \\ 1 & 6 & 4 \end{vmatrix}$ b. $\begin{vmatrix} 1 & 2 & 7 \\ 2 & 3 & 8 \\ -1 & -2 & -7 \end{vmatrix}$

c. $\begin{vmatrix} 2 & 1 & 3 \\ -2 & 1 & 3 \\ 0 & 2 & 5 \end{vmatrix}$ d. $\begin{vmatrix} 1 & 0 & 0 \\ 0 & -1 & 0 \\ 0 & 0 & 4 \end{vmatrix}$

In Problems 1–17 calculate the determinant.

1. $\begin{vmatrix} 1 & 2 \\ 3 & 4 \end{vmatrix}$

2. $\begin{vmatrix} 3 & -5 \\ 6 & 2 \end{vmatrix}$

3. $\begin{vmatrix} 3 & 8 \\ 4 & 2 \end{vmatrix}$

4. $\begin{vmatrix} -1 & -2 \\ -7 & -3 \end{vmatrix}$

5. $\begin{vmatrix} 2 & 6 \\ 1 & 3 \end{vmatrix}$

6. $\begin{vmatrix} 1 & 0 & 0 \\ 0 & -1 & 0 \\ 0 & 0 & 2 \end{vmatrix}$

7. $\begin{vmatrix} 3 & 0 & 0 \\ 0 & 2 & 5 \\ 0 & 0 & -6 \end{vmatrix}$

8. $\begin{vmatrix} 3 & -1 & 2 \\ 0 & 4 & 5 \\ 0 & 0 & 1 \end{vmatrix}$

9. $\begin{vmatrix} -1 & 0 & 0 \\ -2 & 5 & 0 \\ 26 & 49 & 0 \end{vmatrix}$

10. $\begin{vmatrix} 0 & 0 & a \\ 0 & b & 0 \\ c & 0 & 0 \end{vmatrix}$

11. $\begin{vmatrix} a & 0 & 0 \\ 0 & b & 0 \\ 0 & 0 & c \end{vmatrix}$

12. $\begin{vmatrix} 2 & 1 & -1 \\ 3 & -2 & 0 \\ 5 & 1 & 6 \end{vmatrix}$

13. $\begin{vmatrix} 0 & 0 & 0 & a \\ 0 & 0 & b & 0 \\ 0 & c & 0 & 0 \\ d & 0 & 0 & 0 \end{vmatrix}$

14. $\begin{vmatrix} 0 & -2 & 3 \\ 1 & 2 & -3 \\ 4 & 0 & 5 \end{vmatrix}$

15. $\begin{vmatrix} 2 & 3 & -1 \\ 2 & 4 & 2 \\ -1 & 3 & -2 \end{vmatrix}$

16. $\begin{vmatrix} 2 & -1 & 3 \\ 4 & 0 & 6 \\ 5 & -2 & 3 \end{vmatrix}$

17. $\begin{vmatrix} 2 & 3 & 2 \\ 4 & -1 & 5 \\ 6 & 2 & 7 \end{vmatrix}$

Answers to Readiness Check

I. a II. b III. b

In Problems 18–21 solve the given system using Cramer's rule.

18. $3x + y - z = -2$
$x - 2y + 3z = 13$
$4x + 2y - 2z = -6$

19. $x + y + z = 5$
$3x + 5z = -3$
$-2x + 3y - z = 2$

20. $2x + 2y + z = 7$
$x + 2y - z = 0$
$-x + y + 3z = 1$

21. $-3x + 2y + 5z = 8$
$4x - y + 2z = -3$
$-2x + 5y - 3z = 4$

22. (a) Show that the slopes of the two lines whose equations are given in system (1) are $-a/b$ and $-c/d$ if $b \neq 0$ and $d \neq 0$.
(b) Show that the two lines are parallel if and only if $ad - bc = 0$.
(c) Show that if $ad - bc = 0$, then system (1) has no solutions if $ed - bf \neq 0$ and an infinite number of solutions if $ed - bf = 0$.

In Problems 23–28 determine whether each matrix is invertible by computing its determinant. Do not compute the inverse if it is invertible.

23. $\begin{pmatrix} 1 & 0 & 0 \\ 8 & 0 & 0 \\ 13 & 9 & -5 \end{pmatrix}$

24. $\begin{pmatrix} 3 & 1 & 6 \\ 0 & 2 & 5 \\ 0 & 0 & 3 \end{pmatrix}$

25. $\begin{pmatrix} 1 & -1 & 1 \\ 2 & 1 & 5 \\ 3 & 1 & 2 \end{pmatrix}$

26. $\begin{pmatrix} 7 & 1 & 2 \\ 3 & -1 & 2 \\ 1 & 3 & -2 \end{pmatrix}$

27. $\begin{pmatrix} 3 & 2 & 4 \\ -1 & 6 & 8 \\ 5 & 10 & 16 \end{pmatrix}$

28. $\begin{pmatrix} 3 & 7 & 1 \\ 4 & -1 & 2 \\ 5 & 6 & 2 \end{pmatrix}$

Summary Outline of Chapter 8

- An **m × n matrix** is a rectangular array of mn numbers arranged in m rows and n columns. If $m = n$, the matrix is a **square matrix.** p. 439
- A **row vector** is a matrix with one row. p. 440
- A **column vector** is a matrix with one column. p. 440
- **Addition of Matrices:** If A and B are $m \times n$ matrices, then $A + B$ is the $m \times n$ matrix obtained by adding the corresponding components of A and B. p. 442
- **Dot Product:** If $\mathbf{a} = \begin{pmatrix} a_1 \\ a_2 \\ \vdots \\ a_n \end{pmatrix}$ and $\mathbf{b} = (b_1, b_2, \ldots, b_n)$, then $\mathbf{a} \cdot \mathbf{b} = a_1b_1 + a_2b_2 + \cdots + a_nb_n$. p. 447
- **Matrix Product:** If A is an $m \times n$ matrix and B is an $n \times p$ matrix, then AB is an $m \times p$ matrix whose ijth component is the dot product of the ith row of A and the jth column of B. p. 448
- An **Identity Matrix** is an $n \times n$ matrix, denoted by I, with 1's down the main diagonal and 0's everywhere else. For every $n \times n$ matrix A, $AI = IA = A$. p. 452
- **Matrix Inverse:** If A is an $n \times n$ matrix, then the **inverse** of A, if it exists, is a matrix A^{-1}, where $AA^{-1} = A^{-1}A = I$, the identity matrix. p. 456
- If $A = \begin{pmatrix} a & b \\ c & d \end{pmatrix}$ and $D = ad - bc \neq 0$, then A is invertible and $A^{-1} = \begin{pmatrix} d/D & -b/D \\ -c/D & a/D \end{pmatrix}$. p. 461
- The **determinant** of the 2×2 matrix $A = \begin{pmatrix} a & b \\ c & d \end{pmatrix}$, denoted by $\det A$, is defined by $\det A = ad - bc$. p. 465
- The ijth **minor** of an $n \times n$ matrix is the $(n - 1) \times (n - 1)$ matrix obtained by deleting the ith row and jth column of A. p. 466

- The ijth **cofactor** of A is the determinant of the ijth minor of A multiplied by $+1$ if $i+j$ is even and -1 if $i+j$ is odd. p. 466
- $\det A = a_{11}A_{11} + a_{12}A_{12} + \cdots + a_{1n}A_{1n}$, where A_{ij} is the ijth cofactor of A. p. 467
- If A is a square $(n \times n)$ matrix, then A is invertible if and only if $\det A \neq 0$. p. 470

Review Exercises for Chapter 8

In Exercises 1–8 perform the indicated computations.

1. $3\begin{pmatrix} -2 & 1 \\ 0 & 4 \\ 2 & 3 \end{pmatrix}$

2. $\begin{pmatrix} 1 & 0 & 3 \\ 2 & -1 & 6 \end{pmatrix} + \begin{pmatrix} 2 & 0 & 4 \\ -2 & 5 & 8 \end{pmatrix}$

3. $-3\begin{pmatrix} 4 & 6 & -2 \\ 2 & 1 & 0 \\ -1 & 5 & 2 \end{pmatrix} + 2\begin{pmatrix} 5 & 1 & 0 \\ -3 & 4 & 2 \\ 4 & -2 & 1 \end{pmatrix}$

4. $\begin{pmatrix} 2 & 3 \\ -1 & 4 \end{pmatrix}\begin{pmatrix} 5 & -1 \\ 2 & 7 \end{pmatrix}$

5. $\begin{pmatrix} 2 & 3 & 1 & 5 \\ 0 & 6 & 2 & 4 \end{pmatrix}\begin{pmatrix} 5 & 7 & 1 \\ 2 & 0 & 3 \\ 1 & 0 & 0 \\ 0 & 5 & 6 \end{pmatrix}$

6. $\begin{pmatrix} 2 & 3 & 5 \\ -1 & 6 & 4 \\ 1 & 0 & 6 \end{pmatrix}\begin{pmatrix} 0 & -1 & 2 \\ 3 & 1 & 2 \\ -7 & 3 & 5 \end{pmatrix}$

7. $\begin{pmatrix} 1 & 0 & 3 & -1 & 5 \\ 2 & 1 & 6 & 2 & 5 \end{pmatrix}\begin{pmatrix} 7 & 1 \\ 2 & 3 \\ -1 & 0 \\ 5 & 6 \\ 2 & 3 \end{pmatrix}$

8. $\begin{pmatrix} 1 & -1 & 2 \\ 3 & 5 & 6 \\ 2 & 4 & -1 \end{pmatrix}\begin{pmatrix} 2 \\ 1 \\ 3 \end{pmatrix}$

In Exercises 9–14 calculate the determinant.

9. $\begin{vmatrix} -1 & 2 \\ 0 & 4 \end{vmatrix}$

10. $\begin{vmatrix} -3 & 5 \\ -7 & 4 \end{vmatrix}$

11. $\begin{vmatrix} 1 & -2 & 3 \\ 0 & 4 & 5 \\ 0 & 0 & 6 \end{vmatrix}$

12. $\begin{vmatrix} 5 & 0 & 0 \\ 6 & 2 & 0 \\ 10 & 100 & 6 \end{vmatrix}$

13. $\begin{vmatrix} 1 & -1 & 2 \\ 3 & 4 & 2 \\ -2 & 3 & 4 \end{vmatrix}$

14. $\begin{vmatrix} 3 & 1 & -2 \\ 4 & 0 & 5 \\ -6 & 1 & 3 \end{vmatrix}$

In Exercises 15 and 16 solve the system by using Cramer's rule.

15. $\begin{aligned} x - y + z &= 7 \\ 2x \phantom{{}-y} - 5z &= 4 \\ 3y - z &= 2 \end{aligned}$

16. $\begin{aligned} 2x + 3y - z &= 5 \\ -x + 2y + 3z &= 0 \\ 4x - y + z &= -1 \end{aligned}$

In Exercises 17–21 calculate the inverse of the given matrix (if the inverse exists).

17. $\begin{pmatrix} 4 & 7 \\ 2 & 5 \end{pmatrix}$

18. $\begin{pmatrix} -1 & 2 \\ 2 & -4 \end{pmatrix}$

19. $\begin{pmatrix} 1 & 2 & 0 \\ 2 & 1 & -1 \\ 3 & 1 & 1 \end{pmatrix}$

20. $\begin{pmatrix} -1 & 2 & 0 \\ 4 & 1 & -3 \\ 2 & 5 & -3 \end{pmatrix}$

21. $\begin{pmatrix} 2 & 1 & 0 \\ -1 & 2 & 3 \\ 0 & 2 & 1 \end{pmatrix}$

In Exercises 22–24 first write the system in the form $A\mathbf{x} = \mathbf{b}$, then calculate A^{-1}, and, finally, use matrix multiplication to obtain the solution vector.

22. $\begin{aligned} x - 3y &= 4 \\ 2x + 5y &= 7 \end{aligned}$

23. $\begin{aligned} x + 2y \phantom{{}-z} &= 3 \\ 2x + y - z &= -1 \\ 3x + y + z &= 7 \end{aligned}$

24. $\begin{aligned} 2x \phantom{{}+3y} + 4z &= 7 \\ -x + 3y + z &= -4 \\ y + 2z &= 5 \end{aligned}$

Chapter 9

Introduction to Discrete Mathematics

Discrete mathematics is concerned primarily with the analysis of both *finite* collections of objects and infinite collections in which there is one object for every positive integer. Almost every student of computer science studies discrete mathematics. But some of the topics in this vast subject are important to virtually everyone who studies mathematics.

The topics in this chapter are less connected than those in earlier chapters. Each section, except for Sections 9.1 and 9.8, is intended as a brief introduction to an important topic. We begin with a discussion of mathematical induction — a very convenient way to prove statements involving positive integers.

9.1 Mathematical Induction

Mathematical induction is the name given to an elementary logical principle that can be used to prove a certain type of mathematical statement. Typically, we use mathematical induction to prove that a certain statement or equation holds for every positive integer. For example, we may need to prove that $2^n > n$ for all integers $n \geq 1$. To do this, we proceed in two steps:

The Two Steps of Mathematical Induction

Step 1 We prove that the statement is true for some integer N (usually $N = 1$).

Step 2 We *assume* that the statement is true for an integer k and then *prove* that it is true for the integer $k + 1$.

If we can complete these two steps, then we will have demonstrated the validity of the statement for *all* positive integers greater than or equal to N. To convince you of this fact, we reason as follows: Since the statement is true for N (by step (1)), it is true for the integer $N + 1$ (by step (2)). Then it is also true for the integer $(N + 1) + 1 = N + 2$ (again by step (2)), and so on. We now demonstrate the procedure with some examples.

EXAMPLE 1 *A Formula for the Sum of the First n Integers*

Prove that the sum of the first n positive integers is equal to $n(n + 1)/2$.

SOLUTION We are asked to show that

$$1 + 2 + 3 + \cdots + n = \frac{n(n + 1)}{2} \tag{1}$$

You may first wish to try a few examples to illustrate that formula (1) really works. For example

$$1 + 2 + 3 + 4 + 5 + 6 + 7 + 8 + 9 + 10 = \frac{10(11)}{2} = 55$$

Step 1 If $n = 1$, then the sum of the first 1 integer is 1. But $(1)(1 + 1)/2 = 1$ so that equation (1) holds in the case $n = 1$.

Step 2 Assume that (1) holds for $n = k$; that is,

$$1 + 2 + 3 + \cdots + k = \frac{k(k + 1)}{2}$$

We must now show that it holds for $n = k + 1$. That is, we must show that

$$1 + 2 + 3 + \cdots + k + (k + 1) = \frac{(k + 1)(k + 2)}{2}$$

But

$$\begin{aligned}
1 + 2 + 3 + \cdots + k + (k + 1) &= \overbrace{(1 + 2 + 3 + \cdots + k)}^{= k(k+1)/2 \text{ by assumption}} + (k + 1) \\
&= \frac{k(k + 1)}{2} + (k + 1) \\
&= \frac{k(k + 1) + 2(k + 1)}{2} \\
&= \frac{k^2 + 3k + 2}{2} \\
&= \frac{(k + 1)(k + 2)}{2}
\end{aligned}$$

and the proof is complete.

Where the Difficulty Lies

Mathematical induction is sometimes difficult at first sight because of Step 2. Step 1 is usually easy to carry out. In Example 1, for instance, we inserted the value $n = 1$ on both sides of the equation (1) and verified that $1 = 1(1 + 1)/2$. Step 2 was much more difficult. Let us look at it again.

We *assumed* that equation (1) was valid for $n = k$. We did not prove it. That assumption is called the **induction hypothesis.** We then used the induction hypothesis to show that equation (1) holds for $n = k + 1$. Perhaps this will be clearer if we look at a particular value for k, say $k = 10$. Then we have

Assumption $1 + 2 + 3 + 4 + 5 + 6 + 7 + 8 + 9 + 10$

$$= \frac{10(10 + 1)}{2} = \frac{10(11)}{2} = 55 \tag{2}$$

To prove $1 + 2 + 3 + 4 + 5 + 6 + 7 + 8 + 9 + 10 + 11$

$$= \frac{11(11 + 1)}{2} = \frac{11(12)}{2} = 66 \tag{3}$$

The actual proof $(1 + 2 + 3 + 4 + 5 + 6 + 7 + 8 + 9 + 10) + 11$

By the induction hypothesis (2)

$$\downarrow$$

$$= \frac{10(11)}{2} + 11 = \frac{10(11)}{2} + \frac{2(11)}{2}$$

$$= \frac{11(10 + 2)}{2} = \frac{11(12)}{2}$$

which is equation (3). Thus, *if* (2) is true, then (3) is true.

The beauty of the method of mathematical induction is that we do not have to prove each case separately, as we did in this illustration. Rather, we prove it for a first case, *assume* it for a general case, and then prove it for the general case plus 1. Two steps take care of an infinite number of cases. It's really quite a remarkable idea.

EXAMPLE 2 *A Formula for the Sum of the First n Squares*

Prove that the sum of the squares of the first n positive integers is $n(n + 1)(2n + 1)/6$.

SOLUTION We must prove that

$$1^2 + 2^2 + 3^2 + \cdots + n^2 = \frac{n(n + 1)(2n + 1)}{6} \tag{4}$$

Step 1 Since $\dfrac{1(1 + 1)(2 \cdot 1 + 1)}{6} = 1 = 1^2$, equation (4) is valid for $n = 1$.

Step 2 Suppose that equation (4) holds for $n = k$; that is

Induction Hypothesis $$1^2 + 2^2 + 3^2 + \cdots + k^2 = \frac{k(k + 1)(2k + 1)}{6}$$

Then to prove that (4) is true for $n = k + 1$, we have

$$
\begin{aligned}
1^2 + 2^2 + 3^2 + \cdots + k^2 + (k+1)^2 &= (1^2 + 2^2 + 3^2 + \cdots + k^2) + (k+1)^2 \\
&\overset{\text{Induction Hypothesis}}{=} \frac{k(k+1)(2k+1)}{6} + (k+1)^2 \\
&= \frac{k(k+1)(2k+1) + 6(k+1)^2}{6} \\
&= \frac{k+1}{6}[k(2k+1) + 6(k+1)] \\
&= \frac{k+1}{6}[2k^2 + 7k + 6] \\
&= \frac{k+1}{6}[(k+2)(2k+3)] \\
&= \frac{(k+1)(k+2)[2(k+1)+1]}{6}
\end{aligned}
$$

which is equation (4) for $n = k + 1$, and the proof is complete. Again, you may wish to experiment with this formula. For example,

$$
\begin{aligned}
1^2 + 2^2 + 3^2 + 4^2 + 5^2 + 6^2 + 7^2 &= \frac{7(7+1)(2 \cdot 7 + 1)}{6} \\
&= \frac{7 \cdot 8 \cdot 15}{6} = 140 \quad \blacksquare
\end{aligned}
$$

EXAMPLE 3 *The Sum of a Geometric Progression*

For $a \neq 1$, use mathematical induction to prove the formula for the sum of a geometric progression:

$$1 + a + a^2 + \cdots + a^n = \frac{1 - a^{n+1}}{1 - a} \tag{5}$$

SOLUTION

Step 1 If $n = 0$ (the first integer in this case), then

$$\frac{1 - a^{0+1}}{1 - a} = \frac{1 - a}{1 - a} = 1$$

Thus equation (5) holds for $n = 0$ because the left side of (5) begins and stops at 1 when $n = 0$. (We use $n = 0$ instead of $n = 1$ since $a^0 = 1$ is the first term.)

Step 2 Assume that (5) holds for $n = k$; that is,

Induction Hypothesis
$$1 + a + a^2 + \cdots + a^k = \frac{1 - a^{k+1}}{1 - a}$$

Then

$$1 + a + a^2 + \cdots + a^k + a^{k+1} = (1 + a + a^2 + \cdots + a^k) + a^{k+1}$$

Induction Hypothesis
$$\downarrow$$
$$= \frac{1 - a^{k+1}}{1 - a} + a^{k+1}$$

$$= \frac{1 - a^{k+1} + (1 - a)a^{k+1}}{1 - a} = \frac{1 - a^{k+1} + a^{k+1} - a \cdot a^{k+1}}{1 - a} = \frac{1 - a^{k+2}}{1 - a}$$

so that equation (5) also holds for $n = k + 1$, and the proof is complete. ■

EXAMPLE 4 *A Proof That $2n + n^3$ Is Divisible by 3*

Use mathematical induction to prove that $2n + n^3$ is divisible by 3 for every positive integer n.

SOLUTION

Step 1 If $n = 1$, then $2n + n^3 = 2 \cdot 1 + 1^3 = 2 + 1 = 3$, which *is* divisible by 3.

Step 2 Assume that $2k + k^3$ is divisible by 3. Induction Hypothesis

This means that $\frac{2k + k^3}{3} = m$, an integer. Then, expanding $(k + 1)^3$, we obtain

Use formula (7) on p. 42
$$\downarrow$$
$$\begin{aligned} 2(k + 1) + (k + 1)^3 &= 2k + 2 + (k^3 + 3k^2 + 3k + 1) \\ &= 2k + k^3 + 3k^2 + 3k + 3 \\ &= 2k + k^3 + 3(k^2 + k + 1) \end{aligned}$$

Then

$$\begin{aligned} \frac{2(k + 1) + (k + 1)^3}{3} &= \frac{2k + k^3}{3} + \frac{3(k^2 + k + 1)}{3} \\ &= m + k^2 + k + 1 = \text{an integer} \end{aligned}$$

Thus, $2(k + 1) + (k + 1)^3$ is divisible by 3. This shows that the statement is true for $n = k + 1$.

FOCUS ON
Mathematical Induction

The first mathematician to give a formal proof by the explicit use of mathematical induction was the Italian clergyman Franciscus Maurolicus (1494–1575), who was the abbot of Messina in Sicily, and is considered the greatest geometer of the sixteenth century. In his *Arithmetic,* published in 1575, Maurolicus used mathematical induction to prove, among other things, that, for every positive integer n,

$$1 + 3 + 5 + \cdots + (2n - 1) = n^2$$

You are asked to prove this in Problem 4.

The induction proofs of Maurolicus were given in a sketchy style that is difficult to follow. A clearer exposition of the method was given by the French mathematician Blaise Pascal (1623–1662). In his *Traité du Triangle Arithmétique,* published in 1662, Pascal proved a formula for the sum of binomial coefficients. He used his formula to develop what is today called the *Pascal triangle.* We shall discuss this triangle in Section 9.8.

Although the method of mathematical induction was used formally in 1575, the term *mathematical induction* was not used until 1838. In that year, one of the originators of set theory, Augustus de Morgan (1806–1871), published an article in the *Penny Cyclopedia* (London) entitled "Induction (Mathematics)." At the end of that article, he used the term we use today. However, the term did not enjoy widespread use until the twentieth century.

Problems 9.1

Readiness Check

I. Which of the following might be the first step in the mathematical induction process?
  a. Assume the statement is true for $n = 1$.
  b. Prove the statement is true for $n = k + 1$.
  c. Assume the statement is true for $n = k$.
  d. Prove the statement is true for $n = 1$.

II. Which of the following is true about mathematical induction?
  a. It is a method of proof by contradiction.
  b. It proves a statement is true for $n = 1$ and $n = k$.
  c. It proves a statement is true for every integer greater than or equal to a given "first" integer.
  d. It is an indirect method of proof.

III. Which of the following is true in the proof that $1 + 2n \le 3n$ by mathematical induction?
  a. Assume $1 + 2(1) \le 3$.
  b. $1 + 2(k + 1) \le 3(k + 1)$ must be proved.
  c. $1 + 2k \le 3^k$ must be proved.
  d. $1 + 2(5) \le 3^5$ proves the statement.

In Problems 1–20 prove the given formula using mathematical induction. Unless otherwise stated, assume that n is a positive or nonnegative integer.

1. $2 + 4 + 6 + \cdots + 2n = n(n + 1)$
2. $1 + 4 + 7 + \cdots + (3n - 2) = \dfrac{n(3n - 1)}{2}$
3. $2 + 5 + 8 + \cdots + (3n - 1) = \dfrac{n(3n + 1)}{2}$
4. $1 + 3 + 5 + \cdots + (2n - 1) = n^2$
5. $2^n > n$
6. $2^n < n!$ for $n = 4, 5, 6, \ldots$ where $n! = 1 \cdot 2 \cdot 3 \cdots (n - 1) \cdot n$
7. $1 + 2 + 4 + 8 + \cdots + 2^n = 2^{n+1} - 1$
8. $1 + 3 + 9 + 27 + \cdots + 3^n = \dfrac{3^{n+1} - 1}{2}$
9. $1 + \dfrac{1}{2} + \dfrac{1}{4} + \cdots + \dfrac{1}{2^n} = 2 - \dfrac{1}{2^n}$
10. $1 - \dfrac{1}{3} + \dfrac{1}{9} - \cdots + \left(-\dfrac{1}{3}\right)^n = \dfrac{3}{4}\left[1 - \left(-\dfrac{1}{3}\right)^{n+1}\right]$
11. $1^3 + 2^3 + 3^3 + \cdots + n^3 = \dfrac{n^2(n + 1)^2}{4}$
12. $1 \cdot 2 + 2 \cdot 3 + 3 \cdot 4 + \cdots + n(n + 1) = \dfrac{n(n + 1)(n + 2)}{3}$

Answers to Readiness Check

I. d II. c III. b

13. $1 \cdot 2 + 3 \cdot 4 + 5 \cdot 6 + \cdots + (2n-1)(2n) = \dfrac{n(n+1)(4n-1)}{3}$

14. $\dfrac{1}{2^2-1} + \dfrac{1}{3^2-1} + \dfrac{1}{4^2-1} + \cdots + \dfrac{1}{(n+1)^2-1} = \dfrac{3}{4} - \dfrac{1}{2(n+1)} - \dfrac{1}{2(n+2)}$

15. $n + n^2$ is even

16. $n < \dfrac{n^2-n}{12} + 2$ if $n > 10$

17. $n(n^2+5)$ is divisible by 6

* 18. $3n^5 + 5n^3 + 7n$ is divisible by 15

* 19. $x^n - 1$ is divisible by $x - 1$

* 20. $x^n - y^n$ is divisible by $x - y$

In Problems 21–30† find the smallest positive integer k for which the given statement holds. Then use mathematical induction to prove that the statement is true for $n \geq k$.

21. $20n < n^2$

22. $10 + 50n^2 < n^3$

* 23. $100 \log n < n$

* 24. $n^2 < e^n$

* 25. $n^3 < e^n$

* 26. $25n^{10} < e^n$

27. $e^n < n!$

28. $n^2 < 2^n$

* 29. Give a formal proof that $(ab)^n = a^n b^n$ for every integer n.

* 30. Prove that there are exactly 2^n subsets of a set containing n elements.

31. Prove that if $2k - 1$ is an even integer for some integer k, then $2(k+1) - 1 = 2k + 2 - 1 = 2k + 1$ is also an even integer. What, if anything, can you conclude by the proof?

* 32. What is wrong with the following proof that each horse in a set of n horses has the same color as every other horse in the set?

Step 1. It is true for $n = 1$ since there is only one horse in the set and it obviously has the same color as itself.

Step 2. Suppose it is true for $n = k$. That is, each horse in a set containing k horses is the same color as every other horse in the set. Let $h_1, h_2, \ldots, h_k, h_{k+1}$ denote the $k + 1$ horses in a set S. Let $S_1 = \{h_1, h_2, \ldots, h_k\}$ and $S_2 = \{h_2, h_3, \ldots, h_k, h_{k+1}\}$. Then both S_1 and S_2 contain k horses, so the horses in each set are of the same color. We write $h_i = h_j$ to indicate that horse i has the same color as horse j. Then we have

$$h_1 = h_2 = h_3 = \cdots = h_k$$

and

$$h_2 = h_3 = h_4 = \cdots = h_k = h_{k+1}$$

This means that

$$h_1 = h_2 = h_3 = \cdots = h_k = h_{k+1}$$

so all the horses in S have the same color. This proves the statement in the case $n = k + 1$, so the statement is true for all n.

9.2 Sequences of Real Numbers and the Summation Notation

According to a popular dictionary,‡ a *sequence* is "the following of one thing after another." In mathematics, we could define a sequence intuitively as a succession of numbers that never terminates. The numbers in the sequence are called the *terms* of the sequence. In a sequence, there is one term for each positive integer.

EXAMPLE 1 *The Sequence of Powers of $\frac{1}{2}$*

Consider the sequence

1st term	2nd term	3rd term	4th term	5th term		nth term	
↓	↓	↓	↓	↓		↓	
$\dfrac{1}{2}$,	$\dfrac{1}{4}$,	$\dfrac{1}{8}$,	$\dfrac{1}{16}$,	$\dfrac{1}{32}$,	. . . ,	$\dfrac{1}{2^n}$,	. . .

† Problems 23–28 use material from Chapter 6. These problems should be omitted if Chapter 6 has not yet been covered.

‡ *The Random House Dictionary,* Ballantine Books, New York, 1978.

We see that there is one term for each positive integer. The terms in this sequence form an infinite ordered set of real numbers, which we write as

$$A = \left\{\frac{1}{2}, \frac{1}{4}, \frac{1}{8}, \ldots, \frac{1}{2^n}, \ldots\right\} \tag{1}$$

That is, the set A consists of all numbers of the form $1/2^n$, where n is a positive integer. There is another way to describe this set. We define the function f by the rule $f(n) = 1/2^n$, where the domain of f is the set of positive integers. Then the set A is precisely the set of values taken by the function f.

In general, we have the following formal definition.

Definition of a Sequence

A **sequence** of real numbers is a function whose domain is the set of positive or nonnegative integers. The values taken by the function are called **terms** of the sequence.

NOTATION We will often denote the terms of a sequence by a_n. Thus, if the function given in the definition is f, then $a_n = f(n)$. With this notation, *we can denote the set of values taken by the sequence by* $\{a_n\}$. Also, we will use n, m, and so on, as integer variables and x, y, and so on, as real variables.

EXAMPLE 2 *Five Sequences*

The following are sequences of real numbers:

(a) $\{a_n\} = \left\{\frac{1}{n}\right\}$ (b) $\{a_n\} = \{\sqrt{n}\}$ (c) $\{a_n\} = \left\{\frac{1}{n!}\right\}$

(d) $\{a_n\} = \left\{\frac{e^n}{n!}\right\}$ (e) $\{a_n\} = \left\{\frac{n-1}{n}\right\}$

We sometimes denote a sequence by writing out the values $\{a_1, a_2, a_3, \ldots\}$.

EXAMPLE 3 *The Terms of Five Sequences*

We write out the values of the sequences in Example 2:

(a) $\left\{1, \frac{1}{2}, \frac{1}{3}, \frac{1}{4}, \ldots, \frac{1}{n}, \ldots\right\}$

(b) $\{\sqrt{1}, \sqrt{2}, \sqrt{3}, \sqrt{4}, \ldots, \sqrt{n}, \ldots\}$

(c) $\left\{1, \frac{1}{2}, \frac{1}{6}, \frac{1}{24}, \ldots, \frac{1}{n!}, \ldots\right\}$

(d) $\left\{e, \frac{e^2}{2}, \frac{e^3}{6}, \frac{e^4}{24}, \ldots, \frac{e^n}{n!}, \ldots\right\}$

(e) $\left\{0, \frac{1}{2}, \frac{2}{3}, \frac{3}{4}, \ldots, \frac{n-1}{n}, \ldots\right\}$ ■

EXAMPLE 4 ***Finding the General Term of a Sequence***

Find the general term a_n of the sequence $\{-1, 1, -1, 1, -1, 1, -1, \ldots\}$.

SOLUTION We see that $a_1 = -1$, $a_2 = 1$, $a_3 = -1$, $a_4 = 1, \ldots$. Hence,

$$a_n = \begin{cases} -1, & \text{if } n \text{ is odd} \\ 1, & \text{if } n \text{ is even} \end{cases}$$

A more concise way to write this term is $\quad a_n = (-1)^n$

In a number of situations, it is useful to write the sum of the first N terms of a sequence. If the sequence is $\{a_n\}$, then the sum of the first N terms, denoted by S_N, is given by

$$S_N = a_1 + a_2 + a_3 + \cdots + a_N \tag{2}$$

We now introduce a shorthand notation for writing a sum of terms.

The Σ Notation

A sum can be written† with sigma notation.

Sigma Notation

$$S_N = \sum_{k=M}^{N} a_k = a_M + a_{M+1} + \cdots + a_N \tag{3}$$

which is read "the sum of the terms a_k as k goes from M to N." In this context, Σ is called the **summation sign,** and k is called the **index of summation.**

EXAMPLE 5 ***Writing Out a Sum***

Write out the sum $\sum_{k=1}^{5} b_k$.

SOLUTION Starting at $k = 1$ and ending at $k = 5$, we obtain

$$\sum_{k=1}^{5} b_k = b_1 + b_2 + b_3 + b_4 + b_5$$

† The Greek letter Σ (sigma) was first used to denote a sum by the Swiss mathematician Leonhard Euler (1707–1783).

EXAMPLE 6 *Writing Out a Sum*

Write out the sum $\Sigma_{k=3}^{6} c_k$.

SOLUTION Starting at $k = 3$ and ending at $k = 6$, we obtain

$$\sum_{k=3}^{6} c_k = c_3 + c_4 + c_5 + c_6 \quad ■$$

EXAMPLE 7 *Evaluating a Sum*

Calculate $\Sigma_{k=0}^{3} (k + 1)$.

SOLUTION Here $a_k = k + 1$, so that

$$\sum_{k=0}^{3} (k + 1) = (0 + 1) + (1 + 1) + (2 + 1) + (3 + 1)$$

$$= 1 + 2 + 3 + 4 = 10 \quad ■$$

EXAMPLE 8 *Evaluating a Sum*

Calculate $\Sigma_{k=-2}^{3} k^2$.

SOLUTION Here $a_k = k^2$, and k ranges from -2 to 3.

$$\sum_{k=-2}^{3} k^2 = (-2)^2 + (-1)^2 + (0)^2 + 1^2 + 2^2 + 3^2$$

$$= 4 + 1 + 0 + 1 + 4 + 9 = 19$$

NOTE As in Examples 7 or 8, the index of summation can take on negative integer values or zero.

EXAMPLE 9 *Writing a Sum Using Sigma Notation*

Write the sum $S_8 = 1 - 2 + 3 - 4 + 5 - 6 + 7 - 8$ by using the summation sign.

SOLUTION Since $1 = (-1)^2$, $-2 = (-1)^3 \cdot 2$, $3 = (-1)^4 \cdot 3, \ldots$, we have

$$S_8 = \sum_{k=1}^{8} (-1)^{k+1} k \quad ■$$

EXAMPLE 10 *A Sum That Comes Up in Calculus*

Write the following sum by using the summation sign:

$$S = (\tfrac{1}{8})^2\tfrac{1}{8} + (\tfrac{2}{8})^2\tfrac{1}{8} + \cdots + (\tfrac{7}{8})^2\tfrac{1}{8} + (\tfrac{8}{8})^2\tfrac{1}{8}$$

SOLUTION We have

$$S = \sum_{k=1}^{8} \left(\frac{k}{8}\right)^2 \frac{1}{8} = \sum_{k=1}^{8} \left(\frac{1}{8^3}\right) k^2 = \frac{1}{8^3} \sum_{k=1}^{8} k^2$$

In Example 10, we used the following fact:

$$\sum_{k=1}^{n} ca_k = c \sum_{k=1}^{n} a_k$$

where c is a constant. To see why this is true, observe that

$$\begin{aligned}\sum_{k=1}^{n} ca_k &= ca_1 + ca_2 + ca_3 + \cdots + ca_n \\ &= c(a_1 + a_2 + a_3 + \cdots + a_n) = c \sum_{k=1}^{n} a_k\end{aligned}$$

This and other facts are summarized below.

Facts About the Sigma Notation

Let $\{a_n\}$ and $\{b_n\}$ be real sequences, and let c be a real number. Then

$$\sum_{k=M}^{N} ca_k = c \sum_{k=M}^{N} a_k \tag{4}$$

$$\sum_{k=M}^{N} (a_k + b_k) = \sum_{k=M}^{N} a_k + \sum_{k=M}^{N} b_k \tag{5}$$

$$\sum_{k=M}^{N} (a_k - b_k) = \sum_{k=M}^{N} a_k - \sum_{k=M}^{N} b_k \tag{6}$$

$$\sum_{k=M}^{N} a_k = \sum_{k=M}^{m} a_k + \sum_{k=m+1}^{N} a_k \quad \text{if } M < m < N \tag{7}$$

The proofs of these facts are left as exercises (see Problems 71–73).

Problems 9.2

Readiness Check

I. Which of the following is a general term for the sequence $\{\frac{3}{4}, -\frac{4}{5}, \frac{5}{6}, -\frac{6}{7}, \frac{7}{8}, -\frac{8}{9}, \ldots\}$? [Starting at $n = 1$]

a. $\frac{n-1}{n}$

b. $\sum_{k=3}^{9} (-1)^k \frac{k}{k+1}$

c. $\frac{n+2}{n+3}(-1)^{n+1}$

d. $(-1)^n \frac{n}{n+1}$

II. Which of the following is $\{(-1)^{n+1}(2n+1)^{-1}\}$ [Starting at $n = 1$]?

a. $\{3, 5, 7, 9, \ldots\}$

b. $\{\frac{1}{3}, -\frac{1}{5}, \frac{1}{7}, -\frac{1}{9}, \ldots\}$

c. $\{3, -5, 7, -9, \ldots\}$

d. $\{\frac{1}{3}, \frac{1}{5}, \frac{1}{7}, \frac{1}{9}, \ldots\}$

III. Which of the following is the same as $\frac{1}{3} + \frac{4}{4} + \frac{9}{5} + \frac{16}{6}$?

a. $\sum_{k=1}^{3} \frac{k^2}{k+2}$ b. $\sum_{i=0}^{3} \frac{(i+1)^2}{i+3}$

c. $\sum_{j=-1}^{2} \frac{j^2}{j+2}$ d. $\sum_{k=0}^{2} \frac{(k+1)^2}{k+3}$

IV. Which of the following is true about sigma notation?

a. $\sum_{k=1}^{n} ck = \sum_{k=1}^{n} c + \sum_{k=1}^{n} k$

b. $\sum_{k=1}^{n} (a_k \cdot b_k) = \sum_{k=1}^{n} a_k + \sum_{k=1}^{n} b_k$

c. $\sum_{k=1}^{n} (a_k - b_k) = \sum_{k=1}^{n} a_k + (-1)^n \sum_{k=1}^{n} b_k$

d. $\sum_{k=1}^{n} ck = c \sum_{k=1}^{n} k$

V. Which of the following is *not* true?

a. $\sum_{k=1}^{n} a_k + \sum_{k=1}^{n} b_k = \sum_{k=1}^{n} (a_k + b_k)$

b. $\sum_{k=1}^{n} a_k \cdot \sum_{k=1}^{n} b_k = \sum_{k=1}^{n} a_k b_k$

c. $\sum_{k=1}^{n} a_k = \sum_{k=1}^{n-1} a_k + a_n$

d. $\sum_{k=1}^{n} a_k - \sum_{k=1}^{n} b_k = \sum_{k=1}^{n} (a_k - b_k)$

In Problems 1–25 find the first five terms of the given sequence. Start with $n = 1$.

1. $\{n\}$
2. $\{n + 4\}$
3. $\{4n - 2\}$
4. $\left\{\frac{1}{n}\right\}$
5. $\left\{\frac{1}{n+2}\right\}$
6. $\left\{\frac{1}{n^2+1}\right\}$
7. $\{n(n+1)\}$
8. $\{3n(5n-2)\}$
9. $\left\{\frac{n+1}{n}\right\}$
10. $\left\{\frac{1}{3^n}\right\}$
11. $\left\{2 - \frac{1}{2^n}\right\}$
12. $\{\sqrt{n}\}$
13. $\{n^4\}$
14. $\{3\sqrt{n}\}$
15. $\left\{\frac{(-1)^n}{n+3}\right\}$
16. $\{(-1)^n 2^{-n}\}$
17. $\{e^n\}$
18. $\left\{\frac{1}{n!}\right\}$
19. $\{e^{1/n}\}$
20. $\{(-1)^n\}$
21. $\{(-1)^{2n}\}$
22. $\left\{\left(\frac{7}{5}\right)^n\right\}$
23. $\left\{\left(-\frac{2}{3}\right)^n\right\}$
24. $\left\{\frac{n+1}{n^{3/2}}\right\}$
25. $\{(-1)^{2n+1} n\}$

In Problems 26–40 find the general term, a_n, of the given sequence.

26. $\{1, 3, 5, 7, \ldots\}$
27. $\{1, -2, 3, -4, 5, -6, \ldots\}$
28. $\{1, 4, 9, 16, 25, \ldots\}$
29. $\{1, 8, 27, 64, \ldots\}$

Answers to Readiness Check

I. c II. b III. b IV. d V. b

30. $\left\{1, \frac{1}{2}, \frac{1}{3}, \frac{1}{4}, \frac{1}{5}, \frac{1}{6}, \ldots\right\}$
31. $\left\{\frac{1}{4}, -\frac{1}{9}, \frac{1}{16}, -\frac{1}{25}, \ldots\right\}$
32. $\left\{\frac{1}{2}, \frac{2}{3}, \frac{3}{4}, \frac{4}{5}, \frac{5}{6}, \ldots\right\}$
33. $\left\{\frac{5}{2}, -\frac{9}{4}, \frac{17}{8}, -\frac{33}{16}, \frac{65}{32}, \ldots\right\}$
34. $\{1, 2 \cdot 5, 3 \cdot 5^2, 4 \cdot 5^3, 5 \cdot 5^4, \ldots\}$
* 35. $\left\{\frac{1}{3}, \frac{2}{5}, \frac{3}{7}, \frac{4}{9}, \frac{5}{11}, \ldots\right\}$
36. $\{e, e^2, e^3, e^4, e^5, \ldots\}$
37. $\left\{\frac{1}{3}, -\frac{1}{9}, \frac{1}{27}, -\frac{1}{81}, \frac{1}{243}, \ldots\right\}$
* 38. $\{2, 6, 12, 20, 30, 42, \ldots\}$
39. $\left\{\frac{1}{2}, \frac{4}{3}, \frac{9}{4}, \frac{16}{5}, \frac{25}{6}, \ldots\right\}$
* 40. $\left\{2, \frac{5}{8}, \frac{10}{27}, \frac{17}{64}, \frac{26}{125}, \ldots\right\}$

In Problems 41–46 evaluate the given sums.

41. $\sum_{k=1}^{5} 3k$ 42. $\sum_{i=1}^{3} i^3$

43. $\sum_{k=0}^{6} 1$ 44. $\sum_{k=1}^{8} 3^k$

45. $\sum_{k=-2}^{4} k^2$ 46. $\sum_{j=5}^{7} \frac{2j+3}{j-2}$

In Problems 47–61 write each sum by using the Σ notation.

47. $1 + 2 + 4 + 8 + 16$
48. $1 - 3 + 9 - 27 + 81 - 243$
49. $\frac{2}{3} + \frac{3}{4} + \frac{4}{5} + \frac{5}{6} + \frac{6}{7} + \frac{7}{8} + \cdots + \frac{n}{n+1}$
50. $1 - \frac{1}{2!} + \frac{1}{3!} - \frac{1}{4!} + \frac{1}{5!} - \frac{1}{6!} + \frac{1}{7!}$
51. $1 + 2^{1/2} + 3^{1/3} + 4^{1/4} + 5^{1/5} + \cdots + n^{1/n}$
52. $1 - x^2 + x^4 - x^6 + x^8 - x^{10} + x^{12}$
53. $x^6 + x^{10} + x^{14} + x^{18} + x^{22}$
54. $-1 + \frac{1}{a} - \frac{1}{a^2} + \frac{1}{a^3} - \frac{1}{a^4} + \frac{1}{a^5} - \frac{1}{a^6} + \frac{1}{a^7} - \frac{1}{a^8}$
55. $1 \cdot 3 + 3 \cdot 5 + 5 \cdot 7 + 7 \cdot 9 + 9 \cdot 11 + 11 \cdot 13 + 13 \cdot 15 + 15 \cdot 17$
56. $2^2 \cdot 4 + 3^2 \cdot 6 + 4^2 \cdot 8 + 5^2 \cdot 10 + 6^2 \cdot 12 + 7^2 \cdot 14$
57. $\frac{1}{32}(\frac{1}{32})^2 + \frac{1}{32}(\frac{2}{32})^2 + \frac{1}{32}(\frac{3}{32})^2 + \cdots + \frac{1}{32}(\frac{32}{32})^2$
58. $\frac{1}{2^n}\left(\frac{1}{2^n}\right)^2 + \frac{1}{2^n}\left(\frac{2}{2^n}\right)^2 + \cdots + \frac{1}{2^n}\left(\frac{2^{n-1}}{2^n}\right)^2 + \frac{1}{2^n}\left(\frac{2^n}{2^n}\right)^2$
59. $\frac{1}{64}(\frac{1}{64})^3 + \frac{1}{64}(\frac{2}{64})^3 + \frac{1}{64}(\frac{3}{64})^3 + \cdots + \frac{1}{64}(\frac{64}{64})^3$
* 60. $0.1e^{0.1} + 0.2e^{0.2} + 0.3e^{0.3} + \cdots + e^1$
61. $\frac{1}{50}\sqrt{\frac{1}{50}} + \frac{1}{50}\sqrt{\frac{2}{50}} + \frac{1}{50}\sqrt{\frac{3}{50}} + \cdots + \frac{1}{50}\sqrt{\frac{49}{50}} + \frac{1}{50}\sqrt{\frac{50}{50}}$

Observe that there is some flexibility in the Σ notation. For instance, $\Sigma_{i=3}^{8} i$, $\Sigma_{j=3}^{8} j$, $\Sigma_{k=5}^{10} (k-2)$, and $\Sigma_{L=0}^{5} (L+3)$ each equals $3 + 4 + 5 + 6 + 7 + 8$. In Problems 62–64 you are given three expressions; two give the same sum, and one is different. Identify the one that does not equal the other two.

62. $\sum_{k=0}^{7} (2k+1)$, $\sum_{j=1}^{15} j$, $\sum_{i=2}^{9} (2i-3)$

63. $\sum_{k=1}^{7} k^2$, $\sum_{j=0}^{6} (7-j)^2$, $\sum_{i=1}^{7} (7-i)^2$

64. $\left(\sum_{k=7}^{11} k\right)^4$, $\sum_{m=-11}^{-7} m^4$, $\sum_{n=7}^{11} n^4$

* 65. Suppose that g is some function whose domain includes the integers.
  (a) ***(Telescoping sum)*** Prove that $\Sigma_{k=1}^{n} [g(k) - g(k-1)] = g(n) - g(0)$.
  (b) If we let $g(k) = \frac{1}{2}k \cdot (k+1)$, show that $g(k) - g(k-1) = k$, and part (a) yields
$$\sum_{k=1}^{n} k = \frac{n(n+1)}{2}.$$
  Similarly, if we use $G(k) = \frac{1}{3}k \cdot (k+1) \cdot (k+2)$, show that $G(k) - G(k-1) = k(k+1)$.
* 66. Show that $\Sigma_{k=1}^{n} k^2 = n(n+1)(2n+1)/6$. [Hint: Find $\Sigma_{k=1}^{n} k(k+1)$ by using the result of Problem 65(b).]
* 67. Find a short formula with which to compute $\Sigma_{k=1}^{n} k^3$. [Hint: First look at a telescoping sum with $g(k) = \frac{1}{4}k \cdot (k+1) \cdot (k+2) \cdot (k+3)$ and then use formulas for $\Sigma_{k=1}^{n} k$ and $\Sigma_{k=1}^{n} k^2$.]
* 68. Find a short formula for computing the sum of the first n odd integers. Show that your formula is correct for all n. [Hint: Try a telescoping sum with $g(k) = k^2$.]
* 69. Find a short formula by which to compute
$$\frac{1}{3} + \frac{1}{8} + \frac{1}{15} + \frac{1}{24} + \frac{1}{35} + \cdots + \frac{1}{n(n+2)} = \frac{1}{1 \cdot 3} + \frac{1}{2 \cdot 4} + \frac{1}{3 \cdot 5} + \frac{1}{4 \cdot 6} + \cdots + \frac{1}{n^2 + 2n}$$

* 70. Find a short formula for computing

$$\frac{1}{2}+\frac{1}{6}+\frac{1}{12}+\frac{1}{20}+\cdots+\frac{1}{n^2+n}=\frac{1}{1\cdot 2}+\frac{1}{2\cdot 3}+\frac{1}{3\cdot 4}+\frac{1}{4\cdot 5}+\cdots+\frac{1}{(n+1)\cdot n}$$

$\left[\text{Hint: } \frac{1}{k\cdot(k+1)}=\frac{1}{k}-\frac{1}{k+1}.\right]$

71. Prove formula (5) by writing out the terms in $\sum_{k=M}^{N}(a_k+b_k)$.

72. Prove formula (6).

$\left[\text{Hint: Use (4) to show that } \sum_{k=M}^{N}(-a_k)=-\sum_{k=M}^{N}a_k. \text{ Then use (5).}\right]$

73. Prove formula (7).

9.3 First-Order Recursion Relations: Arithmetic and Geometric Sequences

Definition

A **first-order recursion relation** for the sequence $\{a_0, a_1, a_2, \ldots, a_n, \ldots\}$ is an equation that relates a_{n+1} to a_n. The first term in the sequence is called the **initial value** of the sequence.

EXAMPLE 1 *Three First-Order Recursions*

The following are first-order recursion relations:

(a) $a_{n+1}=a_n+5$
(b) $p_{n+1}=1.1p_n$
(c) $b_{n+1}=b_n^2+5b_n-6$

NOTE Recursion relations are often called **difference equations.**

A first-order recursion relation together with a given term generates a sequence.

EXAMPLE 2 *Finding the General Term of a Sequence*

Find the sequence generated by the recursion relation

$$a_{n+1}=a_n+5 \qquad \text{if } a_1=2 \tag{1}$$

SOLUTION Using (1) repeatedly, we have

$$\begin{aligned}
n=1 &\quad a_2=a_1+5=2+5=7\\
n=2 &\quad a_3=a_2+5=7+5=12=2+5\cdot 2\\
n=3 &\quad a_4=a_3+5=12+5=17=2+5\cdot 3\\
n=4 &\quad a_5=a_4+5=17+5=22=2+5\cdot 4
\end{aligned}$$

Do you see the pattern? We will prove in a moment that

$$a_n = 2 + 5(n - 1)$$

If we write out the terms in the sequence, we obtain

$$\{a_1, a_2, a_3, \ldots\} = \{2, 7, 12, 17, 22, \ldots\}$$

Arithmetic Sequences

Definition

An **arithmetic sequence** is a sequence $\{a_n\}$ generated by a first-order recursion relation. An arithmetic sequence takes one of the forms

$$a_{n+1} = a_n + d \qquad \text{or} \qquad a_{n+1} - a_n = d \tag{2}$$

The number d is called the **common difference** between a_{n+1} and a_n.

Theorem

For an arithmetic sequence,

$$a_n = a_1 + (n - 1)d \tag{3}$$

Proof

We prove this theorem by mathematical induction.

Step 1 If $n = 1$, (3) becomes

$$a_1 = a_1 + (1 - 1)d = a_1$$

so (3) holds for $n = 1$.

Step 2 Assume that (3) holds for $n = k > 1$. That is,

Induction Hypothesis $\quad a_k = a_1 + (k - 1)d$

Then, from (2),

$$\begin{aligned} a_{k+1} &= a_k + d = \overbrace{a_1 + (k - 1)d}^{= a_k} + d = a_1 + [(k - 1) + 1]d \\ &= a_1 + kd = a_1 + [(k + 1) - 1]d \end{aligned}$$

which is equation (3) for $n = k + 1$. This proves the theorem. ■

EXAMPLE 3 *Finding a Particular Term of an Arithmetic Sequence*

Find the 20th term of the sequence $\{5, 9, 13, 17, 21, \ldots\}$.

SOLUTION The difference between successive terms is 4 and $a_1 = 5$, so, from (3),

$$a_n = 5 + 4(n - 1)$$

When $n = 20$, we obtain

$$a_{20} = 5 + 4(20 - 1) = 5 + 4(19) = 5 + 76 = 81$$

Sums of Terms of Arithmetic Sequences

Let $\{a_n\}$ be an arithmetic sequence with common difference d and define

$$S_N = \sum_{k=1}^{N} a_k$$

That is, S_N is the sum of the first N terms of the arithmetic sequence.

Theorem: Finite Sum of an Arithmetic Sequence

$$S_N = Na_1 + N(N - 1)\frac{d}{2} \tag{4}$$

Proof

We use (3) repeatedly. We have

$$\begin{aligned} a_2 &= a_1 + d \\ a_3 &= a_1 + 2d \\ a_4 &= a_1 + 3d \\ &\vdots \\ a_N &= a_1 + (N - 1)d \end{aligned}$$

So

$$\begin{aligned} S_N = \sum_{k=1}^{N} a_k &= a_1 + a_2 + a_3 + \cdots + a_N \\ &= \underbrace{a_1+(a_1+d)+(a_1+2d)+(a_1+3d)+\cdots+[a_1+(N-1)d]}_{\text{N terms}} \\ &= Na_1 + d + 2d + 3d + \cdots + (N - 1)d \\ &= Na_1 + [1 + 2 + 3 + \cdots + (N - 1)]d \end{aligned} \tag{5}$$

But, from the result of Example 1 on p. 476,

$$1 + 2 + 3 + \cdots + n = \frac{n(n + 1)}{2} \tag{6}$$

Since there are $N - 1$ terms between the brackets in (5), we set $n = N - 1$ in (6) to obtain

$$1 + 2 + 3 + \cdots + (N - 1) = \frac{(N - 1)[(N - 1) + 1]}{2} = \frac{N(N - 1)}{2}$$

Inserting this value into (5) gives us

$$S_N = Na_1 + d\left[\frac{N(N - 1)}{2}\right] = Na_1 + N(N - 1)\frac{d}{2} \quad ■$$

EXAMPLE 4 ***Finding the Sum of the First 20 Terms of a Sequence***

Find the sum of the first 20 terms of the sequence $\{5, 9, 13, 17, 21, \ldots\}$.

SOLUTION In Example 3, we found that $a_n = 5 + 4(n - 1)$. Thus $a_1 = 5$, $d = 4$, and, from (4), we obtain

$$S_{20} = 20a_1 + 20(19)\frac{d}{2} = 20(5) + (380)\left(\frac{4}{2}\right)$$
$$= 100 + (380)(2) = 100 + 760 = 860$$

We now turn to a different type of example. To make the formulas we develop simpler, we will consider a_0 (where $n = 0$) to be the first term.

EXAMPLE 5 ***Finding the General Term of a Geometric Sequence***

Find the sequence generated by the recursion relation

$$a_{n+1} = 3a_n, \qquad a_0 = 2 \tag{7}$$

SOLUTION Using (7) repeatedly, we have

$$\begin{aligned}
n = 0 \quad & a_1 = 3a_0 = 3 \cdot 2 = 6 \\
n = 1 \quad & a_2 = 3a_1 = 3 \cdot 6 = 18 = 2 \cdot 3^2 \\
n = 2 \quad & a_3 = 3a_2 = 3 \cdot 18 = 54 = 2 \cdot 3^3 \\
n = 3 \quad & a_4 = 3a_3 = 3 \cdot 54 = 162 = 2 \cdot 3^4
\end{aligned}$$

Do you see the pattern? We will shortly indicate that

$$a_n = 2 \cdot 3^n$$

If we write out the terms in the sequence, we obtain

$$\{a_0, a_1, a_2, a_3, \ldots\} = \{2, 6, 18, 54, 162, 486, \ldots\}$$

This sequence is called a geometric sequence.

Geometric Sequences

Definition

A **geometric sequence** is a sequence $\{a_n\}$ generated by the first-order recursion relation

$$a_{n+1} = ra_n, \qquad r \neq 0, 1 \tag{6}$$

The number r is called the **common ratio** of a_{n+1} and a_n.

Theorem

For a geometric sequence

$$a_n = a_0 r^n \tag{9}$$

The proof, using mathematical induction, is left as an exercise (see Problem 86). We can use this theorem to help solve compound interest problems.

Compound Interest

Suppose P_0 dollars are invested in an enterprise (which may be, for instance, a bank, bonds, or a common stock) with an annual interest rate of r. **Simple interest** is the amount earned on the P dollars over a period of time. If the P_0 dollars are invested for n years, then the simple interest I is given by

Simple Interest Formula

$$I = P_0 rn \tag{10}$$

EXAMPLE 6 ***Computing Simple Interest***

If \$1000 is invested for 5 years with an interest rate of 6%, then $r = 0.06$ and the simple interest earned is

$$I = (\$1000)(0.06)(5) = \$300$$

Compound interest is interest paid on the interest previously earned as well as on the original investment. Suppose that interest is paid annually. Then if P_0 dollars are invested, the interest after one year is rP_0 dollars, and the original investment is now worth

$$P_0 + rP_0 = P_0(1 + r) \tag{11}$$

Now let P_n denote the value of the investment after n years. Then, using the reasoning that led to equation (11), P_{n+1}, the value of the investment after $n + 1$ years (one year after year n) is given by

$$P_{n+1} = P_n(1 + r) \tag{12}$$

That is, P_n satisfies a first-order recursion relation of the form (8), so $\{P_n\}$ is a geometric sequence. Thus, from (9)

Compound Interest Formula (Annual Compounding)

$$P_n = P_0(1 + r)^n \tag{13}$$

is the value of P_0 dollars invested for n years at an interest rate of r compounded annually.

EXAMPLE 7 *Computing Compound Interest*

If the interest in Example 6 is compounded annually, then after 5 years the investment is worth (with $P_0 = \$1000$, $r = 0.06$, and $n = 5$)

$$P_5 = 1000(1 + 0.06)^5 = 1000(1.06)^5$$
$$\approx \$1338.23$$

The following result is sometimes useful. You are asked to prove it in Problem 85.

Theorem: The Solution to a Certain Recursion Relation

If $r \neq 0, 1$, then the solution to

$$a_{n+1} = ra_n + b \tag{14}$$

is

$$a_n = \frac{b}{1 - r} + \left(a_0 - \frac{b}{1 - r}\right)r^n \tag{15}$$

There are many other types of first-order recursion relations. Unfortunately, it is usually impossible to find a nice formula like (3), (9), or (15) to represent the general term. This often presents little problem, however, because it is easy to generate the terms recursively on a computer.

Problems 9.3

Readiness Check

I. Which of the following is an arithmetic sequence?
a. $\{7, -7, 7, -7, 7, -7, \ldots\}$
b. $\{12, 4, \frac{4}{3}, \frac{4}{9}, \ldots\}$
c. $\{8, 5, 2, -1, -4, \ldots\}$
d. $\{-3, -6, -12, -24, \ldots\}$

II. Which of the following is true of $\{1, 3, 6, 10, \ldots\}$?
a. It is an arithmetic sequence and $d = n$.
b. It is a geometric sequence and $r = 2$.
c. It is a first-order recursion relation having 4 terms.
d. It is neither a geometric nor an arithmetic sequence.

III. Which of the following is true about $\{1, -3, 9, -27, \ldots\}$?
a. $d = 3$ b. $d = -3$
c. $r = 3$ d. $r = -3$

IV. Which of the following is true of $\{(\frac{2}{3})^{-n}\}$?
a. It is a geometric sequence with $r = \frac{3}{2}$.
b. It is an arithmetic sequence with $d = \frac{2}{3}$.
c. It is a geometric sequence with $r = \frac{2}{3}$.
d. It is neither a geometric nor an arithmetic sequence.

V. Which of the following is an arithmetic sequence?
a. $\left\{\frac{2^{n-1}}{4}\right\}$
b. $2, 4, 8, 16, \ldots$
c. $\frac{n-2}{5}$
d. $2, -4, 6, -8, \ldots$

In Problems 1–6 the first five terms of an arithmetic sequence are given. Write out the general term in the form $a_n = a_1 + (n - 1)d$.

1. $\{1, 4, 7, 10, 13, \ldots\}$
2. $\{6, 11, 16, 21, 26, \ldots\}$
3. $\{4, 1, -2, -5, -8, \ldots\}$
4. $\left\{2, \frac{5}{2}, 3, \frac{7}{2}, 4, \ldots\right\}$
5. $\{-12, -6, 0, 6, 12, \ldots\}$
6. $\{0, 5, 10, 15, 20, \ldots\}$

In Problems 7–12 an arithmetic sequence is given. Write out the first five terms. Start with $n = 1$.

7. $\{4 + n\}$
8. $\{3 + 2n\}$
9. $\left\{6 - \frac{1}{3}n\right\}$
10. $\{50 + 80n\}$
11. $\left\{\frac{1}{2} + 6n\right\}$
12. $\{25n\}$

In Problems 13–18 the first five terms of a geometric sequence are given. Write out the general term in the form $a_n = a_0 r^n$.

13. $\{1, 2, 4, 8, 16, \ldots\}$
14. $\{2, 6, 18, 54, 162, \ldots\}$
15. $\{6, -6, 6, -6, 6, \ldots\}$
16. $\{80, 40, 20, 10, 5, \ldots\}$
17. $\left\{7, \frac{7}{3}, \frac{7}{9}, \frac{7}{27}, \frac{7}{81}, \ldots\right\}$
18. $\{1, -2, 4, -8, 16, \ldots\}$

In Problems 19–24 a geometric sequence is given. Write out the first six terms. Start with $n = 0$.

19. $\{5 \cdot 2^n\}$
20. $\left\{10\left(\frac{1}{2}\right)^n\right\}$
21. $\{3(-1)^n\}$
22. $\{-3(2)^n\}$
23. $\{2^{2n}\}$
24. $\left\{\frac{5^n}{10}\right\}$

In Problems 25–40 a sequence is given. Determine whether it is arithmetic, geometric, or neither. If arithmetic, give the common difference. If geometric, determine the common ratio.

25. $\{6n - 3\}$
26. $\{6^n - 3\}$
27. $\{3 \cdot 6^n\}$
28. $\left\{\frac{4 + 3n^2}{5}\right\}$
29. $\left\{\frac{4 + 3n}{5}\right\}$
30. $\left\{\frac{3^n}{5}\right\}$
31. $\left\{\frac{4 + 3^n}{5}\right\}$
32. $\{-1 - n\}$
33. $\{(-1)^n\}$
34. $\{n^{-1}\}$
35. $\{2 + 10n\}$
36. $\{2 \cdot 10^n\}$
37. $\{1, -1, 1, -1, 1, \ldots\}$

Answers to Readiness Check

I. c II. d III. d IV. a V. c

38. $\{2, 3, 5, 8, 13, \ldots\}$
39. $\{3, -6, 12, -24, 48, \ldots\}$
40. $\{2, 12, 22, 32, 42, \ldots\}$
41. Find the 10th term of the arithmetic sequence $\{3, 7, 11, \ldots\}$.
42. Find the 50th term of the arithmetic sequence $\{4, \frac{9}{2}, 5, \ldots\}$.
43. Find the 100th term of the arithmetic sequence $\{12, 9, 6, \ldots\}$.
44. Find the eighth term of the geometric sequence $\{2, 6, 18, \ldots\}$.
45. Find the eighth term of the geometric sequence $\{4, 2, 1, \ldots\}$.
46. Find the seventh term of the geometric sequence $\{3, 4, \frac{16}{3}, \ldots\}$.
47. If the sixth term of an arithmetic sequence is 120 and the common difference is 5, find the first term a_1.
48. If the 12th term of an arithmetic sequence is 5 and the common difference is $-\frac{1}{2}$, find the first term a_1.

In Problems 49–52 two terms of an arithmetic sequence are given. Find the common difference.

49. $a_4 = 10$, $a_8 = 26$
50. $a_1 = 6$, $a_{10} = 33$
51. $a_8 = 8$, $a_{24} = 4$
52. $a_4 = 3$, $a_{10} = -21$
53. If the fifth term of a geometric sequence is 80 and the common ratio is 2, find the first term a_0.
54. If the sixth term of a geometric sequence is $\frac{1}{8}$ and the common ratio is $-\frac{1}{2}$, find the first term.

In Problems 55–58 two terms of a geometric sequence are given. Find the common ratio.

55. $a_2 = 4$, $a_7 = 972$
56. $a_2 = -9$, $a_6 = -\dfrac{1}{9}$
57. $a_3 = 4$, $a_7 = 15$ [Round to 4 decimal places.]
58. $a_6 = 12$, $a_{11} = -5$

In Problems 59–62 find the value of an investment after n years if P_0 dollars is invested at an interest rate of r, compounded annually.

59. $P_0 = \$5{,}000$, $n = 5$, $r = 8\%$ $(= 0.08)$
60. $P_0 = \$1{,}500$, $n = 8$, $r = 12\%$
61. $P_0 = \$10{,}000$, $n = 10$, $r = 6\frac{1}{2}\%$
62. $P_0 = \$6{,}250$, $n = 20$, $r = 7\frac{1}{2}\%$

* 63. Suppose that P_0 dollars is invested for n years at an interest rate of r, and that interest is compounded m times a year. If P_n denotes the value of the investment after n years, show that

$$P_n = P_0\left(1 + \frac{r}{m}\right)^{mn}$$

64. A sum of \$5000 is invested for 8 years at a return of 7% per year. How much simple interest is paid over that period of time, assuming that interest is not compounded?
65. If the money in Problem 64 is compounded annually, what is the investment worth after 8 years?
66. If the money in Problem 64 is compounded monthly, what is the investment worth after 8 years? [Hint: Use the result of Problem 63.]
67. Calculate the percentage increase in return on investment if P_0 dollars are invested for 10 years at 6% compounded annually and quarterly.
68. As a gimmick to lure depositors, some banks offer 5% interest compounded daily in comparison with their competitors who offer $5\frac{1}{8}\%$ compounded annually. Which bank would you choose?
69. Suppose a competitor in Problem 68 now compounds $5\frac{1}{8}\%$ semiannually. Which bank would you choose?
70. Radium transmutes (decays) at a rate of 1 percent every 25 years. Consider a sample of r_0 grams of radium. If r_n is the amount of radium remaining in the sample after $25n$ years, obtain a difference equation for r_n and find its solution. How much radium is left after 100 years? [Hint: The amount of radium present at any time is 99% of what it was 25 years earlier.]

* 71. A fair coin is marked "1" on one side and "2" on the other side. The coin is tossed repeatedly, and a cumulative sum of the outcomes is recorded. Define P_n to be the probability that at some time the cumulative sum takes on the value n, where n is the number of the toss. Prove that $P_n = \frac{1}{2}P_{n-1}$. Assuming that $P_0 = 1$, derive the formula for P_n. [Hint: the probability that either side turns up on one toss is $\frac{1}{2}$]

In Problems 72–76 find an expression for a_n.

72. $a_{n+1} = 3a_n + 5$; $a_0 = 1$
73. $a_{n+1} = \frac{1}{2}a_n - 3$; $a_0 = 4$
74. $a_{n+1} = -a_n + 2$; $a_0 = 2$
75. $a_{n+1} = -2a_n + 10$; $a_0 = 5$
76. $a_{n+1} = (1.08)a_n - 10$; $a_0 = 0$
77. An individual invests \$10,000 in an account that pays 8% interest compounded annually. At the end of each year, she removes \$300.
    (a) Find a recurrence relation for the amount of money in the account after n years.
    (b) Solve this equation using $P_0 = \$10{,}000$.
    (c) How much is in the account after 5 years? after 10 years?

In Problems 78–80 find a general expression for a_n.

78. $a_{n+1} = (n + 1)a_n$, $a_1 = 1$

79. $a_{n+1} = \dfrac{n+2}{n+1} a_n,\ a_1 = 2$

80. $a_{n+1} = \dfrac{n+5}{n+3} a_n,\ a_1 = 1$

In Problems 81–84 find the first six terms of the sequence generated by the given recursion relation.

81. $a_{n+1} = a_n + n^2,\ a_1 = 2$
82. $a_{n+1} = \sqrt{a_n + 1},\ a_0 = 0$
83. $a_{n+1} = \dfrac{a_n}{2 + a_n},\ a_1 = 4$
84. $a_{n+1} - na_n = n!,\ a_0 = 5$

* 85. Show that if $r \neq 0, 1$ then the solution to $a_{n+1} = ra_n + b$ is

$$a_n = \frac{b}{1-r} + \left(a_0 - \frac{b}{1-r}\right) r^n$$

[Hint: Use mathematical induction. The result is true for $n = 0$ because $r^0 = 1$.]

86. If $a_{n+1} = ra_n$, show that $a_n = a_0 r^n$. [Hint: Use mathematical induction starting with $n = 0$].

9.4 Geometric Progressions and Series

In Section 9.3, we considered the geometric sequence whose general term was

$$a_n = a_0 r^n, \qquad r \neq 0, 1 \tag{1}$$

In this section, we will take both finite and infinite sums of terms from a geometric sequence.

An infinite sum is called an **infinite series.** Infinite series are important in a wide variety of advanced applications. Most calculus books devote an entire chapter to them. In this section, we will look only at one kind of infinite series. We begin with finite sums.

Consider the sum

$$S_7 = 1 + 2 + 4 + 8 + 16 + 32 + 64 + 128$$

This can be written as

$$S_7 = 1 + 2 + 2^2 + 2^3 + 2^4 + 2^5 + 2^6 + 2^7 = \sum_{k=0}^{7} 2^k$$

Definition of the Sum of a Geometric Progression

The **sum of a geometric progression** is a finite sum of the form

$$S_n = 1 + r + r^2 + r^3 + \cdots + r^{n-1} + r^n = \sum_{k=0}^{n} r^k$$

where r is a real number and n is a fixed positive integer.

We now obtain a formula for the sum of a geometric progression. The following result was proved in Example 3 in Section 9.1.

Theorem

If $r \neq 0$ or 1, the sum of a geometric progression $S_n = \Sigma_{k=0}^{n} r^k$ is given by

$$S_n = \frac{1 - r^{n+1}}{1 - r} \tag{2}$$

NOTE If $r = 1$, we obtain

$$S_n = \overbrace{1 + 1 + \cdots + 1}^{n+1 \text{ terms}} = n + 1$$

EXAMPLE 1 ***Calculating the Sum of a Geometric Progression***

Calculate $S_7 = 1 + 2 + 4 + 8 + 16 + 32 + 64 + 128$, using formula (2).

SOLUTION Here $r = 2$ and $n = 7$, so

$$S_7 = \frac{1 - 2^8}{1 - 2} = 2^8 - 1 = 256 - 1 = 255 \quad \blacksquare$$

EXAMPLE 2 ***Calculating the Sum of a Geometric Progression***

Calculate $\Sigma_{k=0}^{10} (\frac{1}{2})^k$.

SOLUTION Here $r = \frac{1}{2}$ and $n = 10$, so

$$S_{10} = \frac{1 - (\frac{1}{2})^{11}}{1 - \frac{1}{2}} = \frac{1 - \frac{1}{2048}}{\frac{1}{2}} = 2\left(\frac{2047}{2048}\right) = \frac{2047}{1024} \quad \blacksquare$$

EXAMPLE 3 ***Calculating the Sum of a Geometric Progression***

Calculate

$$S_6 = 1 - \frac{2}{3} + \left(\frac{2}{3}\right)^2 - \left(\frac{2}{3}\right)^3 + \left(\frac{2}{3}\right)^4 - \left(\frac{2}{3}\right)^5 + \left(\frac{2}{3}\right)^6$$

$$= \sum_{k=0}^{6} \left(-\frac{2}{3}\right)^k$$

SOLUTION Here $r = -\frac{2}{3}$ and $n = 6$, so

$$S_6 = \frac{1 - (-\frac{2}{3})^7}{1 - (-\frac{2}{3})} = \frac{1 + \frac{128}{2187}}{\frac{5}{3}} = \frac{3}{5}\left(\frac{2315}{2187}\right) = \frac{463}{729}$$

EXAMPLE 4 *Calculating the Sum of a Geometric Progression*

Calculate the sum $1 + b^2 + b^4 + b^6 + \cdots + b^{20} = \sum_{k=0}^{10} b^{2k}$ for $b \neq \pm 1$.

SOLUTION Note that the sum can be written $1 + b^2 + (b^2)^2 + (b^2)^3 + \cdots + (b^2)^{10}$. Here $r = b^2 \neq 1$ and $n = 10$, so

$$S_{10} = \frac{1 - (b^2)^{11}}{1 - b^2} = \frac{1 - b^{22}}{1 - b^2}$$

The sum of a geometric progression is the sum of a finite number of terms. We now show what happens if the number of terms is infinite. Consider the sum

$$S = 1 + \frac{1}{2} + \frac{1}{4} + \frac{1}{8} + \frac{1}{16} + \cdots = \sum_{k=0}^{\infty} \left(\frac{1}{2}\right)^k$$

What can such a sum mean? We will give a formal definition in a moment. For now, let us show why it is reasonable to say that $S = 2$. Let $S_n = \sum_{k=0}^{n} (\frac{1}{2})^k = 1 + \frac{1}{2} + \frac{1}{4} + \cdots + (\frac{1}{2})^n$. Then

$$S_n = \frac{1 - (\frac{1}{2})^{n+1}}{1 - \frac{1}{2}} = 2\left[1 - \left(\frac{1}{2}\right)^{n+1}\right]$$

Thus for any n (no matter how large), $1 \leq S_n < 2$. Also,

$$2 - S_n = 2(\tfrac{1}{2})^{n+1} = (\tfrac{1}{2})^n$$

In Table 1, we give values of $(\frac{1}{2})^n$ for various values of n.

Table 1

n	1	2	4	8	16	25	50	100	200
$(\frac{1}{2})^n$	$\frac{1}{2} = 0.5$	0.25	0.0625	0.00390625	0.000015258	0.000000029	8.9×10^{-16}	7.9×10^{-31}	6.2×10^{-61}

We see that $(\frac{1}{2})^n$ gets arbitrarily small as n increases. In the notation of Section 4.2 (p. 237), we have

$$(\tfrac{1}{2})^n \longrightarrow 0 \qquad \text{as } n \to \infty$$

Thus, since $2 - S_n = (\frac{1}{2})^n$,

$$2 - S_n \longrightarrow 0 \qquad \text{as } n \to \infty$$

or

$$S_n \longrightarrow 2 \qquad \text{as } n \to \infty$$

This means that

$$S_n = 1 + \tfrac{1}{2} + (\tfrac{1}{2})^2 + \cdots + (\tfrac{1}{2})^n \longrightarrow 2 \qquad \text{as } n \to \infty$$

But, as $n \to \infty$, $S_n = \sum_{k=0}^{n} (\frac{1}{2})^k \to \sum_{k=0}^{\infty} (\frac{1}{2})^k$.

Thus, it makes sense to *define*

$$S = \sum_{k=0}^{\infty} (\tfrac{1}{2})^k = 2$$

Geometric Series

Definition

The infinite sum $\sum_{k=0}^{\infty} (\frac{1}{2})^k$ is called a **geometric series.** In general, a **geometric series** is an infinite sum of the form

$$S = \sum_{k=0}^{\infty} r^k = 1 + r + r^2 + r^3 + \cdots \tag{3}$$

Now, consider the sum, for $-1 < r < 1$,

$$S_n = \sum_{k=0}^{n} r^k = \frac{1 - r^{n+1}}{1 - r} = \frac{1}{1 - r} - \frac{r^{n+1}}{1 - r} \tag{4}$$

For any value of r between -1 and 1, we can construct a table like Table 1 to show that

$$r^n \longrightarrow 0 \qquad \text{as } n \to \infty \quad \text{if } -1 < r < 1$$

Since r is fixed as n varies, this shows that if $-1 < r < 1$,

$$\frac{r^{n+1}}{1 - r} \longrightarrow 0 \qquad \text{as } n \to \infty$$

Thus, from (4), we see that

$$S_n \longrightarrow \frac{1}{1 - r} \qquad \text{as } n \to \infty \quad \text{if } -1 < r < 1$$

Finally, we have

Theorem: Value of a Geometric Series

If $-1 < r < 1$,

$$S = \sum_{k=0}^{\infty} r^k = \frac{1}{1 - r}. \tag{5}$$

EXAMPLE 5 ***The Values of Three Geometric Series***

(a) $1 - \frac{2}{3} + (\frac{2}{3})^2 - \cdots = \sum_{k=0}^{\infty} (-\frac{2}{3})^k = \dfrac{1}{1-\left(-\frac{2}{3}\right)} = \dfrac{1}{\frac{5}{3}} = \dfrac{3}{5}$

(b) $$1 + \frac{\pi}{4} + \left(\frac{\pi}{4}\right)^2 + \left(\frac{\pi}{4}\right)^3 + \cdots = \sum_{k=0}^{\infty} \left(\frac{\pi}{4}\right)^k = \frac{1}{1-\frac{\pi}{4}} = \frac{4}{4-\pi} \approx 4.66$$

(c) We can write the number $\frac{1}{3}$ as

$$\frac{1}{3} = 0.33333\ldots = \frac{3}{10} + \frac{3}{100} + \frac{3}{1000} + \cdots + \frac{3}{10^n} + \cdots$$
$$= \frac{3}{10}\left(1 + \frac{1}{10} + \frac{1}{100} + \frac{1}{1000} + \cdots\right)$$
$$= \frac{3}{10}\left[\sum_{k=0}^{\infty} \left(\frac{1}{10}\right)^k\right]$$

That is, we can write $\frac{1}{3}$ as a constant times a geometric series. ■

EXAMPLE 6 ***Writing a Repeating Decimal as a Rational Number***

Express the **repeating decimal** 0.123123123 . . . as a rational number (the quotient of two integers).

SOLUTION

$$0.123123123\ldots$$
$$= 0.123 + 0.000123 + 0.000000123 + \cdots$$
$$= \frac{123}{10^3} + \frac{123}{10^6} + \frac{123}{10^9} + \cdots = \frac{123}{10^3}\left[1 + \frac{1}{10^3} + \frac{1}{(10^3)^2} + \cdots\right]$$

Equation (5) ↓

$$= \frac{123}{1000}\sum_{k=0}^{\infty}\left(\frac{1}{1000}\right)^k = \frac{123}{1000}\left[\frac{1}{1-\frac{1}{1000}}\right] = \frac{123}{1000} \cdot \frac{1}{\frac{999}{1000}}$$
$$= \frac{123}{1000} \cdot \frac{1000}{999} = \frac{123}{999} = \frac{41}{333}$$

In general, we can use the geometric series to write any repeating decimal in the form of a fraction by using the technique of Example 6. In fact, *the rational numbers are exactly those real numbers that can be written as repeating decimals*. Repeating decimals include numbers like $3 = 3.00000\ldots$ and $\frac{1}{4} = 0.25 = 0.25000000\ldots$

Annuities

Geometric progressions and series are useful in a number of business applications. One of them is described here.

In Section 9.3 we discussed the compound interest formula

$$P_n = P_0(1 + r)^n$$

where

P_0 is the original principal
r is the annual interest rate
n is the number of years the investment is held
P_n is the amount (in dollars) after n years.

Moreover, if interest is compounded m times a year, we have (see Problem 63 in Section 9.3)

$$P_n = P_0\left(1 + \frac{r}{m}\right)^{mn}$$

Often, people (and corporations) do not deposit large sums of money and then sit back and watch them grow. Rather, money is invested in smaller amounts at periodic intervals (for example, monthly deposits in a bank, annual life insurance premiums, and installment loan payments).

An **annuity** is a fixed amount of money that is paid or received at regular intervals. The time between successive payments of an annuity is its **payment interval,** and the time from the beginning of the first interval to the end of the last interval is called its **term.** The value of the annuity after the last payment has been made or received is called the **future value** of the annuity. If payment is made and interest is computed at the *end* of each time period, then the annuity is called an **ordinary annuity.**

EXAMPLE 7 *Calculating the Future Value of an Annuity*

Suppose \$1000 is invested in a savings plan at the end of each year and 7% interest is paid, compounded annually. How much will be in the account after 4 years?

SOLUTION To find the value of the annuity after 4 years, we compute the value of each of the 4 payments after 4 years and then find the sum of these payments. First, \$1000 deposited at the end of the first year will be earning interest for 3 years. Thus the \$1000 will be worth

$$1000(1 + 0.08)^3 = \$1259.71$$

Similarly, \$1000 deposited at the end of the second year will be earning interest for 2 years. Thus it will be worth

$$1000(1.08)^2 = \$1166.40$$

Continuing, we see that, after the fourth year, the \$1000 invested at the end of the third year will be worth

$$1000(1.08) = \$1080$$

Finally, the \$1000 invested at the end of the fourth year will not earn any interest and so will be worth \$1000. Summing, we have

$$\begin{aligned}\text{future value of the annuity} &= \$1259.71 + \$1166.40 + \$1080 + \$1000 \\ &= \$4506.11\end{aligned}$$

As Example 7 suggests, computing the future value of an annuity by calculating the future value of each payment separately can be a tedious undertaking. Imagine trying to compute the future value of annuity consisting of 360 payments (as in a 30-year mortgage). Fortunately, there is a much easier way to do it.

Suppose that B dollars are deposited or received at the end of each time period. An interest rate of r is paid at the same time. Let A_n denote the amount in the account after n time periods. Then after one period, we deposit B dollars; after two periods, the B dollars have now become $B(1 + r)$, and we deposit another B dollars to obtain $A_2 = B + B(1 + r)$. After three periods, we have $A_3 = B + B(1 + r) + B(1 + r)^2$, and, after n periods,

$$\begin{aligned}A_n &= B + B(1 + r) + B(1 + r)^2 + \cdots + B(1 + r)^{n-1} \\ &= B[1 + (1 + r) + (1 + r)^2 + \cdots + (1 + r)^{n-1}]\end{aligned}$$

From equation (2)
↓

$$= B\left[\frac{1 - (1 + r)^n}{1 - (1 + r)}\right] = B\left[\frac{1 - (1 + r)^n}{-r}\right] = B\left[\frac{(1 + r)^n - 1}{r}\right]$$

Thus we have the following:

Future Value, A_n, of an Annuity with Interest Compounded Annually

$$A_n = \frac{B[(1 + r)^n - 1]}{r} \tag{6}$$

where B is the amount deposited at the end of each year, r is the annual interest rate, n is the number of years, and A_n is the future value (in dollars) of the annuity after n years.

EXAMPLE 7 (CONTINUED)

We can solve Example 7 much more quickly if we use equation (6). We have

$$A_4 = \frac{1000[(1.08)^4 - 1]}{0.08} = \$4506.11$$

If interest is compounded m times a year, a formula similar to (6) holds.

Future Value, A_n, of an Annuity with Interest Compounded m Times in the Interval of Deposit

$$A_n = B\left\{\frac{\left[1+\frac{r}{m}\right]^{mn}-1}{\left[1+\frac{r}{m}\right]^{m}-1}\right\} \tag{7}$$

where n is the number of deposits of the amount B and r is the rate of interest for the interval of deposit.

EXAMPLE 8 *Calculating the Future Value of an Annuity*

If a man deposits \$500 every 6 months and this is compounded quarterly at 6%, how much does he have after 10 years?

SOLUTION Here $B = 500$ and $r = 0.03$ since the interval of deposit is $\frac{1}{2}$ year. Then $m = 2$ (interest is compounded twice every $\frac{1}{2}$ year), $n = 20$ (there are 20 semiannual deposits), and

$$A_{20} = 500\left\{\frac{\left[1+\frac{0.03}{2}\right]^{2(20)}-1}{\left[1+\frac{0.03}{2}\right]^{2}-1}\right\}$$

$$= 500\left(\frac{1.015^{40}-1}{1.015^{2}-1}\right) \approx 500\left(\frac{1.814018-1}{1.030225-1}\right) = 13{,}465.98$$

FOCUS ON
Zeno's Paradox

The idea of an infinite number of numbers having a finite sum is a natural one now, but it was not always the case. Some of the early work on limits was motivated by unresolved questions that had been posed by Greek mathematicians. For example, the fifth-century-B.C. philosopher and mathematician Zeno (ca. 495–435 B.C.) posed four problems that came to be known as **Zeno's paradoxes.** In the second of these, Zeno argued that the legendary Greek hero Achilles could never overtake a tortoise. Suppose that the tortoise starts 100 yd ahead and that Achilles can run 10 times as fast as the tortoise. Then when Achilles has run 100 yd, the tortoise has run 10 yd, and when Achilles has covered this distance, the tortoise is still a yard ahead; and so on. It seems that the tortoise will stay ahead!

We can view this seeming paradox in another way, which is equally contradictory to common sense. Suppose that a man is standing a certain distance, say 10 ft, from a door (see Figure 1).

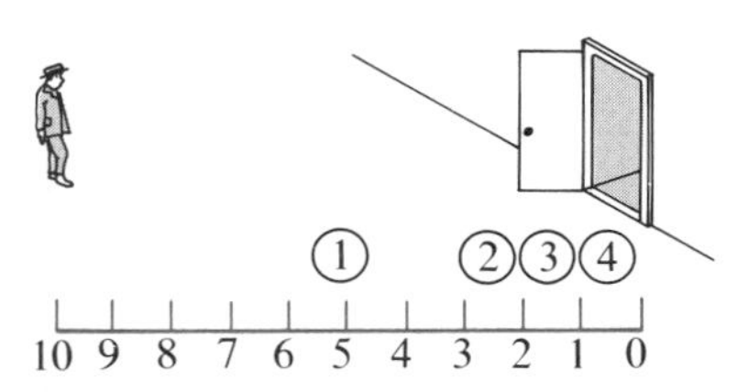

Figure 1 The distance from a man to a door.

Using Zeno's reasoning, we may claim that it is impossible for the man to walk to the door. In order to reach the door, the man must walk half the distance (5 ft) to the door. He then reaches point ① on Figure 1. From point ①, 5 ft from the door, he must again walk halfway ($2\frac{1}{2}$ ft) to the door, to point ②. Continuing in this manner, no matter how close he comes to the door, he must walk halfway to the door and halfway from there and halfway from there and so on. Thus, no matter how close the man gets to the door, he still has half of some remaining distance to cover. It seems that the man will never actually reach the door. Of course, this contradicts our common sense. But where is the flaw in Zeno's reasoning?

It took more than 2000 years for mathematicians to provide a satisfactory answer to this question, and in order to do so, they had to use the notion of a geometric series. Intuitively, we sense that Zeno's man is indeed covering an infinite number of intervals in his walk toward the door, but each interval is over a shorter and shorter distance and, therefore, takes less and less time. Indeed, the time necessary to walk over each succeeding interval "approaches" the limit zero, thus allowing the man to reach the door. Let us prove that the man can indeed reach the door in finite time.

Suppose the man in Figure 1 starts walking toward the door at the fixed velocity of 5 ft/sec. Let us calculate the time it takes him to walk to the door, using Zeno's argument. Since velocity × time = distance, we have t = distance/velocity, where t stands for time. Thus it takes the man $t = (5 \text{ ft})/(5 \text{ ft/sec}) = 1$ sec to walk to the point 5 ft from the door (recall that he starts 10 ft from the door). To walk to the next point, $2\frac{1}{2}$ ft from the door, takes $(2\frac{1}{2} \text{ ft})/(5 \text{ ft/sec}) = \frac{1}{2}$ sec. The next point takes $(1\frac{1}{4} \text{ ft})/(5 \text{ ft/sec}) = \frac{1}{4}$ sec to reach. It is clear that to reach succeeding points, each half the distance to the door from the preceding point, the man will take $\frac{1}{8}$ sec, $\frac{1}{16}$ sec, . . . , $(\frac{1}{2})^n$ sec, ... Thus the total time he takes to walk to the door is

$$t = 1 + \tfrac{1}{2} + \tfrac{1}{4} + \tfrac{1}{8} + \tfrac{1}{16} + \cdots = 2 \text{ seconds}$$

since this is nothing but the sum of a geometric series with $r = \frac{1}{2}$. Hence, the man will reach the door in 2 seconds, which is certainly not surprising. Therefore we see that with the concept of an infinite sum, Zeno's paradox is really no paradox at all.

In Problem 27 you are asked to "explain" the seeming paradox in the original version of Zeno's paradox: the race between Achilles and the tortoise.

Problems 9.4

Readiness Check

I. Which of the following is true about the sum of a geometric progression?
   a. It is a sum of the form $\Sigma_{k=0}^{n}\ n$.
   b. It is a sum of the form $\Sigma_{k=0}^{n}\ k$.
   c. It is a sum of the form $\Sigma_{k=0}^{n}\ r^k$.
   d. It is a sum of the form $\Sigma_{k=0}^{n}\ k^2$.

II. Which of the following is true about an infinite geometric series $S = \Sigma_{k=0}^{\infty}\ r^k$?
   a. The sum exists if $r > 1$ or $r < -1$.
   b. The sum exists if $-1 < r < 1$.
   c. The sum exists if $r = 1$ or $r = -1$.
   d. The sum never exists.

III. Which of the following is true about

$$1 - \tfrac{3}{4} + \tfrac{9}{16} - \tfrac{27}{64} + \cdots?$$

   a. The sum is $\frac{4}{7}$.
   b. It is a finite geometric series with $r = \frac{3}{4}$.
   c. It is an infinite arithmetic series.
   d. There is no sum for this series.

IV. Which of the following is the sum of

$$1 - 3 + 9 - 27 + \cdots?$$

   a. 4 b. 3 c. -20
   d. The sum does not exist.

In Problems 1–11 calculate the sum of the given geometric progression.

1. $1 + 4 + 16 + 64 + 256$
2. $1 + \dfrac{1}{4} + \dfrac{1}{16} + \cdots + \dfrac{1}{4^8}$
3. $1 - 5 + 25 - 125 + 625 - 3125$
4. $0.2 + 0.2^2 + 0.2^3 + \cdots + 0.2^9$
5. $0.3^2 - 0.3^3 + 0.3^4 - 0.3^5 + 0.3^6 - 0.3^7 + 0.3^8$
6. $1 + b^4 + b^8 + b^{12} + b^{16} + b^{20}$
7. $1 - \dfrac{1}{b^2} + \dfrac{1}{b^4} - \dfrac{1}{b^6} + \dfrac{1}{b^8} - \dfrac{1}{b^{10}} + \dfrac{1}{b^{12}} - \dfrac{1}{b^{14}}$
8. $\pi - \pi^3 + \pi^5 - \pi^7 + \pi^9 - \pi^{11} + \pi^{13}$
9. $1 + \sqrt{2} + 2 + 2^{3/2} + 4 + 2^{5/2} + 8 + 2^{7/2} + 16$

Answers to Readiness Check

I. c II. b III. a IV. d

10. $1 - \frac{1}{\sqrt{3}} + \frac{1}{3} - \frac{1}{3\sqrt{3}} + \frac{1}{9} - \frac{1}{9\sqrt{3}} + \frac{1}{27} - \frac{1}{27\sqrt{3}} + \frac{1}{81}$
11. $-16 + 64 - 256 + 1024 - 4096$
12. A bacteria population initially contains 1000 organisms and each bacterium produces 2 live bacteria every 2 hr. How many organisms will be alive after 12 hr if none of the bacteria dies during the growing period?

In Problems 13–22 calculate the sum of the given geometric series.

13. $1 + \frac{1}{3} + \frac{1}{3^2} + \frac{1}{3^3} + \cdots$
14. $1 - \frac{1}{2} + \frac{1}{4} - \frac{1}{8} + \frac{1}{16} - \cdots$
15. $1 + \frac{1}{10} + \frac{1}{100} + \frac{1}{1000} + \cdots$
16. $1 - \frac{1}{10} + \frac{1}{100} - \frac{1}{1000} + \cdots$
17. $1 + \frac{1}{e} + \frac{1}{e^2} + \frac{1}{e^3} + \cdots$
18. $1 + 0.7 + 0.7^2 + 0.7^3 + \cdots$
19. $1 - 0.31 + 0.31^2 - 0.31^3 + \cdots$
20. $\frac{1}{4} + \frac{1}{16} + \frac{1}{64} + \cdots$ [Hint: Factor out the term $\frac{1}{4}$.]
21. $\frac{3}{5} - \frac{3}{25} + \frac{3}{125} - \cdots$
22. $\frac{1}{9} + \frac{1}{27} + \frac{1}{81} + \cdots$
23. How large must n be in order that $(\frac{1}{2})^n < 0.01$? [Hint: Use logarithms.]
24. How large must n be in order that $(0.8)^n < 0.01$?
25. How large must n be in order that $(0.99)^n < 0.01$?
26. Show that if $x > 1$,

$$1 + \frac{1}{x} + \frac{1}{x^2} + \frac{1}{x^3} + \cdots = \frac{x}{x-1}$$

* 27. Suppose that in the original version of Zeno's paradox, the tortoise is moving at a rate of 1 km/hr while Achilles is running at a rate of 201 km/hr. Give the tortoise a 40-km head start.
(a) Show, using the arguments of this section, that Achilles will really overtake the tortoise.
(b) How long will it take?

In Problems 28–37 find the future value of an ordinary annuity with payments of B dollars, over n periods, and at an interest rate of r per period.

28. $B = \$500$, $n = 10$, $r = 0.03$
29. $B = \$500$, $n = 10$, $r = 0.002$
30. $B = \$500$, $n = 8$, $r = 0.03$
31. $B = \$500$, $n = 10$, $r = 0.10$
32. $B = \$625$, $n = 30$, $r = 0.025$
33. $B = \$4000$, $n = 60$, $r = 0.08$
34. $B = \$3$, $n = 104$, $r = 0.06/52$
35. $B = \$200$, $n = 16$, $r = 0.064$
36. $B = \$3785$, $n = 27$, $r = 0.0375$
37. $B = \$10$, $n = 520$, $r = 0.08/52$

In Problems 38–47 write the repeating decimals as rational numbers.

38. 0.666 . . .
39. 0.424242 . . .
40. 0.282828 . . .
41. 0.717171 . . .
42. 0.214214214 . . .
43. 0.312312 . . .
* 44. 0.124242424 . . .
* 45. 0.11362362362 . . .
46. 0.513651365136 . . .
47. 19.1919 . . .
* 48. At what time between 1 P.M. and 2 P.M. is the minute hand of a clock exactly over the hour hand? [Hint: The minute hand moves 12 times as fast as the hour hand. Start at 1:00 P.M. When the minute hand has reached 1, the hour hand points to $1 + \frac{1}{12}$; when the minute hand has reached $1 + \frac{1}{12}$, the hour hand has reached $1 + \frac{1}{12} + \frac{1}{12} \cdot \frac{1}{12}$; and so on. Now add up the geometric series.]
* 49. At what time between 7 A.M. and 8 A.M. is the minute hand exactly over the hour hand?
* 50. A ball is dropped from a height of 8 m. Each time it hits the ground, it rebounds to a height of two-thirds the height from which it fell. Find the total distance traveled by the ball until it comes to rest (that is, until it stops bouncing).

9.5 The Mathematics of Counting I: The Fundamental Principle of Counting

Suppose I flip a coin and ask, "What is the probability that it comes out heads?" You can answer that question only if you assume that I am flipping a *fair* coin; that is, one for which a head and a tail are equally likely to occur. If you suspect that I am flipping the coin in a careful manner in order to affect the outcome, or if you suspect that the coin is weighted to make heads more likely than tails, you will be unable to answer the question. But if you assume that the coin is fair, you can reason as follows: There are two possible outcomes, and each is equally likely. Thus, the outcome is a head approximately half the time, so the probability of a head is $\frac{1}{2}$, or 50%. At this point, you do not know precisely what the word *probability* means. Nevertheless, you have obtained the correct answer. That is, you computed a probability based on known information.

Now let us ask a similar but more difficult question. In poker, a *flush* consists of five cards of the same suit (that is, five spades, five hearts, five diamonds, or five clubs). If you are dealt five cards from a standard deck of 52 cards, what is the probability of obtaining a flush? We will defer the answer to this question until Section 9.7. However, let us see how we could approach the problem even before we define the word *probability*. We could reason as in the coin problem. Suppose that there are x ways to deal 5 cards from a deck of 52 and that there are y ways to obtain a flush. Then, assuming that all deals are equally likely (that is, no one is cheating), we could reason that there are y chances of obtaining a flush out of x possible hands—so the probability of obtaining a flush is y/x.

The reasoning here is the same as before, but now we are faced with the problem of computing x and y. That is, we must *count* the number of poker hands and the number of flushes. It turns out (see Examples 12 and 13 in Section 9.6) that $x = 2{,}598{,}960$ (the number of possible poker hands) and $y = 5148$ (the number of flushes) so that the probability of being dealt a flush is

$$\frac{5148}{2{,}598{,}960} \approx 0.00198 \approx \frac{1}{505}$$

That is, a flush will be dealt, on the average, in 1 out of every 505 hands.

These examples illustrate that in order to solve problems in probability, it is necessary to be able to count. In this section and in Section 9.6, we discuss techniques for counting in many different situations.

Before stating our first general principle, we begin an example.

EXAMPLE 1 *Counting the Number of Ways to Choose a Brand of Toothpaste and a Brand of Soap*

A large hotel chain supplies its guests with toothpaste and soap. The chain has been offered substantial discounts on 3 brands of toothpaste and 4 brands of soap. In how many ways can it choose 1 brand of toothpaste and 1 brand of soap?

SOLUTION Let us represent the brands of toothpaste by A, B, and C and the brands of soap by 1, 2, 3, and 4. The possibilities are listed below.

$$\begin{array}{ccc} A1 & B1 & C1 \\ A2 & B2 & C2 \end{array}$$

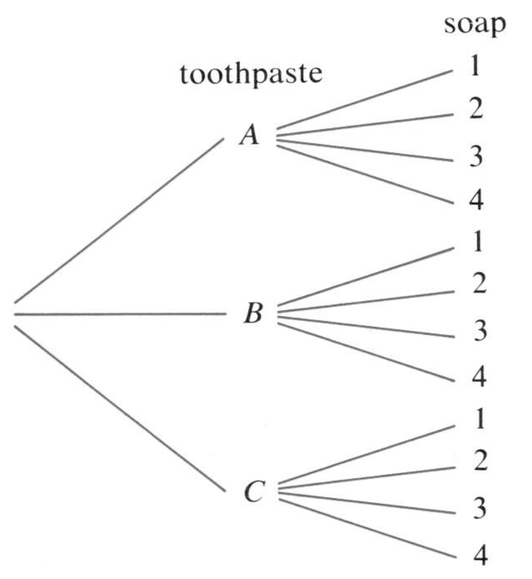

Figure 1 Tree diagram of toothpaste-soap possibilities.

$A3 \quad B3 \quad C3$

$A4 \quad B4 \quad C4$

That is, there are 12 possibilities in all. We note that we could pair each of the 3 brands of toothpaste with one of 4 brands of soap. This tells us that there are $4 + 4 + 4$, or $3 \times 4 = 12$, possibilities.

Another way to see this is to use a **tree diagram.** In Figure 1, the first set of branches in the tree contains the brands of toothpaste and the second, or auxiliary, set of branches lists the brands of soap. From the figure, we can see that there are $3 \times 4 = 12$ possibilities. In order to obtain the number of ways of pairing a brand of toothpaste with a brand of soap, we *multiplied* the number of each kind together.

This example illustrates the **fundamental principle of counting.**

Fundamental Principle of Counting

Two Activities. If activity 1 can be performed in n_1 ways and activity 2 can be performed in n_2 ways, then the two activities can be performed in $n_1 \cdot n_2$ ways.

Three Activities. If activities 1, 2, and 3 can be performed in n_1, n_2, and n_3 ways, respectively, then the three activities can be performed in $n_1 \cdot n_2 \cdot n_3$ ways.

k Activities. If activities 1, 2, 3, . . . , k can be performed in n_1, n_2, n_3, . . . , n_k ways, respectively, then the k activities can be performed in $n_1 \cdot n_2 \cdot n_3 \cdots n_k$ ways.

NOTE In Example 1, there were two activities; $n_1 = 3$ and $n_2 = 4$, so $n_1 n_2 = 3 \cdot 4 = 12$.

EXAMPLE 2 *Counting the Number of Ways of Choosing Three Insects*

In a certain environment there are 14 species of fruit flies, 17 species of moths, and 13 species of mosquitoes. A biologist wishes to choose one species of each type for an experiment. In how many ways can this be done?

SOLUTION There are 3 activities with $n_1 = 14$, $n_2 = 17$, and $n_3 = 13$; by the fundamental principle of counting, there are $14 \cdot 17 \cdot 13 = 3094$ different ways of selecting one species of each type. In this example, it would be very time-consuming to provide a list of the possibilities without the aid of a computer.

EXAMPLE 3 *Counting the Possible Outcomes of a Coin Flip*

A coin is flipped 6 times. How many possible outcomes of heads and tails can be obtained?

SOLUTION Here we are performing 6 identical activities, each of which has 2 possible outcomes—H or T. Thus, by the fundamental principle of counting, there are $2 \cdot 2 \cdot 2 \cdot 2 \cdot 2 \cdot 2 = 2^6 = 64$ possible outcomes. Four of them are *HHTHTT*, *THTHTH*, *HTTTHH*, and *TTTTTT*. Note that here we are concerned with the order of the outcomes. Thus, for example, *HHHTTT* is different from *HTHTHT*.

Problems 9.5

Readiness Check

I. Which of the following activities would employ the following tree diagram?

a. Tossing a coin and choosing a side of a triangle
b. Tossing a coin and rolling a die
c. Choosing a side of a rectangle and choosing a vowel
d. Tossing a coin twice

II. Which of the following is true of the fundamental principle of counting if 3 activities can be performed?

a. 3 is a factor in the product that yields the number of outcomes.
b. The sum of the number of ways these activities can be performed will yield the number of outcomes.
c. There will be 3 factors needed to find the number of outcomes.
d. 3 will be in the sum that yields the number of outcomes.

III. Which of the following are true?

a. There are 8 possible outcomes when a coin is tossed 3 times. [Assume that *HTH* is one of the outcomes.]
b. There are 36 possible outcomes when a die is rolled.
c. There are 104 possible outcomes when a letter and 4 digits are chosen.
d. There are 5 possible outcomes in choosing a vowel and an even number between 3 and 5. [Assume that *y* is not a vowel.]

IV. Which of the following would have six possible outcomes?

a. Flipping a coin three times
b. Throwing two dice
c. Flipping a coin and choosing a letter from the word *CAR*
d. Throwing a die and choosing a letter from the word *WE*

In Problems 1–12 a procedure or activity is described. In each problem, determine the number of possible outcomes. Solve first by using a tree diagram and then by using the fundamental principle of counting.

1. A coin is flipped twice. Assume that *HT* and *TH* are different.
2. A coin is flipped three times. Assume that *HTH* and *THH*, for example, are different.
3. Two dice are thrown (a die is a cube whose sides are numbered 1 through 6). Assume, for example, that (2, 5) and (5, 2) are different.
4. A student committee selects a chairperson and a vice chairperson from among its 5 members (no student can be both).
5. A coin is flipped and a die is thrown. [Hint: Two possible outcomes are *H*5 and *T*2.]

Answers to Readiness Check

I. a II. c III. a, d IV. c

6. A clinic stocks one of 5 brands of aspirin and one of 4 brands of antiseptic cream.
7. A businessman with a large fleet of automobiles must choose, from bids, a type of car from among 3 possibilities, a type of tire from between 2 brands, and a brand of gasoline from among 4 brands.
8. Four applicants take an aptitude test for a job and are ranked according to their performances on the test (assume that there are no ties).
9. Seven politicians compete in a municipal election in which the first-place winner is elected mayor, the second-place winner is elected city council chairperson, and the third-place winner is elected city clerk and recorder.
10. A family has 2 children (assume that *GB* and *BG* are different).
11. A family has 3 children.
12. A family has 4 children.
13. A biologist attempts to classify 46,200 species of insects by assigning to each species 3 letters from the alphabet. Will the classification be completed? If not, determine the number of letters that should be used. [Hint: *ABC* and *CBA* are different.]
14. It is estimated that there are 2 million species of insects, 1 million species of plants, 20,000 species of fish, and 8700 species of birds. If one species from each of these 4 categories is to be chosen for a comparative study, in how many ways can this be done?
15. How many different "words" of 3 letters can be formed from the 4 letters *A*, *U*, *G*, and *C* if repetitions are allowed? (The words of the genetic code are formed from triplets or codons of 4 bases: adenine, uracil, guanine, and cytosine.)
16. In a primitive religion, a professional diviner appeals to his god by reciting a verse. To choose the appropriate verse, 16 smooth palm nuts are grasped by the diviner between both hands. The diviner then attempts to grasp them all in his right hand. Since palm nuts are relatively large, they are difficult to grasp in one hand. If 1 or 2 nuts remain in his left hand, this number is recorded. If any other number of nuts remains, then the result is ignored. This procedure is repeated 8 times, producing a sequence of 8 numbers, each equal to 1 or 2, which determines the verse that the diviner recites. How many verses must the diviner know?
17. In a comparative study of digital watches, 5 characteristics are to be examined. If there are 6 recognizable differences in each of 3 characteristics and 8 recognizable differences in each of the remaining two characteristics, what is the maximum number of different brands of digital watches that could be distinguished by these 5 characteristics?
18. A license tag in a certain state consists of 1 letter followed by 4 numbers. How many possible license plates are there?
19. In Problem 18, how many license plates are there if the number 0 cannot be used?
20. In Problem 18, how many license plates are there if no number can be used more than once?
21. Eight horses compete at a race track. Assuming no ties, how many possible outcomes are there (only first-, second-, and third-place finishes are considered)?
22. ** In Problem 21, how many outcomes are there if ties are allowed?
23. Four classes of students each choose one representative on a committee. If the classes contain 47, 51, 54, and 55 students, in how many ways can the representatives be chosen?

9.6 The Mathematics of Counting II: Permutations and Combinations

In problems involving probability, it is often necessary to count the number of ways in which a set of objects can be arranged. For example, let us count the number of ways the letters *A*, *B*, and *C* can be arranged, without using any letter more than once; they are as follows:

$$\begin{array}{ccc} ABC & BAC & CAB \\ ACB & BCA & CBA \end{array}$$

There are six possible orders in which the three letters can be written. Each such ordering, or rearrangement, is called a *permutation*.

Definition of Permutation of n Objects

A **permutation** of n objects is an arrangement or ordering of the n objects.

NOTE When we use the word *permutation,* we imply that the order in which things are written or done is important.

In problems involving permutations, *order is important.*

Thus, the permutations ABC and ACB are *different* permutations because their orderings are different.

EXAMPLE 1 *The Number of Permutations of Five Objects*

How many permutations are there of the letters A, B, C, D, and E?

SOLUTION We could solve this problem by writing out all the permutations. This would take a lot of space and time and is unnecessary. Let us think of the 5 letters as ''sitting'' in chairs: There are 5 choices for the letter in the first chair. Suppose it is a B. Then there are 4 choices for the letter in the second chair, 3 for the third chair, and 2 for the fourth chair. Finally, there is 1 letter that remains to be placed in the fifth chair. Thus, by the fundamental principal of counting, there are $5 \cdot 4 \cdot 3 \cdot 2 \cdot 1 = 120$ permutations of the 5 letters.

NOTE In Example 1, we could have been talking about any set of five objects. Thus we have shown that *there are* 120 *permutations of five objects.*

Before continuing, it is useful to introduce the *factorial* notation.

The product of the first n positive integers is called ***n* factorial** and is denoted by $n!$ By convention, zero factorial is equal to 1. We have

Illustration of Factorials

$$0! = 1$$
$$1! = 1$$
$$2! = 2 \cdot 1 = 2$$
$$3! = 3 \cdot 2 \cdot 1 = 6$$
$$4! = 4 \cdot 3 \cdot 2 \cdot 1 = 24$$
$$\vdots$$
$$n! = n(n-1)(n-2)\cdots 3 \cdot 2 \cdot 1 \tag{1}$$

Note that

$$4! = 4 \cdot 3 \cdot 2 \cdot 1 = 4 \cdot 3!$$
$$5! = 5 \cdot 4 \cdot 3 \cdot 2 \cdot 1 = 5 \cdot 4!$$

and, in general, for $n \geq 1$,

$$n! = n \cdot (n - 1)! \tag{2}$$

We can now restate the result of Example 1 in a more compact notation: *There are* 5! *permutations of five objects.*

We can extend this result. If, in Example 1, there were n objects instead of five letters, then we could place them in n chairs. There are n choices for the first chair, $n - 1$ choices for the second chair, and so on. Using the same reasoning as in Example 1, we conclude that

There are $n!$ permutations of n objects.

CALCULATOR NOTE Many hand-held calculators have factorial buttons that produce values of $n!$ for n ranging from 0 to 69. The reason that 69! is the largest allowable value is that $69! \approx 1.7 \times 10^{98}$ whereas $70! \approx 1.2 \times 10^{100}$, and many calculators do not carry numbers bigger than 10^{100}.

EXAMPLE 2 *The Number of Rankings of Eight Refrigerators*

Eight brands of refrigerators are ranked according to fixed criteria. Assuming no ties, how many rankings are possible?

SOLUTION Here the order in which the brands are ranked is important, and so we can rephrase the question as, "How many permutations are there of 8 objects?" The answer is $8! = 8 \cdot 7 \cdot 6 \cdot 5 \cdot 4 \cdot 3 \cdot 2 \cdot 1 = 40{,}320$.

Very often, we are not interested in all possible orderings of n objects but in the possible orders of some subset of the n objects.

EXAMPLE 3 *Counting the Number of Possible Outcomes in a Horse Race*

At a race track horses compete for first place (win), second place (place), and third place (show). If 5 horses are running, how many different outcomes are possible, assuming that there are no ties?

SOLUTION Any one of the 5 horses can win, so there are 5 ways to choose the winner. The winning horse cannot also come in second, so once the winner is chosen, there are 4 choices for second place. Similarly, once the first and second choices are determined, there are 3 possibilities for third place. Using the fundamental principle of counting, there are $5 \cdot 4 \cdot 3 = 60$ possible outcomes.

Permutation of n Objects Taken k at a Time

A **permutation of n objects taken k at a time** (with $0 \leq k \leq n$) is any selection of k objects in a definite order from the n objects. We denote the number of permutations of n objects taken k at a time by $P_{n,\,k}$. By definition, $P_{n,\,0} = 1$.

EXAMPLE 4

How many 4-letter "words" (not necessarily English) can be formed from the letters of the word *STRANGE* if no repetitions are permitted?

SOLUTION First we note that we are asking the question, "How many permutations are there of 7 objects taken 4 at a time?" Some of these permutations are *STAR*, *RTAS*, *RATS*, *GEAR*, and *GRAE*. The order in which the letters are chosen is important, which is why we have a permutation. Reasoning as before, we note that there are 7 ways to choose the first letter, 6 ways to choose the second letter, 5 ways to choose the third letter, and 4 ways to choose the fourth letter. Then, using the fundamental principle of counting, we conclude that $7 \cdot 6 \cdot 5 \cdot 4 = 840$ four-letter words can be made from the letters of the word *STRANGE*.

Before going further, let us derive a general formula for $P_{n,\,k}$. If we wish to choose k objects from n objects where order is important, then there are n ways to choose the first object, $n - 1$ ways to choose the second, $n - 2$ ways to choose the third, and so on; there are $n - (k - 1) = n - k + 1$ ways to choose the kth object. That is,

Number of Permutations of *n* Objects Taken *k* at a Time

$$P_{n,\,k} = n(n-1)(n-2)(n-3)\cdots(n-k+1), \qquad (3)$$

$$\text{for } 1 \le k \le n$$

EXAMPLE 5 ***Evaluating*** $P_{n,\,k}$

Compute

(a) $P_{8,\,2}$ $\quad P_{11,\,4}$

(a) $n - k + 1 = 8 - 2 + 1 = 7$, so, in (3), we stop at 7 and $P_{8,\,2} = 8 \cdot 7 = 56$
(b) $n - k + 1 = 11 - 4 + 1 = 8$, so we stop at 8 and $P_{11,\,4} = 11 \cdot 10 \cdot 9 \cdot 8 = 7920$ ■

EXAMPLE 6

First, second, and third prizes are to be awarded in a competition among 20 persons. In how many ways can the prizes be distributed?

SOLUTION Since order counts here (John getting first prize and Susan getting second prize is a different outcome from Susan first and John second), we recognize this as a permutation of 20 objects taken 3 at a time:

$$P_{20,\,3} = 20 \cdot 19 \cdot 18 = 6840$$

We can write $P_{n,k}$ (formula (3)) in another way that is more convenient to use if you have a calculator with a factorial key. For example, from Example 5 we find that

$$P_{11,4} = 11 \cdot 10 \cdot 9 \cdot 8$$

We multiply and divide this expression by 7!

$$P_{11,4} = \frac{11 \cdot 10 \cdot 9 \cdot 8 \cdot 7 \cdot 6 \cdot 5 \cdot 4 \cdot 3 \cdot 2 \cdot 1}{7!} = \frac{11!}{7!}$$

In general

$$P_{n,k} = n(n-1)(n-2)\cdots(n-k+1)$$

Multiply and divide by $(n-k)!$

$$= \frac{n(n-1)(n-2)\cdots(n-k+1)(n-k)(n-k-1)\cdots 2 \cdot 1}{(n-k)!}$$

$$= \frac{n!}{(n-k)!}$$

That is,

Number of Permutations of *n* Things Taken *k* at a Time

$$P_{n,k} = \frac{n!}{(n-k)!}$$

$P_{n,k}$ *on a Calculator*

Many hand-held calculators have function keys for computing $P_{n,k}$ (usually for $n \le 69$). On such a calculator, $P_{n,k}$ is often denoted by ${}_nP_r$.

EXAMPLE 7 *Evaluating $P_{n,k}$ on a Calculator*

Evaluate the following on a calculator:

(a) $P_{11,4}$ (b) $P_{56,12}$

SOLUTION To obtain each answer, press the given keys.

(a) [1] [1] [${}_nP_r$] [4] [=] 7920 is displayed.
(b) [5] [6] [${}_nP_r$] [1] [2] [=] $2.674664976 \times 10^{20}$ is displayed.

In the counting problems so far encountered, we were interested in the number of orderings of a set of objects. There is a type of counting problem in which the ordering of objects is not relevant. For example, from a committee of five people, we may wish to choose three people for a subcommittee. The order in which the three people are chosen is not of interest. Instead, we are interested in the number of ways that a group of three people may be chosen.

EXAMPLE 8 ***Counting the Number of Ways of Choosing a Subcommittee***

From 5 people on a committee, 3 are to be chosen for a subcommittee. In how many ways can this be done?

SOLUTION Let us, for simplicity, label the 5 people as A, B, C, D, and E. Then we can list the possible subcommittees.

$$\begin{matrix} ABC & ABE & ACE & BCD & BDE \\ ABD & ACD & ADE & BCE & CDE \end{matrix} \tag{4}$$

This is all! If you think that this list is too small, then look again. Since the order in which the subcommittee members are chosen is *not* relevant, we see, for example, that ABC (listed) and CAB (unlisted) represented the *same* subcommittee. Thus we can conclude that 10 subcommittees can be formed.

Let us solve the problem in a different way. We know that the number of permutations of 5 objects taken 3 at a time is $5 \cdot 4 \cdot 3 = 60$. These 60 permutations include ABC, ACB, BAC, BCA, CAB, and CBA. As permutations, these 6 are different. But they all represent the *same* subcommittee. The list of 60 permutations contains many duplications if order does not matter. More precisely, each subcommittee listed in (4) contains 3 objects that can be permuted in $3! = 6$ ways. We see that for every subcommittee listed above there are $3! = 6$ permutations of 3 of the letters $ABCDE$. Thus there are 6 times as many permutations as there are subcommittees; putting it another way, to obtain the number of subcommittees, we must divide the number of permutations by 6. Hence

$$\text{Number of subcommittees} = \frac{P_{5,\,3}}{3!} = \frac{60}{6} = 10$$

Before continuing, we have an important definition.

Combination of n Objects Taken k at a Time

Assume that $0 \le k \le n$. A **combination of n objects taken k at a time** is any selection of k of the n objects without regard to order. The symbol $\binom{n}{k}$, read "n choose k," is used to denote the number of combinations of n objects taken k at a time.

The number $\binom{n}{k}$ is called a **binomial coefficient.** In Example 8 we found that $\binom{5}{3} = 10$

NOTE In some books the symbol $C_{n,\,k}$ or ${}_nC_k$ is used instead of $\binom{n}{k}$ to denote the number of combinations of n objects taken k at a time.

How many combinations of n objects taken k at a time are there? Reasoning as in Example 8, we observe that there are

$$P_{n,\,k} = n(n-1)(n-2)\cdots(n-k+1)$$

permutations of the n objects taken k at a time. But any set of k objects can be permuted in $k!$ ways. Therefore there are $k!$ permutations for every combination. Thus, in order to determine the number of combinations of n objects taken k at a time, we must divide the number of permutations of n objects taken k at a time by $k!$. That is,

Number of Combinations of *n* Objects Taken *k* at a Time

$$\binom{n}{k} = \frac{P_{n,\,k}}{k!} = \frac{n(n-1)(n-2)\cdots(n-k+1)}{k!}, \quad 0 \le k \le n \qquad (5)$$

To obtain a more compact expression, we multiply the numerator and denominator of the last expression by $(n-k)!$

$$\begin{aligned}\binom{n}{k} &= \frac{n(n-1)(n-2)(n-3)\cdots(n-k+1)(n-k)!}{(n-k)!k!} \\ &= \frac{n(n-1)(n-2)\cdots(n-k+1)(n-k)(n-k-1)(n-k-2)\cdots 3\cdot 2\cdot 1}{(n-k)!k!} \\ &= \frac{n!}{(n-k)!k!}\end{aligned}$$

Thus,

Number of Combinations of *n* Objects Taken *k* at a Time

$$\binom{n}{k} = \frac{n!}{(n-k)!k!}, \quad 0 \le k \le n \qquad (6)$$

NOTE Formula (5) is easier to use than formula (6), unless you have a calculator with a factorial key.

EXAMPLE 9 *Computing Two Binomial Coefficients*

Compute

(a) $\binom{8}{3}$ (b) $\binom{10}{6}$

SOLUTION We use formula (5):

(a) $\binom{8}{3} = \frac{8\cdot 7\cdot 6}{3!} = \frac{8\cdot 7\cdot 6}{6} = 8\cdot 7 = 56$

(b) $\binom{10}{6} = \frac{10 \cdot 9 \cdot 8 \cdot 7 \cdot 6 \cdot 5}{6!} = \frac{10 \cdot 9 \cdot 8 \cdot 7 \cdot 6 \cdot 5}{6 \cdot 5 \cdot 4 \cdot 3 \cdot 2} \overset{4 \cdot 2 = 8}{\downarrow =} \frac{10 \cdot 9 \cdot 7}{3}$

$= 10 \cdot 3 \cdot 7 = 210$

A Good Suggestion

When computing $\binom{n}{k}$, always divide out common factors in the numerator and denominator before doing any other computations — unless you are carrying out the computations on a calculator with a factorial key.

Before going further, we stress that if $0 \le k \le n$, then $\binom{n}{k}$ is a positive integer. This follows from the fact that $\binom{n}{k}$ denotes the number of ways of choosing k objects from n objects. It is also possible to prove directly that $\binom{n}{k} = \frac{n!}{(n-k)!k!}$ must be an integer.

$\binom{n}{k}$ *on a Calculator*

Many calculators have function keys for computing $\binom{n}{k}$ (usually for $n \le 69$). On such a calculator $\binom{n}{k}$ is often denoted by ${}_nC_r$.

EXAMPLE 10 *Evaluating* $\binom{n}{k}$ *on a Calculator*

Evaluate the following on a calculator:

(a) $\binom{11}{4}$ (b) $\binom{56}{12}$

SOLUTION To obtain each answer, press the given keys.

(a) [1] [1] [${}_nC_r$] [4] [=] 330 is displayed.
(b) [5] [6] [${}_nC_r$] [1] [2] [=] $5.583833073 \times 10^{11}$ is displayed. ■

EXAMPLE 11 ***Counting the Number of Ways of Choosing 3 Razors from 12***

A common procedure in quality control is to take a sample of a manufactured product and test it for defects. From a collection of 12 electric razors, a manufacturer wishes to select 3 for extensive testing. In how many ways can this be done?

SOLUTION Since the order in which the razors are chosen is not important, we are being asked to compute the number of combinations of 12 objects taken 3 at a time. The answer is therefore

$$\binom{12}{3} = \frac{12 \cdot 11 \cdot 10}{3!} = \frac{12 \cdot 11 \cdot 10}{3 \cdot 2} = 2 \cdot 11 \cdot 10 = 220 \quad \blacksquare$$

EXAMPLE 12 ***Counting the Number of Poker Hands***

A poker hand consists of 5 cards selected from a standard deck of 52 cards. How many poker hands are there?

SOLUTION We discussed this question briefly at the beginning of Section 9.5. Now we can give a complete answer by noting that since the order in which the cards are dealt is irrelevant, we must compute the number of combinations of 52 objects taken 5 at a time.

$$\binom{52}{5} = \frac{52 \cdot 51 \cdot 50 \cdot 49 \cdot 48}{5!} = \frac{52 \cdot 51 \cdot 50 \cdot 49 \cdot 48}{5 \cdot 4 \cdot 3 \cdot 2}$$

$$\frac{48}{4 \cdot 3} = 4 \text{ and } \frac{50}{5 \cdot 2} = 5 \longrightarrow = 52 \cdot 51 \cdot 5 \cdot 49 \cdot 4 = 2{,}598{,}960 \quad \blacksquare$$

EXAMPLE 13 ***Counting the Number of Flushes***

A flush in poker consists of 5 cards of the same suit. How many flushes are there?

SOLUTION In order to make a flush, we must do two things: First, we must choose a suit, and second, we must choose 5 cards in that suit. There are 4 ways to choose a suit (since there are 4 suits: spades, hearts, diamonds, and clubs). Because each suit contains 13 cards, there are

$$\binom{13}{5} = \frac{13 \cdot 12 \cdot 11 \cdot 10 \cdot 9}{5!} = \frac{13 \cdot 12 \cdot 11 \cdot 10 \cdot 9}{5 \cdot 4 \cdot 3 \cdot 2} = 13 \cdot 11 \cdot 9 = 1287$$

ways to choose 5 cards from 13 cards. Thus, from the fundamental principle of counting, there are $4 \cdot 1287 = 5148$ possible flushes in a standard deck of 52.† ■

EXAMPLE 14 *Counting the Number of Ways of Selecting 2 Boys and 3 Girls for Testing*

Six boys and 11 girls in a class are suspected of having an infectious disease. Blood samples are to be taken from 2 of the boys and 3 of the girls to test for the disease. In how many ways can this be done?

SOLUTION There are

$$\binom{6}{2} = \frac{6 \cdot 5}{2!} = 15$$

ways to choose the 2 boys and

$$\binom{11}{3} = \frac{11 \cdot 10 \cdot 9}{3!} = 165$$

ways to choose the 3 girls. By the fundamental principle of counting, there are $15 \cdot 165 = 2475$ ways to choose 2 boys and 3 girls.

NOTE In the last example, we solved the problem the way we did by assuming that the orders in which the boys and girls were chosen did not matter so that the problem involved combinations.

TO THE STUDENT The topics with which many students have the most difficulty involve permutations and combinations. The formulas in this section can easily be memorized. It's learning when to use each formula that is difficult. Facility comes only with *practice*!

To help you with this difficulty, the problem set is divided into three parts. Part 1 includes only problems involving permutations. Part 2 contains only problems involving combinations. In Part 3, both types of problems are included.

In the real world, problems don't jump at you and say, "We are permutation problems." You must determine that for yourself. Remember, if you are unsure whether to use a formula for a permutation or a combination, ask yourself the question, "Does order count?" That is, if I take a solution to the problem and rearrange the items in the solution, do I get a different solution? If the answer is yes, you have a permutation problem. If the answer is no, you are dealing with a problem involving combinations. Finally, in some problems no formula will be helpful. In those cases, it is necessary to think hard and use common sense.

---

†These include straight flushes, which are 5 cards of the same suit that are in order. Thus ♡ 4 ♡ 5 ♡ 6 ♡ 7 ♡ 8 is a straight flush.

Problems 9.6

Readiness Check

I. Which of the following is equal to $P_{5,3}$?
 a. $5 \cdot 4$
 b. 60
 c. 30
 d. $5 \cdot 4 \cdot 3 \cdot 2 \cdot 1$

II. Which of the following is true about permutations?
 a. Repetition of choices is possible.
 b. $P_{n,n} = 1$
 c. Order is not important.
 d. $P_{n,n} = n!$

III. Which of the following is true?
 a. $0! = 0$
 b. $n! = n(n-1)(n-2)(n-3)\cdots 3 \cdot 2 \cdot 1$
 c. $n! = n(n-1)(n-2)\cdots(n-k)$
 d. $n!$ is the sum of all the integers from n down to 1.

IV. Which of the following is equal to $\binom{4}{4}$?
 a. 0
 b. 1
 c. 12
 d. 24

V. Which of the following is true?
 a. $\binom{n}{n-2} > \binom{n}{2}$
 b. $P_{n,2} = P_{n,n-2}$
 c. $\binom{n}{n-2} = \binom{n}{2}$
 d. It cannot be determined unless n is given.

Part 1 Permutation Problems

1. How many permutations are there of 7 objects?
2. How many ways can 6 children be lined up?
3. How many ways can the letters of the word *CATNIP* be arranged?
4. How many 4-letter words (not necessarily in English) can be made from the letters of the word *CATNIP*?

In Problems 5–18 compute the number of permutations.

5. $P_{6,3}$ 6. $P_{6,1}$ 7. $P_{8,4}$
8. $P_{9,5}$ 9. $P_{12,8}$ 10. $P_{13,12}$
11. $P_{9,8}$ 12. $P_{4,0}$ 13. $P_{n,0}$
14. $P_{6,6}$ 15. $P_{n,n}$ 16. $P_{n,n-1}$
17. $P_{n,n-2}$ 18. $P_{n,1}$

19. A scrabble player has 7 distinct letters in front of her and wishes to play a 4-letter word. If she chooses to test each possible 4-letter permutation before playing, how many words must she test?
20. If the scrabble player of Problem 19 tests all 6-letter words and it takes her 2 seconds to test each word, how long will it take?
21. An environmental group ranks the 50 members of the state legislature according to their actions on certain key issues. Assuming no ties, how many rankings are possible?
22. The group of Problem 21 finds the 12 legislators with the worst environmental records and accords them the title "the dirty dozen." In how many ways can a dirty-dozen list be compiled? Here the dozen are listed in order with the worst record listed first. Assume no ties.
23. The environmentalists of Problem 21 determine the 10 legislators with the best records and put their names on an environmental honor roll. In how many ways can the honor roll be determined? Here the members of the honor roll are listed in order with the best record listed first.

* 24. How many distinct permutations are there of the letters of the word *RABBIT*?

* 25. How many distinct permutations are there of the letters of the word *ERROR*?

* 26. How many distinct permutations are there of the letters of the word *BARBAROUS*?

Part 2 Combination Problems

In Problems 27–47 determine the number of combinations.

27. $\binom{6}{2}$ 28. $\binom{7}{6}$ 29. $\binom{5}{2}$
30. $\binom{8}{1}$ 31. $\binom{11}{3}$ 32. $\binom{10}{7}$

Answers to Readiness Check

I. b II. d III. b IV. b V. c

33. $\binom{9}{5}$ 34. $\binom{9}{4}$ 35. $\binom{11}{7}$

36. $\binom{13}{1}$ 37. $\binom{13}{12}$ 38. $\binom{12}{4}$

39. $\binom{8}{7}$ 40. $\binom{7}{0}$ 41. $\binom{5}{5}$

42. $\binom{n}{0}$ 43. $\binom{n}{n}$ 44. $\binom{n}{n-1}$

45. $\binom{n}{1}$ 46. $\binom{n}{2}$ 47. $\binom{n}{n-2}$

48. Show that $\binom{n}{k} = \binom{n}{n-k}$. Explain why this must be true.
49. From a 10-person committee, how many different 4-person subcommittees can be formed?
50. An exam contains 10 questions, and a student must answer 6 of them. In how many ways can this be done?
51. A bridge hand contains 13 cards. From a standard deck of 52 cards, how many bridge hands can be dealt?
52. The Supreme Court has 9 members. In how many ways can a 5-to-4 decision be reached?
53. In Problem 52 in how many ways can a 6-to-3 decision be reached?
54. A salesman carries 14 brands of shirts, but can show only 6 brands at any one sales call. In how many ways can these 6 brands be chosen?
55. A certain course covers 10 topics in probability and 8 topics in matrix theory. The final exam will have 5 questions with at most 1 from any topic. Three questions will be on probability, and 2 will be on matrix theory. In how many ways can the topics examined be chosen?
56. Six persons are to be chosen from a group of 10 men and 10 women.
    (a) What is the number of ways that the 6 persons can be chosen?
    (b) What is the number of ways that more men than women can be chosen?

* 57. How many poker hands contain a full house (3 tens and 2 kings, for example)? [Hint: Do this problem in several steps. First, compute the number of ways of choosing a denomination to give you 3 of a kind. In how many ways can the 3 be chosen out of the 4 in the denomination (3 tens out of 4 tens, for example)? Then determine the number of ways to choose a denomination giving a pair and compute the number of ways of, for instance, choosing 2 kings from among 4 kings. Finally, use the fundamental principle of counting.]

Part 3 Assorted Counting Problems

58. A 10-member committee chooses a chairperson, vice chairperson, secretary and treasurer. In how many ways can this be done?
59. In how many ways can a 4-person subcommittee be chosen from the committee of Problem 58?
60. In a genetics experiment, 4 white peas, 7 red peas, and 5 pink peas are chosen for pollination from a sample of 10 white, 10 red, and 10 pink peas. (The color of the peas refers to the color of their flowers.) In how many ways can this be done?
61. How many "words" can be formed from the symbols X and Y if each word must contain at least one X and if the maximum length of the words is 3 letters, with the order of the letters not being relevant? (The X and Y chromosomes determine sex. Normal females and males are XX and XY, but nondisjunction of the sex chromosomes may give rise to X, XXX, XXY, and XYY chromosomes.)
62. Five drugs are used in the treatment of a disease. It is believed that the sequence in which the drugs are used may have a considerable effect on the result of treatment. In how many different orders can the drugs be administered?
63. How many poker hands contain a straight (5 cards in ascending order, such as 5 6 7 8 9 or 9 10 J Q K)?

* 64. How many poker hands contain exactly 3 of a kind?

* 65. How many poker hands contain 2 pairs?

* 66. How many poker hands contain only 1 pair?

67. Suppose that there are 5 highways joining cities A and B and 3 highways joining cities B and C. How many different routes join cities A and C?
68. In Problem 67, in how many ways can the round trip from city A to city C be made
    (a) without traveling the same route twice?
    (b) without traveling the same highway twice?
69. How many distinct groups of 5 letters can be chosen from the letters A, B, C, D, E, F, G, and H?
70. How many distinct 5-letter "words" can be made from the letters A, B, C, D, E, F, G, and H?
71. Among 20 teachers in a mathematics department, students give prizes to the best teacher and the worst teacher. In how many ways can the prizes be awarded?
72. In the same math department, two faculty members are chosen to represent the department in a university committee. In how many ways can this be done?
73. On an accounting test, there are 20 multiple-choice questions, each with four possible answers. If a student guesses on each question, how many possible sets of answers are there?
74. If, on the test of Problem 73, the student must first select 8 questions and then guess, how many possible sets of answers are there?

75. Of 100 selected stocks on the New York Stock Exchange, 37 advanced and 51 declined. In how many different ways could this happen?
76. A committee of the House of Representatives has 9 Democrats and 7 Republicans. In how many ways can a subcommittee consisting of 3 Democrats and 2 Republicans be chosen?
77. In Problem 76 in how many ways can a majority (Democratic) chairman and vice chairman and a minority (Republican) chairman and vice chairman be chosen?
78. A football team contains (offensively) 7 backs, 11 linemen, 5 ends, and 3 quarterbacks. In how many ways can the coach select a starting team of 3 backs, 5 linemen, 2 ends, and 1 quarterback?
* 79. Fifteen children must be placed on 4 teams containing 4, 3, 5, and 3 members, respectively. In how many ways can this be done?
80. A laboratory cage contains 8 white mice and 6 brown mice. Find the number of ways of choosing 5 mice from the cage if
    (a) they can be of either color.
    (b) three must be white and 2 must be brown.
    (c) they must be of the same color.
81. A committee contains 12 members. A minimum quorum for meetings of the committee consists of 8 members. In how many ways can a quorum occur?
* 82. Three types of bacteria are cultured in 9 test tubes. Three test tubes contain bacteria of the first type, 4 contain bacteria of the second type, and 2 contain bacteria of the third type. In how many distinct ways can the test tubes be arranged in a row in a test tube rack, assuming that we cannot distinguish among test tubes containing the same types of bacteria?
83. Mr. and Mrs. Smith and 4 other people are seated in 6 chairs (in a row). In how many seating arrangements will the Smiths end up sitting together?
84. The regions of the country are divided into area-code zones, and in each zone the telephone numbers start with a 3-digit prefix followed by a 4-digit number. Each 3-digit area code has a 0 or a 1 as its second digit. Assume that each 3-digit prefix contains no 0.
    (a) What is the maximum number of area codes?
    (b) What is the maximum number of distinct telephone numbers in a given area-code zone?
    (c) What is the maximum number of distinct telephone numbers in the country?
** 85. In Problem 84 how many telephone numbers in a zone will have exactly 1 repeated digit?
* 86. In Problem 84 how many telephone numbers in a zone will have at least 1 repeated digit?
87. In Problem 84 how many telephone numbers in a zone start with the number 23?
* 88. In Problem 84 how many telephone numbers in a zone contain the numbers 2 and 3?
* 89. Five people, labeled P_1, P_2, P_3, P_4, and P_5, are to be tested for blood types A, B, AB, and O.
    (a) How many different distributions of blood types are possible among these 5 people?
    (b) In how many of these distributions is there at least 1 person of each blood type?
90. In the computer diagnosis of a certain disease, the computer is programmed to take into account 2 primary and 6 secondary symptoms. The disease is diagnosed positively if at least 1 of the primary symptoms and at least 2 of the secondary symptoms are present.
    (a) How many combinations of the symptoms can occur?
    (b) How many combinations of the symptoms lead to a negative diagnosis?
    (c) How many combinations of the symptoms lead to a positive diagnosis?
91. For the purposes of this problem, two families are said to have the *same structure* if the number of children is the same and the sexes of the children taken in order of birth are the same. We restrict our attention to families of 5 children.
    (a) How many different structures are possible for such families?
    (b) In how many of these structures are both the oldest and the youngest children girls?
    (c) In how many structures is the fifth child the second girl of the family?

9.7 Introduction to Probability: Equiprobable Spaces

The common feature of every situation involving probabilities is an action, or occurrence, which can take place in several ways. It may rain tomorrow, or it may not. You may collect on your insurance policy, or you may not. You may win at roulette, or you may not. We analyze these situations by comparing the likelihood of occurrences of the various possibilities.

The theory of probability is developed as a study of the outcomes of trials of an experiment. An **experiment** is a phenomenon to be observed according to a clearly defined procedure. It may be as simple as tossing a coin and observing the outcome or as complex as polling 1000 people from a large population to determine their preferences on a variety of social, economic, and political issues. A **trial** of an experiment is a single performance of the experiment.

The **sample space** S of an experiment is the *set* of all possible **outcomes** of the experiment. If the experiment has a finite number of outcomes, the sample space is said to be **finite.**

EXAMPLE 1 *The Sample Space of a Coin-Tossing Experiment*

Consider the experiment of tossing a coin once and observing the outcome. The sample space is $S = \{H, T\}$. The possible outcomes are H and T. ■

EXAMPLE 2 *The Sample Space of Another Coin-Tossing Experiment*

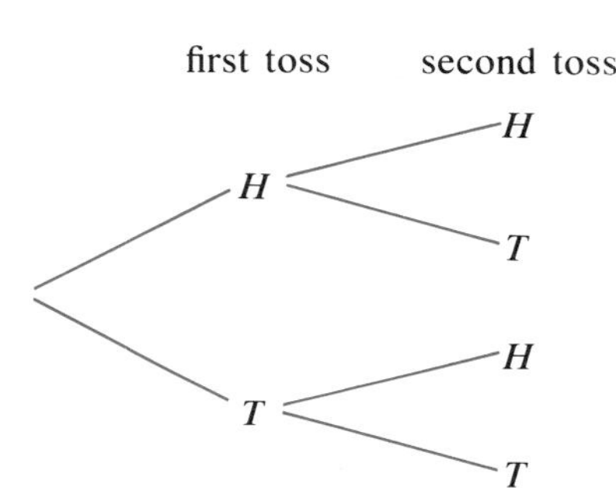

Figure 1 Tree diagram of the coin-tossing experiment.

Consider the experiment of tossing a coin twice and observing the outcome. Then

$$S = \{HH, HT, TH, TT\}$$

The possible outcomes are HH, HT, TH, and TT. It is often useful to illustrate the possibilities using a tree diagram. This is done in Figure 1. ■

EXAMPLE 3 *The Sample Space of a Dice-Throwing Experiment*

Two dice are tossed. The two numbers that appear are recorded. The sample space contains 36 pairs of numbers:

$$S = \{(1, 1), (1, 2), \ldots, (1, 6), (2, 1), \ldots, (2, 6), \ldots, (6, 1), \ldots, (6, 6)\}$$ ■

EXAMPLE 4 *The Sample Space of Another Dice-Throwing Experiment*

Two dice are tossed and the sum of the numbers that turn up is recorded. The sample space is given by

$$S = \{2, 3, 4, 5, 6, 7, 8, 9, 10, 11, 12\}$$

NOTE Examples 3 and 4 illustrate the important point that the sample space of an experiment depends not only on what is being done but also on what question is asked.

EXAMPLE 5 *The Sample Space of a Poll*

One thousand people are polled and asked whether they are for or against the peacetime draft. The number of people for the draft is recorded. The sample space is

$$S = \{0, 1, 2, \ldots, 1000\}$$

As these examples illustrate, sample spaces are all around us. In order to define the probability that something happens, we need first to define an event.

An **event** E is a subset of the sample space S. That is, an event is a set of possible outcomes chosen from S. The event E is called a **simple event** if it contains exactly one of the outcomes. The empty set ϕ, which is a subset of S, is called the **impossible event,** whereas S itself is called the **certain event.**

EXAMPLE 6 *The Events in a Coin-Tossing Experiment*

As in Example 2, a coin is tossed twice and the outcome is observed. What are the events?

SOLUTION The sample space is $\{HH, HT, TH, TT\}$. There are 4 simple events:

$$\{HH\}, \quad \{HT\}, \quad \{TH\}, \quad \{TT\}$$

There are 6 events containing two outcomes:

$$\underbrace{\{HH, HT\}}_{\text{Head on first toss}}, \quad \underbrace{\{HH, TH\}}_{\text{Head on second toss}}, \quad \underbrace{\{HH, TT\}}_{\text{Both tosses the same}}, \quad \underbrace{\{HT, TH\}}_{\text{Both tosses different}}, \quad \underbrace{\{HT, TT\}}_{\text{Tail on second toss}}, \quad \underbrace{\{TH, TT\}}_{\text{Tail on first toss}}$$

There are 4 events containing 3 outcomes:

$$\underbrace{\{HH, HT, TH\}}_{\text{At least one head}}, \quad \{HH, HT, TT\}, \quad \{HH, TH, TT\}, \quad \underbrace{\{HT, TH, TT\}}_{\text{At least one tail}}$$

There are two other subsets:

$$\phi \quad \text{The impossible event}$$

and

$$S = \{HH, HT, TH, TT\} \quad \text{The certain event}$$

Counting, we find that our sample space contains 16 events.

EXAMPLE 7 ***Two Events in a Dice-Throwing Experiment***

Two dice are tossed, and (as in Example 4) the sum of the numbers that turn up is recorded. Describe each of the following events:

(a) The sum is odd.
(b) The sum is greater than 8.

SOLUTION The sample space is given by

$$S = \{2, 3, 4, 5, 6, 7, 8, 9, 10, 11, 12\}$$

(a) $E = \{\text{sum is odd}\} = \{3, 5, 7, 9, 11\}$
(b) $E = \{\text{sum} > 8\} = \{9, 10, 11, 12\}$

Our next step is to define what we mean by the probability of an event in a sample space. We will do this only in the case that each simple event is as likely to occur as any other. The more general situation must await a course in probability or statistics. Our definition of probability is the intuitive definition already used in Section 9.5.

Suppose that in an experiment all the outcomes (simple events) in its sample space S are equally likely to occur. Then S is said to be an **equiprobable** (or **uniform**) **space.**

Suppose that S is a finite equiprobable space. The **probability** of an event E in S, written $P(E)$, is defined to be the number of outcomes in E divided by the number of outcomes in S.

Probability of Events in a Finite Equiprobable Space

If $n(E)$ denotes the number of outcomes in E and $n(S)$ denotes the number of outcomes in S, then the probability of an event in a finite equiprobable space is given by

$$P(E) = \frac{n(E)}{n(S)} = \frac{\text{number of outcomes in } E}{\text{number of outcomes in } S} \tag{1}$$

With the definition of probability given in (1), we can compute an astonishingly large number of probabilities. To do so, we need to do two things. First, we must make sure that we are dealing with an equiprobable space. This is not always obvious (or correct). Second, we must, if necessary, use the counting techniques described in Sections 9.5 and 9.6.

EXAMPLE 8 ***Calculating the Probability of a Head***

A fair coin is tossed. What is the probability that the result is a head?

SOLUTION The word *fair* means that H and T are equally likely. Thus with $E = \{H\}$ and $S = \{H, T\}$, we have

$$P(E) = \frac{n(E)}{n(S)} = \frac{1}{2} \quad \blacksquare$$

EXAMPLE 9 ***Calculating the Probability of a Sum of 7 When Two Dice Are Thrown***

Two fair dice are thrown. What is the probability that the sum of the numbers showing is 7?

SOLUTION Look at Examples 3 and 4. The experiment has at least 2 different sample spaces. Which one should we use? Since the dice are fair, every number is just as likely as every other number to turn up on any one die. That means that each of the 36 pairs in Example 3 is equally likely to occur. In Example 3, the event $E = \{\text{sum of } 7\}$ is the subset of S given by

$$E = \{(1, 6), (6, 1), (2, 5), (5, 2), (4, 3), (3, 4)\}$$

Thus

$$P(\text{sum of } 7) = P(E) = \frac{n(E)}{n(S)} = \frac{6}{36} = \frac{1}{6}$$

WARNING If we make the mistake of using the sample space of Example 4, we obtain the wrong answer. Here $E = \{7\}$ contains only one outcome (that is, it is a simple event), suggesting that $P(E) = n(E)/n(S) = \frac{1}{11}$. The problem here is that the space $S = \{2, 3, \ldots, 12\}$ is *not* an equiprobable space because the outcomes in it are *not* equally likely. Certainly 2, which can occur in only one way ((1, 1)) is less likely to occur than 7, which can occur in six ways. ■

A key to determining that a sample space is an equiprobable space is the use of the words *at random* in the problem.

> If people or objects are chosen at random, then each person or object is equally likely to be chosen.

EXAMPLE 10 ***Calculating Probabilities When 2 People Are Chosen from a Group of 10***

Suppose that in a group of 10 persons, 4 are male. If 2 are chosen at random, what is the probability that (a) both are male, (b) both are female, and (c) 1 is male and 1 is female?

SOLUTION Let A be the event that both are male, B the event that both are female, and C the event that there is 1 of each. The sample space S consists of pairs of people and, since the order in which the people are chosen is irrelevant, S contains $\binom{10}{2} = 45$ outcomes.

(a) There are $\binom{4}{2} = 6$ ways to choose 2 males from among the 4 present, so

$$P(A) = \frac{6}{45} = \frac{2}{15}$$

(b) There are $\binom{6}{2} = 15$ ways to choose 2 females, so

$$P(B) = \frac{15}{45} = \frac{1}{3}$$

(c) There are 4 ways to choose a male and 6 ways to choose a female, so, by the fundamental principle of counting, there are $4 \cdot 6 = 24$ outcomes in C. Thus

$$P(C) = \frac{24}{45} = \frac{8}{15} \quad \blacksquare$$

EXAMPLE 11 *Calculating the Probability of a Flush*

What is the probability of obtaining a flush when a 5-card poker hand is dealt?

SOLUTION Before answering this question, we must first assume that all poker hands are equally likely to occur; that is, we assume that the poker game is honest and the cards are well shuffled. Then $E = \{\text{flush}\}$ and, from Example 13 in Section 9.6, $n(E) = 5148$. Similarly, from Example 12 in Section 9.6, $n(S) = 2{,}598{,}960$. Thus

$$P(E) = \frac{5148}{2{,}598{,}960} \approx 0.0019807923 \approx \frac{1}{505} \quad \blacksquare$$

EXAMPLE 12 *Calculating the Probability of Vehicle Repairs*

Of 20 commercial vehicles that break down at the same time, 15 have been repaired within 3 days. Suppose that 5 vehicles were chosen at random from the 20. What is the probability that exactly 3 were repaired within 3 days?

SOLUTION The event E whose probability we seek is given by

$$E = \{3 \text{ are repaired, } 2 \text{ are not}\}$$

The sample space here consists of all sets of 5 vehicles chosen from the 20.

To obtain $n(E)$, we must compute the number of ways of choosing 3 vehicles from the 15 repaired vehicles and 2 vehicles from the remaining group. Thus, from the fundamental principle of counting,

$$n(E) = \binom{15}{3}\binom{5}{2} = 455 \cdot 10 = 4550$$

$$n(S) = \binom{20}{5} = 15{,}504$$

so that

$$P(E) = \frac{n(E)}{n(S)} = \frac{4550}{15{,}504} \approx 0.29$$

We could solve this problem because the words *at random* suggest that all combinations of five vehicles were equally likely to be chosen.

Problems 9.7

Readiness Check

I. Which of the following is true about a simple event E?
  a. It has exactly one of the outcomes from S.
  b. It is equal to $n(S)$.
  c. It is certain to occur.
  d. $E = \{1\}$.

II. Which of the following is the event E of getting an even number when choosing one number from the positive integers less than 11?
  a. $\{2, 4, 6, 8, 10\}$
  b. 5
  c. $\{2\}$
  d. 6

III. Which of the following is $n(E)$ for the event of getting 3 heads when 3 coins are tossed?
  a. 1/8 b. $\binom{3}{3}$ c. 8 d. 1

IV. For which of the following events is $n(E) = 6$?
  a. Getting a sum less than five when two dice are rolled
  b. Getting six heads when six coins are tossed [Hint: Order doesn't matter.]
  c. Getting an even digit when a digit is chosen from the set $\{1, 2, 3, 4, 5, 6\}$
  d. Getting a vowel when a letter is chosen from the word *SCHISM*

V. Which of the following is true of the sample space $S = \{x$: x is a letter in the word *TIRED*$\}$?
  a. The sample space is equiprobable.
  b. $n(S) = 6$
  c. The event of getting a vowel is $\{a, e, i, o, u\}$.
  d. The probability of getting a consonant is 1/3.

In Problems 1–5, (a) describe the sample space of the experiment and (b) find the probability of a simple event in the sample space.

1. Draw a card at random from a standard 52-card deck.
2. Choose at random an integer from 1 to 10.
3. Choose 5 persons at random from a group of 10.
4. Dial a 7-digit number at random.
5. Choose a chairperson and co-chairperson at random from a 6-person committee.
6. The numbers 1 to 100 are written on slips of paper and placed in a bowl. After the bowl is thoroughly shaken, one of the slips of paper is drawn at random.
   (a) What is the probability that the number drawn is greater than 75?

Answers to Readiness Check

I. a II. a III. d IV. a V. a

(b) What is the probability that the number drawn is divisible by 3?
(c) What is the probability that the number drawn is divisible by 15?

7. A professor assigns 12 different grades to the 12 students in his class. Because of a computer error, the grades are distributed at random on the transcripts of his students.
(a) What is the probability that every student receives his or her correct grade?
(b) What is the probability that at least one receives an incorrect grade?
(c) What is the probability that exactly 11 students receive their correct grades?

8. A cage contains 6 white mice and 4 brown mice. Consider the experiment of drawing 3 mice at random from the cage and observing the colors of those drawn.
(a) Describe the sample space of the experiment.
(b) Calculate the probabilities of the 4 possible distributions of color (3 white, 2 white and 1 brown, and so on).

9. Six persons are chosen at random from a group of 20 men and 8 women.
(a) What is the probability that all 6 chosen are men?
(b) What is the probability that 3 men and 3 women will be selected?

10. A chimpanzee is placed at a toy typewriter with the letters A, B, C, D, and E. The chimpanzee types 4 keys at random.
(a) What is the probability that the word *BEAD* is typed?
(b) What is the probability that all typed letters are the same?

In Problems 11–15 a 5-card poker hand is dealt at random from a standard 52-card deck. Find the probability of the given hand.

* 11. A full house
12. A straight
* 13. 3 of a kind only
* 14. Two pairs only
* 15. One pair

16. A bridge hand consists of 13 cards from a deck of 52 cards. If a bridge hand is dealt at random, what is the probability that all 13 cards will be of the same suit?

17. An accounting quiz contains 5 multiple-choice questions with 4 possible answers for each question. If a student guesses on all 5 questions, what is the probability that he will get a score of 100%?

18. A local union has 8 members, 2 of whom are women. Two of the members are chosen by a lottery to represent the union on a bargaining council. Find the probability of each event.
(a) Both women are chosen.
(b) One man and 1 woman are chosen.
(c) Two men are chosen.

* 19. Of a group of blood donors, 3 have type A blood, 8 have type O blood, 3 have type B blood, and 2 have type AB blood. If 3 people are chosen at random, what is the probability that exactly 2 have the same blood type?

* 20. Suppose you flip a coin 8 times in a row.
(a) What is the probability of getting the outcome *HHTTHHTT*?
(b) What is the probability of getting all heads?
(c) What is the probability of getting at least 1 head?
(d) What is the probability that the first toss is a tail?

* 21. A family is known to have 4 children.
(a) What is the probability that all the children are of the same sex?
(b) What is the probability that there is no girl in this family older than any boy in the family? [Hint: Assume that boys and girls are equally likely at birth.]

22. In a group of 12 people, 4 people are under 20 years of age, 5 are between ages 20 and 40, and 3 are over 40 years old. Six people are chosen at random from this group.
(a) What is the sample space of this experiment? What is the probability of a simple event?
(b) What is the probability that the 3 people over 40 are chosen?
(c) What is the probability that the 6 youngest people are chosen?

23. In a group of 15 people, 10 are right-handed, 4 are left-handed, and one is ambidextrous. Four people are to be chosen at random from this group.
(a) What is the sample space of this experiment?
(b) What is the probability of a simple event in this experiment?
(c) What is the probability that the 4 left-handed people are chosen?

24. What are the sample spaces of the following experiments? Determine the number of simple events for each experiment and the probability of each simple event.
(a) Choose 3 integers at random from 1 to 100 without repetition.
(b) Choose 2 blue objects and 3 red objects at random from a set consisting of 5 blue objects and 4 red objects.

9.8 The Binomial Theorem

The binomial theorem provides a useful device for evaluating expressions of the form $(x + y)^n$, where n is a positive integer. You are familiar with the expression

$$(x + y)^2 = x^2 + 2xy + y^2$$

In addition, it is not difficult to show that

$$\begin{aligned}
(x + y)^3 &= x^3 + 3x^2y + 3xy^2 + y^3\\
(x + y)^4 &= x^4 + 4x^3y + 6x^2y^2 + 4xy^3 + y^4\\
(x + y)^5 &= x^5 + 5x^4y + 10x^3y^2 + 10x^2y^3 + 5xy^4 + y^5\\
(x + y)^6 &= x^6 + 6x^5y + 15x^4y^2 + 20x^3y^3 + 15x^2y^4 + 6xy^5 + y^6
\end{aligned}$$

We add to these expressions

$$(x + y)^0 = 1 \quad \text{and} \quad (x + y)^1 = x + y$$

The goal of this section is to obtain an expansion for $(x + y)^n$. To help you see a pattern, we arrange the coefficients of $(x + y)^n$ for $n = 0, 1, 2, 3, 4, 5$, and 6 in a triangular form:

This triangle is called **Pascal's triangle,** named for the great French mathematician Blaise Pascal (1623–1662). In his 1662 work *Traité du Triangle Arithmétique,* Pascal used mathematical induction to prove that every number in the triangle, except the 1's at the end of each row, is the sum of the two numbers diagonally above it. For example, in the shaded minitriangle in Figure 1, we see that $15 = 10 + 5$.

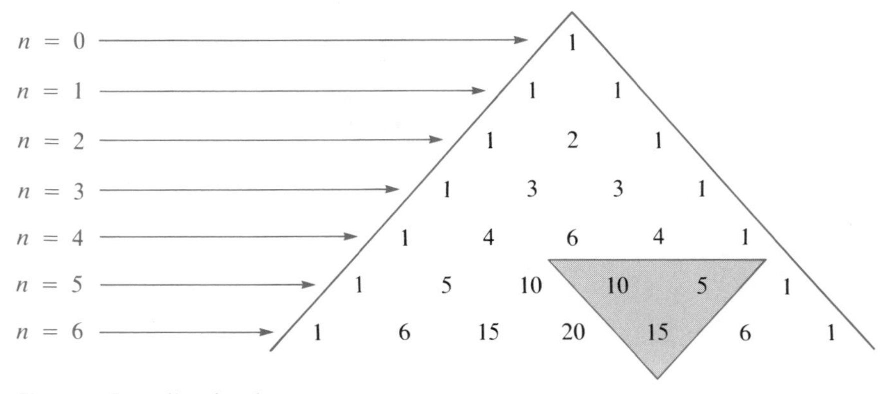

Figure 1 Pascal's triangle.

Before we go on to a statement of the binomial theorem, we reintroduce a notation first seen in Section 9.6.

The Binomial Coefficient

Let n and k denote nonnegative integers. Then

$$\binom{n}{k} = \frac{n(n-1)(n-2)\cdots(n-k+1)}{k(k-1)\cdots 3\cdot 2\cdot 1} = \frac{n!}{k!(n-k)!} \tag{1}$$

where $n! = n(n-1)(n-2)\cdots 3\cdot 2\cdot 1$, and, by convention, $0! = 1$.

The number $\binom{n}{k}$ is called a **binomial coefficient.**

We computed several binomial coefficients in Section 9.6 (see Example 9 in Section 9.6).

EXAMPLE 1 *Evaluating Two Binomial Coefficients*

Compute (a) $\binom{6}{0}$, (b) $\binom{8}{8}$.

SOLUTION

$$\text{(a)}\ \binom{6}{0} = \frac{6!}{0!6!} = \frac{6!}{1(6!)} = 1$$

$$\text{(b)}\ \binom{8}{8} = \frac{8!}{8!0!} = \frac{8!}{(8!)1} = 1$$

Generalizing Example 1, we see that

$$\binom{n}{0} = 1 = \binom{n}{n} \qquad \text{for any positive integer } n$$

We now write the binomial coefficients (up to $n = 6$) in a triangular form (see Figure 2). It turns out that the numbers in Figure 2 are the same as the

$$\binom{0}{0}$$

$$\binom{1}{0} \quad \binom{1}{1}$$

$$\binom{2}{0} \quad \binom{2}{1} \quad \binom{2}{2}$$

$$\binom{3}{0} \quad \binom{3}{1} \quad \binom{3}{2} \quad \binom{3}{3}$$

$$\binom{4}{0} \quad \binom{4}{1} \quad \binom{4}{2} \quad \binom{4}{3} \quad \binom{4}{4}$$

$$\binom{5}{0} \quad \binom{5}{1} \quad \binom{5}{2} \quad \binom{5}{3} \quad \binom{5}{4} \quad \binom{5}{5}$$

$$\binom{6}{0} \quad \binom{6}{1} \quad \binom{6}{2} \quad \binom{6}{3} \quad \binom{6}{4} \quad \binom{6}{5} \quad \binom{6}{6}$$

Figure 2 Pascal's triangle using binomial coefficients.

numbers in Figure 1. For example, in the circled row,

$$\binom{5}{0} = 1 \quad \binom{5}{1} = 5 \quad \binom{5}{2} = 10 \quad \binom{5}{3} = 10 \quad \binom{5}{4} = 5 \quad \binom{5}{5} = 1$$

This is the sixth row in Figure 1. Thus, we see that the triangular table of binomial coefficients is the Pascal triangle.

There is a formula that is useful for writing Pascal's triangle in terms of the binomial coefficients. We have

$$\binom{n}{k-1} + \binom{n}{k} = \frac{n!}{[n-(k-1)]!(k-1)!} + \frac{n!}{(n-k!)k!}$$
$$= \frac{n!}{(n+1-k)!(k-1)!} + \frac{n!}{(n-k)!k!}$$

But

$$\frac{n!}{(n+1-k)!(k-1)!} \overset{\substack{\text{Multiply and} \\ \text{divide by } k}}{=} \frac{n!k}{(n+1-k)!k(k-1)!} \overset{k(k-1)!=k!}{=} k\left[\frac{n!}{(n+1-k)!k!}\right]$$

and

$$\frac{n!}{(n-k)!k!} \overset{\substack{\text{Multiply and} \\ \text{divide by } n+1-k}}{=} \frac{n!(n+1-k)}{(n+1-k)(n-k)!k!} \overset{(n+1-k)(n-k)!=(n+1-k)!}{=} (n+1-k)\left[\frac{n!}{(n+1-k)!k!}\right]$$

Thus

$$\binom{n}{k-1} + \binom{n}{k} = \frac{n!}{(n+1-k)!k!}[k+(n+1-k)] = \frac{n!(n+1)}{(n+1-k)!k!}$$
$$\overset{(n+1)n!=(n+1)!}{=} \frac{(n+1)!}{(n+1-k)!k!} = \binom{n+1}{k}$$

That is,

$$\binom{n}{k-1} + \binom{n}{k} = \binom{n+1}{k} \tag{2}$$

EXAMPLE 2 ***The Sum of Successive Binomial Coefficients***

From (2) we have, for example,

$$\binom{5}{2} + \binom{5}{3} = \binom{6}{3} \quad \text{and} \quad \binom{6}{1} + \binom{6}{2} = \binom{7}{2}$$

This shows how to get successive rows in Pascal's triangle.

The binomial theorem enables us to compute $(x + y)^n$ without writing down the rows of the Pascal triangle. To see what we should get, observe that the sixth (circled) row of the triangle in Figure 2 is equal to

$$1 \quad 5 \quad 10 \quad 10 \quad 5 \quad 1$$

But, as we have seen, these numbers are the coefficients of the expansion of $(x + y)^5$. Thus, we have

$$(x + y)^5 = \underbrace{\binom{5}{0}}_{=1} x^5 + \binom{5}{1} x^4 y + \binom{5}{2} x^3 y^2 + \binom{5}{3} x^2 y^3 + \binom{5}{4} x y^4 + \underbrace{\binom{5}{5}}_{=1} y^5$$

The binomial theorem generalizes this result.

The Binomial Theorem

Let n be a positive integer. Then

$$(x + y)^n = x^n + \binom{n}{1} x^{n-1} y + \binom{n}{2} x^{n-2} y^2 + \cdots + \binom{n}{j} x^{n-j} y^j + \cdots + \binom{n}{n-1} x y^{n-1} + y^n \tag{3}$$

The proof of this theorem will be given after we have done some examples. Using the summation notation discussed in Section 9.2, page 483, we may write the binomial theorem in the following form:

$$(x + y)^n = \sum_{j=0}^{n} \binom{n}{j} x^{n-j} y^j \tag{3'}$$

Note that in (3′) the first term ($j = 0$) is $\binom{n}{0} x^{n-0} y^0 = x^n$ because $\binom{n}{0} = 1$ and $y^0 = 1$. Also, the last term ($j = n$) is $\binom{n}{n} x^{n-n} y^n = x^0 y^n = y^n$ because $\binom{n}{n} = 1$.

EXAMPLE 3 *Using the Binomial Theorem*

Compute $(x + y)^7$.

SOLUTION

$$\begin{aligned}(x + y)^7 &= x^7 + \binom{7}{1} x^6 y + \binom{7}{2} x^5 y^2 + \binom{7}{3} x^4 y^3 + \binom{7}{4} x^3 y^4 \\ &\quad + \binom{7}{5} x^2 y^5 + \binom{7}{6} x y^6 + y^7 \\ &= x^7 + 7x^6 y + 21x^5 y^2 + 35x^4 y^3 + 35x^3 y^4 + 21x^2 y^5 + 7xy^6 + y^7 \quad \blacksquare\end{aligned}$$

EXAMPLE 4 ***Finding a Coefficient Using the Binomial Theorem***

Find the coefficient of the term containing x^3y^6 in the expansion of $(x + y)^9$.

SOLUTION In (3), we obtain the term x^3y^6 by setting $j = 6$ (so that $9 - j = 3$). The coefficient is

$$\binom{9}{6} = \frac{9!}{6!3!} = \frac{9 \cdot 8 \cdot 7}{3!} = \frac{9 \cdot 8 \cdot 7}{3 \cdot 2} = 84 \quad \blacksquare$$

EXAMPLE 5 ***Using the Binomial Theorem***

Calculate $(2x - 3y)^4$.

SOLUTION

$$\begin{aligned}(2x - 3y)^4 &= (2x)^4 + \binom{4}{1}(2x)^3(-3y)^1 + \binom{4}{2}(2x)^2(-3y)^2 \\ &\quad + \binom{4}{3}(2x)^1(-3y)^3 + (-3y)^4 \\ &= 16x^4 + 4(8x^3)(-3y) + 6(4x^2)(9y^2) + 4(2x)(-27y^3) + 81y^4 \\ &= 16x^4 - 96x^3y + 216x^2y^2 - 216xy^3 + 81y^4\end{aligned}$$

NOTE Binomial coefficients can be obtained on some hand-held calculators, as we discussed in Section 9.6 (see p. 516).

EXAMPLE 6 ***Finding a Coefficient Using the Binomial Theorem***

Find the coefficient of the term containing y^6 in the expansion $(2x + y^2)^9$.

SOLUTION The term containing y^6 will be of the form $(2x)^6(y^2)^3$ since $6 + 3 = 9$. This term is then

$$\binom{9}{3}(2x)^6(y^2)^3 = 84 \cdot 2^6x^6y^6 = 84 \cdot 64x^6y^6$$

and the coefficient is $84 \cdot 64 = 5376$.

Proof of the Binomial Theorem

Consider the product

$$(x + y)^n = \underbrace{(x + y)(x + y)(x + y) \cdots (x + y)}_{n \text{ factors}} \tag{4}$$

From the material in Section 1.7, we know that the product (4) is a sum of terms. In each term, we take either an x or y from each of the n $(x + y)$'s in (4). Thus each term in the sum will be of the form

$$x^ky^{n-k}$$

That is, we take the product of k x's and $(n - k)$ y's. What is the coefficient of $x^k y^{n-k}$? The answer is not difficult if you look at it the right way. There are n x's—one in each factor $x + y$. We must choose k of them. From Section 9.6, we know that the number of ways to choose k x's from n x's is $\binom{n}{k}$. Thus, there are $\binom{n}{k}$ ways to get the term $x^k y^{n-k}$, so the coefficient of $x^k y^{n-k}$ is $\binom{n}{k}$. This is true for $k = 0, 1, \ldots, n$, and the theorem is proved. ■

NOTE The binomial theorem also can be proved using mathematical induction. However, the induction proof is much more detailed and so is omitted.

Problems 9.8

Readiness Check

I. Which of the following is the binomial coefficient $\binom{8}{3}$?
a. 24
b. 6720
c. 336
d. 56

II. Which of the following represents the sixth row of Pascal's triangle?
a. 1 6 15 15 6 1
b. The coefficients of $(x - y)^5$
c. $\binom{5}{0}$ $\binom{5}{1}$ $\binom{5}{2}$ $\binom{5}{3}$ $\binom{5}{4}$ $\binom{5}{5}$
d. $P_{6,0}$ $P_{6,1}$ $P_{6,2}$ $P_{6,3}$ $P_{6,4}$ $P_{6,5}$ $P_{6,6}$

III. Which of the following is the binomial coefficient of the sixth term of $(x + y)^9$?
a. $\binom{9}{5}$
b. 84
c. $\binom{8}{5} + \binom{8}{6}$
d. 36

IV. Which of the following is true of a binomial expansion of the form $(x + y)^n$ that contains an x^3y^7 term?
a. The binomial coefficient is $\binom{7}{3}$.
b. The binomial coefficient is the third term in the seventh row of Pascal's triangle.
c. The binomial is raised to the tenth power.
d. The expansion also contains an x^4y^5 term.

In Problems 1–8 calculate the binomial coefficients.

1. $\binom{5}{3}$
2. $\binom{7}{4}$
3. $\binom{7}{3}$
4. $\binom{10}{5}$
5. $\binom{11}{3}$
6. $\binom{35}{35}$
7. $\binom{8}{0}$
8. $\binom{12}{7}$

In Problems 9–43 carry out the indicated binomial expansion.

9. $(x - y)^5$
10. $(x - 2y)^3$
11. $(x - 2y)^4$
12. $(4x + 5y)^3$
13. $(a + b)^8$
14. $(u - w)^4$
15. $(2a - 3b)^5$
16. $\left(\frac{u}{2} + \frac{v}{3}\right)^3$
17. $\left(\frac{u}{3} - \frac{v}{4}\right)^3$
18. $\left(\frac{v}{2} + u\right)^5$
19. $(x^2 + 2y)^4$
20. $(d^2 + d^4)^3$
21. $(a^2 + b^3)^4$
22. $(2a^3 - 3b^2)^4$
23. $(\sqrt{x} + \sqrt{y})^6$
24. $(\sqrt{x} - \sqrt{y})^6$
25. $(3\sqrt{x} + 3\sqrt{y})^4$
26. $(xy^2 + z)^5$
27. $(ab - cd)^3$
28. $\left(\frac{u}{v} + w\right)^5$
29. $\left(w - \frac{u}{v}\right)^4$
30. $(1 + x)^8$

Answers to Readiness Check

I. d II. c III. a IV. c

31. $(1-a)^{10}$
32. $(x+z)^3$
33. $(1+\sqrt{x})^3$
34. $(\sqrt{2}-y)^4$
35. $(\sqrt{5}+\sqrt{7})^4$
36. $(1-c^2)^4$
37. $(1+z^6)^4$
* 38. $(a+b+1)^4 = ((a+b)+1)^4$
* 39. $(u+v-2)^4$
* 40. $(x+y+z)^3$
* 41. $(x+y+z)^4$
* 42. $(x+y+z)^6$
43. $(x^n+y^n)^5$
44. Find the coefficient of x^5y^7 in the expansion of $(x+y)^{12}$.
45. Find the coefficient of a^7b^3 in the expansion of $(a-b)^{10}$.
46. Find the coefficient of u^4v^2 in the expansion of $(2u-3v)^6$.
47. Find the coefficient of x^8y^{12} in the expansion of $(x^2+y^3)^8$.
48. Show that in the expansion of $(x+y)^n$ the coefficient of x^ky^{n-k} is equal to the coefficient of $x^{n-k}y^k$.
49. Show that $\binom{n}{k} = \binom{n}{n-k}$ and explain why this answers the question in Problem 48.
50. Show that $\binom{n}{1} = \binom{n}{n-1} = n$.
51. Show that for any integer n,
$$\binom{n}{0} + \binom{n}{1} + \binom{n}{2} + \cdots + \binom{n}{n} = 2^n$$
[Hint: Expand $(1+1)^n$.]
52. Show that for any positive integer n,
$$\binom{n}{0} - \binom{n}{1} + \binom{n}{2} - \binom{n}{3} + \cdots + (-1)^n\binom{n}{n} = 0$$
[Hint: Expand $(1-1)^n$.]
53. According to **Stirling's formula,** when n is large,
$$n! \approx \sqrt{2\pi n}\left(\frac{n}{e}\right)^n \qquad \text{where } e \approx 2.718281828$$
Use Stirling's formula to estimate (a) 100! (b) 200! [Hint: Use common logarithms.]
* 54. Use the result of Problem 51 to prove that each set containing n elements has precisely 2^n subsets.
55. Find the eighth row of Pascal's triangle from the seventh row in Figure 1.
56. Find the ninth and tenth rows of Pascal's triangle.

Summary Outline of Chapter 9

- **Mathematical Induction:** To prove that something is true for each positive integer, prove it for the first integer. Then assume it true for the integer k and prove it true for the integer $k+1$. p. 475
- **The Sigma Notation:**
$$\sum_{k=M}^{n} a_k = a_M + a_{M+1} + a_{M+2} + \cdots + a_n$$ p. 483
- The sequence $a_1, a_2, a_3, \ldots$ defined by $a_{n+1} = a_n + d$ is called an **arithmetic sequence.** For an arithmetic sequence, $a_n = a_1 + (n-1)d$. p. 489
- The sequence defined by $a_{n+1} = ra_n$, $r \neq 0, 1$, is called a **geometric sequence.** For a geometric sequence, $a_n = a_0r^n$. p. 492
- The **sum of a geometric progression** is a sum of the form
$$S_n = 1 + r + r^2 + r^3 + \cdots + r^n = \sum_{k=0}^{n} r^k = \frac{1-r^{n+1}}{1-r} \qquad \text{for } r \neq 1$$ p. 497
- A **geometric series** is an infinite sum of the form
$$S = 1 + r + r^2 + r^3 + \cdots = \sum_{k=0}^{\infty} r^k = \frac{1}{1-r}, \qquad -1 < r < 1$$ p. 499
- A **permutation** of n objects is an arrangement of the objects. p. 510

- **Factorial:** n factorial, denoted by $n!$, is the product of the first n integers:

$$n! = n(n-1)(n-2)\cdots(3)(2)(1)$$

By convention, $0! = 1$. p. 510

- The number of permutations of n objects is $n!$ p. 511
- The number of permutations of n objects taken k at a time, denoted by $P_{n,k}$, is given by

$$P_{n,k} = n(n-1)(n-2)\cdots(n-k+1) = \frac{n!}{(n-k)!}$$

pp. 512, 513

- A **combination** of n objects taken k at a time is any selection of the n objects without regard to order. p. 514
- The number of combinations of n objects taken k at a time, denoted by $\binom{n}{k}$, is given by

$$\binom{n}{k} = \frac{P_{n,k}}{k!} = \frac{n!}{(n-k)!k!} = \frac{n(n-1)(n-2)\cdots(n-k+1)}{k!} \quad \text{for } 0 \le k \le n$$

p. 515

- The **sample space,** S, of an **experiment** is the set of all possible outcomes. p. 522
- An **event** E is a subset of the sample space. An event containing one outcome is called a **simple event.** p. 523
- A sample space is an **equiprobable space** if all simple events are equally likely. p. 524
- **Probability in an Equiprobable Space:**

$$P(E) = \frac{\text{number of outcomes in } E}{\text{number of outcomes in } S}$$

p. 524

- **The Binomial Theorem:**

$$(x+y)^n = x^n + \binom{n}{1}x^{n-1}y + \binom{n}{2}x^{n-2}y^2 + \cdots + \binom{n}{j}x^{n-j}y^j + \cdots + \binom{n}{n-1}xy^{n-1} + y^n$$

$$= \sum_{j=0}^{n} \binom{n}{j} x^{n-j}y^j$$

p. 532

Review Exercises for Chapter 9

In Exercises 1–4 use mathematical induction to prove the given formula.

1. $1 + 3 + 3^2 + \cdots + 3^{n-1} = \dfrac{3^n - 1}{2}$
2. $1^3 + 2^3 + 3^3 + \cdots + n^3 = \dfrac{n^2(n+1)^2}{4}$
3. $1 + 5 + 9 + \cdots + (4n - 3) = n(2n - 1)$
4. $x^n - 2^n$ is divisible by $x - 2$.

In Exercises 5 and 6 find the first 5 terms of each sequence. Start with $n = 1$.

5. $\{3n + 7\}$ 6. $\left\{\dfrac{n+2}{n+5}\right\}$

In Exercises 7 and 8 find the general term of each sequence.

7. $\{4, 7, 10, 13, \ldots\}$ 8. $\{2, 5, 10, 17, 26, \ldots\}$

In Exercises 9 and 10 evaluate the given sum.

9. $\displaystyle\sum_{k=1}^{6} (2k + 1)$ 10. $\displaystyle\sum_{k=1}^{4} \frac{k}{k^2 + 1}$

In Exercises 11–13 write each sum by using the Σ notation.

11. $1 + 3 + 9 + 27 + 81 + \cdots + 3^m$
12. $1 + 2^{1/5} + 3^{1/5} + 4^{1/5} + \cdots + n^{1/5}$
13. $x^3 - x^7 + x^{11} - x^{15} + x^{19} - x^{23}$

In Exercises 14–17 write the general term for each sequence.

14. $\{3, 9, 15, 21, 27, \ldots\}$
15. $\left\{6, \dfrac{11}{2}, 5, \dfrac{9}{2}, 4, \dfrac{7}{2}, \ldots\right\}$
16. $\{2, 6, 18, 54, 162, \ldots\}$
17. $\left\{-27, 9, -3, 1, -\dfrac{1}{3}, \ldots\right\}$

In Exercises 18–21 write out the first six terms of each sequence. Start with $n = 0$.

18. $\{3 + 4n\}$ 19. $\{2 \cdot 3^n\}$

20. $\{\frac{1}{2} \cdot 3^n\}$ 21. $\left\{2 - \frac{1}{3^n}\right\}$

22. Find the 25th term of the arithmetic sequence $\{5, 9, 13, 17, 21, \ldots\}$.
23. Find the 10th term of the geometric sequence $\{1, 4, 16, 64, \ldots\}$.
24. If \$5000 is invested at $6\frac{1}{2}\%$ compounded annually, what is it worth after 8 years?
25. Find an expression for a_n if $a_{n+1} = 3a_n - 4$; $a_0 = 1$.

In Exercises 26–29 find each sum.

26. $1 + \frac{1}{4} + (\frac{1}{4})^2 + \cdots + (\frac{1}{4})^7$
27. $3 + 3^2 + 3^3 + \cdots + 3^8$
28. $1 + \frac{1}{4} + (\frac{1}{4})^2 + (\frac{1}{4})^3 + \cdots$
29. $1 - \frac{1}{2} + \frac{1}{2^2} - \frac{1}{2^3} + \frac{1}{2^4} - \cdots$

In Exercises 30–36 a procedure or activity is described. In each exercise determine the number of possible outcomes.

30. A coin is flipped and a die is thrown.
31. A coin is flipped 4 times.
32. Three dice are thrown.
33. A drugstore purchases one each of 5 brands of toothpaste, 6 brands of shampoo, and 8 brands of deodorant.
34. Nine people run for 3 positions in the city council.
35. Nine people run for chairperson, vice chairperson, and secretary of the city council.
36. A family has 6 children.
37. How many 5-letter "words" can be made from the letters of the word *PRIMATE*?
38. Evaluate
    (a) $P_{7,3}$ (b) $P_{8,4}$ (c) $P_{9,8}$ (d) $P_{9,1}$ (e) $P_{10,6}$
39. Evaluate
    (a) $\binom{9}{4}$ (b) $\binom{7}{5}$ (c) $\binom{9}{8}$ (d) $\binom{9}{1}$ (e) $\binom{10}{6}$
40. A scrabble player has 7 letters and wishes to play a 5-letter word. If she chooses to test each possible 5-letter permutation and if each test takes $1\frac{1}{2}$ seconds, how long will it take her to complete her play?
41. Eleven heavyweight contenders are ranked by the World Boxing Association (WBA). How many rankings are possible? Assume no ties.
42. If in Exercise 41, the WBA ranks only the top 6 contenders among the 12, how many rankings are possible?
43. How many distinct permutations are there of the letters of the word *BANANAS*?
44. A textbook has 4 chapters on probability and 6 on statistics. If an instructor wishes to cover 2 probability chapters and 3 statistics chapters, how many choices does she have?
45. How many poker hands contain a flush?
46. What is the probability of being dealt a flush?
47. What is the probability of being dealt a straight?
48. John and Mary are on a 6-person committee that is selecting a chairperson and vice chairperson at random. What is the probability that Mary will be chosen chairperson and John vice chairperson?
49. In Exercise 48, what is the probability of the Mary-John outcome if it is known that Aaron and Kerstin, two other members of the committee, have not been selected for either position?
50. In Exercise 48, the committee selects instead 2 members for a subcommittee. What is the probability that Mary and John will be selected?
51. Of 50 stocks on the American Stock Exchange, 18 advanced and 13 declined. In how many ways could this happen?
52. A chimpanzee is placed in front of a toy typewriter with the 26 letters of the alphabet in front of him. If he punches 5 letters at random (without repetition) what is the probability that he will spell out the word *CHIMP*?
53. If a student guesses on a 10-question, true-false test, what is the probability of each event?
    (a) He gets 100%.
    (b) He gets exactly 3 right.
54. What is the probability of getting a sum divisible by 3 if 2 dice are thrown?
55. Four people, labeled P_1, P_2, P_3, and P_4, are to be tested for blood type (A, B, AB, O).
    (a) How many different distributions of blood type are possible among these 4 people?
    (b) In how many of these distributions is there exactly 1 person of each type?

In Exercises 56–60 use the binomial theorem.

56. Expand $(x + y)^3$
57. Expand $(a + b)^6$ 58. Expand $(u - w)^4$
59. Expand $(3x - 5y)^5$
60. Find the coefficient of a^5b^8 in the expansion of $(a + b)^{13}$.

Appendix A

Graphing Using a Calculator

It has long been possible to generate graphs of a wide variety of functions on a computer. Recently, hand-held calculators with graphing capabilities have become available. In this appendix, we shall discuss some techniques that will help you use the graphing calculator more effectively. We shall also discuss some ways that graphing calculators can be used to solve algebraic problems like approximating the zeros of functions, finding where two curves intersect, and solving inequalities.

The graphing calculators currently available are produced by (in alphabetical order) Casio, Hewlett-Packard, Sharp, and Texas Instruments. This appendix is generic; that is, it is intended for use with any graphing calculator. All applications cited in this appendix can be carried out with any of the calculators currently available. Therefore, our discussion will focus on graphing techniques rather than on specific keystrokes; that is, we will not tell you which buttons to push. For that reason:

> It is essential that you read the instruction manual that accompanies your graphing calculator before you read any further.

Problems that will require you to use a graphing calculator can be found in selected sections of this text or in the graphing calculator supplement that is available from the publisher.

I. Obtaining a Graph on a Calculator

In order to obtain a graph on a calculator, two things must be done (not necessarily in the order given here).

A. Enter the Function to Be Graphed

Read your manual to learn the procedure that must be used to enter functions. There will be a special way to enter the function variable, which is most often denoted by X.

B. Determine the Range and Scale

You must tell the calculator the range of values over which you wish the function to be graphed and specify the scale on the x- and y-axes.

The range is given by entering the smallest and largest values to be taken for x and y. The scale is the length represented by each tick mark on an axis. If the x-scale is 2, for example, then the distance between two successive tick marks represents a length of 2 units. In this appendix, we use the following notation:

x_{min} = minimum value of x
x_{max} = maximum value of x
x_{scl} = the scale on the x-axis

y_{min} = minimum value of y
y_{max} = maximum value of y
y_{scl} = the scale on the y-axis

Most calculators have a key (sometimes labeled RANGE) that must be pressed in order to enter range and scale values.

NOTE If you graph a function that you have entered but do not specify range and scale values, then one of three things will happen. First, the calculator may use the range and scale values that were entered for the previous graph that was sketched. Second, the calculator may use some "standard" built-in range and scale values. These are called "standard defaults." On one TI calculator, for example, the standard defaults for both the x- and y-axes are the intervals $[-10, 10]$ with a scale of 1. Third, if the calculator has built-in graphs, then preset ranges and scales will be used whenever one of these graphs is sketched. For example, on one Casio calculator the graph of $y = \ln x$ is built in. When this function is sketched, the calculator uses the following range and scale values:

$$x_{min} = -1 \qquad y_{min} = -1.6$$
$$x_{max} = 8.4 \qquad y_{max} = 2.368$$
$$x_{scl} = 2 \qquad y_{scl} = 1$$

Before pressing a graphing key, check the range and scale values. Unexpected things can happen if you do not enter these values yourself.

The hardest part about graphing on a calculator is choosing appropriate range values. Our first example illustrates why care in choosing these values is essential.

EXAMPLE 1 ***Finding an Appropriate Range in Order to Generate a Graph***

Sketch the graph of $y = -x^3 + 2x^2 + 5x - 6 = -(x + 2)(x - 1)(x - 3)$.

SOLUTION Suppose that we do not notice that the cubic can be factored and we arbitrarily choose the following range and scale values:

Ranges: $-2 \leq x \leq 2;\ -3 \leq y \leq 3$

Scales: Each x-axis tick represents 1 unit, and each y-axis tick represents 2 units. That is,

$$x_{min}: -2 \qquad y_{min}: -3$$
$$x_{max}: 2 \qquad y_{max}: 3$$
$$x_{scl}: 1 \qquad y_{scl}: 2$$

After the function is entered, the graph in Figure 1 is obtained.

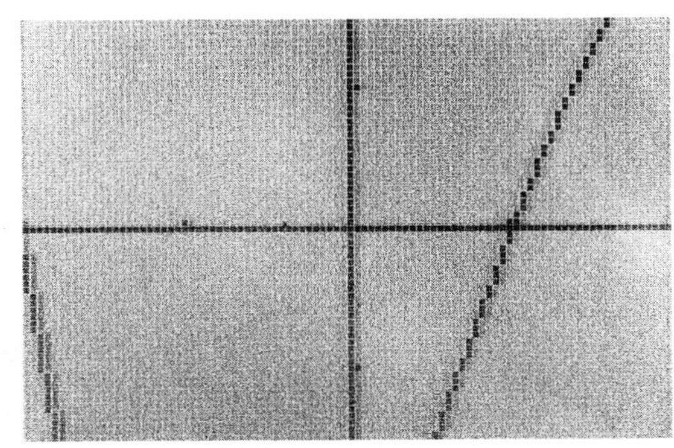

Figure 1 Graph of $y = -x^3 + 2x^2 + 5x - 6$ for $-2 \le x \le 2$, $-3 \le y \le 3$

This graph is accurate but not very useful. We need to see what happens outside of our rather limited range. Let us greatly expand the ranges:

$$-20 \le x \le 20;\ -200 \le y \le 200;\ \text{with } x_{\text{scl}} = 1 \text{ and } y_{\text{scl}} = 20$$

The graph now appears as in Figure 2:

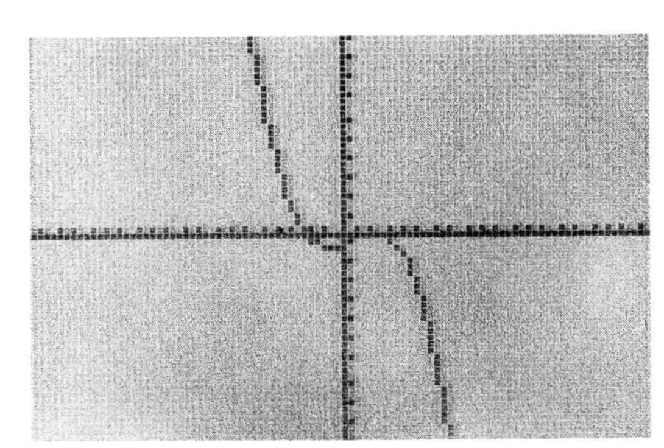

Figure 2 Graph of $y = -x^3 + 2x^2 + 5x - 6$ for $-20 \le x \le 20$, $-200 \le y \le 200$

This graph looks different but is still not what we want. Its appearance should not be surprising. For $|x|$ large, $-x^3 + 2x^2 + 5x - 6 \approx -x^3$, so since we used a large range of x-values, we have obtained a graph that looks like the graph of $-x^3$.

We can get a much more revealing graph by thinking a bit before entering range values. If $f(x) = -x^3 + 2x^2 + 5x - 6$, then, for example, $f(-4) = 70$ and $f(4) = -18$. It is not hard to see that if $x < -4$, then $f(x) > 70$, and if $x > 4$, then $f(x) < -18$. Thus, most of the interesting behavior of this function occurs for $-4 \le x \le 4$ and $-18 \le y \le 70$. Setting these range values and letting $x_{\text{scl}} = 1$ and $y_{\text{scl}} = 2$, we obtain the graph in Figure 3.

This is the type of graph we want. We can clearly see the zeros at -2, 1, and 3. Other interesting behavior, like intervals over which the function is increasing or decreasing, is plainly shown.

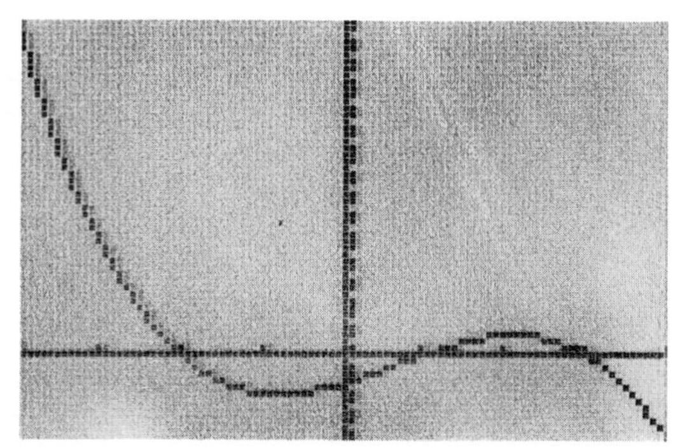

Figure 3 Graph of $y = -x^3 + 2x^2 + 5x - 6$ for $-4 \le x \le 4$, $-18 \le y \le 70$

NOTE On some calculators, there is a ''zoom'' or ''factor'' feature. This feature allows you to zoom in or zoom out by a factor you set. If you zoom out in Figure 1, for example, then you get a more global picture of the graph. If you zoom out too much, you might get a picture like the one in Figure 2. Then you could zoom in to obtain the more accurate graph in Figure 3. You can experiment with the zoom feature, zooming in and out, until you get a graph that looks accurate. You can save a lot of confusion, however, if you first think about what reasonable range values should be. ■

EXAMPLE 2 *Sketching the Graph of a Rational Function Having a Vertical Asymptote*

Sketch the graph of $y = f(x) = \dfrac{x^2 - 2x + 5}{x + 2}$.

SOLUTION We first note that the function is not defined at $x = -2$. As x gets close to -2 from either side, $|y|$ gets large. Also, $|y|$ gets large as $|x|$ gets large. To see this, we divide to obtain

$$\frac{x^2 - 2x + 5}{x + 2} = x - 4 + \frac{13}{x + 2}.$$

Since $\dfrac{13}{x + 2} \to 0$ as $x \to \pm\infty$, we see that, for x large,

$$\frac{x^2 - 2x + 5}{x + 2} \approx x - 4.$$

(The symbol $x \to \infty$ is explained on page 237.)

You must be very careful when you enter functions in your calculator. Without parentheses around the numerator and denominator of this function, you would not be graphing the function you want to graph. Specifically, if you enter

$$y = x^2 - 2x + 5 \div x + 2,$$

you will obtain the graph of

$$y = x^2 - 2x + \frac{5}{x} + 2$$

This is very different from the correctly entered function

$$y = (x^2 - 2x + 5) \div (x + 2).$$

There are many ranges of values that will give us a suitable graph. Each one must include $x = -2$ and allow for reasonably large values for $|y|$. Here is one set of range values:

$$-12 \le x \le 10, \ -25 \le y \le 15$$

We use scales of 1 on each axis.

The graph is given in Figure 4.

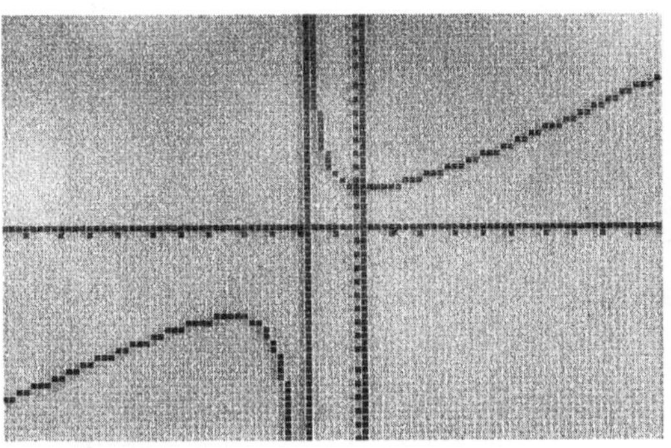

(a) Graph showing the vertical asymptote

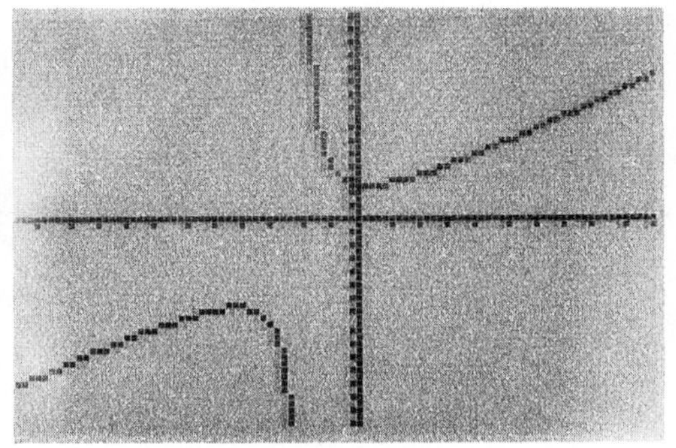

(b) Graph not showing the vertical asymptote

Figure 4 Graph of $y = \dfrac{x^2 - 2x + 5}{x + 2}$ for $-12 \le x \le 10, \ -25 \le y \le 15$

If your calculator shows a vertical line at $x = -2$, it is because the calculator is attempting to connect the points on the graph. Thus, some calculators may show the vertical asymptote (Figure 4(a)) at $x = -2$ (see page 280 for a discussion of vertical asymptotes) and some may not (Figure 4(b)).† ■

EXAMPLE 3 ***Sketching the Graph of a Rational Function with Horizontal and Vertical Asymptotes***

Sketch the graph of $y = \dfrac{2x^2 - 3x + 5}{x^2 - 1}$.

---

† One TI calculator draws the asymptote if the range values for x are set to $[-12, 10]$ but does not draw it if the range values are set to $[-10, 10]$. Thus the appearance or nonappearance of asymptotes depends both on the calculator used and the range settings.

SOLUTION We first note that $x^2 - 1 = 0$ when $x = 1$ and $x = -1$. Since the numerator is not zero at either of these values, the lines $x = 1$ and $x = -1$ are vertical asymptotes (see page 280). If we divide numerator and denominator by x^2, we obtain

$$\frac{2x^2 - 3x + 5}{x^2 - 1} = \frac{2 - \dfrac{3}{x} + \dfrac{5}{x^2}}{1 - \dfrac{1}{x^2}}.$$

The terms $\frac{3}{x}$, $\frac{5}{x^2}$, and $\frac{1}{x^2} \to 0$ as $x \to \pm\infty$, so $\frac{2x^2 - 3x + 5}{x^2 - 1} \to 2$ as $x \to \pm\infty$. Thus $y = 2$ is a horizontal asymptote to the graph (see page 280). This means that for x large in either direction, y is close to 2. On the other hand, $|y| \to \infty$ as x gets near 1 or -1 so we must allow for large values of y.

In Figure 5, we provide the graph for the values $-3 \le x \le 3$, $-20 \le y \le 20$.

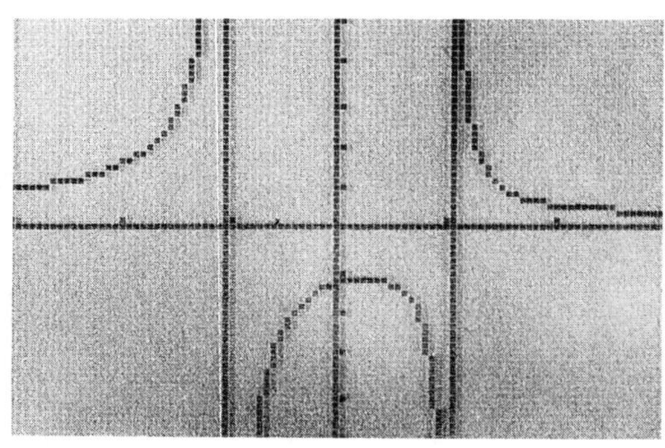

(a) Graph showing vertical asymptotes

(b) Graph not showing vertical asymptotes

Figure 5 Graph of $y = \dfrac{2x^2 - 3x + 5}{x^2 - 1}$ for $-3 \le x \le 3$, $-20 \le y \le 20$, $x_{\text{scl}} = 1$, $y_{\text{scl}} = 5$

This graph gives us an accurate picture. However, if we want to see what happens as $|x|$ gets larger (to see the approach to the asymptote $y = 2$), then we can increase the x-values and decrease the y-values.

In Figure 6, we give the graph for $-12 \le x \le 12$ and $-10 \le y \le 10$.

Now we see clearly that the graph approaches a horizontal line as $|x|$ gets large. However, we lose an accurate picture in the interval $-1 < x < 1$.

Which of the graphs in Figures 5 and 6 is better? The answer here is not clear. If you want to see what happens to the function as $|x|$ becomes large, then Figure 6 is better. However, Figure 5 provides a clearer picture of the shape of the curve and is closer to the one that would appear in a textbook. A

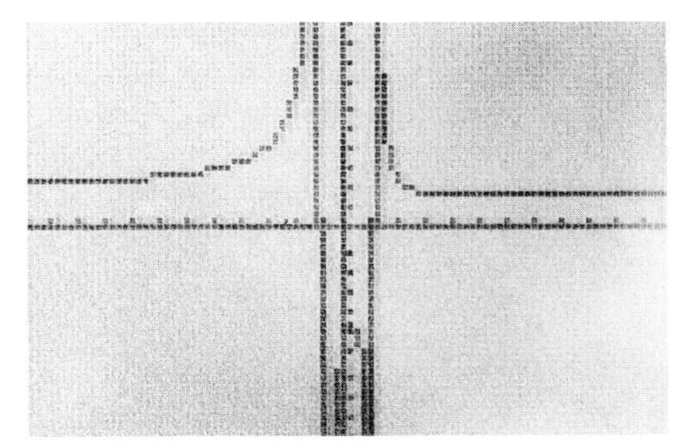

Figure 6 Graph of $y = \dfrac{2x^2 - 3x + 5}{x^2 - 1}$ for $-12 \le x \le 12,\ -10 \le y \le 10$

nice feature of a graphing calculator, especially one with a "zoom" feature, is that it allows you to experiment with different range values in order to obtain a graph that best suits your needs. ■

EXAMPLE 4 *Sketching the Graph of an Exponential Function*

Sketch the graph of $y = f(x) = e^{x^2+1}$.

SOLUTION $x^2 + 1 \ge 1$ so $e^{x^2+1} \ge e^1 = e \approx 2.718$. Thus the minimum value for y is 2.718. Also $f(-2) = f(2) = e^5 \approx 148$. One suitable set of ranges is

$$-2 \le x \le 2;\ 0 \le y \le 20.$$

We allow the value $y = 0$ so that the graph will clearly exhibit the minimum value at $x = 0$. We use the scale values $x_{\text{scl}} = 0.25$ and $y_{\text{scl}} = 1$.

Consult your calculator manual to see how to enter the function e^{x^2+1}. (A discussion of how to obtain exponential values on a calculator appears on page 335.) When e^{x^2+1} is entered, the graph in Figure 7 appears.

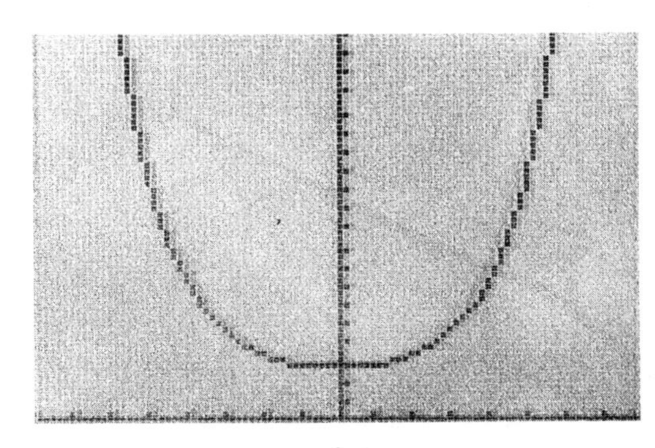

Figure 7 Graph of $y = e^{x^2+1}$ for $-2 \le x \le 2,\ 0 \le y \le 20$ ■

II. Other Uses of Calculator Graphing

There are many types of problems that can be solved on a calculator by using suitable graphs. Among these are (1) finding the zeros of a function and (2) finding points of intersections of graphs. The two problems are really the same because finding where $f(x) = g(x)$ is equivalent to finding the zeros of $f(x) - g(x)$. We will, however, treat these as distinct problems and give examples of each.

EXAMPLE 5 ***Using Graphs to Find the Zeros of a Polynomial***

Find all zeros of $f(x) = x^3 - 3x^2 + 7x - 8$.

SOLUTION $f(0) = -8, f(x) < 0$ if $x < 0, f(1) = -3, f(2) = 2$, and $f(3) = 13$. Reasonable ranges are $-2 \leq x \leq 3$ and $-10 \leq y \leq 15$. Using a scale of 1 on each axis, we obtain the graph in Figure 8. (The numbers on the x-axis were added to make things clearer.)

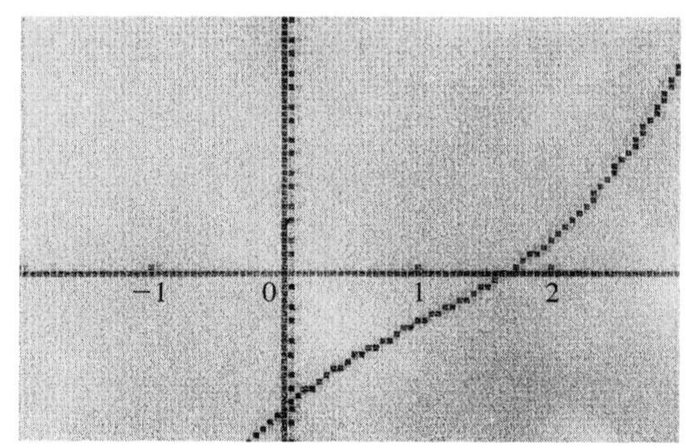

Figure 8 Graph of $y = x^3 - 3x^2 + 7x - 8$ for $-2 \leq x \leq 3$, $-10 \leq y \leq 15$

The graph suggests that there is one and only one zero between 1 and 2. (We already knew that there was at least one zero because $f(1) < 0$ and $f(2) > 0$.)

There are several ways to approximate this zero more closely. If your calculator has a zoom feature, you can zoom in on the part of the graph where the curve crosses the x-axis and find the zero to at least 4 or 5 decimal place accuracy with little difficulty. Whether or not you have this feature, or any similar one, you can improve your estimate by changing the range values along the x-axis. This is what we do now.

Here are some new range and scale values:

$$1 \leq x \leq 2;\ x_{scl} = 0.05;\ -3 \leq y \leq 2;\ y_{scl} = 0.1$$

From the graph in Figure 9, we see that the graph crosses the x-axis in the interval $1.6 < x < 1.7$.

To get more precision, we can change range and scale again:

$$1.6 \leq x \leq 1.7;\ x_{scl} = 0.01$$
$$-0.5 \leq y \leq 0.5;\ y_{scl} = 0.1$$

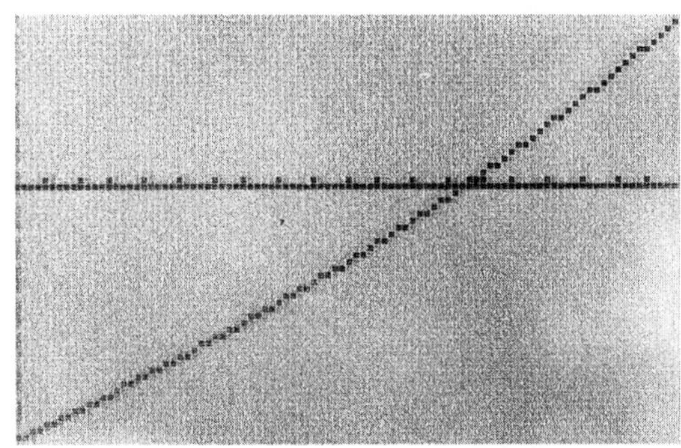

Figure 9 Graph of $y = x^3 - 3x^2 + 7x - 8$ for $1 \le x \le 2$, $-3 \le y \le 2$, $x_{\text{scl}} = 0.05$, $y_{\text{scl}} = 0.1$

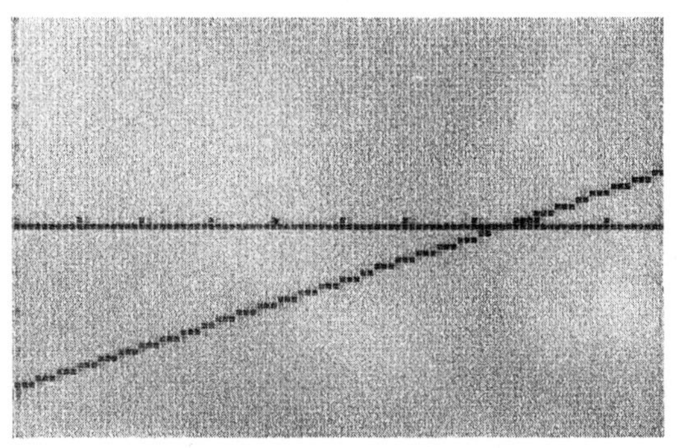

Figure 10 Graph of $y = x^3 - 3x^2 + 7x - 8$ for $1.6 \le x \le 1.7$, $-0.5 \le y \le 0.5$, $x_{\text{scl}} = 0.01$, $y_{\text{scl}} = 0.1$

Now we see, in Figure 10, that the zero falls between 1.67 and 1.68. We stop here because there are faster ways to get more precision if more precision is needed.

A NOTE ON ACCURACY We say that a number x_A approximates a number x with k *decimal places of accuracy* if

$$|x - x_A| < \frac{1}{2} \times 10^{-k}.$$

For example, suppose we wish to approximate $\pi \approx 3.14159265359$.

A First Approximation

$$\pi \approx 3.1$$

Then

$$|\pi - 3.1| \approx 0.04159265359$$

Now $10^{-1} = 0.1$ and $\dfrac{1}{2} \times 10^{-1} = 0.05$.

So

$$|\pi - 3.1| < 0.05 = \frac{1}{2} \times 10^{-1}$$

So 3.1 approximates π with one decimal place accuracy.

A Second Approximation

$$\pi \approx 3.1415$$

Then

$$|\pi - 3.1415| \approx 0.00009265359$$

Now

$$10^{-4} = 0.0001 \quad \text{and} \quad \frac{1}{2} \times 10^{-4} = 0.00005$$

$$10^{-3} = 0.001 \quad \text{and} \quad \frac{1}{2} \times 10^{-3} = 0.0005$$

So since $0.00005 < 0.00009265359 < 0.0005$, we see that 3.1415 approximates π with three (but not four) decimal place accuracy.

ANOTHER NOTE ON THE ZOOM FEATURE On a calculator with a zoom feature, the zero in Example 5 can be approximated quicker than by rescaling manually. If we zoom in on the graph in Figure 8 by a scale of 10 in each axis, with the zooming centered at the place where the curve crosses the x-axis, then we obtain the graph in Figure 11.

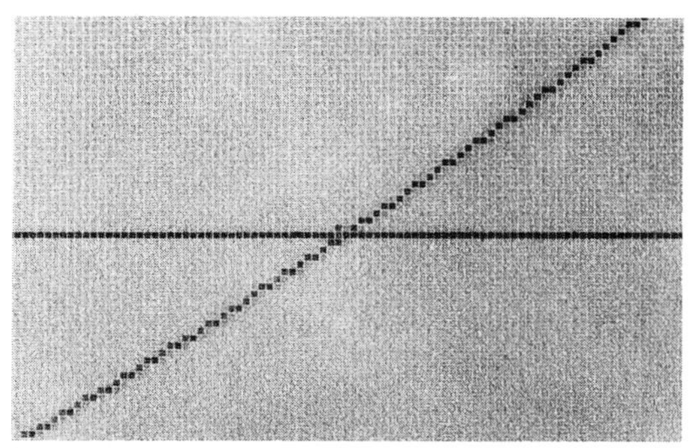

Figure 11 Graph of $y = x^3 - 3x^2 + 7x - 8$ (zoomed in from Figure 8)

Using the trace feature on the calculator, we can move along the curve to locate two points near where the curve crosses the x-axis. One point should be to the left of the zero, and the other should be to the right. We can find, in Figure 11, that $1.65 <$ the zero < 1.69. If we zoom in again by a factor of 10, we get the graph in Figure 12.

Then we find that the zero is between $x = 1.671$ and $x = 1.674$. Thus we know that, to two decimal places, $x \approx 1.67$.

The reason that there are no tick marks in Figures 11 and 12 is that, in zooming in, the calculator we used did not change the scale values. There-

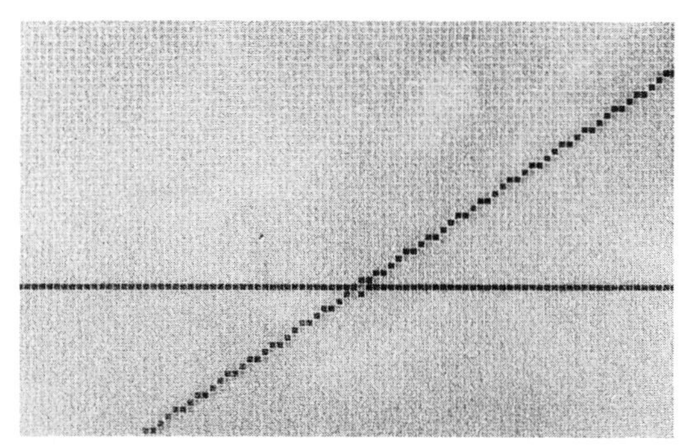

Figure 12 Graph of $y = x^3 - 3x^2 + 7x - 8$ (zoomed in from Figure 11)

fore, the scales stayed at 1 unit on each axis. Since the range of x-values is far less than 1 unit, no tick marks appear.

Alternatively, if your calculator has a zoom feature, it may also have a "box" feature. This allows you to draw a rectangle on the screen and then zoom in on the rectangle within. If you blow up a small rectangle around the zero, you can see more precisely where the curve crosses the x-axis. By doing this repeatedly, you can approximate the zero quickly and accurately.

We will say no more about zoom and box features in this appendix, but you should use them if they are available on your calculator. ■

EXAMPLE 6 *Finding the Zeros of a Polynomial*

Find all real zeros of $p(x) = 2x^4 + 1.5x^3 - 9x^2 - x + 5$.

SOLUTION We first observe that the $2x^4$ term dominates the other terms for x large. For example, if $x = 5$, then $2x^4 = 1250$ while $1.5x^3 - 9x^2 - x + 5 = -37.5$. The range values $-5 \le x \le 5$ include all the "interesting" behavior in the graph. If we set the y values at $-10 \le y \le 10$ with $x_{\text{scl}} = 1$ and $y_{\text{scl}} = 1$, we obtain the graph in Figure 13.

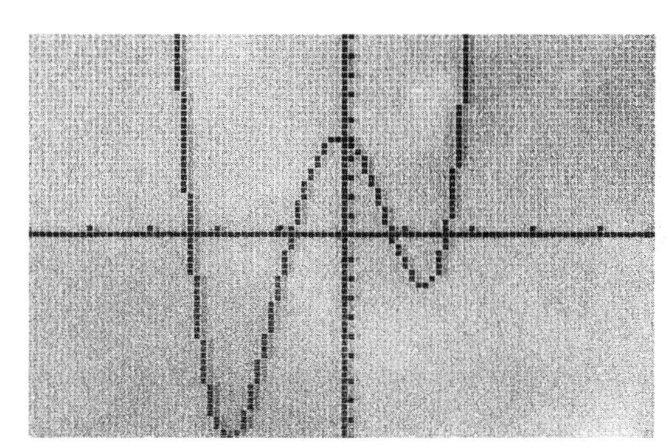

Figure 13 Graph of $y = 2x^4 + 1.5x^3 - 9x^2 - x + 5$ for $-5 \le x \le 5$, $-10 \le y \le 10$

Evidently $p(x)$ has four zeros. There is one between -3 and -2, one between -1 and $-\frac{1}{2}$, one between $\frac{1}{2}$ and 1, and one between 1 and 2. We can obtain each zero by choosing x-values near the zero.

First zero We set $-3 \le x \le -2$, $-1 \le y \le 1$ and $x_{\text{scl}} = y_{\text{scl}} = 0.1$ to obtain the graph in Figure 14.

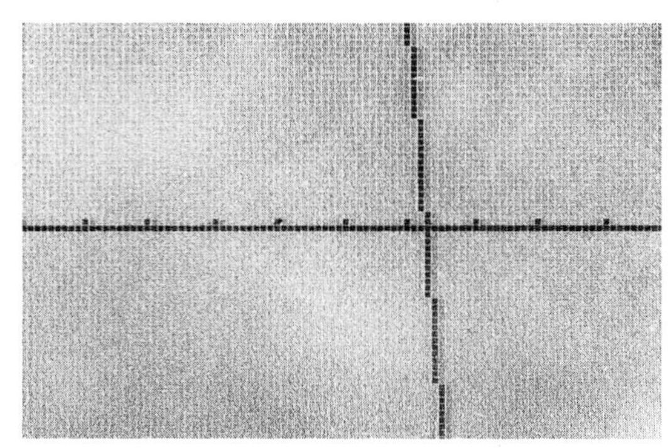

Figure 14 Graph of $y = 2x^4 + 1.5x^3 - 9x^2 - x + 5$, $-3 \le x \le -2$, $-1 \le y \le 1$, $x_{\text{scl}} = 0.1$

The zero is between -2.4 and -2.3. Rescaling again for $-2.4 \le x \le -2.3$, $-1 \le y \le 1$, $x_{\text{scl}} = y_{\text{scl}} = 0.01$, we obtain the graph in Figure 15.

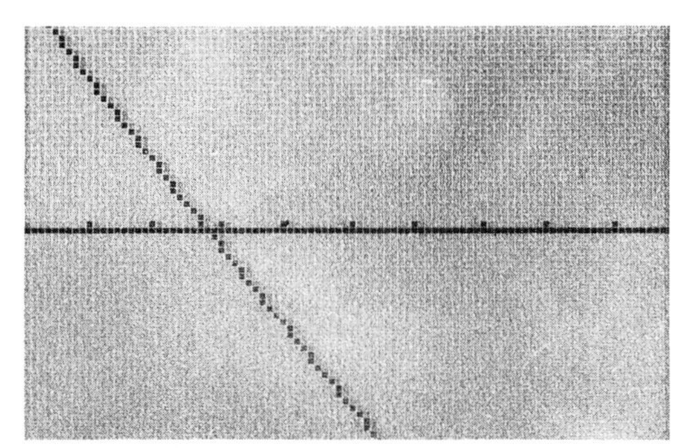

Figure 15 Graph of $y = 2x^4 + 1.5x^3 - 9x^2 - x + 5$, $-2.4 \le x \le -2.3$, $x_{\text{scl}} = 0.01$

Now we see that the zero is between -2.38 and -2.37. Continuing in this manner, we find, to three decimal place accuracy, $x \approx -2.371$.

Second zero Since the zero is between -1 and -0.5, we rescale with $-1.0 \le x \le -0.5$, $-1 \le y \le 1$, $x_{\text{scl}} = y_{\text{scl}} = 0.1$. The result is sketched in Figure 16.
The zero is very close to -0.8. We can continue to find that, to three decimal places, $x \approx -0.807$.

Third zero This zero is between $\frac{1}{2}$ and 1, so in Figure 17, we sketch the curve for $0.5 \le x \le 1.0$ and $x_{\text{scl}} = 0.1$.

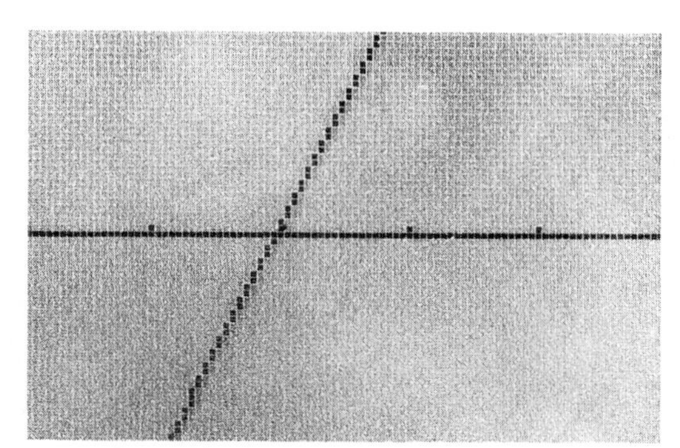

Figure 16 Graph of $y = 2x^4 + 1.5x^3 - 9x^2 - x + 5$, $-1.0 \leq x \leq -0.5$, $x_{scl} = 0.1$

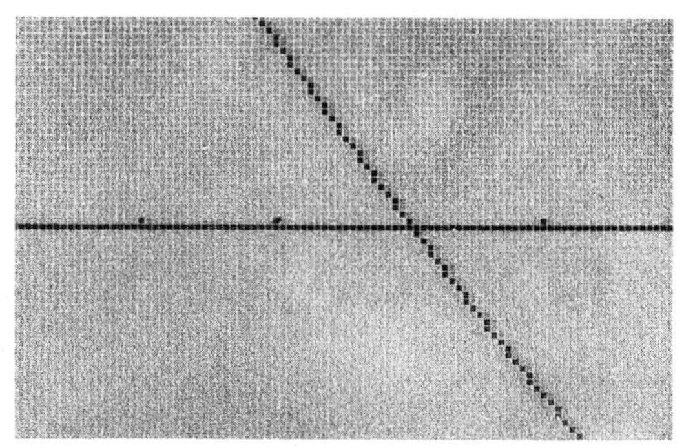

Figure 17 Graph of $y = 2x^4 + 1.5x^3 - 9x^2 - x + 5$, $0.5 \leq x \leq 1.0$, $x_{scl} = 0.1$

The curve seems to cross the x-axis near $x = 0.8$. To three decimal places, $x = 0.803$.

FOURTH ZERO Since this zero is between 1 and 2, we set $1 \leq x \leq 2$ with $x_{scl} = 0.1$ to obtain the graph in Figure 18.

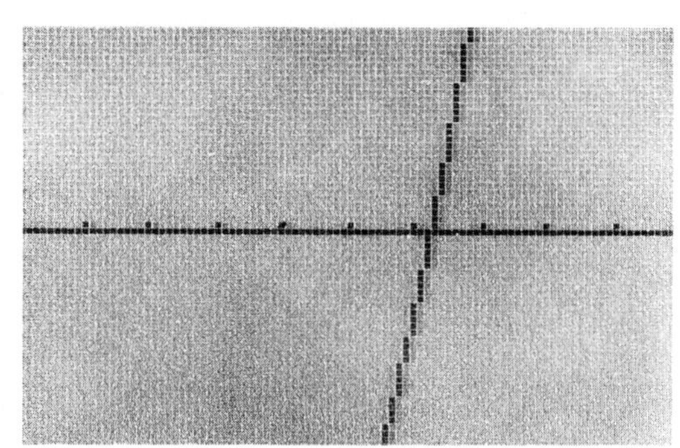

Figure 18 Graph of $y = 2x^4 + 1.5x^3 - 9x^2 - x + 5$, $1 \leq x \leq 2$, $x_{scl} = 0.1$

The last zero is near 1.6. We find that, to three decimal places, $x \approx 1.626$.

Thus, to three decimal places, the four zeros of $p(x) = 2x^4 + 1.5x^3 - 9x^2 - x + 5$ are -2.371, -0.807, 0.803, and 1.626. ■

EXAMPLE 7 ***Finding Points of Intersection of Two Graphs***

Find, to two decimal place accuracy, all points of intersection of the graphs of $y = 2 - x$ and $y = \ln x$.

SOLUTION $\ln x$ is defined only when $x > 0$ (see page 346). Also, if $x > 2$, $2 - x < 0$ and $\ln x > \ln 2 \approx 0.6931 > 0$, so there are no points of intersection for $x > 2$. We use the following ranges and scales on our initial graphs:

$x_{\min}$: 0 $\quad y_{\min}$: -2

$x_{\max}$: 2 $\quad y_{\max}$: 2

x_{scl}: 0.5 $\quad y_{\text{scl}}$: 1

Depending on the calculator, we can sketch both curves on the same screen in one of two ways.

ON SOME CALCULATORS You can enter several functions. When the "graph" key is pressed, the functions will be sketched, one after the other. (It is most common that four functions can be entered.)

ON OTHER CALCULATORS Simply sketch one function and do not clear it. Then, sketch the second function. As long as the range and scale values are not changed, both functions will appear on the same graph.
The graphs of both functions are given in Figure 19.

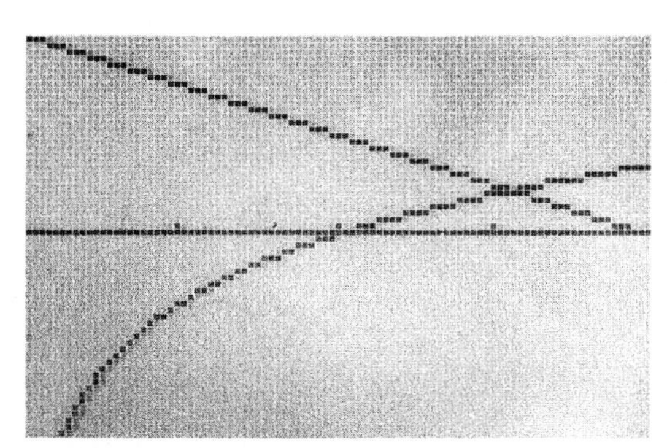

Figure 19 Graphs of $y = 2 - x$ and $y = \ln x$ for $0 \le x \le 2$, $-2 \le y \le 2$, $x_{\text{scl}} = 0.5$, $y_{\text{scl}} = 1$

It appears that the graphs intersect between 1.4 and 1.6 and probably closer to 1.6.

In Figure 20, we draw the graphs for x in the interval $[1.5, 1.6]$:

$x_{\min}$: 1.5 $\quad y_{\min}$: 0.4 ($\ln 1.5 \approx 0.405$)

$x_{\max}$: 1.6 $\quad y_{\max}$: 0.5 ($\ln 1.6 \approx 0.47$)

x_{scl}: 0.01 $\quad y_{\text{scl}}$: 0.01

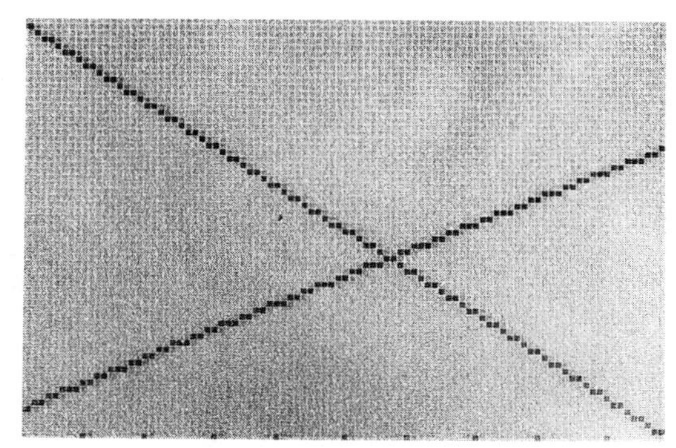

Figure 20 Graphs of $y = 2 - x$ and $y = \ln x$ for $1.5 \le x \le 1.6$, $0.4 \le y \le 0.5$, $x_{scl} = y_{scl} = 0.01$

The graphs seem to cross between 1.55 and 1.56. To be more precise, we sketch the graphs for x in the interval [1.55, 1.56] (see Figure 21).

x_{min}: 1.55 y_{min}: .43 ($\ln 1.55 \approx 0.438$)

x_{max}: 1.56 y_{max}: .45 ($\ln 1.56 \approx 0.445$)

x_{scl}: 0.001 y_{scl}: 0.002

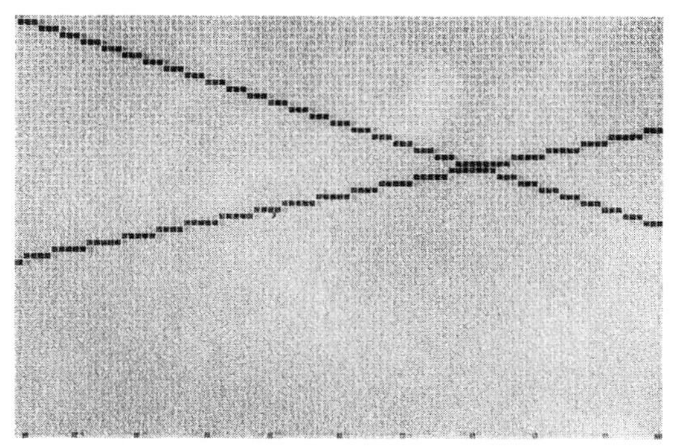

Figure 21 Graphs of $y = 2 - x$ and $y = \ln x$ for $1.55 \le x \le 1.56$, $0.43 \le y \le 0.45$, $x_{scl} = 0.001$, $y_{scl} = 0.002$

We see that the graphs intersect near the value 1.557. Therefore, to two decimal places, the graphs intersect at $x = 1.56$.

NOTE 1 Using Newton's method (see Section 4.6), we can show that, to 9 decimal places, the curves intersect at $x = 1.557145599$.

NOTE 2 This problem seemed more difficult than problems involving finding zeros because it is hard to see where the graphs intersect in relation to tick marks along the x-axis. This difficulty can be avoided in one of two ways: First, if your calculator has a trace function, and most do, then you can use it to find, approximately, the x- and y-coordinates of the point of intersection. If you do this, you don't have to worry about the tick marks.

Second, you can change the problem. If $2 - x = \ln x$, then

$$f(x) = 2 - x - \ln x = 0.$$

That is, you can change the problem into one of finding a zero of a function. This, as we have seen, is easier.

EXAMPLE 8 *Using Graphs to Solve a Quadratic Inequality*

Solve the quadratic inequality

$$4 + 4x - x^2 > 1 - 2x.$$

SOLUTION We observe that

$$f(x) > g(x) \text{ is equivalent to } f(x) - g(x) > 0.$$

It is simpler to solve the equivalent inequality

$$(4 + 4x - x^2) - (1 - 2x) > 0$$

or

$$4 + 4x - x^2 - 1 + 2x > 0$$

or

$$3 + 6x - x^2 > 0.$$

Before continuing, we observe that $y = f(x) > 0$ for precisely those values of x at which the graph of f lies above the x-axis.

We therefore graph $f(x) = 3 + 6x - x^2$ and see where it lies above the x-axis.

We use the range and scale values:

$x_{\min}$: -2 $\quad y_{\min}$: -4

$x_{\max}$: 10 $\quad y_{\max}$: 12

x_{scl}: 1 $\quad y_{\text{scl}}$: 1

The graph is given in Figure 22.

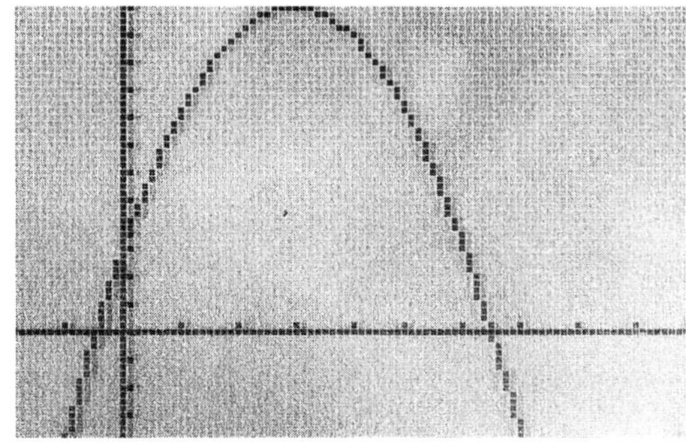

Figure 22 Graph of $y = 3 + 6x - x^2$ for $-2 \le x \le 10$, $-4 \le y \le 12$, $x_{\text{scl}} = y_{\text{scl}} = 1$

We see that $3 + 6x - x^2$ has two zeros, one between -1 and 0 and the other between 6 and 7. Moreover, $3 + 6x - x^2$ is above the x-axis between these two zeros.

We can find these zeros more precisely by using the method of Examples 5 and 6. To two decimal places, the zeros are -0.46 and 6.46. Thus the solution set of the inequality is, approximately,

$$-0.46 < x < 6.46.$$

We can check this answer by obtaining the points of intersection algebraically:

$$3 + 6x - x^2 = 0$$

$$x^2 - 6x - 3 = 0 \qquad \text{Multiply both sides by } -1$$

$$x = \frac{6 \pm \sqrt{36 - 4(1)(-3)}}{2} \qquad \text{Quadratic formula (see page 97)}$$

$$x = \frac{6 \pm \sqrt{48}}{2} = \frac{6 \pm 4\sqrt{3}}{2} = 3 \pm 2\sqrt{3}$$

The points of intersection are

$$x = 3 - 2\sqrt{3} \approx -0.464101615 \quad \text{and} \quad x \approx 3 + 2\sqrt{3} = 6.464101615.$$

This illustrates that the graphical technique is useful for obtaining an approximate solution to a problem. To obtain very accurate answers, other techniques may be more efficient.

III. Graphing Conic Sections

As long as your calculator can plot two functions on the same screen, you can use it to obtain the graphs of conic sections. We illustrate this first with a simple example.

EXAMPLE 9 *Graphing a Circle*

Graph the circle $x^2 + y^2 = 4$.

SOLUTION This is the circle of radius 2 centered at the origin. We cannot graph this directly since we must enter a function in the form $y = f(x)$.† Solving the equation of the circle for y, we have

$$y^2 = 4 - x^2$$

$$y = \pm\sqrt{4 - x^2} \qquad (1)$$

Equation (1) is not the equation of a function. In fact, it is the equation of *two* functions:

$$y_1 = \sqrt{4 - x^2} \qquad \text{The upper half of the circle}$$

$$y_2 = -\sqrt{4 - x^2} \qquad \text{The lower half of the circle}$$

---

† Some calculators can sketch graphs given in parametric form. On such a calculator, we could graph the circle in another way. We will not discuss this method here.

If we plot these two functions together, we obtain the entire circle. Using the range values $-2 \leq x \leq 2$ and $-2 \leq y \leq 2$ with $x_{\text{scl}} = y_{\text{scl}} = 1$, we obtain the graph in Figure 23.

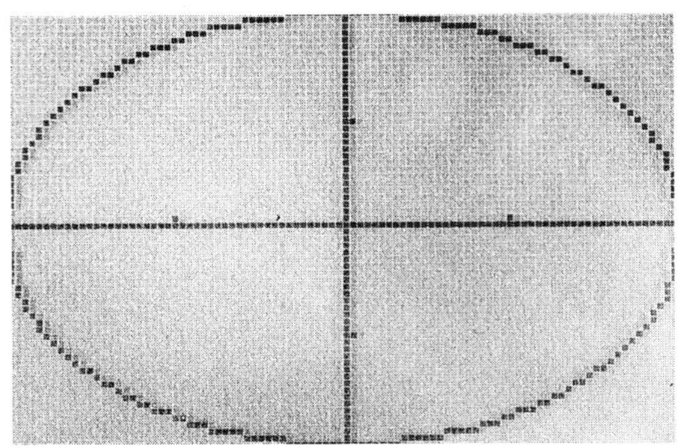

Figure 23 Graphs of the two functions $y_1 = \sqrt{4 - x^2}$ and $y_2 = -\sqrt{4 - x^2}$, $-2 \leq x \leq 2$, $-2 \leq y \leq 2$

This figure appears as an ellipse since the built-in scales on the x- and y-axes are not the same. (The rectangular screen is longer along the x-axis than along the y-axis.) There are two ways to fix this. The easier way can be used if your calculator has a function key labeled SQUARE. By using this function, the calculator will automatically rescale along the x-axis or the y-axis to make units along both axes have approximately the same length. (That is, a tick mark corresponding to a scale value of 1 will have approximately the same length on each axis.) When this is done, the graph appears as in Figure 24.

Alternatively, rescale the screen yourself. In this case, the circle in Figure 24 is obtained if the x-range is expanded to $-3 \leq x \leq 3$.

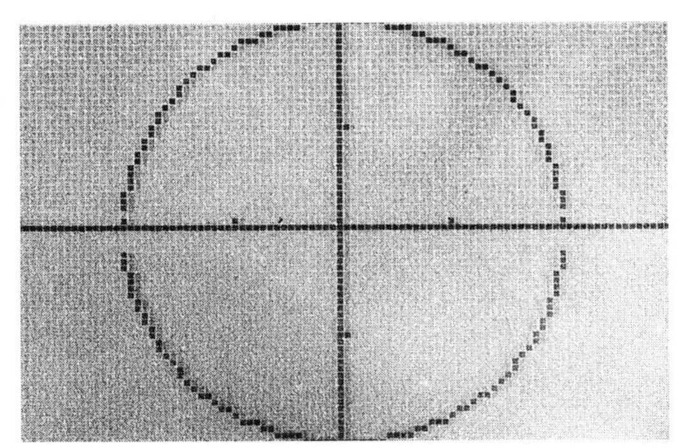

Figure 24 Graphs of $y_1 = \sqrt{4 - x^2}$ and $y_2 = -\sqrt{4 - x^2}$, rescaled ■

EXAMPLE 10 *Graphing a Hyperbola*

Graph the hyperbola

$$\frac{y^2}{9} - \frac{x^2}{16} = 1 \qquad (2)$$

SOLUTION We discuss hyperbolas in Section 5.5. To graph this hyperbola, we first solve equation (2) for y:

$$\frac{y^2}{9} = 1 + \frac{x^2}{16}$$

$$y^2 = 9 + \frac{9}{16}x^2$$

$$y = \pm\sqrt{9 + \frac{9}{16}x^2}$$

Since $x^2 \geq 0$, $\sqrt{9 + \frac{9}{16}x^2} \geq \sqrt{9} = 3$. Thus either $y \geq 3$ or $y \leq -3$. x can take on any real value. As in Example 9, we graph the two functions

$$y_1 = \sqrt{9 + \frac{9}{16}x^2} \quad \text{and} \quad y_2 = -\sqrt{9 + \frac{9}{16}x^2}$$

with range values, $-20 \leq x \leq 20$, $-6 \leq y \leq 6$ and $x_{\text{scl}} = y_{\text{scl}} = 1$. The graph is given in Figure 25.

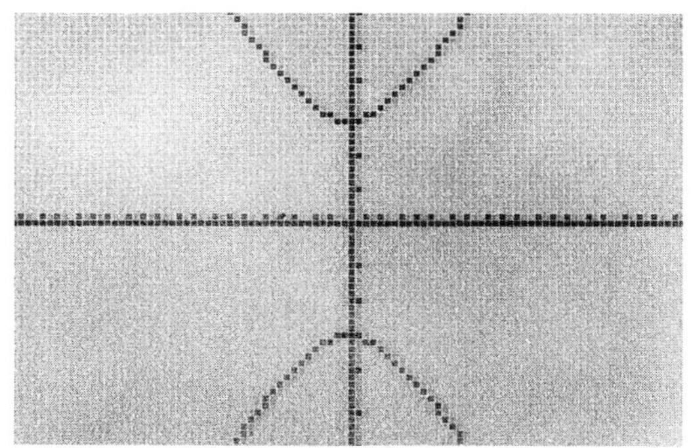

Figure 25 Graphs of $y_1 = \sqrt{9 + \frac{9}{16}x^2}$ and $y_2 = -\sqrt{9 + \frac{9}{16}x^2}$ for $-20 \leq x \leq 20$, $-6 \leq y \leq 6$.

If we "square" the graph as in Example 9, we obtain the more accurate sketch in Figure 26.

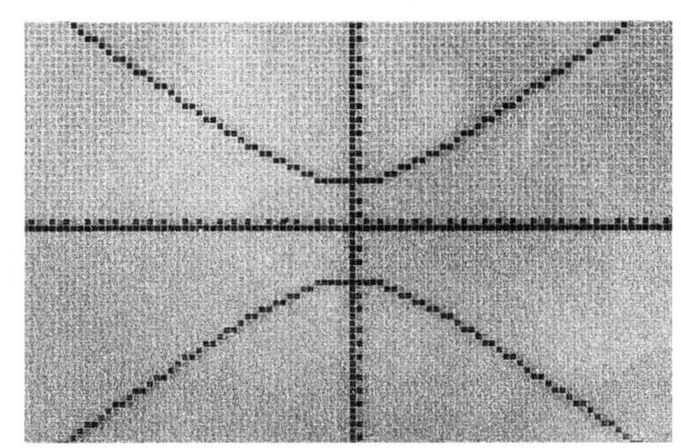

Figure 26 Graph of the hyperbola rescaled: $-20 \leq x \leq 20$, $-13.33 \leq y \leq 13.33$ ■

EXAMPLE 11 *Graphing an Ellipse*

Sketch the graph of the second-degree equation

$$9x^2 + 36x + 4y^2 - 8y + 4 = 0. \tag{3}$$

SOLUTION We graphed this curve in Example 6 in Section 5.3 (see p. 299). To write equation (3) in a form that can be entered on a calculator, we first complete the squares (see page 196).

$$\begin{aligned}
9x^2 + 36x + 4y^2 - 8y + 4 &= 0\\
9(x^2 + 4x) + 4(y^2 - 2y) &= -4\\
9(x^2 + 4x + 4) - 9 \cdot 4 + 4(y^2 - 2y + 1) - 4 \cdot 1 &= -4\\
9(x + 2)^2 - 36 + 4(y - 1)^2 - 4 &= -4\\
9(x + 2)^2 + 4(y - 1)^2 &= 36\\
4(y - 1)^2 &= 36 - 9(x + 2)^2\\
(y - 1)^2 &= 9 - \frac{9}{4}(x + 2)^2\\
y - 1 &= \pm\sqrt{9 - \frac{9}{4}(x + 2)^2}\\
y &= 1 \pm \sqrt{9 - \frac{9}{4}(x + 2)^2}
\end{aligned}$$

Before sketching these two curves, we observe that we must have

$$\begin{aligned}
9 - \frac{9}{4}(x + 2)^2 &\geq 0\\
\frac{9}{4}(x + 2)^2 &\leq 9\\
(x + 2)^2 &\leq 4\\
|x + 2| &\leq 2\\
-2 \leq x + 2 &\leq 2 \quad \text{See page 134}\\
-4 \leq x &\leq 0
\end{aligned}$$

This is our range of values for x.

Also,

$$0 \leq (x + 2)^2 \leq 4$$

If $(x + 2)^2 = 4$, then $\sqrt{9 - \frac{9}{4}(x + 2)^2} = 0$.

If $(x + 2)^2 = 0$, then $\sqrt{9 - \frac{9}{4}(x + 2)^2} = \sqrt{9} = 3$.

Thus

$$0 \leq \sqrt{9 + \frac{9}{4}(x + 2)^2} \leq 3$$

so

$$1 \le 1 + \sqrt{9 + \frac{9}{4}(x + 2)^2} \le 4$$

and

$$-2 \le 1 - \sqrt{9 + \frac{9}{4}(x + 2)^2} \le 1.$$

Thus a suitable range of y-values is $-2 \le y \le 4$. Using these values, we obtain the ellipse in Figure 27.

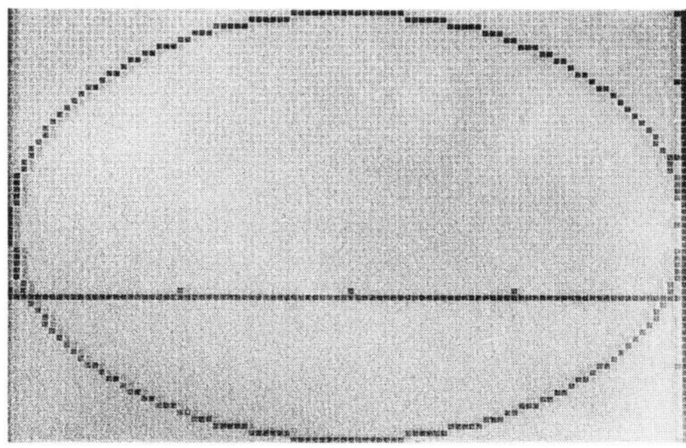

Figure 27 Graphs of the curves $y_1 = 1 + \sqrt{9 - \frac{9}{4}(x + 2)^2}$ and $y_2 = 1 - \sqrt{9 - \frac{9}{4}(x + 2)^2}$, $-4 \le x \le 0$, $-2 \le y \le 4$, $x_{\text{scl}} = y_{\text{scl}} = 1$

Finally, if we square to equalize the spacing along the axes, we obtain the curve in Figure 28.

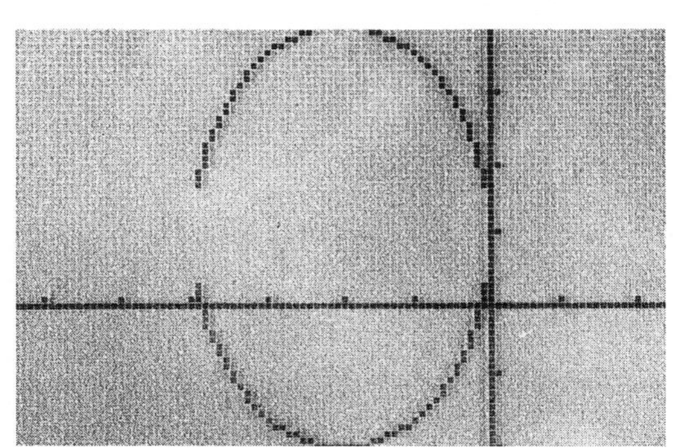

Figure 28 Graph of the ellipse rescaled: $-6.5 \le x \le 2.5$, $-2 \le y \le 4$

IV. Graphing Calculators Cannot Graph All Functions Accurately

We have seen that a remarkable amount of information can be obtained by graphing a function on a calculator. However, as our final example illustrates, graphing calculators do have their limitations.

EXAMPLE 12 ***A Polynomial Whose Graph Cannot Be Obtained Accurately on a Graphing Calculator***

Graph the polynomial $p(x) = -\dfrac{x^5}{40} + x^4 + 3x^3 - 5x^2 - 10x + 4$, and find each of its zeros to one decimal place accuracy.

SOLUTION We set our initial range and scale values as follows:

$$\begin{aligned} x_{\min} &= -5 & y_{\min} &= -15 \\ x_{\max} &= 5 & y_{\max} &= 15 \\ x_{\text{scl}} &= 1 & y_{\text{scl}} &= 5 \end{aligned}$$

The graph in Figure 29 is obtained.

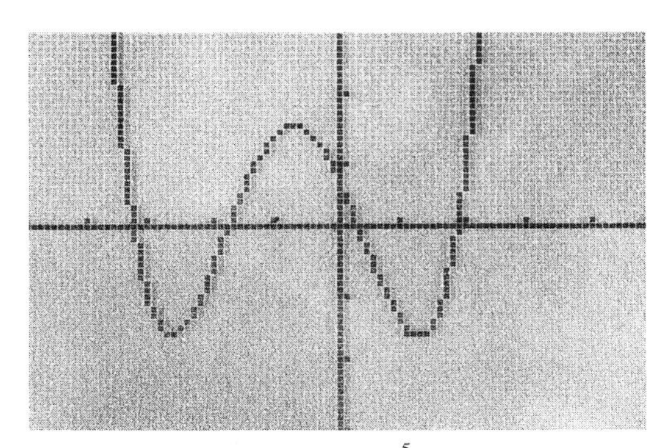

Figure 29 Graph of $p(x) = \dfrac{-x^5}{40} + x^4 + 3x^3 - 5x^2 - 10x + 4$ for $-5 \le x \le 5$, $-15 \le y \le 15$

We can see four zeros quite clearly. To one decimal place, they are -3.2, -1.7, 0.4, and 1.9.

But something is clearly wrong. If x is positive and large, then the term $\dfrac{-x^5}{40}$ will dominate the other terms. For example, if $x = 100$, then $\dfrac{-x^5}{40} = -250{,}000{,}000$ while $x^4 + 3x^3 - 5x^2 - 10x + 4 = 102{,}949{,}004$ so $p(100) = -147{,}050{,}996$. We see that the graph of $p(x)$ must eventually turn downward and become (and stay) negative as x increases. Thus $p(x)$ has a fifth zero. To find it, we change our range and scale values as follows (after some trial and error):

$$\begin{aligned} x_{\min} &= -5 & y_{\min} &= -100{,}000 \\ x_{\max} &= 45 & y_{\max} &= 350{,}000 \\ x_{\text{scl}} &= 5 & y_{\text{scl}} &= 100{,}000 \end{aligned}$$

We obtain the sketch in Figure 30.

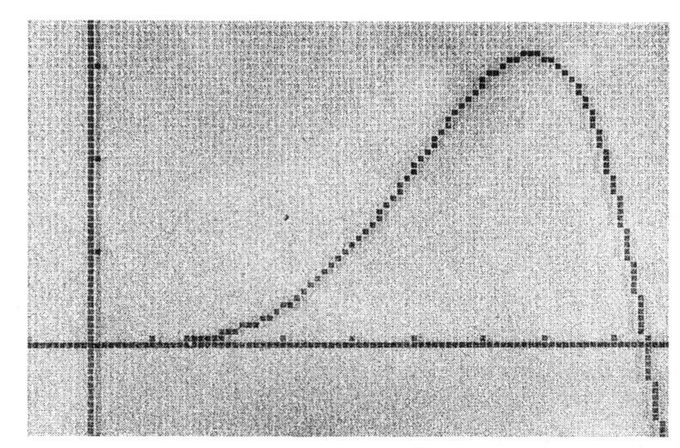

Figure 30 Graph of $p(x) = \dfrac{-x^5}{40} + x^4 + 3x^3 - 5x^2 - 10x + 4$ for $-5 \le x \le 45$, $-100{,}000 \le y \le 350{,}000$, $x_{\text{scl}} = 5$, $y_{\text{scl}} = 100{,}000$

From this figure, we can determine that $p(x)$ reaches a maximum height of approximately 312,000 when x is near 34. It then decreases rapidly and is zero at $x = 42.7$ (rounded to 1 decimal place).

However, in obtaining a graph that showed the behavior of $p(x)$ for x greater than 30, we completely lost the picture obtained in Figure 29. The reason for this should be clear. In the graph in Figure 29, $p(x)$ reaches an approximate minimum value of -8 at $x \approx -3$ and $x \approx 1.3$ and a maximum value of approximately 8 at $x \approx -0.7$. But the number 8 is negligible compared to the maximum height of 312,000 at $x \approx 34$. It is simply impossible to scale the x- and y-axes to show the behavior for $-4 \le x \le 4$ and $30 \le x \le 50$ simultaneously on your calculator screen. If we show one, we lose the other. In fact, an accurate sketch drawn *to scale* that portrayed the graph of $p(x)$ accurately would have to be drawn on a very large piece of paper indeed. For example, if we used a scale of 1 mm = 1 unit (this is very small—there are 25 mm in an inch), then we would need a positive y-axis that extended 312,000 mm = 312 m $\approx$ 1024 ft, which is almost a fifth of a mile! And if we use a scale much smaller than 1 mm per unit, then as in Figure 30, we simply would not see how $p(x)$ behaved for $-4 \le x \le 4$. The graph in this interval would appear as part of the x-axis.

This example illustrates the fact that a bit of thought is needed when using even the best available graphing device. You should, among other things, think about what happens to the graph of a function as $|x|$ gets large. It may be that no range setting will give you an accurate picture.

Appendix A Problems

In Problems 1–17 obtain sketches of each graph by first setting the appropriate calculator range and scale.

1. $y = x^2 + 2$
2. $y = -x^2 + 4x + 2$
3. $y = x^3 - x$
4. $y = 2x^3 - 3x^2 + 4x - 5$
5. $y = -x^4 + 2x^2 - x + 3$
6. $y = x^5 - x^4 - x^3 + x^2 + x - 2$
7. $y = \dfrac{x}{x+1}$
8. $y = \dfrac{x^2 - 3}{x}$

9. $y = \dfrac{x+2}{x-4}$

10. $y = \dfrac{x^3 - x^2 + 1}{x - 3}$

11. $y = \dfrac{x-2}{x^2-1}$

12. $y = e^{3x-2}$

13. $y = \ln(x^2 - 4)$

14. $3 - 2\ln(1 - x) + 4e^x$

15. $\dfrac{e^{-1/x}}{x+2}$

16. $\dfrac{1 - \ln x}{x^2 + 4}$

17. $e^{x - \ln x}$

In Problems 18–27 approximate, to two decimal place accuracy, all zeros of each polynomial by sketching appropriate graphs. [*Note:* To be sure that you have two decimal place accuracy, you should approximate each zero to three decimal places.]

18. $x^2 - 3x - 10$

19. $6x^2 + 5x - 56$

20. $x^3 + x^2 + 5$

21. $x^3 - x^2 - 1$

22. $3x^5 + 2x^4 + x^2 + 3$

23. $x^7 - x^2 - 12$

24. $6x^4 - 2x^3 + 3x^2 - 4x - 2$

25. $x^3 - 6x^2 - 15x + 4$

26. $x^3 + 14x^2 + 60x + 78$

27. $4x^4 - 4x^3 - 23x^2 + x + 10$

In Problems 28–31 find, by graphing, the number of solutions to each equation, and approximate each solution to two decimal place accuracy.

28. $\ln x = x - 4$

29. $e^{-x} = \ln x$

30. $\dfrac{x^2 - 4}{x^2 - 9} = \dfrac{x}{4}$

31. $x^2 - 3x + 5 = \ln(x^4 + x + 5)$

In Problems 32–40 find an approximate solution to each inequality.

32. $2x + 3 > 5$

33. $|3x - 5| < 4$ [*Hint:* Use the Abs key or its equivalent on your calculator.]

34. $\left|\dfrac{x}{2} - 5\right| < 8$

35. $x^2 - 2 > x + 5$

36. $\dfrac{1}{x+2} > \dfrac{x}{x+1}$

37. $\dfrac{x}{x^2 - 3} < \dfrac{1}{2 - x}$

38. $2x^2 + 3x - 5 < 3 - 7x$

39. $x^3 - 2x^2 + 7x - 5 > 2x^2 - x + 4$

40. $3x^3 + 7 < 1 - 2x + x^4$

In Problems 41–55 sketch the graph of each conic section.

41. $x^2 + y^2 = 25$

42. $(x + 2)^2 + (y - 3)^2 = 12$

43. $x^2 + 2x + y^2 - 4y = 20$

44. $2x^2 + 3y^2 = 12$

45. $7x^2 - 11y^2 = 47$

46. $19y^2 - 43x^2 = 6$

47. $\dfrac{x^2}{37} + \dfrac{y^2}{121} = 1$

48. $7x^2 - 3y = 4$

49. $3y^2 + 12x = 6$

50. $\dfrac{x}{2} - \dfrac{y^2}{9} = 3$

51. $2x^2 + 7x + 5y^2 - 2y = 8$

52. $3x^2 + 6x + 9y - 7 = 0$

53. $2y^2 - 8y - 12x^2 + 12x = 16$

54. $17x^2 - 32x - 43y - 11y^2 = 33$

55. $\dfrac{1}{2}x^2 + 4x + \dfrac{1}{3}y^2 - 3y = 5$

Table 1

Exponential Functions

x	e^x	e^{-x}	x	e^x	e^{-x}
0.00	1.0000	1.0000	3.0	20.086	0.0498
0.05	1.0513	0.9512	3.1	22.198	0.0450
0.10	1.1052	0.9048	3.2	24.533	0.0408
0.15	1.1618	0.8607	3.3	27.113	0.0369
0.20	1.2214	0.8187	3.4	29.964	0.0334
0.25	1.2840	0.7788	3.5	33.115	0.0302
0.30	1.3499	0.7408	3.6	36.598	0.0273
0.35	1.4191	0.7047	3.7	40.447	0.0247
0.40	1.4918	0.6703	3.8	44.701	0.0224
0.45	1.5683	0.6376	3.9	49.402	0.0202
0.50	1.6487	0.6065	4.0	54.598	0.0183
0.55	1.7333	0.5769	4.1	60.340	0.0166
0.60	1.8221	0.5488	4.2	66.686	0.0150
0.65	1.9155	0.5220	4.3	73.700	0.0136
0.70	2.0138	0.4966	4.4	81.451	0.0123
0.75	2.1170	0.4724	4.5	90.017	0.0111
0.80	2.2255	0.4493	4.6	99.484	0.0101
0.85	2.3396	0.4274	4.7	109.95	0.0091
0.90	2.4596	0.4066	4.8	121.51	0.0082
0.95	2.5857	0.3867	4.9	134.29	0.0074
1.0	2.7183	0.3679	5.0	148.41	0.0067
1.1	3.0042	0.3329	5.1	164.02	0.0061
1.2	3.3201	0.3012	5.2	181.27	0.0055
1.3	3.6693	0.2725	5.3	200.34	0.0050
1.4	4.0552	0.2466	5.4	221.41	0.0045
1.5	4.4817	0.2231	5.5	244.69	0.0041
1.6	4.9530	0.2019	5.6	270.43	0.0037
1.7	5.4739	0.1827	5.7	298.87	0.0033
1.8	6.0496	0.1653	5.8	330.30	0.0030
1.9	6.6859	0.1496	5.9	365.04	0.0027
2.0	7.3891	0.1353	6.0	403.43	0.0025
2.1	8.1662	0.1225	6.5	665.14	0.0015
2.2	9.0250	0.1108	7.0	1096.6	0.0009
2.3	9.9742	0.1003	7.5	1808.0	0.0006
2.4	11.023	0.0907	8.0	2981.0	0.0003
2.5	12.182	0.0821	8.5	4914.8	0.0002
2.6	13.464	0.0743	9.0	8103.1	0.0001
2.7	14.880	0.0672	9.5	13,360	0.00007
2.8	16.445	0.0608	10.0	22,026	0.00005
2.9	18.174	0.0550			

Table 2

Natural Logarithms

n	$\log_e n$	n	$\log_e n$	n	$\log_e n$
0.0	—	4.5	1.5041	9.0	2.1972
0.1	−2.3026	4.6	1.5261	9.1	2.2083
0.2	−1.6094	4.7	1.5476	9.2	2.2192
0.3	−1.2040	4.8	1.5686	9.3	2.2300
0.4	−0.9163	4.9	1.5892	9.4	2.2407
0.5	−0.6931	5.0	1.6094	9.5	2.2513
0.6	−0.5108	5.1	1.6292	9.6	2.2618
0.7	−0.3567	5.2	1.6487	9.7	2.2721
0.8	−0.2231	5.3	1.6677	9.8	2.2824
0.9	−0.1054	5.4	1.6864	9.9	2.2925
1.0	0.0000	5.5	1.7047	10	2.3026
1.1	0.0953	5.6	1.7228	11	2.3979
1.2	0.1823	5.7	1.7405	12	2.4849
1.3	0.2624	5.8	1.7579	13	2.5649
1.4	0.3365	5.9	1.7750	14	2.6391
1.5	0.4055	6.0	1.7918	15	2.7081
1.6	0.4700	6.1	1.8083	16	2.7726
1.7	0.5306	6.2	1.8245	17	2.8332
1.8	0.5878	6.3	1.8405	18	2.8904
1.9	0.6419	6.4	1.8563	19	2.9444
2.0	0.6931	6.5	1.8718	20	2.9957
2.1	0.7419	6.6	1.8871	25	3.2189
2.2	0.7885	6.7	1.9021	30	3.4012
2.3	0.8329	6.8	1.9169	35	3.5553
2.4	0.8755	6.9	1.9315	40	3.6889
2.5	0.9163	7.0	1.9459	45	3.8067
2.6	0.9555	7.1	1.9601	50	3.9120
2.7	0.9933	7.2	1.9741	55	4.0073
2.8	1.0296	7.3	1.9879	60	4.0943
2.9	1.0647	7.4	2.0015	65	4.1744
3.0	1.0986	7.5	2.0149	70	4.2485
3.1	1.1314	7.6	2.0281	75	4.3175
3.2	1.1632	7.7	2.0412	80	4.3820
3.3	1.1939	7.8	2.0541	85	4.4427
3.4	1.2238	7.9	2.0669	90	4.4998
3.5	1.2528	8.0	2.0794	95	4.5539
3.6	1.2809	8.1	2.0919	100	4.6052
3.7	1.3083	8.2	2.1041	200	5.2983
3.8	1.3350	8.3	2.1163	300	5.7038
3.9	1.3610	8.4	2.1282	400	5.9915
4.0	1.3863	8.5	2.1401	500	6.2146
4.1	1.4110	8.6	2.1518	600	6.3969
4.2	1.4351	8.7	2.1633	700	6.5511
4.3	1.4586	8.8	2.1748	800	6.6846
4.4	1.4816	8.9	2.1861	900	6.8024

Using the Common Logarithm Tables (Table 3)

Table 3 contains four-decimal-place approximations for common logarithms of numbers between 1.00 and 9.99 in intervals of 0.01. The use of these tables is illustrated in the following example.

Example

Approximate each of the following:

(a) log 33.4 (b) log 33,400 (c) log 0.0334

Solution

(a) Express the number in scientific notation

$$33.4 = 3.34 \times 10^1$$

Therefore, $\log 33.4 = \log (3.34 \times 10^1)$
$= \log 3.34 + \log 10^1$

To find the log of 3.34 in Table 3, look down the left-hand column to $n = 3.3$ and over to the right to $n = 4$. The log of 3.34 is shown as .5237. The log of 10^1 is, of course, 1.

Therefore, $\log 33.4 = \log 3.34 + \log 10^1$
$= .5237 + 1$
$= 1.5237$

(b) $\log 33{,}400 = \log (3.34 \times 10^4)$
$= \log 3.34 + \log 10^4$
$= .5237 + 4$
$= 4.5237$

(c) $\log 0.0334 = \log 3.334 + \log 10^{-2}$
$= 0.5237 + (-2) = -1.4763$

Table 3

Common Logarithms

n	0	1	2	3	4	5	6	7	8	9
1.0	.0000	.0043	.0086	.0128	.0170	.0212	.0253	.0294	.0334	.0374
1.1	.0414	.0453	.0492	.0531	.0569	.0607	.0645	.0682	.0719	.0755
1.2	.0792	.0828	.0864	.0899	.0934	.0969	.1004	.1038	.1072	.1106
1.3	.1139	.1173	.1206	.1239	.1271	.1303	.1335	.1367	.1399	.1430
1.4	.1461	.1492	.1523	.1553	.1584	.1614	.1644	.1673	.1703	.1732
1.5	.1761	.1790	.1818	.1847	.1875	.1903	.1931	.1959	.1987	.2014
1.6	.2041	.2068	.2095	.2122	.2148	.2175	.2201	.2227	.2253	.2279
1.7	.2304	.2330	.2355	.2380	.2405	.2430	.2455	.2480	.2504	.2529
1.8	.2553	.2577	.2601	.2625	.2648	.2672	.2695	.2718	.2742	.2765
1.9	.2788	.2810	.2833	.2856	.2878	.2900	.2923	.2945	.2967	.2989
2.0	.3010	.3032	.3054	.3075	.3096	.3118	.3139	.3160	.3181	.3201
2.1	.3222	.3243	.3263	.3284	.3304	.3324	.3345	.3365	.3385	.3404
2.2	.3424	.3444	.3464	.3483	.3502	.3522	.3541	.3560	.3579	.3598
2.3	.3617	.3636	.3655	.3674	.3692	.3711	.3729	.3747	.3766	.3784
2.4	.3802	.3820	.3838	.3856	.3874	.3892	.3909	.3927	.3945	.3962
2.5	.3979	.3997	.4014	.4031	.4048	.4065	.4082	.4099	.4116	.4133
2.6	.4150	.4166	.4183	.4200	.4216	.4232	.4249	.4265	.4281	.4298
2.7	.4314	.4330	.4346	.4362	.4378	.4393	.4409	.4425	.4440	.4456
2.8	.4472	.4487	.4502	.4518	.4533	.4548	.4564	.4579	.4594	.4609
2.9	.4624	.4639	.4654	.4669	.4683	.4698	.4713	.4728	.4742	.4757
3.0	.4771	.4786	.4800	.4814	.4829	.4843	.4857	.4871	.4886	.4900
3.1	.4914	.4928	.4942	.4955	.4969	.4983	.4997	.5011	.5024	.5038
3.2	.5051	.5065	.5079	.5092	.5105	.5119	.5132	.5145	.5159	.5172
3.3	.5185	.5198	.5211	.5224	.5237	.5250	.5263	.5276	.5289	.5302
3.4	.5315	.5328	.5340	.5353	.5366	.5378	.5391	.5403	.5416	.5428
3.5	.5441	.5453	.5465	.5478	.5490	.5502	.5514	.5527	.5539	.5551
3.6	.5563	.5575	.5587	.5599	.5611	.5623	.5635	.5647	.5658	.5670
3.7	.5682	.5694	.5705	.5717	.5729	.5740	.5752	.5763	.5775	.5786
3.8	.5798	.5809	.5821	.5832	.5843	.5855	.5866	.5877	.5888	.5899
3.9	.5911	.5922	.5933	.5944	.5955	.5966	.5977	.5988	.5999	.6010
4.0	.6021	.6031	.6042	.6053	.6064	.6075	.6085	.6096	.6107	.6117
4.1	.6128	.6138	.6149	.6160	.6170	.6180	.6191	.6201	.6212	.6222
4.2	.6232	.6243	.6253	.6263	.6274	.6284	.6294	.6304	.6314	.6325
4.3	.6335	.6345	.6355	.6365	.6375	.6385	.6395	.6405	.6415	.6425
4.4	.6435	.6444	.6454	.6464	.6474	.6484	.6493	.6503	.6513	.6522
4.5	.6532	.6542	.6551	.6561	.6571	.6580	.6590	.6599	.6609	.6618
4.6	.6628	.6637	.6646	.6656	.6665	.6675	.6684	.6693	.6702	.6712
4.7	.6721	.6730	.6739	.6749	.6758	.6767	.6776	.6785	.6794	.6803
4.8	.6812	.6821	.6830	.6839	.6848	.6857	.6866	.6875	.6884	.6893
4.9	.6902	.6911	.6920	.6928	.6937	.6946	.6955	.6964	.6972	.6981
5.0	.6990	.6998	.7007	.7016	.7024	.7033	.7042	.7050	.7059	.7067
5.1	.7076	.7084	.7093	.7101	.7110	.7118	.7126	.7135	.7143	.7152
5.2	.7160	.7168	.7177	.7185	.7193	.7202	.7210	.7218	.7226	.7235
5.3	.7243	.7251	.7259	.7267	.7275	.7284	.7292	.7300	.7308	.7316
5.4	.7324	.7332	.7340	.7348	.7356	.7364	.7372	.7380	.7388	.7396

Common Logarithms

n	0	1	2	3	4	5	6	7	8	9
5.5	.7404	.7412	.7419	.7427	.7435	.7443	.7451	.7459	.7466	.7474
5.6	.7482	.7490	.7497	.7505	.7513	.7520	.7528	.7536	.7543	.7551
5.7	.7559	.7566	.7574	.7582	.7589	.7597	.7604	.7612	.7619	.7627
5.8	.7634	.7642	.7649	.7657	.7664	.7672	.7679	.7686	.7694	.7701
5.9	.7709	.7716	.7723	.7731	.7738	.7745	.7752	.7760	.7767	.7774
6.0	.7782	.7789	.7796	.7803	.7810	.7818	.7825	.7832	.7839	.7846
6.1	.7853	.7860	.7868	.7875	.7882	.7889	.7896	.7903	.7910	.7917
6.2	.7924	.7931	.7938	.7945	.7952	.7959	.7966	.7973	.7980	.7987
6.3	.7993	.8000	.8007	.8014	.8021	.8028	.8035	.8041	.8048	.8055
6.4	.8062	.8069	.8075	.8082	.8089	.8096	.8102	.8109	.8116	.8122
6.5	.8129	.8136	.8142	.8149	.8156	.8162	.8169	.8176	.8182	.8189
6.6	.8195	.8202	.8209	.8215	.8222	.8228	.8235	.8241	.8248	.8254
6.7	.8261	.8267	.8274	.8280	.8287	.8293	.8299	.8306	.8312	.8319
6.8	.8325	.8331	.8338	.8344	.8351	.8357	.8363	.8370	.8376	.8382
6.9	.8388	.8395	.8401	.8407	.8414	.8420	.8426	.8432	.8439	.8445
7.0	.8451	.8457	.8463	.8470	.8476	.8482	.8488	.8494	.8500	.8506
7.1	.8513	.8519	.8525	.8531	.8537	.8543	.8549	.8555	.8561	.8567
7.2	.8573	.8579	.8585	.8591	.8597	.8603	.8609	.8615	.8621	.8627
7.3	.8633	.8639	.8645	.8651	.8657	.8663	.8669	.8675	.8681	.8686
7.4	.8692	.8698	.8704	.8710	.8716	.8722	.8727	.8733	.8739	.8745
7.5	.8751	.8756	.8762	.8768	.8774	.8779	.8785	.8791	.8797	.8802
7.6	.8808	.8814	.8820	.8825	.8831	.8837	.8842	.8848	.8854	.8859
7.7	.8865	.8871	.8876	.8882	.8887	.8893	.8899	.8904	.8910	.8915
7.8	.8921	.8927	.8932	.8938	.8943	.8949	.8954	.8960	.8965	.8971
7.9	.8976	.8982	.8987	.8993	.8998	.9004	.9009	.9015	.9020	.9025
8.0	.9031	.9036	.9042	.9047	.9053	.9058	.9063	.9069	.9074	.9079
8.1	.9085	.9090	.9096	.9101	.9106	.9112	.9117	.9122	.9128	.9133
8.2	.9138	.9143	.9149	.9154	.9159	.9165	.9170	.9175	.9180	.9186
8.3	.9191	.9196	.9201	.9206	.9212	.9217	.9222	.9227	.9232	.9238
8.4	.9243	.9248	.9253	.9258	.9263	.9269	.9274	.9279	.9284	.9289
8.5	.9294	.9299	.9304	.9309	.9315	.9320	.9325	.9330	.9335	.9340
8.6	.9345	.9350	.9355	.9360	.9365	.9370	.9375	.9380	.9385	.9390
8.7	.9395	.9400	.9405	.9410	.9415	.9420	.9425	.9430	.9435	.9440
8.8	.9445	.9450	.9455	.9460	.9465	.9469	.9474	.9479	.9484	.9489
8.9	.9494	.9499	.9504	.9509	.9513	.9518	.9523	.9528	.9533	.9538
9.0	.9542	.9547	.9552	.9557	.9562	.9566	.9571	.9576	.9581	.9586
9.1	.9590	.9595	.9600	.9605	.9609	.9614	.9619	.9624	.9628	.9633
9.2	.9638	.9643	.9647	.9652	.9657	.9661	.9666	.9671	.9675	.9680
9.3	.9685	.9689	.9694	.9699	.9703	.9708	.9713	.9717	.9722	.9727
9.4	.9731	.9736	.9741	.9745	.9750	.9754	.9759	.9763	.9768	.9773
9.5	.9777	.9782	.9786	.9791	.9795	.9800	.9805	.9809	.9814	.9818
9.6	.9823	.9827	.9832	.9836	.9841	.9845	.9850	.9854	.9859	.9863
9.7	.9868	.9872	.9877	.9881	.9886	.9890	.9894	.9899	.9903	.9908
9.8	.9912	.9917	.9921	.9926	.9930	.9934	.9939	.9943	.9948	.9952
9.9	.9956	.9961	.9965	.9969	.9974	.9978	.9983	.9987	.9991	.9996

Answers to Selected Odd-Numbered Problems

Chapter 1

Problems 1.2, page 6

1. $\frac{2}{9}$ **3.** $\frac{147}{999} = \frac{49}{333}$ **5.** $\frac{2121}{999} = \frac{707}{333}$
7. $0.83333\ldots = 0.8\overline{3}$ **9.** $0.41666\ldots = 0.41\overline{6}$
11. $0.12871287\ldots = 0.\overline{1287}$ **13.** 1.78125
15. One simple answer is $\sqrt{5}$ and $-\sqrt{5}$.

Problems 1.3, page 13

1. Commutative Law for Addition
3. Associative Law for Multiplication
5. Multiplicative Identity
7. Associative Law for Addition
9. Multiplicative Inverse
11. Left Distributive Law
13. Additive Substitution Law
15. Multiplicative Substitution Law
17. Additive Reduction Law
19. Additive Substitution Law
21. $\frac{7}{3}$ **23.** $\frac{7}{12}$ **25.** $\frac{2}{7}$ **27.** $\frac{1}{20}$ **29.** $\frac{21}{4}$
31. 14 **33.** $\frac{3}{5}$ **35.** 2 **37.** $\frac{15}{28}$ **39.** $\frac{3}{10}$
41. $\frac{21}{20}$ **43.** $\frac{15}{8}$ **45.** $\frac{13}{12}$ **47.** $\frac{19}{60}$ **49.** $\frac{55}{26}$
63. Any choice except where $a = b$
65. $\frac{1+2}{3+4} = \frac{3}{7}$ but $\frac{1}{3} + \frac{2}{4} = \frac{1}{3} + \frac{1}{2} = \frac{5}{6}$
67. Example: $-4 - (-9) = 5$ not closed

Problems 1.4, page 19

1. $<$ **3.** $>$ **5.** $>$ **7.** $>$ **9.** $<$
11. 4 **13.** -5 **15.** $4 - \sqrt{2} \approx 2.586$ **17.** e
19. c, f **21.** a, d, e **23.** Inequality property (a)
25. Inequality property (c)
27. Inequality property (d)
29. Inequality property (d)
31. Inequality property (e) **33.** 5 **35.** 9
37. 5 **39.** 2.2 **41.** 5.4

Problems 1.5, page 27

1. 9 **3.** $\frac{1}{64}$ **5.** $\frac{1}{9}$ **7.** 16 **9.** 2
11. -1 **13.** 128 **15.** $\frac{4}{9}$ **17.** $\frac{1}{64}$ **19.** $\frac{1}{5}$
21. -1 **23.** -1 **25.** -10 **27.** -1000
29. $\frac{1049}{257}$ **31.** $\frac{1}{x^2}$ **33.** y^2 **35.** $\frac{4}{x^2}$
37. $\frac{3}{2b}$ **39.** $\frac{25e^2}{49d^2}$ **41.** $\frac{8y^3}{125x^3}$ **43.** $\frac{a}{2b}$
45. $\frac{u}{v}$ **47.** $\frac{x^4w^2}{z^2}$ **49.** $\frac{1}{a^4b^3}$ **51.** $a^2b^2c^2$
53. xy^3 **55.** w^2z **57.** ab **59.** $\frac{y^4}{8x^3}$
61. $\frac{25xy^4}{24}$ **63.** $\frac{729}{10a^3x}$ **65.** 1 **67.** $\frac{y^2}{x}$
69. $\frac{w}{v}$ **71.** $x^4y^2z^{12}$ **73.** 3.65×10^2
75. 5.21236×10^2 **77.** 1.0×10^{-7}
79. 2.9028×10^4 ft
81. $5.87849983012 \times 10^{12}$ miles
83. 9.11×10^{-31} kg **85.** 9.29×10^7 miles
87. 5.55×10^{-7} m **89.** -1.6387064×10^4
91. 3.278902863×10^2 **93.** $6.586374179 \times 10^{31}$
95. -3.38346718×10^0 **97.** $-6.775052713 \times 10^{-8}$
99. $-7.97305921 \times 10^{-1}$ **101.** 4.495092941×10^5
103. 2.704813829×10^0
105. 10 $[(\frac{1}{2})^9 \approx 0.001953$ and $(\frac{1}{2})^{10} \approx 0.000977]$
107. 38 $[(1.2)^{37} \approx 850.6$ and $(1.2)^{38} \approx 1020.7]$
109. $4^5 = 1024$ $(5^4 = 625)$ **111.** 1000^{1001}

Problems 1.6, page 35

1. 32 **3.** 512 **5.** $\frac{1}{2}$ **7.** 4 **9.** $\frac{1}{4}$
11. 16 **13.** 3 **15.** $\frac{1}{9}$ **17.** 10
19. not a real number **21.** 243 **23.** 100,000
25. not a real number **27.** -10 **29.** -10
31. 2 **33.** 32 **35.** 64 **37.** $\frac{1}{2}$ **39.** 16
41. 100 **43.** 121 **45.** $\frac{1}{27}$ **47.** $\frac{1}{8}$
49. $\frac{1}{256}$ **51.** $6\sqrt{3}$ **53.** $-10\sqrt[3]{5}$
55. $100\sqrt{3}/3$ **57.** $\frac{1}{2}$ **59.** $\frac{5 \cdot 3^{2/3}}{3} = \frac{5}{3}\sqrt[3]{9}$
61. 1 **63.** 1 **65.** xy^2 **67.** $\frac{1}{4}$ **69.** $\sqrt{b/a}$
71. $z^{3/4}/y^{2/3}$ **73.** $1/y^{1.9} = y^{-1.9}$ **75.** $\sqrt{xy}/y$
77. $x^{2/3}y^2$ **79.** $x^{3/2}y^{3/2}$ **81.** y^4 **83.** $w^{7/12}$
85. $(x-1)^{5/2}y$ **87.** $\frac{x^6\sqrt{y}}{y^5} = x^6y^{-9/2}$ **89.** $5\sqrt{2}$
91. x^{3n} **93.** $y^{3k/2}$ **95.** -2 **97.** $\frac{1}{4}$
99. $-\frac{7}{3}$ **101.** $-\frac{8}{3}$ **103.** $\frac{2}{3}$ **105.** 2.15443469
107. -3.981071706 **109.** 0.152893846
111. 3.899419867 **113.** 0.120714399
115. (a) $1338.23 (b) $1548.08 (c) $1983.10
117. $16,501.93 **119.** 76.67

Problems 1.7, page 45

1. 4 **3.** 0 **5.** 6 **7.** $4x^2 - 6x + 8$
9. $3x^3 + x^2 + 2x + 1$ **11.** $3x^3 - 3x^2 + 8x - 7$

13. $9x^3 - 11x^2 + 27x - 25$
15. $6x^5 - 11x^4 + 25x^3 - 25x^2 + 29x - 12$
17. $3x^7 - 6x + 9$ **19.** $x^7 - x^4 - 2x + 6$ **21.** 11
23. $x^2 + 6x + 8$; 2 **25.** $x^2 - 25$; 2
27. $-15x^2 + 31x - 10$; 2 **29.** $12x^2 - 5x - 2$; 2
31. $49x^2 + 21x + 2$; 2 **33.** $24x^2 + 35x + 4$; 2
35. $2x^2 - 15x + 27$; 2 **37.** $2x^2 - 9x + 10$; 2
39. $x^8 - 16$; 8
41. $-18x^4 - 15x^3 + 18x^2 - 14x + 4$; 4
43. $x^5 + x^3 - x^2 - 1$; 5
45. $4x^5 - 12x^4 + 9x^3 + 8x^2 - 12x$; 5
47. $acx^6 + (ad + bc)x^4 + aex^3 + bdx^2 + bex$; 6
49. $x^{30} - 2x^{10}$; 30 **51.** $x^8 - 4x^4 + 12x^2 - 9$; 8
53. $x^4 - 16$; 4 **55.** $9x^2 - 12x + 4$; 2
57. $x^3 - 3x^2 + 3x - 1$; 3 **59.** $x^3 + 2x^2 - 5x - 6$; 3
61. $x^4 + 12x^3 + 54x^2 + 108x + 81$; 4
63. (a) 7 (b) 6 (c) 12 **65.** (a) 9 (b) 1 (c) 0
67. (a) 5 (b) 4 (c) -1125
69. (a) 8 (b) 16 (c) 1296
71. (a) $n + m$ (b) 1 (c) 1 **73.** 8 **75.** 4
77. 9 **79.** $4a^2 - 4ab + b^2$
81. $27w^3 - 108w^2z + 144wz^2 - 64z^3$
83. $\dfrac{1}{x^2} - \dfrac{2}{xy} + \dfrac{1}{y^2}$
85. $x^2 + 2xy + y^2 + z^2 - 2yz - 2xz$
87. $x^2 + y^2 + z^2 - 2xy - 2yz + 2xz$
89. $x^2 + 2xy + y^2 - z^2$
91. $x^4 - 2x^2y^2 + 2x^2z^2 + y^4 - 2y^2z^2 + z^4$
93. $x^2 + 2 + 2xy + \dfrac{1}{x^2} + \dfrac{2y}{x} + y^2$
95. $x^2 + y^2 + z^2 + w^2 + 2xy + 2xz + 2xw + 2yz + 2yw + 2zw$
97. $x^2 + 2xy + y^2 - z^2 - 2zw - w^2$
99. $x - 2\sqrt{xy} + y$ **101.** $w^6 - 3w^2 + \dfrac{3}{w^2} - \dfrac{1}{w^6}$
103. $30x^5 - 120x^3y^2 + 12x^2y - 48y^3$
105. $ac + ad + bc + bd$ **107.** $z^2 - \dfrac{1}{z^2}$
109. $4x^4 - x^3y^3 + 5x^3z^2 + 8xy - 2y^4 + 10yz^2 + 12xz^4 - 3y^3z^4 + 15z^6$
111. $\dfrac{8}{x^3} - \dfrac{36y}{x^2} + \dfrac{54y^2}{x} - 27y^3$
113. (a) 8184 (b) 8160

Problems 1.8, page 54

1. $2(x + 2)$ **3.** $5(2x + 5)$ **5.** $4(2x + 3y)$
7. $a(b + 2c)$ **9.** $2x^2(4x + 7)$
11. $(x + 2)[(x - 3) + (x^2 + 1)] = (x + 2)^2(x - 1)$
13. $(x - 1)(x + 1)$ **15.** $(z - w)(z + w)$
17. $[(x - 2) + 3][(x - 2) - 3] = (x - 5)(x + 1)$
19. $[(y + 4) + (z - 2)][(y + 4) - (z - 2)] = (y + z + 2)(y - z + 6)$
21. $4(4 + 3b^2)(4 - 3b^2)$ **23.** $(2z + 5)(2z - 5)$
25. $(w^3 - 3u^3)(w^3 + 3u^3)$ **27.** $(w - 1)^2$
29. $(q - 3)^2$ **31.** $(4w^2 - 3z)^2$ **33.** $-(x - 2)^2$
35. $(x^2 - 3)^2$ **37.** $(4w^2 - 1)^2$ **39.** $3(r - 5)^2$
41. $(a^3 + b^3 - 3)(a^3 + b^3 + 3)$ **43.** $(x + 2)(x + 5)$
45. not possible **47.** $(x - 2)(x + 3)$
49. $(x - 14)(x + 2)$ **51.** $(x - 7)(x - 6)$
53. not possible **55.** $-2(x - 1)(x - 2)$
57. $(5x + 2)(x - 1)$ **59.** $(3x - 2)(4x + 7)$
61. $(x - 2y)(x - y)$ **63.** $(3x + y)(2x - 5y)$
65. $(x + m)(x - n)$ **67.** $(x - 3)(x^2 + 3x + 9)$
69. $(x + 4)(x^2 - 4x + 16)$
71. $(2x + 1)(4x^2 - 2x + 1)$
73. $(x + 3y)(x^2 - 3xy + 9y^2)$
75. $2y^3(x + z)(x^2 - xz + z^2)$
77. $y(x + y) - 1(x + y) = (x + y)(y - 1)$
79. $(2x + 9y^2)(xy + 2)$
81. $(x - 2y - 2)(x - 2y + 2)$ **83.** $2(x - 1)(x + 1)$
85. $-6(x + 1)^2$ **87.** $-3(x - 2)(x - 5)$
89. $x^2(x - 2)(x^2 + 2x + 4)$ **91.** $xy(x - 3y)(x + 2y)$
93. $(x - y)^2$ **95.** $(x^2 + y^2)^2$
97. $(x + 1)^2(2x + 1)$
99. $x^{-5}(x^3 - 1) = x^{-5}(x - 1)(x^2 + x + 1)$
101. $3x^{-5}(2 + 5x^2)$ **103.** $3x^{-2}(x^4 - 2 + 4x)$
105. $(x - x^{-1})(x + x^{-1})$ **107.** $(x^{-2} + x^3)^2$
109. $(5x - 2x^{-3})^2$ **111.** $(y^{-1} - 1)(y^{-1} - 2)$
113. $(u^{-2} - 5)(u^{-2} + 1)$ **115.** $(2z^{-1} + 1)(z^{-1} - 4)$

Problems 1.9, page 64

1. $2x$ **3.** $\dfrac{x + 2}{3x^2 + 4}$ **5.** $1/2y^3$ **7.** $1/xy$
9. $\frac{1}{2}$ **11.** $\dfrac{16x^2}{3}$ **13.** $(x + 1)/x^3$
15. $(x + 2)/(x + 1)$ **17.** $z^2/(z + 1)^2 = (z/(z + 1))^2$
19. $(w^2 + 1)/(w + 1)$ **21.** $(x^2 + 1)/(x + 1)^2$
23. $\dfrac{x + 1}{x + 4}$ **25.** $(z - 2)/(z - 9)$
27. $(3x - 1)/(2x - 1)$ **29.** $(6z + 1)/(6z - 7)$
31. $(x^2 + xy + y^2)/(x - y)^2$
33. $\dfrac{(x + 3)(x - 3)}{(x - 4)(x - 1)} = \dfrac{x^2 - 9}{x^2 - 5x + 4}$
35. $4(z + 3)/5(z - 1)$ **37.** $(4 + y)/2$
39. $(1 + 4x)/2$ **41.** $(3x - 8)/x^2$
43. $(x - 2)/(2x - 2) = (x - 2)/2(x - 1)$
45. $\dfrac{20 - z^2}{4z}$ **47.** $-2/15s$ **49.** $37/6y^2$
51. $\dfrac{15x - 12a - 3b}{(x - a)(x - b)}$ **53.** $\dfrac{-2(x - 11)}{(x - 3)(x + 5)}$
55. $\dfrac{4(x + 4)}{(x + 2)(x - 2)} = \dfrac{4(x + 4)}{x^2 - 4}$
57. $\dfrac{7x - y + 2}{(x + y)(x - y)} = \dfrac{7x - y + 2}{x^2 - y^2}$ **59.** $\dfrac{-x + 8}{(x - 2)^2}$
61. $\dfrac{-2(y - 18)}{(y - 3)(y + 6)} = \dfrac{-2(y - 18)}{y^2 + 3y - 18}$ **63.** $\dfrac{2x + 1}{3x - 1}$

65. $\dfrac{3y-5x}{2(y+3x)}$ **67.** $\dfrac{x^2+4x}{2x^2-7x+2}$

69. $\dfrac{-2x-h}{x^2(x+h)^2}$ **71.** $\dfrac{x^2+1}{x^3}$ **73.** $\dfrac{4x^2+25x+8}{(x+5)^{4/3}}$

75. $\dfrac{x^3+x^2-x-1}{x^6}$ **77.** $\dfrac{3\sqrt{x}}{x}$

79. $\dfrac{\sqrt{5}(x+1)^{3/4}}{x+1}$ **81.** $\dfrac{\sqrt{5}-\sqrt{3}}{2}$

83. $\dfrac{-(2+\sqrt{2x})}{4-2x}=\dfrac{2+\sqrt{2x}}{2(x-2)}$ **85.** $\dfrac{x-2\sqrt{xy}+y}{x-y}$

87. $\dfrac{2(1+x^{1/3}+x^{2/3})}{1-x}$ **89.** $\dfrac{1}{\sqrt{x+h+4}+\sqrt{x+4}}$

91. $\dfrac{-1}{\sqrt{x}\sqrt{x+2}(\sqrt{x}+\sqrt{x+2})}$

93. $\dfrac{-2x-h}{x\sqrt{(x+h)^2}(x+\sqrt{(x+h)^2})}$ **95.** 4×10^{-4}

Review Exercises for Chapter 1, page 68

1. $\frac{42}{99}=\frac{14}{33}$ **3.** $0.8333\ldots=0.8\overline{3}$ **5.** $-\frac{3}{35}$
7. $\frac{5}{4}$ **9.** $-\frac{3}{16}$ **11.** 7 **13.** 25 **15.** $\frac{1}{16}$
17. 243 **19.** $\frac{1}{225}$ **21.** w^2 **23.** $5d^2/4$
25. $\dfrac{9x}{32y^3}$ **27.** $3x/4y^2z$ **29.** 3.729×10^{-4}
31. 3.6661×10^9
33. $4.344923458\times10^{-4}\approx4.345\times10^{-4}$
35. $2.197009257\times10^{-3}\approx2.197\times10^{-3}$ **37.** $\frac{1}{27}$
39. 4 **41.** 289 **43.** 15,625
45. -2.410142264 **47.** 0.533061446
49. $3\sqrt[3]{10}$ **51.** 4 **53.** $x^{7/20}$ **55.** $\frac{16}{81}$
57. $1/x^{2.2}=x^{-2.2}$ **59.** $-x^3+6x^2-10x+4$
61. $-4x^3+20x^2-35x+13$
63. $4x^5-13x^4+33x^3-35x^2+20x-3$
65. $x^2-14x+49$; degree 2 **67.** x^2-64; degree 2
69. $x^3-6x^2+10x-8$; degree 3
71. $6x^3+5x^2+2x+12$; degree 3
73. $x^3+6x^2+12x+8$; degree 3
75. $16x^2-8xy+y^2$
77. $4/x^2-12/xy+9/y^2=(2y-3x)^2/x^2y^2$
79. $x^2+4xy-6xz-12yz+4y^2+9z^2$
81. $x-x^{1/4}y^{1/2}+x^{3/4}y^{3/2}-y^2$ **83.** $3(x^2+4)$
85. $4(x-2y)(x+2y)$ **87.** $(2x-3w)(2x+3w)$
89. $(z+7)^2$ **91.** $(2r-3s)^2$ **93.** $(x+5)(x-2)$
95. $(4z+5)(z-2)$ **97.** $(x-10)(x^2+10x+100)$
99. $(xy-2)(x^2y^2+2xy+4)$
101. $(x+3y+4)(x+3y-4)$ **103.** $x(x-2)(x+7)$
105. $(x+2)^2(2x-5)$ **107.** x/y^2
109. $(x+3)(x+2)/(x+4)^2$ **111.** $\dfrac{3x+5}{x+2}$
113. $(x+1)(x-1)/(x+4)(x-3)$ **115.** $(1+3x)/3$
117. $(25+x^4)/5x^2$ **119.** $\dfrac{3x^2+13x-8}{(x-2)(x+3)}$
121. $(6x^2-11x-5)/x(x+1)(x-5)$
123. $(-x-22)/(2x+3)$
125. $(2x+5\sqrt{xy}+3y)/(4x-9y)$

Chapter 2

Problems 2.1, page 76

1. $z=\frac{5}{4}$ **3.** $x=0$ **5.** $y=\frac{3}{10}$ **7.** $z=\frac{1}{2}$
9. $x=\frac{1}{2}$ **11.** $v=\frac{1}{24}$ **13.** $z=-4$
15. $x=1$ **17.** $w=-\frac{3}{2}$ **19.** $y=1$
21. $m=-11$ **23.** $s=\frac{4}{17}$ **25.** $x=-\frac{13}{9}$
27. $x=\frac{1}{2}$ **29.** $a=-\frac{15}{7}$ **31.** $x\approx3.399$
33. $z\approx0.2375$ **35.** $p\approx6.0\times10^{-3}$
37. $x\approx-0.398$ **39.** $w\approx0.036$
41. You obtain $6=-1$ on simplifying.
43. You obtain $-15=1$ on simplifying.
45. You obtain $6=5$ on simplifying.
47. $5/(y+2)=0$ can never be true
49. You obtain $-6=2$ on simplifying.
57. $w=y/4z^3$ **59.** $d=a$
61. $y=\dfrac{\dfrac{z}{4}+\dfrac{3}{x}-10}{x}=\dfrac{xz+12-40x}{4x^2}$
63. $z=(c-bc-b)/a$
65. $a=\dfrac{-3x^2+3bx+x-3b}{3x-3b+2}$ **67.** $q=-b/2$
69. -1 is a solution of $x^2=1$ but not of $x=1$.
71. Yes, $x=2$ is the only real solution of each equation. [However, the equation $x^5=32$ has 4 complex solutions, whereas $x^3=8$ has only 2 complex solutions, so if you include complex solutions, the equations are not equivalent. We discuss complex numbers in Section 2.4.]
73. One cannot cancel $(x-5)$ from both sides of the equation; that is dividing by zero.

Problems 2.2, page 89

1. $r=C/2\pi$ **3.** $h=V/\pi r^2$ **5.** $R=E/I$
7. $m_1=Fr^2/Gm_2$ **9.** $P=A(1+r/n)^{-nt}$
11. \$140 **13.** \$622 **15.** 9.4%
17. 4.5 *years* **19.** \$523,810
21. (a) \$3420 in the $5\frac{1}{2}$% account; \$4580 in the 6% account (b) 5.786%
23. \$11,771 in the mutual fund; \$13,229 in the CD
25. 7.143 m/sec
27. $4\frac{4}{9}$ hours = 4 hours, 26 minutes, 40 seconds
29. (a) $C(q)=35q+1650$ (b) \$9175
31. 0.4 liter of air per breath
33. 15 g of the 30%; 25 g of the 22% **35.** 148 oz.
37. $7\frac{1}{3}$% **39.** 4.60 meters **41.** 4 seconds
43. $26\frac{2}{3}$°C **45.** 130 **47.** $22\frac{2}{9}$ years
49. after a total of 360 at bats **51.** 60 feet by 40 feet

53. 13.54 pounds of \$4.80 per pound coffee; 11.46 pounds of \$3.60 per pound coffee
55. 86
57. $\frac{12}{7}$ hours, or approximately 1 hour, 43 minutes
59. 60 ohms **61.** 57 and 58
63. 23, 25, 27, and 29 **65.** 4 years old
67. $66\frac{2}{3}$ feet, $\frac{40}{9}$ seconds **69.** 100 miles
71. There are many such "tricks." Here's one: pick a number; add 8; square your answer; add 4 times your number; add 36; take the square root; subtract your number.
73. (a) The bird will lose 45 miles each hour, so it will never fly 100 miles forward and will take $2\frac{2}{9}$ hr to be blown 100 miles backward.
(b) There may be a different way to look at a problem that leads to a different and sometimes easier solution.

Problems 2.3, page 102

1. 3 (double root) **3.** -5 (double root)
5. $2, -3$ **7.** $-2, -5$ **9.** $6, -4$ **11.** 1, 2
13. $-\frac{1}{2}, -\frac{3}{2}$ **15.** $\frac{1}{6}, -\frac{3}{2}$ **17.** $(y - \frac{3}{2})^2 - \frac{9}{4}$
19. $(r + \frac{7}{2})^2 - \frac{49}{4}$ **21.** $(x + 1.6)^2 - 2.56$
23. $(u + 7.603)^2 - 57.805609$ **25.** $-1 \pm \sqrt{3}$
27. $-2 \pm \sqrt{7}$ **29.** $(1 \pm \sqrt{5})/2$
31. $(11 \pm \sqrt{93})/2$ **33.** no real roots
35. $y = 0.126510207$, $y = -0.123310207$
37. $D = 5$; two real roots **39.** $D = 0$; one real root
41. $D = -15$; no real roots
43. $D = -56$; no real roots
45. $D = 31.85$; two real roots **47.** $-6, 1$
49. $(-2 \pm \sqrt{34})/3$ **51.** no real roots
53. no real roots **55.** $(-2 \pm \sqrt{10})/3$
57. $(15{,}106 \pm \sqrt{3{,}709{,}576{,}900})/75{,}004 = 1.013443231$ and -0.610638046
59. 2,5
61. $(-8 \pm \sqrt{108})/2 = -4 \pm \sqrt{27} = -4 \pm 3\sqrt{3}$
63. $w = \frac{1}{4}$ (reduces to a linear equation) **65.** 0, 1
67. $y = -2$ (reduces to a linear equation)
69. sum $= -2$, product $= -3$
71. sum $= -7$, product $= -\frac{1}{7}$
73. sum $= 50$, product $= 137$
75. sum $= -8056$, product $= -1137$ **77.** $c = -8$
79. $c = -6$ **81.** $c = -7$ **83.** $c = \frac{4}{5}$
85. $c = 18$ (roots are 3 and 6) **87.** $c = -1$
89. $x^2 + 2x - 3$ **91.** $x^2 + 4x + 3$
93. $x^2 - 12x + 36$ **95.** $x^2 + x + \frac{1}{4}$
97. $x^2 - x - \frac{15}{4}$ **99.** $x^2 + 1.245x - 2.955906$
101. $x^2 + 7053x + 10{,}505{,}492$
103. $2.9049; -1.2049$ **105.** $1.0532; -2.5771$
107. $-1.5701 \times 10^{-4}; -7.6430 \times 10^{-3}$
109. $(x - 4)(2x + 5)$ **111.** $(5x + 4)(2x + 7)$
113. $(5x - 9)(3x + 5)$

Problems 2.4, p. 111

1. $5i$ **3.** $\sqrt{5}\, i$ **5.** $3 + 7i$ **7.** 5
9. $-2 + \frac{1}{3}i$ **11.** $-\sqrt{73}\, i$ **13.** $\dfrac{4}{7} - \dfrac{\sqrt{2}}{7}i$
15. $3 - 8i$ **17.** $-\frac{1}{2} + i$ **19.** $6 + 8i$
21. $\frac{1}{2} - \frac{1}{20}i$ **23.** $2\sqrt{3}\, i$ **25.** $-1 + 3i$
27. $-2i$ **29.** $12 + 16i$ **31.** $31 - 29i$
33. $-\frac{7}{4} - \frac{23}{8}i$ **35.** $-19 + 4i$ **37.** $-i$ **39.** i
41. -4 **43.** $\frac{1}{2} - \frac{1}{2}i$ **45.** $-\frac{3}{2} - 4i$ **47.** i
49. i **51.** $\frac{19}{10} - \frac{3}{10}i = 1.9 - 0.3i$ **53.** $-\frac{1}{2}i$
55. $\frac{1}{2} + \frac{1}{2}i$ **57.** $\frac{2}{29} + \frac{5}{29}i$ **59.** $\frac{1}{13} + \frac{3}{26}i$
61. $\dfrac{a}{a^2 + b^2} + \dfrac{b}{a^2 + b^2}i$ **63.** $u = 6, v = -1$
65. $u = \frac{2}{13}, v = \frac{4}{13}$ **67.** $\pm 2i$ **69.** $-\frac{1}{2} \pm \dfrac{\sqrt{7}}{2}i$
71. $3 \pm i$ **73.** $\dfrac{3}{4} \pm \dfrac{\sqrt{31}}{4}i$ **75.** $-\frac{1}{3} \pm (\sqrt{14}/6)i$
77. $(8.06 \pm 27.90388503i)/25.44 \approx 0.317 \pm 1.097i$
81. $\sqrt{\dfrac{a}{b}} = \dfrac{\sqrt{a}}{\sqrt{b}}$ only if $a \geq 0$ and $b > 0$

Problems 2.5, page 118

1. 19 **3.** 120 **5.** 10 **7.** $\pm 2, \pm\sqrt{3}$
9. $\pm 3, \pm 3i$ **11.** $\pm 3, \pm 4$ **13.** $4, -2 \pm 2\sqrt{3}\, i$
15. $1, 2, -\frac{1}{2} \pm (\sqrt{3}/2)i, -1 \pm \sqrt{3}\, i$
17. $-1, -2, \frac{1}{2} \pm (\sqrt{3}/2)i, 1 \pm \sqrt{3}\, i$
19. $\pm 2, \pm 2i, \pm 1, \pm i$
21. $\pm 125\sqrt{5} = \pm 5^{7/2}$ are the only real solutions; two other solutions are $\pm(-2)^{7/2} = \mp 8\sqrt{2}\, i$.
23. no solutions [note that $(-7)^{10}$ does not work since $(-7)^{10} > 0$]
25. $(2 \pm \sqrt{109})/5$ **27.** 1 **29.** $5, -2$
31. 4 ($z = -1$ is not a solution)
33. 7 (0 is not a solution) **35.** $-1, 2$
37. $x = (7 + \sqrt{97})/8$ ($x = (7 - \sqrt{97})/8$ is not a solution)
39. 6 ($z = 2$ is not a solution) **41.** -1
43. $33 + 6\sqrt{29}$ ($33 - 6\sqrt{29}$ is not a solution)
45. $(3 + 2\sqrt{813})/23$ ($(3 - 2\sqrt{813})/23$ is not a solution)
47. ± 3 (± 2 are not solutions) **49.** 0
51. $-7, \pm 1$ **53.** $1.2290, -2.6617$
55. $2.9367 \times 10^6, 4.4739 \times 10^{-5}$
57. $\pm 0.2608, \pm 1.8494$ **59.** $(1 \pm \sqrt{3}\, i)/2, -1$
61. $\sqrt[3]{5}, \sqrt[3]{5}(-1 \pm \sqrt{3}\, i)/2$ $[\sqrt[3]{5} = 1.709975947]$
63. Hint: -1 has three cube roots

Problems 2.6, page 126

1. 80 items
3. 112 or $721\frac{1}{3}$ (if the answer must be an integer, throw out the answer $721\frac{1}{3}$)
5. (a) $R = 14q$ (b) $P = -0.01q^2 + 11q - 250$ (c) 180 dolls

7. 58 sets (the answer obtained is 57.9985533)
9. (a) $P = -0.01q^2 + 150q - 5000$ (b) 155 sets
11. $\sqrt{12.5} \approx 3.54$ seconds
13. $\sqrt{200/4.9} \approx 6.39$ seconds
15. $\sqrt{2500/4.9} \approx 22.59$ seconds
17. after 4.24 seconds
19. $\sqrt{1000/1.96} \approx 22.59$ seconds
21. approximately 11.11%
23. 7.5% (twice the semiannual interest rate)
25. 11 ft by 25 ft **27.** 27 and 29, or −27 and −29
29. $\frac{8}{5}$ or $-\frac{5}{8}$ **31.** 8 ohms, 10 ohms **33.** 6 cm
35. Jeff: $9 + \sqrt{65} \approx 17.062$ hours, Mary: $7 + \sqrt{65} \approx 15.062$ hours

Problems 2.7, page 132

1. $[-1, 6]$ **3.** $(-3, 6]$ **5.** $[0, \infty)$
7. $(-\infty, 7)$ **9.** $[\frac{1}{2}, \frac{7}{5}]$ **11.** $[39, 429)$
13. $(1, 2)$ **15.** $(0, 8)$ **17.** $[-2, 0]$
19. $(0, 5]$ **21.** $[-1.32, 4.16)$ **23.** $(-\infty, 0)$
25. $(-\infty, 2]$ **27.** $(-\infty, -5)$ **29.** $(-\infty, 11)$
31. $(\frac{15}{2}, \infty)$ **33.** $[-1, \infty)$ **35.** $(-\infty, \frac{9}{2}]$
37. $(-\infty, -2]$ **39.** $(-\infty, \frac{3}{4})$ **41.** $[-1, 2]$
43. $[-4, 5)$ **45.** $[-\frac{1}{10}, \frac{1}{5})$ **47.** $[-3, 1)$
49. $[-7, -1)$ **51.** $(-4, \frac{25}{2}]$
53. $[(ad - c)/b, (de - c)/b)$ **55.** $[15.733, \infty)$
57. $(3.438, \infty)$ **59.** $(-\infty, -15)$
61. $[1.8155 \times 10^{-2}, 2.30375 \times 10^{-2}]$

Problems 2.8, page 142

1. 1 **3.** −1 **5.** −2 **7.** $\pi - 2 \approx 1.1416$
9. $7 - \pi \approx 3.8584$ **11.** $(-1, 1)$
13. $\{x: x \le -4 \text{ or } x \ge 4\}$ **15.** $\mathbb{R}$
17. no solution **19.** (1.3)
21. $\{x: x < 1 \text{ or } x > 5\}$ **23.** $(-\frac{3}{2}, 1)$
25. $\{x: x \le 4 \text{ or } x \ge 6\}$ **27.** $\{x: x < -5 \text{ or } x > 1\}$
29. $\{x: x \le \frac{4}{3} \text{ or } x \ge \frac{8}{5}\}$ **31.** $\{x: x < -\frac{53}{3} \text{ or } x > \frac{19}{3}\}$
33. $\{x: x \le (c - b)/a \text{ or } x \ge (-b - c)/a\}$ **35.** $(\frac{5}{4}, \infty)$
37. $\{x: x < 2 \text{ or } x > 5\}$ **39.** $(-\frac{1}{2}, \frac{1}{2})$
41. $\{w: w \le 0 \text{ or } w \ge 5\}$ **43.** $\{x \in \mathbb{R}: x \ne 2\}$
45. $x = 2$ **47.** no solution **49.** $(-\infty, \infty)$
51. $(2, 7)$ **53.** $(-2, 3)$
55. $[-1 - 2\sqrt{2}, -1 + 2\sqrt{2}]$
57. $\{x: x \le -\frac{7}{3} \text{ or } x \ge 1\}$ **59.** $(0, \frac{6}{5})$
61. $(-\infty, 1)$ **63.** $\{x: x \le -5 \text{ or } 0 \le x \le 3\}$
65. $(\frac{9}{2}, 5)$ **67.** $\{x: x < -3 \text{ or } 2 < x < 7\}$
69. $\{x: x < \frac{11}{5} \text{ or } x > 3\}$ **71.** $\{x: x < -1 \text{ or } x \ge 2\}$
73. $\{x: x < -4 \text{ or } 3 < x < 4\}$
81. (a) $[-5, 5]$ (b) $\{x: x < -3 \text{ or } x > 3\}$ (c) $[-2, 2]$ (d) no solution
83. (a) $x > \frac{5}{2}$ (b) $y < 0$ **87.** $s < 0$
89. no; [1.5625, 1.8225]
91. $1.25 \text{ sec} \le t \le 6.25 \text{ sec}$ **93.** $[-1.125, 4.5]$
95. $\{x: x < -0.752833078 \text{ or } x > -0.729555895\}$

97. $(2.999, 3.001)$ **99.** $(-3.51, -3.49)$
101. $\{x: \sqrt{0.99} < x < \sqrt{1.01} \text{ or } -\sqrt{1.01} < x < -\sqrt{0.99}\}$
109. *c* **111.** *i* **113.** *a* **115.** *f* **117.** *d*
119. $\{x: x < -2.4 \text{ or } x > 0.9\}$ **121.** $(2, 2.4)$
123. $(-1.6, \infty)$ **125.** $\{x: x < 1 \text{ or } x > 2\}$
127. $\{x: x < -1.6 \text{ or } x > 0.6\}$
129. $\{x: x < -3 \text{ or } -2.8 < x < 0.7 \text{ or } x > 0.8\}$

Review Exercises for Chapter 2, page 146

1. −5 **3.** $-\frac{9}{8}$ **5.** $-\frac{13}{5}$ **7.** $-\frac{3}{7}$ **9.** $-\frac{11}{8}$
11. $(a - d)/b$ **13.** $r = \sqrt{Gm_1m_2/F}$ **15.** \$4800
17. (a) $C = 60q + 2300$ (b) \$10,400 **19.** 96
21. ±3 **23.** 2, −7 **25.** 1, 6
27. $(-5 \pm \sqrt{17})/2$ **29.** $-5 \pm 4\sqrt{2}$
31. $(-3 \pm 2\sqrt{6})/5$ **33.** $(-1 \pm \sqrt{3}\, i)/2$
35. $-1 \pm \sqrt{2}\, i$ **37.** sum $= -\frac{10}{3}$, product $= -\frac{4}{3}$
39. $D = 60$; two roots **41.** $D = 0$; 1 double root
43. $D = 16 - 4\sqrt{2} \approx 10.34$; two roots
45. $-5 + 3i$ **47.** $7 - 8i$ **49.** $27 + 24i$
51. $-i$ **53.** $\frac{3}{5} + \frac{4}{5}i$ **55.** 47 **57.** ±2
59. $\pm 2, \pm\sqrt{2}\, i$ **61.** $(7 \pm \sqrt{249})/4$ **63.** 0, 4
65. approximately 225
67. (a) $P = -0.01q^2 + 15q - 2000$ (b) $q = 320$
69. 28.22 seconds **71.** $[3, 7]$ **73.** $(0, 5]$
75. $[4, \infty)$ **77.** $(-\infty, 7)$ **79.** $(-\infty, \frac{5}{2}]$
81. $[-2, 5)$ **83.** $(1, 7)$
85. $\{x: x < -1 \text{ or } x > 9\}$ **87.** $(1, 4)$
89. no solution **91.** $\{x: x < \frac{5}{2} \text{ or } x > \frac{7}{2}\}$
93. $\{x: x < -2 \text{ or } x > \frac{26}{5}\}$ **95.** $\{x: x < 1 \text{ or } x > 6\}$
97. $(1, 6)$ **99.** $\{x: x \le -5 \text{ or } x \ge 2\}$
101. $[-3, 4]$ **103.** $\{x: x < -5 \text{ or } -2 < x < 1\}$

Chapter 3

Problems 3.1, page 155

1. IV

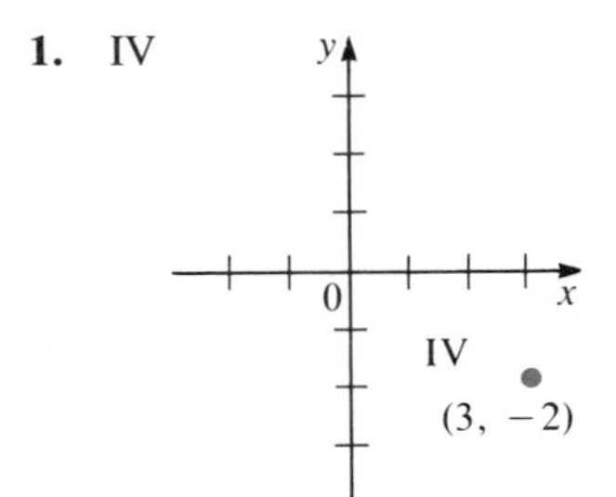

3. on x-axis

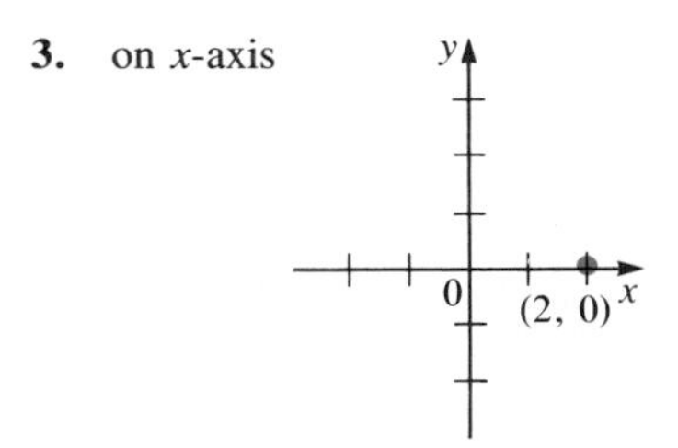

5. III

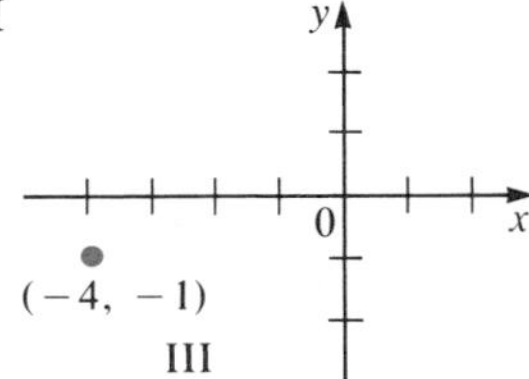

7. I

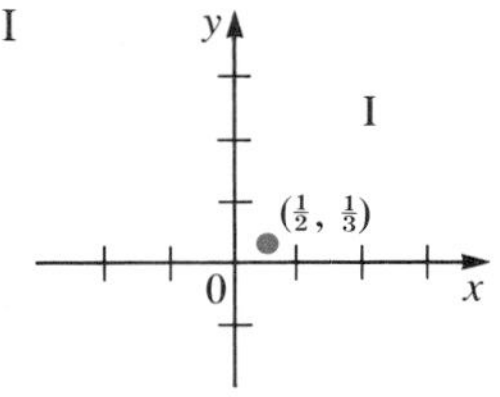

9. on y-axis

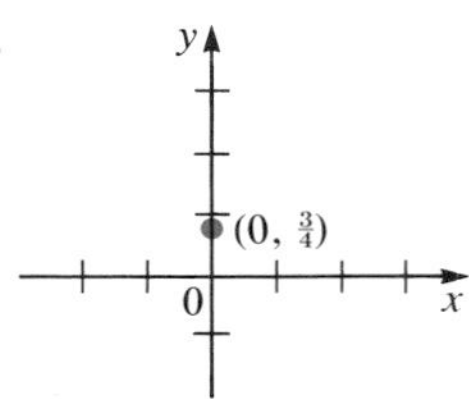

11. 5 **13.** $\sqrt{202}$ **15.** $\sqrt{29}$ **17.** $\sqrt{c^2 + d^2}$
19. $\sqrt{70.5805} \approx 8.40$ **21.** yes **23.** yes
25. $4, 4, 4\sqrt{2}$; yes
27. $\sqrt{10}, \sqrt{40} = 2\sqrt{10}, \sqrt{50} = 5\sqrt{2}$; yes
29. $4, \frac{1}{2}\sqrt{65}, \frac{1}{2}\sqrt{113}$; no **31.** 1 or −15
33. (6, −1) **35.** $(\frac{3}{2}, 7)$ **37.** (3.9, 0.8)
41. (0, 3)
43. $x^2 + (y - 2)^2 = 1$

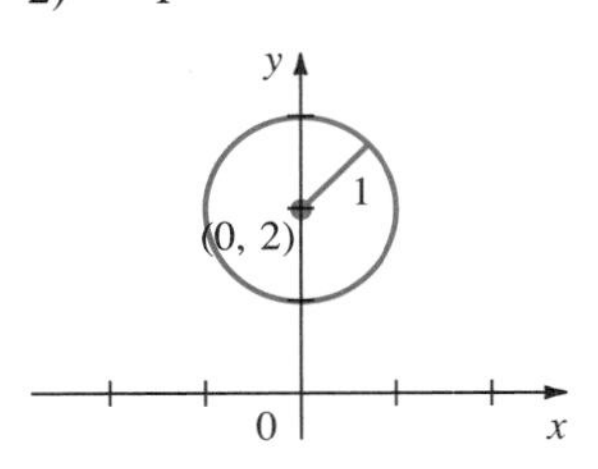

45. $(x - 1)^2 + (y - 1)^2 = 2$

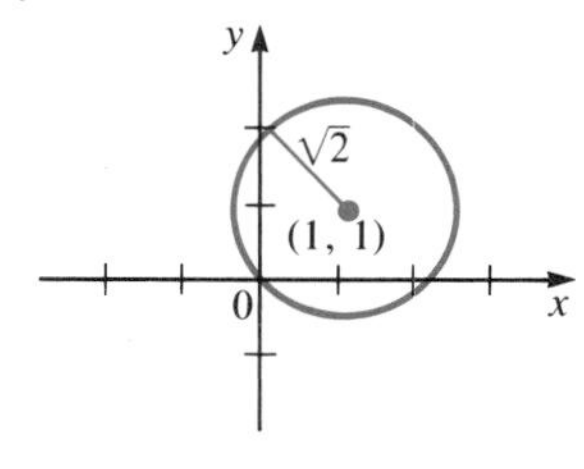

47. $(x + 1)^2 + (y - 4)^2 = 25$

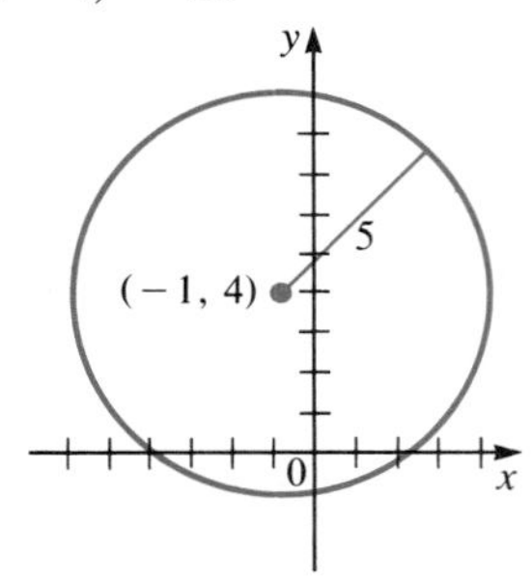

49. $(x - \pi)^2 + (y - 2\pi)^2 = \pi$

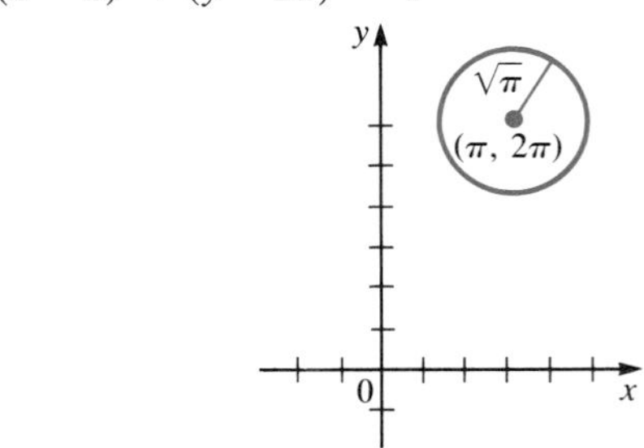

51. $(x - 3)^2 + (y + 2)^2 = 16$

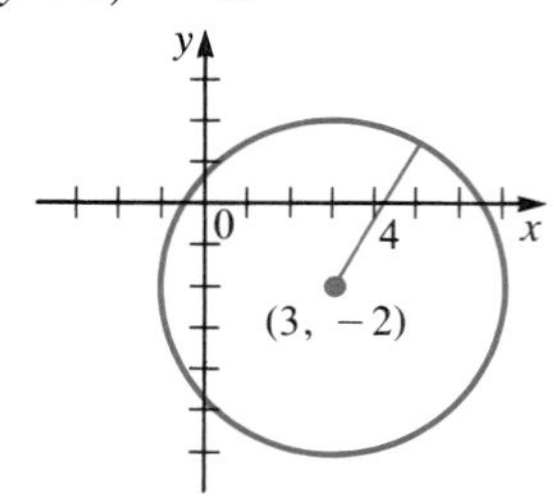

53. center (0, 3), radius $\sqrt{6}$
55. center $(-\frac{1}{2}, \frac{1}{4})$, radius = $\sqrt{66}/4$
57. $(x - 3)^2 + (y + 7)^2 = 34$ **59.** (*a*) $\frac{11}{2}$ (b) 8

Problems 3.2, page 168

1. −2

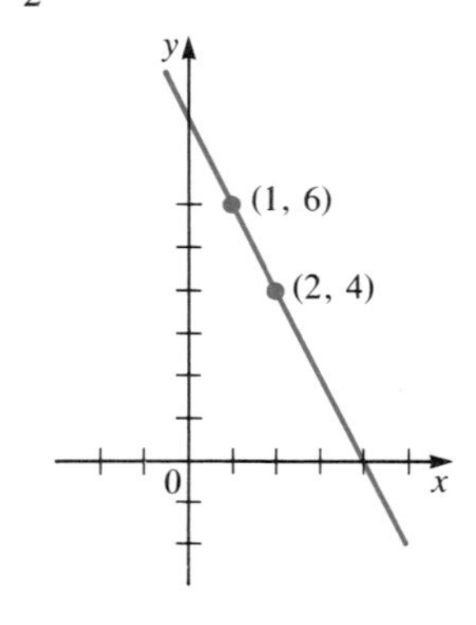

3. 1

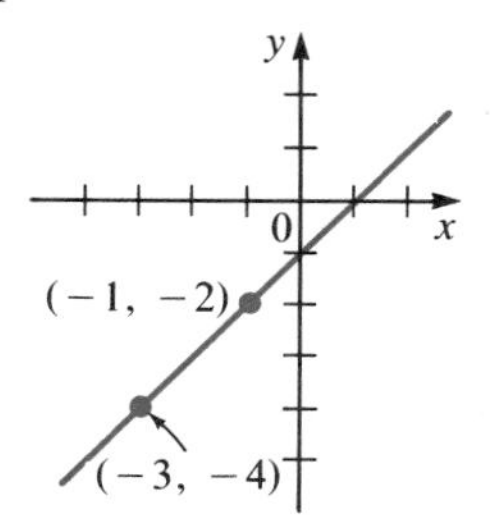

5. $-\frac{5}{12}$

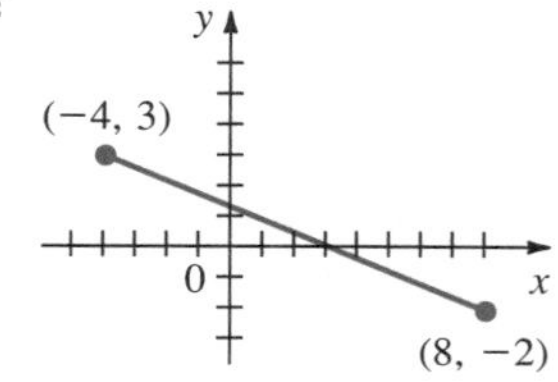

7. 0

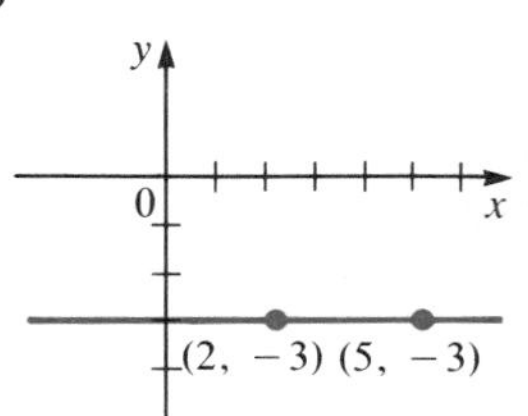

9. −1

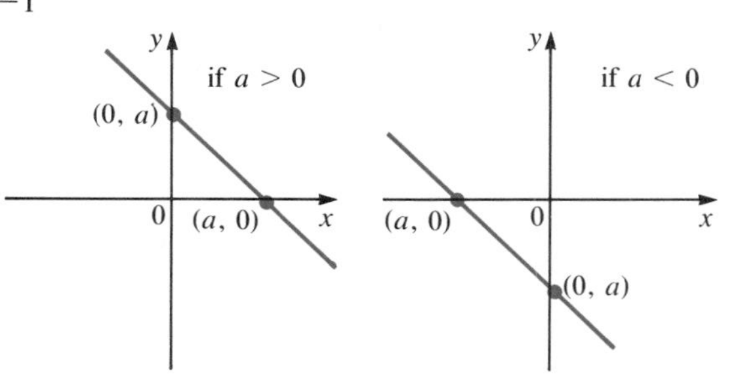

11. $\dfrac{b-d}{a-c}$

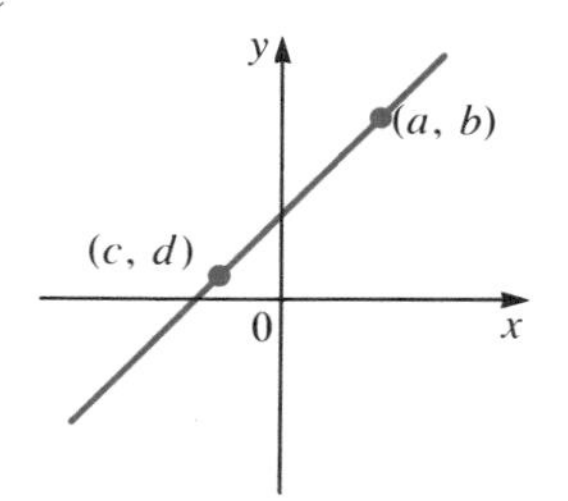

13. $-\frac{6}{5}$ **15.** 0 **17.** $-\frac{9}{2}$ **19.** $\frac{1}{6}$

21. perpendicular

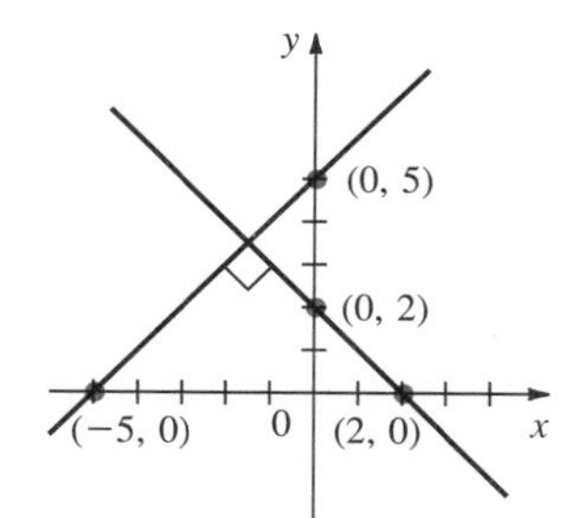

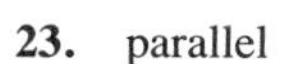
23. parallel

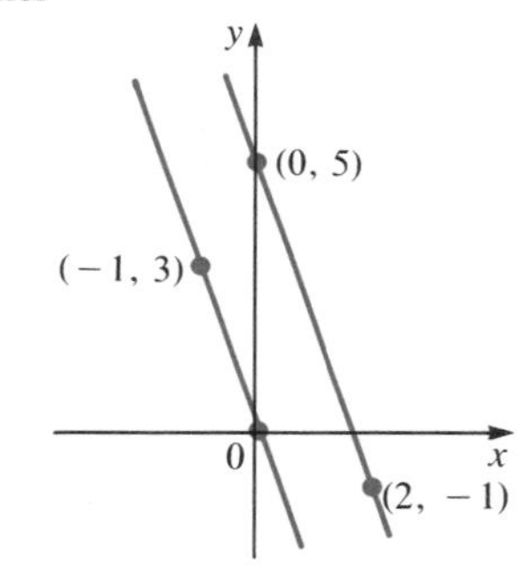

25. perpendicular

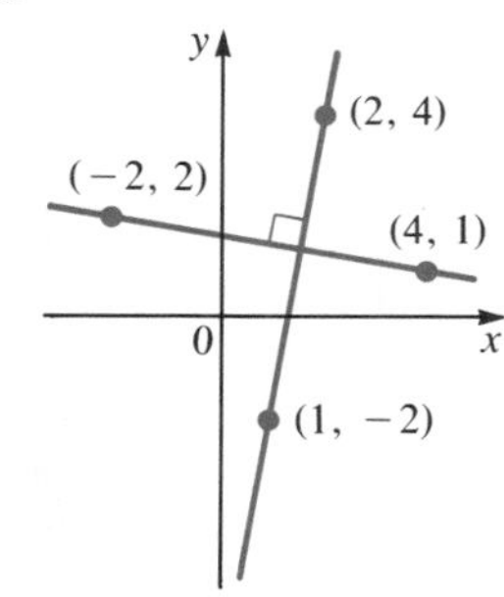

27. perpendicular

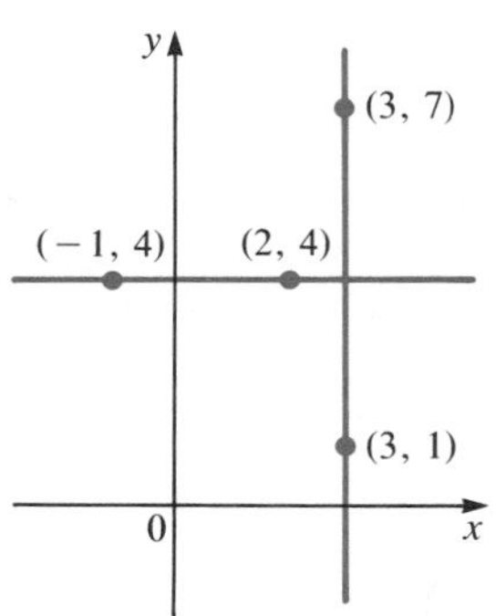

29. 4.5 **33.** −0.2 **37.** $\sqrt{410} - 20 \approx 0.248457$

41. $y + 1 = 3(x - 3)$
$y = 3x - 10$
$3x - y = 10$

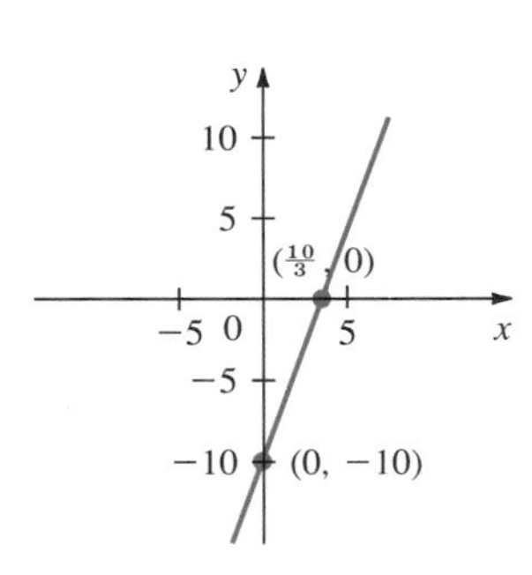

43. $y - 3 = 0(x - 8)$

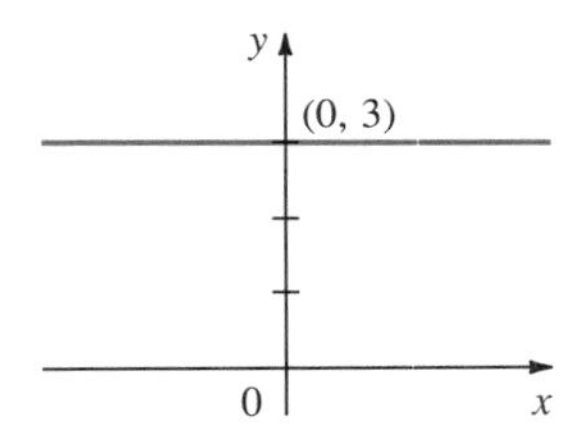

45. (a) $y + \frac{1}{2} = -\frac{3}{16}(x - 3)$ (b) $y = -\frac{3}{16}x + \frac{1}{16}$
(c) $3x + 16y = 1$

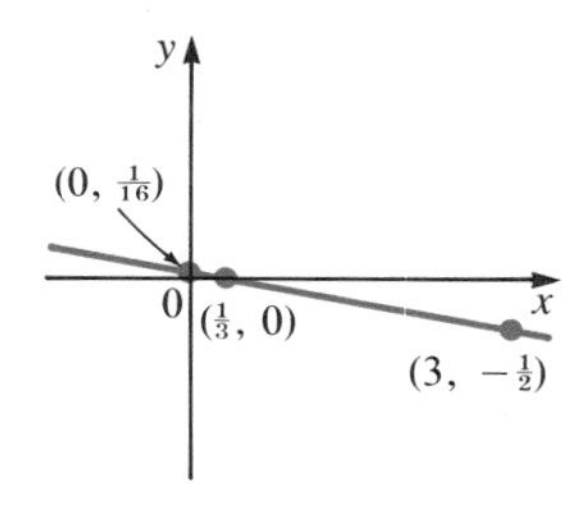

47. (a) $y - 2 = 1(x - 8)$ (b) $y = x - 6$ (c) $x - y = 6$

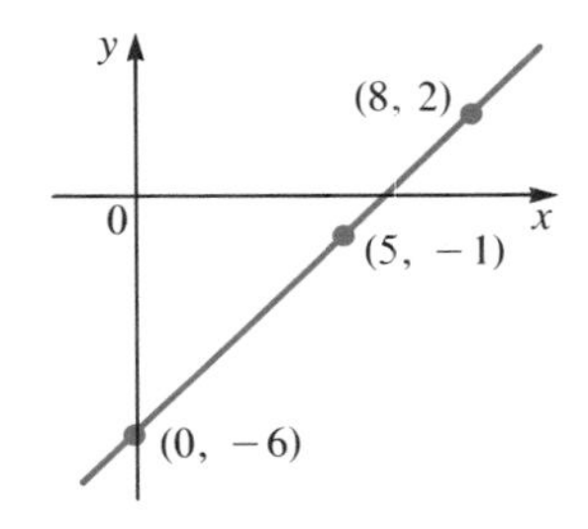

49. (a) $y - 1 = \frac{3}{7}(x + 5)$ (b) $y = \frac{3}{7}x + \frac{22}{7}$
(c) $3x - 7y = -22$

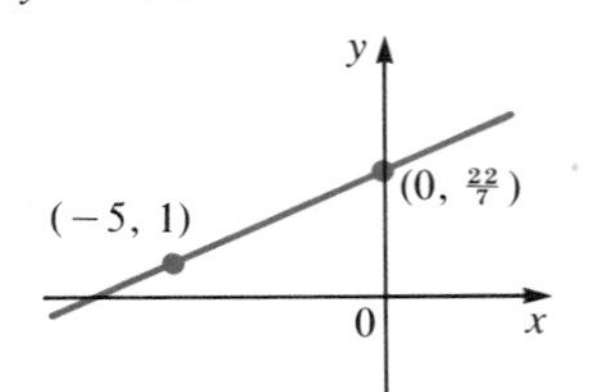

51. (a) $y - b = c(x - a)$ (b) $y = cx + (b - ac)$
(c) $cx - y = ac - b$

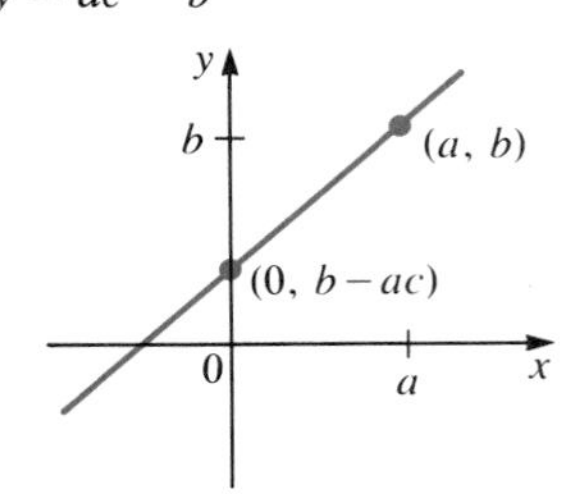

53. (a) $y - 0.0058 = 12.611(x - 0.0146)$
(b) $y = 12.611x - 0.1783206$
(c) $12.611x - y = 0.1783206$

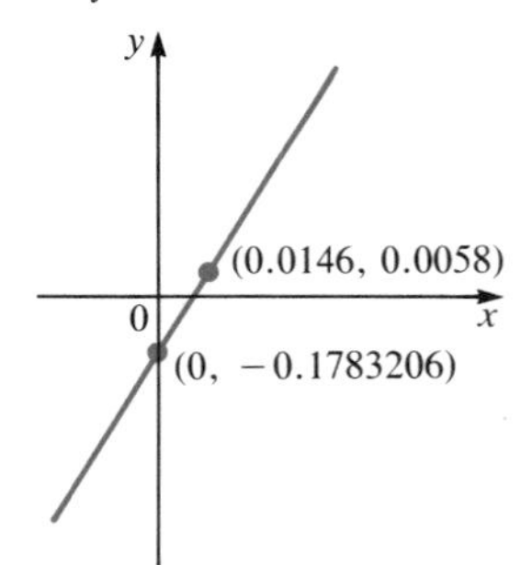

55. $y = \frac{5}{7}x + \frac{25}{7}$ **57.** $y = -\frac{3}{4}x + \frac{13}{4}$

59. (4, 1)

61. infinite number of points of intersection of the form $(x, \frac{3}{2} - 2x)$

63. $(\frac{50}{17}, -\frac{5}{17})$ **65.** (123.0560092, 173.0156701)

69.

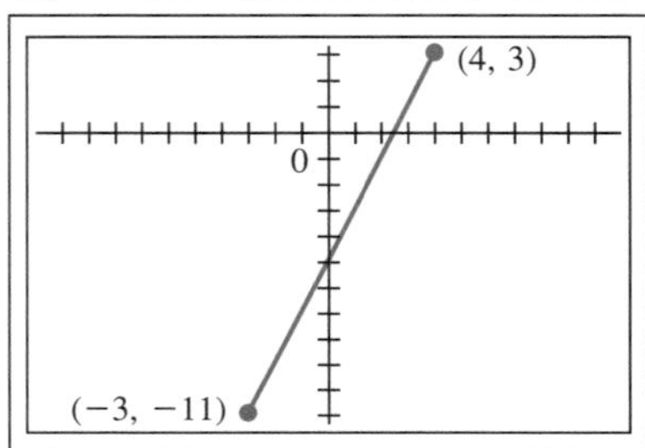

71.

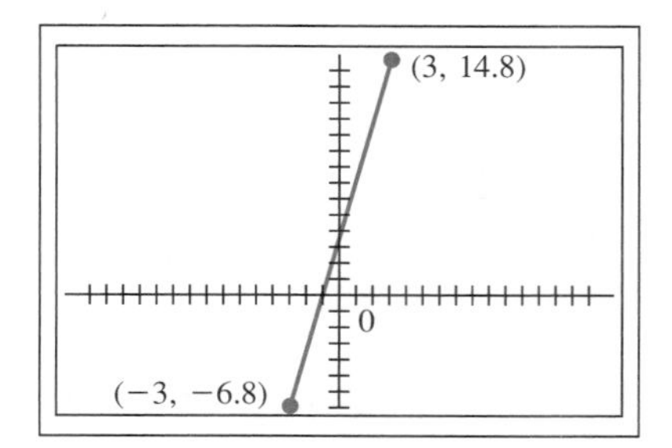

73.

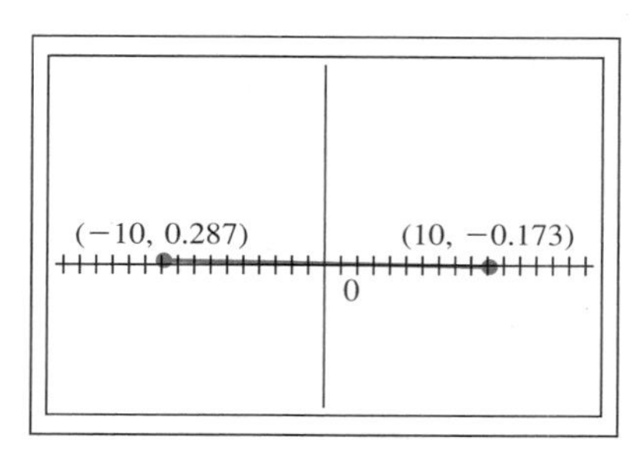

75.

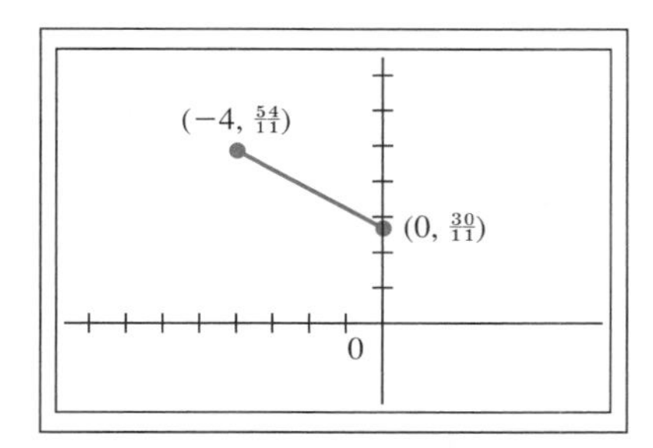

77.

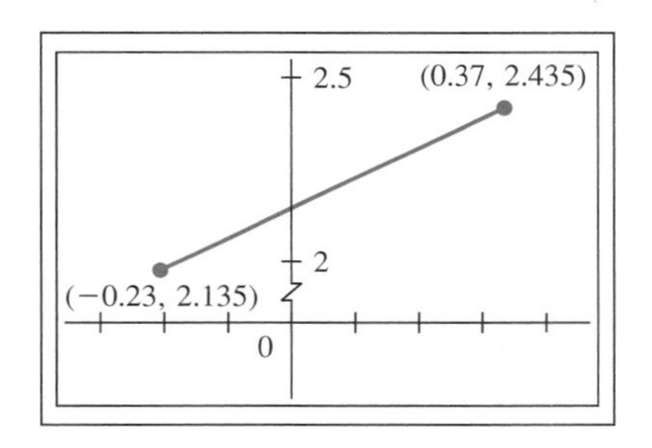

Problems 3.3, page 179

1. $1, \frac{1}{2}, -1, -\frac{1}{4}, \frac{1}{1+x^2}, \frac{1}{1+\sqrt{x}} = \frac{1-\sqrt{x}}{1-x}$
3. $-1, 7, 17, -\frac{1}{2}, 2w-1, 2w^{10}-1$
5. $0, 16, 16, 25, s^{4/5}, s^4 - 4s^3 + 6s^2 - 4s + 1$
7. $1, 0, 2, 2\sqrt{2}, n^{3/2}, \sqrt{\frac{1+w}{w}}$
9. $1, 3, 91, 31, n^4 - n^2 + 1, \frac{1}{n^6} - \frac{1}{n^3} + 1$
11. function
13. not a function because $g(a)$ is not unique
15. function
17. not a function with domain D because it is not defined at w
19. function **21.** yes **23.** yes **25.** no
27. yes **29.** yes **31.** yes
33. $x^n = a$ has two solutions if n is even **35.** yes
37. yes **39.** yes **41.** no
43. no (the y-axis passes through two points on the graph)
45. domain: all real numbers; range: all real numbers
47. domain: all real numbers except 0; range: $(0, \infty)$
49. domain: all real numbers except -1; range: all real numbers except 0
51. domain: $[1, \infty)$; range: $[0, \infty)$
53. domain: all real numbers except 0; range: $(0, \infty)$
55. domain: all real numbers; range: $[0, \infty)$
57. domain: all real numbers; range: $[0, \infty)$
59. domain: all real numbers; range: all integers
61. 1.419936, 54.372225, −1676.384304
63. −0.980796045, 0.227833604
65. $x^3 + 3x^2\Delta x + 3x(\Delta x)^2 + (\Delta x)^3, 3x^2 + 3x\Delta x + (\Delta x)^2$
67. domain: all real numbers except 0; range: $\{1, -1\}$
69. $A(W) = 25W - W^2$; domain: $(0, 25)$; range: $\left(0, \frac{625}{4}\right)$
71.
$$d(t) = \begin{cases} 30\sqrt{t^2 - 6t + 18}, & 0 \le t < 3 \\ 180 - 30t, & 3 \le t \le 6 \\ 30t - 180, & 6 < t \le 9 \\ 30\sqrt{t^2 - 18t + 90}, & 9 < t \le 12 \end{cases}$$
73. (a) 1183 (b) 1187 (c) 1203 (d) 1248 (e) 1204
75. even **77.** even **79.** neither **81.** even
83. even **85.** even **87.** odd **89.** odd
91. neither

Problems 3.4, page 188

1. $m = 0.12$ represents the rate of interest

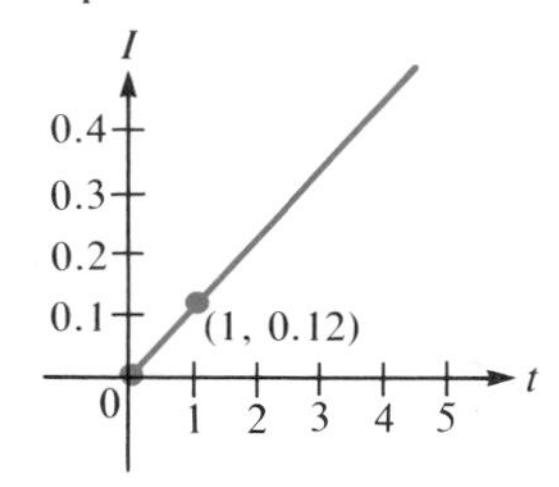

3. $m = 130$ represents annual simple interest on a \$1000 investment at 13% interest

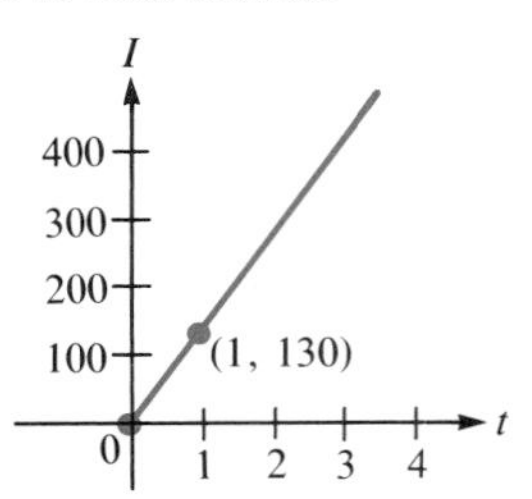

5. $m = 70$ represents velocity

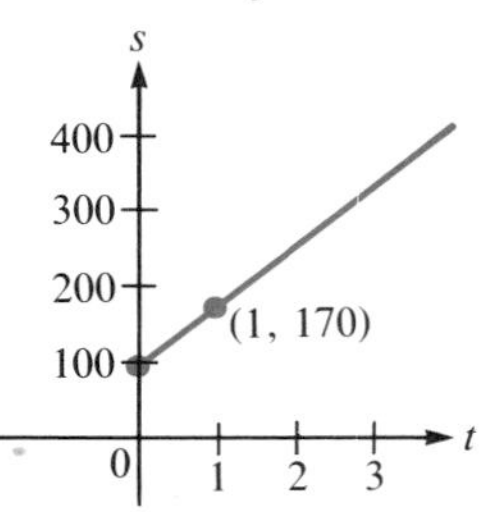

7. $m = -0.025$ represents the amount the price is reduced for each unit sold

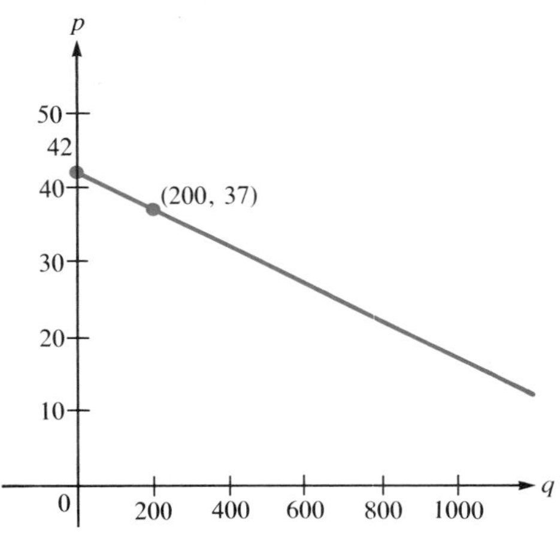

9. $m = 0.015$ represents how the price increases as the number of units (q) produced increases

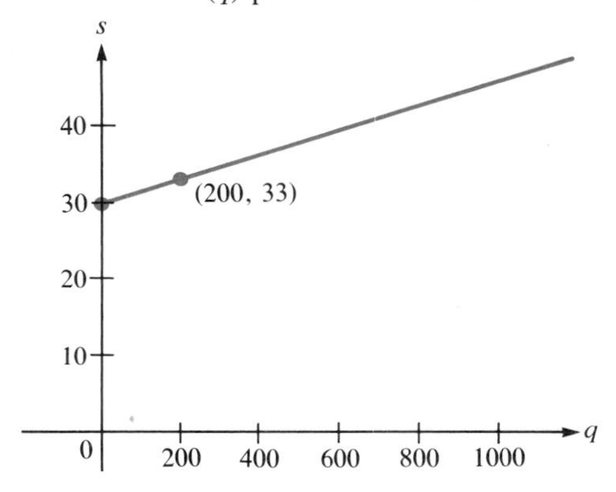

11. \$34.50 **13.** \$485

15. endurance line slope is $-\frac{21}{4}$; sprint line slope is $-\frac{5}{2}$

17. (a) -4 (b) \$4 per bushel

19. (a) $C = 6 + 0.07(h - 30) = 0.07h + 3.90,\ h \geq 30$

(b)

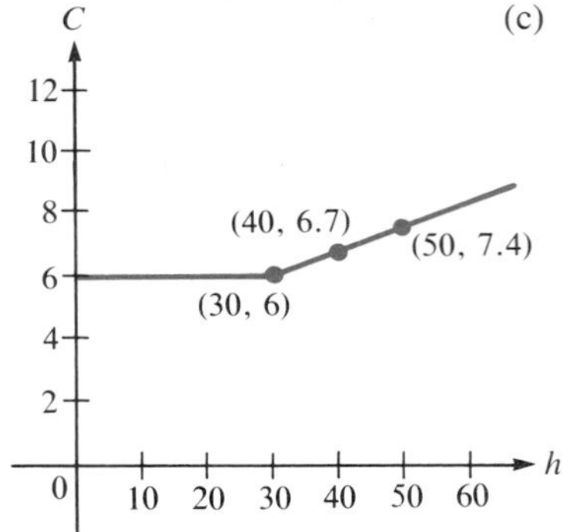

(c) \$9.15

21. (a) $T = 7434 + 0.44(I - 28{,}800) = 0.44I - 5238$

(b)

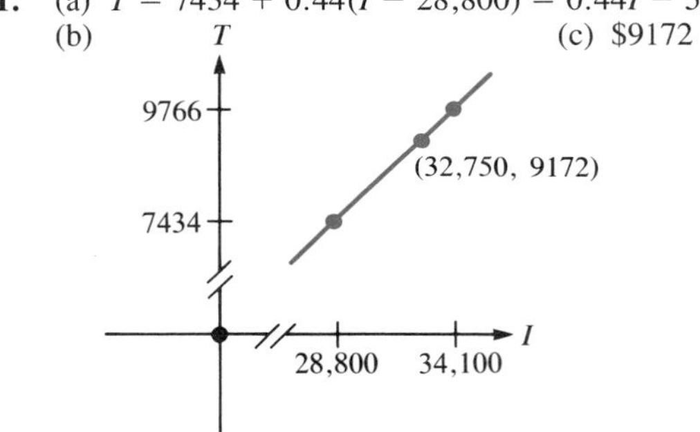

(c) \$9172

23. 0.34 **25.** \$1.56

Problems 3.5, page 199

1.

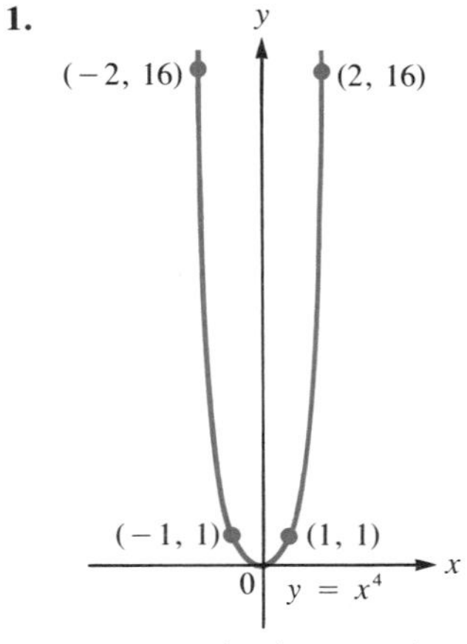

symmetric about y-axis

3.

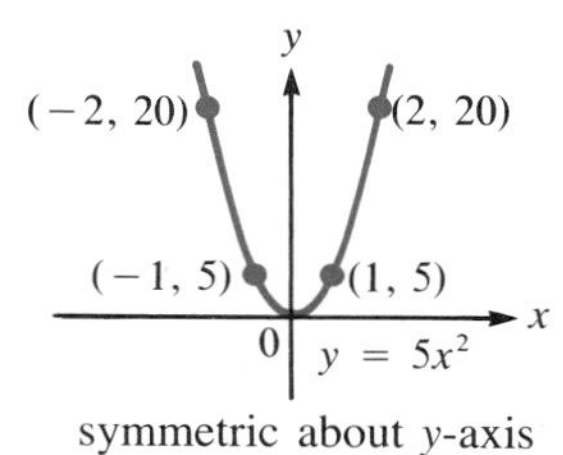

symmetric about y-axis

5.

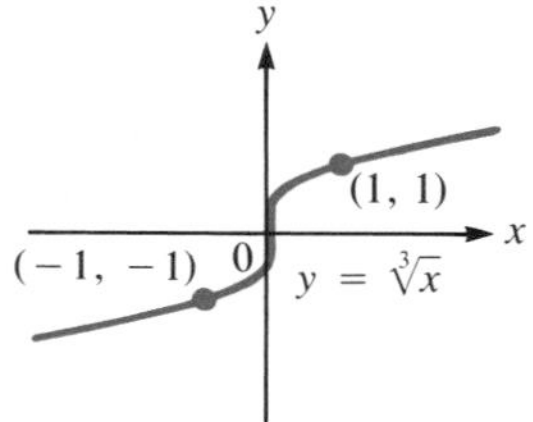

symmetric about origin

7.

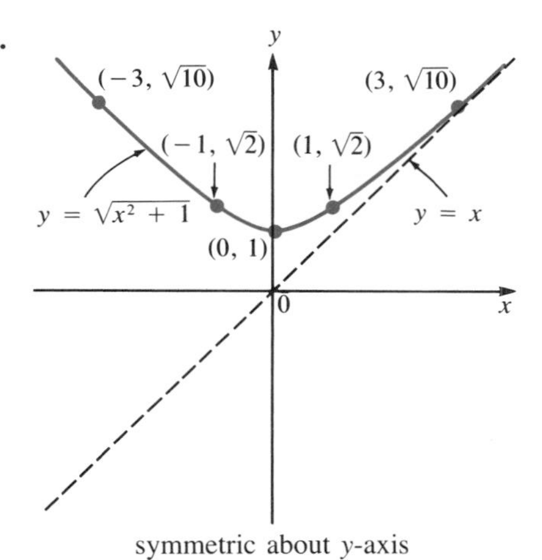

symmetric about y-axis

9.

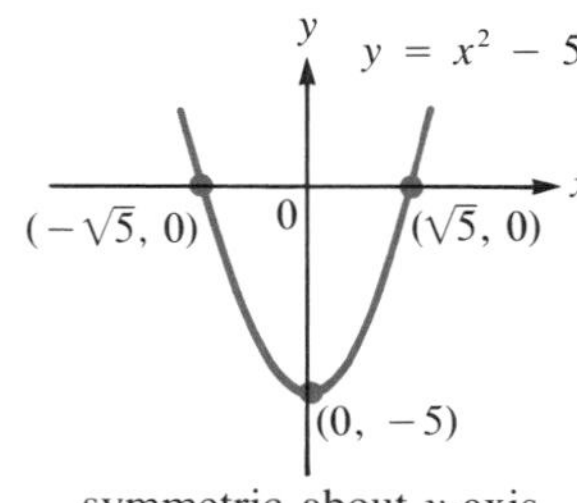

symmetric about y-axis

11.

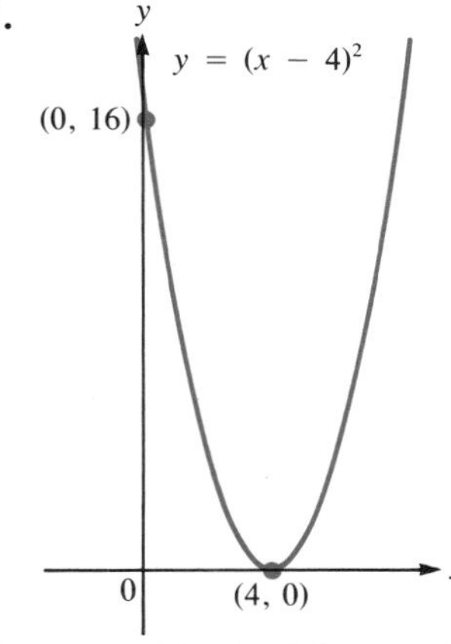

symmetric about line $x = 4$

13.

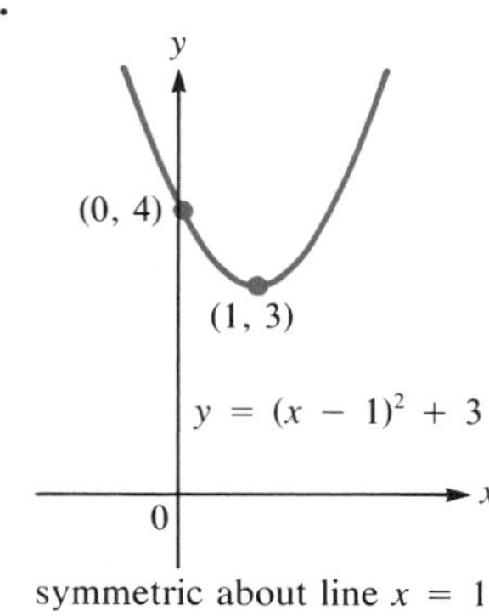

symmetric about line $x = 1$

15.

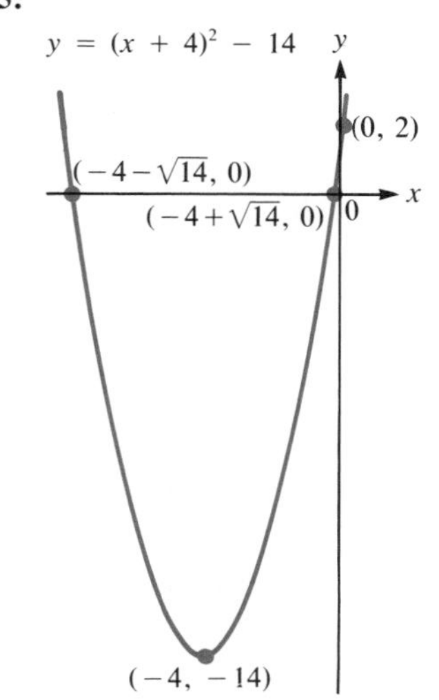

17.

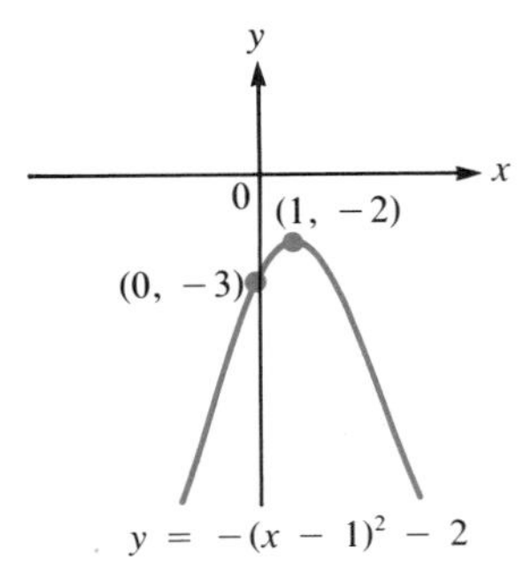

19.

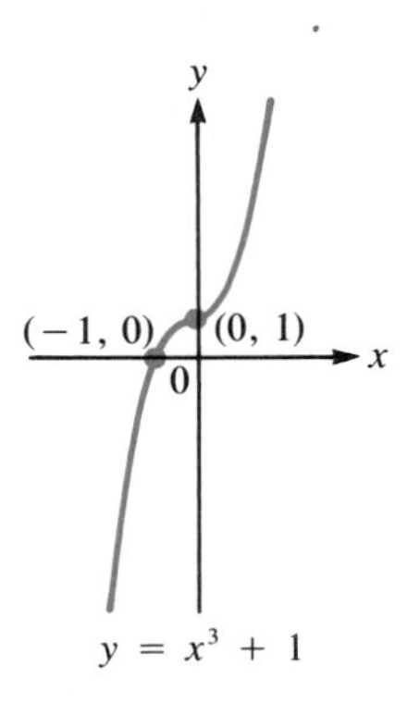

21.

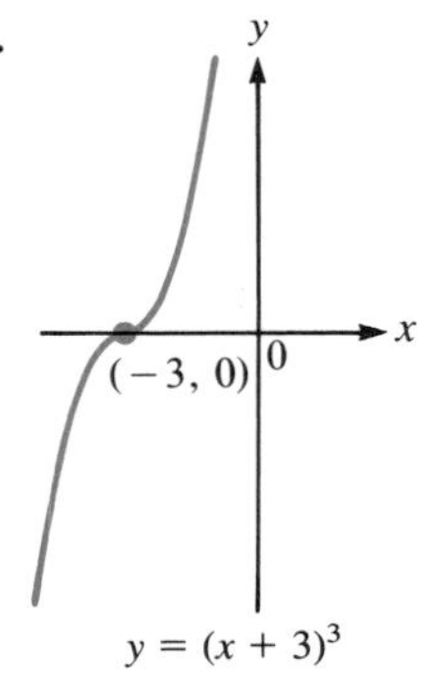

23.

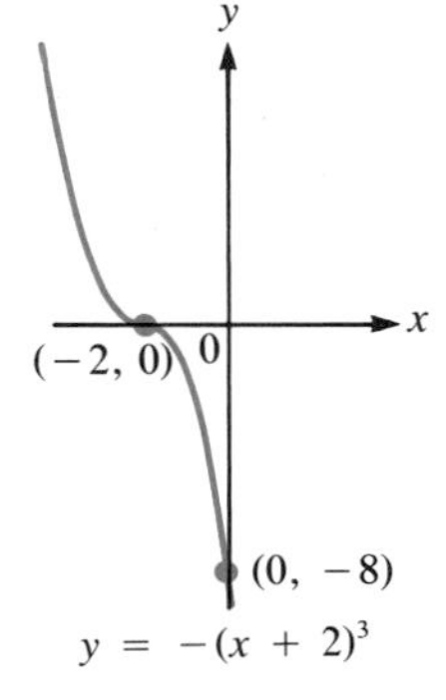

25. (a)

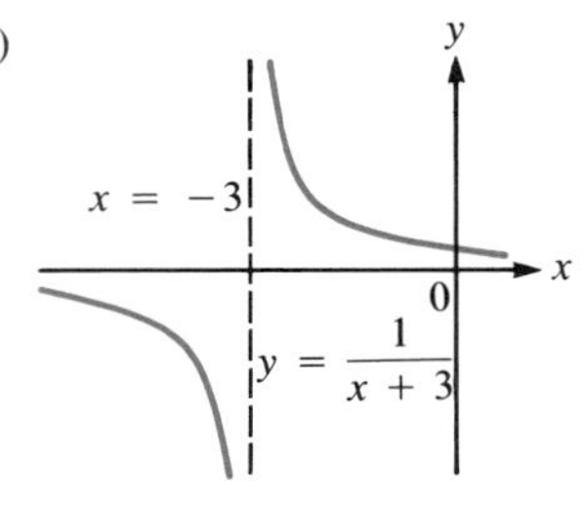

(b)

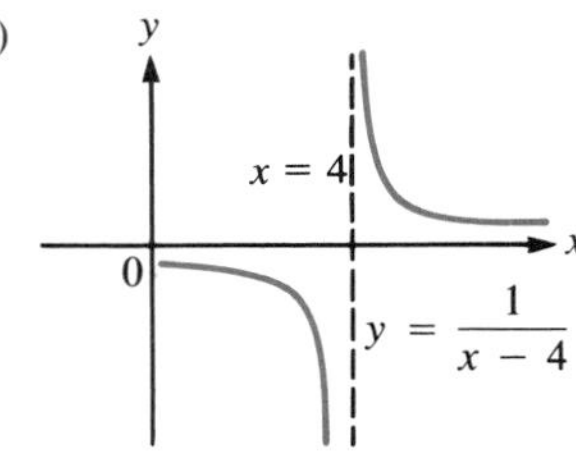

(c)

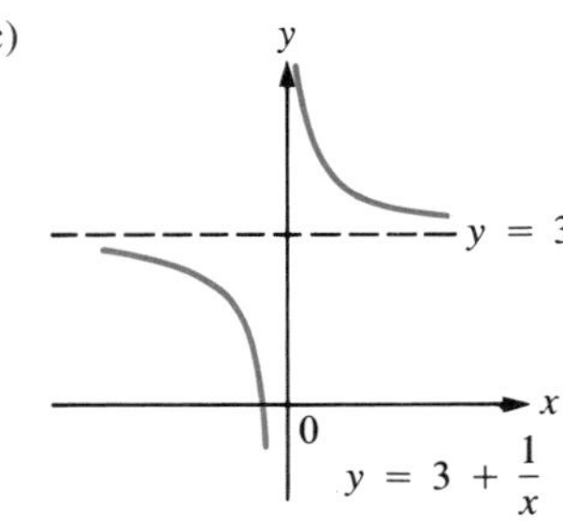

(d)

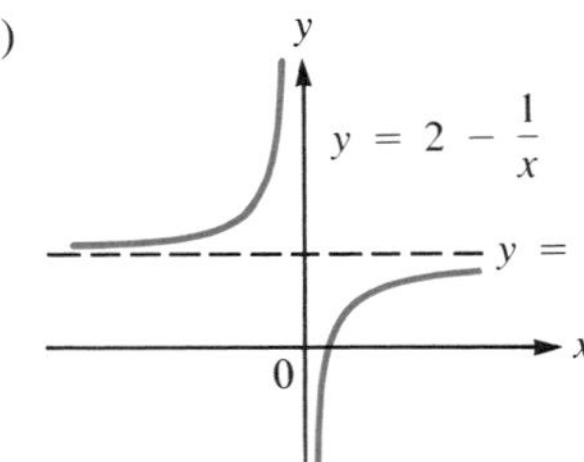

(e)

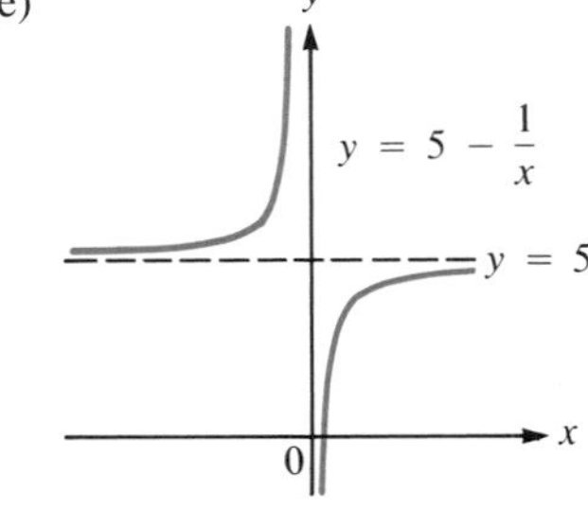

27.

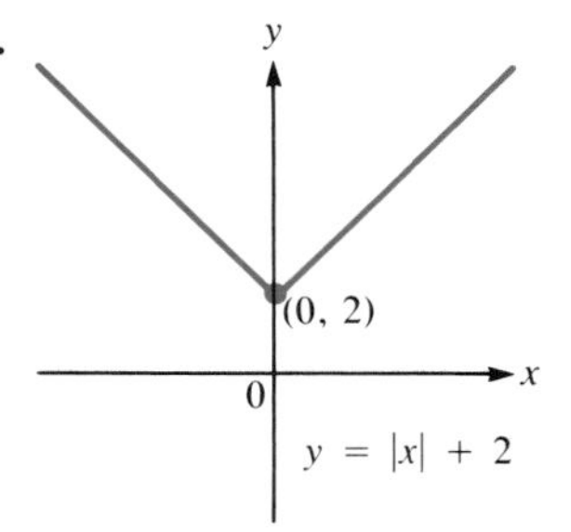

29.

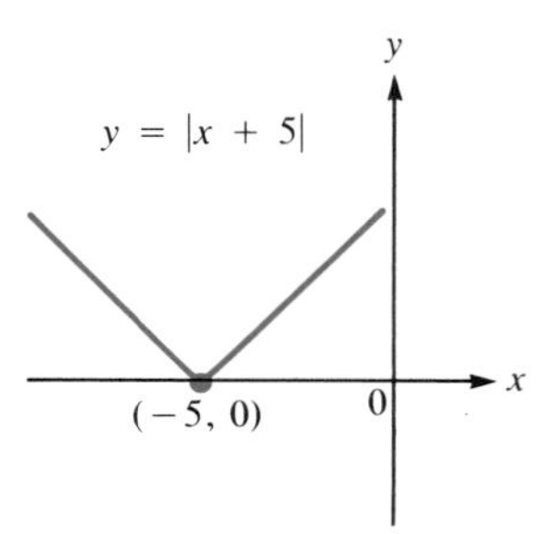

31.

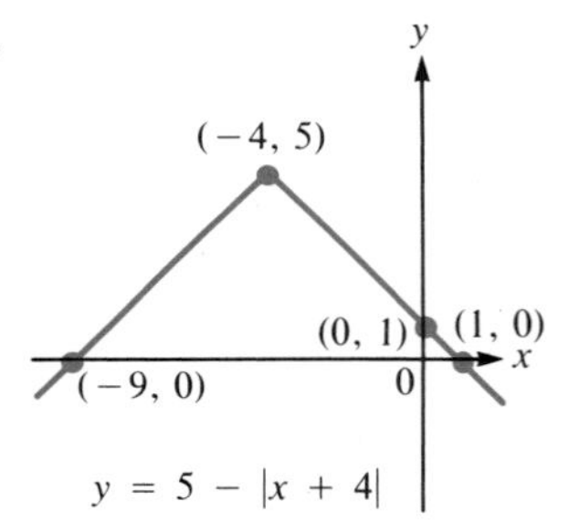

33.

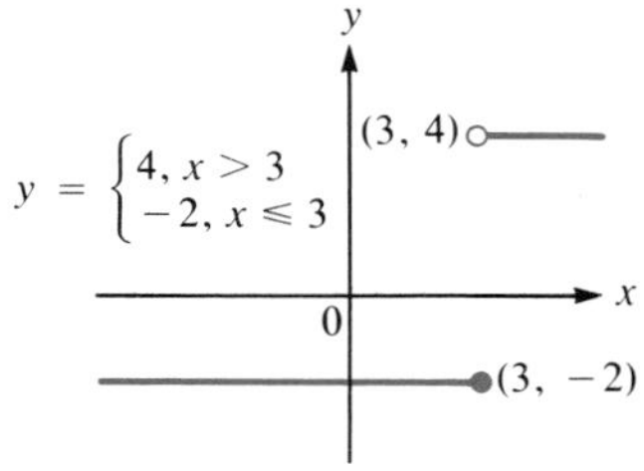

35.

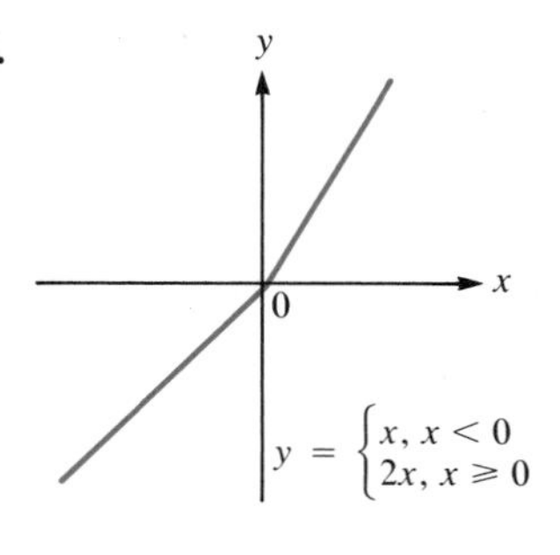

37.

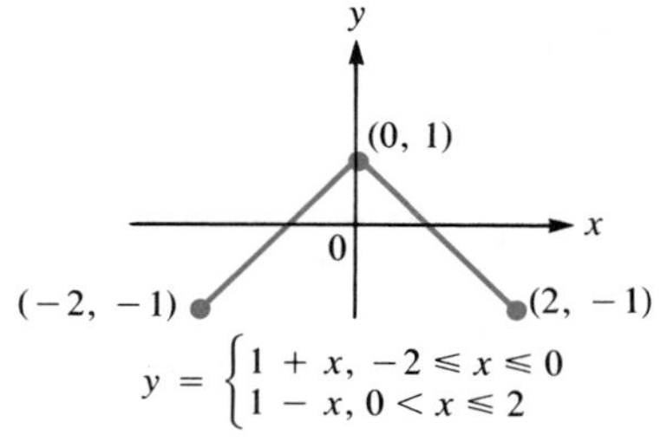

39.

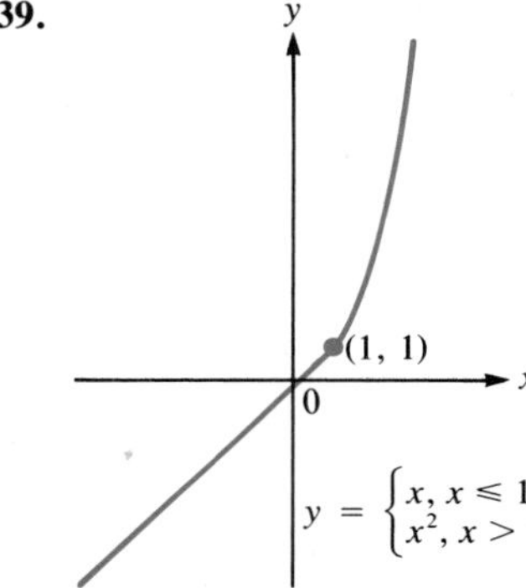

41.

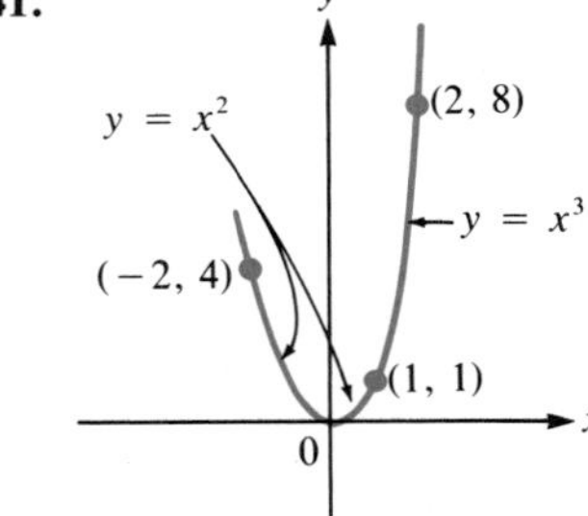

43.

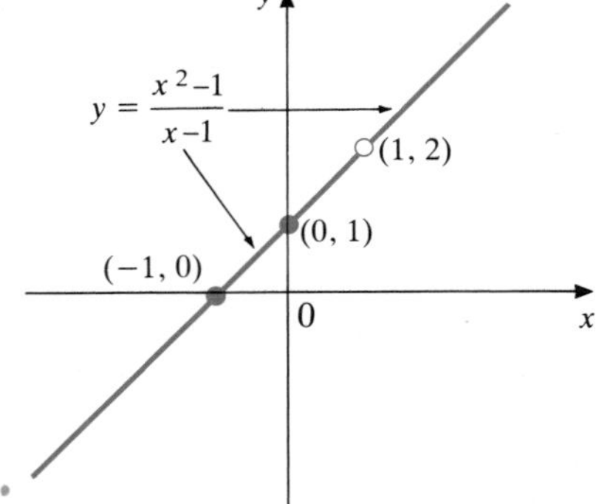

45.

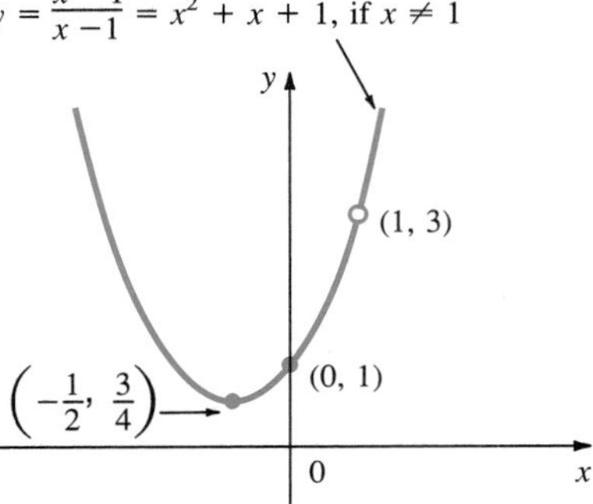

47. (a)

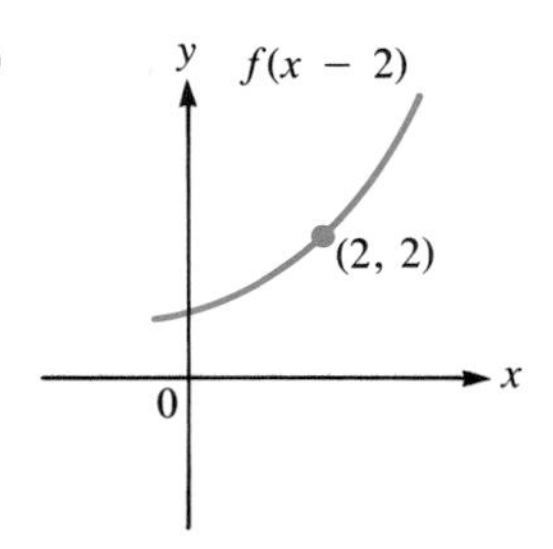

(b)

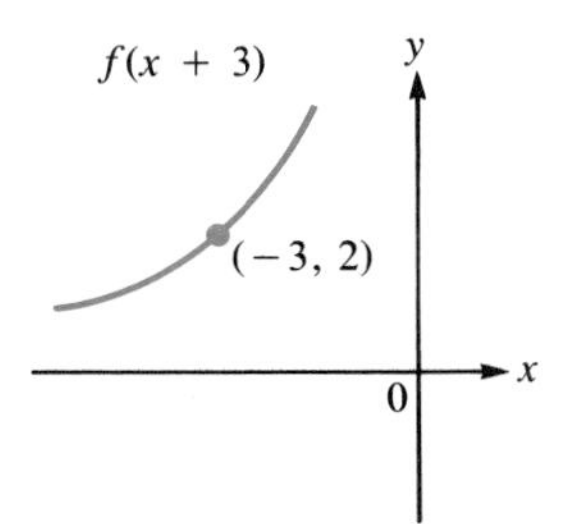

(c)

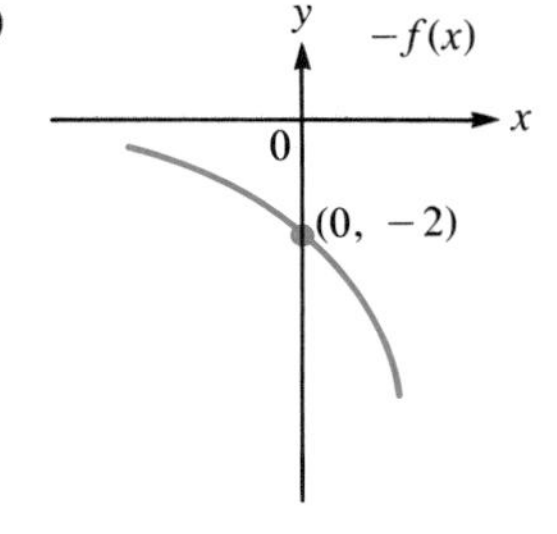

(d)

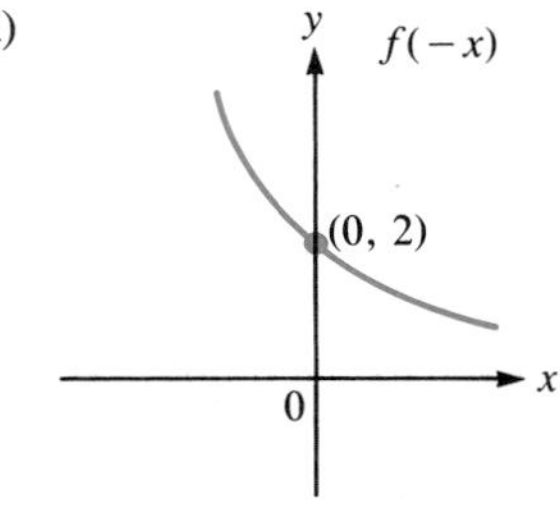

(e)

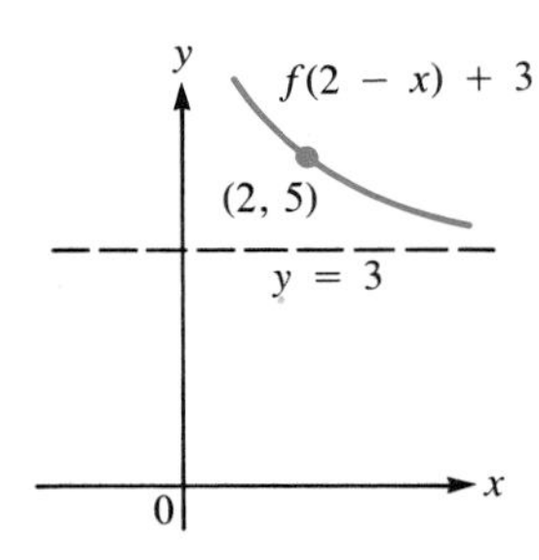

49. (a)

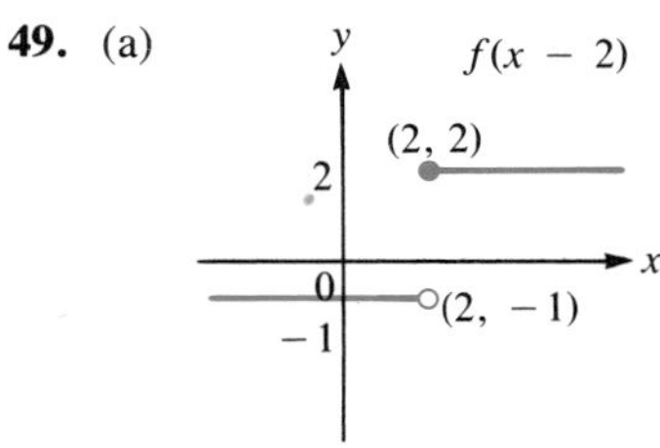

(b)

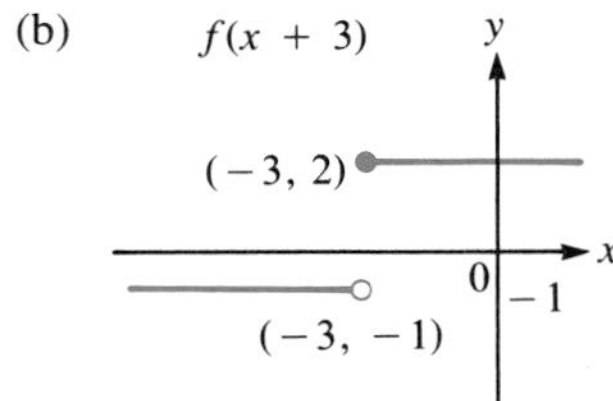

(c)

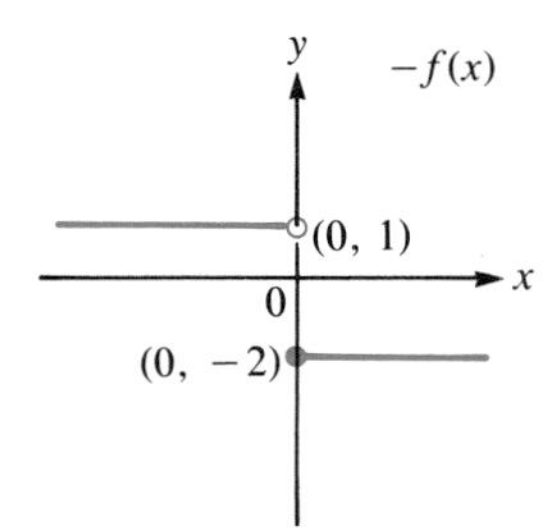

(d)

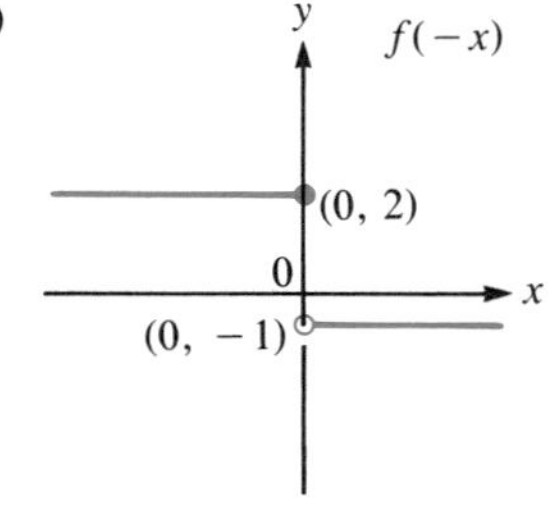

(e)

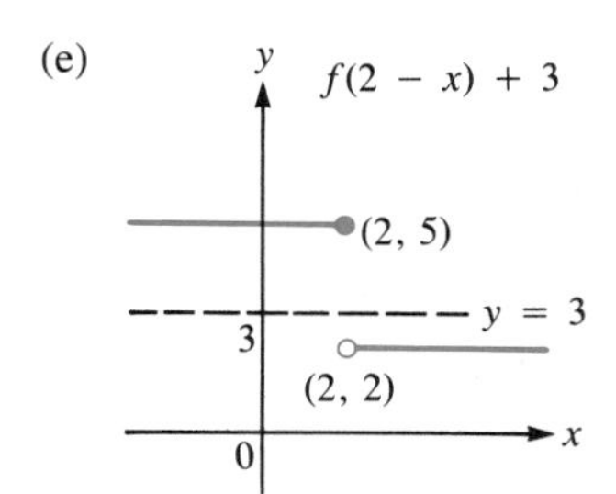

51. (a)

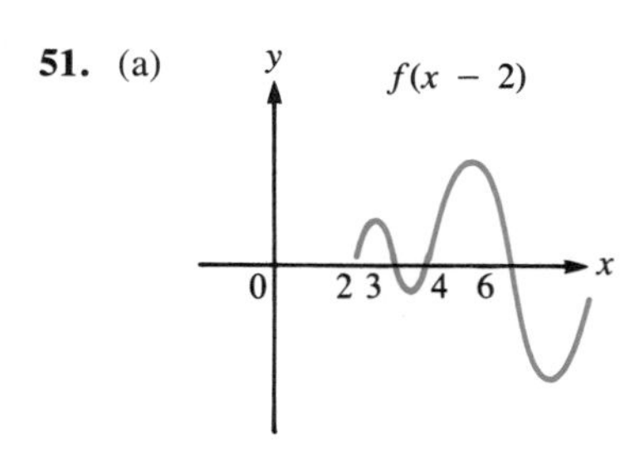

(b)

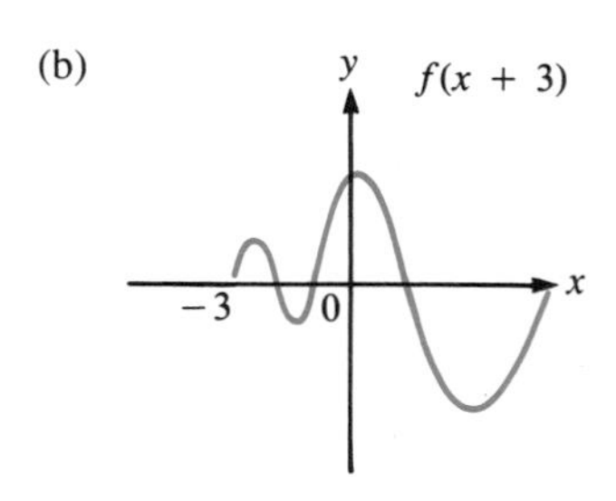

(c)

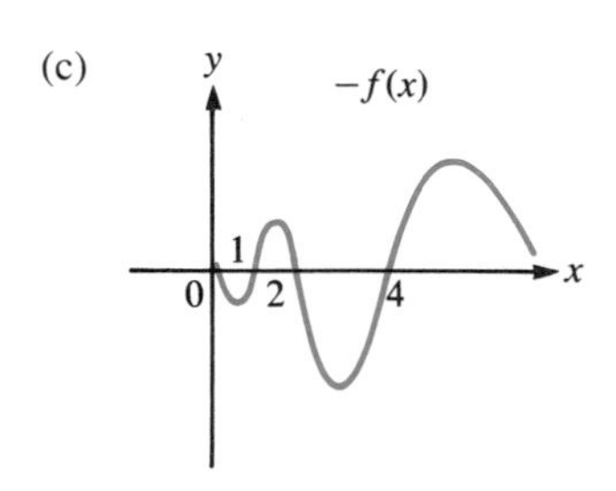

(d)

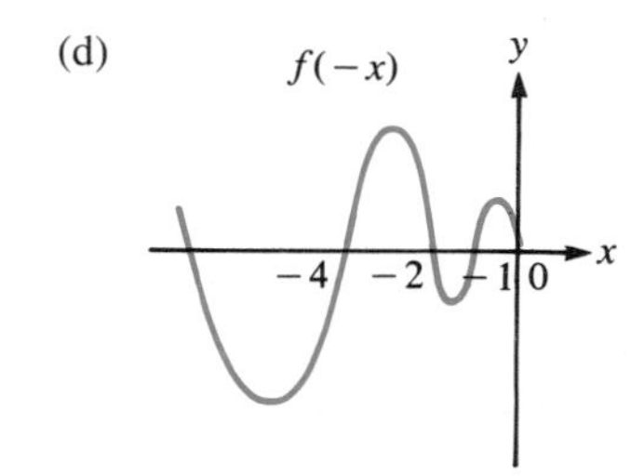

(e)

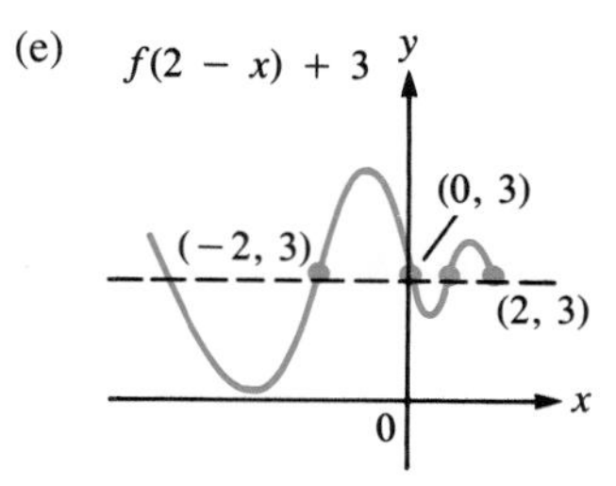

53. (a)

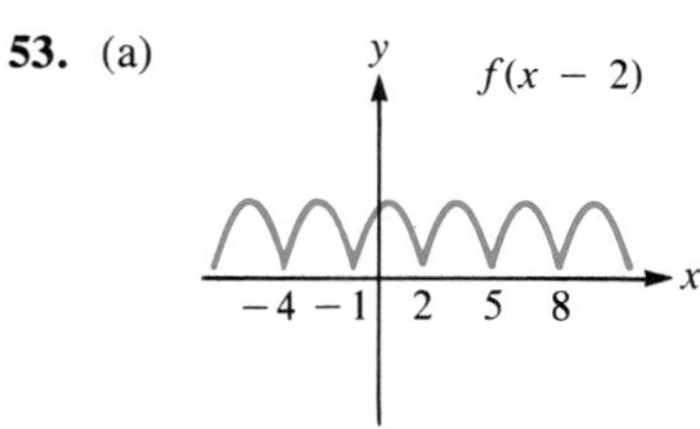

(b)

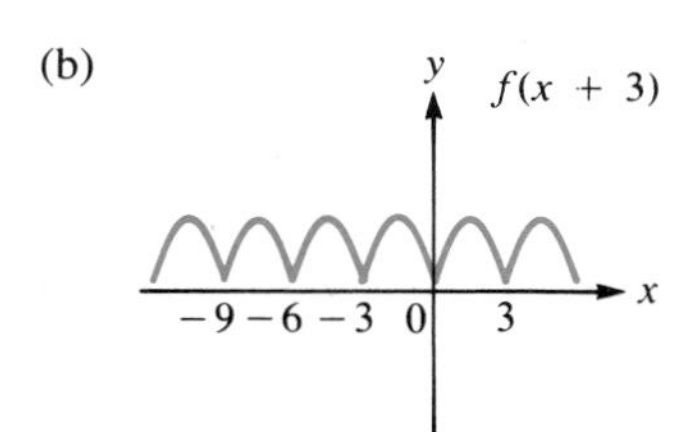

(c)

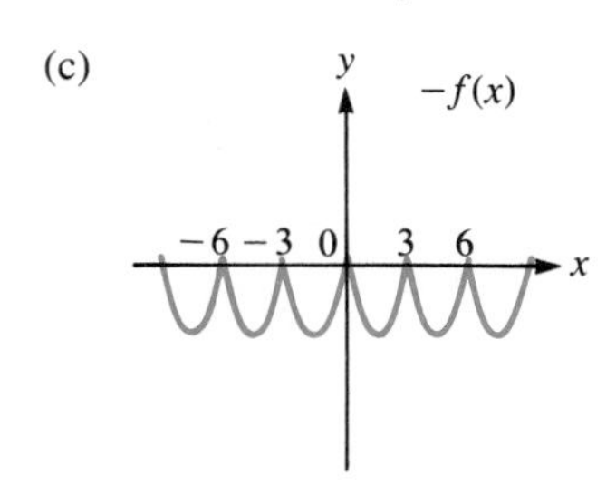

(d)

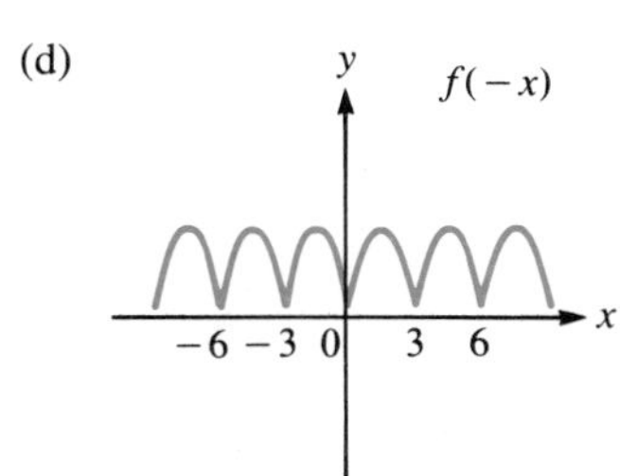

(e)

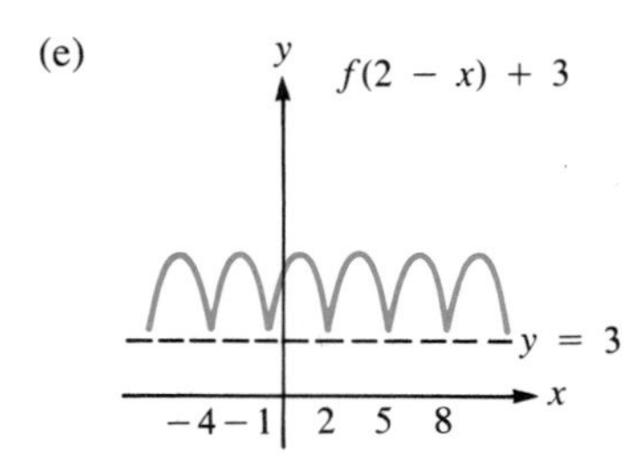

55.

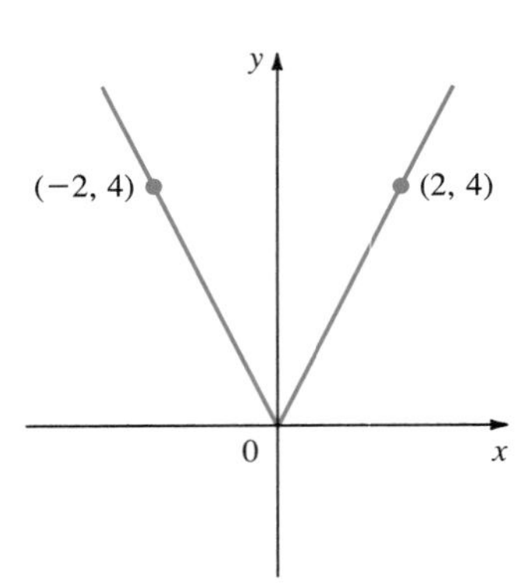

57.

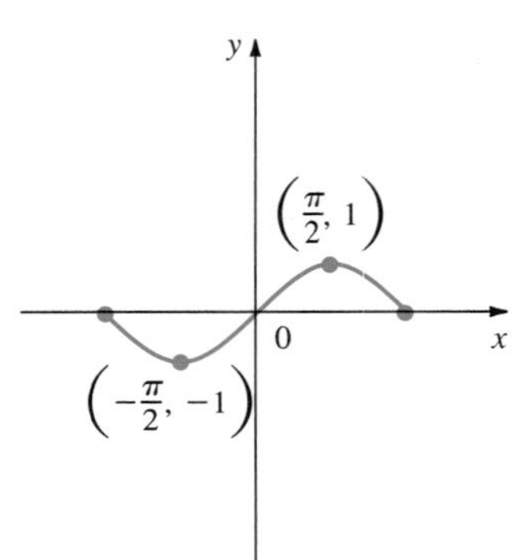

59.

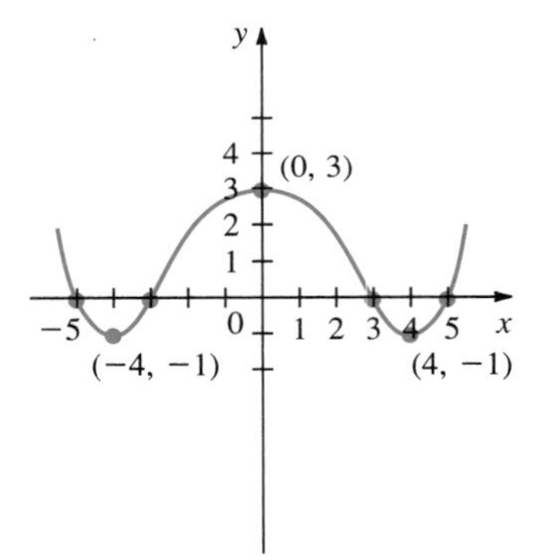

61.

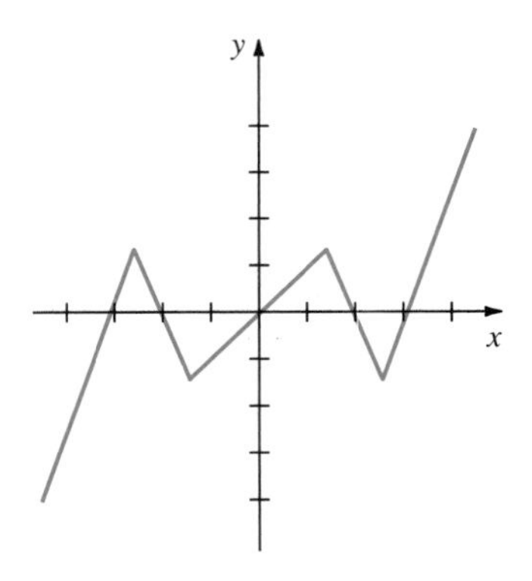

63. $y = \sqrt[3]{x} \quad -9 \le x \le 9,\ -2.5 \le y \le 2.5$

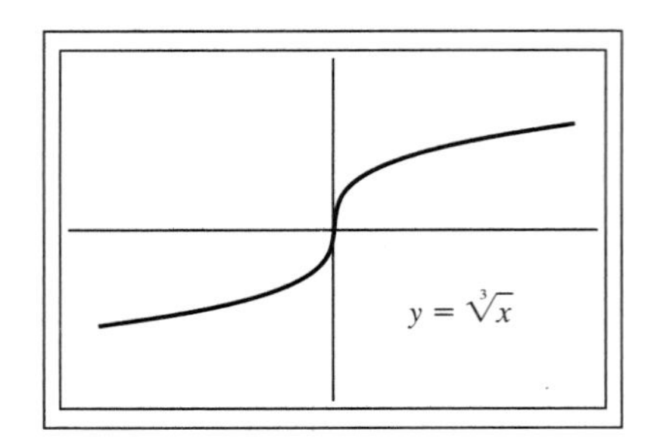

65.

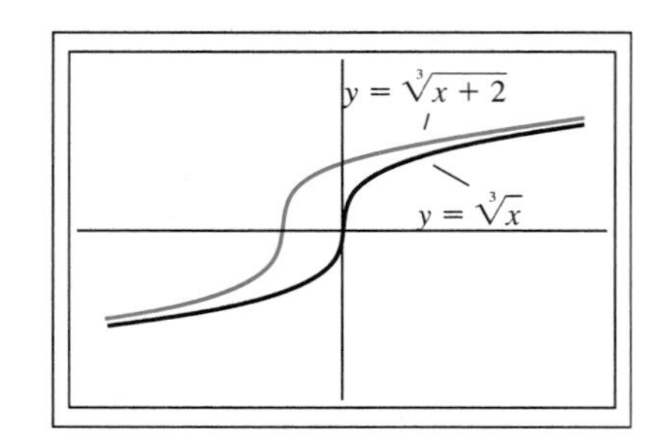

67.

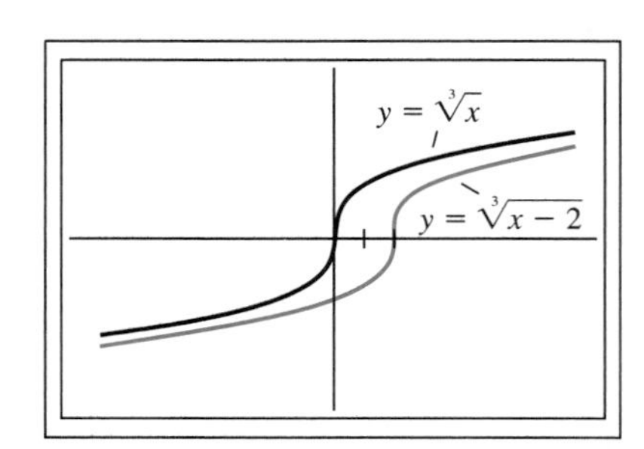

69.

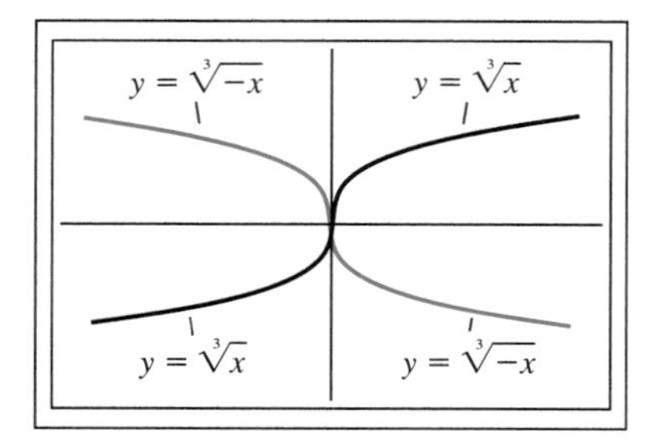

71.

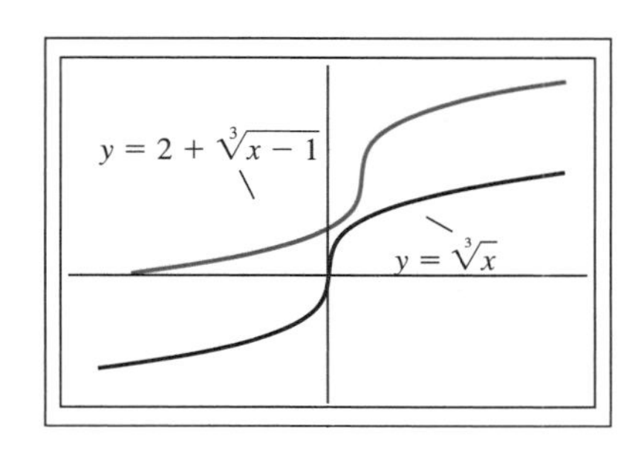

73. $y = -(x-2)^3 + 2(x-2)^2 + 5(x-2) - 6$
$-2 \le x \le 6,\ -18 \le y \le 70$

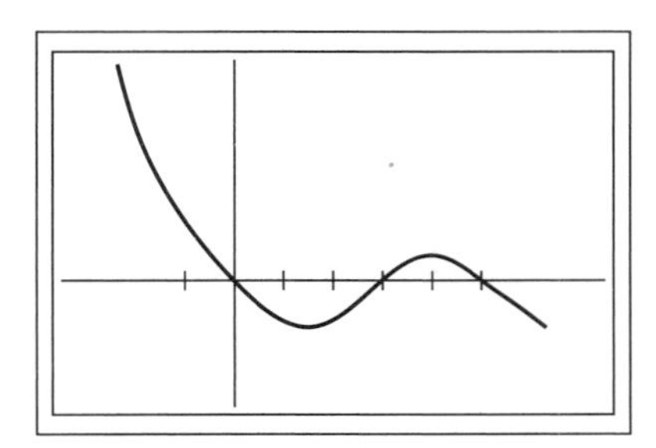

75. $y = -x^3 + 2x^2 + 5x - 5 \qquad -4 \le x \le 4,\ -17 \le y \le 71$

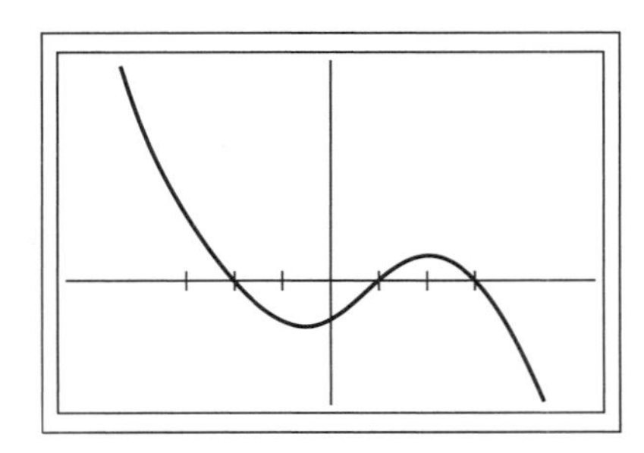

77. $y = x^3 - 2x^2 - 5x + 6 \qquad -4 \le x \le 4,\ -70 \le y \le 18$

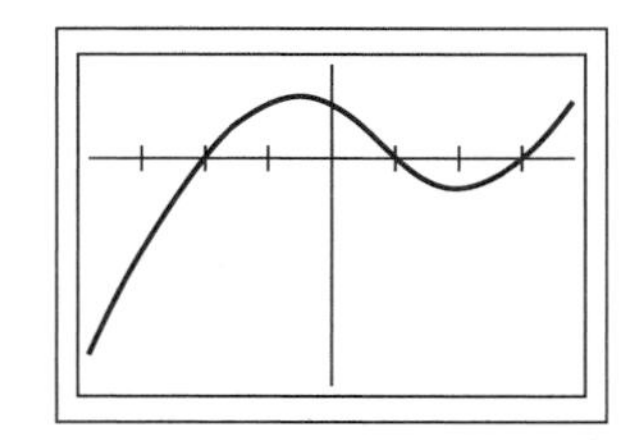

79. $y = -(2-x)^3 + 2(2-x)^2 + 5(2-x) - 6$
$-2 \le x \le 6,\ -18 \le y \le 70$

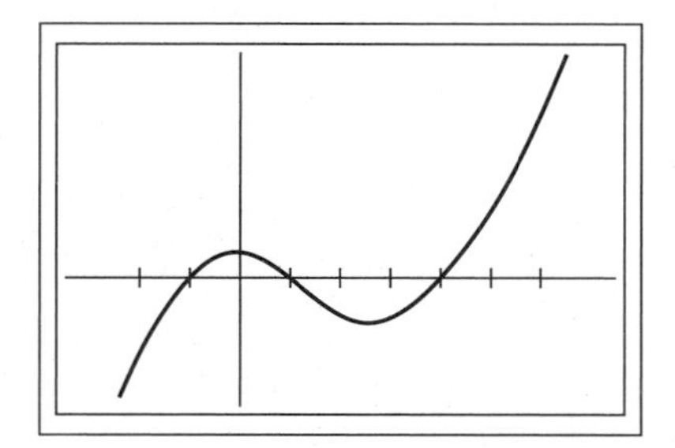

81. $y = -(1+x)^3 + 2(1+x)^2 + 5(1+x) - 10$
$-5 \le x \le 3,\ -22 \le y \le 66$

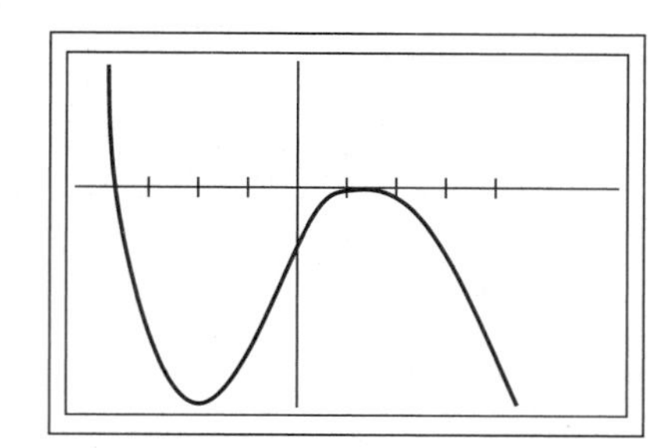

Problems 3.6, page 204

1. $f + g$: $-x + 3$; domain: all real numbers
$f - g$: $5x + 3$; domain: all real numbers
$f \cdot g$: $-6x^2 - 9x$; domain: all real numbers
$\frac{f}{g}$: $-\frac{2x+3}{3x}$; domain: all real numbers except 0

3. $f + g$: 14; domain: all real numbers
$f - g$: -6; domain: all real numbers
fg: 40; domain: all real numbers
f/g: 0.4; domain: all real numbers

5. $f + g$: $x + 1/x$; domain: all real numbers except 0
$f - g$: $x - 1/x$; domain: all real numbers except 0
fg: 1; domain: all real numbers except 0
f/g: x^2; domain: all real numbers except 0

7. $f + g$: $\sqrt{x+1} + \sqrt{1-x}$; domain: $[-1, 1]$
$f - g$: $\sqrt{x+1} - \sqrt{1-x}$; domain: $[-1, 1]$
$f \cdot g$: $\sqrt{1-x^2}$; domain: $[-1, 1]$
$\frac{f}{g}$: $\frac{\sqrt{1-x^2}}{1-x}$; domain $[-1, 1)$

9. $f + g$: $x^5 + 2 - |x|$; domain: all real numbers
$f - g$: $x^5 + |x|$; domain: all real numbers
fg: $x^5 - x^5|x| - |x| + 1$; domain: all real numbers
f/g: $(x^5 + 1)/(1 - |x|)$; domain: all real numbers except 1 and -1

11. $f+g$: $\sqrt[5]{x+2}+\sqrt[4]{x-3}$; domain: $[3, \infty)$
$f-g$: $\sqrt[5]{x+2}-\sqrt[4]{x-3}$; domain: $[3, \infty)$
fg: $\sqrt[5]{x+2}\sqrt[4]{x-3}$; domain: $[3, \infty)$
f/g: $\sqrt[5]{x+2}/\sqrt[4]{x-3}$; domain: $(3, \infty)$
13. $f \circ g$: $3x-6$; domain: all real numbers
$g \circ f$: $3x-2$; domain: all real numbers
15. $f \circ g$: 5; domain: all real numbers
$g \circ f$: 8; domain: all real numbers
17. $f \circ g$: $1/2x$; domain: all real numbers except 0
$g \circ f$: $1/2x$; domain: all real numbers except 0
19. $f \circ g$: $2x+2$; domain: all real numbers
$g \circ f$: $2x-1$; domain: all real numbers
21. $f \circ g$: $\dfrac{x+1}{x-1}$; domain: all real numbers except 0, 1
$g \circ f$: $\dfrac{2}{x}$; domain: all real numbers except 0, 2
23. $f \circ g$: $\sqrt{1-\sqrt{x-1}}$; domain: $[1, 2]$
$g \circ f$: $\sqrt{\sqrt{1-x}-1}$; domain: $(-\infty, 0]$
27. $g_1(x)=x-5$; $g_2(x)=5-x$
29. Domain of k is $[0, \infty)$; $h(x)=\sqrt{x}$, $g(x)=x+1$, $f(x)=x^{5/7}$
31. $ad+b=bc+d$
33. (a) $C(p)=168{,}000-3200p$
(b) $R(p)=20{,}000p-400p^2$
(c) $P(p)=-400p^2+23{,}200p-168{,}000$
(d) \$29.00; \$168,400

53.

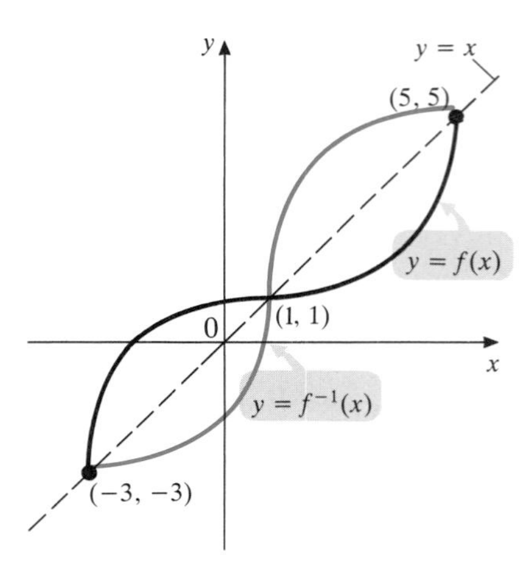

55.

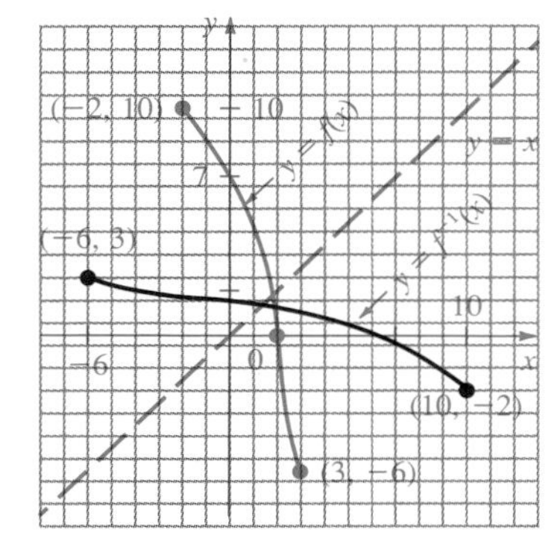

Problems 3.7, page 211

1. $\dfrac{x+1}{2}=\dfrac{1}{2}x+\dfrac{1}{2}$ **3.** $\dfrac{12x+3}{8}=\dfrac{3}{2}x+\dfrac{3}{8}$
5. $-\frac{11}{2}x+\frac{3}{2}$ **7.** $3/x$ **9.** $\dfrac{1-x}{x}=\dfrac{1}{x}-1$
11. $4-3/x$ **13.** $\sqrt[3]{4/x-2}$ **15.** x^2-2; $x \geq 0$
17. $(1-x^2)/2$, $x \geq 0$ **19.** $1/x^3+7$
21. $x/(1-x)$ **23.** $\sqrt[3]{x-3}$
25. $-\sqrt{x-1}$ on $(-\infty, 0]$; $\sqrt{x-1}$ on $[0, \infty)$
27. $4-\sqrt{x}$ on $(-\infty, 4]$; $4+\sqrt{x}$ on $[4, \infty)$
29. $-\sqrt[4]{x}$ on $(-\infty, 0]$; $\sqrt[4]{x}$ on $[0, \infty)$
31. $-\sqrt[4]{x+5}$ on $(-\infty, 0]$; $\sqrt[4]{x+5}$ on $[0, \infty)$
33. $\frac{7}{2}-\sqrt{x+\frac{25}{4}}$ on $(-\infty, \frac{7}{2}]$; $\frac{7}{2}+\sqrt{x+\frac{25}{4}}$ on $[\frac{7}{2}, \infty)$
35. $-\sqrt{-2+\sqrt{x}}$ on $(-\infty, 0]$; $\sqrt{-2+\sqrt{x}}$ on $[0, \infty]$
37. $-x$ on $(-\infty, 0]$; x on $[0, \infty)$ **43.** yes **45.** no
47. no **49.** yes **51.** yes

Problems 3.8, page 216

1. $y=10x$ **3.** $z=3.3/w$ **5.** $x=kz^3$
7. $w=k\sqrt{z}$ **9.** $y=k/v^{1/4}$ **11.** $R=kx^3\sqrt{z}$
13. $w=ku^2/v^3$ **15.** (a) $C=0.002q^2$ (b) \$80
17. (a) $S=4\pi r^2$ (b) 36π in^2
19. is multiplied by 27 **21.** quadruples
23. 87.91 days (the true value is 87.96 days)
25. increases to 75 lb/in^2
27. (a) 625 ft-candles (b) 25 ft-candles
29. $50\sqrt{5}$ ft/sec $\approx$ 111.8 ft/sec or 76 mph
31. is multiplied by $\frac{4}{3}$ **33.** nothing

Review Exercises for Chapter 3, page 219

1. $\sqrt{10}$ **3.** $2\sqrt{2}$ **5.** $\left(-2, \dfrac{9}{2}\right)$
7. center $(-2, 4)$, radius 2
9. (a) 5 (b) x-intercept = y-intercept = 2

(c)

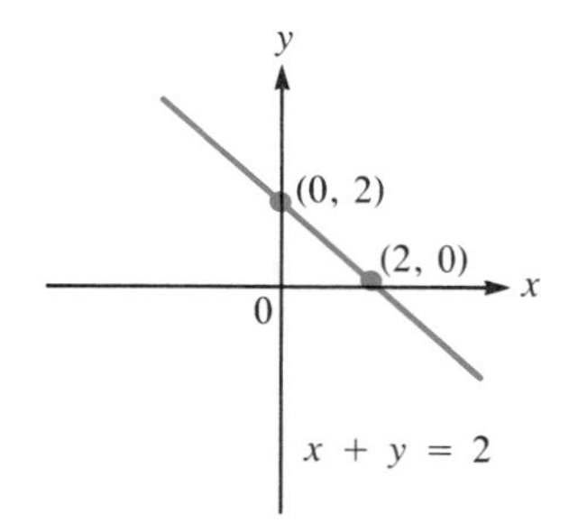

11. (a) $\frac{21}{5}$ (b) x-intercept 5, y-intercept 3
(c)

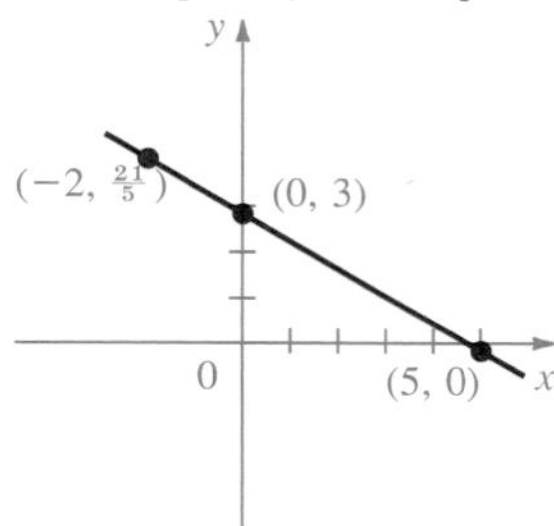

In Exercises 13–17 the equations are given in the order (a) slope-intercept equation (b) standard equation, (c) point-slope equation.

13. (a) $y = \frac{2}{3}x + \frac{11}{3}$ (b) $2x - 3y = -11$
(c) $y - 5 = \frac{2}{3}(x - 2)$

15. $y = x - 6$; $x - y = 6$; $y + 2 = 1(x - 4)$

17. (a) $x = 1$ (vertical line) (b) $x = 1$ (c) $0 = x - 1$

19. $y = -\frac{2}{5}x - \frac{11}{5}$

21. The slope -0.04 indicates that the price decreases 4 cents for every increase of 1 in the total quantity sold

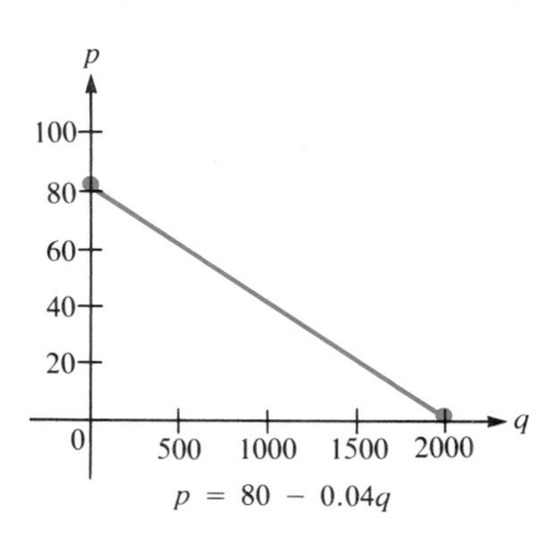

23. \$60 ($q = 500$)

25. yes; all real numbers; all real numbers

27. yes; all real numbers except 0; all real numbers except 0

29. yes; domain: $[-3, \infty)$; range: $[0, \infty)$

31. yes; all real numbers; $[-\frac{1}{2}, \frac{1}{2}]$

33. yes; domain: $\{x: x \le -\sqrt{6} \text{ or } x \ge \sqrt{6}\}$; range $[0, \infty)$

37. even **39.** even

41.

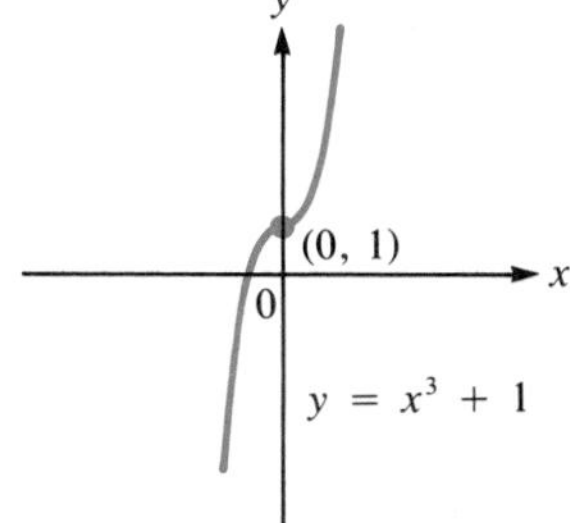

43.

(−2, 1) (0, 1) (−1, 0) 0

$y = (x + 1)^2$

45.

(0, 7) (2, 3) 0

$y = (x - 2)^2 + 3$

47.

(1, 5) 0 $(1 + \sqrt[3]{5}, 0)$

$y = 5 - (x - 1)^3$

49.

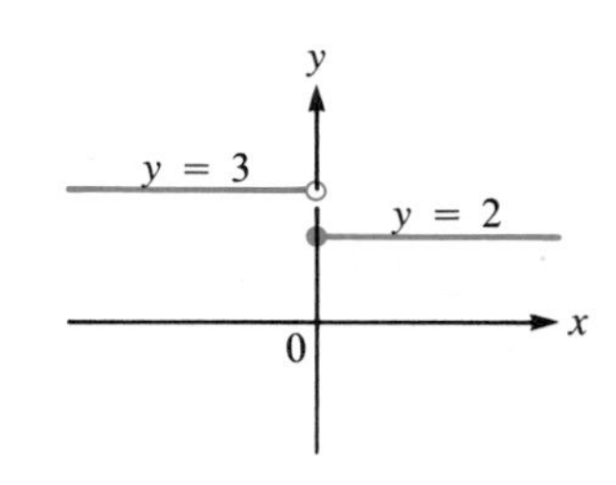

51. (a)

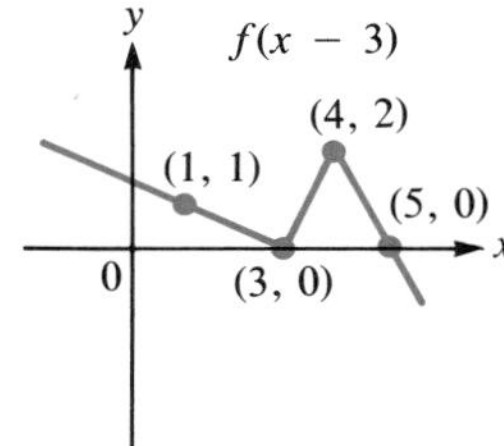

(b)

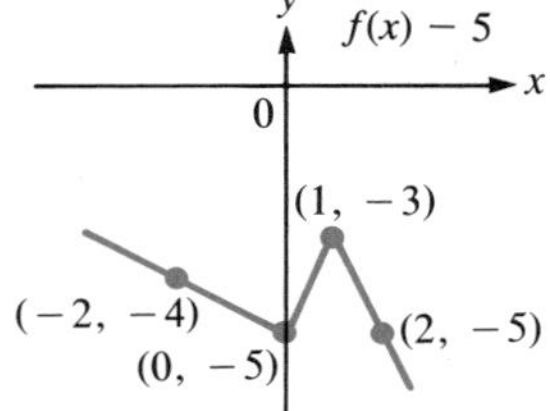

(c)

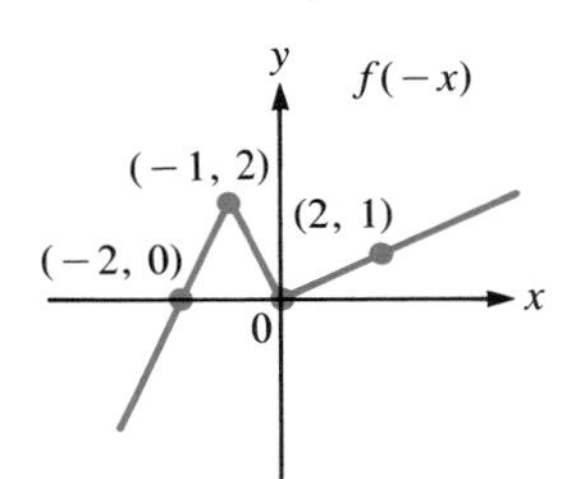

(d)

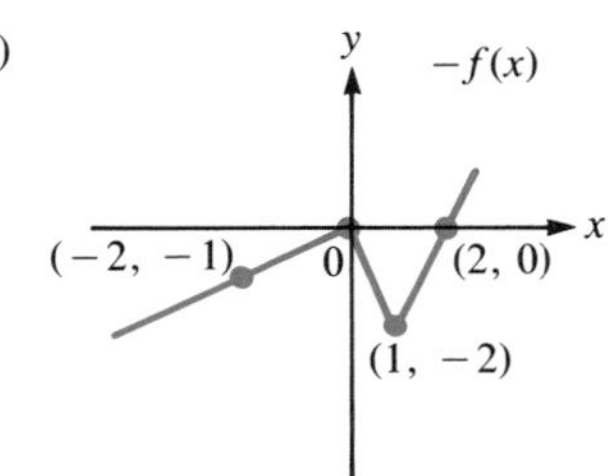

(e)

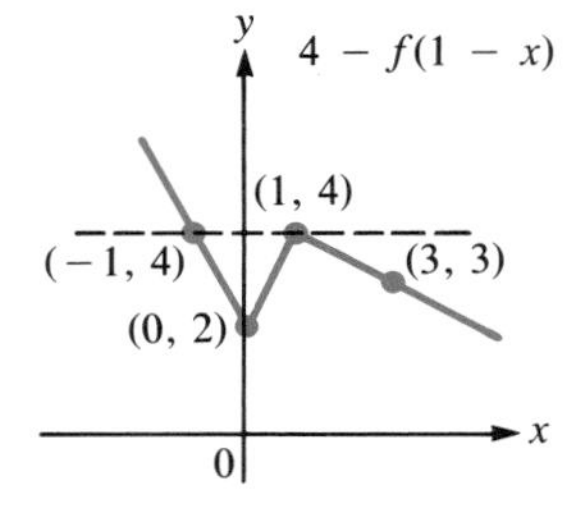

53. $f+g$: $\sqrt{x+1}+x^3$ domain: $[-1, \infty)$
$f-g$: $\sqrt{x+1}-x^3$ domain: $[-1, \infty)$
fg: $x^3\sqrt{x+1}$ domain: $[-1, \infty)$
g/f: $x^3/\sqrt{x+1}$ domain: $(-1, \infty)$
$f \circ g$: $\sqrt{x^3+1}$ domain: $[-1, \infty)$
$g \circ f$: $(x+1)^{3/2}$ domain: $[-1, \infty)$

55. $(x+1)/4$ **57.** $2/x$ **59.** $x^2+2,\ x \geq 0$

61. $f^{-1}(x) = -\sqrt{(x-3)/2}$ on $(-\infty, 0]$
$f^{-1}(x) = \sqrt{(x-3)/2}$ on $[0, \infty)$

63.

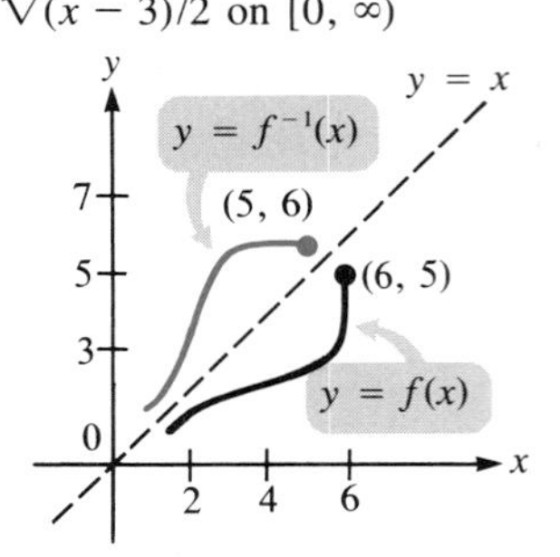

65. $v = k/u$ **67.** $P = ks/t^3$ **69.** 2.5 lb

Chapter 4

Problems 4.1, page 229

1.

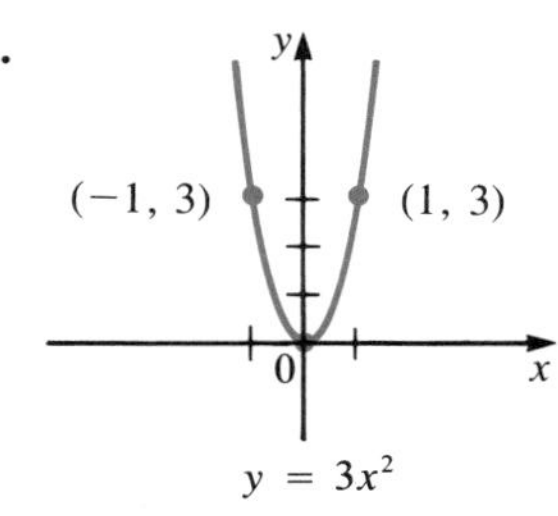

3.

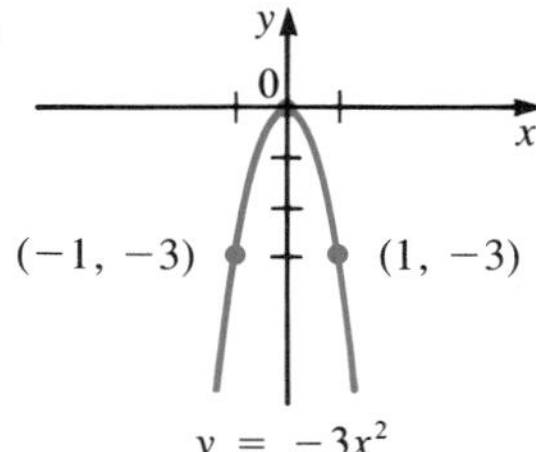

5.

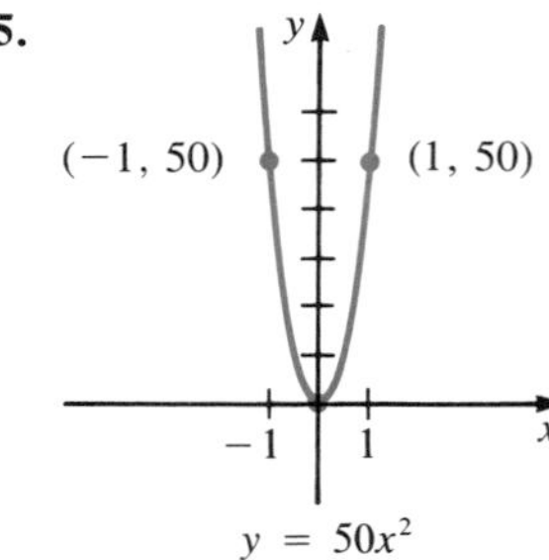

7. $y = (x+1)^2 - 1$; vertex $(-1, -1)$

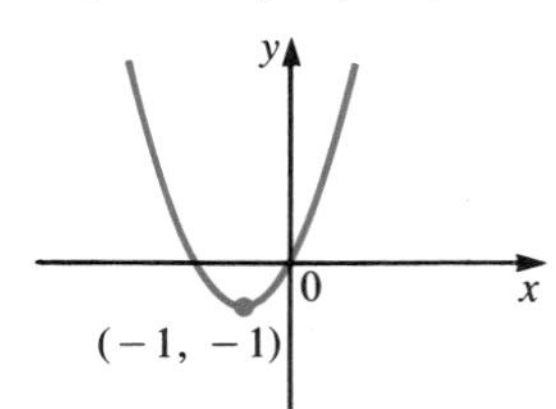

9. $y = \left(x + \frac{5}{2}\right)^2 - \frac{9}{4}$; vertex $\left(-\frac{5}{2}, -\frac{9}{4}\right)$

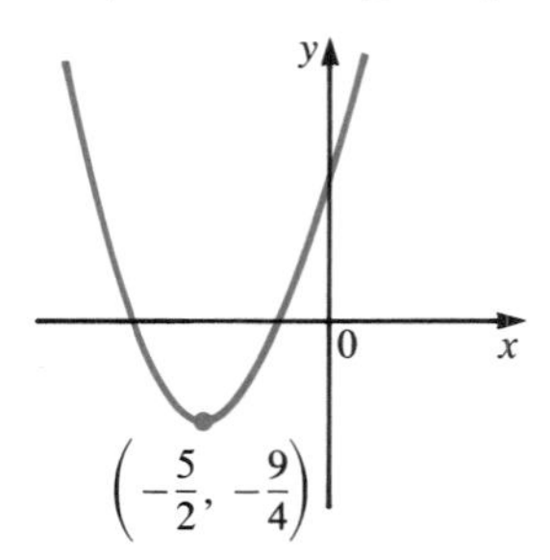

11. $y = 2(x+1)^2 - 7$; vertex $(-1, -7)$

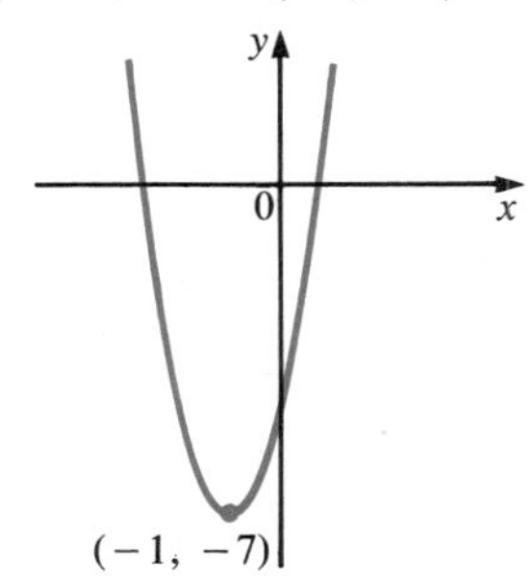

13. $y = -3\left(x - \frac{3}{2}\right)^2 + \frac{75}{4}$; vertex $\left(\frac{3}{2}, \frac{75}{4}\right)$

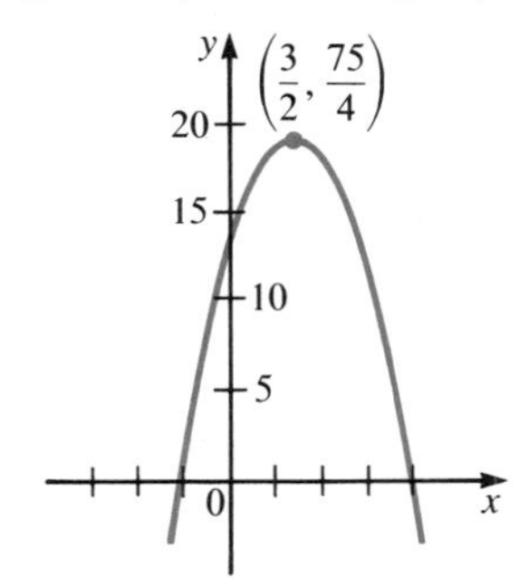

15. $y = -7\left(x - \frac{3}{14}\right)^2 - \frac{47}{28}$; vertex $\left(\frac{3}{14}, -\frac{47}{28}\right)$

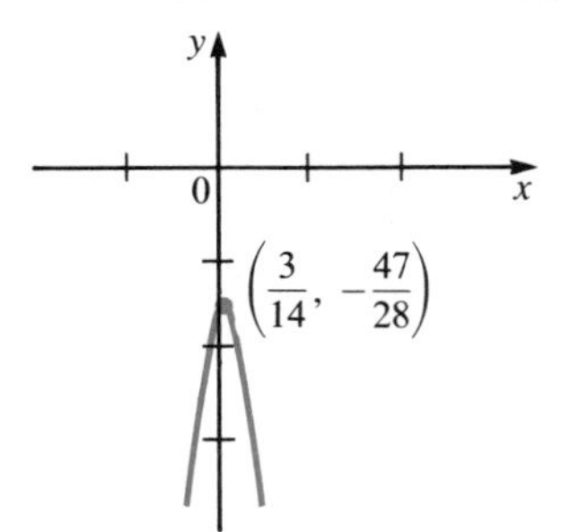

17. $y = \frac{1}{10}(x+1)^2 - \frac{11}{10}$; vertex $\left(-1, -\frac{11}{10}\right)$

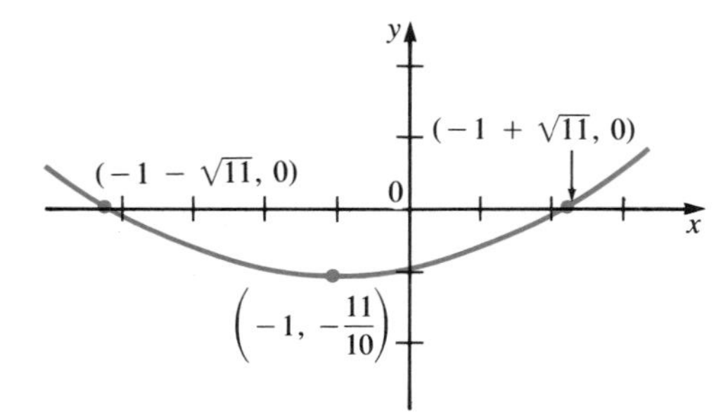

19. axis: $x = 0$
minimum: -4

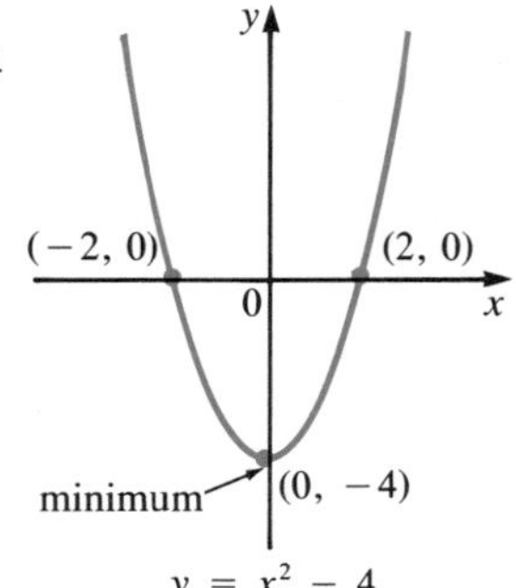

21. axis: $x = \frac{5}{2}$
minimum: $-\frac{9}{4}$

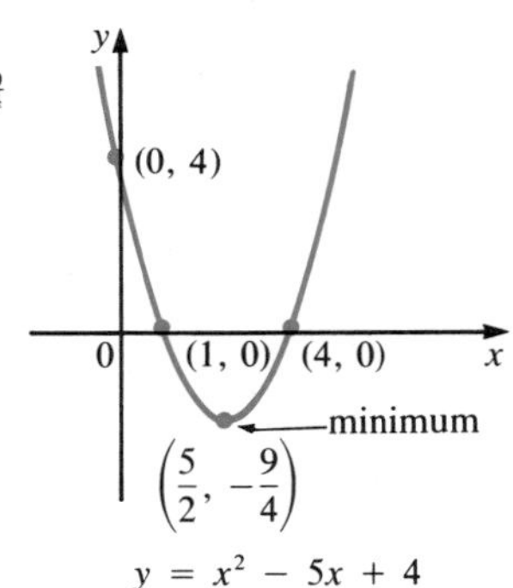

23. axis: $x = 0$
maximum: 1

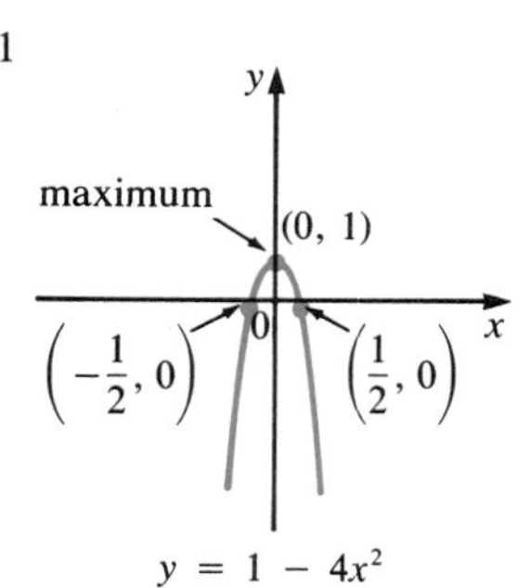

25. axis: $t = -1$
minimum: -8

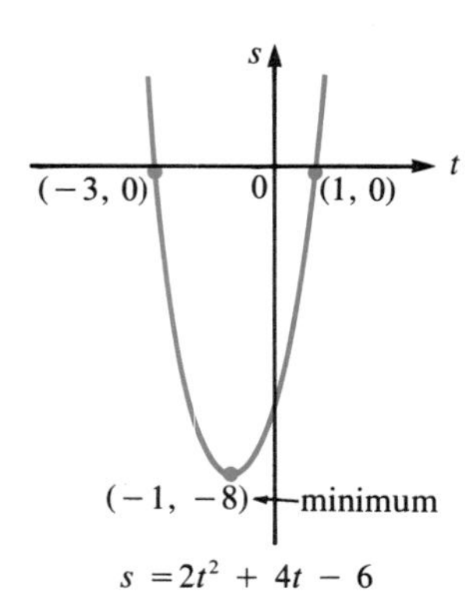

27. axis: $u = -1$
minimum: 0

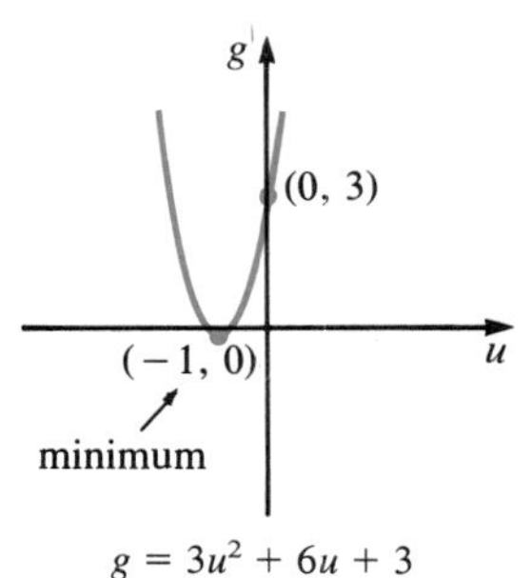

29. axis: $x = -2$
minimum: -2

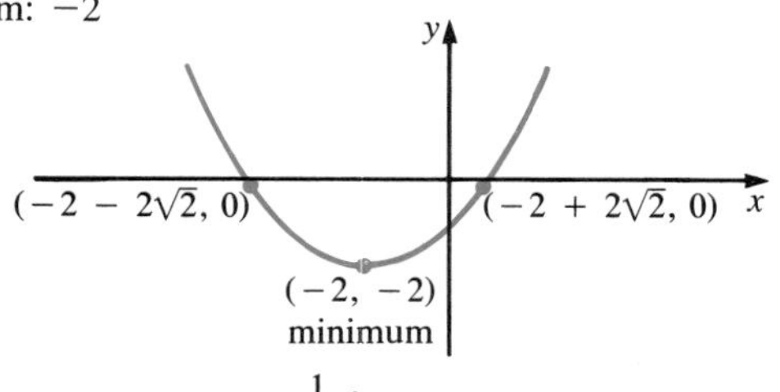

31. $y = \frac{1}{3}x^2 - \frac{8}{3}x + 4 = \frac{1}{3}(x - 4)^2 - \frac{4}{3}$
33. $y = -\frac{1}{3}x^2 + 3$
35. $y = -2x^2 + 6x = -2(x - \frac{3}{2})^2 + \frac{9}{2}$
37. (a) 20 (b) \$1125

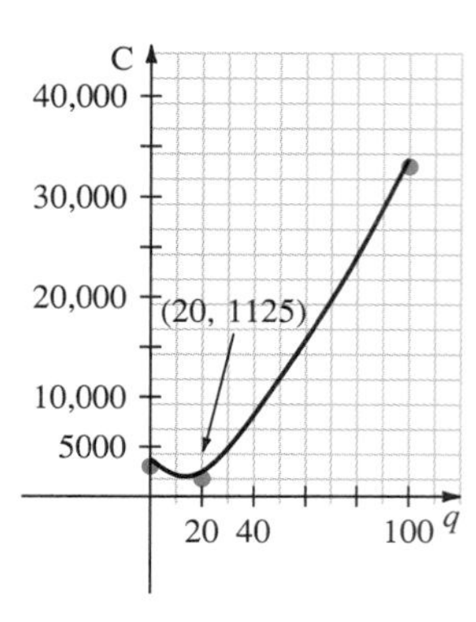

39. (a) 75 (b) \$3875

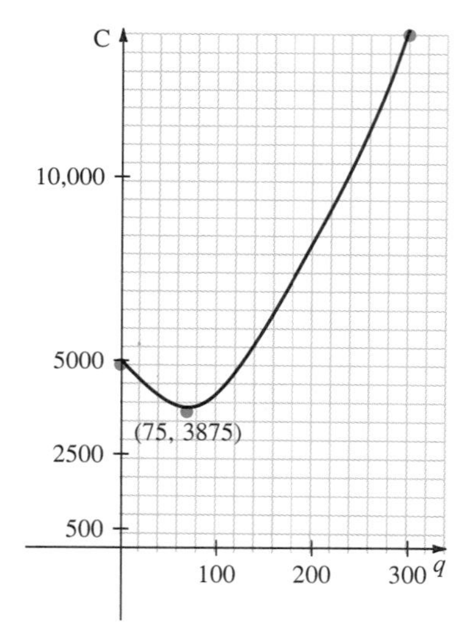

41. (a) 2.5 seconds (b) 25 feet
43. (a) 330 seconds = 5.5 min
(b) 435,600 feet = 82.5 miles
45. (a) $200/4.9 \approx 40.8$ seconds (b) 2040.8 m
47. 1275.51 m
49. Minimum is -31.36303797 when $x = 1.08/4.74 = 0.227848101$
51. Minimum is 0.00199975 when $x = 0.05$

53. $y = 5x^2$
$-5 \le x \le 5,\ -1 \le y \le 125$; one zero

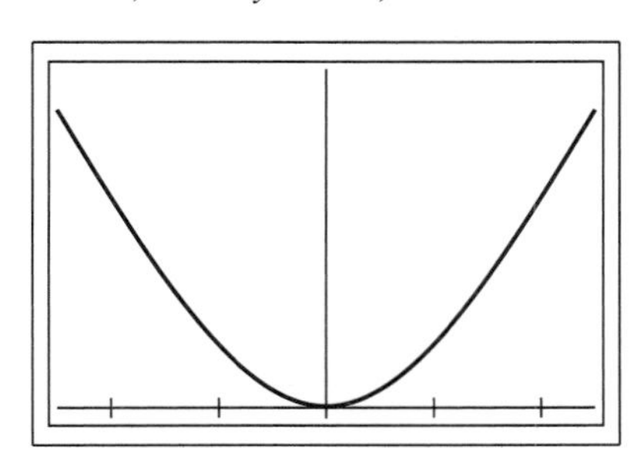

55. $y = 5x^2 + 8x + 9$
$-5 \le x \le 3,\ -1 \le y \le 80$; no real zeros

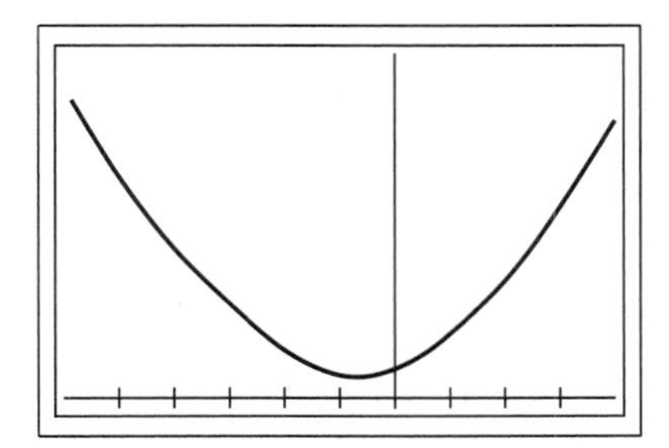

57. $y = 0.02x^2 - 0.036x - 0.071$
$-5 \le x \le 5,\ -0.2 \le y \le 0.2$; two real zeros

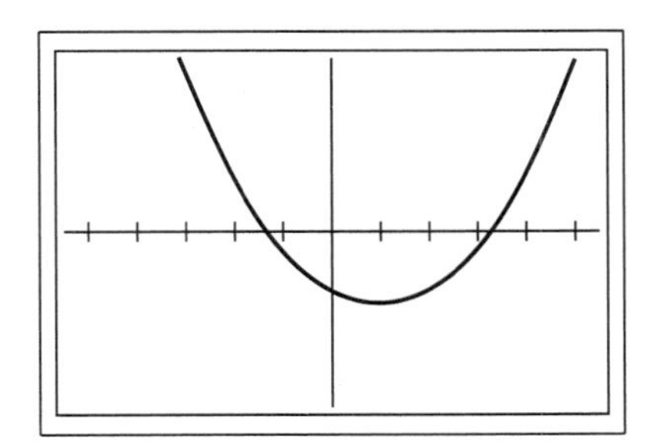

59. $y = -12.6x^2 - 8.9x + 47.5$
$-4 \le x \le 4,\ -50 \le y \le 50$; two real zeros

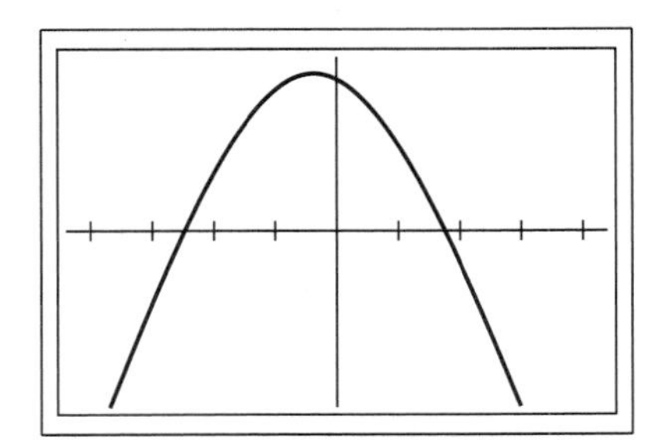

Problems 4.2, page 240

1.

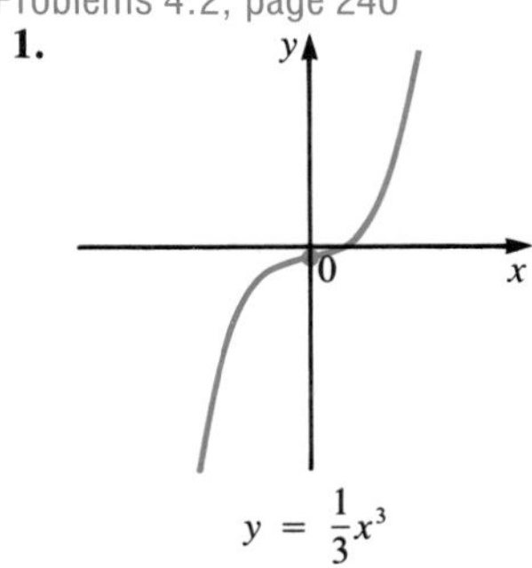

$y = \frac{1}{3}x^3$

3.

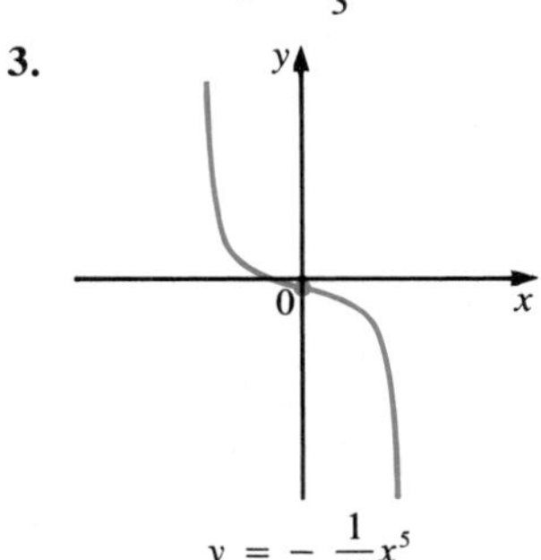

$y = -\frac{1}{10}x^5$

5.

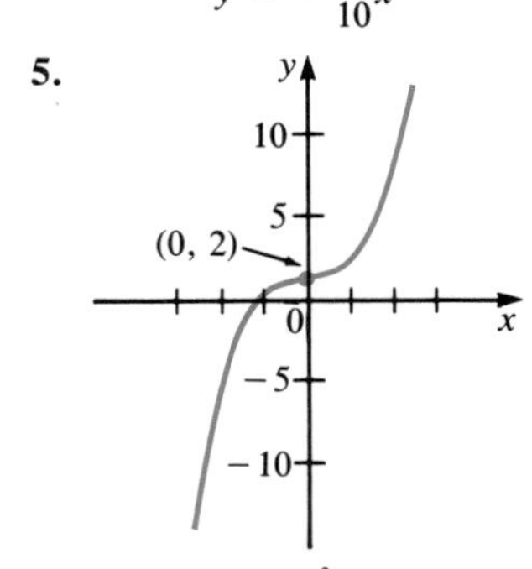

$y = x^3 + 2$

7.

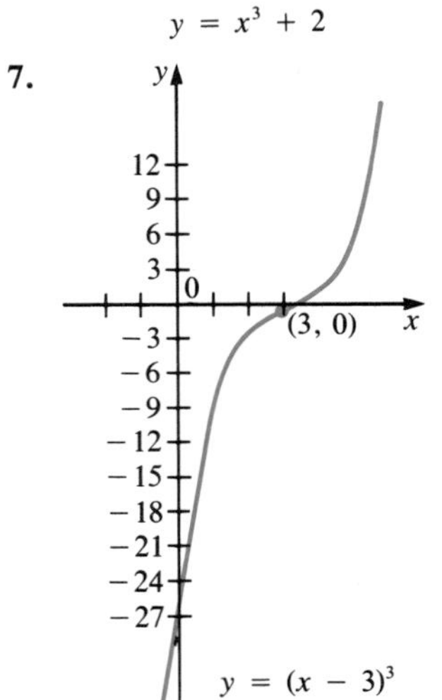

$y = (x - 3)^3$

9.

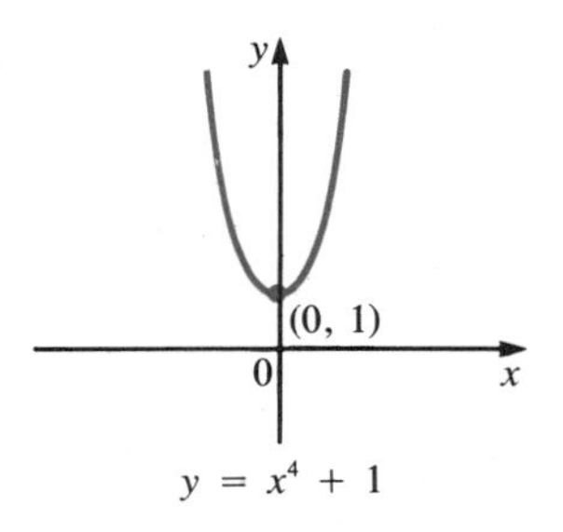

$y = x^4 + 1$

11.

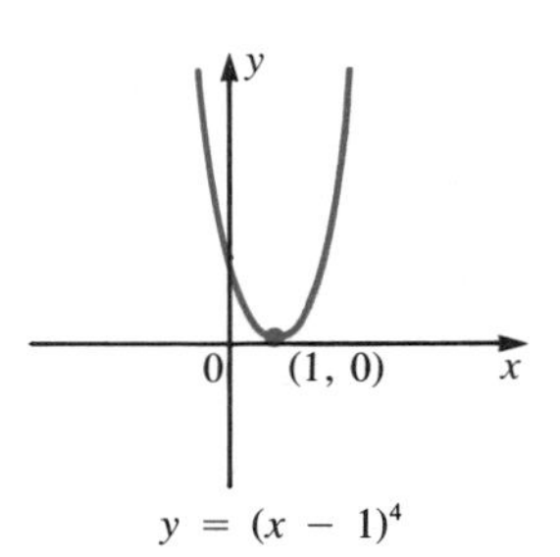

$y = (x - 1)^4$

13.

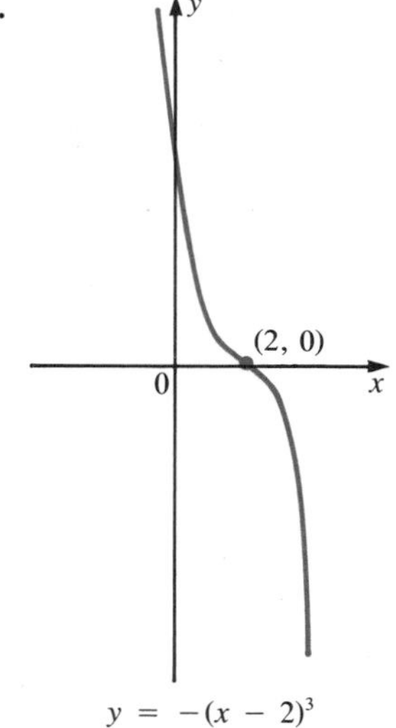

$y = -(x - 2)^3$

15.

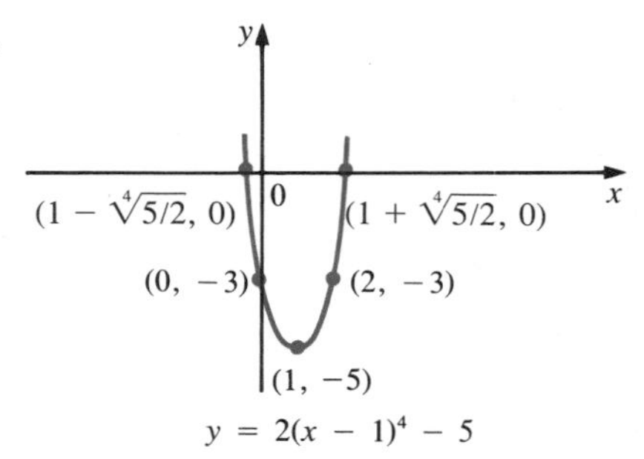

$y = 2(x - 1)^4 - 5$

17.

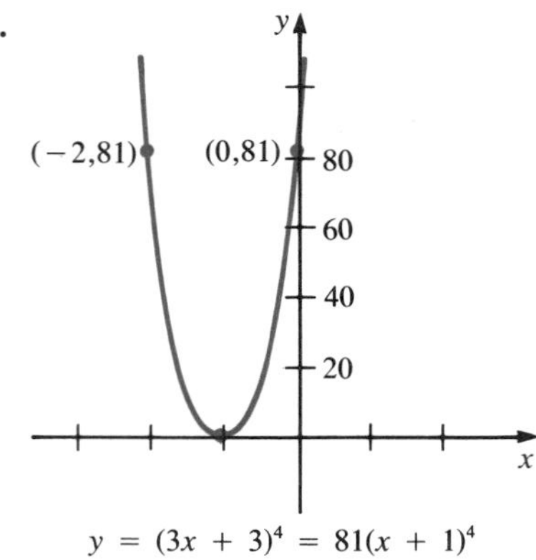

$y = (3x + 3)^4 = 81(x + 1)^4$

19. (a) 1, 2, 3 (b) positive $(1, 2)$ and $(3, \infty)$; negative $(-\infty, 1)$ and $(2, 3)$
(c)

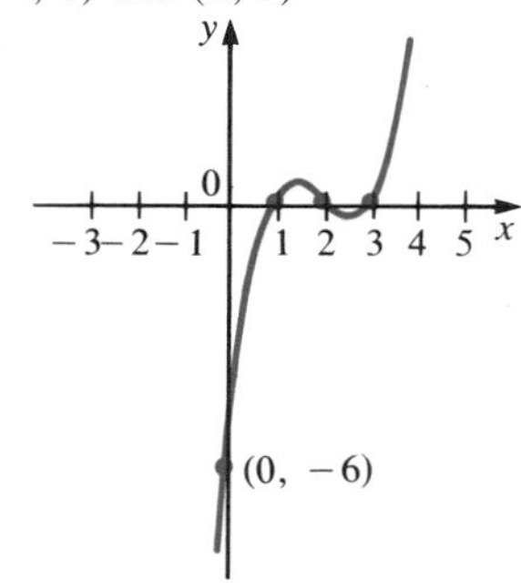

$y = (x - 1)(x - 2)(x - 3)$

21. (a) −5, 4, 6 (b) positive $(-5, 4)$ and $(6, \infty)$; negative $(-\infty, -5)$ and $(4, 6)$
(c)

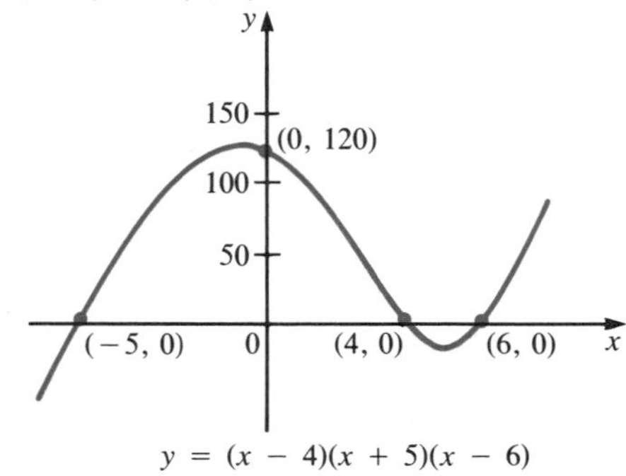

$y = (x - 4)(x + 5)(x - 6)$

23. (a) −6, −5, −2, 2 (b) positive $(-6, -5)$ and $(-2, 2)$; negative $(-\infty, -6)$, $(-5, -2)$ and $(2, \infty)$

(c)

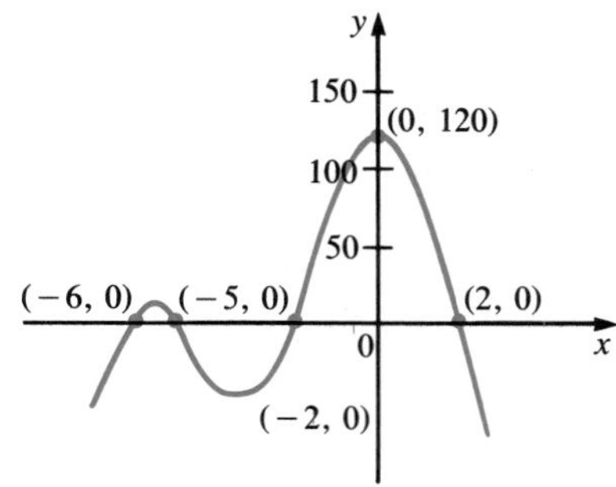

$y = -(x + 2)(x + 6)(x - 2)(x + 5)$

25. (a) −2, −1, 1, 2 (b) positive $(-\infty, -2)$, $(-1, 1)$ and $(2, \infty)$; negative $(-2, -1)$ and $(1, 2)$
(c)

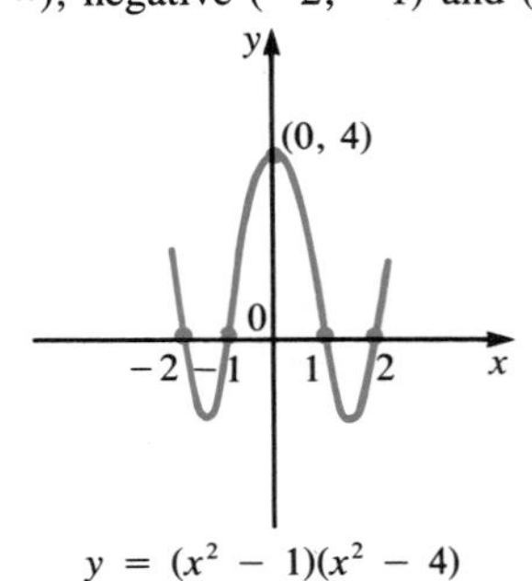

$y = (x^2 - 1)(x^2 - 4)$

27. (a) 0, 9 (b) positive $(9, \infty)$; negative $(-\infty, 0)$ and $(0, 9)$
(c)

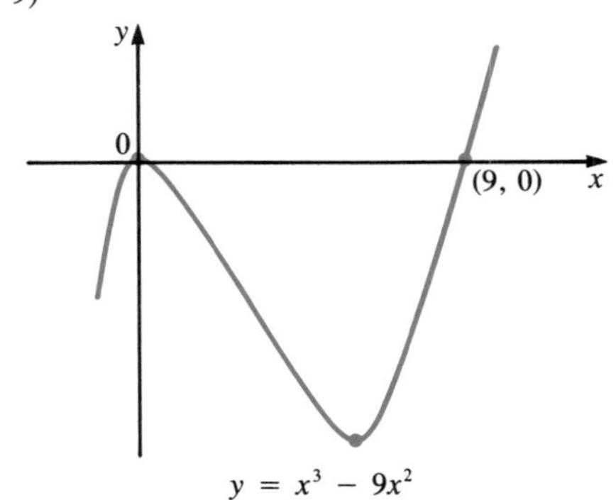

$y = x^3 - 9x^2$

29. (a) −4, −2, 0 (b) positive $(-4, -2)$ and $(0, \infty)$; negative $(-\infty, -4)$ and $(-2, 0)$
(c)

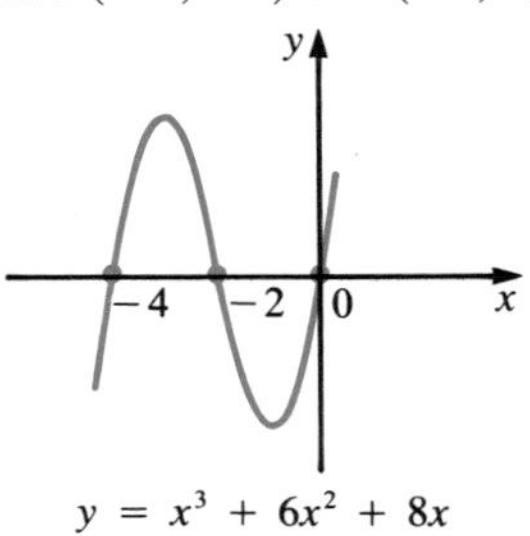

$y = x^3 + 6x^2 + 8x$

31. (a) 0, 1, 2 (b) positive $(-\infty, 0)$, $(0, 1)$ and $(2, \infty)$; negative $(1, 2)$
(c)

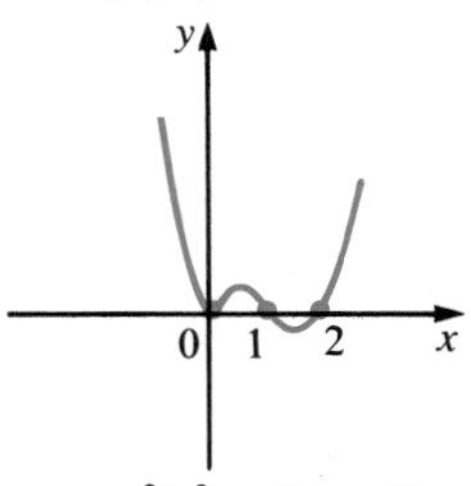

$y = x^2(x^2 - 3x + 2)$

33. $x(x^2 + 1)$

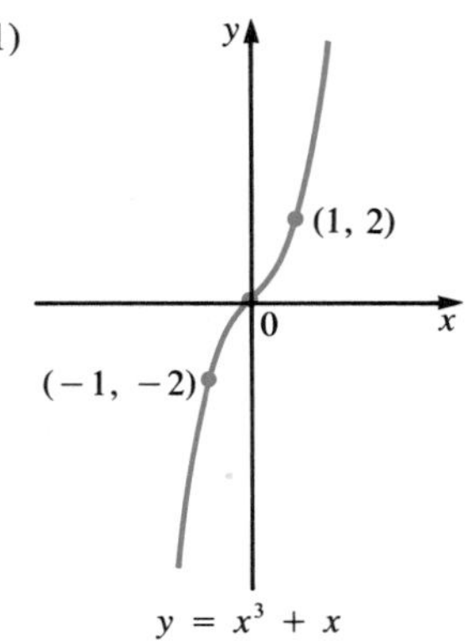

$y = x^3 + x$

35. $x(x^2 + x + 5)$

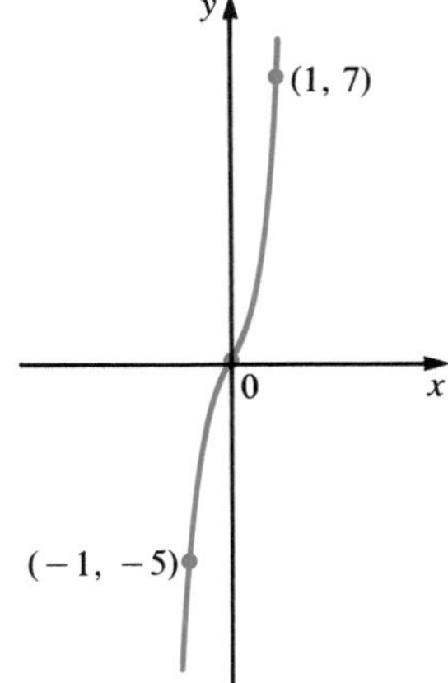

$y = x^3 + x^2 + 5x$

37. $(x^2 + 1)(x^2 + 2)$

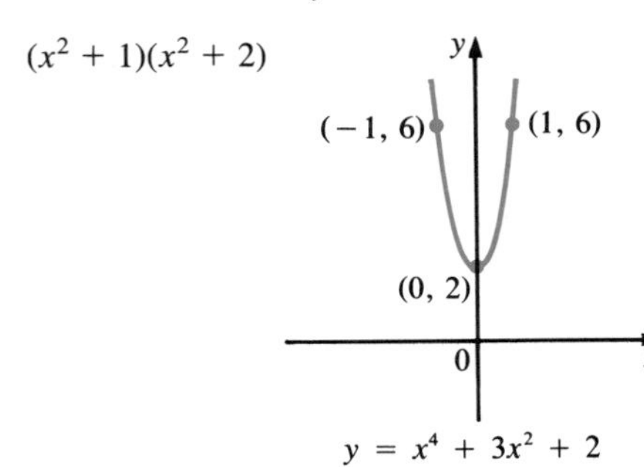

$y = x^4 + 3x^2 + 2$

39. $(x + 1)(x^2 - x + 1)$

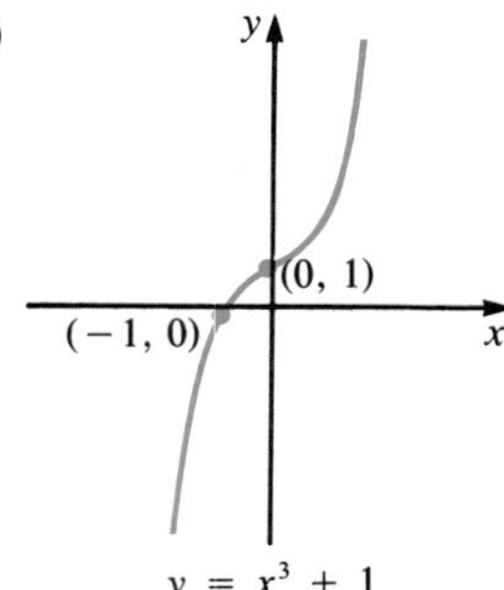

41. 1.42 **43.** 1.34 **45.** $x^2 - 3x - 28$
47. $x^3 - 3x^2 + 2x$ **49.** $x^4 - 2x^3 - 3x^2 + 2x + 2$
51. $x^6 - 10x^4 + 31x^2 - 30$ **53.** $(x - k)^n$
55. $-\frac{1}{4}x^3 - \frac{1}{4}x^2 + \frac{3}{2}x$ **57.** \$20.77
59. Positive for $x > \sqrt{\frac{3}{5}}$ and in $(-\sqrt{\frac{3}{5}}, 0)$
negative for $x < -\sqrt{\frac{3}{5}}$ and in $(0, \sqrt{\frac{3}{5}})$
61. $P_2(x) = \frac{3}{2}x^2 - \frac{1}{2}$

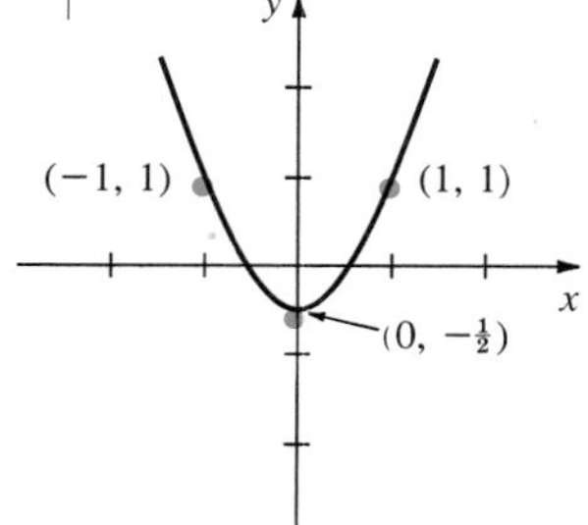

63. (a) $V(s) = s^2(s + 2) = s^3 + 2s^2,\ s \geq 0$
(b)

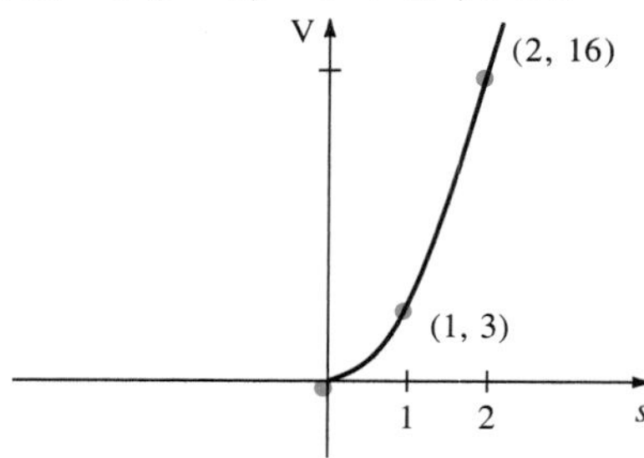

65. $y = x^3 + 2x^2 + x + 1$
$-3 \leq x \leq 2,\ -5 \leq y \leq 8$; one real zero

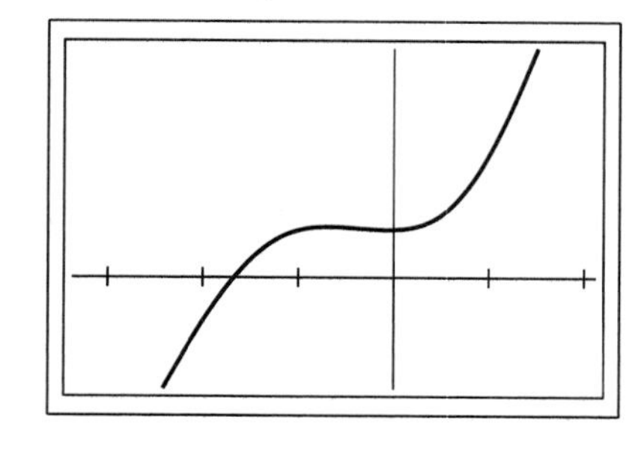

67. $y = -467x^3 - 506x^2 + 288x + 143$
$-2 \leq x \leq 2,\ -500 \leq y \leq 500$; one real zero

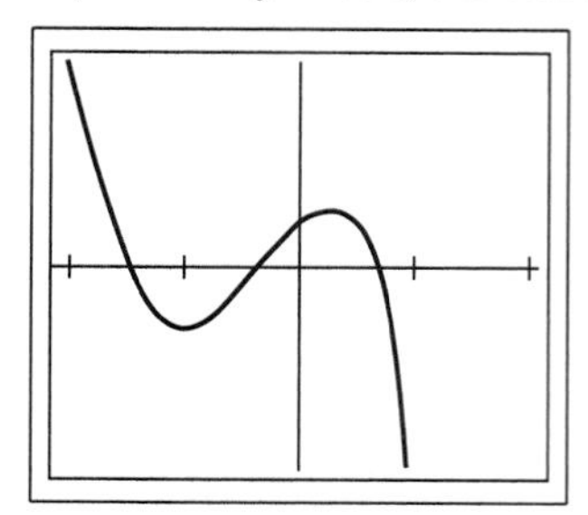

69. $y = x^4 + x^3 + x^2 + x + 1$
$-2 \leq x \leq 2,\ 0 \leq y \leq 10$; no real zero

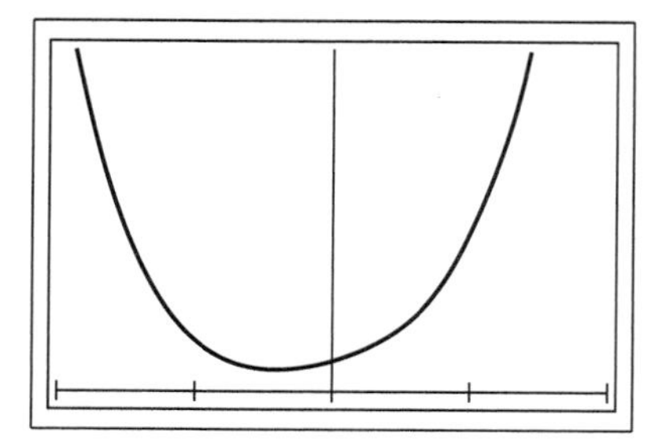

71. $y = -2x^4 + 3x^3 - x + 5$
$-2 \leq x \leq 3,\ -6 \leq y \leq 6$; two real zeros

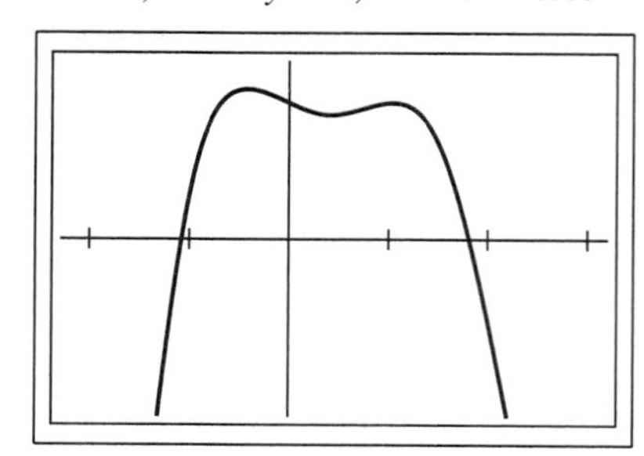

73. $y = -x^5 - 10x^4 + 6x^3 + 8x^2 - 12x + 1$
$-2 \leq x \leq 2,\ -50 \leq y \leq 10$; three real zeros (see second graph)

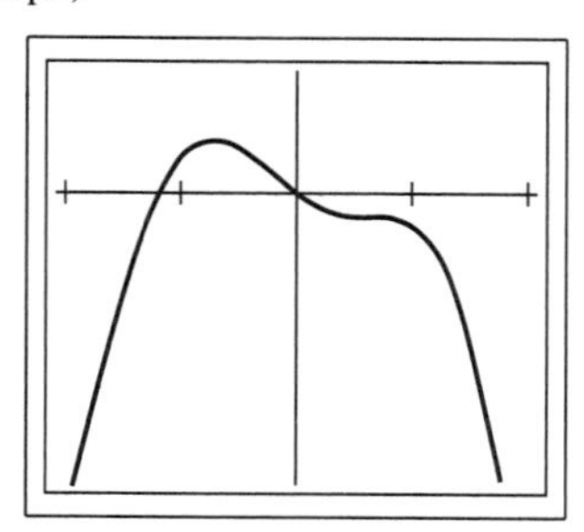

Graph showing behavior near the origin.

This graph doesn't show how the curve looks for x large in the negative direction

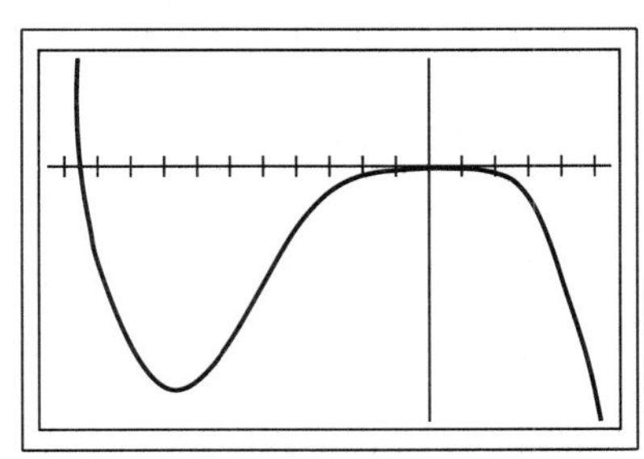

Second graph for $-11 \le x \le 6$, $-11{,}000 \le y \le 3000$

75. $y = 3x^5 - 50x^3 + 134x + 60$
$-4 \le x \le 5$, $-200 \le y \le 400$

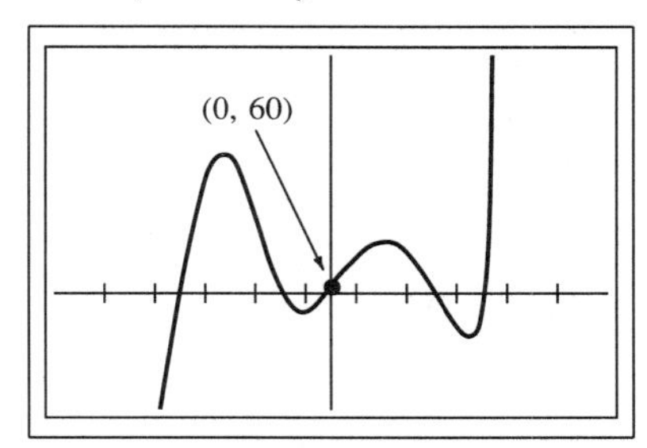

Problems 4.3, page 253

1. $q(x) = x - 2;\ r(x) = 0$
3. $q(x) = 3x + 4;\ r(x) = 0$
5. $q(x) = x + 3;\ r(x) = -7$
7. $q(x) = 4x + 8;\ r(x) = 80$
9. $q(x) = 2x;\ r(x) = -1$
11. $q(x) = x^2 + x + 4;\ r(x) = 0$
13. $q(x) = 2x^2 + 9x + 21;\ r(x) = 77$
15. $q(x) = x^2 - \frac{1}{2}x + \frac{3}{4};\ r(x) = \frac{9}{4}$
17. $q(x) = x;\ r(x) = x^2 - x$
19. $q(x) = 4x^2 + 2x - 17;\ r(x) = -28x + 89$
21. $q(x) = x - 2;\ r(x) = 0$
23. $q(x) = 5x - 29;\ r(x) = 148$
25. $q(x) = x^3 - 2x^2 + 4x - 8;\ r(x) = 32$
27. $q(x) = 2x^2 + x + 6;\ r(x) = 14$
29. $q(x) = 2x^4 + 2x^3 + 2x^2 + 2x + 2;\ r(x) = 1$
31. $q(x) = -3x^3 - \frac{1}{2}x^2 + \frac{1}{4}x - \frac{1}{8},\ r(x) = \frac{49}{16}$
33. $q(x) = x^4 + \frac{3}{2}x^3 + \frac{7}{4}x^2 + \frac{15}{8}x + \frac{31}{16};\ r(x) = \frac{63}{32}$
35. $q(x) = x^9 + x^8 + x^7 + x^6 + x^5 + x^4 + x^3 + x^2 + x + 1;$
$r(x) = 0$
37. $q(x) = x^3 - \frac{3}{4}x^2 + \frac{25}{16}x - \frac{75}{64};\ r(x) = -\frac{543}{256}$
39. $q(x) = -3x^3 + 9x^2 - 20x + 38;\ r(x) = -71$
41. $q(x) = x^{199} - x^{198} + x^{197} - x^{196} + \cdots + x^5 - x^4$
$+ x^3 - x^2 + x - 1;\ r(x) = 2$
43. 5 **45.** -156 **47.** 39 **49.** $\frac{15}{16}$
51. $2x^3 - 8x^2 + 2x + 12$ **53.** $-\frac{1}{8}x^3 + \frac{1}{4}x^2 - \frac{1}{8}x + \frac{1}{4}$
55. $-x^3 + x^2 + 6x$
57. $\frac{1}{12}x^4 + \frac{1}{6}x^3 - \frac{13}{12}x^2 - \frac{7}{6}x + 2 =$
$\frac{1}{12}(x - 1)(x + 2)(x - 3)(x + 4)$
59. $2x^3 - 5x^2 + x + 2 = 2(x - 1)(x - 2)(x + \frac{1}{2})$
61. -2.4475 **63.** 249.842936
65. -5.835997094 **67.** 49.928
69. 8.2283296×10^{37} **71.** -4
73. -7.779950048 **75.** -9 **77.** $-1 \pm \sqrt{5}$
79. $\frac{1}{2} \pm i$ **81.** $(5 \pm \sqrt{23}i)/4$ **83.** $1 \pm \sqrt{7}i$
85. $(1.45 \pm \sqrt{24.8085})/2 = 3.215406593,$
-1.765406593
87. $-4, 3$ **89.** $1, -\frac{5}{2}$ **91.** -6 **93.** 4
101. (a) \$52,000
(b) 487.2983346 rounds to 487 tons

Problems 4.4, page 266

1. -5 of 1, 4 of 2 **3.** -1 of 1, -2 of 1, 7 of 1
5. $\pm 2i$ of 5 **7.** ± 2 of 3, 6 of 5, -3 of 5
9. 0 positive, 2 or 0 negative
11. 0 positive, 1 negative
13. 3 or 1 positive, 0 negative
15. 0 positive, 1 or 3 negative
17. 0 positive, 1 negative
19. 2 or 0 positive, 3 or 1 negative
21. no changes of sign for $f(x)$ or $f(-x)$
23. $(x - 1)(x - 2)(x + 2);\ 1, 2, -2$
25. $(x - 1)(x + 1)(x - 2);\ 1, -1, 2$
27. $(2x + 1)(x^2 + 5x + 2);\ -\frac{1}{2}, (-5 \pm \sqrt{17})/2$
29. $(6x + 1)(x^2 + 2x + 5);\ -\frac{1}{6}$ is the only real zero
31.

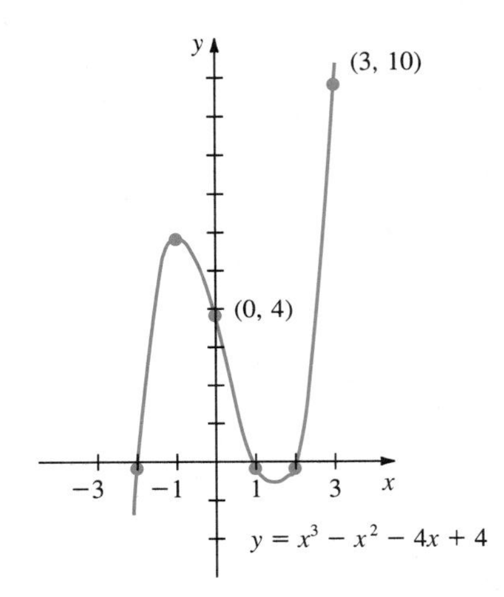

33.

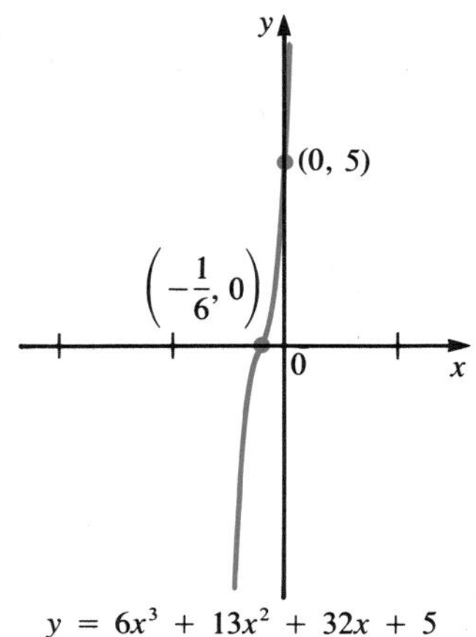

$y = 6x^3 + 13x^2 + 32x + 5$

35. $x^3 + 2x^2 + x + 2$ **37.** $x^4 + x^2$
39. $4x^4 + 8x^3 + 16x^2 - 8x - 20$
41. upper bound 1; lower bound -3; 1, -3
43. upper bound 2; lower bound -3; 2, -2
45. upper bound 4; lower bound -4; 1, -1, 3, -3 *Note:* The smallest upper and largest lower bounds are 3 and -3, respectively, but the upper and lower bound theorem ensures only that 4 and -4 are upper and lower bounds.
47. upper bound 1, lower bound -1; $\frac{1}{3}$; $-\frac{1}{3}$ [These are the best upper and lower bounds.]
49. upper bound 1, lower bound -2; no rational zeros
51. upper bound 2, lower bound -2; 0 (double zero), 1 (double zero), -2
53. upper bound 1, lower bound -2; $-\frac{1}{2}$, $\frac{1}{3}$ [These are the best upper and lower bounds]
71. $\sqrt[3]{\sqrt{17} + 4} - \sqrt[3]{\sqrt{17} - 4} - 1 \approx 0.5127453266$
73. -1.75 **75.** -1.37, -0.35, 0.64
77. no zeros **79.** -1.06, 1.78
81. -10.49, -1.24, 0.70
83. -3.72, -1.46, -0.49, 2.10, 3.57

Problems 4.5, page 270

1. -0.68 **3.** -0.40 **5.** 0.48
7. -2.09, 0.24, 7.85 **9.** 1.466
11. -1.265 **13.** 0.945
15. 7.97 is the only zero

Problems 4.6, page 274

1. 5 **3.** -0.33333333 or $-\frac{1}{3}$
5. $0.618033988 = (-1 + \sqrt{5})/2$
7. The zeros are 0.807417596 and 6.192582404 (or $(7 \pm \sqrt{29})/2$).
9. 1.465571232 **11.** -1.265189523
13. 0.945228112
15. -7.086130198, -4.428006732; -2.485863071
17. The program does not terminate; nothing is approached. The quadratic equation $x^2 + 5x + 7 = 0$ has no real solutions.
19. $3.35 \times 3.35 \times 5.35$
21. 397.24 tons or 644.51 tons

Review Exercises for Chapter 4, page 276

1.

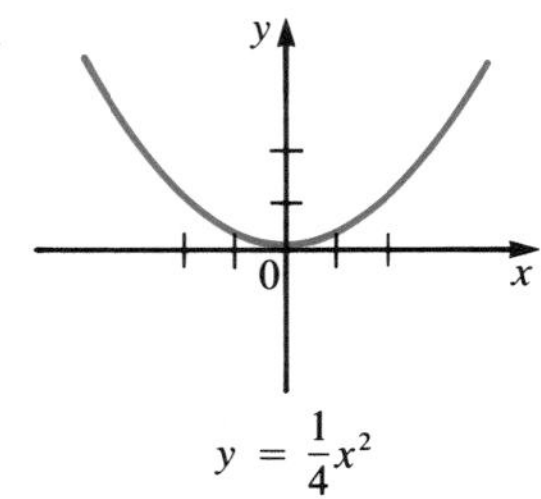

$y = \frac{1}{4}x^2$

3. $y = (x + 2)^2 - 4$; vertex $(-2, -4)$
5. $y = 2(x + \frac{3}{2})^2 - \frac{25}{2}$; vertex $(-\frac{3}{2}, -\frac{25}{2})$
7.

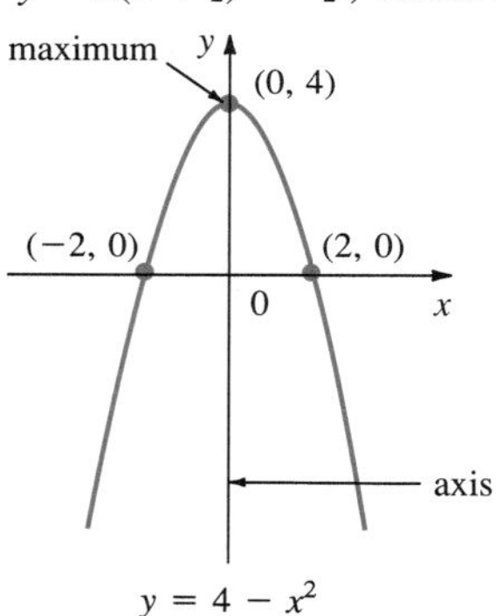

$y = 4 - x^2$

9. (a) 5 seconds (b) 100 feet
11.

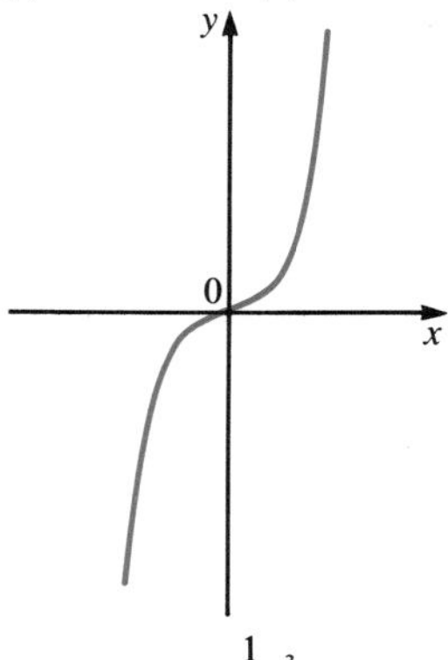

$y = \frac{1}{2}x^3$

13.

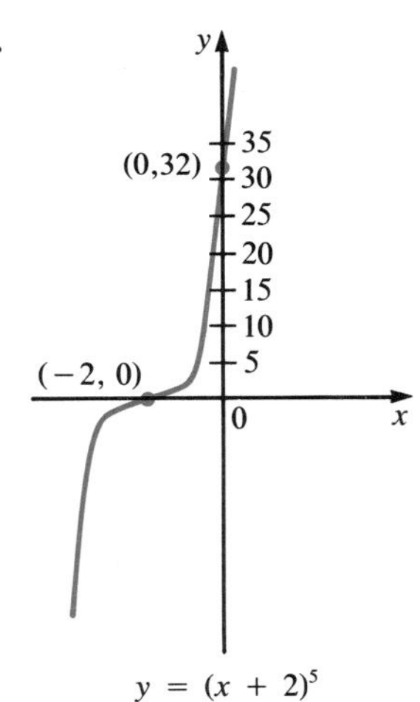

$y = (x + 2)^5$

15.

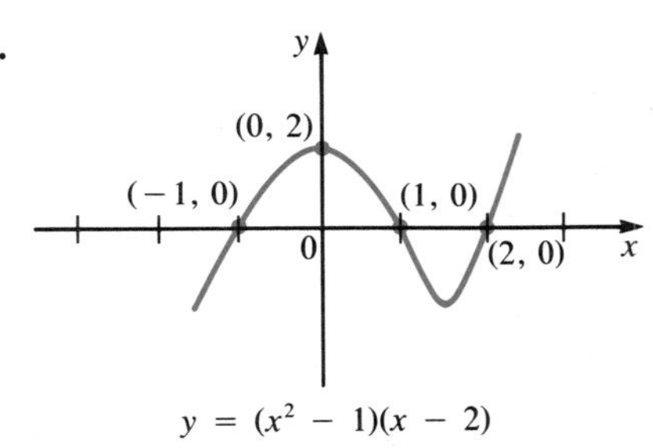

$y = (x^2 - 1)(x - 2)$

17.

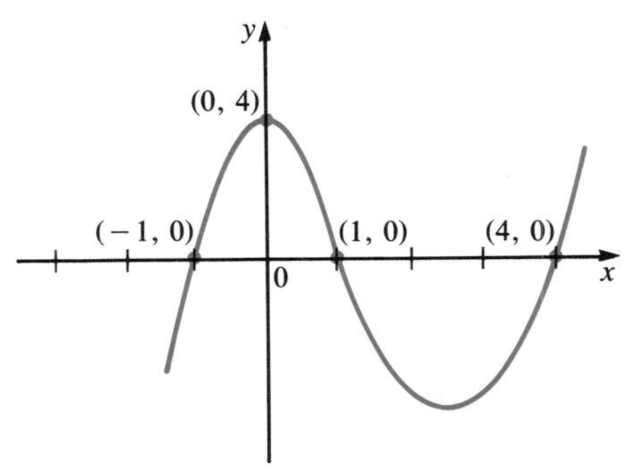

$y = (x - 1)(x^2 - 3x - 4)$
$= (x - 1)(x + 1)(x - 4)$

19.

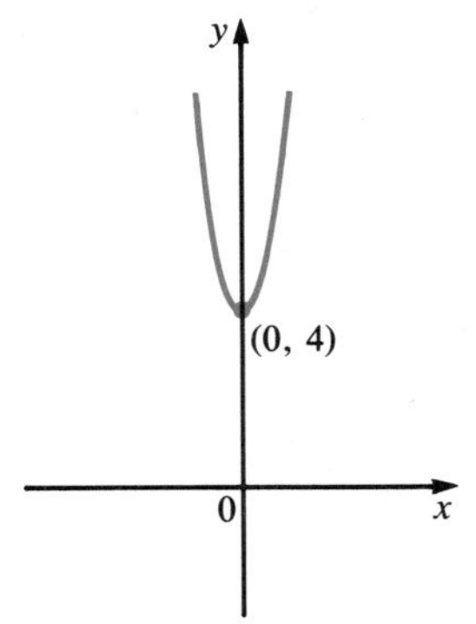

$y = x^4 + 4x^2 + 4 = (x^2 + 2)^2$

21. $q(x) = x - 3;\ r(x) = 7x + 1$
23. $q(x) = x^2 + 4x + 15;\ r(x) = 55x^2 - 18x - 72$
25. $q(x) = 3x^3 - x + 5;\ r(x) = -14$
27. $q(x) = 4x^4 - 8x^3 + 15x^2 - 30x + 62;\ r(x) = -125$
29. 2123 **31.** 1 positive, 2 or 0 negative
33. 2 or 0 positive, 2 or 0 negative **35.** 1, 2, 4
37. $-\frac{1}{2}, i, -i$ **39.** no rational zeros **41.** $-1, 1, 4$
43. 0.77 **45.** 1.24 **47.** 0.77091700
49. 1.24405130

Chapter 5

Problems 5.1, page 288

1. vertical $x = 1, x = -1$; horizontal $y = 0$ (the $x -$ axis)
3. vertical $x = -2, x = -3$; horizontal $y = 0$
5. vertical $x = -1, x = -6$; horizontal $y = 0$
7. vertical $x = 9$; horizontal $y = 1$
9. vertical $x = 2, x = 0$; horizontal $y = 0$
11. vertical $x = 2$; horizontal $y = 0$
13. vertical $x = -\frac{4}{3}$; no horizontal **15.** no asymptotes
17. vertical $x = 1, x = -2, x = 4$; horizontal $y = 1$
19. no asymptotes **21.** $y = x$ **23.** $y = -\frac{2}{5}x$
25. $y = \frac{1}{2}x$ **27.** e **29.** a **31.** h **33.** i
35. f
37.

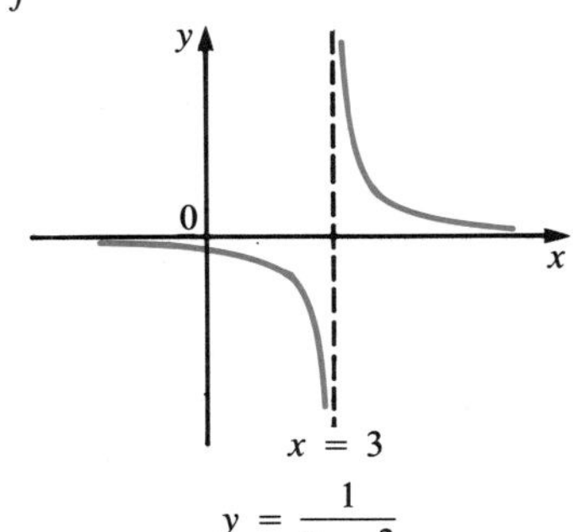

$y = \dfrac{1}{x - 3}$

39.

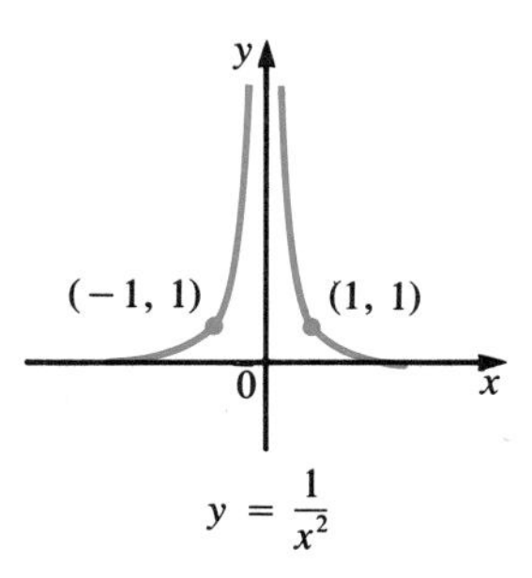

41.

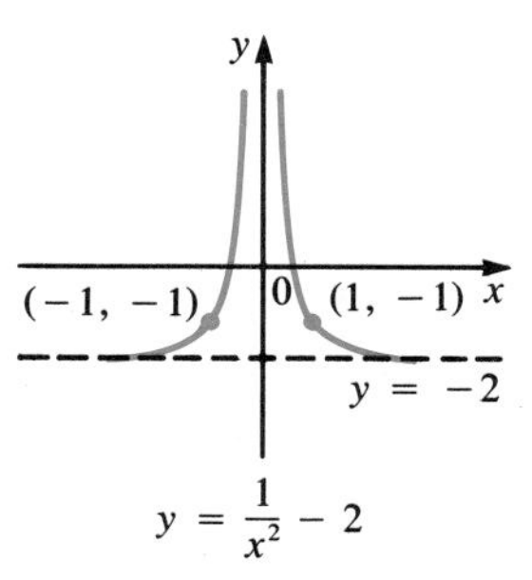

43.

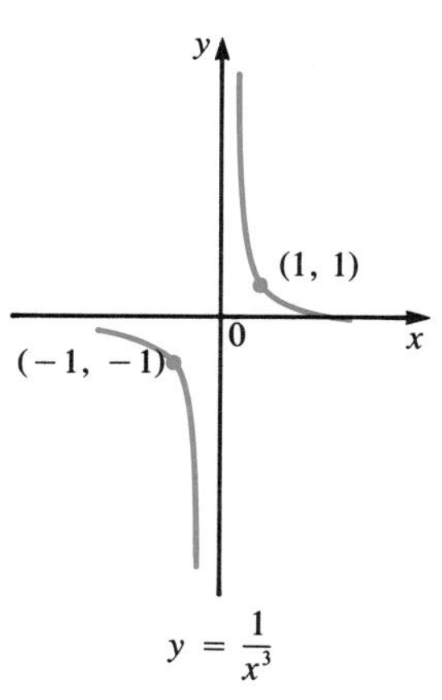

45.

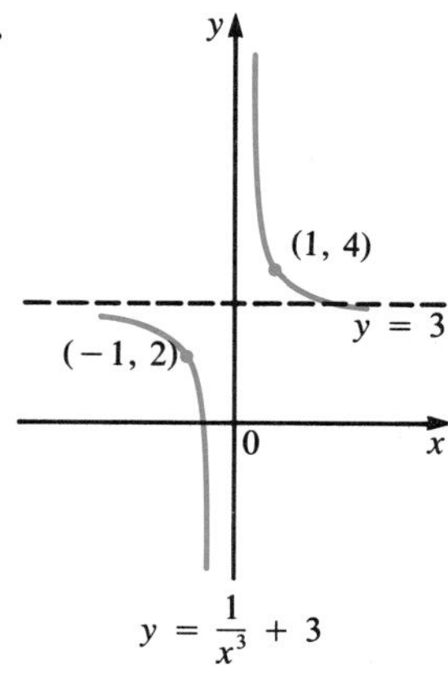

47.

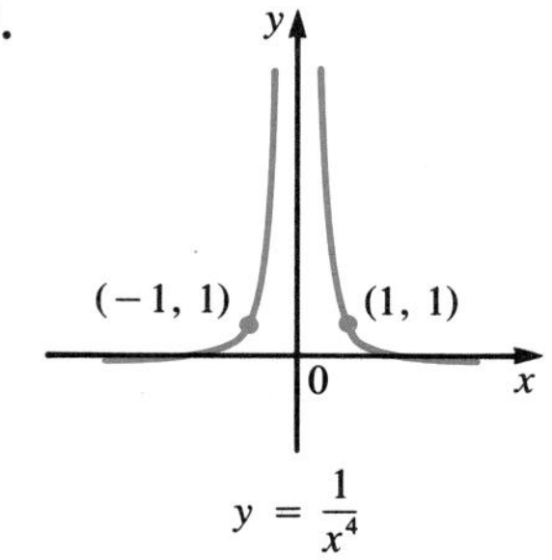

49.

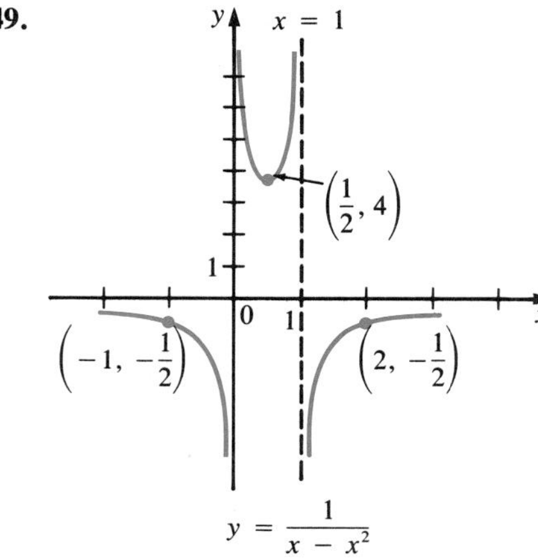

51.

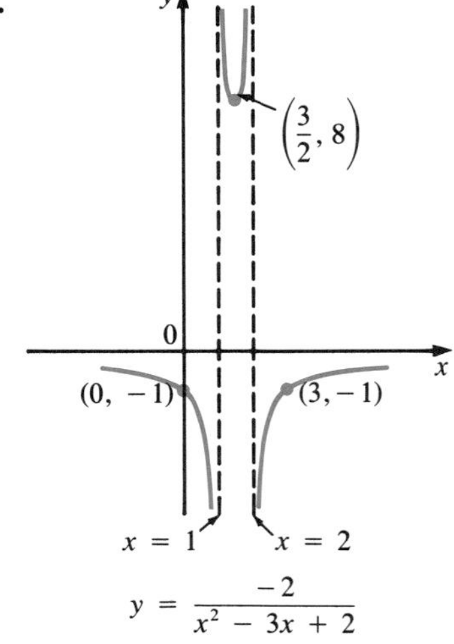

53.

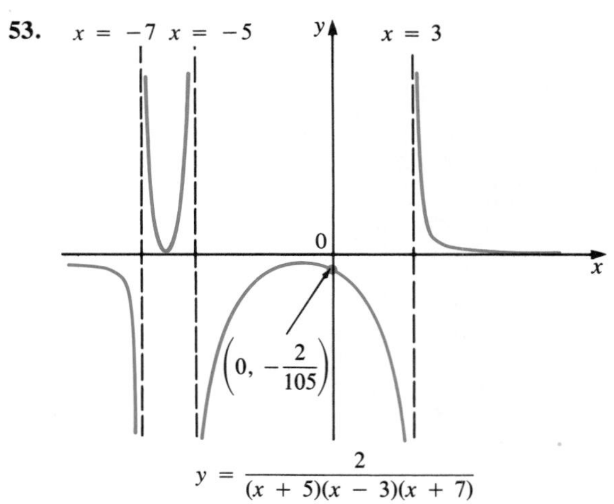

$$y = \frac{2}{(x + 5)(x - 3)(x + 7)}$$

55.

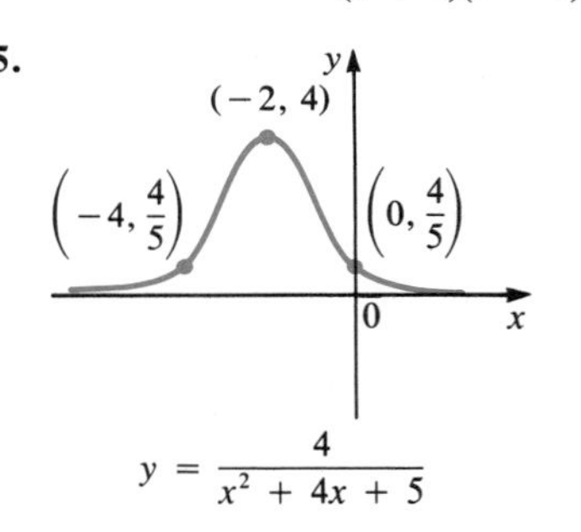

$$y = \frac{4}{x^2 + 4x + 5}$$

57.

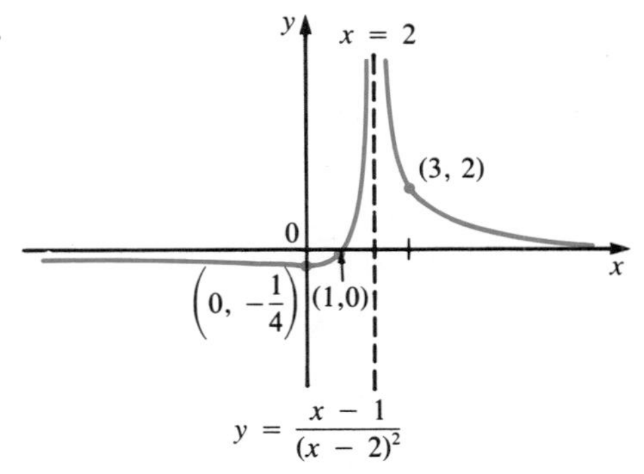

$$y = \frac{x - 1}{(x - 2)^2}$$

59.

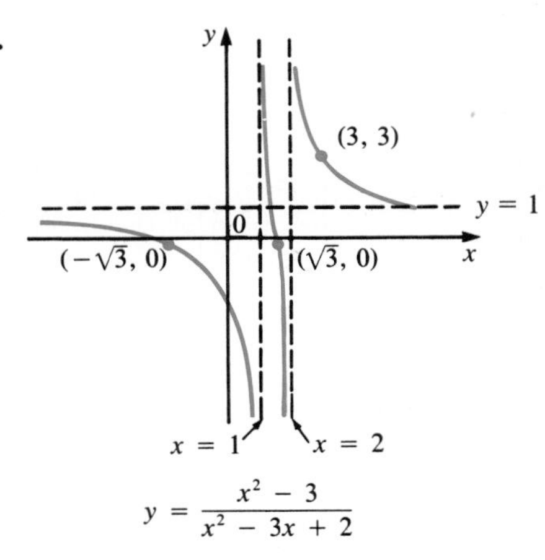

$$y = \frac{x^2 - 3}{x^2 - 3x + 2}$$

61.

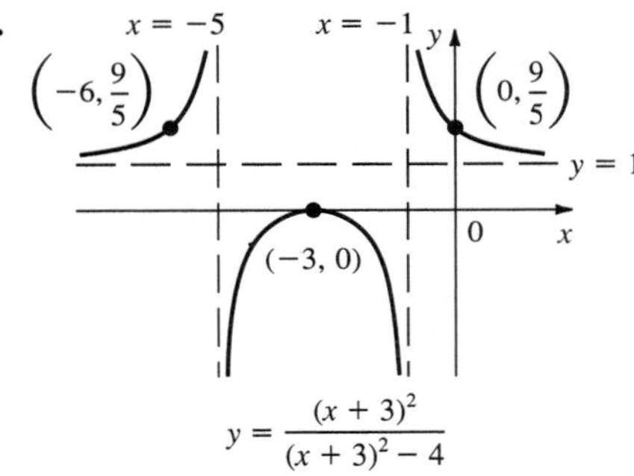

$$y = \frac{(x + 3)^2}{(x + 3)^2 - 4}$$

63.

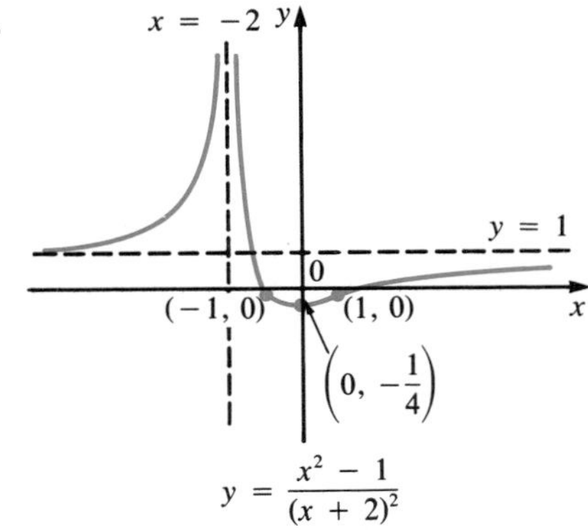

$$y = \frac{x^2 - 1}{(x + 2)^2}$$

65.

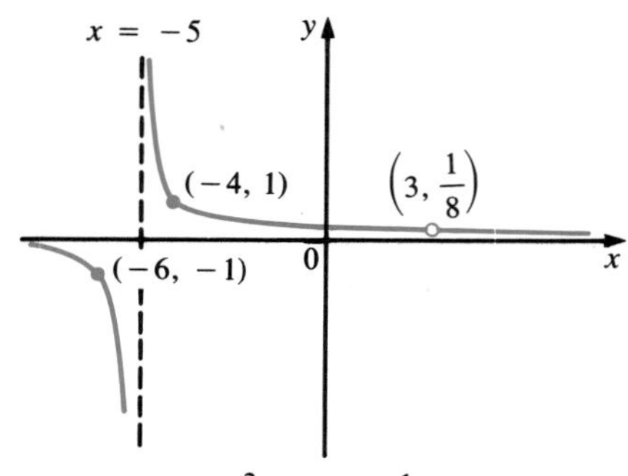

$$y = \frac{x - 3}{x^2 + 2x - 15} = \frac{1}{x + 5}, \quad x \neq 3$$

67.

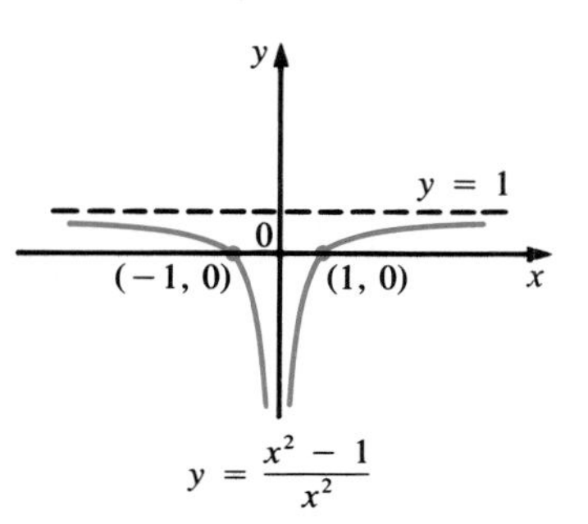

$$y = \frac{x^2 - 1}{x^2}$$

69.

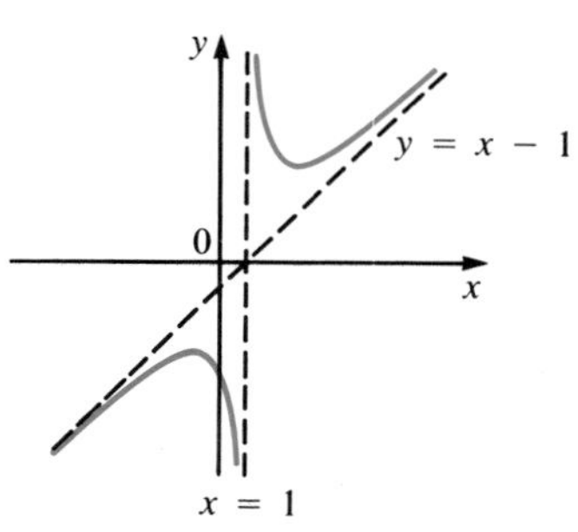

$$y = \frac{x^2 - 2x + 4}{x - 1}$$

71.

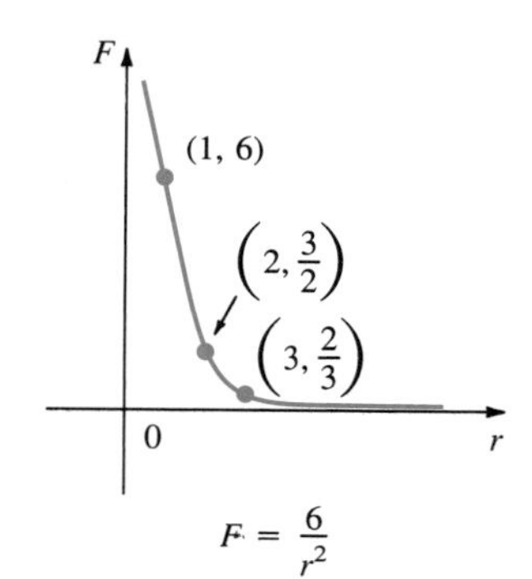

$$F = \frac{6}{r^2}$$

73.

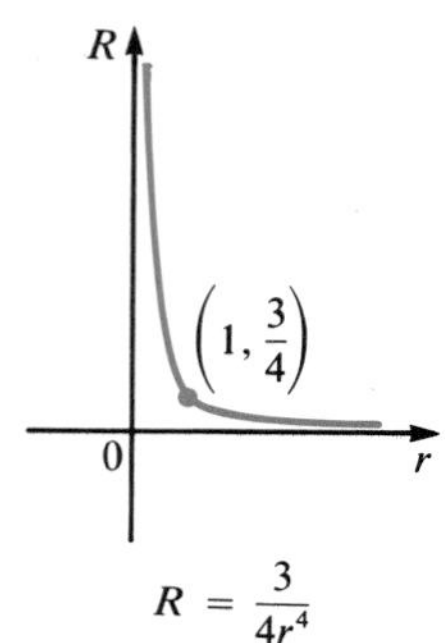

$$R = \frac{3}{4r^4}$$

75.

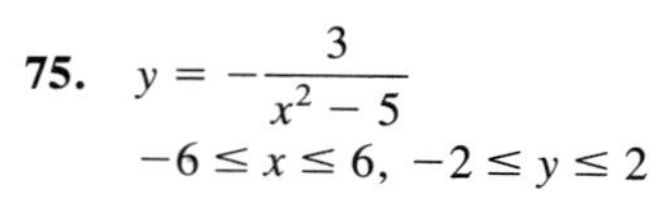

$$y = -\frac{3}{x^2 - 5}$$

$-6 \leq x \leq 6, \ -2 \leq y \leq 2$

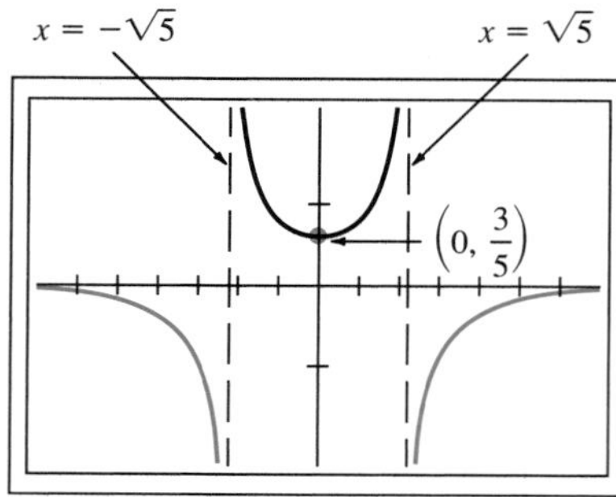

77.

$$y = \frac{x^2 - 3}{x^2 + 3}$$

$-10 \leq x \leq 10, \ -1 \leq y \leq 1$

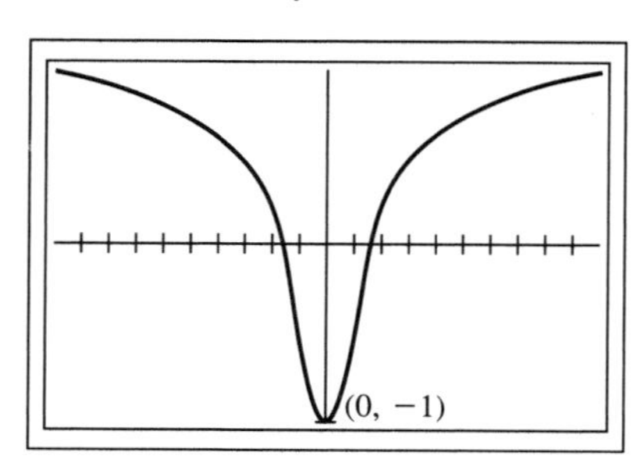

79. $y = \dfrac{15 - 27x}{32x - 18}$
$0.5 \le x \le 0.65,\ -3 \le y \le 1$
$x_{scl} = 0.025$

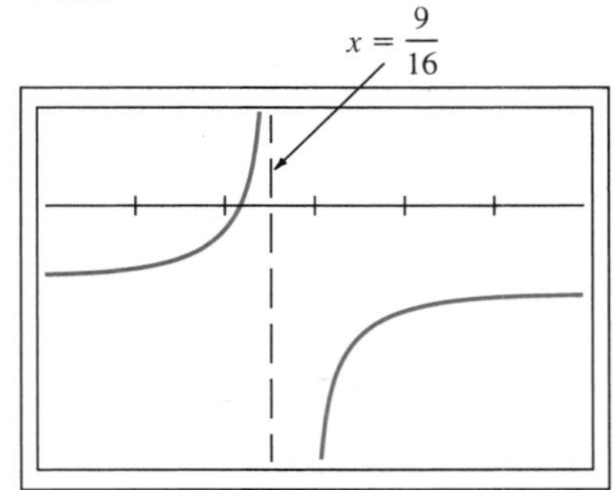

81. $y = \dfrac{x^2 - 2x + 7}{x^2 - 4x - 12}$
$-6 \le x \le 15,\ -4 \le y \le 6$

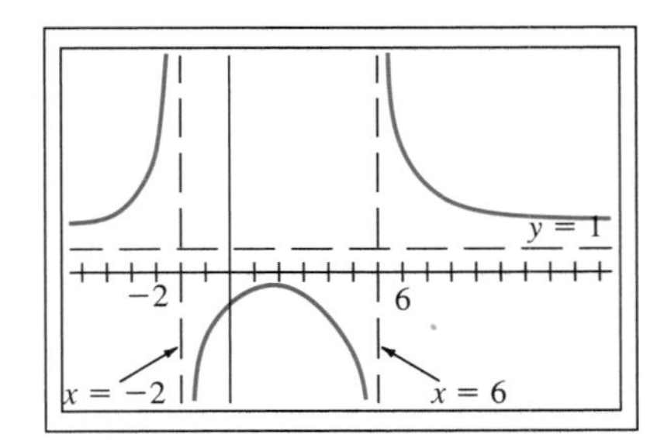

83. $y = \dfrac{x^3 + x - 4}{x^2 - 15}$
$-15 \le x \le 15,\ -25 \le y \le 25$

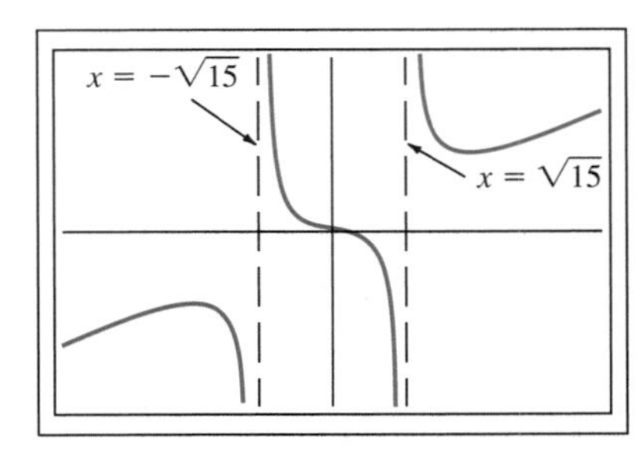

85. $y = \dfrac{2x^3 - 2x^2 + x + 5}{x^3 + x^2 + x + 3}$
$-6 \le x \le 5,\ -3 \le y \le 10$

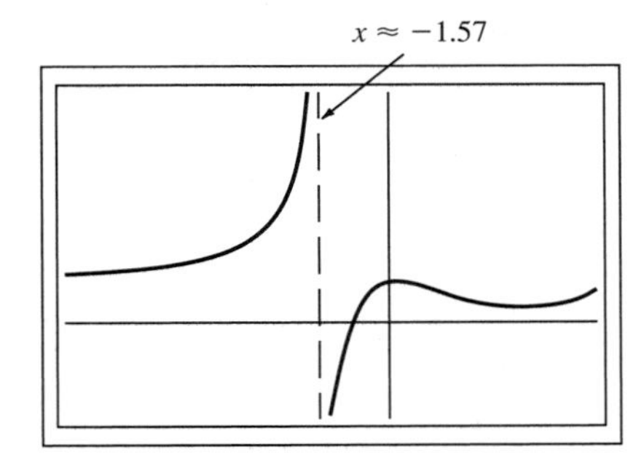

Problems 5.3, page 301

1. center: (0, 0)
foci: (0, ±3)
vertices: (0, ±5)
major axis: line segment between (0, −5) and (0, 5)
minor axis: line segment between (−4, 0) and (4, 0)

$e = \dfrac{3}{5} = 0.6$

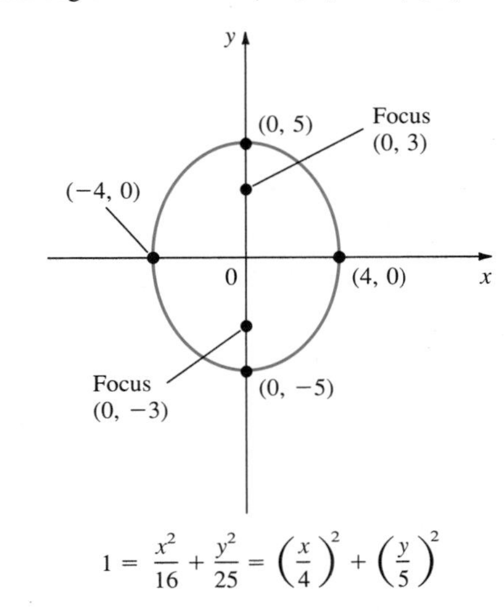

Note: In the following answers, $\overline{(a, b)(c, d)}$ denotes the line segment between (a, b) and (c, d).

3. center: (0, 0)
foci: $(0, \pm 2\sqrt{2})$
vertices: $(0, \pm 3)$
major axis: $\overline{(0, -3)(0, 3)}$
minor axis: $\overline{(-1, 0)(1, 0)}$

$e = \dfrac{2\sqrt{2}}{3} \approx 0.94$

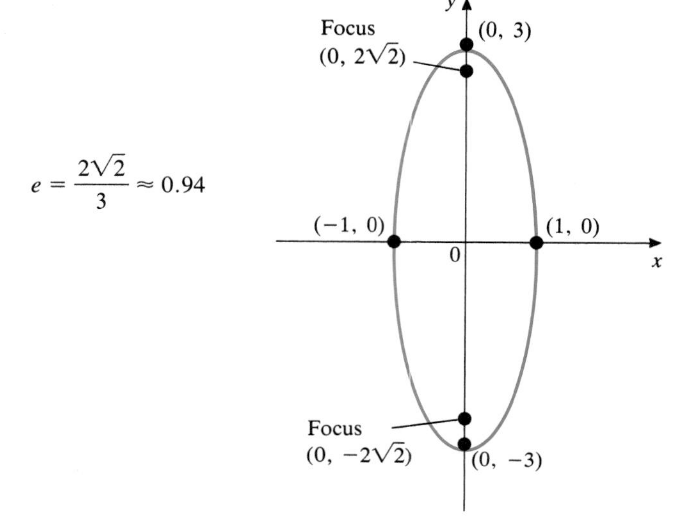

$x^2 + (\frac{y}{3})^2 = 1$

5. center: (0, 0)
foci: $(\pm 2\sqrt{3}, 0)$
vertices: $(\pm 4, 0)$
major axis: $\overline{(-4, 0)(4, 0)}$
minor axis: $\overline{(0, -2)(0, 2)}$

$e = \dfrac{\sqrt{3}}{2} \approx 0.87$

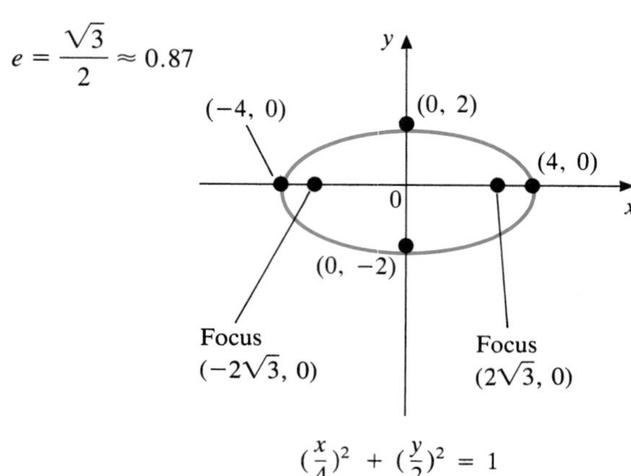

$(\frac{x}{4})^2 + (\frac{y}{2})^2 = 1$

7. center: (1, −3)
foci: (1, −6), (1, 0)
vertices: (1, −8), (1, 2)
major axis: $\overline{(1, -8)(1, 2)}$
minor axis: $\overline{(-3, -3)(5, -3)}$

$e = \dfrac{3}{5} = 0.6$

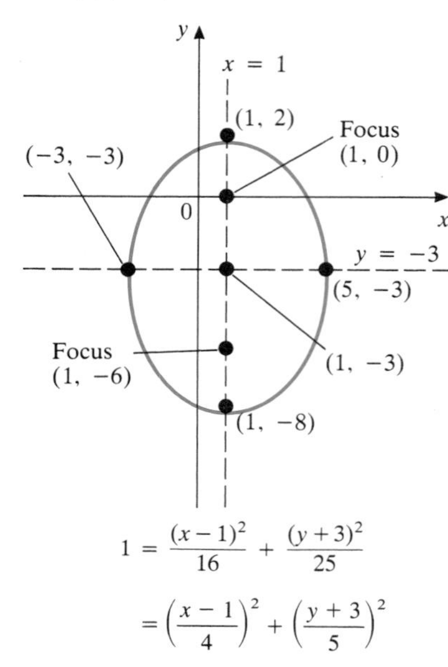

$$1 = \frac{(x-1)^2}{16} + \frac{(y+3)^2}{25} = \left(\frac{x-1}{4}\right)^2 + \left(\frac{y+3}{5}\right)^2$$

9. The graph is a circle, centered at (0, 0) with radius 1 (the unit circle).

11. center: (0, 0)
foci: $(\pm 3\sqrt{3}/2, 0)$
vertices: $(\pm 3, 0)$
major axis: $\overline{(-3, 0)(3, 0)}$
minor axis: $\overline{(0, -\frac{3}{2})(0, \frac{3}{2})}$

$e = \dfrac{\sqrt{3}}{2} \approx 0.87$

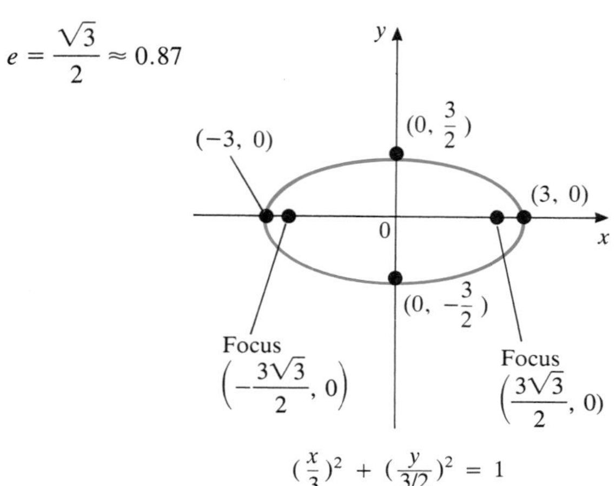

$(\frac{x}{3})^2 + (\frac{y}{3/2})^2 = 1$

13. center: $(-1, -3)$
foci: $(-1, -3 \pm 2\sqrt{3})$
vertices: $(-1, -7), (-1, 1)$
major axis: $\overline{(-1, -7)(-1, 1)}$
minor axis: $\overline{(-3, -3)(1, -3)}$

$e = \dfrac{\sqrt{3}}{2} \approx 0.87$

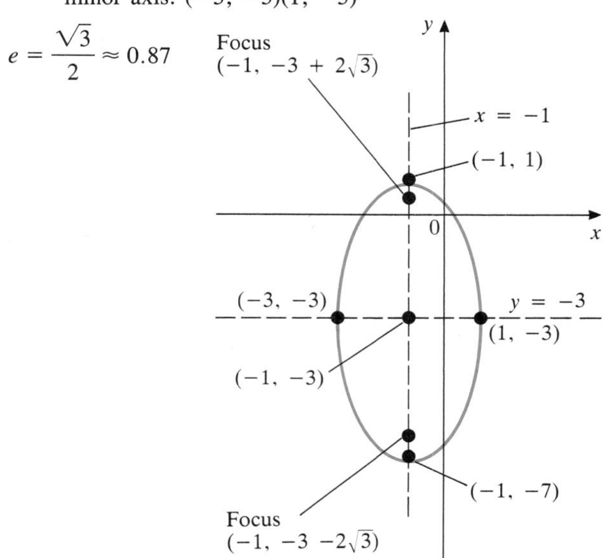

$$4(x+1)^2 + (y+3)^2 = 16$$

$$\left(\frac{x+1}{2}\right)^2 + \left(\frac{y+3}{4}\right)^2 = 1$$

15. center: $(-1, 3)$
foci: $(-1, 3 \pm 2\sqrt{3})$
vertices: $(-1, -1), (-1, 7)$
major axis: $\overline{(-1, -1)(-1, 7)}$
minor axis: $\overline{(-3, 3)(1, 3)}$

$e = \dfrac{\sqrt{3}}{2} \approx 0.87$

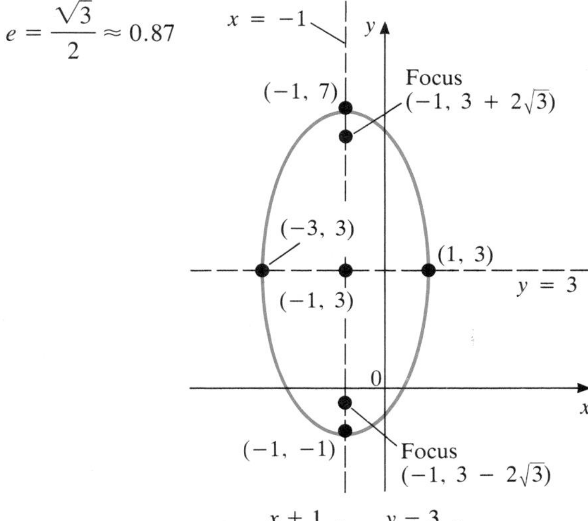

$$\left(\frac{x+1}{2}\right)^2 + \left(\frac{y-3}{4}\right)^2 = 1$$

17. center: $(-2, \frac{1}{4})$
foci: $(-2 \pm \sqrt{\frac{325}{48}}, \frac{1}{4})$
vertices: $(-2 \pm \sqrt{\frac{65}{6}}, \frac{1}{4})$
major axis: $\overline{(-2 - \sqrt{\frac{65}{6}}, \frac{1}{4})(-2 + \sqrt{\frac{65}{6}}, \frac{1}{4})}$
minor axis: $\overline{(-2, (1 - \sqrt{65})/4)(-2, (1 + \sqrt{65})/4)}$

$e = \dfrac{\sqrt{10}}{4} \approx 0.79$

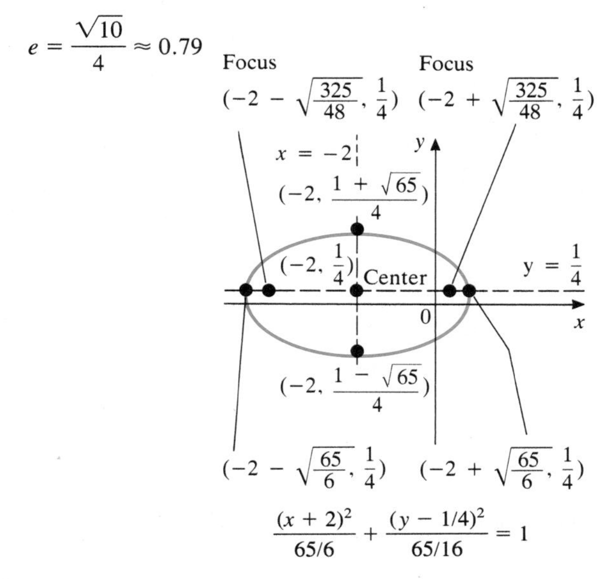

$$\frac{(x+2)^2}{65/6} + \frac{(y-1/4)^2}{65/16} = 1$$

19. $\dfrac{x^2}{9} + \dfrac{y^2}{25} = 1$ **21.** $\dfrac{(x+1)^2}{9} + \dfrac{(y-4)^2}{25} = 1$

25. 0.9679 **27.** 8.3 AU

29. Lanes 1 and 6 have $5\sqrt{7} \approx 13.2$ ft, lanes 2 and 5 have $10\sqrt{3} \approx 17.3$ ft, lanes 3 and 4 have $5\sqrt{15} \approx 19.4$ ft.

31. (b); $(-\sqrt{21}, 0)$ and $(\sqrt{21}, 0)$

33. (h); $(-2, 4)$ and $(-2, -2)$

35. (c); $(0, \sqrt{13})$ and $(0, -\sqrt{13})$

37. (e); $(3, 1 + \sqrt{5})$ and $(3, 1 - \sqrt{5})$

39.

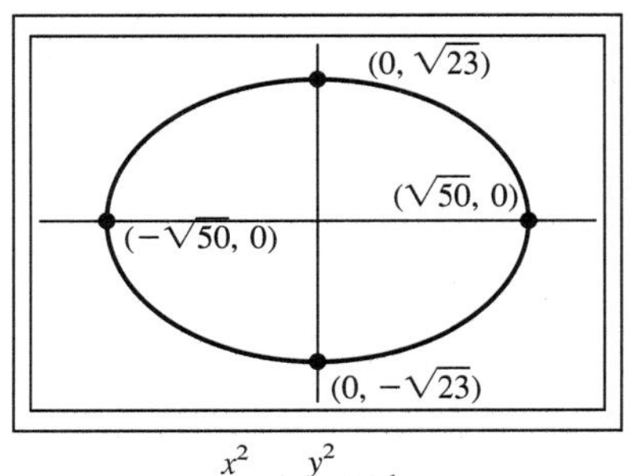

$$\frac{x^2}{50} + \frac{y^2}{23} = 1$$

41.

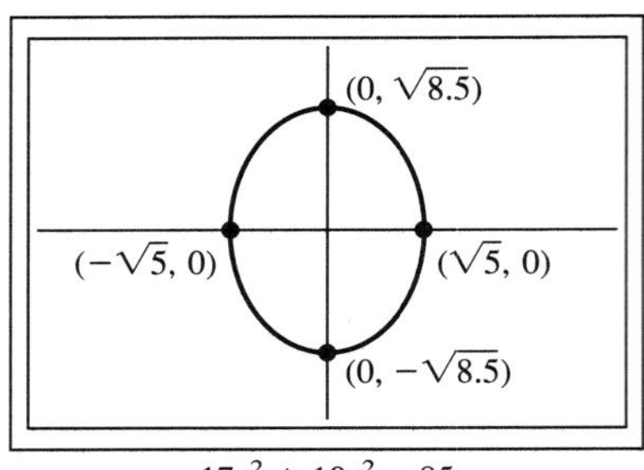

$17x^2 + 10y^2 = 85$

43.

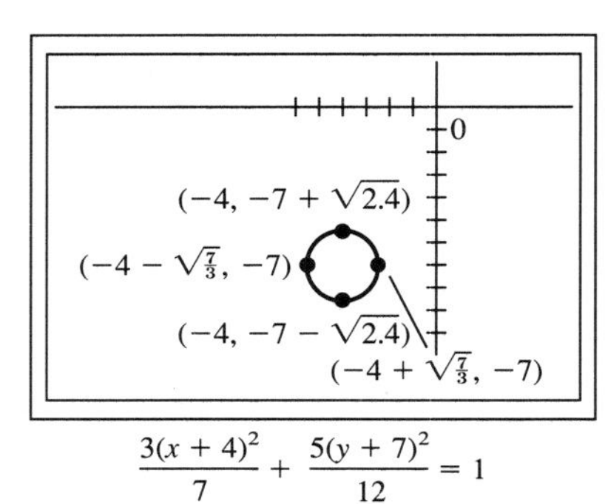

$$\frac{3(x+4)^2}{7} + \frac{5(y+7)^2}{12} = 1$$

Note: This is almost, but not quite, a circle. $\frac{3}{7} \approx 0.429$ and $\frac{5}{12} \approx 0.417$. If they were equal, it would be a circle.

45.

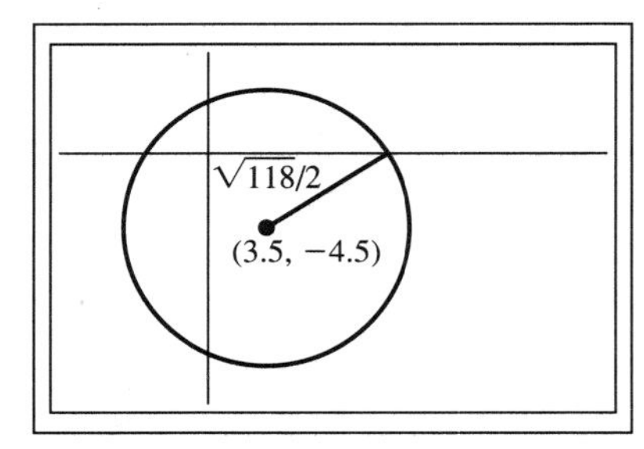

The circle $(x - \frac{7}{2})^2 + (y + \frac{9}{2})^2 = \frac{118}{4}$

47.

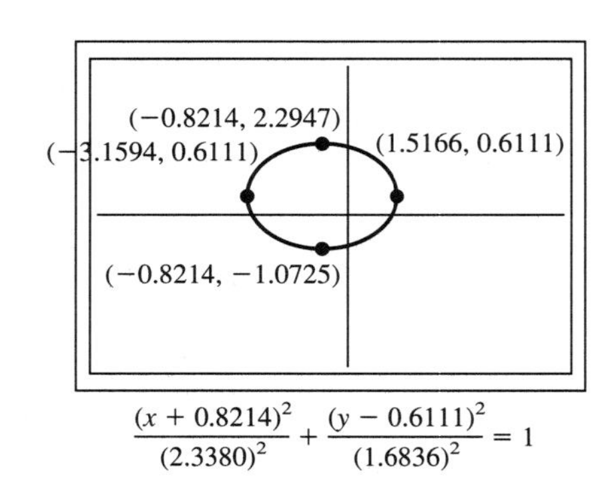

$$\frac{(x+0.8214)^2}{(2.3380)^2} + \frac{(y-0.6111)^2}{(1.6836)^2} = 1$$

Problems 5.4, page 309

1. focus: (0, 4)
directrix: $y = -4$
axis: y-axis
vertex: (0, 0)

y Focus (0, 4) 0 x Directrix $y = -4$

$x^2 = 16y$

3. focus: (0, −4)
directrix: $y = 4$
axis: y-axis
vertex: (0, 0)

y Directrix $y = 4$ 0 x Focus (0, −4)

$x^2 = -16y$

5. focus: $(0, \frac{3}{8})$
directrix: $y = -\frac{3}{8}$
axis: y-axis
vertex: (0, 0)

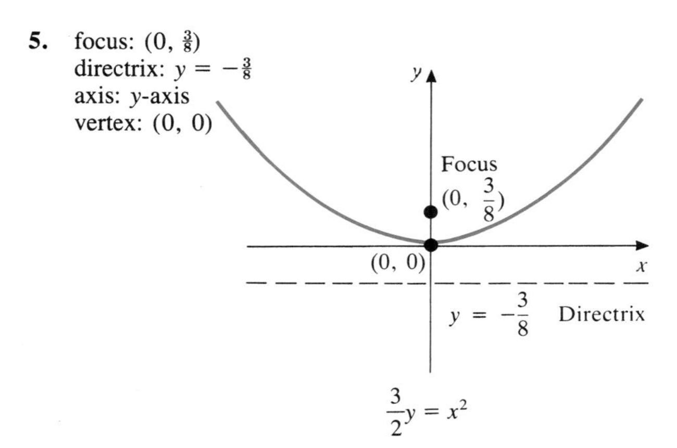

$\frac{3}{2}y = x^2$

7. focus: $(0, -\frac{9}{16})$
directrix: $y = \frac{9}{16}$
axis: y-axis
vertex: $(0, 0)$

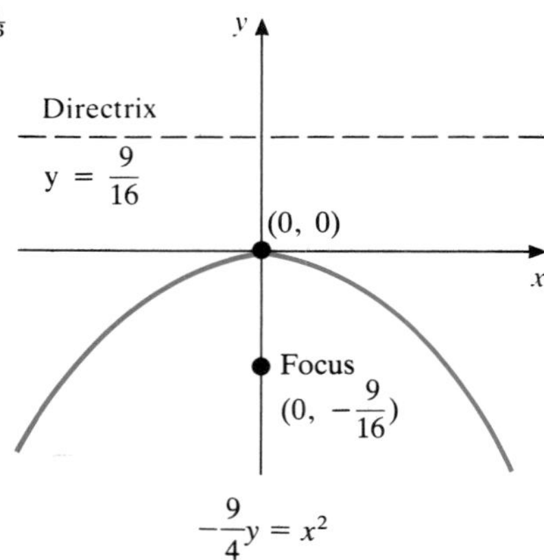

9. vertex: $(1, -3)$
focus: $(1, -7)$
directrix: $y = 1$
axis: $x = 1$

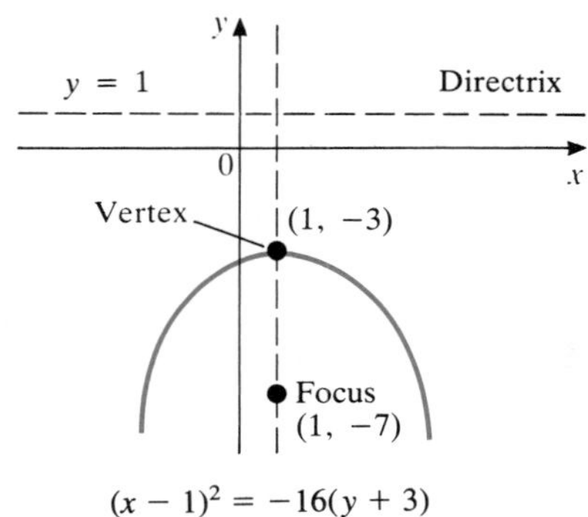

11. vertex: $(0, \frac{9}{4})$
focus: $(0, \frac{5}{4})$
directrix: $y = \frac{13}{4}$
axis: $x = 0$

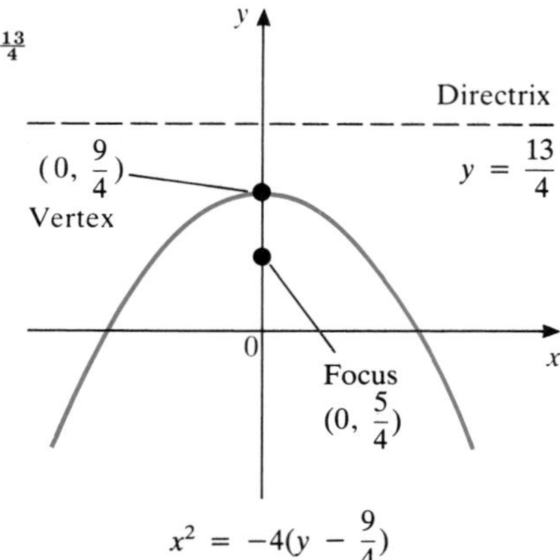

13. vertex: $(-1, 0)$
focus: $(-1, -\frac{1}{4})$
directrix: $y = \frac{1}{4}$
axis: $x = -1$

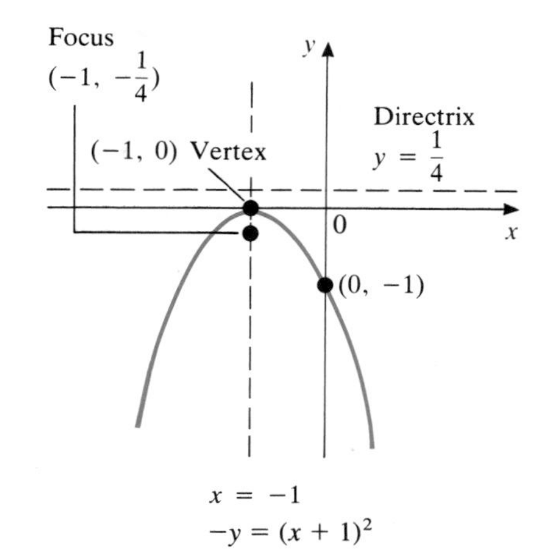

15. vertex: $(-2, 4)$
focus: $(-2, \frac{15}{4})$
directrix: $y = \frac{17}{4}$
axis: $x = -2$

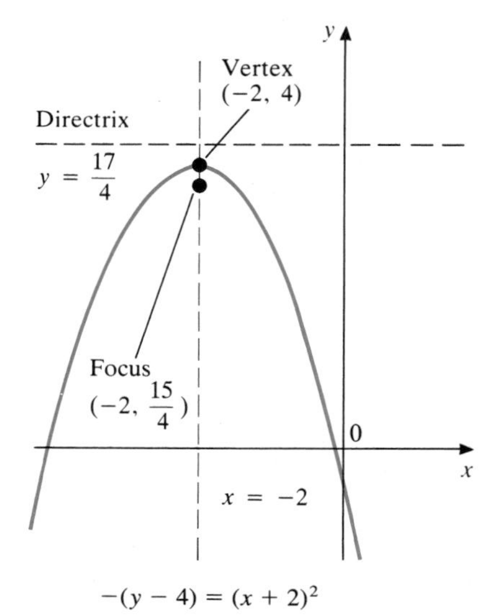

17. vertex: $(-2, -4)$
focus: $(-2, -\frac{15}{4})$
directrix: $y = -\frac{17}{4}$
axis: $x = -2$

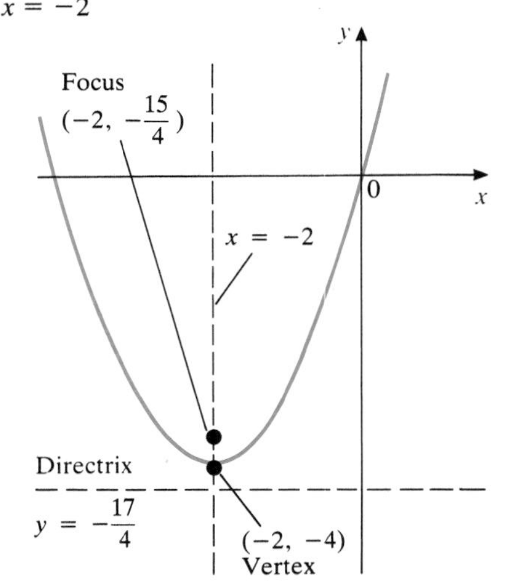

$y + 4 = (x + 2)^2$

19. $x^2 = 16y$ **21.** $(y - 5)^2 = -12(x + 2)$

23. focus $(0, -1)$; directrix $x = 6$ **27.** $x^2 = 8y$

29. (a)

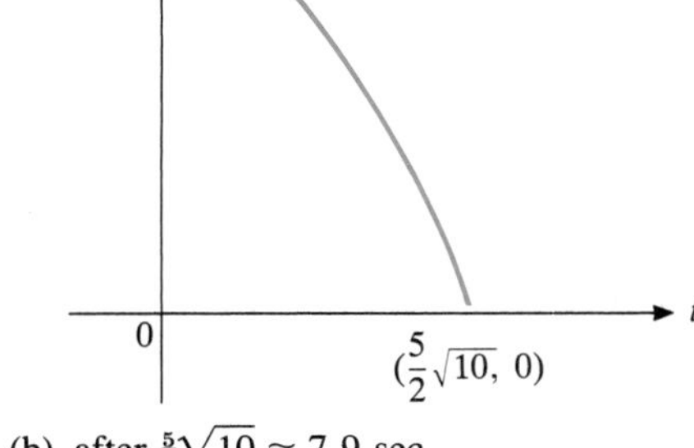

(b) after $\frac{5}{2}\sqrt{10} \approx 7.9$ sec

31. f **33.** b **35.** c **37.** e

39.

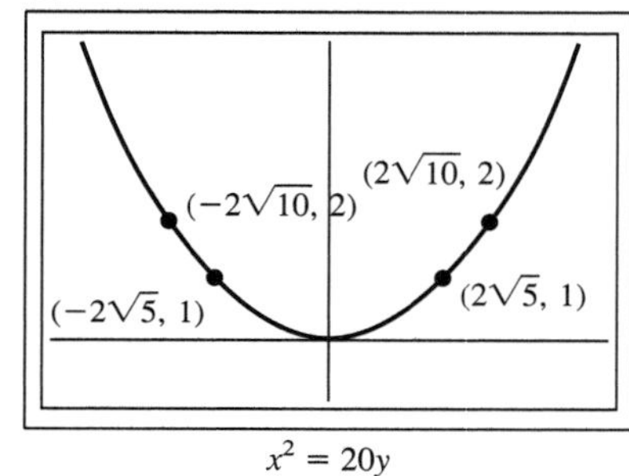

$x^2 = 20y$

41.

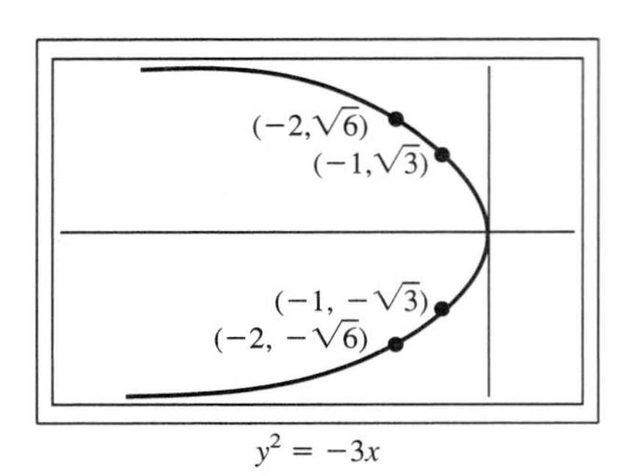

$y^2 = -3x$

43.

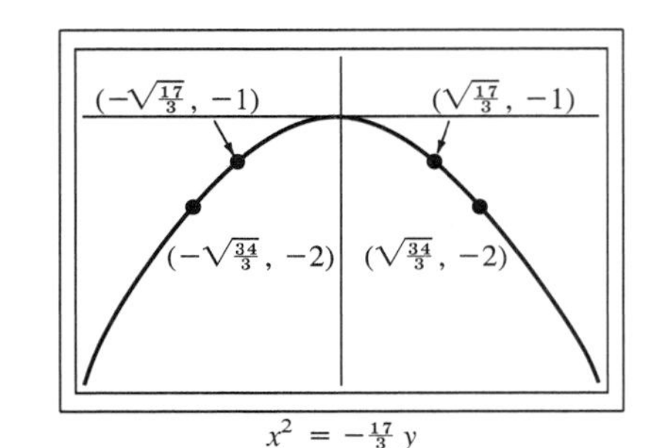

$x^2 = -\frac{17}{3}y$

45.

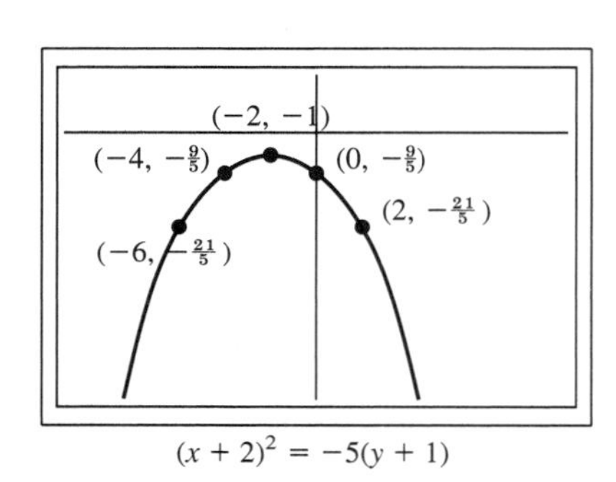

$(x + 2)^2 = -5(y + 1)$

47.

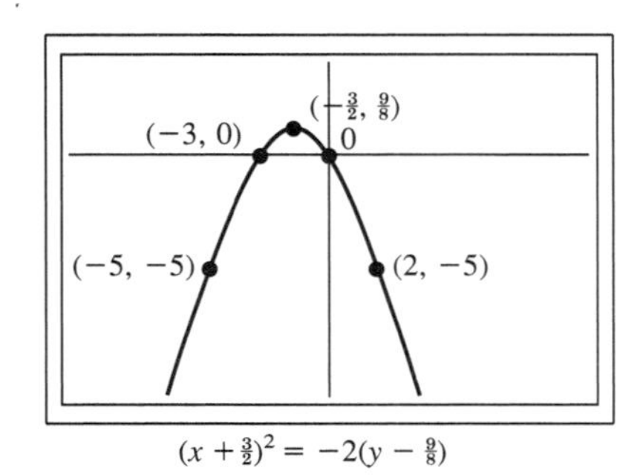

$(x + \frac{3}{2})^2 = -2(y - \frac{9}{8})$

Problems 5.5, page 320

1. center: (0. 0)
foci: $(\pm\sqrt{41}, 0)$
vertices: $(\pm 4, 0)$
transverse axis: $\overline{(-4, 0)(4, 0)}$
asymptotes: $y = \pm\frac{5}{4}x$
conjugate axis: $\overline{(0,5)(0, -5)}$

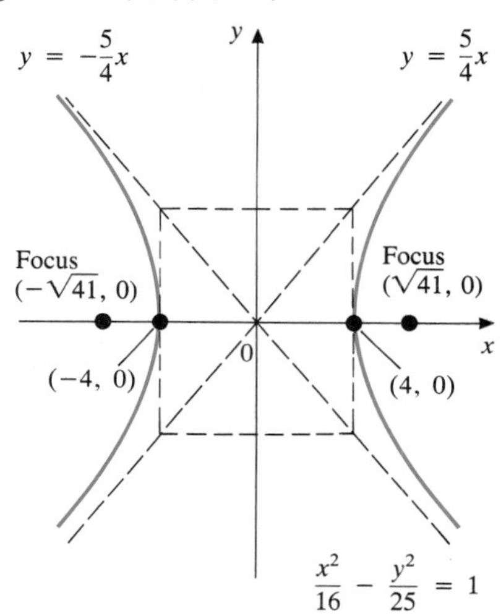

3. center: (0, 0)
foci: $(0, \pm\sqrt{41})$
vertices: $(0, \pm 5)$
transverse axis: $\overline{(0, -5)(0,5)}$
asymptotes: $y = \pm\frac{5}{4}x (x = \pm\frac{4}{5}y)$
conjugate axis: $\overline{(-4, 0)(4,0)}$

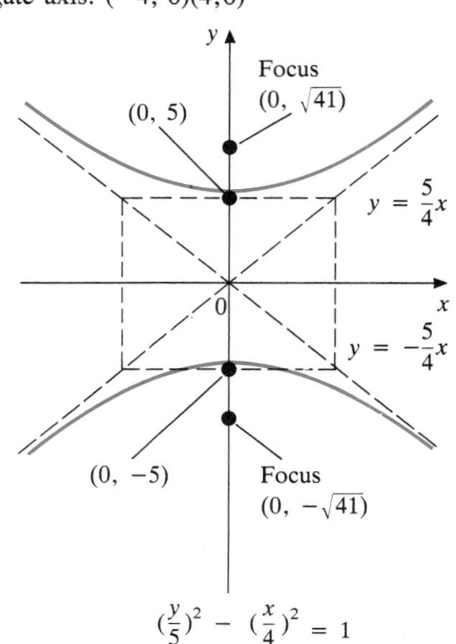

5. center: (0, 0)
foci: $(0, \pm\sqrt{2})$
vertices: $(0, \pm 1)$
transverse axis: $\overline{(0, -1)(0,1)}$
asymptotes: $y = \pm x$
conjugate axis: $\overline{(-1, 0)(1, 0)}$

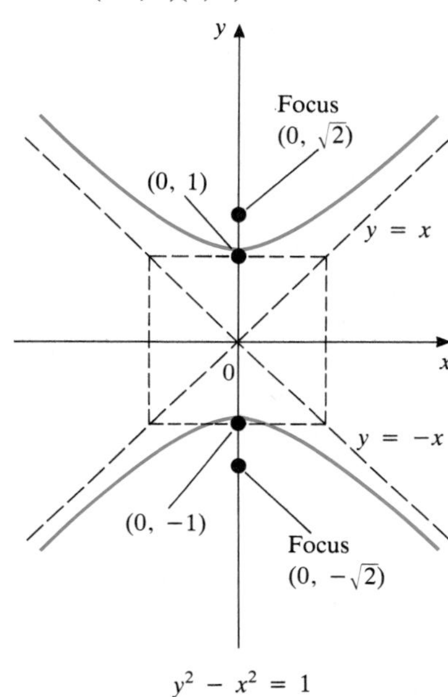

7. center: (0, 0)
foci: $(\pm 3\sqrt{5}/2, 0)$
vertices: $(\pm 3, 0)$
transverse axis: $\overline{(-3, 0)(3, 0)}$
asymptotes: $y = \pm x/2$
conjugate axis: $\overline{(0, -\frac{3}{2})(0, \frac{3}{2})}$

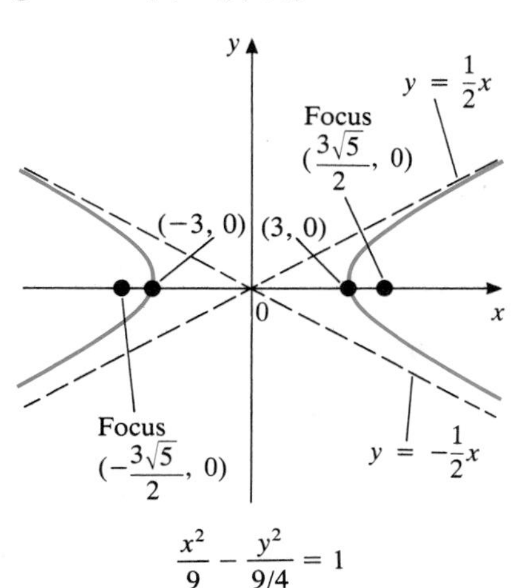

9. center: $(0, 0)$
foci: $(0, \pm 3\sqrt{5}/2)$
vertices: $(0, \pm 3)$
transverse axis: $\overline{(0, -3)(0, 3)}$
asymptotes: $y = \pm 2x$
conjugate axis: $\overline{(-\frac{3}{2}, 0)(\frac{3}{2}, 0)}$

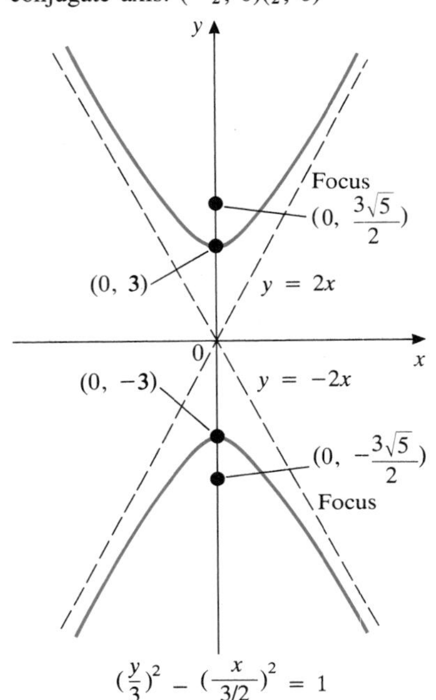

$$\left(\frac{y}{3}\right)^2 - \left(\frac{x}{3/2}\right)^2 = 1$$

11. center: $(0, 0)$
foci: $(\pm\sqrt{\frac{10}{3}}, 0)$
vertices: $(\pm\sqrt{2}, 0)$
transverse axis: $\overline{(-\sqrt{2}, 0)(\sqrt{2}, 0)}$
asymptotes: $y = \pm\sqrt{\frac{2}{3}}x$
conjugate axis: $\overline{\left(0, -\frac{2}{\sqrt{3}}\right)\left(0, \frac{2}{\sqrt{3}}\right)}$

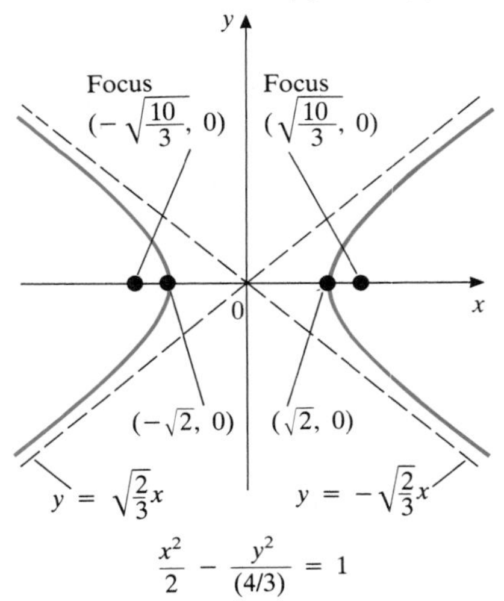

$$\frac{x^2}{2} - \frac{y^2}{(4/3)} = 1$$

13. center: $(0, 0)$
foci: $(0, \pm\sqrt{\frac{10}{3}})$
vertices: $(0, \pm\sqrt{2})$
transverse axis: $\overline{(0, -\sqrt{2})(0, \sqrt{2})}$
asymptotes: $y = \pm\sqrt{\frac{3}{2}}x$
conjugate axis: $\overline{\left(-\frac{2}{\sqrt{3}}, 0\right)\left(\frac{2}{\sqrt{3}}, 0\right)}$

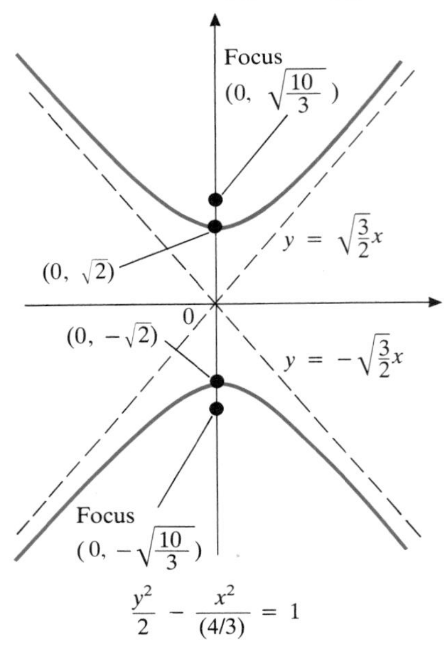

$$\frac{y^2}{2} - \frac{x^2}{(4/3)} = 1$$

15. center: $(1, -2)$
foci: $(1 - \sqrt{5}, -2)$, $(1 + \sqrt{5}, -2)$
vertices: $(-1, -2)$, $(3, -2)$
transverse axis: $\overline{(-1, -2)(3, -2)}$
asymptotes: $y = \pm(\frac{1}{2})(x - 1) - 2$
conjugate axis: $\overline{(1, -1)(1, -3)}$

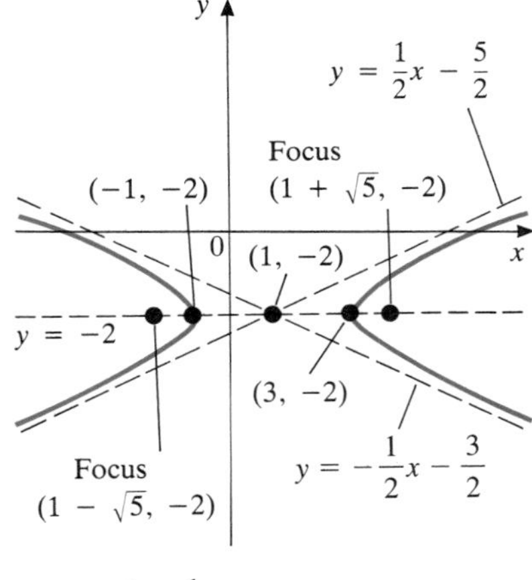

$$\left(\frac{x-1}{2}\right)^2 - (y+2)^2 = 1$$

17. center: $(-1, -3)$
foci: $(-1 \pm 2\sqrt{5}, -3)$
vertices: $(-3, -3), (1, -3)$
transverse axis: $\overline{(-3, -3)(1, -3)}$
asymptotes: $y = \pm 2(x + 1) - 3$
conjugate axis: $\overline{(-1, 1)(-1, -7)}$

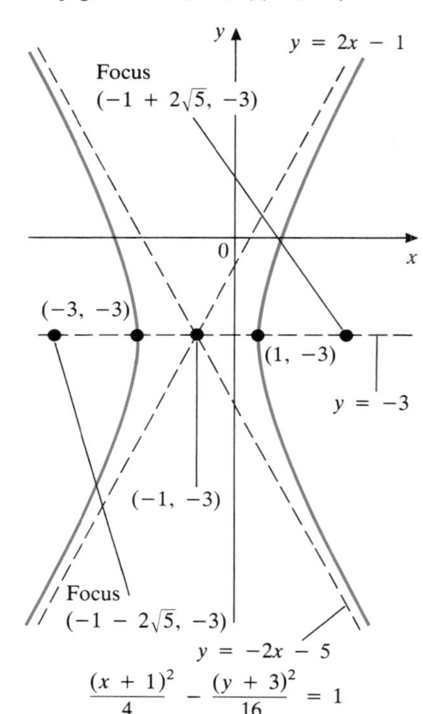

$$\frac{(x+1)^2}{4} - \frac{(y+3)^2}{16} = 1$$

19. center: $(4, 2)$
foci: $(4 - \sqrt{\tfrac{325}{6}}, 2)(4 + \sqrt{\tfrac{325}{6}}, 2)$
vertices: $(4 - \sqrt{\tfrac{65}{2}}, 2), (4 + \sqrt{\tfrac{65}{2}}, 2)$
transverse axis: $\overline{(4 - \sqrt{\tfrac{65}{2}}, 2)(4 + \sqrt{\tfrac{65}{2}}, 2)}$
asymptotes: $y = \pm\sqrt{\tfrac{2}{3}}(x - 4) + 2$
conjugate axis: $\overline{(4, 2 - \sqrt{\tfrac{65}{3}})(4, 2 + \sqrt{\tfrac{65}{3}})}$

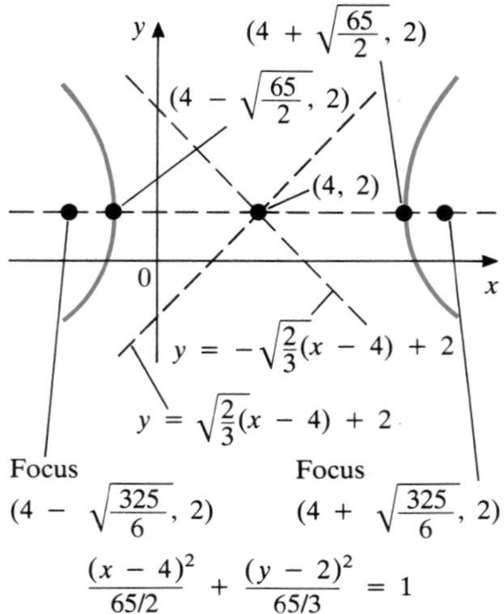

$$\frac{(x-4)^2}{65/2} + \frac{(y-2)^2}{65/3} = 1$$

21. $\frac{x^2}{9} - \frac{y^2}{7} = 1$ **23.** $\frac{y^2}{4} - \frac{x^2}{5} = 1$

25. $\frac{(x-2)^2}{4} - \frac{(y-1)^2}{5} = 1$ **27.** $\frac{x^2}{4} - \frac{y^2}{4} = 1$

29. $\frac{(x-3)^2}{4} - \frac{(y-1)^2}{16} = 1$

31. $\frac{(x-4)^2}{9} - \frac{(y+3)^2}{7} = 1$ **35.** $\sqrt{41}/4$

37. $\sqrt{2}$ **39.** $\sqrt{5}/2$ **41.** $\sqrt{5/3} = \sqrt{15}/3$

43. $\sqrt{5/3} = \sqrt{15}/3$ **45.** $\frac{x^2}{\frac{11}{35}} - \frac{y^2}{\frac{11}{6}} = 1$

47. (c) $x + \sqrt{3}y = -2 + \sqrt{3}$ and $x - \sqrt{3}y = -2 - \sqrt{3}$

49. $\frac{x^2}{2{,}350{,}089} - \frac{y^2}{1{,}649{,}911} = 1$, with coordinates given in meters. Note that $1533^2 = 2{,}350{,}089$ and $1{,}649{,}911 = 2000^2 - 1533^2$

51.

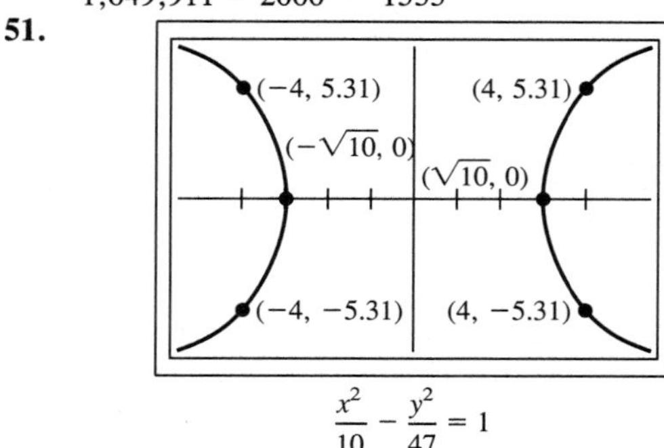

$$\frac{x^2}{10} - \frac{y^2}{47} = 1$$

53.

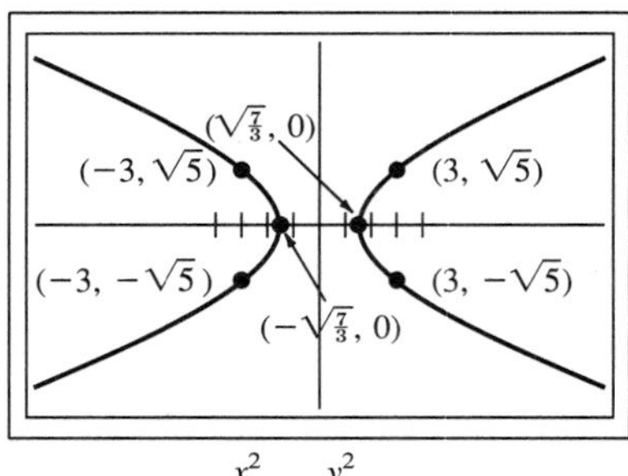

$$\frac{x^2}{7/3} - \frac{y^2}{7/4} = 1$$

55.

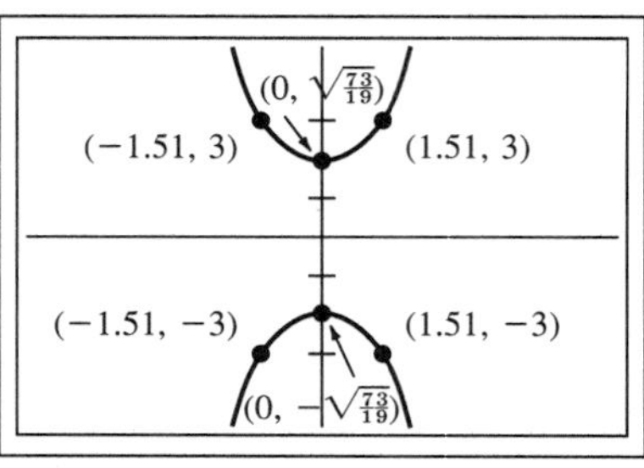

$$\frac{y^2}{73/19} - \frac{x^2}{73/43} = 1$$

57.

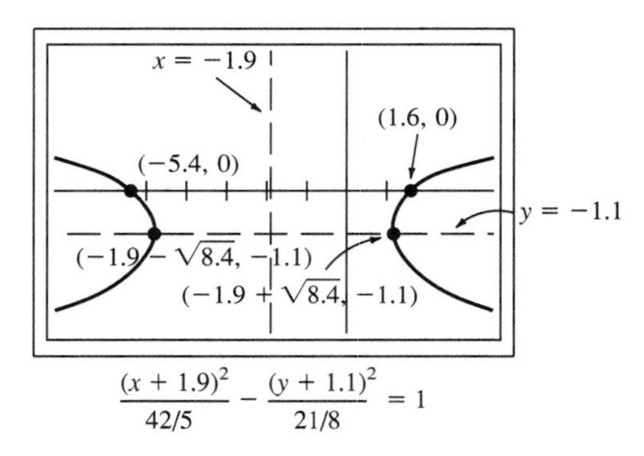

$$\frac{(x + 1.9)^2}{42/5} - \frac{(y + 1.1)^2}{21/8} = 1$$

59.

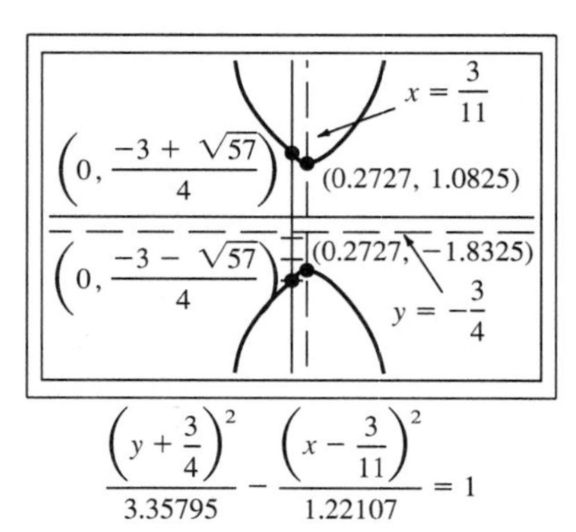

$$\frac{\left(y + \frac{3}{4}\right)^2}{3.35795} - \frac{\left(x - \frac{3}{11}\right)^2}{1.22107} = 1$$

Review Exercises for Chapter 5, page 323

1. vertical $x = 2$, $x = -2$; horizontal $y = 0$

3. vertical $x = \dfrac{5 + \sqrt{41}}{2}$, $x = \dfrac{5 - \sqrt{41}}{2}$; horizontal $y = 3$

5.

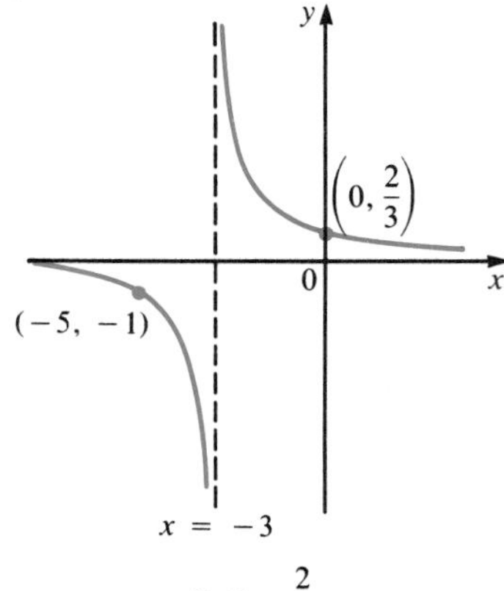

$$y = \frac{2}{x + 3}$$

7.

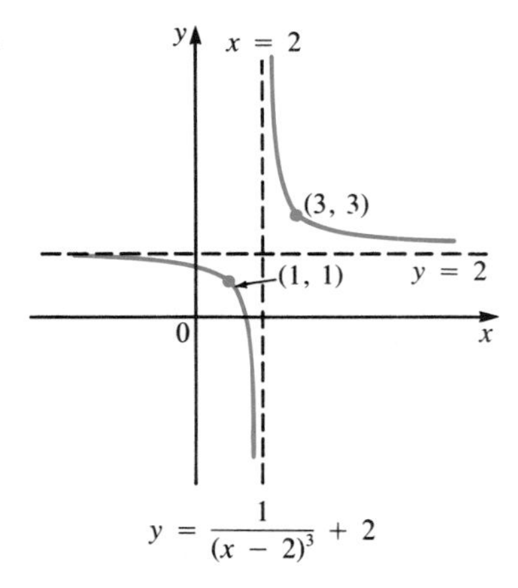

$$y = \frac{1}{(x - 2)^3} + 2$$

9.

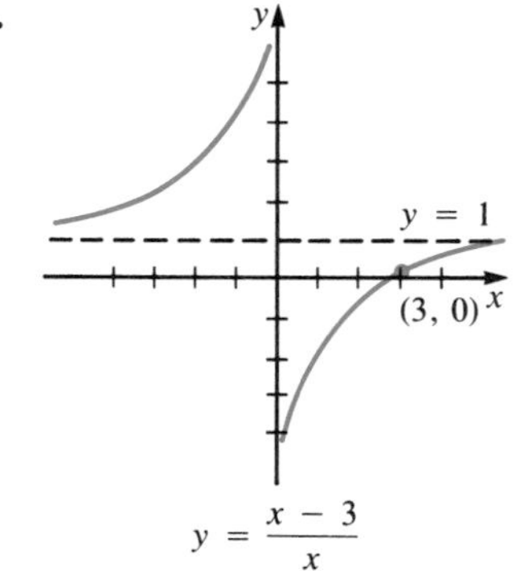

$$y = \frac{x - 3}{x}$$

11.

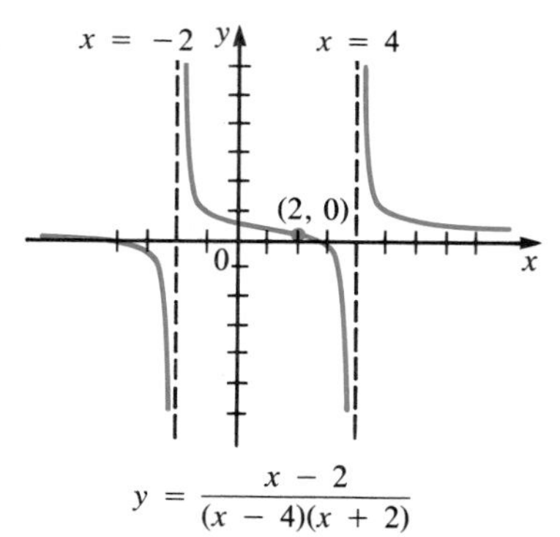

13. Ellipse
center: (0, 0)
foci: $(0, \pm\sqrt{7})$
vertices: $(0, \pm 4)$
major axis: $\overline{(0, -4)(0, 4)}$
minor axis: $\overline{(-3, 0)(3, 0)}$
eccentricity: $\sqrt{7}/4$

$$\left(\frac{x}{3}\right)^2 + \left(\frac{y}{4}\right)^2 = 1$$

15. Parabola
focus: $(0, \frac{9}{64})$
directrix: $y = -\frac{9}{64}$
axis: $x = 0$
vertex: (0, 0)

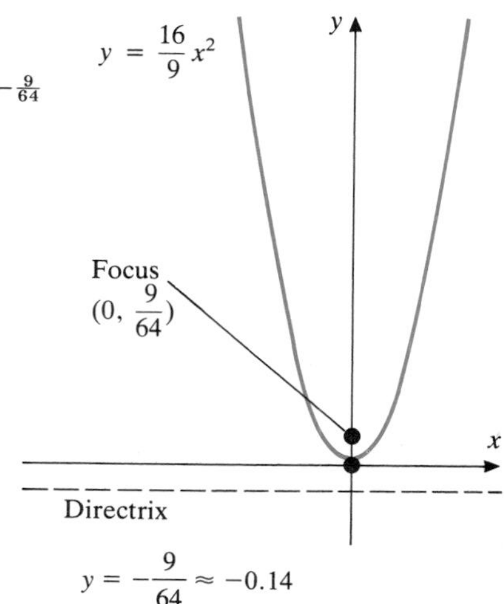

17. Hyperbola
center: (0, 0)
foci: $(0, \pm 5)$
vertices: $(0, \pm 3)$
eccentricity: $\frac{5}{3}$
transverse axis: $\overline{(0, -3)(0, 3)}$
asymptotes: $y = \pm\frac{3}{4}x$
conjugate axis: $\overline{(-4, 0)(4, 0)}$

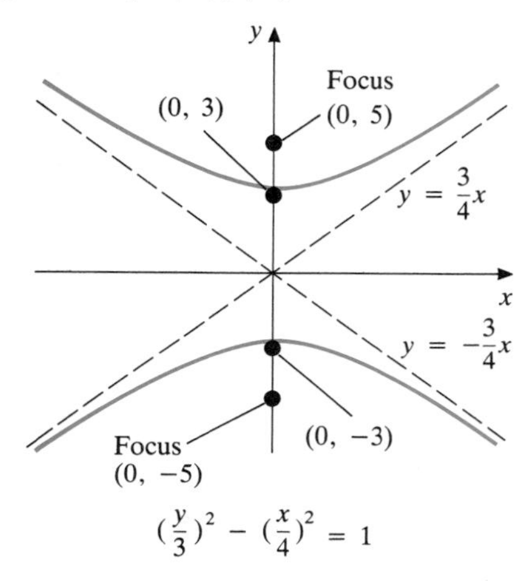

$$\left(\frac{y}{3}\right)^2 - \left(\frac{x}{4}\right)^2 = 1$$

19. Ellipse
center: $(1, -1)$
foci: $(1, -1-\sqrt{5})$, $(1, -1+\sqrt{5})$
vertices: $(1, -4)$, $(1, 2)$
major axis: $\overline{(1, -4)(1, 2)}$
minor axis: $\overline{(-1, -1)(3, -1)}$
eccentricity: $\sqrt{5}/3$

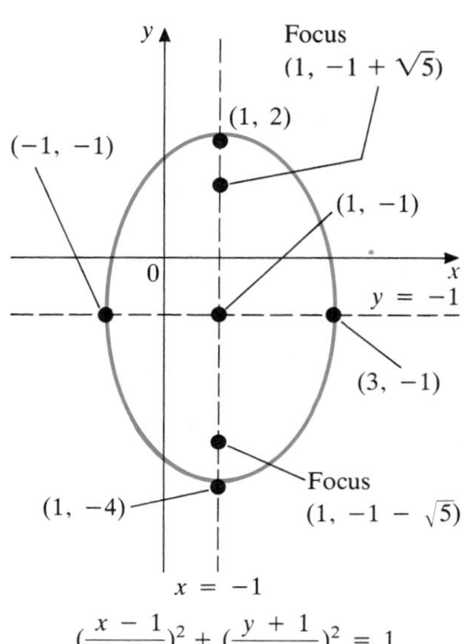

21. single point $(-2, 5)$

23. two straight lines: i.e., a degenerate hyperbola

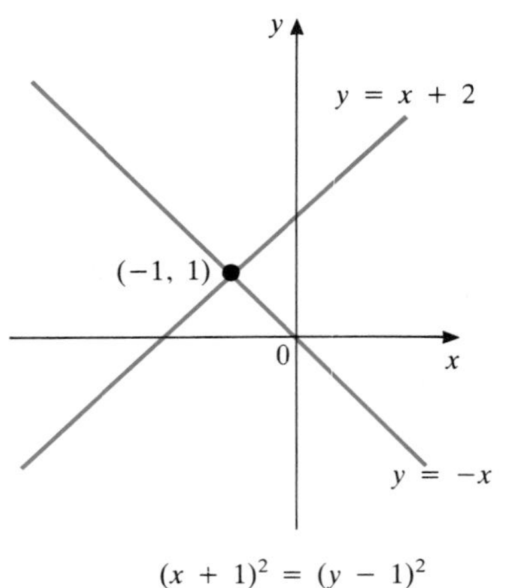

$$(x + 1)^2 = (y - 1)^2$$

25. Ellipse
center: $(-\frac{1}{2}, -4)$
foci: $(-\frac{1}{2}, -4 - 3/\sqrt{2}), (-\frac{1}{2}, -4 + 3/\sqrt{2})$
vertices: $(-\frac{1}{2}, -4 - 3\sqrt{2}), (-\frac{1}{2}, -4 + 3\sqrt{2})$
major axis: $(-\frac{1}{2}, -4 - 3\sqrt{2})(-\frac{1}{2}, -4 + 3\sqrt{2})$
minor axis: $(-\frac{1}{2} - 3\sqrt{\frac{3}{2}}, -4)(-\frac{1}{2} + 3\sqrt{\frac{3}{2}}, -4)$
eccentricity: $\frac{1}{2}$

$$\frac{(x + \frac{1}{2})^2}{27/2} + \frac{(y + 4)^2}{18} = 1$$

27. unique ellipse: $\dfrac{x^2}{25} + \dfrac{y^2}{16} = 1$

29. $\dfrac{y^2}{4} - \dfrac{x^2}{5} = 1$

Chapter 6

Problems 6.1, page 332

1.

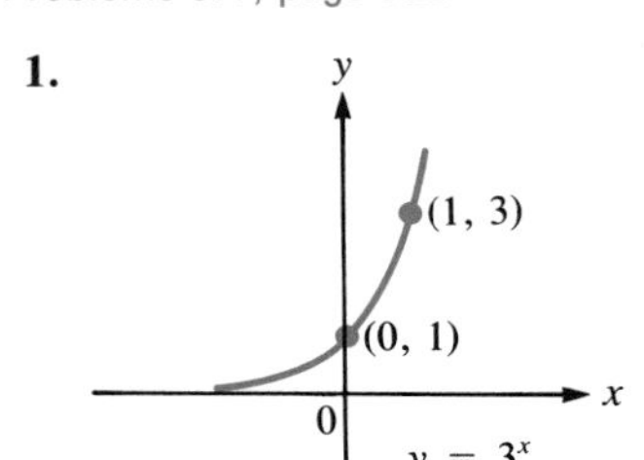

3.

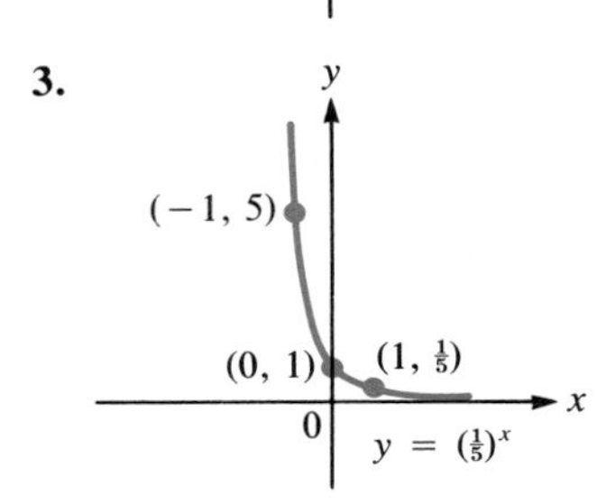

5.

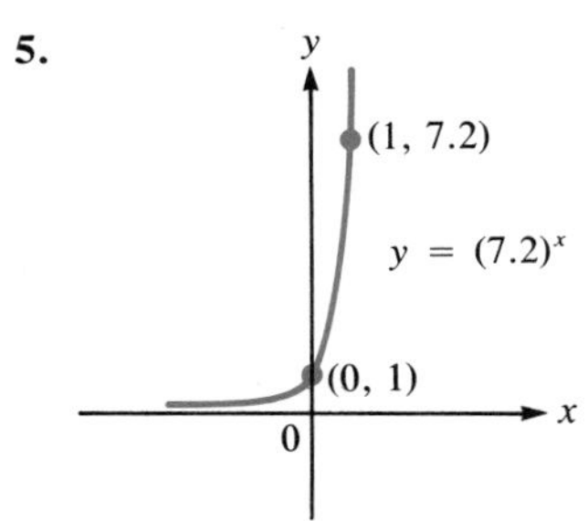

7.

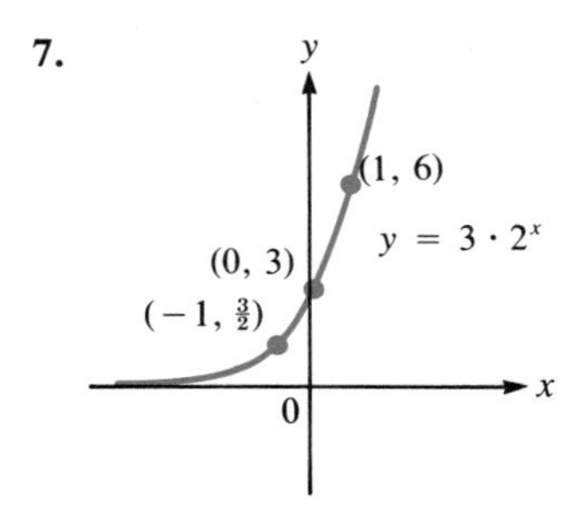

9.

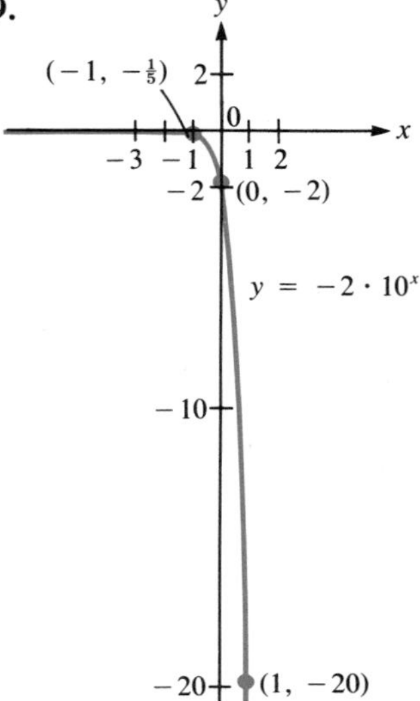

11.

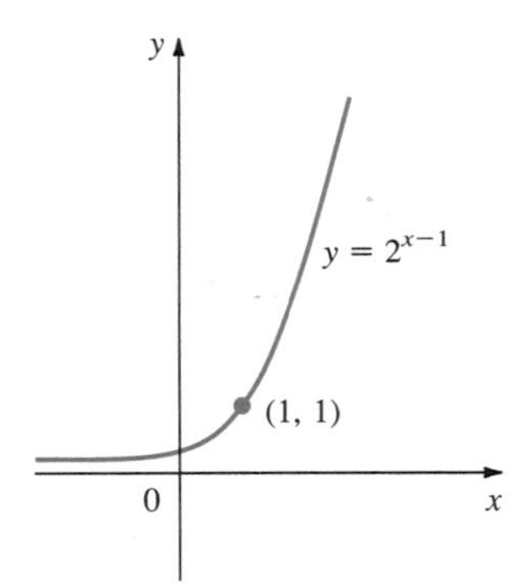

13.

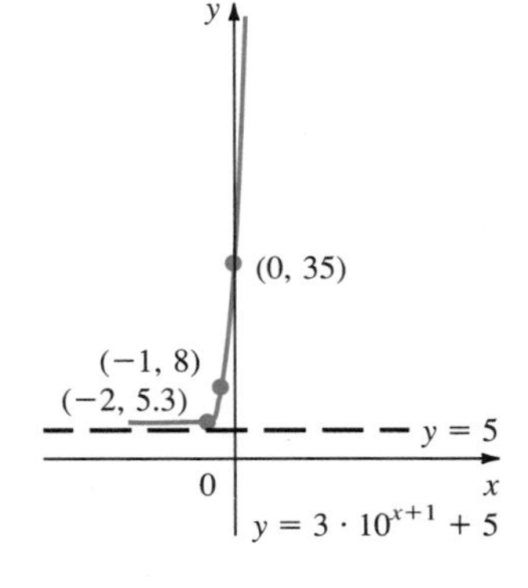

15. 158.4893192 **17.** 0.10881882
19. 0.643440775 **21.** 4.728804388
23. \$6312.38; interest is \$1312.38
25. \$6352.45; interest is \$1352.45
27. \$12,348.08; interest is \$4348.08
29. \$12,418.65; interest is \$4418.65
31. \$16,564.17; interest is \$6564.17
33. 79.08% annually; 81.40% quarterly so percentage difference is 2.32%
35. $5\frac{1}{8}$% semiannually (5.191% > 5.127%)
37. 3.61224%
39. (a) 12.36% for *A* and 11.57% for *B* (b) Account *A* by \$24.77
41. *d* **43.** *a*
45.

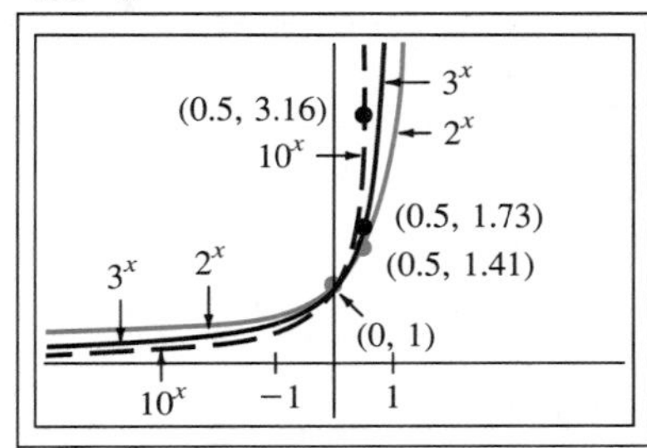

47.

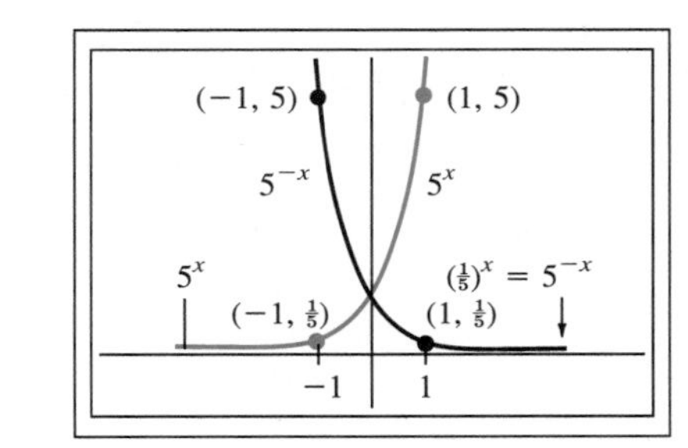

49.

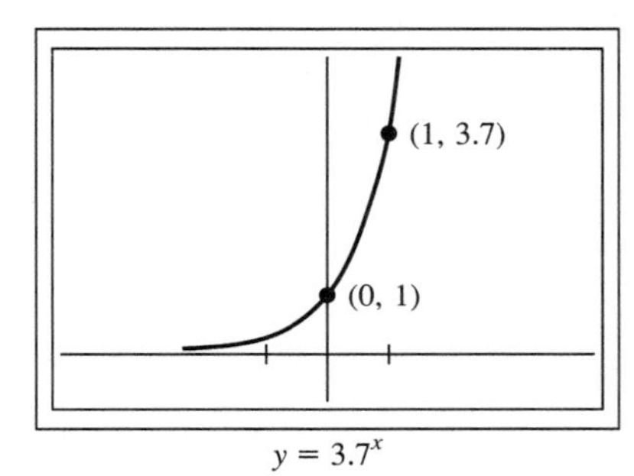

$y = 3.7^x$

51.

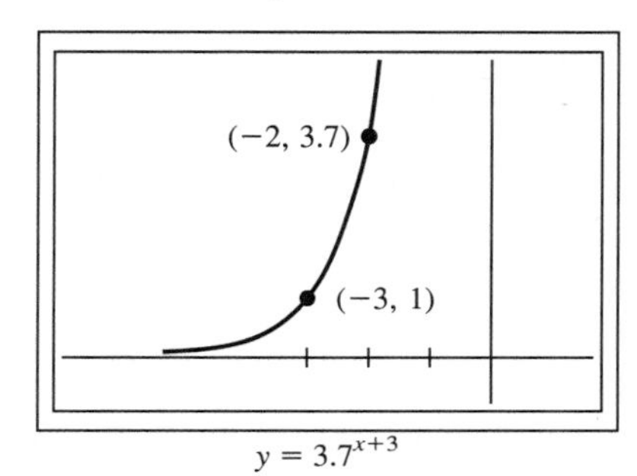

$y = 3.7^{x+3}$

53.

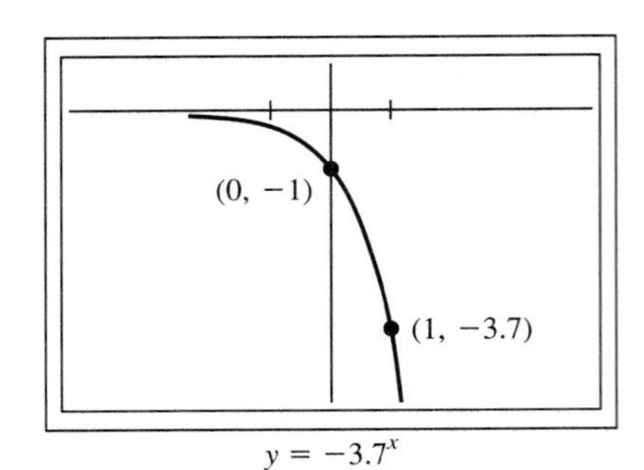

$y = -3.7^x$

55.

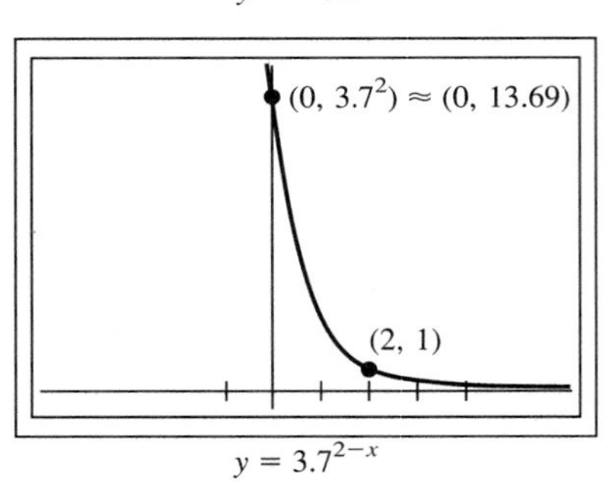

$y = 3.7^{2-x}$

57.

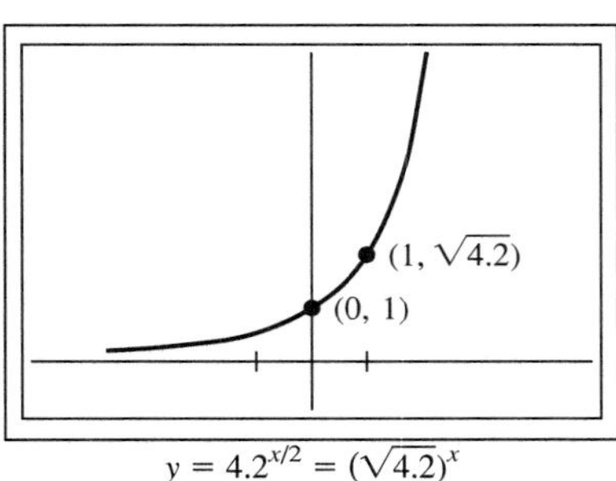

$y = 4.2^{x/2} = (\sqrt{4.2})^x$

59.

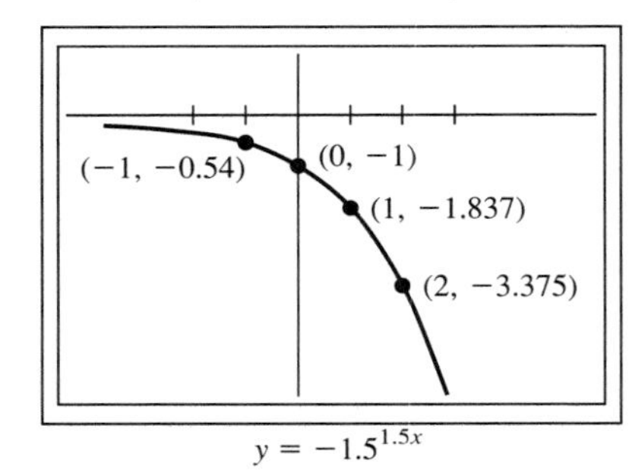

$y = -1.5^{1.5x}$

61.

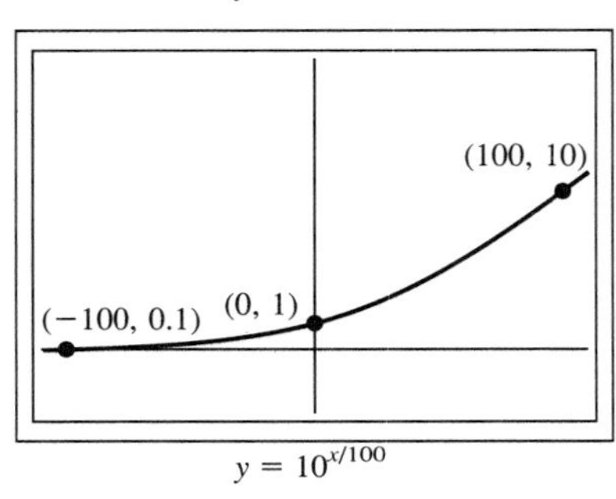

$y = 10^{x/100}$

Problems 6.2, page 342

1. 23.3361 **3.** 5.65223 **5.** 5.86486×10^{12} **7.** 1.24371×10^{-7} **9.** 23.1407 **11.** *f* **13.** *a* **15.** *c* **17.** *d* **19.** *i*

21.

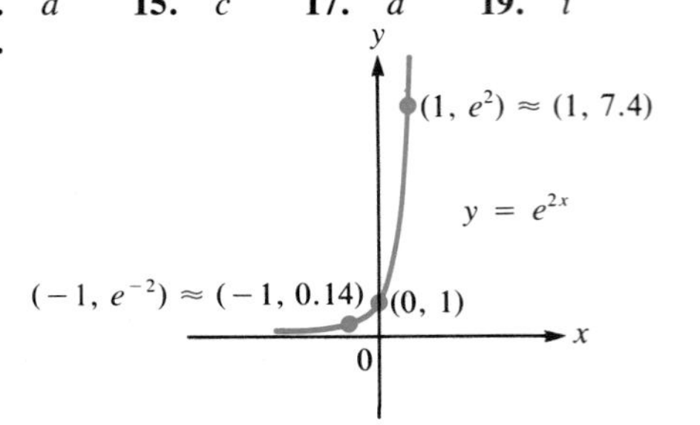

23.

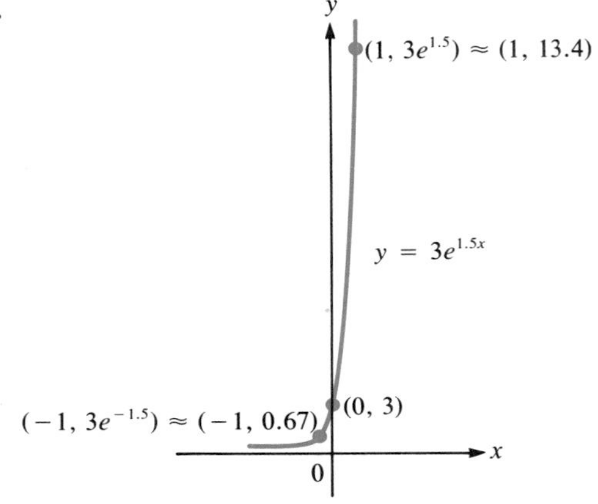

25.

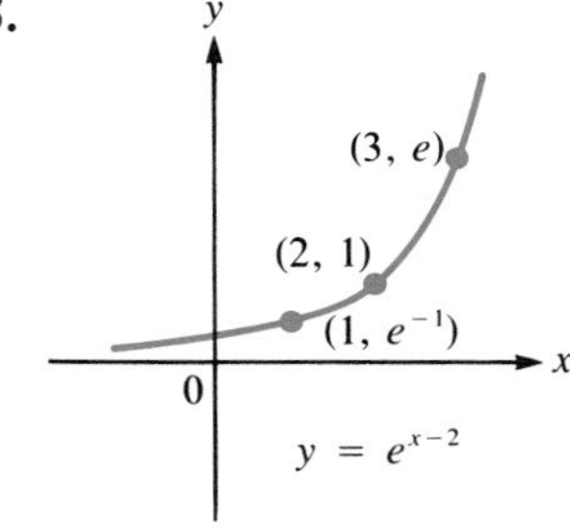

27.

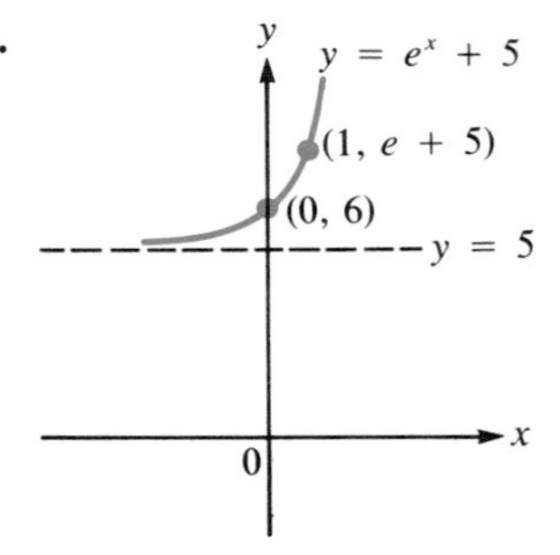

29. (a) 1.1388 (b) 0.69073
31. \$14,838.49; interest \$5838.49
33. \$28,292.17; interest \$18,292.17
35. \$20,544.33 **37.** the first (5.127% > 5.125%)
39. 5.8 years
41. (a) 12.36%, 11.63% (b) the first, by \$23.04
43. (a) 5,042,728,201 (b) 5,573,076,555 (c) 6,159,202,133
45. (a) 121.35°F (b) 71.49°F (c) 53.55°F
47. (c)

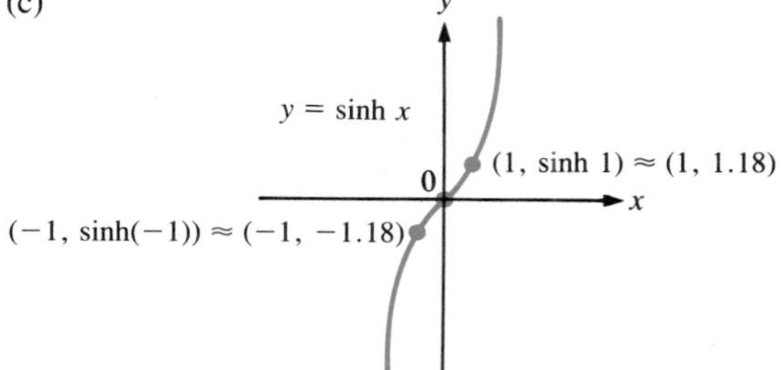

49. (d)

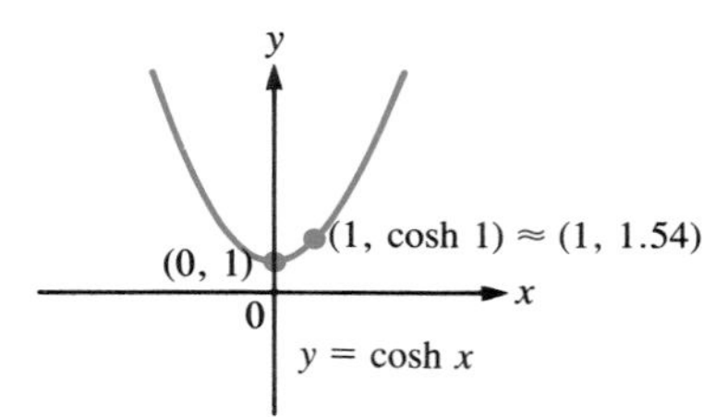

51.

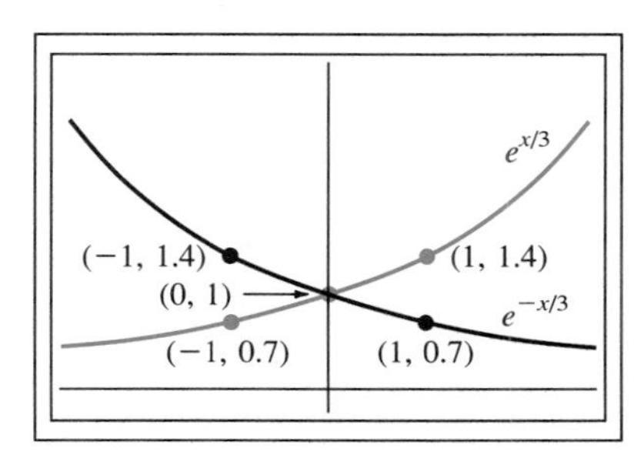

53.

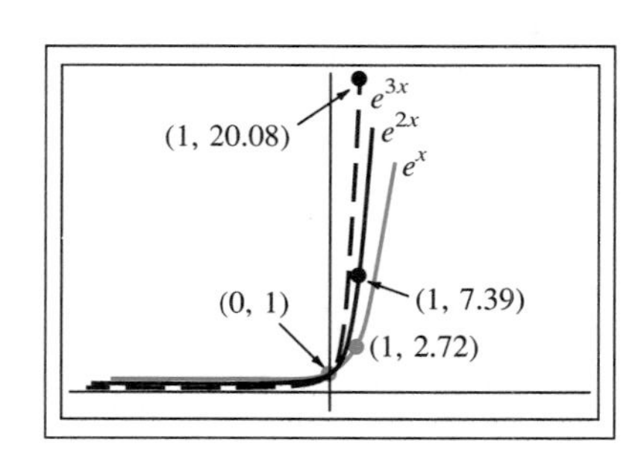

55.

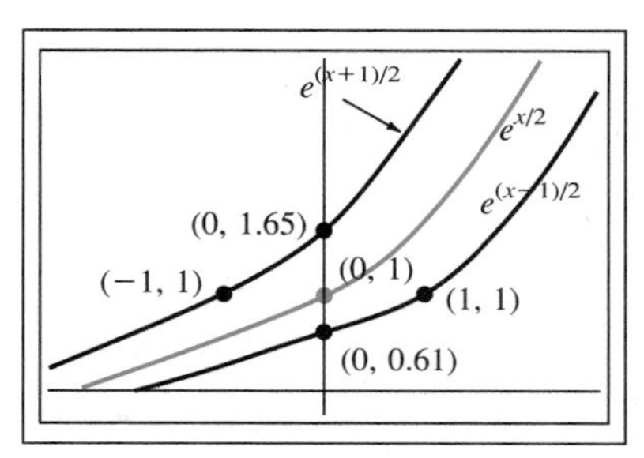

57.

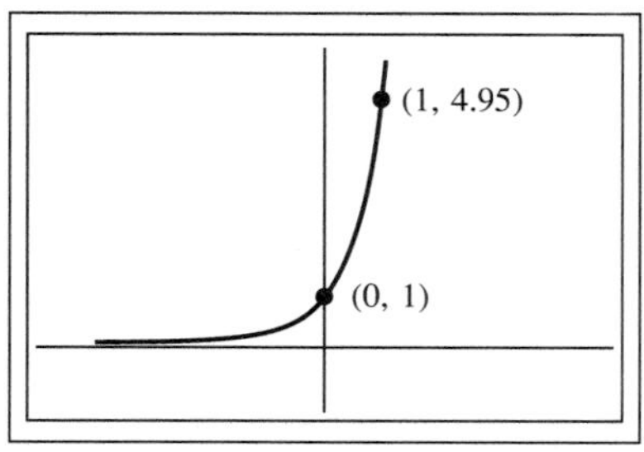

$y = e^{1.6x}$

59.

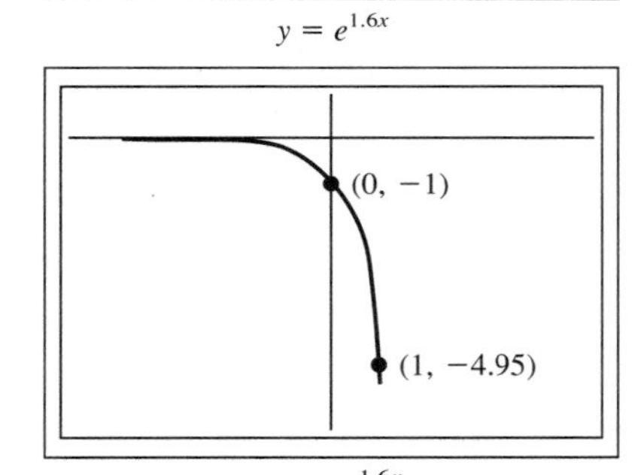

$y = -e^{1.6x}$

61.

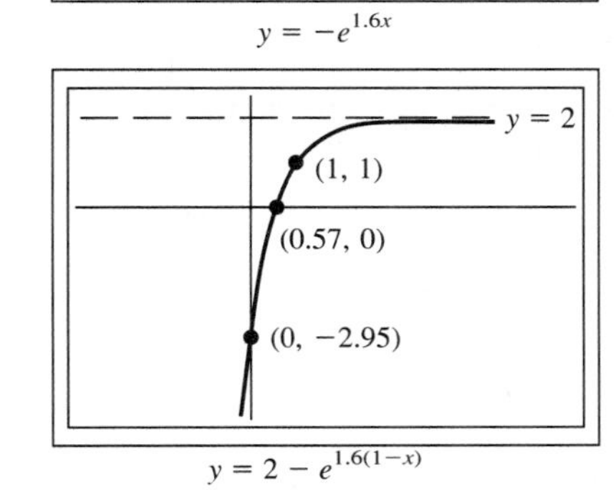

$y = 2 - e^{1.6(1-x)}$

63.

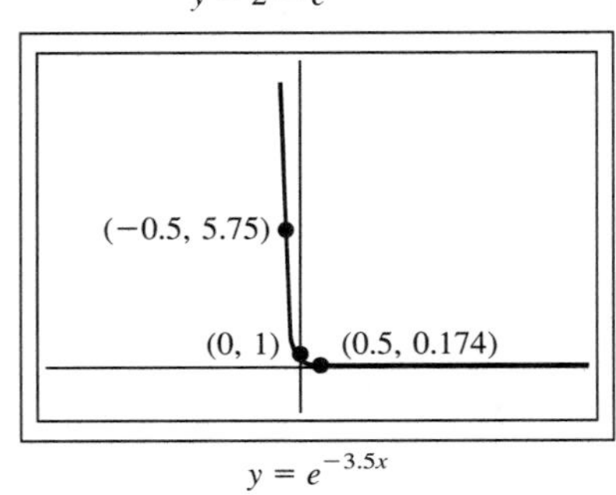

$y = e^{-3.5x}$

65.

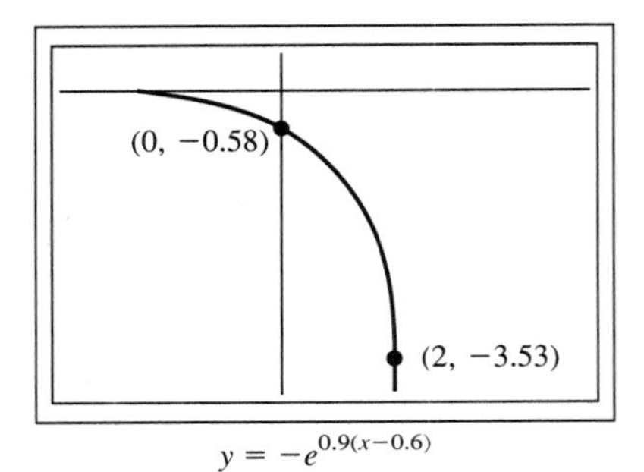

$y = -e^{0.9(x-0.6)}$

67.

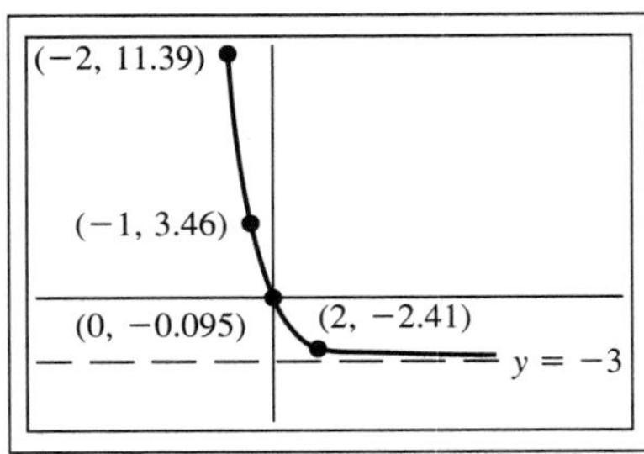

$y = -3 + 4e^{-0.8(x+0.4)}$

69.

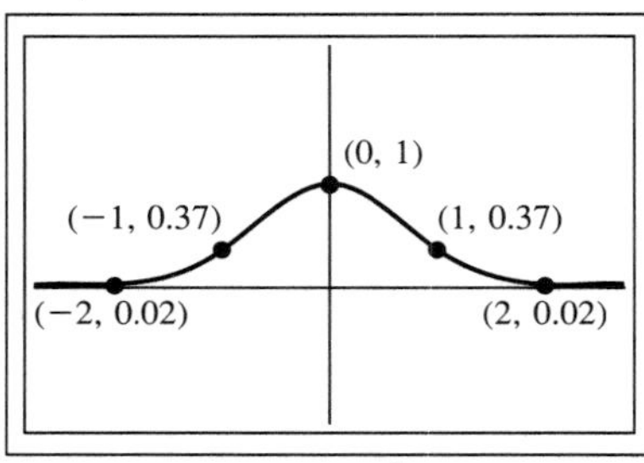

$y = e^{-x^2}$

71.

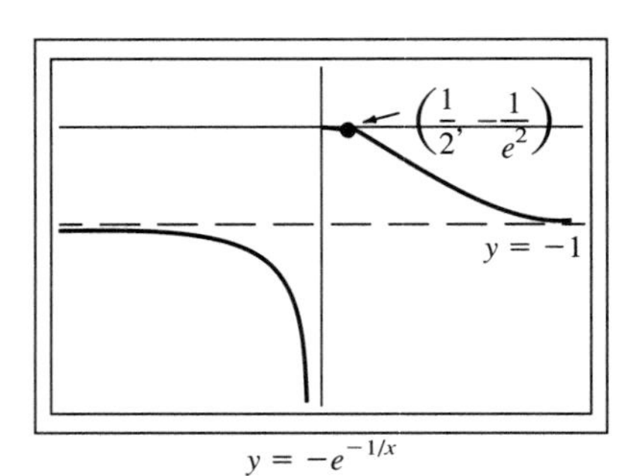

$y = -e^{-1/x}$

73.

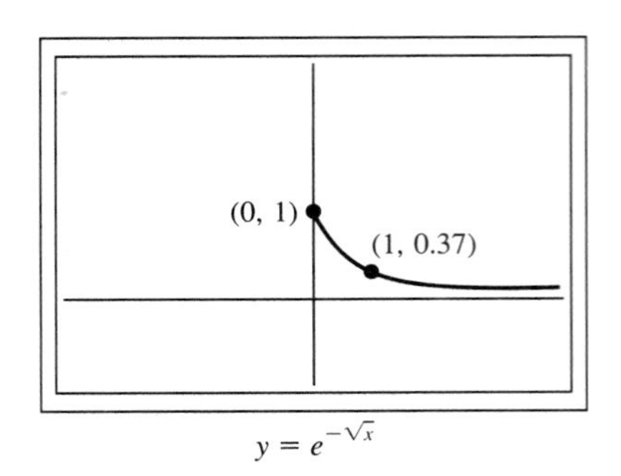

$y = e^{-\sqrt{x}}$

Problems 6.3, page 350

1. $16^{1/2} = 4$ **3.** $(\frac{1}{2})^{-4} = 16$ **5.** $12^0 = 1$
7. $10^{-3} = 0.001$ **9.** $e^2 = e^2$ **11.** $\log_2 32 = 5$
13. $\log_{1/3} 9 = -2$ **15.** $\log_{10} 2 = 0.301029995$
17. 3 **19.** −2 **21.** $-\frac{1}{2}$ **23.** 5 **25.** −9
27. −4.5 **29.** b **31.** c **33.** 2.0793
35. 3.5836 **37.** 2.5378 **39.** −0.4622
41. 0.2689 **43.** −1.7689 **45.** −4.6134
47. $4 \log_a x + 3 \log_a y$ **49.** $2 \log_a x + \frac{3}{4} \log_a y$
51. $\frac{1}{2} \log_a x + \frac{1}{3} \log_a y - \frac{1}{5} \log_a z$
53. $\frac{4}{5} \log_a x + \frac{4}{5} \log_a y - \frac{4}{5} \log_a z$ **55.** $\log_a 1 = 0$
57. $\log_a xy^2$ **59.** $\log_a 20wx$ **61.** $\log_a xy^2z^3$
63. $\log_a (x - 2)$ **65.** $\log_a (x^2y - 2y^2 + 2x^2 - 4y)$
67. $\log_a \{(x + y)^4/(x - y)^3\}$

Problems 6.4, page 359

1. 3 **3.** 20 **5.** 3.4 **7.** 0.235 **9.** π
11. 6.4 **13.** $\ln (x^2 + 3x)$ **15.** $\ln xyz$
17. $\ln (z^{1/2}w^{3/4}/(x^2 + 1)^2)$ **19.** ln 2
21. $\ln \dfrac{2^3 3^4}{w^3 z^4} = \ln \dfrac{648}{w^3 z^4}$ **23.** $(-1, \infty)$
25. $(-\infty, 1)$ **27.** $\{x: x \neq 0\}$
29. $\{x: x < -1 \text{ or } x > 1\}$ **31.** $\{x: x < -3 \text{ or } x > 4\}$
33. $\mathbb{R}$ **35.** $(0, \infty)$ **37.** f **39.** i **41.** e
43. b **45.** 144 **47.** $b^a \cdot a^b$ **49.** $\frac{10}{21}$
51. $54^3 = 157{,}464$ **53.** 9
55. (a) 2.8351 (b) 6.8351 (c) −1.1649 (d) −6.1649
57. (a) 1.9786 (b) 4.9786 (c) −1.0214 (d) −4.0214
59. 2.0214 **61.** 2.2364 **63.** −0.91755
65. 5.2865 **67.** 5.2131
69. (a) increases by $\ln 27 = 3 \ln 3$ (b) decreases by $\ln 1000 = 3 \ln 10$
71. 1.757699625 **73.** 4.698987376
75. 6.907755279 **77.** −3.42651519
79. 0.497149872 **81.** $\pi = 3.141592654$
83. 1.561842388 **85.** 1.351393465
87. 1.261859507 **89.** 1.796488803
91. 0.79248125 **93.** 1.10091231
95. (a) 0.4307 (b) 0.8614 (c) −1.2920

97. (a) $\ln 0.8 \approx -0.223143491$ and $\ln 1.2 \approx 0.182321542$
(b) 0.693143053
(c) $\ln 3 \approx 1.098604387$ and $\ln 8 \approx 2.079429159$

99. $100! \approx 9.32 \times 10^{157}$; $200! \approx 7.88 \times 10^{374}$

101. (c) $2 \pm \sqrt{5}$ (e) $\ln(2 + \sqrt{5}) \approx 1.4436$

103. $\ln(4 + \sqrt{17}) \approx 2.0947$

105. $\ln(2 \pm \sqrt{3}) \approx \pm 1.316957897$

107.

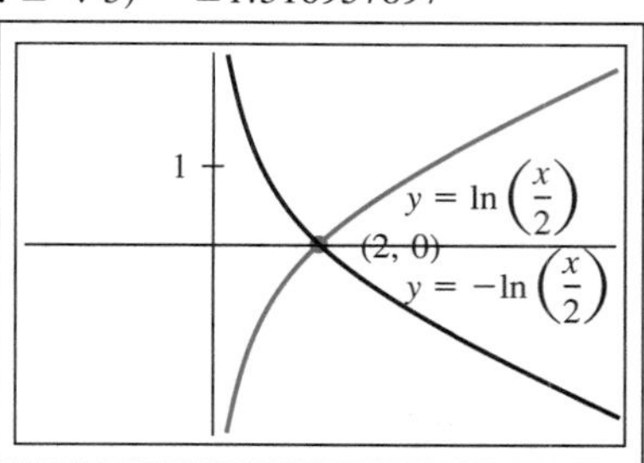

109.

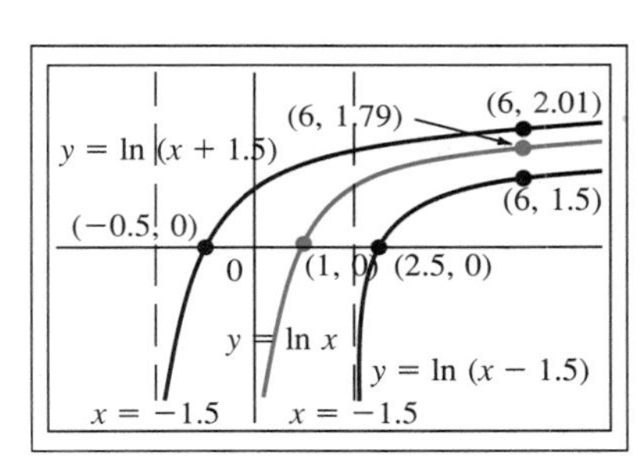

111.

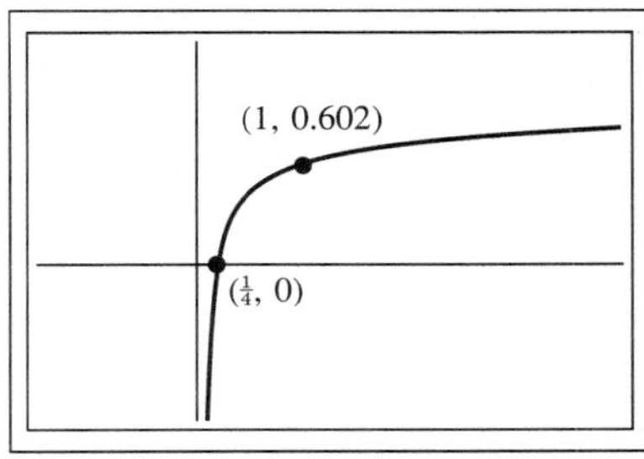

$y = \log 4x = \log x + \log 4$

113.

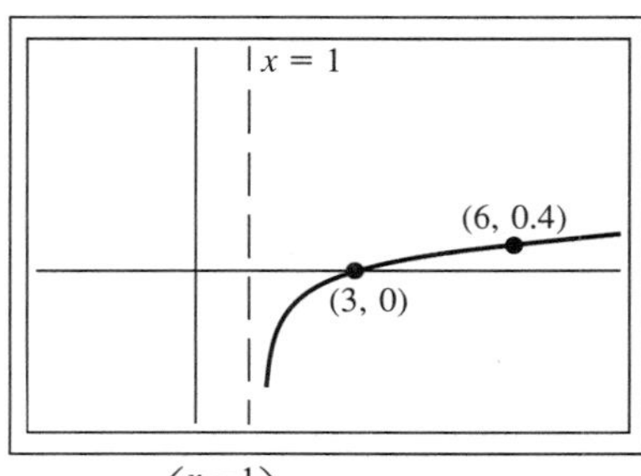

$y = \log\left(\frac{x-1}{2}\right) = \log(x - 1) - \log 2$

115.

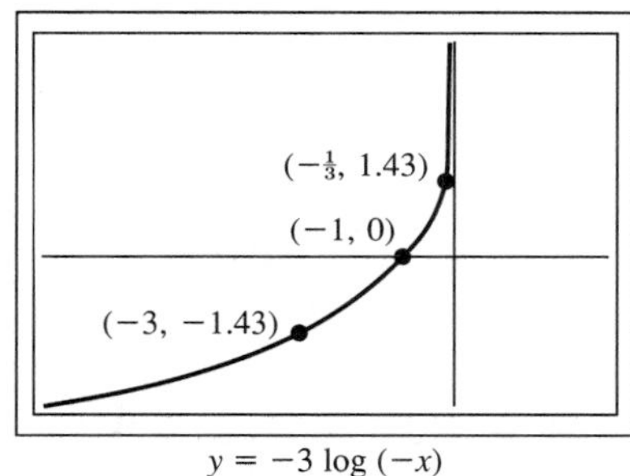

$y = -3 \log(-x)$

117.

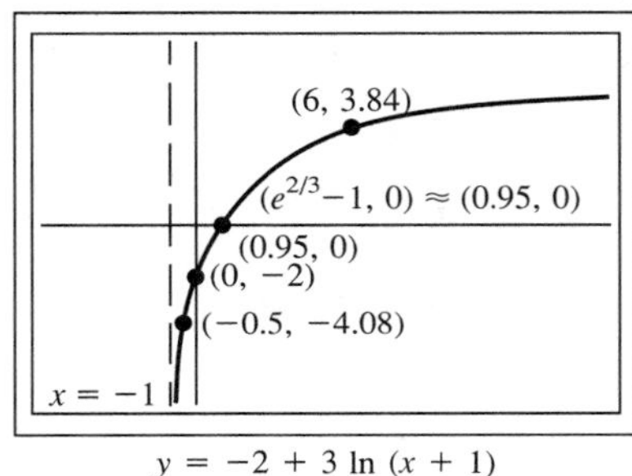

$y = -2 + 3 \ln(x + 1)$

119.

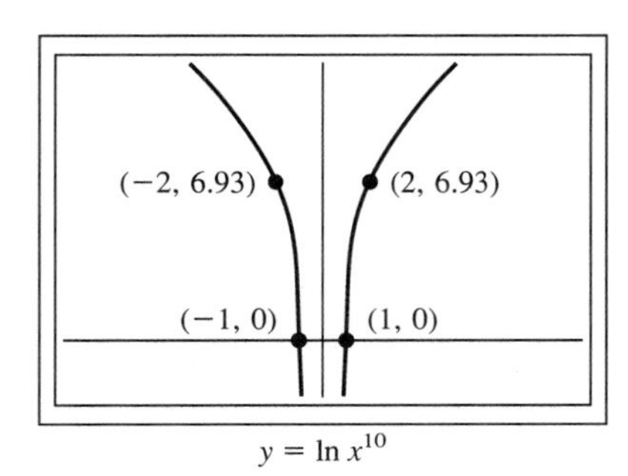

$y = \ln x^{10}$

121.

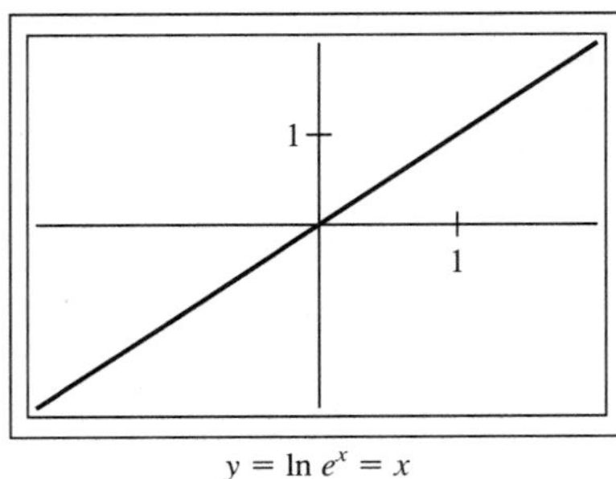

$y = \ln e^x = x$

123.

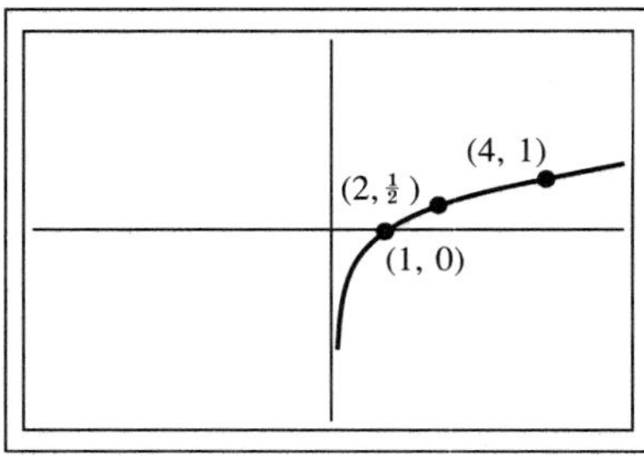

$y = \log_4 x = \dfrac{\ln x}{\ln 4}$

125.

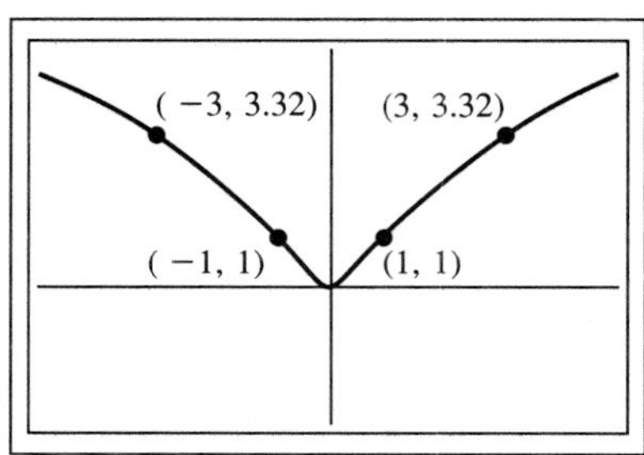

$$y = \log_2 (x^2 + 1) = \frac{\ln (x^2 + 1)}{\ln 2} \approx 1.44 \ln (x^2 + 1)$$

Problems 6.5, page 369

1. 2 **3.** −1 **5.** $2^8 = 256$ **7.** $e^{-1} \approx 0.368$
9. 10 **11.** 1 **13.** $\ln 10 \approx 2.303$
15. no solution **17.** $\dfrac{\log 14}{\log 7} = \dfrac{\ln 14}{\ln 7} \approx 1.356$
19. $\dfrac{\log 8}{\log 231} = \dfrac{\ln 8}{\ln 231} \approx 0.382$ **21.** $e^{1.6} \approx 4.953$
23. $10^{-1.57} \approx 0.027$ **25.** $1 + \ln 2 \approx 1.693$
27. $2 \ln 4 \approx 2.773$ **29.** $-5 \ln 5 \approx -8.047$
31. $\frac{1}{3}e \approx 0.906$ **33.** 4 **35.** 8.920
37. no solution [$z = -11$ is wrong because $\ln(-11 + 4)$ is undefined]
39. $\ln 4 \approx 1.386$ or $\ln 2 \approx 0.693$
41. $2 \ln 2 \approx 1.386$
43. no solution (e^x cannot equal −1 or −2)
45. 48 **47.** 0.0023 **49.** 4 **51.** $\frac{1}{10}$
53. $\frac{1}{2}\sqrt[3]{5} \approx 0.855$ **55.** ±2 **57.** $\frac{1}{5}$ **59.** 3
61. $\frac{1}{8}$ **63.** $-\frac{4999}{1000} = -4.999$ **65.** $\frac{1}{3}$ **67.** −4
69. no solution [$\log_4(-\frac{4}{3})$ is undefined]
71. 8 **73.** 0 **75.** 0.46, 5.36 **77.** 1.52
79. −0.86, 1.20 **81.** 3.82 **83.** −3.98, 282.12

Problems 6.6, page 380

1. 10.15 years
3. (a) 55.07 years (b) 22.11 years (c) 13.87 years (d) 11.12 years (e) 7.46 years
5. (a) 54.93 years (b) 21.97 years (c) 13.73 years (d) 10.99 years (e) 7.32 years
7. \$9185.12 **9.** \$6499.31 **11.** 3.596%
13. 7.324% **15.** in 2012 (26.39 years later)
17. (a) 273,037,968 (b) in 2065 (84.8 years after 1980)
19. (a) 19,057,278 (b) in 2004 (29.56 years after 1974)
21. in 2003 (22.74 years after 1980)
23. 13.76% **25.** 2.71% **27.** 141.46
29. (a) −4.138°C (b) 1.677 minutes = 100.62 seconds
31. (a) 135.69 years (b) 450.76 years
33. 4,620,981 years
35. (a) 625.53 mbar (b) 303.42 mbar (c) 429.03 mbar (d) 348.63 mbar (e) 57,396.3 meters
37. (a) 10,000 (b) 7 (c) after about 214.2 hours = 8 days, 22.2 hours
39. 7.22 **41.** 3.162×10^{-5} moles per liter
43. (a) 15.849 times as bright (b) 1.75 (c) 1.99526 times as bright (d) 11.633 (e) 1.6004

Review Exercises for Chapter 6, page 383

1.

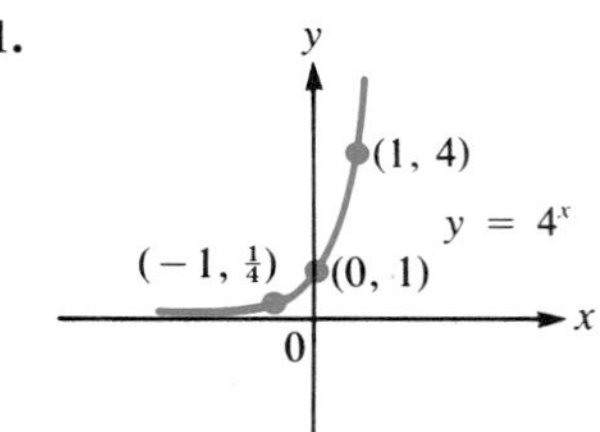

3.

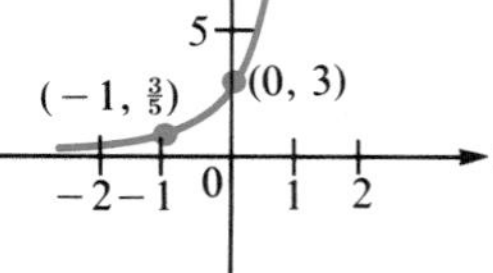

5.

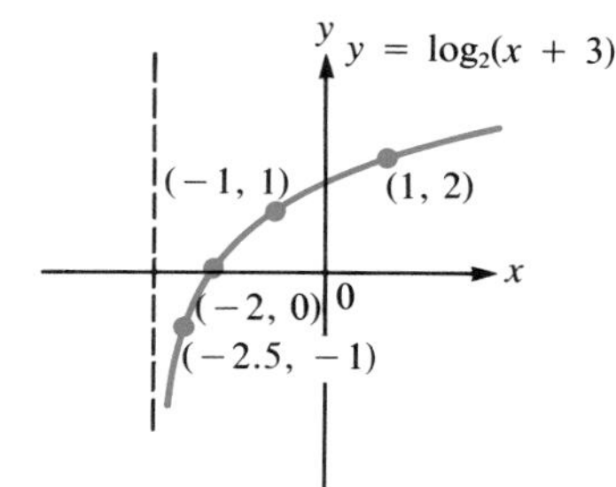

7.

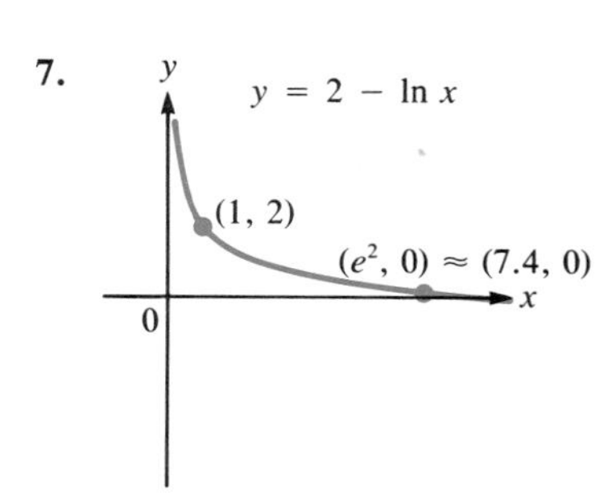

9. 5.473947392 **11.** 0.079913677
13. 1.45331834 **15.** 13.81551056 **17.** 4
19. 25 **21.** $10^{10^{-9}} \approx 1$
23. $e^2 - 3 = 4.389056099$
25. no solution [ln (−2) is not defined]
27. 1 (−3 is incorrect)
29. increases by $\ln 8 = 3 \ln 2$
31. (a) $9^{3/2} = 27$ (b) $(\frac{1}{2})^{-3} = 8$
33. $\log_3 (x^2 - 2x)$ **35.** $\log ((x + 1)z^2/(x + 3))$
37. $\ln \dfrac{(x + 1)^3(z + 12)^{1/3}}{(y + 4)^{1/2}}$ **39.** 2.807354922
41. 2.604726683 **43.** \$2100
45. \$9910.60; interest = \$1910.60
47. \$9892.33; interest = \$3892.33
49. 8.32871%
51. (a) 23.19 years (b) 10.75 years (c) 8.75 years (c) 5.86 years
53. 21,170; 44,817
55. (a) 54.85°C (b) 67.566 minutes
57. 3.1063 weeks ≈ 21.744 days

Chapter 7

Problems 7.1, page 388

1. (1, 2) **3.** (−2, 0) **5.** $(\frac{1}{2}, -\frac{3}{2})$
7. infinite number of solutions of form $(x, \frac{2}{3}x)$, where x is a real number.
9. (−1, 2) **11.** $(c/(b + a), c/(b + a))$ if $b \pm a \neq 0$
13. unique solution if $a \neq 0$ and $b \neq 0$
15. no solution if $a = 0$, $b = 0$, and either c or d is nonzero.
17. (1, −3) **19.** no point of intersection
21. $(\frac{38}{25}, -\frac{7}{25})$ **23.** \$3,500,000 in A, \$500,000 in B
25. No solution will use all materials.
27. 50 acres for corn, 450 acres for soybeans

Problems 7.2, page 401

1. $(\frac{8}{9}, -\frac{2}{3}, -\frac{7}{9})$ **3.** (0, 2, 0) **5.** $(\frac{26}{15}, \frac{2}{15}, -\frac{1}{3})$
7. (10, 14, −26)
9. $(-\frac{4}{5}z, \frac{9}{5}z, z)$, where z is any real number
11. no solution
13. $(z + 16, -z - 6, z)$ where z is any real number
15. $(\frac{20}{13} - \frac{4}{13}w, -\frac{28}{13} + \frac{3}{13}w, -\frac{45}{13} + \frac{9}{13}w, w)$, where w is any real number
17. $(18 - 4w, -\frac{15}{2} + 2w, -31 + 7w, w)$, where w is any real number
19. (3, −2) **21.** yes **23.** yes **25.** no
27. yes **29.** no
31. 1100 units of product 1, 1450 units of product 2, 840 units of product 3
33. 6 days in England, 4 days in France, 4 days in Spain
35. no unique solution (two equations in three unknowns); if 200 shares of McDonald's, then 300 shares in Eastern and 100 shares in Hilton
37. no unique solution; graze $\frac{72}{7} - \frac{2}{7}R$, $R \geq 6$; move $\frac{96}{7} - \frac{5}{7}R$, $R \geq 6$; Rest (R) at least 6 hours
39. 20 units of each
41. 1.5 batches love potion, 1 batch of cold remedy
43. 32 sodas, 128 shakes
45. 6 units of A, 2.5 units of B
47. 4 mechanics, 3 attendants **49.** $b = 2a - c$
51. (1.90081, 4.19411, −11.34852)
53. $(-2y, y)$, where y is any number
55. $(5y, y)$, where y is any real number
57. $(\frac{1}{6}z, \frac{5}{6}z, z)$, where z is a real number
59. $(-\frac{4}{7}z, \frac{5}{7}z, z)$, where z is a real number
61. $(-\frac{13}{5}z + \frac{2}{5}w, \frac{22}{5}z - \frac{3}{5}w, z, w)$, where z and w are real numbers
63. (0, 0, 0, 0) **65.** $(3y, y)$, where y is a real number
67. If $a_{11}a_{22} - a_{12}a_{21} = 0$, the two equations are equations of the same line.
69. $k = \frac{161}{11}$

Problems 7.3, page 410

1. $(-\sqrt{2}, -\sqrt{2})$, $(\sqrt{2}, \sqrt{2})$

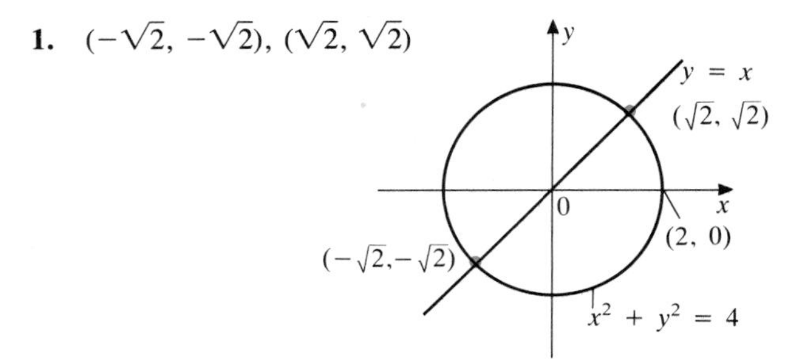

3. $(-3, -4), (4, 3)$

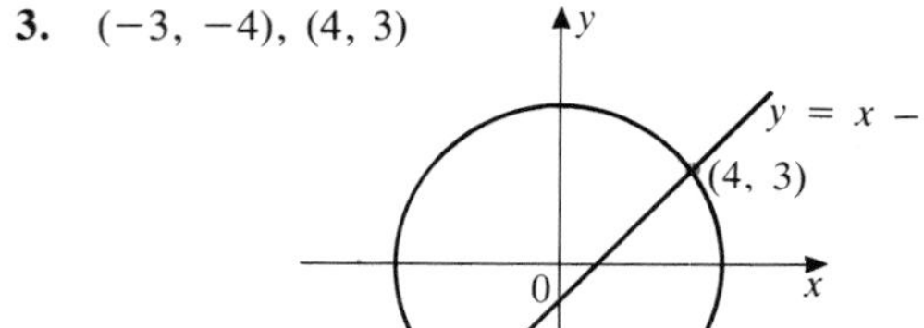

5. $(0, 0), (2, 4)$

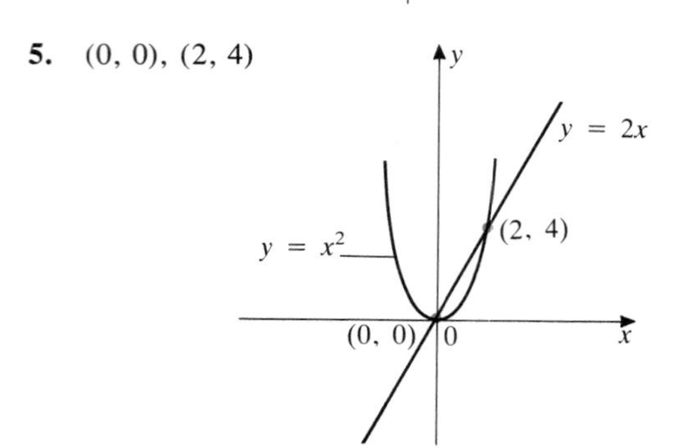

7. $(-3, -2), (-3, 2), (3, -2), (3, 2)$

9. $(-2, 0), (2, 0)$

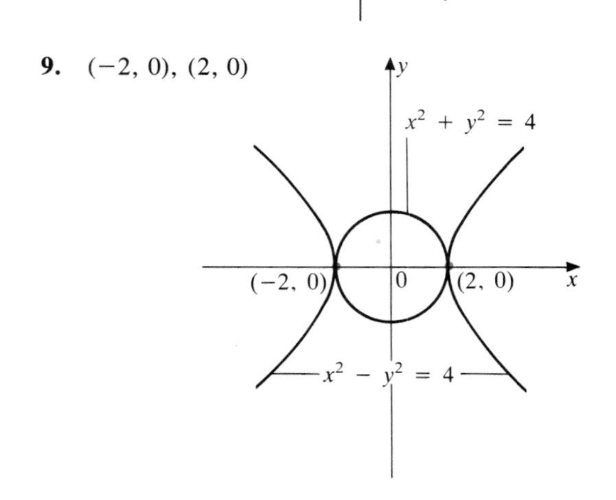

11. $(-2\sqrt{5}, 1 - \sqrt{5}), (2\sqrt{5}, 1 + \sqrt{5})$

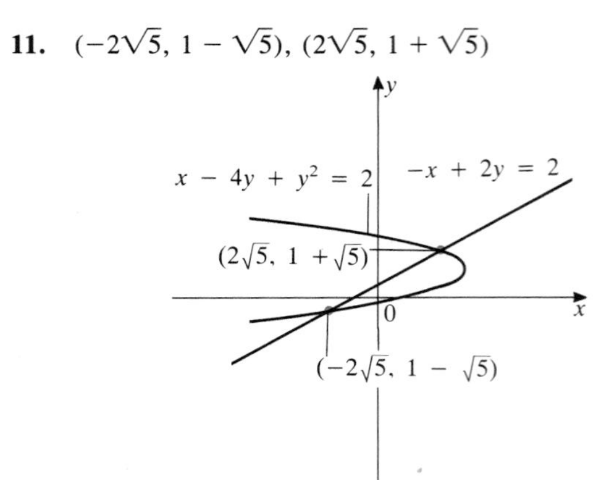

13. $(0, 0)$

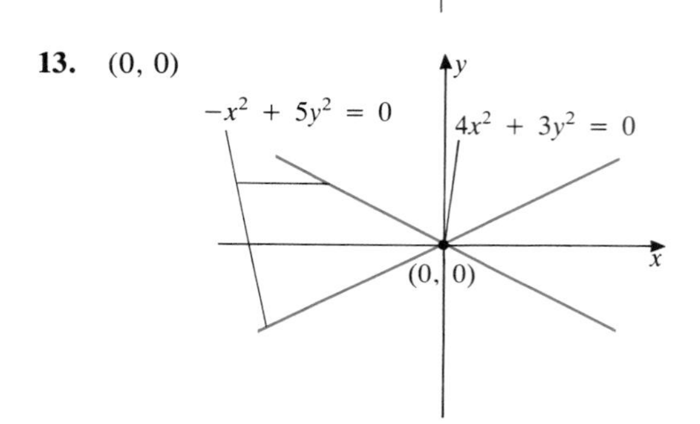

15. $\left(-4, \frac{15}{4}\right), (5, -3)$

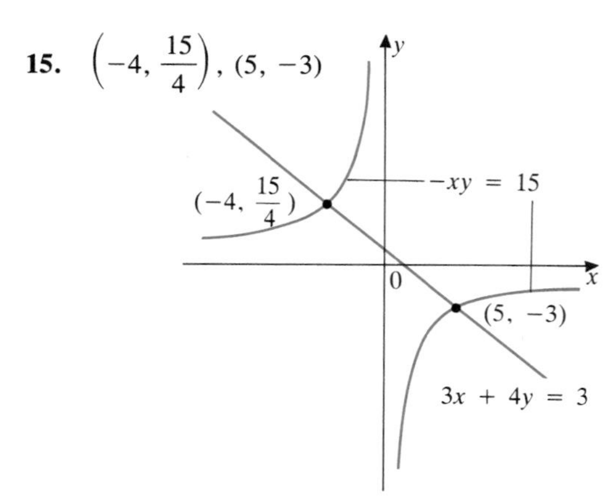

17. $\left(\pm\sqrt{\frac{\sqrt{65} - 1}{8}}, \frac{\sqrt{65} - 1}{4}\right) \approx (\pm 0.94, 1.77)$

19. $(3 + \sqrt{5}, 3 - \sqrt{5}), (3 - \sqrt{5}, 3 + \sqrt{5})$

21. $\left(-\frac{1}{3}, -\frac{1}{3}\right)$ **23.** no real solutions

25. no real solutions **27.** no real solutions

29. $(-3, 0), (2, -5)$ **31.** $\left(\frac{1}{2}, -\frac{1}{3}, 1\right)$

33. $(0, 1)$ **35.** $(\ln 4, 5)$ **37.** $(e^2, 5)$

39. $(0, 3)$

41. $(i, 2), (-i, 2)$

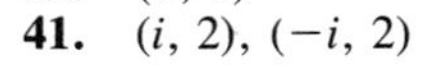

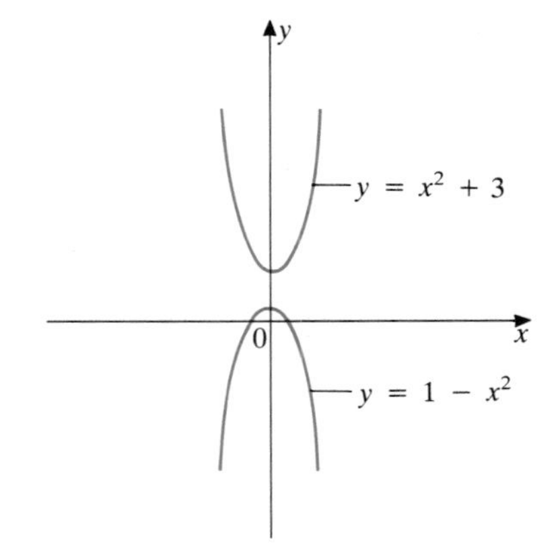

43. $\left(-\sqrt{\frac{5}{2}}, -\sqrt{\frac{3}{2}}i\right), \left(-\sqrt{\frac{5}{2}}, \sqrt{\frac{3}{2}}i\right), \left(\sqrt{\frac{5}{2}}, -\sqrt{\frac{3}{2}}i\right), \left(\sqrt{\frac{5}{2}}, \sqrt{\frac{3}{2}}i\right)$

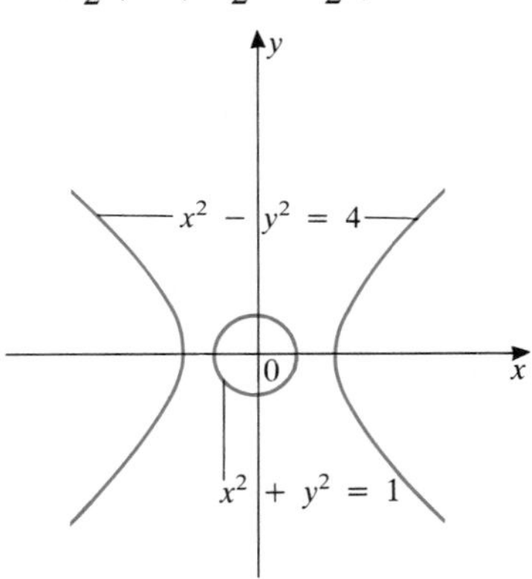

45. $\left(-\sqrt{\frac{48}{7}}, -\sqrt{\frac{5}{7}}i\right), \left(-\sqrt{\frac{48}{7}}, \sqrt{\frac{5}{7}}i\right), \left(\sqrt{\frac{48}{7}}, -\sqrt{\frac{5}{7}}i\right), \left(\sqrt{\frac{48}{7}}, \sqrt{\frac{5}{7}}i\right)$

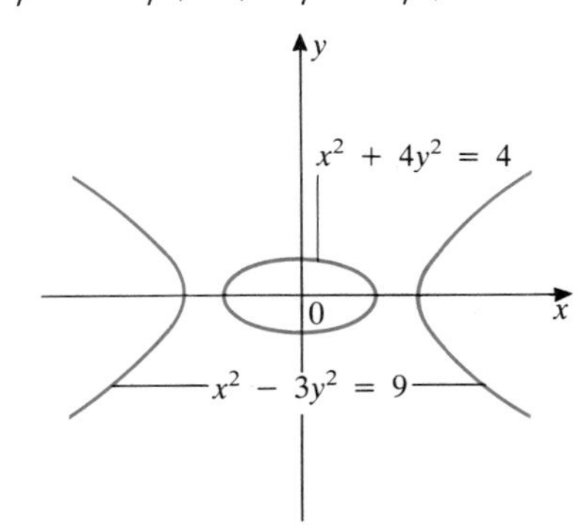

47. 8 and 12 **49.** 4 and 12 **51.** $\frac{1}{2}$ cm, $\frac{7}{2}$ cm

53. 144 people at \$250 per person

Problems 7.4, page 417

1. $\frac{\frac{1}{2}}{x-1} + \frac{-\frac{1}{2}}{x+1}$ **3.** $\frac{2}{x+1} + \frac{5}{x-3}$

5. $2 + \frac{13}{x-1} - \frac{8}{x-3}$ **7.** $\frac{7}{18(x-5)} - \frac{1}{18(x+1)}$

9. $\frac{1}{12(x+2)} + \frac{1}{4(x-2)} - \frac{1}{3(x-1)}$

11. $\frac{3}{x+1} - \frac{2}{x+2} - \frac{4}{x+3}$

13. $\frac{-2}{x} + \frac{\frac{1}{2}}{x+1} + \frac{\frac{3}{2}}{x-1}$

15. $\frac{3}{x-2} + \frac{4}{(x-2)^2} + \frac{2}{(x-2)^3}$

17. $x^2 + 4 + \frac{2}{x^2+1}$ **19.** $\frac{2}{x+1} + \frac{3}{(x+1)^2}$

21. $-\frac{1}{(1+x)^2} + \frac{2}{(1+x)^3}$

23. $\frac{-\frac{1}{35}}{x+3} + \frac{\frac{6}{35}}{x-2} + \frac{-\frac{1}{7}x - \frac{4}{7}}{x^2+x+1}$

25. $\frac{-\frac{1}{12}}{x+2} + \frac{\frac{1}{12}}{x-2} + \frac{\frac{1}{6}}{x+1} + \frac{-\frac{1}{6}}{x-1}$

27. $\frac{-\frac{1}{12}}{(x+3)^2} + \frac{\frac{1}{12}}{(x-3)^2}$ **29.** $-x - \frac{8}{x-4} - \frac{8}{x+4}$

31. $\frac{2}{x} - \frac{\frac{5}{4}}{x-1} + \frac{\frac{3}{2}}{(x-1)^2} - \frac{\frac{3}{4}}{x+1}$

33. $\frac{2x-1}{x^2+2} + \frac{-3x+2}{x^2+3}$ **35.** $-\frac{1}{x} + \frac{1}{x^3} + \frac{2x}{x^2+1}$

37. $\frac{23x+6}{5(x^2+x+1)} + \frac{2x+139}{5(x^2-4x-9)}$

39. $\frac{x-4}{x^2+4x+13} + \frac{3x+53}{(x^2+4x+13)^2}$

Problems 7.5, page 423

In the graphs below each solution set is screened (i.e., shaded).

1.

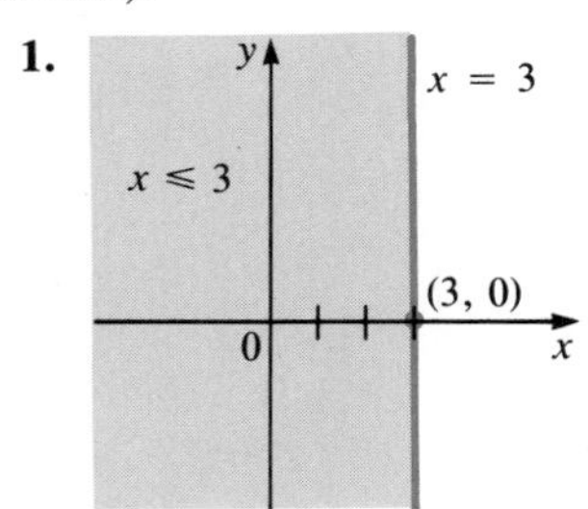

3.

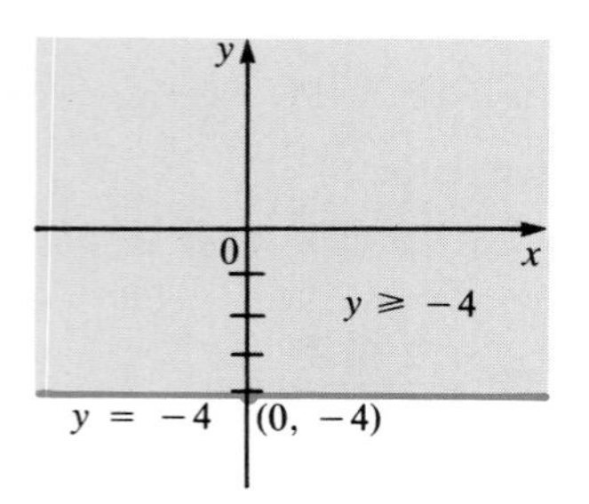

5.

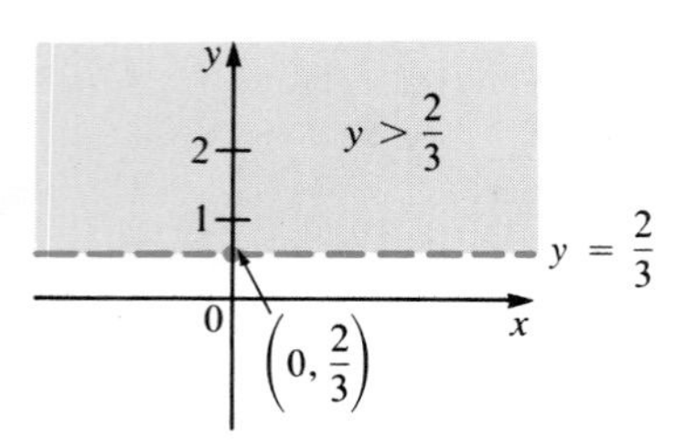

7.

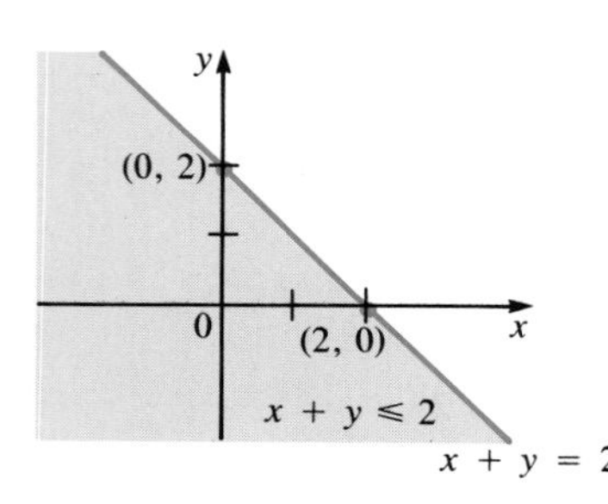

9.

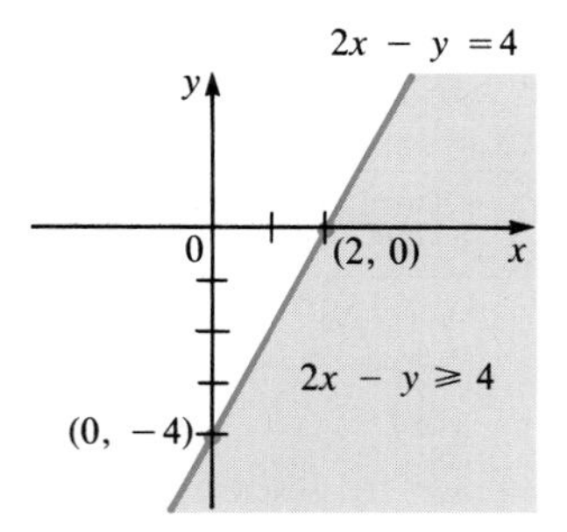

11.

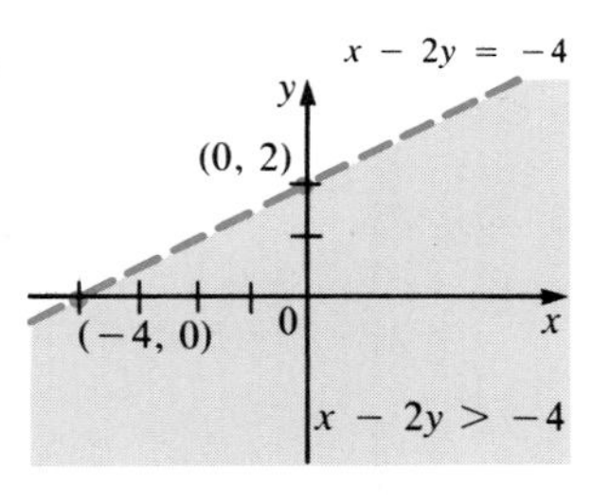

13.

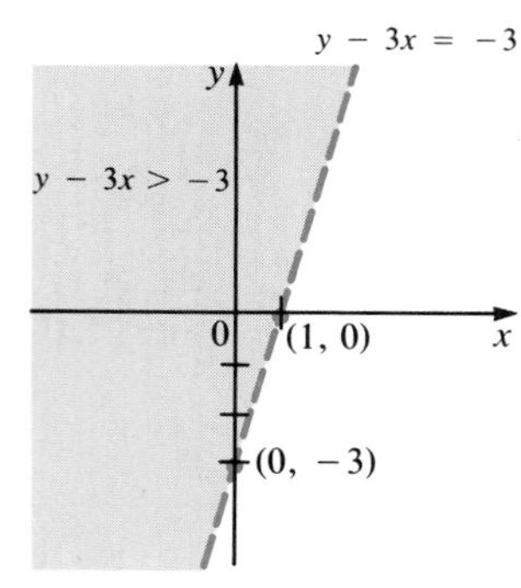

15.

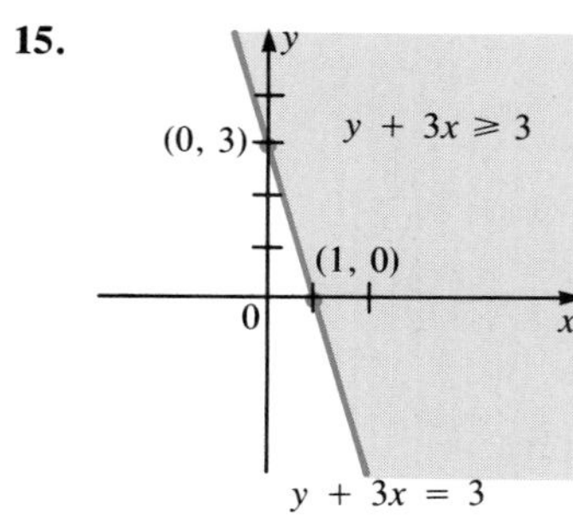

17.

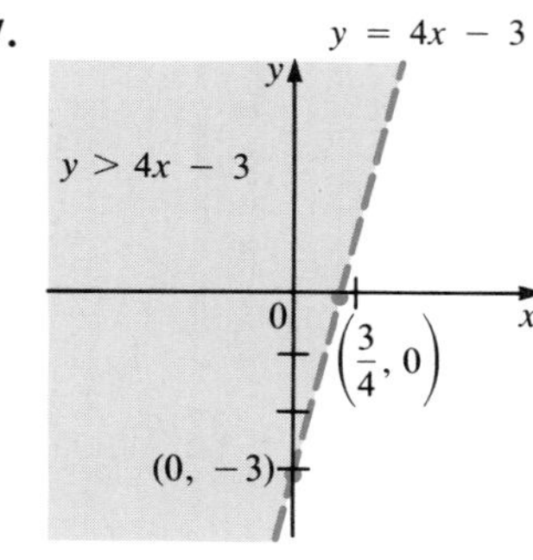

19.

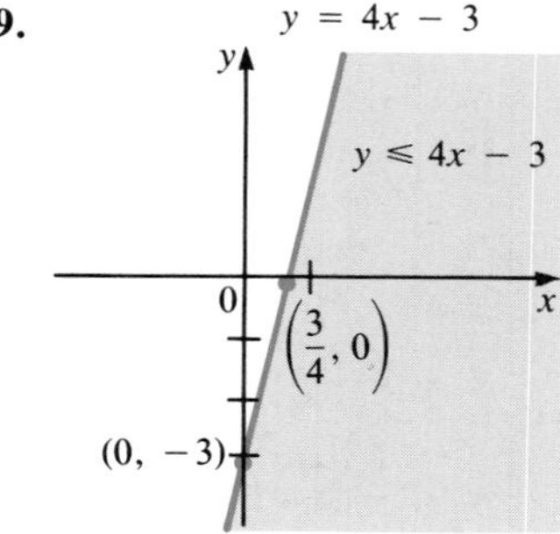

21.

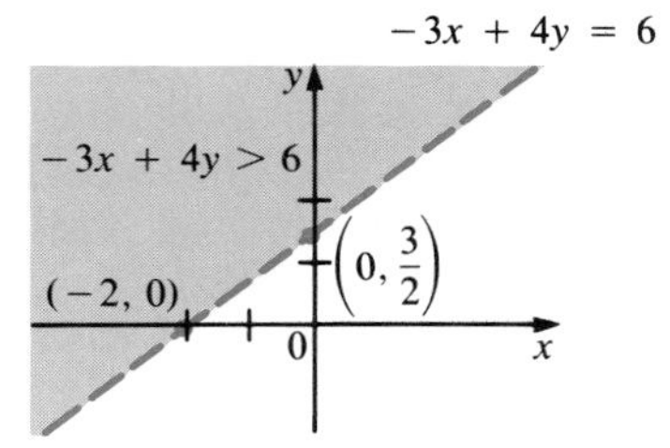

23.

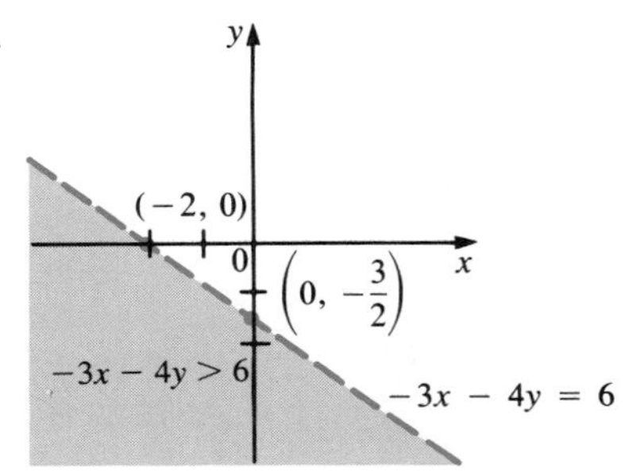

25.

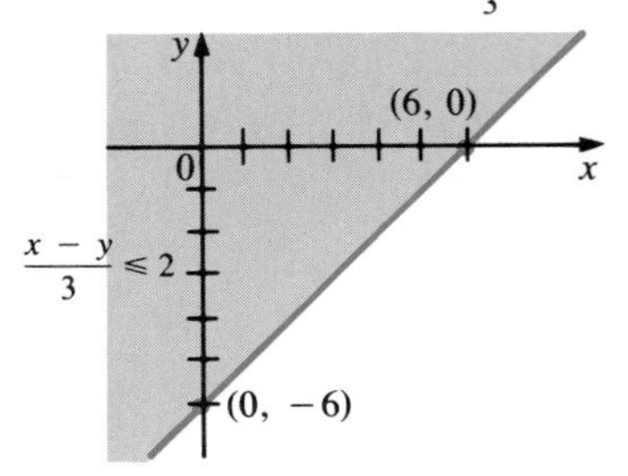

27.

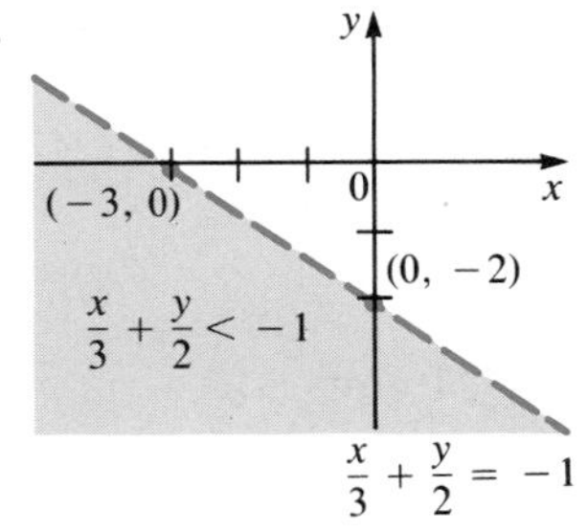

29.

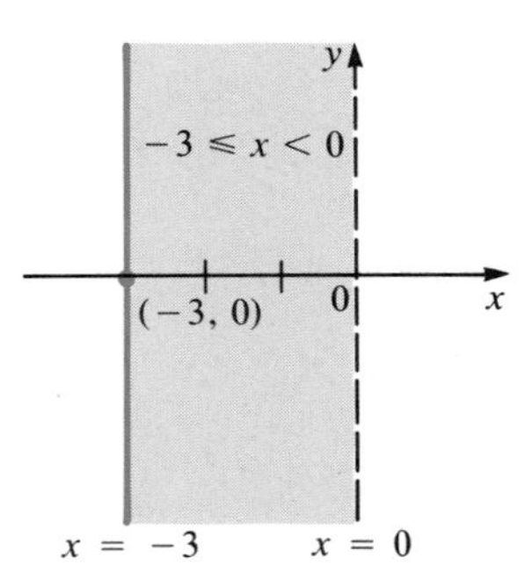

31.

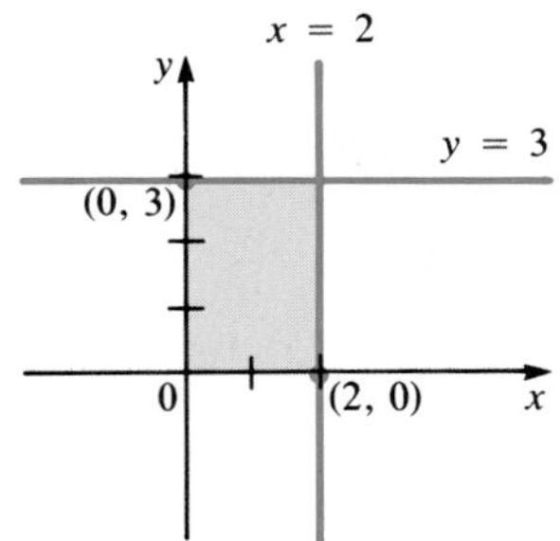

33.

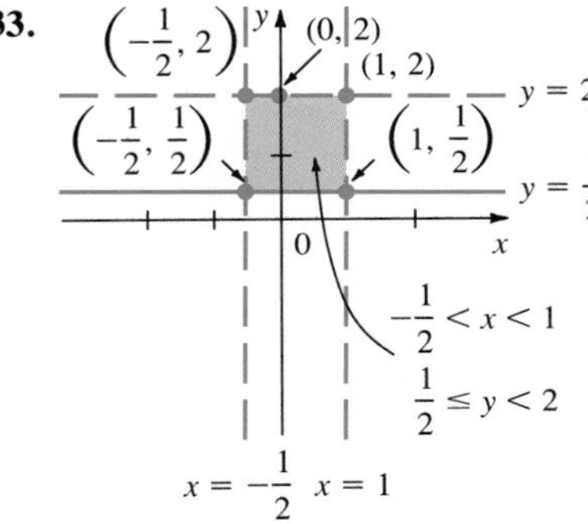

35.

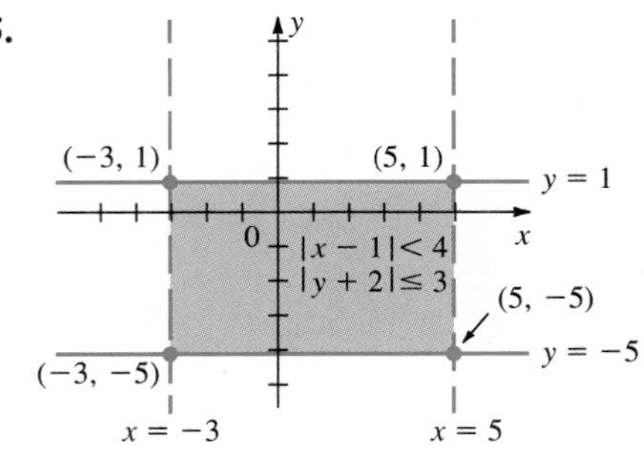

37.

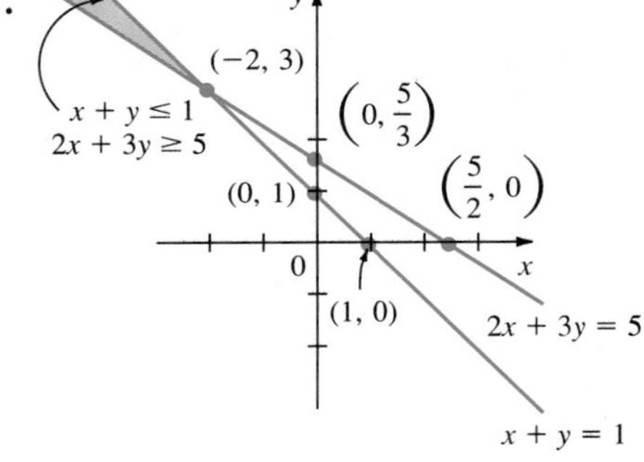

39.

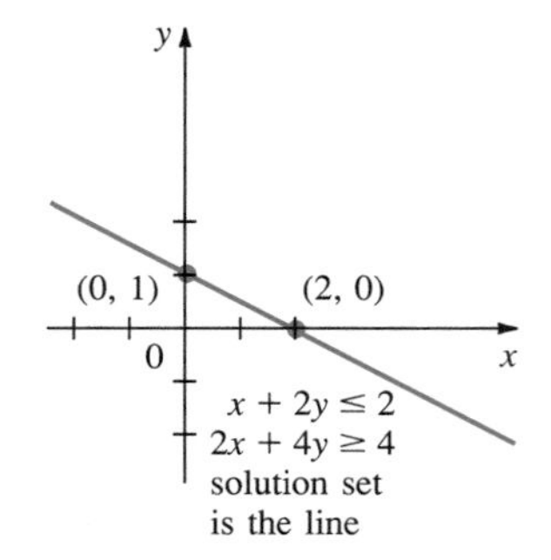

41.

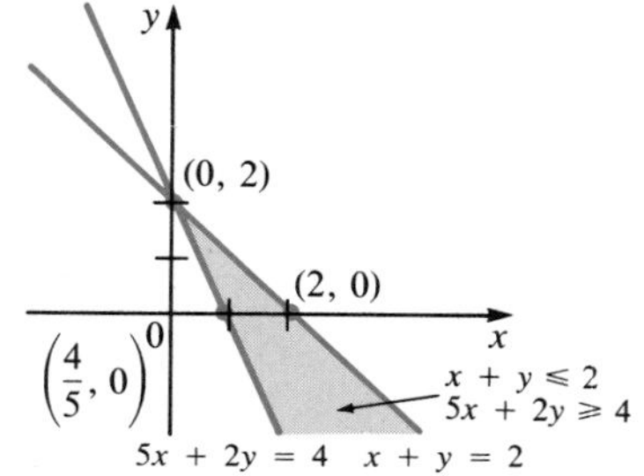

43.

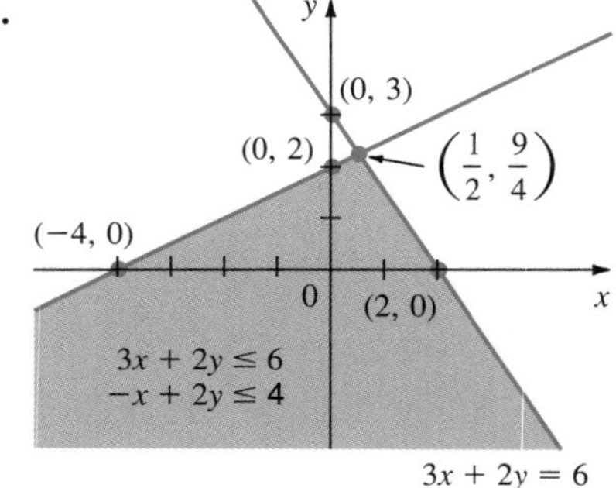

45.

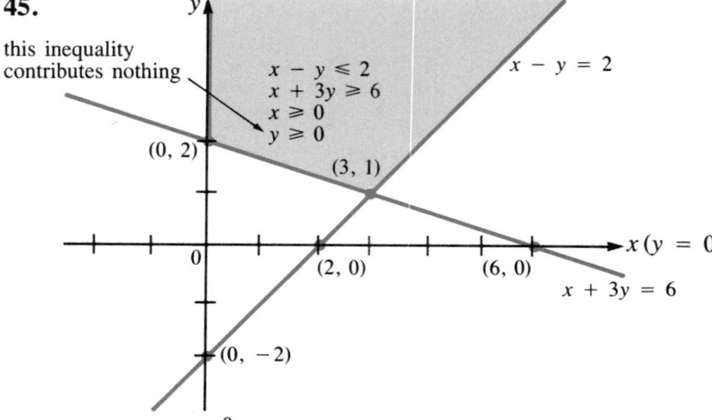

Problems 7.6, page 432

1. 65 chairs; 40 tables; Profit = \$525
3. 380 chairs; no tables; Profit = \$1900
5. 380 chairs; no tables; Profit = \$3040
7. 5 mgd from each; Cost = \$4000 per day
9. $(1, 3); f = 13$
11. Any point on the line $2x + 3y = 10$ between $\left(0, \frac{10}{3}\right)$ and $(2, 2)$ will yield the maximum value $f = 10$.
13. $\left(\frac{10}{11}, \frac{10}{11}\right); f = \frac{80}{11}$ **15.** $(0, 1); f = 12$
17. $(2, 1); g = 13$ **19.** $(1, 0); g = 3$
21. Food I $= \frac{2}{3}$ lb; Food II $= \frac{7}{3}$ lb; Cost per lb $= \$\frac{8}{9} \approx 89¢$
23. $s_1 = 1$; $s_2 = 5$; Energy = 13 units

Review Exercises for Chapter 7, page 436

1. $(\frac{1}{7}, \frac{10}{7})$ **3.** no solution **5.** $(0, 0, 0)$
7. $(2, -1, 3)$
9. $(\frac{1}{3}z, \frac{7}{3}z, z)$, where z is a real number
11. no solution **13.** $(0, 0, 0, 0)$
15. $(-z, w + 4, z, w)$, where z and w are real numbers
17. yes **19.** no
21. $\begin{pmatrix} 1 & 0 & -6 \\ 0 & 1 & \frac{5}{4} \end{pmatrix}$ or $\begin{pmatrix} 1 & 4 & -1 \\ 0 & 1 & \frac{5}{4} \end{pmatrix}$
23. $\frac{\frac{1}{5}}{x-2} - \frac{\frac{1}{5}}{x+3} = \frac{1}{5(x-2)} - \frac{1}{5(x+3)}$
25. $\frac{-\frac{2}{3}}{x+1} + \frac{\frac{5}{4}}{x+2} + \frac{\frac{5}{12}}{x-2} = -\frac{2}{3(x+1)} + \frac{5}{4(x+2)} + \frac{5}{12(x-2)}$
27. $\frac{\frac{2}{9}x + \frac{10}{9}}{x^2+2} + \frac{-\frac{2}{9}x + \frac{1}{3}}{x^2 - 2x + 3} = \frac{2x+10}{9(x^2+2)} + \frac{-2x+3}{9(x^2-2x+3)}$
29. $(-3/\sqrt{2}, 3/\sqrt{2}), (3/\sqrt{2}, -3/\sqrt{2})$
31. $(-\sqrt{6}, -2), (-\sqrt{6}, 2), (\sqrt{6}, -2), (\sqrt{6}, 2)$
33. $(1 - \sqrt{2}, -2 - 2\sqrt{2}), (1 + \sqrt{2}, -2 + 2\sqrt{2})$
35. $\left(\frac{19}{23}, -\frac{19}{17}\right)$ **37.** $(0, 1)$
39.

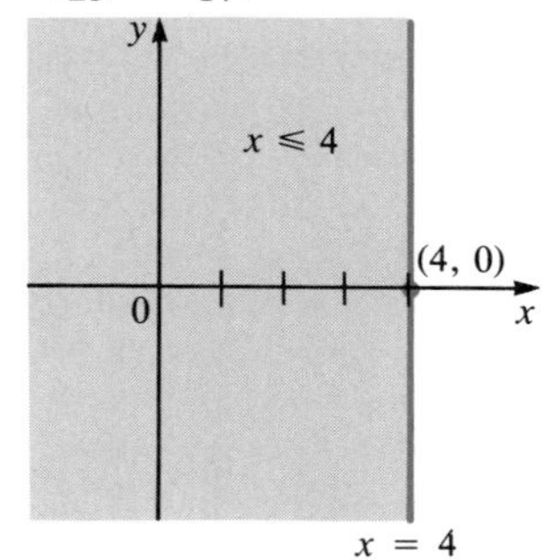

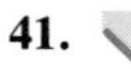

41.

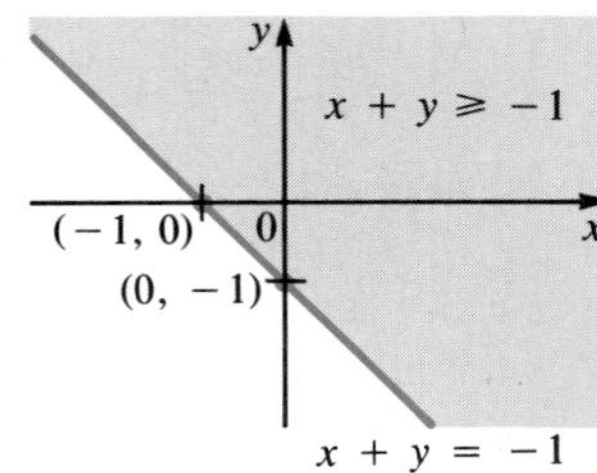

43.

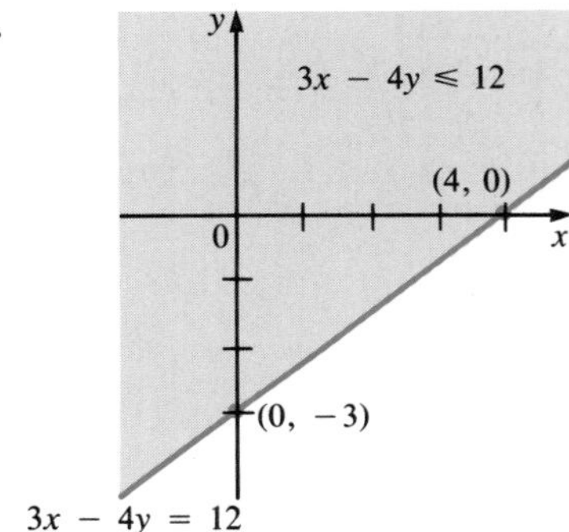

45.

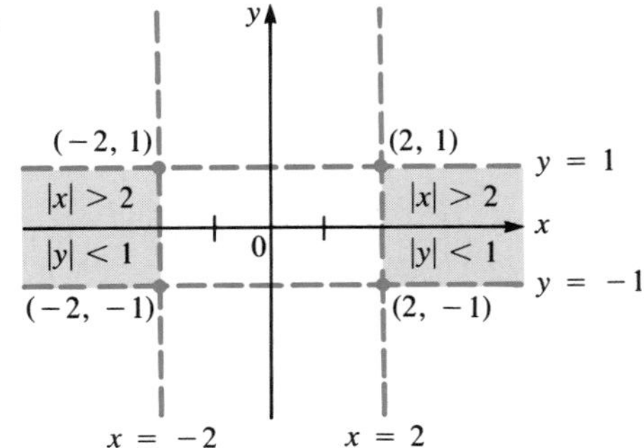

47.

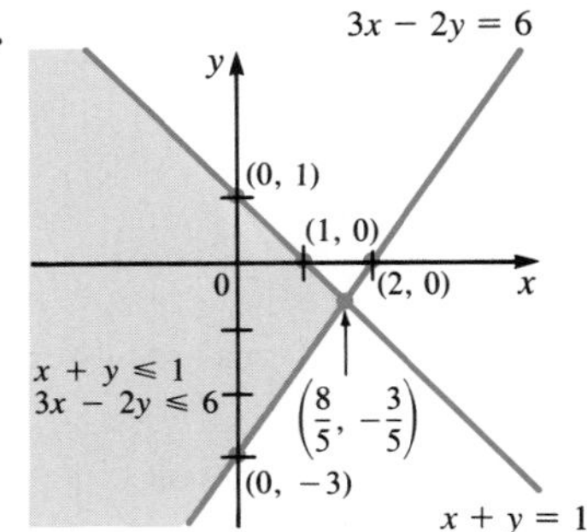

49. $(\frac{4}{5}, \frac{12}{5})$; $f = \frac{68}{5}$ **51.** $(\frac{16}{9}, \frac{10}{9})$; $g = \frac{68}{9}$
53. unbounded (no maximum)

Chapter 8

Problems 8.1, page 444

1. 2×2, square **3.** 2×2, square **5.** 3×2
7. 2×4 **9.** 3×1 **11.** 3×3, square
13. 1×4 **15.** no **17.** no **19.** yes

21. $\begin{pmatrix} 8 & 20 \\ 12 & -4 \end{pmatrix}$ **23.** $\begin{pmatrix} -1 & -2 & -5 \\ -7 & 2 & 0 \end{pmatrix}$

25. not possible **27.** $\begin{pmatrix} -12 & 30 & 23 \\ 35 & -30 & 3 \\ -29 & 35 & 6 \end{pmatrix}$

29. $\begin{pmatrix} 5 & 6 & 2 \\ 3 & 4 & 1 \\ 0 & -7 & 2 \end{pmatrix}$ **31.** $\begin{pmatrix} 2 & 6 \\ 4 & 10 \\ -2 & 4 \end{pmatrix}$

33. $\begin{pmatrix} 0 & 4 \\ 6 & 11 \\ -8 & 5 \end{pmatrix}$ **35.** $\begin{pmatrix} 0 & 0 \\ 0 & 0 \\ 0 & 0 \end{pmatrix}$ **37.** $\begin{pmatrix} -2 & 4 \\ 7 & 15 \\ -15 & 10 \end{pmatrix}$

39. $\begin{pmatrix} 4 & 10 \\ 17 & 22 \\ -9 & 1 \end{pmatrix}$ **41.** $\begin{pmatrix} 0 & 6 \\ 5 & 14 \\ -9 & 9 \end{pmatrix}$

43. $\begin{pmatrix} -2 & 2 & -2 \\ -1 & -7 & -10 \\ 0 & -4 & 6 \end{pmatrix}$ **45.** $\begin{pmatrix} 1 & 1 & 5 \\ 8 & 5 & 10 \\ 7 & -7 & 3 \end{pmatrix}$

47. $\begin{pmatrix} -1 & -1 & -1 \\ -2 & -3 & -10 \\ -7 & 3 & 5 \end{pmatrix}$ **49.** $\begin{pmatrix} -1 & -1 & -5 \\ -8 & -5 & -10 \\ -7 & 7 & -3 \end{pmatrix}$

51. $1.25S = \begin{pmatrix} 7.5 & 23.75 & 17.5 & 57.5 \\ 10 & 35 & 15 & 50 \\ 5 & 32.5 & 21.25 & 68.75 \end{pmatrix}$

53. $1.3A + 0.8B = \begin{pmatrix} 13.9 & 32.4 & 9 & 69.2 \\ 14.5 & 33.4 & 21.6 & 48.5 \\ 9.4 & 32.5 & 10.5 & 65.3 \end{pmatrix}$

Problems 8.2, page 453

1. 2 **3.** 1 **5.** 32 **9.** \$1535

11. $\begin{pmatrix} -6 & 15 \\ -8 & 1 \end{pmatrix}$ **13.** $\begin{pmatrix} -3 & -3 \\ 1 & 3 \end{pmatrix}$ **15.** $\begin{pmatrix} 9 & 8 \\ 7 & -2 \end{pmatrix}$

17. not possible **19.** $\begin{pmatrix} -1 & 2 & -9 \\ -1 & 0 & 1 \\ 16 & 3 & -5 \end{pmatrix}$ **21.** $(7 \quad 16)$

23. $\begin{pmatrix} 3 & -2 & 1 \\ 4 & 0 & 6 \\ 5 & 1 & 9 \end{pmatrix}$ **25.** $\begin{pmatrix} a & b & c \\ d & e & f \\ g & h & j \end{pmatrix}$

27. (a) 5 (b) 4 (c) $\begin{pmatrix} 1 & 3 & 1 & 2 \\ 0 & 1 & 2 & 2 \end{pmatrix}$

29. (a) $(100 \quad 200 \quad 400 \quad 100)$ (b) $\begin{pmatrix} 46 \\ 34 \\ 15 \\ 10 \end{pmatrix}$ (c) \$18,400

31. (a) $\begin{pmatrix} 80{,}000 & 45{,}000 & 40{,}000 \\ 50 & 20 & 10 \end{pmatrix}$ (b) $\begin{pmatrix} 1 \\ 3 \\ 1 \end{pmatrix}$
(c) money \$255,000, shares 120

33. (a) $\begin{pmatrix} \frac{1}{5} & \frac{2}{5} & \frac{1}{5} \\ 0 & 1 & \frac{1}{2} \\ \frac{1}{10} & 0 & 0 \\ \frac{1}{8} & \frac{1}{8} & \frac{1}{16} \end{pmatrix}$ (b) $\begin{pmatrix} 2 \\ 1 \\ 3 \end{pmatrix}$
(c) bread $\frac{7}{5}$ loaves, milk $\frac{5}{2}$ quarts, coffee $\frac{1}{5}$ pound, cheese $\frac{9}{16}$ pound

35. $\begin{pmatrix} 2 & -1 \\ 4 & 5 \end{pmatrix}\begin{pmatrix} x \\ y \end{pmatrix} = \begin{pmatrix} 3 \\ 7 \end{pmatrix}$

37. $\begin{pmatrix} 3 & 6 & -7 \\ 2 & -1 & 3 \end{pmatrix}\begin{pmatrix} x \\ y \\ z \end{pmatrix} = \begin{pmatrix} 0 \\ 1 \end{pmatrix}$

39. $\begin{pmatrix} 2 & 0 & 4 \\ 3 & 2 & 0 \\ 0 & -4 & -9 \end{pmatrix}\begin{pmatrix} x \\ y \\ z \end{pmatrix} = \begin{pmatrix} 9 \\ 6 \\ 3 \end{pmatrix}$

Problems 8.3, page 464

1. $\begin{pmatrix} -5 & 3 \\ 2 & -1 \end{pmatrix}$ **3.** $\begin{pmatrix} 0 & 1 \\ 1 & 0 \end{pmatrix}$ **5.** not invertible

7. $\begin{pmatrix} -\frac{1}{2} & 0 & \frac{1}{2} \\ \frac{9}{10} & \frac{3}{5} & -\frac{7}{10} \\ -\frac{1}{10} & -\frac{2}{5} & \frac{3}{10} \end{pmatrix}$ **9.** not invertible

11. $\begin{pmatrix} \frac{7}{20} & \frac{1}{10} & -\frac{3}{20} \\ \frac{3}{10} & -\frac{1}{5} & \frac{3}{10} \\ -\frac{1}{10} & \frac{2}{5} & -\frac{1}{10} \end{pmatrix}$ **13.** $\begin{pmatrix} -\frac{3}{4} & 1 & \frac{3}{2} & -\frac{5}{4} \\ \frac{1}{4} & 0 & -\frac{1}{2} & \frac{3}{4} \\ \frac{1}{4} & -1 & -\frac{1}{2} & \frac{3}{4} \\ \frac{3}{4} & 0 & -\frac{1}{2} & \frac{1}{4} \end{pmatrix}$

15. not invertible

Problems 8.4, page 472

1. -2 **3.** -26 **5.** 0 **7.** -36 **9.** 0
11. abc **13.** $abcd$ **15.** -32 **17.** 0
19. $(\frac{77}{13}, \frac{42}{13}, -\frac{54}{13})$ **21.** $(-\frac{113}{127}, \frac{111}{127}, \frac{91}{127})$
23. det = 0; not invertible **25.** det = -15; invertible
27. det = 0; not invertible

Review Exercises for Chapter 8, page 474

1. $\begin{pmatrix} -6 & 3 \\ 0 & 12 \\ 6 & 9 \end{pmatrix}$ **3.** $\begin{pmatrix} -2 & -16 & 6 \\ -12 & 5 & 4 \\ 11 & -19 & -4 \end{pmatrix}$

5. $\begin{pmatrix} 17 & 39 & 41 \\ 14 & 20 & 42 \end{pmatrix}$ **7.** $\begin{pmatrix} 9 & 10 \\ 30 & 32 \end{pmatrix}$ **9.** -4

11. 24 **13.** 60 **15.** $(\frac{123}{19}, \frac{24}{19}, \frac{34}{19})$

17. $\begin{pmatrix} \frac{5}{6} & -\frac{7}{6} \\ -\frac{1}{3} & \frac{2}{3} \end{pmatrix}$ **19.** $\begin{pmatrix} -\frac{1}{4} & \frac{1}{4} & \frac{1}{4} \\ \frac{5}{8} & -\frac{1}{8} & -\frac{1}{8} \\ \frac{1}{8} & -\frac{5}{8} & \frac{3}{8} \end{pmatrix}$

21. $\begin{pmatrix} \frac{4}{7} & \frac{1}{7} & -\frac{3}{7} \\ -\frac{1}{7} & -\frac{2}{7} & \frac{6}{7} \\ \frac{2}{7} & \frac{4}{7} & -\frac{5}{7} \end{pmatrix}$

23. $\begin{pmatrix} 1 & 2 & 0 \\ 2 & 1 & -1 \\ 3 & 1 & 1 \end{pmatrix}\begin{pmatrix} x \\ y \\ z \end{pmatrix} = \begin{pmatrix} 3 \\ -1 \\ 7 \end{pmatrix}$; inverse given in answer to Exercise 19; solution is $(\frac{3}{4}, \frac{9}{8}, \frac{29}{8})$

Chapter 9

Problems 9.1, page 480

1. For $n = 1$, $2 = 1(1 + 1)$ is true. Assume that $2 + 4 + 6 + \cdots + 2k = k(k + 1)$.
Then $2 + 4 + 6 + \cdots + 2k + 2(k + 1)$
$= k(k + 1) + 2(k + 1)$
$= (k + 1)(k + 2)$ so true for $n = k + 1$.

3. For $n = 1$, $2 = \frac{1(3 + 1)}{2} = \frac{4}{2}$ is true. Assume true for $n = k$.
$2 + 5 + 8 + \cdots + (3k - 1) + (3k + 2)$
$= \frac{k(3k + 1)}{2} + (3k + 2)$
$= \frac{3k^2 + k + 6k + 4}{2}$
$= \frac{3k^2 + 7k + 4}{2}$
$= \frac{(k + 1)(3k + 4)}{2}$

5. For $n = 1$, $2^1 > 1$ is true
Assume $2^k > k$.
$2^k \cdot 2 > k(2)$
$2^{k+1} > 2k$ but $2k > k + 1$ for $k > 1$
$2^{k+1} > k + 1$

7. For $n = 0$, $1 = 2^0 = 2^1 - 1 = 1$ is true. Assume true for $n = k$.
$1 + 2 + 4 + 8 + \cdots + 2^k + 2^{k+1} = 2^{k+1} - 1 + 2^{k+1}$
$= 2 \cdot 2^{k+1} - 1$
$= 2^{k+2} - 1$

9. For $n = 0$, $1 = \frac{1}{2^0} = 2 - \frac{1}{2^0} = 2 - 1 = 1$ is true.
Assume true for $n = k$.

$$1 + \frac{1}{2} + \frac{1}{4} + \cdots + \frac{1}{2^k} + \frac{1}{2^{k+1}} = 2 - \frac{1}{2^k} + \frac{1}{2^{k+1}}$$
$$= 2 - \frac{2}{2^{k+1}} + \frac{1}{2^{k+1}}$$
$$= 2 - \frac{1}{2^{k+1}}$$

11. For $n = 1$, $1^3 = \frac{1^2(1+1)^2}{4} = \frac{1(2)^2}{4} = 1$ is true.
Assume true for $n = k$.
$1^3 + 2^3 + 3^3 + \cdots + k^3 + (k+1)^3$
$$= \frac{k^2(k+1)^2}{4} + (k+1)^3$$
$$= (k+1)^2\left[\frac{k^2}{4} + (k+1)\right]$$
$$= (k+1)^2 \cdot \frac{k^2+4k+4}{4} = \frac{(k+1)^2(k+2)^2}{4}$$

13. For $n = 1$, $1 \cdot 2 = (2-1)(2) = \frac{1(2)(3)}{3} = 2$ is true.
Assume true for $n = k$.
$1 \cdot 2 + 3 \cdot 4 + 5 \cdot 6 + \cdots + (2k-1)(2k) + (2k+1)(2k+2)$
$$= \frac{k(k+1)(4k-1)}{3} + (2k+1)(2k+2)$$
$$= \frac{(k+1)[k(4k-1) + (2k+1)(6)]}{3}$$
$$= \frac{(k+1)[4k^2 + 11k + 6]}{3}$$
$$= \frac{(k+1)(k+2)[4(k+1) - 1]}{3}$$

15. For $n = 1$, $n + n^2 = 1 + 1 = 2$ is even is true.
Assume $k + k^2$ is even.
$(k+1) + (k+1)^2 = k + 1 + k^2 + 2k + 1$
$= k^2 + 3k + 2$
$= (k + k^2) + (2k + 2)$
$= (k + k^2) + 2(k+1)$
Both $(k + k^2)$ and $2(k+1)$ are even so the sum is even.

17. For $n = 1$, $n(n^2 + 5) = 1(1 + 5) = 6$ is divisible by 6 is true. Assume $k(k^2 + 5)$ is divisible by 6.
$(k+1)((k+1)^2 + 5) = (k+1)(k^2 + 2k + 6)$
$= k^3 + 3k^2 + 8k + 6$
$= (k^3 + 5k) + (3k^2 + 3k) + 6$
$= k(k^2 + 5) + 3(k)(k+1) + 6$
$k(k^2 + 5)$ is divisible by 6 by assumption; $3(k)(k+1)$ is divisible by 6 since either k or $(k+1)$ is even. The sum of three numbers, each divisible by 6, is divisible by 6.

19. For $n = 1$, $(x^1 - 1) \div (x - 1) = 1$ is true. Assume $(x^k - 1) \div (x - 1) = q(x)$, a polynomial.
$x^{k+1} - 1 = x^{k+1} - x^k + x^k - 1$
$= x^k(x-1) + (x^k - 1)$
Then $(x^{k+1} - 1) \div (x - 1) = x^k + q(x)$

21. For $k = 21$, $20(21) < 21^2$ is true since $420 < 441$.
Assume $20k < k^2$ for $k > 21$.
Show $20(k+1) < (k+1)^1$
$20k + 20 < k^2 + 2k + 1$ is true because
$20k < k^2$ and $20 < 2k + 1$ holds for $k > \frac{19}{2}$.

23. For $k = 238$, $100 \log 238 \approx 237.6577 < 238$ is true.
Assume $100 \log k < k$ for $k > 238$.
Show $100 \log (k+1) < k + 1$.
But $100 \log (k+1) = 100[\log (k+1) - \log k + \log k]$
$$= 100\left[\log\left(\frac{k+1}{k}\right) + \log k\right] = 100\left[\log\left(1 + \frac{1}{k}\right) + \log k\right] < 100\left[\log\left(1 + \frac{1}{238}\right) + \log k\right]$$
$\approx 100[0.00182 + \log k] < 100[0.01 + \log k]$
$= 100 \log k + 1 < k + 1$

25. For $k = 5$, $5^3 = 125 < e^5 \approx 148.4$
Assume $k^3 < e^k$ for $k > 5$.
Show $(k+1)^3 < e^{k+1}$.
$$(k+1)^3 = k^3\left(1 + \frac{1}{k}\right)^3 < e^k\left(1 + \frac{1}{k}\right)^3 \underset{k \geq 6}{\leq} e^k\left(1 + \frac{1}{6}\right)^3$$
$\approx 1.59e^k < e \cdot e^k = e^{k+1}$

27. For $k = 6$, $e^6 < 6!$ is true since $403.43 < 720$.
Assume $e^k < k!$
Show $e^{k+1} < (k+1)!$ for $k > 6$.
Given $e^k < k!$ multiply by $e < k + 1$
$e^k \cdot e < (k!)(k+1)$
$e^{k+1} < (k+1)!$

29. To prove for all *integers*
For $n = 1$, $(ab)^1 = ab = a^1b^1$ so is true
Assume $(ab)^k = a^kb^k$
Show $(ab)^{k+1} = a^{k+1}b^{k+1}$
$(ab)^{k+1} = (ab)^k(ab) = a^kb^kab = a^kab^kb$
$= a^{k+1}b^{k+1}$

31. Assume $2k - 1$ is even
Then $(2k - 1) + 2$ is even
But $(2k - 1) + 2 = 2k + 1 = 2(k+1) - 1$.
It appears that we have proved that $2k - 1$ is even.
But $2k - 1$ is not even for any integer. Conclusion: Step 1 of an induction proof is essential.

Problems 9.2, page 486

1. 1, 2, 3, 4, 5 **3.** 2, 6, 10, 14, 18
5. $\frac{1}{3}, \frac{1}{4}, \frac{1}{5}, \frac{1}{6}, \frac{1}{7}$ **7.** 2, 6, 12, 20, 30
9. $2, \frac{3}{2}, \frac{4}{3}, \frac{5}{4}, \frac{6}{5}$ **11.** $\frac{3}{2}, \frac{7}{4}, \frac{15}{8}, \frac{31}{16}, \frac{63}{32}$
13. 1, 16, 81, 256, 625 **15.** $-\frac{1}{4}, \frac{1}{5}, -\frac{1}{6}, \frac{1}{7}, -\frac{1}{8}$

17. e, e^2, e^3, e^4, e^5 **19.** $e, e^{1/2}, e^{1/3}, e^{1/4}, e^{1/5}$
21. 1, 1, 1, 1, 1 **23.** $-\frac{2}{3}, \frac{4}{9}, -\frac{8}{27}, \frac{16}{81}, -\frac{32}{243}$
25. $-1, -2, -3, -4, -5$ **27.** $(-1)^{n+1}n$ **29.** n^3
31. $(-1)^{n+1}/(n+1)^2$ **33.** $\dfrac{(-1)^{n+1}(2^{n+1}+1)}{2^n}$
35. $n/(2n+1)$ **37.** $(-1)^{n+1}/3^n$ **39.** $n^2/(n+1)$
41. 45 **43.** 7 **45.** 35 **47.** $\sum_{k=0}^{4} 2^k$
49. $\sum_{k=2}^{n} \dfrac{k}{k+1} = \sum_{k=1}^{n-1} \dfrac{k+1}{k+2}$ **51.** $\sum_{k=1}^{n} k^{1/k}$
53. $\sum_{k=1}^{5} x^{4k+2}$ **55.** $\sum_{k=1}^{8} (2k-1)(2k+1)$
57. $\sum_{k=1}^{32} \dfrac{1}{32}\left(\dfrac{k}{32}\right)^2 = \dfrac{1}{32^3}\sum_{k=1}^{32} k^2$
59. $\sum_{k=1}^{64} \dfrac{1}{64}\left(\dfrac{k}{64}\right)^3 = \dfrac{1}{64^4}\sum_{k=1}^{64} k^3$
61. $\sum_{k=1}^{50} \dfrac{1}{50}\sqrt{\dfrac{k}{50}} = \dfrac{1}{50^{3/2}}\sum_{k=1}^{50} \sqrt{k}$ **63.** $\sum_{i=1}^{7} (7-i)^2$
67. $n^2(n+1)^2/4$ **69.** $\frac{1}{2}[\frac{3}{2} - 1/(n+1) - 1/(n+2)]$

Problems 9.3, page 494

1. $a_n = 1 + 3(n-1) = -2 + 3n$
3. $a_n = 4 - 3(n-1) = 7 - 3n$
5. $a_n = -12 + 6(n-1) = -18 + 6n$
7. 5, 6, 7, 8, 9 **9.** $\frac{17}{3}, \frac{16}{3}, 5, \frac{14}{3}, \frac{13}{3}$
11. $\frac{13}{2}, \frac{25}{2}, \frac{37}{2}, \frac{49}{2}, \frac{61}{2}$ **13.** 2^n **15.** $6(-1)^n$
17. $7(\frac{1}{3})^n$ **19.** 5, 10, 20, 40, 80, 160
21. 3, −3, 3, −3, 3, −3
23. 1, 4, 16, 64, 256, 1024 **25.** arithmetic, 6
27. geometric, 6 **29.** arithmetic, $\frac{3}{5}$ **31.** neither
33. geometric, −1 **35.** arithmetic, 10
37. geometric, −1 **39.** geometric, −2 **41.** 39
43. −285 **45.** $\frac{1}{32}$ **47.** 95 **49.** 4
51. $-\frac{1}{4}$ **53.** 5 **55.** 3 **57.** $(\frac{15}{4})^{1/4} \approx 1.3916$
59. \$7346.64 **61.** \$18,771.37 **65.** \$8590.93
67. 79.085%, 81.402%
69. $5\frac{1}{8}$% semiannually [5.19% is better than 5.13%]
71. $P_n = 1/2^n$ **73.** $a_n = -6 + 10/2^n$
75. $a_n = \frac{10}{3} + \frac{5}{3}(-2)^n$
77. (a) $P_{n+1} = P_n(1.08) - 300$
(b) $P_n = 3750 + 6250(1.08)^n$
(c) \$12,933.30, \$17,243.28
79. $a_n = n + 1$ **81.** 2, 3, 7, 16, 32, 57
83. $4, \frac{2}{3}, \frac{1}{4}, \frac{1}{9}, \frac{1}{19}, \frac{1}{39}$

Problems 9.4, page 504

1. 341 **3.** −2604
5. $0.3^2((1 + 0.3^7)/1.3) = 0.06924591$
7. $(b^{16} - 1)/(b^{16} + b^{14})$
9. $(16\sqrt{2} - 1)/(\sqrt{2} - 1) = 31 + 15\sqrt{2}$ **11.** −3280
13. $\frac{3}{2}$ **15.** $\frac{10}{9}$ **17.** $\dfrac{e}{e-1}$
19. $\dfrac{1}{1.31} \approx 0.763358778$ **21.** $\frac{1}{2}$ **23.** 7
25. 459 **27.** (b) 12 minutes **29.** \$5045.24
31. \$7968.71 **33.** \$5,012,853.18 **35.** \$5306.72
37. \$7957.13 **39.** $\frac{14}{33}$ **41.** $\frac{71}{99}$ **43.** $\frac{104}{333}$
45. 11,351/99,900 **47.** $\frac{1900}{99}$
49. $38\frac{2}{11} \approx 38.1818$ minutes

Problems 9.5, page 508

1. 4 **3.** 36 **5.** 12 **7.** 24 **9.** 210
11. 8
13. No; three letters can be chosen in $26^3 = 17{,}576$ ways. Four letters are needed, which yield $26^4 = 456{,}976$ possible classifications.
15. 64 **17.** 13,824 **19.** 170,586 **21.** 336
23. 7,119,090

Problems 9.6, page 519

1. 5040 **3.** 720 **5.** 120 **7.** 1680
9. 19,958,400 **11.** 362,880 **13.** 1 **15.** $n!$
17. $n!/2$ **19.** 840 **21.** $50! \approx 3.0414 \times 10^{64}$
23. $50!/40! \approx 3.7276 \times 10^{16}$ **25.** 20 **27.** 15
29. 10 **31.** 165 **33.** 126 **35.** 330
37. 13 **39.** 8 **41.** 1 **43.** 1 **45.** n
47. $n(n-1)/2$ **49.** 210
51. $\binom{52}{13} = 635{,}013{,}559{,}680$ **53.** 84 **55.** 3360
57. 3744 **59.** 210 **61.** 6
63. 10,240 (including the 40 straight flushes)
65. 123,552 **67.** 15 **69.** 56 **71.** 380
73. $4^{20} \approx 1.0995 \times 10^{12}$
75. $\binom{100}{37} \cdot \binom{63}{51} \approx 9.1261 \times 10^{39}$ **77.** 3024
79. $\binom{15}{4} \cdot \binom{11}{3} \cdot \binom{8}{5} \cdot \binom{3}{3} = 12{,}612{,}600$
81. $\binom{12}{8} + \binom{12}{9} + \binom{12}{10} + \binom{12}{11} + \binom{12}{12} = 794$
83. 240
85. $9\left[\binom{3}{2} \cdot 8 \cdot 8 \cdot 7 \cdot 6 \cdot 5 + \binom{3}{1} \cdot 8 \cdot 7 \cdot \binom{4}{1} \cdot 7 \cdot 6 \cdot 5 + 8 \cdot 7 \cdot 6 \cdot \binom{4}{2} \cdot 6 \cdot 5\right] + \left[9 \cdot 8 \cdot 7 \cdot \binom{4}{2} \cdot 6 \cdot 5\right] = 2{,}268{,}000$
87. 90,000 **89.** (a) 1024 (b) 240
91. (a) 32 (b) 8 (c) 4

Problems 9.7, page 527

1. (a) The set consisting of each of the 52 cards (b) $\frac{1}{52}$
3. (a) The set consisting of all subsets containing exactly five of the people. (b) $\frac{1}{252}$
5. (a) The set consisting of the ordered pairs containing two distinct people (b) $\frac{1}{30}$
7. (a) $1/12! \approx 2.0877 \times 10^{-9}$ (b) $1 - 1/12!$ (c) 0
9. (a) $\binom{20}{6}\Big/\binom{28}{6} \approx 0.1029$
(b) $\binom{20}{3}\binom{8}{3}\Big/\binom{28}{6} \approx 0.1695$
11. $3744/2{,}598{,}960 \approx 0.00144 \approx 1/694$
13. $13\binom{4}{3}\binom{12}{2}4^2\Big/\binom{52}{5} = 54{,}912/2{,}598{,}960 \approx 0.02113 \approx 1/47.3$
15. $13\binom{4}{2}\binom{12}{3}4^3\Big/\binom{52}{5} = 1{,}098{,}240/2{,}598{,}960 \approx 0.4227 \approx 1/2.37$
17. $\frac{1}{1024}$
19. $\left(\binom{3}{2}13 + \binom{8}{2}8 + \binom{3}{2}13 + 14\right)\Big/\binom{16}{3} = 316/560 \approx 0.5643$
21. (a) $\frac{1}{8}$ (b) $\frac{5}{16}$
23. (a) All groups of four people from among the 15. (b) $\frac{1}{1365}$ (c) $\frac{1}{1365}$

Problems 9.8, page 534

1. 10 **3.** 35 **5.** 165 **7.** 1
9. $x^5 - 5x^4y + 10x^3y^2 - 10x^2y^3 + 5xy^4 - y^5$
11. $x^4 - 8x^3y + 24x^2y^2 - 32xy^3 + 16y^4$
13. $a^8 + 8a^7b + 28a^6b^2 + 56a^5b^3 + 70a^4b^4 + 56a^3b^5 + 28a^2b^6 + 8ab^7 + b^8$
15. $32a^5 - 240a^4b + 720a^3b^2 - 1080a^2b^3 + 810ab^4 - 243b^5$
17. $\frac{u^3}{27} - \frac{u^2v}{12} + \frac{uv^2}{16} - \frac{v^3}{64}$
19. $x^8 + 8x^6y + 24x^4y^2 + 32x^2y^3 + 16y^4$
21. $a^8 + 4a^6b^3 + 6a^4b^6 + 4a^2b^9 + b^{12}$
23. $x^3 + 6x^{5/2}y^{1/2} + 15x^2y + 20x^{3/2}y^{3/2} + 15xy^2 + 6x^{1/2}y^{5/2} + y^3$
25. $81x^2 + 324x^{3/2}y^{1/2} + 486xy + 324x^{1/2}y^{3/2} + 81y^2$
27. $a^3b^3 - 3a^2b^2cd + 3abc^2d^2 - c^3d^3$
29. $w^4 - 4w^3u/v + 6w^2u^2/v^2 - 4wu^3/v^3 + u^4/v^4$
31. $1 - 10a + 45a^2 - 120a^3 + 210a^4 - 252a^5 + 210a^6 - 120a^7 + 45a^8 - 10a^9 + a^{10}$
33. $1 + 3\sqrt{x} + 3x + x^{3/2}$
35. $25 + 20\sqrt{35} + 210 + 28\sqrt{35} + 49 = 284 + 48\sqrt{35}$
37. $1 + 4z^6 + 6z^{12} + 4z^{18} + z^{24}$
39. $u^4 + 4u^3v + 6u^2v^2 + 4uv^3 + v^4 - 8u^3 - 24u^2v - 24uv^2 - 8v^3 + 24u^2 + 48uv + 24v^2 - 32u - 32v + 16$
41. $x^4 + 4x^3y + 6x^2y^2 + 4xy^3 + y^4 + 4x^3z + 12x^2yz + 12xy^2z + 4y^3z + 6x^2z^2 + 12xyz^2 + 6y^2z^2 + 4xz^3 + 4yz^3 + z^4$
43. $x^{5n} + 5x^{4n}y^n + 10x^{3n}y^{2n} + 10x^{2n}y^{3n} + 5x^ny^{4n} + y^{5n}$
45. -120 **47.** 70
53. (a) $100! \approx 9.3248476 \times 10^{157}$
(b) $200! \approx 7.883293288 \times 10^{374}$
55. 1 7 21 35 35 21 7 1

Review Exercises for Chapter 9, page 536

1. For $k = 1$, $3^{1-1} = 3^0 = 1 = \frac{3-1}{2} = 1$ is true.
Assume $1 + 3 + 3^2 + \cdots + 3^{k-1} = \frac{3^k - 1}{2}$
Show $1 + 3 + 3^2 + \cdots + 3^{k-1} + 3^k = \frac{3^{k+1} - 1}{2}$
$[1 + 3 + 3^2 + \cdots + 3^{k-1}] + 3^k = \frac{3^k - 1}{2} + 3^k$
$= \frac{3^k - 1 + 2 \cdot 3^k}{2} = \frac{3 \cdot 3^k - 1}{2} = \frac{3^{k+1} - 1}{2}$
3. For $k = 1$, $1 = [4(1) - 3] = 1(2 - 1) = 1$ is true
Assume $1 + 5 + 9 + \cdots + (4k - 3) = k(2k - 1)$
Show $1 + 5 + 9 + \cdots + (4k - 3) + [4(k + 1) - 3] = (k + 1)[2(k + 1) - 1]$
$[1 + 5 + 9 + \cdots + (4k - 3)] + (4k + 1) = k(2k - 1) + 4k + 1 = 2k^2 + 3k + 1 = (k + 1)(2k + 1) = (k + 1)[2(k + 1) - 1]$
5. 10, 13, 16, 19, 22
7. $3n + 1$, starting with $n = 1$ **9.** 48 **11.** $\sum_{k=0}^{m} 3^k$
13. $\sum_{k=0}^{5} (-1)^k x^{4k+3}$ **15.** $-\frac{1}{2}n + \frac{13}{2}$, starting with $n = 1$
17. $(-3)^{3-n}$, starting with $n = 0$
19. 2, 6, 18, 54, 162, 486 **21.** 1, $\frac{5}{3}$, $\frac{17}{9}$, $\frac{53}{27}$, $\frac{161}{81}$, $\frac{485}{243}$
23. 262, 144 **25.** $a_n = 2 - 3^n$ **27.** 9840
29. $\frac{2}{3}$ **31.** 16 **33.** 240 **35.** 504
37. 2520 **39.** (a) 126 (b) 21 (c) 9 (d) 9 (e) 210
41. $11! = 39{,}916{,}800$ **43.** 420
45. $4\binom{13}{5} = 5148$ (including straight flushes)
47. $\frac{10 \cdot 4^5}{\binom{52}{5}} = \frac{10{,}240}{2{,}598{,}960} \approx 0.00394 \approx \frac{1}{253.8}$ (including straight flushes)
49. $\frac{1}{12}$ **51.** $\binom{50}{18} \cdot \binom{32}{13} \approx 6.27 \times 10^{21}$
53. (a) $\frac{1}{1024}$ (b) $\frac{15}{128}$ **55.** (a) 256 (b) 24
57. $a^6 + 6a^5b + 15a^4b^2 + 20a^3b^3 + 15a^2b^4 + 6ab^5 + b^6$
59. $243x^5 - 2025x^4y + 6750x^3y^2 - 11{,}250x^2y^3 + 9375xy^4 - 3125y^5$

1.

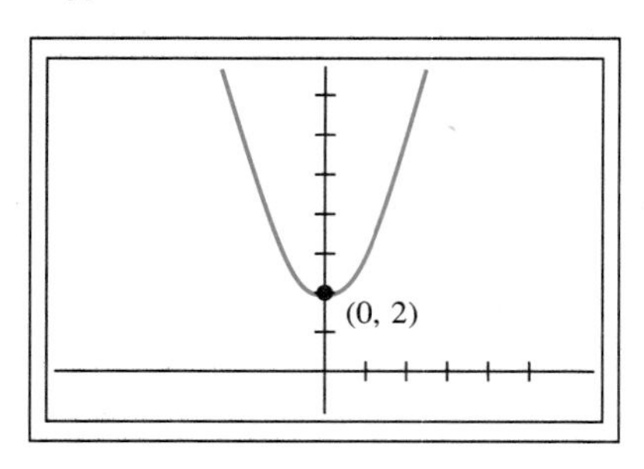

$y = x^2 + 2$

3.

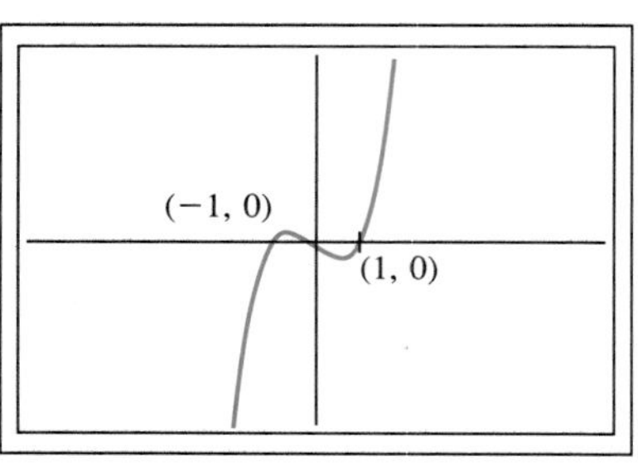

$y = x^3 - x$

5.

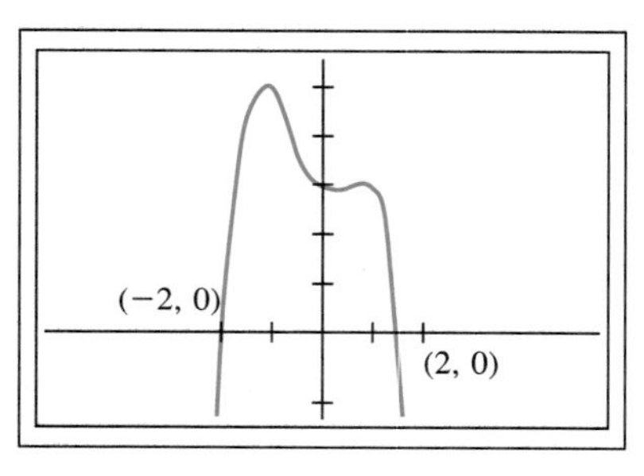

$y = -x^4 + 2x^2 - x + 3$

7.

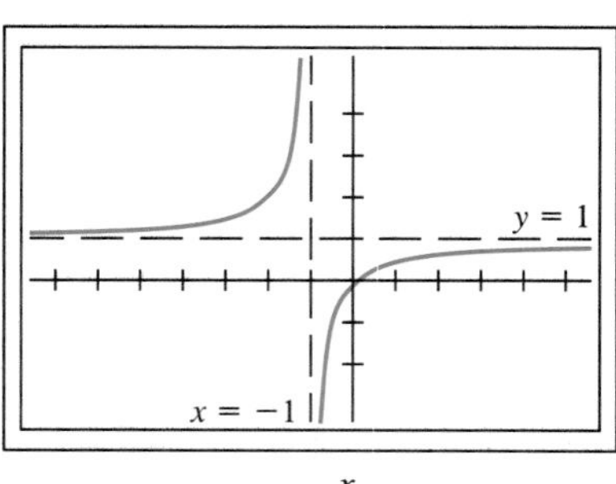

$y = \dfrac{x}{x + 1}$

9.

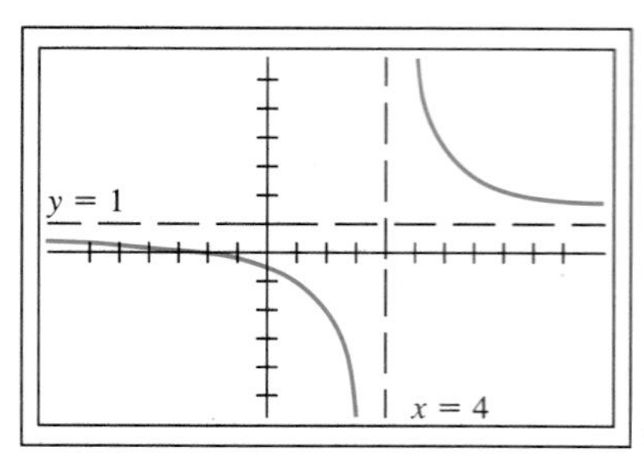

$y = \dfrac{x + 2}{x - 4}$

11.

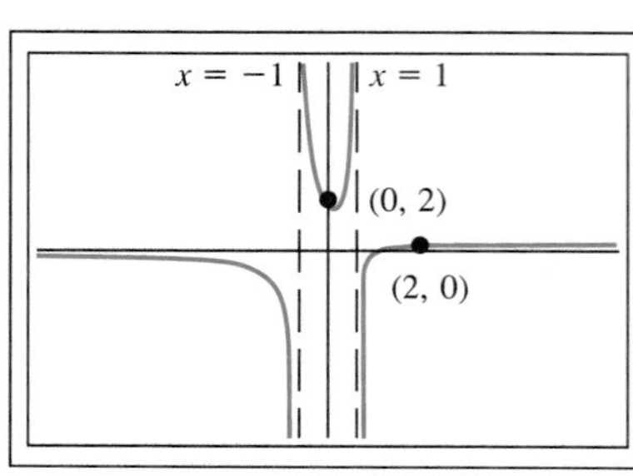

$y = \dfrac{x - 2}{x^2 - 1}$

13.

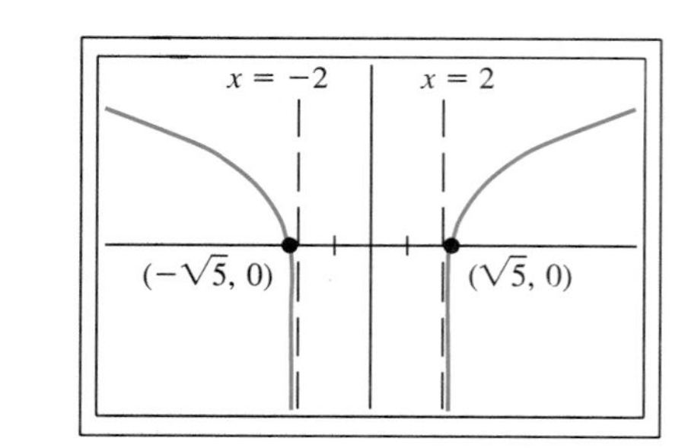

$y = \ln\,(x^2 - 4)$

15.

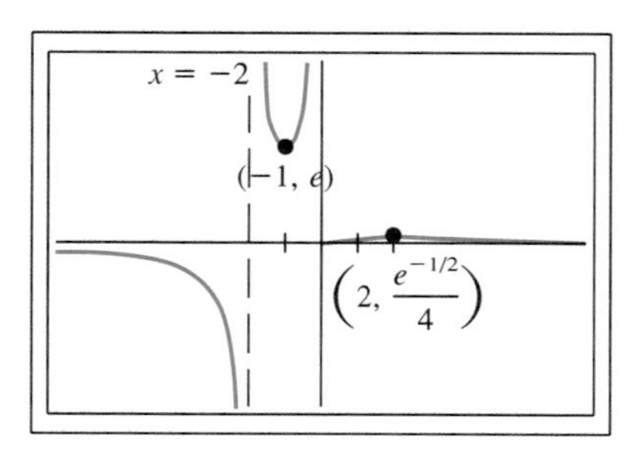

$y = \dfrac{e^{-1/x}}{x + 2}$

17.

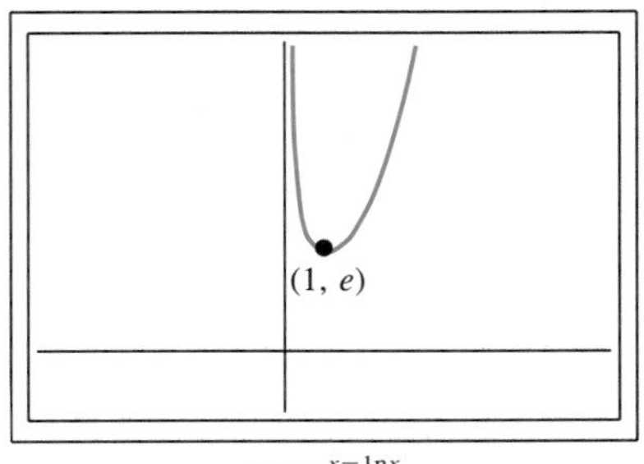

$y = e^{x - \ln x}$

19. 2.67, −3.50 **21.** 1.47 **23.** 1.46
25. −2.09, 0.24, 7.85
27. −1.82, −0.72, 0.67, 2.87 **29.** 1.31
31. 1.79, 2.62 **33.** $\{x: 0.33 < x < 3.00\}$
35. $\{x: x < -2.19 \text{ or } x > 3.19\}$
37. $\{x: x < -1.73 \text{ or } -0.82 < x < 1.73 \text{ or } 1.82 < x < 2\}$
39. $\{x: x > 2.22\}$
41.

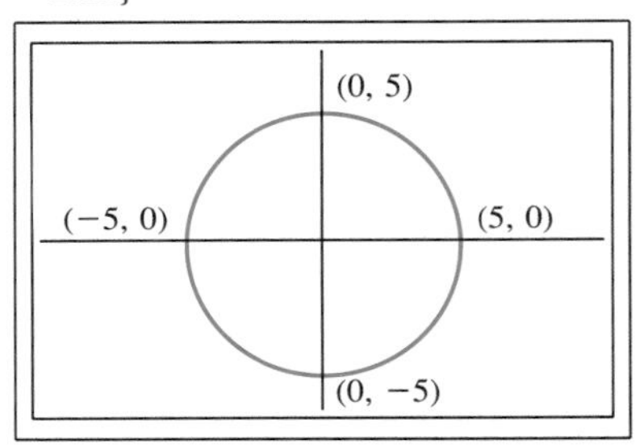

$x^2 + y^2 = 25$ (circle)

43.

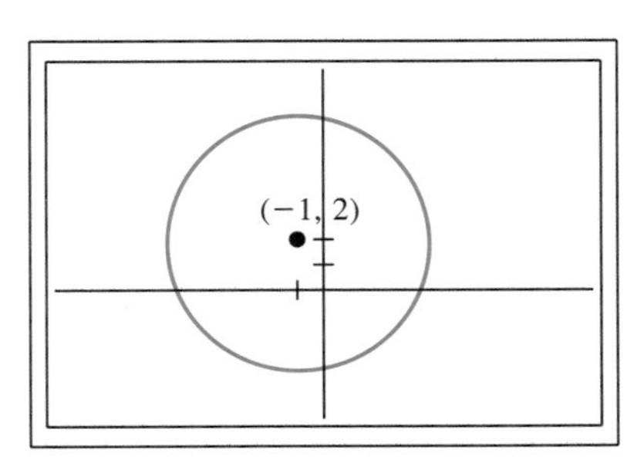

$x^2 + 2x + y^2 - 4y = 20$ (circle)
$(x + 1)^2 + (y - 2)^2 = 25$
$y = 2 \pm \sqrt{25 - (x + 1)^2}$

45.

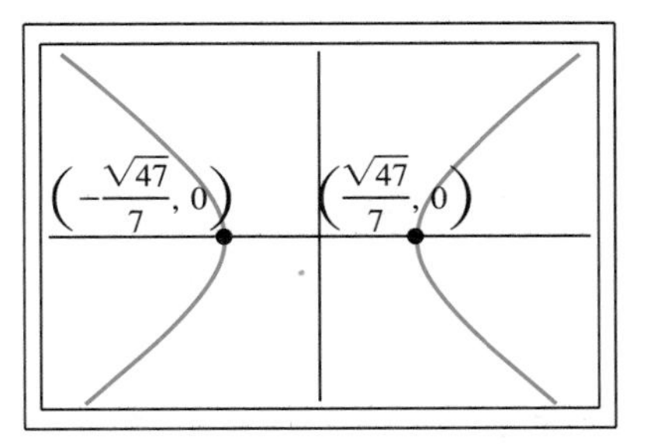

$7x^2 - 11y^2 = 47$ (hyperbola)
$y = \pm\sqrt{(7x^2 - 47)/11}$

47.

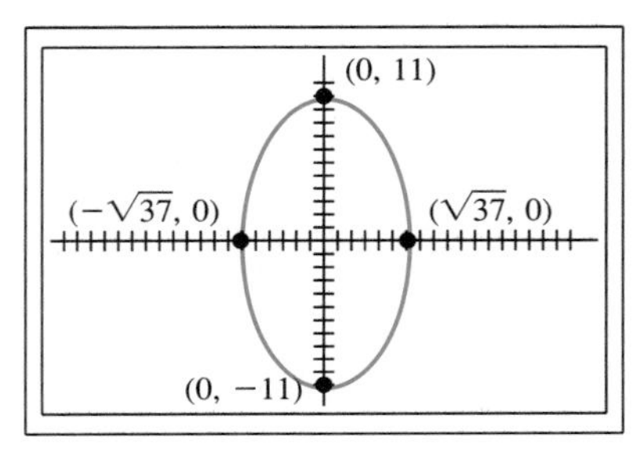

$\frac{x^2}{37} + \frac{y^2}{121} = 1$ (ellipse)
$y = \pm 11\sqrt{\left(1 - \frac{x^2}{37}\right)}$

49.

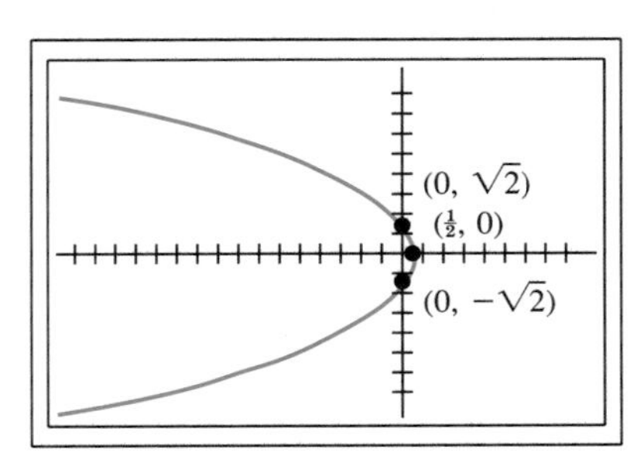

$3y^2 + 12x = 6$ (parabola)
$y = \pm\sqrt{2 - 4x}$

51.

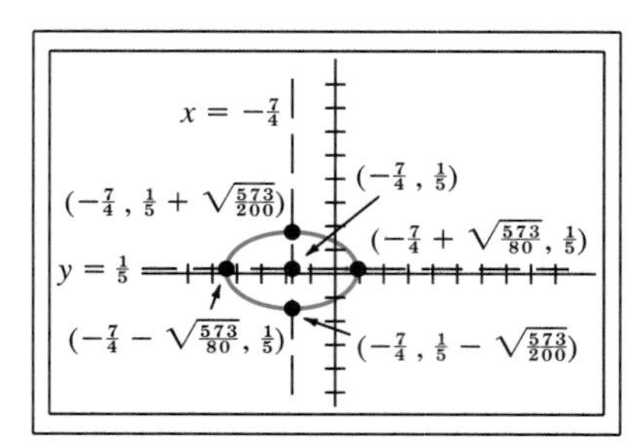

$2x^2 + 7x + 5y^2 - 2y = 8$ (ellipse)

$$\frac{(x + \frac{7}{4})^2}{573/80} + \frac{(y - \frac{1}{5})^2}{573/200} = 1$$

$$y = \tfrac{1}{5} \pm \sqrt{\tfrac{573}{200}[1 - \tfrac{80}{573}(x + \tfrac{7}{4})^2]}$$

53.

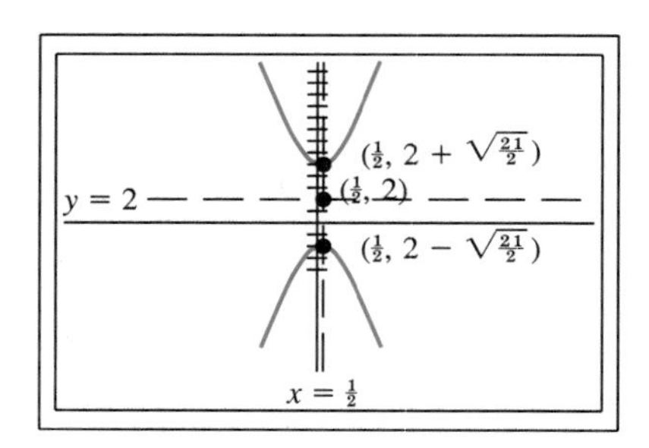

$2y^2 - 8y - 12x^2 + 12x = 16$ (hyperbola)

$$\frac{(y - 2)^2}{21/2} - \frac{(x - \frac{1}{2})^2}{7/4} = 1$$

$$y = 2 \pm \sqrt{\tfrac{21}{2}[1 + \tfrac{4}{7}(x - \tfrac{1}{2})^2]}$$

55.

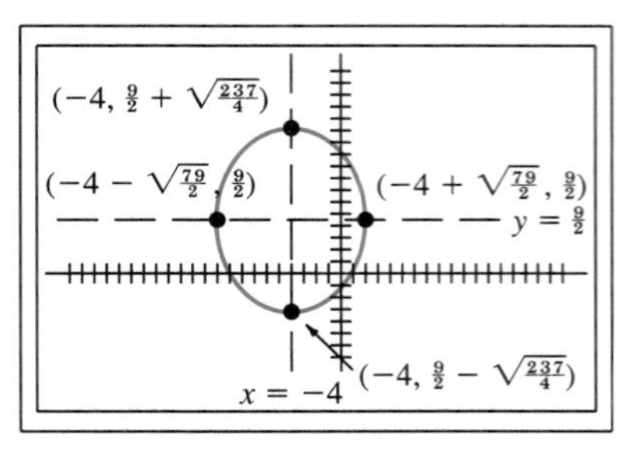

$\frac{1}{2}x^2 + 4x + \frac{1}{3}y^2 - 3y = 5$ (ellipse)

$$\frac{(x + 4)^2}{79/2} + \frac{(y - \frac{9}{2})^2}{237/4} = 1$$

$$y = \frac{9}{2} \pm \sqrt{\frac{237}{4}\left[1 - \frac{2(x + 4)^2}{79}\right]}$$

Index

Page numbers followed by "n" signify information included in a footnote on that page.

I.4 Index

B
C 2
D 3
E 4
F 5
G 6
H 7
I 8
J 9